MW00838177

ADAPTIVE FILTERS

ADAPTIVE FILTERS

ALI H. SAYED
University of California at Los Angeles

IEEE Press

A JOHN WILEY & SONS, INC., PUBLICATION

Cover design by Michael Rutkowski.

Published by John Wiley & Sons, Inc., Hoboken, New Jersey
Published simultaneously in Canada

Library of Congress Cataloging-in-Publication Data:

Sayed, Ali H.
Adaptive filters / Ali H. Sayed.
p. cm.
Includes bibliographical references and index.
ISBN 978-0-470-25388-5 (cloth)
1. Adaptive filters. I. Title.
TK7872.F5S285 2008
621.3815'324--dc22

2008003731

To my parents

Contents

PART VI: BLOCK ADAPTIVE FILTERS

PART VII: LEAST-SQUARES METHODS

PART XI: ROBUST FILTERS

REFERENCES AND INDICES

Preface

I am indebted to the many readers and colleagues who have written to me on several occasions with encouraging feedback on my earlier textbook *Fundamentals of Adaptive Filtering* (Wiley, NJ, 2003). Their enthusiastic comments encouraged me to pursue this second project in an effort to create a revised version for teaching purposes. During this exercise, I decided to remove some advanced material and move select topics to the problems. I also opted to fundamentally restructure the entire text into eleven consecutive *parts* with each part consisting of a series of focused lectures and ending with bibliographic comments, problems, and computer projects. I believe this restructuring into a sequence of lectures will provide readers and instructors with more flexibility in designing and managing their courses. I also collected most background material on random variables and linear algebra into three chapters at the beginning of the book. Students and readers have found this material of independent interest in its own right. At the same time, I decided to maintain the same general style and features of the earlier publication in terms of presentation and exposition, motivation, problems, computer projects, summary, and bibliographic notes. These features have been well received by our readers.

AREA OF STUDY

Adaptive filtering is a topic of immense practical relevance and deep theoretical challenges that persist even to this date. There are several notable texts on the subject that describe many of the features that have marveled students and researchers over the years. In this textbook, we choose to step back and to take a broad look at the field. In so doing, we feel that we are able to bring forth, to the benefit of the reader, extensive commonalities that exist among different classes of adaptive algorithms and even among different filtering theories. We are also able to provide a uniform treatment of the subject in a manner that addresses some existing limitations, provides additional insights, and allows for extensions of current theory.

We do not have any illusions about the difficulties that arise in any attempt at understanding adaptive filters more fully. This is because adaptive filters are, by design, time-variant, nonlinear, and stochastic systems. Any one of these qualifications alone would have resulted in a formidable system to study. Put them together and you face an almost impossible task. It is no wonder then that current practice tends to study different adaptive schemes separately, with techniques and assumptions that are usually more suitable for one adaptation form over another. It is also no surprise that most treatments of adaptive filters, including the one adopted in this textbook, need to rely on some simplifying assumptions in order to make filter analysis and design a more tractable objective.

Still, in our view, three desirable features of any study of adaptive filters would be (1) to attempt to keep the number of simplifying assumptions to a minimum, (2) to delay their use until necessary, and (3) to apply similar assumptions uniformly across different classes of adaptive algorithms. This last feature enables us to evaluate and compare the performance of adaptive schemes under similar assumptions on the data, while delaying the use of assumptions enables us to extract the most information possible about actual filter performance. In our discussions in this book we pay particular attention to these three features throughout the presentation.

In addition, we share the conviction that a thorough understanding of the performance and limitations of adaptive filters requires a solid grasp of the fundamentals of least-mean-squares estimation theory. These fundamentals help the designer understand what it is that an adaptive filter is trying to accomplish and how well it performs in this regard. For this reason, Parts I (*Optimal Estimation*) and II (*Linear Estimation*) of the book are designed to provide the reader with a self-contained and easy-to-follow exposition of estimation theory, with a focus on topics that are relevant to the subject matter of the book. In these initial parts, special emphasis is placed on geometric interpretations of several fundamental results. The reader is advised to pay close attention to these interpretations since it will become clear, time and again, that cumbersome algebraic manipulations can often be simplified by recourse to geometric constructions. These constructions not only provide a more lasting appreciation for the results of the book, but they also expose the reader to powerful tools that can be useful in other contexts as well, other than adaptive filtering and estimation theory.

The reader is further advised to master the convenience of the vector notation, which is used extensively throughout this book. Besides allowing a compact exposition of ideas and a compact representation of results, the vector notation also allows us to exploit to great effect several important results from linear algebra and matrix theory and to capture, in elegant ways, many revealing characteristics of adaptive filters. We cannot emphasize strongly enough the importance of linear algebraic and matrix tools in our presentation, as well as the elegance that they bring to the subject. The combined power of the geometric point of view and the vector notation is perhaps best exemplified by our detailed treatment later in this book of least-squares theory and its algorithmic variants. Of course, the reader is exposed to geometric and vector formulations in the early chapters of the book.

STRUCTURE OF THE BOOK

The book is divided into eleven core parts, in addition to a leading part on *Background Material* and a trailing part on *References and Indices*. Table P.1 lists the various parts. Each of the core parts, numbered I through XI, consists of four distinctive elements in the following order: (i) a series of lectures where the concepts are introduced, (ii) a summary of all lectures combined, (iii) bibliographic commentary, and (iv) problems and computer projects.

Lectures and Concepts. In the early parts of the book, each concept is motivated from first principles; starting from the obvious and ending with the more advanced. We follow this route of presentation until the reader develops enough maturity in the field. As the book progresses, we expect the reader to become more sophisticated and, therefore, we cut back on the "obvious."

Summaries. For ease of reference, at the end of each part, we collect a summary of the key concepts and results introduced in the respective lectures.

Bibliographic Commentaries. In the remarks at the end of each part we provide a wealth of references on the main contributors to the results discussed in the respective lectures. Rather than scatter references throughout the lectures, we find it useful to collect all references at the end of the part in the form of a narrative. We believe that this way of presentation gives the reader a more focused perspective on how the references and the contributions relate to each other both in time and context.

Problems. The book contains a significant number of problems, some more challenging than others and some more applied than others. The problems should be viewed as an *integral* part of the text, especially since additional results appear in them. It is for this reason, and also for the benefit of the reader, that we have chosen to formulate and design most problems in a guided manner. Usually, and especially in the more challenging cases, a problem starts by stating its objective followed by a sequence of guided steps until the final answer is attained. In most cases, the answer to each step appears stated in the body of the problem. In this way, a reader would know what the answer should be, even if the reader fails to solve the problem. Thus rather than ask the reader to "find an expression for x,", we would generally ask instead to "show that x is given by $x = \ldots$" and then give the expression for x.

All instructors can request copies of a free solutions manual from the publisher.

Moreover, several problems in the book have been designed to introduce readers to useful topics from related fields, such as multi-antenna receivers, cyclic-prefixing, maximal ratio combining, OFDM receivers, and so forth. Students are usually surprised to learn how classical concepts and ideas form the underpinnings of seemingly advanced techniques.

Computer Projects. We have included several computer projects (see the listing in Table P.2) to show students, and also practitioners, how the results developed in the book can be useful in situations of practical interest (e.g., linear equalization, decision feedback equalization, channel estimation, beamforming, tracking fading channels, line echo cancellation, acoustic echo cancellation, active noise control, OFDM receivers). In designing these projects, we have made an effort at choosing topics that are relevant to practitioners. We have also made an effort at illustrating to students how a solid theoretical understanding can guide them in challenging situations. MATLAB[1] programs are available for solving all computer projects in the book, in addition to a solutions manual. The programs are offered without any guarantees. While we have found them to be effective for the instructional purposes of this textbook, the programs are not intended to be examples of full-blown or optimized designs; practitioners should use them at their own risk. For example, in order to keep the codes at a level that is easy to understand by students, we have often decided to sacrifice performance in lieu of simplicity.

MATLAB programs that solve all computer projects in the book, in addition to a solutions manual for the projects with extensive commentary and typical performance plots, can be downloaded for free by all readers (including students and instructors) from either the publisher's website or the author's website.

Background Material. We provide three self-contained chapters that explain all the required background material on random variables and linear algebra for the purposes of this book. Actually, after progressing sufficiently enough in the book, students will be able to master many useful concepts from linear algebra and matrix theory, in addition to adaptive filtering.

COVERAGE AND TOPICS

The material in the book can be categorized into five broad areas of study (A through E), as listed in Table P.3. Area A covers the fundamentals of least-mean-squares estimation theory with several application examples. Areas B and C deal mainly with LMS-type adaptive filters, while areas D and E deal with least-squares-type adaptive filters. If an instructor wishes to focus mostly on LMS-type filters, then the instructor can do so by covering only material from within areas B and C. Even in this case, students will still be exposed to the recursive-least-squares (RLS) algorithm and its performance results from the discussions in Chapter 14 and Area C. However, for a more-in-depth treatment of RLS and its many variants, instructors will need to select chapters from within Area D as well.

DEPENDENCIES AMONG THE CORE PARTS

Figure P.1 illustrates the dependencies among the eleven core parts in the book. In the figure, the material in a part that is at the receiving end of an arrow requires some (but not necessarily all) of the material from the part at the origin of the arrow. A dashed arrow indicates that the dependency between the respective parts is weak and, if desired, the parts can be covered independently of each other. For example, in order to cover Part III (*Stochastic Gradient Methods*), the instructor would need to cover Part II (*Linear Estimation*). The material in Part I (*Optimal Estimation*) is not necessary for Part II (*Linear Estimation*) but it is useful for a better understanding of it. Figure P.1 can be

[1] MATLAB is a registered trademark of the MathWorks Inc., 24 Prime Park Way, Natick, MA 01760-1500, http://www.mathworks.com.

TABLE P.1 A breakdown of the book structure into eleven core parts.

Parts	Chapters
Background Material	A. Random Variables B. Linear Algebra C. Complex Gradients
I. Optimal Estimation	1. Scalar-Valued Data 2. Vector-Valued Data
II. Linear Estimation	3. Normal Equations 4. Orthogonality Principle 5. Linear Models 6. Constrained Estimation 7. Kalman Filter
III. Stochastic Gradient Methods	8. Steepest-Descent Technique 9. Transient Behavior 10. LMS Algorithm 11. Normalized LMS Algorithm 12. Other LMS-Type Algorithms 13. Affine Projection Algorithm 14. RLS Algorithm
IV. Mean-Square Performance	15. Energy Conservation 16. Performance of LMS 17. Performance of NLMS 18. Performance of Sign-Error LMS 19. Performance of RLS and Other Filters 20. Nonstationary Environments 21. Tracking Performance
V. Transient Performance	22. Weighted Energy Conservation 23. LMS with Gaussian Regressors 24. LMS with Non-Gaussian Regressors 25. Data-Normalized Filters
VI. Block Adaptive Filters	26. Transform-Domain Adaptive Filters 27. Efficient Block Convolution 28. Block and Subband Adaptive Filters
VII. Least-Squares Methods	29. Least-Squares Criterion 30. Recursive Least-Squares 31. Kalman Filtering and RLS 32. Order and Time-Update Relations
VIII. Array Algorithms	33. Norm and Angle Preservation 34. Unitary Transformations 35. QR and Inverse QR Algorithms
IX. Fast RLS Algorithms	36. Hyperbolic Rotations 37. Fast Array Algorithm 38. Regularized Prediction Problems 39. Fast Fixed-Order Filters
X. Lattice Filters	40. Three Basic Estimation Problems 41. Lattice Filter Algorithms 42. Error-Feedback Lattice Filters 43. Array Lattice Filters
XI. Robust Filters	44. Indefinite Least-Squares 45. Robust Adaptive Filters 46. Robustness Properties
References and Indices	

TABLE P.2 A listing of all computer projects in the book. MATLAB programs that solve these projects can be downloaded by all readers from the publisher's or author's websites, in addition to a solutions manual.

Computer project	Topic
I.1	Comparing optimal and suboptimal estimators
II.1	Linear equalization and decision devices
II.2	Beamforming.
II.3	Decision-feedback equalization
III.1	Constant-modulus criterion
III.2	Constant-modulus algorithm
III.3	Adaptive channel equalization
III.4	Blind adaptive equalization
IV.1	Line echo cancellation
IV.2	Tracking Rayleigh fading channels
V.1	Transient behavior of LMS
VI.1	Acoustic echo cancellation
VII.1	OFDM receiver
VII.2	Tracking Rayleigh fading channels
VIII.1	Performance of array implementations in finite precision
IX.1	Stability issues in fast least-squares
X.1	Performance of lattice filters in finite precision
XI.1	Active noise control

TABLE P.3 A breakdown of the book structure into five broad topic areas.

Category	Parts
A. Introduction and Foundations	Part I: Optimal Estimation Part II: Linear Estimation
B. Stochastic–Gradient Methods	Part III: Stochastic–Gradient Methods Part VI: Block Adaptive Filters
C. Performance Analyses	Part IV: Mean-Square Performance Part V: Transient Performance
D. Least-Squares Methods	Part VII: Least-Squares Methods Part VIII: Array Algorithms Part IX: Fast RLS Algorithms Part X: Lattice Filters
E. Indefinite Least–Squares	Part XI: Robust Filters

used by instructors to design different course sequences according to their needs and interests. For example, if the instructor is interested in covering only LMS-type adaptive filters and in studying their performance, then one possibility is to cover material from within Parts II, III, IV, and V.

AUDIENCE

The book is intended for a graduate-level course on adaptive filtering. Although it is beneficial that students have some familiarity with basic concepts from matrix theory, linear algebra, and random variables, the book includes three chapters on background material in these areas. The review is done in a motivated manner and is tailored to the needs of the presentation. From our experience,

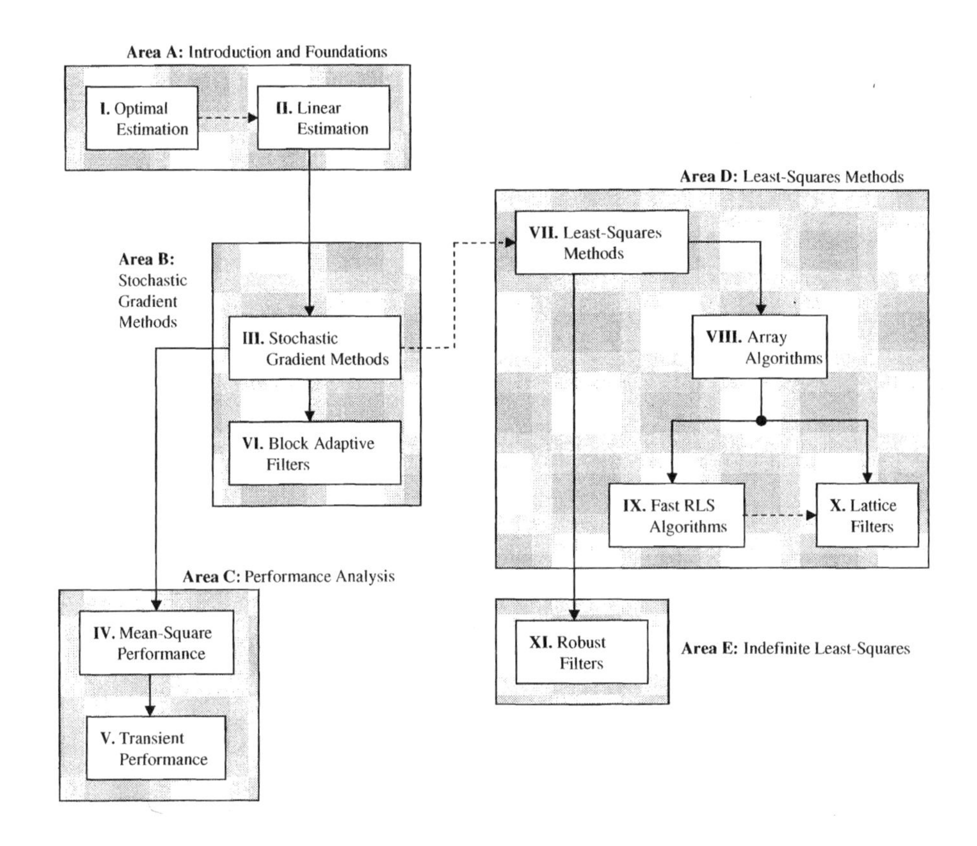

FIGURE P.1 Dependencies among the chapters. Instructors can design different course sequences in accordance with their needs and interests.

these reviews are sufficient for a thorough understanding of the discussions in the book. In addition, several of the problems reinforce the linear algebraic and matrix concepts, so much so that students will get valuable training in linear algebra and matrix theory, in addition to adaptive filtering, from reading (and understanding) this book.

The book is also intended to be a reference for researchers, which explains why we have chosen to include some advanced topics in a handful of places. As a result, the book contains ample material for instructors to design courses according to their interests. Clearly, we do not expect instructors to cover all the material in the book in a typical course offering; such an objective would be counterproductive. In our own teaching of the material, we instead *focus on some key sections and chapters and request that students complement the discussions by means of reading and problem solving*. As explained below, several key sections have been designed to convey the main concepts; while the remaining sections tend to include more advanced material and also illustrative examples. Once students understand the basic principles, you will be amazed at how well they can follow the other lectures on their own and even solve the pertinent problems.

To facilitate course planning, Table P.4 lists the key chapters or sections from the various core parts of the book for both lecturing and reading purposes. For example, Part V (*Transient Performance*) studies the transient behavior of a large family of adaptive filters in a uniform manner. The main idea is captured by the transient analysis of the LMS algorithm in Chapters 23 and 24; these chapters rely on the machinery developed in Chapter 22. Once students understand the framework as applied to LMS, they will be able to study the transient analysis of other filters on their own. This is one key advantage of adopting and emphasizing a uniform treatment of adaptive filter performance throughout our presentation. Similar remarks hold for the steady-state and tracking analyses of Part IV (*Mean-Square Performance*). It is sufficient to illustrate how the methodology applies to the special case of LMS, for example, by covering Chapters 15 and 16, as well as Sec. 21.1. Students would then do well in studying the extensions on their own if desired.

TABLE P.4 A suggested list of key chapters and sections.

Part	Key chapters for lecturing	Key chapters for reading
Part I	Chapters 1 and 2	
Part II	Chapters 3, 4	Sections 5.2, 5.3, 5.5, 6.3, 6.5
	Sections 5.1, 5.4, 6.1, 6.2, 6.4	Chapter 7
Part III	Chapters 8, 9, 10, 11, 14	Chapters 12, 13
Part IV	Chapters 15, 16, 20, 21.1	Chapters 17, 18, 19, 21.2–21.8
Part V	Chapters 22, 23, 24	Chapter 25
Part VI	Chapters 26, 27, 28	
Part VII	Chapters 29, 30	Chapters 31, 32
Part VIII	Chapters 33, 34, 35	
Part IX	Chapters 36, 37	Chapters 38, 39
Part X	Chapters 40, 41	Chapters 42, 43
Part XI	Chapters 44, 45	Chapter 46

SOME FEATURES OF OUR TREATMENT

There are some distinctive features in our treatment of adaptive filtering. Among other elements, experts will be able to notice the following contributions:

(a) We treat a large variety of adaptive algorithms.

(b) Parts IV and V study the mean-square performance of adaptive filters by resorting to energy-conservation arguments. While the performance of different adaptive filters is usually studied separately in the literature, the framework adopted in this book applies uniformly across different classes of adaptive filters. In addition, the same framework is used for steady-state analysis, transient analysis, tracking analysis, and robustness analysis (in Part XI).

(c) Part VI studies block adaptive filters, and the related class of subband adaptive filters, in a manner that clarifies the connections between these two families more directly than prior treatments. Our presentation also indicates how to move beyond DFT-based transforms and how to use other classes of orthogonal transforms for block adaptive filtering (as explained in Chapter 10 of Sayed (2003)).

(d) Parts VII-IX provide a detailed treatment of least-squares adaptive filters that is distinct from prevailing approaches in a handful of respects. First, we focus on regularized least-squares problems from the onset and take the regularization factor into account in all derivations. Second, we insist on deriving time- and order-update relations independent of any structure in the regression data (e.g., we do not require the regressors to arise from a tapped-delay-line implementation). In this way, one can pursue efficient least-squares filtering even for some non-FIR structures (as explained in Chapter 16 of Sayed (2003)). Third, we emphasize the role and benefits of array-based schemes. And, finally, we highlight the role of geometric constructions and the insights they bring into least-squares theory.

(e) Part XI develops the theory of robust adaptive filters by studying indefinite least-squares problems and by relying on energy arguments as well. In the process, the robustness and optimality properties of several adaptive filters are clarified. The presentation in this part is developed in a manner that parallels our treatment of least-squares problems in Chapters 29 and 30 so that readers can appreciate the similarities and distinctions between both theories (classical least-squares versus indefinite least-squares).

Notation

NOTATION

The objective of this book is not limited to providing a treatment of adaptive filters, but also to bring forth connections between adaptive filtering and other filtering theories. To do so, it becomes necessary to adopt a notation that captures with relative ease the similarities and connections that exist among different filtering theories. One main reason behind our choice of notation is that in our treatment of adaptive filters we need to distinguish between *four* types of variables:

random, scalar, vector, and matrix variables

While in many treatments of adaptive filters, no notational distinction is usually made between random quantities and their realizations, it is nevertheless important in our treatment of the subject to distinguish between the stochastic and nonstochastic domains. This distinction allows us, among other results, to describe with more transparency the connections that exist between filters that are derived in the stochastic domain (e.g., Kalman filters) and filters that are motivated by working with signal realizations (e.g., least-squares filters).

Once the reader becomes familiar with our convention, it will be straightforward to deduce the nature of a variable appearing in an equation by simply recalling the rules listed below. Whenever exceptions to these rules are used in the text, they will be obvious from the context. In some rare instances, rather than insist on following a strict convention, we may opt to relax our notation in a manner that best suits the discussion at hand. In general, however, the notation is consistent and motivated, and it should not be a hurdle to any attentive reader.

The following is a list of the notational conventions used in the textbook:

(a) We use **boldface letters** to denote random variables and normal font letters to denote their realizations (i.e., deterministic values), like $\boldsymbol{x}$ and x, respectively. In other words, we reserve the boldface notation to random quantities.

(b) We use CAPITAL LETTERS for matrices and small letters for *both* vectors and scalars, for example, X and x, respectively. In view of the first convention, $\boldsymbol{X}$ would denote a matrix with random entries, while X would denote a matrix realization (i.e., a matrix with deterministic values). Likewise, $\boldsymbol{x}$ would denote a vector with random entries, while x would denote a vector realization (or a vector with deterministic values). One notable *exception* to the capital letter notation is the use of such letters to refer to filter orders and to the total number of data points. For example, we usually write M to denote the number of taps in a filter and N to denote the total number of observations. These exceptions will be obvious from the context.

(c) We use *parentheses* to denote the time dependency of a scalar quantity and *subscripts* to denote the time dependency of a vector or matrix quantity; for example, if d is a scalar then $d(i)$ refers to its value at time (or iteration) i. On the other hand, if u is a vector, then u_i denotes its value at time (or iteration) i. Thus by looking at $d(i)$ and u_i, it is easy in this manner to distinguish between which one is a scalar and which one is a vector: $d(i)$ is a scalar and u_i is a vector. The time-dependency notation is useful to distinguish between scalars and vectors.

(d) We use the superscript T to denote transposition and the superscript $*$ to denote transposition with complex conjugation; for example, if

$$A = \begin{bmatrix} \alpha & \beta \\ \gamma & \lambda \end{bmatrix}$$

then

$$A^{\mathsf{T}} = \begin{bmatrix} \alpha & \gamma \\ \beta & \lambda \end{bmatrix} \quad \text{and} \quad A^* = \begin{bmatrix} \alpha^* & \gamma^* \\ \beta^* & \lambda^* \end{bmatrix}$$

(e) All vectors in our presentation are *column vectors* with one notable exception (starting from Chapter 10 onwards). We choose to represent the regression vector u_i as a *row* vector. Although this convention for the regressor is not essential, we have found it to be convenient for the following reasons:

(e.1) We shall frequently encounter in our discussions the inner product between the regressor u_i and some weight column vector w. In this way, we can simply write the inner product as $u_i w$ without the need for a transposition symbol.

(e.2) Usually, the regressor u_i arises as the state vector of a tapped delay line, as is shown in Fig. N.1 for a finite-impulse-response channel, say, of order M and weight vector w,

$$u_i = \begin{bmatrix} u(i) & u(i-1) & \dots & u(i-M+1) \end{bmatrix} \quad \text{(adopted convention)}$$

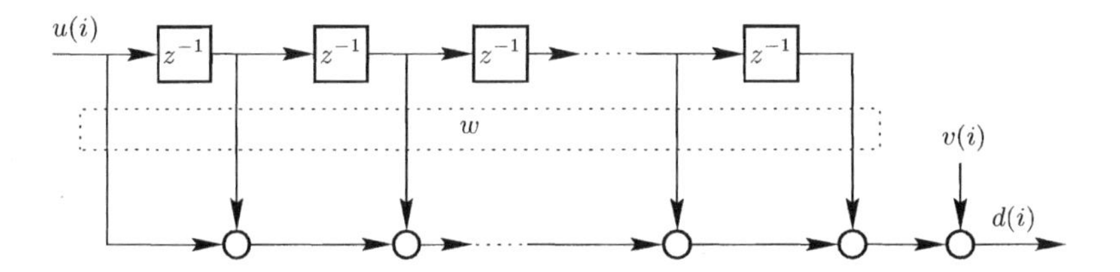

FIGURE N.1 A tapped delay line structure. The state of the channel is defined as the vector that contains the entries at the outputs of the delay elements, in addition to the input signal.

With u_i defined as a row vector, the noisy output of the channel will be described by

$$d(i) = u_i w + v(i)$$

Had we defined u_i as a column vector instead, say, as

$$u_i = \begin{bmatrix} u(i) \\ u(i-1) \\ \vdots \\ u(i-M+1) \end{bmatrix}$$

then the output of the channel would have been described by

$$d(i) = u_i^{\mathsf{T}} w + v(i)$$

with the transposition symbol used. Of course, the use of the transposition symbol is not of any consequence in its own right, especially when dealing with real data. However, we shall often deal with complex data (e.g., in channel equalization applications). In such situations, in addition to u_i^{T}, we shall also encounter terms involving u_i^* in our developments such as when defining the covariance matrix of the regression data. In this way, both transposition superscripts, $*$ and T, will appear in the discussions. We prefer to stick to a single symbol for transposition. By defining u_i as a row vector, we avoid such situations and use almost exclusively the complex-conjugation symbol $*$ throughout the presentation in the book.

(e.3) Alternatively, we could have defined the regressor as the following column vector:

$$u_i = \begin{bmatrix} u^*(i) \\ u^*(i-1) \\ \vdots \\ u^*(i-M+1) \end{bmatrix}$$

with conjugated entries $\{u^*(j)\}$ instead of the actual entries $\{u(j)\}$. In this case, the output of the channel would have been described by

$$d(i) = u_i^* w + v(i)$$

for both cases of real and complex data and, therefore, only terms involving u_i^* will appear in our discussions (and not both u_i^* and u_i^{T}). Nevertheless, if we define u_i as above, then the entries of u_i will not relate directly to the data stored in the channel (i.e., to the state of the channel) but to their conjugate values, and the designer will need to keep this fact in mind during simulation and algorithm coding.

(e.4) Finally, the notation $u_i w$ is convenient for MATLAB simulations of adaptive filters (e.g., in the computer projects used throughout the book). The designer would need to be more careful in coding $u_i^* w$ as opposed to $u_i w$.

For these reasons, we have opted to define the regressor u_i as a row vector. All other vectors in our treatment are column vectors. In our teaching of the subject over the years, we have found that students adapt very well to this convention. We have made every effort to make the notation consistent and coherent for the benefit of the reader. Please excuse imperfections.

SYMBOLS

We collect here, for ease of reference, a list of the main symbols used throughout the text.

$\mathbb{R}$	set of real numbers
$\mathbb{R}^+$	set of positive real numbers
$\cdot^T$	matrix transposition
$\cdot^*$	complex conjugation for scalars; Hermitian transposition for matrices
$\boldsymbol{x}$	boldface letter denotes a random scalar or vector variable
$\boldsymbol{X}$	boldface capital letter denotes a random matrix
x	letter in normal font denotes a vector or a scalar in Euclidean space
X	capital letter in normal font denotes a matrix in Euclidean space
$\mathsf{E}\,\boldsymbol{x}$	expected value of the random variable $\boldsymbol{x}$
$\boldsymbol{x} \perp \boldsymbol{y}$	orthogonal random variables $\boldsymbol{x}$ and $\boldsymbol{y}$ (i.e., $\mathsf{E}\,\boldsymbol{x}\boldsymbol{y}^* = 0$)
$x \perp y$	orthogonal vectors x and y (i.e., $x^*y = 0$)
$\|x\|^2$	x^*x for a column vector x; squared Euclidean norm of x
$\|x\|$	$\sqrt{x^*x}$ for a column vector x; Euclidean norm of x
$\|x\|_W^2$	x^*Wx for a column vector x and positive-definite matrix W
$\|A\|_2$	maximum singular value of A; also the spectral norm of A
$\|A\|_F$	Frobenius norm of A
$a \triangleq b$	quantity a is defined as b
$\mathsf{Re}(x)$	real part of x
$\mathsf{Im}(x)$	imaginary part of x
0	zero scalar, vector, or matrix
I_n	identify matrix of size $n \times n$
$\mathrm{col}\{a, b\}$	column vector with entries a and b
$\mathrm{vec}\{A\}$	column vector formed by stacking the columns of A
$\mathrm{vec}^{-1}\{a\}$	square matrix formed by unstacking its columns from a
$\mathrm{diag}\{A\}$	column vector with the diagonal entries of A
$\mathrm{diag}\{a\}$	diagonal matrix with entries read from the column a
$\mathrm{diag}\{a, b\}$	diagonal matrix with diagonal entries a and b
$a \oplus b$	same as $\mathsf{diag}\{a, b\}$
$A \otimes B$	Kronecker product of A and B
$A^\dagger$	pseudo-inverse of A
$P > 0$	positive-definite matrix P
$P \geq 0$	positive-semidefinite matrix P
$P^{1/2}$	square-root factor of a matrix $P \geq 0$, usually lower triangular
$A > B$	means that $A - B$ is positive-definite
$A \geq B$	means that $A - B$ is positive-semidefinite
$\det A$	determinant of the matrix A
$\mathsf{Tr}(A)$	trace of the matrix A
$O(M)$	constant multiple of M, or of the order of M
$\log a$	logarithm of a relative to base 10
$\ln a$	natural logarithm of a
$\diamond$	end of a theorem/lemma/proof/example/remark
Thm.	theorem
Def.	definition
Prob.	problem
Ex.	example
Fig.	figure

QR	QR factorization of a matrix
LDU	lower-diagonal-upper decomposition of a matrix
UDL	upper-diagonal-lower decomposition of a matrix
LDL*	LDU decomposition of a Hermitian matrix
UDU*	UDL decomposition of a Hermitian matrix
LTI	linear time-invariant
l.l.m.s.e.	linear least-mean-squares estimation/estimator
m.m.s.e.	minimum mean-square error
i.i.d.	independent and identically distributed
pdf	probability density function
a.e.	almost everywhere
AR	autoregressive model
MA	moving average model
ARMA	autoregressive moving average model
FIR	finite impulse response filter
IIR	infinite impulse response filter
SNR	signal to noise ratio
z^{-1}	unit-time delay operator
$X(z)$	bilateral z-transform of a scalar sequence $\{x(i)\}$
$X(e^{j\omega})$	discrete-time Fourier transform of $\{x(i)\}$
$\mathcal{X}(z)$	bilateral z-transform of a vector sequence $\{x(i)\}$
$\hat{\boldsymbol{x}}_{i\|j}$	l.l.m.s. estimator of $\boldsymbol{x}_i$ given observations $\{\boldsymbol{y}_k\}$ up to time j
$\hat{\boldsymbol{x}}_i$ or $\hat{\boldsymbol{x}}_{i\|i-1}$	l.l.m.s. estimator of $\boldsymbol{x}_i$ given observations $\{\boldsymbol{y}_k\}$ up to time $i-1$
$\tilde{\boldsymbol{x}}_{i\|j}$	estimation error $\boldsymbol{x}_i - \hat{\boldsymbol{x}}_{i\|j}$
$\tilde{\boldsymbol{x}}_i$ or $\tilde{\boldsymbol{x}}_{i\|i-1}$	estimation error $\boldsymbol{x}_i - \hat{\boldsymbol{x}}_i$
w_i	weight estimate at time or iteration i (a column vector)
$\tilde{w}_i$	weight error vector at time or iteration i (a column vector)
u_i	regressor at time or iteration i (a row vector)
$e(i)$	output estimation error at time or iteration i
$e_a(i)$	*a priori* estimation error at time or iteration i
$e_p(i)$	*a posteriori* estimation error at time or iteration i
R_u	covariance matrix of the regression data
σ_v^2	noise variance
$d(i)$	value of a *scalar* variable d at time or iteration i
d_i	value of a *vector* variable d at time or iteration i

Acknowledgments

I continue to be indebted to my students and associates who have deepened my understanding of adaptive filtering over the years. I wish to acknowledge M. Rupp, V. H. Nascimento, J. Mai, N. R. Yousef, R. Merched, T. Y. Al-Naffouri, H. Shin, A. Tarighat, C. G. Lopes, Q. Zou, and F. Cattivelli, all of whom have contributed directly or indirectly to the material in this book. I am also grateful to the many friends and colleagues who have provided me with valuable feedback on my earlier textbook, and whose comments continue to be reflected in this project. I wish to mention N. Al-Dhahir (University of Texas, Dallas), J. C. M. Bermudez (Federal University of Santa Catarina, Brazil), E. Eweda (Ajman University, United Arab Emirates), T. Kailath (Stanford University), D. G. Manolakis (MIT), D. Morgan (Lucent Technologies), V. H. Nascimento (University of São Paulo, Brazil), M. R. Petraglia (Federal University of Rio de Janeiro, Brazil), P. A. Regalia (The Catholic University of America), M. Rupp (Tech. Unisersitaet Wien, Austria), S. Theodoridis (University of Athens, Greece), and M. Verhaegen (Delft University, The Netherlands).

I am also thankful to the many colleagues whose kind words of praise on the earlier publication have encouraged me to pursue the current project. I wish to thank L. Cheded (King Fahd University of Petroleum and Minerals, Saudi Arabia), P. De Leon (New Mexico State University), M. Kaveh (University of Minnesota), L. Ljung (Linkoping University, Sweden), G. Maalouli (General Dynamics Corporation), S. K. Mitra (University of California at Santa Barbara), B. Ottersten (Royal Institute of Technology, Sweden), A. Petropulu (Drexel University), H. Sakai (Kyoto University, Japan), T. Soderstrom (Uppsala University, Sweden), P. Stoica (Uppsala University, Sweden), P. P. Vaidyanathan (California Institute of Technology), M. Vetterli (EPFL, Switzerland), and B. Wahlberg (Royal Institute of Technology, Sweden). I would like to thank the National Science Foundation for its support of the research that led to this textbook. I would also like to thank my publisher, George Telecki, from Wiley for his help with this project and Amy Hendrickson for her assistance with Latex and formatting.

My wife, Laila, and daughters, Faten and Samiya, have always provided me with their utmost support and encouragement. My parents, Soumaya and Hussein, have been overwhelming in their support. For all the sacrifices they have endured to guarantee their son's education, I dedicate this volume to them knowing very well that this tiny gift can never match their gift.

ALI H. SAYED

Westwood, California
December 2007

BACKGROUND MATERIAL

Random Variables

The presentation in the book relies on some basic concepts from probability theory and random variables. For the benefit of the reader, we shall motivate these concepts whenever needed as well as highlight their relevance in the estimation context. In this way, readers will be introduced to the necessary concepts in a gradual and motivated manner, and they will come to appreciate their significance away from unnecessary abstractions. In this initial chapter, we collect several concepts of general interest. These concepts complement well the material in future chapters and will be called upon at different stages of the discussion.

A.1 VARIANCE OF A RANDOM VARIABLE

We initiate our presentation with an intuitive explanation for what the variance of a random variable means. The explanation will help the reader appreciate the value of the least-mean-squares criterion, which is used extensively in later chapters.

Consider a scalar *real-valued* random variable $\boldsymbol{x}$ with mean value $\bar{x}$ and variance σ_x^2, i.e.,

$$\boxed{\bar{x} \triangleq \mathsf{E}\,\boldsymbol{x}, \qquad \sigma_x^2 \triangleq \mathsf{E}\,(\boldsymbol{x}-\bar{x})^2 = \mathsf{E}\,\boldsymbol{x}^2 - \bar{x}^2} \tag{A.1}$$

where the symbol E denotes the expectation operator. Observe that we are using boldface letters to denote random variables, which will be our convention in this book. When $\boldsymbol{x}$ has zero mean, its variance is simply $\sigma_x^2 = \mathsf{E}\,\boldsymbol{x}^2$. Intuitively, the variance of $\boldsymbol{x}$ defines an interval on the real axis around $\bar{x}$ where the values of $\boldsymbol{x}$ are most likely to occur:

1. A small σ_x^2 indicates that $\boldsymbol{x}$ is more likely to assume values that are close to its mean, $\bar{x}$.

2. A large σ_x^2 indicates that $\boldsymbol{x}$ can assume values over a wider interval around its mean.

For this reason, it is customary to regard the variance of a random variable as a measure of *uncertainty* about the value it can assume in a given experiment. A small variance indicates that we are more certain about what values to expect for $\boldsymbol{x}$ (namely, values that are close to its mean), while a large variance indicates that we are less certain about what values to expect. These two situations are illustrated in Figs. A.1 and A.2 for two different probability density functions.

Figure A.1 plots the probability density function (pdf) of a Gaussian random variable $\boldsymbol{x}$ for two different variances. In both cases, the mean of the random variable is fixed at $\bar{x} = 20$ while the variance is $\sigma_x^2 = 225$ in one case and $\sigma_x^2 = 4$ in the other. Recall that the pdf of a Gaussian random variable is defined in terms of $(\bar{x}, \sigma_x^2)$ by the expression

$$\boxed{f_{\boldsymbol{x}}(x) = \frac{1}{\sqrt{2\pi}\,\sigma_x}\, e^{-(x-\bar{x})^2/2\sigma_x^2}, \quad x \in (-\infty, \infty)} \tag{A.2}$$

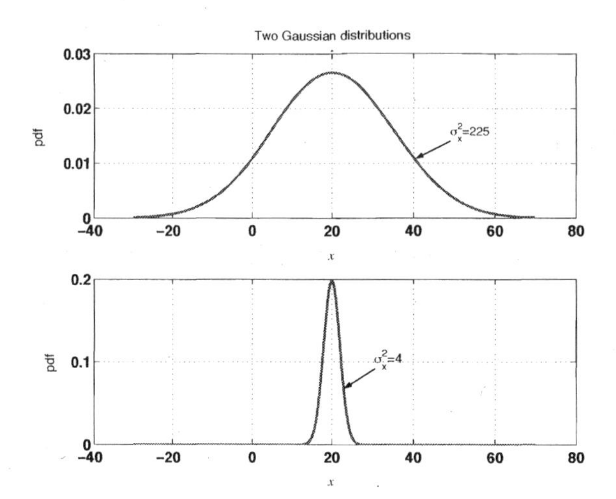

FIGURE A.1 The figure shows the plots of the probability density functions of a Gaussian random variable $\boldsymbol{x}$ with mean $\bar{x} = 20$, variance $\sigma_x^2 = 225$ in the top plot, and variance $\sigma_x^2 = 4$ in the bottom plot.

where σ_x is called the *standard deviation* of $\boldsymbol{x}$. Recall further that the pdf of a random variable is useful in several respects. In particular, it allows us to evaluate probabilities of events of the form

$$P(a \leq \boldsymbol{x} \leq b) \;=\; \int_a^b f_{\boldsymbol{x}}(x)\mathrm{d}x$$

i.e., the probability of $\boldsymbol{x}$ assuming values inside the interval $[a, b]$. From Fig. A.1 we observe that the smaller the variance of $\boldsymbol{x}$, the more concentrated its pdf is around its mean.

Figure A.2 provides similar plots for a random variable $\boldsymbol{x}$ with a Rayleigh distribution, namely, with a pdf given by

$$\boxed{f_{\boldsymbol{x}}(x) \;=\; \frac{x}{\alpha^2}\, e^{-x^2/2\alpha^2}, \qquad x \geq 0, \;\; \alpha > 0} \tag{A.3}$$

where α is a positive parameter that determines the mean and variance of $\boldsymbol{x}$ according to the expressions (see Prob. I.1):

$$\bar{x} = \alpha\sqrt{\frac{\pi}{2}}, \qquad \sigma_x^2 = \left(2 - \frac{\pi}{2}\right)\alpha^2 \tag{A.4}$$

Observe, in particular, and in contrast to the Gaussian case, that the mean and variance of a Rayleigh-distributed random variable cannot be chosen independently of each other since they are linked through the parameter α. In Fig. A.2, the top plot corresponds to $\bar{x} = 1$ and $\sigma_x^2 = 0.2732$, while the bottom plot corresponds to $\bar{x} = 3$ and $\sigma_x^2 = 2.4592$.

These remarks on the variance of a random variable can be further qualified by invoking a well-known result from probability theory known as Chebyshev's inequality — see Probs. I.2 and I.3. The result states that for a random variable $\boldsymbol{x}$ with mean $\bar{x}$ and variance σ_x^2, and for any given scalar $\delta > 0$, it holds that

$$\boxed{P(|\boldsymbol{x} - \bar{x}| \geq \delta) \;\leq\; \sigma_x^2/\delta^2} \tag{A.5}$$

That is, the probability that $\boldsymbol{x}$ assumes values outside the interval $(\bar{x} - \delta, \bar{x} + \delta)$ does not exceed σ_x^2/δ^2, with the bound being proportional to the variance of $\boldsymbol{x}$. Hence, for a fixed δ, the smaller the variance of $\boldsymbol{x}$ the smaller the probability that $\boldsymbol{x}$ will assume values outside the interval $(\bar{x} - \delta, \bar{x} + \delta)$. Choose, for instance, $\delta = 5\sigma_x$. Then (A.5) gives

$$P(|\boldsymbol{x} - \bar{x}| \geq 5\sigma_x) \;\leq\; 1/25 = 4\%$$

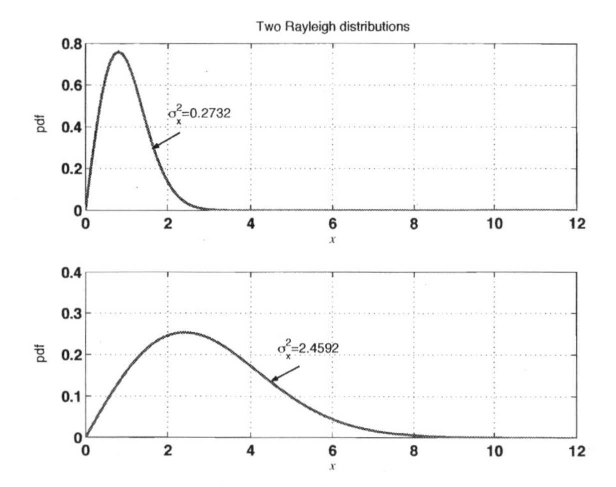

FIGURE A.2 The figure shows the plots of the probability density functions of a Rayleigh random variable $\boldsymbol{x}$ with mean $\bar{x} = 1$ and variance $\sigma_x^2 = 0.2732$ in the top plot, and mean $\bar{x} = 3$ and variance $\sigma_x^2 = 2.4592$ in the bottom plot.

In other words, there is at most 4% chance that $\boldsymbol{x}$ will assume values outside the interval $(\bar{x} - 5\sigma_x, \bar{x} + 5\sigma_x)$.

Actually, the bound that is provided by Chebyshev's inequality is generally not tight. Consider, for example, a zero-mean Gaussian random variable $\boldsymbol{x}$ with variance σ_x^2 and choose $\delta = 2\sigma_x$. Then, from Chebyshev's inequality (A.5) we would obtain

$$P(|\boldsymbol{x}| \geq 2\sigma_x) \;\leq\; 1/4 = 25\%$$

whereas direct evaluation of the integral

$$P(|\boldsymbol{x}| \geq 2\sigma_x) \;\stackrel{\Delta}{=}\; 1 - 2\left(\frac{1}{\sqrt{2\pi}\,\sigma_x}\int_0^{2\sigma_x} e^{-x^2/2\sigma_x^2}\,dx\right)$$

yields

$$P(|\boldsymbol{x}| \geq 2\sigma_x) \;\approx\; 4.56\%$$

Remark A.1 (Zero-variance random variables) One useful consequence of Chebyshev's inequality is the following. It allows us to interpret a zero-variance random variable as one that is equal to its mean with probability one. That is,

$$\boxed{\sigma_x^2 = 0 \;\Longrightarrow\; \boldsymbol{x} = \bar{x} \;\text{ with probability one}}$$

This is because, for any small $\delta > 0$, we obtain from (A.5) that

$$P(|\boldsymbol{x} - \bar{x}| \geq \delta) \;\leq\; 0$$

But since the probability of any event is necessarily a nonnegative number, we conclude that $P(|\boldsymbol{x}-\bar{x}| \geq \delta) = 0$, for any $\delta > 0$, so that $\boldsymbol{x} = \bar{x}$ with probability one. We shall call upon this result on several occasions (see, e.g., the proof of Thm. 1.2).

A.2 DEPENDENT RANDOM VARIABLES

In estimation theory, it is generally the case that information about one variable is extracted from observations of another variable. The relevance of the information extracted from the observations is a function of how closely related the two random variables are, as measured by relations of dependency or correlation between them.

The dependency between two real-valued random variables $\{\boldsymbol{x}, \boldsymbol{y}\}$ is characterized by their *joint* probability density function (pdf). Thus, let $f_{\boldsymbol{x},\boldsymbol{y}}(x, y)$ denote the joint (continuous) pdf of $\boldsymbol{x}$ and $\boldsymbol{y}$; this function allows us to evaluate probabilities of events of the form:

$$P(a \leq \boldsymbol{x} \leq b, c \leq \boldsymbol{y} \leq d) = \int_c^d \int_a^b f_{\boldsymbol{x},\boldsymbol{y}}(x, y) \mathrm{d}x \mathrm{d}y$$

namely, the probability that $\boldsymbol{x}$ and $\boldsymbol{y}$ assume values inside the intervals $[a, b]$ and $[c, d]$, respectively. Let also $f_{\boldsymbol{x}|\boldsymbol{y}}(x|y)$ denote the *conditional* pdf of $\boldsymbol{x}$ given $\boldsymbol{y}$; this function allows us to evaluate probabilities of events of the form

$$P(a \leq \boldsymbol{x} \leq b \mid \boldsymbol{y} = y) = \int_a^b f_{\boldsymbol{x}|\boldsymbol{y}}(x|y) \mathrm{d}x$$

namely, the probability that $\boldsymbol{x}$ assumes values inside the interval $[a, b]$ given that $\boldsymbol{y}$ is fixed at the value y. It is known that the joint and conditional pdfs of two random variables are related via Bayes' rule, which states that

$$\boxed{f_{\boldsymbol{x},\boldsymbol{y}}(x, y) \;=\; f_{\boldsymbol{y}}(y)\, f_{\boldsymbol{x}|\boldsymbol{y}}(x|y) \;=\; f_{\boldsymbol{x}}(x)\, f_{\boldsymbol{y}|\boldsymbol{x}}(y|x)} \tag{A.6}$$

in terms of the probability density functions of the individual random variables $\boldsymbol{x}$ and $\boldsymbol{y}$.

The variables $\{\boldsymbol{x}, \boldsymbol{y}\}$ are said to be *independent* if

$$f_{\boldsymbol{x}|\boldsymbol{y}}(x|y) = f_{\boldsymbol{x}}(x) \quad \text{and} \quad f_{\boldsymbol{y}|\boldsymbol{x}}(y|x) = f_{\boldsymbol{y}}(y)$$

in which case the pdfs of $\boldsymbol{x}$ and $\boldsymbol{y}$ are not modified by conditioning on $\boldsymbol{y}$ and $\boldsymbol{x}$, respectively. Otherwise, the variables are said to be *dependent*. In particular, when the variables are independent, it follows that $\mathsf{E}\,\boldsymbol{x}\boldsymbol{y} = \mathsf{E}\,\boldsymbol{x}\mathsf{E}\,\boldsymbol{y}$. It also follows that independent random variables are uncorrelated, meaning that their cross-correlation is zero as can be verified from the definition of cross-correlation:

$$\begin{aligned}
\sigma_{xy} &\overset{\Delta}{=} \mathsf{E}\,(\boldsymbol{x} - \bar{x})(\boldsymbol{y} - \bar{y}) \\
&= \mathsf{E}\,\boldsymbol{x}\boldsymbol{y} - \bar{x}\bar{y} \\
&= \mathsf{E}\,\boldsymbol{x}\;\mathsf{E}\,\boldsymbol{y} - \bar{x}\bar{y} \\
&= 0
\end{aligned}$$

The converse statement is not true: uncorrelated random variables can be dependent. Consider the following example. Let $\boldsymbol{\theta}$ be a random variable that is uniformly distributed over the interval $[0, 2\pi]$. Define the zero-mean random variables $\boldsymbol{x} = \cos\boldsymbol{\theta}$ and $\boldsymbol{y} = \sin\boldsymbol{\theta}$. Then $\boldsymbol{x}^2 + \boldsymbol{y}^2 = 1$ so that $\boldsymbol{x}$ and $\boldsymbol{y}$ are dependent. However, $\mathsf{E}\,\boldsymbol{x}\boldsymbol{y} = \mathsf{E}\,\cos\boldsymbol{\theta}\sin\boldsymbol{\theta} = 0.5\mathsf{E}\,\sin 2\boldsymbol{\theta} = 0$, so that $\boldsymbol{x}$ and $\boldsymbol{y}$ are uncorrelated.

A.3 COMPLEX-VALUED RANDOM VARIABLES

Although we have focused so far on real-valued random variables, we shall often encounter applications that deal with random variables that assume complex values. Accordingly, a complex-valued random variable is defined as one whose real and imaginary parts are *real*-valued random variables, say,

$$\boxed{\boldsymbol{x} = \boldsymbol{x}_{\mathsf{r}} + j\boldsymbol{x}_{\mathsf{i}}, \quad j \overset{\Delta}{=} \sqrt{-1}}$$

where $\boldsymbol{x}_{\mathsf{r}}$ and $\boldsymbol{x}_{\mathsf{i}}$ denote the real and imaginary parts of $\boldsymbol{x}$. Therefore, the pdf of a complex-valued random variable $\boldsymbol{x}$ can be characterized in terms of the joint pdf, $f_{\boldsymbol{x}_{\mathsf{r}},\boldsymbol{x}_{\mathsf{i}}}(\cdot,\cdot)$, of its real and imaginary parts. This means that we can regard a complex random variable as a function of two real random variables. The mean of $\boldsymbol{x}$ is

$$\begin{aligned} \mathsf{E}\boldsymbol{x} &\stackrel{\Delta}{=} \mathsf{E}\boldsymbol{x}_r + j\mathsf{E}\boldsymbol{x}_i \\ &= \bar{x}_r + j\bar{x}_i \end{aligned}$$

in terms of the means of its real and imaginary parts. The variance of $\boldsymbol{x}$, on the other hand, is defined as

$$\boxed{\sigma_x^2 \stackrel{\Delta}{=} \mathsf{E}(\boldsymbol{x}-\bar{x})(\boldsymbol{x}-\bar{x})^* = \mathsf{E}|\boldsymbol{x}-\bar{x}|^2} \tag{A.7}$$

where the symbol $*$ denotes complex conjugation.

Comparing with the definition (A.1) in the real case, we see that the above definition is different because of the use of conjugation (in the real case, the conjugate of $(\boldsymbol{x}-\bar{x})$ is $(\boldsymbol{x}-\bar{x})$ itself and the above definition reduces to (A.1)). The use of the conjugate term in (A.7) is necessary in order to guarantee that σ_x^2 will be a nonnegative real number. In particular, it is immediate to verify from (A.7) that

$$\sigma_x^2 = \sigma_{x_r}^2 + \sigma_{x_i}^2$$

in terms of the individual variances of $\boldsymbol{x}_r$ and $\boldsymbol{x}_i$.

We shall say that two complex-valued random variables $\boldsymbol{x}$ and $\boldsymbol{y}$ are uncorrelated if, and only if, their cross-correlation is zero, i.e.,

$$\sigma_{xy} \stackrel{\Delta}{=} \mathsf{E}(\boldsymbol{x}-\bar{x})(\boldsymbol{y}-\bar{y})^* = 0$$

On the other hand, we shall say that they are *orthogonal* if, and only if,

$$\mathsf{E}\boldsymbol{x}\boldsymbol{y}^* = 0$$

It can be immediately verified that the concepts of orthogonality and uncorrelatedness coincide if at least one of the random variables has zero mean.

Example A.1 (QPSK constellation)

Consider a signal $\boldsymbol{x}$ that is chosen uniformly from a QPSK constellation, i.e., $\boldsymbol{x}$ assumes any of the values $\pm\sqrt{2}/2 \pm j\sqrt{2}/2$ with probability $1/4$ (see Fig. A.3). Clearly, $\boldsymbol{x}$ is a complex-valued random variable; its mean and variance are easily found to be $\bar{x} = 0$ and $\sigma_x^2 = 1$.

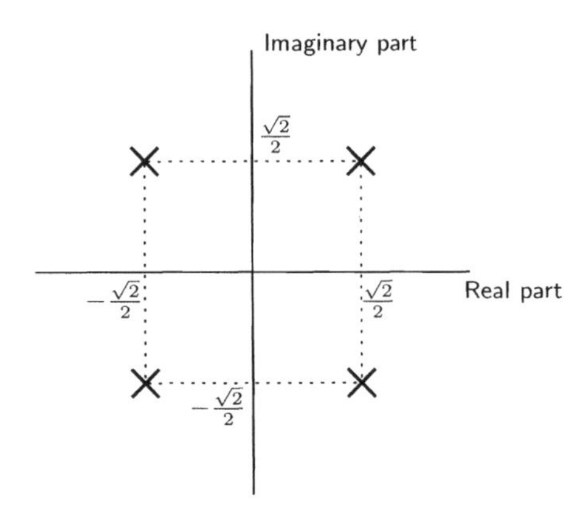

FIGURE A.3 A QPSK constellation.

◇

A.4 VECTOR-VALUED RANDOM VARIABLES

A vector-valued random variable is a collection (in column or row vector forms) of random variables. The individual entries can be real or complex-valued themselves. For example, if $\boldsymbol{x} = \text{col}\{\boldsymbol{x}(0), \boldsymbol{x}(1)\}$ is a random vector with entries $\boldsymbol{x}(0)$ and $\boldsymbol{x}(1)$, then we shall define its mean as the vector of individual means,

$$\bar{x} = \mathsf{E}\boldsymbol{x} = \begin{bmatrix} \bar{x}(0) \\ \bar{x}(1) \end{bmatrix} \stackrel{\Delta}{=} \begin{bmatrix} \mathsf{E}\boldsymbol{x}(0) \\ \mathsf{E}\boldsymbol{x}(1) \end{bmatrix}$$

and its covariance *matrix* as

$$\boxed{R_x \stackrel{\Delta}{=} \mathsf{E}(\boldsymbol{x} - \bar{x})(\boldsymbol{x} - \bar{x})^*} \tag{A.8}$$

where the symbol $*$ now denotes complex-conjugate transposition (i.e., we transpose the vector and then replace each of its entries by the corresponding conjugate value). Note that we are using parentheses to index the scalar entries of a vector, e.g., $\boldsymbol{x}(k)$ denotes the kth entry of $\boldsymbol{x}$. Moreover, if $\boldsymbol{x}$ were a *row* random vector, rather than a *column* random vector, then its covariance matrix would instead be defined as

$$R_x \stackrel{\Delta}{=} \mathsf{E}(\boldsymbol{x} - \bar{x})^*(\boldsymbol{x} - \bar{x})$$

with the conjugate term coming first. This is because, in this case, it is the product $(\boldsymbol{x} - \bar{x})^*(\boldsymbol{x} - \bar{x})$ that yields a matrix, while the product $(\boldsymbol{x} - \bar{x})(\boldsymbol{x} - \bar{x})^*$ would be a scalar.

For the two-element column vector $\boldsymbol{x} = \text{col}\{\boldsymbol{x}(0), \boldsymbol{x}(1)\}$ we obtain

$$R_x = \begin{bmatrix} \mathsf{E}\,|\boldsymbol{x}(0) - \bar{x}(0)|^2 & \mathsf{E}\,[\boldsymbol{x}(0) - \bar{x}(0)][\boldsymbol{x}(1) - \bar{x}(1)]^* \\ \mathsf{E}\,[\boldsymbol{x}(1) - \bar{x}(1)][\boldsymbol{x}(0) - \bar{x}(0)]^* & \mathsf{E}\,|\boldsymbol{x}(1) - \bar{x}(1)|^2 \end{bmatrix}$$

with the individual variances of the variables $\{\boldsymbol{x}(0), \boldsymbol{x}(1)\}$ appearing on the diagonal and the cross-correlations between them appearing on the off-diagonal entries. In the zero-mean case, the definition of R_x, and the above expression, simplify to

$$\boxed{R_x \stackrel{\Delta}{=} \mathsf{E}\boldsymbol{x}\boldsymbol{x}^*}$$

and

$$R_x = \begin{bmatrix} \mathsf{E}\,|\boldsymbol{x}(0)|^2 & \mathsf{E}\boldsymbol{x}(0)\boldsymbol{x}^*(1) \\ \mathsf{E}\boldsymbol{x}(1)\boldsymbol{x}^*(0) & \mathsf{E}\,|\boldsymbol{x}(1)|^2 \end{bmatrix}$$

It should be noted that the covariance matrix R_x is Hermitian, i.e., it satisfies

$$R_x = R_x^*$$

Moreover, R_x is a nonnegative-definite matrix, written as

$$R_x \geq 0$$

By definition, a Hermitian matrix R is said to be nonnegative definite if, and only if, $a^*Ra \geq 0$ for any column vector a (real or complex-valued). In order to verify that $R_x \geq 0$, we introduce the scalar-valued random variable $\boldsymbol{y} = a^*(\boldsymbol{x} - \bar{x})$, where a is an arbitrary column vector. Then $\boldsymbol{y}$ has zero mean and

$$\sigma_y^2 = \mathsf{E}\,|\boldsymbol{y}|^2 = a^*R_xa$$

But since the variance of any scalar-valued random variable is always nonnegative, we conclude that $a^*R_xa \geq 0$ for any a. This means that R_x is nonnegative definite, as claimed.

For real-valued data, the symbol $*$ is replaced by the transposition symbol T, and R_x is defined as

$$\boxed{R_x \stackrel{\Delta}{=} \mathsf{E}(\boldsymbol{x} - \bar{x})(\boldsymbol{x} - \bar{x})^{\mathsf{T}}}$$

In this case, the covariance matrix R_x therefore becomes symmetric rather than *Hermitian*. In other words, it now satisfies

$$R_x = R_x^{\mathsf{T}}$$

Moreover, R_x continues to be nonnegative-definite, which now means that $a^{\mathsf{T}} R_x a \geq 0$ for any real-valued column vector a.

Example A.2 (Transmissions over a noisy channel)

Consider the setting of Fig. A.4. A sequence of independent and identically distributed (i.i.d.) symbols $\{\boldsymbol{s}(i)\}$ is transmitted over an initially relaxed FIR channel with transfer function $C(z) = 1 + 0.5z^{-1}$, where z^{-1} denotes the unit-time delay in the z-transform domain. Each symbol is either $+1$ with probability p or -1 with probability $1 - p$. The output of the channel is corrupted by zero-mean additive white Gaussian noise $\boldsymbol{v}(i)$ of unit variance. The whiteness assumption means that $\mathsf{E}\,\boldsymbol{v}(i)\boldsymbol{v}^*(j) = \delta_{ij}$, where δ_{ij} denotes the Kronecker delta function that is equal to unity when $i = j$ and zero otherwise. In other words, the noise terms are uncorrelated with each other. The noise $\boldsymbol{v}(i)$ and the symbols $\boldsymbol{s}(j)$ are also assumed to be independent of each other for all i and j.

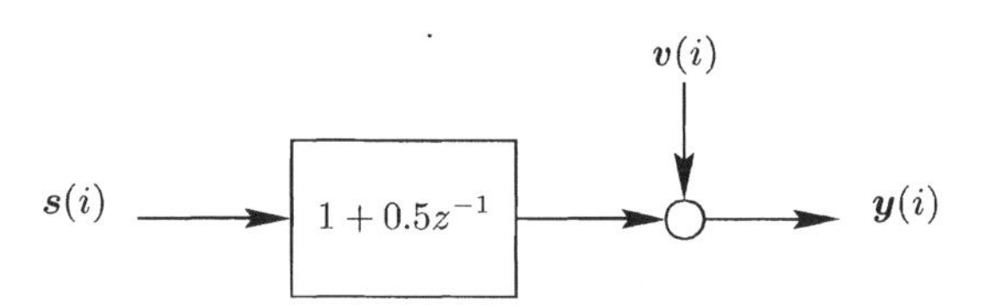

FIGURE A.4 Transmissions over an additive white Gaussian FIR channel.

The output of the channel at time i is given by $\boldsymbol{y}(i) = \boldsymbol{s}(i) + 0.5\boldsymbol{s}(i-1) + \boldsymbol{v}(i)$. Assume we collect $N + 1$ measurements, $\{\boldsymbol{y}(i), i = 0, 1, \ldots, N\}$, into a column vector $\boldsymbol{y}$, and then pose the problem of recovering the transmitted symbols $\{\boldsymbol{s}(i), i = 0, 1, \ldots, N\}$ over the same interval of time. If we collect the symbols $\{\boldsymbol{s}(i)\}$ into a column vector $\boldsymbol{x}$, say $\boldsymbol{x} = \text{col}\{\boldsymbol{s}(0), \boldsymbol{s}(1), \ldots, \boldsymbol{s}(N)\}$, then we are faced with the problem of estimating a vector $\boldsymbol{x}$ from a vector $\boldsymbol{y}$. Here the entries of $\boldsymbol{x}$ and $\boldsymbol{y}$ are all real-valued. If the symbols $\boldsymbol{s}(i)$ were instead chosen from a QPSK constellation, then both $\boldsymbol{x}$ and $\boldsymbol{y}$ will be complex-valued.

◇

A.5 GAUSSIAN RANDOM VECTORS

Gaussian random variables play an important role in many situations, especially when we deal with the sum of a large number of random variables. In this case, a fundamental result in probability theory, known as the *central limit theorem*, states that under conditions often reasonable in applications, the probability density function (pdf) of the sum of independent random variables approaches that of a Gaussian distribution. Specifically, if $\{\boldsymbol{x}(i), i = 1, 2, \ldots, N\}$ are independent real-valued Gaussian random variables with means $\{\bar{x}(i)\}$ and variances $\{\sigma_x^2(i)\}$ each, then the pdf of the normalized variable

$$\boldsymbol{y} \;\triangleq\; \frac{\sum_{i=1}^{N}[\boldsymbol{x}(i) - \bar{x}(i)]}{\sqrt{\sum_{i=1}^{N}\sigma_x^2(i)}}$$

approaches that of a Gaussian distribution with zero mean and unit variance, i.e.,

$$f_{\boldsymbol{y}}(y) = \frac{1}{\sqrt{2\pi}}e^{-y^2/2} \quad \text{as } N \longrightarrow \infty$$

or, equivalently,

$$\lim_{N\to\infty} P(\boldsymbol{y} \leq a) = \frac{1}{\sqrt{2\pi}} \int_{-\infty}^{a} e^{-y^2/2} dy$$

It is for this reason that the term "Gaussian noise" generally refers to the combined effect of many independent disturbances.

In this section we describe the general form of the pdf of a vector Gaussian random variable. However, as the discussion will show, we need to distinguish between two cases depending on whether the random variable is real or complex. In the complex case, the random variable will need to satisfy a certain *circularity* assumption in order for the given form of the pdf to be valid.

Real-Valued Gaussian Random Variables

We start with the real case. Thus, consider a $p \times 1$ random vector $\boldsymbol{x}$ with mean $\bar{x}$ and a nonsingular covariance matrix

$$R_x = \mathsf{E}\,(\boldsymbol{x} - \bar{x})(\boldsymbol{x} - \bar{x})^{\mathsf{T}}$$

We say that $\boldsymbol{x}$ has a Gaussian distribution if its pdf has the form

$$f_{\boldsymbol{x}}(x) = \frac{1}{\sqrt{(2\pi)^p}} \frac{1}{\sqrt{\det R_x}} \exp\left\{-\tfrac{1}{2}(x - \bar{x})^{\mathsf{T}} R_x^{-1} (x - \bar{x})\right\} \qquad \text{(A.9)}$$

in terms of the determinant of R_x. Of course, when $p = 1$, the above expression reduces to the pdf considered in the text in (A.2) with R_x replaced by σ_x^2.

Now consider a second $q \times 1$ Gaussian random vector $\boldsymbol{y}$ with mean $\bar{y}$ and covariance matrix

$$R_y = \mathsf{E}\,(\boldsymbol{y} - \bar{y})(\boldsymbol{y} - \bar{y})^{\mathsf{T}}$$

so that its pdf is given by

$$f_{\boldsymbol{y}}(y) = \frac{1}{\sqrt{(2\pi)^q}} \frac{1}{\sqrt{\det R_y}} \exp\left\{-\tfrac{1}{2}(y - \bar{y})^{\mathsf{T}} R_y^{-1} (y - \bar{y})\right\}$$

Let R_{xy} denote the cross-covariance matrix between $\boldsymbol{x}$ and $\boldsymbol{y}$, i.e.,

$$R_{xy} = \mathsf{E}\,(\boldsymbol{x} - \bar{x})(\boldsymbol{y} - \bar{y})^{\mathsf{T}}$$

We then say that the random variables $\{\boldsymbol{x}, \boldsymbol{y}\}$ have a joint Gaussian distribution if their joint pdf has the form

$$f_{\boldsymbol{x},\boldsymbol{y}}(x, y) = \frac{1}{\sqrt{(2\pi)^{p+q}}} \frac{1}{\sqrt{\det R}} \exp\left\{-\tfrac{1}{2}\begin{bmatrix} (x - \bar{x})^{\mathsf{T}} & (y - \bar{y})^{\mathsf{T}} \end{bmatrix} R^{-1} \begin{bmatrix} x - \bar{x} \\ y - \bar{y} \end{bmatrix}\right\} \qquad \text{(A.10)}$$

in terms of the covariance matrix R of $\text{col}\{\boldsymbol{x}, \boldsymbol{y}\}$, namely,

$$R = \mathsf{E}\left(\begin{bmatrix} \boldsymbol{x} \\ \boldsymbol{y} \end{bmatrix} - \begin{bmatrix} \bar{x} \\ \bar{y} \end{bmatrix}\right)\left(\begin{bmatrix} \boldsymbol{x} \\ \boldsymbol{y} \end{bmatrix} - \begin{bmatrix} \bar{x} \\ \bar{y} \end{bmatrix}\right)^{\mathsf{T}} = \begin{bmatrix} R_x & R_{xy} \\ R_{xy}^{\mathsf{T}} & R_y \end{bmatrix}$$

It can be seen from (A.10) that the joint pdf of $\{\boldsymbol{x}, \boldsymbol{y}\}$ is completely determined by the mean, covariances, and cross-covariance of $\{\boldsymbol{x}, \boldsymbol{y}\}$, i.e., by the first- and second-order moments $\{\bar{x}, \bar{y}, R_x, R_y, R_{xy}\}$.

Complex-Valued Random Variables and Circularity

Let us now examine the case of complex-valued random vectors. We consider again two real random vectors $\{\boldsymbol{x}, \boldsymbol{y}\}$, both assumed of size $p \times 1$, with joint pdf given by (cf. (A.10)):

$$f_{\boldsymbol{x},\boldsymbol{y}}(x, y) = \frac{1}{(2\pi)^p} \frac{1}{\sqrt{\det R}} \exp\left\{-\frac{1}{2}\begin{bmatrix} (x - \bar{x})^{\mathsf{T}} & (y - \bar{y})^{\mathsf{T}} \end{bmatrix} R^{-1} \begin{bmatrix} x - \bar{x} \\ y - \bar{y} \end{bmatrix}\right\} \qquad \text{(A.11)}$$

Let $z = x + jy$, where $j = \sqrt{-1}$, denote a complex-valued random variable defined in terms of $\{x, y\}$. Its mean is simply

$$\bar{z} = \mathsf{E}\, z = \bar{x} + j\bar{y}$$

while its covariance matrix is

$$R_z \triangleq \mathsf{E}\,(z-\bar{z})(z-\bar{z})^* = (R_x + R_y) + j(R_{yx} - R_{xy}) \tag{A.12}$$

which is expressed in terms of the covariances and cross-covariance of $\{x, y\}$.

We shall say that the complex variable z has a Gaussian distribution if its real and imaginary parts $\{x, y\}$ are jointly Gaussian. Since z is a function of $\{x, y\}$, its pdf is characterized by the joint pdf of $\{x, y\}$ as in (A.11), i.e., in terms of $\{\bar{x}, \bar{y}, R_x, R_y, R_{xy}\}$. However, we would like to express the pdf of z in terms of its own first- and second-order moments, i.e., in terms of $\{\bar{z}, R_z\}$. It turns out that this is not always possible. This is because knowledge of $\{\bar{z}, R_z\}$ alone is not enough to recover the moments $\{\bar{x}, \bar{y}, R_x, R_y, R_{xy}\}$. More information is needed in the form of a circularity condition.

To see this, assume we only know $\{\bar{z}, R_z\}$. Then this information is enough to recover $\{\bar{x}, \bar{y}\}$ since $\bar{z} = \bar{x} + j\bar{y}$. However, the information is not enough to recover $\{R_x, R_y, R_{xy}\}$. This is because, as we see from (A.12), knowledge of R_z allows us to recover the values of $(R_x + R_y)$ and $(R_{yx} - R_{xy})$ via

$$R_x + R_y = \mathsf{Re}(R_z), \qquad R_{yx} - R_{xy} = \mathsf{Im}(R_z) \tag{A.13}$$

in terms of the real and imaginary parts of R_z. This information is not sufficient to determine the individual covariances (R_x, R_y, R_{xy}).

In order to be able to uniquely recover $\{R_x, R_y, R_{xy}\}$ from R_z, it is generally further assumed that the random variable z satisfies a *circularity* condition. This means that z should satisfy

$$\boxed{\mathsf{E}\,(z-\bar{z})(z-\bar{z})^{\mathsf{T}} = 0} \qquad \text{(circularity condition)}$$

with the transposition symbol T used instead of Hermitian conjugation. Knowledge of R_z and this circularity condition are enough to recover $\{R_x, R_y, R_{xy}\}$ from R_z. Indeed, using the fact that

$$\mathsf{E}\,(z-\bar{z})(z-\bar{z})^{\mathsf{T}} = (R_x - R_y) + j(R_{yx} + R_{xy})$$

we find that, in view of the circularity assumption, it must hold that $R_x = R_y$ and $R_{xy} = -R_{yx}$. Consequently, combining with (A.13), we can solve for $\{R_x, R_y, R_{xy}\}$ to get

$$R_x = R_y = \frac{1}{2}\,\mathsf{Re}(R_z) \quad \text{and} \quad R_{xy} = -R_{yx} = -\frac{1}{2}\,\mathsf{Im}(R_z) \tag{A.14}$$

It follows that the covariance matrix of $\text{col}\{x, y\}$ can be expressed in terms of R_z as

$$R = \frac{1}{2}\begin{bmatrix} \mathsf{Re}(R_z) & -\mathsf{Im}(R_z) \\ \mathsf{Im}(R_z) & \mathsf{Re}(R_z) \end{bmatrix}$$

Actually, it also follows that R should have the symmetry structure

$$R = \begin{bmatrix} R_x & R_{xy} \\ -R_{xy} & R_x \end{bmatrix} \tag{A.15}$$

with the same matrix R_x appearing on the diagonal, and with R_{xy} and its negative appearing at the off-diagonal locations. Observe further that when z happens to be scalar-valued, then R_{xy} becomes a scalar, say, σ_{xy}, and the condition $R_{xy} = -R_{yx}$ can only hold if $\sigma_{xy} = 0$. That is, the real and imaginary parts of z will need to be independent in the scalar case.

Using the result (A.15), we can now verify that the joint pdf of $\{x, y\}$ in (A.11) can be rewritten in terms of $\{\bar{z}, R_z\}$ as shown below — compare with (A.9) in the real case. Observe in particular that the factors of 2, as well as the square-roots, disappear from the pdf expression in the complex case.

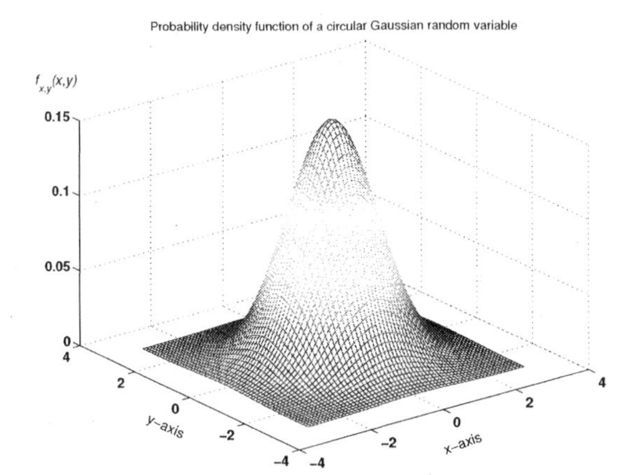

FIGURE A.5 A typical plot of the probability density function of a zero-mean scalar and circular Gaussian random variable.

Lemma A.1 (Circular Gaussian random variables) The pdf of a complex valued circular (or spherically invariant) Gaussian random variable z of dimension p is given by

$$f_z(z) = \frac{1}{\pi^p}\frac{1}{\det R_z}\exp^{\{-(z-\bar{z})^* R_z^{-1}(z-\bar{z})\}} \tag{A.16}$$

Proof: Using (A.15) we have

$$\det R = \det(R_x)\cdot\det(R_x + R_{xy}R_x^{-1}R_{xy})$$

Likewise, using the expression $R_z = 2[R_x - jR_{xy}]$, we obtain

$$\begin{aligned}[\det R_z]^2 &= \det(R_z)\cdot\det(R_z^{\mathsf{T}}) \\ &= 2^{2p}\det[R_x(I - jR_x^{-1}R_{xy})]\cdot\det(R_x - jR_{xy}^{\mathsf{T}})\end{aligned}$$

But

$$R_{xy}^{\mathsf{T}} = R_{yx} = -R_{xy}$$

and, for matrices A and B of compatible dimensions, $\det(AB) = \det(BA)$. Hence,

$$\begin{aligned}[\det R_z]^2 &= 2^{2p}\det R_x \det[(R_x + jR_{xy})(I - jR_x^{-1}R_{xy})] \\ &= 2^{2p}\det(R_x)\cdot\det(R_x + R_{xy}R_x^{-1}R_{xy})\end{aligned}$$

We conclude that $\det R = 2^{-2p}(\det R_z)^2$. Finally, some algebra will show that the exponents in (A.11) and (A.16) are identical.

◇

Figure A.5 plots the pdf of a scalar zero-mean complex-valued and circular Gaussian random variable z using $R = \mathrm{I}_2$, i.e., $\sigma_x^2 = \sigma_y^2 = 1$ and $\sigma_{xy} = 0$ so that $\sigma_z^2 = 2$. Therefore, in this example, the real and imaginary parts of z are independent Gaussian random variables with identical variances

When (A.16) holds, we can check that uncorrelated jointly Gaussian random variables will also be independent; this is one of the main reasons for the assumption of circularity.

Two Fourth-Order Moment Results

We establish below two useful results concerning the evaluation of fourth-order moments of Gaussian random variables, in both cases of real and complex-valued data. Although these results will only be used later in Parts IV (*Mean-Square Performance*) and V (*Transient Performance*) when we study the performance of adaptive filters, we list them here because their proofs relate to the earlier discussion on Gaussian random variables.

Lemma A.2 (Fourth-moment of real Gaussian variables) Let $\boldsymbol{x}$ be a real-valued Gaussian random column vector with zero-mean and a diagonal covariance matrix, say, $\mathsf{E}\,\boldsymbol{x}\boldsymbol{x}^{\mathsf{T}} = \Lambda$. Then for any symmetric matrix W of compatible dimensions it holds that

$$\mathsf{E}\left\{\boldsymbol{x}\boldsymbol{x}^{\mathsf{T}} W \boldsymbol{x}\boldsymbol{x}^{\mathsf{T}}\right\} = \Lambda \mathsf{Tr}(W\Lambda) + 2\Lambda W \Lambda \tag{A.17}$$

Proof: The argument is based on the fact that uncorrelated Gaussian random variables are also independent, so that if $\boldsymbol{x}(i)$ is the ith element of $\boldsymbol{x}$, then $\boldsymbol{x}(i)$ is independent of $\boldsymbol{x}(j)$ for $i \neq j$. Now let S denote the desired matrix, i.e., $S = \mathsf{E}\left\{\boldsymbol{x}\boldsymbol{x}^{\mathsf{T}} W \boldsymbol{x}\boldsymbol{x}^{\mathsf{T}}\right\}$, and let S_{ij} denote its (i,j)th element. Assume also that $\boldsymbol{x}$ is p-dimensional. Then

$$S_{ij} = \mathsf{E}\left\{\boldsymbol{x}(i)\boldsymbol{x}(j)\left(\sum_{m=0}^{p-1}\sum_{n=0}^{p-1} \boldsymbol{x}(m) W_{mn} \boldsymbol{x}(n)\right)\right\}$$

The right-hand side is nonzero only when there are two pairs of equal indices $\{i = j,\ m = n\}$ or $\{i = m,\ j = n\}$ or $\{i = n,\ j = m\}$. Assume first that $i = j$ (which corresponds to the diagonal elements of S). Then the expectation is nonzero only for $m = n$, i.e.,

$$S_{ii} = \mathsf{E}\left\{\boldsymbol{x}^2(i)\sum_{m=0}^{p-1} W_{mm}\boldsymbol{x}^2(m)\right\} = \sum_{m=0}^{p-1} W_{mm}\mathsf{E}\left\{\boldsymbol{x}^2(i)\boldsymbol{x}^2(m)\right\} = \lambda_i \mathsf{Tr}(W\Lambda) + 2W_{ii}\lambda_i^2$$

where we used the fact that for a zero-mean *real* scalar-valued Gaussian random variable $\boldsymbol{a}$ we have $\mathsf{E}\,\boldsymbol{a}^4 = 3\left(\mathsf{E}\,\boldsymbol{a}^2\right)^2 = 3\sigma_a^4$, where σ_a^2 denotes the variance of $\boldsymbol{a}$, $\sigma_a^2 = \mathsf{E}\,\boldsymbol{a}^2$. We are also denoting the diagonal entries of Λ by $\{\lambda_i\}$.

For the off-diagonal elements of S (i.e., for $i \neq j$), we must have either $i = n, j = m$, or $i = m, j = n$, so that

$$\begin{aligned} S_{ij} &= \mathsf{E}\left\{\boldsymbol{x}(i)\boldsymbol{x}(j)\left(\boldsymbol{x}(i)W_{ij}\boldsymbol{x}(j)\right)\right\} + \mathsf{E}\left\{\boldsymbol{x}(i)\boldsymbol{x}(j)\left(\boldsymbol{x}(j)W_{ji}\boldsymbol{x}(i)\right)\right\} \\ &= \left(W_{ij} + W_{ji}\right)\mathsf{E}\left\{\boldsymbol{x}^2(i)\boldsymbol{x}^2(j)\right\} = \left(W_{ij} + W_{ji}\right)\lambda_i\lambda_j \end{aligned}$$

Using the fact that W is symmetric, so that $W_{ij} = W_{ji}$, and collecting the expressions for S_{ij}, in both cases of $i = j$ and $i \neq j$, into matrix form we get the desired result (A.17).

◇

The equivalent result for complex-valued circular Gaussian random variables is the following. The only difference is the factor of 2 in (A.17). This is because the fourth-order moment of a zero-mean *complex* scalar-valued circular random variable $\boldsymbol{a}$, of variance $\sigma_a^2 = \mathsf{E}\,|\boldsymbol{a}|^2$, is now given by $\mathsf{E}\,|\boldsymbol{a}|^4 = 2\left(\mathsf{E}\,|\boldsymbol{a}|^2\right)^2 = 2\sigma_a^4$ — see, e.g., App. 1.B of Sayed (2003).

Lemma A.3 (Fourth-moment of complex Gaussian variables) Let $\boldsymbol{z}$ be a circular complex-valued Gaussian random column vector with zero-mean and a diagonal covariance matrix, say, $\mathsf{E}\,\boldsymbol{z}\boldsymbol{z}^* = \Lambda$. Then for any Hermitian matrix W of compatible dimensions it holds that

$$\mathsf{E}\left\{\boldsymbol{z}\boldsymbol{z}^* W \boldsymbol{z}\boldsymbol{z}^*\right\} = \Lambda \mathsf{Tr}(W\Lambda) + \Lambda W \Lambda \tag{A.18}$$

Linear Algebra

Throughout the book, the reader will be exposed to a variety of concepts from linear algebra and matrix theory in a motivated manner. In this way, after progressing sufficiently enough into the book, readers will be able to master many of the useful concepts described herein.

B.1 HERMITIAN AND POSITIVE-DEFINITE MATRICES

Hermitian matrices. The Hermitian conjugate, A^*, of a matrix A is the complex conjugate of its transpose, e.g.,

$$\text{if} \quad A = \begin{bmatrix} 1 & -j \\ 2+j & 1-j \end{bmatrix} \quad \text{then} \quad A^* = \begin{bmatrix} 1 & 2-j \\ j & 1+j \end{bmatrix}, \quad \text{where} \quad j = \sqrt{-1}$$

A Hermitian matrix is a square matrix satisfying $A^* = A$, e.g.,

$$\text{if} \quad A = \begin{bmatrix} 1 & 1+j \\ 1-j & 1 \end{bmatrix} \quad \text{then} \quad A^* = \begin{bmatrix} 1 & 1+j \\ 1-j & 1 \end{bmatrix} = A$$

so that A is Hermitian.

Spectral decomposition. Hermitian matrices can only have *real* eigenvalues. To see this, assume u_i is an eigenvector of A corresponding to an eigenvalue λ_i, i.e., $Au_i = \lambda_i u_i$. Multiplying from the left by u_i^* we get $u_i^* A u_i = \lambda_i \|u_i\|^2$, where $\|\cdot\|$ denotes the Euclidean norm of its argument. Now the scalar quantity on the left-hand side of this equality is real since it coincides with its complex conjugate, namely $(u_i^* A u_i)^* = u_i^* A^* u_i = u_i^* A u_i$. Therefore, λ_i must be real too.

Another important property of Hermitian matrices, whose proof requires a more involved argument, is that such matrices always have a *full* set of orthonormal eigenvectors. That is, if A is $n \times n$ Hermitian, then there will exist n orthonormal eigenvectors u_i satisfying

$$Au_i = \lambda_i u_i, \quad \|u_i\|^2 = 1, \quad u_i^* u_j = 0 \ \text{ for } i \neq j$$

In compact matrix notation we can write this so-called spectral (or modal or eigen-) decomposition of A as

$$\boxed{A = U \Lambda U^*}$$

where $\Lambda = \text{diag}\{\lambda_1, \lambda_2, \ldots, \lambda_n\}$, $U = \begin{bmatrix} u_1 & u_2 & \ldots & u_n \end{bmatrix}$, and U satisfies

$$\boxed{UU^* = U^*U = \text{I}}$$

We say that U is a unitary matrix. Here, the notation $\text{diag}\{a, b\}$ denotes a diagonal matrix with diagonal entries a and b.

Adaptive Filters, by Ali H. Sayed

Lemma B.1 (Rayleigh-Ritz characterization of eigenvalues) If A is an $n \times n$ Hermitian matrix, then it holds that for all vectors x,

$$\lambda_{\min}\|x\|^2 \leq x^*Ax \leq \lambda_{\max}\|x\|^2$$

as well as

$$\lambda_{\min} = \min_{x \neq 0}\left(\frac{x^*Ax}{x^*x}\right) = \min_{\|x\|=1} x^*Ax, \quad \lambda_{\max} = \max_{x \neq 0}\left(\frac{x^*Ax}{x^*x}\right) = \max_{\|x\|=1} x^*Ax$$

where $\{\lambda_{\min}, \lambda_{\max}\}$ denote the smallest and largest eigenvalues of A. The ratio x^*Ax/x^*x is called the Rayleigh-Ritz ratio.

Proof: Let $A = U\Lambda U^*$, where U is unitary and Λ has real entries, and define $y = U^*x$ for any vector x. Then

$$x^*Ax = x^*U\Lambda U^*x = y^*\Lambda y = \sum_{k=1}^{n}\lambda_k|y(k)|^2$$

with the $\{y(k)\}$ denoting the individual entries of y. Now since the squared terms $\{|y(k)|^2\}$ are nonnegative, we get

$$\lambda_{\min}\sum_{k=1}^{n}|y(k)|^2 \leq \sum_{k=1}^{n}\lambda_k|y(k)|^2 \leq \lambda_{\max}\sum_{k=1}^{n}|y(k)|^2$$

or, equivalently, $\lambda_{\min}\|y\|^2 \leq x^*Ax \leq \lambda_{\max}\|y\|^2$. Using the fact that U is unitary and, hence,

$$\|y\|^2 = y^*y = x\underbrace{UU^*}_{=\mathrm{I}}x = \|x\|^2$$

we conclude that $\lambda_{\min}\|x\|^2 \leq x^*Ax \leq \lambda_{\max}\|x\|^2$. The lower and upper bounds are achieved when x is chosen as the eigenvector corresponding to $\lambda_{\min}$ or $\lambda_{\max}$, respectively.

◇

Positive-definite matrices. An $n \times n$ Hermitian matrix A is positive-semidefinite (also called nonnegative definite) if it satisfies $x^*Ax \geq 0$ for all column vectors x. It is positive-definite if $x^*Ax > 0$ except when $x = 0$. We denote a positive-definite matrix by writing $A > 0$ and a positive-semidefinite matrix by writing $A \geq 0$. Among the several characterizations of positive-definite matrices, we note the following.

Lemma B.2 (Eigenvalues of positive-definite matrices) An $n \times n$ Hermitian matrix A is positive-definite if, and only if, all its eigenvalues are positive.

Proof: Let $A = U\Lambda U^*$ denote the spectral decomposition of A. Let also u_i be the ith column of U with λ_i the corresponding eigenvalue, i.e., $Au_i = \lambda_i u_i$ with $\|u_i\|^2 = 1$. If we multiply this equality from the left by u_i^* we get

$$u_i^*Au_i = \lambda_i\|u_i\|^2 = \lambda_i > 0$$

where the last inequality follows from the fact that $x^*Ax > 0$ for any nonzero vector x. Therefore, $A > 0$ implies $\lambda_i > 0$. Conversely, assume all $\lambda_i > 0$ and multiply the equality $A = U\Lambda U^*$ by any nonzero vector x and its conjugate transpose, from right and left, to get $x^*Ax = x^*U\Lambda U^*x$. Now define the matrix $\Lambda^{1/2} = \text{diag}\{\sqrt{\lambda_1}, \sqrt{\lambda_2}, \ldots, \sqrt{\lambda_n}\}$ and the vector $y = \Lambda^{1/2}U^*x$. The vector y is nonzero since U and $\Lambda^{1/2}$ are nonsingular matrices and, therefore, the product $\Lambda^{1/2}U^*$ cannot map a nonzero vector x to 0. Then the above equality becomes $x^*Ax = \|y\|^2 > 0$, which establishes that $A > 0$.

In a similar vein, we can show that

$$\boxed{A \geq 0 \iff \lambda_i \geq 0}$$

Note further that since $\det A = (\det U)(\det \Lambda)(\det U^*)$ and $(\det U)(\det U^*) = 1$, we find that $\det A = \det \Lambda = \prod_{i=1}^{n} \lambda_i$. Therefore, the determinant of a positive-definite matrix is positive,

$$\boxed{A > 0 \implies \det A > 0}$$

B.2 RANGE SPACES AND NULLSPACES OF MATRICES

Let A denote an $m \times n$ matrix without any constraint on the relative sizes of n and m.

Range spaces. The *column span* or the *range space* of A is defined as the set of all $m \times 1$ vectors q that can be generated by Ap, for all $n \times 1$ vectors p. We denote the column span of A by

$$\boxed{\mathcal{R}(A) \stackrel{\Delta}{=} \{\text{set of all } \; q \; \text{ such that } \; q = Ap \; \text{ for some } \; p\}}$$

Nullspaces. The nullspace of A is the set of all $n \times 1$ vectors p that are annihilated by A, namely, that satisfy $Ap = 0$. We denote the nullspace of A by

$$\boxed{\mathcal{N}(A) \stackrel{\Delta}{=} \{\text{set of all } \; p \; \text{ such that } \; Ap = 0\}}$$

Properties. A useful property that follows from the definitions of range spaces and nullspaces is that any vector z from the nullspace of A^* (not A) is orthogonal to any vector p in the range space of A, i.e.,

$$\boxed{z \in \mathcal{N}(A^*), \;\; q \in \mathcal{R}(A) \quad \Longrightarrow \quad z^* q = 0}$$

Indeed, $z \in \mathcal{N}(A^*)$ implies that $A^* z = 0$ or, equivalently, $z^* A = 0$. Now write $q = Ap$ for some p. Then $z^* q = z^* Ap = 0$, as desired. Another useful property is that the matrices $A^* A$ and A^* have the *same* range space (i.e., they span the same space). Also, A and $A^* A$ have the same nullspace.

> **Lemma B.3 (Range and nullspaces)** For any $m \times n$ matrix A, it holds that $\mathcal{R}(A^*) = \mathcal{R}(A^* A)$ and $\mathcal{N}(A) = \mathcal{N}(A^* A)$.

Proof: One direction is immediate. Take a vector $q \in \mathcal{R}(A^* A)$, i.e., $q = A^* Ap$ for some p. Define $r = Ap$, then $q = A^* r$. This shows that $q \in \mathcal{R}(A^*)$ and we conclude that $\mathcal{R}(A^* A) \subset \mathcal{R}(A^*)$. The proof of the converse statement requires more effort.

Take a vector $q \in \mathcal{R}(A^*)$ and let us show that $q \in \mathcal{R}(A^* A)$. Assume, to the contrary, that q does not lie in $\mathcal{R}(A^* A)$. This implies that there exists a vector z in the nullspace of $A^* A$ that is not orthogonal to q, i.e., $A^* Az = 0$ and $z^* q \neq 0$. Now, if we multiply the equality $A^* Az = 0$ by z^* from the left we obtain that $z^* A^* Az = 0$ or, equivalently, $\|Az\|^2 = 0$, where $\|\cdot\|$ denotes the Euclidean norm of its vector argument. Therefore, Az is necessarily the zero vector, $Az = 0$. But from $q \in \mathcal{R}(A^*)$ we have that $q = A^* p$ for some p. Then $z^* q = z^* A^* p = 0$, which contradicts $z^* q \neq 0$. Therefore, we must have $q \in \mathcal{R}(A^* A)$ and we conclude that $\mathcal{R}(A^*) \subset \mathcal{R}(A^* A)$.

The second assertion in the lemma is more immediate. If $Ap = 0$ then $A^* Ap = 0$ so that $\mathcal{N}(A) \subset \mathcal{N}(A^* A)$. Conversely, if $A^* Ap = 0$ then $p^* A^* Ap = 0$ and we must have $Ap = 0$. That is, $\mathcal{N}(A^* A) \subset \mathcal{N}(A)$. Combining both facts we conclude that $\mathcal{N}(A) = \mathcal{N}(A^* A)$.

◇

A useful consequence of Lemma B.3 is that linear systems of equations of the form $A^* Ax = A^* b$ always have a solution x for any vector b. This is because $A^* b$ belongs to $\mathcal{R}(A^*)$ and, therefore, also belongs to $\mathcal{R}(A^*A)$.

Rank. The rank of a matrix A is defined as the number of linearly independent columns (or rows) of A. It holds that

$$\boxed{\text{rank}(A) \leq \min\{m, n\}}$$

That is, the rank of a matrix never exceeds its smallest dimension. A matrix is said to have *full rank* if

$$\text{rank}(A) = \min\{m, n\}$$

Otherwise, the matrix is said to be *rank deficient*.

If A is a square matrix (i.e., $m = n$), then rank deficiency is also equivalent to a zero determinant, $\det A = 0$. Indeed, if A is rank deficient, then there exists a nonzero p such that $Ap = 0$. This means that 0 is an eigenvalue of A so that its determinant must be zero — recall that the determinant of a square matrix is equal to the product of its eigenvalues (see Prob. II.2). A useful result is the following, the proof of which is instructive.

Lemma B.4 (Invertible product) Let A be $N \times n$, with $N \geq n$. Then

$$A \text{ has full rank } \iff A^*A \text{ is positive-definite}$$

That is, every tall full rank matrix is such that the square matrix A^*A is invertible (actually, positive-definite).

Proof: Let us first show that A has full rank only if A^*A is invertible. Thus, assume A has full rank but that A^*A is not invertible. This means that there exists a nonzero vector p such that $A^*Ap = 0$, which implies $p^*A^*Ap = 0$ or, equivalently, $\|Ap\|^2 = 0$. That is, $Ap = 0$. This shows that the nullspace of A is nontrivial so that A cannot have full rank, which is a contradiction. Therefore a full rank A implies an invertible matrix A^*A.

Conversely, assume A^*A is invertible and let us show that A has to have full rank. Assume not. Then there exists a nonzero vector p such that $Ap = 0$. It follows that $A^*Ap = 0$, which contradicts the invertibility of A^*A. This is because $A^*Ap = 0$ implies that p is an eigenvector of A^*A corresponding to the zero eigenvalue. Hence, the determinant of A^*A is necessarily zero.

Finally, let us show that A^*A is positive-definite. For this purpose, take any nonzero vector x and consider the product x^*A^*Ax, which evaluates to $\|Ax\|^2$. Then, the product x^*A^*Ax is necessarily positive; it cannot be zero since the nullspace of A, in view of A being full rank, contains only the zero vector.

◇

In fact, when A has full rank, not only A^*A is positive-definite, but any product of the form A^*BA for any Hermitian positive-definite matrix B:

$$\boxed{A: \ N \times n, \ N \geq n, \quad \text{full rank} \quad \iff \quad A^*BA > 0, \quad \text{for any } B > 0} \tag{B.1}$$

To see this, recall from Sec. B.1 that every Hermitian matrix B admits an eigen-decomposition of the form

$$B = U\Lambda U^* \tag{B.2}$$

where Λ is diagonal with the eigenvalues of B, and U is a unitary matrix with the orthonormal eigenvectors of B. Define the matrices

$$\Lambda^{1/2} \triangleq \text{diag}\left\{\sqrt{\lambda_1}, \sqrt{\lambda_2}, \ldots, \sqrt{\lambda_n}\right\}, \quad \bar{A} \triangleq \Lambda^{1/2}U^*A$$

Then $A^*BA = \bar{A}^*\bar{A}$. Now the matrix $\bar{A}$ has full rank if, and only if, A has full rank and, in view of the result of the previous lemma, the full rank property of $\bar{A}$ is equivalent to the positive-definiteness of $\bar{A}^*\bar{A}$, as desired.

B.3 SCHUR COMPLEMENTS

In this section we assume inverses exist as needed. Thus, consider a block matrix

$$M = \begin{bmatrix} A & B \\ C & D \end{bmatrix}$$

The Schur complement of A in M is denoted by Δ_A and is defined as

$$\boxed{\Delta_A \triangleq D - CA^{-1}B}$$

Likewise, the Schur complement of D in M is denoted by Δ_D and is defined as

$$\boxed{\Delta_D \triangleq A - BD^{-1}C}$$

Block factorizations. In terms of these Schur complements, it is easy to verify by direct calculation that the matrix M can be factored in either of the following two useful forms:

$$\begin{aligned} \begin{bmatrix} A & B \\ C & D \end{bmatrix} &= \begin{bmatrix} \mathrm{I} & 0 \\ CA^{-1} & \mathrm{I} \end{bmatrix} \begin{bmatrix} A & 0 \\ 0 & \Delta_A \end{bmatrix} \begin{bmatrix} \mathrm{I} & A^{-1}B \\ 0 & \mathrm{I} \end{bmatrix} \\ &= \begin{bmatrix} \mathrm{I} & BD^{-1} \\ 0 & \mathrm{I} \end{bmatrix} \begin{bmatrix} \Delta_D & 0 \\ 0 & D \end{bmatrix} \begin{bmatrix} \mathrm{I} & 0 \\ D^{-1}C & \mathrm{I} \end{bmatrix} \end{aligned}$$

Inertia. When M is Hermitian (i.e., $M = M^*$), its eigenvalues are real (cf. Sec. B.1). We then define its inertia as the triplet $\mathrm{In}\{M\} = \{I_+, I_-, I_0\}$, where

$$\begin{aligned} I_+(M) &= \text{the number of positive eigenvalues of } M \\ I_-(M) &= \text{the number of negative eigenvalues of } M \\ I_0(M) &= \text{the number of zero eigenvalues of } M \end{aligned}$$

Congruence. Given a Hermitian matrix M, the matrices M and QMQ^* are said to be congruent for any invertible matrix Q. An important result regarding congruent matrices is the following, which states that congruence preserves inertia.

Lemma B.5 (Sylvester's law of inertia) Congruent matrices have identical inertia, i.e., for any Hermitian M and invertible Q, it holds that $\mathrm{In}\{M\} = \mathrm{In}\{QMQ^*\}$.

Thus, assume that M has the block Hermitian form

$$M = \begin{bmatrix} A & B \\ B^* & D \end{bmatrix} \quad \text{with} \quad A = A^*, \qquad D = D^*$$

and consider the corresponding block factorizations

$$\begin{aligned} \begin{bmatrix} A & B \\ B^* & D \end{bmatrix} &= \begin{bmatrix} \mathrm{I} & 0 \\ B^*A^{-1} & \mathrm{I} \end{bmatrix} \begin{bmatrix} A & 0 \\ 0 & \Delta_A \end{bmatrix} \begin{bmatrix} \mathrm{I} & A^{-1}B \\ 0 & \mathrm{I} \end{bmatrix} \\ &= \begin{bmatrix} \mathrm{I} & BD^{-1} \\ 0 & \mathrm{I} \end{bmatrix} \begin{bmatrix} \Delta_D & 0 \\ 0 & D \end{bmatrix} \begin{bmatrix} \mathrm{I} & 0 \\ D^{-1}B^* & \mathrm{I} \end{bmatrix} \end{aligned}$$

where now

$$\Delta_A = D - B^*A^{-1}B, \qquad \Delta_D = A - BD^{-1}B^*$$

The above factorizations have the form of congruence relations so that we must have

$$\text{In}(M) = \text{In}\left(\begin{bmatrix} A & 0 \\ 0 & \Delta_A \end{bmatrix}\right) \quad \text{and} \quad \text{In}(M) = \text{In}\left(\begin{bmatrix} \Delta_D & 0 \\ 0 & D \end{bmatrix}\right)$$

Positive-definite matrices. When M is positive-definite, all its eigenvalues will be positive (cf. Sec. B.1). Then, from the above inertia equalities, it will follow that $\{A, \Delta_A, \Delta_D, D\}$ can only have positive eigenvalues as well. In other words, it is easy to conclude that

$$\boxed{M > 0 \iff A > 0 \text{ and } \Delta_A > 0}$$

Likewise,

$$\boxed{M > 0 \iff D > 0 \text{ and } \Delta_D > 0}$$

B.4 CHOLESKY FACTORIZATION

Let us continue to assume that M is positive-definite and let us decompose it now as

$$M = \begin{bmatrix} a & b^* \\ b & D \end{bmatrix}$$

where a is its $(0, 0)$ entry, so that b is a column vector and a is a scalar. The positive-definiteness of M guarantees $a > 0$ and $D > 0$. Let us further consider the first block factorization shown above for M and write

$$M = \begin{bmatrix} 1 & 0 \\ b/a & \text{I} \end{bmatrix} \begin{bmatrix} a & 0 \\ 0 & \Delta_a \end{bmatrix} \begin{bmatrix} 1 & b^*/a \\ 0 & \text{I} \end{bmatrix}$$

We can rewrite this factorization more compactly in the form

$$M = \mathcal{L}_0 \begin{bmatrix} a(0) & \\ & \Delta_0 \end{bmatrix} \mathcal{L}_0^* \tag{B.3}$$

where $\mathcal{L}_0$ is the lower-triangular matrix

$$\mathcal{L}_0 \triangleq \begin{bmatrix} 1 & 0 \\ b/a & \text{I} \end{bmatrix}$$

and $a(0) = a$, $\Delta_0 = \Delta_a = D - bb^*/a$. Observe that the first column of $\mathcal{L}_0$ is the first column of M normalized by the inverse of the $(0, 0)$ entry of M. Moreover, the positive-definiteness of M guarantees $a(0) > 0$ and $\Delta_0 > 0$. Note further that if M is $n \times n$ then Δ_0 is $(n-1) \times (n-1)$.

Expression (B.3) provides a factorization for M that consists of a lower-triangular matrix followed by a *block-diagonal* matrix and by an upper-triangular matrix. Now since Δ_0 is in turn positive-definite, we can introduce a similar factorization for it, which we denote by

$$\Delta_0 = L_1 \begin{bmatrix} a(1) & \\ & \Delta_1 \end{bmatrix} L_1^*$$

for some lower-triangular matrix L_1 and where $a(1)$ is the $(0, 0)$ entry of Δ_0. Moreover, Δ_1 is the Schur complement of $a(1)$ in Δ_0, and its dimensions are $(n-2) \times (n-2)$. In addition, the first column of L_1 coincides with the first column of Δ_0 normalized by the inverse of the $(0, 0)$ entry of Δ_0. Also, the positive-definiteness of Δ_0 guarantees $a(1) > 0$ and $\Delta_1 > 0$. Substituting the above factorization of Δ_0 into the factorization for M we get

$$M = \mathcal{L}_0 \begin{bmatrix} 1 & \\ & L_1 \end{bmatrix} \begin{bmatrix} a(0) & & \\ & a(1) & \\ & & \Delta_1 \end{bmatrix} \begin{bmatrix} 1 & \\ & L_1^* \end{bmatrix} \mathcal{L}_0^*$$

But since the product of two lower-triangular matrices is also lower-triangular, we conclude that

$$\mathcal{L}_0 \begin{bmatrix} 1 & \\ & L_1 \end{bmatrix}$$

is lower-triangular, so that we can denote it by $\mathcal{L}_1$ and write instead

$$M = \mathcal{L}_1 \begin{bmatrix} a(0) & & \\ & a(1) & \\ & & \Delta_1 \end{bmatrix} \mathcal{L}_1^*$$

Clearly, the first column of $\mathcal{L}_1$ is the first column of $\mathcal{L}_0$ and the second column of $\mathcal{L}_1$ is formed from the first column of L_1.

We can proceed to factor Δ_1, which would lead to an expression of the form

$$M = \mathcal{L}_2 \begin{bmatrix} a(0) & & & \\ & a(1) & & \\ & & a(2) & \\ & & & \Delta_2 \end{bmatrix} \mathcal{L}_2^*$$

where $a(2) > 0$ is the $(0,0)$ entry of Δ_1 and $\Delta_2 > 0$ is the Schur complement of $a(2)$ in Δ_1. Continuing in this fashion we arrive (after $(n-1)$ Schur complementation steps) at a factorization for M of the form $M = \mathcal{L}_{n-1}\mathcal{D}\mathcal{L}_{n-1}^*$, where $\mathcal{L}_{n-1}$ is $n \times n$ lower-triangular and $\mathcal{D}$ is $n \times n$ diagonal with positive entries $\{a(i)\}$. The columns of $\mathcal{L}_{n-1}$ are the successive leading columns of the Schur complements $\{\Delta_i\}$, normalized by the inverses of the $(0,0)$ entries of $\{\Delta_i\}$. The diagonal entries of $\mathcal{D}$ coincide with these $(0,0)$ entries of the $\{\Delta_i\}$.

If we define $\bar{L} \triangleq \mathcal{L}_{n-1}\mathcal{D}^{1/2}$, where $\mathcal{D}^{1/2}$ is a diagonal matrix with the positive square-roots of the $\{a(i)\}$, we obtain

$$\boxed{M = \bar{L}\bar{L}^*} \quad \text{(lower-upper triangular factorization)}$$

In summary, this argument shows that every positive-definite matrix can be factored as the product of a lower-triangular matrix with positive diagonal entries by its conjugate transpose. This factorization is called the *Cholesky* factorization of M.

Had we instead partitioned M as

$$M = \begin{bmatrix} A & b \\ b^* & d \end{bmatrix}$$

where d is now a scalar, and had we used the block factorization

$$M \quad = \quad \begin{bmatrix} \mathrm{I} & b/d \\ 0 & 1 \end{bmatrix} \begin{bmatrix} \Delta_d & 0 \\ 0 & d \end{bmatrix} \begin{bmatrix} \mathrm{I} & 0 \\ b^*/d & 1 \end{bmatrix}$$

we would have arrived at a similar factorization for M of the form

$$\boxed{M = \bar{U}\bar{U}^*} \quad \text{(upper-lower triangular factorization)}$$

where $\bar{U}$ is an upper-triangular matrix with positive diagonal entries.

Lemma B.6 (Cholesky factorization) Every positive-definite matrix M admits a unique factorization of either form $M = \bar{L}\bar{L}^* = \bar{U}\bar{U}^*$, where $\bar{L}$ ($\bar{U}$) is a lower-triangular (upper-triangular) matrix with positive entries along its diagonal.

Proof: The existence of the factorizations was proved prior to the statement of the lemma. It remains to establish uniqueness. We show this for one of the factorizations. A similar argument applies to the other factorization. Thus, assume that

$$M = \bar{L}_1\bar{L}_1^* = \bar{L}_2\bar{L}_2^* \tag{B.4}$$

are two Cholesky factorizations for M. Then

$$\bar{L}_2^{-1}\bar{L}_1 = \bar{L}_2^*\bar{L}_1^{-*} \tag{B.5}$$

where the compact notation A^{-*} stands for $A^{-*} = [A^{-1}]^* = [A^*]^{-1}$. But since the inverse of a lower-triangular matrix is lower-triangular, and since the product of two lower-triangular matrices is also lower-triangular, we conclude that $\bar{L}_2^{-1}\bar{L}_1$ is lower-triangular. Likewise, we can verify that the product $\bar{L}_2^*\bar{L}_1^{-*}$ is upper-triangular. Therefore, the equality (B.5) will hold if, and only if, $\bar{L}_2^{-1}\bar{L}_1$ is a diagonal matrix, which means that

$$\bar{L}_1 = \bar{L}_2 D \tag{B.6}$$

for some diagonal matrix D. We want to show that D is the identity matrix. Indeed, it is easy to see from (B.4) that the $(0,0)$ entries of $\bar{L}_1$ and $\bar{L}_2$ must coincide so that the leading entry of D must be unity. This further implies from (B.6) that the first column of $\bar{L}_1$ should coincide with the first column of $\bar{L}_2$, so that using (B.4) again we conclude that the $(1,1)$ entries of $\bar{L}_1$ and $\bar{L}_2$ also coincide. Hence, the second entry of D is also unity. Proceeding in this fashion we get $D = \mathrm{I}$.

◇

Remark B.1 (Evaluation of Cholesky factors) While we obtained the Cholesky factorization of a positive-definite matrix by performing a sequence of successive Schur complements, this need not be the preferred method numerically for the evaluation of the Cholesky factor. Later in Ch. 34 we shall see that the same Cholesky factor can be obtained by applying to the matrix a sequence of unitary rotations, which are better conditioned numerically (so that this alternative method of computation is less sensitive to roundoff errors).

Remark B.2 (Alternative factorization) We also conclude from the discussion in this section that every positive-definite matrix M admits a unique factorization of either form $M = LDL^* = UD_uU^*$, where L (U) is a lower-triangular (upper-triangular) matrix with unit diagonal entries, and D and D_u are diagonal matrices with positive entries. [Actually, $L = \mathcal{L}_{n-1}$ and $D = \mathcal{D}$.]

B.5 QR DECOMPOSITION

The QR decomposition of a matrix is a very useful tool and we shall comment on its value and convenience on a handful of occasions. It can be motivated as follows.

Consider a collection of n column vectors, $\{h_i,\ i = 0, 1, \ldots, n-1\}$, of dimensions $N \times 1$ each with $N \geq n$. These vectors can be converted into an equivalent set of orthonormal vectors $\{q_i,\ i = 0, 1, \ldots, n-1\}$, which span the same linear subspace as the $\{h_i\}$, by appealing to the classical Gram-Schmidt procedure. This is an iterative procedure that operates as follows. It starts with $q_0 = h_0/\|h_0\|$, where $\|h_0\|$ denotes the Euclidean norm of h_0, and then repeats for $i > 0$:

$$r_i \quad = \quad h_i - \sum_{j=0}^{i-1} (q_j^* h_i)\, q_j, \qquad q_i = r_i/\|r_i\|$$

Thus, observe, for example, that $r_1 = h_1 - (q_0^* h_1)\, q_0$, which is simply the residual vector that results from projecting h_1 onto q_0. Clearly, by the orthogonality property of least-squares solutions, this residual vector is orthogonal to q_0. By normalizing its norm to unity and by defining $q_1 = r_1/\|r_1\|$, we end up with a unit-norm vector that is orthogonal to q_0. In this way, we would have replaced the original vectors $\{h_0, h_1\}$ by two orthonormal vectors $\{q_0, q_1\}$. More generally, for any i, the vector r_i is the residual that results from projecting h_i onto the successive orthonormal vectors $\{q_0, q_1, \ldots, q_{i-1}\}$.

Now we can express each column h_i as a linear combination of $\{q_0, q_1, \ldots, q_i\}$ as follows:

$$h_i = (q_0^* h_i) q_0 + (q_1^* h_i) q_1 + \ldots + (q_{i-1}^* h_i) q_{i-1} + \|r_i\|\; q_i \tag{B.7}$$

If we collect the $\{h_i\}$ into a matrix $H = [h_0 \; h_1 \; \ldots \; h_{n-1}]$, and if we collect the coefficients of the linear combinations (B.7), for all $i = 0, 1, \ldots, n-1$, into an $n \times n$ upper-triangular matrix R,

$$R = \begin{bmatrix} \|r_0\| & q_0^* h_1 & q_0^* h_2 & \ldots & q_0^* h_{n-1} \\ & \|r_1\| & q_1^* h_2 & \ldots & q_1^* h_{n-1} \\ & & \ddots & & \vdots \\ & & & \ddots & \vdots \\ & & & & \|r_{n-1}\| \end{bmatrix}$$

we find that we can express the decomposition (B.7) in matrix form as follows:

$$\boxed{H = \widehat{Q} R} \tag{B.8}$$

where $\widehat{Q}$ is $N \times n$ with orthonormal columns $\{q_i\}$, i.e., $\widehat{Q} = [q_0 \; q_1 \; \ldots \; q_{n-1}]$ with $q_j^* q_i = \delta_{ij}$. When H has full column-rank, i.e., when H has rank n, all the diagonal entries of R will be positive. The factorization $H = \widehat{Q} R$ is referred to as the *reduced* QR decomposition of the matrix H; in effect, the QR decomposition of a matrix simply amounts to the orthonormalization of its columns.

> **Definition B.1 (Reduced QR decomposition)** Given an $N \times n$, $n \leq N$, matrix $H = \begin{bmatrix} h_0 & h_1 & \ldots & h_{n-1} \end{bmatrix}$. It can be decomposed as $H = \widehat{Q} R$, where R is $n \times n$ and upper triangular and $\widehat{Q}$ is $N \times n$ with orthonormal columns.

It is often convenient to employ the *full* QR decomposition of H, as opposed to its reduced QR decomposition. The full decomposition is obtained by appending $N - n$ orthonormal columns to $\widehat{Q}$ so that it becomes a unitary $N \times N$ (square) matrix Q. Correspondingly, we also append rows of zeros to R so that (B.8) becomes

$$\boxed{H = Q \begin{bmatrix} R \\ 0 \end{bmatrix}} \tag{B.9}$$

where $Q = [\widehat{Q} \; q_n \; \ldots \; q_{N-1}]$.

> **Definition B.2 (Full QR decomposition)** Given an $N \times n$, $n \leq N$, matrix $H = \begin{bmatrix} h_0 & h_1 & \ldots & h_{n-1} \end{bmatrix}$. It can be decomposed as in (B.9), where R is $n \times n$ and upper triangular and Q is $N \times N$ and unitary.

B.6 SINGULAR VALUE DECOMPOSITION

The singular value decomposition of a matrix (SVD, for short) is another powerful tool that is useful for both analytical and numerical purposes. It enables us to represent any matrix (square or not, invertible or not, Hermitian or not) as the product of three matrices with special and desirable properties: two of the matrices are unitary (orthogonal in the real case) and the third matrix is composed of a diagonal matrix and a zero block.

Specifically, the SVD of a matrix A states that if A is $n \times m$ then there exist an $n \times n$ unitary matrix U ($UU^* = \mathrm{I}$), an $m \times m$ unitary matrix V ($VV^* = \mathrm{I}$), and a diagonal matrix Σ with nonnegative entries such that:

(i) If $n \leq m$, then Σ is $n \times n$ and

$$\boxed{A = U \begin{bmatrix} \Sigma & 0 \end{bmatrix} V^*, \quad A : n \times m, \quad n \leq m}$$

(ii) If $n \geq m$, then Σ is $m \times m$ and

$$\boxed{A = U \begin{bmatrix} \Sigma \\ 0 \end{bmatrix} V^*, \quad A : n \times m, \quad n \geq m}$$

Observe that U and V are square matrices, while the central matrix has the dimensions of A. The diagonal entries of Σ are called the singular values of A and are usually ordered in decreasing order, say, $\Sigma = \text{diag}\{\sigma_1, \sigma_2, \ldots, \sigma_p, 0, \ldots, 0\}$ with $\sigma_1 \geq \sigma_2 \geq \ldots \geq \sigma_p > 0$. If Σ has p nonzero diagonal entries then A has rank p. The columns of U and V are called the left- and right-singular vectors of A, respectively.

Constructive Proof of the SVD

One proof of the SVD decomposition follows from the eigen-decomposition of a Hermitian nonnegative-definite matrix. The argument given here assumes $n \leq m$, but it can be adjusted to handle the case $n \geq m$.

Note that AA^* is a Hermitian nonnegative-definite matrix and, consequently, there exists an $n \times n$ unitary matrix U and an $n \times n$ diagonal matrix Σ^2, with nonnegative entries, such that $AA^* = U\Sigma^2U^*$. This representation simply corresponds to the eigen-decomposition of AA^*. The diagonal entries of Σ^2 are the eigenvalues of AA^*, which are nonnegative (and, hence, the notation Σ^2); they are also equal to the nonzero eigenvalues of A^*A. The columns of U are the corresponding orthonormal eigenvectors. By proper reordering, we can always arrange the diagonal entries of Σ^2 in decreasing order so that Σ^2 can be put into the form $\Sigma^2 = \text{diagonal } \{\sigma_1^2, \sigma_2^2, \ldots, \sigma_p^2, 0, \ldots, 0\}$, where $p = \text{rank}(AA^*)$ and $\sigma_1^2 \geq \sigma_2^2 \geq \ldots \geq \sigma_p^2 > 0$. If we define the $m \times m$ matrix

$$V_1 = A^*U \text{ diag}\{\sigma_1^{-1}, \ldots, \sigma_p^{-1}, 0, \ldots, 0\}$$

then it is immediate to conclude that $A = U\Sigma {V_1}^*$, and that V_1 satisfies

$$V_1^* V_1 = \begin{bmatrix} \mathrm{I} & 0 \\ 0 & 0 \end{bmatrix}$$

But the columns of V_1 can be completed to a full unitary basis V of an m-dimensional space, say, $V = \begin{bmatrix} V_1 & V_2 \end{bmatrix}$, such that $VV^* = V^*V = \mathrm{I}$. This allows us to conclude that we can write $A = U \begin{bmatrix} \Sigma & 0 \end{bmatrix} V^*$, which is the desired SVD of A in the $n \leq m$ case. A similar argument establishes the SVD decomposition of A in the $n > m$ case.

Spectral Norm of a Matrix

Assume $n \geq m$ and consider the $m \times m$ square matrix A^*A. Using the Rayleigh-Ritz characterization of the eigenvalues of a matrix from Sec. B.1, we have that

$$\lambda_{\max}(A^*A) = \max_{x \neq 0} \left(\frac{x^*A^*Ax}{x^*x} \right) = \max_{x \neq 0} \left(\frac{\|Ax\|^2}{\|x\|^2} \right)$$

But since $\sigma_1^2 = \lambda_{\max}(A^*A)$, we conclude that the largest singular value of A satisfies:

$$\sigma_1 = \max_{x \neq 0} \left(\|Ax\|/\|x\|\right)$$

In other words, the square of the maximum singular value, σ_1^2, measures the maximum energy gain from x to Ax. The same conclusion holds when $n \leq m$ since $\sigma_1^2 = \lambda_{\max}(A^*A)$ as well, and the argument can be repeated. In addition, if we select $x = v_1$ (i.e., as the right singular vector corresponding to σ_1), then this choice for x achieves the maximum gain since $Av_1 = \sigma u_1$ and, therefore,

$$\frac{\|Av_1\|^2}{\|v_1\|^2} = \frac{\|\sigma_1 u_1\|^2}{\|v_1\|^2} = \sigma_1^2$$

The maximum singular value of a matrix is called its spectral norm or its 2-induced norm, written as

$$\|A\|_2 \stackrel{\Delta}{=} \sigma_1 = \max_{x \neq 0} \left(\frac{\|Ax\|}{\|x\|}\right) = \max_{\|x\|=1} \|Ax\|$$

This definition can be taken as a matrix norm because it can be verified to satisfy the properties of a norm. Specifically, a matrix norm $\|\cdot\|$ should satisfy the following properties, for any matrices A and B and for any complex scalar α:

1. $\|A\| \geq 0$ always and $\|A\| = 0$ if, and only if, $A = 0$.
2. $\|\alpha A\| = |\alpha| \cdot \|A\|$.
3. Triangle inequality: $\|A + B\| \leq \|A\| + \|B\|$.
4. Submultiplicative property: $\|AB\| \leq \|A\| \cdot \|B\|$.

Pseudo-inverses

The pseudo-inverse of a matrix is a generalization of the concept of inverses for square invertible matrices; it is defined for matrices that need not be invertible or even square.

Given an $n \times m$ matrix A, its pseudo-inverse is defined as the $m \times n$ matrix $A^\dagger$ that satisfies the following four requirements:

$$\text{(i) } AA^\dagger A = A \quad \text{(ii) } A^\dagger AA^\dagger = A^\dagger \quad \text{(iii) } (AA^\dagger)^* = AA^\dagger \quad \text{(iv) } (A^\dagger A)^* = A^\dagger A$$

The SVD of A can be used to determine its pseudo-inverse as follows. Introduce the matrix $\Sigma^\dagger =$ diagonal $\{\sigma_1^{-1}, \sigma_2^{-1}, \ldots, \sigma_p^{-1}, 0, \ldots, 0\}$. That is, we invert the nonzero entries of Σ and keep the zero entries unchanged.

(i) When $n \leq m$, we define

$$A^\dagger = V \begin{bmatrix} \Sigma^\dagger \\ 0 \end{bmatrix} U^* \qquad (A : n \times m, \;\; A^\dagger : m \times n, \;\; n \leq m)$$

(ii) When $n \geq m$, we define

$$A^\dagger = V \begin{bmatrix} \Sigma^\dagger & 0 \end{bmatrix} U^* \qquad (A : n \times m, \;\; A^\dagger : m \times n, \;\; n \geq m)$$

It can be verified that this $A^\dagger$ satisfies the four defining properties (i)–(iv) listed above. In addition, it can be verified, by replacing A by its SVD in the expressions below, that when A has *full rank*, its pseudo-inverse is given by

1. when $n \leq m$, $A^\dagger = A^*(AA^*)^{-1}$.
2. when $n \geq m$, $A^\dagger = (A^*A)^{-1}A^*$.

Minimum Norm Solution of Least-Squares Problems

A useful application of the pseudo-inverse of a matrix is in the characterization of the minimum norm solution of a least-squares problem with infinitely many solutions.

Lemma B.7 (Minimum norm solution) Consider a least-squares problem of the form

$$\min_{w} \|y - Hw\|^2$$

with possibly an infinite number of solutions $\{\widehat{w}\}$. The least-norm solution among these (i.e., the solution $\widehat{w}$ with the smallest Euclidean norm) is unique and is given by $\widehat{w} = H^\dagger y$. In other words, this particular $\widehat{w}$ is the unique solution to the following optimization problem:

$$\min_{w \in \mathcal{W}} \|w\| \quad \text{where } \mathcal{W} = \{w \text{ such that } \|y - Hw\|^2 \text{ is minimum}\}$$

Proof: Let $N \times M$ denote the dimensions of H. We establish the result for the case $N \geq M$ (i.e., for the over-determined case) by using the convenience of the SVD representation. A similar argument applies to the under-determined case. Thus, let $p \leq M$ denote the rank of H and introduce its SVD, say,

$$H = U \begin{bmatrix} \Sigma \\ 0 \end{bmatrix} V^*$$

with $\Sigma = \text{diag}\{\sigma_1, \ldots, \sigma_p, 0, \ldots, 0\}$. Then

$$\|y - Hw\|^2 \quad = \quad \|U^*y - U^*HVV^*w\|^2 = \left\| f - \begin{bmatrix} \Sigma \\ 0 \end{bmatrix} z \right\|^2$$

where we introduced the vectors $z = V^*w$ and $f = U^*y$. Therefore, the problem of minimizing $\|y - Hw\|^2$ over w is equivalent to the problem of minimizing the rightmost term over z. Let $\{z(i), f(i)\}$ denote the individual entries of $\{z, f\}$. Then

$$\left\| f - \begin{bmatrix} \Sigma \\ 0 \end{bmatrix} z \right\|^2 = \sum_{i=1}^{p} |f(i) - \sigma_i z(i)|^2 + \sum_{i=p+1}^{N+1} |f(i)|^2$$

The second term is independent of z. Hence, any solution z has to satisfy $z(i) = f(i)/\sigma_i$ for $i = 1$ to p and $z(i)$ arbitrary for $i = p + 1$ to $i = M$. The one with the smallest Euclidean norm requires that these latter values be chosen as zero. In this case, the minimum norm solution becomes

$$\widehat{w} \quad = \quad V \,\text{col}\{f(1)/\sigma_1, \ldots, f(p)/\sigma_p, 0, \ldots, 0\} \; = \; V \begin{bmatrix} \Sigma^\dagger & 0 \end{bmatrix} U^*y \; = \; H^\dagger y$$

as desired.

◇

B.7 KRONECKER PRODUCTS

Let $A = [a_{ij}]_{i,j=1}^{m}$ and $B = [b_{ij}]_{i,j=1}^{n}$ be $m \times m$ and $n \times n$ matrices, respectively. Their Kronecker product (also called their tensor product) is denoted by $A \otimes B$ and is defined as the $mn \times mn$ matrix whose entries are given by (see, e.g., Horn and Johnson (1994)):

$$A \otimes B \; = \; \begin{bmatrix} a_{11}B & a_{12}B & \ldots & a_{1m}B \\ a_{21}B & a_{22}B & \ldots & a_{2m}B \\ \vdots & & & \vdots \\ a_{m1}B & a_{m2}B & \ldots & a_{mm}B \end{bmatrix}$$

In other words, each entry of A is replaced by a scaled multiple of B. In particular, if A is the identity matrix, I_m, then $\mathrm{I}_m \otimes B$ is a block diagonal matrix with B repeated along its diagonal:

$$\mathrm{I}_m \otimes B = \text{diag}\underbrace{\{B, B, \ldots, B\}}_{m \text{ times}}$$

On the other hand, $A \otimes I_n$ is not a block diagonal matrix. For example, if $m = 2 = n$ and $B = I_2$ then

$$A \otimes I_2 = \left[\begin{array}{cc|cc} a_{11} & & a_{12} & \\ & a_{11} & & a_{12} \\ \hline a_{21} & & a_{22} & \\ & a_{21} & & a_{22} \end{array}\right]$$

One of the main uses of Kronecker products, at least for our purposes, is that they allow us to replace matrix operations by vector operations, as we saw repeatedly in the body of the chapter. The following is a list of some of the useful properties of Kronecker products, with property (vi) being the one that we used the most in our development.

Lemma B.8 (Useful properties) Consider $m \times m$ and $n \times n$ matrices A and B and let $\{\alpha_i, i = 1, \ldots, m\}$ and $\{\beta_j, j = 1, \ldots, n\}$ denote their eigenvalues, respectively. The matrices may be real or complex-valued. Then it holds that:

(i) $(A \otimes B)(C \otimes D) = AC \otimes BD$.

(ii) If A and B are invertible matrices, then $(A \otimes B)^{-1} = A^{-1} \otimes B^{-1}$; observe that the order of the matrices A and B is not switched.

(iii) $(A \otimes B)$ has mn eigenvalues and they are equal to all combinations $\{\alpha_i\beta_j\}$, for $i = 1, \ldots, m$ and $j = 1, \ldots, n$.

(iv) $\text{Tr}(A \otimes B) = \text{Tr}(A)\text{Tr}(B)$.

(v) $\det(A \otimes B) = (\det A)^n (\det B)^m$.

(vi) For any matrices $\{A, B, C, X\}$ of compatible dimensions, if $C = AXB$, then $\text{vec}\{C\} = (B^{\mathsf{T}} \otimes A)\text{vec}\{X\}$.

(vii) $(A \otimes B)^{\mathsf{T}} = A^{\mathsf{T}} \otimes B^{\mathsf{T}}$ as well as $(A \otimes B)^* = A^* \otimes B^*$.

Proof: Part (i) follows by direct calculation from the definition of Kronecker products. Part (ii) follows by using part (i) to note that $(A \otimes B)(A^{-1} \otimes B^{-1}) = I \otimes I = I$. Part (iii) follows from part (i) by choosing C as a right eigenvector for A and D as a right eigenvector for B, say, $C = q_i$ and $D = p_j$ where $Aq_i = \alpha_i q_i$ and $Bp_j = \beta_j p_j$. Then

$$(A \otimes B)(q_i \otimes p_j) = \alpha_i\beta_j(q_i \otimes p_j)$$

which shows that $(q_i \otimes p_j)$ is an eigenvector of $(A \otimes B)$ with eigenvalue $\alpha_i\beta_j$. Part (iv) follows from part (iii) since

$$\text{Tr}(A) = \sum_{i=1}^{m} \alpha_i, \qquad \text{Tr}(B) = \sum_{j=1}^{n} \beta_j$$

and, therefore,

$$\text{Tr}(A)\text{Tr}(B) = \left(\sum_{i=1}^{m} \alpha_i\right)\left(\sum_{j=1}^{n} \beta_j\right) = \sum_{i=1}^{m}\sum_{j=1}^{n} \alpha_i\beta_j$$

Part (v) also follows from part (iii) since

$$\det(A \otimes B) = \prod_{i=1}^{m}\prod_{j=1}^{n} \alpha_i\beta_j = \left(\prod_{i=1}^{m} \alpha_i\right)^n \cdot \left(\prod_{i=j}^{n} \beta_j\right)^m = (\det A)^n \cdot (\det B)^m$$

Part (vi) follows from the definition of Kronecker products and from noting that the vec representation of the rank one matrix ab^{T} is simply $b \otimes a$, i.e., $\text{vec}(ab^{\mathsf{T}}) = b \otimes a$. Finally, part (vii) follows from the definition of Kronecker products.

◇

Complex Gradients

In this chapter we explain how to differentiate a scalar-valued function $g(z)$ with respect to a complex-valued argument, z, and its complex conjugate, z^*. The argument z could be either a scalar or a vector.

C.1 CAUCHY-RIEMANN CONDITIONS

We start with a scalar argument $z = x + jy$, where $j = \sqrt{-1}$. In this case, we can regard $g(z)$ as a function of the two real scalar variables, x and y, say,

$$g(z) = u(x,y) \; + \; jv(x,y) \tag{C.1}$$

with $u(\cdot,\cdot)$ denoting its real part and $v(\cdot,\cdot)$ denoting its imaginary part. Now, from complex function theory, the derivative of $g(z)$ at a point $z_o = x_o + jy_o$ is defined as (see, e.g., Ahlfors (1979)):

$$\frac{dg}{dz} \stackrel{\Delta}{=} \lim_{\Delta z \to 0} \frac{g(x_o + \Delta x, y_o + \Delta y) \; - \; g(x_o, y_o)}{\Delta x + j\Delta y}$$

where $\Delta z = \Delta x + j\Delta y$. For $g(z)$ to be differentiable at z_o, in which case it is also said to be *analytic* at z_o, the above limit should exist regardless of the direction from which z approaches z_o. In particular, if we assume $\Delta y = 0$ and $\Delta x \to 0$, then the above definition gives

$$\frac{dg}{dz} = \frac{\partial u}{\partial x} + j\frac{\partial v}{\partial x} \tag{C.2}$$

If, on the other hand, we assume that $\Delta x = 0$ and $\Delta y \to 0$ so that $\Delta z = j\Delta y$, then the definition gives

$$\frac{dg}{dz} = \frac{\partial v}{\partial y} - j\frac{\partial u}{\partial y} \tag{C.3}$$

The expressions (C.2) and (C.3) should coincide. Therefore, by adding them we get

$$\frac{dg}{dz} = \frac{1}{2}\left(\frac{\partial u}{\partial x} + j\frac{\partial v}{\partial x} + \frac{\partial v}{\partial y} - j\frac{\partial u}{\partial y}\right)$$

or, more compactly,

$$\boxed{\frac{dg}{dz} \stackrel{\Delta}{=} \frac{1}{2}\left\{\frac{\partial g}{\partial x} - j\frac{\partial g}{\partial y}\right\}} \tag{C.4}$$

Observe that the equality of expressions (C.2) and (C.3) implies that the real and imaginary parts of $g(\cdot)$ should satisfy the conditions

$$\frac{\partial u}{\partial x} = \frac{\partial v}{\partial y} \quad \text{and} \quad \frac{\partial u}{\partial y} = -\frac{\partial v}{\partial x}$$

which are known as the *Cauchy-Riemann* conditions. It can be shown that these conditions are not only necessary for a complex function $g(z)$ to be differentiable at $z = z_o$, but if the partial derivatives of $u(\cdot, \cdot)$ and $v(\cdot, \cdot)$ are continuous, then they are also sufficient.

C.2 SCALAR ARGUMENTS

More generally, if g is a function of both z and z^*, we define its partial derivatives with respect to z and z^* as follows:

$$\boxed{\frac{\partial g}{\partial z} = \frac{1}{2}\left\{\frac{\partial g}{\partial x} - j\frac{\partial g}{\partial y}\right\}, \qquad \frac{\partial g}{\partial z^*} = \frac{1}{2}\left\{\frac{\partial g}{\partial x} + j\frac{\partial g}{\partial y}\right\}} \tag{C.5}$$

Note, in particular, from the Cauchy-Riemann conditions, that if $g(z)$ is an analytic function of z then it must necessarily hold that $\partial g/\partial z^* = 0$.

Examples

We illustrate the definitions (C.4)–(C.5) considering several examples.

1. Let $g(z) = z = x + jy$. Then

$$\frac{\partial g}{\partial z} = (1 - j^2)/2 = 1, \qquad \frac{\partial g}{\partial z^*} = (1 + j^2)/2 = 0$$

2. Let $g(z) = z^2 = (x + jy)(x + jy) = (x^2 - y^2) + j2xy$. Then

$$\frac{\partial g}{\partial z} = 2(x + jy) = 2z, \qquad \frac{\partial g}{\partial z^*} = 0$$

In Examples 1 and 2, since g is a function of z alone, it holds that $\partial g/\partial z = dg/dz$.

3. Let $g(z) = |z|^2 = zz^* = (x + jy)(x - jy) = x^2 + y^2$. Then

$$\frac{\partial g}{\partial z} = (x - jy) = z^*, \qquad \frac{\partial g}{\partial z^*} = (x + jy) = z$$

4. Let $g(z) = \lambda + \alpha z + \beta z^* + \gamma z z^*$, where $(\lambda, \alpha, \beta, \gamma)$ are complex constants. That is,

$$g(z) = \left[\lambda + \alpha x + \beta x + \gamma(x^2 + y^2)\right] + j\left[\alpha y - \beta y\right]$$

Then

$$\frac{\partial g}{\partial z} = \alpha + \gamma z^*, \qquad \frac{\partial g}{\partial z^*} = \beta + \gamma z$$

C.3 VECTOR ARGUMENTS

Now assume that z is a *column* vector, say,

$$z = \text{col}\{z_1, z_2, \ldots, z_n\}, \quad z_i = x_i + jy_i$$

The *complex gradient* of g with respect to z is denoted by $\nabla_z g(z)$, or simply $\nabla_z g$, and is defined as the *row* vector

$$\boxed{\nabla_z g \triangleq \left[\begin{array}{cccc} \partial g/\partial z_1 & \partial g/\partial z_2 & \ldots & \partial g/\partial z_n \end{array}\right]} \qquad \left\{\begin{array}{l} z \text{ is a column} \\ \nabla_z g \text{ is a row} \end{array}\right.$$

Likewise, the complex gradient of g with respect to z^* is defined as the *column* vector

$$\nabla_{z^*} g \triangleq \begin{bmatrix} \partial g/\partial z_1^* \\ \partial g/\partial z_2^* \\ \vdots \\ \partial g/\partial z_n^* \end{bmatrix} \qquad \left\{ \begin{array}{l} z^* \text{ is a row} \\ \nabla_z g \text{ is a column} \end{array} \right.$$

The reason why we choose to define $\nabla_z g$ as a row vector and $\nabla_{z^*} g$ as a column vector is because the subsequent differentiation results will be consistent with what we are used to from the standard differentiation of functions of real-valued arguments. Let us again consider a few examples:

1. Let $g(z) = \alpha^* z$, where $\{\alpha, z\}$ are column vectors. Then $\nabla_z g = \alpha^*$ and $\nabla_{z^*} g = 0$.
2. Let $g(z) = z^* \beta$, where $\{\beta, z\}$ are column vectors. Then $\nabla_z g = 0$ and $\nabla_{z^*} g = \beta$.
3. Let $g(z) = z^* z$, where z is a column vector. Then $\nabla_z g = z^*$ and $\nabla_{z^*} g = z$.
4. Let $g(z) = \lambda + \alpha^* z + z^* \beta + z^* \Gamma z$, where λ is a scalar, $\{\alpha, \beta\}$ are column vectors and Γ is a matrix. Then $\nabla_z g = \alpha^* + z^* \Gamma$ and $\nabla_{z^*} g = \beta + \Gamma z$.

Hessian Matrix

The complex gradient of $\nabla_z g$ with respect to z^* is called the *Hessian* matrix of g, and it is denoted by

$$\nabla_z^2 g \triangleq \nabla_{z^*} [\nabla_z g] = \begin{bmatrix} \frac{\partial^2 g}{\partial z_1^* \partial z_1} & \frac{\partial^2 g}{\partial z_1^* \partial z_2} & \cdots & \frac{\partial^2 g}{\partial z_1^* \partial z_n} \\ \frac{\partial^2 g}{\partial z_2^* \partial z_1} & \frac{\partial^2 g}{\partial z_2^* \partial z_2} & \cdots & \frac{\partial^2 g}{\partial z_2^* \partial z_n} \\ \cdot & \cdot & & \cdot \\ \frac{\partial^2 g}{\partial z_n^* \partial z_1} & \frac{\partial^2 g}{\partial z_n^* \partial z_2} & \cdots & \frac{\partial^2 g}{\partial z_n^* \partial z_n} \end{bmatrix}$$

It holds that $\nabla_{z^*} [\nabla_z g] = \nabla_z [\nabla_{z^*} g]$. Consider again the example $g(z) = \lambda + \alpha^* z + z^* \beta + z^* \Gamma z$, then its Hessian matrix is $\nabla_z^2 g = \Gamma$.

Real-Valued Arguments

When z is real, say, $z = x$, and $g(x) = \lambda + \alpha^{\mathsf{T}} x + x^{\mathsf{T}} \beta + x^{\mathsf{T}} \Gamma x$, with $x = \text{col}\{x_1, x_2, \ldots, x_n\}$ and a symmetric matrix Γ, then the gradient of g with respect to x is again defined as

$$\nabla_x g \triangleq \begin{bmatrix} \partial g/\partial x_1 & \partial g/\partial x_2 & \ldots & \partial g/\partial x_n \end{bmatrix}$$

so that we now obtain $\nabla_x g = \alpha^{\mathsf{T}} + \beta^{\mathsf{T}} + 2x^{\mathsf{T}} \Gamma$. Likewise, the Hessian matrix becomes $\nabla_x^2 g = 2\Gamma$. Observe the difference from the complex case (in terms of an additional scaling factor that is equal to 2). This is because, in the complex case, the symbols $\{z, z^*\}$ are treated as separate variables.

PART I

OPTIMAL ESTIMATION

CHAPTER 1

Scalar-Valued Data

In this first part of the book we focus on the basic, yet fundamental, problem of estimating an unobservable quantity from a collection of measurements in the least-mean-squares sense. The estimation task is made more or less difficult depending on how much information the measured data convey about the unobservable quantity. We shall study this estimation problem with increasing degrees of complexity, starting from a simple scenario and building up to more sophisticated cases.

The material is developed initially at a slow pace. This is done deliberately in order to familiarize readers (and especially students) with the basic concepts of estimation theory for both real- *and* complex-valued random variables, as well as for scalar- *and* vector-valued random variables. We hope that, by the end of our exposition, the reader will be convinced that these different scenarios (of real vs. complex and scalar vs. vector) can be masked by adopting a uniform vector and complex-conjugation notation. The notation is introduced gradually in the two initial chapters and will be used throughout the book thereafter.

Before plunging into a study of least-mean-squares estimation theory, and the reasons for its widespread use, the reader is advised to consult the review material in Secs. A.1–A.4. These sections provide an intuitive explanation for what the variance of a random variable means. The sections also introduce several useful concepts such as complex- and vector-valued random variables and the notions of independence and uncorrelatedness between two random variables. The explanations will help the reader appreciate the value of the least-mean-squares criterion, which is used extensively in later sections and chapters.

1.1 ESTIMATION WITHOUT OBSERVATIONS

We initiate our discussions of estimation theory by posing and solving a simple (almost trivial) estimation problem. Thus, suppose that all we know about a real-valued random variable $\boldsymbol{x}$ is its mean $\bar{x}$ and its variance σ_x^2, and that we wish to estimate the value that $\boldsymbol{x}$ will assume in a given experiment. We shall denote the *estimate* of $\boldsymbol{x}$ by $\hat{x}$; it is a deterministic quantity (i.e., a number). But how do we come up with a value for $\hat{x}$? And how do we decide whether this value is optimal or not? And if optimal, in what sense? These inquiries are at the heart of every estimation problem.

To answer these questions, we first need to choose a cost function to penalize the estimation error. The resulting estimate $\hat{x}$ will be optimal only in the sense that it leads to the smallest cost value. Different choices for the cost function will generally lead to different choices for $\hat{x}$, each of which will be optimal in its own way.

The design criterion we shall adopt is the *mean-square-error* criterion. It is based on introducing the error signal

$$\tilde{\boldsymbol{x}} \overset{\Delta}{=} \boldsymbol{x} - \hat{x}$$

and then determining $\hat{x}$ by minimizing the mean-square-error (m.s.e.), which is defined as the expected value of $\tilde{\boldsymbol{x}}^2$, i.e.,

$$\min_{\hat{x}} \mathsf{E}\, \tilde{\boldsymbol{x}}^2 \tag{1.1}$$

The error $\tilde{\boldsymbol{x}}$ is a random variable since $\boldsymbol{x}$ is random. The resulting estimate, $\hat{x}$, will be called the *least-mean-squares estimate* of $\boldsymbol{x}$. The following result is immediate (and, in fact, intuitively obvious as we explain below).

Lemma 1.1 (Lack of observations) The least-mean-squares estimate of $\boldsymbol{x}$ given knowledge of $(\bar{x}, \sigma_x^2)$ is $\hat{x} = \bar{x}$. The resulting minimum cost is $\mathsf{E}\,\tilde{\boldsymbol{x}}^2 = \sigma_x^2$.

Proof: Expand the mean-square error by subtracting and adding $\bar{x}$ as follows:

$$\mathsf{E}\,\tilde{\boldsymbol{x}}^2 = \mathsf{E}\,(\boldsymbol{x} - \hat{x})^2 = \mathsf{E}\,[(\boldsymbol{x} - \bar{x}) + (\bar{x} - \hat{x})]^2 = \sigma_x^2 + (\bar{x} - \hat{x})^2$$

The choice of $\hat{x}$ that minimizes the m.s.e. is now evident. Only the term $(\bar{x} - \hat{x})^2$ is dependent on $\hat{x}$ and this term can be annihilated by choosing $\hat{x} = \bar{x}$. The resulting minimum mean-square error (m.m.s.e.) is then

$$\text{m.m.s.e.} \stackrel{\Delta}{=} \mathsf{E}\,\tilde{\boldsymbol{x}}^2 = \sigma_x^2$$

An alternative derivation would be to expand the cost function as

$$\mathsf{E}\,(\boldsymbol{x} - \hat{x})^2 = \mathsf{E}\,\boldsymbol{x}^2 - 2\bar{x}\hat{x} + \hat{x}^2$$

and to differentiate it with respect to $\hat{x}$. By setting the derivative equal to zero we arrive at the same conclusion, namely, $\hat{x} = \bar{x}$.

◇

There are several good reasons for choosing the mean-square-error criterion (1.1). The simplest one perhaps is that the criterion is amenable to mathematical manipulations, more so than any other criterion. In addition, the criterion is essentially attempting to force the estimation error to assume values close to its mean, which happens to be zero. This is because

$$\mathsf{E}\,\tilde{\boldsymbol{x}} = \mathsf{E}\,(\boldsymbol{x} - \hat{x}) = \mathsf{E}\,(\boldsymbol{x} - \bar{x}) = \bar{x} - \bar{x} = 0$$

and, by minimizing $\mathsf{E}\,\tilde{\boldsymbol{x}}^2$, we are in effect minimizing the variance of the error, $\tilde{\boldsymbol{x}}$. In view of the discussion in Sec. A.1 regarding the interpretation of the variance of a random variable, we find that the mean-square-error criterion is therefore attempting to increase the likelihood of small errors.

The effectiveness of the estimation procedure (1.1) can be measured by examining the value of the minimum cost, which is the variance of the resulting estimation error. The above lemma tells us that the minimum cost is equal to σ_x^2. That is,

$$\sigma_{\tilde{x}}^2 = \sigma_x^2$$

so that the estimate $\hat{x} = \bar{x}$ does not reduce our initial uncertainty about $\boldsymbol{x}$ since the error variable still has the same variance as $\boldsymbol{x}$ itself! We thus find that the performance of the mean-square-error design procedure is limited in this case. Clearly, we are more interested in estimation procedures that result in error variances that are smaller than the original signal variance. We shall discuss one such procedure in the next section.

The reason for the poor performance of the estimate $\hat{x} = \bar{x}$ lies in the lack of more sophisticated prior information about $\boldsymbol{x}$. Note that Lemma 1.1 simply tells us that the best we can do, in the absence of any other information about a random variable $\boldsymbol{x}$, other than its mean and variance, is to use the mean value of $\boldsymbol{x}$ as our estimate. This statement is, in a sense, intuitive. After all, the mean value of a random variable is, by definition, an indication of the value that we would expect to occur on average in repeated experiments. Hence, in answer to the question: what is the best guess for $\boldsymbol{x}$?, the analysis tells us that the best guess is what we would expect for $\boldsymbol{x}$ on average! This is a circular answer, but one that is at least consistent with intuition.

Example 1.1 (Binary signal)

Assume $\boldsymbol{x}$ represents a BPSK (binary phase-shift keying) signal that is equal to ± 1 with probability $1/2$ each. Then

$$\bar{x} = \frac{1}{2} \cdot (1) + \frac{1}{2} \cdot (-1) = 0$$

and

$$\sigma_x^2 = \mathsf{E}\,\boldsymbol{x}^2 = 1$$

Now given knowledge of $\{\bar{x}, \sigma_x^2\}$ alone, the best estimate of $\boldsymbol{x}$ in the least-mean-squares sense is $\hat{x} = \bar{x} = 0$. This example shows that the least-mean-squares (and, hence, optimal) estimate does not always lead to a meaningful solution! In this case, $\hat{x} = 0$ is not useful in guessing whether $\boldsymbol{x}$ is 1 or -1 in a given realization. If we could incorporate into the design of the estimator the knowledge that $\boldsymbol{x}$ is a BPSK signal, or some other related information, then we could perhaps come up with a better estimate for $\boldsymbol{x}$.

1.2 ESTIMATION GIVEN DEPENDENT OBSERVATIONS

So let us examine the case in which more is known about a random variable $\boldsymbol{x}$, other than its mean and variance. Specifically, let us assume that we have access to an observation of a second random variable $\boldsymbol{y}$ that is related to $\boldsymbol{x}$ in some way. For example, $\boldsymbol{y}$ could be a noisy measurement of $\boldsymbol{x}$, say, $\boldsymbol{y} = \boldsymbol{x} + \boldsymbol{v}$, where $\boldsymbol{v}$ denotes the disturbance, or $\boldsymbol{y}$ could be the sign of $\boldsymbol{x}$, or dependent on $\boldsymbol{x}$ in some other manner.

Given two dependent random variables $\{\boldsymbol{x}, \boldsymbol{y}\}$, we therefore pose the problem of determining the least-mean-squares *estimator* of $\boldsymbol{x}$ given $\boldsymbol{y}$. Observe that we are now employing the terminology *estimator* of $\boldsymbol{x}$ as opposed to *estimate* of $\boldsymbol{x}$. In order to highlight this distinction, we denote the estimator of $\boldsymbol{x}$ by the boldface notation $\hat{\boldsymbol{x}}$; it is a random variable that is defined as a function of $\boldsymbol{y}$, say,

$$\hat{\boldsymbol{x}} = h(\boldsymbol{y})$$

for some function $h(\cdot)$ to be determined. Once the function $h(\cdot)$ has been determined, evaluating it at a particular occurrence of $\boldsymbol{y}$, say, for $\boldsymbol{y} = y$, will result in an estimate for $\boldsymbol{x}$, i.e.,

$$\hat{x} = h(\boldsymbol{y})\big|_{\boldsymbol{y}=y} = h(y)$$

Different occurrences for $\boldsymbol{y}$ lead to different estimates $\hat{x}$. In Sec. 1.1 we did not need to make this distinction between an estimator $\hat{\boldsymbol{x}}$ and an estimate $\hat{x}$. There we sought directly an estimate $\hat{x}$ for $\boldsymbol{x}$ since we did not have access to a random variable $\boldsymbol{y}$; we only had access to the deterministic quantities $\{\bar{x}, \sigma_x^2\}$.

The criterion we shall use to determine the estimator $\hat{\boldsymbol{x}}$ is still the mean-square-error criterion. We define the error signal

$$\boxed{\tilde{\boldsymbol{x}} \stackrel{\Delta}{=} \boldsymbol{x} - \hat{\boldsymbol{x}}} \tag{1.2}$$

and then determine $\hat{\boldsymbol{x}}$ by minimizing the mean-square-error over all possible functions $h(\cdot)$:

$$\boxed{\min_{h(\cdot)} \ \mathsf{E}\, \tilde{\boldsymbol{x}}^2} \tag{1.3}$$

The solution is given by the following statement.

Theorem 1.1 (Optimal mean-square-error estimator) The least-mean-squares estimator (l.m.s.e.) of $\boldsymbol{x}$ given $\boldsymbol{y}$ is the conditional expectation of $\boldsymbol{x}$ given $\boldsymbol{y}$, i.e., $\hat{\boldsymbol{x}} = \mathsf{E}\,(\boldsymbol{x}|\boldsymbol{y})$. The resulting estimate is

$$\hat{x} = \mathsf{E}\,(\boldsymbol{x}|\boldsymbol{y} = y) \ = \ \int_{S_x} x f_{\boldsymbol{x}|\boldsymbol{y}}(x|y)\mathrm{d}x$$

where S_x denotes the support (or domain) of the random variable $\boldsymbol{x}$. Moreover, the estimator is unbiased, i.e., $\mathsf{E}\,\hat{\boldsymbol{x}} = \bar{x}$, and the resulting minimum cost is $\mathsf{E}\,\tilde{\boldsymbol{x}}^2 \ = \ \sigma_x^2 \ - \ \sigma_{\hat{x}}^2$.

Proof: There are several ways to establish the result. Our argument is based on recalling that for any two random variables $\boldsymbol{x}$ and $\boldsymbol{y}$, it holds that (see Prob. I.4):

$$\boxed{\mathsf{E}\,\boldsymbol{x} \ = \ \mathsf{E}\,[\mathsf{E}\,(\boldsymbol{x}|\boldsymbol{y})]} \tag{1.4}$$

where the outermost expectation on the right-hand side is with respect to $\boldsymbol{y}$, while the innermost expectation is with respect to $\boldsymbol{x}$. We shall indicate these facts explicitly by showing the variables with respect to which the expectations are performed, so that (1.4) is rewritten as

$$\mathsf{E}\,\boldsymbol{x} \ = \ \mathsf{E}_y[\mathsf{E}_x(\boldsymbol{x}|\boldsymbol{y})]$$

It now follows that, for any function of $\boldsymbol{y}$, say, $g(\boldsymbol{y})$, it holds that

$$\mathsf{E}_{x,y}\ \boldsymbol{x}g(\boldsymbol{y}) = \mathsf{E}_y\left[\mathsf{E}_x(\boldsymbol{x}g(\boldsymbol{y})|\boldsymbol{y})\right] = \mathsf{E}_y\left[\mathsf{E}_x(\boldsymbol{x}|\boldsymbol{y})g(\boldsymbol{y})\right] = \mathsf{E}_{x,y}\left[\mathsf{E}_x(\boldsymbol{x}|\boldsymbol{y})\right] g(\boldsymbol{y})$$

This means that, for any $g(\boldsymbol{y})$, it holds that $\mathsf{E}_{x,y}\left[\boldsymbol{x} - \mathsf{E}_x(\boldsymbol{x}|\boldsymbol{y})\right] g(\boldsymbol{y}) = 0$, which we write more compactly as

$$\boxed{\mathsf{E}\,\left[\boldsymbol{x} - \mathsf{E}\,(\boldsymbol{x}|\boldsymbol{y})\right] g(\boldsymbol{y}) = 0} \tag{1.5}$$

Expression (1.5) states that the random variable $\boldsymbol{x} - \mathsf{E}\,(\boldsymbol{x}|\boldsymbol{y})$ is uncorrelated with any function $g(\cdot)$ of $\boldsymbol{y}$. Indeed, as mentioned before in Sec. A.2, two random variables $\boldsymbol{a}$ and $\boldsymbol{b}$ are uncorrelated if, and only if, their cross-correlation is zero, i.e., $\mathsf{E}\,(\boldsymbol{a} - \bar{a})(\boldsymbol{b} - \bar{b}) = 0$. On the other hand, the random variables are said to be *orthogonal* if, and only if, $\mathsf{E}\,\boldsymbol{ab} = 0$. It is easy to verify that the concepts of orthogonality and uncorrelatedness coincide if at least one of the random variables is zero mean. From equation (1.5) we conclude that the variables $\boldsymbol{x} - \mathsf{E}\,(\boldsymbol{x}|\boldsymbol{y})$ and $g(\boldsymbol{y})$ are orthogonal. However, since $\boldsymbol{x} - \mathsf{E}\,(\boldsymbol{x}|\boldsymbol{y})$ is zero mean, then we can also say that they are uncorrelated.

Using this intermediate result, we return to the cost function (1.3), add and subtract $\mathsf{E}(\boldsymbol{x}|\boldsymbol{y})$ to its argument, and express it as

$$\mathsf{E}(\boldsymbol{x}-\hat{\boldsymbol{x}})^2 = \mathsf{E}\left[\boldsymbol{x}-\mathsf{E}(\boldsymbol{x}|\boldsymbol{y})+\mathsf{E}(\boldsymbol{x}|\boldsymbol{y})-\hat{\boldsymbol{x}}\right]^2$$

The term $\mathsf{E}(\boldsymbol{x}|\boldsymbol{y})-\hat{\boldsymbol{x}}$ is a function of $\boldsymbol{y}$. Therefore, if we choose $g(\boldsymbol{y}) = \mathsf{E}(\boldsymbol{x}|\boldsymbol{y})-\hat{\boldsymbol{x}}$, then from the orthogonality property (1.5) we conclude that

$$\mathsf{E}(\boldsymbol{x}-\hat{\boldsymbol{x}})^2 = \mathsf{E}\left[\boldsymbol{x}-\mathsf{E}(\boldsymbol{x}|\boldsymbol{y})\right]^2 \;+\; \mathsf{E}\left[\mathsf{E}(\boldsymbol{x}|\boldsymbol{y})-\hat{\boldsymbol{x}}\right]^2$$

Now only the second term on the right-hand side is dependent on $\hat{\boldsymbol{x}}$ and the m.s.e. is minimized by choosing $\hat{\boldsymbol{x}} = \mathsf{E}(\boldsymbol{x}|\boldsymbol{y})$. To evaluate the resulting m.m.s.e. we first note that the optimal estimator is unbiased since

$$\mathsf{E}\,\hat{\boldsymbol{x}} \;=\; \mathsf{E}\left[\mathsf{E}(\boldsymbol{x}|\boldsymbol{y})\right] \;=\; \mathsf{E}\,\boldsymbol{x} \;=\; \bar{x}$$

and its variance is therefore given by $\sigma_{\hat{x}}^2 = \mathsf{E}\,\hat{\boldsymbol{x}}^2 - \bar{x}^2$. Moreover, in view of the orthogonality property (1.5), and in view of the fact that $\hat{\boldsymbol{x}} = \mathsf{E}(\boldsymbol{x}|\boldsymbol{y})$ is itself a function of $\boldsymbol{y}$, we have

$$\boxed{\mathsf{E}(\boldsymbol{x}-\hat{\boldsymbol{x}})\hat{\boldsymbol{x}} = 0} \tag{1.6}$$

In other words, the estimation error, $\tilde{\boldsymbol{x}}$, is uncorrelated with the optimal estimator. Using this result, we can evaluate the m.m.s.e. as follows:

$$\begin{aligned}
\mathsf{E}\,\tilde{\boldsymbol{x}}^2 &= \mathsf{E}\left[\boldsymbol{x}-\hat{\boldsymbol{x}}\right]\left[\boldsymbol{x}-\hat{\boldsymbol{x}}\right] \;=\; \mathsf{E}\left[\boldsymbol{x}-\hat{\boldsymbol{x}}\right]\boldsymbol{x} \qquad \text{(because of (1.6))}\\
&= \mathsf{E}\,\boldsymbol{x}^2 \;-\; \mathsf{E}\,\hat{\boldsymbol{x}}[\tilde{\boldsymbol{x}}+\hat{\boldsymbol{x}}]\\
&= \mathsf{E}\,\boldsymbol{x}^2 \;-\; \mathsf{E}\,\hat{\boldsymbol{x}}^2 \qquad \text{(because of (1.6))}\\
&= \left(\mathsf{E}\,\boldsymbol{x}^2 \;-\; \bar{x}^2\right) + \left(\bar{x}^2 \;-\; \mathsf{E}\,\hat{\boldsymbol{x}}^2\right) \;=\; \sigma_x^2 - \sigma_{\hat{x}}^2
\end{aligned}$$

◇

Theorem 1.1 tells us that the least-mean-squares estimator of $\boldsymbol{x}$ is its conditional expectation given $\boldsymbol{y}$. This result is again intuitive. In answer to the question: what is the best guess for $\boldsymbol{x}$ given that we observed $\boldsymbol{y}$?, the analysis tells us that the best guess is what we would expect for $\boldsymbol{x}$ given the occurrence of $\boldsymbol{y}$!

Example 1.2 (Noisy measurement of a binary signal)

Let us return to Ex. 1.1, where $\boldsymbol{x}$ is a BPSK signal that assumes the values ± 1 with probability $1/2$. Assume now that in addition to the mean and variance of $\boldsymbol{x}$, we also have access to a noisy observation of $\boldsymbol{x}$, say,

$$\boldsymbol{y} = \boldsymbol{x} + \boldsymbol{v}$$

Assume further that the signal $\boldsymbol{x}$ and the disturbance $\boldsymbol{v}$ are independent, with $\boldsymbol{v}$ being a zero-mean Gaussian random variable of unit variance, i.e., its pdf is given by

$$f_{\boldsymbol{v}}(v) = \frac{1}{\sqrt{2\pi}}\,e^{-v^2/2}$$

Our intuition tells us that we should be able to do better here than in Ex. 1.1. But beware, even here, we shall make some interesting observations.

According to Thm. 1.1, the optimal estimate of $\boldsymbol{x}$ given an observation of $\boldsymbol{y}$ is

$$\hat{x} = \mathsf{E}(\boldsymbol{x}|\boldsymbol{y}=y) \;=\; \int_{-\infty}^{\infty} x f_{\boldsymbol{x}|\boldsymbol{y}}(x|y)dx \tag{1.7}$$

We therefore need to determine the conditional pdf, $f_{\boldsymbol{x}|\boldsymbol{y}}(x|y)$, and evaluate the integral (1.7). For this purpose, we start by noting, from probability theory, that the pdf of the sum of two independent random variables, namely, $\boldsymbol{y} = \boldsymbol{x} + \boldsymbol{v}$, is equal to the convolution of their individual pdfs, i.e.,

$$f_{\boldsymbol{y}}(y) = \int_{-\infty}^{\infty} f_{\boldsymbol{x}}(x) f_{\boldsymbol{v}}(y-x) \mathrm{d}x$$

In this example, we have

$$f_{\boldsymbol{x}}(x) = \frac{1}{2}\delta(x-1) \;+\; \frac{1}{2}\delta(x+1)$$

where $\delta(\cdot)$ is the Dirac-delta function, so that $f_{\boldsymbol{y}}(y)$ is given by

$$f_{\boldsymbol{y}}(y) = \frac{1}{2} f_{\boldsymbol{v}}(y+1) + \frac{1}{2} f_{\boldsymbol{v}}(y-1) \tag{1.8}$$

Moreover, the joint pdf of $\{\boldsymbol{x}, \boldsymbol{y}\}$ is given by

$$\begin{aligned} f_{\boldsymbol{x},\boldsymbol{y}}(x,y) &= f_{\boldsymbol{x}}(x) \cdot f_{\boldsymbol{y}|\boldsymbol{x}}(y|x) \\ &= \left[\frac{1}{2}\delta(x-1) \;+\; \frac{1}{2}\delta(x+1)\right] \cdot f_{\boldsymbol{v}}(y-x) \\ &= \frac{1}{2} f_{\boldsymbol{v}}(y-1)\delta(x-1) + \frac{1}{2} f_{\boldsymbol{v}}(y+1)\delta(x+1) \end{aligned}$$

Using (A.6) we get

$$f_{\boldsymbol{x}|\boldsymbol{y}}(x|y) \;=\; \frac{f_{\boldsymbol{x},\boldsymbol{y}}(x,y)}{f_{\boldsymbol{y}}(y)} = \frac{f_{\boldsymbol{v}}(y-1)\delta(x-1)}{f_{\boldsymbol{v}}(y+1) + f_{\boldsymbol{v}}(y-1)} + \frac{f_{\boldsymbol{v}}(y+1)\delta(x+1)}{f_{\boldsymbol{v}}(y+1) + f_{\boldsymbol{v}}(y-1)}$$

Substituting into expression (1.7) for $\hat{x}$ and integrating we obtain

$$\begin{aligned} \hat{x} &= \frac{f_{\boldsymbol{v}}(y-1)}{f_{\boldsymbol{v}}(y+1) + f_{\boldsymbol{v}}(y-1)} - \frac{f_{\boldsymbol{v}}(y+1)}{f_{\boldsymbol{v}}(y+1) + f_{\boldsymbol{v}}(y-1)} \\ &= \frac{1}{\left(\dfrac{e^{-(y+1)^2/2}}{e^{-(y-1)^2/2}}\right) + 1} - \frac{1}{\left(\dfrac{e^{-(y-1)^2/2}}{e^{-(y+1)^2/2}}\right) + 1} = \frac{e^{y} - e^{-y}}{e^{y} + e^{-y}} \stackrel{\Delta}{=} \mathsf{tanh}\, y \end{aligned}$$

In other words, the least-mean-squares estimator of $\boldsymbol{x}$ is the hyperbolic tangent function,

$$\boxed{\hat{\boldsymbol{x}} = \mathsf{tanh}(\boldsymbol{y})} \tag{1.9}$$

The result is represented schematically in Fig. 1.1.

Figure 1.2 plots the function $\mathsf{tanh}(y)$. We see that it tends to ± 1 as $y \longrightarrow \pm\infty$. For other values of y, the function assumes real values that are distinct from ± 1. This is a bit puzzling from the designer's perspective. The designer is interested in knowing whether the symbol $\boldsymbol{x}$ is $+1$ or -1 based on the observed value of $\boldsymbol{y}$. The above construction tells the designer to estimate $\boldsymbol{x}$ by computing $\mathsf{tanh}(y)$. But this value will never be exactly $+1$ or -1; it will be a real number inside

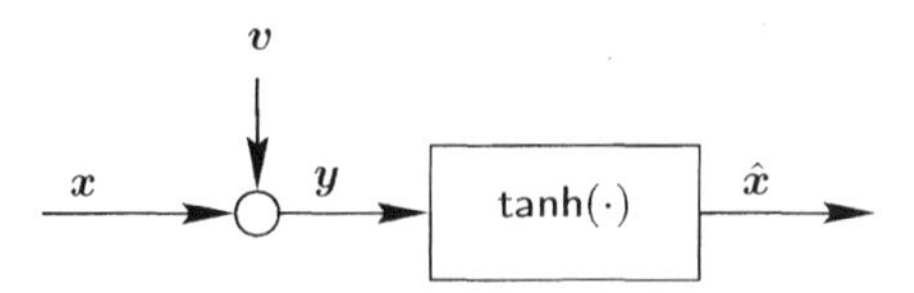

FIGURE 1.1 Optimal estimation of a BPSK signal embedded in unit-variance additive Gaussian noise.

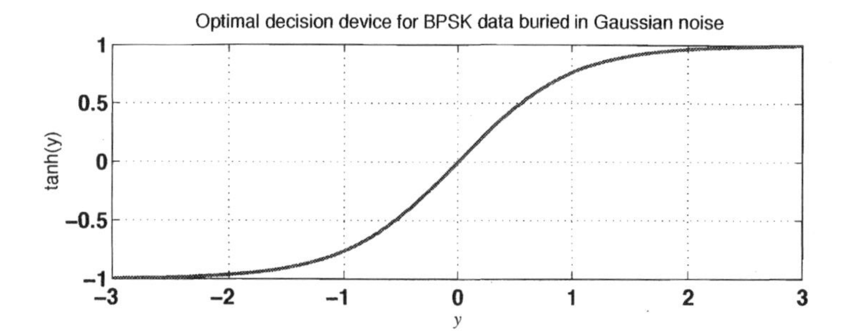

FIGURE 1.2 A plot of the function $\mathsf{tanh}(y)$.

the interval $(-1, 1)$. The designer will then be induced to make a hard decision of the form:

$$\text{decide in favor of} \begin{cases} +1 & \text{if } \hat{x} \text{ is nonnegative} \\ -1 & \text{if } \hat{x} \text{ is negative} \end{cases}$$

In effect, the designer ends up implementing the alternative estimator:

$$\hat{\boldsymbol{x}} = \mathsf{sign}[\mathsf{tanh}(\boldsymbol{y})] \tag{1.10}$$

where $\mathsf{sign}(\cdot)$ denotes the sign of its argument; it is equal to $+1$ if the argument is nonnegative and -1 otherwise.

We therefore have a situation where the optimal estimator, although known in closed form, does not solve the original problem of recovering the symbols ± 1 directly. Instead, the designer is forced to implement a suboptimal solution; it is suboptimal from a least-mean-squares point of view. Even more puzzling, the designer could consider implementing the alternative (and simpler) suboptimal estimator:

$$\hat{\boldsymbol{x}} = \mathsf{sign}(\boldsymbol{y}) \tag{1.11}$$

where the $\mathsf{sign}(\cdot)$ function operates directly on $\boldsymbol{y}$ rather than on $\mathsf{tanh}(\boldsymbol{y})$ — see Fig. 1.3. Both suboptimal implementations (1.10) and (1.11) lead to the same result since, as is evident from Fig. 1.2, $\mathsf{sign}[\mathsf{tanh}(y)] = \mathsf{sign}(y)$. In the computer project at the end of this part we shall compare the performance of the optimal and suboptimal estimators (1.9)–(1.11).

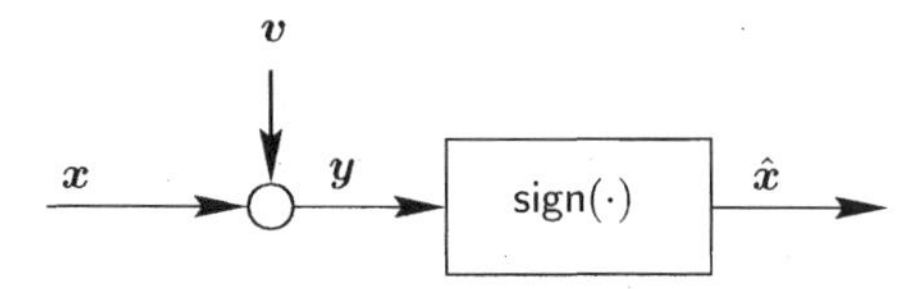

FIGURE 1.3 Suboptimal estimation of a BPSK signal embedded in unit-variance additive Gaussian noise.

We may mention that in the digital communications literature, especially in studies on equalization methods, an implementation using (1.11) is usually said to be based on *hard decisions*, while an implementation using (1.9) is said to be based on *soft decisions*.

◇

Remark 1.1 (Complexity of optimal estimation) Example 1.2 highlights one of the inconveniences of working with the optimal estimator of Thm. 1.1. Although the form of the optimal solution is given explicitly by $\hat{\boldsymbol{x}} = \mathsf{E}(\boldsymbol{x}|\boldsymbol{y})$, in general it is not an easy task to find a closed-form expression for the conditional expectation of two random variables (especially for other choices of probability density functions). Moreover, even when a closed-form expression can be found, one is usually led to a nonlinear estimator whose implementation may not be practical or may even be costly. For this reason, from Part II (*Linear Estimation*) onwards, we shall restrict the class of estimators to *linear* estimators, and study the capabilities of these estimators.

◇

The purpose of Exs. 1.1 and 1.2 is not to confuse the reader, but rather to stress the fact that an optimal estimator is optimal only in the sense that it satisfies a certain optimality criterion. One should not confuse an optimal guess with a perfect guess. One should also not confuse an optimal guess with a practical one; an optimal guess need not be perfect or even practical, though it can suggest good practical solutions.

1.3 ORTHOGONALITY PRINCIPLE

There are two important conclusions that follow from the proof of Thm. 1.1, namely, the orthogonality properties (1.5) and (1.6). The first one states that the difference

$$\boldsymbol{x} - \mathsf{E}\,(\boldsymbol{x}|\boldsymbol{y})$$

is orthogonal to any function $g(\cdot)$ of $\boldsymbol{y}$. Now since we already know that the conditional expectation, $\mathsf{E}\,(\boldsymbol{x}|\boldsymbol{y})$, is the optimal least-mean-squares estimator of $\boldsymbol{x}$, we can restate this result by saying that the estimation error $\tilde{\boldsymbol{x}}$ is orthogonal to any function of $\boldsymbol{y}$,

$$\boxed{\mathsf{E}\,\tilde{\boldsymbol{x}}\;g(\boldsymbol{y}) = 0} \tag{1.12}$$

We shall sometimes use a geometric notation to refer to this result and write instead

$$\boxed{\tilde{\boldsymbol{x}} \perp g(\boldsymbol{y})} \tag{1.13}$$

where the symbol $\perp$ is used to signify that the two random variables are orthogonal; a schematic representation of this orthogonality property is shown in Fig. 1.4.
Relation (1.13) admits the following interpretation. It states that the optimal estimator $\hat{\boldsymbol{x}} = \mathsf{E}\,(\boldsymbol{x}|\boldsymbol{y})$ is such that the resulting error, $\tilde{\boldsymbol{x}}$, is orthogonal to (and, in fact, also uncorrelated with) any transformation of the data $\boldsymbol{y}$. In other words, the optimal estimator is such that no matter how we modify the data $\boldsymbol{y}$, there is no way we can extract additional information from the data in order to reduce the variance of $\tilde{\boldsymbol{x}}$ any further. This is because any additional processing of $\boldsymbol{y}$ will remain uncorrelated with $\tilde{\boldsymbol{x}}$.

The second orthogonality property (1.6) is a special case of (1.13). It states that

$$\boxed{\tilde{\boldsymbol{x}} \perp \hat{\boldsymbol{x}}}$$

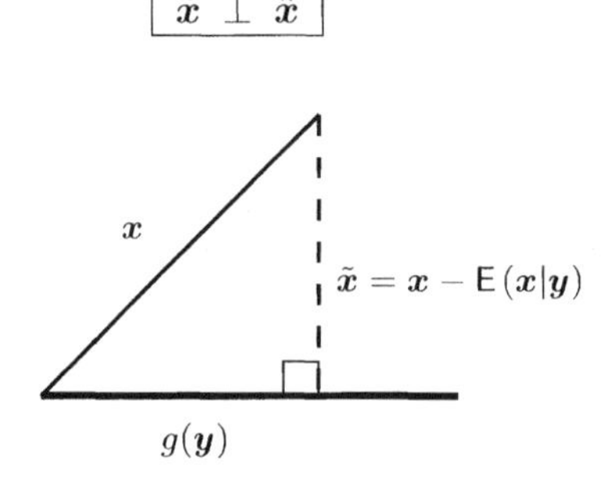

FIGURE 1.4 The orthogonality condition: $\tilde{\boldsymbol{x}} \perp g(\boldsymbol{y})$.

That is, the estimation error is orthogonal to (or uncorrelated with) the estimator itself. This is a special case of (1.13) since $\hat{\boldsymbol{x}}$ is a function of $\boldsymbol{y}$ by virtue of the result $\hat{\boldsymbol{x}} = \mathsf{E}\,(\boldsymbol{x}|\boldsymbol{y})$.

In summary, the optimal least-mean-squares estimator is such that the estimation error is orthogonal to the estimator and, more generally, to any function of the observation. It turns out that the converse statement is also true so that the orthogonality condition (1.13) is in fact a *defining* property of optimality in the least-mean-squares sense.

Theorem 1.2 (Orthogonality condition) Given two random variables $\boldsymbol{x}$ and $\boldsymbol{y}$, an estimator $\hat{\boldsymbol{x}} = h(\boldsymbol{y})$ is optimal in the least-mean-squares sense (1.3) if, and only if, $\hat{\boldsymbol{x}}$ is unbiased (i.e., $\mathsf{E}\,\hat{\boldsymbol{x}} = \bar{x}$) and $\boldsymbol{x} - \hat{\boldsymbol{x}} \perp g(\boldsymbol{y})$ for any function $g(\cdot)$.

Proof: One direction has already been proven prior to the statement of the theorem, namely, if $\hat{\boldsymbol{x}}$ is the optimal estimator and hence, $\hat{\boldsymbol{x}} = \mathsf{E}\,(\boldsymbol{x}|\boldsymbol{y})$, then we already know from (1.13) that $\tilde{\boldsymbol{x}} \perp g(\boldsymbol{y})$, for any $g(\cdot)$. Moreover, we know from Thm. 1.1 that this estimator is unbiased.

Conversely, assume $\hat{\boldsymbol{x}}$ is some unbiased estimator for $\boldsymbol{x}$ and that it satisfies $\boldsymbol{x} - \hat{\boldsymbol{x}} \perp g(\boldsymbol{y})$, for any $g(\cdot)$. Define the random variable $\boldsymbol{z} = \hat{\boldsymbol{x}} - \mathsf{E}\,(\boldsymbol{x}|\boldsymbol{y})$ and let us show that it is the zero variable with probability one. For this purpose, we note first that $\boldsymbol{z}$ is zero mean since

$$\mathsf{E}\,\boldsymbol{z} = \mathsf{E}\,\hat{\boldsymbol{x}} \;-\; \mathsf{E}\,(\mathsf{E}\,(\boldsymbol{x}|\boldsymbol{y})) \;=\; \bar{x} - \bar{x} \;=\; 0$$

Moreover, from (1.5) we have $\boldsymbol{x} - \mathsf{E}\,(\boldsymbol{x}|\boldsymbol{y}) \perp g(\boldsymbol{y})$ and, by assumption, we have $\boldsymbol{x} - \hat{\boldsymbol{x}} \perp g(\boldsymbol{y})$ for any $g(\cdot)$. Subtracting these two conditions we conclude that $\boldsymbol{z} \;\perp g(\boldsymbol{y})$, which is the same as $\mathsf{E}\,\boldsymbol{z}g(\boldsymbol{y}) = 0$. Now since the variable $\boldsymbol{z}$ itself is a function of $\boldsymbol{y}$, we may choose $g(\boldsymbol{y}) = \boldsymbol{z}$ to get $\mathsf{E}\,\boldsymbol{z}^2 = 0$. We thus find that $\boldsymbol{z}$ is zero mean and has zero variance, so that, from Remark A.1, we conclude that $\boldsymbol{z} = 0$, or equivalently, $\hat{\boldsymbol{x}} = \mathsf{E}\,(\boldsymbol{x}|\boldsymbol{y})$, with probability one.

◇

Example 1.3 (Suboptimal estimator for a binary signal)

Consider again Ex. 1.2, where $\boldsymbol{x}$ is a BPSK signal that assumes the values ± 1 with probability $1/2$. Let us verify that the estimator $\hat{\boldsymbol{x}} = \mathsf{sign}(\boldsymbol{y})$ is not optimal in the least-mean squares sense. We already know that this is the case because we found in Ex. 1.2 that the optimal estimator is $\mathsf{tanh}(\boldsymbol{y})$. Here we wish to verify the sub-optimality of $\mathsf{sign}(\boldsymbol{y})$ without assuming prior knowledge of the optimal estimator, and by relying solely on the orthogonality condition (1.12).

According to Thm. 1.2, we need to verify that the estimator $\mathsf{sign}(\boldsymbol{y})$ fails the orthogonality test. In particular, we shall exhibit a function $g(\boldsymbol{y})$ such that the difference $\boldsymbol{x} - \mathsf{sign}(\boldsymbol{y})$ is correlated with it. Actually, we shall choose $g(\boldsymbol{y}) = \mathsf{sign}(\boldsymbol{y})$ and verify that

$$\mathsf{E}\,[\boldsymbol{x} - \mathsf{sign}(\boldsymbol{y})]\mathsf{sign}(\boldsymbol{y}) \;\neq\; 0 \tag{1.14}$$

Let us first check whether the estimator $\hat{\boldsymbol{x}} = \mathsf{sign}(\boldsymbol{y})$ is biased or not. For this purpose we recall that $\boldsymbol{y} = \boldsymbol{x} + \boldsymbol{v}$ and that

$$\mathsf{sign}(\boldsymbol{x} + \boldsymbol{v}) \;=\; \begin{cases} +1 & \text{if } \boldsymbol{x} + \boldsymbol{v} \geq 0 \\ -1 & \text{if } \boldsymbol{x} + \boldsymbol{v} < 0 \end{cases}$$

We therefore need to evaluate the probability of the events $\boldsymbol{x} + \boldsymbol{v} \geq 0$ and $\boldsymbol{x} + \boldsymbol{v} < 0$. For the first case we have

$$\boldsymbol{x} + \boldsymbol{v} \geq 0 \Longleftrightarrow \;(\boldsymbol{x} = +1 \text{ and } \boldsymbol{v} \geq -1) \;\text{ or }\; (\boldsymbol{x} = -1 \text{ and } \boldsymbol{v} \geq 1)$$

Now recall that $\boldsymbol{x}$ and $\boldsymbol{v}$ are independent and that $\boldsymbol{v}$ is a zero-mean unit-variance Gaussian random variable. Thus, let

$$P(\boldsymbol{v} \geq 1) \stackrel{\Delta}{=} \alpha \tag{1.15}$$

Then

$$P(\boldsymbol{v} \geq -1) = 1 - P(\boldsymbol{v} \leq -1) = 1 - P(\boldsymbol{v} \geq 1) = 1 - \alpha$$

and we obtain

$$P(\boldsymbol{x} + \boldsymbol{v} \geq 0) = (1-\alpha)/2 + \alpha/2 = 1/2$$

Consequently, $P(\boldsymbol{x} + \boldsymbol{v} < 0) = 1/2$ and $\mathsf{E}\,\mathsf{sign}(\boldsymbol{x} + \boldsymbol{v}) = 0$. This means that the estimator $\hat{\boldsymbol{x}} = \mathsf{sign}(\boldsymbol{y})$ is unbiased. We now return to (1.14) and note that

$$\mathsf{E}\,[\boldsymbol{x} - \mathsf{sign}(\boldsymbol{y})]\mathsf{sign}(\boldsymbol{y}) = \mathsf{E}\;\boldsymbol{x}\mathsf{sign}(\boldsymbol{y}) - 1$$

Therefore, all we need to do in order to verify that (1.14) holds is to check that $\mathsf{E}\,\boldsymbol{x}\mathsf{sign}(\boldsymbol{y})$ does not evaluate to one. To do this, we introduce the random variable $\boldsymbol{z} = \boldsymbol{x}\mathsf{sign}(\boldsymbol{y})$ and proceed to evaluate its mean. It is clear from the definition of $\boldsymbol{z}$ that

$$\boldsymbol{z} = \begin{cases} +1 & \text{if } (\boldsymbol{x} = +1 \text{ and } \boldsymbol{v} \geq -1) \text{ or } (\boldsymbol{x} = -1 \text{ and } \boldsymbol{v} < 1) \\ -1 & \text{if } (\boldsymbol{x} = +1 \text{ and } \boldsymbol{v} < -1) \text{ or } (\boldsymbol{x} = -1 \text{ and } \boldsymbol{v} \geq 1) \end{cases}$$

The events

$$(\boldsymbol{x} = +1 \text{ and } \boldsymbol{v} \geq -1) \quad \text{or} \quad (\boldsymbol{x} = -1 \text{ and } \boldsymbol{v} < 1)$$

each has probability $0.5(1-\alpha)$. Likewise, the events

$$(\boldsymbol{x} = +1 \text{ and } \boldsymbol{v} < -1) \quad \text{or} \quad (\boldsymbol{x} = -1 \text{ and } \boldsymbol{v} \geq 1)$$

each has probability 0.5α. It then follows that $\mathsf{E}\,\boldsymbol{z} = 1 - 2\alpha \neq 1$, so that $\boldsymbol{x} - \mathsf{sign}(\boldsymbol{y})$ is correlated with $\mathsf{sign}(\boldsymbol{y})$. Hence, the estimator $\mathsf{sign}(\boldsymbol{y})$ does not satisfy the orthogonality condition (1.12) and, therefore, it cannot be the optimal least-mean-squares estimator.

◇

1.4 GAUSSIAN RANDOM VARIABLES

We mentioned earlier in Remark 1.1 that it is not always possible to determine a closed form expression for the optimal estimator $\mathsf{E}\,(\boldsymbol{x}|\boldsymbol{y})$. Only in some special cases this calculation can be carried out to completion (as we did in Ex. 1.2 and as we shall do in another example below). This difficulty will motivate us to limit ourselves in Part II (*Linear Estimation*) to the subclass of *linear (or affine) estimators*, namely, to choices of $h(\cdot)$ in (1.3) that are *affine* functions of the observation, say, $h(\boldsymbol{y}) = k\boldsymbol{y} + b$ for some constants k and b to be determined. Despite its apparent narrowness, this class of estimators performs reasonably well in many applications.

There is an important special case for which the *optimal* estimator of Thm. 1.1 turns out to be affine in $\boldsymbol{y}$. This scenario happens when the random variables $\boldsymbol{x}$ and $\boldsymbol{y}$ are jointly Gaussian. To see this, let us introduce the matrix

$$R \stackrel{\Delta}{=} \begin{bmatrix} \sigma_x^2 & \sigma_{xy} \\ \sigma_{xy} & \sigma_y^2 \end{bmatrix}$$

where $\{\sigma_x^2, \sigma_y^2, \sigma_{xy}\}$ denote the variances and cross-correlation of $\boldsymbol{x}$ and $\boldsymbol{y}$, respectively,

$$\sigma_x^2 = \mathsf{E}\,(\boldsymbol{x} - \bar{x})^2, \qquad \sigma_y^2 = \mathsf{E}\,(\boldsymbol{y} - \bar{y})^2, \qquad \sigma_{xy} = \mathsf{E}\,(\boldsymbol{x} - \bar{x})(\boldsymbol{y} - \bar{y})$$

The matrix R can be regarded as the *covariance* matrix of the column vector col$\{\boldsymbol{x}, \boldsymbol{y}\}$, namely,

$$R = \mathsf{E}\left(\begin{bmatrix} \boldsymbol{x} \\ \boldsymbol{y} \end{bmatrix} - \begin{bmatrix} \bar{x} \\ \bar{y} \end{bmatrix}\right)\left(\begin{bmatrix} \boldsymbol{x} \\ \boldsymbol{y} \end{bmatrix} - \begin{bmatrix} \bar{x} \\ \bar{y} \end{bmatrix}\right)^{\mathsf{T}} \tag{1.16}$$

where the symbol T denotes vector transposition, and the notation col$\{\alpha, \beta\}$ denotes a column vector whose entries are α and β. As explained in Sec. A.4, every such covariance matrix is necessarily symmetric, $R = R^{\mathsf{T}}$. Moreover, R is also nonnegative-definite, written as $R \geq 0$. To proceed with the analysis, we are going to assume that the covariance matrix R is positive-definite and, hence, invertible — see Prob. I.6.

Now the joint pdf of two jointly Gaussian random variables $\{\boldsymbol{x}, \boldsymbol{y}\}$ is given by (see Sec. A.5 for a review of Gaussian random variables and their probability density functions):

$$\boxed{f_{\boldsymbol{x},\boldsymbol{y}}(x, y) = \frac{1}{2\pi} \frac{1}{\sqrt{\det R}} \exp\left\{-\tfrac{1}{2} \begin{bmatrix} x - \bar{x} & y - \bar{y} \end{bmatrix} R^{-1} \begin{bmatrix} x - \bar{x} \\ y - \bar{y} \end{bmatrix}\right\}} \tag{1.17}$$

Also, the individual probability density functions of $\boldsymbol{x}$ and $\boldsymbol{y}$ are given by

$$\begin{aligned} f_{\boldsymbol{x}}(x) &= \frac{1}{\sqrt{2\pi}} \frac{1}{\sigma_x} \exp\left\{-(x - \bar{x})^2 / 2\sigma_x^2\right\} \\ f_{\boldsymbol{y}}(y) &= \frac{1}{\sqrt{2\pi}} \frac{1}{\sigma_y} \exp\left\{-(y - \bar{y})^2 / 2\sigma_y^2\right\} \end{aligned}$$

According to Thm. 1.1, the least-mean-squares estimator of $\boldsymbol{x}$ given $\boldsymbol{y}$ is $\hat{\boldsymbol{x}} = \mathsf{E}(\boldsymbol{x}|\boldsymbol{y})$, which requires that we determine the conditional pdf $f_{\boldsymbol{x}|\boldsymbol{y}}(x|y)$. This pdf can be obtained from the calculation:

$$\begin{aligned} f_{\boldsymbol{x}|\boldsymbol{y}}(x|y) &= \frac{f_{\boldsymbol{x},\boldsymbol{y}}(x, y)}{f_{\boldsymbol{y}}(y)} \\ &= \frac{\dfrac{1}{2\pi} \dfrac{1}{\sqrt{\det R}} \exp\left\{-\tfrac{1}{2} \begin{bmatrix} x - \bar{x} & y - \bar{y} \end{bmatrix} R^{-1} \begin{bmatrix} x - \bar{x} \\ y - \bar{y} \end{bmatrix}\right\}}{\dfrac{1}{\sqrt{2\pi}} \dfrac{1}{\sigma_y} \exp\left\{-(y - \bar{y})^2 / 2\sigma_y^2\right\}} \end{aligned} \tag{1.18}$$

In order to simplify the above ratio, we shall use the fact that R can be factored into a product of an upper-triangular, diagonal, and lower-triangular matrices, as follows (this can be checked by straightforward algebra):

$$\boxed{R = \begin{bmatrix} 1 & \sigma_{xy}/\sigma_y^2 \\ 0 & 1 \end{bmatrix} \begin{bmatrix} \sigma^2 & 0 \\ 0 & \sigma_y^2 \end{bmatrix} \begin{bmatrix} 1 & 0 \\ \sigma_{xy}/\sigma_y^2 & 1 \end{bmatrix}} \tag{1.19}$$

where we introduced the scalar

$$\sigma^2 \triangleq \sigma_x^2 - \sigma_{xy}^2 / \sigma_y^2$$

which is called the *Schur complement* of σ_y^2 in R; it is guaranteed to be positive in view of the assumed positive-definiteness of R itself. Indeed, and more generally, let

$$R = \begin{bmatrix} A & B \\ B^{\mathsf{T}} & C \end{bmatrix}$$

be any symmetric matrix with possibly matrix-valued entries $\{A, B, C\}$ satisfying $A = A^{\mathsf{T}}$ and $C = C^{\mathsf{T}}$. Assume further that C is invertible. Then it is easy to verify by direct calculation that every such matrix can be factored in the form

$$\begin{bmatrix} A & B \\ B^{\mathsf{T}} & C \end{bmatrix} = \begin{bmatrix} \mathrm{I} & BC^{-1} \\ 0 & \mathrm{I} \end{bmatrix} \begin{bmatrix} \Sigma & 0 \\ 0 & C \end{bmatrix} \begin{bmatrix} \mathrm{I} & 0 \\ C^{-1}B^{\mathsf{T}} & \mathrm{I} \end{bmatrix}$$

where

$$\Sigma = A - BC^{-1}B^{\mathsf{T}}$$

is called the Schur complement of R with respect to C. The factorization (1.19) is a special case of this result where the entries $\{A, B, C\}$ are scalars: $A = \sigma_x^2$, $B = \sigma_{xy}$, and $C = \sigma_y^2$. Moreover, the determinant of a positive-definite matrix is always positive — see Sec. B.1. We see from (1.19) that $\det R = \sigma^2\sigma_y^2$, so that σ^2 is necessarily positive since $\det R > 0$.

Now, by inverting both sides of (1.19), we find that the inverse of R can be factored as

$$R^{-1} = \begin{bmatrix} 1 & 0 \\ -\sigma_{xy}/\sigma_y^2 & 1 \end{bmatrix} \begin{bmatrix} 1/\sigma^2 & 0 \\ 0 & 1/\sigma_y^2 \end{bmatrix} \begin{bmatrix} 1 & -\sigma_{xy}/\sigma_y^2 \\ 0 & 1 \end{bmatrix} \tag{1.20}$$

where we used the simple fact that for any scalar a,

$$\begin{bmatrix} 1 & 0 \\ a & 1 \end{bmatrix}^{-1} = \begin{bmatrix} 1 & 0 \\ -a & 1 \end{bmatrix}, \quad \begin{bmatrix} 1 & a \\ 0 & 1 \end{bmatrix}^{-1} = \begin{bmatrix} 1 & -a \\ 0 & 1 \end{bmatrix}$$

Then

$$\begin{bmatrix} x - \bar{x} & y - \bar{y} \end{bmatrix} R^{-1} \begin{bmatrix} x - \bar{x} \\ y - \bar{y} \end{bmatrix} = \frac{[(x - \bar{x}) - \sigma_{xy}\sigma_y^{-2}(y - \bar{y})]^2}{\sigma^2} + \frac{(y - \bar{y})^2}{\sigma_y^2}$$

where the right-hand side is expressed as the sum of two quadratic terms. It follows that

$$\exp\left\{-\frac{1}{2}\begin{bmatrix} x - \bar{x} & y - \bar{y} \end{bmatrix} R^{-1} \begin{bmatrix} x - \bar{x} \\ y - \bar{y} \end{bmatrix}\right\} =$$
$$\exp\left\{-[(x - \bar{x}) - \sigma_{xy}\sigma_y^{-2}(y - \bar{y})]^2/2\sigma^2\right\} \exp\left\{-(y - \bar{y})^2/2\sigma_y^2\right\}$$

This equality, along with $\det R = \sigma^2\sigma_y^2$, allows us to simplify expression (1.18) for $f_{\boldsymbol{x}|\boldsymbol{y}}(x|y)$ to

$$f_{\boldsymbol{x}|\boldsymbol{y}}(x|y) = \frac{1}{\sqrt{2\pi}} \frac{1}{\sqrt{\sigma^2}} \exp\left\{-[(x - \bar{x}) - \sigma_{xy}\sigma_y^{-2}(y - \bar{y})]^2/2\sigma^2\right\}$$

This expression has the form of the pdf of a Gaussian random variable with variance σ^2 and mean value $\bar{x} + \sigma_{xy}\sigma_y^{-2}(y - \bar{y})$. Consequently, the optimal estimator is given by the *affine* relation:

$$\boxed{\hat{\boldsymbol{x}} = \mathsf{E}(\boldsymbol{x}|\boldsymbol{y}) = \bar{x} + \frac{\sigma_{xy}}{\sigma_y^2}(\boldsymbol{y} - \bar{y})} \tag{1.21}$$

Moreover, the resulting m.m.s.e., which is the variance of $\tilde{\boldsymbol{x}} = \boldsymbol{x} - \hat{\boldsymbol{x}}$, is given by

$$\boxed{\text{m.m.s.e.} \overset{\Delta}{=} \sigma_{\tilde{x}}^2 = \sigma_x^2 - \sigma_{\hat{x}}^2 = \sigma_x^2 - \frac{\sigma_{xy}^2}{\sigma_y^2} = \sigma^2} \tag{1.22}$$

Observe that, in this Gaussian case, the m.m.s.e. is completely specified by the second-order statistics of the random variables $\{\boldsymbol{x}, \boldsymbol{y}\}$ (namely, σ_x^2, σ_y^2, and σ_{xy}). Note also that the m.m.s.e. is smaller than σ_x^2.

Example 1.4 (Correlation coefficient)

A measure of the correlation between two random variables is their correlation coefficient, defined by

$$\rho_{xy} \triangleq \sigma_{xy}/\sigma_x\sigma_y$$

It is shown in Prob. I.6 that ρ_{xy} always lies in the interval $[-1, 1]$. As ρ_{xy} moves closer to zero, the variables $\boldsymbol{x}$ and $\boldsymbol{y}$ become more uncorrelated (in the Gaussian case, this also means that the variables become less dependent). We see from (1.22) that the m.m.s.e. in the Gaussian case can be rewritten in the form

$$\text{m.m.s.e.} = \sigma_x^2(1 - \rho_{xy}^2)$$

This shows that when $\rho_{xy} = 0$, which occurs when $\sigma_{xy} = 0$, the resulting m.m.s.e. is σ_x^2. Also, from (1.21), the estimator collapses to $\hat{\boldsymbol{x}} = \bar{x}$. That is, we are reduced to the simple estimator studied in Sec. 1.1. This is expected since in the Gaussian case, a zero cross-correlation means that the random variables $\boldsymbol{x}$ and $\boldsymbol{y}$ are independent so there is no additional information available that we can use to estimate $\boldsymbol{x}$, besides its mean and variance.

Example 1.5 (Gaussian noise)

Let $\boldsymbol{x}$ denote a Gaussian random variable with mean $\bar{x} = 1$ and variance $\sigma_x^2 = 2$. Similarly, let $\boldsymbol{v}$ denote a Gaussian random variable independent of $\boldsymbol{x}$, with mean $\bar{v} = 2$ and variance σ_v^2. Now consider the noisy measurement

$$\boldsymbol{y} = 2\boldsymbol{x} + \boldsymbol{v}$$

and let us estimate $\boldsymbol{x}$ from $\boldsymbol{y}$. According to (1.21), we need to determine the quantities $\{\bar{y}, \sigma_{xy}, \sigma_y^2\}$. From the above equation we find that

$$\bar{y} = 2\bar{x} + \bar{v} = 4$$

The independence of $\boldsymbol{x}$ and $\boldsymbol{v}$ implies that

$$\sigma_y^2 = 4\sigma_x^2 + \sigma_v^2 = 8 + \sigma_v^2$$

Finally, the cross-correlation σ_{xy} is given by

$$\sigma_{xy} = \mathsf{E}\,(\boldsymbol{x} - \bar{x})(\boldsymbol{y} - \bar{y}) = \mathsf{E}\,(\boldsymbol{x} - 1)(2\boldsymbol{x} + \boldsymbol{v} - 4) = 4$$

where we used

$$\mathsf{E}\boldsymbol{x}^2 = \sigma_x^2 + \bar{x}^2 = 3$$

and

$$\mathsf{E}\boldsymbol{x}\boldsymbol{v} = \mathsf{E}\boldsymbol{x}\mathsf{E}\boldsymbol{v} = 2$$

Using (1.21) and (1.22) we obtain

$$\hat{\boldsymbol{x}} = 1 + \frac{4}{8 + \sigma_v^2}(\boldsymbol{y} - 4) \quad \text{and} \quad \sigma_{\tilde{x}}^2 = 2 - \frac{16}{8 + \sigma_v^2} = \frac{2\sigma_v^2}{8 + \sigma_v^2}$$

Moreover, since $\sigma_{\tilde{x}}^2 = \sigma_x^2 - \sigma_{\hat{x}}^2$, we also find that

$$\sigma_{\hat{x}}^2 = \frac{16}{8 + \sigma_v^2}$$

CHAPTER 2

Vector-Valued Data

We have focused so far on *scalar real-valued* random variables. The results however can be extended in a straightforward manner, by using the convenience and power of the vector notation, to the cases of vector-valued and complex-valued random variables.

These two situations are common in applications. For example, in channel estimation problems, the quantities to be estimated are the samples of the impulse response sequence of a supposedly finite-impulse-response (FIR) channel. If we group these samples into a vector x, then we are faced with the problem of estimating a vector rather than a scalar quantity. Likewise, in quadrature amplitude modulation (QAM) or in quadrature phase-shift keying (QPSK) transmissions over a communications channel, the transmitted symbols are complex-valued. The recovery of these symbols at the receiver requires that we solve an estimation problem that involves estimating complex-valued quantities.

2.1 OPTIMAL ESTIMATOR IN THE VECTOR CASE

It turns out that the optimal estimator in the general vector and complex-valued case is still given by the conditional expectation of $\boldsymbol{x}$ given $\boldsymbol{y}$. To see this, let us start with a special case. Assume $\boldsymbol{x}$ and $\boldsymbol{y}$ are both real-valued with $\boldsymbol{x}$ a *scalar* and $\boldsymbol{y}$ a vector, say,

$$\boldsymbol{y} = \text{col}\{\boldsymbol{y}(0), \boldsymbol{y}(1), \ldots, \boldsymbol{y}(q-1)\}$$

As before, let $\hat{\boldsymbol{x}} = h(\boldsymbol{y})$ denote an estimator for $\boldsymbol{x}$. Since $\boldsymbol{y}$ is vector-valued, the function $h(\cdot)$ operates on the entries of $\boldsymbol{y}$ and provides a real scalar quantity as a result. More explicitly, we write

$$\hat{\boldsymbol{x}} = h\left(\boldsymbol{y}(0), \boldsymbol{y}(1), \ldots, \boldsymbol{y}(q-1)\right)$$

The function $h(\cdot)$ is to be chosen optimally by minimizing the variance of the error $\tilde{\boldsymbol{x}} = \boldsymbol{x} - \hat{\boldsymbol{x}}$, i.e., by solving

$$\min_{h(\cdot)} \mathsf{E}\, \tilde{\boldsymbol{x}}^2$$

The same argument that we used to establish Thm. 1.1 can be repeated here to verify that the optimal estimator is still given by

$$\hat{\boldsymbol{x}} \;=\; \mathsf{E}\,(\boldsymbol{x}|\boldsymbol{y}) \;=\; \mathsf{E}\,[\boldsymbol{x}|\boldsymbol{y}(0), \boldsymbol{y}(1), \ldots, \boldsymbol{y}(q-1)] \quad . \tag{2.1}$$

The only difference between this result and that of Thm. 1.1 is that the conditional expectation is now computed relative to a collection of random variables $\{\boldsymbol{y}(i)\}$, rather than a single random variable. Moreover, for any function $g(\cdot)$ of $\boldsymbol{y}$, the orthogonality condition (1.5) extends to this case and is still given by

$$\mathsf{E}\,[\boldsymbol{x} - \mathsf{E}\,(\boldsymbol{x}|\boldsymbol{y})]g(\boldsymbol{y}) = 0 \tag{2.2}$$

Let us return to Ex. 1.2, where x is a BPSK signal that is either $+1$ or -1 with probability $1/2$ each. Assume that we collect two noisy measurements $y(0)$ and $y(1)$ of x, say, $y(0) = x + v(0)$ and $y(1) = x + v(1)$, where $\{v(0), v(1)\}$ are zero-mean unit-variance Gaussian random variables that are independent of each other and of x. The value of x is the same in both measurements (i.e., if it is $+1$ in the measurement $y(0)$, it is also $+1$ in the measurement $y(1)$, and similarly for -1.) We may interpret $\{y(0), y(1)\}$ as the noisy signals measured at two antennas as a result of transmitting x over two additive Gaussian-noise channels — see Fig. 2.1.

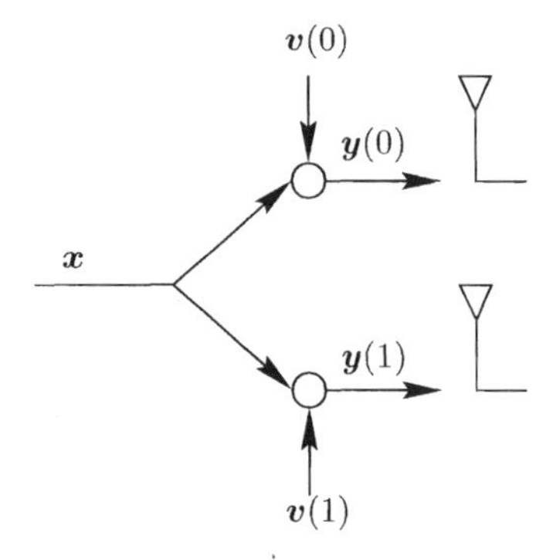

FIGURE 2.1 Reception by two antennas of a symbol x transmitted over two additive Gaussian-noise channels.

We can then pose the problem of estimating x given *both* measurements $\{y(0), y(1)\}$. According to (2.1), the solution is given by

$$\hat{x} = \mathsf{E}\,[x|y(0), y(1)]$$

The evaluation of the conditional expectation in this case is a trivial extension of the derivation given in Ex. 1.2, and it is left as an exercise to the reader — see Prob. I.13, where the more general case of multiple measurements is treated. The result of that problem shows that

$$\hat{x} = \tanh[y(0) + y(1)]$$

In the context of the two-antenna example of Fig. 2.1, this result leads to the optimal receiver structure shown in Fig. 2.2.

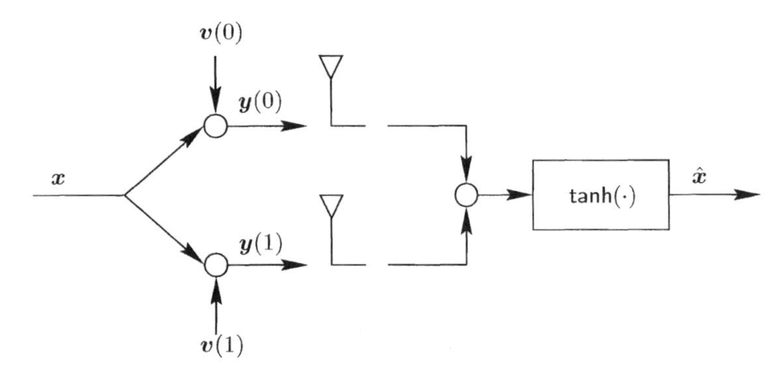

FIGURE 2.2 Optimal receiver structure for recovering a symbol x from two separate measurements over additive Gaussian-noise channels.

◇

Let us now study the general case and determine the form of the optimal estimator for a *vector-valued* random variable $\boldsymbol{x}$ given another vector-valued random variable $\boldsymbol{y}$, with both variables allowed to be complex-valued as well. Thus, assume that $\boldsymbol{x}$ is $p-$dimensional while $\boldsymbol{y}$ is $q-$dimensional.

Again, let $\hat{\boldsymbol{x}} = h(\boldsymbol{y})$ denote an estimator for $\boldsymbol{x}$. Since $\boldsymbol{x}$ and $\boldsymbol{y}$ are vector-valued, the function $h(\cdot)$ operates on the entries of $\boldsymbol{y}$ and provides a vector quantity as a result. More explicitly, we can write for the individual entries of $\hat{\boldsymbol{x}}$ and $\boldsymbol{y}$,

$$\begin{bmatrix} \hat{\boldsymbol{x}}(0) \\ \hat{\boldsymbol{x}}(1) \\ \hat{\boldsymbol{x}}(2) \\ \vdots \\ \hat{\boldsymbol{x}}(p-1) \end{bmatrix} = \begin{bmatrix} h_0[\boldsymbol{y}(0), \boldsymbol{y}(1), \ldots, \boldsymbol{y}(q-1)] \\ h_1[\boldsymbol{y}(0), \boldsymbol{y}(1), \ldots, \boldsymbol{y}(q-1)] \\ h_2[\boldsymbol{y}(0), \boldsymbol{y}(1), \ldots, \boldsymbol{y}(q-1)] \\ \vdots \\ h_{p-1}[\boldsymbol{y}(0), \boldsymbol{y}(1), \ldots, \boldsymbol{y}(q-1)] \end{bmatrix}$$

where the $\{h_k(\cdot)\}$ represent the individual mappings from the observation vector $\boldsymbol{y}$ to the estimators $\{\hat{\boldsymbol{x}}(k)\}$. We can then seek optimal functions $\{h_k(\cdot)\}$ that minimize the variance of the error in each component of $\boldsymbol{x}$, namely, each $h_k(\cdot)$ is determined by solving

$$\boxed{\min_{h_k(\cdot)} \quad \mathsf{E}\, |\tilde{\boldsymbol{x}}(k)|^2} \tag{2.3}$$

where

$$\tilde{\boldsymbol{x}}(k) \stackrel{\Delta}{=} \boldsymbol{x}(k) - h_k(\boldsymbol{y})$$

Actually, this formulation is equivalent to solving over all $\{h_k(\cdot)\}$ the following problem:

$$\boxed{\min_{\{h_k(\cdot)\}} \quad \mathsf{E}\, \tilde{\boldsymbol{x}}^* \tilde{\boldsymbol{x}}} \tag{2.4}$$

This is because the quantity $\mathsf{E}\,\tilde{\boldsymbol{x}}^*\tilde{\boldsymbol{x}}$ in (2.4) is the sum of the individual terms $\mathsf{E}\,|\tilde{\boldsymbol{x}}(k)|^2$,

$$\mathsf{E}\,\tilde{\boldsymbol{x}}^*\tilde{\boldsymbol{x}} = \mathsf{E}\,|\tilde{\boldsymbol{x}}(0)|^2 + \mathsf{E}\,|\tilde{\boldsymbol{x}}(1)|^2 + \ldots + \mathsf{E}\,|\tilde{\boldsymbol{x}}(p-1)|^2$$

with each term $\mathsf{E}\,|\tilde{\boldsymbol{x}}(k)|^2$ depending only on the corresponding function $h_k(\cdot)$. In this way, minimizing the sum $\mathsf{E}\,\tilde{\boldsymbol{x}}^*\tilde{\boldsymbol{x}}$ over all $\{h_k(\cdot)\}$ is equivalent to minimizing each individual term, $\mathsf{E}\,|\tilde{\boldsymbol{x}}(k)|^2$, over its $h_k(\cdot)$. Note further that

$$\mathsf{E}\,\tilde{\boldsymbol{x}}^*\tilde{\boldsymbol{x}} = \mathsf{Tr}\,(\mathsf{E}\,\tilde{\boldsymbol{x}}\tilde{\boldsymbol{x}}^*) = \mathsf{Tr}\,(R_{\tilde{x}})$$

That is, the scalar quantity $\mathsf{E}\,\tilde{\boldsymbol{x}}^*\tilde{\boldsymbol{x}}$ in (2.4) is equal to the trace of the error covariance matrix $R_{\tilde{x}}$. This is because the trace of a matrix is equal to the sum of its diagonal elements and, therefore, for any column vector a, it holds that $a^*a = \mathsf{Tr}(aa^*)$. Then problem (2.4) is also equivalent to solving over all $\{h_k(\cdot)\}$:

$$\boxed{\min_{\{h_k(\cdot)\}} \quad \mathsf{Tr}(R_{\tilde{x}})} \tag{2.5}$$

Now the solution to the general problem (2.3) follows from the special case discussed at the beginning of this section. Indeed, if we express $\boldsymbol{x}(k)$ and $h_k(\cdot)$ in terms of their real and imaginary parts, say,

$$\boldsymbol{x}(k) \stackrel{\Delta}{=} \boldsymbol{x}_{\mathsf{r}}(k) + j\,\boldsymbol{x}_{\mathsf{i}}(k), \qquad h_k(\boldsymbol{y}) \stackrel{\Delta}{=} h_{\mathsf{r},k}(\boldsymbol{y}) + j\,h_{\mathsf{i},k}(\boldsymbol{y})$$

then we can expand the error criterion as

$$\mathsf{E}\,|\boldsymbol{x}(k) - h_k(\boldsymbol{y})|^2 = \mathsf{E}\,[\boldsymbol{x}_{\mathsf{r}}(k) - h_{\mathsf{r},k}(\boldsymbol{y})]^2 \;+\; \mathsf{E}\,[\boldsymbol{x}_{\mathsf{i}}(k) - h_{\mathsf{i},k}(\boldsymbol{y})]^2$$

and we are reduced to minimizing the sum of two nonnegative quantities over the unknowns $\{h_{\mathsf{r},k}(\cdot), h_{\mathsf{i},k}(\cdot)\}$. This is equivalent to minimizing each term separately,

$$\min_{h_{\mathsf{r},k}(\cdot)} \mathsf{E}\,[\boldsymbol{x}_{\mathsf{r}}(k) - h_{\mathsf{r},k}(\boldsymbol{y})]^2, \qquad\qquad \min_{h_{\mathsf{i},k}(\cdot)} \mathsf{E}\,[\boldsymbol{x}_{\mathsf{i}}(k) - h_{\mathsf{i},k}(\boldsymbol{y})]^2$$

and the solution we already know from (2.1) to be given by

$$\begin{aligned}
\hat{\boldsymbol{x}}_{\mathsf{r}}(k) &= \mathsf{E}\,[\boldsymbol{x}_{\mathsf{r}}(k)|\boldsymbol{y}_{\mathsf{r}}(0), \boldsymbol{y}_{\mathsf{i}}(0), \boldsymbol{y}_{\mathsf{r}}(1), \boldsymbol{y}_{\mathsf{i}}(1), \ldots, \boldsymbol{y}_{\mathsf{r}}(q-1), \boldsymbol{y}_{\mathsf{i}}(q-1)] \\
\hat{\boldsymbol{x}}_{\mathsf{i}}(k) &= \mathsf{E}\,[\boldsymbol{x}_{\mathsf{i}}(k)|\boldsymbol{y}_{\mathsf{r}}(0), \boldsymbol{y}_{\mathsf{i}}(0), \boldsymbol{y}_{\mathsf{r}}(1), \boldsymbol{y}_{\mathsf{i}}(1), \ldots, \boldsymbol{y}_{\mathsf{r}}(q-1), \boldsymbol{y}_{\mathsf{i}}(q-1))
\end{aligned}$$

Therefore, the optimal choice for $h_k(\cdot)$ is

$$\hat{\boldsymbol{x}}(k) = \mathsf{E}\,[\boldsymbol{x}(k)|\boldsymbol{y}]$$

so that the optimal estimator that minimizes the variances of the individual errors $\{\tilde{\boldsymbol{x}}(k)\}$ is

$$\boxed{\hat{\boldsymbol{x}} = \mathsf{E}\,(\boldsymbol{x}|\boldsymbol{y}) \;\triangleq\; \begin{bmatrix} \mathsf{E}\,[\boldsymbol{x}(0)|\boldsymbol{y}] \\ \mathsf{E}\,[\boldsymbol{x}(1)|\boldsymbol{y}] \\ \vdots \\ \mathsf{E}\,[\boldsymbol{x}(p-1)|\boldsymbol{y}] \end{bmatrix}} \tag{2.6}$$

Likewise, using the property (1.5) of conditional expectations, we conclude that the orthogonality condition in this case is still given by

$$\mathsf{E}\,[\boldsymbol{x} - \mathsf{E}\,(\boldsymbol{x}|\boldsymbol{y})]g(\boldsymbol{y}) = 0 \tag{2.7}$$

for any function $g(\cdot)$ of the observation vector $\boldsymbol{y}$.

Theorem 2.1 (Optimal estimation in the vector case) The least-mean-squares estimator of a (possibly complex-valued) vector $\boldsymbol{x}$ given another (possibly complex-valued) vector $\boldsymbol{y}$ is the conditional expectation of $\boldsymbol{x}$ given $\boldsymbol{y}$, i.e., $\hat{\boldsymbol{x}} = \mathsf{E}\,(\boldsymbol{x}|\boldsymbol{y})$. This estimator solves

$$\min_{\hat{\boldsymbol{x}}} \mathsf{Tr}(R_{\tilde{x}})$$

where $R_{\tilde{x}} = \mathsf{E}\,\tilde{\boldsymbol{x}}\tilde{\boldsymbol{x}}^*$ and $\tilde{\boldsymbol{x}} = \boldsymbol{x} - \hat{\boldsymbol{x}}$.

Example 2.2 (Estimation of transmitted symbols)

Consider again the setting of Ex. A.2 and assume $N = 2$, so that we are interested in estimating the vector $\boldsymbol{x} = \text{col}\{\boldsymbol{s}(0), \boldsymbol{s}(1)\}$ from the observation $\boldsymbol{y} = \text{col}\{\boldsymbol{y}(0), \boldsymbol{y}(1)\}$, where

$$\boldsymbol{y}(0) = \boldsymbol{s}(0) + \boldsymbol{v}(0) \quad \text{and} \quad \boldsymbol{y}(1) = \boldsymbol{s}(1) + 0.5\boldsymbol{s}(0) + \boldsymbol{v}(1)$$

Here we are assuming that transmissions start at time 0 so that $s(-1) = 0$. If we introduce the 2×2 matrix

$$H = \begin{bmatrix} 1 & 0 \\ 0.5 & 1 \end{bmatrix}$$

and the vector $\boldsymbol{v} = \text{col}\{\boldsymbol{v}(0), \boldsymbol{v}(1)\}$, then the above equations can be written more compactly in matrix form as

$$\boldsymbol{y} = H\boldsymbol{x} + \boldsymbol{v}$$

We are therefore faced with the problem of estimating $\boldsymbol{x}$ from $\boldsymbol{y}$, with the noise term $\boldsymbol{v}$ assumed independent of $\boldsymbol{x}$. Now recall that the symbols $s(i)$ are either $+1$ or -1 with probabilities p or $1-p$, respectively. Hence, the vector $\boldsymbol{x}$ can assume one of four values:

$$\boldsymbol{x} \in \left\{ \begin{bmatrix} +1 \\ +1 \end{bmatrix}, \begin{bmatrix} -1 \\ -1 \end{bmatrix}, \begin{bmatrix} +1 \\ -1 \end{bmatrix}, \begin{bmatrix} -1 \\ +1 \end{bmatrix} \right\} \triangleq \{m_0, m_1, m_2, m_3\}$$

with probabilities $\{p^2,\ (1-p)^2,\ p(1-p),\ p(1-p)\}$, respectively. Observe that we are denoting the four possibilities for $\boldsymbol{x}$ by $\{m_i, i = 0, \ldots, 3\}$ for compactness of notation. Let also $q = 1 - p$. Moreover, the pdf of $\boldsymbol{v}$ is Gaussian and given by

$$f_{\boldsymbol{v}}(v) = \frac{1}{2\pi} \exp\left\{-v^{\mathsf{T}} v/2\right\}$$

since the covariance matrix of $\boldsymbol{v}$ is assumed to be the identity matrix. It then follows that the pdf of $\boldsymbol{y}$ is given by

$$f_{\boldsymbol{y}}(y) = p^2 f_{\boldsymbol{v}}\,(y - Hm_0) \;+\; q^2 f_{\boldsymbol{v}}\,(y - Hm_1) \;+\; pqf_{\boldsymbol{v}}\,(y - Hm_2) \;+\; pqf_{\boldsymbol{v}}\,(y - Hm_3)$$

Similarly, we obtain, as in Ex. 1.2, that

$$\begin{aligned} f_{\boldsymbol{x},\boldsymbol{y}}(x, y) &= f_{\boldsymbol{x}}(x) \cdot f_{\boldsymbol{v}}(y - Hx) \\ &= p^2 f_{\boldsymbol{v}}(y - Hm_0)\delta(x - m_0) \;+\; q^2 f_{\boldsymbol{v}}(y - Hm_1)\delta(x - m_1) \\ &\quad + pqf_{\boldsymbol{v}}(y - Hm_2)\delta(x - m_2) \;+\; pqf_{\boldsymbol{v}}(y - Hm_3)\delta(x - m_3) \end{aligned}$$

The expressions so derived for $f_{\boldsymbol{y}}(y)$ and $f_{\boldsymbol{x},\boldsymbol{y}}(x, y)$ allow us to evaluate $f_{\boldsymbol{x}|\boldsymbol{y}}(x|y)$, from which we can evaluate the desired conditional expectation $\mathsf{E}\,(\boldsymbol{x}|\boldsymbol{y})$ and, consequently, $\{\hat{s}(0), \hat{s}(1)\}$. This final computation is left as an exercise to the reader — see Prob. I.15.

◇

2.2 SPHERICALLY INVARIANT GAUSSIAN VARIABLES

We saw earlier in Sec. 1.4 that for *scalar* real-valued Gaussian random variables $\{\boldsymbol{x}, \boldsymbol{y}\}$, the optimal estimator of $\boldsymbol{x}$ given $\boldsymbol{y}$ depends in an affine manner on the observation $\boldsymbol{y}$. The same conclusion holds in the general *vector* complex-valued case.

So assume that $\boldsymbol{x}$ and $\boldsymbol{y}$ are jointly Gaussian random *vector* variables with a nonsingular covariance matrix

$$R \triangleq \begin{bmatrix} R_x & R_{xy} \\ R_{yx} & R_y \end{bmatrix}$$

where

$$\begin{aligned} R_x &= \mathsf{E}\,(\boldsymbol{x} - \bar{x})(\boldsymbol{x} - \bar{x})^* \\ R_y &= \mathsf{E}\,(\boldsymbol{y} - \bar{y})(\boldsymbol{y} - \bar{y})^* \\ R_{xy} &= \mathsf{E}\,(\boldsymbol{x} - \bar{x})(\boldsymbol{y} - \bar{y})^* = R_{yx}^* \end{aligned}$$

The variables $\{\boldsymbol{x}, \boldsymbol{y}\}$ are assumed to be complex-valued with dimensions $p \times 1$ for $\boldsymbol{x}$ and $q \times 1$ for $\boldsymbol{y}$.

If $\boldsymbol{x}$ and $\boldsymbol{y}$ were *real-valued*, then their individual probability density functions, as well as their joint pdf, would be given by (see Sec. A.5):

$$f_{\boldsymbol{x}}(x) = \frac{1}{\sqrt{(2\pi)^p}} \frac{1}{\sqrt{\det R_x}} \exp\left\{-\frac{1}{2}(x-\bar{x})^{\mathsf{T}} R_x^{-1}(x-\bar{x})\right\}$$

$$f_{\boldsymbol{y}}(y) = \frac{1}{\sqrt{(2\pi)^q}} \frac{1}{\sqrt{\det R_y}} \exp\left\{-\frac{1}{2}(y-\bar{y})^{\mathsf{T}} R_y^{-1}(y-\bar{y})\right\}$$

$$f_{\boldsymbol{x},\boldsymbol{y}}(x,y) = \frac{1}{\sqrt{(2\pi)^{p+q}}} \frac{1}{\sqrt{\det R}} \exp\left\{-\frac{1}{2}\left[\begin{array}{cc}(x-\bar{x})^{\mathsf{T}} & (y-\bar{y})^{\mathsf{T}}\end{array}\right] R^{-1} \left[\begin{array}{c} x-\bar{x} \\ y-\bar{y}\end{array}\right]\right\}$$

In particular, observe that if $\boldsymbol{x}$ and $\boldsymbol{y}$ were *uncorrelated*, i.e., if $R_{xy} = 0$, then the covariance matrix R becomes block diagonal, with entries $\{R_x, R_y\}$, and it is straightforward to verify from the above pdf expressions that in this case $f_{\boldsymbol{x},\boldsymbol{y}}(x,y) = f_{\boldsymbol{x}}(x) f_{\boldsymbol{y}}(y)$. In other words, uncorrelated real-valued Gaussian random variables are also independent.

When, on the other hand, $\boldsymbol{x}$ and $\boldsymbol{y}$ are *complex-valued*, they need to satisfy two conditions in order for their individual and joint pdfs to have forms similar to the above in the Gaussian case. These conditions are known as *circularity* assumptions, and the need for them is explained in Sec. A.5. The conditions are as follows. Each variable is required to be *circular*, meaning that $\{\boldsymbol{x}, \boldsymbol{y}\}$ should satisfy

$$\mathsf{E}\,(\boldsymbol{y}-\bar{y})(\boldsymbol{y}-\bar{y})^{\mathsf{T}} = 0 \quad \text{and} \quad \mathsf{E}\,(\boldsymbol{x}-\bar{x})(\boldsymbol{x}-\bar{x})^{\mathsf{T}} = 0$$

with the transposition symbol T used instead of the conjugation symbol $*$. The variables are also required to be second-order circular, i.e.,

$$\mathsf{E}\,(\boldsymbol{x}-\bar{x})(\boldsymbol{y}-\bar{y})^{\mathsf{T}} = 0$$

These circularity assumptions are not needed when the variables $\{\boldsymbol{x}, \boldsymbol{y}\}$ are real-valued. The circularity of $\boldsymbol{x}$ in the complex case guarantees that its pdf in the Gaussian case will have the form

$$f_{\boldsymbol{x}}(x) = \frac{1}{\pi^p} \frac{1}{\det R_x} \exp\left\{-(x-\bar{x})^* R_x^{-1}(x-\bar{x})\right\}$$

Likewise, the circularity of $\boldsymbol{y}$ guarantees that its pdf will have the form

$$f_{\boldsymbol{y}}(y) = \frac{1}{\pi^q} \frac{1}{\det R_y} \exp\left\{-(y-\bar{y})^* R_y^{-1}(y-\bar{y})\right\}$$

The second-order circularity of $\boldsymbol{x}$ and $\boldsymbol{y}$ guarantees that the joint pdf of $\{\boldsymbol{x}, \boldsymbol{y}\}$ will have the form

$$f_{\boldsymbol{x},\boldsymbol{y}}(x,y) = \frac{1}{\pi^{p+q}} \frac{1}{\det R} \exp\left\{-\left[\begin{array}{cc}(x-\bar{x})^* & (y-\bar{y})^*\end{array}\right] R^{-1} \left[\begin{array}{c} x-\bar{x} \\ y-\bar{y}\end{array}\right]\right\}$$

Thus, observe again that if $\boldsymbol{x}$ and $\boldsymbol{y}$ were *uncorrelated*, then the above pdf expressions lead to

$$f_{\boldsymbol{x},\boldsymbol{y}}(x,y) = f_{\boldsymbol{x}}(x) \cdot f_{\boldsymbol{y}}(y)$$

which shows that uncorrelated circular Gaussian random variables are also independent. This conclusion would not have held without the circularity assumptions in the complex case. We may add that circular Gaussian random variables are also called spherically-invariant Gaussian random variables.

Now the least-mean-squares estimator of $\boldsymbol{x}$ given $\boldsymbol{y}$ requires that we determine the conditional pdf $f_{\boldsymbol{x}|\boldsymbol{y}}(x|y)$. This can be obtained from the calculation

$$f_{\boldsymbol{x}|\boldsymbol{y}}(x|y) = \frac{f_{\boldsymbol{x},\boldsymbol{y}}(x,y)}{f_{\boldsymbol{y}}(y)} = \frac{1}{\pi^p}\frac{\det R_y}{\det R}\frac{\exp\left\{-\left[\begin{array}{cc}(x-\bar{x})^* & (y-\bar{y})^*\end{array}\right] R^{-1}\left[\begin{array}{c}x-\bar{x}\\ y-\bar{y}\end{array}\right]\right\}}{\exp\left\{-(y-\bar{y})^* R_y^{-1}(y-\bar{y})\right\}}$$

Following the same argument that we used earlier in Sec. 1.4, we can simplify the above expression by introducing the *block* upper-diagonal-lower triangular factorization (whose validity can again be verified, e.g., by direct calculation):

$$R \triangleq \left[\begin{array}{cc} R_x & R_{xy} \\ R_{yx} & R_y \end{array}\right] = \left[\begin{array}{cc} \mathrm{I} & R_{xy}R_y^{-1} \\ 0 & \mathrm{I} \end{array}\right]\left[\begin{array}{cc} \Sigma & 0 \\ 0 & R_y \end{array}\right]\left[\begin{array}{cc} \mathrm{I} & 0 \\ R_y^{-1}R_{yx} & \mathrm{I} \end{array}\right]$$

where Σ is the Schur complement of R_y in R, namely,

$$\Sigma = R_x - R_{xy}R_y^{-1}R_{yx}$$

Inverting both sides of the above factorization for R we get

$$R^{-1} = \left[\begin{array}{cc} \mathrm{I} & 0 \\ -R_y^{-1}R_{yx} & \mathrm{I} \end{array}\right]\left[\begin{array}{cc} \Sigma^{-1} & 0 \\ 0 & R_y^{-1} \end{array}\right]\left[\begin{array}{cc} \mathrm{I} & -R_{xy}R_y^{-1} \\ 0 & \mathrm{I} \end{array}\right]$$

which allows us to express the term

$$\left[\begin{array}{cc}(x-\bar{x})^* & (y-\bar{y})^*\end{array}\right] R^{-1}\left[\begin{array}{c}x-\bar{x}\\ y-\bar{y}\end{array}\right]$$

which appears in the expression for $f_{\boldsymbol{x},\boldsymbol{y}}(x,y)$, as the following separable sum of two quadratic terms,

$$[(x-\bar{x}) - R_{xy}R_y^{-1}(y-\bar{y})]^*\Sigma^{-1}[(x-\bar{x}) - R_{xy}R_y^{-1}(y-\bar{y})] \; + \; (y-\bar{y})^*R_y^{-1}(y-\bar{y})$$

Substituting this equality into the expression for $f_{\boldsymbol{x}|\boldsymbol{y}}(x|y)$, and using

$$\det R = \det\Sigma\cdot\det R_y$$

we conclude that

$$f_{\boldsymbol{x}|\boldsymbol{y}}(x|y) \; =$$

$$\frac{1}{\pi^p}\frac{1}{\det\Sigma}\exp\left\{-[(x-\bar{x}) - R_{xy}R_y^{-1}(y-\bar{y})]^*\Sigma^{-1}[(x-\bar{x}) - R_{xy}R_y^{-1}(y-\bar{y})]\right\}$$

which can be interpreted as the pdf of a circular Gaussian random variable with covariance matrix Σ and mean value $\bar{x} + R_{xy}R_y^{-1}(y-\bar{y})$. We therefore conclude that

$$\boxed{\hat{\boldsymbol{x}} \triangleq \mathsf{E}(\boldsymbol{x}|\boldsymbol{y}) \; = \; \bar{x} + R_{xy}R_y^{-1}(\boldsymbol{y}-\bar{y})}$$

and the resulting m.m.s.e. matrix is

$$\boxed{\text{m.m.s.e.} \triangleq R_{\tilde{x}} \; = \; R_x - R_{\hat{x}} = R_x - R_{xy}R_y^{-1}R_{yx}}$$

These are the extensions to the vector case of expressions (1.21) and (1.22) in the scalar case. Note further that in the zero-mean case we obtain

$$\boxed{\hat{\boldsymbol{x}} = R_{xy}R_y^{-1}\boldsymbol{y}}$$

with $\{R_x, R_y, R_{xy}\}$ defined accordingly,

$$R_x = \mathsf{E}\,\boldsymbol{x}\boldsymbol{x}^*, \quad R_y = \mathsf{E}\,\boldsymbol{y}\boldsymbol{y}^*, \quad R_{xy} = \mathsf{E}\,\boldsymbol{x}\boldsymbol{y}^*$$

Observe from the above expressions that the solution of the optimal estimation problem in the Gaussian case is completely determined by the second-order moments of the variables $\{\boldsymbol{x}, \boldsymbol{y}\}$ (i.e., by R_x, R_y, and R_{xy}). This means that, in the Gaussian case, the m.m.s.e. matrix can be evaluated *beforehand* by the designer (i.e., prior to the collection of the observations); a step that provides a mechanism for checking whether the least-mean-squares estimator will be an acceptable solution.

Lemma 2.1 (Circular Gaussian variables) If $\boldsymbol{x}$ and $\boldsymbol{y}$ are two circular and jointly Gaussian random variables with means $\{\bar{x}, \bar{y}\}$ and covariance matrices $\{R_x, R_y, R_{xy}\}$, then the least-mean-squares estimator of $\boldsymbol{x}$ given $\boldsymbol{y}$ is

$$\hat{\boldsymbol{x}} \;=\; \bar{x} + R_{xy}R_y^{-1}(\boldsymbol{y} - \bar{y})$$

and the resulting minimum cost is m.m.s.e. $= R_x - R_{xy}R_y^{-1}R_{yx}$.

2.3 EQUIVALENT OPTIMIZATION CRITERION

A useful fact to highlight at the end of this chapter is that the optimal estimator $\mathsf{E}\,(\boldsymbol{x}|\boldsymbol{y})$ defined by (2.6), which solves problems (2.3)–(2.5), is also the optimal solution of another related *matrix-valued* error criterion (cf. (2.9) further ahead), as we now explain.

Thus, consider the following alternative formulation. Assume that we pose the problem of estimating $\boldsymbol{x}$ from $\boldsymbol{y}$ by requiring that the functions $\{h_k(\cdot)\}$ be such that they minimize the variance of any *arbitrary* linear combination of the entries of the error vector, say, $a^*(\boldsymbol{x} - h(\boldsymbol{y}))$ for any a. That is, assume we replace the optimization problem (2.3) by the alternative problem

$$\min_{\{h_k(\cdot)\}} \; \mathsf{E}\,|a^*\tilde{\boldsymbol{x}}|^2, \quad \text{for any column vector } a \tag{2.8}$$

The error vector $\tilde{\boldsymbol{x}}$ is dependent on the choice of h and, therefore, the covariance matrix $\mathsf{E}\,\tilde{\boldsymbol{x}}\tilde{\boldsymbol{x}}^*$ is also dependent on h. Let us indicate this fact explicitly by writing

$$R_{\tilde{x}}(h) \;\triangleq\; \mathsf{E}\,\tilde{\boldsymbol{x}}\tilde{\boldsymbol{x}}^*$$

Now note that

$$\mathsf{E}\,|a^*\tilde{\boldsymbol{x}}|^2 = a^*R_{\tilde{x}}(h)a$$

so that problem (2.8) is in effect seeking an optimal function h^o such that, for any vector a and for any other h,

$$a^*R_{\tilde{x}}(h)a \;\geq\; a^*R_{\tilde{x}}(h^o)a$$

That is, the difference matrix $R_{\tilde{x}}(h) - R_{\tilde{x}}(h^o)$ should be nonnegative-definite for all h. For this reason, we can equivalently interpret (2.8) as the problem of minimizing the error covariance matrix $R_{\tilde{x}}$ itself, written as

$$\boxed{\min_{h(\cdot)} \ \mathsf{E}\,\tilde{\boldsymbol{x}}\tilde{\boldsymbol{x}}^*} \tag{2.9}$$

Comparing with (2.4) we see that we are replacing the scalar $\mathsf{E}\,\tilde{\boldsymbol{x}}^*\tilde{\boldsymbol{x}}$ by the matrix $\mathsf{E}\,\tilde{\boldsymbol{x}}\tilde{\boldsymbol{x}}^*$.

Let us now verify that the solution to (2.8), or equivalently (2.9), is again $h^o(\boldsymbol{y}) = \mathsf{E}\,(\boldsymbol{x}|\boldsymbol{y})$. For this purpose, we recall that, for any $h(\boldsymbol{y})$,

$$\tilde{\boldsymbol{x}} = \boldsymbol{x} - \hat{\boldsymbol{x}} = \boldsymbol{x} - h(\boldsymbol{y})$$

so that the covariance matrix $R_{\tilde{x}}(h)$ is given by

$$\begin{aligned} R_{\tilde{x}}(h) &= \mathsf{E}\,[\boldsymbol{x} - h(\boldsymbol{y})][\boldsymbol{x} - h(\boldsymbol{y})]^* \\ &= \mathsf{E}\,\boldsymbol{x}\boldsymbol{x}^* - \mathsf{E}\,\boldsymbol{x}h^*(\boldsymbol{y}) - \mathsf{E}\,h(\boldsymbol{y})\boldsymbol{x}^* + \mathsf{E}\,h(\boldsymbol{y})h^*(\boldsymbol{y}) \end{aligned}$$

We now verify that

$$R_{\tilde{x}}(h) - R_{\tilde{x}}(h^o) \geq 0$$

for any choice of h. Indeed, from the orthogonality property (2.7), we have that $\boldsymbol{x} - h^o(\boldsymbol{y})$ is uncorrelated with any function of $\boldsymbol{y}$. Hence,

$$\begin{aligned} R_{\tilde{x}}(h^o) &= \mathsf{E}\,[\boldsymbol{x} - h^o(\boldsymbol{y})][\boldsymbol{x} - h^o(\boldsymbol{y})]^* \\ &= \mathsf{E}\,[\boldsymbol{x} - h^o(\boldsymbol{y})]\boldsymbol{x}^* \\ &= \mathsf{E}\,\boldsymbol{x}\boldsymbol{x}^* - \mathsf{E}\,h^o(\boldsymbol{y})\boldsymbol{x}^* \end{aligned}$$

Subtracting from $R_{\tilde{x}}(h)$ leads to

$$R_{\tilde{x}}(h) - R_{\tilde{x}}(h^o) \;=\; -\mathsf{E}\,\boldsymbol{x}h^*(\boldsymbol{y}) - \mathsf{E}\,h(\boldsymbol{y})\boldsymbol{x}^* + \mathsf{E}\,h(\boldsymbol{y})h^*(\boldsymbol{y}) + \mathsf{E}\,h^o(\boldsymbol{y})\boldsymbol{x}^*$$

From the orthogonality property (2.7) we again have that

$$\mathsf{E}\,[\boldsymbol{x} - h^o(\boldsymbol{y})]h^{o*}(\boldsymbol{y}) = 0, \quad \mathsf{E}\,[\boldsymbol{x} - h^o(\boldsymbol{y})]h^*(\boldsymbol{y}) = 0$$

so that

$$\mathsf{E}\,\boldsymbol{x}h^{o*}(\boldsymbol{y}) = \mathsf{E}\,h^o(\boldsymbol{y})h^{o*}(\boldsymbol{y}) \quad \text{and} \quad \mathsf{E}\,\boldsymbol{x}h^*(\boldsymbol{y}) = \mathsf{E}\,h^o(\boldsymbol{y})h^*(\boldsymbol{y})$$

These two equalities allow us to rewrite the difference $R_{\tilde{x}}(h) - R_{\tilde{x}}(h^o)$ as a perfect square:

$$R_{\tilde{x}}(h) - R_{\tilde{x}}(h^o) = \mathsf{E}\,[h^o(\boldsymbol{y}) - h(\boldsymbol{y})][h^o(\boldsymbol{y}) - h(\boldsymbol{y})]^*$$

The right-hand side is nonnegative-definite for all h, as desired. Finally, since the cost used in (2.4) is simply the trace of the error covariance matrix, we conclude that minimizing the error covariance matrix is equivalent to minimizing its trace.

Lemma 2.2 (Cost function) The conditional expectation of $\boldsymbol{x}$ given $\boldsymbol{y}$ is optimal relative to either cost

$$\min_{\hat{\boldsymbol{x}}} \mathsf{Tr}(R_{\tilde{x}}) \quad \text{or} \quad \min_{\hat{\boldsymbol{x}}} R_{\tilde{x}}$$

where $R_{\tilde{x}} = \mathsf{E}\,\tilde{\boldsymbol{x}}\tilde{\boldsymbol{x}}^*$ and $\tilde{\boldsymbol{x}} = \boldsymbol{x} - \hat{\boldsymbol{x}}$.

Summary and Notes

In this initial part we highlighted several concepts and results in least-mean-squares estimation theory. Some of these concepts are reproduced below in a less technical language in order to reinforce their importance.

SUMMARY OF MAIN RESULTS

1. The variance of a random variable serves as a measure of the amount of uncertainty about the variable: the larger the variance the less certain we are about the value it may assume in an experiment.

2. The least-mean-squares error criterion is useful in that it leads to tractable mathematical solutions. The criterion is also intuitively appealing. By seeking to minimize the variance of the estimation error we are in effect attempting to force this error to assume values close to its mean and, hence, to assume small values since the mean is zero.

3. The least-mean-squares estimator of a random variable $\boldsymbol{x}$ given another random variable $\boldsymbol{y}$ is the conditional expectation estimator, namely, $\hat{\boldsymbol{x}} = \mathsf{E}(\boldsymbol{x}|\boldsymbol{y})$. This estimator is optimal in the sense that it minimizes the covariance matrix of the error vector (or, equivalently, its trace), i.e., it solves
$$\min_{h(\cdot)} \mathsf{E}\tilde{\boldsymbol{x}}\tilde{\boldsymbol{x}}^* \quad \text{or} \quad \min_{h(\cdot)} \mathsf{Tr}(R_{\tilde{x}})$$

4. A defining property of the least-mean-squares estimator is that the resulting estimation error is uncorrelated with any function of the observations, namely,
$$\mathsf{E}(\boldsymbol{x} - \hat{\boldsymbol{x}})g(\boldsymbol{y}) = 0 \quad \text{for any function } g(\cdot) \text{ of } \boldsymbol{y}$$
In particular, $\tilde{\boldsymbol{x}} \perp \hat{\boldsymbol{x}}$ and $\tilde{\boldsymbol{x}} \perp \boldsymbol{y}$.

5. The evaluation of the conditional expectation, $\mathsf{E}(\boldsymbol{x}|\boldsymbol{y})$, is a formidable task in most cases. However, for circular Gaussian random variables the estimator $\hat{\boldsymbol{x}}$ is related to the observation $\boldsymbol{y}$ in an affine manner, namely,
$$\hat{\boldsymbol{x}} = \bar{x} + R_{xy}R_y^{-1}(\boldsymbol{y} - \bar{y})$$
where
$$R_x = \mathsf{E}(\boldsymbol{x} - \bar{x})(\boldsymbol{x} - \bar{x})^*, \quad R_y = \mathsf{E}(\boldsymbol{y} - \bar{y})(\boldsymbol{y} - \bar{y})^* \quad \text{and} \quad R_{xy} = \mathsf{E}(\boldsymbol{x} - \bar{x})(\boldsymbol{y} - \bar{y})^*$$
In particular, the estimator is completely determined from knowledge of the first and second-order moments of $\{\boldsymbol{x}, \boldsymbol{y}\}$, namely, their means, covariances and cross-covariance.

BIBLIOGRAPHIC NOTES

Probability theory. The exposition in these two initial chapters assumes some basic knowledge of probability theory; mainly with regards to the concepts of mean, variance, probability density function, and vector-random variables. Most of these ideas were defined and introduced from first principles. If additional help is needed, some accessible references on probability theory and basic random variable concepts are Papoulis (1991), Picinbono (1993), Leon-Garcia (1994), Stark and Woods (1994), and Durrett (1996). The textbook by Leon-Garcia (1994) is rich in examples, and is particularly suited to an engineering audience.

Mean-square-error performance. The squared-error criterion, whereby the square of the estimation error is used as a measure of performance, has a distinguished history. It dates back to C. F. Gauss (1795), who developed a deterministic least-squares-error criterion as opposed to the stochastic least-mean-squares criterion of this chapter. Gauss' formulation was motivated by his work on celestial orbits, and we shall comment on it more fully in the remarks to Part VII (*Least-Squares Methods*) when we study the least-squares criterion. A distinctive feature of the square-error criterion is that it penalizes large errors more than small errors. In this way, it is more sensitive to the presence of outliers in the data. This is in contrast, for example, to Laplace's proposition to use the absolute error criterion as a performance measure (see Sheynin (1977)). Gauss was very much aware of the distinction between both design criteria and this is how he commented on his squared-error criterion in relation to Laplace's absolute-error criterion:

> "Laplace has also considered the problem in a similar manner, but he adopted the absolute value of the error as the measure of this loss. Now if I am not mistaken, this convention is no less arbitrary than mine. Should an error of double size be considered as tolerable as a single error twice repeated or worse? Is it better to assign only twice as much influence to a double error or more? The answers are not self-evident, and the problem cannot be resolved by mathematical proofs, but only by an arbitrary decision."
>
> Extracted from the translation by Stewart (1995).

Besides Gauss' motivation, there are many good reasons for using the mean-square-error criterion, not the least of which is the fact that it leads to a closed-form characterization of the solution as a conditional mean. In addition, for Gaussian random variables, it can be argued that the least-mean-squares error estimator is practically optimal for any other choice of the error cost function (quadratic or otherwise) — see, for example, Pugachev (1958) and Zakai (1964).

Statistical theory. There is extensive work on the least-mean-squares error criterion in the statistical literature. For instance, the result of Thm. 1.1 on the conditional mean estimator is related to the so-called Rao-Blackwell theorem from statistics (see, e.g., Caines (1988) and Scharf (1991)). However, in statistics, there is often a distinction between the classical approach and the Bayesian approach to estimation. In the classical approach, the unknown quantity to be estimated is modeled as a deterministic but unknown constant; we shall encounter this situation in Chapter 6 while studying the Gauss-Markov theorem. The Bayesian approach, on the other hand, models the unknown quantity as a random variable, which is the point of view we adopted in this chapter. Such Bayesian formulations allow us to incorporate prior knowledge about the unknown variable itself into the solution, such as information about its probability density function. This fact helps explain why Bayesian techniques are dominant in many successful filtering and estimation designs; still the Bayesian approach has not been immune to controversies along its history (see Box and Tiao (1973)).

Complex random variables. Complex variables, as well as complex random variables, are frequent in electrical engineering (and perhaps more so than in other disciplines). One notable example arises in digital communications where symbols are often selected at random from a complex constellation (or even in the complex representation of bandpass signals). Since complex random variables will play a prominent role throughout this textbook, we have chosen to motivate them from first principles in the body of the chapters. In Sec. A.5 we pursue their study more closely and focus, in particular, on the important class of complex-valued *Gaussian* random variables. It is explained in that section that a certain circularity assumption needs to be satisfied if the resulting pdf in the complex case is to be uniquely determined by the first and second-order moments of the complex random variable, as happens in the real case. The main conclusion appears in the statement of Lemma A.1, which shows the form of a complex Gaussian distribution under the circularity assumption. The original derivation of this form is due to Wooding (1956) — see also Goodman (1963) and Miller (1974). It is for this reason that, in future discussions, whenever we refer to a complex Gaussian distribution we shall often attach the qualification "circular" to it and refer instead to a circular Gaussian distribution.

Linear algebra. Throughout the book, the reader will be exposed to a variety of concepts from linear algebra and matrix theory in a self-contained and motivated manner. In this way, after progressing sufficiently enough into the book, students will be able to master many useful concepts. Several of these concepts are summarized in the background Chapter B. If additional help is needed, some accessible references on matrix theory are the two volumes by Gantmacher (1959), the book by Bellman (1970), and the two volumes by Horn and Johnson (1987,1994). Accessible references on linear algebra are, for example, the books by Strang (1988,1993), Lay (1994), and Lax (1997).

Problems and Computer Projects

PROBLEMS

Problem I.1 (Rayleigh distribution) Consider a Rayleigh-distributed random variable $\boldsymbol{x}$ with pdf given by (A.3). Show that its mean and variance are given by (A.4).

Problem I.2 (Markov's inequality) Suppose $\boldsymbol{x}$ is a scalar *nonnegative* real-valued random variable with probability density function $f_{\boldsymbol{x}}(x)$. Show that $P[\boldsymbol{x} \geq \alpha] \leq \mathsf{E}\,\boldsymbol{x}/\alpha$.

Problem I.3 (Chebyshev's inequality) Consider a scalar real-valued random variable $\boldsymbol{x}$ with mean $\bar{x}$ and variance σ_x^2. Let $\boldsymbol{y} = (\boldsymbol{x} - \bar{x})^2$. Apply Markov's inequality to $\boldsymbol{y}$ to establish Chebyshev's inequality (A.5).

Problem I.4 (Conditional expectation) Consider two real-valued random variables $\boldsymbol{x}$ and $\boldsymbol{y}$. Establish that $\mathsf{E}\left[\mathsf{E}\left(\boldsymbol{x}|\boldsymbol{y}\right)\right] = \mathsf{E}\,\boldsymbol{x}$. That is, show that

$$\int_{\mathcal{S}_x} x f_{\boldsymbol{x}}(x)\mathrm{d}x = \int_{\mathcal{S}_y} f_{\boldsymbol{y}}(y)\left[\int_{\mathcal{S}_x} x f_{\boldsymbol{x}|\boldsymbol{y}}(x|y)\mathrm{d}x\right]\mathrm{d}y$$

where $\mathcal{S}_x$ and $\mathcal{S}_y$ denote the supports of the variables $\boldsymbol{x}$ and $\boldsymbol{y}$, respectively.

Problem I.5 (Numerical example) Assume $\boldsymbol{y}$ is a random variable that is red with probability $1/3$ and blue with probability $2/3$. Likewise, $\boldsymbol{x}$ is a random variable that is Gaussian with mean 1 and variance 2 if $\boldsymbol{y}$ is red, and uniformly distributed between -1 and 1 if $\boldsymbol{y}$ is blue. Find the individual and joint pdfs of $\{\boldsymbol{x}, \boldsymbol{y}\}$. Find also $\mathsf{E}\,\boldsymbol{x}$ and $\mathsf{E}\left(\boldsymbol{x}|\boldsymbol{y}\right)$.

Problem I.6 (Correlation coefficient) Consider two scalar random variables $\{\boldsymbol{x}, \boldsymbol{y}\}$ with means $\{\bar{x}, \bar{y}\}$, variances $\{\sigma_x^2, \sigma_y^2\}$, and cross-correlation σ_{xy}. Define the correlation coefficient $\rho_{xy} = \sigma_{xy}/(\sigma_x\sigma_y)$. Show that $|\rho_{xy}| \leq 1$.

Problem I.7 (Fully correlated random variables) Consider two scalar real-valued random variables $\boldsymbol{x}$ and $\boldsymbol{y}$ with correlation coefficient ρ_{xy}, means $\{\bar{x}, \bar{y}\}$, and variances $\{\sigma_x^2, \sigma_y^2\}$. Show that $|\rho_{xy}| = 1$ if, and only if, $\boldsymbol{x} - \bar{x} = \pm\dfrac{\sigma_x}{\sigma_y}(\boldsymbol{y} - \bar{y})$.

Problem I.8 (Chi-square distribution) Let $\boldsymbol{x}$ be a real-valued random variable with pdf $f_{\boldsymbol{x}}(x)$. Define $\boldsymbol{y} = \boldsymbol{x}^2$.

(a) Use the fact that for any nonnegative $\boldsymbol{y}$, the event $\{\boldsymbol{y} \leq y\}$ occurs whenever $\{-\sqrt{y} \leq \boldsymbol{x} \leq \sqrt{y}\}$ to conclude that the pdf of $\boldsymbol{y}$ is given by

$$f_{\boldsymbol{y}}(y) = \frac{1}{2}\frac{f_{\boldsymbol{x}}(\sqrt{y})}{\sqrt{y}} + \frac{1}{2}\frac{f_{\boldsymbol{x}}(-\sqrt{y})}{\sqrt{y}}, \quad y > 0$$

(b) Assume $\boldsymbol{x}$ is Gaussian with zero mean and unit variance. Conclude that $f_{\boldsymbol{y}}(y) = \frac{1}{\sqrt{2\pi y}}e^{-y/2}$ for $y > 0$.

Remark. The above pdf is known as the Chi-square distribution with one degree of freedom. A Chi-square distribution with k degrees of freedom is characterized by the pdf:

$$f_{\boldsymbol{y}}(y) = \frac{1}{2^{k/2}\Gamma(k/2)}\, y^{(k-2)/2}e^{-y/2}, \quad y > 0$$

where $\Gamma(\cdot)$ is the so-called Gamma function; it is defined by the integral $\Gamma(z) = \int_0^\infty s^{z-1}e^{-s}ds$ for $z > 0$. The function $\Gamma(\cdot)$ has the following useful properties: $\Gamma(1/2) = \sqrt{\pi}$, $\Gamma(z+1) = z\Gamma(z)$ for any $z > 0$, and $\Gamma(n+1) = n!$ for any integer $n \geq 0$.

Problem I.9 (Chi and Rayleigh distributions) Consider an FIR channel with two real-valued taps, $\boldsymbol{x}(1)$ and $\boldsymbol{x}(2)$. The taps are assumed to be independent zero-mean unit-variance Gaussian random variables.

(a) Use the result of part (b) of Prob. I.8 to show that the random variable $\boldsymbol{w} = \boldsymbol{x}^2(1) + \boldsymbol{x}^2(2)$ has a Chi-square distribution with two degrees of freedom, i.e., $f_{\boldsymbol{w}}(w) = \frac{1}{2}e^{-w/2}$ for $w \geq 0$.

(b) Conclude that the random variable $\boldsymbol{z} = \sqrt{\boldsymbol{x}^2(1) + \boldsymbol{x}^2(2)}$ has a Rayleigh distribution, namely, show that $f_{\boldsymbol{z}}(z) = ze^{-z^2/2}$ for $z \geq 0$, with $\bar{z} = \sqrt{\pi/2}$ and $\sigma_z^2 = (2 - \pi/2)$.

Problem I.10 (Uniform noise) Consider Ex. 1.2 but assume now that the noise is uniformly distributed between $-1/2$ and $1/2$. Show that $\hat{\boldsymbol{x}} = 1$ when $\boldsymbol{y} \in [1/2, 3/2]$ and $\hat{\boldsymbol{x}} = -1$ when $\boldsymbol{y} \in [-3/2, -1/2]$.

Problem I.11 (Estimator for a binary signal) Consider the same setting of Ex. 1.2 but assume now that the noise $\boldsymbol{v}$ has a generic variance σ_v^2.

(a) Show that the optimal least-mean-squares estimator of $\boldsymbol{x}$ given $\boldsymbol{y}$ is $\hat{\boldsymbol{x}} = \tanh(\boldsymbol{y}/\sigma_v^2)$. Plot the estimate $\hat{x}$ as a function of y for the values $\sigma_v^2 = 0.5, 1, 2$.

(b) Argue that $\hat{\boldsymbol{x}} = \text{sign}(\boldsymbol{y})$ can be chosen as a suboptimal estimator.

Problem I.12 (Biased measurements) Consider the same setting of Ex. 1.2 but assume now that the noise $\boldsymbol{v}$ has mean $\bar{v}$ and unit variance.

(a) Show that the optimal least-mean-squares estimator of $\boldsymbol{x}$ given $\boldsymbol{y}$ is $\hat{\boldsymbol{x}} = \tanh(\boldsymbol{y} - \bar{v})$. Plot the estimate $\hat{x}$ as a function of y for the values $\bar{v} = -0.5, 0, 0.5$.

(b) Argue that $\hat{\boldsymbol{x}} = \text{sign}(\boldsymbol{y} - \bar{v})$ can be chosen as a suboptimal estimator.

Problem I.13 (BPSK signal) Consider noisy observations $\boldsymbol{y}(i) = \boldsymbol{x} + \boldsymbol{v}(i)$, where $\boldsymbol{x}$ and $\boldsymbol{v}(i)$ are independent real-valued random variables, $\boldsymbol{v}(i)$ is a white-noise Gaussian random process with zero mean and variance σ_v^2, and $\boldsymbol{x}$ takes the values ± 1 with equal probability. The value of $\boldsymbol{x}$ is either $+1$ or -1 for all measurements $\{\boldsymbol{y}(i)\}$. The whiteness assumption on $\boldsymbol{v}(i)$ means that $\mathsf{E}\,\boldsymbol{v}(i)\boldsymbol{v}(j) = 0$ for $i \neq j$.

(a) Show that the least-mean-squares estimate of $\boldsymbol{x}$ given $\{y(0), \ldots, y(N-1)\}$ is

$$\hat{x}_N = \tanh\left(\sum_{i=0}^{N-1} y(i)/\sigma_v^2\right)$$

(b) Assume $\boldsymbol{x}$ takes the value 1 with probability p and the value -1 with probability $1-p$. Show that the least-mean-squares estimate of $\boldsymbol{x}$ given $\{y(0), \ldots, y(N-1)\}$ is

$$\hat{x}_N = \tanh\left[\frac{1}{2}\ln\left(\frac{p}{1-p}\right) + \sum_{i=0}^{N-1} y(i)/\sigma_v^2\right]$$

in terms of the natural logarithm of $p/(1-p)$.

(c) Assume the noise is correlated. Let $R_v = \mathsf{E}\,\boldsymbol{v}\boldsymbol{v}^*$ where $\boldsymbol{v} = \text{col}\{\boldsymbol{v}(0), \boldsymbol{v}(1), \ldots, \boldsymbol{v}(N-1)\}$. Show that the least-mean-squares estimate of $\boldsymbol{x}$ given $\{y(0), \ldots, y(N-1)\}$ is now

$$\hat{x}_N = \tanh\left[\frac{1}{2}\ln\left(\frac{p}{1-p}\right) + y^{\mathsf{T}}R_v^{-1}h\right]$$

where $y = \text{col}\{y(0), y(1), \ldots, y(N-1)\}$ and $h = \text{col}\{1, 1, \ldots, 1\}$.

Problem I.14 (Interfering signals) Two random variables $\boldsymbol{x}_1$ and $\boldsymbol{x}_2$ are transmitted over known channels with gains α_1 and α_2, as indicated in Fig. I.1. The channels simply scale the random variables by $\{\alpha_1, \alpha_2\}$. Zero-mean Gaussian noise is added to each transmission and the result is $\boldsymbol{y} = \alpha_1\boldsymbol{x}_1 + \alpha_2\boldsymbol{x}_2 + \boldsymbol{v}_1 + \boldsymbol{v}_2$, where $\boldsymbol{v}_1$ and $\boldsymbol{v}_2$ are independent of each other and of $\boldsymbol{x}_1$ and $\boldsymbol{x}_2$. The variances of the Gaussian noises $\boldsymbol{v}_1$ and $\boldsymbol{v}_2$ are σ_{v1}^2 and σ_{v2}^2, respectively.

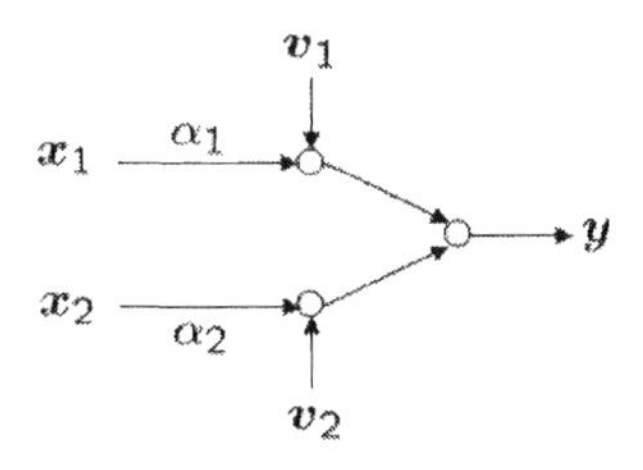

FIGURE I.1 Two interfering signals transmitted over flat channels.

The random variable $\boldsymbol{x}_1$ assumes the value $+1$ with probability p and the value -1 with probability $1-p$. The random variable $\boldsymbol{x}_2$ is distributed as follows:

$$\text{if } \boldsymbol{x}_1 = +1 \text{ then} \quad \boldsymbol{x}_2 = \begin{cases} +2 \text{ with probability } q \\ -2 \text{ with probability } 1-q \end{cases}$$

$$\text{if } \boldsymbol{x}_1 = -1 \text{ then} \quad \boldsymbol{x}_2 = \begin{cases} +3 \text{ with probability } r \\ -3 \text{ with probability } 1-r \end{cases}$$

(a) Find the means of $\boldsymbol{x}_1$, $\boldsymbol{x}_2$, and $\boldsymbol{y}$.

(b) Find the minimum-mean-square-error estimator of $\boldsymbol{x}_1$ given $\boldsymbol{y}$.

(c) Specialize the results to the case $\alpha_1 = \alpha_2 = \alpha$, $\sigma_{v_1}^2 = \sigma_{v_2}^2 = 1/4$.

Problem I.15 (Optimal receiver) Complete the derivation of Ex. 2.1.

(a) Verify that

$$\hat{x} = \frac{1}{f_y(y)}\left(\sum_{k=0}^{3} a(k) m_k f_v(y - Hm_k)\right)$$

where $a(0) = p^2$, $a(1) = q^2$ and $a(2) = a(3) = pq$.

(b) Let $a = p^2 \cdot e^{-\frac{1}{2}(-2y(0)-3y(1)+2)}$, $b = q^2 \cdot e^{-\frac{1}{2}(-2y(0)+3y(1)+2)}$, $c = pq \cdot e^{-\frac{1}{2}(-2y(0)+y(1))}$, and $d = pq \cdot e^{-\frac{1}{2}(2y(0)-y(1))}$. Show that the expression in part (a) leads to

$$\begin{bmatrix} \hat{s}(0) \\ \hat{s}(1) \end{bmatrix} = \frac{1}{a+b+c+d}\begin{bmatrix} a-b+c-d \\ a-b-c+d \end{bmatrix}$$

Problem I.16 (Exponential distribution) Suppose we observe $\boldsymbol{y} = \boldsymbol{x} + \boldsymbol{v}$, where $\boldsymbol{x}$ and $\boldsymbol{v}$ are independent real-valued random variables with exponential distributions with parameters λ_1 and λ_2 $(\lambda_1 \neq \lambda_2)$. That is, the pdfs of $\boldsymbol{x}$ and $\boldsymbol{v}$ are $f_{\boldsymbol{x}}(x) = \lambda_1 e^{-\lambda_1 x}$ for $x \geq 0$ and $f_{\boldsymbol{v}}(v) = \lambda_2 e^{-\lambda_2 v}$ for $v \geq 0$, respectively.

(a) Using the fact that the pdf of the sum of two independent random variables is the convolution of the individual pdfs, show that

$$f_{\boldsymbol{y}}(y) = \frac{\lambda_1\lambda_2}{\lambda_2 - \lambda_1} \cdot e^{-\lambda_2 y} \cdot \left[e^{(\lambda_2-\lambda_1)y} - 1\right], \quad y \geq 0$$

(b) Establish that $f_{\boldsymbol{x},\boldsymbol{y}}(x,y) = \lambda_1\lambda_2 e^{(\lambda_2-\lambda_1)x-\lambda_2 y}$, for $x \geq 0$ and $y \geq 0$.

(c) Show that the least-mean-squares estimate of $\boldsymbol{x}$ given $\boldsymbol{y} = y$ is

$$\hat{x} = \frac{1}{\lambda_1 - \lambda_2} - \frac{e^{-\lambda_1 y}}{e^{-\lambda_2 y} - e^{-\lambda_1 y}}\, y$$

Problem I.17 (Equivalent criterion) Show that the least-mean-squares estimator $\hat{\boldsymbol{x}} = \mathsf{E}(\boldsymbol{x}|\boldsymbol{y})$ also minimizes $\mathsf{E}\,\tilde{\boldsymbol{x}}^* W \tilde{\boldsymbol{x}}$, for any Hermitian nonnegative-definite matrix W.

Problem I.18 (Second- and fourth-order moments) Consider an $M \times M$ positive-definite matrix R and introduce its eigen-decomposition (cf. Sec. B.1), $R = \sum_{i=1}^{M} \lambda_i u_i u_i^*$, where the λ_i are the eigenvalues of R (all positive) and the u_i are the eigenvectors of R. The u_i are orthonormal, i.e., $u_j^* u_i = 0$ for all $i \neq j$ and $u_i^* u_i = 1$. Let $\boldsymbol{h}$ be a random vector with probability distribution $P(\boldsymbol{h} = u_j) = \lambda_j / \mathsf{Tr}(R)$.

(a) Show that $\mathsf{E}\,\boldsymbol{h}\boldsymbol{h}^* = R/\mathsf{Tr}(R)$ and $\mathsf{E}\,\boldsymbol{h}\boldsymbol{h}^*\boldsymbol{h}\boldsymbol{h}^* = R/\mathsf{Tr}(R)$.

(b) Show that $\mathsf{E}\,\boldsymbol{h}^* R^{-1}\boldsymbol{h} = M/\mathsf{Tr}(R)$ and $\mathsf{E}\,\boldsymbol{h}\boldsymbol{h}^* R^{-1}\boldsymbol{h}\boldsymbol{h}^* = \mathrm{I}/\mathsf{Tr}(R)$, where I is the identity matrix.

(c) Show that $\mathsf{E}\,\boldsymbol{h}^*\boldsymbol{h} = 1$ and $\mathsf{E}\,\boldsymbol{h} = \frac{1}{\mathsf{Tr}(R)} \sum_{j=1}^{M} \lambda_j u_j$.

Problem I.19 (Independent and Gaussian variables) Consider two independent and zero-mean random variables $\{\boldsymbol{u}, \boldsymbol{w}\}$, where $\boldsymbol{u}$ is a row vector and $\boldsymbol{w}$ is a column vector; both are M-dimensional. The covariance matrices of $\boldsymbol{u}$ and $\boldsymbol{w}$ are defined by $\mathsf{E}\,\boldsymbol{u}^*\boldsymbol{u} = \sigma_u^2 \mathrm{I}$ and $\mathsf{E}\,\boldsymbol{w}\boldsymbol{w}^* = C$. In addition, $\boldsymbol{u}$ is assumed to be circular Gaussian. Let $\boldsymbol{e}_a = \boldsymbol{u}\boldsymbol{w}$.

(a) Show that $\mathsf{E}\,|\boldsymbol{e}_a|^2 = \sigma_u^2 \mathsf{Tr}(C)$.

(b) Use the result of Lemma A.3 to show that $\mathsf{E}\,\|\boldsymbol{u}\|^2 \cdot |\boldsymbol{e}_a|^2 = (M+1)\sigma_u^4 \mathsf{Tr}(C)$, where the notation $\|\cdot\|$ denotes the Euclidean norm of its argument.

Problem I.20 (Fourth-moment) Assume $\boldsymbol{u}$ is a circular Gaussian random row vector with a diagonal covariance matrix Λ. Define $\boldsymbol{z} = \|\boldsymbol{u}\|^2$. What is the variance of $\boldsymbol{z}$?

Problem I.21 (Covariance equation) Consider two column vectors $\{\boldsymbol{w}, \boldsymbol{z}\}$ that are related via $\boldsymbol{z} = \boldsymbol{w} + \mu \boldsymbol{u}^*(\boldsymbol{d} - \boldsymbol{u}\boldsymbol{w})$, where $\boldsymbol{u}$ is a circular Gaussian random variable with a diagonal covariance matrix, $\mathsf{E}\,\boldsymbol{u}^*\boldsymbol{u} = \Lambda$ ($\boldsymbol{u}$ is a row vector). Moreover, μ is a positive constant and $\boldsymbol{d} = \boldsymbol{u}w^o + \boldsymbol{v}$, for some constant vector w^o and random scalar $\boldsymbol{v}$ with variance σ_v^2. The variables $\{\boldsymbol{v}, \boldsymbol{u}, \boldsymbol{w}\}$ are independent of each other. Define $\boldsymbol{e}_a = \boldsymbol{u}(w^o - \boldsymbol{w})$, as well as the error vectors $\tilde{\boldsymbol{z}} = w^o - \boldsymbol{z}$ and $\tilde{\boldsymbol{w}} = w^o - \boldsymbol{w}$, and denote their covariances by $\{R_{\tilde{z}}, R_{\tilde{w}}\}$. Assume $\mathsf{E}\,\boldsymbol{z} = \mathsf{E}\,\boldsymbol{w} = w^o$, while all other random variables are zero-mean.

(a) Verify that $\tilde{\boldsymbol{z}} = \tilde{\boldsymbol{w}} - \mu \boldsymbol{u}^*(\boldsymbol{e}_a + \boldsymbol{v})$.

(b) Use the result of Lemma A.3 to show that

$$R_{\tilde{z}} = R_{\tilde{w}} - \mu R_{\tilde{w}}\Lambda - \mu \Lambda R_{\tilde{w}} + \mu^2\left(\Lambda \mathsf{Tr}(R_{\tilde{w}}\Lambda) + \Lambda R_{\tilde{w}}\Lambda\right) \;+\; \mu^2 \sigma_v^2 \Lambda$$

(c) How would the result of part (b) change if $\boldsymbol{u}$ were real-valued Gaussian and all other variables were also real-valued?

COMPUTER PROJECT

Project I.1 (Comparing optimal and suboptimal estimators) The purpose of this project[2] is to compare the performance of an optimal least-mean-squares estimator with three approximations for it, along the lines discussed in Ex. 1.2. Thus consider the setting of Prob. I.13.

(a) Write a MATLAB program that generates a BPSK random variable $\boldsymbol{x}$ that is equal to $+1$ with probability p and to -1 with probability $1-p$.

(b) Simulate the estimator of part (b) of Prob. I.13 for different number of observations N. Generate observations $\{y(i)\}$ and plot $\hat{x}_N$ as a function of N for $1 \leq N \leq 10$, with all observations assumed generated by the same value of x — either $+1$ or -1, and using zero-mean Gaussian noise with unit variance. Plot $\hat{x}_N$ for the cases $p = 0.1, 0.3, 0.5, 0.8$.

(c) Compare the performance of the optimal estimate $\hat{x}_N$ with the averaged estimate

$$\hat{x}_{N,\text{av}} \triangleq \frac{1}{N}\sum_{i=0}^{N-1} y(i)$$

for several values of N, say, for $1 \leq N \leq 300$, and for the same values of p in part (b). Does it take many more samples N for the averaged estimate $\hat{x}_{N,\text{av}}$ to provide a good result compared with the optimal nonlinear estimate $\hat{x}_N$?

(d) Fix $p = 1/2$, and define the nonlinear decision device:

$$\text{sign}[z] = \begin{cases} +1 & \text{if } z \geq 0 \\ -1 & \text{if } z < 0 \end{cases}$$

Consider also the alternative (sign-of-optimal) estimate $\hat{x}_{\text{dec}} = \text{sign}\,[\hat{x}_N]$. It is clear that $\hat{x}_{\text{dec}}$ assumes the values ± 1, whereas the optimal estimate $\hat{x}_N$ does not. Is $\hat{x}_{\text{dec}}$ a better estimate than $\hat{x}_N$? The answer in the mean-square sense is of course negative since we already know that $\hat{x}_N$ is the best estimate. To verify this fact do the following. Fix the number of observations at $N = 10$. Then perform 1000 experiments, with each experiment i resulting in an optimal estimate $\hat{x}_{10}(i)$ and an estimate $\hat{x}_{\text{dec}}(i)$. For each estimate, the value of x is fixed at either $+1$ or -1. Compute the sample variances

$$\frac{1}{1000}\sum_{i=1}^{1000}|x - \hat{x}_{10}(i)|^2, \qquad \frac{1}{1000}\sum_{i=1}^{1000}|x - \hat{x}_{\text{dec}}(i)|^2$$

Which one is smaller? Repeat for the following (sign-of-average) estimate:

$$\hat{x}_{\text{sign}} = \text{sign}\,[\hat{x}_{N,\text{av}}]$$

That is, apply the decision device to the estimate that is obtained from averaging.

[2] MATLAB programs that solve all computer projects in the book can be downloaded for free by all readers (including students and instructors) from the publisher or author's websites. The purpose of these programs is to allow readers to experiment freely with the concepts covered in the chapters. The readers may also download extensive commentary and typical performance plots for all computer projects from the same websites. Detailed solutions also appear in the textbook by Sayed (2003).

PART II

LINEAR ESTIMATION

CHAPTER 3

Normal Equations

In Secs. 1.2 and 2.1 we studied the problem of determining the optimal function $h(\cdot)$ that minimizes the mean-square error of estimating a random variable $\boldsymbol{x}$ from another random variable $\boldsymbol{y}$. Specifically, we solved

$$\min_{h(\cdot)} \; \mathsf{E}\,\tilde{\boldsymbol{x}}\tilde{\boldsymbol{x}}^* \tag{3.1}$$

over all functions $h(\cdot)$ of $\boldsymbol{y}$. The optimal solution was found to be the conditional expectation of $\boldsymbol{x}$ given $\boldsymbol{y}$, i.e.,

$$\hat{\boldsymbol{x}} = \mathsf{E}\,(\boldsymbol{x}|\boldsymbol{y})$$

Such conditional expectations are generally hard to evaluate in closed-form, except in some special cases. We encountered three such cases in Part I (*Optimal Estimation*), namely the case of a BPSK signal embedded in additive Gaussian noise (Ex. 1.2), the case of jointly Gaussian random variables (studied in Secs. 1.4 and 2.2), and the case of random variables with exponential distributions (studied in Prob. I.16). Due to the difficulty in evaluating $\mathsf{E}\,(\boldsymbol{x}|\boldsymbol{y})$ in general, it is common practice to restrict the choice of $h(\cdot)$ to the subclass of *affine* functions of $\boldsymbol{y}$, i.e., to functions of the form

$$h(\boldsymbol{y}) \;=\; K\boldsymbol{y} + b \tag{3.2}$$

for some matrix K and for some vector b to be determined. For general vector-valued variables $\{\boldsymbol{x}, \boldsymbol{y}\}$, K is a matrix and b is a vector. When $\{\boldsymbol{x}, \boldsymbol{y}\}$ are scalars, then $\{K, b\}$ will also be scalars. If $\boldsymbol{x}$ is a scalar and $\boldsymbol{y}$ is a vector, then K will be a row vector and b a scalar. In some applications, the variables $\{\boldsymbol{x}, \boldsymbol{y}\}$ may be matrix-valued, in which case $\{K, b\}$ will be matrix-valued. In other words, the dimensions of $\{K, b\}$ need to be consistent with the dimensions of $\{\boldsymbol{x}, \boldsymbol{y}\}$, and these dimensions will be obvious from the context.

Affine functions of the form (3.2) are easier to implement than most nonlinear functions. For example, when K is a row vector, the product $K\boldsymbol{y}$ amounts to an inner product. Moreover, by deliberately restricting $h(\cdot)$ to be an affine function of $\boldsymbol{y}$, the evaluation of the optimal $h(\cdot)$ is greatly simplified. In particular, it will be seen that all we need to know in order to determine the optimal parameters $\{K, b\}$ are the first and second-order moments of $\{\boldsymbol{x}, \boldsymbol{y}\}$, namely, $\mathsf{E}\,\boldsymbol{x}, \mathsf{E}\,\boldsymbol{y}, \mathsf{E}\,\boldsymbol{x}\boldsymbol{x}^*, \mathsf{E}\,\boldsymbol{y}\boldsymbol{y}^*$, and $\mathsf{E}\,\boldsymbol{x}\boldsymbol{y}^*$. No other moments are needed. In contrast, evaluation of the optimal estimator, $\hat{\boldsymbol{x}} = \mathsf{E}\,(\boldsymbol{x}|\boldsymbol{y})$, requires full knowledge of the conditional pdf $f_{\boldsymbol{x}|\boldsymbol{y}}(x|y)$.

Of course, the minimum mean-square error (m.m.s.e.) that results from using the affine estimator (3.2) will be larger than the m.m.s.e. that results from the optimal design (3.1). One notable exception is the case of jointly Gaussian random variables $\{\boldsymbol{x}, \boldsymbol{y}\}$. It was verified in Sec. 2.2 that for Gaussian random variables, the optimal estimator is an affine function of the observation (so that, in this case, the affine and optimal estimators achieve the same m.m.s.e.) – see Lemma 2.1. Still, in general and for other signal distributions, the performance of the affine estimator (3.2) is reasonable enough for many important applications, and this fact explains its widespread adoption in current practice.

In this chapter we shall study the linear (affine) estimation problem in some detail. Before proceeding, we would like to remind the reader that the presentation in Part I (*Optimal Estimation*) has shown that all results for scalar real-valued random variables could be extended rather immediately to vector and also complex-valued random variables by relying on the vector and complex conjugation notation. For this reason, we shall deal directly with the vector and complex-valued case in the sequel.

3.1 MEAN-SQUARE ERROR CRITERION

In order to simplify the presentation, we assume first that $\{\boldsymbol{x}, \boldsymbol{y}\}$ are *zero-mean* random variables. Later in Sec. 4.4 we show how the nonzero-mean case can be accommodated through the process of centering the random variables.

Thus, consider two zero-mean vector-valued random variables $\boldsymbol{x}$ and $\boldsymbol{y}$ and let

$$\boxed{\bar{x} \triangleq \mathsf{E}\,\boldsymbol{x} = 0, \quad \bar{y} \triangleq \mathsf{E}\,\boldsymbol{y} = 0, \quad R_x \triangleq \mathsf{E}\,\boldsymbol{x}\boldsymbol{x}^*, \quad R_y \triangleq \mathsf{E}\,\boldsymbol{y}\boldsymbol{y}^*, \quad R_{xy} \triangleq \mathsf{E}\,\boldsymbol{x}\boldsymbol{y}^*}$$

The dimensions of $\boldsymbol{x}$ and $\boldsymbol{y}$ need not be identical, say, $\boldsymbol{x}$ is $p \times 1$ and $\boldsymbol{y}$ is $q \times 1$. The analysis can be adjusted to treat the case of matrix-valued data as well, say, where $\boldsymbol{x}$ is $p \times r$ and $\boldsymbol{y}$ is $q \times s$ with r and s larger than one. However, it is common (and also sufficient) to illustrate the main ideas by focusing on the case of column vectors $\boldsymbol{x}$ and $\boldsymbol{y}$. In this case, $\{R_x, R_y, R_{xy}\}$ are $\{p \times p, q \times q, p \times q\}$ matrices, respectively.

We now seek an affine estimator for $\boldsymbol{x}$, namely, one of the form

$$\hat{\boldsymbol{x}} = K\boldsymbol{y} + b$$

for some constants $\{K, b\}$ to be determined, where K is $p \times q$ and b is $p \times 1$. The determination of $\{K, b\}$ is based on two considerations. First, the estimator should be unbiased, which means that it should satisfy

$$\boxed{\mathsf{E}\,\hat{\boldsymbol{x}} = 0}$$

But since

$$\mathsf{E}\,\hat{\boldsymbol{x}} = K\mathsf{E}\,\boldsymbol{y} + b \;=\; 0 + b = b$$

we find that we must have $b = 0$. This means that the estimator that we are seeking is effectively a linear estimator, i.e., it is one of the form $\hat{\boldsymbol{x}} = K\boldsymbol{y}$. Second, the coefficient matrix K should be chosen optimally so as to minimize the error covariance matrix (or its trace), as we now explain.

Let $\{k_i^*\}$ denote the individual rows of K. Then the estimator for each entry of $\boldsymbol{x}$, say, $\boldsymbol{x}(i)$, is given by the inner product $k_i^*\boldsymbol{y}$,

$$\underbrace{\begin{bmatrix} \hat{\boldsymbol{x}}(0) \\ \hat{\boldsymbol{x}}(1) \\ \vdots \\ \hat{\boldsymbol{x}}(p-1) \end{bmatrix}}_{=\hat{\boldsymbol{x}}} = \begin{bmatrix} k_0^*\boldsymbol{y} \\ k_1^*\boldsymbol{y} \\ \vdots \\ k_{p-1}^*\boldsymbol{y} \end{bmatrix} = \underbrace{\begin{bmatrix} \rule{2em}{0.4pt} \\ \rule{2em}{0.4pt} \\ \vdots \\ \rule{2em}{0.4pt} \end{bmatrix}}_{=K} \boldsymbol{y}$$

The optimal choices for the column vectors $\{k_i\}$ are determined by solving

$$\boxed{\min_{k_i} \ \mathsf{E}\,|\tilde{\boldsymbol{x}}(i)|^2} \qquad \text{for each } i = 0, 1, \ldots, p-1 \tag{3.3}$$

where

$$\tilde{\boldsymbol{x}}(i) = \boldsymbol{x}(i) - \hat{\boldsymbol{x}}(i)$$

is the estimation error for the i-th entry of $\boldsymbol{x}$. That is, each column vector k_i is determined by minimizing the error variance of the corresponding entry in $\boldsymbol{x}$. The optimization problems (3.3) can be grouped together and stated equivalently as the problem of determining the matrix K by solving

$$\boxed{\min_{K} \ \mathsf{E}\,\tilde{\boldsymbol{x}}^*\tilde{\boldsymbol{x}}} \tag{3.4}$$

where $\tilde{\boldsymbol{x}} = \boldsymbol{x} - \hat{\boldsymbol{x}}$. This is because the scalar quantity $\mathsf{E}\,\tilde{\boldsymbol{x}}^*\tilde{\boldsymbol{x}}$ in (3.4) is simply the sum of the individual error variances that appear in (3.3),

$$\mathsf{E}\,\tilde{\boldsymbol{x}}^*\tilde{\boldsymbol{x}} \;=\; \mathsf{E}\,|\tilde{\boldsymbol{x}}(0)|^2 \;+\; \mathsf{E}\,|\tilde{\boldsymbol{x}}(1)|^2 \;+\; \ldots \;+\; \mathsf{E}\,|\tilde{\boldsymbol{x}}(p-1)|^2 \tag{3.5}$$

and each term $\mathsf{E}\,|\tilde{\boldsymbol{x}}(i)|^2$ depends on the corresponding k_i alone. In this way, minimizing $\mathsf{E}\,\tilde{\boldsymbol{x}}^*\tilde{\boldsymbol{x}}$ over K is equivalent to minimizing each term $\mathsf{E}\,|\tilde{\boldsymbol{x}}(i)|^2$ over its k_i, so that problems (3.3) and (3.4) are equivalent. We now proceed slowly and show how to solve (3.3) over the corresponding *vector* arguments $\{k_i\}$. Later in Sec. 3.4, we shall show how to minimize $\mathsf{E}\,\tilde{\boldsymbol{x}}^*\tilde{\boldsymbol{x}}$ directly over the matrix argument K, rather than in steps over the individual $\{k_i\}$.

Therefore, continuing with (3.3), we expand the cost function to obtain a *quadratic* expression in the unknown column vector k_i,

$$\begin{aligned}
\mathsf{E}\,|\tilde{\boldsymbol{x}}(i)|^2 &\stackrel{\Delta}{=} \mathsf{E}\,|\boldsymbol{x}(i) - k_i^*\boldsymbol{y}|^2 \\
&= \mathsf{E}\,[\boldsymbol{x}(i) - k_i^*\boldsymbol{y}][\boldsymbol{x}(i) - k_i^*\boldsymbol{y}]^* \\
&= \mathsf{E}\,|\boldsymbol{x}(i)|^2 - [\mathsf{E}\,\boldsymbol{x}(i)\boldsymbol{y}^*]k_i - k_i^*[\mathsf{E}\,\boldsymbol{y}\boldsymbol{x}^*(i)] + k_i^* R_y k_i
\end{aligned}$$

This is a *scalar-valued* cost function of a possibly *complex-valued vector* quantity, k_i. We denote it by

$$J(k_i) \;\stackrel{\Delta}{=}\; \mathsf{E}\,|\boldsymbol{x}(i)|^2 - [\mathsf{E}\,\boldsymbol{x}(i)\boldsymbol{y}^*]k_i - k_i^*[\mathsf{E}\,\boldsymbol{y}\boldsymbol{x}^*(i)] + k_i^* R_y k_i \tag{3.6}$$

The quantity $\mathsf{E}\,|\boldsymbol{x}(i)|^2$ that appears in the expression for $J(k_i)$ is the variance of $\boldsymbol{x}(i)$ and is therefore equal to the i-th diagonal entry of R_x. We denote it by

$$\sigma_{x,i}^2 = \mathsf{E}\,|\boldsymbol{x}(i)|^2$$

Likewise, the quantity $\mathsf{E}\,\boldsymbol{x}(i)\boldsymbol{y}^*$ is the i-th row of the cross-covariance matrix R_{xy}. We denote it by $R_{xy,i} = \mathsf{E}\,\boldsymbol{x}(i)\boldsymbol{y}^*$. In this way, we can rewrite $J(k_i)$ as

$$\boxed{J(k_i) \;\stackrel{\Delta}{=}\; \sigma_{x,i}^2 - R_{xy,i}k_i - k_i^* R_{yx,i} + k_i^* R_y k_i} \tag{3.7}$$

where $\{\sigma_{x,i}^2, R_{xy,i}, R_y\}$ are known quantities and k_i is the unknown column vector; $\sigma_{x,i}^2$ is a scalar, $R_{xy,i}$ is a row vector, and R_y is a nonnegative-definite matrix. Moreover, $R_{yx,i} = R_{xy,i}^*$. Our objective is to minimize $J(k_i)$ over k_i.

We can proceed in at least two different ways. One way relies on standard differentiation techniques from calculus, while the second way relies on a completion-of-squares argument. While it may be easier for the reader to assimilate the differentiation argument due to familiarity with the concept of derivatives and function minimization, the second argument on completion of squares will serve as a powerful example of the convenience of the vector notation. Since this argument will be useful in other scenarios as well, we choose to include it here for the benefit of the reader, in addition to the differentiation argument. Still, both arguments require careful reasoning, as we explain below. Later, when similar derivations are needed in other chapters, we shall move more swiftly through the presentation.

3.2 MINIMIZATION BY DIFFERENTIATION

We start with the differentiation argument. As the reader may recall from a basic course on calculus, in order to determine the vector k_i that minimizes $J(k_i)$, we need to differentiate $J(k_i)$ with respect to each one of the entries of k_i, namely, $\{k_{ij}, j = 0, 1, \ldots, q-1\}$ and set the derivatives equal to zero. The main complication that arises here, relative to what the reader may be familiar with from a standard course on calculus, is that the entries $\{k_{ij}\}$ are *complex-valued*, and we therefore need to explain what is meant by differentiating $J(k_i)$ with respect to a complex variable. This is done in Chapter. C, where we show how to differentiate a function with respect to a complex scalar and even a complex vector. When the rules derived in that appendix are applied to the cost function $J(k_i)$ in (3.7) we find that the complex gradient vector of $J(k_i)$ with respect to k_i is given by

$$\nabla_{k_i} J(k_i) = -R_{xy,i} + k_i^* R_y$$

Observe that this result is consistent with what we would expect from the standard rules of differentiation for functions of real variables, with the vectors k_i and k_i^* treated as different quantities. Also, if all data were real-valued, in which case

$$J(k_i) = \mathsf{E}\,\boldsymbol{x}^2(i) - [\mathsf{E}\,\boldsymbol{x}(i)\boldsymbol{y}^{\mathsf{T}}]k_i - k_i^{\mathsf{T}}[\mathsf{E}\,\boldsymbol{y}\boldsymbol{x}(i)] + k_i^{\mathsf{T}} R_y k_i$$

then we would have obtained instead $\nabla_{k_i} J(k_i) = -R_{xy,i} + 2k_i^{\mathsf{T}} R_y$, with an additional factor of 2 — see Chapter C.

By setting the complex gradient equal to zero at the optimal choice $k_i = k_i^o$, we find that k_i^o should satisfy the linear equations

$$k_i^{o*} R_y = R_{xy,i}, \qquad i = 0, 1, \ldots, p-1 \tag{3.8}$$

The vector k_i^o so obtained minimizes $J(k_i)$ since the Hessian matrix of $J(k_i)$ with respect to k_i is equal to R_y, which is a nonnegative-definite matrix. Hessian matrices are defined in Chapter C; they are obtained by further differentiating $\nabla_{k_i} J(k_i)$ with respect to k_i^*.

If we collect the row vectors $\{k_i^{o*}\}$ from (3.8) into a matrix K_o we find that this desired solution matrix should satisfy

$$\boxed{K_o R_y = R_{xy}} \tag{3.9}$$

These equations are called the normal equations, for reasons explained later in Remark 4.1 in the next chapter.

3.3 MINIMIZATION BY COMPLETION OF SQUARES

We now re-establish (3.9), from first principles, by using a completion-of-squares argument that avoids the need for dealing with complex gradients. Thus, consider the cost function

(3.7) and note that it can be expressed in matrix form as follows:

$$J(k_i) = \begin{bmatrix} 1 & k_i^* \end{bmatrix} \begin{bmatrix} \sigma_{x,i}^2 & -R_{xy,i} \\ -R_{yx,i} & R_y \end{bmatrix} \begin{bmatrix} 1 \\ k_i \end{bmatrix} \tag{3.10}$$

with a Hermitian center matrix and with the unknown vector k_i, and its conjugate transpose, multiplying from both sides.

Now given any Hermitian matrix of the form

$$M = \begin{bmatrix} A & B \\ B^* & C \end{bmatrix}$$

with $A = A^*$, $C = C^*$, and C invertible, it can be verified by direct calculation that M can be factored into a product of a block upper-triangular, block-diagonal, and block lower-triangular matrices as:

$$\begin{bmatrix} A & B \\ B^* & C \end{bmatrix} = \begin{bmatrix} \mathrm{I} & BC^{-1} \\ 0 & \mathrm{I} \end{bmatrix} \begin{bmatrix} \Sigma & 0 \\ 0 & C \end{bmatrix} \begin{bmatrix} \mathrm{I} & 0 \\ C^{-1}B^* & \mathrm{I} \end{bmatrix} \tag{3.11}$$

where

$$\Sigma \triangleq A - BC^{-1}B^*$$

is called the Schur complement of M with respect to C. This is a general matrix result; it is an immediate extension to Hermitian matrices of a similar factorization result that was described in a footnote in Sec. 1.4 for symmetric matrices. The validity of (3.11) can be verified by expanding the right-hand side. The factorization (3.11) for M is valid as long as the matrix C is invertible, which means that we cannot apply the result directly to the center matrix

$$\begin{bmatrix} \sigma_{x,i}^2 & -R_{xy,i} \\ -R_{yx,i} & R_y \end{bmatrix}$$

that appears in our expression (3.10) for $J(k_i)$. The reason being that the covariance matrix R_y is only required to be nonnegative-definite (and, hence, possibly singular). However, there is a generalization of (3.11) for block matrices M with possibly singular matrices C. Indeed, it is easy to verify, also by direct calculation, that we can alternatively factor any such M as

$$\begin{bmatrix} A & B \\ B^* & C \end{bmatrix} = \begin{bmatrix} \mathrm{I} & D \\ 0 & \mathrm{I} \end{bmatrix} \begin{bmatrix} \Sigma & 0 \\ 0 & C \end{bmatrix} \begin{bmatrix} \mathrm{I} & 0 \\ D^* & \mathrm{I} \end{bmatrix} \tag{3.12}$$

where D is defined as *any* solution to the linear system of equations

$$DC = B \tag{3.13}$$

and

$$\Sigma = A - BD^*$$

Clearly, when C is invertible, the above factorization reduces to (3.11). However, when C is singular, many solutions D may exist for (3.13) and, consequently, many factorizations of the form (3.12) may be possible for M. Our arguments will show that for the matrix M in question, namely, the center matrix in (3.10), a factorization of the form (3.12) always exists since the corresponding equations (3.13) will always have a solution D — see Sec. 4.3.

Applying (3.12) to the center matrix in (3.10) we can write

$$\begin{bmatrix} \sigma_{x,i}^2 & -R_{xy,i} \\ -R_{yx,i} & R_y \end{bmatrix} = \begin{bmatrix} 1 & -k_i^{o*} \\ 0 & \mathrm{I} \end{bmatrix} \begin{bmatrix} \sigma_{x,i}^2 - R_{xy,i}k_i^o & 0 \\ 0 & R_y \end{bmatrix} \begin{bmatrix} 1 & 0 \\ -k_i^o & \mathrm{I} \end{bmatrix} \tag{3.14}$$

where k_i^o is any solution to the linear system of equations

$$k_i^{o*} R_y = R_{xy,i} \tag{3.15}$$

Substituting the factorization (3.14) into expression (3.10) for $J(k_i)$ and expanding the right-hand side, we find that $J(k_i)$ can be expressed in the equivalent form

$$\boxed{J(k_i) = (\sigma_{x,i}^2 - R_{xy,i}k_i^o) \; + \; (k_i - k_i^o)^* R_y (k_i - k_i^o)} \tag{3.16}$$

This is a revealing form for $J(k_i)$ since only the second term depends on the unknown k_i. But since R_y is nonnegative-definite, this second term is always nonnegative and it will be minimized (actually made equal to zero) by choosing $k_i = k_i^o$ (see Prob. II.5). We conclude that the minimizing k_i is given by any solution to (3.15). In addition, the resulting m.m.s.e. is given by any of the following equivalent forms:

$$\boxed{J(k_i^o) = \sigma_{x,i}^2 - R_{xy,i}k_i^o \; = \; \sigma_{x,i}^2 - k_i^{o*} R_y k_i^o \; = \; \sigma_{x,i}^2 - k_i^{o*} R_{yx,i}} \tag{3.17}$$

If we again collect the row vectors $\{k_i^{o*}\}$ into a matrix K_o we find that K_o should satisfy (3.9), namely, $K_o R_y = R_{xy}$. Moreover, the minimum value of the related cost (3.5) can be seen to be (see Probs. II.5 and II.6):

$$\boxed{\mathsf{E}\,\tilde{\boldsymbol{x}}^* \tilde{\boldsymbol{x}} \; = \; \mathsf{Tr}\,(R_x - K_o R_y K_o^*)} \tag{3.18}$$

3.4 MINIMIZATION OF THE ERROR COVARIANCE MATRIX

The same completion-of-squares argument can be applied directly to the solution of (3.4) rather than the solution of the individual problems (3.3). To see this, let $J(K)$ denote the error-covariance matrix,

$$\boxed{J(K) \; \triangleq \; \mathsf{E}\,\tilde{\boldsymbol{x}}\tilde{\boldsymbol{x}}^* \; = \; \mathsf{E}\,(\boldsymbol{x} - K\boldsymbol{y})(\boldsymbol{x} - K\boldsymbol{y})^*}$$

so that, as in (3.10),

$$J(K) = \begin{bmatrix} \mathrm{I} & K \end{bmatrix} \begin{bmatrix} R_x & -R_{xy} \\ -R_{yx} & R_y \end{bmatrix} \begin{bmatrix} \mathrm{I} \\ K^* \end{bmatrix}$$

Following the same arguments as in the previous section, we can factor the center matrix as

$$\begin{bmatrix} R_x & -R_{xy} \\ -R_{yx} & R_y \end{bmatrix} = \begin{bmatrix} \mathrm{I} & -K_o \\ 0 & \mathrm{I} \end{bmatrix} \begin{bmatrix} R_x - R_{xy}K_o^* & 0 \\ 0 & R_y \end{bmatrix} \begin{bmatrix} \mathrm{I} & 0 \\ -K_o^* & \mathrm{I} \end{bmatrix}$$

where K_o is any solution to $K_o R_y = R_{xy}$. It then follows that

$$J(K) \; = \; (R_x - R_{xy}K_o^*) \; + \; (K - K_o)R_y(K - K_o)^* \tag{3.19}$$

where the last term is nonnegative definite for any K since $R_y \geq 0$. Now the criterion in (3.4) is related to $J(K)$ via $\mathsf{E}\,\tilde{\boldsymbol{x}}^*\tilde{\boldsymbol{x}} = \mathsf{Tr}[J(K)]$ so that, from (3.19),

$$\mathsf{E}\,\tilde{\boldsymbol{x}}^*\tilde{\boldsymbol{x}} \geq \mathsf{Tr}(R_x - R_{xy}K_o^*) \qquad \text{for any } K \tag{3.20}$$

This is because the trace of the nonnegative-definite matrix $(K - K_o)R_y(K - K_o)^*$ is nonnegative. This result follows from the fact that the trace of a matrix, which is the sum of its diagonal elements, is also equal to the sum of its eigenvalues — see Prob. II.3. Now since the eigenvalues of a nonnegative matrix are necessarily nonnegative, it follows that $\mathsf{Tr}[(K - K_o)R_y(K - K_o)^*] \geq 0$. Equality in (3.20) is achieved by setting $K = K_o$ in (3.19).

This derivation reveals another aspect of the solution K_o. It not only minimizes the trace of the error covariance matrix, as in (3.4), but it also minimizes the error-covariance matrix itself since we also get $J(K) \geq J(K_o)$ for any K. We can therefore interpret K_o as also the solution to the problem

$$\boxed{\min_K \ \mathsf{E}\,\tilde{\boldsymbol{x}}\tilde{\boldsymbol{x}}^*} \tag{3.21}$$

which is in terms of the error-covariance matrix, rather than its trace. The resulting minimum value is

$$\boxed{J(K_o) = R_x - R_{xy}K_o^*} \tag{3.22}$$

The optimization problem (3.21) is interesting for two reasons. First, the cost function $J(K) = \mathsf{E}\,\tilde{\boldsymbol{x}}\tilde{\boldsymbol{x}}^*$ is matrix-valued. That is, it assumes *matrix values* for each choice of K. Second, the unknown argument, K, is a *matrix* itself. In this way, problem (3.21) involves minimizing a matrix-valued cost function over a matrix-valued argument.

3.5 OPTIMAL LINEAR ESTIMATOR

We shall have more to say about the solution(s) K_o of the normal equations $K_oR_y = R_{xy}$. For now, we summarize the main conclusions of the last two sections.

Theorem 3.1 (Optimal linear estimator) Given zero-mean random variables $\boldsymbol{x}$ and $\boldsymbol{y}$, the linear least-mean-squares estimator (l.l.m.s.e.) of $\boldsymbol{x}$ given $\boldsymbol{y}$ is

$$\hat{\boldsymbol{x}} = K_o\boldsymbol{y}$$

where K_o is any solution to the linear system of equations $K_oR_y = R_{xy}$. This estimator minimizes the following two error measures:

$$\min_K \mathsf{E}\,\tilde{\boldsymbol{x}}^*\tilde{\boldsymbol{x}} \qquad \text{and} \qquad \min_K \mathsf{E}\,\tilde{\boldsymbol{x}}\tilde{\boldsymbol{x}}^*$$

The scalar cost on the left is the trace of the matrix cost on the right. The resulting minimum mean-square errors, as defined by (3.4) and (3.22), are given by

$$\begin{aligned}\min_K \mathsf{E}\,\tilde{\boldsymbol{x}}^*\tilde{\boldsymbol{x}} &= \mathsf{Tr}\,(R_x - K_oR_yK_o^*)\\ \min_K \mathsf{E}\,\tilde{\boldsymbol{x}}\tilde{\boldsymbol{x}}^* &= R_x - K_oR_yK_o^*\end{aligned}$$

CHAPTER 4

Orthogonality Principle

Before examining the properties of the solution(s) K_o of the normal equations $K_oR_y = R_{xy}$, we consider some illustrative examples in the context of symbol estimation and channel equalization. Additional examples and applications are discussed in Chapter 5.

4.1 DESIGN EXAMPLES

Example 4.1 (Noisy measurements of a binary signal)

We reconsider Ex. 1.2 of a BPSK signal $\boldsymbol{x}$ that assumes the values ± 1 with probability $1/2$. The measurement $\boldsymbol{y}$ is $\boldsymbol{y} = \boldsymbol{x} + \boldsymbol{v}$, where $\boldsymbol{x}$ and the disturbance $\boldsymbol{v}$ are independent of each other, with $\boldsymbol{v}$ being zero-mean Gaussian of unit variance.

Both $\boldsymbol{x}$ and $\boldsymbol{y}$ have zero means so that, according to Thm. 3.1, the optimal linear estimator of $\boldsymbol{x}$ is $\hat{\boldsymbol{x}} = k_o\boldsymbol{y}$, where the (now scalar) coefficient k_o is obtained from solving $k_o\sigma_y^2 = \sigma_{xy}$. We therefore need to determine the quantities $\{\sigma_y^2, \sigma_{xy}\}$. Now since $\{\boldsymbol{x}, \boldsymbol{v}\}$ are independent we have

$$\sigma_y^2 \;=\; \sigma_x^2 + \sigma_v^2 = 1 + 1 = 2$$

Moreover,

$$\sigma_{xy} = \mathsf{E}\,\boldsymbol{x}\boldsymbol{y} = \mathsf{E}\,\boldsymbol{x}(\boldsymbol{x}+\boldsymbol{v}) = \mathsf{E}\,\boldsymbol{x}^2 + 0 = \mathsf{E}\,\boldsymbol{x}^2 = 1$$

so that $k_o = 1/2$, and the optimal linear estimator is $\hat{\boldsymbol{x}} = \boldsymbol{y}/2$. That is, we simply scale the received signal by $1/2$. In contrast, the optimal estimator was found in Ex. 1.2 to be given by the nonlinear transformation $\tanh(\boldsymbol{y})$. In addition, observe that the form of the linear estimator, $\hat{\boldsymbol{x}} = \boldsymbol{y}/2$, is valid regardless of whether the noise $\boldsymbol{v}$ is Gaussian or not (i.e., the Gaussian assumption on $\boldsymbol{v}$ is not needed to arrive at $k_o = 1/2$). The form of the optimal estimator, $\hat{\boldsymbol{x}} = \tanh(\boldsymbol{y})$, on the other hand, is very much tied to the Gaussian assumption on $\boldsymbol{v}$.

Let us now reconsider Ex. 2.1, where we collect two noisy measurements $\boldsymbol{y}(0)$ and $\boldsymbol{y}(1)$ of $\boldsymbol{x}$, say,

$$\boldsymbol{y}(0) = \boldsymbol{x} + \boldsymbol{v}(0) \quad \text{and} \quad \boldsymbol{y}(1) = \boldsymbol{x} + \boldsymbol{v}(1)$$

where $\{\boldsymbol{v}(0), \boldsymbol{v}(1)\}$ are zero-mean unit-variance Gaussian random variables that are independent of each other and of $\boldsymbol{x}$. The value of $\boldsymbol{x}$ is the same in both measurements (i.e., if it is $+1$ in the measurement $\boldsymbol{y}(0)$, it is also $+1$ in the measurement $\boldsymbol{y}(1)$, and similarly for -1) — recall Fig. 2.1. Introduce the column vector $\boldsymbol{y} = \text{col}\{\boldsymbol{y}(0), \boldsymbol{y}(1)\}$. Then, according to Thm. 3.1, the optimal linear estimator of $\boldsymbol{x}$ given $\boldsymbol{y}$ is $\hat{\boldsymbol{x}} = k_o^*\boldsymbol{y}$, where k_o^* is now 1×2 and is obtained from the solution of the normal equations $k_o^*R_y = R_{xy}$. To determine $\{R_y, R_{xy}\}$ we proceed as follows. Since $\{\boldsymbol{x}, \boldsymbol{v}(0), \boldsymbol{v}(1)\}$ are independent we get

$$R_y = \mathsf{E}\,\boldsymbol{y}\boldsymbol{y}^* = \begin{bmatrix} \mathsf{E}\,|\boldsymbol{y}(0)|^2 & \mathsf{E}\,\boldsymbol{y}(0)\boldsymbol{y}^*(1) \\ \mathsf{E}\,\boldsymbol{y}(1)\boldsymbol{y}^*(0) & \mathsf{E}\,|\boldsymbol{y}(1)|^2 \end{bmatrix} = \begin{bmatrix} 2 & 1 \\ 1 & 2 \end{bmatrix}$$

where we used the fact that

$$\mathsf{E}\,\boldsymbol{y}(0)\boldsymbol{y}^*(1) = \mathsf{E}\,(\boldsymbol{x}+\boldsymbol{v}(0))(\boldsymbol{x}+\boldsymbol{v}(1))^* = \mathsf{E}\,|\boldsymbol{x}|^2 = 1$$

Likewise,

$$R_{xy} = \mathsf{E}\,\boldsymbol{x}\boldsymbol{y}^* = \begin{bmatrix} \mathsf{E}\,\boldsymbol{x}\boldsymbol{y}^*(0) & \mathsf{E}\,\boldsymbol{x}\boldsymbol{y}^*(1) \end{bmatrix} = \begin{bmatrix} 1 & 1 \end{bmatrix}$$

so that

$$k_o^* = R_{xy}R_y^{-1} = \begin{bmatrix} 1 & 1 \end{bmatrix}\begin{bmatrix} 2 & 1 \\ 1 & 2 \end{bmatrix}^{-1} = \frac{1}{3}\begin{bmatrix} 1 & 1 \end{bmatrix}$$

That is,

$$\hat{\boldsymbol{x}} = \frac{1}{3}\left[\boldsymbol{y}(0)+\boldsymbol{y}(1)\right]$$

Again, this expression for the linear least-mean-squares estimator of $\boldsymbol{x}$ given $\{\boldsymbol{y}(0),\boldsymbol{y}(1)\}$ holds regardless of whether the noises $\{\boldsymbol{v}(0),\boldsymbol{v}(1)\}$ are Gaussian or not. Only the first and second moments of $\{\boldsymbol{v}(0),\boldsymbol{v}(1)\}$, namely, their means and variances, are needed to determine k_o^*. In the context of the two-antenna example of Fig. 2.1, the above result leads to the optimal *linear* receiver structure shown in Fig. 4.1.

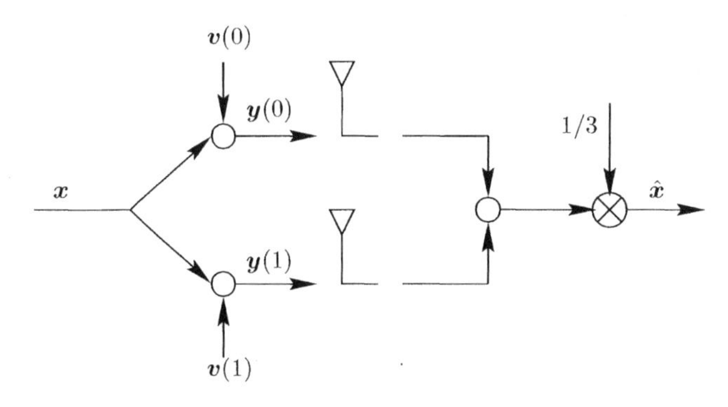

FIGURE 4.1 An optimal linear receiver for recovering a BPSK transmission from two measurements in the presence of additive unit-variance uncorrelated noises.

Example 4.2 (Multiple measurements of a binary signal)

Continuing with Ex. 4.1, let us examine what happens if we increase the number of available measurements from 2 to N, say,

$$\boldsymbol{y}(i) = \boldsymbol{x}+\boldsymbol{v}(i), \qquad i = 0,1,\ldots,N-1$$

Introduce the observation vector $\boldsymbol{y} = \text{col}\{\boldsymbol{y}(0),\boldsymbol{y}(1),\ldots,\boldsymbol{y}(N-1)\}$. Then, say, for $N=5$,

$$R_{xy} = \begin{bmatrix} 1 & 1 & 1 & 1 & 1 \end{bmatrix}, \qquad R_y = \begin{bmatrix} 2 & 1 & 1 & 1 & 1 \\ 1 & 2 & 1 & 1 & 1 \\ 1 & 1 & 2 & 1 & 1 \\ 1 & 1 & 1 & 2 & 1 \\ 1 & 1 & 1 & 1 & 2 \end{bmatrix}$$

so that

$$\hat{\boldsymbol{x}} = R_{xy}R_y^{-1}\boldsymbol{y} = \left(\begin{bmatrix} 1 & 1 & 1 & 1 & 1 \end{bmatrix} \begin{bmatrix} 2 & 1 & 1 & 1 & 1 \\ 1 & 2 & 1 & 1 & 1 \\ 1 & 1 & 2 & 1 & 1 \\ 1 & 1 & 1 & 2 & 1 \\ 1 & 1 & 1 & 1 & 2 \end{bmatrix}^{-1} \right) \cdot \boldsymbol{y}$$

We need to evaluate R_y^{-1}. Due to the special structure of R_y, its inverse can be evaluated in closed form for any N, as we explain below. Later, in Sec. 5.5, when we reconsider this problem, we shall show how to evaluate $\hat{\boldsymbol{x}}$ via a more direct route.

Observe that, for any N, the matrix R_y can be expressed as $R_y = \mathrm{I} + aa^{\mathsf{T}}$, where a is the $N \times 1$ column vector $a = \mathrm{col}\{1, 1, 1, \ldots, 1\}$. In other words, R_y is a rank-one modification of the identity matrix. This is a useful observation since the inverse of every such matrix has a similar form (see Prob. II.1). Specifically,

$$\left(\mathrm{I} + aa^{\mathsf{T}}\right)^{-1} = \mathrm{I} - \frac{aa^{\mathsf{T}}}{1 + \|a\|^2} = \mathrm{I} - \frac{1}{N+1}aa^{\mathsf{T}}$$

where $\|a\|^2$ denotes the squared Euclidean norm of a, $\|a\|^2 = a^{\mathsf{T}}a$. Using this result we find that

$$R_{xy}R_y^{-1} = a^{\mathsf{T}}\left(\mathrm{I} - \frac{aa^{\mathsf{T}}}{N+1}\right) = a^{\mathsf{T}} - \frac{N}{N+1}a^{\mathsf{T}} = \frac{a^{\mathsf{T}}}{N+1}$$

so that

$$\hat{\boldsymbol{x}} = R_{xy}R_y^{-1}\boldsymbol{y} = \frac{a^{\mathsf{T}}}{N+1}\boldsymbol{y} = \frac{1}{N+1}\sum_{k=0}^{N-1}\boldsymbol{y}(k)$$

Example 4.3 (Transmissions over a noisy channel)

Consider again the setting of Exs. A.4 and 2.2, where independent and identically distributed (i.i.d.) symbols $\{\boldsymbol{s}(i)\}$ are transmitted over the FIR channel $C(z) = 1 + 0.5z^{-1}$. Each symbol is either $+1$ with probability p or -1 with probability $1 - p$, and the output of the channel is corrupted by zero-mean additive white Gaussian noise $\boldsymbol{v}(i)$ of unit variance. The noise and the symbols are independent of each other (see Fig. 4.2). We also assume, for simplicity, that $p = 1/2$. We want to estimate the vector $\boldsymbol{x} = \mathrm{col}\{\boldsymbol{s}(0), \boldsymbol{s}(1)\}$ from the observation vector $\boldsymbol{y} = \mathrm{col}\{\boldsymbol{y}(0), \boldsymbol{y}(1)\}$, where

$$\boldsymbol{y}(0) = \boldsymbol{s}(0) + \boldsymbol{v}(0), \quad \boldsymbol{y}(1) = \boldsymbol{s}(1) + 0.5\boldsymbol{s}(0) + \boldsymbol{v}(1) \tag{4.1}$$

We are assuming that transmissions start at time 0 so that $\boldsymbol{s}(-1) = 0$. According to Thm. 3.1, the optimal linear estimator of $\boldsymbol{x}$ is $\hat{\boldsymbol{x}} = K_o\boldsymbol{y}$, where K_o is 2×2 and is obtained by solving the normal equations $K_oR_y = R_{xy}$. We therefore need to determine $\{R_y, R_{xy}\}$.

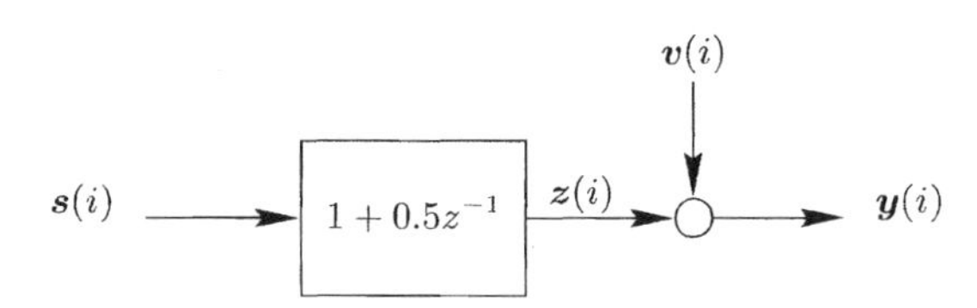

FIGURE 4.2 Data transmissions through an additive Gaussian-noise channel.

It follows from the relations (4.1) that

$$R_{xy} = \mathsf{E}\begin{bmatrix} s(0) \\ s(1) \end{bmatrix}\begin{bmatrix} y^*(0) & y^*(1) \end{bmatrix} = \begin{bmatrix} 1 & 0.5 \\ 0 & 1 \end{bmatrix}$$

Moreover,

$$R_y = \mathsf{E}\begin{bmatrix} y(0) \\ y(1) \end{bmatrix}\begin{bmatrix} y^*(0) & y^*(1) \end{bmatrix} = \begin{bmatrix} 2 & 1/2 \\ 1/2 & 9/4 \end{bmatrix}$$

so that

$$K_o = \begin{bmatrix} 1 & 0.5 \\ 0 & 1 \end{bmatrix}\begin{bmatrix} 2 & 1/2 \\ 1/2 & 9/4 \end{bmatrix}^{-1} = \frac{4}{17}\begin{bmatrix} 2 & 1/2 \\ -1/2 & 2 \end{bmatrix}$$

That is,

$$\hat{s}(0) = 8y(0)/17 + 2y(1)/17 \quad \text{and} \quad \hat{s}(1) = -2y(0)/17 + 8y(1)/17$$

This example is pursued further below and in Probs. II.9 and II.10. Later in Sec. 5.3 we study the general case of estimating a block of data $\{s(\cdot)\}$, of generic length N, from a block of observations $\{y(\cdot)\}$.

Example 4.4 (Linear channel equalization)

Consider again the setting of Ex. 4.1, and assume now that the transmissions started in the remote past (rather than at time 0, i.e., $i > -\infty$) so that all random processes $\{s(i), v(i), y(i)\}$ can be assumed to be wide-sense stationary — the need for this assumption will become evident soon. By a wide-sense stationary process $s(\cdot)$, we mean one with a constant mean and whose auto-correlation sequence is only a function of the time lag, i.e., $r_s(k) = \mathsf{E}\, s(i)s^*(i-k)$ is only a function of k.

Now referring to Fig. 4.3, we see that the output of the channel at any time instant i is a linear combination of the current symbol $s(i)$ and the previous symbol $s(i-1)$, i.e.,

$$z(i) = s(i) + 0.5s(i-1)$$

We therefore say that the channel introduces *inter-symbol interference* or ISI, since a symbol transmitted at a prior time, $s(i-1)$, interferes with the output at the time of the current symbol, $s(i)$. The measurement $y(i)$ that is available at the receiver is a noisy version of $z(i)$, namely,

$$y(i) = s(i) + 0.5s(i-1) + v(i) \tag{4.2}$$

The purpose of this example is to show how to design an equalizer for the channel. The function of the equalizer is to process the received signal $\{y(i)\}$ in order to recover the transmitted symbol $\{s(i)\}$, or a delayed version of it, say, $\{s(i-\Delta)\}$ for some Δ. There are many different structures that can be used for equalization purposes. In Fig. 4.3 we show an FIR equalizer structure that consists of three taps $\{\alpha(0), \alpha(1), \alpha(2)\}$. Its output at any time instant i is given by the linear combination

$$\alpha(0)y(i) + \alpha(1)y(i-1) + \alpha(2)y(i-2)$$

We wish to determine the taps $\{\alpha(0), \alpha(1), \alpha(2)\}$ so that the output of the equalizer is the optimal linear estimator for $s(i)$ (we choose $\Delta = 0$ in this example — see Prob. II.17 for nonzero values of Δ). This procedure is known as linear minimum mean-square-error equalization in communications applications. We discuss it in greater detail in Sec. 5.4 for higher-order channels and equalizers, and also for nonzero values of Δ. The discussion here is meant to illustrate some of the key concepts.

Observe from Fig. 4.3 that at any time instant i, the equalizer uses three observations $\{y(i), y(i-1), y(i-2)\}$ in order to estimate $s(i)$. Therefore, the observation vector is $y = \text{col}\{y(i), y(i-1), y(i-2)\}$ and the variable we wish to estimate is $x = s(i)$. We then know from Thm.3.1 that

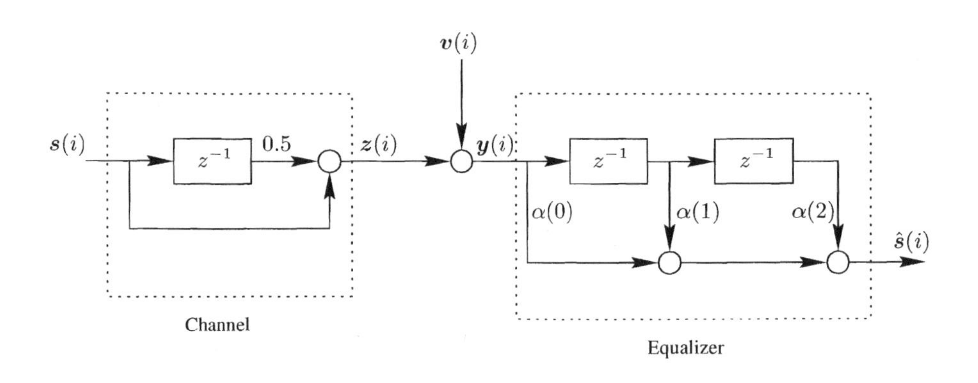

FIGURE 4.3 Linear channel equalization.

the optimal linear estimator for $\boldsymbol{x}$ is given by

$$\hat{\boldsymbol{x}} \stackrel{\Delta}{=} \hat{\boldsymbol{s}}(i) = k_o^* \boldsymbol{y}$$

where the row vector k_o^* is found from solving the normal equations $k_o^* R_y = R_{xy}$. Once k_o^* is found, its entries give the desired tap coefficients $\{\alpha(0), \alpha(1), \alpha(2)\}$, i.e.,

$$k_o^* = \begin{bmatrix} \alpha(0) & \alpha(1) & \alpha(2) \end{bmatrix}$$

In order to find k_o^*, we need to determine $\{R_{xy}, R_y\}$. Thus, let $r_y(k)$ denote the auto-correlation sequence of the stationary process $\{\boldsymbol{y}(i)\}$, i.e., $r_y(k) = \mathsf{E}\,\boldsymbol{y}(i)\boldsymbol{y}^*(i-k)$. Then

$$R_y = \mathsf{E}\,\boldsymbol{y}\boldsymbol{y}^* = \begin{bmatrix} r_y(0) & r_y(1) & r_y(2) \\ r_y^*(1) & r_y(0) & r_y(1) \\ r_y^*(2) & r_y^*(1) & r_y(0) \end{bmatrix}$$

To determine $\{r_y(0), r_y(1), r_y(2)\}$, we use the output equation (4.2). Multiplying it from the right by $\boldsymbol{y}^*(i)$ we get

$$\boldsymbol{y}(i)\boldsymbol{y}^*(i) = \boldsymbol{s}(i)\boldsymbol{y}^*(i) \;+\; 0.5\boldsymbol{s}(i-1)\boldsymbol{y}^*(i) \;+\; \boldsymbol{v}(i)\boldsymbol{y}^*(i)$$

Taking expectations of both sides, and recalling that the variables $\{\boldsymbol{s}(i), \boldsymbol{s}(i-1), \boldsymbol{v}(i)\}$ are independent of each other, we find that

$$\begin{aligned} \mathsf{E}\,\boldsymbol{y}(i)\boldsymbol{y}^*(i) &= r_y(0) \\ \mathsf{E}\,\boldsymbol{s}(i)\boldsymbol{y}^*(i) &= \mathsf{E}\,\boldsymbol{s}(i)[\boldsymbol{s}(i) + 0.5\boldsymbol{s}(i-1) + \boldsymbol{v}(i)]^* \;=\; 1 \\ \mathsf{E}\,\boldsymbol{s}(i-1)\boldsymbol{y}^*(i) &= \mathsf{E}\,\boldsymbol{s}(i-1)[\boldsymbol{s}(i) + 0.5\boldsymbol{s}(i-1) + \boldsymbol{v}(i)]^* \;=\; 1/2 \\ \mathsf{E}\,\boldsymbol{v}(i)\boldsymbol{y}^*(i) &= \mathsf{E}\,\boldsymbol{v}(i)[\boldsymbol{s}(i) + 0.5\boldsymbol{s}(i-1) + \boldsymbol{v}(i)]^* \;=\; 1 \end{aligned}$$

so that

$$r_y(0) = 1 + 1/4 + 1 \;=\; 9/4$$

Likewise, multiplying (4.2) from the right by $\boldsymbol{y}^*(i-1)$ and taking expectations of both sides, we get $r_y(1) = 1/2$. Finally, multiplying (4.2) from the right by $\boldsymbol{y}^*(i-2)$ and taking expectations we get $r_y(2) = 0$. In summary, we find that

$$R_y = \begin{bmatrix} 9/4 & 1/2 & 0 \\ 1/2 & 9/4 & 1/2 \\ 0 & 1/2 & 9/4 \end{bmatrix}$$

In a similar fashion, we can evaluate R_{xy},

$$R_{xy} = \mathsf{E}\,\boldsymbol{x}\boldsymbol{y}^* = \begin{bmatrix} \mathsf{E}\,\boldsymbol{s}(i)\boldsymbol{y}^*(i) & \mathsf{E}\,\boldsymbol{s}(i)\boldsymbol{y}^*(i-1) & \mathsf{E}\,\boldsymbol{s}(i)\boldsymbol{y}^*(i-2) \end{bmatrix}$$

Thus, multiplying (4.2) from the right by $\boldsymbol{s}^*(i)$ and taking expectations we get $\mathsf{E}\,\boldsymbol{y}(i)\boldsymbol{s}^*(i) = 1$. Likewise, multiplying (4.2) from the right by $\boldsymbol{s}^*(i+1)$ and taking expectations we get $\mathsf{E}\,\boldsymbol{y}(i)\boldsymbol{s}^*(i+1) = 0$. Similarly, $\mathsf{E}\,\boldsymbol{y}(i)\boldsymbol{s}^*(i+2) = 0$. It further follows from the assumed stationarity of the processes $\{\boldsymbol{s}(i), \boldsymbol{v}(i)\}$ that

$$\mathsf{E}\,\boldsymbol{y}(i)\boldsymbol{s}^*(i+1) = \mathsf{E}\,\boldsymbol{y}(i-1)\boldsymbol{s}^*(i) = [\mathsf{E}\,\boldsymbol{s}(i)\boldsymbol{y}^*(i-1)]^*$$

and

$$\mathsf{E}\,\boldsymbol{y}(i)\boldsymbol{s}^*(i+2) = \mathsf{E}\,\boldsymbol{y}(i-2)\boldsymbol{s}^*(i) = [\mathsf{E}\,\boldsymbol{s}(i)\boldsymbol{y}^*(i-2)]^*$$

Hence,

$$R_{xy} = \begin{bmatrix} 1 & 0 & 0 \end{bmatrix}$$

Later, in Sec. 5.4, we describe a more general procedure for determining the quantities $\{R_y, R_{xy}\}$ for arbitrary channel and equalizer lengths.

Using the just derived values for $\{R_{xy}, R_y\}$, we are led to

$$k_o^* = R_{xy}R_y^{-1} = \begin{bmatrix} 1 & 0 & 0 \end{bmatrix} \begin{bmatrix} 9/4 & 1/2 & 0 \\ 1/2 & 9/4 & 1/2 \\ 0 & 1/2 & 9/4 \end{bmatrix}^{-1} = \begin{bmatrix} 0.4688 & -0.1096 & 0.0244 \end{bmatrix}$$

That is, $\alpha(0) = 0.4688$, $\alpha(1) = -0.1096$, and $\alpha(2) = 0.0244$. Moreover, the resulting m.m.s.e. is $\text{m.m.s.e.} = \sigma_x^2 - R_{xy}k_o = 0.5312$, where $\sigma_x^2 = \sigma_s^2 = 1$. A computer project at the end of this part illustrates the operation of optimal linear equalizers designed in this manner.

◇

4.2 ORTHOGONALITY CONDITION

The linear least-mean-squares estimator admits an important geometric interpretation in the form of an orthogonality condition. This can be seen by rewriting the normal equations (3.9) as $K_o\mathsf{E}\,\boldsymbol{y}\boldsymbol{y}^* = \mathsf{E}\,\boldsymbol{x}\boldsymbol{y}^*$ or, equivalently,

$$\mathsf{E}\,(\boldsymbol{x} - K_o\boldsymbol{y})\boldsymbol{y}^* = 0 \tag{4.3}$$

The difference $\boldsymbol{x} - K_o\boldsymbol{y}$ is the estimation error, $\tilde{\boldsymbol{x}}$. Therefore, equality (4.3) states that the error is orthogonal to (or uncorrelated with) the observation vector $\boldsymbol{y}$, namely,

$$\boxed{\mathsf{E}\,\tilde{\boldsymbol{x}}\boldsymbol{y}^* = 0}$$

which we also write as (see Fig. 4.4):

$$\boxed{\tilde{\boldsymbol{x}} \perp \boldsymbol{y}} \tag{4.4}$$

We thus conclude that for linear least-mean-squares estimation, the estimation error is orthogonal to the data and, in fact, to any *linear* transformation of the data, say, $A\boldsymbol{y}$ for any matrix A. This fact means that no further linear transformation of $\boldsymbol{y}$ can extract additional information about $\boldsymbol{x}$ in order to further reduce the error covariance matrix. Moreover, since the estimator $\hat{\boldsymbol{x}}$ is itself a linear function of $\boldsymbol{y}$, we obtain, as a special case, that

$$\boxed{\tilde{\boldsymbol{x}} \perp \hat{\boldsymbol{x}}} \tag{4.5}$$

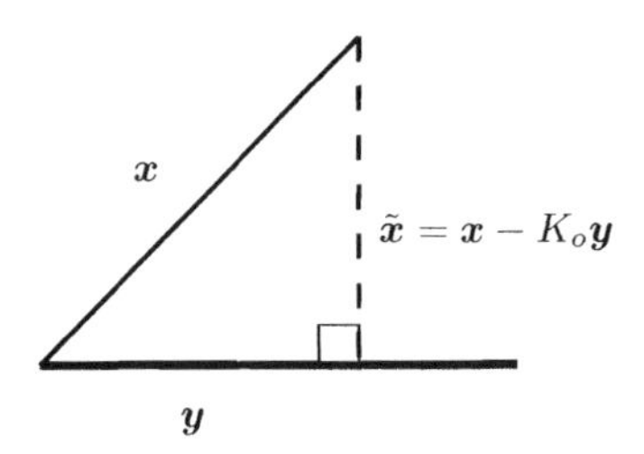

FIGURE 4.4 The orthogonality condition for linear estimation: $\tilde{x} \perp y$.

That is, the estimation error is also orthogonal to the estimator. Actually, the orthogonality condition is the *defining* property of optimality in the linear least-mean-squares sense, as the following result shows.

> **Theorem 4.1 (Orthogonality principle)** Given two zero-mean random variables x and y, a linear estimator $\hat{x} = K_o y$ is optimal in the least-mean-squares sense (3.3) if, and only if, it satisfies $x - \hat{x} \perp y$, i.e., $\mathsf{E}\,(x - \hat{x})y^* = 0$.

Proof: One direction was argued prior to the statement of the theorem. Specifically, if $\hat{x}$ is the optimal linear estimator, then we know from (4.4) that $\tilde{x} \perp y$. Conversely, assume $\hat{x}$ is a linear estimator for x that satisfies $x - \hat{x} \perp y$, and let $\hat{x} = Ky$ for some K. It then follows from $x - \hat{x} \perp y$ that $\mathsf{E}\,(x - Ky)y^* = 0$, so that K satisfies the normal equations $KR_y = R_{xy}$ and, hence, from Thm. 3.1 we conclude that $\hat{x}$ should be the optimal linear estimator.

◇

Remark 4.1 (Terminology) The designation *normal equations* for (3.9) is motivated by the fact that these equations arise from the orthogonality condition (4.3). In the adaptive filtering literature, an optimal solution K_o of the normal equations is often mistakenly called the "Wiener solution." As explained in App. 3.C of Sayed (2003), Wiener solved a more elaborate problem.

◇

Example 4.5 (Signal with exponential auto-correlation)

Consider a scalar zero-mean stationary random process $\{z(t)\}$ with auto-correlation function:

$$R_z(\tau) \triangleq \mathsf{E}\,z(t)z^*(t-\tau) = e^{-\alpha|\tau|} \tag{4.6}$$

That is, the samples of $z(t)$ become less correlated as the time gap between them increases. A so-called *random telegraph* signal has this property — see Prob. II.23. It is claimed that the linear least-mean squares estimator of $z(T_3)$ given $z(T_1)$ and $z(T_2)$ (assuming $T_1 < T_2 < T_3$) is

$$\hat{z}(T_3) = e^{-\alpha(T_3 - T_2)} z(T_2)$$

That is, the estimator of a future value depends only on the most recent observation, $z(T_2)$. We can verify the validity of this claim by checking whether the above estimator satisfies the orthogonality condition.

So define the observation vector $\boldsymbol{y} = \text{col}\{\boldsymbol{z}(T_1), \boldsymbol{z}(T_2)\}$ and let $\boldsymbol{x} = \boldsymbol{z}(T_3)$. We can now evaluate the cross-correlation vector

$$\mathsf{E}\,(\boldsymbol{x} - \hat{\boldsymbol{x}})\boldsymbol{y}^* = \mathsf{E}\,\left[\boldsymbol{z}(T_3) - e^{-\alpha(T_3 - T_2)}\boldsymbol{z}(T_2)\right]\boldsymbol{y}^*$$

If the answer is zero then the orthogonality condition is satisfied and the estimator is optimal in the linear least-mean-squares sense, as claimed. Otherwise, the estimator is not optimal. Using the given auto-correlation function (4.6), it is easy to verify that

$$\begin{aligned}\mathsf{E}\left[\boldsymbol{z}(T_3) - e^{-\alpha(T_3-T_2)}\boldsymbol{z}(T_2)\right]\boldsymbol{y}^* &= \left[\begin{array}{cc} e^{-\alpha(T_3-T_1)} - e^{-\alpha(T_3-T_1)} & e^{-\alpha(T_3-T_2)} - e^{-\alpha(T_3-T_2)} \end{array}\right] \\ &= \left[\begin{array}{cc} 0 & 0 \end{array}\right]\end{aligned}$$

so that the estimator is optimal.

4.3 EXISTENCE OF SOLUTIONS

Consider again the normal equations (3.9), namely,

$$K_o R_y = R_{xy} \tag{4.7}$$

In general, such linear systems of equations can have a unique solution, no solution at all, or an infinite number of solutions. This depends on the rank of the coefficient matrix R_y and on how the right-hand side matrix R_{xy} relates to R_y. We shall explain these facts here and, in the process, provide the reader with an opportunity to get acquainted with some basic concepts from matrix theory and linear algebra. The reader may wish to review the material in Sec. B.2 for an overview of the concepts of nullspace, range space, and rank of a matrix.

Unique solution. The argument will show that a unique solution K_o exists if, and only if, the covariance matrix R_y is positive-definite. Indeed, if $R_y > 0$ then all its eigenvalues are positive and, consequently, R_y is nonsingular. In this case, equation (4.7) will have a unique solution K_o given by $K_o = R_{xy}R_y^{-1}$. Conversely, assume a unique solution K_o exists to the normal equations (4.7) and let us establish that it must hold that $R_y > 0$. Assume, to the contrary, that R_y is singular. Then there should exist a nonzero row vector c^* such that $c^* R_y = 0$ or, equivalently, $R_y c = 0$. The vector c belongs to the *nullspace* of R_y, written as $c \in \mathcal{N}(R_y)$. It is now easy to see that by adding to the rows of K_o any such vector c^*, we obtain a matrix K_o' that satisfies the same equations (4.7), i.e., $K_o' R_y = R_{xy}$. This contradicts the fact that K_o is unique so that R_y must be positive-definite.

Infinitely many solutions. We now show that the normal equations (4.7) will have infinitely many solutions K_o if, and only if, R_y is singular. One direction of the proof is obvious. Assume K_o is one solution and that R_y is singular. Then by adding to the rows of K_o any vector c^* (or any combination of vectors) from the nullspace of R_y, we obtain another solution K_o' (as explained above). Hence, infinitely many solutions exist in this case. Conversely assume that many solutions exist. Let K_o and K_o' denote any two of these solutions. Then subtracting the equations $K_o R_y = R_{xy}$ and $K_o' R_y = R_{xy}$ we obtain $(K_o - K_o')R_y = 0$, which means that R_y is singular, since there is at least one nonzero row in $K_o - K_o'$ and this row annihilates R_y.

Existence of solutions. The only question that remains regarding $K_o R_y = R_{xy}$ is whether solutions always exist. The answer is affirmative. That is, the normal equations (4.7) are always *consistent*. Recall that a linear system of equations $Ax = b$ is said to be *consistent* if the vector b lies in the range space of A, written as $b \in \mathcal{R}(A)$. This is equivalent to saying that a solution vector x exists. To establish this fact for the normal equations $K_o R_y = R_{xy}$, we need to show that for any two random variables $\{\boldsymbol{x}, \boldsymbol{y}\}$, it always holds that the columns of R_{yx} lie in the column span of R_y, i.e., that there exists at least one matrix K_o^* that satisfies $R_y K_o^* = R_{yx}$, which by transposition is equivalent to $K_o R_y = R_{xy}$. This statement is obviously true when R_y is nonsingular since then $K_o = R_{xy} R_y^{-1}$. The argument is more involved when R_y is singular. We first establish a preliminary result that provides an equivalent characterization for checking whether the normal equations $K_o R_y = R_{xy}$ are consistent or not.

Lemma 4.1 (Consistent equations) The equations $K_o R_y = R_{xy}$ are consistent (i.e., there exists at least one solution K_o) if, and only if,

$$c^* R_{yx} = 0 \quad \text{for any} \quad c \in \mathcal{N}(R_y)$$

that is, $c^* R_{yx} = 0$ for any column vector c satisfying $R_y c = 0$.

Proof: We first verify that the condition $c^* R_{yx} = 0$ for all $c \in \mathcal{N}(R_y)$ implies the existence of a matrix K_o satisfying $K_o R_y = R_{xy}$. Let us assume, to the contrary, that the equations $K_o R_y = R_{xy}$ are not consistent. This means that R_{yx} does not lie in the column span of R_y, which in turn means that there should exist a vector $c \in \mathcal{N}(R_y)$ that is not orthogonal to R_{yx}. This conclusion contradicts the assumption that $c^* R_{yx} = 0$ for all $c \in \mathcal{N}(R_y)$.

We now establish the converse statement, namely, that the existence of a K_o satisfying $K_o R_y = R_{xy}$ implies $c^* R_{yx} = 0$ for all $c \in \mathcal{N}(R_y)$. This claim is obvious. For any such c, we have $R_y c = 0$ and, hence, $K_o R_y c = 0$. This implies that $R_{xy} c = 0$ or, equivalently, $c^* R_{yx} = 0$.

◇

Therefore, in order to establish that the equations $K_o R_y = R_{xy}$ are consistent, it is enough to prove that $c^* R_{yx} = 0$ for all $c \in \mathcal{N}(R_y)$. So let us assume that this latter condition does not hold. This means that there exists at least one nonzero vector c such that $c^* R_y = 0$ and $c^* R_{yx} \neq 0$. This statement leads to a contradiction. Indeed, $c^* R_y = 0$ implies that $c^* R_y c = c^* (\mathsf{E}\, \boldsymbol{y}\boldsymbol{y}^*) c = 0$ or, equivalently, $\mathsf{E}\, |c^* \boldsymbol{y}|^2 = 0$. We therefore have a zero-mean random variable $c^* \boldsymbol{y}$ (because $\boldsymbol{y}$ is zero-mean itself) with zero variance. Using Remark A.1, we conclude that $c^* \boldsymbol{y}$ is the zero random variable with probability one. It follows that

$$c^* R_{yx} \stackrel{\Delta}{=} c^* (\mathsf{E}\, \boldsymbol{y}\boldsymbol{x}^*) = \mathsf{E}\, (c^* \boldsymbol{y}) \boldsymbol{x}^* = 0$$

This contradicts the assumption $c^* R_{yx} \neq 0$ and we conclude that the normal equations (4.7) are always consistent.

Uniqueness of estimator. Interesting enough, regardless of which solution K_o we pick (in the case when a multitude of solutions exist), the resulting estimator, $\hat{\boldsymbol{x}} = K_o \boldsymbol{y}$, and the resulting m.m.s.e., $R_x - K_o R_y K_o^*$, will always assume the same values. In other words, their values are independent of the specific choice of K_o.

To see this, let us first establish the result for the m.m.s.e. Thus, let K_o and K_o' be two possible solutions of (4.7), i.e., $K_o R_y = R_{xy}$ and $K_o' R_y = R_{xy}$. Then the difference $K_o - K_o'$ satisfies

$$[K_o - K_o']\, R_y = 0 \tag{4.8}$$

Now denote the minimum mean-square errors by

$$\Delta_1 = R_x - K_o R_y K_o^* = R_x - R_{xy} K_o^* \quad \text{and} \quad \Delta_2 = R_x - K_o' R_y K_o'^* = R_x - R_{xy} K_o'^*$$

Subtracting the expressions for $\{\Delta_1, \Delta_2\}$ we obtain

$$\Delta_2 - \Delta_1 \;=\; R_{xy}\left[K_o - K_o'\right]^* \;=\; K_o R_y \left[K_o - K_o'\right]^* \;=\; 0$$

where in the second equality we substituted R_{xy} by $K_o R_y$, and in the third equality we used (4.8). Therefore, $\Delta_1 = \Delta_2$. This means, as desired, that the value of the m.m.s.e. is independent of K_o.

Let us now verify that no matter which K_o we pick, the corresponding estimator $\hat{\boldsymbol{x}}$ will be the same. So let again K_o and K_o' be two possible solutions and define $C = K_o' - K_o$. Then from (4.8) we have $CR_y = 0$. Let further $\hat{\boldsymbol{x}} = K_o \boldsymbol{y}$ and $\hat{\boldsymbol{x}}' = K_o' \boldsymbol{y}$. We want to verify that $\hat{\boldsymbol{x}} = \hat{\boldsymbol{x}}'$ with probability one. For this purpose, note that the condition $CR_y = 0$ implies $CR_yC^* = 0$ or, equivalently, $\mathsf{E}\,(C\boldsymbol{y})(\boldsymbol{y}^* C^*) = 0$. We therefore have a zero-mean random variable $C\boldsymbol{y}$ (because $\boldsymbol{y}$ is zero-mean itself) with a zero covariance matrix. Using Remark A.1, we again conclude that $C\boldsymbol{y}$ is the zero random variable with probability one. It then follows from

$$\hat{\boldsymbol{x}}' = K_o' \boldsymbol{y} = (K_o + C)\boldsymbol{y} = \hat{\boldsymbol{x}} + C\boldsymbol{y}$$

that $\hat{\boldsymbol{x}} = \hat{\boldsymbol{x}}'$ with probability one, as desired. We summarize our discussions in the following statement.

Theorem 4.2 (Properties of the linear estimator) Consider the same setting of Thm. 3.1. Then the normal equations $K_o R_y = R_{xy}$ that define the linear least-mean-squares estimator have the following properties:

1. They are always consistent, i.e., a solution K_o always exists.
2. The solution K_o is unique if, and only if, $R_y > 0$.
3. Infinitely many solutions K_o exist if, and only if, R_y is singular.

In case 3, regardless of which solution K_o is chosen, the values of the estimator, $\hat{\boldsymbol{x}} = K_o \boldsymbol{y}$, and the m.m.s.e., $(R_x - K_o R_y K_o^*)$, remain invariant.

4.4 NONZERO-MEAN VARIABLES

Starting from Sec. 3.1, the discussion has focused so far on zero-mean random variables $\boldsymbol{x}$ and $\boldsymbol{y}$. When the means are nonzero, we should seek an unbiased estimator for $\boldsymbol{x}$ of the form

$$\hat{\boldsymbol{x}} = K\boldsymbol{y} + b \tag{4.9}$$

for some matrix K and some vector b. As before, the optimal values for $\{K, b\}$ are determined through the minimization of the mean-square error,

$$\boxed{\min_{K,b} \; \mathsf{E}\,\tilde{\boldsymbol{x}}^* \tilde{\boldsymbol{x}}} \tag{4.10}$$

where $\tilde{\boldsymbol{x}} = \boldsymbol{x} - \hat{\boldsymbol{x}}$.

To solve this problem, we start by noting that since the estimator should be unbiased we must enforce $\mathsf{E}\,\hat{\boldsymbol{x}} = \bar{x}$. Taking expectations of both sides of (4.9) shows that the vector b must satisfy $\bar{x} = K\bar{y} + b$. Using this expression for b, we can eliminate it from the expression for $\hat{\boldsymbol{x}}$, which becomes $\hat{\boldsymbol{x}} = K\boldsymbol{y} + (\bar{x} - K\bar{y})$ or, equivalently,

$$(\hat{\boldsymbol{x}} - \bar{x}) = K(\boldsymbol{y} - \bar{y}) \tag{4.11}$$

This expression shows that the desired gain matrix K should map the now zero-mean variable $(\boldsymbol{y} - \bar{y})$ to another zero-mean variable $(\hat{\boldsymbol{x}} - \bar{x})$. In other words, we are reduced to solving the problem of estimating the zero-mean random variable $\boldsymbol{x} - \bar{x}$ from the also zero-mean random variable $\boldsymbol{y} - \bar{y}$.

We already know that the solution K_o is found by solving

$$K_o R_y = R_{xy} \tag{4.12}$$

in terms of the covariance and cross-covariance matrices $\{R_y, R_{xy}\}$ of the zero-mean variables $\{\boldsymbol{x} - \bar{x}, \boldsymbol{y} - \bar{y}\}$, i.e.,

$$R_y \stackrel{\Delta}{=} \mathsf{E}\,(\boldsymbol{y} - \bar{y})(\boldsymbol{y} - \bar{y})^*, \qquad R_{xy} \stackrel{\Delta}{=} \mathsf{E}\,(\boldsymbol{x} - \bar{x})(\boldsymbol{y} - \bar{y})^*$$

Therefore, the optimal solution in the nonzero-mean case is given by

$$\boxed{\hat{\boldsymbol{x}} = \bar{x} + K_o[\boldsymbol{y} - \bar{y}]} \tag{4.13}$$

with K_o obtained from solving (4.12).

Comparing (4.13) to the zero-mean case from Thm. 3.1, we see that the solution to the nonzero-mean case simply amounts to replacing $\boldsymbol{x}$ and $\boldsymbol{y}$ by the centered variables $(\boldsymbol{x} - \bar{x})$ and $(\boldsymbol{y} - \bar{y})$, respectively, and then solving a linear estimation problem with these centered (zero-mean) variables. For this reason, there is no loss of generality, for linear estimation purposes, to assume that all random variables are zero-mean; the results for the nonzero-mean case can be deduced via centering.

It further follows from (4.12) that

$$K_o \mathsf{E}\,(\boldsymbol{y} - \bar{y})(\boldsymbol{y} - \bar{y})^* \;=\; \mathsf{E}\,(\boldsymbol{x} - \bar{x})(\boldsymbol{y} - \bar{y})^*$$

or, equivalently,

$$\mathsf{E}\,[(\boldsymbol{x} - \bar{x}) - K_o(\boldsymbol{y} - \bar{y})](\boldsymbol{y} - \bar{y})^* \;=\; 0$$

so that the orthogonality condition (4.4) in the nonzero-mean case becomes

$$\boxed{\mathsf{E}\,\tilde{\boldsymbol{x}}(\boldsymbol{y} - \bar{y})^* = 0 \qquad \text{or, equivalently,} \qquad \tilde{\boldsymbol{x}} \perp (\boldsymbol{y} - \bar{y})}$$

where $\tilde{\boldsymbol{x}} = (\boldsymbol{x} - \bar{x}) - (\hat{\boldsymbol{x}} - \bar{x}) = \boldsymbol{x} - \hat{\boldsymbol{x}}$. Moreover, the resulting m.m.s.e. matrix $\mathsf{E}\,\tilde{\boldsymbol{x}}\tilde{\boldsymbol{x}}^*$ is equal to

$$\boxed{\text{m.m.s.e.} = R_x - K_o R_y K_o^*}$$

with $R_x = \mathsf{E}\,(\boldsymbol{x} - \bar{x})(\boldsymbol{x} - \bar{x})^*$.

CHAPTER 5

Linear Models

We apply the linear estimation theory of the previous two chapters to the important special case of linear models, which arises often in applications. Specifically, we now assume that the zero-mean random vectors $\{\boldsymbol{x}, \boldsymbol{y}\}$ are related via a linear model of the form

$$\boxed{\boldsymbol{y} = H\boldsymbol{x} + \boldsymbol{v}} \tag{5.1}$$

for some $q \times p$ matrix H. Here $\boldsymbol{v}$ denotes a zero-mean random noise vector with known covariance matrix, $R_v = \mathsf{E}\,\boldsymbol{v}\boldsymbol{v}^*$. The covariance matrix of $\boldsymbol{x}$ is also assumed to be known, say, $\mathsf{E}\,\boldsymbol{x}\boldsymbol{x}^* = R_x$. Both $\{\boldsymbol{x}, \boldsymbol{v}\}$ are uncorrelated, i.e., $\mathsf{E}\,\boldsymbol{x}\boldsymbol{v}^* = 0$, and we further assume that $R_x > 0$ and $R_v > 0$.

5.1 ESTIMATION USING LINEAR RELATIONS

According to Thm. 3.1, when $R_y > 0$, the linear least-mean-squares estimator of $\boldsymbol{x}$ given $\boldsymbol{y}$ is

$$\hat{\boldsymbol{x}} = R_{xy}R_y^{-1}\boldsymbol{y} \tag{5.2}$$

Because of (5.1), the covariances $\{R_{xy}, R_y\}$ can be determined in terms of the given matrices $\{H, R_x, R_v\}$. Indeed, the uncorrelatedness of $\{\boldsymbol{x}, \boldsymbol{v}\}$ gives

$$\begin{aligned} R_y &= \mathsf{E}\,\boldsymbol{y}\boldsymbol{y}^* = \mathsf{E}\,(H\boldsymbol{x}+\boldsymbol{v})(H\boldsymbol{x}+\boldsymbol{v})^* &= HR_xH^* + R_v \\ R_{xy} &= \mathsf{E}\,\boldsymbol{x}\boldsymbol{y}^* = \mathsf{E}\,\boldsymbol{x}(H\boldsymbol{x}+\boldsymbol{v})^* &= R_xH^* \end{aligned}$$

Moreover, since $R_v > 0$ we get $R_y > 0$. The expression (5.2) for $\hat{\boldsymbol{x}}$ then becomes

$$\boxed{\hat{\boldsymbol{x}} = R_xH^*\left[R_v + HR_xH^*\right]^{-1}\boldsymbol{y}} \tag{5.3}$$

This expression can be rewritten in an equivalent form by using the so-called *matrix inversion formula* or *lemma*. This formula is a very useful matrix theory result and it will be called upon several times throughout this book. The result states that for arbitrary matrices $\{A, B, C, D\}$ of compatible dimensions, if A and C are invertible, then

$$\boxed{(A + BCD)^{-1} = A^{-1} - A^{-1}B(C^{-1} + DA^{-1}B)^{-1}DA^{-1}} \tag{5.4}$$

The identity can be verified algebraically; it essentially shows how the inverse of the sum $A + BCD$ is related to the inverse of A.

Applying (5.4) to the matrix $\left[R_v + HR_xH^*\right]^{-1}$ in (5.3), with the identifications

$$A = R_v, \quad B = H, \quad C = R_x, \quad D = H^*$$

we obtain

$$\begin{aligned}\hat{\boldsymbol{x}} &= R_xH^*\left\{R_v^{-1} - R_v^{-1}H(R_x^{-1} + H^*R_v^{-1}H)^{-1}H^*R_v^{-1}\right\}\boldsymbol{y} \\ &= \left\{R_x(R_x^{-1} + H^*R_v^{-1}H) - R_xH^*R_v^{-1}H\right\}(R_x^{-1} + H^*R_v^{-1}H)^{-1}H^*R_v^{-1}\boldsymbol{y} \\ &= \left[R_x^{-1} + H^*R_v^{-1}H\right]^{-1}H^*R_v^{-1}\boldsymbol{y}\end{aligned}$$

where in the second equality we factored out the term $(R_x^{-1} + H^*R_v^{-1}H)^{-1}H^*R_v^{-1}\boldsymbol{y}$ from the right. Hence,

$$\boxed{\hat{\boldsymbol{x}} = \left[R_x^{-1} + H^*R_v^{-1}H\right]^{-1}H^*R_v^{-1}\boldsymbol{y}} \tag{5.5}$$

This alternative form can be useful in several contexts. Observe, for example, that when H is a column vector, the quantity $(R_v + HR_xH^*)$ that appears in (5.3) is a matrix, while the quantity $(R_x^{-1} + H^*R_v^{-1}H)$ that appears in (5.5) is a scalar. In this case, the representation (5.5) leads to a simpler expression for $\hat{\boldsymbol{x}}$. In general, the convenience of using (5.3) or (5.5) depends on the situation at hand.

It further follows that the m.m.s.e. matrix is given by

$$\begin{aligned}\text{m.m.s.e.} = \mathsf{E}\,\tilde{\boldsymbol{x}}\tilde{\boldsymbol{x}}^* &= \mathsf{E}\,(\boldsymbol{x} - \hat{\boldsymbol{x}})\boldsymbol{x}^*, \quad \text{since } \tilde{\boldsymbol{x}} \perp \hat{\boldsymbol{x}} \\ &= R_x - \left[R_x^{-1} + H^*R_v^{-1}H\right]^{-1}H^*R_v^{-1}HR_x \\ &= \left[R_x^{-1} + H^*R_v^{-1}H\right]^{-1}\end{aligned}$$

where in the last equality we used the matrix inversion lemma again. That is,

$$\boxed{\text{m.m.s.e.} = \left[R_x^{-1} + H^*R_v^{-1}H\right]^{-1}}$$

Theorem 5.1 (Linear estimator for linear models) Let $\{\boldsymbol{y}, \boldsymbol{x}, \boldsymbol{v}\}$ be zero-mean random variables that are related via the linear model $\boldsymbol{y} = H\boldsymbol{x} + \boldsymbol{v}$, for some data matrix H of compatible dimensions. Both $\boldsymbol{x}$ and $\boldsymbol{v}$ are assumed uncorrelated with invertible covariance matrices, $R_v = \mathsf{E}\,\boldsymbol{v}\boldsymbol{v}^*$ and $R_x = \mathsf{E}\,\boldsymbol{x}\boldsymbol{x}^*$. The linear least-mean-squares estimator of $\boldsymbol{x}$ given $\boldsymbol{y}$ can be evaluated by either expression:

$$\hat{\boldsymbol{x}} = R_xH^*\left[R_v + HR_xH^*\right]^{-1}\boldsymbol{y} = \left[R_x^{-1} + H^*R_v^{-1}H\right]^{-1}H^*R_v^{-1}\boldsymbol{y}$$

and the resulting minimum mean-square error matrix is

$$\text{m.m.s.e.} = \left[R_x^{-1} + H^*R_v^{-1}H\right]^{-1}$$

Remark 5.1 (Centering for linear models) If the variables $\{\boldsymbol{x}, \boldsymbol{v}\}$ in (5.1) were not zero-mean, say, $\mathsf{E}\,\boldsymbol{x} = \bar{x}$ and $\mathsf{E}\,\boldsymbol{v} = \bar{v}$, then the above results will still hold with $\hat{\boldsymbol{x}}$ replaced by $(\hat{\boldsymbol{x}} - \bar{x})$ and $\boldsymbol{y}$ replaced by $(\boldsymbol{y} - \bar{y}) = (\boldsymbol{y} - H\bar{x} - \bar{v})$. Indeed, the covariance matrices $\{R_x, R_v, R_y\}$ will need to be defined accordingly as

$$\begin{aligned}R_x &= \mathsf{E}\,(\boldsymbol{x} - \bar{x})(\boldsymbol{x} - \bar{x})^*, \quad R_v = \mathsf{E}\,(\boldsymbol{v} - \bar{v})(\boldsymbol{v} - \bar{v})^* \\ R_y &= \mathsf{E}\,(\boldsymbol{y} - \bar{y})(\boldsymbol{y} - \bar{y})^*, \quad R_{xy} = \mathsf{E}\,(\boldsymbol{x} - \bar{x})(\boldsymbol{y} - \bar{y})^*\end{aligned}$$

and the uncorrelatedness of $\{\boldsymbol{x}, \boldsymbol{v}\}$ will now amount to requiring $\mathsf{E}(\boldsymbol{x}-\bar{x})(\boldsymbol{v}-\bar{v})^* = 0$. Under these conditions, it will still hold that $R_{xy} = R_x H^*$ and $R_y = R_v + HR_xH^*$, and the expressions for $\hat{\boldsymbol{x}}$ will become (cf. (4.13)):

$$\begin{aligned}\hat{\boldsymbol{x}} &= \bar{x} + \left[R_x^{-1} + H^* R_v^{-1} H\right]^{-1} H^* R_v^{-1}(\boldsymbol{y} - H\bar{x} - \bar{v}) \\ &= \bar{x} + R_x H^* \left[HR_xH^* + R_v\right]^{-1}(\boldsymbol{y} - H\bar{x} - \bar{v})\end{aligned}$$

◇

We now illustrate the application of Thm. 5.1 to several important examples including channel estimation, channel equalization, and block data estimation.

5.2 APPLICATION: CHANNEL ESTIMATION

Consider an FIR channel whose tap vector $\boldsymbol{c}$ is unknown; it is modeled as a zero-mean random vector with a known covariance matrix, $R_c = \mathsf{E}\,\boldsymbol{c}\boldsymbol{c}^*$. The following experiment is performed with the purpose of estimating $\boldsymbol{c}$, assumed of length M. The channel is assumed initially at rest (i.e., no initial conditions in its delay elements) and a known input sequence $\{s(i)\}$, also called a *training* sequence, is applied to the channel. The resulting output sequence $\{\boldsymbol{z}(i)\}$ is measured in the presence of additive noise, $\boldsymbol{v}(i)$, as shown in Fig. 5.1. The available measurements are

$$\boldsymbol{y}(i) = \boldsymbol{z}(i) + \boldsymbol{v}(i) \tag{5.6}$$

where $\boldsymbol{v}(i)$ is a zero-mean noise sequence that is uncorrelated with $\boldsymbol{c}$.

Assume we collect a block of measurements $\{s(\cdot), \boldsymbol{y}(\cdot)\}$ over the interval $0 \le i \le N$. Then we can write in matrix form, say, for $M = 3$ and $N = 6$,

$$\underbrace{\begin{bmatrix} \boldsymbol{y}(0) \\ \boldsymbol{y}(1) \\ \boldsymbol{y}(2) \\ \boldsymbol{y}(3) \\ \boldsymbol{y}(4) \\ \boldsymbol{y}(5) \\ \boldsymbol{y}(6) \end{bmatrix}}_{\boldsymbol{y}:(N+1)\times 1} = \underbrace{\begin{bmatrix} s(0) & & \\ s(1) & s(0) & \\ s(2) & s(1) & s(0) \\ s(3) & s(2) & s(1) \\ s(4) & s(3) & s(2) \\ s(5) & s(4) & s(3) \\ s(6) & s(5) & s(4) \end{bmatrix}}_{H:(N+1)\times M} \boldsymbol{c} + \underbrace{\begin{bmatrix} \boldsymbol{v}(0) \\ \boldsymbol{v}(1) \\ \boldsymbol{v}(2) \\ \boldsymbol{v}(3) \\ \boldsymbol{v}(4) \\ \boldsymbol{v}(5) \\ \boldsymbol{v}(6) \end{bmatrix}}_{\boldsymbol{v}:(N+1)\times 1} \tag{5.7}$$

where we are further defining the quantities $\{\boldsymbol{y}, H, \boldsymbol{v}\}$. Note that the *data matrix* H has a rectangular Toeplitz structure, i.e., it has constant entries along its diagonals. In addition, each row of H amounts to a state vector (also called a regressor) of the FIR channel.

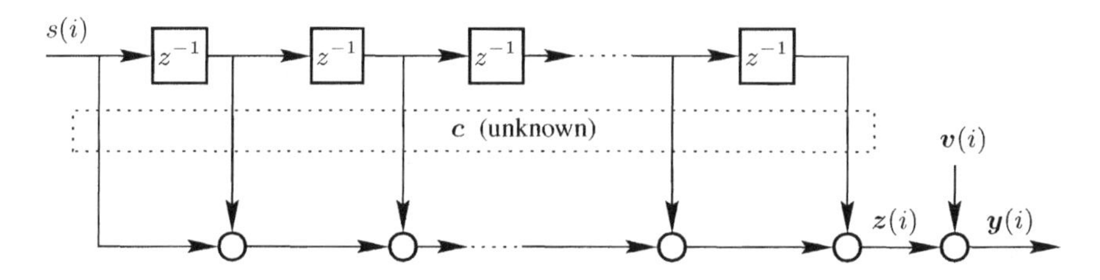

FIGURE 5.1 Channel estimation in the presence of additive noise.

Specifically, the $i-$th row of H has the form

$$\begin{bmatrix} s(i) & s(i-1) & \ldots & s(i-M+1) \end{bmatrix}$$

which contains the input at time i, $s(i)$, as well as the outputs of all delay elements in the channel. Thus, it is common to refer to the $i-$th row of H as the state vector or the regressor of the channel at the time instant i.

The quantities $\{\boldsymbol{y}, H\}$ so defined are both available to the designer, in addition to the covariance matrices $\{R_c = \mathsf{E}\,\boldsymbol{c}\boldsymbol{c}^*, R_v = \mathsf{E}\,\boldsymbol{v}\boldsymbol{v}^*\}$ (by assumption). In particular, if the noise sequence $\{\boldsymbol{v}(i)\}$ is assumed white with variance σ_v^2, then $R_v = \mathsf{E}\,\boldsymbol{v}\boldsymbol{v}^* = \sigma_v^2 \mathrm{I}$. With this information, we can estimate the channel as follows. Since we have a linear model relating $\boldsymbol{y}$ to $\boldsymbol{c}$, as indicated by (5.7), then according to Thm. 5.1, the optimal linear estimator for $\boldsymbol{c}$ can be obtained from either expression:

$$\boxed{\hat{\boldsymbol{c}} = R_c H^* \left[H R_c H^* + R_v\right]^{-1} \boldsymbol{y} = \left[R_c^{-1} + H^* R_v^{-1} H\right]^{-1} H^* R_v^{-1} \boldsymbol{y}} \tag{5.8}$$

5.3 APPLICATION: BLOCK DATA ESTIMATION

Our second application is in the context of data (or symbol) recovery. We consider the same FIR channel as in Fig. 5.1, except that now we assume that its tap vector is known. For example, it could have been estimated via a prior training procedure as explained in the previous section. We denote this tap vector by c, with individual entries

$$c \triangleq \text{col}\{c(0), c(1), \ldots, c(M-1)\}$$

The channel is initially at rest and its output sequence, $\{\boldsymbol{z}(i)\}$, is again measured in the presence of additive noise, $\boldsymbol{v}(i)$, as shown in Fig. 5.2. The signals $\{\boldsymbol{v}(\cdot), \boldsymbol{s}(\cdot)\}$ are assumed uncorrelated. What we would like to estimate now are the symbols $\{\boldsymbol{s}(\cdot)\}$ that are being transmitted through the channel. Observe that, to be consistent with our notation, since c is deterministic and $\boldsymbol{s}(\cdot)$ is random, we are now using a boldface letter for the latter and a normal font for the former.

Suppose we collect a block of measurements $\{\boldsymbol{y}(\cdot)\}$, say, $(N+1)$ measurements over the interval $0 \le i \le N$. Rather than relate the $\{\boldsymbol{y}(\cdot)\}$ to the channel tap vector c through a data matrix, as we did in (5.7), we now relate the $\{\boldsymbol{y}(\cdot)\}$ to the $\{\boldsymbol{s}(\cdot)\}$ through a *channel matrix*. More specifically, assume again that $M = 3$ and $N = 6$ for illustration purposes.

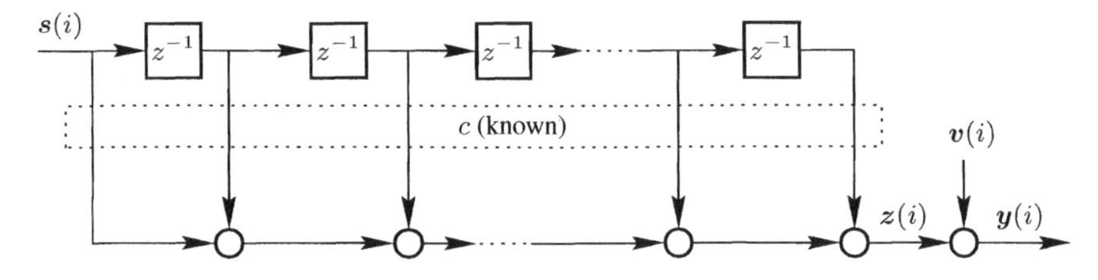

FIGURE 5.2 Block data estimation in the presence of additive noise.

Then we can write

$$\underbrace{\begin{bmatrix} y(0) \\ y(1) \\ y(2) \\ y(3) \\ y(4) \\ y(5) \\ y(6) \end{bmatrix}}_{y:(N+1)\times 1} = \underbrace{\begin{bmatrix} c(0) & & & & & & \\ c(1) & c(0) & & & & & \\ c(2) & c(1) & c(0) & & & & \\ & c(2) & c(1) & c(0) & & & \\ & & c(2) & c(1) & c(0) & & \\ & & & c(2) & c(1) & c(0) & \\ & & & & c(2) & c(1) & c(0) \end{bmatrix}}_{H:(N+1)\times(N+1)} \underbrace{\begin{bmatrix} s(0) \\ s(1) \\ s(2) \\ s(3) \\ s(4) \\ s(5) \\ s(6) \end{bmatrix}}_{s:(N+1)\times 1} + \underbrace{\begin{bmatrix} v(0) \\ v(1) \\ v(2) \\ v(3) \\ v(4) \\ v(5) \\ v(6) \end{bmatrix}}_{v:(N+1)\times 1}$$

Note that the channel matrix H is now square Toeplitz of size $(N+1)\times(N+1)$; it also has a banded structure. The quantities $\{y, H\}$ so defined are available to the designer, in addition to the covariance matrices $\{R_s, R_v\}$ (by assumption). In particular, if the data and noise sequences $\{s(\cdot), v(\cdot)\}$ are white with variances $\{\sigma_s^2, \sigma_v^2\}$, then $R_s = \mathsf{E}\, ss^* = \sigma_s^2 \mathrm{I}$ and $R_v = \mathsf{E}\, vv^* = \sigma_v^2 \mathrm{I}$. With this information, we can estimate the symbols in the vector s as follows. Observe again that we have a linear model relating y to the unknown symbol vector s. According to Thm. 5.1, the optimal linear estimator for s can then be found from either expression:

$$\boxed{\hat{s} = R_s H^* \left[H R_s H^* + R_v\right]^{-1} y = \left[R_s^{-1} + H^* R_v^{-1} H\right]^{-1} H^* R_v^{-1} y} \tag{5.9}$$

5.4 APPLICATION: LINEAR CHANNEL EQUALIZATION

Our third application is in the context of linear channel equalization. More specifically, we generalize the discussion of Ex. 4.1.

Consider again an FIR channel as shown in Fig. 5.2, with a known tap vector c of length M. Data symbols $\{s(\cdot)\}$ are transmitted through the channel and the output sequence, $\{z(i)\}$, is measured in the presence of additive noise, $v(i)$. The signals $\{v(\cdot), s(\cdot)\}$ are assumed uncorrelated. Due to channel memory, each measurement $y(i)$ contains contributions not only from $s(i)$ but also from prior symbols since

$$y(i) = c(0)s(i) + \underbrace{\sum_{k=1}^{M-1} c(k)s(i-k)}_{\text{ISI}} + v(i) \tag{5.10}$$

The second term on the right-hand side is termed *inter-symbol-interference* (ISI); it refers to the interference that is caused by prior symbols in $y(i)$. The purpose of an equalizer is to recover $s(i)$. To achieve this task, an equalizer does not only rely on the most recent measurement $y(i)$, but it also employs several prior measurements $\{y(i-k)\}$, say, for $k = 1, 2, \ldots, L-1$. These prior measurements contain information that is correlated with the ISI term in $y(i)$. For example, the expression for $y(i-1)$ is

$$y(i-1) = c(0)s(i-1) + \underbrace{\sum_{k=1}^{M-1} c(k)s(i-1-k)}_{\text{ISI}} + v(i-1)$$

and the ISI term in it shares several data symbols with the ISI term in $y(i)$. It is for this reason that prior measurements are useful in eliminating (or reducing) ISI. In this section, we are interested in an equalizer structure of the form shown in Fig. 5.3. The equalizer is chosen to have an FIR structure with L coefficients, so that for each time instant i, it would employ the L observations

$$\boldsymbol{y}_i \triangleq \begin{bmatrix} \boldsymbol{y}(i) \\ \boldsymbol{y}(i-1) \\ \boldsymbol{y}(i-2) \\ \vdots \\ \boldsymbol{y}(i-L+1) \end{bmatrix}$$

in order to estimate $\boldsymbol{s}(i-\Delta)$ for some integer delay $\Delta \geq 0$. We remark that we shall frequently deal with *time sequences* in this book. So when we write $\boldsymbol{y}(i)$ we are referring to the value of the time sequence $\boldsymbol{y}(\cdot)$ at time i. Not only that, but the notation $\boldsymbol{y}(\cdot)$, with parentheses, also means that $\boldsymbol{y}(\cdot)$ is a scalar. This is because for vector-valued time-sequences, we shall instead write $\boldsymbol{y}_i$, with a subscript rather than parentheses, to refer to its value at time i. In other words, we shall use parenthesis for *time-indexing* in the scalar case, e.g., $\{\boldsymbol{y}(i)\}$, and subscripts in the vector case, e.g., $\{\boldsymbol{y}_i\}$.

It is useful to express the observation vector $\boldsymbol{y}_i$ in terms of the transmitted data as follows. Assume for the sake of illustration that $L=5$ (i.e., an equalizer with 5 taps) and $M=4$ (a channel with four taps). Then we have

$$\underbrace{\begin{bmatrix} \boldsymbol{y}(i) \\ \boldsymbol{y}(i-1) \\ \boldsymbol{y}(i-2) \\ \boldsymbol{y}(i-3) \\ \boldsymbol{y}(i-4) \end{bmatrix}}_{\boldsymbol{y}_i:L\times 1} = \underbrace{\begin{bmatrix} c(0) & c(1) & c(2) & c(3) & & & & \\ & c(0) & c(1) & c(2) & c(3) & & & \\ & & c(0) & c(1) & c(2) & c(3) & & \\ & & & c(0) & c(1) & c(2) & c(3) & \\ & & & & c(0) & c(1) & c(2) & c(3) \end{bmatrix}}_{H:L\times(L+M-1)} \underbrace{\begin{bmatrix} \boldsymbol{s}(i) \\ \boldsymbol{s}(i-1) \\ \boldsymbol{s}(i-2) \\ \boldsymbol{s}(i-3) \\ \boldsymbol{s}(i-4) \\ \boldsymbol{s}(i-5) \\ \boldsymbol{s}(i-6) \\ \boldsymbol{s}(i-7) \end{bmatrix}}_{\boldsymbol{s}_i:(L+M)-1\times 1}$$

$$+ \underbrace{\begin{bmatrix} \boldsymbol{v}(i) \\ \boldsymbol{v}(i-1) \\ \boldsymbol{v}(i-2) \\ \boldsymbol{v}(i-3) \\ \boldsymbol{v}(i-4) \end{bmatrix}}_{\boldsymbol{v}_i:L\times 1}$$

Note that the observation vector $\boldsymbol{y}_i$ has L entries, the data vector $\boldsymbol{s}_i$ has $L+M-1$ entries, and the channel matrix is now $L\times(L+M-1)$. We thus find that there is a linear relation between the vectors $\{\boldsymbol{y}_i, \boldsymbol{s}_i\}$ and this relation can be used to evaluate the covariance and cross-covariance quantities that are needed to estimate $\boldsymbol{s}(i-\Delta)$ from $\boldsymbol{y}_i$ in the linear least-mean-squares sense. Specifically, let us write

$$\hat{\boldsymbol{s}}(i-\Delta) = w^* \boldsymbol{y}_i \tag{5.11}$$

for some column vector w to be determined. We denote the optimal choice for w by w^o; the entries of w^{o*} will correspond to the optimal tap coefficients for the equalizer. According to Thm. 3.1, w^{o*} is given by

$$w^{o*} = R_{sy} R_y^{-1} \tag{5.12}$$

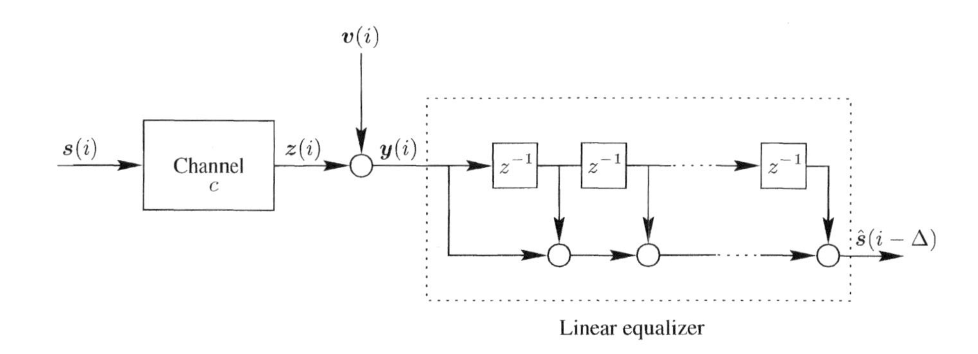

FIGURE 5.3 Linear equalization of an FIR channel in the presence of additive noise.

where

$$\boxed{R_{sy} \triangleq \mathsf{E}\, s(i-\Delta)\boldsymbol{y}_i^* \qquad (1\times L)} \tag{5.13}$$

denotes the cross-covariance vector between $s(i-\Delta)$ and $\boldsymbol{y}_i$, and

$$\boxed{R_y \triangleq \mathsf{E}\, \boldsymbol{y}_i\boldsymbol{y}_i^* \qquad (L\times L)} \tag{5.14}$$

denotes the covariance matrix of the observation vector, $\boldsymbol{y}_i$. Observe that since the processes $\{s(\cdot), \boldsymbol{y}(\cdot)\}$ are jointly wide-sense stationary, the quantities $\{R_{sy}, R_y\}$ are independent of i. In order to determine $\{R_{sy}, R_y\}$ we resort to the aforementioned linear model relating $\{\boldsymbol{y}_i, \boldsymbol{s}_i, \boldsymbol{v}_i, H\}$. To begin with,

$$R_y = \mathsf{E}\,(H\boldsymbol{s}_i + \boldsymbol{v}_i)(H\boldsymbol{s}_i + \boldsymbol{v}_i)^* \;=\; HR_sH^* + R_v$$

where

$$\boxed{R_s \triangleq \mathsf{E}\, \boldsymbol{s}_i\boldsymbol{s}_i^* \quad ((L+M-1)\times(L+M-1)) \quad \text{and} \quad R_v \triangleq \mathsf{E}\, \boldsymbol{v}_i\boldsymbol{v}_i^* \quad (L\times L)}$$

and

$$R_{sy} = \mathsf{E}\, s(i-\Delta)(H\boldsymbol{s}_i + \boldsymbol{v}_i)^* \;=\; \left(\mathsf{E}\, s(i-\Delta)\boldsymbol{s}_i^*\right)H^*$$

since $\{v(\cdot), s(\cdot)\}$ are uncorrelated. The value of $\mathsf{E}\, s(i-\Delta)\boldsymbol{s}_i^*$ depends on the assumed correlation between the transmitted symbols. If the $\{s(\cdot)\}$ are independent and identically distributed with variance σ_s^2, then $R_s = \sigma_s^2\mathrm{I}$ and

$$\begin{aligned} \mathsf{E}\, s(i-\Delta)\boldsymbol{s}_i^* &= \mathsf{E}\, s(i-\Delta)\left[\begin{array}{cccc} s^*(i) & s^*(i-1) & \ldots & s^*(i-7) \end{array}\right] \\ &= \left[\begin{array}{ccccccc} 0 & \ldots & 0 & \sigma_s^2 & 0 & \ldots & 0 \end{array}\right] \end{aligned}$$

with Δ leading zeros. We continue with this assumption on the data $\{s(\cdot)\}$ for simplicity of presentation. But it should be clear that the above derivation applies even for correlated (but stationary) data. In a similar vein, we assume that the noise sequence $\{v(\cdot)\}$ is white with variance σ_v^2 so that $R_v = \sigma_v^2\mathrm{I}$. Again, the development applies even for correlated (but stationary) noise. We therefore find that

$$\boxed{R_y = \sigma_s^2 HH^* + \sigma_v^2\mathrm{I} \quad \text{and} \quad R_{sy} = \left[\begin{array}{ccccccc} 0 & \ldots & 0 & \sigma_s^2 & 0 & \ldots & 0 \end{array}\right] H^*} \tag{5.15}$$

Observe that R_{sy} is proportional to the $(\Delta+1)$–th row of H^*. Substituting (5.15) into (5.12) we arrive at the following expression for the equalizer tap vector:

$$\boxed{w^{o*} = \sigma_s^2 e_\Delta^* H^* \left(\sigma_s^2 HH^* + \sigma_v^2 \mathrm{I}\right)^{-1}} \tag{5.16}$$

where e_Δ denotes the basis vector with Δ leading zeros,

$$e_\Delta \triangleq \begin{bmatrix} 0 & \dots & 0 & 1 & 0 & \dots & 0 \end{bmatrix}^{\mathsf{T}} \quad \text{(with } \Delta \text{ leading zeros)}$$

Moreover, according to Thm. 3.1, the resulting minimum mean-square error is given by $\text{m.m.s.e} = \sigma_s^2 - R_{sy}R_y^{-1}R_{ys}$ and, hence,

$$\boxed{\text{m.m.s.e} \;=\; \sigma_s^2\left(1 - \sigma_s^2 e_\Delta^* H^* R_y^{-1} H e_\Delta\right)} \tag{5.17}$$

An intuitive way to understand the usefulness of using a nonzero delay Δ is to recall that channels have group delays. Loosely speaking, the group delay of a channel is a measure of the amount of delay that a signal undergoes when transmitted through the channel. For this reason, the channel output, $\boldsymbol{z}(i)$, would be more correlated with a delayed version of $\boldsymbol{s}(i)$, than with $\boldsymbol{s}(i)$ itself. It therefore makes sense to use the channel output to estimate a delayed replica of the input.

Example 5.1 (Numerical illustration)

Let us use the above results to re-examine Ex. 4.4, where $L=3$, $M=2$, $\Delta=0$, $\{c(0),c(1)\}=\{1,0.5\}$, and $\{\sigma_s^2,\sigma_v^2\}=\{1,1\}$. Therefore, for this case, we have

$$H = \begin{bmatrix} 1 & 0.5 & & \\ & 1 & 0.5 & \\ & & 1 & 0.5 \end{bmatrix}$$

so that from (5.15),

$$R_y = \begin{bmatrix} 9/4 & 1/2 & 0 \\ 1/2 & 9/4 & 1/2 \\ 0 & 1/2 & 9/4 \end{bmatrix} \quad \text{and} \quad R_{sy} = \begin{bmatrix} 1 & 0 & 0 \end{bmatrix}$$

Using (5.16) and (5.17) we get

$$w^{o*} = \begin{bmatrix} 0.4688 & -0.1096 & 0.0244 \end{bmatrix} \quad \text{and} \quad \text{m.m.s.e} = 0.5312$$

which are the same results from Ex. 4.1.

◇

5.5 APPLICATION: MULTIPLE-ANTENNA RECEIVERS

Let us reconsider Ex. 4.1 with N noisy measurements,

$$\boldsymbol{y}(i) = \boldsymbol{x} + \boldsymbol{v}(i), \qquad i = 0,1,\dots,N-1$$

of some zero-mean random variable $\boldsymbol{x}$. Let $\boldsymbol{y} = \text{col}\{\boldsymbol{y}(0),\boldsymbol{y}(1),\dots,\boldsymbol{y}(N-1)\}$ denote the observation vector. In Ex. 4.1 we evaluated the linear least-mean-squares estimator of $\boldsymbol{x}$ given $\boldsymbol{y}$ by computing $R_{xy}R_y^{-1}$ explicitly. Here we evaluate $\hat{\boldsymbol{x}}$ by showing first that $\boldsymbol{y}$

and x are related through a linear model as in (5.1), and then using (5.5). For generality, we shall assume that the variances of x and $v(i)$ are σ_x^2 and σ_v^2, respectively. In Ex. 4.1, we used $\sigma_x^2 = \sigma_v^2 = 1$.

Introduce the $N \times 1$ column vectors:

$$v \triangleq \text{col}\{v(0), v(1), \ldots, v(N-1)\}, \qquad h \triangleq \text{col}\{1, 1, \ldots, 1\}$$

Then $y = hx + v$ and the covariance matrix of v is $\sigma_v^2 \mathrm{I}$. We now obtain from (5.5) that

$$\hat{x} = \left[1/\sigma_x^2 \ + \ h^* h/\sigma_v^2\right]^{-1} h^* y/\sigma_v^2 \ = \ \frac{1}{N + \frac{1}{\mathsf{SNR}}} \sum_{i=0}^{N-1} y(i)$$

where $\mathsf{SNR} = \sigma_x^2/\sigma_v^2$. Observe that we are not dividing by the number of observations (which is N) but by $N + 1/\mathsf{SNR}$. We shall comment on the significance of this observation in the next chapter.

Compared with the solution of Ex. 4.1, observe how the expression for $\hat{x}$ is obtained here more immediately. We did not explicitly form the $N \times N$ covariance matrix R_y and the $1 \times N$ cross-covariance vector R_{xy}, and then evaluate $R_{xy} R_y^{-1}$. Instead, we used the linear relation $y = hx + v$ and the formula (5.5). In this formula, the term $(R_x^{-1} + H^* R_v^{-1} H)$ is a scalar and its inversion is trivialized. This is in contrast to the term $(R_v + H R_x H^*)$, whose inverse appears in the alternative formula (5.3) — this term is equal to R_y. Again, if we interpret the $\{y(k)\}$ as noisy measurements that are collected at multiple antennas as a result of transmitting a signal x over additive noise channels, then we find that the expression for $\hat{x}$ suggests the optimal linear receiver structure shown in Fig. 5.4.

In Probs. II.25 and II.26 we extend this result further by showing how to incorporate channel gains into the design procedure. Specifically, we pursue receiver structures that are optimal according to two criteria: in one criterion, the SNR at the output of the receiver is maximized (resulting in the so-called maximal-ratio-combining technique), while in the second criterion the same minimum mean-square-error design of the current example is used.

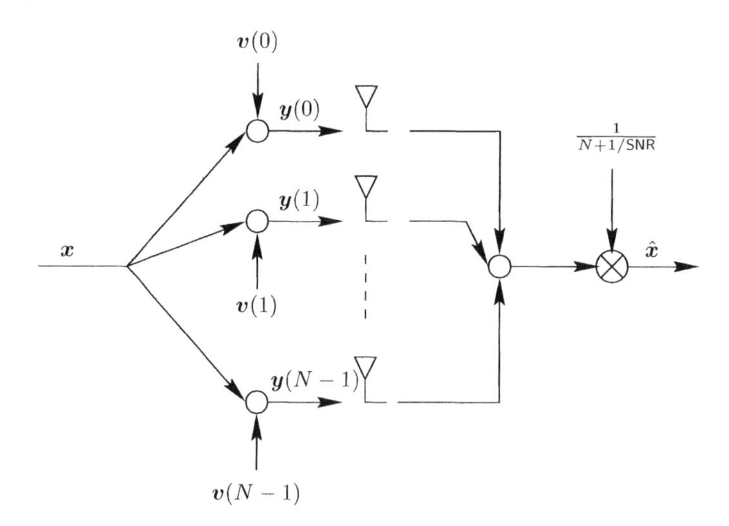

FIGURE 5.4 An optimal linear receiver for recovering a symbol x transmitted over additive-noise channels from multiple-antenna measurements.

CHAPTER 6

Constrained Estimation

In Sec. 5.1 we studied the problem of estimating a random variable $\boldsymbol{x}$ from a noisy observation $\boldsymbol{y}$ that is related to $\boldsymbol{x}$ via the linear model

$$\boldsymbol{y} = H\boldsymbol{x} + \boldsymbol{v} \tag{6.1}$$

where H is a known data matrix and $\boldsymbol{v}$ is some disturbance, with $\boldsymbol{x}$ and $\boldsymbol{v}$ satisfying

$$\mathsf{E}\,\boldsymbol{x} = 0, \quad \mathsf{E}\,\boldsymbol{v} = 0, \quad \mathsf{E}\,\boldsymbol{x}\boldsymbol{x}^* = R_x, \quad \mathsf{E}\,\boldsymbol{v}\boldsymbol{v}^* = R_v, \quad \mathsf{E}\,\boldsymbol{x}\boldsymbol{v}^* = 0 \tag{6.2}$$

The linear least-mean-squares estimator of $\boldsymbol{x}$ given $\boldsymbol{y}$ was found to be given by either expression

$$\hat{\boldsymbol{x}} = R_x H^* \left[R_v + H R_x H^*\right]^{-1} \boldsymbol{y} \;=\; \left[R_x^{-1} + H^* R_v^{-1} H\right]^{-1} H^* R_v^{-1} \boldsymbol{y} \tag{6.3}$$

with the right-most expression valid whenever $R_x > 0$ and $R_v > 0$.

In Sec. 5.5, we applied these results to a simple, yet revealing example. Given N measurements $\{\boldsymbol{y}(0), \boldsymbol{y}(1), \ldots, \boldsymbol{y}(N-1)\}$ of a random variable $\boldsymbol{x}$ with variance σ_x^2,

$$\boldsymbol{y}(i) = \boldsymbol{x} + \boldsymbol{v}(i), \qquad i = 0, 1, \ldots, N-1$$

i.e., given

$$\begin{bmatrix} \boldsymbol{y}(0) \\ \boldsymbol{y}(1) \\ \vdots \\ \boldsymbol{y}(N-1) \end{bmatrix} = \begin{bmatrix} 1 \\ 1 \\ \vdots \\ 1 \end{bmatrix} \boldsymbol{x} + \begin{bmatrix} \boldsymbol{v}(0) \\ \boldsymbol{v}(1) \\ \vdots \\ \boldsymbol{v}(N-1) \end{bmatrix} \tag{6.4}$$

we estimated $\boldsymbol{x}$ from the $\{\boldsymbol{y}(i)\}$ and found that

$$\hat{\boldsymbol{x}} = \frac{1}{N + \frac{1}{\mathsf{SNR}}} \sum_{i=0}^{N-1} \boldsymbol{y}(i) \tag{6.5}$$

where $\mathsf{SNR} = \sigma_x^2/\sigma_v^2$. In (6.4), the variable $\boldsymbol{x}$ is assumed to have been initially selected at random and then N noisy measurements of this same value are made — see Fig. 2.9. The observations are subsequently used to estimate $\boldsymbol{x}$ according to (6.5).

But what if we consider a different model for $\boldsymbol{x}$, whereby it is assumed to be a *constant* of unknown value, say, x, rather than a random quantity? How will the expression for $\hat{\boldsymbol{x}}$ change? The purpose of this chapter is to study such estimators. Specifically, we shall now consider linear models of the form

$$\boldsymbol{y} = Hx + \boldsymbol{v} \tag{6.6}$$

where, compared with (6.1), we are replacing the boldface letter $\boldsymbol{x}$ by the normal letter x (remember that we reserve the boldface notation to random variables throughout this book). The observation vector $\boldsymbol{y}$ in (6.6) continues to be random since the disturbance $\boldsymbol{v}$ is random. Moreover, any estimator for x that is based on $\boldsymbol{y}$ will also be a random variable itself. Given (6.6), we shall then study the problem of designing an optimal linear estimator for x of the form $\hat{\boldsymbol{x}} = K\boldsymbol{y}$, for some K to be determined. It will turn out that, for such problems, K is found by solving a *constrained* least-mean-squares estimation problem, as opposed to the unconstrained estimation problem (3.21). Once the estimation problem is solved, we shall then apply it in Secs. 6.3–6.5 to three examples: channel and noise estimation, decision-feedback equalization and antenna beamforming.

6.1 MINIMUM-VARIANCE UNBIASED ESTIMATION

Thus consider a zero-mean random noise variable $\boldsymbol{v}$ with a positive-definite covariance matrix $R_v = \mathsf{E}\,\boldsymbol{v}\boldsymbol{v}^* > 0$, and let $\boldsymbol{y}$ be a noisy measurement of Hx,

$$\boxed{\boldsymbol{y} = Hx + \boldsymbol{v}} \tag{6.7}$$

where x is the unknown constant vector that we wish to estimate. The dimensions of the data matrix H are denoted by $N \times n$ and it is further assumed that $N \geq n$,

$$\boxed{H : \; N \times n, \quad N \geq n} \tag{6.8}$$

That is, H is assumed to be a tall matrix so that the number of available entries in $\boldsymbol{y}$ is at least as many as the number of unknown entries in x. Note that we use the capital letter N for the larger dimension and the small letter n for the smaller dimension. We shall use this convention in the book whenever it is relevant to indicate how the row and column dimensions of a matrix compare to each other.

We also assume that the matrix H in (6.7) has *full rank*, i.e., that all its columns are linearly independent and, hence,

$$\boxed{\text{rank}(H) \;=\; n} \tag{6.9}$$

This condition guarantees that the matrix product H^*H is invertible (in fact, positive-definite — recall Lemma B.4). It also guarantees that the product $H^*R_v^{-1}H$ is positive-definite — see expression (B.1). For the benefit of the reader, Sec. B.2 reviews several basic concepts regarding range spaces, nullspaces, and ranks of matrices.

Problem Formulation

We are interested in determining a linear estimator for x of the form $\hat{\boldsymbol{x}} = K\boldsymbol{y}$, for some $n \times N$ matrix K. The choice of K should satisfy two conditions:

1. **Unbiasedness**. First, the estimator $\hat{\boldsymbol{x}}$ should be unbiased. That is, the choice of K should guarantee $\mathsf{E}\,\hat{\boldsymbol{x}} = x$, which is the same as $K\mathsf{E}\,\boldsymbol{y} = x$. But from (6.7) we have $\mathsf{E}\,\boldsymbol{y} = Hx$ so that K should satisfy $KHx = x$, no matter what the value of x is. This condition means that K should satisfy

$$\boxed{KH = \mathrm{I}} \tag{6.10}$$

   Note that KH is $n \times n$ and is therefore a square matrix.

2. **Optimality**. Second, the choice of K should minimize the covariance matrix of the estimation error, $\tilde{x} = x - \hat{x}$. Using the condition $KH = \mathrm{I}$, we find that

$$\hat{x} = Ky = K(Hx + v) = KHx + Kv = x + Kv$$

so that $\tilde{x} = -Kv$. This means that the error covariance matrix, as a function of K, is given by

$$\boxed{\mathsf{E}\,\tilde{x}\tilde{x}^* = \mathsf{E}\,(Kvv^*K^*) \;=\; KR_vK^*} \tag{6.11}$$

Combining (6.10) and (6.11), we conclude that the desired K is found by solving the following constrained optimization problem:

$$\boxed{\min_K \quad KR_vK^* \qquad \text{subject to } KH = \mathrm{I}} \tag{6.12}$$

The estimator $\hat{x} = K_o y$ that results from the solution of (6.12) is known as the *minimum-variance-unbiased estimator*, or m.v.u.e. for short. It is also sometimes called the best linear unbiased estimator (BLUE).

Interpretation and Solution

Let $\mathcal{J}(K)$ denote the cost function that appears in (6.12), i.e.,

$$\mathcal{J}(K) \;\stackrel{\Delta}{=}\; KR_vK^*$$

Then problem (6.12) means the following. We seek a matrix K_o satisfying $K_oH = \mathrm{I}$ such that

$$\mathcal{J}(K) - \mathcal{J}(K_o) \geq 0 \quad \text{for all } \; K \; \text{ satisfying } \; KH = \mathrm{I}$$

There are several ways of determining K_o. We choose to use the already known solution of the linear estimation problem (cf. Sec. 5.1) in order to guess what the solution K_o for (6.12) should be. Once this is done, we shall then provide an independent verification of the result.

Thus recall, as mentioned in the introduction of this chapter, that for two zero-mean random variables $\{x, y\}$ that are related as in (6.1), the linear least-mean-squares estimator of x given y is (cf. the second expression in (6.3)):

$$\hat{x} = (R_x^{-1} + H^*R_v^{-1}H)^{-1}H^*R_v^{-1}y$$

Now assume that the covariance matrix of x has the particular form $R_x = \alpha\mathrm{I}$, with a sufficiently large positive scalar α (i.e., $\alpha \to \infty$). That is, assume that the variance of each of the entries of x is infinitely large. In this way, x can be "interpreted" as playing the role of some unknown constant vector, x. Then the above expression for $\hat{x}$ reduces to

$$\hat{x} = (H^*R_v^{-1}H)^{-1}H^*R_v^{-1}y$$

This conclusion suggests that the choice

$$K_o = (H^*R_v^{-1}H)^{-1}H^*R_v^{-1}$$

solves the problem of estimating the unknown vector x from model (6.7). We shall now establish this result more directly; the result is known as the Gauss-Markov theorem.

Theorem 6.1 (Gauss-Markov Theorem) Consider the linear model $\boldsymbol{y} = Hx + \boldsymbol{v}$, where $\boldsymbol{v}$ is a zero-mean random variable with positive-definite covariance matrix R_v, and x is an unknown constant vector. Assume further that H is a full-rank $N \times n$ matrix with $N \geq n$. Then the minimum-variance-unbiased linear estimator of x given $\boldsymbol{y}$ is $\hat{\boldsymbol{x}} = K_o\boldsymbol{y}$, where

$$K_o = (H^* R_v^{-1} H)^{-1} H^* R_v^{-1}$$

Moreover, the resulting cost is m.m.s.e. $= (H^* R_v^{-1} H)^{-1}$.

Proof: For any matrix K that satisfies $KH = \mathrm{I}$, it is easy to verify that

$$\mathcal{J}(K) = KR_vK^* \;=\; (K - K_o)R_v(K - K_o)^* \;+\; K_oR_vK_o^* \tag{6.13}$$

This is because

$$KR_vK_o^* \;=\; KR_v[R_v^{-1}H(H^*R_v^{-1}H)^{-1}] \;=\; KH(H^*R_v^{-1}H)^{-1} = (H^*R_v^{-1}H)^{-1}$$

Likewise, $K_oR_vK_o^* \;=\; (H^*R_v^{-1}H)^{-1}$. Relation (6.13) expresses the cost $\mathcal{J}(K)$ as the sum of two nonnegative-definite terms: one is independent of K and is equal to $K_oR_vK_o^*$, while the other is dependent on K. It is then clear, since $R_v > 0$, that the cost is minimized by choosing $K = K_o$, and that the resulting minimum cost is

$$\mathcal{J}(K_o) = (H^*R_v^{-1}H)^{-1}$$

Note further that the matrix K_o in the statement of the theorem satisfies the constraint $K_oH = \mathrm{I}$.

Remark 6.1 (Constrained optimization) Sometimes in applications (see Secs. 6.4 and 6.5), optimization problems of the form (6.12) arise without being explicitly related to a minimum-variance-unbiased estimation problem (as in the statement of Thm. 6.1). For this reason, we also state the following conclusion here for later reference. The solution of a generic constrained optimization problem of the form

$$\boxed{\min_K \; KR_vK^* \qquad \text{subject to } KH = \mathrm{I} \text{ and } R_v > 0} \tag{6.14}$$

is given by

$$\boxed{K_o = (H^*R_v^{-1}H)^{-1}H^*R_v^{-1}}$$

with the resulting minimum cost equal to

$$\text{minimum cost} = (H^*R_v^{-1}H)^{-1} \tag{6.15}$$

6.2 EXAMPLE: MEAN ESTIMATION

Let us reconsider the example of Sec. 5.5, where we assumed that we are given N measurements

$$\boldsymbol{y}(i) = \boldsymbol{x} + \boldsymbol{v}(i), \qquad i = 0, 1, \ldots, N-1$$

of the same random variable $\boldsymbol{x}$ with variance σ_x^2. The noise sequence $\boldsymbol{v}(i)$ was further assumed to be white with zero mean and variance σ_v^2. The linear least-mean-squares estimator (l.l.m.s.e.) of $\boldsymbol{x}$ given the $\{\boldsymbol{y}(i)\}$ was found to be (cf. (6.5)):

$$\boxed{\hat{\boldsymbol{x}}_{\mathsf{llmse}} = \frac{1}{N + \frac{1}{\mathsf{SNR}}} \sum_{i=0}^{N-1} \boldsymbol{y}(i)}$$

where $\mathsf{SNR} = \sigma_x^2/\sigma_v^2$.

Now assume instead that we model x as an unknown constant, rather than a random variable, say,

$$\boldsymbol{y}(i) = x + \boldsymbol{v}(i), \qquad i = 0, 1, \ldots, N-1 \tag{6.16}$$

In this case, the value of x can be regarded as the mean value of each $\boldsymbol{y}(i)$. We collect the measurements and the noises into vector form,

$$\boldsymbol{y} \stackrel{\Delta}{=} \text{col}\{\boldsymbol{y}(0), \boldsymbol{y}(1), \ldots, \boldsymbol{y}(N-1)\}, \qquad \boldsymbol{v} \stackrel{\Delta}{=} \text{col}\{\boldsymbol{v}(0), \boldsymbol{v}(1), \ldots, \boldsymbol{v}(N-1)\}$$

and define the data vector

$$h = \text{col}\{1, 1, \ldots, 1\}$$

Then

$$\boldsymbol{y} = hx + \boldsymbol{v}$$

with $R_v = \mathsf{E}\,\boldsymbol{v}\boldsymbol{v}^* = \sigma_v^2 \mathrm{I}$. Invoking the result of Thm. 6.1 with $H = h$, we conclude that the optimal linear estimator, or the m.v.u.e., of x is

$$\boxed{\hat{\boldsymbol{x}}_{\mathsf{mvue}} = \frac{1}{N} \sum_{i=0}^{N-1} \boldsymbol{y}(i)} \tag{6.17}$$

This result is simply the sample-mean estimator that the reader may be familiar with from an introductory course on statistics. Comparing the expressions for $\hat{\boldsymbol{x}}_{\mathsf{llmse}}$ and $\hat{\boldsymbol{x}}_{\mathsf{mvue}}$ we see that we are now dividing the sum of the observations $\{\boldsymbol{y}(i)\}$ by N, and not by $N + 1/\mathsf{SNR}$. This modification guarantees that the estimator $\hat{\boldsymbol{x}}_{\mathsf{mvue}}$ is truly unbiased.

6.3 APPLICATION: CHANNEL AND NOISE ESTIMATION

We reconsider the channel estimation problem of Sec. 5.2, except that now the channel tap vector is modeled as an unknown *constant* vector, c, rather than a random vector, $\boldsymbol{c}$, as shown in Fig. 6.1.

By repeating the construction of Sec. 5.2 we again obtain (cf. (5.7)):

$$\underbrace{\begin{bmatrix} \boldsymbol{y}(0) \\ \boldsymbol{y}(1) \\ \boldsymbol{y}(2) \\ \boldsymbol{y}(3) \\ \boldsymbol{y}(4) \\ \boldsymbol{y}(5) \\ \boldsymbol{y}(6) \end{bmatrix}}_{\boldsymbol{y}:(N+1)\times 1} = \underbrace{\begin{bmatrix} s(0) & & \\ s(1) & s(0) & \\ s(2) & s(1) & s(0) \\ s(3) & s(2) & s(1) \\ s(4) & s(3) & s(2) \\ s(5) & s(4) & s(3) \\ s(6) & s(5) & s(4) \end{bmatrix}}_{H:(N+1)\times M} c + \underbrace{\begin{bmatrix} \boldsymbol{v}(0) \\ \boldsymbol{v}(1) \\ \boldsymbol{v}(2) \\ \boldsymbol{v}(3) \\ \boldsymbol{v}(4) \\ \boldsymbol{v}(5) \\ \boldsymbol{v}(6) \end{bmatrix}}_{\boldsymbol{v}:(N+1)\times 1}$$

where we are defining the quantities $\{\boldsymbol{y}, H, \boldsymbol{v}\}$ and where H is $(N+1) \times M$. Using the result of Thm. 6.1, we find that the optimal estimator of c is now given by

$$\boxed{\hat{\boldsymbol{c}}_{\text{mvue}} = (H^* R_v^{-1} H)^{-1} H^* R_v^{-1} \boldsymbol{y}} \tag{6.18}$$

where R_v is the covariance matrix of $\boldsymbol{v}$. This result is different from the linear least-mean-squares estimator (l.l.m.s.e.) found in Sec. 5.2 (see (5.8)), namely,

$$\hat{\boldsymbol{c}}_{\text{llmse}} = \left[R_c^{-1} + H^* R_v^{-1} H\right]^{-1} H^* R_v^{-1} \boldsymbol{y}$$

which requires knowledge of the covariance matrix $R_c = \mathsf{E}\,\boldsymbol{c}\boldsymbol{c}^*$ when $\boldsymbol{c}$ is modeled as a random variable. The estimator (6.18) requires knowledge of only $\{H, R_v, \boldsymbol{y}\}$. Actually, if the noise sequence $\{\boldsymbol{v}(i)\}$ is modeled as white with variance σ_v^2, then $R_v = \sigma_v^2 \mathrm{I}$ and R_v would end up disappearing from the expression for $\hat{\boldsymbol{c}}_{\text{mvue}}$. Specifically, (6.18) would become

$$\boxed{\hat{\boldsymbol{c}}_{\text{mvue}} = (H^* H)^{-1} H^* \boldsymbol{y}} \tag{6.19}$$

It is worth remarking that expression (6.19) has the form of a least-squares solution; which we shall study in great detail in Part VII (*Least-Squares Methods*).

Note from (6.19) that we do not need to know σ_v^2; the estimator is now only dependent on the available data (namely, the measurements $\{\boldsymbol{y}(i)\}$ and the data matrix H). If desired, we can estimate σ_v^2 itself as follows. Since

$$\boldsymbol{v}(i) = \boldsymbol{y}(i) - \left[\begin{array}{cccc} s(i) & s(i-1) & \ldots & s(i-M+1) \end{array}\right] c$$

an estimator for σ_v^2 would be

$$\widehat{\boldsymbol{\sigma}_v^2} = \frac{1}{N+1} \sum_{i=0}^{N} \left| \boldsymbol{y}_i - \left[\begin{array}{cccc} s(i) & s(i-1) & \ldots & s(i-M+1) \end{array}\right] \hat{\boldsymbol{c}}_{\text{mvue}} \right|^2$$

i.e.,

$$\boxed{\widehat{\boldsymbol{\sigma}_v^2} = \frac{1}{N+1} \|\boldsymbol{y} - H\hat{\boldsymbol{c}}_{\text{mvue}}\|^2} \tag{6.20}$$

Expressions (6.19) and (6.20) are often used in practice to perform channel and noise variance estimation. Later in Prob. VII.19 we shall show that the alternative estimator

$$\widehat{\boldsymbol{\sigma}_v^2} = \frac{1}{N+1-M} \|\boldsymbol{y} - H\hat{\boldsymbol{c}}_{\text{mvue}}\|^2$$

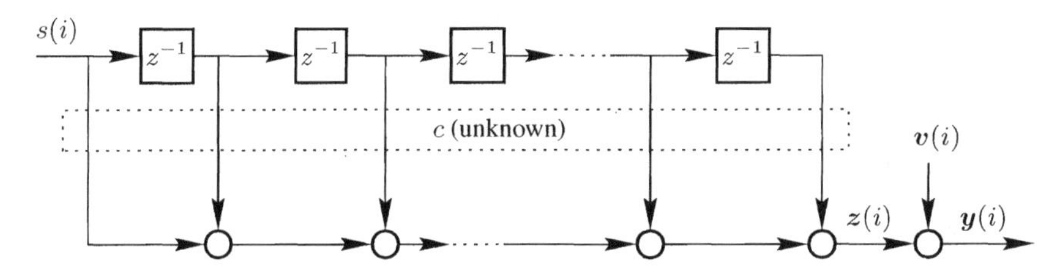

FIGURE 6.1 Channel and noise estimation.

with $(N+1-M)$ instead of $(N+1)$, is an unbiased estimator for σ_v^2.

6.4 APPLICATION: DECISION FEEDBACK EQUALIZATION

Our second application is in the context of channel equalization, which we already encountered in Sec. 5.4 while studying *linear* equalizers. In this section, we extend the discussion to decision-feedback equalizers.

Thus consider an FIR channel with a known column tap vector c of length M (i.e., with M taps), say, with transfer function

$$\boxed{C(z) = c(0) + c(1)z^{-1} + \ldots + c(M-1)z^{-(M-1)}}$$

Data symbols $\{\boldsymbol{s}(\cdot)\}$ are transmitted through the channel and the output sequence is measured in the presence of additive noise, $\boldsymbol{v}(i)$. The signals $\{\boldsymbol{v}(\cdot), \boldsymbol{s}(\cdot)\}$ are assumed uncorrelated. Due to the channel memory, each measurement $\boldsymbol{y}(i)$ contains contributions not only from $\boldsymbol{s}(i)$ but also from prior symbols, since

$$\boldsymbol{y}(i) = c(0)\boldsymbol{s}(i) \; + \; \underbrace{\sum_{k=1}^{M-1} c(k)\boldsymbol{s}(i-k)}_{\text{ISI}} \; + \boldsymbol{v}(i)$$

The second term on the right-hand side describes the *inter-symbol-interference* (ISI); it refers to the interference that is caused by prior symbols. The purpose of an equalizer is to combat ISI and to recover $\boldsymbol{s}(i)$ from measurements of the output sequence.

As was discussed in Sec. 5.4, in order to achieve this task, a linear equalizer employs current and prior measurements $\{\boldsymbol{y}(i-k)\}$, say, for $k = 0, 1, \ldots, L-1$. This is because prior measurements contain information that is correlated with the ISI term in $\boldsymbol{y}(i)$ and, therefore, they can help in estimating the interference term and removing its effect. Of course, if possible, it would be preferable to use the prior symbols $\{\boldsymbol{s}(i-1), \boldsymbol{s}(i-2), \ldots\}$ themselves in order to cancel their effect from $\boldsymbol{y}(i)$ rather than rely on the prior measurements $\{\boldsymbol{y}(i-1), \boldsymbol{y}(i-2), \ldots\}$.

Decision-feedback equalizers (DFE) attempt to implement this strategy and are therefore better suited for channels with pronounced ISI. In addition to using an FIR filter in the feedforward path, as in linear equalization, a DFE employs a *feedback* filter in order to feed back previous decisions and use them to reduce ISI. The DFE structure is shown in Fig. 6.2

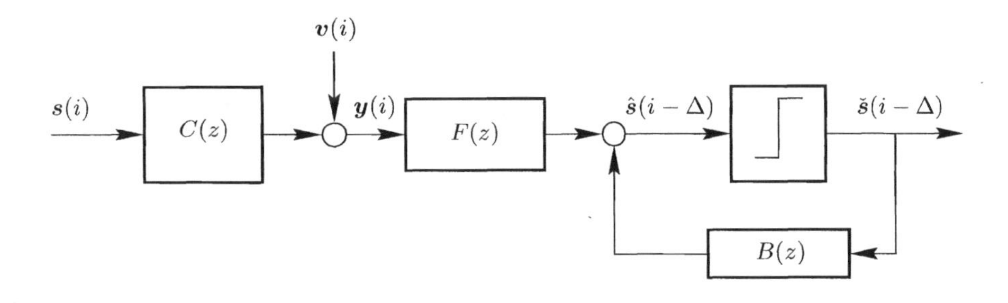

FIGURE 6.2 A decision-feedback equalizer. It consists of a feedforward filter, a feedback filter, and a decision device.

for estimating a delayed version of $s(i)$, with the transfer functions of the feedforward and feedback filters denoted by $\{F(z), B(z)\}$, respectively. It is seen from the figure that the input to the feedback filter comes from the output of the decision device, denoted by $\check{s}(i-\Delta)$. The purpose of this device is to map the estimator $\hat{s}(i-\Delta)$, which is obtained by combining the outputs of the feedforward and feedback filters, to the closest point in the symbol constellation. Now in linear equalization, the feedforward filter reduces ISI by attempting to force the combined system $C(z)F(z)$ to be close to

$$C(z)F(z) \approx z^{-\Delta}$$

In general, this objective is difficult to attain, especially for channels with pronounced ISI, and $C(z)F(z)$ will have a nontrivial impulse response sequence (we say that $C(z)F(z)$ will have trailing inter-symbol interference). The purpose of the feedback filter in a DFE implementation is to use prior decisions in order to cancel this trailing ISI.

Equalizer Design

Assume the feedforward filter has L taps and denote its transfer function by

$$\boxed{F(z) = f(0) + f(1)z^{-1} + \ldots + f(L-1)z^{-(L-1)}}$$

with coefficients $\{f(i)\}$. Likewise, assume the feedback filter has Q taps with a transfer function of the form

$$\boxed{B(z) = -b(1)z^{-1} - b(2)z^{-2} - \ldots - b(Q)z^{-Q}}$$

with coefficients denoted by $\{-b(i)\}$ for convenience. Note that this filter is strictly causal in that it does not have a direct path from its input to its output (i.e., $b(0) = 0$). This is because *previous* decisions are being fed back through $B(z)$.

The criterion for designing the equalizer coefficients $\{f(i), b(i)\}$ is, as usual, to minimize the variance of the error signal,

$$\tilde{s}(i-\Delta) = s(i-\Delta) - \hat{s}(i-\Delta)$$

In so doing, the designer expects that $\hat{s}(i-\Delta)$ will be sufficiently close to $s(i-\Delta)$ so that the decision device would be able to map $\hat{s}(i-\Delta)$ to the correct symbol in the signal constellation. Therefore, the $\{f(i), b(i)\}$ will be determined by solving

$$\boxed{\min_{\left\{\begin{array}{c} f(0), f(1), \ldots, f(L-1) \\ b(1), b(2), \ldots, b(Q) \end{array}\right\}} \mathsf{E}\,|\tilde{s}(i-\Delta)|^2} \tag{6.21}$$

The presence of the decision device makes (6.21) a *nonlinear* optimization problem. This is because $\hat{s}(i-\Delta)$ will be a nonlinear function of the measured data $\{y(i)\}$. In order to facilitate the design of $\{F(z), B(z)\}$, it is customary to assume that

$$\boxed{\text{The decisions } \{\check{s}(i-\Delta)\} \text{ are correct and equal to } \{s(i-\Delta)\}} \tag{6.22}$$

That is, we assume that the decision device gives correct decisions. This assumption highlights one difficulty with decision-feedback equalization. Erroneous decisions can happen and they get fed back into the equalizer through the feedback filter. Longer feedback filters tend to keep these errors longer within the equalizer and can cause performance degradation especially at low signal-to-noise ratios. Still, in general, decision-feedback equalizers tend to outperform linear equalizers (see Prob. II.40).

To solve (6.21) we first examine the dependence of the error variance on the unknown coefficients $\{f(i), b(i)\}$. From Fig. 6.2 we have

$$\begin{aligned}\hat{s}(i-\Delta) &= [\,f(0)\boldsymbol{y}(i)+f(1)\boldsymbol{y}(i-1)\ldots+f(L-1)\boldsymbol{y}(i-L+1)\,]\\ &\quad -[\,b(1)\boldsymbol{s}(i-\Delta-1)+b(2)\boldsymbol{s}(i-\Delta-2)+\ldots+b(Q)\boldsymbol{s}(i-\Delta-Q)\,]\end{aligned}$$

where we used assumption (6.22) to replace $\check{s}(j)$ by $\boldsymbol{s}(j)$. We can rewrite this expression more compactly in vector form as follows. We collect the coefficients of $F(z)$ into a row vector:

$$\boxed{f^* \triangleq [\; f(0) \quad f(1) \quad \ldots \quad f(L-1)\;]}$$

and the coefficients of $-B(z)$ into another row vector with a leading entry that is equal to one,

$$\boxed{b^* \triangleq [\; 1 \quad b(1) \quad b(2) \quad \ldots \quad b(Q)\;]}$$

We also define the following column vectors of observations and data symbols:

$$\boldsymbol{y}_i \triangleq \underbrace{\begin{bmatrix} \boldsymbol{y}(i) \\ \boldsymbol{y}(i-1) \\ \boldsymbol{y}(i-2) \\ \vdots \\ \boldsymbol{y}(i-L+1) \end{bmatrix}}_{L\times 1}, \qquad \boldsymbol{s}_\Delta = \underbrace{\begin{bmatrix} \boldsymbol{s}(i-\Delta) \\ \boldsymbol{s}(i-\Delta-1) \\ \boldsymbol{s}(i-\Delta-2) \\ \vdots \\ \boldsymbol{s}(i-\Delta-Q) \end{bmatrix}}_{(Q+1)\times 1} \tag{6.23}$$

and denote their covariances and cross-covariances by

$$\boxed{R \triangleq \mathsf{E}\begin{bmatrix} \boldsymbol{s}_\Delta \\ \boldsymbol{y}_i \end{bmatrix}\begin{bmatrix} \boldsymbol{s}_\Delta \\ \boldsymbol{y}_i \end{bmatrix}^* \triangleq \begin{bmatrix} R_s & R_{sy} \\ R_{ys} & R_y \end{bmatrix}}$$

where R_s is $(Q+1)\times(Q+1)$ and R_{sy} is $(Q+1)\times L$. We assume that the processes $\{\boldsymbol{s}(\cdot), \boldsymbol{y}(\cdot)\}$ are jointly wide-sense stationary so that the quantities $\{R_{sy}, R_y, R_s\}$ are independent of i. We further assume that the covariance matrix R is positive-definite and, hence, invertible. The positive-definiteness of R guarantees that both R_y and the Schur complement of R with respect to R_y are positive-definite matrices as well (see Sec. B.3), i.e.,

$$\boxed{R_y > 0 \qquad \text{and} \qquad R_\delta \triangleq R_s - R_{sy}R_y^{-1}R_{ys} > 0} \tag{6.24}$$

where we are denoting the Schur complement by R_δ. Hence, $\{R_y, R_\delta\}$ are also invertible.

With these definitions, the error signal $\tilde{s}(i-\Delta)$ can be written as

$$\tilde{s}(i-\Delta) = \boldsymbol{s}(i-\Delta) - \hat{s}(i-\Delta) = b^*\boldsymbol{s}_\Delta - f^*\boldsymbol{y}_i$$

so that the optimization problem (6.21) becomes

$$\min_{f,b} \mathsf{E}\,|b^*\boldsymbol{s}_\Delta - f^*\boldsymbol{y}_i|^2 \tag{6.25}$$

We shall denote the optimal vector solutions by f^*_{opt} and b^*_{opt}. Rather than minimize the variance of $b^*\boldsymbol{s}_\Delta - f^*\boldsymbol{y}_i$ simultaneously over $\{f, b\}$, we shall minimize it over one vector at a time. Thus assume that we *fix* the vector b and let us minimize the error variance over f. To do so, we introduce the scalar $\alpha = b^* s_\Delta$ so that the error signal becomes $\tilde{s}(i-\Delta) = \alpha - f^* y_i$. In this way, $\tilde{s}(i-\Delta)$ can be interpreted as the error that results from estimating α from y_i through the choice of f. In other words, we are reduced to solving

$$\boxed{\min_f \ \mathsf{E}\,|\alpha - f^* y_i|^2 \qquad \text{where} \qquad \alpha = b^* s_\Delta}$$

which is a standard linear least-mean-squares estimation problem. From Thm. 3.1, we know that the optimal choice for f is

$$f^*_{\rm opt} = R_{\alpha y} R_y^{-1} \ = \ b^* R_{sy} R_y^{-1} \tag{6.26}$$

where we used the fact that

$$R_{\alpha y} \ \stackrel{\Delta}{=} \ \mathsf{E}\,\alpha y_i^* = \mathsf{E}\,b^* s_\Delta y_i^* = b^* R_{sy}$$

The resulting minimum mean-square error is, again from Thm. 3.1,

$$\begin{aligned} \text{m.m.s.e.} \ &\stackrel{\Delta}{=} \ \mathsf{E}\,|\alpha - f^*_{\rm opt} y_i|^2 \\ &= \ R_\alpha - R_{\alpha y} R_y^{-1} R_{y\alpha} \\ &= \ b^* R_s b - b^* R_{sy} R_y^{-1} R_{ys} b \\ &= \ b^*[R_s - R_{sy} R_y^{-1} R_{ys}] b \\ &= \ b^* R_\delta b \end{aligned} \tag{6.27}$$

Substituting this expression into (6.25), we find that we now need to solve

$$\min_b \ b^* R_\delta b \tag{6.28}$$

But recall that the leading entry of b is unity, so that (6.28) is actually a constrained problem of the form

$$\boxed{\min_b \ b^* R_\delta b \qquad \text{subject to} \ \ b^* e_0 = 1}$$

where e_0 is the first basis vector, of dimension $(Q+1) \times 1$,

$$e_0 \ \stackrel{\Delta}{=} \ \text{col}\{1, 0, 0 \ldots, 0\}$$

Using the result stated in Remark 6.1, we find that the optimal choice of b is

$$b^*_{\rm opt} = \frac{e_0^{\mathsf T} R_\delta^{-1}}{e_0^{\mathsf T} R_\delta^{-1} e_0}$$

The term that appears in the denominator is the $(0,0)$ entry of R_δ^{-1}, while the term in the numerator is the first row of R_δ^{-1}. This means that the optimal vector $b^*_{\rm opt}$ is obtained by normalizing the first row of R_δ^{-1} to have a unit leading entry. Substituting the above expression for $b^*_{\rm opt}$ into $b^* R_\delta b$ we find that the resulting m.m.s.e. of the original optimization problem (6.21) is

$$\boxed{\text{m.m.s.e.} \ = \ \frac{1}{e_0^{\mathsf T} R_\delta^{-1} e_0}} \tag{6.29}$$

In summary, under assumption (6.22) that the decisions $\{\check{s}(i-\Delta)\}$ are correct, the optimal coefficients $\{f(i), b(i)\}$ of the DFE can be found as follows:

$$\boxed{b^*_{\text{opt}} = \frac{e_0^{\mathsf{T}} R_\delta^{-1}}{e_0^{\mathsf{T}} R_\delta^{-1} e_0} \quad \text{and} \quad f^*_{\text{opt}} = b^*_{\text{opt}} R_{sy} R_y^{-1}} \tag{6.30}$$

The entries of $\{f^*_{\text{opt}}, b^*_{\text{opt}}\}$ provide the desired tap coefficients $\{b(i), f(i)\}$.

Using the Channel Model

The expressions (6.29)–(6.30) for $\{b^*_{\text{opt}}, f^*_{\text{opt}}, \text{m.m.s.e.}\}$ are in terms of the covariance and cross-covariance matrices $\{R_s, R_{sy}, R_y\}$, which can be evaluated from the channel model $C(z)$ and from the given statistical information about $\{\boldsymbol{s}(\cdot), \boldsymbol{v}(\cdot)\}$. To do so, we proceed as in Sec. 5.4.

We first express the observation vector $\boldsymbol{y}_i$ in terms of the transmitted data. Assume for the sake of illustration that $L = 5$ (i.e., a feedforward filter with 5 taps) and $M = 4$ (a channel with four taps). Then we can write

$$\underbrace{\begin{bmatrix} \boldsymbol{y}(i) \\ \boldsymbol{y}(i-1) \\ \boldsymbol{y}(i-2) \\ \boldsymbol{y}(i-3) \\ \boldsymbol{y}(i-4) \end{bmatrix}}_{\boldsymbol{y}_i : L\times 1} = \underbrace{\begin{bmatrix} c(0) & c(1) & c(2) & c(3) & & & & \\ & c(0) & c(1) & c(2) & c(3) & & & \\ & & c(0) & c(1) & c(2) & c(3) & & \\ & & & c(0) & c(1) & c(2) & c(3) & \\ & & & & c(0) & c(1) & c(2) & c(3) \end{bmatrix}}_{H : L\times(L+M-1)} \underbrace{\begin{bmatrix} \boldsymbol{s}(i) \\ \boldsymbol{s}(i-1) \\ \boldsymbol{s}(i-2) \\ \boldsymbol{s}(i-3) \\ \boldsymbol{s}(i-4) \\ \boldsymbol{s}(i-5) \\ \boldsymbol{s}(i-6) \\ \boldsymbol{s}(i-7) \end{bmatrix}}_{\underline{\boldsymbol{s}}_i : (L+M)-1\times 1}$$

$$+ \underbrace{\begin{bmatrix} \boldsymbol{v}(i) \\ \boldsymbol{v}(i-1) \\ \boldsymbol{v}(i-2) \\ \boldsymbol{v}(i-3) \\ \boldsymbol{v}(i-4) \end{bmatrix}}_{\boldsymbol{v}_i : L\times 1}$$

That is,

$$\boxed{\boldsymbol{y}_i = H\underline{\boldsymbol{s}}_i + \boldsymbol{v}_i}$$

where, for general $\{L, M\}$,

$$\underline{\boldsymbol{s}}_i \triangleq \begin{bmatrix} \boldsymbol{s}(i) \\ \boldsymbol{s}(i-1) \\ \boldsymbol{s}(i-2) \\ \vdots \\ \boldsymbol{s}(i-L-M+2) \end{bmatrix}, \quad \boldsymbol{v}_i \triangleq \begin{bmatrix} \boldsymbol{v}(i) \\ \boldsymbol{v}(i-1) \\ \boldsymbol{v}(i-2) \\ \vdots \\ \boldsymbol{v}(i-L+1) \end{bmatrix} \tag{6.31}$$

and H is the $L \times (L+M-1)$ channel matrix. We therefore find that there is a linear relation between the vectors $\{\boldsymbol{y}_i, \underline{\boldsymbol{s}}_i\}$ and this relation can be used to evaluate R_y as

$$R_y = \mathsf{E}\,(H\underline{\boldsymbol{s}}_i + \boldsymbol{v}_i)(H\underline{\boldsymbol{s}}_i + \boldsymbol{v}_i)^* = HR_{\underline{s}}H^* + R_v$$

where

$$R_{\underline{s}} \stackrel{\Delta}{=} \mathsf{E}\,\underline{s}_i\underline{s}_i^* \quad ((L+M-1)\times(L+M-1)) \quad \text{and} \quad R_v \stackrel{\Delta}{=} \mathsf{E}\,v_iv_i^* \quad (L\times L)$$

Likewise,

$$R_{sy} = \mathsf{E}\,s_\Delta(H\underline{s}_i + v_i)^* \;=\; (\mathsf{E}\,s_\Delta\underline{s}_i^*)\,H^*$$

since $\{v(\cdot), s(\cdot)\}$ are uncorrelated. We still need to evaluate $\mathsf{E}\,s_\Delta\underline{s}_i^*$, where $\{s_\Delta, \underline{s}_i\}$ are defined by (6.23) and (6.31) in terms of the transmitted symbols. Of course, the value of $\mathsf{E}\,s_\Delta\underline{s}_i^*$ depends on the assumed correlation between the transmitted symbols.

It is common that Δ be chosen such that all the entries of s_Δ fall within the entries of $\underline{s}_i$. This condition requires the channel and filter lengths, as well as the delay Δ, to satisfy

$$\boxed{\Delta + Q \leq L + M - 2} \tag{6.32}$$

With this condition, if the $\{s(\cdot)\}$ are assumed to be independent and identically distributed with variance σ_s^2, then it can be seen that

$$\mathsf{E}\,s_\Delta\underline{s}_i^* \;=\; \begin{bmatrix} 0 & \ldots & 0 & \sigma_s^2\mathrm{I}_{Q+1} & 0 & \ldots & 0 \end{bmatrix} \quad ((Q+1)\times(L+M-1))$$

with Δ leading zero columns, followed by a $(Q+1)\times(Q+1)$ identity matrix scaled by σ_s^2, followed (or not) by zero columns. Even if Δ exceeds the bound in (6.32), we can still evaluate $\mathsf{E}\,s_\Delta\underline{s}_i^*$ and complete the calculations. We can express the above $\mathsf{E}\,s_\Delta\underline{s}_i^*$ more compactly as

$$\mathsf{E}\,s_\Delta\underline{s}_i^* \;=\; \begin{bmatrix} 0_{(Q+1)\times\Delta} & \sigma_s^2\mathrm{I}_{Q+1} & 0 \end{bmatrix}$$

Likewise,

$$R_s = \sigma_s^2\mathrm{I}_{Q+1} \qquad \text{and} \qquad R_{\underline{s}} = \sigma_s^2\mathrm{I}_{L+M-1}$$

We continue with the assumption of i.i.d. symbols $\{s(\cdot)\}$ for simplicity of presentation. But it should be noted that the derivation applies even for correlated (but stationary) data. In a similar vein, we assume that the noise sequence $\{v(\cdot)\}$ is white with variance σ_v^2 so that $R_v = \sigma_v^2\mathrm{I}$. Again, the development applies even for correlated (but stationary) noise. We thus find that

$$\boxed{R_y = \sigma_s^2HH^* + \sigma_v^2\mathrm{I}_L \qquad \text{and} \qquad R_{sy} = \begin{bmatrix} 0_{(Q+1)\times\Delta} & \sigma_s^2\mathrm{I}_{Q+1} & 0 \end{bmatrix} H^*} \tag{6.33}$$

and

$$R_\delta \;=\; \sigma_s^2\mathrm{I} - R_{sy}(\sigma_s^2HH^* + \sigma_v^2\mathrm{I}_L)^{-1}R_{ys}$$

This latter expression can be rewritten, by virtue of the matrix inversion lemma (5.4), as

$$\boxed{R_\delta = \Phi\left(\frac{1}{\sigma_s^2}\mathrm{I} + \frac{1}{\sigma_v^2}H^*H\right)^{-1}\Phi^*} \tag{6.34}$$

where

$$\Phi = \begin{bmatrix} 0_{(Q+1)\times\Delta} & \mathrm{I}_{Q+1} & 0 \end{bmatrix}$$

Expressions (6.33) and (6.34) can now be used with (6.29)–(6.30) to determine the optimal equalizer coefficients and the resulting m.m.s.e.

Example 6.1 (Numerical illustration)

Let us reconsider Ex. 4.4 and design a DFE equalizer rather than a linear equalizer for the channel $C(z) = 1 + 0.5z^{-1}$, for which $M = 2$. We select a feedforward filter with 3 taps (i.e., $L = 3$) and a feedback filter with one tap (i.e., $Q = 1$). We also select $\Delta = 1$. The resulting structure is shown in Fig. 6.3. For this example, $\sigma_s^2 = 1$, $\sigma_v^2 = 1$, and

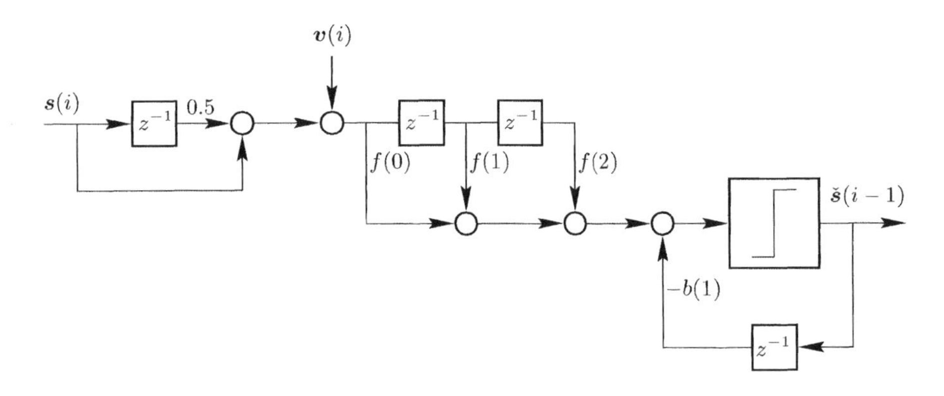

FIGURE 6.3 A DFE structure for the channel $1 + 0.5z^{-1}$.

$$H = \begin{bmatrix} 1 & 0.5 & & \\ & 1 & 0.5 & \\ & & 1 & 0.5 \end{bmatrix}$$

so that from (6.33) and (6.34)

$$R_y = \begin{bmatrix} 9/4 & 1/2 & 0 \\ 1/2 & 9/4 & 1/2 \\ 0 & 1/2 & 9/4 \end{bmatrix}, \quad R_{sy} = \begin{bmatrix} 0.5 & 1 & 0 \\ 0 & 0.5 & 1 \end{bmatrix}, \quad R_\delta = \begin{bmatrix} 0.4992 & -0.1218 \\ -0.1218 & 0.5175 \end{bmatrix}$$

Using (6.30) we obtain

$$b_{\text{opt}}^* = \begin{bmatrix} 1.0000 & 0.2354 \end{bmatrix}, \qquad f_{\text{opt}}^* = \begin{bmatrix} 0.1176 & 0.4706 & 0.0000 \end{bmatrix}$$

and the resulting m.m.s.e. is

$$\text{m.m.s.e.} = b_{\text{opt}} R_\delta b_{\text{opt}}^* = 0.4705$$

That is,

$$B_{\text{opt}}(z) = -0.2354z^{-1} \quad \text{and} \quad F_{\text{opt}}(z) = 0.1176 + 0.4706z^{-1}$$

Had we selected instead $\Delta = 0$, as we did in Ex. 4.1, then the only quantity that changes is the cross-covariance R_{sy}, which becomes

$$R_{sy} = \begin{bmatrix} 1 & 0 & 0 \\ 0.5 & 1 & 0 \end{bmatrix} \quad \text{so that} \quad R_\delta = \begin{bmatrix} 0.5312 & -0.1248 \\ -0.1248 & 0.4992 \end{bmatrix}$$

and, therefore,

$$b_{\text{opt}}^* = \begin{bmatrix} 1 & 0.2500 \end{bmatrix}, \qquad f_{\text{opt}}^* = \begin{bmatrix} 0.5000 & 0.0000 & 0.0000 \end{bmatrix}$$

That is, $B_{\text{opt}}(z) = -0.25z^{-1}$ and $F_{\text{opt}}(z) = 0.5$. The resulting m.m.s.e. in this case is

$$\text{m.m.s.e.} = b_{\text{opt}}^* R_\delta b_{\text{opt}} = 0.5000$$

We see that for this example with $\Delta = 0$, the DFE results in a smaller mean-square error than the linear equalizer designed in Ex. 4.1, which resulted in m.m.s.e. $= 0.5312$. A more noticeable difference in performance between decision-feedback equalizers and linear equalizers can be observed for channels with more pronounced inter-symbol interference. The performance of DFEs is examined in greater detail in a computer project at the end of this part.

◇

Formulation as Linear Estimation

The design of a DFE can alternatively be pursued by formulating an *unconstrained* linear estimation problem, of the same form studied in Thm. 3.1, as opposed to splitting its solution into two steps: unconstrained estimation for determining the feedforward filter and constrained estimation for determining the feedback filter, as was done above.

To see this, we introduce the extended vector

$$\boldsymbol{r} \triangleq \underbrace{\left[\begin{array}{c} \boldsymbol{s}(i-\Delta-1) \\ \boldsymbol{s}(i-\Delta-2) \\ \vdots \\ \boldsymbol{s}(i-\Delta-Q) \\ \hline \boldsymbol{y}(i) \\ \boldsymbol{y}(i-1) \\ \vdots \\ \boldsymbol{y}(i-L+1) \end{array}\right]}_{(L+Q)\times 1}$$

which, under the assumption (6.22) of correct decisions, contains all the observations that are used by the feedforward and feedback filters in order to estimate the variable $\boldsymbol{x} = \boldsymbol{s}(i-\Delta)$. We also define the $1 \times (L+Q)$ row vector

$$\boxed{k^* = \left[\begin{array}{cccc|cccc} -b(1) & -b(2) & \ldots & -b(Q) & f(0) & f(1) & \ldots & f(L-1) \end{array}\right]}$$

which contains the equalizer coefficients $\{f(i), b(i)\}$ that we wish to determine. In this way,

$$\hat{\boldsymbol{s}}(i-\Delta) = k^* \boldsymbol{r}$$

and the estimation error, $\tilde{\boldsymbol{x}} = \tilde{\boldsymbol{s}}(i-\Delta)$, is given by

$$\tilde{\boldsymbol{x}} = \boldsymbol{x} - k^* \boldsymbol{r}$$

so that problem (6.21) becomes equivalent to solving

$$\min_k \mathsf{E}\, |\boldsymbol{x} - k^* \boldsymbol{r}|^2 \tag{6.35}$$

This is a standard linear least-mean-squares formulation, and its solution is given by (cf. Thm. 3.1):

$$\boxed{k_o^* = R_{xr} R_r^{-1}} \tag{6.36}$$

where

$$R_{xr} = \mathsf{E}\, \boldsymbol{s}(i-\Delta)\boldsymbol{r}^* \quad \text{and} \quad R_r = \mathsf{E}\, \boldsymbol{r}\boldsymbol{r}^*$$

and that the resulting m.m.s.e. is

$$\boxed{\text{m.m.s.e.} = \sigma_s^2 - R_{xr}R_r^{-1}R_{rx}} \tag{6.37}$$

Some algebra will show that expressions (6.36) and (6.37) lead to the same solutions (6.29)–(6.30) — see Prob. II.41.

6.5 APPLICATION: ANTENNA BEAMFORMING

Our third application is in the context of antenna beamforming. In this application, we desire to combine the measurements of an antenna array in order to maximize its gain along a particular direction.

Consider the diagram of Fig. 6.4, which shows a linear array of sensors or antennas, assumed uniformly spaced, with the separation between two adjacent elements denoted by d. The antenna array is assumed to be far from a source radiating an electromagnetic wave of the form

$$r(t) = s(t)e^{j\omega_c t}, \qquad j \triangleq \sqrt{-1}$$

where ω_c denotes the carrier frequency and $s(t)$ denotes the envelope, also called the baseband signal. Since the source is sufficiently distant from the antenna array, the arriving wavefronts can be assumed to be planar at the array. The output of the array is obtained by linearly combining the measurements at the antennas. These measurements are subjected to noise with the noise at antenna j at time t denoted by $v_j(t)$.

Let θ denote the direction of arrival of the wavefront relative to the plane of the antennas, as indicated in the figure, and consider the two leftmost antennas, labelled 0 and 1. The distance that separates the planar waves arriving at these two elements is equal to $d\cos\theta$.

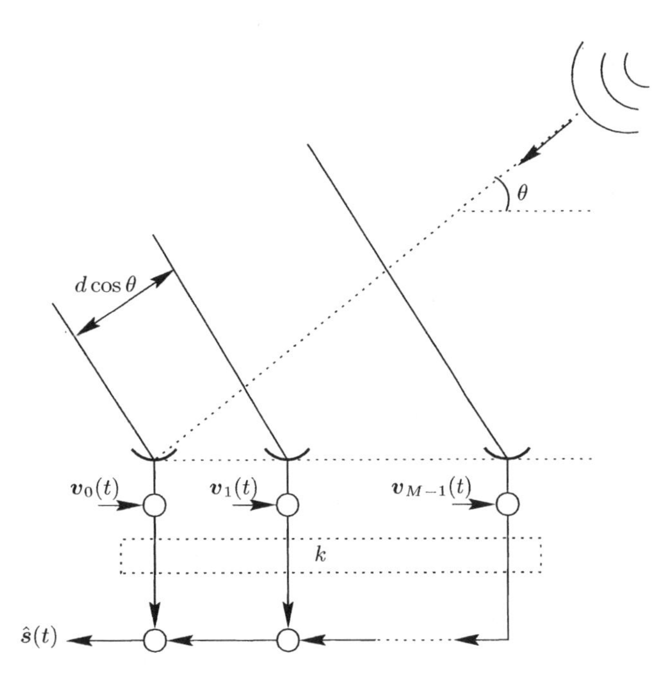

FIGURE 6.4 A uniformly-spaced array of antennas.

Moreover, the interval of time Δt that is needed for the wave to propagate from antenna 1 to antenna 0 is $d\cos\theta/c$, where c is the speed of propagation or, equivalently,

$$\Delta t = (2\pi d\cos\theta)/(\omega_c\lambda)$$

where λ is the wavelength of the wavefront; it is related to c via

$$\boxed{c = \frac{\omega_c\lambda}{2\pi}}$$

Now if $s(t)e^{j\omega_c t}$ is the signal at antenna 0 at time t, then $s(t+\Delta t)e^{j\omega_c(t+\Delta t)}$ is the signal at that *same* time instant at antenna 1. Therefore, if we assume a slowly-varying envelope, i.e., if $s(t+\Delta t) \approx s(t)$, then the signal at antenna 1 at time t is

$$s(t)e^{j\omega_c\left(t+\frac{2\pi d\cos\theta}{\omega_c\lambda}\right)} = s(t)e^{j\omega_c t}e^{j\frac{2\pi}{\lambda}d\cos\theta}$$

More generally, by following a similar argument, the signal arriving at the n-th antenna at time t is of the form

$$s(t)e^{j\omega_c t}e^{j\frac{2\pi n}{\lambda}d\cos\theta}, \qquad n = 0, 1, \ldots, M-1$$

If at time t we take a snapshot of the values of the signals at the M antenna elements we obtain, from left to right,

$$s(t)e^{j\omega_c t}\begin{bmatrix} 1 & e^{j\frac{2\pi}{\lambda}d\cos\theta} & e^{j\frac{4\pi}{\lambda}d\cos\theta} & \ldots & e^{j\frac{2\pi(M-1)}{\lambda}d\cos\theta} \end{bmatrix}$$

Usually, before processing by the antenna array, the incident signals are first converted to baseband, which means that the carrier component $e^{j\omega_c t}$ is removed. In this way, the signals received by the antennas at time t can be assumed to be of the form:

$$\boxed{s(t)e^{j\frac{2\pi n}{\lambda}d\cos\theta}, \qquad n = 0, 1, \ldots, M-1}$$

Now let $\boldsymbol{y}$ denote a noisy snapshot of the baseband signals at the antennas, i.e.,

$$\boldsymbol{y} = \underbrace{\begin{bmatrix} 1 \\ e^{j\frac{2\pi}{\lambda}d\cos\theta} \\ e^{j\frac{4\pi}{\lambda}d\cos\theta} \\ \vdots \\ e^{j\frac{2\pi(M-1)}{\lambda}d\cos\theta} \end{bmatrix}}_{h:M\times 1} s(t) + \underbrace{\begin{bmatrix} \boldsymbol{v}_0(t) \\ \boldsymbol{v}_1(t) \\ \boldsymbol{v}_2(t) \\ \vdots \\ \boldsymbol{v}_{M-1}(t) \end{bmatrix}}_{\boldsymbol{v}:M\times 1}$$

where $\boldsymbol{v}$ is a column vector whose entries correspond to the noise components at the individual antennas. Moreover, the column vector h is dependent on the direction of arrival θ, the wavelength λ, and the geometry of the array as defined by $\{M, d\}$. Observe that the entries of $\boldsymbol{y}$ correspond to measurements at different points in space rather than at different points in time, i.e., we are now dealing with an application that involves *spatial* sampling as opposed to *time* sampling.

The purpose of a beamformer is to combine the entries of the snapshot, say, as $k^*\boldsymbol{y}$ for some column vector k, in order to achieve two objectives.

1. **Directionality**. The vector k must satisfy $k^*h = 1$. This condition guarantees that when the incident direction of arrival is θ, the response of the beamformer, in the absence of any noise, will be

$$k^*\boldsymbol{y} = s(t)k^*h = s(t)$$

In other words, an incident wave along the direction θ will pass through the beamformer undistorted.

2. **Interference attenuation**. When the snapshot $\boldsymbol{y}$ is corrupted by additive noise (including the effects of interferences caused by signals not originating from the direction θ), we would like the output of the beamformer to provide an estimate of $s(t)$. For this purpose, we should choose k in order to minimize the error variance, i.e., by solving

$$\min_k \ \mathsf{E}|\tilde{\boldsymbol{s}}(t)|^2$$

where

$$\tilde{\boldsymbol{s}}(t) = s(t) - \hat{\boldsymbol{s}}(t) \ = \ s(t) - k^*\boldsymbol{y} = s(t) - k^*[hs(t) + \boldsymbol{v}(t)] = -k^*\boldsymbol{v}$$

since $k^*h = 1$. Therefore, the second requirement on k is to minimize the variance k^*R_vk, where R_v is the covariance matrix of $\boldsymbol{v}$.

The beamforming problem, also known as the *linearly-constrained minimum-variance* problem, or as the *minimum-variance distortionless-response* problem, is then to solve

$$\boxed{\min_k \ k^*R_vk \quad \text{subject to} \quad k^*h = 1}$$

Using the result in Remark 6.1, we find that the optimal choice for k is

$$\boxed{k_o^* = \frac{h^*R_v^{-1}}{h^*R_v^{-1}h}}$$

in which case the beamformer output is

$$\boxed{\hat{\boldsymbol{s}}(t) = \frac{h^*R_v^{-1}}{h^*R_v^{-1}h} \cdot \boldsymbol{y}}$$

and the resulting m.m.s.e. is

$$\boxed{\text{m.m.s.e.} \ = \ \frac{1}{h^*R_v^{-1}h}}$$

If the noise at the antennas is spatially white, i.e., if the noise at each sensor is uncorrelated with the noises at the other sensors, and if the variance of all noises is σ_v^2, then $R_v = \sigma_v^2 \mathrm{I}$. In this case, the output of the beamformer becomes

$$\hat{\boldsymbol{s}}(t) = \frac{h^*R_v^{-1}}{h^*R_v^{-1}h} \cdot \boldsymbol{y} \ = \ \frac{1}{M} \sum_{n=0}^{M-1} \boldsymbol{y}_n(t) e^{-j\frac{2\pi n}{\lambda} d\cos\theta}$$

where we have denoted the individual entries of $\boldsymbol{y}$ by $\boldsymbol{y}_n(t)$. This expression shows that, in the case of white spatial noise, the beamformer first aligns the phases of the signals it receives and then averages them.

CHAPTER 7

Kalman Filter

The theory developed in Chapters 3–5 on linear estimation can be used to introduce one of the most celebrated tools in linear least-mean-squares estimation theory, namely the Kalman filter. The filter has an intimate relation with adaptive filter theory, so much so that a solid understanding of its functionality can suggest extensions of classical adaptive schemes. A demonstration to this effect will be given later in Chapter 31, after we have progressed sufficiently enough in our treatment of adaptive filters. At that stage, we shall tie up the Kalman filter with adaptive least-squares theory and show how it can motivate useful extensions. For the time being, it suffices to treat the material in this chapter as simply an application of linear least-mean-squares estimation theory.

7.1 INNOVATIONS PROCESS

Consider two zero-mean random variables $\{\boldsymbol{x}, \boldsymbol{y}\}$. We already know from Thm. 3.1 that the linear least-mean-squares estimator of $\boldsymbol{x}$ given $\boldsymbol{y}$ is $\hat{\boldsymbol{x}} = K_o\boldsymbol{y}$, where K_o is any solution to the normal equations

$$K_o R_y = R_{xy} \tag{7.1}$$

In the sequel we assume that R_y is positive-definite so that K_o is uniquely defined as $K_o = R_{xy}R_y^{-1}$.

Usually, the variable $\boldsymbol{y}$ is vector-valued, say, $\boldsymbol{y} = \text{col}\{\boldsymbol{y}_0, \boldsymbol{y}_1, \ldots, \boldsymbol{y}_N\}$, where each $\boldsymbol{y}_i$ is also possibly a vector. Now assume that we could somehow replace $\boldsymbol{y}$ by another vector $\boldsymbol{e}$ of similar dimensions, say,

$$\boldsymbol{e} = A\boldsymbol{y} \tag{7.2}$$

for some lower triangular invertible matrix A. Assume further that the transformation A could be chosen such that the entries of $\boldsymbol{e}$, denoted by $\boldsymbol{e} = \text{col}\{\boldsymbol{e}_0, \boldsymbol{e}_1, \ldots, \boldsymbol{e}_N\}$, are uncorrelated with each other, i.e.,

$$\mathsf{E}\,\boldsymbol{e}_i\boldsymbol{e}_j^* \overset{\Delta}{=} R_{e,i}\delta_{ij}$$

where δ_{ij} denotes the Kronecker delta function that is unity when $i = j$ and zero otherwise, and $R_{e,i}$ denotes the covariance matrix of $\boldsymbol{e}_i$. Then the covariance matrix of $\boldsymbol{e}$ will be block diagonal,

$$R_e \overset{\Delta}{=} \mathsf{E}\,\boldsymbol{e}\boldsymbol{e}^* = \text{diag}\{R_{e,0}, R_{e,1}, \ldots, R_{e,N}\}$$

and, in addition, the problem of estimating $\boldsymbol{x}$ from $\boldsymbol{y}$ would be equivalent to the problem of estimating $\boldsymbol{x}$ from $\boldsymbol{e}$. To see this, let $\hat{\boldsymbol{x}}_{|\boldsymbol{e}}$ denote the linear least-mean-squares estimator of $\boldsymbol{x}$ given $\boldsymbol{e}$, i.e.,

$$\hat{\boldsymbol{x}}_{|\boldsymbol{e}} = R_{xe}R_e^{-1}\boldsymbol{e} \tag{7.3}$$

Likewise, let $\hat{x}_{|y}$ denote the estimator of x given y,

$$\hat{x}_{|y} = R_{xy}R_y^{-1}y \tag{7.4}$$

Then since

$$R_e = \mathsf{E}ee^* = A(\mathsf{E}yy^*)A^* = AR_yA^*$$

and

$$R_{xe} = \mathsf{E}xe^* = (\mathsf{E}xy^*)A^* = R_{xy}A^*$$

we find that

$$\hat{x}_{|e} = R_{xe}R_e^{-1}e = R_{xy}A^*(AR_yA^*)^{-1}e = R_{xy}R_y^{-1}A^{-1}e = R_{xy}R_y^{-1}y$$

That is,

$$\boxed{\hat{x}_{|e} = \hat{x}_{|y}} \tag{7.5}$$

as claimed. This means that we can replace the problem of estimating x from y by the problem of estimating x from e. The key advantage of working with e instead of y is that R_e in (7.3) is block-diagonal and, hence, the estimator $\hat{x}_{|e}$ can be evaluated as the combined sum of individual estimators. Specifically, expression (7.3) gives

$$\hat{x}_{|e} = \sum_{i=0}^{N} (\mathsf{E}xe_i^*)R_{e,i}^{-1}e_i = \sum_{i=0}^{N} \hat{x}_{|e_i}$$

This result shows that we can estimate x from y by estimating x individually from each e_i and then combining the resulting estimators. In particular, if we replace the notations $\hat{x}_{|e}$ and $\hat{x}_{|y}$ by the more suggestive notation $\hat{x}_{|N}$, in order to indicate that the estimator of x is based on the observations y_0 through y_N, then the above expression shows that

$$\hat{x}_{|N} = \sum_{i=0}^{N} \hat{x}_{|e_i}, = \hat{x}_{|e_N} + \sum_{i=0}^{N-1} \hat{x}_{|e_i}$$

where the last sum on the right-hand side is simply the estimator of x using the observations y_0 through y_{N-1}. It follows that

$$\hat{x}_{|N} = \hat{x}_{|N-1} + \hat{x}_{|e_N}$$

i.e.,

$$\boxed{\hat{x}_{|N} = \hat{x}_{|N-1} + (\mathsf{E}xe_N^*)R_{e,N}^{-1}e_N} \tag{7.6}$$

This is a useful recursive formula; it shows how the estimator of x can be updated recursively by adding the contribution of the most recent variable e_N.

The question now is how to generate the variables $\{e_i\}$ from the $\{y_i\}$. One possible transformation is the so-called Gram-Schmidt procedure. Let $\hat{y}_{i|i-1}$ denote the estimator of y_i that is based on the observations up to time $i-1$, i.e., on $\{y_0, y_1, \ldots, y_{i-1}\}$. The same argument that led to (7.5) shows that $\hat{y}_{i|i-1}$ can be alternatively calculated by estimating y_i from $\{e_0, \ldots, e_{i-1}\}$. Then we can construct e_i as

$$\boxed{e_i \overset{\Delta}{=} y_i - \hat{y}_{i|i-1}} \tag{7.7}$$

That is, we can choose e_i as the estimation error that results from estimating y_i from the observations $\{y_0, y_1 \ldots, y_{i-1}\}$. In order to verify that the resulting $\{e_i\}$ are uncorrelated with each other, we recall that, by virtue of the orthogonality condition of linear least-mean-squares estimation (cf. Thm. 4.1),

$$e_i \perp \{y_0, y_1, \ldots, y_{i-1}\}$$

That is, e_i is uncorrelated with the observations $\{y_0, y_1 \ldots, y_{i-1}\}$. It then follows that e_i should be uncorrelated with any e_j for $j < i$ since, by definition, e_j is a linear combination of the observations $\{y_0, y_1 \ldots, y_j\}$ and, moreover,

$$\{y_0, y_1, \ldots, y_j\} \subset \{y_0, y_1, \ldots, y_{i-1}\} \qquad \text{for} \quad j < i$$

By the same token, e_i is uncorrelated with any e_j for $j > i$.

It is instructive to see what choice of a transformation A in (7.2) corresponds to the use of the Gram-Schmidt procedure. Assume, for illustration purposes, that $N = 2$. Then writing (7.7) for $i = 0, 1, 2$ we get

$$\begin{bmatrix} e_0 \\ e_1 \\ e_2 \end{bmatrix} = \begin{bmatrix} \mathrm{I} & & \\ -(\mathsf{E}\, y_1 y_0^*)(\mathsf{E}\, y_0 y_0^*)^{-1} & \mathrm{I} & \\ \times & \times & \mathrm{I} \end{bmatrix} \begin{bmatrix} y_0 \\ y_1 \\ y_2 \end{bmatrix}$$

where the entries $\times$ arise from the calculation

$$\begin{bmatrix} \times & \times \end{bmatrix} = -\left(\mathsf{E}\, y_2 \begin{bmatrix} y_0^* & y_1^* \end{bmatrix}\right) \left(\mathsf{E} \begin{bmatrix} y_0 \\ y_1 \end{bmatrix} \begin{bmatrix} y_0 \\ y_1 \end{bmatrix}^*\right)^{-1}$$

We thus find that A is a lower triangular transformation with unit entries along its diagonal. The lower triangularity of A is relevant since it translates into a causal relationship between the $\{e_i\}$ and the $\{y_i\}$. By causality we mean that each e_i can be computed from $\{y_j, j \le i\}$ and, similarly, each y_i can be recovered from $\{e_j, j \le i\}$. We also see from the construction (7.7) that we can regard e_i as the "new information" in y_i given $\{y_0, \ldots, y_{i-1}\}$. Therefore, it is customary to refer to the $\{e_i\}$ as the innovations process associated with the $\{y_i\}$.

7.2 STATE-SPACE MODEL

As we now proceed to show, the Kalman filter is an efficient procedure for determining the innovations when the observation process $\{y_i\}$ arises from a finite-dimensional linear state-space model.

What we mean by a state-space model for $\{y_i\}$ is the following. We assume that y_i satisfies an equation of the form

$$\boxed{y_i = H_i x_i + v_i, \quad i \ge 0} \tag{7.8}$$

in terms of an $n \times 1$ so-called state-vector x_i, which in turn obeys a recursion of the form

$$\boxed{x_{i+1} = F_i x_i + G_i n_i, \quad i \ge 0} \tag{7.9}$$

The processes v_i and n_i are assumed to be $p \times 1$ and $m \times 1$ zero-mean white noise processes, respectively, with covariances and cross-covariances denoted by

$$\mathsf{E} \begin{bmatrix} n_i \\ v_i \end{bmatrix} \begin{bmatrix} n_j \\ v_j \end{bmatrix}^* \triangleq \begin{bmatrix} Q_i & S_i \\ S_i^* & R_i \end{bmatrix} \delta_{ij}$$

whereas the initial state $\boldsymbol{x}_0$ is assumed to have zero mean, covariance matrix Π_0, and to be uncorrelated with $\{\boldsymbol{n}_i\}$ and $\{\boldsymbol{v}_i\}$, i.e.,

$$\mathsf{E}\,\boldsymbol{x}_0\boldsymbol{x}_0^* = \Pi_0, \quad \mathsf{E}\,\boldsymbol{n}_i\boldsymbol{x}_0^* = 0, \quad \text{and} \quad \mathsf{E}\,\boldsymbol{v}_i\boldsymbol{x}_0^* = 0 \quad \text{for all } i \geq 0$$

The assumptions on $\{\boldsymbol{x}_0, \boldsymbol{n}_i, \boldsymbol{v}_i\}$ can be compactly restated as

$$\boxed{\mathsf{E}\begin{bmatrix}\boldsymbol{n}_i\\ \boldsymbol{v}_i\\ \boldsymbol{x}_0\\ 1\end{bmatrix}\begin{bmatrix}\boldsymbol{n}_j\\ \boldsymbol{v}_j\\ \boldsymbol{x}_0\end{bmatrix}^* = \begin{bmatrix}\begin{bmatrix}Q_i & S_i\\ S_i^* & R_i\end{bmatrix}\delta_{ij} & \begin{matrix}0\\0\end{matrix}\\ \begin{matrix}0 & 0\end{matrix} & \Pi_0\\ \begin{matrix}0 & 0\end{matrix} & 0\end{bmatrix}} \tag{7.10}$$

It is also assumed that the matrices

$$F_i\ (n{\times}n),\quad G_i\ (n{\times}m),\quad H_i\ (p{\times}n),\quad Q_i\ (m{\times}m),\quad R_i\ (p{\times}p),\quad S_i\ (m{\times}p),\quad \Pi_0\ (n{\times}n)$$

are known *a priori*. The process $\boldsymbol{v}_i$ is called measurement noise and the process $\boldsymbol{n}_i$ is called process noise. We now examine how the innovations $\{\boldsymbol{e}_i\}$ of a process $\{\boldsymbol{y}_i\}$ satisfying a state-space model of the form (7.8)–(7.10) can be evaluated.

7.3 RECURSION FOR THE STATE ESTIMATOR

Let $\{\hat{\boldsymbol{y}}_{i|i-1}, \hat{\boldsymbol{x}}_{i|i-1}, \hat{\boldsymbol{v}}_{i|i-1}\}$ denote the estimators of the variables $\{\boldsymbol{y}_i, \boldsymbol{x}_i, \boldsymbol{v}_i\}$ from the observations $\{\boldsymbol{y}_0, \boldsymbol{y}_1, \ldots, \boldsymbol{y}_{i-1}\}$, respectively. Then using $\boldsymbol{y}_i = H_i\boldsymbol{x}_i + \boldsymbol{v}_i$, and appealing to linearity, we have

$$\hat{\boldsymbol{y}}_{i|i-1} = H_i\hat{\boldsymbol{x}}_{i|i-1} + \hat{\boldsymbol{v}}_{i|i-1} \tag{7.11}$$

Now the assumptions on our state-space model imply that

$$\boldsymbol{v}_i \perp \boldsymbol{y}_j \quad \text{for} \quad j \leq i-1$$

i.e., $\boldsymbol{v}_i$ is uncorrelated with the observations $\{\boldsymbol{y}_j, j \leq i-1\}$, so that

$$\hat{\boldsymbol{v}}_{i|i-1} = 0$$

This is because from the model (7.8)–(7.9), $\boldsymbol{y}_j$ is a linear combination of the variables $\{\boldsymbol{v}_j, \boldsymbol{n}_{j-1}, \ldots, \boldsymbol{n}_0, \boldsymbol{x}_0\}$, all of which are uncorrelated with $\boldsymbol{v}_i$ for $j \leq i-1$. Consequently,

$$\boldsymbol{e}_i = \boldsymbol{y}_i - \hat{\boldsymbol{y}}_{i|i-1} = \boldsymbol{y}_i - H_i\hat{\boldsymbol{x}}_{i|i-1} \tag{7.12}$$

Therefore, the problem of finding the innovations reduces to one of finding $\hat{\boldsymbol{x}}_{i|i-1}$. For this purpose, we can use (7.6) to write

$$\begin{aligned}\hat{\boldsymbol{x}}_{i+1|i} &= \hat{\boldsymbol{x}}_{i+1|i-1} + (\mathsf{E}\,\boldsymbol{x}_{i+1}\boldsymbol{e}_i^*)\,R_{e,i}^{-1}\boldsymbol{e}_i\\ &= \hat{\boldsymbol{x}}_{i+1|i-1} + (\mathsf{E}\,\boldsymbol{x}_{i+1}\boldsymbol{e}_i^*)\,R_{e,i}^{-1}(\boldsymbol{y}_i - H_i\hat{\boldsymbol{x}}_{i|i-1})\end{aligned} \tag{7.13}$$

where

$$\boxed{R_{e,i} \triangleq \mathsf{E}\,\boldsymbol{e}_i\boldsymbol{e}_i^*} \tag{7.14}$$

But since $\boldsymbol{x}_{i+1}$ obeys the state equation $\boldsymbol{x}_{i+1} = F_i\boldsymbol{x}_i + G_i\boldsymbol{n}_i$, we also obtain, again by linearity, that

$$\hat{\boldsymbol{x}}_{i+1|i-1} = F_i\hat{\boldsymbol{x}}_{i|i-1} + G_i\hat{\boldsymbol{n}}_{i|i-1} = F_i\hat{\boldsymbol{x}}_{i|i-1} + 0 \tag{7.15}$$

since $n_i \perp y_j, j \leq i-1$. By combining Eqs. (7.12)–(7.15) we arrive at the following recursive equations for determining the innovations:

$$\boxed{\begin{array}{rcl} e_i & = & y_i - H_i \hat{x}_{i|i-1} \\ \hat{x}_{i+1|i} & = & F_i \hat{x}_{i|i-1} + K_{p,i} e_i, \quad i \geq 0 \end{array}} \tag{7.16}$$

with initial conditions

$$\hat{x}_{0|-1} = 0, \qquad e_0 = y_0 \tag{7.17}$$

and where we have defined the gain matrix

$$\boxed{K_{p,i} \stackrel{\Delta}{=} (\mathsf{E}\, x_{i+1} e_i^*)\, R_{e,i}^{-1}} \tag{7.18}$$

The subscript "p" indicates that $K_{p,i}$ is used to update a predicted estimator of the state vector. By combining the equations in (7.16) we also find that

$$\boxed{\hat{x}_{i+1|i} = F_{p,i} \hat{x}_{i|i-1} + K_{p,i} y_i, \quad F_{p,i} \stackrel{\Delta}{=} F_i - K_{p,i} H_i, \quad \hat{x}_{0|-1} = 0,\ i \geq 0} \tag{7.19}$$

which shows that in finding the innovations, we actually also have a complete recursion for the state-estimator $\{\hat{x}_{i|i-1}\}$.

7.4 COMPUTING THE GAIN MATRIX

We still need to evaluate $K_{p,i}$ and $R_{e,i}$. To do so, we introduce the state-estimation error $\tilde{x}_{i|i-1} = x_i - \hat{x}_{i|i-1}$, and denote its covariance matrix by

$$\boxed{P_{i|i-1} \stackrel{\Delta}{=} \mathsf{E}\, \tilde{x}_{i|i-1} \tilde{x}_{i|i-1}^*} \tag{7.20}$$

Then, as we are going to see, the $\{K_{p,i}, R_{e,i}\}$ can be expressed in terms of $P_{i|i-1}$ and, in addition, the evaluation of $P_{i|i-1}$ will require propagating a so-called Riccati recursion.

To see this, note first that

$$e_i = y_i - H_i \hat{x}_{i|i-1} = H_i x_i - H_i \hat{x}_{i|i-1} + v_i = H_i \tilde{x}_{i|i-1} + v_i \tag{7.21}$$

Moreover, $v_i \perp \tilde{x}_{i|i-1}$. This is because $\tilde{x}_{i|i-1}$ is a linear combination of the variables $\{v_0, \ldots, v_{i-1}, x_0, n_0, \ldots, n_{i-1}\}$, all of which are uncorrelated with v_i. This claim follows from the definition $\tilde{x}_{i|i-1} = x_i - \hat{x}_{i|i-1}$ and from the fact that $\hat{x}_{i|i-1}$ is a linear combination of $\{y_0, \ldots, y_{i-1}\}$ and x_i is a linear combination of $\{x_0, n_0, \ldots, n_{i-1}\}$. Therefore, we get

$$\boxed{R_{e,i} = \mathsf{E}\, e_i e_i^* = R_i + H_i P_{i|i-1} H_i^*} \tag{7.22}$$

Likewise, since

$$\mathsf{E}\, x_{i+1} e_i^* = F_i (\mathsf{E}\, x_i e_i^*) + G_i (\mathsf{E}\, n_i e_i^*) \tag{7.23}$$

with the terms $\mathsf{E}\, x_i e_i^*$ and $\mathsf{E}\, n_i e_i^*$ given by

$$\begin{array}{rcl} \mathsf{E}\, x_i e_i^* & = & \mathsf{E}\left(\hat{x}_{i|i-1} + \tilde{x}_{i|i-1}\right) e_i^* \\ & = & \mathsf{E}\, \tilde{x}_{i|i-1} e_i^*, \quad \text{since } e_i \perp \hat{x}_{i|i-1} \\ & = & \mathsf{E}\, \tilde{x}_{i|i-1} (H_i \tilde{x}_{i|i-1} + v_i)^* \\ & = & \mathsf{E}\, \tilde{x}_{i|i-1} (H_i \tilde{x}_{i|i-1})^* + 0, \quad \text{since } v_i \perp \tilde{x}_{i|i-1} \\ & = & P_{i|i-1} H_i^* \end{array}$$

and

$$\begin{aligned} \mathsf{E}\,\boldsymbol{n}_i\boldsymbol{e}_i^* &= \mathsf{E}\,\boldsymbol{n}_i(H_i\tilde{\boldsymbol{x}}_{i|i-1}+\boldsymbol{v}_i)^* \\ &= 0+\mathsf{E}\,\boldsymbol{n}_i\boldsymbol{v}_i^*, \quad \text{since } \boldsymbol{n}_i \perp \tilde{\boldsymbol{x}}_{i|i-1} \\ &= S_i \end{aligned}$$

we get

$$\boxed{K_{p,i} = (\mathsf{E}\,\boldsymbol{x}_{i+1}\boldsymbol{e}_i^*)\,R_{e,i}^{-1} = (F_iP_{i|i-1}H_i^* + G_iS_i)R_{e,i}^{-1}} \tag{7.24}$$

7.5 RICCATI RECURSION

Since $\boldsymbol{n}_i \perp \boldsymbol{x}_i$, it can be easily seen from $\boldsymbol{x}_{i+1} = F_i\boldsymbol{x}_i + G_i\boldsymbol{n}_i$ that the covariance matrix of $\boldsymbol{x}_i$ obeys the recursion

$$\Pi_{i+1} = F_i\Pi_iF_i^* + G_iQ_iG_i^*, \quad \Pi_i \triangleq \mathsf{E}\,\boldsymbol{x}_i\boldsymbol{x}_i^* \tag{7.25}$$

Likewise, since $\boldsymbol{e}_i \perp \hat{\boldsymbol{x}}_{i|i-1}$, it can be seen from $\hat{\boldsymbol{x}}_{i+1|i} = F_i\hat{\boldsymbol{x}}_{i|i-1} + K_{p,i}\boldsymbol{e}_i$ that the covariance matrix of $\hat{\boldsymbol{x}}_{i|i-1}$ satisfies the recursion

$$\Sigma_{i+1} = F_i\Sigma_iF_i^* + K_{p,i}R_{e,i}K_{p,i}^*, \quad \Sigma_i \triangleq \mathsf{E}\,\hat{\boldsymbol{x}}_{i|i-1}\hat{\boldsymbol{x}}_{i|i-1}^* \tag{7.26}$$

with initial condition $\Sigma_0 = 0$. Now the orthogonal decomposition

$$\boldsymbol{x}_i = \hat{\boldsymbol{x}}_{i|i-1} + \tilde{\boldsymbol{x}}_{i|i-1} \quad \text{with} \quad \hat{\boldsymbol{x}}_{i|i-1} \perp \tilde{\boldsymbol{x}}_{i|i-1}$$

shows that $\Pi_i = \Sigma_i + P_{i|i-1}$. It is then immediate to conclude that the matrix $P_{i+1|i} = \Pi_{i+1} - \Sigma_{i+1}$ satisfies the recursion

$$\boxed{P_{i+1|i} = F_iP_{i|i-1}F_i^* + G_iQ_iG_i^* - K_{p,i}R_{e,i}K_{p,i}^*, \quad P_{0|-1} = \Pi_0} \tag{7.27}$$

which is known as the Riccati recursion.

7.6 COVARIANCE FORM

In summary, we arrive at the following statement of the Kalman filter, also known as the covariance form of the filter.

Algorithm 7.1 (The Kalman filter) Given observations $\{\boldsymbol{y}_i\}$ that satisfy the state-space model (7.8)–(7.10), the innovations process $\{\boldsymbol{e}_i\}$ can be recursively computed as follows. Start with $\hat{\boldsymbol{x}}_{0|-1} = 0$, $P_{0|-1} = \Pi_0$, and repeat for $i \geq 0$:

$$\begin{aligned} R_{e,i} &= R_i + H_iP_{i|i-1}H_i^* \\ K_{p,i} &= (F_iP_{i|i-1}H_i^* + G_iS_i)R_{e,i}^{-1} \\ \boldsymbol{e}_i &= \boldsymbol{y}_i - H_i\hat{\boldsymbol{x}}_{i|i-1} \\ \hat{\boldsymbol{x}}_{i+1|i} &= F_i\hat{\boldsymbol{x}}_{i|i-1} + K_{p,i}\boldsymbol{e}_i \\ P_{i+1|i} &= F_iP_{i|i-1}F_i^* + G_iQ_iG_i^* - K_{p,i}R_{e,i}K_{p,i}^* \end{aligned}$$

7.7 MEASUREMENT AND TIME-UPDATE FORM

The implementation described in Alg. 7.1 is known as the prediction form of the Kalman filter since it relies on propagating the one-step prediction $\{\hat{\boldsymbol{x}}_{i|i-1}\}$. There is an alternative implementation of the Kalman filter, known as the time- and measurement-update form. It relies on going from $\hat{\boldsymbol{x}}_{i|i-1}$ to $\hat{\boldsymbol{x}}_{i|i}$ (a *measurement-update* step), and on going from $\hat{\boldsymbol{x}}_{i|i}$ to $\hat{\boldsymbol{x}}_{i+1|i}$ (a *time-update* step).

For the measurement-update step, it can be verified by arguments similar to the ones used in deriving the prediction form, that

$$\hat{\boldsymbol{x}}_{i|i} = \hat{\boldsymbol{x}}_{i|i-1} + K_{f,i}\boldsymbol{e}_i, \qquad K_{f,i} \stackrel{\Delta}{=} P_{i|i-1}H_i^* R_{e,i}^{-1} \tag{7.28}$$

with error covariance matrix

$$P_{i|i} \stackrel{\Delta}{=} \mathsf{E}\,\tilde{\boldsymbol{x}}_{i|i}\tilde{\boldsymbol{x}}_{i|i}^* = P_{i|i-1} - P_{i|i-1}H_i^* R_{e,i}^{-1} H_i P_{i|i-1} \tag{7.29}$$

Likewise, for the time-update step we get

$$\hat{\boldsymbol{x}}_{i+1|i} = F_i\hat{\boldsymbol{x}}_{i|i} + G_i S_i R_{e,i}^{-1}\boldsymbol{e}_i \tag{7.30}$$

with

$$P_{i+1|i} = F_i P_{i|i} F_i^* + G_i(Q_i - S_i R_{e,i}^{-1} S_i^*)G_i^* - F_i K_{f,i} S_i^* G_i^* - G_i S_i K_{f,i}^* F_i^* \tag{7.31}$$

When $S_i = 0$, this latter recursion simplifies to $P_{i+1|i} = F_i P_{i|i} F_i^* + G_i Q_i G_i^*$.

Algorithm 7.2 (Time- and measurement-update forms) Given observations $\{\boldsymbol{y}_i\}$ that satisfy the state-space model (7.8), (7.9), and (7.10), the innovations process $\{\boldsymbol{e}_i\}$ can be recursively computed as follows. Start with $\hat{\boldsymbol{x}}_{0|-1} = 0$, $P_{0|-1} = \Pi_0$, and repeat for $i \geq 0$:

$$\begin{aligned}
R_{e,i} &= R_i + H_i P_{i|i-1} H_i^* \\
K_{f,i} &= P_{i|i-1} H_i^* R_{e,i}^{-1} \\
\boldsymbol{e}_i &= \boldsymbol{y}_i - H_i \hat{\boldsymbol{x}}_{i|i-1} \\
\hat{\boldsymbol{x}}_{i|i} &= \hat{\boldsymbol{x}}_{i|i-1} + K_{f,i}\boldsymbol{e}_i \\
\hat{\boldsymbol{x}}_{i+1|i} &= F_i \hat{\boldsymbol{x}}_{i|i} + G_i S_i R_{e,i}^{-1}\boldsymbol{e}_i \\
P_{i|i} &= P_{i|i-1} - P_{i|i-1} H_i^* R_{e,i}^{-1} H_i P_{i|i-1} \\
P_{i+1|i} &= F_i P_{i|i} F_i^* + G_i(Q_i - S_i R_{e,i}^{-1} S_i^*)G_i^* - F_i K_{f,i} S_i^* G_i^* - G_i S_i K_{f,i}^* F_i^*
\end{aligned}$$

Summary and Notes

The chapters in Part II introduce the basic principles underlying (constrained and unconstrained) linear least-mean-squares estimation with several design examples. The most relevant results are summarized below.

SUMMARY OF MAIN RESULTS

1. Given two zero-mean random variables $\{\boldsymbol{x}, \boldsymbol{y}\}$, the optimal linear estimator (optimal in the least-mean-squares sense) of $\boldsymbol{x}$ given $\boldsymbol{y}$ is $\hat{\boldsymbol{x}} = K_o\boldsymbol{y}$, where K_o is any solution to the normal equations $K_oR_y = R_{xy}$. This construction minimizes the error covariance matrix (or, equivalently, its trace), and the resulting minimum mean-square error matrix is m.m.s.e. $= R_x - K_oR_{yx}$.
2. The normal equations are always consistent, i.e., they always admit a solution K_o. The solution is unique only when R_y is positive-definite; otherwise there are infinitely many solutions.
3. No matter which solution we pick for K_o (when many solutions exist), the resulting estimator and minimum mean-square error values remain invariant.
4. The optimal linear estimator satisfies the orthogonality condition $\tilde{\boldsymbol{x}} \perp \boldsymbol{y}$. That is, the error is orthogonal to the observations, and to any linear transformation of the observations for that matter. In particular, $\tilde{\boldsymbol{x}} \perp \hat{\boldsymbol{x}}$.
5. If the variables $\{\boldsymbol{x}, \boldsymbol{y}\}$ do not have zero means, we can replace them by their centered versions, $\{\boldsymbol{x} - \bar{x}, \boldsymbol{y} - \bar{y}\}$, and then find $\hat{\boldsymbol{x}}$ from $\hat{\boldsymbol{x}} - \bar{x} = K_o(\boldsymbol{y} - \bar{y})$.
6. When $\{\boldsymbol{x}, \boldsymbol{y}\}$ are related by a linear model, say $\boldsymbol{y} = H\boldsymbol{x} + \boldsymbol{v}$, then the optimal linear estimator of $\boldsymbol{x}$ given $\boldsymbol{y}$ can be determined from either expression

$$\hat{\boldsymbol{x}} \quad = \quad R_xH^*\left[R_v + HR_xH^*\right]^{-1}\boldsymbol{y} \quad = \quad \left[R_x^{-1} + H^*R_v^{-1}H\right]^{-1}H^*R_v^{-1}\boldsymbol{y}$$

   where it is assumed that $R_v > 0$ and $R_x > 0$. The resulting minimum mean-square error is m.m.s.e. $= \left[R_x^{-1} + H^*R_v^{-1}H\right]^{-1}$.
7. Given $\boldsymbol{y} = Hx + \boldsymbol{v}$, where H is a tall full rank matrix and x is an unknown deterministic quantity that we wish to estimate, the minimum variance unbiased estimator of x is given by $\hat{\boldsymbol{x}} = (H^*R_v^{-1}H)^{-1}H^*R_v^{-1}\boldsymbol{y}$. The corresponding minimum mean-square error is m.m.s.e. $= (H^*R_v^{-1}H)^{-1}$.
8. The solution to a generic constrained optimization problem of the form

$$\min_K \ KR_vK^* \quad \text{subject to} \quad KH = \mathrm{I} \ \text{ and } \ R_v > 0$$

   is given by $K_o = (H^*R_v^{-1}H)^{-1}H^*R_v^{-1}$, with the minimum cost equal to $(H^*R_v^{-1}H)^{-1}$.
9. Such constrained estimation problems are useful in the design of decision-feedback equalizers and antenna beamformers. The expressions for the feedforward and feedback filters in a decision-feedback equalizer are given in Sec. 6.4, while the expression for the gain vector of an antenna beamformer is given in Sec. 6.5.

BIBLIOGRAPHIC NOTES

Linear estimation. In this part we covered the basics of linear estimation theory, and highlighted those concepts that are most relevant to the subject matter of the book. However, it should be noted that linear least-mean-squares-error estimation is a rich field and it has had a distinguished history. The pioneering work in the area was done independently by Kolmogorov (1939,1941a,b) and Wiener (1942,1949). Kolmogorov was motivated by the work of Wold (1938) on stationary processes and solved a so-called linear prediction problem for discrete-time stationary random processes (see, e.g., Probs. II.31 and II.32 for some discussion on prediction problems). Wiener, on the other hand, solved a continuous-time prediction problem under causality constraints by means of an elegant technique now known as the Wiener-Hopf technique (Wiener and Hopf (1931)). Appendix 3.C of Sayed (2003) describes one particular causal estimation problem that exemplifies some of the elegance involved in the solution of Wiener and Hopf; the appendix focuses on a discrete finite-time horizon estimation problem, whereas Wiener and Hopf (1931) studied the more demanding continuous-time infinite horizon case. Unfortunately, in the literature of adaptive filtering, there seems to exist a persistent confusion with regards to Wiener's contribution. Most authors often mistakenly refer to the normal equations (3.9) as Wiener's solution. As is shown in App. 3.C of Sayed (2003), Wiener solved a deeper and more elaborate problem. Readers interested in more details about Wiener's contribution may consult the textbook by Kailath, Sayed, and Hassibi (2000), which offers a detailed treatment of the subject.

Wiener and Kalman filters. The studies of Kolmogorov and Wiener laid the foundations for most of the subsequent developments in linear estimation theory. In particular, their investigations were fundamental to the development of the Kalman filter two decades later in elegant articles by Kalman (1960) and Kalman and Bucy (1961). While the works of Wiener and Kolmogorov focused on stationary random processes, Kalman's filter had the powerful feature of being the optimal filter for both stationary *and* nonstationary processes. For this reason, the Kalman filter is considered in many respects to be one of the best successes of linear estimation theory. The filter also plays a prominent role in the context of adaptive filtering, as we shall indicate later in Parts VII–X on least-squares theory. In preparation for these discussions, we provided a derivation of the Kalman filter in Chapter 7 for discrete-time models by using the innovations approach of Kailath (1968); this approach exploits to great advantage the orthogonality condition of linear least-mean-squares estimation. As is shown in Chapter 7, rather than determine the optimal steady-state filter for stationary processes, as was the case in Wiener's work, Kalman devised a recursive algorithm that is optimal during all stages of adaptation and which, in steady-state, tends to Wiener's solution. In Kalman's filter, recursion (7.27) plays a prominent role in propagating the error covariance matrix, and we shall encounter a special case of it later in Chapter 30 when we study the recursive least-squares algorithm (RLS). Kalman termed this recursion the Riccati recursion in analogy to a differential equation attributed to Riccati (1724), and later used by Legendre in the calculus of variations and by Bellman (1957) in optimal control theory.

Stochastic modeling. What is particularly significant about the works of Kolmogorov, Wiener, and Kalman is the insight that the problem of separating signals from noise, as well as prediction and filtering problems, can be approached *statistically*. In other words, they can be approached by formulating statistical performance measures and by modelling the variables involved as *random* rather than deterministic quantities (as was the case with our formulation of the mean-square-error criterion in this chapter). This point of view is in clear contrast, for example, to deterministic (least-squares-based) estimation techniques studied earlier by Gauss (1809) and Legendre (1805,1810), and which we shall study in some detail starting in Chapter 29. [An overview of the historical progress from Gauss' work to Kalman's work can be found in Sorenson (1970) and in the edited volume by Sorenson (1985) — see also Kailath (1974).] The insight of formulating estimation problems *stochastically* in this manner has had a significant impact on the fields of signal processing, communications, and control. In particular, it has led to the development of several disciplines that nowadays go by the names of statistical signal processing, statistical communications theory, optimal stochastic control, and stochastic system theory, as can be seen, for instance, by examining the titles of some of the references in these fields (e.g., Lee (1960), Aström (1970), Caines (1988), Scharf (1991), and Kay (1993)).

Linear models. In this part we did not treat the linear least-mean-squares estimation problem in its generality; we only focused on the part of the material that is relevant to the development of adaptive filters in subsequent chapters. In particular, we emphasized the role played by the orthogonality principle (Sec. 4.2), as well as the simplifications that result when the underlying variables are related via a linear model (Sec. 5.1). Such linear models will play a crucial role throughout our development of adaptive filtering in this book. For instance, in Sec. 5.4 we already showed how the linear model can be exploited to derive compact closed-form expressions for the design of finite-length linear equalizers.

Matrix inversion formula. The matrix inversion formula (5.4) is often attributed to Woodbury (1950). One of the first uses of the formula in the context of filtering theory is due to Kailath (1960a) and Ho (1963) — see the article by Henderson and Searle (1981) for an account on the origin of the formula.

Space-time coding. Linear least-mean-squares estimation is useful in many applications. To illustrate this fact, in Probs. II.24 and II.38, we show its relevance to a transmit diversity signaling technique devised by Alamouti (1998). The scheme exploits spatial diversity and results in a simple receiver structure. Due to its many attractive features, this signaling technique has been adopted in several wireless standards for code division multiple access (CDMA) communications such as WCDMA and CDMA2000. A brief overview of CDMA systems is provided in Computer Project 11.3 of Sayed (2003).

Single-carrier-frequency-domain equalization. In Prob. II.28 we also show how linear least-mean-squares estimation is useful in the context of single-carrier-frequency-domain equalization, which involves the addition of a cyclic prefix to the data before transmission in order to simplify the structure of the receiver. One of the first works on single-carrier-frequency-domain equalization is that of Walzman and Schwartz (1973). More recent works appear in Sari, Karam, and Jeanclaude (1995), Clark (1998), Al-Dhahir (2001), and Younis, Sayed, and Al-Dhahir (2003). Cyclic prefixing is also useful in the context of orthogonal frequency-division multiplexing (OFDM), where the data are first Fourier transformed before the inclusion of a cyclic prefix. OFDM is studied later in Computer Project VII.1.

Unbiased estimators. In all least-mean-squares estimation problems studied in Chapters 1–7, the estimators were required to be unbiased for both cases of deterministic and stochastic unknown variables. Sometimes, the condition of unbiasedness can be a hurdle to minimizing the mean-square error. This is because, as is well known from the statistical literature, there are estimators that are biased but that can achieve smaller error variances than unbiased estimators (see, e.g., Rao (1973), Cox and Hinkley (1974), and Kendall and Stuart (1976–1979)).

Two interesting examples to this effect are the following (the first example is from Kay (1993, pp. 310–311) while the second example is from Rao (1973)). In Sec. 6.2 we studied the problem of estimating the mean value, x, of N measurements $\{\boldsymbol{y}(i)\}$. The minimum-variance unbiased estimator for x was seen to be given by the sample mean estimator:

$$\hat{\boldsymbol{x}}_{\textsf{mvue}} = \frac{1}{N}\sum_{i=0}^{N-1} \boldsymbol{y}(i)$$

The value of x was not restricted in any way; it was only assumed to be an unknown constant and that it could assume any value in the interval $(-\infty, \infty)$ (assuming real-valued data). But what if we know beforehand that x is limited to some interval, say $[-\alpha, \alpha]$ for some finite $\alpha > 0$? One way to incorporate this piece of information into the design of an estimator for x is to perhaps consider the following alternative construction:

$$\breve{x} = \begin{cases} -\alpha & \text{if } \hat{x}_{\mathsf{mvue}} < -\alpha \\ \hat{x}_{\mathsf{mvue}} & \text{if } -\alpha \leq \hat{x}_{\mathsf{mvue}} \leq \alpha \\ \alpha & \text{if } \hat{x}_{\mathsf{mvue}} > \alpha \end{cases}$$

in terms of a realization for $\hat{\boldsymbol{x}}_{\mathsf{mvue}}$. In this way, $\breve{\boldsymbol{x}}$ will always assume values within $[-\alpha, \alpha]$. A calculation in Kay (1993) shows that although the (truncated mean) estimator $\breve{\boldsymbol{x}}$ is biased, it nevertheless satisfies $\mathsf{E}\,(x - \breve{\boldsymbol{x}})^2 < \mathsf{E}\,(x - \hat{\boldsymbol{x}})^2$. In other words, the truncated mean estimator results in a smaller mean-square error.

A second classical example from the realm of statistics is the variance estimator. In this case, the parameter to be estimated is the variance of a random variable $\boldsymbol{y}$ given access to several observations of it, say $\{\boldsymbol{y}(i)\}$. Let σ_y^2 denote the variance of $\boldsymbol{y}$. Two well-known estimators for σ_y^2 are

$$\widehat{\boldsymbol{\sigma}_y^2} = \frac{1}{N}\sum_{i=0}^{N-1}(\boldsymbol{y}(i) - \bar{y})^2 \quad \text{and} \quad \breve{\boldsymbol{\sigma}_y^2} = \frac{1}{N+1}\sum_{i=0}^{N-1}(\boldsymbol{y}(i) - \bar{y})^2$$

where $\bar{y} = \mathsf{E}\,\boldsymbol{y}$. The first one is unbiased while the second one is biased. However, it is shown in Rao (1973) that $\mathsf{E}\,(\sigma_y^2 - \breve{\boldsymbol{\sigma}_y^2})^2 < \mathsf{E}\,(\sigma_y^2 - \widehat{\boldsymbol{\sigma}_y^2})^2$. We therefore see that biased estimators can result in smaller mean-square errors. However, unbiasedness is often a desirable property in practice since it guarantees that, on average, the estimator agrees with the unknown quantity that we seek to estimate.

Channel equalization. The earliest work on equalization methods for digital communications applications dates back to the mid 1960s. It was first proposed by Lucky (1965) in the design of a linear equalizer. Lucky did not use the mean-square-error criterion, as we did in Secs. 5.4 and 6.4, but rather the so-called peak-distortion criterion. Lucky's solution led to what is known as the *zero-forcing* equalizer whereby the equalizer is essentially the inverse of the channel. This design method ignores the presence of noise and, as a result, it can lead to noise amplification and performance degradation. Another early work on equalization is that of DiToro (1965).

Soon afterwards, Widrow (1966), followed by Gersho (1969) and Proakis and Miller (1969), proposed using the mean-square-error criterion instead of the peak-distortion criterion. The mean-square-error criterion takes the noise into account and generally leads to superior performance. Since then, the criterion has been used extensively in the design of optimal equalizer and receiver structures, including decision-feedback equalizers and fractionally-spaced equalizers. The first work on decision-feedback-equalization was that of Austin (1967), followed by Monsen (1971), George, Bowen, and Storey (1971), Price (1972), and Salz (1973). Some of the earliest references on fractionally-spaced equalizers were those by Brady (1970), Ungerboeck (1976), Qureshi and Forney (1977), and Gitlin and Weinstein (1981).

Appendix 3.B of Sayed (2003) motivates the need for equalization methods and explains their role in compensating for the time dispersion introduced by communications channels and in combating inter-symbol interference. Among other results, the appendix explains the origin of the discrete-time channel model (5.10), as well as the concepts of symbol-spaced and fractionally-spaced equalization. The material in the appendix complements the discussions in Secs. 5.4 and 6.4 on the design of linear and decision-feedback equalizers. Further details on equalization techniques can be found in the textbooks by Gitlin, Hayes, and Weinstein (1992) and Proakis (2000). Another accessible reference is the survey article by Qureshi (1985), which deals primarily with adaptive equalization.

Finite-length DFE. A finite-length formulation of the decision-feedback equalizer that is similar to the one presented in Sec. 6.4 can be found in Al-Dhahir and Cioffi (1995). A similar design of decision-feedback equalizers for MIMO systems (e.g., systems that involve multiple transmit and receive antennas) is studied in Probs. II.42–II.45 following the work of Al-Dhahir and Sayed (2000).

Beamforming. Beamforming has applications in several areas including radar, sonar, and communications (see, e.g., the book by Johnson and Dudgeon (1993) and the articles by Van Veen and Buckley (1988) and Krim and Viberg (1996); the last article discusses several other issues in array signal processing and subspace methods).

Problems and Computer Projects

PROBLEMS

Problem II.1 (Rank-one modification of the identity matrix) Consider any matrix of the form $\mathrm{I} + xy^{\mathsf{T}}$, where x and y are column vectors. Use the matrix inversion formula (5.4) to show that its inverse is also a rank-one modification of the identity. More specifically, show that

$$\boxed{\left(\mathrm{I} + xy^{\mathsf{T}}\right)^{-1} = \mathrm{I} - \frac{xy^{\mathsf{T}}}{1 + y^{\mathsf{T}}x}}$$

Problem II.2 (Determinant of a matrix) Consider a square matrix A. A fundamental result in matrix theory is that every such matrix admits a so-called *canonical Jordan decomposition*, which is of the form $A = UJU^{-1}$, where $J = \text{diag}\{J_1, \ldots, J_r\}$ is a block diagonal matrix, say with r blocks. Each J_i is bi-diagonal with identical diagonal entries λ_i, and with unit entries along the lower diagonal, namely,

$$J_i \triangleq \begin{bmatrix} \lambda_i & & & \\ 1 & \lambda_i & & \\ & \ddots & \ddots & \\ & & 1 & \lambda_i \end{bmatrix}$$

The size of each J_i, say $n_i \times n_i$, indicates the multiplicity of the eigenvalue λ_i. Show that $\det A = \prod_{i=1}^{r} \lambda_i^{n_i}$.

Remark. When A is Hermitian, it can be shown that J is necessarily diagonal (rather than block diagonal with bi-diagonal blocks).

Problem II.3 (Trace of a matrix) Use the canonical Jordan factorization of Prob. II.2, and the fact that $\text{Tr}(XY) = \text{Tr}(YX)$ for any matrices $\{X, Y\}$ of compatible dimensions, to show that the trace of a matrix, which is the sum of its diagonal elements, is also equal to the sum of its eigenvalues. What about the determinant of a matrix?

Problem II.4 (Matrix norm) Let A be an $n \times n$ matrix with eigenvalues $\{\lambda_i\}$ and introduce its spectral radius

$$\boxed{\rho(A) \triangleq \max_{1 \leq i \leq n} |\lambda_i|}$$

Let further $A = TJT^{-1}$ denote the Jordan canonical form of A (cf. Prob. II.2), and define the $n \times n$ diagonal matrix $D = \text{diag}\{\epsilon, \epsilon^2, \ldots, \epsilon^n\}$, for any given positive scalar ϵ.

(a) Show that DJD^{-1} has the same form as J with the unit entries of J replaced by ϵ. Conclude that the one norm of DJD^{-1} is equal to $\rho(A) + \epsilon$.

Remark. The one norm of an $n \times n$ matrix B is denoted by $\|B\|_1$ and is defined as

$$\|B\|_1 \triangleq \max_{1 \leq j \leq n} \sum_{i=1}^{n} |b_{ij}|$$

That is, it is equal to the maximum absolute column sum of B.

(b) For any $n \times n$ matrix B, define the function $\|B\|_\rho \triangleq \|DT^{-1}BTD^{-1}\|_1$. Show that the function $\|\cdot\|_\rho$ so defined is a matrix norm (i.e., show that it satisfies the properties of a norm, as defined in Sec. B.6.)

(c) Verify that $\rho(A) \leq \|A\|_\rho \leq \rho(A) + \epsilon$.

Problem II.5 (Minimum of a quadratic form) Consider the quadratic cost function $J(x) = (x-c)^*A(x-c)$, where x and c are column vectors and A is a Hermitian nonnegative-definite matrix. Argue that the minimum value of $J(x)$ is zero and it is achieved at $x = c + d$ for any d satisfying $Ad = 0$.

Problem II.6 (Trace of error covariance matrix) Use (3.5), and expression (3.17) for the optimal value of $\mathsf{E}\,|\tilde{\boldsymbol{x}}(i)|^2$, to justify (3.18).

Problem II.7 (Determinant of error covariance matrix) For any square matrices A and B satisfying $A \geq B \geq 0$, show that $\det A \geq \det B$. Conclude that the linear least-mean-squares estimator of Thm. 3.1 also minimizes the determinant of the error covariance matrix, $\det\,(\mathsf{E}\,\tilde{\boldsymbol{x}}\tilde{\boldsymbol{x}}^*)$.

Remark. We therefore see that the linear least-mean-squares estimator minimizes both the trace and the determinant of the error covariance matrix. This fact is behind the use of the terminologies *arithmetic* SNR and *geometric* SNR as performance measures, which are defined by

$$\text{ASNR} \triangleq \frac{\text{Tr}(R_x)}{\text{Tr}(R_{\tilde{x}})}, \qquad \text{GSNR} \triangleq \left(\frac{\det(R_x)}{\det(R_{\tilde{x}})}\right)^{1/p}$$

where p is the dimension of $\boldsymbol{x}$. The covariance matrix $R_{\tilde{x}}$ denotes the m.m.s.e. of the estimation problem, $R_{\tilde{x}} = R_x - K_o R_{yx}$.

Problem II.8 (Weighted error cost) Show that the estimator of Thm. 3.1 also minimizes $\mathsf{E}\,\tilde{\boldsymbol{x}}^*W\tilde{\boldsymbol{x}}$ for any $W \geq 0$.

Problem II.9 (Independence vs. uncorrelatedness) How would the answer to Ex. 4.4 change if the noise sequence $\{\boldsymbol{v}(i)\}$ were only assumed to be uncorrelated with, rather than independent of, the data $\{\boldsymbol{s}(i)\}$?

Problem II.10 (Second-order statistics) How would the answer to Ex. 4.4 change if the symbols $\{\boldsymbol{s}(i)\}$ were instead chosen uniformly from a QPSK constellation?

Problem II.11 (Nonzero means) Refer again to Ex. 4.4 and assume a generic value for p, $0 \leq p \leq 1$. [In the example, we used $p = 1/2$.] In this case, the random variables $\{\boldsymbol{s}(0), \boldsymbol{s}(1)\}$ do not have zero means anymore, and the linear least-mean-squares estimator of $\boldsymbol{x} = \text{col}\{\boldsymbol{s}(0), \boldsymbol{s}(1)\}$ given $\boldsymbol{y} = \text{col}\{\boldsymbol{y}(0), \boldsymbol{y}(1)\}$ becomes $\hat{\boldsymbol{x}} = \bar{x} + R_{xy}R_y^{-1}(\boldsymbol{y} - \bar{y})$, where

$$\bar{x} = \mathsf{E}\,\boldsymbol{x}, \qquad \bar{y} = \mathsf{E}\,\boldsymbol{y}, \qquad R_{xy} = \mathsf{E}\,(\boldsymbol{x} - \bar{x})(\boldsymbol{y} - \bar{y})^*, \qquad R_y = \mathsf{E}\,(\boldsymbol{y} - \bar{y})(\boldsymbol{y} - \bar{y})^*$$

(a) Find $\{\bar{x}, \bar{y}, R_y, R_{xy}\}$.

(b) Determine $\hat{\boldsymbol{s}}(0)$ and $\hat{\boldsymbol{s}}(1)$.

(c) Simplify the results for $p = 1/2$.

Problem II.12 (Perfect estimation) Consider a linear model of the form $\boldsymbol{d} = w^{o*}\boldsymbol{u} + \boldsymbol{v}$, where $\{w^o, \boldsymbol{u}\}$ are column vectors and $\boldsymbol{v}$ is a scalar. Both $\{\boldsymbol{u}, \boldsymbol{v}\}$ are zero-mean uncorrelated random variables with $R_u = \mathsf{E}\,\boldsymbol{u}\boldsymbol{u}^* > 0$. Moreover, w^o is unknown. Let k^{o*} denote the row vector that defines the linear least-mean-squares estimator of $\boldsymbol{d}$ given $\boldsymbol{u}$, i.e., $\hat{\boldsymbol{d}} = k^{o*}\boldsymbol{u}$. Show that $k^o = w^o$.

Remark. This result has the following useful interpretation. If an observation $\boldsymbol{d}$ happens to be related linearly to some data $\boldsymbol{u}$, then the linear least-mean-squares estimator of $\boldsymbol{d}$ given $\boldsymbol{u}$ identifies *exactly* the unknown vector w^o that relates $\boldsymbol{d}$ to $\boldsymbol{u}$. This result is useful in channel estimation applications, and also in the study of convergence properties of adaptive schemes, as will be shown later in this book — see, e.g., Secs. 10.5 and 15.2.

Problem II.13 (Correlated component) Assume a zero-mean random variable $\boldsymbol{x}$ consists of two components, $\boldsymbol{x} = \boldsymbol{x}_c + \boldsymbol{z}$, and that only $\boldsymbol{x}_c$ is correlated with the observation vector $\boldsymbol{y}$. Show that the linear least-mean-squares estimator of $\boldsymbol{x}$ given $\boldsymbol{y}$ is simply the linear least-mean-squares estimator of $\boldsymbol{x}_c$ given $\boldsymbol{y}$.

Remark. The result shows that the linear least-mean-squares estimator can only estimate that part of $\boldsymbol{x}$ that is correlated with the observation. This remark sounds obvious but it forms the basis of some useful applications — see, e.g., Prob. III.46.

Problem II.14 (Estimation of x^2) Let $\boldsymbol{y} = \boldsymbol{x} + \boldsymbol{v}$, where $\boldsymbol{x}$ and $\boldsymbol{v}$ are independent zero-mean Gaussian real-valued random variables with variances σ_x^2 and σ_v^2, respectively. Show that the linear least-mean-squares estimator of $\boldsymbol{x}^2$ using $\{\boldsymbol{y}, \boldsymbol{y}^2\}$ is

$$\widehat{\boldsymbol{x}^2} = \sigma_x^2 + \frac{\sigma_x^4}{\sigma_x^4 + 2\sigma_x^2\sigma_v^2 + \sigma_v^4} \cdot (\boldsymbol{y}^2 - \sigma_x^2 - \sigma_v^2)$$

Hint: If $\boldsymbol{s}$ is a zero-mean real-valued Gaussian random variable with variance σ_s^2, then it holds that $\mathsf{E}\,\boldsymbol{s}^3 = 0$ and $\mathsf{E}\,\boldsymbol{s}^4 = 3(\mathsf{E}\,\boldsymbol{s}^2)^2 = 3\sigma_s^4$.

Problem II.15 (Interfering signals) Refer to Prob. I.14. Find the linear least-mean-square-error estimator of $\boldsymbol{x}_1$ given $\boldsymbol{y}$.

Problem II.16 (Power distortion through a channel) A zero-mean independent and identically distributed sequence $\{\boldsymbol{s}(\cdot)\}$, with variance σ_s^2, is applied to an FIR channel with impulse response vector c (i.e., c contains the samples of the channel impulse response). Let $\{\boldsymbol{z}(\cdot)\}$ denote the channel output sequence. Verify that the variance of $\boldsymbol{z}(\cdot)$ is given by $\sigma_z^2 = \sigma_s^2\|c\|^2$, in terms of the squared Euclidean norm of c.

Problem II.17 (Linear equalization and delayed estimation) Refer to Ex. 4.4 on linear equalization. Assume now that we wish to determine the optimal coefficients $\{\alpha(0), \alpha(1), \alpha(2)\}$ of the linear equalizer in order to estimate $\boldsymbol{s}(i-\Delta)$, for some Δ. That is, the output of the equalizer in Fig. 4.3 should be changed from $\hat{\boldsymbol{s}}(i)$ to $\hat{\boldsymbol{s}}(i-\Delta)$.

(a) Find R_{xy} for $\Delta = 0, 1, 2, 3$.

(b) Find the optimal equalizer coefficients, and the corresponding minimum mean-square errors, for $\Delta = 0, 1, 2, 3$,

(c) Would it be useful to use values of Δ larger than 3? Why?

(d) Which value of Δ results in the smallest m.m.s.e.? Can you explain why?

(e) The above calculations assume $\sigma_v^2 = 1$. For an arbitrary noise variance σ_v^2, verify that

$$R_y = \begin{bmatrix} \frac{5}{4} + \sigma_v^2 & 1/2 & 0 \\ 1/2 & \frac{5}{4} + \sigma_v^2 & 1/2 \\ 0 & 1/2 & \frac{5}{4} + \sigma_v^2 \end{bmatrix}$$

Does the value of R_{xy} depend on σ_v^2? What about the m.m.s.e.? What about the value of Δ that results in the smallest m.m.s.e.?

(f) The variance of the transmitted data $\{\boldsymbol{s}(i)\}$ is unity. Verify that the variance of the channel output, $\{\boldsymbol{z}(i)\}$, is $\sigma_z^2 = 5/4$.

(g) Compute the SNR at the output of the equalizer, $\text{SNR} = \mathsf{E}\,|\boldsymbol{s}(i-\Delta)|^2/\mathsf{E}\,|\tilde{\boldsymbol{s}}(i-\Delta)|^2$, for the case $\Delta = 1$ and $\sigma_v^2 = 0.05$.

Problem II.18 (Equalizer performance) Refer again to Ex. 4.4 on linear equalization. The SNR at the output of the equalizer is defined as $\text{SNR} = \mathsf{E}\,|\boldsymbol{s}(i)|^2/\mathsf{E}\,|\tilde{\boldsymbol{s}}(i)|^2$. We want to examine the performance of the equalizer as a function of its number of taps, say L. Let σ_v^2 denote the variance of the white noise $\boldsymbol{v}(i)$.

(a) Define the observation vector $\boldsymbol{y} \triangleq \text{col}\{\boldsymbol{y}(i), \boldsymbol{y}(i-1), \ldots, \boldsymbol{y}(i-L+1)\}$, and show that its $L \times L$ covariance matrix is Toeplitz, namely,

$$R_y \triangleq \begin{bmatrix} r_y(0) & r_y(1) & r_y(2) & \ldots \\ r_y(1) & r_y(0) & r_y(1) & \ldots \\ r_y(2) & r_y(1) & r_y(0) & \ldots \\ \vdots & \vdots & \vdots & \ddots \end{bmatrix}_{L \times L}$$

with entries

$$r_y(k) = \begin{cases} 5/4 + \sigma_v^2 & k = 0 \\ 1/2 & k = 1 \\ 0 & 2 \leq k \leq L-1 \end{cases}$$

We shall denote this covariance matrix by $R_y^{(L)}$, with the superscript L indicating the size of the vector $\boldsymbol{y}$.

(b) Let $\boldsymbol{x} = \boldsymbol{s}(i)$. Verify that $R_{xy} = \begin{bmatrix} 1 & 0 & \ldots & 0 \end{bmatrix}_{1 \times L}$.

(c) Show that the m.m.s.e. pertaining to the problem of estimating $\boldsymbol{x}$ from $\boldsymbol{y}$, and the SNR at the output of the equalizer, are given by

$$\text{m.m.s.e.} = 1 - \frac{\det R_y^{(L-1)}}{\det R_y^{(L)}}, \qquad \text{SNR} = \frac{1}{1 - \frac{\det R_y^{(L-1)}}{\det R_y^{(L)}}}$$

Problem II.19 (Useful identity) Let $\{\boldsymbol{x}, \boldsymbol{y}\}$ denote two zero-mean random variables with positive-definite covariance matrices $\{R_x, R_y\}$. Let $\hat{\boldsymbol{x}}$ denote the linear least-mean-squares estimator of $\boldsymbol{x}$ given $\boldsymbol{y}$. Likewise, let $\hat{\boldsymbol{y}}$ denote the linear least-mean-squares estimator of $\boldsymbol{y}$ given $\boldsymbol{x}$. Introduce the estimation errors $\tilde{\boldsymbol{x}} = \boldsymbol{x} - \hat{\boldsymbol{x}}$ and $\tilde{\boldsymbol{y}} = \boldsymbol{y} - \hat{\boldsymbol{y}}$, and denote their covariance matrices by $R_{\tilde{x}}$ and $R_{\tilde{y}}$, respectively.

(a) Show that $R_x R_{\tilde{x}}^{-1} \hat{\boldsymbol{x}} = R_{xy} R_{\tilde{y}}^{-1} \boldsymbol{y}$.

(b) Assume $\{\boldsymbol{y}, \boldsymbol{x}\}$ are related via a linear model of the form $\boldsymbol{y} = H\boldsymbol{x} + \boldsymbol{v}$, where $\boldsymbol{v}$ is zero-mean with covariance matrix R_v and is uncorrelated with $\boldsymbol{x}$. Verify that the identity of part (a) reduces to $R_{\tilde{x}}^{-1} \hat{\boldsymbol{x}} = H^* R_v^{-1} \boldsymbol{y}$.

Problem II.20 (Combining estimators) Let $\boldsymbol{x}$ be a zero-mean random variable with an $M \times M$ positive-definite covariance matrix R_x. Let $\hat{\boldsymbol{x}}_1$ denote the linear least-mean-squares estimator of $\boldsymbol{x}$ given a zero-mean observation $\boldsymbol{y}_1$. Likewise, let $\hat{\boldsymbol{x}}_2$ denote the linear least-mean-squares estimator of the same variable $\boldsymbol{x}$ given a second zero-mean observation $\boldsymbol{y}_2$. That is, we have two separate estimators for $\boldsymbol{x}$ from two separate sources. Let P_1 and P_2 denote the corresponding error covariance matrices, i.e., $P_1 = \mathsf{E}\,\tilde{\boldsymbol{x}}_1 \tilde{\boldsymbol{x}}_1^*$, $P_2 = \mathsf{E}\,\tilde{\boldsymbol{x}}_2 \tilde{\boldsymbol{x}}_2^*$, where $\tilde{\boldsymbol{x}}_j = \boldsymbol{x} - \hat{\boldsymbol{x}}_j$, and assume $P_1 > 0$ and $P_2 > 0$. Assume further that the cross-covariance matrix

$$\mathsf{E}\left(\begin{bmatrix} \boldsymbol{x} \\ \boldsymbol{y}_1 \end{bmatrix} \begin{bmatrix} \boldsymbol{x} \\ \boldsymbol{y}_2 \end{bmatrix}^* \right)$$

has rank M.

(a) Show that the linear least-mean-squares estimator of $\boldsymbol{x}$ given *both* observations $\{\boldsymbol{y}_1, \boldsymbol{y}_2\}$, denoted by $\hat{\boldsymbol{x}}$, satisfies $P^{-1}\hat{\boldsymbol{x}} = P_1^{-1}\hat{\boldsymbol{x}}_1 + P_2^{-1}\hat{\boldsymbol{x}}_2$, where P denotes the resulting error covariance matrix and is given by $P^{-1} = P_1^{-1} + P_2^{-1} - R_x^{-1}$.

Remark. This useful result tells us that the estimators $\{\hat{\boldsymbol{x}}_1, \hat{\boldsymbol{x}}_2\}$ and their error covariances $\{P_1, P_2\}$ can be fused together without the need to access the original measurements $\{\boldsymbol{y}_1, \boldsymbol{y}_2\}$.

(b) Assume $\{\boldsymbol{y}_1, \boldsymbol{x}\}$ and $\{\boldsymbol{y}_2, \boldsymbol{x}\}$ are related via linear models of the form, $\boldsymbol{y}_1 = H_1\boldsymbol{x} + \boldsymbol{v}_1$ and $\boldsymbol{y}_2 = H_2\boldsymbol{x} + \boldsymbol{v}_2$, where $\{\boldsymbol{v}_1, \boldsymbol{v}_2\}$ are zero-mean with covariance matrices $\{R_{v_1}, R_{v_2}\}$ and are uncorrelated with each other and with $\boldsymbol{x}$. Verify that this situation satisfies the required rank-deficiency condition and conclude that the estimator of x given $\{y_1, y_2\}$ is given by the expression in part (a).

Problem II.21 (Linear and optimal estimators) A random variable z is defined as follows

$$z = \begin{cases} -x & \text{with probability } p \\ Hx + v & \text{with probability } 1-p \end{cases}$$

where x and v are zero-mean uncorrelated random variables. Assume we know the linear least-mean-squares estimator of x given y, namely, $\hat{x}_{|y}$, where y is a zero-mean random random variable that is also uncorrelated with v.

(a) Find an expression for $\hat{z}_{|y}$ in terms of $\hat{x}_{|y}$.

(b) More generally, what can you say about the relation between $\mathsf{E}\,(z|y)$ and $\mathsf{E}\,(x|y)$?

(c) Find an expression for the linear least-mean-squares estimator $\hat{x}_{|z}$ and the corresponding m.m.s.e.

(d) How would your answers to parts (a) and (b) change if the random variables $\{v, y\}$ were jointly circular and Gaussian?

Problem II.22 (Distributed processing) Consider a distributed network with m nodes as shown in Fig. II.1. Each node k observes a zero-mean measurement y_k that is related to an unknown zero-mean variable x via a linear model of the form $y_k = H_k x + v_k$, where the data matrix H_k is known, and the noise v_k is zero mean and uncorrelated with x. The noises across all nodes are uncorrelated with each other. Let $\{R_x, R_{v_k}\}$ denote the positive-definite covariance matrices of $\{x, v_k\}$, respectively. Introduce the following notation:

- At each node k, the notation $\hat{x}_k$ denotes the linear least-mean-squares estimator of x that is based on the observation y_k. Likewise, P_k denotes the resulting error covariance matrix, $P_k = \mathsf{E}\,\tilde{x}_k\tilde{x}_k^*$.
- At each node k, the notation $\hat{x}_{1:k}$ denotes the linear least-mean-squares estimator of x that is based on the observations $\{y_1, y_2, \ldots, y_k\}$, i.e., on the observations collected at nodes 1 through k. Likewise, $P_{1:k}$ denotes the resulting error covariance matrix, $P_{1:k} = \mathsf{E}\,\tilde{x}_{1:k}\tilde{x}_{1:k}^*$.

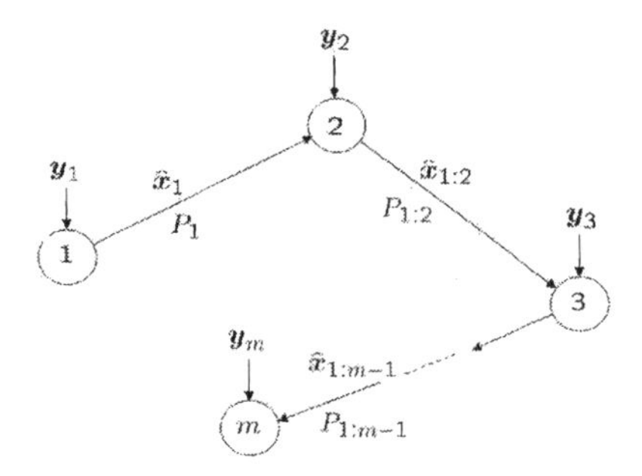

FIGURE II.1 A distributed estimation network.

The network functions as follows. Node 1 uses y_1 to estimate x. The resulting estimator, $\hat{x}_1$, and the corresponding error covariance matrix, $P_1 = \mathsf{E}\,\tilde{x}_1\tilde{x}_1^*$, are transmitted to node 2. Node 2 in turn uses its measurement y_2 and the data $\{\hat{x}_1, P_1\}$ received from node 1 to compute the estimator of x that is based on both observations $\{y_1, y_2\}$. Note that node 2 *does not* have access to y_1 but only to y_2 and the information received from node 1. The estimator computed by node 2, $\hat{x}_{1:2}$, and the corresponding error covariance matrix, $P_{1:2}$, are then transmitted to node 3. Node 3 evaluates $\{\hat{x}_{1:3}, P_{1:3}\}$ using $\{y_3, \hat{x}_{1:2}, P_{1:2}\}$ and transmits $\{\hat{x}_{1:3}, P_{1:3}\}$ to node 4 and so forth.

(a) Find an expression for $\hat{x}_{1:m}$ in terms of $\hat{x}_{1:m-1}$ and $\hat{x}_m$.

(b) Find an expression for $P_{1:m}^{-1}$ in terms of $\{P_{1:m-1}^{-1}, P_m^{-1}, R_x^{-1}\}$.

(c) Find a recursion relating $P_{1:m}$ to $P_{1:m-1}$.

(d) Show that $P_{1:m}$ is a non-increasing sequence as a function of m.

(e) Assume $H_k = H$ for all k and $R_{v_k} = R_v > 0$. Assume further that H is tall and has full column rank. Find

$$\lim_{m\to\infty} P_{1:m}$$

Problem II.23 (Random telegraph signal) The probability distribution function of a Poisson process with rate λ is defined by

$$P(N=k) = \frac{(\lambda t)^k}{k!}e^{-\lambda t}, \qquad k = 0,1,2,\ldots$$

where λ is the average number of events per second and N is the number of events that can occur in the interval of time $[0,t]$. Now consider a random process $\boldsymbol{x}(t)$ that is generated as follows. Its initial value $\boldsymbol{x}(0)$ is ± 1 with probability $1/2$, and it changes polarity thereafter with each occurrence of an event in a Poisson process.

(a) Show that the probability that $\boldsymbol{x}(t) = +1$ given that $\boldsymbol{x}(0) = +1$ is given by the expression $P[\boldsymbol{x}(t) = +1|\boldsymbol{x}(0) = +1] = \left(1 + e^{-2\lambda t}\right)/2$. Repeat for $P[\boldsymbol{x}(t) = -1|\boldsymbol{x}(0) = -1]$.

(b) Show also that $P[\boldsymbol{x}(t) = -1|\boldsymbol{x}(0) = +1] = \left(1 - e^{-2\lambda t}\right)/2$. Repeat for $P[\boldsymbol{x}(t) = +1|\boldsymbol{x}(0) = -1]$.

(c) Show that, for any t, $P[\boldsymbol{x}(t) = +1] = P[x(t) = -1] = 1/2$.

(d) Show that $\boldsymbol{x}(t)$ is a zero-mean stationary random process with auto-correlation function $R_x(\tau) = e^{-2\lambda|\tau|}$.

(e) Show that the linear least-mean-squares estimator of $\boldsymbol{x}(0)$ given both $\boldsymbol{x}(T)$ and $\boldsymbol{x}(2T)$ is

$$\widehat{\boldsymbol{x}(0)} = \frac{e^{-2\lambda T} - e^{-6\lambda T}}{1 - e^{-4\lambda T}}\boldsymbol{x}(T)$$

Problem II.24 (Space-time coding) Consider a two-transmit one-receive antenna system, as shown in Fig. II.2. Let α denote the channel gain from transmit antenna one to the receiver and let β denote the channel gain from the transmit antenna two to the same receiver. The channels between the two transmitters and the receiver simply scale the transmitted data by the scalar gains $\{\alpha, \beta\}$; we say that they are flat channels since their frequency responses will be flat. Both gains $\{\alpha, \beta\}$ are assumed known. In space-time coding, symbols $\{\boldsymbol{s}_1(i), \boldsymbol{s}_2(i)\}$ are transmitted at time i from the two antennas followed by the symbols $\{-\boldsymbol{s}_2^*(i), \boldsymbol{s}_1^*(i)\}$ at time $i+1$. The corresponding received signals will therefore be

$$\boldsymbol{r}(i) = \alpha\boldsymbol{s}_1(i) + \beta\boldsymbol{s}_2(i) + \boldsymbol{v}(i), \qquad \boldsymbol{r}(i+1) = -\alpha\boldsymbol{s}_2^*(i) + \beta\boldsymbol{s}_1^*(i) + \boldsymbol{v}(i+1)$$

The noise sequence $\{\boldsymbol{v}(\cdot)\}$ is white with variance σ_v^2 and uncorrelated with the data $\{\boldsymbol{s}_1(\cdot), \boldsymbol{s}_2(\cdot)\}$. The symbols $\{\boldsymbol{s}_1(\cdot), \boldsymbol{s}_2(\cdot)\}$ are assumed independent and of variance σ_s^2 each.

(a) Verify that

$$\begin{bmatrix} \boldsymbol{r}(i) \\ \boldsymbol{r}^*(i+1) \end{bmatrix} = \underbrace{\begin{bmatrix} \alpha & \beta \\ \beta^* & -\alpha^* \end{bmatrix}}_{H} \begin{bmatrix} \boldsymbol{s}_1(i) \\ \boldsymbol{s}_2(i) \end{bmatrix} + \begin{bmatrix} \boldsymbol{v}(i) \\ \boldsymbol{v}^*(i+1) \end{bmatrix}$$

and that the matrix H satisfies $H^*H = (|\alpha|^2 + |\beta|^2)\mathrm{I}$.

(b) Show that the linear least-mean-squares estimators of $\{\boldsymbol{s}_1(i), \boldsymbol{s}_2(i)\}$ given both measurements $\{\boldsymbol{r}(i), \boldsymbol{r}(i+1)\}$ are given by

$$\begin{bmatrix} \hat{\boldsymbol{s}}_1(i) \\ \hat{\boldsymbol{s}}_2(i) \end{bmatrix} = \frac{\sigma_s^2}{\sigma_s^2(|\alpha|^2 + |\beta|^2) + \sigma_v^2} \cdot H^* \begin{bmatrix} \boldsymbol{r}(i) \\ \boldsymbol{r}^*(i+1) \end{bmatrix}$$

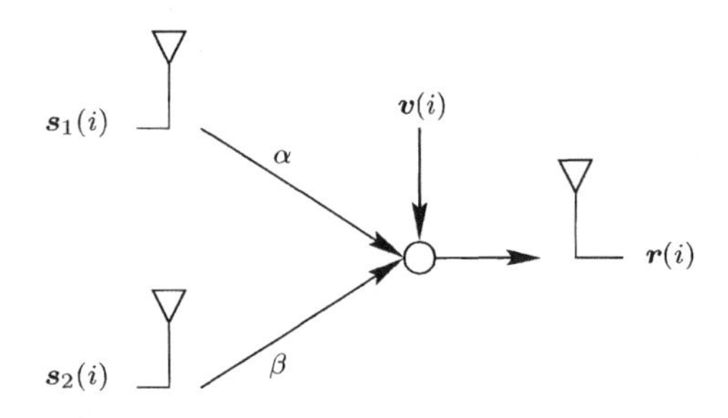

FIGURE II.2 A two-transmit one-receive antenna system.

Remark. The idea of coding the data in the manner described in this problem in order to exploit transmit diversity, and subsequently simplify the structure of the receiver, is due to Alamouti (1998).

Problem II.25 (Maximal ratio combining) Consider a one-transmit two-receive antenna system, as shown in Fig. II.3. Let α denote the channel gain from the transmit antenna to the first receiver and let β denote the channel gain from the same transmit antenna to the second receiver. Both channel gains are possibly complex-valued. A symbol s is transmitted and the received signals by both antennas are $y = \alpha s + v$ and $z = \beta s + w$, where $\{v, w\}$ denote zero-mean noise components that are uncorrelated with each other and with s and have the same variance σ_v^2. On the receiver side, the scalar signals $\{y, z\}$ are to be combined linearly, say as $\hat{s} = ay + bz$, in order to generate an enhanced signal $\hat{s}$ with maximal signal-to-noise ratio. We wish to determine the values of $\{a, b\}$.

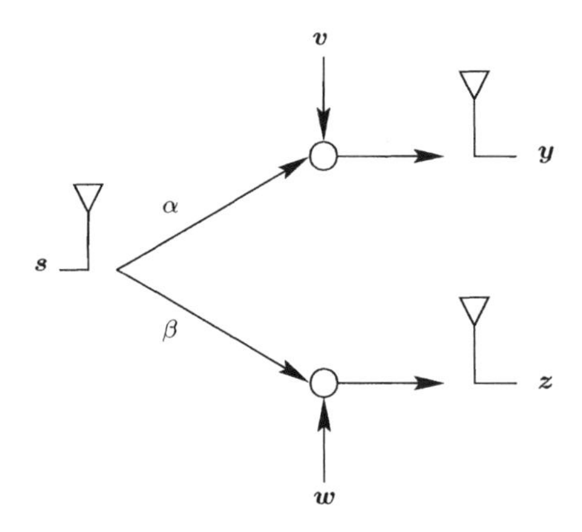

FIGURE II.3 A one-transmit two-receive antenna system.

(a) Collect the measurements into vector form,

$$\begin{bmatrix} y \\ z \end{bmatrix} = \underbrace{\begin{bmatrix} \alpha \\ \beta \end{bmatrix}}_{h} s + \begin{bmatrix} v \\ w \end{bmatrix}$$

and let $r^* = [a \;\; b]$ denote the desired combination vector. Argue that the SNR after processing by r is $\text{SNR} = \sigma_s^2|r^*h|^2/\sigma_v^2\|r\|^2$, where r^*h is the inner product between r and h.

(b) Use the Cauchy-Schwartz inequality to conclude that a choice for r that maximizes the SNR is $r = h$ and that the resulting maximum SNR is $\text{SNR}_{\max} = \sigma_s^2(|\alpha|^2 + |\beta|^2)/\sigma_v^2$. Conclude that an optimal linear combination is $\hat{s} = \alpha^* y + \beta^* z$.

Remark. Maximal ratio combining is a classical technique due to Brennan (1959). As can be seen from the result of this problem, MRC is an optimal spatial diversity receiver that maximizes the output SNR and thereby reduces signal distortions caused by multipath propagation. This technique is employed in RAKE receivers, as discussed in Computer Project 11.3 of Sayed (2003).

Problem II.26 (MMSE linear combining) Consider the same setting as Prob. II.25, except that now it is desired to determine the coefficients $\{a, b\}$ by minimizing the mean-square error in estimating s from z,

$$\min_{a,b} \mathsf{E}\left| s - \begin{bmatrix} a & b \end{bmatrix} \begin{bmatrix} y \\ z \end{bmatrix} \right|^2$$

Use the result of Thm. 5.1 to show that the linear least-mean-squares estimator of s given $\{y, z\}$ is given by

$$\hat{s} = \frac{1}{\sigma_v^2/\sigma_s^2 + |\alpha|^2 + |\beta|^2}(\alpha^* y + \beta^* z) \quad \text{with} \quad \text{m.m.s.e.} = \frac{\sigma_v^2}{\sigma_v^2/\sigma_s^2 + |\alpha|^2 + |\beta|^2}$$

Problem II.27 (Cyclic prefixing) Assume 9 data points, $\{s(0), s(1), s(2), \ldots, s(8)\}$, are transmitted through a channel with transfer function $H(z) = h(0) + h(1)z^{-1} + h(2)z^{-2}$, and collect the resulting channel outputs $\{z(2), z(3), \ldots, z(8)\}$.

(a) Verify that the transmissions and measurements are related via

$$\underbrace{\begin{bmatrix} z(8) \\ z(7) \\ z(6) \\ z(5) \\ z(4) \\ z(3) \\ z(2) \end{bmatrix}}_{z:7\times 1} = \underbrace{\begin{bmatrix} h(0) & h(1) & h(2) & & & & & & \\ & h(0) & h(1) & h(2) & & & & & \\ & & h(0) & h(1) & h(2) & & & & \\ & & & h(0) & h(1) & h(2) & & & \\ & & & & h(0) & h(1) & h(2) & & \\ & & & & & h(0) & h(1) & h(2) & \\ & & & & & & h(0) & h(1) & h(2) \end{bmatrix}}_{H:7\times 9} \underbrace{\begin{bmatrix} s(8) \\ s(7) \\ s(6) \\ s(5) \\ s(4) \\ s(3) \\ s(2) \\ \hline s(1) \\ s(0) \end{bmatrix}}_{s:9\times 1}$$

Remark. The channel matrix H defined above has a Toeplitz structure, i.e., it has constant elements along its diagonals.

(b) Now assume that the transmitted data is designed such that $s(8) = s(1)$ and $s(7) = s(0)$. That is, after 7 transmissions $\{s(0), \ldots, s(6)\}$, we repeat the first 2 symbols $\{s(0), s(1)\}$. Show that the transmissions and measurements are now related as

$$\underbrace{\begin{bmatrix} z(8) \\ z(7) \\ z(6) \\ z(5) \\ z(4) \\ z(3) \\ z(2) \end{bmatrix}}_{z:7\times 1} = \underbrace{\begin{bmatrix} h(0) & h(1) & h(2) & & & & \\ & h(0) & h(1) & h(2) & & & \\ & & h(0) & h(1) & h(2) & & \\ & & & h(0) & h(1) & h(2) & \\ & & & & h(0) & h(1) & h(2) \\ h(2) & & & & & h(0) & h(1) \\ h(1) & h(2) & & & & & h(0) \end{bmatrix}}_{H:7\times 7} \underbrace{\begin{bmatrix} s(8) \\ s(7) \\ s(6) \\ s(5) \\ s(4) \\ s(3) \\ s(2) \end{bmatrix}}_{s:7\times 1}$$

Remark. Observe that the channel matrix H is now square. It still has a Toeplitz structure. However, in addition, it also has a circulant structure. This means that each of its rows can be obtained from the previous one by circularly shifting it to the right by one entry.

(c) More generally, consider a channel with memory M, i.e., $H(z) = h(0) + h(1)z^{-1} + \ldots + h(M)z^{-M}$, and assume that we collect $N+1$ measurements starting from some time instant i, i.e., $\boldsymbol{z}$ is now $(N+1) \times 1$ with data $\boldsymbol{z} = \text{col}\{\boldsymbol{z}(i+N), \boldsymbol{z}(i+N-1), \ldots, \boldsymbol{z}(i+1), \boldsymbol{z}(i)\}$. Assume further that the transmissions are designed with a cyclic prefix, namely, the last M values of $\boldsymbol{s}(\cdot)$ coincide with the M values of $\boldsymbol{s}(\cdot)$ prior to time i,

$$\boldsymbol{s}(i+N-k) = \boldsymbol{s}(i-1-k), \quad k = 0, 1, \ldots, M-1$$

Let $\boldsymbol{s} = \text{col}\{\boldsymbol{s}(i+N), \boldsymbol{s}(i+N-1), \ldots, \boldsymbol{s}(i+1), \boldsymbol{s}(i)\}$ and assume $N > M$. Verify that $\boldsymbol{z}$ and $\boldsymbol{s}$ are related via an $(N+1) \times (N+1)$ circulant channel matrix H.

Problem II.28 (Frequency-domain equalization) Independent and identically distributed data $\boldsymbol{s}(i)$, with variance σ_s^2, are transmitted through a known channel with transfer function

$$H(z) = h(0) + h(1)z^{-1} + h(2)z^{-2} + \ldots + h(M)z^{-M}$$

A cyclic prefix is incorporated into the data, as explained in part (c) of Prob. II.27. The output of the channel is observed under additive white noise, $\boldsymbol{v}(i)$, with variance σ_v^2, i.e., $\boldsymbol{y}(i) = \boldsymbol{z}(i) + \boldsymbol{v}(i)$. Collect a block of $(N+1)$ measurements, starting from some time instant i, $\boldsymbol{y} = \text{col}\{\boldsymbol{y}(i+N), \boldsymbol{y}(i+N-1), \ldots, \boldsymbol{y}(i+1), \boldsymbol{y}(i)\}$, and define the corresponding data vector $\boldsymbol{s} = \text{col}\{\boldsymbol{s}(i+N), \boldsymbol{s}(i+N-1), \ldots, \boldsymbol{s}(i+1), \boldsymbol{s}(i)\}$. We know from Prob. II.27 that $\boldsymbol{y}$ and $\boldsymbol{s}$ are related via $\boldsymbol{y} = H\boldsymbol{s} + \boldsymbol{v}$, where H is $(N+1) \times (N+1)$ circulant, and $\boldsymbol{v} = \text{col}\{\boldsymbol{v}(i+N), \ldots, \boldsymbol{v}(i+1), \boldsymbol{v}(i)\}$ denotes the noise vector.

(a) Every circulant matrix H can be diagonalized by the discrete Fourier transform (DFT), i.e., it holds that $\Lambda = FHF^*$, where F is the DFT matrix of size $(N+1)$:

$$\boxed{[F]_{ik} \triangleq \frac{1}{\sqrt{N+1}} e^{\frac{-j2\pi ik}{(N+1)}}, \quad i, k = 0, 1, \ldots, N, \quad j = \sqrt{-1}}$$

and Λ is diagonal with eigenvalues denoted by $\{\lambda_i\}$ — see, e.g., Prob. VI.6. Define the frequency-transformed vectors $\overline{\boldsymbol{y}} = F\boldsymbol{y}$, $\overline{\boldsymbol{s}} = F\boldsymbol{s}$, and $\overline{\boldsymbol{v}} = F\boldsymbol{v}$. That is, $\{\overline{\boldsymbol{y}}, \overline{\boldsymbol{s}}, \overline{\boldsymbol{v}}\}$ are the DFTs of the vectors $\{\boldsymbol{y}, \boldsymbol{s}, \boldsymbol{v}\}$. Show that $\overline{\boldsymbol{y}} = \Lambda\overline{\boldsymbol{s}} + \overline{\boldsymbol{v}}$ and that the covariance matrices of $\{\overline{\boldsymbol{v}}, \overline{\boldsymbol{s}}\}$ are $\{\sigma_v^2 \mathrm{I}, \sigma_s^2 \mathrm{I}\}$.

(b) Show that the linear least-mean-squares estimator of $\overline{\boldsymbol{s}}$ is $\widehat{\overline{\boldsymbol{s}}} = \sigma_s^2 \Lambda^* \left(\sigma_v^2 \mathrm{I} + \sigma_s^2 \Lambda\Lambda^*\right)^{-1} \overline{\boldsymbol{y}}$. Conclude that the entries of $\{\overline{\boldsymbol{y}}, \widehat{\overline{\boldsymbol{s}}}\}$ are related to each other via a simple scaling operation, namely, for the k-th entries:

$$\boxed{\widehat{\overline{\boldsymbol{s}}}(k) = \frac{\lambda_k^* \sigma_s^2}{\sigma_v^2 + \sigma_s^2 |\lambda_k|^2} \cdot \overline{\boldsymbol{y}}(k)}$$

Remark. The result of this problem explains the concept behind single-carrier-frequency-domain equalization (for more details see, e.g., Sari, Karam, and Jeanclaude (1995), Clark (1998), Al-Dhahir (2001), and Younis, Sayed, and Al-Dhahir (2003)). The data $\{\boldsymbol{s}(i)\}$ are transmitted with cyclic prefixing, and a block of measurements $\boldsymbol{y}$ is collected. The DFT of $\boldsymbol{y}$ is computed and its entries scaled according to the expression in part (b) above, by using the diagonal elements $\{\lambda_k\}$ of the frequency-transformed channel matrix H. The estimators $\widehat{\overline{\boldsymbol{s}}}$ are then transformed back to the time domain by undoing the DFT operation, i.e., $\widehat{\boldsymbol{s}} = F^* \widehat{\overline{\boldsymbol{s}}}$. The use of a cyclic prefix adds redundancy to the transmitted data and results in overhead costs. Therefore, there is a compromise to be balanced between the design of a simple receiver structure and the loss in information capacity. Cyclic prefixing is also useful in the context of orthogonal frequency-division multiplexing (OFDM), where the receiver structure can be derived in much the same manner as in this problem. We leave the details of OFDM to Computer Project VII.1, where it will be shown how least-mean-squares (as well as least-squares) techniques are useful in the design of OFDM receivers.

Problem II.29 (Multi-transmit multi-receive antenna system) Independent and identically distributed data $\{\boldsymbol{s}_1(\cdot), \boldsymbol{s}_2(\cdot)\}$ are transmitted from two antennas and received by two antennas, after travelling through a multi-channel environment and being corrupted by additive white noises. Transmissions from antenna 1 travel through channels $\{h_{11}, h_{12}\}$, which refer to the impulse response sequences of the channels between transmit antenna 1 and receive antennas 1 and 2. Likewise, transmissions from antenna 2 travel through channels $\{h_{21}, h_{22}\}$, which refer to the impulse response sequences of the channels between transmit antenna 2 and receive antennas 1 and 2. This scenario corresponds to transmissions through a MIMO (multi-input multi-output) channel — see Fig. II.4. The signals $\{\boldsymbol{s}_1(\cdot), \boldsymbol{s}_2(\cdot), \boldsymbol{v}_1(\cdot), \boldsymbol{v}_2(\cdot)\}$ are assumed uncorrelated with each other, with the signals having variances $\{\sigma_{s1}^2, \sigma_{s2}^2\}$ and the noises having variances $\{\sigma_{v1}^2, \sigma_{v2}^2\}$.

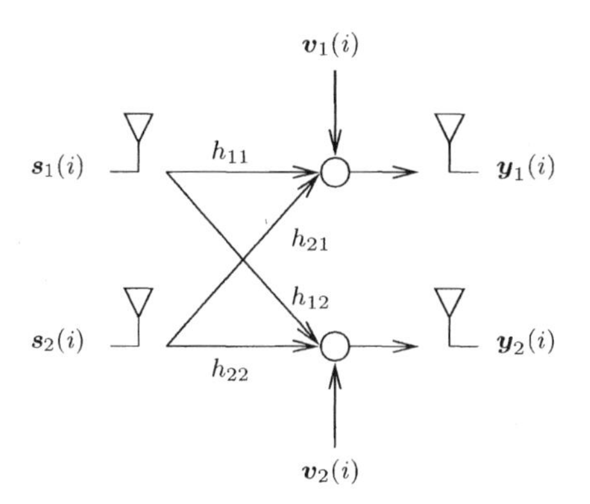

FIGURE II.4 A two-transmit two-receive antenna system.

The transmissions from both users are not only distorted by the channels but they also interfere with each other. We wish to design a linear equalizer (receiver) in order to recover the transmitted data $\{\boldsymbol{s}_1(\cdot), \boldsymbol{s}_2(\cdot)\}$ from the corrupted measurements $\{\boldsymbol{y}_1(\cdot), \boldsymbol{y}_2(\cdot)\}$. Assume in this problem that the transfer functions of the channels are

$$h_{11}(z) = 1 + \alpha z^{-1}, \qquad h_{22}(z) = 1 + \beta z^{-1}, \qquad h_{12}(z) = az^{-1}, \qquad h_{21}(z) = bz^{-1}$$

in terms of $\{\alpha, \beta, a, b\}$. Define the vectors of measurements and data at time i,

$$\boldsymbol{y}_i \triangleq \begin{bmatrix} \boldsymbol{y}_1(i) \\ \boldsymbol{y}_2(i) \end{bmatrix}, \qquad \boldsymbol{s}_i \triangleq \begin{bmatrix} \boldsymbol{s}_1(i) \\ \boldsymbol{s}_2(i) \end{bmatrix}$$

and choose the receiver to be of length 3, i.e., it generates an estimate for $\boldsymbol{s}_i$ as follows:

$$\hat{\boldsymbol{s}}_i = W_0 \boldsymbol{y}_i + W_1 \boldsymbol{y}_{i-1} + W_2 \boldsymbol{y}_{i-2}$$

where the $\{W_0, W_1, W_2\}$ are 2×2 coefficient matrices to be determined optimally in the least-mean-squares sense.

(a) Verify that

$$\underbrace{\begin{bmatrix} \boldsymbol{y}_i \\ \boldsymbol{y}_{i-1} \\ \boldsymbol{y}_{i-2} \end{bmatrix}}_{\boldsymbol{y}} = \underbrace{\begin{bmatrix} \mathrm{I}_2 & P & & \\ & \mathrm{I}_2 & P & \\ & & \mathrm{I}_2 & P \end{bmatrix}}_{H} \underbrace{\begin{bmatrix} \boldsymbol{s}_i \\ \boldsymbol{s}_{i-1} \\ \boldsymbol{s}_{i-2} \\ \boldsymbol{s}_{i-3} \end{bmatrix}}_{\boldsymbol{s}} + \underbrace{\begin{bmatrix} \boldsymbol{v}_i \\ \boldsymbol{v}_{i-1} \\ \boldsymbol{v}_{i-2} \end{bmatrix}}_{\boldsymbol{v}}$$

where P is the 2×2 matrix

$$P \triangleq \begin{bmatrix} \alpha & b \\ a & \beta \end{bmatrix}$$

Observe that the channel matrix H has a block Toeplitz structure with 2×2 block entries.

(b) Let $\{R_y, R_s, R_v\}$ denote the covariances of $\{\boldsymbol{y}, \boldsymbol{s}, \boldsymbol{v}\}$. Show that

$$R_y = HR_sH^* + R_v, \qquad \mathsf{E}\,\boldsymbol{s}_i\boldsymbol{y}^* = (\mathsf{E}\,\boldsymbol{s}_i\boldsymbol{s}^*)H^* = \begin{bmatrix} D_s & 0 & 0 & 0 \end{bmatrix} H^*$$

where

$$R_v = \begin{bmatrix} D_v & & \\ & D_v & \\ & & D_v \end{bmatrix}, \quad D_v = \begin{bmatrix} \sigma_{v1}^2 & \\ & \sigma_{v2}^2 \end{bmatrix}, \quad D_s = \begin{bmatrix} \sigma_{s1}^2 & \\ & \sigma_{s2}^2 \end{bmatrix}$$

(c) Determine the receiver coefficients $\{W_0, W_1, W_2\}$ and the resulting m.m.s.e. in terms of $\{H, D_v, D_s\}$.

(d) Draw a block diagram representation for the receiver when $\{\alpha = 1/2, \beta = -1/2, a = 1, b = -1, \sigma_{s1}^2 = \sigma_{s2}^2 = 1, \sigma_{v1}^2 = \sigma_{v2}^2 = 1/2\}$.

(e) Determine the receiver coefficients in the following situations:

(c.1) $a = 0, b = 0$, and all other values as in part (d).

(c.2) $a = 0$ and all other values as in part (d).

(c.3) $b = 0$ and all other values as in part (d).

(c.4) $h_{21} = 0, h_{22} = 0$ and all other values as in part (d).

Problem II.30 (MIMO channel estimation) Consider a scenario similar to Prob. II.29, except that now *known* symbols $\{s_1(\cdot), s_2(\cdot)\}$ are transmitted from two antennas and received by two antennas, after travelling through a multi-channel environment and being corrupted by uncorrelated additive white noises with variances $\{\sigma_{v1}^2, \sigma_{v2}^2\}$. Transmissions from antenna 1 travel through channels $\{h_{11}, h_{12}\}$, while transmissions from antenna 2 travel through channels $\{h_{22}, h_{21}\}$. The impulse response sequences are now modeled as zero-mean random variables, $\{\boldsymbol{h}_{11}, \boldsymbol{h}_{12}, \boldsymbol{h}_{21}, \boldsymbol{h}_{22}\}$, and they are assumed to be of first-order for simplicity (i.e., each has two taps at most). Let $\boldsymbol{h} = \text{col}\{\boldsymbol{h}_{11}, \boldsymbol{h}_{12}, \boldsymbol{h}_{21}, \boldsymbol{h}_{22}\}$ denote a column vector with the entries of the individual channels stacked on top of each other; its covariance matrix is taken as $R_h = \sigma_h^2 \mathrm{I}$ (assumed known).

Define again

$$\boldsymbol{y}_i \triangleq \begin{bmatrix} \boldsymbol{y}_1(i) \\ \boldsymbol{y}_2(i) \end{bmatrix}, \quad s_i \triangleq \begin{bmatrix} s_1(i) \\ s_2(i) \end{bmatrix}$$

where $\boldsymbol{y}_i$ is random while s_i is deterministic, and assume $N+1$ measurement vectors $\boldsymbol{y}_i$ are collected, say $\boldsymbol{y} = \text{col}\{\boldsymbol{y}_0, \boldsymbol{y}_1, \ldots, \boldsymbol{y}_N\}$.

(a) Verify that $\boldsymbol{y}$ is related to $\boldsymbol{h}$ via $\boldsymbol{y} = S\boldsymbol{h} + \boldsymbol{v}$ for some $\{S, \boldsymbol{v}\}$ to be determined.

(b) Determine an expression for the linear least-mean-squares estimator of $\boldsymbol{h}$ given $\boldsymbol{y}$. Find also the resulting m.m.s.e.

Problem II.31 (Lattice recursions) Consider a zero-mean scalar random process $\{\boldsymbol{y}(i)\}$. For each time instant i, let

$$\begin{aligned} \hat{\boldsymbol{y}}(i|i-1:i-m) &= \text{l.l.m.s. estimator of } \boldsymbol{y}(i) \text{ given the } m \text{ past observations} \\ &\quad \{\boldsymbol{y}(i-1), \boldsymbol{y}(i-2), \ldots, \boldsymbol{y}(i-m)\} \\ \hat{\boldsymbol{y}}(i-m-1|i-1:i-m) &= \text{l.l.m.s. estimator of } \boldsymbol{y}(i-m-1) \text{ given the same observations} \\ &\quad \{\boldsymbol{y}(i-1), \boldsymbol{y}(i-2), \ldots, \boldsymbol{y}(i-m)\} \end{aligned}$$

We refer to $\hat{\boldsymbol{y}}(i|i-1:i-m)$ as the *forward* prediction of $\boldsymbol{y}(i)$ that is based on the previous observations. Likewise, we refer to $\hat{\boldsymbol{y}}(i-m-1|i-1:i-m)$ as the *backward* prediction of $\boldsymbol{y}(i-m-1)$ that is based on the same observations, as indicated below:

$$\boldsymbol{y}(i-m-1)\ \boxed{\boldsymbol{y}(i-m) \ \ \ldots \ \ \boldsymbol{y}(i-2) \ \ \boldsymbol{y}(i-1)}\ \boldsymbol{y}(i)$$

In one case, we are using the boxed observations to predict the future (hence, the designation *forwards prediction*), while in the second case we are using the same observations to "predict" the past (and, hence, the designation *backwards prediction*).

Let $\{\boldsymbol{f}_m(i), \boldsymbol{b}_m(i-1)\}$ represent the corresponding residual errors, also known as forward and backward estimation errors,

$$\begin{aligned} \boldsymbol{f}_m(i) &= \boldsymbol{y}(i) - \hat{\boldsymbol{y}}(i|i-1:i-m) \\ \boldsymbol{b}_m(i-1) &= \boldsymbol{y}(i-m-1) - \hat{\boldsymbol{y}}(i-m-1|i-1:i-m) \end{aligned}$$

and denote their variances and cross-correlation by

$$\zeta_m^f(i) = \mathsf{E}\,|\boldsymbol{f}_m(i)|^2, \qquad \zeta_m^b(i-1) = \mathsf{E}\,|\boldsymbol{b}_m(i-1)|^2, \qquad \delta_m(i) = \mathsf{E}\,\boldsymbol{b}_m(i-1)\boldsymbol{f}_m^*(i)$$

(a) Show that the forward and backward errors, as well as their variances, satisfy the order-update relations:

$$\boxed{\begin{aligned} \kappa_m^f(i) &= \delta_m^*(i)/\zeta_m^b(i-1) \\ \kappa_m^b(i) &= \delta_m(i)/\zeta_m^f(i) \\ \boldsymbol{f}_{m+1}(i) &= \boldsymbol{f}_m(i) - \kappa_m^f(i)\boldsymbol{b}_m(i-1) \\ \boldsymbol{b}_{m+1}(i) &= \boldsymbol{b}_m(i-1) - \kappa_m^b(i)\boldsymbol{f}_m(i) \\ \zeta_{m+1}^f(i) &= \zeta_m^f(i) - |\delta_m(i)|^2/\zeta_m^b(i-1) \\ \zeta_{m+1}^b(i) &= \zeta_m^b(i-1) - |\delta_m(i)|^2/\zeta_m^f(i) \end{aligned}}$$

in terms of the so-called reflection coefficients $\{\kappa_m^f(i), \kappa_m^b(i)\}$. *Remark.* Later, in Part X (*Lattice Filters*), we study order-recursive least-squares (as opposed to least-mean-squares) lattice filters. It is instructive to compare the above lattice recursions with the recursions appearing in Tables 40.1 and 42.1.

(b) Define the normalized reflection coefficient $\kappa_m^a(i) = \delta_m^*(i)/\zeta_m^{b/2}(i-1)\zeta_m^{f/2}(i)$. Show that $|\kappa_m^a(i)| \leq 1$.

(c) Assume $|\kappa_m^a(i)| < 1$ for all i and m, and define the normalized errors $\boldsymbol{b}_m''(i) = \boldsymbol{b}_m(i)/\zeta_m^{b/2}(i)$ and $\boldsymbol{f}_m''(i) = \boldsymbol{f}_m(i)/\zeta_m^{f/2}(i)$. Show that the normalized errors satisfy the recursions:

$$\begin{aligned} \boldsymbol{f}_{m+1}''(i) &= \frac{1}{\sqrt{1-|\kappa_m^a(i)|^2}}\left[\boldsymbol{f}_m''(i) - \kappa_m^a(i)\boldsymbol{b}_m''(i-1)\right] \\ \boldsymbol{b}_{m+1}''(i) &= \frac{1}{\sqrt{1-|\kappa_m^a(i)|^2}}\left[\boldsymbol{b}_m''(i-1) - \kappa_m^{a*}(i)\boldsymbol{f}_m''(i)\right] \end{aligned}$$

in terms of the single reflection coefficient $\kappa_m^a(i)$.

(d) Assume now that $\{\boldsymbol{y}(i)\}$ is a wide-sense stationary random process so that $\mathsf{E}\,\boldsymbol{y}(i)\boldsymbol{y}^*(j) = \mathsf{E}\,\boldsymbol{y}(i-k)\boldsymbol{y}^*(j-k)$ for all k. Show that, in this case, the quantities $\{\zeta_m^f(i), \zeta_m^b(i), \kappa_m^a(i)\}$ become independent of i and that the normalized errors $\{\boldsymbol{f}_m''(i), \boldsymbol{b}_m''(i)\}$ become related as shown in Fig. II.5.

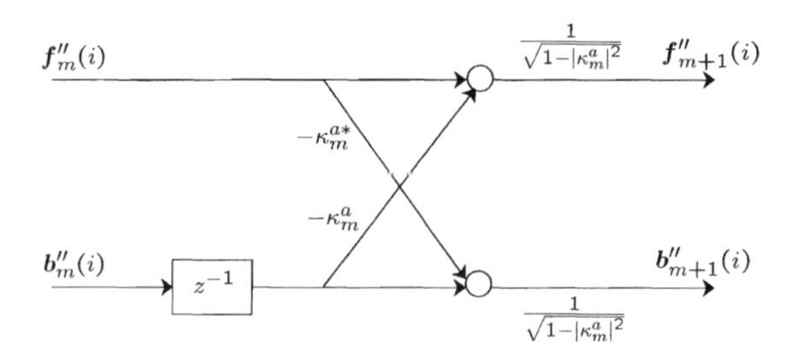

FIGURE II.5 A lattice section showing how the normalized forward and backward estimation errors for stationary processes are related in terms of the normalized reflection coefficient.

Problem II.32 (Levinson-Durbin algorithm) Refer to Prob. II.31 and assume, as in part (d) of that problem, that the process $\{\boldsymbol{y}(i)\}$ is stationary. Express the forward residual error as

$$\boldsymbol{f}_m(i) = \boldsymbol{y}(i) + a_m(1)\boldsymbol{y}(i-1) + a_m(2)\boldsymbol{y}(i-2) + \ldots + a_m(m)\boldsymbol{y}(i-m)$$

for some coefficients $\{a_m(j), j = 1, \ldots, m\}$. Let $c(k) = \mathsf{E}\,\boldsymbol{y}(i)\boldsymbol{y}^*(i-k)$.

(a) Show that the backward residual error is given by a similar expression:

$$\boldsymbol{b}_m(i-1) = \boldsymbol{y}(i-m-1) + a_m^*(m)\boldsymbol{y}(i-1) + a_m^*(m-1)\boldsymbol{y}(i-2) + \ldots + a_m^*(1)\boldsymbol{y}(i-m)$$

in terms of the conjugated coefficients $\{a_m^*(j),\ j = 1, \ldots, m\}$.

(b) Collect the $\{a_m(j)\}$ into a row vector as $a_m = [a_m(m)\ a_m(m-1)\ \ldots\ a_m(1)\ 1]$. Show that a_m satisfies the so-called Yule-Walker equations

$$\boxed{a_m T_m = \begin{bmatrix} 0 & \ldots & 0 & \zeta_m^f \end{bmatrix}}$$

where $\zeta_m^f = \mathsf{E}\,|\boldsymbol{f}_m(i)|^2$ and T_m is an $(m+1) \times (m+1)$ Toeplitz matrix whose first column is $\text{col}\{c(0), c(1), \ldots, c(m)\}$,

$$T_m \triangleq \begin{bmatrix} c(0) & c^*(1) & c^*(2) & \ldots & c^*(m) \\ c(1) & c(0) & c^*(1) & \ldots & c^*(m-1) \\ c(2) & c(1) & c(0) & \ldots & c^*(m-2) \\ \vdots & \vdots & \vdots & \ddots & \vdots \\ c(m) & c(m-1) & c(m-2) & \ldots & c(0) \end{bmatrix}$$

Show further that $\zeta_m^b = \zeta_m^f$ where $\zeta_m^b = \mathsf{E}\,|\boldsymbol{b}_m(i)|^2$.

(c) Refer to the recursions of part (a) of Prob. II.31. Argue that the reflection coefficients are now related as $\kappa_m^f = \kappa_m^{b*}$. Let $\kappa_m^f = \kappa_m$. Deduce the following so-called Levinson-Durbin recursions for the prediction vectors $\{a_m\}$:

$$\boxed{\begin{bmatrix} a_{m+1} \\ a_{m+1}^{\#} \end{bmatrix} = \begin{bmatrix} 1 & -\kappa_m \\ -\kappa_m^* & 1 \end{bmatrix} \begin{bmatrix} 0 & a_m \\ a_m^{\#} & 0 \end{bmatrix}, \quad \begin{bmatrix} a_0 \\ a_0^{\#} \end{bmatrix} = \begin{bmatrix} 1 \\ 1 \end{bmatrix}}$$

where $a_m^{\#} \triangleq [\ 1\ a_m^*(1)\ \ldots\ a_m^*(m)\ ]$.

(d) Show also that $\zeta_{m+1}^f = \zeta_m^f(1 - |\kappa_m|^2)$ and

$$\kappa_m = \frac{c(m+1) + \sum_{i=1}^{m} a_m(m+1-i)c(i)}{\zeta_m^f}$$

Remark. The Yule-Walker equations were introduced by Yule (1927) in his studies on fitting an autoregressive model to sunspot data. The efficient algorithm of parts (c) and (d) for order-updating the solution of the Yule-Walker equations was derived by Durbin (1960) and earlier (in a more general context) by Levinson (1947).

Problem II.33 (Block signal processing) Consider a 6–th order FIR filter with transfer function

$$K_o(z) \triangleq \alpha(0) + \alpha(1)z^{-1} + \alpha(2)z^{-2} + \alpha(3)z^{-3} + \alpha(4)z^{-4} + \alpha(5)z^{-5}$$

Its so-called 3rd-order polyphase components $\{E_i(z)\}$ are defined via the identity

$$K_o(z) = E_0(z^3) + z^{-1}E_1(z^3) + z^{-2}E_2(z^3)$$

where each $E_i(z)$ is a polynomial in z^{-1} of degree one.

(a) Verify that $E_0(z) = \alpha(0) + \alpha(3)z^{-1}$, $E_1(z) = \alpha(1) + \alpha(4)z^{-1}$, and $E_2(z) = \alpha(2) + \alpha(5)z^{-1}$.

(b) Let $\{x(n), y(n)\}$ denote the input and output sequences of the fullband filter $K_o(z)$. Define the vectors of size 3 each,

$$x_n \triangleq \begin{bmatrix} x(3n) \\ x(3n-1) \\ x(3n-2) \end{bmatrix}, \quad y_n \triangleq \begin{bmatrix} y(3n) \\ y(3n-1) \\ y(3n-2) \end{bmatrix}$$

Verify that the *matrix* transfer function that maps x_n to y_n is given by

$$\mathcal{K}_o(z) = \begin{bmatrix} E_0(z) & E_1(z) & E_2(z) \\ z^{-1}E_2(z) & E_0(z) & E_1(z) \\ z^{-1}E_1(z) & z^{-1}E_2(z) & E_0(z) \end{bmatrix}$$

Remark. The matrix transfer function $\mathcal{K}_o(z)$ allows us to evaluate the output signal on a block-by-block basis. The terminology of fullband filters and polyphase components is common in multirate digital signal processing. The transfer matrix $\mathcal{K}_o(z)$ shown above has a so-called pseudo-circulant structure. A pseudo-circulant matrix is a circulant matrix with the exception that all entries below the main diagonal are further multiplied by the same factor z^{-1}. A circulant matrix is a Toeplitz matrix (i.e., one with identical entries along the diagonals) with the additional property that its first row is circularly shifted to the right in order to form the other rows. We shall return to the result of this problem later in Chapter 26.

Problem II.34 (Oblique projection) Consider zero mean-random variables $\{\boldsymbol{x}, \boldsymbol{y}, \boldsymbol{z}, \boldsymbol{d}\}$ and let $R_{yd} = \mathsf{E}\,\boldsymbol{y}\boldsymbol{d}^*$, $R_{zy} = \mathsf{E}\,\boldsymbol{z}\boldsymbol{y}^*$, and $R_{xd} = \mathsf{E}\,\boldsymbol{x}\boldsymbol{d}^*$. We want to determine estimators for $\boldsymbol{x}$ and $\boldsymbol{z}$ such that:

1. $\hat{\boldsymbol{x}} = K_x^o\boldsymbol{y}$, for some coefficient matrix K_x^o to be determined.
2. $\hat{\boldsymbol{z}} = K_z^o\boldsymbol{d}$, for some coefficient matrix K_z^o to be determined.
3. The estimation error $\tilde{\boldsymbol{x}} = \boldsymbol{x} - \hat{\boldsymbol{x}}$ is orthogonal to $\boldsymbol{d}$, i.e., $\mathsf{E}\,\tilde{\boldsymbol{x}}\boldsymbol{d}^* = 0$.
4. The estimation error $\tilde{\boldsymbol{z}} = \boldsymbol{z} - \hat{\boldsymbol{z}}$ is orthogonal to $\boldsymbol{y}$, i.e., $\mathsf{E}\,\tilde{\boldsymbol{z}}\boldsymbol{y}^* = 0$.

We refer to $\hat{\boldsymbol{z}}$ as the *oblique projection* of $\boldsymbol{z}$ onto $\boldsymbol{d}$ because the resulting error, $\tilde{\boldsymbol{z}}$, is not orthogonal to $\boldsymbol{d}$ but rather to $\boldsymbol{y}$. In other words, we are projecting $\boldsymbol{z}$ onto $\boldsymbol{d}$ along a direction that is orthogonal to $\boldsymbol{y}$. Likewise, we say that $\hat{\boldsymbol{x}}$ is the oblique projection of $\boldsymbol{x}$ onto $\boldsymbol{y}$.

(a) Assuming R_{dy} is invertible, show that $K_z^o = R_{zy}R_{dy}^{-1}$ and $K_x^o = R_{xd}R_{yd}^{-1}$.

(b) Assume that $\{\boldsymbol{y}, \boldsymbol{d}, \boldsymbol{x}, \boldsymbol{z}\}$ are related via $\boldsymbol{y} = H\boldsymbol{x} + \boldsymbol{v}$ and $\boldsymbol{d} = A\boldsymbol{z} + \boldsymbol{w}$, with $\mathsf{E}\,\boldsymbol{v}\boldsymbol{w}^* = W$, $\mathsf{E}\,\boldsymbol{z}\boldsymbol{x}^* = \Pi$, $\mathsf{E}\,\boldsymbol{v}\boldsymbol{z}^* = 0$ and $\mathsf{E}\,\boldsymbol{w}\boldsymbol{x}^* = 0$. Determine $\hat{\boldsymbol{z}}$ and $\hat{\boldsymbol{x}}$ in terms of $\{H, A, \Pi, W, \boldsymbol{y}, \boldsymbol{d}\}$.

Remark. Oblique projections arise in the study of instrumental variable methods in system identification (see, e.g., Söderström and Stoica (1983)). They also arise in some communications and array processing applications and in higher-order spectral (HOS) analysis methods (see, e.g., Kayalar and Weinert (1989) and Behrens and Scharf (1994). See also Prob. VII.23). A Kalman-type filter for performing oblique state-space estimation was developed by Sayed and Kailath (1995).

Problem II.35 (Valve pressure) Let p_0 denote the initial pressure in a valve and assume that it is decreasing exponentially. Three noisy measurements of the pressure in the valve are available at time instants $\{t_0, t_1, t_2\}$, say $\boldsymbol{y}(i) = p_0e^{-t_i} + \boldsymbol{v}(i)$ for $i = 0, 1, 2$, where the $\{\boldsymbol{v}(i)\}$ are uncorrelated zero-mean random variables with variances $\sigma_{v,i}^2$. Let $\hat{\boldsymbol{p}}_0$ denote the minimum-variance-unbiased estimator of p_0 given $\{\boldsymbol{y}(0), \boldsymbol{y}(1), \boldsymbol{y}(2)\}$. Show that

$$\hat{\boldsymbol{p}}_0 = \left(\frac{e^{-2t_0}}{\sigma_{v,0}^2} + \frac{e^{-2t_1}}{\sigma_{v,1}^2} + \frac{e^{-2t_2}}{\sigma_{v,2}^2}\right)^{-1}\left(\frac{e^{-t_0}}{\sigma_{v,0}^2}\boldsymbol{y}(0) + \frac{e^{-t_1}}{\sigma_{v,1}^2}\boldsymbol{y}(1) + \frac{e^{-t_2}}{\sigma_{v,2}^2}\boldsymbol{y}(2)\right)$$

and that the resulting m.m.s.e. is

$$\text{m.m.s.e.} = \left(\frac{e^{-2t_0}}{\sigma_{v,0}^2} + \frac{e^{-2t_1}}{\sigma_{v,1}^2} + \frac{e^{-2t_2}}{\sigma_{v,2}^2}\right)^{-1}$$

Problem II.36 (Constrained mean-square error) Let $\boldsymbol{d}$ denote a scalar zero-mean random variable with variance σ_d^2, and let $\boldsymbol{u}$ denote a $1 \times M$ zero-mean random vector with covariance matrix $R_u = \mathsf{E}\,\boldsymbol{u}^*\boldsymbol{u} > 0$. Consider the constrained optimization problem

$$\boxed{\min_w \mathsf{E}\,|\boldsymbol{d} - \boldsymbol{u}w|^2 \quad \text{subject to } c^*w = \alpha}$$

where c is a known $M \times 1$ vector and α is a known real scalar.

(a) Let $z = w - R_u^{-1}R_{du}$ and $R_{du} = \mathsf{E}\,\boldsymbol{d}\boldsymbol{u}^*$. Show that the above optimization problem is equivalent to the following:

$$\min_z \left[(\sigma_d^2 - R_{ud}R_u^{-1}R_{du}) + z^*R_uz \right] \quad \text{subject to } c^*z = \alpha - c^*R_u^{-1}R_{du}$$

(b) Use Remark 6.1 to conclude that the desired optimal solution, w^o, of the constrained optimization problem is given by

$$\boxed{w^o = R_u^{-1}R_{du} - \left(\frac{c^*R_u^{-1}R_{du} - \alpha}{c^*R_u^{-1}c}\right)R_u^{-1}c}$$

Verify that this solution satisfies the constraint $c^*w^o = \alpha$.

(c) Alternatively, arrive at the same expression for w^o by using a Lagrange multiplier argument. Specifically, introduce a complex Lagrange multiplier λ and consider the extended cost function $J_e(w, \lambda) = \mathsf{E}\,|\boldsymbol{d} - \boldsymbol{u}w|^2 + 2\text{Re}[\lambda(c^*w - \alpha)]$, in terms of the real part of $\lambda(c^*w - \alpha)$. Set the individual gradients of $J_e(w, \lambda)$ with respect to w and with respect to λ equal to zero and determine w^o.

Problem II.37 (Eigenfilters) Let $\boldsymbol{y} = \boldsymbol{s} + \boldsymbol{v}$ be a vector of measurements, where $\boldsymbol{v}$ is noise and $\boldsymbol{s}$ is the desired signal. Both $\boldsymbol{v}$ and $\boldsymbol{s}$ are zero-mean uncorrelated random vectors with covariance matrices $\{R_v, R_s\}$, respectively. We wish to determine a unit-norm column vector, w, such that the signal-to-noise ratio in the output signal, $w^*\boldsymbol{y}$, is maximized.

(a) Verify that the covariance matrices of the signal and noise components in $w^*\boldsymbol{y}$ are equal to w^*R_sw and w^*R_vw, respectively.

(b) Use the Rayleigh-Ritz characterization of eigenvalues from Sec. B.1 to conclude that the solution of $\max_{\|w\|=1} w^*R_sw$ is given by the unit-norm eigenvector that corresponds to the maximum eigenvalue of R_s, written as $w^o = q_{\max}$, where $R_sq_{\max} = \lambda_{\max}q_{\max}$. Assume that $R_v = \sigma_v^2\mathrm{I}$. Verify further that the resulting maximum SNR is equal to $\lambda_{\max}/\sigma_v^2$.

(c) Assume now that $\boldsymbol{v}$ is colored noise and introduce the Cholesky factorization $R_v = \bar{L}\bar{L}^*$ (the Cholesky factorization of a positive-definite matrix is described in Sec. B.3). Repeat the argument of part (b) to show that the solution of

$$\max_{\|w\|=1} \left(\frac{w^*R_sw}{w^*R_vw}\right)$$

is now related to the unit-norm eigenvector that corresponds to the maximum eigenvalue of $\bar{L}^{-1}R_s\bar{L}^{-*}$.

Problem II.38 (Space-time coding) Consider the same setting as in Prob. II.24, except that now we are interested in estimating the channel gains $\{\alpha, \beta\}$, which are assumed to be unknown constants. For this purpose, known symbols $\{s_1(i), s_2(i)\}$ are transmitted at time i from the two antennas followed by the symbols $\{-s_2^*(i), s_1^*(i)\}$ at time $i+1$. Assume the symbols have unit magnitude, i.e., $|s_k(i)|^2 = 1$ for $k = 1, 2$.

(a) Verify that the received signals are given by

$$\begin{bmatrix} \boldsymbol{r}(i) \\ \boldsymbol{r}(i+1) \end{bmatrix} = \underbrace{\begin{bmatrix} s_1(i) & s_2(i) \\ -s_2^*(i) & s_1^*(i) \end{bmatrix}}_{S} \begin{bmatrix} \alpha \\ \beta \end{bmatrix} + \begin{bmatrix} \boldsymbol{v}(i) \\ \boldsymbol{v}(i+1) \end{bmatrix}$$

where the matrix S satisfies $S^*S = (|s_1(i)|^2 + |s_2(i)|^2)\mathrm{I} = 2\mathrm{I}$.

(b) Find the minimum-variance-unbiased estimator of $\{\alpha, \beta\}$ given $\{\boldsymbol{r}(i), \boldsymbol{r}(i+1)\}$. Find also the resulting minimum mean-square error.

Problem II.39 (Constrained optimization) Consider the more general optimization problem

$$\min_K \; KR_vK^* \quad \text{subject to} \quad KH = A \quad \text{and} \quad R_v > 0$$

where K is $m \times N$, H is $N \times n$, A is $m \times n$, $n < N$, $m < N$, and H has full rank. In the text we assumed A is square and equal to the identity matrix (see (6.14)). Show that the optimal solution is given by $K_o = A(H^*R_v^{-1}H)^{-1}H^*R_v^{-1}$ and that the resulting minimum cost is equal to $A(H^*R_v^{-1}H)^{-1}A^*$.

Problem II.40 (Comparing linear and decision-feedback equalization) Refer to Sec. 6.4 and set $B(z) = 0$. The resulting structure would become a linear equalizer. The purpose of this problem is to show that the m.m.s.e. when $B(z) = 0$ is larger than or equal to the m.m.s.e. for nontrivial feedback filters.

(a) When $B(z) = 0$, and using (6.27), argue that the m.m.s.e. of the resulting linear equalizer is equal to $e_0^{\mathsf{T}} R_\delta e_0$ where $e_0 = \text{col}\{1, 0, 0 \ldots, 0\}$. That is, the m.m.s.e. is equal to the $(0, 0)$ entry of R_δ.

(b) We know from (6.29) that the m.m.s.e. of the DFE design is given by $1/(e_0^{\mathsf{T}} R_\delta^{-1} e_0)$. We would like to show that $1/e_0^{\mathsf{T}} R_\delta^{-1} e_0 \leq e_0^{\mathsf{T}} R_\delta e_0$. Let us establish a more general result. Let A denote any positive-definite Hermitian matrix and partition it as

$$A = \begin{bmatrix} a & x^* \\ x & B \end{bmatrix}$$

where a is a positive scalar, x is a column vector and B is also Hermitian positive-definite, by virtue of the positive-definiteness of A — see Sec. B.3. Show that the $(0, 0)$ entry of A^{-1} is positive and given by $[A^{-1}]_{00} = (a - x^*B^{-1}x)^{-1}$. Conclude that $[A]_{00} \geq 1/[A^{-1}]_{00}$.

Problem II.41 (DFE as linear estimation) Refer to the discussion at the end of Sec. 6.4. We wish to show that expressions (6.36) and (6.37) lead to the same solutions (6.29)–(6.30).

(a) Partition the vector $\boldsymbol{r}$ into $\boldsymbol{r} = \text{col}\{\bar{\boldsymbol{s}}, \boldsymbol{y}_i\}$, with $\bar{\boldsymbol{s}}$ denoting the top entries of $\boldsymbol{r}$ that are dependent on $\boldsymbol{s}(\cdot)$, and $\boldsymbol{y}_i$ is as in (6.23). Using (6.23), verify that

$$\boldsymbol{s}_\Delta = \begin{bmatrix} \boldsymbol{s}(i-\Delta) \\ \bar{\boldsymbol{s}} \end{bmatrix}, \qquad R_r \triangleq \mathsf{E}\boldsymbol{r}\boldsymbol{r}^* = \begin{bmatrix} R_{\bar{s}} & R_{\bar{s}y} \\ R_{y\bar{s}} & R_y \end{bmatrix}$$

where $R_{\bar{s}} = \mathsf{E}\,\bar{\boldsymbol{s}}\bar{\boldsymbol{s}}^*$ and $R_{\bar{s}y} = \mathsf{E}\,\bar{\boldsymbol{s}}\boldsymbol{y}^*$. Check also that

$$R_{sr} = \mathsf{E}\,\boldsymbol{s}(i-\Delta)\boldsymbol{r}^* = \begin{bmatrix} \mathsf{E}\,\boldsymbol{s}(i-\Delta)\bar{\boldsymbol{s}}^* & \mathsf{E}\,\boldsymbol{s}(i-\Delta)\boldsymbol{y}^* \end{bmatrix}$$

(b) Let $R_z = R_{\bar{s}} - R_{\bar{s}y}R_y^{-1}R_{y\bar{s}}$. Verify that

$$R_r^{-1} = \begin{bmatrix} \mathrm{I} & 0 \\ -R_y^{-1}R_{y\bar{s}} & \mathrm{I} \end{bmatrix} \begin{bmatrix} R_z^{-1} & \\ & R_y^{-1} \end{bmatrix} \begin{bmatrix} \mathrm{I} & -R_{\bar{s}y}R_y^{-1} \\ & \mathrm{I} \end{bmatrix}$$

(c) Substitute the expressions for R_{sr} and R_r^{-1} into (6.36) and show that k_o^* evaluates to $k_o^* = [-\beta^* \;\; \alpha^*]$, where the row vectors $\{\alpha^*, \beta^*\}$ are given by

$$\beta^* = -\left[R_{s(i-\Delta)\bar{s}} - R_{s(i-\Delta)y}R_y^{-1}R_{y\bar{s}}\right]R_z^{-1}, \quad \alpha^* = \beta^* R_{\bar{s}y}R_y^{-1} + R_{s(i-\Delta)y}R_y^{-1}$$

where $R_{s(i-\Delta)\bar{s}} = \mathsf{E}\,\boldsymbol{s}(i-\Delta)\bar{\boldsymbol{s}}^*$ and $R_{s(i-\Delta)y} = \mathsf{E}\,\boldsymbol{s}(i-\Delta)\boldsymbol{y}^*$.

(d) Verify that the matrix R_δ in (6.24) is given by

$$R_\delta = \begin{bmatrix} \sigma_s^2 - R_{s(i-\Delta)y}R_y^{-1}R_{ys(i-\Delta)} & R_{s(i-\Delta)\overline{s}} - R_{s(i-\Delta)y}R_y^{-1}R_{y\overline{s}} \\ R_{\overline{s}s(i-\Delta)} - R_{\overline{s}y}R_y^{-1}R_{ys(i-\Delta)} & R_z \end{bmatrix}$$

Evaluate the first row of R_δ^{-1} and show that

$$\frac{e_0^{\mathsf{T}}R_\delta^{-1}}{e_0^{\mathsf{T}}R_\delta^{-1}e_0} = \begin{bmatrix} 1 & \beta^* \end{bmatrix}$$

Conclude that $b_{\text{opt}}^* = [1 \;\; \beta^*]$ and $f_{\text{opt}}^* = \alpha^*$.

(e) Show that expression (6.37) for the m.m.s.e. coincides with (6.29).

Problem II.42 (DFE for a multi-transmit multi-receive antenna system) Refer to Prob. II.29. We reconsider the two-transmit two-receive antenna system of that problem and proceed to design a decision-feedback equalizer for it, as opposed to a linear equalizer. The arguments extend the derivation of Sec. 6.4 to the MIMO case. Although we focus on a 2×2 channel model, the results can be extended to multi-antenna systems in a straightforward manner.

At each time instant i, define the vectors of received samples, transmitted symbols, and delayed decisions,

$$\boldsymbol{y}_i \triangleq \begin{bmatrix} \boldsymbol{y}_1(i) \\ \boldsymbol{y}_2(i) \end{bmatrix}, \qquad \boldsymbol{s}_i = \begin{bmatrix} \boldsymbol{s}_1(i) \\ \boldsymbol{s}_2(i) \end{bmatrix}, \qquad \check{\boldsymbol{s}}_{i-\Delta} = \begin{bmatrix} \check{\boldsymbol{s}}_1(i-\Delta) \\ \check{\boldsymbol{s}}_2(i-\Delta) \end{bmatrix}$$

The measurement vectors $\{\boldsymbol{y}_i\}$ are fed into a feedforward FIR filter with 3 matrix taps (i.e., $L = 3$) of dimensions 2×2 each. Likewise, the delayed decisions are fed into a feedback FIR filter with 3 matrix taps, including B_0, of dimensions 2×2 each, so that $Q = 2$ (in Sec. 6.4 we defined Q as the number of taps excluding the direct path coefficient B_0). The input of the decision device is given by

$$\hat{\boldsymbol{s}}_{i-\Delta} = F_0\boldsymbol{y}_i + F_1\boldsymbol{y}_{i-1} + F_2\boldsymbol{y}_{i-2} - B_0\check{\boldsymbol{s}}_{i-\Delta} - B_1\check{\boldsymbol{s}}_{i-\Delta-1} - B_2\check{\boldsymbol{s}}_{i-\Delta-2}$$

where the $\{F_k\}$ and the $\{-B_k\}$ denote the matrix coefficients of the feedforward and feedback filters. It is assumed in the sequel that the decisions that are fed back are correct so that $\check{\boldsymbol{s}}_{i-\Delta-k} = \boldsymbol{s}_{i-\Delta-k}$ for $k = 0, 1, 2$. Our objective is to select the $\{F_k, B_k\}$ in order to minimize the covariance matrix of the estimation error, namely,

$$\min_{F_k, B_k} \; \mathsf{E}\, \tilde{\boldsymbol{s}}_{i-\Delta}\tilde{\boldsymbol{s}}_{i-\Delta}^*$$

(a) Collect the signals within the feedforward and feedback filters into the following vectors:

$$\boldsymbol{y}_{i:i-2} \triangleq \begin{bmatrix} \boldsymbol{y}_i \\ \boldsymbol{y}_{i-1} \\ \boldsymbol{y}_{i-2} \end{bmatrix}, \qquad \boldsymbol{s}_\Delta \triangleq \begin{bmatrix} \boldsymbol{s}_{i-\Delta} \\ \boldsymbol{s}_{i-\Delta-1} \\ \boldsymbol{s}_{i-\Delta-2} \end{bmatrix}$$

and define the corresponding covariance matrix, assumed positive-definite,

$$R \triangleq \mathsf{E}\begin{bmatrix} \boldsymbol{s}_\Delta \\ \boldsymbol{y}_{i:i-2} \end{bmatrix}\begin{bmatrix} \boldsymbol{s}_\Delta \\ \boldsymbol{y}_{i:i-2} \end{bmatrix}^* \triangleq \begin{bmatrix} R_s & R_{sy} \\ R_{ys} & R_y \end{bmatrix}$$

Define the filter matrices $F = \begin{bmatrix} F_0 & F_1 & F_2 \end{bmatrix}$ and $B = \begin{bmatrix} \mathrm{I}_2 + B_0 & B_1 & B_2 \end{bmatrix}$. Verify that $\tilde{\boldsymbol{s}}_{i-\Delta} = B\boldsymbol{s}_\Delta - F\boldsymbol{y}_{i:i-2}$ and show that the optimal coefficient matrices $\{F, B\}$ are related via

$$\boxed{F_{\text{opt}} = B_{\text{opt}}R_{sy}R_y^{-1}}$$

(b) Consider the situation in which only previous decisions from users 1 and 2 are available for feedback. That is, set $B_0 = 0$. Show that the optimal choice of B is obtained by solving

$$\min_B \; BR_\delta B^* \quad \text{subject to } B\Psi = \mathrm{I}_2, \quad \Psi \stackrel{\Delta}{=} \begin{bmatrix} \mathrm{I}_2 \\ 0 \end{bmatrix}$$

where $R_\delta = R_s - R_{sy}R_y^{-1}R_{ys}$. Use the result of Prob. II.39 to conclude that $B_{\text{opt}} = (\Psi^* R_\delta^{-1}\Psi)^{-1}\Psi^* R_\delta^{-1}$, and that the resulting minimum mean-square error is m.m.s.e. $= (\Psi^* R_\delta^{-1}\Psi)^{-1}$.

(c) Introduce the vectors

$$\boldsymbol{s}_{i:i-3} = \begin{bmatrix} \boldsymbol{s}_i \\ \boldsymbol{s}_{i-1} \\ \boldsymbol{s}_{i-2} \\ \boldsymbol{s}_{i-3} \end{bmatrix}, \qquad \boldsymbol{v}_{i:i-2} \stackrel{\Delta}{=} \begin{bmatrix} \boldsymbol{v}_i \\ \boldsymbol{v}_{i-1} \\ \boldsymbol{v}_{i-2} \end{bmatrix}, \qquad \boldsymbol{v}_i \stackrel{\Delta}{=} \begin{bmatrix} \boldsymbol{v}_1(i) \\ \boldsymbol{v}_2(i) \end{bmatrix}$$

Verify that $\boldsymbol{y}_{i:i-2} = H\boldsymbol{s}_{i:i-3} + \boldsymbol{v}_{i:i-2}$, where the channel matrix H has a block Toeplitz structure and is given by

$$H \stackrel{\Delta}{=} \begin{bmatrix} \mathrm{I}_2 & P & & \\ & \mathrm{I}_2 & P & \\ & & \mathrm{I}_2 & P \end{bmatrix}$$

for some 2×2 matrix P.

(d) Let $R_v = \mathsf{E}\,\boldsymbol{v}_{i:i-2}\boldsymbol{v}_{i:i-2}^*$ and $R_{\underline{s}} = \mathsf{E}\,\boldsymbol{s}_{i:i-3}\boldsymbol{s}_{i:i-3}^*$. Show that

$$R_y = HR_{\underline{s}}H^* + R_v \quad \text{and} \quad R_{sy} = (\mathsf{E}\,\boldsymbol{s}_\Delta \boldsymbol{s}_{i:i-3}^*)\, H^*$$

If the maximum number of taps among the channels $\{h_{11}, h_{12}, h_{21}, h_{22}\}$ is denoted by M, show that Δ should satisfy

$$\Delta + Q < L + M - 2$$

in order for the term $\mathsf{E}\,\boldsymbol{s}_\Delta\boldsymbol{s}_{i:i-3}^*$ to have the form $\mathsf{E}\,\boldsymbol{s}_\Delta\boldsymbol{s}_{i:i-3}^* = \begin{bmatrix} 0 & R_s & 0 \end{bmatrix}$. Verify that the leading zero block has dimensions $2(Q+1) \times 2\Delta$, and that $\mathsf{E}\,\boldsymbol{s}_\Delta\boldsymbol{s}_{i:i-3}^*$ is $2(Q+1) \times 2(L+M-1)$.

(e) Assume $R_s = \sigma_s^2\mathrm{I}$, which arises when the sequences $\{s_1(\cdot), s_2(\cdot)\}$ have the same variances. Verify that

$$R_\delta = \Phi\left(\frac{1}{\sigma_s^2}\mathrm{I} + H^* R_v^{-1} H\right)^{-1}\Phi^*$$

where $\Phi = \begin{bmatrix} 0 & \mathrm{I} & 0 \end{bmatrix}$. Identify the sizes of the blocks in Φ.

Remark. The results of Probs. II.42–II.45 on MIMO DFE are based on the work by Al-Dhahir and Sayed (2000).

Problem II.43 (Selection of the delay) Consider the same scenario as in Prob. II.42 but assume that $\Delta + Q = L + M - 2$.

(a) Show that $R_\delta = \Phi\left(R_{\underline{s}}^{-1} + H^* R_v^{-1} H\right)^{-1}\Phi^*$, where $\Phi = \left[0_{2(Q+1)\times 2\Delta} \;\; \mathrm{I}_{2(Q+1)\times 2(Q+1)}\right]$.

(b) Let $X = \left(R_{\underline{s}}^{-1} + H^* R_v^{-1} H\right)$; it does not depend on Δ while Φ does. This observation suggests a way for selecting the values of $\{\Delta, Q\}$. Use the result of part (b) of Prob. II.42 to verify that the m.m.s.e. is given by m.m.s.e. $= (\Psi^*(\Phi X^{-1}\Phi^*)^{-1}\Psi)^{-1}$, where $\Psi = \text{col}\{\mathrm{I}_2, 0\}$. Introduce the triangular factorization $X = LDL^*$, where L is lower-triangular with unit diagonal entries and D is diagonal with entries $\{d_i\}$ (cf. Sec. B.3). Express the m.m.s.e. in terms of the $\{d_i\}$. Which choice of $\{Q, \Delta\}$ minimizes the trace of the m.m.s.e.?

Problem II.44 (Ordered decisions) Consider the same scenario as in Prob. II.42, but now assume that only decisions from user 1 (current and past) are available for use by user 2, and that only past decisions of user 1 are available for his use. In other words, only decisions from the lower-indexed user are available for use by the higher-indexed user. This case corresponds to restricting the coefficient B_0 to being strictly lower-triangular, so that the determination of B_{opt} reduces to a problem of the form

$$\min_{B_0,B_1,B_2} \quad BR_\delta B^* \quad \text{subject to } B\Psi = \text{I}_2 + B_0 \text{ and } B_0 \text{ strictly lower triangular}$$

(a) Let $C = \text{I} + B_0$. Use the result of Prob. II.39 to argue that the solution to the above problem is given by $B_{\text{opt}} = C_{\text{opt}}(\Psi^* R_\delta^{-1}\Psi)^{-1}\Psi^* R_\delta^{-1}$, where C_{opt} is obtained by solving the optimization problem

$$\min_C \; C(\Psi^* R_\delta^{-1}\Psi)^{-1}C^*$$

subject to C being a lower triangular matrix with unit diagonal entries.

(b) Let $X = \Psi^* R_\delta^{-1}\Psi$. That is, X is the leading block of R_δ^{-1}. Introduce the triangular factorization of X, say $X = U_x D_x U_x^*$ where U_x is upper-triangular with unit entries (cf. Sec. B.3). Show that $C_{\text{opt}} = U_x^*$. Conclude that the resulting m.m.s.e. is equal to D_x^{-1}.

(c) Show that the trace of this value for the m.m.s.e. is lower than the trace of the m.m.s.e. found in part (b) of Prob. II.42 (i.e., show that $\text{Tr}(D_x^{-1}) \leq \text{Tr}(X^{-1})$).

Problem II.45 (Multistage detection) Consider the same scenario as in Prob. II.42, but assume now that current and past decisions from user 1 are available for use by user 2 and, likewise, current and past decisions from user 2 are available for use by user 1. This corresponds to a situation in which current decisions for all users are obtained from some previous detection stage. This case restricts the coefficient B_0 to having zero diagonal elements.

(a) Follow the arguments of Prob. II.44 to show that

$$B_{\text{opt}} = C_{\text{opt}}(\Psi^* R_\delta^{-1}\Psi)^{-1}\Psi^* R_\delta^{-1}$$

where $C_{\text{opt}} = \text{I} + B_{0,\text{opt}}$ is obtained by solving the optimization problem

$$\min_C \; CX^{-1}C^* \quad \text{subject to } e_i^{\mathsf{T}} C e_i = 1 \quad \text{for } i = 0, 1$$

where $X = \Psi^* R_\delta^{-1}\Psi$, $e_0 = \text{col}\{1, 0\}$, and $e_1 = \text{col}\{0, 1\}$.

(b) Show that the two rows of C_{opt} are given by

$$e_i^{\mathsf{T}} C_{\text{opt}} = \frac{e_i^{\mathsf{T}} X}{e_i^{\mathsf{T}} X e_i}, \quad i = 0, 1$$

(c) Find the resulting m.m.s.e. Show that its trace is smaller than the traces of the m.m.s.e. values of Probs. II.42 and II.44.

Problem II.46 (Moving average process) Let $\boldsymbol{y}_k = \boldsymbol{v}_k + \boldsymbol{v}_{k-1}$, $k \geq 0$, where $\{\boldsymbol{v}_j, j \geq -1\}$ is a zero-mean stationary white-noise scalar process with unit variance. Show that

$$\hat{\boldsymbol{y}}_{k+1|k} = \frac{k+1}{k+2}(\boldsymbol{y}_k - \hat{\boldsymbol{y}}_{k|k-1})$$

Remark. For Probs. II.46–II.48, refer to the material in Chapter 7.

Problem II.47 (Correlated signal and noise processes) Consider the model $\boldsymbol{y}_i = \boldsymbol{z}_i + \boldsymbol{v}_i$, with $\mathsf{E}\,\boldsymbol{v}_i\boldsymbol{v}_j^* = R_i\delta_{ij}$, $\mathsf{E}\,\boldsymbol{v}_i\boldsymbol{z}_j^* = 0$ for $i > j$, and $\mathsf{E}\,\boldsymbol{v}_i\boldsymbol{z}_i^* = D_i$. All random variables are zero-mean. Let $R_{e,i}$ denote the covariance matrix of the innovations of the process $\{\boldsymbol{y}_i\}$. Let also $\hat{\boldsymbol{z}}_{i|i}$ denote the l.l.m.s.e. of $\boldsymbol{z}_i$ given $\{\boldsymbol{y}_j, 0 \leq j \leq i\}$. Show that $\hat{\boldsymbol{z}}_{i|i} = \boldsymbol{y}_i - (D_i + R_i)R_{e,i}^{-1}\boldsymbol{e}_i$.

Problem II.48 (Filtered residuals) Consider a process $\boldsymbol{y}_i = H_i\boldsymbol{x}_i + \boldsymbol{v}_i$, where $\{\boldsymbol{v}_i\}$ is a white-noise zero-mean process with covariance matrix R_i and uncorrelated with the zero-mean process

$\{\boldsymbol{x}_i\}$. The filtered residuals of $\{\boldsymbol{y}_i\}$ are defined as $\boldsymbol{\nu}_i = \boldsymbol{y}_i - H_i\hat{\boldsymbol{x}}_{i|i}$. Show that the $\{\boldsymbol{\nu}_i\}$ form a white-noise sequence with covariance matrix $R_{\nu,i} = R_i - H_i P_{i|i} H_i^*$.

Problem II.49 (State-space estimation) Consider the state-space model $\boldsymbol{x}_{i+1} = F\boldsymbol{x}_i + G_1\boldsymbol{u}_i + G_2\boldsymbol{u}_{i+1}$, $\boldsymbol{y}_i = H\boldsymbol{x}_i + \boldsymbol{v}_i$ for $i \geq 0$, with zero-mean uncorrelated random variables $\{\boldsymbol{x}_0, \boldsymbol{u}_i, \boldsymbol{v}_i\}$ such that $\mathsf{E}\,\boldsymbol{u}_i\boldsymbol{u}_j^* = Q_i\delta_{ij}$, $\mathsf{E}\,\boldsymbol{v}_i\boldsymbol{v}_j^* = R_i\delta_{ij}$, $\mathsf{E}\,\boldsymbol{x}_0\boldsymbol{x}_0^* = \Pi_0$. Find recursive equations for $\hat{\boldsymbol{x}}_{i|i-1}$ and $P_{i|i-1}$.

COMPUTER PROJECTS

Project II.1 (Linear equalization and decision devices) Consider the three-tap linear equalizer discussed in Ex. 4.1, and studied further in Prob. II.17 where the optimal equalizer coefficients were determined for several values of Δ. The equalizer structure is shown again in Fig. II.6. Symbols $\{\boldsymbol{s}(i)\}$ are transmitted through an FIR channel and corrupted by additive white noise $\{\boldsymbol{v}(i)\}$. The received signal $\{\boldsymbol{y}(i)\}$ is processed by a linear equalizer to generate estimators $\{\hat{\boldsymbol{s}}(i-\Delta)\}$, which are further fed into a decision device, as explained in parts (c) and (d) below. The purpose of this device is to map each $\hat{\boldsymbol{s}}(i-\Delta)$ to the closest symbol in the constellation; the result is denoted by $\check{\boldsymbol{s}}(i-\Delta)$. Choose initially $\sigma_v^2 = 0.004$ so that, according to part (f) of Prob. II.17, the signal-to-noise ratio at the input of the equalizer is approximately 25 dB (the dB value is obtained by computing $10\log(\cdot)$). The purpose of this project is to examine the performance and operation of this three-tap equalizer.

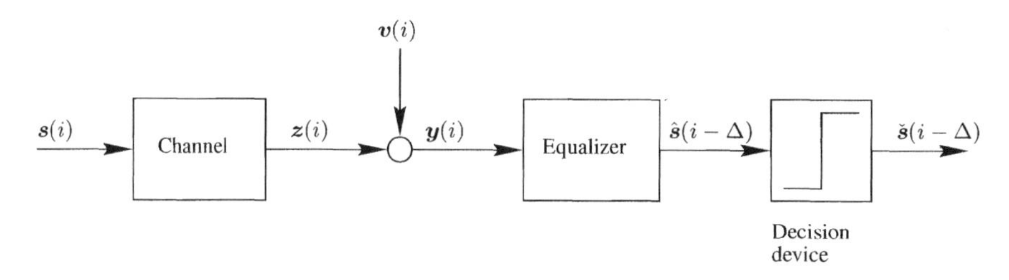

FIGURE II.6 Linear equalization of an FIR channel.

(a) Write a program that evaluates the optimal equalizer coefficients for $\Delta = 0, 1, 2, 3$ and for arbitrary noise variance σ_v^2. Use the program to feed 2000 BPSK symbols $\{\boldsymbol{s}(i)\}$ into the channel $1 + 0.5z^{-1}$ and generate the corresponding equalizer outputs $\{\hat{\boldsymbol{s}}(i-\Delta)\}$, for $\Delta = 0, 1, 2, 3$. Plot the scatter diagrams of $\{\boldsymbol{s}(i), \boldsymbol{y}(i), \hat{\boldsymbol{s}}(i-\Delta)\}$. For each Δ, estimate the m.m.s.e. by computing

$$\frac{1}{N-\Delta}\sum_{i=\Delta+1}^{N} |\boldsymbol{s}(i) - \hat{\boldsymbol{s}}(i)|^2$$

Compare the resulting values with those obtained from theory (cf. Prob. II.17). Plot also the scatter diagrams of $\{\boldsymbol{y}(i), \hat{\boldsymbol{s}}(i-\Delta)\}$ for $\Delta = 0$ when $\sigma_v^2 = 0.05$, which corresponds to SNR$=14$ dB. Compare with the scatter diagrams at 25 dB.

(b) How would the equalizer coefficients $\{\alpha(0), \alpha(1), \alpha(2)\}$ change if the input signal $\{\boldsymbol{s}(i)\}$ were instead chosen uniformly from a QPSK constellation, i.e., $\boldsymbol{s}(i) \in \{\sqrt{2}(\pm 1 \pm j)/2\}$? Repeat the simulations of part (a) for QPSK data with $\sigma_v^2 = 0.004$ (and also $\sigma_v^2 = 0.05$). Now, however, the white noise sequence $\{\boldsymbol{v}(i)\}$ needs to be complex-valued. In order to generate such a sequence with variance σ_v^2, simply generate two separate real-valued white noise sequences $\{\boldsymbol{a}(i), \boldsymbol{b}(i)\}$ with variance $\sigma_v^2/2$ each. Then set $\boldsymbol{v}(i) = \boldsymbol{a}(i) + j\boldsymbol{b}(i)$, where $j = \sqrt{-1}$.

(c) For all cases in part (a), assume now that the output of the equalizer is applied to the nonlinear decision device:

$$\text{sign}[\hat{s}(i-\Delta)] = \begin{cases} +1 & \text{if } \hat{s}(i-\Delta) \geq 0 \\ -1 & \text{if } \hat{s}(i-\Delta) < 0 \end{cases}$$

Determine the number of erroneous decisions for each Δ. Which Δ results in the smallest number of errors?

(d) For all cases in part (b), assume that the output of the equalizer is applied to the nonlinear decision device:

$$\text{dec}[\hat{s}(i-\Delta)] = \frac{\sqrt{2}}{2}\{\ \text{sign}[\ \text{Re}(\hat{s}(i-\Delta))] + j\text{sign}[\ \text{Im}(\hat{s}(i-\Delta))]\ \}$$

which assigns $\hat{s}(i-\Delta)$ to the closest symbol in the QPSK constellation. Determine the number of erroneous decisions for each Δ. Which Δ results in the smallest number of errors?

(e) Fix $\Delta = 1$ and vary the value of σ_v^2 between 0.004 and 0.2 in increments of 0.001, so that the SNR is varied between 8 and 25 dB. Write a program that generates a plot showing how the symbol error rate (SER) varies with SNR. [The SER is defined as the number of erroneous symbols relative to the total number of transmissions.]

(f) Repeat part (e) for QPSK data, by repeating the simulations of parts (b) and (d).

(g) In order to visualize the improvement that is provided by the presence of the linear equalizer, assume that the received signal $\{\boldsymbol{y}(i)\}$ is applied directly to the nonlinear decision device (i.e., let us remove the equalizer). The output of the decision device is then taken as $\hat{s}(i-\Delta)$. For both cases of BPSK and QPSK data, generate plots that show how the symbol error rate varies with the SNR. Compare these plots with the ones obtained in parts (e) and (f) for small and large values of σ_v^2.

Project II.2 (Beamforming) Refer to the discussion in Sec. 6.5 on antenna beamforming, and consider a uniform linear antenna array consisting of 4 elements spaced by $d = \lambda/5$.

(a) Assume first that the noise signal at each antenna is white with real and imaginary parts that are Gaussian with variances 0.1 each (hence, $\sigma_v^2 = 0.2$). Assume also that the noise signals across the antennas are uncorrelated with each other (i.e., assume spatial whiteness in addition to temporal whiteness). Design an optimal beamformer with unit response along the direction 30° and find the theoretical m.m.s.e. Simulate the operation of the beamformer by using as baseband signal the sinusoid $s(t) = \cos(2\pi t)$. Plot portions of the baseband signal $s(\cdot)$ and the signals received at the antennas (sampled at the rate of 100 samples per second). Plot also $s(\cdot)$ along with the real part of the output of the beamformer. Estimate the m.m.s.e. and compare it with the theoretical value.

(b) Assume now that the input signal is the sum of two sinusoids arriving along different directions; one impinges on the antennas at 30°, while the other arrives at 45°. Simulate again the operation of the beamformer by using as input signal $s(t) = \cos(2\pi t) + 0.2\sin(4\pi t)$, where the latter sinusoid is the one arriving at 45°. Assume that the sampling rate is still 100 samples per second. Estimate again the m.m.s.e. and compare it with the theoretical value. Can you explain the difference?

(c) Assume now that the noise signals across the antennas are spatially correlated with an exponential auto-correlation function equal to $0.85^{|k|}$, so that

$$R_v = \begin{bmatrix} 1.0000 & 0.8500 & 0.7225 & 0.6141 \\ 0.8500 & 1.0000 & 0.8500 & 0.7225 \\ 0.7225 & 0.8500 & 1.0000 & 0.8500 \\ 0.6141 & 0.7225 & 0.8500 & 1.0000 \end{bmatrix}$$

In order to generate a noise vector $\boldsymbol{v}$ with this spatial auto-correlation function, we proceed as follows. Use the command chol of MATLAB to determine the 4×4 lower triangular matrix $\bar{L}$ such that $\bar{L}\bar{L}^* = R_v$ (every positive-definite matrix R_v admits a unique factorization of this form — see Sec. B.3). Then generate a complex-valued $M \times 1$ vector e with unit-diagonal covariance matrix by using, say, the command randn, and let $v = \bar{L}e$. Show that the vector v so generated has covariance matrix R_v. Now repeat the simulations of parts (a) and (b) for this case.

(d) Consider again the case of spatially white noise, as in part (a), and assume that the beamformer has been designed optimally for 30°. Now let $s(t) = 1$ be the input signal and perform a simulation that varies the direction of arrival of $s(t)$ between 1° and 180° in increments of 1°. For each direction θ, determine the output of the beamformer, $\hat{s}(t)$, and estimate its average power,

$$\hat{P}(\theta) \triangleq \frac{1}{N}\sum_{i=0}^{N-1} |\hat{s}(i)|^2$$

where N is the number of samples of $\hat{s}(t)$, and $\hat{s}(i)$ its i-th sample. The power gain of the beamformer at direction θ is given by $10\log \hat{P}(\theta)$. Use the command polar of MATLAB to generate a plot of the power gain of the antenna as a function of θ. Simulate two cases: a beamformer with 4 antennas, as above, and a beamformer with 25 antennas.

Project II.3 (Decision feedback equalization) In this project we study the performance of decision feedback equalization for the channel

$$C(z) = 0.5 + 1.2z^{-1} + 1.5z^{-2} - z^{-3}$$

The symbols $\{s(i)\}$ that are transmitted through $C(z)$ are i.i.d. and chosen from a QPSK constellation, i.e., each $s(i)$ is selected randomly from the set

$$\left\{\pm\frac{\sqrt{2}}{2} \pm j\frac{\sqrt{2}}{2}\right\}, \quad j = \sqrt{-1}$$

The noise sequence $\{v(i)\}$ is assumed i.i.d. and complex-valued; its real and imaginary parts are uncorrelated Gaussian random variables with variances 0.039 each, so that $\sigma_v^2 = 0.078$ and the SNR ratio at the input of the equalizer is approximately 18 dB. We start with $L = 13$ and $Q = 2$.

(a) Plot the impulse response, as well as the magnitude of the frequency response, of the channel. Is the channel minimum phase?

(b) Generate $N = 2000$ QPSK data points $\{s(i)\}$ and transmit them through the channel. Plot the scatter diagrams of the transmitted sequence $\{s(i)\}$ and the received sequence $\{y(i)\}$.

(c) Compute the optimal filters $\{f_{\text{opt}}^*, b_{\text{opt}}^*\}$ for values of Δ in the interval $0 \leq \Delta \leq 15$ and generate the sequences $\{\hat{s}(i-\Delta)\}$ and $\{\check{s}(i-\Delta)\}$ at the input and output of the decision device, which is defined by the equation

$$\text{dec}[x] = \frac{\sqrt{2}}{2}\{\text{sign}[\ \text{Re}(x)\ ] + j\text{sign}[\ \text{Im}(x)\ ]\}$$

Plot the number of erroneous decisions as a function of Δ. For $\Delta = 5$, plot the scatter diagrams of the received sequence $\{y(i)\}$ and of the input to the decision device, $\{\hat{s}(i-5)\}$.

(d) For each Δ, compute the theoretical m.m.s.e. by using m.m.s.e. $= 1/e_1^{\mathsf{T}}R_\delta^{-1}e_1$, and plot its value as a function of Δ. Using the actual data, estimate the m.m.s.e. by computing

$$\frac{1}{N-\Delta}\sum_{i=\Delta+1}^{N} |s(i) - \hat{s}(i)|^2$$

Compare the resulting values with the theoretical values. Can you explain why there is a bad fit between theory and practice for smaller values of Δ? Plot also for $\Delta = 5$, the following sequences on three separate subplots:

(i) The channel impulse response sequence.

(ii) The impulse response sequence of the cascade combination of the channel and the feedforward filter.

(iii) The impulse response sequence of the feedback filter delayed by the value of Δ.

You will observe that the sequence in part (ii) has an almost unit-magnitude sample at time instant 5, followed by two nonzero samples that correspond to what we call *post inter-symbol interference*. This interference should be cancelled by the feedback filter. Any residual ISI prior to the peak sample at time instant 5 will not be equalized. Compare the coefficients of the feedback filter in (iii) to the values of the post ISI.

(e) Fix $\Delta = 5$ and let us now examine the effect of changing the length of the feedforward filter. Generate a plot showing the number of erroneous decisions as a function of L, for L varying between 1 and 15. Keep Q fixed at $Q = 2$. Which value of L results in the smallest number of errors?

(f) Now fix $\Delta = 5$ and $L = 6$, and let us vary Q. Generate a plot showing the number of erroneous decisions as a function of Q, for Q varying between 1 and 6. Which value of Q results in the smallest number of errors?

(g) Now fix $\Delta = 5$, $L = 6$, and $Q = 1$. That is, the feedforward filter has 6 taps and the feedback filter has a single tap. In all derivations and simulations so far we assumed $\sigma_v^2 = 0.078$. Now let σ_v^2 vary between 0.12 and 0.78, say in increments of 0.001. Write a program that generates a plot showing how the symbol error rate (SER) varies with SNR.

(h) Let us now compare the performance of the DFE with that of a linear equalizer for the same channel. Recall that we studied linear equalizers in Computer Project II.1. Write a program that determines the optimal linear equalizer for L varying between 1 and 10. The output of the equalizer is fed into the decision device. Generate a plot that shows the number of erroneous decisions as a function of L. Use $\sigma_v^2 = 0.078$ and $\Delta = 4$ for the linear equalizer. Fix $L = 4$ for the linear equalizer and plot the scatter diagrams of the received sequence $\{\boldsymbol{y}(i)\}$ and of the input to the decision device, $\{\hat{\boldsymbol{s}}(i-4)\}$. For this particular channel, do you see any advantage in using the DFE structure over the linear structure?

(i) Now assume the channel $C(z)$ and the noise variance σ_v^2 are not known beforehand but that we know the first 200 transmitted symbols $\{s(i)\}$, in addition to the entire received data record $\{\boldsymbol{y}(i)\}$. Use the initial 200 data $\{s(i), \boldsymbol{y}(i)\}$ to estimate $C(z)$ and σ_v^2, as explained in Sec. 6.3. Note that while the coefficients of the actual channel $C(z)$ are real-valued, the estimated coefficients will in general be complex-valued. You may use the complex-valued estimates, or you may keep only their real parts. If the estimates are good enough, their imaginary parts should be small compared to the real parts. Plot the impulse and frequency responses of the estimated channel and compare them with that of the actual channel. Repeat the design of the DFE equalizer by using the estimates of $C(z)$ and σ_v^2 instead. Use $\sigma_v^2 = 0.078$, $L = 6, Q = 1$, and $\Delta = 5$. Compare the number of errors in this case with the one obtained in part (g) using the exact channel model and the exact noise variance.

(j) Now repeat part (i) using a linear equalizer of length 4, followed by the nonlinear decision device. Compare the number of errors you get in this case with that obtained in part (h) and also with the DFE.

PART III

STOCHASTIC GRADIENT ALGORITHMS

CHAPTER 8

Steepest–Descent Technique

The earlier chapters discussed in some detail the theory of least-mean-squares estimation and highlighted several applications in the context of channel equalization, channel estimation, and antenna beamforming. While the chapters in Part I (*Optimal Estimation*) studied optimal estimators, which are generally nonlinear functions of the observations, the chapters in Part II (*Linear Estimation*) focused on linear estimators with and without constraints. In all cases, the estimators were optimal in the sense that they minimized the mean-square error.

Now there are situations where a designer may be interested in other performance criteria, other than the mean-square error criterion. Several examples to this effect are provided in the problems at the end of this part (e.g., Probs. III.12–III.18). In most of these cases it is generally not possible to describe the optimal solution $\hat{x}$ in *closed-form* in terms of the moments of the underlying variables, and it often becomes necessary to approximate the optimal solution iteratively. The iterative procedure would start from an initial guess for the solution and then improve upon it from one iteration to another. The purpose of this chapter is to describe one class of iterative schemes known as steepest-descent methods, which is at the core of most adaptive filtering techniques.

The steepest-descent methods will be initially motivated by showing how they apply to the *already-studied* case of linear least-mean-squares estimation. By focusing on a situation that is familiar to the reader, and one for which the optimal solution is already known, we will be able to highlight some of the abilities (and deficiencies) of iterative schemes. In particular, we will be able to show, even for the linear estimation problem, that steepest-descent methods are of independent value in their own right. For instance, they will help us avoid the need to invert R_y in order to determine K_o in the solution of the normal equations $K_o R_y = R_{xy}$. Such matrix inversions are challenging from a complexity point of view (requiring of the order of N^3 computations for an $N \times N$ matrix R_y); they are also challenging for ill-conditioned matrices R_y, namely, for matrices that are close to singular and that have a large ratio of largest to smallest eigenvalues. Once the main idea of steepest-descent has been examined in the context of linear estimation, we shall then show how to extend the technique to other estimation problems, with more involved performance criteria.

Steepest-descent methods are not studied in this chapter only because they provide a mechanism for solving more involved estimation problems. In addition to this useful objective, these methods are also important because they will serve as the launching pad for the development of adaptive filters in Chapters 10–14. It is because of this latter objective that, from now on and until the end of this textbook, we shall adopt a notation that is more specific, and also more suited, to the study of adaptive filters.

Notation. In Parts I (*Optimal Estimation*) and II (*Linear Estimation*) of this book, we adopted the $\{\boldsymbol{x}, \boldsymbol{y}\}$ notation, as is common in estimation theory, for the variable to be estimated and for the observation vector. The variables $\{\boldsymbol{x}, \boldsymbol{y}\}$ were general and they could refer to scalars or vectors. The results of the earlier chapters are of broad interest and they are not exclusive to the study of adaptive filters. However, from now on, we shall develop the theory of adaptive filters in greater detail. In this context, we will be mostly interested in the case in which $\boldsymbol{x}$ is a *scalar* and $\boldsymbol{y}$ is a *row* vector. Moreover, the $\{\boldsymbol{x}, \boldsymbol{y}\}$ variables will have specific meanings attached to them. For instance, $\boldsymbol{x}$ will denote the so-called "desired signal" and we shall replace it by the letter $\boldsymbol{d}$, which will be a scalar. The observation vector $\boldsymbol{y}$, on the other hand, will be a row vector and it will be denoted by $\boldsymbol{u}$. In this way, we are now interested in estimating $\boldsymbol{d}$ from $\boldsymbol{u}$. Some motivation for our choice of the *row* vector notation for $\boldsymbol{u}$ appears in the Notation section in the opening pages of this book.

◇

8.1 LINEAR ESTIMATION PROBLEM

So let $\boldsymbol{d}$ be a zero-mean *scalar-valued* random variable with variance σ_d^2,

$$\boxed{\mathsf{E}\,\boldsymbol{d} = 0, \qquad \sigma_d^2 = \mathsf{E}\,|\boldsymbol{d}|^2}$$

and let $\boldsymbol{u}$ be a $1 \times M$ zero-mean random row vector with a positive-definite covariance matrix denoted by R_u

$$\boxed{R_u \stackrel{\Delta}{=} \mathsf{E}\,\boldsymbol{u}^*\boldsymbol{u}} \qquad \text{(a square matrix)}$$

The variables $\{\boldsymbol{d}, \boldsymbol{u}\}$ are allowed to be complex-valued for generality, which, as we saw in several examples in Chapters 1–6, is usually a necessity in digital communications applications. The $M \times 1$ cross-covariance vector of $\{\boldsymbol{d}, \boldsymbol{u}\}$ is denoted by

$$\boxed{R_{du} \stackrel{\Delta}{=} \mathsf{E}\,\boldsymbol{d}\boldsymbol{u}^*} \qquad \text{(a column vector)}$$

We then consider the problem of estimating $\boldsymbol{d}$ from $\boldsymbol{u}$ in the linear least-mean-squares sense as follows:

$$\boxed{\min_{w} \ \mathsf{E}\,|\boldsymbol{d} - \boldsymbol{u}w|^2} \tag{8.1}$$

where w is $M \times 1$ and is known as the weight vector.

Remark 8.1 (Row vector notation) Observe that since we choose $\boldsymbol{u}$ to be a row vector and the unknown w to be a column vector, the inner-product between $\boldsymbol{u}$ and w is simply written as $\boldsymbol{u}w$ with no transposition or conjugation symbols needed.

> We adopt this convention throughout our treatment of adaptive filters in this and subsequent chapters.

All vectors, from this chapter onwards, will be column vectors with the notable exception of $\boldsymbol{u}$, which will be a row vector.

We can proceed to solve (8.1) either afresh (i.e., from first principles) or by invoking the solution of the linear least-mean-squares estimation problem from Chapter 3. We shall argue in both ways in order to reinforce the main ideas.

Using Linear Estimation Theory

One way to determine the solution of (8.1) is to observe that it is a special case of the linear least-mean-squares estimation problem studied in Chapter 3, namely,

$$\min_{K} \mathsf{E}\,(\boldsymbol{x} - K\boldsymbol{y})(\boldsymbol{x} - K\boldsymbol{y})^* \tag{8.2}$$

whose solution was seen to be

$$K_o = R_{xy}R_y^{-1}$$

with minimum cost given by

$$\text{m.m.s.e.} = R_x - R_{xy}R_y^{-1}R_{yx}$$

and where $R_{xy} = \mathsf{E}\,\boldsymbol{x}\boldsymbol{y}^*$ and $R_y = \mathsf{E}\,\boldsymbol{y}\boldsymbol{y}^*$. The statement (8.1) is a slight variation of (8.2) with the unknown w multiplying the observation $\boldsymbol{u}$ from the right. However, if we replace $\boldsymbol{d} - \boldsymbol{u}w$ by its conjugate, we can restate (8.1) as

$$\min_{w^*} \mathsf{E}\,|\boldsymbol{d}^* - w^*\boldsymbol{u}^*|^2 \tag{8.3}$$

which is now of the form (8.2). In particular, we can make the identifications:

$$\begin{array}{cccccccc} \boldsymbol{d}^* & \longleftarrow & \boldsymbol{x}, & \boldsymbol{u}^* \longleftarrow \boldsymbol{y}, & w^* \longleftarrow K \\ \sigma_d^2 & \longleftarrow & R_x, & R_u \longleftarrow R_y, & R_{d^*u^*} \longleftarrow R_{xy} \end{array}$$

Using the already known solution to (8.2), we find that the solution $(w^o)^*$ of (8.3) is given by

$$(w^o)^* = R_{d^*u^*}R_u^{-1} = R_{du}^*R_u^{-1}$$

so that, by transposition,

$$\boxed{w^o = R_u^{-1}R_{du}} \tag{8.4}$$

and the resulting m.m.s.e. is

$$\boxed{\text{m.m.s.e.} = \sigma_d^2 - R_{ud}R_u^{-1}R_{du}} \tag{8.5}$$

where $R_{ud} = \mathsf{E}\,\boldsymbol{u}\boldsymbol{d}^* = R_{du}^*$. We refer to w^o as the optimal weight vector; the terminology "weight vector" refers to the fact that $\boldsymbol{u}w^o$ is a weighted combination of the entries of $\boldsymbol{u}$.

Using the Orthogonality Principle

Alternatively, we can solve (8.1) more directly by invoking the orthogonality principle of linear least-mean-squares estimation. Specifically, from Thm. 4.1, we know that the optimal weight vector w^o should lead to an error variable, $\boldsymbol{d} - \boldsymbol{u}w^o$, that is orthogonal to the observation vector $\boldsymbol{u}$, i.e., it must hold that $\boldsymbol{d} - \boldsymbol{u}w^o \perp \boldsymbol{u}$ or, equivalently,

$$\mathsf{E}\,\boldsymbol{u}^*(\boldsymbol{d} - \boldsymbol{u}w^o) = 0 \tag{8.6}$$

which means that w^o should satisfy the normal equations $R_{du} - R_u w^o = 0$, and we are back to (8.4). Likewise, the resulting m.m.s.e. can be obtained from the orthogonality condition as follows:

$$\begin{aligned} \text{m.m.s.e.} &= \mathsf{E}\,|\boldsymbol{d} - \boldsymbol{u}w^o|^2 \\ &= \mathsf{E}\,(\boldsymbol{d} - \boldsymbol{u}w^o)(\boldsymbol{d} - \boldsymbol{u}w^o)^* \\ &= \mathsf{E}\,(\boldsymbol{d} - \boldsymbol{u}w^o)\boldsymbol{d}^* \qquad \text{(because of (8.6))} \\ &= \sigma_d^2 - R_{ud}R_u^{-1}R_{du} \end{aligned}$$

as in (8.5).

Using Completion-of-Squares

A third way to solve (8.1) is to proceed afresh, from first principles, by using a completion-of-squares argument similar to the one we employed in Sec. 3.3. So let

$$J(w) \;\stackrel{\Delta}{=}\; \mathsf{E}\,|\boldsymbol{d} - \boldsymbol{u}w|^2 \;=\; \mathsf{E}\,(\boldsymbol{d} - \boldsymbol{u}w)(\boldsymbol{d} - \boldsymbol{u}w)^* \tag{8.7}$$

denote the cost function in (8.1). Expanding $J(w)$ we get

$$\boxed{J(w) \;=\; \sigma_d^2 - R_{du}^* w - w^* R_{du} + w^* R_u w} \tag{8.8}$$

which can be expressed in vector form as

$$J(w) = \begin{bmatrix} 1 & w^* \end{bmatrix} \begin{bmatrix} \sigma_d^2 & -R_{ud} \\ -R_{du} & R_u \end{bmatrix} \begin{bmatrix} 1 \\ w \end{bmatrix} \tag{8.9}$$

We can now factor the center matrix into a product of upper-triangular, diagonal, and lower-triangular block matrices:

$$\begin{bmatrix} \sigma_d^2 & -R_{ud} \\ -R_{du} & R_u \end{bmatrix} = \begin{bmatrix} 1 & -R_{ud}R_u^{-1} \\ 0 & 1 \end{bmatrix} \begin{bmatrix} \sigma_d^2 - R_{ud}R_u^{-1}R_{du} & \\ & R_u \end{bmatrix} \begin{bmatrix} 1 & 0 \\ -R_u^{-1}R_{du} & 1 \end{bmatrix}$$

Substituting into (8.9) we obtain

$$J(w) = (\sigma_d^2 - R_{ud}R_u^{-1}R_{du}) \;+\; (w - R_u^{-1}R_{du})^* R_u (w - R_u^{-1}R_{du}) \tag{8.10}$$

from which it is clear, since $R_u > 0$, that $J(w)$ is minimized by choosing w as $w^o = R_u^{-1}R_{du}$, with the resulting minimum value given by

$$\boxed{J_{\min} \;\stackrel{\Delta}{=}\; J(w^o) \;=\; \text{m.m.s.e.} \;=\; \sigma_d^2 \;-\; R_{ud}R_u^{-1}R_{du}} \tag{8.11}$$

In summary, we arrive at the following statement.

> **Theorem 8.1 (Optimal linear estimator)** All random variables are zero-mean. Consider a scalar variable $\boldsymbol{d}$ and a row vector $\boldsymbol{u}$ with $R_u = \mathsf{E}\,\boldsymbol{u}^*\boldsymbol{u} > 0$. The linear least-mean-squares estimator of $\boldsymbol{d}$ given $\boldsymbol{u}$ is $\hat{\boldsymbol{d}} = \boldsymbol{u}w^o$ where
>
> $$w^o = R_u^{-1}R_{du}$$
>
> The resulting minimum mean-square error is m.m.s.e. $= \sigma_d^2 - R_{ud}R_u^{-1}R_{du}$.

8.2 STEEPEST-DESCENT METHOD

The solution w^o of (8.1) is given in closed-form by (8.4). However, as mentioned in the introduction, such closed-form solutions are generally not possible for more elaborate performance criteria, other than the mean-square-error criterion (8.1). In such situations, it becomes necessary to resort to an iterative procedure in order to approximate the solution w^o. In this section we explain how one such iterative procedure can be devised for the

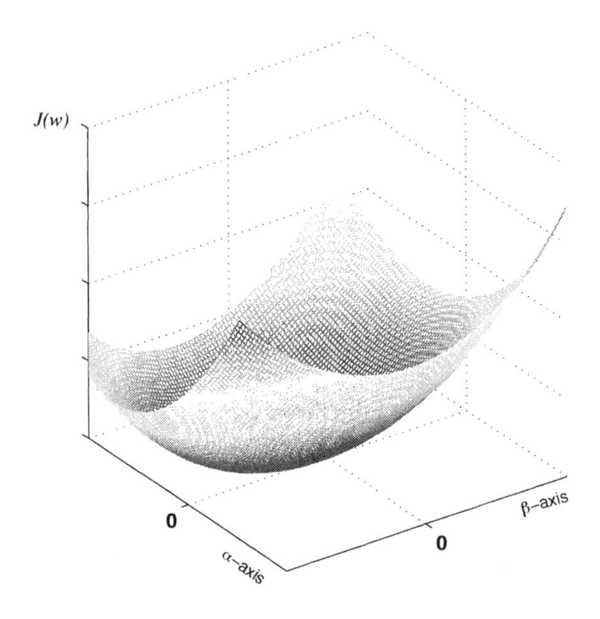

FIGURE 8.1 A typical plot of the quadratic cost function $J(w)$ when w is two-dimensional and real-valued, say, $w = \text{col}\{\alpha, \beta\}$.

already-solved problem (8.1). The ensuing discussion will be used later to motivate similar iterative procedures for more general cost functions.

Consider again the cost function $J(w)$ in (8.8), which is a scalar-valued quadratic function of w. We already know that $J(w)$ has a unique global minimum at $w^o = R_u^{-1}R_{du}$ with minimum value given by (8.11). Figure 4.1 shows a typical plot of $J(w)$ for the case in which w is two-dimensional and real-valued.

Choice of the Search Direction

Now given $J(w)$, and without assuming any prior knowledge about the location of its minimizing argument w^o, we wish to devise a procedure that starts from an initial guess for w^o and then improves upon it in a recursive manner until ultimately converging to w^o. The procedure that we seek is one of the form

$$\text{(new guess)} \;=\; \text{(old guess)} \;+\; \text{(a correction term)}$$

or, more explicitly,

$$\boxed{w_i = w_{i-1} + \mu\, p, \quad i \geq 0} \tag{8.12}$$

where we are writing w_{i-1} to denote a guess for w^o at iteration $(i-1)$, and w_i to denote the updated guess at iteration i. The vector p is an *update direction* vector that we should choose adequately, along with the positive scalar μ, in order to guarantee convergence of w_i to w^o. The scalar μ is called the *step-size* parameter since it affects how small or how large the correction term is. In (8.12), and in all future developments in this book, it is assumed that the index i runs from 0 onwards, so that the initial condition is specified at $i = -1$. Usually, but not always, the initial condition w_{-1} is taken to be zero.

The criterion for selecting μ and p is to enforce, if possible, the condition $J(w_i) < J(w_{i-1})$. In this way, the value of the cost function at the successive iterations will be monotonically decreasing. To show how this condition can be enforced, we start by relating $J(w_i)$ to $J(w_{i-1})$. Evaluating $J(w)$ at $w_i = w_{i-1} + \mu p$ and expanding we get

$$\begin{aligned} J(w_i) &= \sigma_d^2 - R_{du}^*(w_{i-1}+\mu p) - (w_{i-1}+\mu p)^* R_{du} + (w_{i-1}+\mu p)^* R_u (w_{i-1}+\mu p) \\ &= J(w_{i-1}) + \mu(w_{i-1}^* R_u - R_{du}^*)p + \mu p^*(R_u w_{i-1} - R_{du}) + \mu^2 p^* R_u p \quad (8.13) \end{aligned}$$

We can rewrite this equality more compactly by observing from expression (8.8) that the gradient vector of $J(w)$ with respect to w is equal to

$$\nabla_w J(w) = w^* R_u - R_{du}^* \tag{8.14}$$

This means that the term $(w_{i-1}^* R_u - R_{du}^*)$ that appears in (8.13) is simply the value of the gradient vector at $w = w_{i-1}$, i.e.,

$$w_{i-1}^* R_u - R_{du}^* = \nabla_w J(w_{i-1})$$

Similarly, the matrix R_u appearing in $\mu^2 p^* R_u p$ is equal to the Hessian matrix of $J(w)$, i.e.,

$$\mu^2 p^* R_u p = \mu^2 p^* \left[\nabla_w^2 J(w_{i-1})\right] p$$

We can then rewrite (8.13) as

$$\boxed{J(w_i) \;=\; J(w_{i-1}) \;+\; 2\mu \, \mathrm{Re}\left[\nabla_w J(w_{i-1})p\right] \;+\; \mu^2 p^* R_u \, p} \tag{8.15}$$

in terms of the real part of the inner product $\nabla_w J(w_{i-1})p$.

Now the last term on the right-hand side of (8.13) is positive for all nonzero p since $R_u > 0$. Therefore, a *necessary* condition for

$$J(w_i) \; < \; J(w_{i-1}) \tag{8.16}$$

is to require the update direction p to satisfy

$$\boxed{\mathrm{Re}\left[\nabla_w J(w_{i-1})p\right] \; < \; 0} \tag{8.17}$$

This condition guarantees that the second term on the right-hand side of (8.15) is strictly negative. The selection of a vector p according to (8.17) will depend on whether $\nabla_w J(w_{i-1})$ is zero or not. If the gradient vector is zero, then $R_u w_{i-1} = R_{du}$, and thus w_{i-1} already coincides with the desired solution w^o. In this situation, recursion (8.12) would have attained w^o and p should be selected as $p = 0$.

When, on the other hand, the gradient vector at w_{i-1} is nonzero, there are many choices of vectors p that satisfy (8.17). For example, any p of the form

$$\boxed{p \;=\; -B\left[\nabla_w J(w_{i-1})\right]^*} \tag{8.18}$$

for any Hermitian positive-definite matrix B will do (this choice will also give $p = 0$ when $\nabla_w J(w_{i-1}) = 0$). To see this, note that for any such p, the inner product in (8.17) is real-valued and evaluates to

$$\nabla_w \, J(w_{i-1})p \;=\; -\left[\nabla_w \, J(w_{i-1})\right] B \left[\nabla_w \, J(w_{i-1})\right]^*$$

which is negative in view of the positive-definiteness of B. The special choice $B = \mathrm{I}$ is very common and it corresponds to the update direction

$$\boxed{p \;=\; -\;\left[\nabla_w J(w_{i-1})\right]^* \;=\; R_{du} - R_u w_{i-1}} \tag{8.19}$$

This choice for p reduces (8.12) to the recursion

$$\boxed{w_i = w_{i-1} + \mu[R_{du} - R_u w_{i-1}], \quad i \geq 0, \quad w_{-1} = \text{initial guess}} \tag{8.20}$$

The update direction (8.19) has a useful and intuitive interpretation. Recall that the gradient vector at any point of a cost function points toward the direction in which the function is increasing. Now (8.19) is such that, at each iteration, it chooses the update direction p to point in the *opposite* direction of the (conjugate) gradient vector. For this reason, we refer to (8.20) as a steepest-descent method; the successive weight vectors $\{w_i\}$ are obtained by descending along a path of decreasing cost values. The choice of the step-size μ is crucial and, if not chosen with care, it can destroy this desirable behavior. Choosing p according to (8.19) is only a necessary condition for (8.16) to hold; it is not sufficient as we still need to choose μ properly, as we proceed to explain.

Condition on the Step-Size for Convergence

Introduce the weight-error vector

$$\boxed{\tilde{w}_i \stackrel{\Delta}{=} w^o - w_i}$$

It measures the difference between the weight estimate at time i and the optimal weight vector, w^o, which we are attempting to reach.

Subtracting both sides of the steepest-descent recursion (8.20) from w^o we obtain

$$\tilde{w}_i = \tilde{w}_{i-1} - \mu[R_{du} - R_u w_{i-1}]$$

with initial weight-error vector $\tilde{w}_{-1} = w^o - w_{-1}$. Using the fact that w^o satisfies the normal equations $R_u w^o = R_{du}$, we replace R_{du} in the above recursion by $R_u w^o$ and arrive at the weight-error recursion:

$$\boxed{\tilde{w}_i = [\mathrm{I} - \mu R_u]\tilde{w}_{i-1}\,, \quad i \geq 0, \quad \tilde{w}_{-1} = \text{initial condition}} \tag{8.21}$$

This is a homogeneous difference equation with coefficient matrix $(\mathrm{I} - \mu R_u)$. Therefore, a necessary and sufficient condition for the error vector $\tilde{w}_i$ to tend to zero, regardless of the initial condition $\tilde{w}_{-1}$, is to require that all of the eigenvalues of the matrix $(\mathrm{I} - \mu R_u)$ be strictly less than one in magnitude. That is, $(\mathrm{I} - \mu R_u)$ must be a *stable* matrix. This conclusion is a special case of a general result. For any homogeneous recursion of the form $y_i = Ay_{i-1}$, it is well-known that the successive vectors y_i will tend to zero regardless of the initial condition y_{-1} if, and only if, all eigenvalues of A are strictly inside the unit disc. The argument that we give below establishes the result for the special case $A = \mathrm{I} - \mu R_u$. For generic matrices A, the proof is left as an exercise to the reader; see Prob. III.23.

One way to establish that $(\mathrm{I} - \mu R_u)$ must be a stable matrix is the following. Since R_u is a positive-definite Hermitian matrix, its eigen-decomposition has the form (cf. Sec. B.1):

$$\boxed{R_u = U\Lambda U^*} \tag{8.22}$$

where Λ is diagonal with positive entries, $\Lambda = \text{diag}\{\lambda_k\}$, and U is unitary, i.e., it satisfies $UU^* = U^*U = \mathrm{I}$. The columns of U, say, $\{q_k\}$, are the orthonormal eigenvectors of R_u, namely, each q_k satisfies

$$R_u q_k = \lambda_k q_k, \quad \|q_k\|^2 = 1$$

Now define the transformed weight-error vector

$$\boxed{x_i \triangleq U^* \tilde{w}_i} \tag{8.23}$$

Since U is unitary and, hence, invertible, x_i and $\tilde{w}_i$ determine each other uniquely. The vectors $\{x_i, \tilde{w}_i\}$ also have equal Euclidean norms since

$$\|x_i\|^2 = x_i^* x_i = \tilde{w}_i^* \underbrace{UU^*}_{\mathrm{I}} \tilde{w}_i = \tilde{w}_i^* \tilde{w}_i = \|\tilde{w}_i\|^2$$

Therefore, if x_i tends to zero then $\tilde{w}_i$ tends to zero and vice-versa. This means that we can instead seek a condition on μ to force x_i to tend to zero. It is more convenient to work with x_i because it satisfies a difference equation similar to (8.21), albeit one with a *diagonal* coefficient matrix. To see this, we multiply (8.21) by U^* from the left, and replace R_u by $U\Lambda U^*$ and I by UU^*, to get

$$\boxed{x_i = [\mathrm{I} - \mu\Lambda] x_{i-1}, \quad x_{-1} = U^* \tilde{w}_{-1} = \text{initial condition}} \tag{8.24}$$

The coefficient matrix for this difference equation is now diagonal and equal to $(\mathrm{I} - \mu\Lambda)$. It follows that the evolution of the individual entries of x_i are decoupled. Specifically, if we denote these individual entries by

$$x_i = \mathrm{col}\{x_1(i), x_2(i), \ldots, x_M(i)\}$$

then (8.24) shows that the k-th entry of x_i satisfies

$$x_k(i) = (1 - \mu\lambda_k) x_k(i-1)$$

Iterating this recursion from time -1 up to time i gives

$$\boxed{x_k(i) = (1 - \mu\lambda_k)^{i+1} x_k(-1), \quad i \geq 0} \tag{8.25}$$

where $x_k(-1)$ denotes the k-th entry of the initial condition x_{-1}. We refer to the coefficient $(1 - \mu\lambda_k)$ as the *mode* associated with $x_k(i)$. Now in order for $x_k(i)$ to tend to zero regardless of $x_k(-1)$, the mode $(1 - \mu\lambda_k)$ must have less than unit magnitude. This condition is both necessary and sufficient. Therefore, in order for all the entries of the transformed vector x_i to tend to zero, the step-size μ must satisfy

$$|1 - \mu\lambda_k| < 1, \quad \text{for all } k = 1, 2, \ldots, M \tag{8.26}$$

The modes $\{1 - \mu\lambda_k\}$ are the eigenvalues of the coefficient matrix $(\mathrm{I} - \mu R_u)$ in (8.21), and we have therefore established our initial claim that all eigenvalues of this matrix must be less than one in magnitude in order for $\tilde{w}_i$ to converge to zero. The condition (8.26) is of course equivalent to choosing μ such that

$$\boxed{0 < \mu < 2/\lambda_{\max}}$$

where $\lambda_{\max}$ denotes the largest eigenvalue of R_u.

Theorem 8.2 (Steepest-descent algorithm) Consider a zero-mean random variable $\boldsymbol{d}$ with variance σ_d^2 and a zero-mean random row vector $\boldsymbol{u}$ with $R_u = \mathsf{E}\,\boldsymbol{u}^*\boldsymbol{u} > 0$. Let $\lambda_{\max}$ denote the largest eigenvalue of R_u. The solution w^o of the linear least-mean-squares estimation problem

$$\min_w \; \mathsf{E}\,|\boldsymbol{d} - \boldsymbol{u}w|^2$$

can be obtained recursively as follows. Start with any initial guess w_{-1}, choose any step-size μ that satisfies $0 < \mu < 2/\lambda_{\max}$, and iterate for $i \geq 0$:

$$w_i = w_{i-1} + \mu[R_{du} - R_u w_{i-1}]$$

Then $w_i \rightarrow w^o$ as $i \rightarrow \infty$.

8.3 MORE GENERAL COST FUNCTIONS

With the above statement, we have achieved our original goal of deriving an iterative procedure for solving the least-mean-squares estimation problem

$$\min_w \; \mathsf{E}\,|\boldsymbol{d} - \boldsymbol{u}w|^2 \tag{8.27}$$

The ideas developed for this case can be applied to more general optimization problems, say,

$$\min_w J(w)$$

with cost functions $J(w)$ that are *not* necessarily quadratic in w (see, e.g., Probs. III.15–III.18). The update recursion in these cases would continue to be of the form

$$\boxed{w_i = w_{i-1} - \mu\left[\nabla_w J(w_{i-1})\right]^*} \tag{8.28}$$

in terms of the gradient vector of $J(\cdot)$, and using sufficiently small step-sizes. Clearly, the expression for the gradient vector will be different for different cost functions. Now, however, since $J(w)$ may have both local and global minima, the successive iterates w_i from (8.28) need not approach a global minimum of $J(w)$ and, subsequently, recursion (8.28) may end up converging to a local minimum.

For this reason, convergence difficulties can arise for general cost functions and it is usually difficult to predict beforehand whether (8.28) will converge to a local or global minimum. The ultimate convergence behavior will depend on the value of the step-size parameter and on the location of the initial condition w_{-1} relative to the local and global minima of $J(w)$. In the problems, we shall use (8.28) to develop steepest-descent algorithms for some cost functions with multiple minima. The problems, as well as the computer project, will illustrate these difficulties. We did not encounter any of these difficulties with the mean-square-error criterion (8.27) because the cost function in that case has a unique global minimum.

CHAPTER 9

Transient Behavior

In order to gain further insight into the workings of steepest-descent methods, we shall continue to examine recursion (8.20), namely,

$$\boxed{w_i = w_{i-1} + \mu[R_{du} - R_u w_{i-1}], \quad i \geq 0, \quad w_{-1} = \text{initial guess}} \tag{9.1}$$

which pertains to the quadratic cost function (8.8). In particular, we shall now study more closely the manner by which the weight-error vector $\tilde{w}_i$ of (8.21) tends to zero. We repeat the weight-error vector recursion here for ease of reference,

$$\boxed{\tilde{w}_i = [\mathrm{I} - \mu R_u]\tilde{w}_{i-1}\,, \quad i \geq 0, \quad \tilde{w}_{-1} = \text{initial condition}} \tag{9.2}$$

along with its transformed version (8.24):

$$\boxed{x_i = [\mathrm{I} - \mu\Lambda]x_{i-1}, \quad x_{-1} = U^*\tilde{w}_{-1} = \text{initial condition}} \tag{9.3}$$

9.1 MODES OF CONVERGENCE

To begin with, it is clear from (9.3) that the form of the exponential decay of the $k-$th entry of x_i, namely, $x_k(i)$, to zero depends on the value of the mode $1-\mu\lambda_k$. For instance, the sign of $1-\mu\lambda_k$ determines whether the convergence of $x_k(i)$ to zero occurs with or without *oscillation*. When $0 \leq 1-\mu\lambda_k < 1$ the decay of $x_k(i)$ to zero is monotonic. On the other hand, when $-1 < 1-\mu\lambda_k < 0$ the decay of $x_k(i)$ to zero is oscillatory.

Example 9.1 (Exponential decay)

Consider a two-dimensional data vector $\boldsymbol{u}$, i.e., $M = 2$ and R_u is 2×2. Assume the eigenvalues of R_u are $\lambda_{\min} = 1$ and $\lambda_{\max} = 4$. Then μ must satisfy $\mu < 2/\lambda_{\max} = 1/2$ for convergence of the steepest-descent method (9.1) to be guaranteed. If we choose $\mu = 2/5$, then the resulting modes $\{1-\mu\lambda_k\}$ will be $1-\mu\lambda_{\max} = -3/5 < 0$ and $1-\mu\lambda_{\min} = 3/5 > 0$. In this case, both entries of the transformed vector x_i will tend to zero; however, one entry will converge monotonically (the one associated with $\lambda_{\min}$) while the other entry will converge in an oscillatory manner (the one associated with $\lambda_{\max}$). This situation is illustrated in Fig. 9.1.

◇

It is also clear from (8.25) that the mode $(1-\mu\lambda_k)$ with the smallest magnitude determines the entry of x_i that decays to zero at the fastest rate. Likewise, the mode $(1-\mu\lambda_k)$

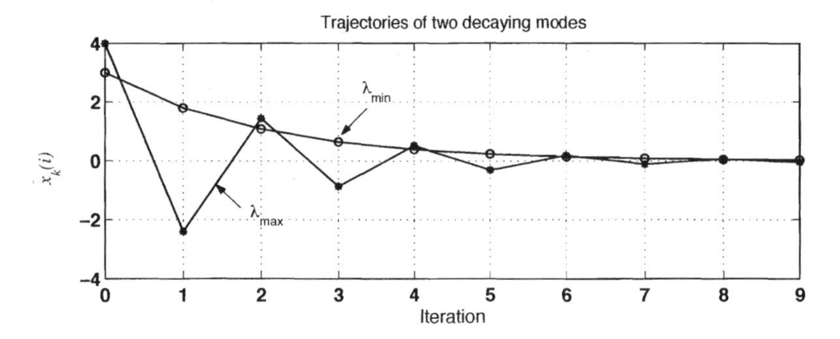

FIGURE 9.1 Two exponentially decaying modes from Ex. 9.1.

with the largest magnitude determines the entry of x_i that decays to zero at the slowest rate. The above example shows that the fastest and slowest rates of convergence are *not* necessarily the ones that are associated with the largest and smallest eigenvalues of R_u, respectively. For the numerical values used in the example, both $\lambda_{\rm min}$ and $\lambda_{\rm max}$ lead to modes $\{1 - \mu\lambda_k\}$ with identical magnitudes (equal to $3/5$). Consider the following alternative example.

Example 9.2 (Fastest rate of decay)

Assume again that $M = 2$ and that $\lambda_{\rm min} = 1$ and $\lambda_{\rm max} = 3$. Then μ must satisfy $\mu < 2/\lambda_{\rm max} = 2/3$. Choose $\mu = 7/12$. Then $1 - \mu\lambda_{\rm max} = -9/12 < 0$ and $1 - \mu\lambda_{\rm min} = 5/12 > 0$. This shows that the entry of x_i that is associated with $\lambda_{\rm min}$ (rather than $\lambda_{\rm max}$) will decay at the fastest rate.

◇

9.2 OPTIMAL STEP-SIZE

In general, for each value of μ, there are M modes, $\{1-\mu\lambda_k, k = 1, 2, \ldots, M\}$. Among these modes there will be at least one with largest magnitude. This largest-magnitude mode exhibits the slowest rate of convergence among the entries of x_i, and we therefore say that it is the one that ultimately determines the convergence rate of the algorithm (9.1). Now, different choices for μ will lead to different slowest modes. This fact suggests that we could select μ optimally by minimizing the magnitude of the slowest mode, i.e., by forcing the magnitude of the slowest mode to be as far away from one as possible. More specifically, we could choose μ optimally by solving the min-max problem:

$$\min_{\mu} \max_{\left(\begin{array}{c} k = 1, \ldots, M \\ \text{subject to} \\ |1 - \mu\lambda_k| < 1 \end{array}\right)} |1 - \mu\lambda_k| \tag{9.4}$$

Figure 9.2 plots the curves $|1 - \mu\lambda_k|$ for different $\lambda_k's$; only four curves are shown corresponding to $\lambda_{\rm max}$ (the left-most curve), $\lambda_{\rm min}$ (the right-most curve), and two other eigenvalues. It is clear from the figure that the choice of μ for which the largest-magnitude mode is furthest from one is the point at which the curves $|1 - \mu\lambda_{\rm max}|$ and $|1 - \mu\lambda_{\rm min}|$

intersect. If we denote this optimal value by μ^o, then μ^o should satisfy

$$1 - \mu^o \lambda_{\min} = -(1 - \mu^o \lambda_{\max})$$

which leads to

$$\boxed{\mu^o = \frac{2}{\lambda_{\max} + \lambda_{\min}}} \tag{9.5}$$

The figure further indicates that there are actually two optimal slowest modes, with identical magnitudes but opposite signs; they are obtained by evaluating $(1 - \mu\lambda_{\min})$ and $(1 - \mu\lambda_{\max})$ at $\mu = \mu^o$:

$$\text{optimal slowest modes} = \left\{ \pm \frac{\lambda_{\max} - \lambda_{\min}}{\lambda_{\max} + \lambda_{\min}} \right.$$

If we define the *eigenvalue spread* of the covariance matrix R_u as $\rho = \lambda_{\max}/\lambda_{\min}$, then we can also write

$$\boxed{\text{optimal slowest modes} = \pm \frac{\rho - 1}{\rho + 1}}$$

These values can be interpreted as corresponding to the most favorable slowest convergence modes that we can expect. Observe that they are dependent on the eigenvalue spread of R_u.

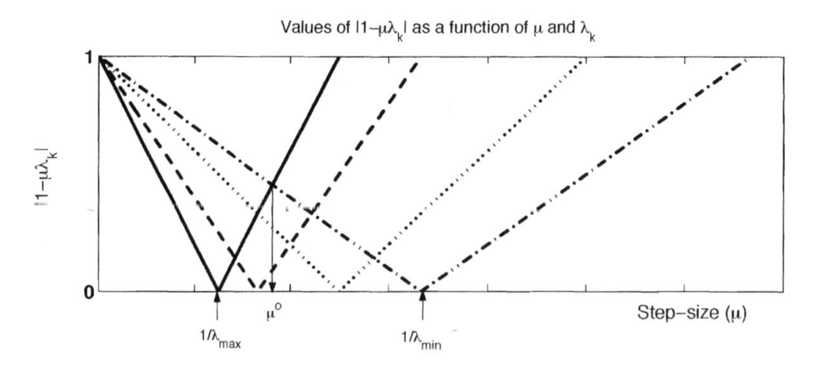

FIGURE 9.2 The plot shows the curves $\{|1 - \mu\lambda_k|\}$ restricted to the vertical interval $[0, 1)$. The optimal step-size occurs at the point of intersection of the curves $|1 - \mu\lambda_{\max}|$ and $|1 - \mu\lambda_{\min}|$.

Lemma 9.1 (Optimal step-size) Consider the setting of Thm. 8.2. The optimal selection for the step-size is

$$\mu^o = \frac{2}{\lambda_{\max} + \lambda_{\min}}$$

where $\{\lambda_{\max}, \lambda_{\min}\}$ are the maximum and minimum eigenvalues of the covariance matrix R_u. This step-size guarantees fastest convergence speed in the sense that the magnitude of the slowest modes is minimized, and this magnitude is given by

$$\text{magnitude of the optimal slowest mode(s)} = \frac{\rho - 1}{\rho + 1}$$

where ρ denotes the eigenvalue spread of R_u, $\rho = \lambda_{\max}/\lambda_{\min}$. There are two optimal slowest modes; one negative and one positive.

9.3 WEIGHT-ERROR VECTOR CONVERGENCE

Let us now examine the convergence behavior of the weight-error vector. Since, as indicated by (8.23), $\tilde{w}_i = Ux_i$, it follows that $\tilde{w}_i$ is a linear combination of the columns of U, and the coefficients of this linear combination are the entries of x_i. Using (8.25) we then get

$$\tilde{w}_i = \sum_{k=1}^{M} q_k \; x_k(i) = \sum_{k=1}^{M} (1 - \mu\lambda_k)^{i+1} \; q_k \; x_k(-1) \tag{9.6}$$

This expression shows that the convergence of $\tilde{w}_i$ to zero is also determined by the slowest converging mode among the $\{1 - \mu\lambda_k\}$; once the faster modes have died out relative to the slowest mode, it is the slowest mode that ultimately determines the convergence rate of $\tilde{w}_i$ to zero. Assume that this slowest mode of convergence corresponds to an eigenvalue λ_{k_o}. Then (9.6) shows that in the limit, as $i \longrightarrow \infty$, $\tilde{w}_i$ tends to zero along the direction of the associated eigenvector, q_{k_o}:

$$\tilde{w}_i \longrightarrow (1 - \mu\lambda_{k_o})^{i+1} x_{k_o}(-1) \cdot q_{k_o} \quad \text{as} \quad i \longrightarrow \infty$$

Moreover, since the evolution of each entry of $\tilde{w}_i$ is governed by a combination of the modes $\{1 - \mu\lambda_k\}$, it does *not* necessarily follow that the choice $\mu = \mu^o$ would result in $\tilde{w}_i$ going to zero at the fastest rate possible. There are instances in which the weight-error vector may converge slower *initially* compared to other choices of the step-size. The optimal step-size μ^o guarantees that in the later stages of convergence, when the slowest mode becomes dominant, the convergence of $\tilde{w}_i$ will be the fastest relative to any other choice of the step-size. We illustrate this behavior in the next example.

Example 9.3 (Rates of convergence)

Figure 9.3 shows six runs of the steepest-descent algorithm (9.1) with different choices of the step-size parameter μ.

For the top figure, the matrix R_u is 5×5 with largest and smallest eigenvalues given by $\lambda_{\max} = 7.7852$ and $\lambda_{\min} = 0.3674$. The optimum step-size in this case is $\mu^o = 0.2453$. The simulation

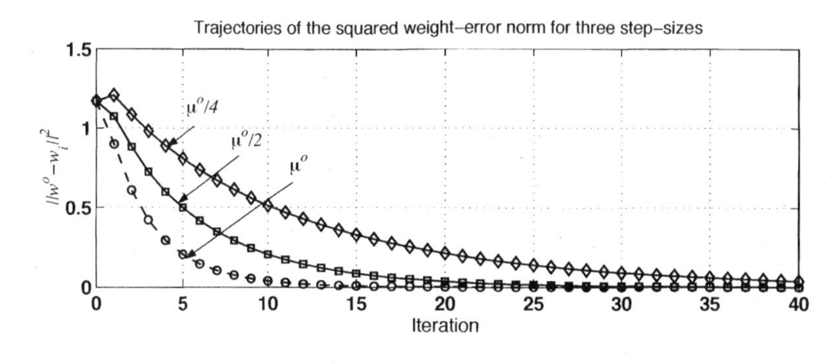

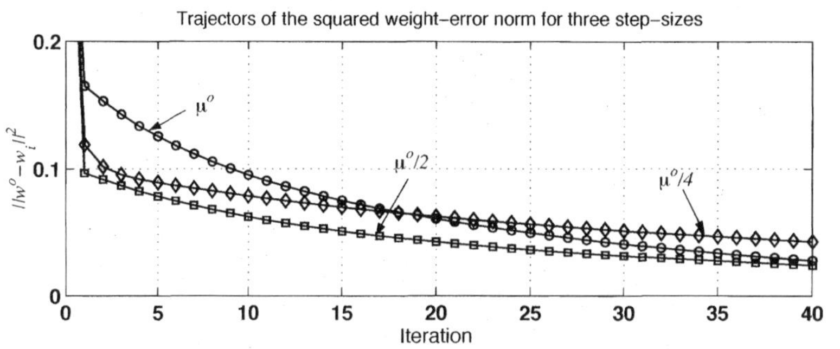

FIGURE 9.3 Plots of the evolution of $\|\tilde{w}_i\|^2$ for three different choices of the step-size; the covariance matrices R_u are different for both plots. The optimal choice μ^o results in fastest convergence in the top plot right from the initial iterations while, in the bottom plot, it results in a slower decay initially but catches up as the iterations progress.

uses the step-sizes $\{\mu^o, \mu^o/2, \mu^o/4\}$, so that the resulting modes are

$$\begin{aligned}
\mu = \mu^o &= 0.2453 \Longrightarrow \{\boxed{0.9099}, 0.5755, 0.6129, 0.8325, \boxed{-0.9099}\} \\
\mu = \mu^o/2 &= 0.1227 \Longrightarrow \{\boxed{0.9549}, 0.9163, 0.8065, 0.7877, 0.0451\} \\
\mu = \mu^o/4 &= 0.0613 \Longrightarrow \{\boxed{0.9775}, 0.9032, 0.8939, 0.8345, 0.5225\}
\end{aligned}$$

Observe that the slowest modes are at $\{\pm 0.9099, 0.9549, 0.9775\}$ with, of course, the smallest among them in magnitude corresponding to $\mu = \mu^o$. We see from the curves in the top figure that the choice μ^o leads to the fastest convergence of $\|\tilde{w}_i\|^2$ to zero.

The bottom plot in Fig. 9.3 illustrates a situation for which convergence is initially slower when the step-size is chosen as $\mu = \mu^o$. The example corresponds to a 5×5 matrix R_u with largest and smallest eigenvalues at $\lambda_{\max} = 7.4724$ and $\lambda_{\min} = 0.0691$. The optimum step-size now is $\mu^o = 0.2652$. In the simulation we again use $\{\mu^o, \mu^o/2, \mu^o/4\}$ so that the resulting modes are:

$$\begin{aligned}
\mu = \mu^o &= 0.2652 \Longrightarrow \{\boxed{-0.9817}, 0.6864, 0.8898, 0.9345, \boxed{0.9817}\} \\
\mu = \mu^o/2 &= 0.1326 \Longrightarrow \{0.0092, 0.84320.9449, 0.9672, \boxed{0.9908}\} \\
\mu = \mu^o/4 &= 0.0663 \Longrightarrow \{0.5046, 0.9216, 0.9725, 0.9836, \boxed{0.9954}\}
\end{aligned}$$

The slowest modes are at $\{\pm 0.9817, 0.9908, 0.9954\}$ with the smallest among them in magnitude corresponding to $\mu = \mu^o$. However, we now see from the bottom plot in Fig. 9.3 that the curve that corresponds to $\mu = \mu^o/2$ decays faster initially than the one that corresponds to $\mu = \mu^o$. Observe however that, as i increases, the curve that corresponds to $\mu = \mu^o$ ultimately converges faster; this is because for larger values of i, it is the slowest mode that determines the rate of convergence of $\tilde{w}_i$.

◇

9.4 TIME CONSTANTS

It is customary to describe the rate of convergence of a steepest-descent algorithm in terms of its time constants, which are defined as follows.

Recall that for an exponential function $f(t) = e^{-t/\tau}$, the time constant is τ and it corresponds to the time required for the value of the function to decay by a factor of e since

$$f(t+\tau) = e^{-(t+\tau)/\tau} = f(t)/e$$

Now, for an exponential discrete-time sequence of the form (cf. (8.25)):[3]

$$|x_k(i)|^2 = (1-\mu\lambda_k)^2\,|x_k(i-1)|^2, \quad i \geq 0$$

the value of $|x_k(i)|^2$ decays by $(1-\mu\lambda_k)^2$ at each iteration. Let T denote the time interval between one iteration and another, and let us fit a decaying exponential function through the points of the sequence $\{|x_k(i)|^2\}$. Denote the function by $f(t) = e^{-t/\tau_k}$, with a time constant τ_k to be determined. Then we must have

$$\begin{aligned} f(t)|_{t=(i-1)T} &= |x_k(i-1)|^2 = e^{-(i-1)T/\tau_k} \\ f(t)|_{t=iT} &= (1-\mu\lambda_k)^2\,|x_k(i-1)|^2 = e^{-iT/\tau_k} \end{aligned}$$

Dividing one expression by the other leads to $e^{-T/\tau_k} = (1-\mu\lambda_k)^2$ or, equivalently,

$$\tau_k \stackrel{\Delta}{=} \frac{-T}{2\ln|1-\mu\lambda_k|} \qquad \text{(measured in units of time)}$$

This value measures the time that is needed for the value of $|x_k(i)|^2$ to decay by a factor of e, which corresponds to a decrease of the order of $10\log e \approx 4.4$ dB. It is common to normalize the value of τ_k to be independent of T. Thus, let $\bar{\tau}_k = \tau_k/T$. Then

$$\boxed{\bar{\tau}_k \stackrel{\Delta}{=} \frac{-1}{2\ln|1-\mu\lambda_k|}} \qquad \text{(measured in iterations)} \tag{9.7}$$

This normalized value measures the approximate number of *iterations* that is needed for the value of $|x_k(i)|^2$ to decay by approximately 4.4 dB. For sufficiently small step-sizes (say, for $\mu\lambda_k \ll 1$), we have $\ln|1-\mu\lambda_k| \approx -\mu\lambda_k$ and we can approximate the expression for $\bar{\tau}_k$ by

$$\bar{\tau}_k \approx \frac{1}{2\mu\lambda_k} \qquad \text{(iterations)}$$

Usually, the largest $\{\bar{\tau}_k, k = 1, 2, \ldots, M\}$ is taken as indicative of the time constant of the steepest-descent method.

[3]Note that we are using the squared quantity $|x_k(i)|^2$ instead of $x_k(i)$ in order to obtain a sequence of decaying nonnegative real numbers.

9.5 LEARNING CURVE

Besides modes and time constants, it is also customary to characterize the convergence performance of a steepest-descent method in terms of its learning curve. Recall that our original problem is to determine the vector w that minimizes

$$J(w) = \mathsf{E}\,|\boldsymbol{d} - \boldsymbol{u}w|^2$$

The steepest-descent recursion (9.1) provides successive iterates w_i with cost values

$$J(w_i) = \mathsf{E}\,|\boldsymbol{d} - \boldsymbol{u}w_i|^2$$

Since, by choosing the step-size μ such that $\mu < 2/\lambda_{\max}$, we are guaranteed a sequence $\{w_i\}$ that converges to the optimal solution w^o, the same condition on μ also guarantees that the successive values $J(w_i)$ will converge to the minimum value of $J(w)$, namely (cf. (8.11)):

$$J(w_i) \;\longrightarrow\; J_{\min} \;=\; \sigma_d^2 \;-\; R_{ud}R_u^{-1}R_{du} \quad \text{as } i \longrightarrow \infty$$

It turns out, as we now verify, that the decay of $J(w_i)$ to $J_{\min}$ is always monotonic. To see this, we recall from (8.10) that

$$J(w) \;=\; J_{\min} \;+\; (w - w^o)^* R_u (w - w^o) \tag{9.8}$$

i.e.,

$$\boxed{J(w_i) = J_{\min} \;+\; \tilde{w}_i^* R_u \tilde{w}_i} \tag{9.9}$$

The term $\tilde{w}_i^* R_u \tilde{w}_i$ represents the *excess mean-square error* at iteration i and it will be denoted by

$$\boxed{\xi(w_i) \;\stackrel{\Delta}{=}\; J(w_i) - J_{\min} \;=\; \tilde{w}_i^* R_u \tilde{w}_i} \tag{9.10}$$

It measures how far the cost at iteration i is from the minimum cost, $J_{\min}$.

If we replace $\tilde{w}_i$ by Ux_i, and use the eigen-decomposition (8.22), we obtain

$$J(w_i) \;=\; J_{\min} + \sum_{k=1}^{M} \lambda_k |x_k(i)|^2 \;=\; J_{\min} \;+\; \sum_{k=1}^{M} \lambda_k (1 - \mu\lambda_k)^{2(i+1)}\; |x_k(-1)|^2$$

which confirms, under the requirement $0 < \mu < 2/\lambda_{\max}$, that $J(w_i) \rightarrow J_{\min}$ as $i \rightarrow \infty$, irrespective of the initial weight-error vector $\tilde{w}_{-1}$. Moreover, the convergence is both exponential *and* monotonic; it is monotonic since, for any k, the coefficient $\lambda_k(1 - \mu\lambda_k)^2$ is positive.

The evolution of $J(w_i)$ as a function of i provides useful information about the learning behavior of a steepest-descent algorithm. For future reference, we shall adopt the following definition.

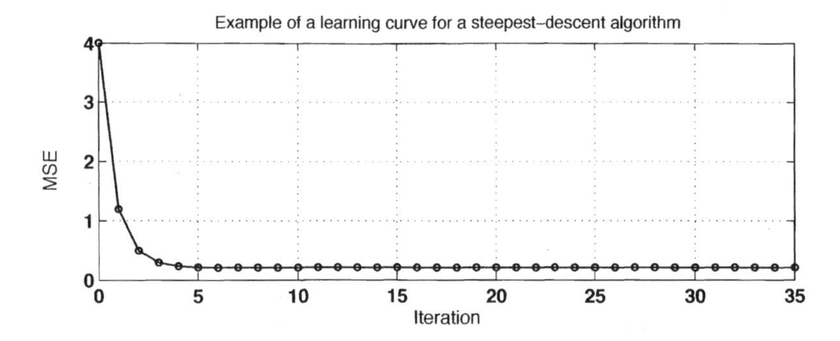

FIGURE 9.4 A typical learning curve $J(i)$ for algorithm (9.1).

> **Definition 9.1 (Learning curve)** The learning curve of a steepest-descent method associated with a cost function $J(w)$ is denoted by $J(i)$ and defined as $J(i) = J(w_{i-1})$ for $i \geq 0$. In particular, for the quadratic cost function $J(w)$ in (8.7), we obtain that its learning curve is given by
>
> $$J(i) = \mathsf{E}\,|e(i)|^2 \quad \text{where} \quad e(i) = d - uw_{i-1}$$
>
> is the so-called *a priori* output estimation error. In this case, the learning curve is also called the *mean-square-error* (MSE) curve.

Observe that the initial value of $J(i)$ is

$$J(0) = J(w_{-1})$$

In other words, it is the value of the cost function evaluated at the initial condition w_{-1}.

In general, the value of the learning curve at an iteration i is a measure of the cost that would result if we freeze the weight estimate at the value obtained at the prior iteration. Correspondingly, in the mean-square-error case, the learning curve is defined in terms of the variance of the *a priori* error $e(i)$ (which uses w_{i-1} and not w_i). Figure 9.4 shows a typical learning curve for the steepest-descent algorithm (9.1) with $M = 3$, $\lambda_{\text{min}} = 0.3$, $\lambda_{\text{max}} = 1$, and $\mu = 1.5385$. The modes $\{1 - \mu\lambda_k\}$ for this simulation are at $\{0.5385, 0.0769, -0.5385\}$.

9.6 CONTOUR CURVES OF THE ERROR SURFACE

Another useful way to examine the performance of a steepest-descent method is by examining the contours of constant value of its cost function, $J(w)$. These contour curves are more easily characterized if we perform a change of coordinates. For any w, we define $z = U^*(w - w^o)$ or, equivalently,

$$\boxed{w = w^o + Uz} \tag{9.11}$$

where U is obtained from the eigen-decompostion (8.22) of R_u. In other words, we replace the $w-$coordinate system by a $z-$coordinate system. The origin of the new system, $z = 0$,

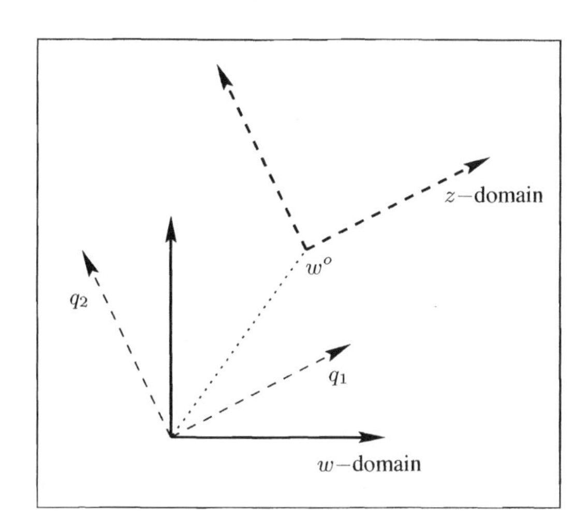

FIGURE 9.5 Change of coordinates from the $w-$domain to the $z-$domain, defined by $z = U^*(w - w^o)$, for the case $M = 2$.

occurs at the point $w = w^o$ in the $w-$coordinate system. Likewise, the first basis vector in the $z-$coordinate system, namely,

$$z_1 = \text{col}\{1, 0, \ldots, 0\}$$

corresponds to the vector

$$w = w^o + q_1$$

in the $w-$coordinate system, where q_1 is the first column of U. This means that the first basis vector in the $z-$domain is obtained by shifting q_1 to w^o in the $w-$domain. A similar construction holds for the other basis vectors of the $z-$domain. This change of basis is illustrated in Fig. 9.5 for the case $M = 2$. We therefore say that the new coordinate system is obtained by shifting the origin to w^o and then rotating the w-axes by U^*. The minimum value of $J(w)$, which occurs at $w = w^o$ in the $w-$domain, will now occur at $z = 0$ in the $z-$domain.

Using (9.8), and the eigen-decomposition $R_u = U\Lambda U^*$, we can express the cost function as

$$J(z) \;=\; J_{\min} + z^*\Lambda z \;=\; J_{\min} + \sum_{k=1}^{M} \lambda_k |z(k)|^2$$

where the $\{z(k)\}$ denote the entries of z. The contour curves of $J(z)$ (and, correspondingly, of $J(w)$), are the curves for which

$$J(z) \;=\; \text{a constant} \tag{9.12}$$

for different constant values. Equation (9.12) defines a hyper-ellipse in $M-$dimensions. The hyper-ellipse is centered at w^o and it has M principal axes. The principal axes are, by definition, the lines that pass through the origin and are normal to the hyper-ellipse. For $J(z)$, its principal axes coincide with the basis vectors of the $z-$coordinate system. To see this, note first that the gradient of $J(z)$ with respect to z^* is equal to Λz. Moreover, any line passing through the origin has the form λz for some scalar λ. Therefore, for any

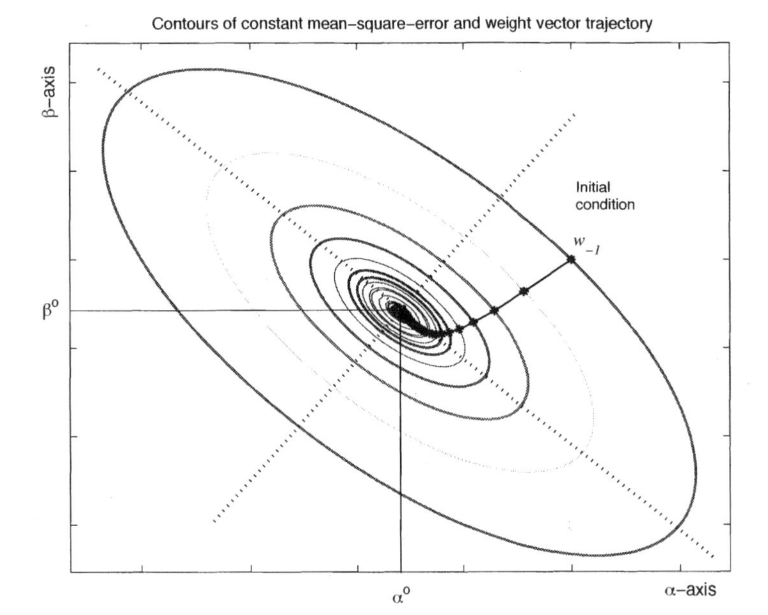

FIGURE 9.6 Elliptic contours of constant mean-square error in two dimensions, where the entries of w are denoted by $w = \text{col}\{\alpha, \beta\}$ and the entries of w^o are $\{\alpha^o, \beta^o\}$. The figure also indicates a typical trajectory starting from some initial condition w_{-1}.

such line to be normal to the hyper-ellipse it should satisfy $\lambda z = \Lambda z$. This equality is possible only if λ is an eigenvalue of Λ and z the corresponding eigenvector. But since Λ is diagonal, this conclusion requires z to be one of the basis vectors. Therefore, the basis vectors of the $z-$coordinate system are normal to the hyper-ellipse and, consequently, they are the principal axes of the hyper-ellipse. We therefore find that the eigenvectors of R_u, when shifted to w^o, are the principal axes of the elliptic contours of $J(w)$.

These conclusions are illustrated in Fig. 9.6 in the 2-dimensional case ($M = 2$). As the figure indicates, the contour curves are ellipses centered at w^o and with principal axes along the directions of the eigenvectors of R_u. Moreover, since the steepest-descent algorithm (9.1) updates w_{i-1} along the direction of the negative (conjugate) gradient vector at w_{i-1}, then the steepest-descent algorithm will be such that it moves from one contour level to another along a direction that is normal to the elliptic contour at w_{i-1}. This process continues until convergence is attained.

9.7 ITERATION-DEPENDENT STEP-SIZES

The steepest-descent algorithm of Thm. 8.2 uses a constant step-size μ. In many instances, it may be desirable to vary the value of the step-size in order to obtain better control over the speed of convergence of the algorithm.

Condition for Convergence

Starting with (8.12), the arguments of Sec. 8.2 would still hold if we replace μ by an iteration-dependent positive step-size $\mu(i)$. In this case, recursion (9.1) would be replaced by

$$\boxed{w_i = w_{i-1} + \mu(i)[R_{du} - R_u w_{i-1}], \quad w_{-1} = \text{initial guess}} \tag{9.13}$$

Of course, not every choice of the step-size sequence $\{\mu(i)\}$ will guarantee convergence of w_i to w^o. For example, one might be tempted to extrapolate the arguments of Sec. 8.2 and conclude that by choosing $\mu(i)$ such that $\mu(i) < 2/\lambda_{\max}$ for all i, the weight-error vector $\tilde{w}_i$ will converge to zero. This conclusion is generally *false*. Consider for illustration purposes, a scalar recursion of the form $x(i) = a(i)x(i-1)$ for $i \geq 0$. Then

$$x(i) = \left(\prod_{j=0}^{i} a(j)\right) x(-1)$$

If the $\{a(j)\}$ are such that $|a(j)| < 1$ for all finite j, it does *not* necessarily follow that

$$\prod_{j=0}^{i} a(j) \longrightarrow 0 \quad \text{as} \quad i \longrightarrow \infty$$

That is, the product of infinitely many numbers that are all less than one in magnitude is not necessarily zero (see Prob. III.2). The product would tend to zero if all the $\{a(j)\}$ have their magnitudes *uniformly* bounded away from one, say, $|a(j)| < \alpha < 1$ for all j and for some $\alpha > 0$.

The following statement provides one necessary condition on $\mu(i)$ in (9.13) for convergence; the proof is given in Probs. III.9 and III.10. As explained after the statement of the theorem, other conditions are possible.

Theorem 9.1 (Convergence condition) Given a zero-mean random variable $\boldsymbol{d}$ with variance σ_d^2 and a zero-mean random row vector $\boldsymbol{u}$ with $R_u = \mathsf{E}\,\boldsymbol{u}^*\boldsymbol{u} > 0$, the solution of the linear least-mean-squares estimation problem

$$\min_{w} \mathsf{E}\,|\boldsymbol{d} - \boldsymbol{u}w|^2$$

can be obtained recursively as follows. Start with any initial guess w_{-1}, choose a bounded step-size sequence $\mu(i) > 0$ that tends to zero, i.e., $\mu(i) \to 0$ as $i \to \infty$, and iterate:

$$w_i = w_{i-1} + \mu(i)[R_{du} - R_u w_{i-1}], \quad i \geq 0$$

Then $w_i \to w^o$ as $i \to \infty$ if, and only if, the step-size sequence satisfies $\sum_{i=0}^{\infty} \mu(i) = \infty$. That is, if and only if, $\{\mu(i)\}$ is a divergent series.

The result of Thm. 9.1 requires the sequence $\mu(i)$ to tend to zero but not too fast since the sequence has to diverge as well. A typical sequence that satisfies the conditions of the theorem is

$$\mu(i) = \frac{\alpha}{i+\beta}, \quad \alpha > 0, \quad \beta > 0, \quad i \geq 0$$

Other examples are any bounded step-size sequences that satisfy both conditions

$$\sum_{i=0}^{\infty} \mu^2(i) < \infty \quad \text{and} \quad \sum_{i=0}^{\infty} \mu(i) = \infty$$

This is because the finite-energy condition on the sequence $\{\mu(i)\}$ guarantees $\mu(i) \to 0$. Still, convergence can occur even if the conditions of the theorem are violated. For example, in Prob. III.4 it is shown that with $\mu(i) \longrightarrow \alpha > 0$ as $i \longrightarrow \infty$, i.e., even with $\mu(i)$ tending to a nonzero limit, but as long as $\alpha < 2/\lambda_{\max}$, then w_i is guaranteed to converge to w^o.

Optimal Iteration-Dependent Step-Size

We saw earlier in Sec. 9.1 that the choice $\mu^o = 2/(\lambda_{\max} + \lambda_{\min})$ results in fastest convergence to steady-state in the constant step-size case. We can also seek an optimal choice for fastest convergence in the iteration-dependent step-size case. To do so, we start from the weight-error recursion

$$\tilde{w}_i = [\mathrm{I} - \mu(i)R_u]\,\tilde{w}_{i-1} \tag{9.14}$$

and use expression (9.10) for the excess mean-square error to write

$$\begin{aligned}
\xi(w_i) &= \tilde{w}_i^* R_u \tilde{w}_i \\
&= \tilde{w}_{i-1}^*[\mathrm{I} - \mu(i)R_u]\; R_u\; [\mathrm{I} - \mu(i)R_u]\tilde{w}_{i-1} \\
&= \tilde{w}_{i-1}^*\left[R_u - 2\mu(i)R_u^2 + \mu^2(i)R_u^3\right]\tilde{w}_{i-1} \\
&= \xi(w_{i-1}) \;-\; 2\mu(i)\tilde{w}_{i-1}^* R_u^2 \tilde{w}_{i-1} \;+\; \mu^2(i)\tilde{w}_{i-1}^* R_u^3 \tilde{w}_{i-1} \\
&= \xi(w_{i-1}) \;-\; \left[2\mu(i)\tilde{w}_{i-1}^* R_u^2 \tilde{w}_{i-1} \;-\; \mu^2(i)\tilde{w}_{i-1}^* R_u^3 \tilde{w}_{i-1}\right]
\end{aligned} \tag{9.15}$$

By maximizing the term

$$2\mu(i)\tilde{w}_{i-1}^* R_u^2 \tilde{w}_{i-1} \;-\; \mu^2(i)\tilde{w}_{i-1}^* R_u^3 \tilde{w}_{i-1}$$

with respect to $\mu(i)$ we can guarantee largest decay in the excess mean-square error at iteration i. This maximization step leads to the optimal choice

$$\boxed{\mu^o(i) = \frac{\tilde{w}_{i-1}^*\; R_u^2\; \tilde{w}_{i-1}}{\tilde{w}_{i-1}^*\; R_u^3\; \tilde{w}_{i-1}}} \tag{9.16}$$

This argument is possible here since the step-size is allowed to vary from one iteration to another and, therefore, we can minimize $\xi(w_i)$ with respect to $\mu(i)$ at each i. The same argument is not possible in the constant step-size case. There we selected μ^o by solving the min-max problem (9.4) instead. We still need to argue that the choice (9.16) leads to a convergent algorithm; in the process we shall also verify that $\{\mu^o(i)\}$ leads to a monotonically decreasing excess mean-square error.

First note that the choice (9.16) does not satisfy the convergence conditions of Thm. 9.1. For instance, assume $\tilde{w}_{i-1}$ and R_u are scalars. Then, expression (9.16) collapses to $\mu^o(i) = 1/R_u$, which does not tend to zero as required by the statement of Thm. 9.1 — see also Prob. III.5. However, as mentioned in the comments following the statement of the theorem, the conditions on $\mu(i)$ in the theorem are only sufficient. There can exist step-size sequences that do not satisfy the conditions of the theorem but nevertheless result in convergence. The sequence $\{\mu^o(i)\}$ is one such example.

To see this, we substitute (9.16) into expression (9.15) for $\xi(w_i)$ and get

$$\xi^o(w_i) \;=\; \xi^o(w_{i-1}) \;-\; \frac{(\tilde{w}_{i-1}^*\; R_u^2\; \tilde{w}_{i-1})^2}{\tilde{w}_{i-1}^*\; R_u^3\; \tilde{w}_{i-1}}$$

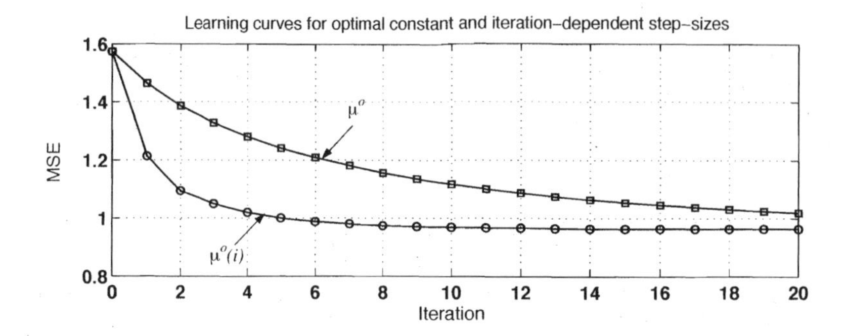

FIGURE 9.7 Typical plots of learning curves corresponding to the cases of an optimal constant step-size and an optimal iteration-dependent step-size.

If $\tilde{w}_{i-1} = 0$, then $w_{i-1} = w^o$ and algorithm (9.13) would have converged since $w_i = w_{i-1}$ for all i. So assume $\tilde{w}_{i-1} \neq 0$. Then we necessarily have, since $R_u > 0$,

$$\xi^o(w_i) \ < \ \xi^o(w_{i-1})$$

That is the sequence $\{\xi^o(w_i)\}$ is monotonically decreasing. Combining this result with the fact that $\{\xi^o(w_i)\}$ is bounded from below (since $\xi^o(w_i) \geq 0$), we conclude that the sequence $\{\xi^o(w_i)\}$ is convergent. The question that remains is whether its limit is zero.

We showed earlier in Sec. 8.2 that there exists a constant step-size that guarantees convergence of $\tilde{w}_i$ to zero (namely, any μ satisfying $\mu < 2/\lambda_{\max}$). Choose any such step-size and let $\xi^\mu(w_i^\mu)$ denote the excess mean-square error that results from the recursion

$$w_i^\mu = w_{i-1}^\mu + \mu[R_{du} - R_u w_{i-1}^\mu]$$

We are writing w_i^μ to distinguish it from the weight estimate obtained from the recursion $w_i = w_{i-1} + \mu^o(i)[R_{du} - R_u w_{i-1}]$ with the optimal iteration-dependent step-size. Then, starting from the same initial conditions $w_{-1} = w_{-1}^\mu$, it holds that

$$\xi^o(w_i) \ \leq \ \xi^u(w_i^\mu)$$

This is because the optimal iteration-dependent step-size minimizes the excess mean-square error at each iteration. But since $\xi^u(w_i^\mu) \to 0$, we conclude that $\xi^o(w_i) \to 0$, as desired.

Figure 9.7 plots the learning curve $J(i) = J(w_{i-1})$ for both cases of an optimal constant step-size and an optimal iteration-dependent step-size. The result confirms the faster decay of the learning curve in the latter case.

9.8 NEWTON'S METHOD

We mentioned in our derivation of the steepest-descent algorithm in Sec. 8.2 that any choice for the search direction of the form (cf. (8.18)):

$$p = -B\ [\nabla_w J(w_{i-1})]^*$$

for any positive-definite matrix B, can be used to enforce the condition

$$\text{Re}\,[\nabla_w J(w_{i-1})p] < 0$$

We chose $B = \mathrm{I}$ in our earlier discussions, which led to the steepest-descent variants of Thms. 8.2 and 9.1. But other choices for B are possible and they lead to different algorithms with different properties. One useful choice for B is

$$B = \left[\nabla_w^2 J(w_{i-1})\right]^{-1}$$

in terms of the inverse of the Hessian matrix, in which case the search direction becomes

$$p = -\left[\nabla_w^2 J(w_{i-1})\right]^{-1} \left[\nabla_w J(w_{i-1})\right]^* \tag{9.17}$$

The resulting steepest-descent recursion (8.12) would be

$$\boxed{w_i = w_{i-1} - \mu \left[\nabla_w^2 J(w_{i-1})\right]^{-1} [\nabla_w J(w_{i-1})]^*, \quad i \geq 0, \quad w_{-1} = \text{initial guess}} \tag{9.18}$$

This recursive form is known as Newton's method.

For the quadratic cost function $J(w)$ of (8.8), we use (8.14) to find that (9.18) reduces to

$$\boxed{w_i = w_{i-1} + \mu R_u^{-1}\left[R_{du} - R_u w_{i-1}\right]} \tag{9.19}$$

We can examine the properties of this algorithm in much the same way as we did for recursion (9.1). So we shall be brief.

Convergence Properties

Subtracting both sides of (9.19) from w^o, and using the fact that w^o satisfies the normal equations $R_u w^o = R_{du}$, we arrive at the weight-error recursion

$$\tilde{w}_i = (1-\mu)\tilde{w}_{i-1} \tag{9.20}$$

In contrast to (9.2) and (9.14), we find that the covariance matrix R_u does not appear any longer in (9.20). In particular, convergence is now guaranteed for all step-sizes μ that satisfy $0 < \mu < 2$; a condition that is independent of R_u.

Actually, the choice $\mu = 1$ in (9.20) leads to immediate convergence because $\tilde{w}_i = 0$ with no further iteration. This is a well-known property of Newton's method; the method guarantees convergence in a single iteration to the minimizing argument of a *quadratic* cost function by choosing $\mu = 1$. This fact can also be seen from recursion (9.19), which for $\mu = 1$ collapses to

$$w_i = w_{i-1} + R_u^{-1} R_{du} - w_{i-1} = w_{i-1} + w^o - w_{i-1} = w^o$$

Of course, applying Newton's method (9.19) to the solution of the least-mean-squares estimation problem (8.1) has the same complexity as using (8.4) since both schemes require the inversion of R_u. The usefulness of Newton's method relative to the normal equations will become evident later in Chapters 10 and 11 when we use it to devise some useful adaptive variants.

Learning Curve

Recursion (9.19) estimates the vector w that minimizes $J(w) = \mathsf{E}\,|\boldsymbol{d} - \boldsymbol{u}w|^2$. It does so by evaluating successive iterates w_i with cost values $J(w_i) = \mathsf{E}\,|\boldsymbol{d} - \boldsymbol{u}w_i|^2$. Since, by choosing $0 < \mu < 2$, we are guaranteed a sequence $\{w_i\}$ that converges to w^o, this same condition on μ guarantees convergence of $J(w_i)$ to the minimum value of $J(w)$, namely (cf. (8.11)),

$$J(w_i) \longrightarrow J_{\min} = J(w^o) = \sigma_d^2 - R_{ud} R_u^{-1} R_{du} \quad \text{as} \quad i \longrightarrow \infty$$

The decay of $J(w_i)$ to J_{min} is again monotonic. This can be seen as follows. Using the representation (9.9),

$$J(w_i) = J_{\text{min}} + \tilde{w}_i^* R_u \tilde{w}_i$$

replacing $\tilde{w}_i$ by Ux_i, where $\tilde{w}_i$ now evolves according to (9.20), and using the eigendecomposition (8.22) for R_u, we obtain

$$J(w_i) = J_{\text{min}} + \sum_{k=1}^{M} \lambda_k |x_k(i)|^2 = J_{\text{min}} + (1-\mu)^{2(i+1)} \sum_{k=1}^{M} \lambda_k |x_k(-1)|^2$$

This expression confirms that, under the requirement $0 < \mu < 2$,

$$\lim_{i \to \infty} J(w_i) = J_{\text{min}}$$

irrespective of the initial weight-error vector $\tilde{w}_{-1}$. Moreover, the convergence is both exponential *and* monotonic and, in contrast to the steepest-descent analysis of Sec. 9.5, convergence is now governed by a *single* mode at $(1-\mu)^2$. Therefore, with Newton's method, we need only associate a single time constant that is equal to (cf. (9.7)):

$$\boxed{\bar{\tau} = -1/2\ln(1-\mu)} \quad \text{(iterations)}$$

The value of $\bar{\tau}$ is an approximation for the number of iterations that is needed for $\|\tilde{w}_i\|^2$ to decay by approximately 4.4 dB.

With regards to the contour curves of the error surface, they are still the same hyperelliptic curves that were described in Sec. 9.6 (after all we are dealing with the same quadratic cost function $J(w)$ from (8.8)). As shown in that section, the principal axes of the contour curves are the eigenvectors of the covariance matrix R_u shifted to the location of w^o. Now, however, the search direction in Newton's method is *not* along the normal direction to the elliptic curves anymore, but along the line connecting w_{i-1} to w^o. To see this, recall that when $\mu = 1$, convergence of Newton's method occurs in a single step, which is only possible if the search direction is along the line connecting w_{i-1} to w^o. When $\mu \neq 1$, we are still moving along the same direction connecting w_{i-1} to w^o but for a shorter distance since from (9.19),

$$w_i = w_{i-1} + \mu(w^o - w_{i-1})$$

Remark 9.1 (Regularization) When the Hessian matrix in (9.18) is close to singular, it is common to employ regularization, in which case Newton's method is sometimes known as the *Levenberg-Marquardt* method and it becomes

$$\boxed{w_i = w_{i-1} - \mu[\epsilon \mathrm{I} + \nabla_w^2 J(w_{i-1})]^{-1} [\nabla J(w_{i-1})]^*, \quad i \geq 0, \quad w_{-1} = \text{initial guess}}$$

The difference relative to Newton's recursion (9.17) is the addition of the small positive parameter ϵ. This algorithm can still be interpreted as a steepest-descent method of the form (8.12) with B in (8.18) chosen as

$$B = \left[\epsilon \mathrm{I} + \nabla_w^2 J(w_{i-1})\right]^{-1}$$

More generally, we can employ iteration-dependent step-sizes, $\mu(i)$, and iteration-dependent regularization parameters, $\epsilon(i) > 0$, and write instead

$$\boxed{w_i = w_{i-1} - \mu(i)[\epsilon(i)\mathrm{I} + \nabla_w^2 J(w_{i-1})]^{-1} [\nabla J(w_{i-1})]^*, \quad i \geq 0, \quad w_{-1} = \text{initial guess}}$$

◇

CHAPTER **10**

LMS Algorithm

In Chapters 10–14 we start to develop the theory of adaptive algorithms by studying *stochastic-gradient* methods. These methods are obtained from steepest-descent implementations by replacing the required gradient vectors and Hessian matrices by some suitable approximations. Different approximations lead to different algorithms with varied degrees of complexity and performance properties. The resulting methods will be generically called *stochastic-gradient algorithms* since, by employing estimates for the gradient vector, the update directions become subject to random fluctuations that are often referred to as *gradient noise*.

Stochastic-gradient algorithms serve at least two purposes. First, they avoid the need to know the *exact signal statistics* (e.g., covariances and cross-covariances), which are necessary for a successful steepest-descent implementation but are nevertheless rarely available in practice. Stochastic-gradient methods achieve this feature by means of a *learning mechanism* that enables them to estimate the required signal statistics. Second, these methods possess a *tracking mechanism* that enables them to track variations in the signal statistics. The two combined capabilities of learning and tracking are the main reasons behind the widespread use of stochastic-gradient methods (and the corresponding adaptive filters). It is because of these abilities that we tend to describe adaptive filters as "smart systems"; smart in the sense that they can learn the statistics of the underlying signals and adjust their behavior to variations in the "environment" in order to keep the performance level at check.

In the body of this chapter and the subsequent chapters, we describe a handful of stochastic-gradient algorithms, including the least-mean-squares (LMS) algorithm, the normalized least-mean-squares (NLMS) algorithm, the affine projection algorithm (APA), and the recursive least-squares algorithm (RLS). In Probs. III.26–III.38 we devise several other stochastic-gradient methods. By the end of the presentation, the reader will have had sufficient exposition to the procedure that is commonly used to motivate and derive adaptive filters.

10.1 MOTIVATION

We start our discussions by reviewing the linear estimation problem of Sec. 8.1 and the corresponding steepest-descent methods of Chapter 8. Thus, let $\boldsymbol{d}$ be a zero-mean scalar-valued random variable with variance $\sigma_d^2 = \mathsf{E}\,|\boldsymbol{d}|^2$. Let further $\boldsymbol{u}^*$ be a zero-mean $M \times 1$ random variable with a positive-definite covariance matrix, $R_u = \mathsf{E}\,\boldsymbol{u}^*\boldsymbol{u} > 0$. The $M \times 1$ cross-covariance vector of $\boldsymbol{d}$ and $\boldsymbol{u}$ is denoted by $R_{du} = \mathsf{E}\,\boldsymbol{d}\boldsymbol{u}^*$. We know from Sec. 8.1 that the weight vector w that solves

$$\min_{w} \; \mathsf{E}\,|\boldsymbol{d} - \boldsymbol{u}w|^2 \tag{10.1}$$

is given by

$$\boxed{w^o = R_u^{-1}R_{du}} \tag{10.2}$$

and that the resulting minimum mean-square error is

$$\boxed{\text{m.m.s.e.} = \sigma_d^2 - R_{ud}R_u^{-1}R_{du}} \tag{10.3}$$

In Chapter 8 we developed several steepest-descent methods that approximate w^o iteratively, until eventually converging to it. For example, in Sec. 8.1 we studied the following recursion with a constant step-size,

$$\boxed{w_i = w_{i-1} + \mu[R_{du} - R_u w_{i-1}], \quad w_{-1} = \text{initial guess}} \tag{10.4}$$

where the update direction, $R_{du} - R_u w_{i-1}$, was seen to be equal to the negative conjugate-transpose of the gradient vector of the cost function at w_{i-1}, i.e.,

$$R_{du} - R_u w_{i-1} = -\left[\nabla_w J(w_{i-1})\right]^* \tag{10.5}$$

where

$$J(w) \triangleq \mathsf{E}\,|\boldsymbol{d} - \boldsymbol{u}w|^2$$

In Sec. 9.7 we allowed for an iteration-dependent step-size, $\mu(i)$, and studied the recursion

$$\boxed{w_i = w_{i-1} + \mu(i)[R_{du} - R_u w_{i-1}], \quad w_{-1} = \text{initial guess}} \tag{10.6}$$

and in Sec. 9.8 we studied Newton's recursion,

$$\boxed{w_i = w_{i-1} + \mu R_u^{-1}[R_{du} - R_u w_{i-1}], \quad w_{-1} = \text{initial guess}} \tag{10.7}$$

where R_u^{-1} resulted from using the inverse of the Hessian matrix of $J(w)$, namely,

$$R_u = \nabla_w^2 J(w_{i-1}) = \nabla_{w^*}\left[\nabla_w J(w_{i-1})\right]$$

More generally, when regularization is employed and when the step-size is also allowed to be iteration-dependent, the recursion for Newton's method is replaced by

$$\boxed{w_i = w_{i-1} + \mu(i)\left[\epsilon(i)\mathrm{I} + R_u\right]^{-1}\left[R_{du} - R_u w_{i-1}\right]} \tag{10.8}$$

for some positive scalars $\{\epsilon(i)\}$; they could be set to a constant value for all i, say, $\epsilon(i) = \epsilon$.

Now all the steepest-descent formulations described above, i.e., (10.4), (10.6), (10.7) and (10.8), rely explicitly on knowledge of $\{R_{du}, R_u\}$. This fact constitutes a limitation in practice and serves as a motivation for the development of stochastic-gradient algorithms for two reasons:

1. **Lack of statistical information**. First, the quantities $\{R_{du}, R_u\}$ are rarely available in practice. As a result, the true gradient vector, $\nabla_w J(w_{i-1})$, and the true Hessian matrix, $\nabla_w^2 J(w_{i-1})$, cannot be evaluated exactly and a true steepest-descent implementation becomes impossible. Stochastic-gradient algorithms replace the gradient vector and the Hessian matrix by approximations for them. There are several ways for obtaining such approximations. The better the approximation, the closer we expect the performance of the resulting stochastic-gradient algorithm to be to that of the original steepest-descent method. In Parts IV (*Mean-Square Performance*) and V

(*Transient Performance*) we shall study and quantify the degradation in performance that occurs as a result of such approximations.

2. **Variation in the statistical information.** Second, and even more important, the quantities $\{R_{du}, R_u,\}$ tend to vary with time.[4] In this way, the optimal weight vector w^o will also vary with time. It turns out that stochastic-gradient algorithms provide a mechanism for tracking such variations in the signal statistics.

We therefore move on to motivate and develop several stochastic-gradient algorithms.

10.2 INSTANTANEOUS APPROXIMATION

Assume that we have access to several observations of the random variables $\boldsymbol{d}$ and $\boldsymbol{u}$ in (10.1), say, $\{d(0), d(1), d(2), d(3), \ldots\}$ and $\{u_0, u_1, u_2, u_3, \ldots\}$. We refer to the $\{u_i\}$ as regressors. Observe that, in conformity with our notation in this book, we are using the boldface letter $\boldsymbol{d}$ to refer to the random variable, and the normal letter d to refer to observations (or realizations) of it. Likewise, we write $\boldsymbol{u}$ for the random vector and u for observations of it.

One of the simplest approximations for $\{R_{du}, R_u\}$ is to use the instantaneous values

$$\widehat{R}_{du} = d(i)u_i^*, \qquad \widehat{R}_u = u_i^* u_i \tag{10.9}$$

This construction simply amounts to dropping the expectation operator from the actual definitions, $R_{du} = \mathsf{E}\,\boldsymbol{d}\boldsymbol{u}^*$ and $R_u = \mathsf{E}\,\boldsymbol{u}^*\boldsymbol{u}$, and replacing the random variables $\{\boldsymbol{d}, \boldsymbol{u}\}$ by the observations $\{d(i), u_i\}$ at iteration i. In this way, the gradient vector (10.5) is approximated by the instantaneous value

$$-\left[\nabla_w J(w_{i-1})\right]^* \approx d(i)u_i^* - u_i^* u_i w_{i-1} = u_i^*[d(i) - u_i w_{i-1}]$$

and the corresponding steepest-descent recursion (10.4) becomes

$$w_i = w_{i-1} + \mu u_i^*[d(i) - u_i w_{i-1}]\,, \quad w_{-1} = \text{initial guess} \tag{10.10}$$

We continue to write w_i to denote the estimate that is obtained via this approximation procedure although, of course, w_i in (10.10) is different from the w_i that is obtained from the steepest-descent algorithm (10.4): the former is based on using instantaneous approximations whereas the latter is based on using $\{R_{du}, R_u\}$. We do so in order to avoid an explosion of notation; the distinction between both estimates is usually clear from the context.

The stochastic-gradient approximation (10.10) is one of the most widely used adaptive algorithms in current practice due to its striking simplicity. It is widely known as the least-mean-squares (LMS) algorithm, or sometimes as the Widrow-Hoff algorithm in honor of its originators.

[4]We have so far interpreted the index i in a steepest-descent recursion, e.g., as in (10.4), as merely an iteration index. In the adaptive context, however, it will become common to interpret i as a time index in which case w_i is interpreted as the weight estimate at time i.

Algorithm 10.1 (LMS Algorithm) Consider a zero-mean random variable $\boldsymbol{d}$ with realizations $\{d(0), d(1), \ldots\}$, and a zero-mean random row vector $\boldsymbol{u}$ with realizations $\{u_0, u_1, \ldots\}$. The optimal weight vector w^o that solves

$$\min_w \mathsf{E}\,|\boldsymbol{d} - \boldsymbol{u}w|^2$$

can be approximated iteratively via the recursion

$$w_i = w_{i-1} + \mu u_i^*[d(i) - u_i w_{i-1}], \quad i \geq 0, \quad w_{-1} = \text{initial guess}$$

where μ is a positive step-size (usually small).

10.3 COMPUTATIONAL COST

A useful property of LMS is its computational simplicity. The evaluations that follow for the number of computations that are required by LMS are intended to provide an approximate idea of its computational complexity. While there can be different ways to perform specific calculations, the resulting overall filter complexity will be of the same order of magnitude (and often quite similar) to the values we derive in this section for LMS, and in subsequent sections for other adaptive filters.

Note that each step of the LMS algorithm requires a handful of straightforward computations, which are explained below:

1. Each iteration (10.10) requires the evaluation of the inner product $u_i w_{i-1}$, between two vectors of size M each. This calculation requires M complex multiplications and $(M-1)$ complex additions. Using the fact that one complex multiplication requires four real multiplications and two real additions, while one complex addition requires two real additions, we find that the evaluation of this inner product requires $4M$ real multiplications and $4M - 2$ real additions.

2. The algorithm also requires the evaluation of the scalar $d(i) - u_i w_{i-1}$. This calculation amounts to one complex addition, i.e., 2 real additions.

3. Evaluation of the product $\mu[d(i) - u_i w_{i-1}]$, where μ is a real scalar, requires two real multiplications when the data is complex-valued. Usually, μ is chosen as a power of 2^{-1}, say, 2^{-m} for some integer $m > 0$. In this case, multiplying μ by $[d(i) - u_i w_{i-1}]$ can be implemented digitally very efficiently by means of shift registers, and we could therefore ignore the cost of this multiplication. In the text, we choose to account for the case of arbitrary step-sizes.

4. The algorithm further requires multiplying the scalar $\mu[d(i) - u_i w_{i-1}]$ by the vector u_i^*. This requires M complex multiplications and, therefore, $4M$ real multiplications and $2M$ real additions.

5. Finally, the addition of the two vectors w_{i-1} and $\mu u_i^*[d(i) - u_i w_{i-1}]$ requires M complex additions, i.e., $2M$ real additions.

In summary, for general complex-valued signals, LMS requires $8M+2$ real multiplications and $8M$ real additions per iteration. On the other hand, for real-valued data, LMS requires $2M + 1$ real multiplications and $2M$ real additions per iteration.

10.4 LEAST-PERTURBATION PROPERTY

The LMS algorithm was derived in Sec. 10.2 as an approximate iterative solution to the linear least-mean-squares estimation problem (10.1), in the sense that it was obtained by replacing the actual gradient vector in the steepest-descent implementation (10.4) by an instantaneous approximation for it. It turns out that LMS could have been derived in a different manner as the *exact* (not approximate) solution of a well-defined estimation problem; albeit one with a different cost criterion. We shall pursue this derivation later in Chapter 45 — see Sec. 45.4, when we study robust adaptive filters. Until then, we shall continue to treat LMS as a stochastic-gradient algorithm and proceed to highlight some of its properties.

One particular property is that LMS can also be regarded as the exact solution to a *local* (as opposed to a global) optimization problem. To see this, assume that we have available some weight estimate at time $i-1$, say, w_{i-1}, and let $\{d(i), u_i\}$ denote the newly measured data. Let w_i denote a new weight estimate that we wish to compute from the available data $\{d(i), u_i, w_{i-1}\}$ in the following manner.

Let μ denote a given positive number and define two estimation errors: the *a priori output* estimation error

$$\boxed{e(i) \stackrel{\Delta}{=} d(i) - u_i w_{i-1}} \tag{10.11}$$

and the *a posteriori output* estimation error

$$\boxed{r(i) \stackrel{\Delta}{=} d(i) - u_i w_i} \tag{10.12}$$

The former measures the error in estimating $d(i)$ by using $u_i w_{i-1}$, i.e., by using the available weight estimate prior to the update, while the latter measures the error in estimating $d(i)$ by using $u_i w_i$, i.e., by using the new weight estimate. We then seek a w_i that solves the constrained optimization problem:

$$\boxed{\min_{w_i} \ \|w_i - w_{i-1}\|^2 \quad \text{subject to} \quad r(i) = (1 - \mu\|u_i\|^2)e(i)} \tag{10.13}$$

In other words, we seek a solution w_i that is closest to w_{i-1} in the Euclidean norm sense and subject to an equality constraint between $r(i)$ and $e(i)$. The constraint is most relevant when the step-size μ is small enough to satisfy $\left|1 - \mu\|u_i\|^2\right| < 1$, i.e., when

$$0 < \mu\|u_i\|^2 < 2 \qquad \text{for all } i \tag{10.14}$$

This is because, when (10.14) holds, the magnitude of the *a posteriori* error, $r(i)$, will always be less than that of the *a priori* error, $e(i)$, i.e., $|r(i)| < |e(i)|$ or, equivalently, $u_i w_i$ will be a better estimate for $d(i)$ than $u_i w_{i-1}$ (except when $e(i) = 0$, in which case $r(i) = 0$ also). One could of course impose different kinds of constraints on $r(i)$ than (10.13); these will not lead to the LMS algorithm but to other algorithms.

The solution of (10.13) can be obtained from first principles as follows. Let $\delta w = (w_i - w_{i-1})$. Then

$$\begin{aligned} u_i \delta w &= u_i w_i - u_i w_{i-1} \\ &= [u_i w_i - d(i)] + [d(i) - u_i w_{i-1}] \\ &= -r(i) + e(i) \\ &= \mu\|u_i\|^2 e(i) \end{aligned}$$

where in the last equality we used the constraint (10.13). This calculation shows that the optimization problem (10.13) is equivalent to determining a vector δw of smallest Euclidean norm that satisfies $u_i\delta w = \mu\|u_i\|^2 e(i)$, i.e.,

$$\min_{\delta w} \ \|\delta w\|^2 \quad \text{subject to} \quad u_i\delta w = \mu\|u_i\|^2 e(i) \tag{10.15}$$

The constraint $u_i\delta w = \mu\|u_i\|^2 e(i)$ amounts to an under-determined linear system of equations in δw, and it therefore admits infinitely many solutions. Among all solutions δw we desire to determine one that has the smallest Euclidean norm. To do so, we distinguish between two cases:

1. $\|u_i\|^2 = 0$. In this case, $u_i = 0$ and the constraint in (10.15) is satisfied by any δw. The δw with smallest Euclidean norm is then $\delta w = 0$, so that $w_i = w_{i-1}$.

2. $\|u_i\|^2 \neq 0$. In this case, it is easy to find at least one solution to $u_i\delta w = \mu\|u_i\|^2 e(i)$. Indeed,

$$\delta w^o = \mu u_i^* e(i) \tag{10.16}$$

is one such solution since if we multiply δw^o by u_i from the left we get

$$u_i\delta w^o = \mu\|u_i\|^2 e(i)$$

However, there are infinitely many other solutions to this equation. To see this, let $\delta w^o + z$ denote any other possible solution, say,

$$u_i(\delta w^o + z) = \mu\|u_i\|^2 e(i)$$

Now since δw^o is a solution, we find that z must satisfy $u_i z = 0$. This means that all solutions to $u_i\delta w = \mu\|u_i\|^2 e(i)$ can be obtained from δw^o by adding to it any column vector z that is orthogonal to u_i (and there are infinitely many such vectors). Still, among all solutions $\{\delta w^o + z\}$, it can be verified that the δw^o in (10.16) has the smallest Euclidean norm since

$$\begin{aligned}
\|\delta w^o + z\|^2 &= \|\delta w^o\|^2 + \|z\|^2 + (\delta w^o)^* z + z^*\delta w^o \\
&= \|\delta w^o\|^2 + \|z\|^2 + \mu(u_i z)e^*(i) + \mu(u_i z)^* e(i) \\
&= \|\delta w^o\|^2 + \|z\|^2 + 0 + 0 \\
&> \|\delta w^o\|^2
\end{aligned}$$

Finally, using $\delta w^o = w_i - w_{i-1}$ and (10.16) we arrive at the following expression for w_i:

$$w_i = w_{i-1} + \mu u_i^*[d(i) - u_i w_{i-1}]$$

which coincides with the LMS recursion (10.10). We have therefore established that LMS is the solution to the localized optimization problem (10.13).

10.5 APPLICATION: ADAPTIVE CHANNEL ESTIMATION

Before proceeding to the derivation of other similar stochastic-gradient algorithms, we find it instructive to describe how such algorithms are useful in the context of several applications, including channel estimation, linear equalization and decision-feedback equalization. These are the same applications that we studied before in Secs. 5.2, 5.4, 6.3 and 6.4, except that now we are going to solve them in an iterative (i.e., adaptive) manner. We start with the problem of channel estimation.

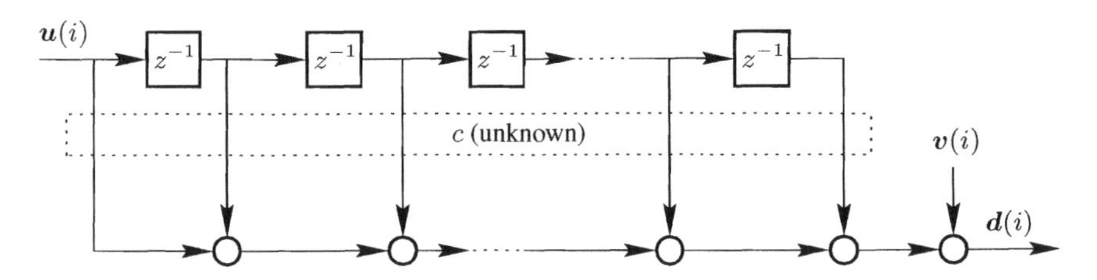

FIGURE 10.1 Noisy measurements of an FIR channel with an impulse response vector c.

Figure 10.1 shows an FIR channel excited by a zero-mean random sequence $\{\boldsymbol{u}(i)\}$. Its output is another zero-mean random sequence $\{\boldsymbol{d}(i)\}$. At any time instant i, the state of the channel is captured by the regressor

$$\boldsymbol{u}_i = \left[\begin{array}{cccc} \boldsymbol{u}(i) & \boldsymbol{u}(i-1) & \ldots & \boldsymbol{u}(i-M+1) \end{array}\right]$$

and its output is given by

$$\boldsymbol{d}(i) = \boldsymbol{u}_i c + \boldsymbol{v}(i) \tag{10.17}$$

where the column vector c denotes the channel impulse response sequence, and $\boldsymbol{v}(i)$ is a zero-mean noise sequence that is uncorrelated with $\boldsymbol{u}_i$. We again remind the reader of our notation for indexing vectors and scalars in this book. We use subscripts as time indices for vectors, e.g., $\boldsymbol{u}_i$, and parenthesis as time indices for scalars, e.g., $\boldsymbol{u}(i)$.

It is further assumed that the moments $R_u = \mathsf{E}\,\boldsymbol{u}_i^*\boldsymbol{u}_i$, $\sigma_d^2 = \mathsf{E}\,|\boldsymbol{d}(i)|^2$, and $R_{du} = \mathsf{E}\,\boldsymbol{d}(i)\boldsymbol{u}_i^*$ are all independent of time. We also let $\{u_i, d(i)\}$ denote observed values for the random variables $\{\boldsymbol{u}_i, \boldsymbol{d}(i)\}$. It is important to distinguish between a measured value $d(i)$ and its stochastic version $\boldsymbol{d}(i)$; similarly for u_i and $\boldsymbol{u}_i$. This distinction is relevant because while an adaptive filtering implementation operates on the measured data $\{d(i), u_i\}$, its derivation and performance are characterized in terms of the statistical properties of the underlying stochastic variables $\{\boldsymbol{d}(i), \boldsymbol{u}_i\}$.

The channel vector c is modeled as an unknown constant vector. This situation is identical to the scenario discussed in Sec. 6.3, where c was estimated by formulating a constrained linear estimation problem — see (6.18). In terms of the notation of the present section, the rows of the data matrix H from Eq. (6.18) would be the $\{\boldsymbol{u}_i\}$, while the entries of the observation vector $\boldsymbol{y}$ in (6.18) would be the $\{\boldsymbol{d}(i)\}$. The solution method (6.18) did not require knowledge of $\{R_{du}, R_u\}$. There is an alternative way to estimate c that does not require knowledge of $\{R_{du}, R_u\}$ either; it relies on using a stochastic-gradient (or adaptive) algorithm. The method can be motivated as follows. Assume we formulate the following linear least-mean-squares estimation problem:

$$\min_w \mathsf{E}\,|\boldsymbol{d}(i) - \boldsymbol{u}_i w|^2$$

whose solution we already know is

$$w^o = R_u^{-1} R_{du} \tag{10.18}$$

Then w^o coincides with the desired unknown c. This is because if we multiply (10.17) by $\boldsymbol{u}_i^*$ from the left and take expectations, we find that

$$\mathsf{E}\,\boldsymbol{u}_i^*\boldsymbol{d}(i) = (\mathsf{E}\,\boldsymbol{u}_i^*\boldsymbol{u}_i)\cdot c + \underbrace{\mathsf{E}\,\boldsymbol{u}_i^*\boldsymbol{v}(i)}_{=0}$$

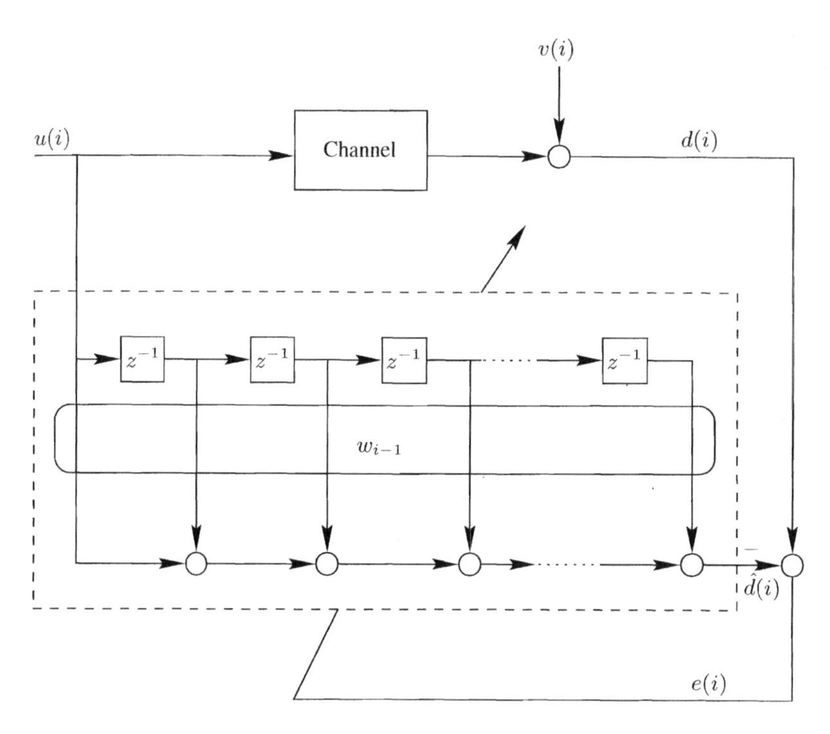

FIGURE 10.2 A structure for adaptive channel estimation.

so that $c = R_u^{-1} R_{du} = w^o$. Therefore, if we can determine w^o then we recover c. However, the moments $\{R_{du}, R_u\}$ are rarely available in practice so that determining w^o via (10.18), or even via a related steepest-descent implementation such as

$$w_i = w_{i-1} \; + \; \mu[R_{du} - R_u w_{i-1}]$$

would not be viable. Instead, we can appeal to a stochastic-gradient approximation for estimating w^o. Using measurements $\{d(i), u_i\}$, we can estimate w^o (and, hence, c) by using, e.g., the LMS recursion:

$$w_i = w_{i-1} \; + \; \mu u_i^* [d(i) - u_i w_{i-1}] \tag{10.19}$$

This discussion suggests the structure shown in Fig. 10.2, which employs an FIR filter with adjustable weights that is connected to the input and output signals $\{d(i), u(i)\}$ of the channel. At each time instant i, the measured output of the channel, $d(i)$, is compared with the output of the adaptive filter, $\hat{d}(i) = u_i w_{i-1}$, and an error signal, $e(i) = d(i) - u_i w_{i-1}$, is generated. The error is then used to adjust the filter coefficients from w_{i-1} to w_i according to (10.19). In steady-state, if the step-size μ is suitably chosen to guarantee filter convergence (usually, small step-sizes will do), then the error signal will assume small values, and the output $\hat{d}(i) = u_i w_{i-1}$ of the adaptive filter will assume values close to $d(i)$. Consequently, from an input/output perspective, the (adaptive filter) mapping from $u(i)$ to $\hat{d}(i)$ will behave similarly to the channel, which maps $u(i)$ to $d(i)$. This construction assumes that the adaptive filter has at least as many taps as the channel itself; otherwise, performance degradation can occur due to under-modeling.

10.6 APPLICATION: ADAPTIVE CHANNEL EQUALIZATION

Our second application is linear channel equalization. In Fig. 10.3, data symbols $\{\boldsymbol{s}(\cdot)\}$ are transmitted through a channel and the output sequence is measured in the presence of additive noise, $\boldsymbol{v}(i)$. The signals $\{\boldsymbol{v}(\cdot), \boldsymbol{s}(\cdot)\}$ are assumed uncorrelated. The noisy output of the channel is denoted by $\boldsymbol{u}(i)$ and is fed into a linear equalizer with L taps. At any particular time instant i, the state of the equalizer is given by

$$\boldsymbol{u}_i = \begin{bmatrix} \boldsymbol{u}(i) & \boldsymbol{u}(i-1) & \ldots & \boldsymbol{u}(i-L+1) \end{bmatrix}$$

It is desired to determine the equalizer tap vector w in order to estimate the signal $\boldsymbol{d}(i) = \boldsymbol{s}(i-\Delta)$ optimally in the least-mean-squares sense. This application coincides with the one described in Sec. 5.4 except that now, in conformity with the notation of this chapter, we are denoting the input to the equalizer by $\boldsymbol{u}(i)$ and the symbol to be estimated by $\boldsymbol{d}(i)$.

Clearly, the equalizer w^o that solves

$$\min_w \mathsf{E}\,|\boldsymbol{d}(i) - \boldsymbol{u}_i w|^2$$

is given by

$$w^o = R_u^{-1} R_{du} \tag{10.20}$$

where $R_u = \mathsf{E}\,\boldsymbol{u}_i^* \boldsymbol{u}_i$ and $R_{du} = \mathsf{E}\,\boldsymbol{d}(i)\boldsymbol{u}_i^*$. In Sec. 5.4 we assumed knowledge of the channel tap vector c (assumed FIR) and used it to evaluate $\{R_{du}, R_u\}$ — refer to equations (5.15), which were defined in terms of a channel matrix H. However, in practice, knowledge of $\{R_{du}, R_u\}$ and even c cannot be taken for granted. Actually, more often than not, these quantities are not available. For this reason, determining w^o via (10.20), or even via a related steepest-descent implementation such as

$$w_i = w_{i-1} \ + \ \mu[R_{du} - R_u w_{i-1}]$$

may not be viable. In such situations, we can appeal to a stochastic-gradient approximation for estimating w^o. Assuming an initial training phase in which transmitted data $\{d(i) = s(i-\Delta)\}$ are known at the receiver (i.e., at the equalizer), we can then use the available measurements $\{d(i), u_i\}$ to estimate w^o iteratively by using, e.g., the LMS recursion:

$$w_i = w_{i-1} \ + \ \mu u_i^*[d(i) - u_i w_{i-1}], \qquad d(i) = s(i-\Delta) \tag{10.21}$$

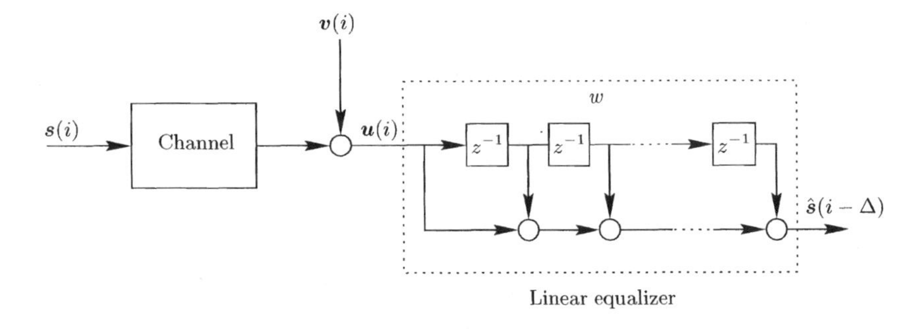

FIGURE 10.3 Linear equalization of an FIR channel in the presence of additive noise.

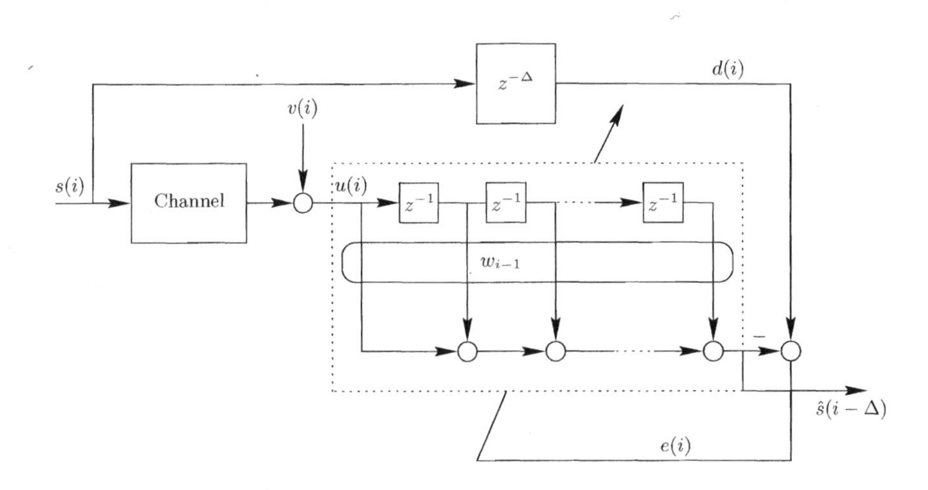

FIGURE 10.4 Adaptive linear equalization of a channel in the presence of additive noise.

This discussion suggests that we consider the structure of Fig. 10.4, with an FIR filter with adjustable weights that is connected in series with the channel. At each time instant i, the symbol $d(i) = s(i-\Delta)$ is compared with the output of the adaptive filter, $\hat{s}(i-\Delta)$, and an error signal, $e(i) = d(i) - u_i w_{i-1}$, is generated. The error is then used to adjust the filter coefficients from w_{i-1} to w_i according to (10.21). In steady-state, the error signal will assume small values and, hence, the output $\hat{s}(i-\Delta)$ of the adaptive filter will assume values close to $s(i-\Delta)$. Consequently, from an input/output perspective, the combination channel/equalizer, which maps $s(i)$ to $\hat{s}(i-\Delta)$, behaves "like" a delay system with transfer function $z^{-\Delta}$. Observe that this scheme for adaptive channel equalization does not require knowledge of the channel.

In practice, following a training phase with a known reference sequence $\{d(i)\}$, an equalizer could continue to operate in one of two modes. In the first mode, its coefficient vector w_i would be frozen and used thereafter to generate future outputs $\{\hat{s}(i-\Delta)\}$. This mode of operation is appropriate when the training phase is successful enough to result in reliable estimates $\hat{s}(i-\Delta)$, namely, estimates that lead to a low probability of error after they are fed into a decision device, which maps $\hat{s}(i-\Delta)$ to the closest point in the symbol constellation, say, $\check{s}(i-\Delta)$. However, if the channel varies slowly with time, it may be necessary to continue to operate the equalizer in a decision-directed mode. In this mode of operation, the weight vector of the equalizer continues to be adapted even after the training phase has ended. Now, however, the output of the decision device replaces the reference sequence $\{d(i)\}$ in the generation of the error sequence $\{e(i)\}$. Figure 10.5 illustrates the operation of an equalizer during the training and decision-directed modes of operation.

10.7 APPLICATION: DECISION-FEEDBACK EQUALIZATION

Our third application is decision-feedback equalization. Again, data symbols $\{\boldsymbol{s}(\cdot)\}$ are transmitted through a channel and the output sequence is measured in the presence of additive noise, $\boldsymbol{v}(i)$. The signals $\{\boldsymbol{v}(\cdot), \boldsymbol{s}(\cdot)\}$ are assumed uncorrelated. The noisy output of the channel is denoted by $\boldsymbol{u}(i)$ and is fed into a feedforward filter with L taps, as indicated in Fig. 10.6 for the case $L = 3$. The output of the decision device is fed into a feedback filter with Q taps; this filter works in conjunction with the feedforward filter in order to supply the decision device with an estimator $\hat{s}(i-\Delta)$. Assuming correct decisions, at any particular time instant i, the state of the equalizer is given by

$$\boldsymbol{u}_i = \left[\begin{array}{ccc|ccc} \boldsymbol{s}(i-\Delta-1) & \ldots & \boldsymbol{s}(i-\Delta-Q) & \boldsymbol{u}(i) & \ldots & \boldsymbol{u}(i-L+1) \end{array} \right]$$

That is, it contains the states of both the feedback and the feedforward filter. It is then desired to determine an equalizer tap vector

$$w \triangleq \text{col}\{-b(1), -b(2), \ldots, -b(Q), f(0), f(1), \ldots, f(L-1)\}$$

that estimates $\boldsymbol{d}(i) = \boldsymbol{s}(i-\Delta)$ optimally in the least-mean-squares sense, where the $\{-b(i)\}$ denote the coefficients of the feedback filter while the $\{f(i)\}$ denote the coefficients of the forward filter. This application coincides with the one described in Sec. 6.4, except that now, in conformity with the notation of this chapter, we are denoting the input to the equalizer by $\boldsymbol{u}(i)$ and the symbol to be estimated by $\boldsymbol{d}(i)$. We are also relying on the formulation of the decision-feedback equalizer as a linear estimation problem (cf. Eq. (6.35)).

Clearly, the equalizer w^o that solves $\min_w \mathsf{E}\,|\boldsymbol{d}(i) - \boldsymbol{u}_i w|^2$ is

$$w^o = R_u^{-1} R_{du} \tag{10.22}$$

where $R_u = \mathsf{E}\,\boldsymbol{u}_i^* \boldsymbol{u}_i$ and $R_{du} = \mathsf{E}\,\boldsymbol{d}(i)\boldsymbol{u}_i^*$. In Sec. 6.4, and also Prob. II.41, we assumed knowledge of the channel tap vector c (assumed FIR) and used it to evaluate $\{R_{du}, R_u\}$ — refer to equations (6.33) and (6.34), which were defined in terms of a channel matrix H. However, in practice, knowledge of $\{R_{du}, R_u, c\}$ is generally unavailable. For this reason, determining w^o via (10.22), or even via a related steepest-descent implementation such as

$$w_i = w_{i-1} + \mu[R_{du} - R_u w_{i-1}]$$

would not be viable. Instead, we can appeal to a stochastic-gradient approximation for estimating w^o. Assuming an initial training phase in which transmitted data $\{d(i) = s(i-\Delta)\}$

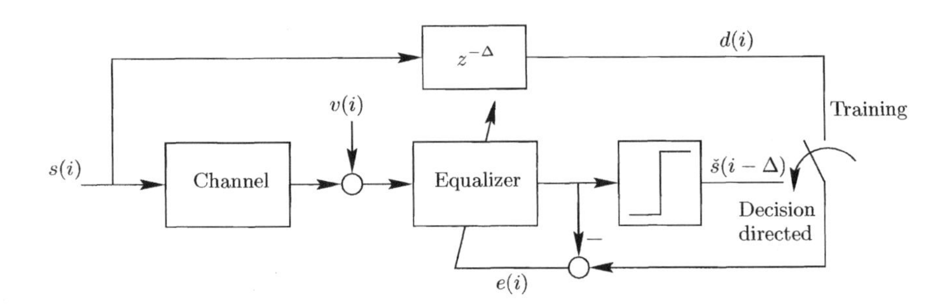

FIGURE 10.5 An adaptive linear equalizer operating in two modes: training mode and decision-direction mode.

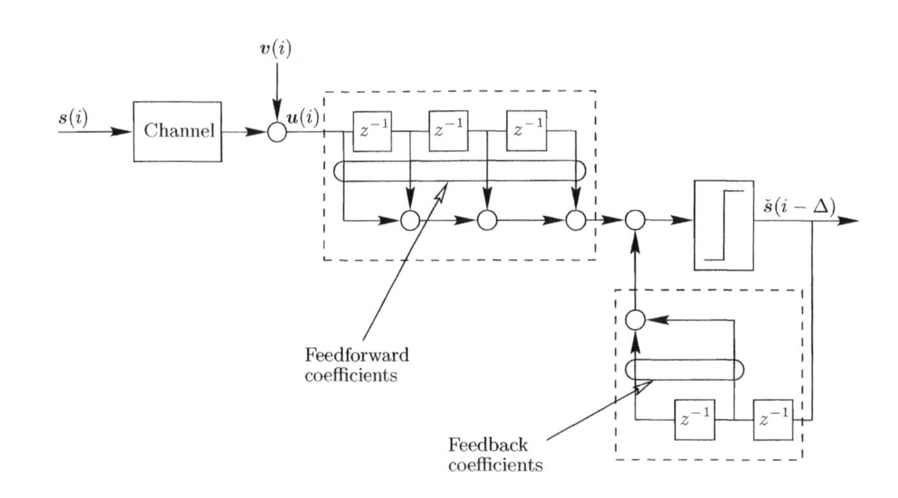

FIGURE 10.6 A decision-feedback equalizer. It consists of a feedforward filter, a feedback filter, and a decision device.

are known at the receiver (i.e., at the equalizer), we can then use the available measurements $\{d(i), u_i\}$ to estimate w^o iteratively by using, e.g., the LMS recursion:

$$w_i = w_{i-1} + \mu u_i^*[d(i) - u_i w_{i-1}], \qquad d(i) = s(i-\Delta) \tag{10.23}$$

Thus, consider the structure shown in Fig. 10.7, with FIR filters with adjustable weights used as feedforward and feedback filters. The top entries of w_i would correspond to the coefficients of the feedback filter, while the bottom entries of w_i would correspond to the coefficients of the feedforward filter. Likewise, the leading entries of the regressor correspond to the state of the feedback filter, while the trailing entries of the regressor correspond to the state of the feedforward filter.

Figure 10.7 depicts two modes of operation: training and decision-directed. During training, delayed symbols are used as a reference sequence. At each time instant i, the symbol $d(i) = s(i-\Delta)$ is compared with the output of the adaptive filter, $\hat{s}(i-\Delta)$ (which is the input to the decision device), and an error signal, $e(i) = d(i) - u_i w_{i-1}$, is generated. The error is then used to adjust the coefficients of the feedback and feedforward filters according to (10.23). During training, the state (or regressor) of the equalizer is given by

$$u_i = \left[\begin{array}{ccc|ccc} s(i-\Delta-1) & \ldots & s(i-\Delta-Q) & u(i) & \ldots & u(i-L+1) \end{array}\right]$$

while during decision-directed operation, the signal $d(i)$ is replaced by the output of the decision device, $\check{s}(i-\Delta)$, so that the state of the equalizer is then given by

$$u_i = \left[\begin{array}{ccc|ccc} \check{s}(i-\Delta-1) & \ldots & \check{s}(i-\Delta-Q) & u(i) & \ldots & u(i-L+1) \end{array}\right]$$

10.8 ENSEMBLE-AVERAGE LEARNING CURVES

It is often necessary to evaluate the performance of a stochastic-gradient algorithm. A common way to do so is to construct its *ensemble-average* learning curve, which is defined below.

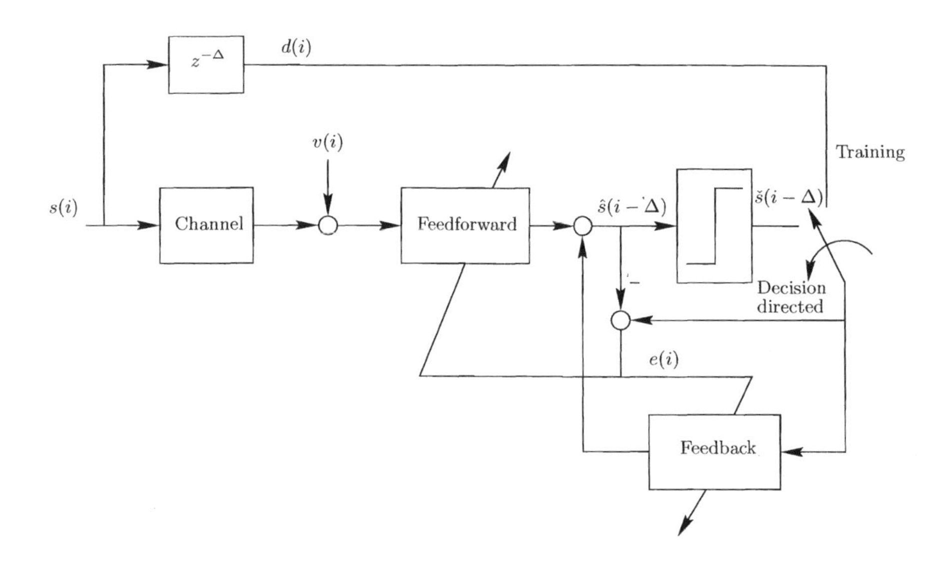

FIGURE 10.7 Adaptive decision-feedback equalization. Both modes of operation are shown: training and decision-directed operation.

Recall that for least-mean-squares estimation, the learning curve of a steepest-descent method was defined in Sec. 9.5 as (cf. (9.1)):

$$\boxed{J(i) \stackrel{\Delta}{=} \mathsf{E}\,|\boldsymbol{d} - \boldsymbol{u}w_{i-1}|^2, \quad i \geq 0}$$

where w_{i-1} is the weight estimate at iteration $i-1$ that is given by the steepest-descent algorithm. Evaluation of $J(i)$ would require knowledge of $\{\sigma_d^2, R_{du}, R_u\}$. However, in a stochastic-gradient implementation, we do not have access to this statistical information but only to observations of the random variables $\boldsymbol{d}$ and $\boldsymbol{u}$, namely $\{d(i), u_i\}$. If we replace $\boldsymbol{d}$ by $d(i)$ and $\boldsymbol{u}$ by u_i in the above expression for $J(i)$, then the difference $\boldsymbol{d} - \boldsymbol{u}w_{i-1}$ becomes $d(i) - u_i w_{i-1}$, with the w_{i-1} now denoting the weight estimate that is obtained from the stochastic-gradient implementation (e.g., LMS). We have denoted the difference $d(i) - u_i w_{i-1}$ by $e(i)$ earlier in this chapter and called it the *a priori* output estimation error,

$$\boxed{e(i) = d(i) - u_i w_{i-1}}$$

We can then estimate the learning curve of an adaptive filter as follows. We run the algorithm for a certain number of iterations, say, for $0 \leq i \leq N$. The duration N is usually chosen large enough so that convergence is observed. We then compute the error sequence $\{e(i)\}$ and the corresponding squared-error curve $\{|e(i)|^2,\ 0 \leq i \leq N\}$. We denote this squared-error curve by

$$\left\{\left|e^{(1)}(i)\right|^2,\ 0 \leq i \leq N\right\}$$

with the superscript $^{(1)}$ used to indicate that this curve is the result of the first experiment. We then run the same stochastic-gradient algorithm a second time, starting from the same initial condition w_{-1} and using data with the same statistical properties as in the first run. After L such experiments, we obtain L squared-error curves,

$$\left\{ \left|e^{(1)}(i)\right|^2, \ \left|e^{(2)}(i)\right|^2, \ \ldots, \ \left|e^{(L)}(i)\right|^2 \right\}, \quad 0 \le i \le N$$

The *ensemble-average* learning curve, over the interval $0 \le i \le N$, is defined as the average over the L experiments:

$$\boxed{\widehat{J}(i) \triangleq \frac{1}{L}\left(\sum_{j=1}^{L} \left|e^{(j)}(i)\right|^2\right), \quad i \ge 0}$$

The averaged curve $\widehat{J}(i)$ so defined is a sample-average approximation of the true learning curve $J(i)$.

Example 10.1 (Learning curves)

We illustrate the construction of learning curves by considering an example in the context of channel estimation, as described in Sec. 10.5. The impulse response sequence of the channel is chosen as $c = \text{col}\{1, 0.5, -1, 2\}$, i.e., its transfer function is

$$C(z) = 1 + 0.5z^{-1} - z^{-2} + 2z^{-3}$$

The channel impulse response, along with its magnitude frequency response, are shown in Fig. 10.8.

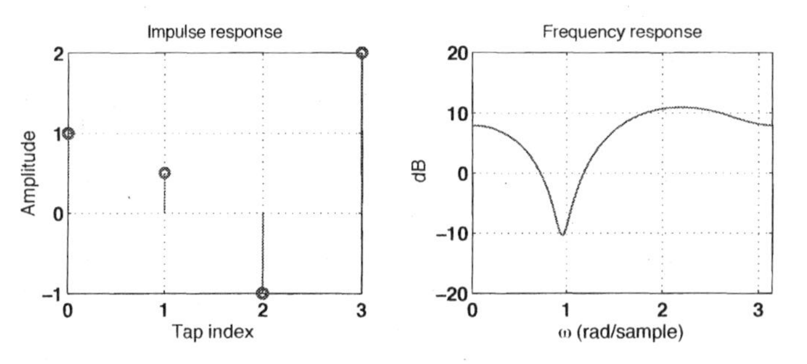

FIGURE 10.8 The impulse-response sequence and the magnitude-frequency response of the channel $C(z) = 1 + 0.5z^{-1} - z^{-2} + 2z^{-2}$.

White input data $\{\boldsymbol{u}(i)\}$ of unit variance is fed into the channel and the output sequence is observed in the presence of white additive noise of variance 0.01. A total of $N = 600$ samples $\{u(i), d(i)\}$ are generated and used to train an adaptive filter with $M = 4$ taps. The filter is trained by using the LMS algorithm of this chapter with step-size $\mu = 0.01$, as well as two other algorithms derived in Chapters 11 and 14 for comparison purposes, namely, the so-called $\epsilon-$NLMS algorthm with step-size $\mu = 0.2$ and $\epsilon = 0.001$, and the RLS algorithm with $\lambda = 0.995$ and the same value of ϵ. All filters start from the same initial condition $w_{-1} = 0$.

Figure 10.9 shows two typical instantaneous squared-error curves for each of the algorithms over the first 200 iterations, i.e., the rows show plots of $|e(i)|^2$ versus time in two random simulations for each algorithm. Observe how the curves die out quicker for $\epsilon-$NLMS and RLS relative to LMS. By averaging $L = 300$ such curves for each algorithm, we obtain the ensemble-averaging learning curves shown in Fig. 10.10. These curves illustrate the fact that the convergence speed increases as we move from LMS to $\epsilon-$NLMS to RLS.

◇

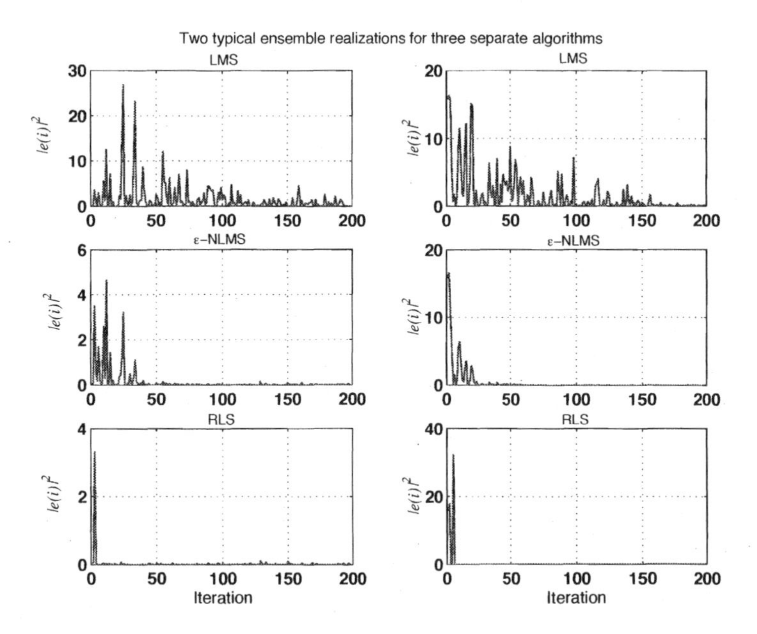

FIGURE 10.9 Typical squared-error curves for LMS, ϵ-NLMS, and RLS.

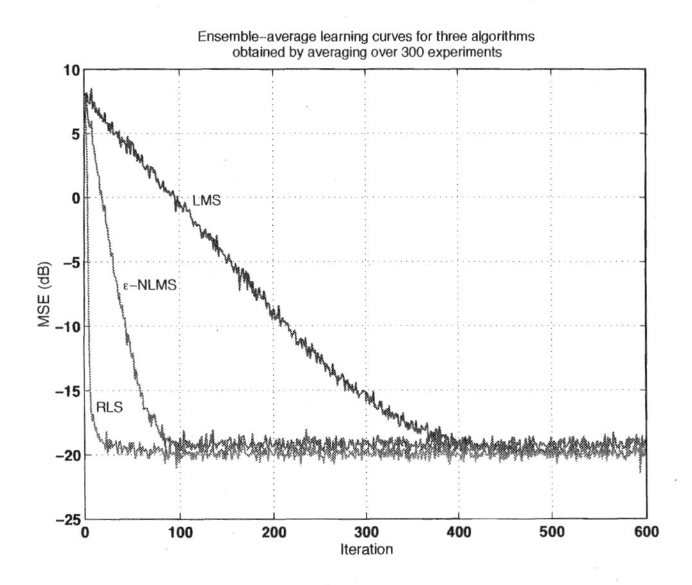

FIGURE 10.10 Ensemble-average learning curves for LMS, ϵ-NLMS, and RLS obtained by averaging over 300 experiments.

CHAPTER 11

Normalized LMS Algorithm

Now that we have illustrated the application of adaptive filters (and specifically LMS) in several contexts, we proceed to the derivation of other similar stochastic-gradient methods. In this chapter we motivate the so-called Normalized LMS algorithm.

11.1 INSTANTANEOUS APPROXIMATION

Assume again that we have access to several observations of the random variables $\boldsymbol{d}$ and $\boldsymbol{u}$ in (10.1), say, $\{d(0), d(1), d(2), d(3), \ldots\}$ and $\{u_0, u_1, u_2, u_3, \ldots\}$. The normalized LMS algorithm can be motivated in much the same way as LMS except that now we start from the regularized Newton's recursion (10.8) and assume that the regularization sequence $\{\epsilon(i)\}$ and the step-size sequence $\mu(i)$ are constants, say, $\epsilon(i) = \epsilon$ and $\mu(i) = \mu$.

Thus, using

$$w_i = w_{i-1} + \mu\,[\epsilon \mathrm{I} + R_u]^{-1}\,[R_{du} - R_u w_{i-1}] \tag{11.1}$$

and replacing the quantities $(\epsilon\mathrm{I} + R_u)$ and $(R_{du} - R_u w_{i-1})$ by the instantaneous approximations $(\epsilon\mathrm{I}+u_i^* u_i)$ and $u_i^*[d(i)-u_i w_{i-1}]$, respectively, we arrive at the stochastic-gradient recursion

$$w_i = w_{i-1} + \mu[\epsilon\mathrm{I} + u_i^* u_i]^{-1}\; u_i^*[d(i) - u_i w_{i-1}] \tag{11.2}$$

This recursion, in its current form, requires the inversion of the matrix $(\epsilon\mathrm{I} + u_i^* u_i)$ at each iteration. This step can be avoided by reworking the recursion into an equivalent simpler form. Thus, note that $(\epsilon\mathrm{I} + u_i^* u_i)$ is a rank-one modification of a multiple of the identity matrix, and the inverse of every such matrix has a similar structure. To see this, we simply apply the matrix inversion formula (5.4) to get

$$[\epsilon\mathrm{I} + u_i^* u_i]^{-1} = \epsilon^{-1}\mathrm{I} - \frac{\epsilon^{-2}}{1+\epsilon^{-1}\|u_i\|^2}\; u_i^* u_i \tag{11.3}$$

where the expression on the right-hand side is a rank-one modification of $\epsilon^{-1}\mathrm{I}$. If we now multiply both sides of (11.3) by u_i^* from the right we obtain

$$[\epsilon\mathrm{I} + u_i^* u_i]^{-1} u_i^* \;=\; \epsilon^{-1} u_i^* - \frac{\epsilon^{-2}}{1+\epsilon^{-1}\|u_i\|^2}\; u_i^*\|u_i\|^2 \;=\; \frac{u_i^*}{\epsilon + \|u_i\|^2}$$

which is a scalar multiple of u_i^*. Substituting into (11.2) we arrive at the $\epsilon-$NLMS recursion:

$$\boxed{w_i = w_{i-1} + \frac{\mu}{\epsilon + \|u_i\|^2}\; u_i^*[d(i) - u_i w_{i-1}]\;,\quad i \geq 0} \tag{11.4}$$

We refer to (11.4) as ϵ-NLMS as opposed to simply NLMS, in order to highlight the presence of the small positive number ϵ; the terminology NLMS will be reserved for $\epsilon = 0$.

Algorithm 11.1 (ϵ-NLMS algorithm) Consider a zero-mean random variable $\boldsymbol{d}$ with realizations $\{d(0), d(1), \ldots\}$, and a zero-mean random row vector $\boldsymbol{u}$ with realizations $\{u_0, u_1, \ldots\}$. The optimal weight vector w^o that solves

$$\min_w \mathsf{E}\,|\boldsymbol{d} - \boldsymbol{u}w|^2$$

can be approximated iteratively via the recursion

$$w_i = w_{i-1} + \frac{\mu}{\epsilon + \|u_i\|^2} u_i^* \left[d(i) - u_i w_{i-1}\right], \quad i \geq 0, \quad w_{-1} = \text{initial guess}$$

where μ is a positive step-size and ϵ is a small positive parameter.

Comparing $\epsilon-$NLMS with the LMS recursion (10.10), we find that the update direction in LMS is a scaled version of the regression vector u_i^*, namely, $\mu u_i^* e(i)$. The "size" of the change to w_{i-1} will therefore be in proportion to the norm of u_i^*. In this way, a vector u_i with a large norm will generally lead to a more substantial change to w_{i-1} than a vector u_i with a small norm. Such behavior can have an adverse effect on the performance of LMS in some applications, e.g., when dealing with speech signals, where intervals of speech activity are usually followed by intervals of pause so that the norm of the regression vector can fluctuate appreciably. In comparison, the correction term that is added to w_{i-1} in the ϵ-NLMS recursion (11.4) is normalized with respect to the squared-norm of the regressor u_i; hence the name normalized LMS. Moreover, the positive constant ϵ in the denominator avoids division by zero or by a small number when the regressor is zero or close to zero.

Moreover, as can be seen from (11.4), $\epsilon-$NLMS employs a time-variant step-size of the form $\mu(i) = \mu/(\epsilon + \|u_i\|^2)$, as opposed to the constant step-size, μ, which is used by LMS. Now since $\epsilon-$NLMS was obtained as a stochastic-gradient approximation to Newton's method, and given the superior convergence speed of Newton's recursion (10.8) as compared to the standard steepest-descent recursion (10.4), we expect ϵ-NLMS to exhibit a faster convergence behavior than LMS. This is indeed the case, and this intuitive argument will be formalized later in Chapters 24 and 25 when we study the transient behavior of LMS and ϵ-NLMS.

11.2 COMPUTATIONAL COST

Compared to LMS, the recursion for ϵ-NLMS requires the evaluation of two additional quantities:

1. One inner product calculation is needed to evaluate $\|u_i\|^2$. This calculation is a special inner product since it involves a vector and its conjugate transpose. It can therefore be evaluated more efficiently than a generic inner product. To see this, let us denote an arbitrary entry of u_i by $a + jb$, then evaluation of $\|u_i\|^2$ requires $M - 1$ real additions of M terms of the form $|a|^2 + |b|^2$. Each such term requires 2 real multiplications and one real addition. Therefore, the computation of $\|u_i\|^2$ requires $2M$ real multiplications and $2M - 1$ real additions.

2. One real addition and one real division is needed in order to evaluate $\mu/(\epsilon + \|u_i\|^2)$.

Therefore, for general complex-valued data, each iteration of ϵ-NLMS requires $10M + 2$ real multiplications, $10M$ real additions, and one real division. On the other hand, for real-valued data, each iteration of ϵ-NLMS requires $3M + 1$ real multiplications, $3M$ real additions, and one real division.

Regressors with Shift-Structure

When the regressor u_i possesses *shift-structure*, the computational cost of ϵ-NLMS can in principle be reduced to a figure that is comparable to that of LMS. So assume

$$u_i = \begin{bmatrix} u(i) & \boxed{u(i-1)\ u(i-2)\ \ldots\ u(i-M+1)} \end{bmatrix} \tag{11.5}$$

for some sequence of values $\{u(i)\}$. Then two successive regressors, u_i and u_{i-1}, will only differ in two entries since

$$u_{i-1} = \begin{bmatrix} \boxed{u(i-1)\ u(i-2)\ \ldots\ u(i-M+1)} & u(i-M) \end{bmatrix} \tag{11.6}$$

The boxed entries in u_i and u_{i-1} are common to both regressors. It follows that

$$\|u_i\|^2 = \|u_{i-1}\|^2 \ - \ |u(i-M)|^2 \ + \ |u(i)|^2 \tag{11.7}$$

so that the update of the successive squared-norms of the regressors can be accomplished as follows. For general complex-valued data, it requires 2 real multiplications and 1 real addition to evaluate each of $|u(i)|^2$ and $|u(i-M)|^2$, and 2 real additions to evaluate the sum $\|u_{i-1}\|^2 \ - \ |u(i-M)|^2 \ + \ |u(i)|^2$. We therefore find that for general complex-valued data and for regressors with shift-structure, each iteration of ϵ-NLMS requires $8M + 6$ real multiplications, $8M + 5$ real additions, and one real division. On the other hand, for real-valued data and for regressors with shift-structure, each iteration of ϵ-NLMS requires $2M + 3$ real multiplications, $2M + 3$ real additions, and one real division. We may remark that, in practice, use of (11.7) is problematic due to the accumulation of rounding errors, which can ultimately destroy the nonnegativity of $\|u_i\|^2$.

11.3 POWER NORMALIZATION

Observe further that the ϵ-NLMS recursion (11.4) can be written as

$$w_i = w_{i-1} + \frac{\mu/M}{\epsilon/M + \|u_i\|^2/M}\, u_i^*[d(i) - u_i w_{i-1}]$$

where we scaled both the numerator and the denominator of the term $\mu/(\epsilon + \|u_i\|^2)$ by M or, more compactly, as

$$w_i = w_{i-1} + \frac{\mu'}{\epsilon' + \|u_i\|^2/M}\, u_i^*[d(i) - u_i w_{i-1}] \tag{11.8}$$

for some smaller step-size μ' and regularization parameter ϵ'. This rewriting of ϵ–NLMS suggests a method for approximating $\|u_i\|^2$. The method replaces $\|u_i\|^2/M$ by a variable $p(i)$ that is updated as follows:

$$\boxed{p(i) \ = \ \beta p(i-1) \ + \ (1-\beta)|u(i)|^2, \qquad p(-1) = 0} \tag{11.9}$$

where β is a positive scalar chosen from within the interval $0 < \beta \leq 1$. In this way, the ϵ-NLMS recursion (11.8) would be replaced by

$$\boxed{w_i = w_{i-1} + \frac{\mu}{\epsilon + p(i)} u_i^*[d(i) - u_i w_{i-1}], \quad i \geq 0} \tag{11.10}$$

where we are restoring the notation $\{\mu, \epsilon\}$ for the step-size and the regularization parameter in order to avoid an explosion of notation; but it is understood that the step-size in the implementation (11.10) is approximately M times smaller than the corresponding step-size in the ϵ-NLMS implementation (11.4). We refer to (11.10) as ϵ-NLMS with power normalization since $p(i)$ can be interpreted as an estimate for the power of the input sequence $\{u(j)\}$ — see Prob. III.24.

The inclusion of the parameter β in (11.9) is meant to introduce *memory* into the recursion for $p(i)$. Indeed, note from (11.9) that

$$p(i) = (1 - \beta) \sum_{j=0}^{i} \beta^{i-j} |u(j)|^2$$

so that input data $\{u(j)\}$ in the remote past are weighted less heavily than most recent data.[5] Now since, for complex data, the evaluation of $p(i)$ requires four real multiplications and three real additions, we find that for general complex-valued data, each iteration of recursion (11.10) requires $8M + 6$ real multiplications, $8M + 4$ real additions, and one real division. On the other hand, for real-valued data, each iteration of recursion (11.10) requires $2M + 4$ real multiplications, $2M + 3$ real additions, and one real division. Usually, β is chosen as a power of 2^{-1} so that multiplications by β and $(1 - \beta)$ could be implemented digitally by means of shift registers. Therefore, in principle, we can ignore the cost of these two required multiplications. In the text, however, we choose to account for more general choices of β.

Algorithm 11.2 (ϵ-NLMS algorithm with power normalization) Consider the same setting of Alg. 11.1. The optimal weight vector w^o can be approximated iteratively via the recursions

$$\begin{aligned} p(i) &= \beta p(i-1) + (1-\beta)|u(i)|^2, \quad p(-1) = 0 \\ w_i &= w_{i-1} + \frac{\mu}{\epsilon + p(i)} u_i^*[d(i) - u_i w_{i-1}], \quad i \geq 0 \end{aligned}$$

where $0 < \beta \leq 1$ (usually close to one), μ is a positive step-size, and the regression vector u_i is assumed to possess shift-structure as in (11.5).

LMS *with Time-Variant Step-Sizes*

More generally, a stochastic-gradient algorithm with a generic time-variant step-size can be obtained from the steepest-descent method (10.6) if we replace $R_{du} - R_u w_{i-1}$ by

$$u_i^*[d(i) - u_i w_{i-1}]$$

[5] In some applications such as speech processing, a constant value of β may not be adequate since speech signals generally undergo significant fluctuations.

which leads to

$$\boxed{w_i = w_{i-1} + \mu(i)u_i^*[d(i) - u_i w_{i-1}], \quad i \geq 0} \tag{11.11}$$

The ϵ-NLMS algorithm, as well as its power-normalized variant, are special cases of (11.11) with particular choices for $\mu(i)$.

11.4 LEAST-PERTURBATION PROPERTY

The ϵ-NLMS algorithm was derived in Sec. 11.1 as an approximate solution to the linear least-mean-squares estimation problem (10.1), in the sense that it was obtained by replacing the gradient vector and Hessian matrix in Newton's recursion (11.1) by instantaneous approximations for them. Later in Sec. 45.2, we shall show that ϵ-NLMS can also be derived as the *exact* (not approximate) solution of a well-defined estimation problem; albeit one with a different cost criterion. Until then, we shall continue to treat ϵ-NLMS as a stochastic-gradient algorithm and proceed to highlight some of its properties.

One particular property is that ϵ-NLMS, just like LMS, can also be regarded as the exact solution to a *local* optimization problem. To see this, we proceed as in Sec. 10.4. Thus, assume that we have available some weight estimate at time $i-1$, say, w_{i-1}, and let $\{d(i), u_i\}$ denote the newly measured data. Let w_i denote a weight estimate that we wish to compute from the available data $\{w_{i-1}, d(i), u_i\}$ in the following manner. Let μ and ϵ be two given positive numbers, and consider the problem of determining w_i by solving the constrained optimization problem:

$$\boxed{\min_{w_i} \ \|w_i - w_{i-1}\|^2 \qquad \text{subject to } r(i) = \left(1 - \frac{\mu\|u_i\|^2}{\epsilon + \|u_i\|^2}\right) e(i)} \tag{11.12}$$

where $\{r(i), e(i)\}$ are defined as in (10.11) and (10.12). In other words, we seek a solution w_i that is closest to w_{i-1} and ensures an equality constraint between $r(i)$ and $e(i)$. This constraint is different from the one we employed earlier in (10.13) for LMS, and it becomes relevant when the step-size μ is small enough to satisfy

$$\left|1 - \frac{\mu\|u_i\|^2}{\epsilon + \|u_i\|^2}\right| < 1$$

i.e., when

$$0 < \mu < \frac{2(\epsilon + \|u_i\|^2)}{\|u_i\|^2} \quad \text{for all } i \tag{11.13}$$

This is because when (11.13) holds, the magnitude of the *a posteriori* error will not exceed that of the *a priori* error, i.e., $|r(i)| < |e(i)|$ except when $e(i) = 0$, in which case $r(i) = 0$ also. In contrast to condition (10.14) in the LMS case, we now find from (11.13) that the bound on μ is more relaxed. In particular, any choice of μ in the interval $(0, 2)$ will do and any such choice is independent of u_i.

The solution of the optimization problem (11.12) can be obtained in precisely the same way as we did in Sec. 10.4 for LMS. The argument will lead us again to the ϵ-NLMS recursion:

$$w_i = w_{i-1} + \frac{\mu}{\epsilon + \|u_i\|^2} u_i^*[d(i) - u_i w_{i-1}], \quad i \geq 0, \quad w_{-1} = \text{initial guess}$$

CHAPTER 12

Other LMS-Type Algorithms

The idea of using instantaneous approximations to devise stochastic-gradient algorithms from steepest-descent implementations is not limited to quadratic cost functions as in (10.1). For instance, in Probs. III.16–III.21 we formulate steepest-descent methods for a variety of other cost functions. If we then employ instantaneous approximations for the associated gradient vectors and Hessian matrices, we would obtain other well-known adaptive algorithms. In this chapter, we list the recursions for various such algorithms of the *blind* and *non-blind* types. Non-blind methods are so-called because they employ a reference sequence $\{d(i)\}$ in their update recursions. On the other hand, blind algorithms do not use a reference sequence.

12.1 NON-BLIND ALGORITHMS

We first list several non-blind methods derived in Probs. III.26–III.31. The derivations in the problems lead to the following statements for the so-called sign-error LMS, leaky-LMS, least-mean-fourth (LMF), and least-mean-mixed-norm (LMMN) algorithms. For all statements below, we consider a zero mean random variable $\boldsymbol{d}$ with realizations $\{d(0), d(1), \ldots\}$, and a zero-mean random row vector $\boldsymbol{u}$ with realizations $\{u_0, u_1, \ldots\}$. Moreover, μ is a positive step-size (usually small).

Algorithm 12.1 (Sign-error LMS algorithm) The weight vector w^o that solves (in terms of the mean of the l_1 norm — cf. Probs. III.14–III.15 and III.29):

$$\min_w \mathsf{E}\,|\boldsymbol{d} - \boldsymbol{u}w|$$

can be approximated iteratively via the recursion

$$w_i = w_{i-1} + \mu u_i^* \mathsf{csgn}[d(i) - u_i w_{i-1}], \qquad i \geq 0$$

In the above statement, the sign of a complex number, $x = x_\mathsf{r} + jx_\mathsf{i}$, is defined as (cf. Prob. III.15):

$$\mathsf{csgn}(x) \triangleq \mathsf{sign}(x_\mathsf{r}) + j\mathsf{sign}(x_\mathsf{i}) \quad \text{where} \quad \mathsf{sign}(a) \triangleq \begin{cases} +1 & \text{if } a > 0 \\ -1 & \text{if } a < 0 \\ 0 & \text{if } a = 0 \end{cases} \tag{12.1}$$

in terms of the sign of a real number and $j = \sqrt{-1}$.

Algorithm 12.2 (Leaky-LMS algorithm) The weight vector w^o that solves (cf. Probs III.12 and III.26):

$$\min_{w} \; [\, \alpha \|w\|^2 \; + \; \mathsf{E}\,|\boldsymbol{d} - \boldsymbol{u}w|^2 \,]$$

for some positive constant α, can be approximated iteratively via the recursion

$$w_i = (1 - \mu\alpha)w_{i-1} + \mu u_i^* [d(i) - u_i w_{i-1}], \quad i \geq 0$$

Algorithm 12.3 (LMF algorithm) The weight vector w^o that solves (cf. Probs. III.16 and III.30):

$$\min_{w} \; \mathsf{E}\,|\boldsymbol{d} - \boldsymbol{u}w|^4$$

can be approximated iteratively via the recursion

$$w_i = w_{i-1} + \mu u_i^* e(i)|e(i)|^2, \quad i \geq 0$$

Algorithm 12.4 (LMMN algorithm) The weight vector w^o that solves (cf. Probs III.17 and III.31):

$$\min_{w} \; \mathsf{E}\left[\, \delta|\boldsymbol{e}|^2 \; + \; \frac{1}{2}(1-\delta)|\boldsymbol{e}|^4 \,\right], \qquad \boldsymbol{e} = \boldsymbol{d} - \boldsymbol{u}w$$

for some constant $0 \leq \delta \leq 1$, can be approximated iteratively via the recursion

$$w_i = w_{i-1} + \mu u_i^* e(i)[\, \delta + (1-\delta)|e(i)|^2 \,], \quad i \geq 0$$

Listing of Non-Blind Algorithms

Table 12.1 summarizes the stochastic-gradient algorithms derived so far, along with some other algorithms that we derive in the sequel. In the table, and in other places in the book, we use the notation $0 \ll \lambda \leq 1$, with the symbol $\ll$, to mean that λ is a positive scalar that is close or equal to one (e.g., $\lambda = 0.995$ or $\lambda = 1$). One way to express this condition more precisely would be to say that λ is a positive scalar inside the interval $(0, 1]$ and such that $1 - \lambda \ll 1$.

Tables 12.2 and 12.3 show the respective *estimated* computational costs per iteration for both cases of real- and complex-valued data. The costs in the second line of Tables 12.2 and 12.3 correspond to the generic recursion (11.11), and they assume that the values of $\mu(i)$ are available; if these values need to be computed then the costs will of course change. For completeness, in the multiplications columns in both tables, the entries between parenthesis

TABLE 12.1 A listing of several stochastic-gradient algorithms for $i \geq 0$; some are derived in the text while others are motivated and derived in the problems. In all of them, the initial weight estimate is specified at $i = -1$ and the signal $e(i)$ denotes the *a priori* output estimation error, $e(i) = d(i) - u_i w_{i-1}$.

Algorithm	Recursion
LMS with constant step-size	$w_i = w_{i-1} + \mu u_i^* e(i)$
LMS with time-variant step-size	$w_i = w_{i-1} + \mu(i) u_i^* e(i)$
ϵ-NLMS	$w_i = w_{i-1} + \dfrac{\mu}{\epsilon + \Vert u_i \Vert^2} u_i^* e(i)$
ϵ-NLMS with power normalization	$p(i) = \beta p(i-1) + (1-\beta)\vert u(i)\vert^2,\ p(-1) = 0$ $w_i = w_{i-1} + \dfrac{\mu}{\epsilon + p(i)} u_i^* e(i)$
sign-error LMS	$w_i = w_{i-1} + \mu u_i^* \text{csgn}[e(i)]$
leaky-LMS	$w_i = (1 - \mu\alpha) w_{i-1} + \mu u_i^* e(i),\ \alpha > 0$
LMF	$w_i = w_{i-1} + \mu u_i^* e(i) \vert e(i) \vert^2$
LMMN	$w_i = w_{i-1} + \mu u_i^* e(i) [\, \delta + (1-\delta)\vert e(i)\vert^2 \,]$ $0 \leq \delta \leq 1$
RLS	$P_i = \lambda^{-1} \left[P_{i-1} - \dfrac{\lambda^{-1} P_{i-1} u_i^* u_i P_{i-1}}{1 + \lambda^{-1} u_i P_{i-1} u_i^*} \right]$ $w_i = w_{i-1} + P_i u_i^* e(i)$ $P_{-1} = \epsilon^{-1} \mathrm{I},\ 0 \ll \lambda \leq 1$
Gauss-Newton (GN)	$P_i = \dfrac{\lambda^{-1}}{1-\alpha} \left[P_{i-1} - \dfrac{\lambda^{-1} P_{i-1} u_i^* u_i P_{i-1}}{\frac{(1-\alpha)}{\alpha} + \lambda^{-1} u_i P_{i-1} u_i^*} \right]$ $w_i = w_{i-1} + \mu P_i\, u_i^* e(i)$ $P_{-1} = \epsilon^{-1} \mathrm{I},\ \alpha > 0$ (small), $0 \ll \lambda \leq 1$

indicate the number of multiplications that would result if the parameters $\{\mu, \beta, \alpha, \delta\}$, whenever meaningful, were chosen as powers of 2^{-1}. Moreover, M is the filter order.

Observe in particular that all LMS-variants require on the order of M floating point operations (multiplications and additions) per iteration. We therefore say that they are $O(M)$ algorithms, with the notation $O(M)$ signifying "on the order of M" or "a multiple of M".

It is worth noting that, in practice, besides addition and multiplication operations, the complexity of an algorithm is also judged in terms of how often it accesses the memory (i.e., in terms of the frequency of its read and write operations).

TABLE 12.2 A comparison of the *estimated* computational cost per iteration for several stochastic-gradient algorithms for *real-valued* data in terms of the number of *real* multiplications, *real* additions, *real* divisions, and comparisons with zero (or sign evaluations).

Algorithm	$\times$	$+$	/	sign
LMS with constant step-size	$2M+1$ $(2M)$	$2M$		
LMS with time-variant step-size	$2M+1$ $(2M)$	$2M$		
ϵ-NLMS	$3M+1$ $(3M)$	$3M$	1	
ϵ-NLMS with power normalization	$2M+4$ $(2M+1)$	$2M+2$	1	
sign-error LMS	$2M$ (M)	$2M$		1
leaky-LMS	$3M+2$ $(2M+1)$	$2M+1$ $(3M+1)$		
LMF	$2M+3$ $(2M+2)$	$2M$		
LMMN	$2M+4$ $(2M+2)$	$2M+2$		
RLS	M^2+5M+1	M^2+3M	1	
GN	M^2+7M+1	M^2+3M+1	3	

12.2 BLIND ALGORITHMS

We now list several blind algorithms derived in Probs. III.33–III.36; such methods are useful in blind channel equalization applications — see Computer Project III.4. In the statements below we consider a zero-mean random row vector $\boldsymbol{u}$ with realizations $\{u_0, u_1, \ldots\}$ and a positive scalar γ. Moreover, μ is a positive step-size (usually small).

Algorithm 12.5 (CMA1-2 and NCMA) A weight vector w^o that solves (cf. Probs. III.21 and III.36):

$$\min_w \mathsf{E}\left(\gamma - |\boldsymbol{u}w|\right)^2$$

can be approximated iteratively via the recursion

$$w_i = w_{i-1} + \mu u_i^* \left[\gamma \frac{z(i)}{|z(i)|} - z(i)\right], \quad z(i) = u_i w_{i-1}, \quad i \geq 0$$

or via the normalized form

$$w_i = w_{i-1} + \frac{\mu u_i^*}{\|u_i\|^2} \left[\gamma \frac{z(i)}{|z(i)|} - z(i)\right], \quad z(i) = u_i w_{i-1}, \quad i \geq 0$$

In both cases, we set $w_i = w_{i-1}$ when $z(i) = 0$.

TABLE 12.3 A comparison of the *estimated* computational cost per iteration for several stochastic-gradient algorithms for the case of *complex-valued* data in terms of the number of *real* multiplications, *real* additions, *real* divisions, and comparisons with zero (or sign evaluations).

Algorithm	$\times$	$+$	$/$	sign
LMS with constant step-size	$8M+2$ $(8M)$	$8M$		
LMS with time-variant step-size	$8M+2$ $(8M)$	$8M$		
ϵ-NLMS	$10M+2$ $(10M)$	$10M$	1	
ϵ-NLMS with power normalization	$8M+6$ $(8M+2)$	$8M+4$	1	
sign-error LMS	$6M$ $(4M)$	$6M$		2
leaky-LMS	$10M+3$ $(8M+2)$	$8M+1$ $(9M+1)$		
LMF	$8M+5$ $(8M+3)$	$8M+1$		
LMMN	$8M+6$ $(8M+3)$	$8M+3$		
RLS	$4M^2+16M+1$	$4M^2+12M-1$	1	
GN	$4M^2+20M+1$	$4M^2+12M$	3	

Algorithm 12.6 (CMA2-2) A weight vector w^o that solves (cf. Probs. III.18 and III.33):

$$\min_w \mathsf{E}\left(\gamma - |\boldsymbol{u}w|^2\right)^2$$

can be approximated iteratively via the recursion

$$w_i = w_{i-1} + \mu u_i^* z(i)[\ \gamma - |z(i)|^2\ ], \quad z(i) = u_i w_{i-1}, \quad i \geq 0$$

Algorithm 12.7 (RCA) A weight vector w^o that solves (cf. Probs. III.19 and III.34):

$$\min_w \mathsf{E}\left|\boldsymbol{u}w \ - \ \gamma \cdot \text{csgn}(\boldsymbol{u}w)\right|^2$$

can be approximated iteratively via the recursion

$$w_i = w_{i-1} + \mu u_i^*[\ \gamma \text{csgn}(z(i)) \ - \ z(i)\ ], \quad z(i) = u_i w_{i-1}, \quad i \geq 0$$

Algorithm 12.8 (MMA) A weight vector w^o that solves (cf. Probs. III.20 and III.35):

$$\min_{w} \mathsf{E}\left\{ \left[(\mathsf{Re}(\boldsymbol{u}w))^2 - \gamma\right]^2 \; + \; \left[(\mathsf{Im}(\boldsymbol{u}w))^2 - \gamma\right]^2 \right\}$$

and where $\mathsf{Re}(\cdot)$ and $\mathsf{Im}(\cdot)$ denote the real and imaginary parts of their arguments, can be approximated iteratively via the recursion

$$\begin{cases} z(i) = u_i w_{i-1} \\ a(i) = \mathsf{Re}[z(i)] \\ b(i) = \mathsf{Im}[z(i)] \\ e(i) = a(i)[\gamma - a^2(i)] \; + \; jb(i)[\gamma - b^2(i)] \\ w_i = w_{i-1} + \mu u_i^* e(i) \end{cases}$$

Listing of Blind Algorithms

Observe that for real-valued data, the RCA recursion reduces to that of CMA1-2. The blind recursions are summarized in Table 12.4.

TABLE 12.4 A listing of several blind stochastic-gradient algorithms derived in the problems. In all of them, the initial weight estimate is specified at iteration $i = -1$ and, clearly, it cannot be chosen as zero since otherwise the recursions do not update the weight estimate.

Algorithm	Recursion
CMA2-2	$w_i = w_{i-1} + \mu u_i^* z(i)[\,\gamma - \|z(i)\|^2\,]$ $z(i) = u_i w_{i-1}$
CMA1-2	$w_i = w_{i-1} + \mu u_i^* \left[\gamma \frac{z(i)}{\|z(i)\|} - z(i)\right]$ $z(i) = u_i w_{i-1}$
NCMA	$w_i = w_{i-1} + \frac{\mu u_i^*}{\|u_i\|^2} \left[\gamma \frac{z(i)}{\|z(i)\|} - z(i)\right]$ $z(i) = u_i w_{i-1}$
RCA	$w_i = w_{i-1} + \mu u_i^* [\,\gamma \mathsf{csgn}(z(i)) \; - \; z(i)\,]$ $z(i) = u_i w_{i-1}$
MMA	$w_i = w_{i-1} + \mu u_i^* e(i)$ $z(i) = u_i w_{i-1}, \;\; a(i) = \mathsf{Re}(z(i)), \;\; b(i) = \mathsf{Im}(z(i))$ $e(i) = a(i)[\gamma - a^2(i)] \; + \; jb(i)[\gamma - b^2(i)]$

Tables 12.5–12.6 summarize the estimated costs for both real- and complex-valued data and for constant step-sizes. For completeness, in the multiplications columns in both tables, the entries between parenthesis indicate the number of multiplications that would result if the step-size μ is chosen as a power of 2^{-1}.

12.3 SOME PROPERTIES

The algorithms described in this chapter serve purposes different from LMS and ϵ-NLMS.

1. Different cost functions correspond to different optimality criteria and the resulting stochastic-gradient algorithms can behave differently than LMS and ϵ-NLMS under

TABLE 12.5 *Estimated* computational cost per iteration for *real-valued* data in terms of the number of *real* multiplications, *real* additions, *real* divisions, and comparisons with zero (or sign evaluations).

Algorithm	$\times$	$+$	$/$	sign
CMA2-2	$2M+3$ $(2M+2)$	$2M$		
CMA1-2	$2M+2$ $(2M+1)$	$2M$	1	
NCMA	$3M+2$ $(3M+1)$	$3M-1$	1	
RCA	$2M+1$ $(2M)$	$2M$		1
MMA	$2M+3$ $(2M+1)$	$2M$		

TABLE 12.6 *Estimated* computational cost per iteration for the case of *complex-valued* data in terms of the number of *real* multiplications, *real* additions, *real* divisions, and comparisons with zero (or sign evaluations).

Algorithm	$\times$	$+$	$/$	sign
CMA2-2	$8M+8$ $(8M+6)$	$8M+2$		
CMA1-2	$8M+4$ $(8M+2)$	$8M$	2	
NCMA	$10M+4$ $(10M+2)$	$10M-1$	3	
RCA	$8M+2$ $(8M)$	$8M$		2
MMA	$8M+6$ $(8M+4)$	$8M$		

varied statistical conditions on the data. We shall see in Parts IV (*Mean-Square Performance*) and V (*Transient Performance*) that this is indeed the case — see, e.g., Sec. 21.5.

2. **Sign-error LMS**. The motivation for introducing sign-error LMS, especially for real-valued data, is due to its computational simplicity. While it may seem from Table 12.2 that the cost per iteration of sign-error LMS is similar to that of LMS, the point is that the step-size μ is usually selected as a power of 2^{-1}, say, $\mu = 2^{-m}$ for some integer $m > 0$. When this is the case, the evaluation of $\mu u_i^{\mathsf{T}}\text{sign}[e(i)]$ in the real case can be implemented digitally very efficiently by means of shift registers, and we can ignore the M multiplications that are needed for a generic μ. In this case, we can replace the $2M$ figure that appears in Table 12.2 by M multiplications. This simplification is not possible for LMS because it uses $e(i)$ instead of its sign. Nevertheless, the simplification in computations for sign-error LMS comes at the expense of slower convergence.

3. **Leaky-LMS**. The LMS algorithm can suffer from a potential instability problem when the covariance matrix R_u is singular or close to singular (an example is given in Prob. IV.40 in the context of fractionally-spaced equalizers). When this happens, the weight estimates w_i can *drift* and grow unbounded — see Prob. III.27 for an example and also Prob. IV.39 for a more detailed explanation. The leaky-LMS algorithm limits the growth of the weight estimates by employing a coefficient, $(1-\mu\alpha)$, in the recursion for the weight vector. We shall study the properties of this algorithm in Part V (*Transient Performance*)– see Probs. V.27–V.32. In particular, we shall see that while leaky-LMS solves the drift problem, it nevertheless introduces a *bias* problem in that the mean value of w_i will not tend to the optimal solution $w^o = R_u^{-1}R_{du}$ of the normal equations.

CHAPTER 13

Affine Projection Algorithm

The LMS and ϵ-NLMS algorithms were obtained by using simple instantaneous approximations for the covariance and cross-covariance quantities $\{R_{du}, R_u\}$. More involved algorithms, with better performance but at increased computational costs, can be obtained by resorting to more sophisticated approximations for $\{R_{du}, R_u\}$. We illustrate this situation by motivating the so-called affine projection algorithm.

13.1 INSTANTANEOUS APPROXIMATION

Just like ϵ-NLMS, we again start from the regularized Newton's recursion (10.8), namely,

$$w_i = w_{i-1} + \mu[\epsilon' \mathrm{I} + R_u]^{-1} \left[R_{du} - R_u w_{i-1}\right] \tag{13.1}$$

albeit with a fixed step-size μ and a fixed regularization parameter ϵ'. Now, however, we shall employ a better approximation for both the covariance matrix, R_u, and the cross-covariance vector, R_{du}. Specifically, we choose a positive integer K (usually $K \leq M$, where $M \times 1$ is the size of the weight vector) and replace $\{R_{du}, R_u\}$ by the following instantaneous approximations:

$$\widehat{R}_u = \frac{1}{K}\left(\sum_{j=i-K+1}^{i} u_j^* u_j\right), \qquad \widehat{R}_{du} = \frac{1}{K}\left(\sum_{j=i-K+1}^{i} d(j)u_j^*\right)$$

In other words, at each iteration i, we use the K most recent regressors and the K most recent observations,

$$\{u_i, u_{i-1}, \ldots, u_{i-K+1}\} \quad \text{and} \quad \{d(i), d(i-1), \ldots, d(i-K+1)\}$$

to compute the approximate values for $\{R_u, R_{du}\}$.

Let $\epsilon = \epsilon'/K$. If we introduce the $K \times M$ block data matrix

$$U_i \triangleq \begin{bmatrix} u_i \\ u_{i-1} \\ \vdots \\ u_{i-K+1} \end{bmatrix} \qquad (K \times M) \tag{13.2}$$

and the $K \times 1$ data vector

$$d_i = \begin{bmatrix} d(i) \\ d(i-1) \\ \vdots \\ d(i-K+1) \end{bmatrix} \qquad (K \times 1) \tag{13.3}$$

then we can express $\{\widehat{R}_u, \widehat{R}_{du}\}$ more compactly as

$$\widehat{R}_u = \frac{1}{K}U_i^*U_i, \qquad \widehat{R}_{du} = \frac{1}{K}U_i^*d_i$$

so that Newton's recursion (13.1) becomes

$$w_i = w_{i-1} + \mu\left(\epsilon\mathrm{I} + U_i^*U_i\right)^{-1} U_i^*[d_i - U_iw_{i-1}] \tag{13.4}$$

Although $U_i^*U_i$ is singular when $K \leq M$, the term $\epsilon\mathrm{I}$ guarantees the invertibility of $\epsilon\mathrm{I} + U_i^*U_i$.

Recursion (13.4) requires the inversion of the $M \times M$ matrix $(\epsilon\mathrm{I} + U_i^*U_i)$ at each iteration. Alternatively, we can invoke the matrix inversion formula (5.4) to verify that

$$\left(\epsilon\mathrm{I} + U_i^*U_i\right)^{-1} U_i^* = U_i^*\left(\epsilon\mathrm{I} + U_iU_i^*\right)^{-1}$$

in which case (13.4) becomes

$$w_i = w_{i-1} + \mu U_i^*\left(\epsilon\mathrm{I} + U_iU_i^*\right)^{-1} [d_i - U_iw_{i-1}] \tag{13.5}$$

This form requires inverting the (usually smaller) $K \times K$ matrix $(\epsilon\mathrm{I} + U_iU_i^*)$ at each iteration; recursion (13.5) is also useful even when $\epsilon = 0$ since $U_iU_i^*$ is generally invertible when $K \leq M$. Algorithm (13.5) is what is known as the *affine projection algorithm* or, more accurately, $\epsilon-$APA, in order to highlight the presence of the regularization factor ϵ. In particular, it is seen that when $K = 1$, ϵ-APA reduces to the $\epsilon-$NLMS recursion of Alg. 11.1. More generally, when compared with ϵ-NLMS, and even with the LMS recursion of Alg. 10.1, we find that ϵ-APA uses a *vector-valued* estimation error, $e_i = d_i - U_iw_{i-1}$, as opposed to the scalar-valued error, $e(i) = d(i) - u_iw_{i-1}$, used by LMS and $\epsilon-$NLMS. This is because the latter algorithms use only the most recent regressor to update the weight vector estimate, whereas ϵ-APA uses the K most recent regressors for this same task. For this reason, affine projection algorithms are sometimes called data-reusing algorithms since they re-use past regressor and reference data. The integer K is referred to as the order of the ϵ-APA filter.

Algorithm 13.1 (ϵ-APA algorithm) Consider a zero-mean random variable $\boldsymbol{d}$ with realizations $\{d(0), d(1), \ldots\}$, and a zero-mean random row vector $\boldsymbol{u}$ with realizations $\{u_0, u_1, \ldots\}$. The optimal weight vector w^o that solves

$$\min_w \mathsf{E}\,|\boldsymbol{d} - \boldsymbol{u}w|^2$$

can be approximated iteratively via the recursion

$$w_i = w_{i-1} + \mu U_i^*\left(\epsilon\mathrm{I} + U_iU_i^*\right)^{-1} [d_i - U_iw_{i-1}]$$

where $\{U_i, d_i\}$ are defined by (13.2)–(13.3) and K is a positive integer that denotes the filter order (usually $K \leq M$).

13.2 COMPUTATIONAL COST

The computational cost of ϵ-APA is a function of its order K. Tables 13.1 and 13.2 show the *estimated* number of real multiplications and real additions that are required in the evaluation of specific terms for both cases of real and complex-valued data. The tables assume that the cost of inverting a $K \times K$ matrix is $O(K^3)$ operations (multiplications and additions). The main conclusion is that the cost of ϵ-APA is $O(K^2M)$ operations per iteration.

TABLE 13.1 Estimated computational cost of ϵ-APA per iteration for *real-valued* data in terms of the number of *real* multiplications and *real* additions.

Term	$\times$	$+$
$U_i w_{i-1}$	KM	$K(M-1)$
$d_i - U_i w_{i-1}$		K
$U_i U_i^*$	K^2M	$K^2(M-1)$
$\epsilon\mathrm{I} + U_i U_i^*$		K
$(\epsilon\mathrm{I} + U_i U_i^*)^{-1}$	K^3	K^3
$(\epsilon\mathrm{I} + U_i U_i^*)^{-1}[d_i - U_i w_{i-1}]$	K^2	$K(K-1)$
$U_i^*(\epsilon\mathrm{I} + U_i U_i^*)^{-1}[d_i - U_i w_{i-1}]$	KM	$(K-1)M$
w_i		M
TOTAL per iteration	$(K^2+2K)M+$ K^3+K	$(K^2+2K)M+$ K^3+K^2

TABLE 13.2 Estimated computational cost of ϵ-APA algorithm per iteration for *complex-valued* data in terms of the number of *real* multiplications and *real* additions.

Term	$\times$	$+$
$U_i w_{i-1}$	$4KM$	$2K(2M-1)$
$d_i - U_i w_{i-1}$		$2K$
$U_i U_i^*$	$4K^2M$	$2K^2(2M-1)$
$\epsilon\mathrm{I} + U_i U_i^*$		$2K$
$(\epsilon\mathrm{I} + U_i U_i^*)^{-1}$	$4K^3$	$4K^3$
$(\epsilon\mathrm{I} + U_i U_i^*)^{-1}[d_i - U_i w_{i-1}]$	$4K^2$	$2K(2K-1)$
$U_i^*(\epsilon\mathrm{I} + U_i U_i^*)^{-1}[d_i - U_i w_{i-1}]$	$4KM$	$2(2K-1)M$
w_i		$2M$
TOTAL per iteration	$4(K^2+2K)M+$ $4K^3+4K$	$4(K^2+2K)M+$ $4K^3+2K^2$

13.3 LEAST-PERTURBATION PROPERTY

In a manner similar to what we did for $\epsilon-$NLMS in Sec. 10.4, we can similarly motivate the affine projection algorithm as the exact solution to a *local* optimization problem. To see this, assume that we have available a weight estimate at time $i-1$, say, w_{i-1}, and let $\{d_i, U_i\}$ denote the data at iteration i (constructed as in (13.2)–(13.3)). Let also μ denote a given positive number and define two estimation error *vectors*: the *a priori output* estimation error

$$\boxed{e_i \ \stackrel{\Delta}{=}\ d_i - U_i w_{i-1}} \tag{13.6}$$

and the *a posteriori output* estimation error

$$\boxed{r_i \ \stackrel{\Delta}{=}\ d_i - U_i w_i} \tag{13.7}$$

We can then seek a w_i that solves the constrained optimization criterion:

$$\boxed{\min_{w_i} \ \ \|w_i - w_{i-1}\|^2 \qquad \text{subject to } \ r_i = \left(\mathrm{I} - \mu U_i U_i^*(\epsilon \mathrm{I} + U_i U_i^*)^{-1}\right) e_i} \tag{13.8}$$

In other words, we seek a w_i that is closest to w_{i-1} in the Euclidean norm sense and subject to an equality constraint between r_i and e_i. As was the case with $\epsilon-$NLMS in Sec. 11.4, we show in Prob. III.41 that because of the constraint (13.8), any step-size μ in the interval $(0,2)$ guarantees the desirable property $\|r_i\|^2 \leq \|e_i\|^2$ (with equality when $e_i = 0$), i.e., it guarantees that $U_i w_i$ will be a better estimate for d_i than $U_i w_{i-1}$. Moreover, the solution of (13.8) can be obtained in the same manner as in Sec. 10.4 for LMS. The argument will lead us again to the ϵ-APA recursion (13.5) — see Prob. III.42.

13.4 AFFINE PROJECTION INTERPRETATION

The formulation (13.8) allows us to explain the reason for the denomination "affine projection". It is because a special case of the affine projection recursion (13.5) admits an interpretation in terms of projections onto affine subspaces. To see this, refer to (13.8) with $K \leq M$ and consider the special choices $\mu = 1$ and $\epsilon = 0$, in which case (13.8) reduces to

$$\min_{w_i} \ \ \|w_i - w_{i-1}\|^2 \qquad \text{subject to } \ r_i = 0$$

or, equivalently,

$$\min_{w_i} \ \ \|w_i - w_{i-1}\|^2 \qquad \text{subject to } \ d_i = U_i w_i \tag{13.9}$$

and the ϵ-APA update (13.5) becomes

$$w_i = w_{i-1} + U_i^*(U_i U_i^*)^{-1}[d_i - U_i w_{i-1}] \tag{13.10}$$

In other words, when $\mu = 1$ and $\epsilon = 0$, APA enforces the equality $d_i = U_i w_i$. In the special case $K = 1$, we recover the $\epsilon-$NLMS scenario:

$$\min_{w_i} \ \ \|w_i - w_{i-1}\|^2 \qquad \text{subject to } \ d(i) = u_i w_i \tag{13.11}$$

whose solution is

$$w_i = w_{i-1} + \frac{u_i^*}{\|u_i\|^2}[d(i) - u_i w_{i-1}] \tag{13.12}$$

In this case, NLMS is enforcing the equality $d(i) = u_i w_i$. This observation admits the following geometric interpretation.

For any given data $\{d(i), u_i\}$, there are infinitely many vectors w that solve $d(i) = u_i w$. The set of all such w is an affine subspace (also called a hyperplane or a manifold), denoted by $\mathcal{M}_i$, whose defining equation is

$$\mathcal{M}_i \ \stackrel{\Delta}{=}\ \{\text{set of all vectors } w \text{ such that } u_i w - d(i) = 0\}$$

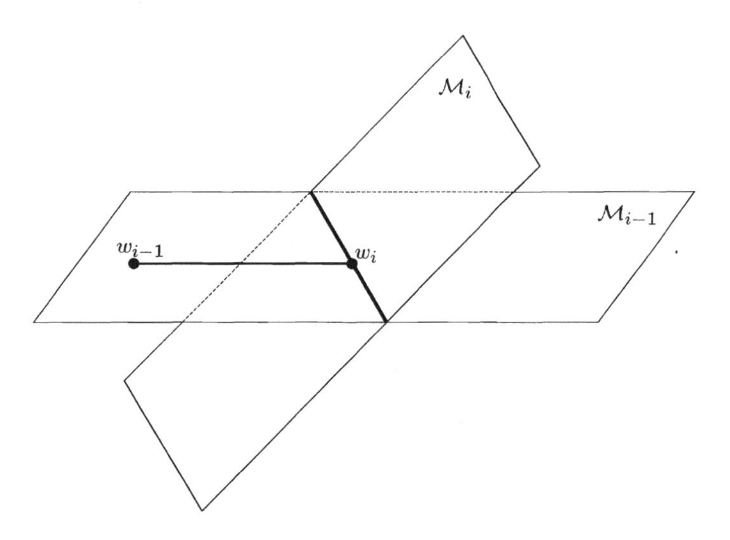

FIGURE 13.1 Manifolds associated with a second-order APA implementation with $\mu = 1, \mathcal{M}_{i-1}$ and $\mathcal{M}_i$. The estimate w_{i-1}, which lies in $\mathcal{M}_{i-1}$, is updated to a point w_i lying in the intersection of both manifolds.

The qualification "affine" is used to indicate that the hyperplane does not necessarily pass through the origin $w = 0$. Given w_{i-1}, and because of (13.11), NLMS selects that particular vector w_i from this subspace that is closest to w_{i-1} in the Euclidean norm sense. We therefore say that w_i is obtained as the projection of w_{i-1} onto the manifold $\mathcal{M}_i$.

When $K > 1$, on the other hand, we find from (13.9) that the APA recursion (13.10) is such that it enforces K equalities (as opposed to a single one):

$$d(i) = u_i w_i, \quad d(i-1) = u_{i-1} w_i, \quad \ldots \quad d(i-K+1) = u_{i-K+1} w_i$$

For each pair of data $\{d(i-j), u_{i-j}\}$, there are infinitely many vectors w that satisfy $d(i-j) = u_{i-j} w$ and which define a manifold $\mathcal{M}_j$. The weight vector w_i that is computed by (13.9)–(13.10) lies in the intersection of the corresponding K manifolds,

$$w_i \in \left(\bigcap_{j=i-K+1}^{i} \mathcal{M}_j \right)$$

We thus say that w_i of APA with $\mu = 1$ is obtained by projecting w_{i-1} onto the intersection of the manifolds $\{\mathcal{M}_j, j = i, i-1, \ldots, i-K+1\}$. Figure 5.8 illustrates this construction for the case $K = 2$. Two manifolds are shown in the figure, $\mathcal{M}_i$ and $\mathcal{M}_{i-1}$. In the figure, the estimate w_{i-1} lies in $\mathcal{M}_{i-1}$ while the updated estimate w_i lies in the intersection of both manifolds, $\mathcal{M}_{i-1} \bigcap \mathcal{M}_i$.

Variations

There are variations of the affine projection algorithm. One such variant is the so-called partial-rank algorithm (PRA), which replaces the update (13.5) by one of the form:

$$w_i = w_{i-K} + \mu U_i^* \left(\epsilon \mathrm{I} + U_i U_i^*\right)^{-1} \left[d_i - U_i w_{i-K}\right] \tag{13.13}$$

where the quantities $\{U_i, d_i\}$ are still defined as in (13.2)–(13.3). The main difference from ϵ-APA is that the weight estimate w_j is kept fixed at w_{i-K} during the time instants $j = i-K+1, i-K+2, \ldots, i-1$ and, hence, the weight vector is updated only once every K iterations. The computational cost of PRA will therefore be less than that of ϵ-APA by a factor of K, namely, $O(KM)$ operations per iteration.

More generally, a family of affine projection algorithms can be defined as follows:

$$w_i = w_{i-1-\alpha(K-1)} + \mu U_i^* \left(\epsilon \mathrm{I} + U_i U_i^*\right)^{-1} \left[d_i - U_i w_{i-1-\alpha(K-1)}\right] \qquad (13.14)$$

where, for example, $\alpha = 0$ corresponds to ϵ-APA form (13.5) while $\alpha = 1$ corresponds to PRA. Moreover, the data $\{U_i, d_i\}$ can also be taken as

$$U_i \stackrel{\Delta}{=} \begin{bmatrix} u_i \\ u_{i-D} \\ \vdots \\ u_{i-(K-1)D} \end{bmatrix} \qquad d_i = \begin{bmatrix} d(i) \\ d(i-D) \\ \vdots \\ d(i-(K-1)D) \end{bmatrix} \qquad (13.15)$$

with their entries separated by multiples of some delay D. The choice $D = 1$ corresponds to ϵ-APA and PRA, but larger values for D can also be employed. One motivation for using $D > 1$ is to increase the separation, and consequently reduce the correlation, among the regressors in U_i. As we are going to see in our studies on the performance of adaptive filters, starting in Chapter 15, the performance of LMS-type filters tends to be sensitive to the correlation of its regression data. And ϵ-APA filters, with $D = 1$ or $D > 1$, address some of these deficiencies albeit at an increased computational cost over NLMS.

Finally, it is worth mentioning that the APA form (13.5), with $\epsilon = 0$, can be rewritten in an equivalent form in terms of *orthogonalized* regressors $\{\tilde{u}_i\}$ that are obtained from the original regressors $\{u_i\}$ via a Gram-Schmidt orthogonalization procedure (see Prob. III.43). The details are carried out in Prob. III.44 and they lead to the following statement.

Algorithm 13.2 (APA with orthogonal update factors) Consider the APA update

$$w_i = w_{i-1} + \mu U_i^* \left(U_i U_i^*\right)^{-1} \left[d_i - U_i w_{i-1}\right]$$

with a full rank regression matrix U_i and $K \leq M$. This update can be equivalently implemented as follows. For each iteration i, perform the following steps. Start with $w_{i-1}^{(0)} = w_{i-1}$ and repeat for $k = 0, 1, \ldots, K-1$:

$$\begin{aligned} \tilde{u}_{i-k} &= u_{i-k}\left(\mathrm{I} - \sum_{j=1}^{k} \frac{\tilde{u}_{i-k+j}^* \tilde{u}_{i-k+j}}{\|\tilde{u}_{i-k+j}\|^2}\right) \\ e^{(k)}(i-k) &= d(i-k) - u_{i-k} w_{i-1}^{(k)} \\ w_{i-1}^{(k+1)} &= w_{i-1}^{(k)} + \frac{\tilde{u}_{i-k}^*}{\|\tilde{u}_{i-k}\|^2} e^{(k)}(i-k) \end{aligned}$$

Then set

$$w_i = (1-\mu) w_{i-1} + \mu w_{i-1}^{(K)}$$

Listing of the Algorithms

Table 13.3 lists the affine projection algorithms described in this section, while Tables 13.4–13.5 compare their computational cost per iteration for real and complex data.

TABLE 13.3 A listing of three affine projection algorithms for $i \geq 0$. In all of them, the initial weight estimate is specified at iteration $i = -1$.

Algorithm	Recursion
APA	$w_i = w_{i-1} + \mu U_i^*(U_iU_i^*)^{-1}[d_i - U_iw_{i-1}]$ $U_i = \begin{bmatrix} u_i \\ \vdots \\ u_{i-K+1} \end{bmatrix}, \quad d_i = \begin{bmatrix} d(i) \\ \vdots \\ d(i-K+1) \end{bmatrix}$
ϵ–APA	$w_i = w_{i-1} + \mu U_i^*(\epsilon I + U_iU_i^*)^{-1}[d_i - U_iw_{i-1}]$
PRA	$w_i = w_{i-K} + \mu U_i^*(\epsilon I + U_iU_i^*)^{-1}[d_i - U_iw_{i-K}]$

TABLE 13.4 A comparison of the *estimated* computational cost per iteration for several affine projection algorithms for *real-valued* data in terms of the number of multiplications and *real* additions.

Algorithm	$\times$	$+$
APA	$(K^2+2K)M + K^3 + K$	$(K^2+2K)M + K^3 + K^2 - K$
ϵ-APA	$(K^2+2K)M + K^3 + K$	$(K^2+2K)M + K^3 + K^2$
PRA	$(K+2)M + K^2 + 1$	$(K+2)M + K^2 + K$

TABLE 13.5 A comparison of the *estimated* computational cost per iteration for several affine projection algorithms for *complex-valued* data in terms of the number of *real* multiplications and *real* additions.

Algorithm	$\times$	$+$
APA	$4(K^2+2K)M + 4K^3 + 4K$	$4(K^2+2K)M + 4K^3 + 2K^2 - K$
ϵ-APA	$4(K^2+2K)M + 4K^3 + 4K$	$4(K^2+2K)M + 4K^3 + 2K^2$
PRA	$4(K+2)M + 4K^2 + 4$	$4(K+2)M + 4K^2 + 2K$

CHAPTER 14

RLS Algorithm

A second example of an algorithm that employs a more sophisticated approximation for R_u is the Recursive-Least-Squares (RLS) algorithm. Although this algorithm can be motivated and derived as the *exact* solution to a well-defined estimation problem with a least-squares cost function, as we shall show in detail later in the book (see, e.g., Sec. 30.6), we shall motivate it here as simply a stochastic-gradient method. In this way, readers can get an early introduction to this important adaptive algorithm.

14.1 INSTANTANEOUS APPROXIMATION

Just like ϵ-NLMS and $\epsilon-$APA, we again start from the regularized Newton's recursion (10.8), namely,

$$w_i = w_{i-1} + \mu(i)\left[\epsilon(i)\mathrm{I} + R_u\right]^{-1}\left[R_{du} - R_u w_{i-1}\right] \tag{14.1}$$

and replace

$$(R_{du} - R_u w_{i-1})$$

by the instantaneous approximation

$$u_i^*[d(i) - u_i w_{i-1}]$$

Now, however, we replace R_u by a better estimate for it, which we choose as the exponentially weighted sample average

$$\widehat{R}_u = \frac{1}{i+1}\sum_{j=0}^{i}\lambda^{i-j}u_j^* u_j$$

for some scalar $0 \ll \lambda \leq 1$. Assume first that $\lambda = 1$. Then the above expression for $\widehat{R}_u$ amounts to averaging all past regressors up to time i, namely,

$$\widehat{R}_u = \frac{1}{i+1}\sum_{j=0}^{i}u_j^* u_j \quad \text{when } \lambda = 1$$

Choosing a value for λ that is less than one introduces *memory* into the estimation of R_u. This is because such a λ would assign larger weights to recent regressors and smaller weights to regressors in the remote past. In this way, the filter will be endowed with a tracking mechanism that enables it to forget data in the remote past and to give more relevance to recent data so that changes in R_u can be better tracked by the resulting algorithm.

We further assume that the step-size in (14.1) is chosen as

$$\mu(i) = 1/(i+1)$$

whereas the regularization factor is chosen as

$$\epsilon(i) = \lambda^{i+1}\epsilon/(i+1)$$

for $i \geq 0$ and for some small positive scalar ϵ. This choice for $\epsilon(i)$ is such that regularization disappears as time progresses. With these approximations and choices, the regularized Newton's recursion (14.1) becomes

$$w_i = w_{i-1} + \left[\lambda^{i+1}\epsilon\mathrm{I} + \sum_{j=0}^{i}\lambda^{i-j}u_j^*u_j\right]^{-1} u_i^*[d(i) - u_iw_{i-1}] \tag{14.2}$$

This recursion is inconvenient in its present form since it requires, at each time instant i, that all previous and present data be combined to form the matrix

$$\Phi_i \triangleq \left(\lambda^{i+1}\epsilon\mathrm{I} + \sum_{j=0}^{i}\lambda^{i-j}u_j^*u_j\right) \tag{14.3}$$

which then needs to be inverted. These two complications (of data storing and matrix inversion) can be alleviated as follows. Observe from the definition of Φ_i that it satisfies the recursion

$$\Phi_i = \lambda\Phi_{i-1} + u_i^*u_i, \qquad \Phi_{-1} = \epsilon\mathrm{I} \tag{14.4}$$

Let $P_i = \Phi_i^{-1}$. Then applying the matrix inversion formula (5.4) to (14.4) gives

$$P_i = \lambda^{-1}\left[P_{i-1} - \frac{\lambda^{-1}P_{i-1}u_i^*u_iP_{i-1}}{1+\lambda^{-1}u_iP_{i-1}u_i^*}\right], \qquad P_{-1} = \epsilon^{-1}\mathrm{I} \tag{14.5}$$

This recursion shows that the update from P_{i-1} to P_i requires only knowledge of the most recent regressor u_i. In this way, at each time instant i, the algorithm only needs to have access to the data $\{w_{i-1}, d(i), u_i, P_{i-1}\}$ in order to determine $\{w_i, P_i\}$. The matrix P_{i-1} essentially summarizes the information from all previous regressors.

Algorithm 14.1 (RLS algorithm) Consider a zero-mean random variable $\boldsymbol{d}$ with realizations $\{d(0), d(1), \ldots\}$, and a zero-mean random row vector $\boldsymbol{u}$ with realizations $\{u_0, u_1, \ldots\}$. The weight vector w^o that solves

$$\min_w \mathsf{E}\,|\boldsymbol{d} - \boldsymbol{u}w|^2$$

can be approximated iteratively via the recursion

$$P_i = \lambda^{-1}\left[P_{i-1} - \frac{\lambda^{-1}P_{i-1}u_i^*u_iP_{i-1}}{1+\lambda^{-1}u_iP_{i-1}u_i^*}\right]$$

$$w_i = w_{i-1} + P_i\, u_i^*[d(i) - u_iw_{i-1}], \quad i \geq 0$$

with initial condition $P_{-1} = \epsilon^{-1}\mathrm{I}$ and where $0 \ll \lambda \leq 1$.

TABLE 14.1 *Estimated* computational cost of the RLS algorithm per iteration for *real-valued* data in terms of the number of *real* multiplications, *real* additions, and *real* divisions.

Term	$\times$	$+$	$/$
$u_i w_{i-1}$	M	$M-1$	
$d(i) - u_i w_{i-1}$		1	
$\lambda^{-1} u_i^*$	M		
$P_{i-1}(\lambda^{-1} u_i^*)$	M^2	$M(M-1)$	
$u_i P_{i-1}(\lambda^{-1} u_i^*)$	M	$M-1$	
$1 + u_i P_{i-1}(\lambda^{-1} u_i^*)$		1	
$1/[1 + u_i P_{i-1}(\lambda^{-1} u_i^*)]$			1
$(\lambda^{-1} u_i P_{i-1} u_i^*) \cdot \dfrac{1}{1 + u_i P_{i-1}(\lambda^{-1} u_i^*)}$	1		
$(\lambda^{-1} P_{i-1} u_i^*) \times \dfrac{\lambda^{-1} u_i P_{i-1} u_i^*}{1 + \lambda^{-1} u_i P_{i-1} u_i^*}$	M		
$P_i u_i^*$		M	
$P_i u_i^* [d(i) - u_i w_{i-1}]$	M		
w_i		M	
TOTAL per iteration	$M^2 + 5M + 1$	$M^2 + 3M$	1

In Chapters 29–30, we shall study RLS in greater detail and show, among other results, that it is in fact the exact solution of a least-squares problem (and, hence, the origin of its name).

14.2 COMPUTATIONAL COST

The computational cost of RLS is one order of magnitude higher than that of LMS and ϵ-NLMS. To see this, we show in Tables 14.1 and 14.2 the *estimated* number of real multiplications and real additions that are required in the evaluation of specific terms for both cases of real and complex-valued data. In particular, the listing in the tables assumes that the RLS calculations are performed in the following manner:

$$\begin{aligned} P_i u_i^* &= \left(P_{i-1}\left(\lambda^{-1} u_i^*\right)\right) - \left(P_{i-1}(\lambda^{-1} u_i^*)\right) \cdot \left(\lambda^{-1} u_i P_{i-1} u_i^*\right) \cdot \frac{1}{1 + \lambda^{-1} u_i P_{i-1} u_i^*} \\ w_i &= w_{i-1} + \left(P_i\, u_i^*\right) \cdot \left(d(i) - \left(u_i w_{i-1}\right)\right) \end{aligned}$$

This manner of calculation is chosen only for convenience of exposition. The order by which the quantities are being computed here need not be the best one in practice; it is certainly not the only one. While other ways of carrying out the calculations may result in a slightly different computational cost, they will all lead to the same order of magnitude, namely, $O(M^2)$. Moreover, in this chapter, we are assuming that all computations are performed in infinite precision. In practice, however, LMS and RLS need to be implemented in finite-precision. In this case, a reliable RLS implementation will usually require a higher precision than a similar LMS implementation. What this means is that, when comparing the computational costs of LMS and RLS in terms of the required number of additions and multiplications, these numbers should be interpreted in light of the fact that the calculations used by RLS should generally be of higher precision.

TABLE 14.2 *Estimated* computational cost of the RLS algorithm per iteration for *complex-valued* data in terms of the number of *real* multiplications, *real* additions, and *real* divisions.

Term	$\times$	$+$	$/$
$u_i w_{i-1}$	$4M$	$4M-2$	
$d(i) - u_i w_{i-1}$		2	
$\lambda^{-1} u_i^*$	$2M$		
$P_{i-1}(\lambda^{-1} u_i^*)$	$4M^2$	$M(4M-2)$	
$u_i P_{i-1}(\lambda^{-1} u_i^*)$	$4M$	$4M-2$	
$1 + u_i P_{i-1}(\lambda^{-1} u_i^*)$		1	
$1/[1 + u_i P_{i-1}(\lambda^{-1} u_i^*)]$			1
$(\lambda^{-1} u_i P_{i-1} u_i^*) \cdot \frac{1}{1 + u_i P_{i-1}(\lambda^{-1} u_i^*)}$	1		
$(\lambda^{-1} P_{i-1} u_i^*) \times \frac{\lambda^{-1} u_i P_{i-1} u_i^*}{1 + \lambda^{-1} u_i P_{i-1} u_i^*}$	$2M$		
$P_i u_i^*$		$2M$	
$P_i u_i^* [d(i) - u_i w_{i-1}]$	$4M$	$2M$	
w_i		$2M$	
TOTAL per iteration	$4M^2 + 16M + 1$	$4M^2 + 12M - 1$	1

Later in the book (e.g., Chapters 37–43) we shall develop more efficient variants of RLS that require the same order of computations as LMS and ϵ-NLMS. These variants will have the same advantage as RLS in terms of faster convergence. However, their algorithmic descriptions will be more involved than what we have seen so far for LMS, ϵ-NLMS, and even RLS.

Summary and Notes

This part of the book deals with the basic principles of steepest-descent methods. The key results are the following.

SUMMARY OF MAIN RESULTS

Steepest-Descent Algorithms

1. Consider a zero-mean random variable $\boldsymbol{d}$ with variance σ_d^2 and a zero-mean random row vector $\boldsymbol{u}$ with $R_u = \mathsf{E}\,\boldsymbol{u}^*\boldsymbol{u} > 0$. The solution of the linear least-mean-squares estimation problem
$$\min_w \mathsf{E}\,|\boldsymbol{d} - \boldsymbol{u}w|^2$$
is given by $w^o = R_u^{-1}R_{du}$. The cost function $J(w) = \mathsf{E}\,|\boldsymbol{d} - \boldsymbol{u}w|^2$ is quadratic in w and has a unique global minimum at w^o.

2. The optimal solution w^o can be determined recursively via a steepest-descent method. Start with any initial guess w_{-1} (e.g., $w_{-1} = 0$) and iterate
$$w_i = w_{i-1} + \mu[R_{du} - R_u w_{i-1}], \quad i \geq 0$$
The successive weight estimates w_i are guaranteed to converge to w^o as long as the step-size μ satisfies $0 < \mu < 2/\lambda_{\max}$, where $\lambda_{\max}$ denotes the maximum eigenvalue of R_u. Fastest convergence is attained by choosing $\mu^o = 2/(\lambda_{\max}+\lambda_{\min})$, where $\lambda_{\min}$ denotes the smallest eigenvalue of R_u. Moreover, the convergence of w_i to w^o is exponential and controlled by the modes $\{1 - \mu\lambda_k\}$ or by the time constants $\{-1/2\ln|1 - \mu\lambda_k|\}$.

3. We can also employ a steepest-descent method with a time-dependent step-size to estimate w^o. Start with any initial guess w_{-1} (e.g., $w_{-1} = 0$) and iterate
$$w_i = w_{i-1} + \mu(i)[R_{du} - R_u w_{i-1}], \quad i \geq 0$$
The successive weight estimates w_i are guaranteed to converge to w^o if the step-size sequence satisfies $\mu(i) \to 0$ and $\sum_{i=0}^{\infty}\mu(i) = \infty$. Other conditions on the step-size can still guarantee convergence. Fastest convergence is attained by choosing
$$\mu^o(i) = \tilde{w}_{i-1}^*\, R_u^2\, \tilde{w}_{i-1} / \tilde{w}_{i-1}^*\, R_u^3\, \tilde{w}_{i-1}$$

4. The learning curve of a steepest-descent method is defined as $J(i) = \mathsf{E}\,|\boldsymbol{d} - \boldsymbol{u}w_{i-1}|^2$. In other words, the value of the learning curve at a particular iteration i is a measure of the cost that would result if we freeze the weight estimate at the value obtained at the prior iteration.

5. The contours of the quadratic cost function $J(w) = \mathsf{E}\,|\boldsymbol{d} - \boldsymbol{u}w|^2$ are elliptic curves centered at w^o and whose principal axes are the eigenvectors of R_u shifted to w^o.

6. The optimal solution w^o can also be determined by employing a Newton-type recursion or a *Levenberg-Marquardt* method,
$$w_i = w_{i-1} - \mu[\epsilon \mathrm{I} + R_u]^{-1}[R_{du} - R_u w_{i-1}], \quad i \geq 0, \quad w_{-1} = \text{initial guess}$$

where ϵ is a small positive number.

7. For more general cost functions $J(w)$ that are not necessarily quadratic in w, steepest-descent methods can be used to attempt to estimate a minimizing argument of $J(w)$ as follows:

$$w_i = w_{i-1} - \mu\left[\nabla_w J(w_{i-1})\right]^*$$

with sufficiently small step-sizes μ, or even

$$w_i = w_{i-1} - \mu[\epsilon \mathrm{I} \ + \ \nabla_w^2 J(w_{i-1})]^{-1}\left[\nabla_w J(w_{i-1})\right]^*$$

in terms of the gradient vector and the Hessian matrix of the cost function. In these more general cases, there is no guarantee beforehand that the weight estimates will converge to a global minimum of the cost function.

Stochastic-Gradient Algorithms

This part also describes the procedure for developing stochastic-gradient approximations from steepest-descent methods. The key step is to replace true gradient vectors and Hessian matrices with instantaneous approximations. These substitutions achieve three objectives:

1. They free the designer from the need to know beforehand the underlying signal statistics.
2. They equip stochastic-gradient algorithms (and the corresponding adaptive filters) with a learning mechanism that allows them to learn the statistics of the underlying signals. Different algorithms learn at different rates and with different accuracies.
3. They also equip stochastic-gradient algorithms with a tracking mechanism that allows them to track variations in the signal statistics. Again, different algorithms track at different rates and with different degrees of success.

We described several stochastic gradient methods and commented on some of their properties:

a) LMS and $\epsilon-$NLMS are the most widely used adaptive filters in current practice due to their simplicity, robustness, and low computational complexity. The $\epsilon-$NLMS algorithm has the advantage of relying on a search direction that is essentially insensitive to the norm of the regression vector. This property is useful when dealing with signals that undergo periods of activity and pause (e.g., speech signals).

b) The RLS algorithm is an order of magnitude costlier than LMS-type algorithms, requiring $O(M^2)$ vs. $O(M)$ operations per iteration. However, RLS converges significantly faster than LMS. Several chapters in this book are devoted to RLS (e.g., Part VII (*Least-Squares Methods*) through Part X (*Lattice Filters*) where several efficient implementations of RLS are described).

c) Although LMS and RLS have been motivated in this chapter as stochastic-gradient algorithms, they can also be derived as *exact* solutions to well-defined optimization problems. These problems are described at length in later stages of the book (e.g., Chapters 30 and 45). In the case of RLS, the algorithm will be shown to be the recursive solution of a regularized least-squares problem, while LMS will be shown to be the recursive solution of an indefinite least-squares problem.

d) Several other LMS-type algorithms are described in the chapter, including sign-error LMS, leaky-LMS, LMF, LMMN, CMA, RCA, and MMA. As will become clear from the analyses in Parts IV (*Mean-Square Performance*) and V (*Transient Performance*), and from the problems therein, some of these algorithms can exhibit superior performance to LMS for some signal statistics. In addition, CMA, RCA, and MMA are particularly suited for applications where reference signals are absent (e.g., in blind equalization applications).

BIBLIOGRAPHIC NOTES

Iterative methods. There is a huge amount of literature on iterative methods for solving linear systems of equations (see, e.g., Faddeev and Faddeeva (1963), Traub (1965), and Golub and

Van Loan (1996)) and also for solving linear and nonlinear optimization methods (see, e.g., Wilde (1964), Wilde and Beightler (1967), Hestenes (1975), and Fletcher (1987)). The steepest-descent and Newton schemes that we described in this chapter are special cases of such iterative methods, when applied to the solution of linear equations or to the minimization of quadratic cost functions. In order to illustrate this connection, we show below how the steepest-descent recursion (8.20) can be alternatively motivated as an iterative method for the solution of the normal equations $R_u w^o = R_{du}$.

Linear equations. Consider an arbitrary linear system of equations $Ax = b$ and assume we can express A as $A = Q - N$ for two matrices Q and N, chosen such that Q is easily invertible and the eigenvalues of $Q^{-1}N$ are all inside the unit circle. Then the equality $Ax = b$ can be written as $Qx = Nx + b$, which leads to $x = Q^{-1}Nx + Q^{-1}b$. An iterative method for finding x then replaces this equation by the *recursion*

$$x_i = Q^{-1}Nx_{i-1} + Q^{-1}b, \quad i \geq 0$$

whose iterates $\{x_i\}$ would converge to x, no matter what the initial condition x_{-1} is, since all the eigenvalues of $Q^{-1}N$ lie inside the unit circle.

To apply this construction to $R_u w^o = R_{du}$, we proceed as follows. We first express R_u as

$$R_u = \underbrace{\mu^{-1}\mathrm{I}}_{Q} - \underbrace{(\mu^{-1}\mathrm{I} - R_u)}_{N} \stackrel{\Delta}{=} Q - N$$

for some positive constant μ to be chosen, and for matrices $\{Q, N\}$ defined as above. Then the normal equations can be rewritten as

$$Qw^o = Nw^o + R_{du}$$

or, equivalently, as

$$\mu^{-1}w^o = (\mu^{-1}\mathrm{I} - R_u)w^o + R_{du}$$

which simplifies to

$$w^o = w^o + \mu[R_{du} - R_u w^o]$$

We can now replace this equality by the recursion

$$w_i = w_{i-1} + \mu[R_{du} - R_u w_{i-1}]$$

and choose μ such that the eigenvalues of $Q^{-1}N = \mathrm{I} - \mu R_u$ all lie inside the unit disc. This is of course the same steepest-descent method that we encountered before in (8.20), with the condition $\mu < 2/\lambda_{\max}$ guaranteeing a stable matrix $Q^{-1}N$.

Stochastic approximation. The idea of using iterative procedures that are based on sample realizations in order to *simultaneously* approximate actual expectations and minimize a certain cost function is at the core of what is known as *stochastic approximation theory*. The theory can be succinctly described as follows. Assume real-valued data and consider the cost function $J(w) = \mathsf{E}\left[f(\boldsymbol{x}, w)\right]$, for some function f and stochastic data $\boldsymbol{x}$. The stochastic-approximation method for minimizing $J(w)$ over w takes the form (see, e.g., Tsypkin (1971, p. 47)):

$$w_i = w_{i-1} - \mu(i)\nabla_w^{\mathsf{T}}\left[f(x_i, w_{i-1}\right]$$

with $f(\cdot,\cdot)$ evaluated at a realization for x at iteration i, and where $\nabla_w f$ denotes the gradient vector of f with respect to w. Moreover, $\mu(i)$ is a step-size sequence, possibly matrix-valued and also time-variant. It is clear that such constructions have a close relation with adaptive filter design. For example, in the adaptive filtering context, we usually have $f(\boldsymbol{x}, w) = (\boldsymbol{d} - \boldsymbol{u}w)^2$ with $\{\boldsymbol{d}, \boldsymbol{u}\}$ playing the role of $\boldsymbol{x}$. Then

$$\nabla_w^{\mathsf{T}}\left[f(x_i, w_{i-1})\right] = -u_i^{\mathsf{T}}[d(i) - u_i w_{i-1}]$$

in terms of realizations $\{d(i), u_i\}$ for $\{\boldsymbol{d}, \boldsymbol{u}\}$; the LMS filter would follow by fixing the step-size at a constant value.

According to Tsypkin (1971, p. 70), the pioneering work in the field of stochastic approximation is that of Robbins and Monro (1951). Although their recursive procedure was a variation of a scheme developed two decades earlier by von Mises and Pollaczek-Geiringer (1929), the work by Robbins and Monro generated tremendous interest and led to many subsequent studies and extensions. While Robbins and Monro's (1951) work dealt primarily with a *scalar* weight, Blum (1954) and Schmetterer (1961) extended the procedure to weight *vectors*. A description of these developments can be found in the book by Wetherhill (1966).

LMS adaptation. During the 1950s, stochastic-approximation theory did not receive much attention in the engineering community until the landmark work by Widrow and Hoff (1960) in which they developed the real form of the LMS algorithm; the complex form of the filter appeared later in Widrow et al. (1975). The LMS filter is considered, in many respects, to be the birthmark of modern adaptive filter theory. Since its inception, the algorithm has been examined and scrutinized from all angles with remarkable resilience to the test of time. Very few algorithms in estimation and filtering theories have found so much success and have been used in so many widespread areas.

Besides LMS, there have also been other works on adaptive structures in the early 1960s. One such example is an adaptive filter that minimizes the mean-square error between an input signal and a reference signal developed by Gabor, Wilby and Woodcock (1961); their filter is described in Tsypkin (1971, p. 156). This latter reference also contains on pages 172–173 commentaries on works on adaptation and learning during the early sixties, including a description of an adaptive filter by Sefl (1960) that is the continuous-time counterpart of LMS; it employs a differential update equation of the form

$$\frac{\mathrm{d}w(t)}{\mathrm{d}t} = \mu(t)u^{\mathsf{T}}(t)\left[d(t) - u(t)w(t)\right]$$

with continuous-time vector variables $\{w(t), u(t)\}$ and scalar variables $\{d(t), \mu(t)\}$.

Other noteworthy works on adaptive structures in the 1960s are those by Applebaum (1966) and Widrow et al. (1967) on adaptive antenna arrays. In Applebaum (1966), an adaptive algorithm is derived that is based on maximizing the signal-to-noise ratio, while Widrow et al. (1967) focus on mean-square error performance and use the LMS algorithm.

NLMS adaptation. The NLMS filter was independently proposed by Nagumo and Noda (1967) and Albert and Gardner (1967); although the actual terminology of normalized least-mean-squares algorithm seems to be due to Bitmead and Anderson (1980). A simplified version of the least-perturbation property (11.12) for NLMS was studied by Goodwin and Sin (1984) assuming $\mu = 1$ and $\epsilon = 0$, in which case the constraint in (11.12) simplifies to $r(i) = 0$. In Nitzberg (1985), an interesting connection is made between NLMS and LMS whereby it is shown that the NLMS algorithm can be obtained from applying LMS repeatedly for every new sample of data.

LMS variations. Since its creation, several modifications have been proposed to the LMS original formulation, such as replacing the error signal or the regressor by their signed versions in lieu of computational simplicity. These ideas seem to have been first suggested by Lucky (1965) in the context of adaptive equalization. Other early works are those by Gersho (1969,1984), Hirsch and Wolf (1970), Claasen and Mecklenbräuker (1981), and Duttweiler (1982). These investigations resulted in the sign-error LMS form (Alg. 12.1), the sign regressor LMS form (see Prob. V.25), and the sign-sign LMS form where both the error signal and the regressor are replaced by their signs, i.e., for real data,

$$w_i = w_{i-1} + \mu\mathsf{sign}[u_i]^{\mathsf{T}}\mathsf{sign}[d(i) - u_i w_{i-1}] \qquad \text{(sign-sign LMS)}$$

Unfortunately, however, the sign-error and sign-sign LMS algorithms converge much slower than LMS. In Duttweiler (1982), it was further suggested that in calculating the filter update of LMS, the quantity $e(i)$ and the entries of u_i could be quantized to the nearest power-of-two-values with the resulting algorithm still having a similar performance to that of LMS.

One variant that has improved convergence performance over sign-error LMS is the dual-sign LMS algorithm (see, e.g., Sethares and Johnson (1989) and Mathews (1991a)). In this variant, rather than only replace the error signal by its sign, the magnitude of the error is also examined and if it is deemed to be large, then the sign of the error is scaled up by a constant larger than one, e.g., for real data,

$$w_i = w_{i-1} + \mu\gamma u_i^{\mathsf{T}}\mathsf{sign}[e(i)], \qquad e(i) = d(i) - u_i w_{i-1} \qquad \text{(dual-sign LMS)}$$

where γ is a positive constant chosen as follows:

$$\begin{cases} \gamma > 1 \text{ and chosen as a power of two} & \text{if } |e(i)| > \text{threshold} \\ \gamma = 1 & \text{if } |e(i)| \leq \text{threshold} \end{cases}$$

Large values of γ tend to increase the convergence speed at the expense of performance degradation. On the other hand, a large threshold tends to improve performance at the expense of slower convergence.

A second variant of sign-error LMS is the power-of-two error LMS algorithm (see Xue and Liu (1986) and Eweda (1992)). In this case, if the magnitude of the error is larger than one, then it is replaced by its sign. Otherwise, its value is quantized to the nearest power-of-two-value in the following manner:

$$w_i = w_{i-1} + \mu\gamma u_i^{\mathsf{T}} \mathsf{sign}[e(i)], \qquad e(i) = d(i) - u_i w_{i-1} \qquad \text{(power-of-two error LMS)}$$

where γ is again a positive number chosen as follows. Let B denote a desired number of bits (excluding the sign bit). Then

$$\gamma = \begin{cases} 1 & \text{if } |e(i)| \geq 1 \\ 2^{\mathsf{fl}[\log_2 |e(i)|]} & \text{if } 2^{-B+1} \leq |e(i)| < 1 \\ 2^{-B} & \text{if } |e(i)| < 2^{-B+1} \end{cases}$$

where the so-called floor function, $\mathsf{fl}[x]$, returns the closest integer that is smaller than its argument. The choice of γ as 2^{-B}, when the error signal is small, guarantees that the filter will not stop adapting. Alternatively, we could have set the value of γ to zero when $|e(i)| < 2^{-B+1}$. In this case, however, the filter update will stop when the error becomes small.

Affine projection algorithms. Although LMS and NLMS are among the most widely used adaptive filters due to their computational simplicity and ease of implementation, colored input signals can deteriorate their convergence speed – see, e.g., App. 9.D of Sayed (2003). To address this problem, Ozeki and Umeda (1984) developed the basic form of the affine projection algorithm (APA). Since then, many variants of APA (also known as data-reusing algorithms) have been devised from different perspectives such as the partial-rank algorithm (PRA) by Kratzer and Morgan (1985), the generalized NLMS algorithm by Morgan and Kratzer (1996), the decorrelating algorithm by Rupp (1998a), the APA with orthogonal correction factors (referred to as NLMS-OCF by Sankaran and Beex (1997,2000)), and the binormalized data-reusing LMS algorithm (which corresponds to the special case $K = 2$ of APA — see Apolinário, Campos, and Diniz (2000)). An early analysis of data-reusing algorithms appears in Roy and Shynk (1989). A more recent study of an APA family appears in Shin and Sayed (2004). For applications of APA to acoustic echo cancellation, see the book by Benetsy et al. (2001).

Variable step-size LMS. As we are going to see in Chapters 16 and 24, the step-size in LMS adaptation controls the rate of convergence and the steady-state performance of the filter. In order to meet the conflicting requirements of fast convergence and good steady-state performance, the step-size needs to be controlled. Various schemes for controlling the step-size of LMS have been proposed in the literature, for instance, by Kwong and Johnston (1992), Mathews and Xie (1993), Aboulnasr and Mayyas (1997), Pazaitis and Constantinides (1999), and Shin, Sayed, and Song (2004a). Kwong and Johnston (1992) use squared instantaneous errors, while Aboulnasr and Mayyas (1997) use the squared autocorrelation of errors at adjacent times, Pazaitis and Constantinides (1999) adopt the fourth-order cumulant of the instantaneous error, and Shin, Sayed, and Song (2004a) attempt to maximize the decrease in the weight-error-vector energy. The step-sizes in these variants are evaluated in the manner shown in Table 14.3 (for the case of real data).

Blind algorithms. Blind algorithms compensate for the lack of an explicit reference signal. They are pointedly referred to as adaptive algorithms for restoring signal properties by Treichler, Johnson, and Larimore (2001). One of the first "blind" algorithms in the context of signal processing is that of Griffiths (1967), who replaced the product $\boldsymbol{u}_i^* d(i)$ in the LMS update (10.10) by R_{du}, in which case

TABLE 14.3 Variable-step-size LMS implementations.

Algorithm	Update
VSS-LMS Kwong and Johnston (1992)	$\mu(i)=\alpha\mu(i-1)+\gamma e^2(i)$
RVS-LMS Aboulnasr and Mayyas (1997)	$\mu(i)=\alpha\mu(i-1)+\gamma p^2(i)$ $p(i)=\beta p(i-1)+(1-\beta)e(i)e(i-1)$
KVS-LMS Pazaitis and Constantinides (1999)	$\mu(i)=\mu_{\max}\left(1-e^{-\alpha C_4^e(i)}\right)$ $C_4^e(i)=f(i)-3p^2(i)$ $f(i)=\beta f(i-1)+(1-\beta)e^4(i)$ $p(i)=\beta p(i-1)+(1-\beta)e^2(i)$
VSS-NLMS Shin, Sayed, and Song (2004a)	$\mu(i)=\dfrac{\mu_{\max}}{\epsilon+\|u_i\|^2}\cdot\dfrac{\|p_i\|^2}{c+\|p_i\|^2}$ $c\approx 1/\mathsf{SNR}$ $p_i=\beta p_{i-1}+(1-\beta)\dfrac{u_i}{\epsilon+\|u_i\|^2}e(i)$

the update for w_i would become

$$w_i = w_{i-1} + \mu R_{du} \;-\; \mu u_i^* z(i)$$

where $z(i) = u_i w_{i-1}$ is the output of the filter. In this implementation, the absence of a reference signal is compensated for by using R_{du}, and there are applications where R_{du} can be computed beforehand — see Treichler, Johnson, and Larimore (2001) for an example in the context of interference suppression.

The first truly blind algorithm is due to Sato (1975), who developed his algorithm for blind equalization purposes. In Sato's algorithm, all signals are real-valued and the data to be recovered belongs to a BPSK constellation; its update equation has the form

$$w_i = w_{i-1} + \mu u_i^{\mathsf{T}}\left[\gamma \mathsf{sign}[z(i)] - z(i)\right], \quad z(i) = u_i w_{i-1}$$

which is seen to be a special case of CMA1-2 and RCA for real-data. Actually, RCA extends Sato's algorithm to the complex domain (see Prob. III.34). Although simple to implement, the Sato and RCA methods tend to face convergence difficulties.

The more general constant-modulus algorithms described in Algs. 12.5 and 12.6 were developed by Godard (1980) and Treichler and Agee (1983). These authors arrived at the same algorithms, and the only minor difference between their formulations is with regards to the interpretation of the constant γ. The motivation of Treichler and Agee (1983) was to equalize constant-envelope signals. For this reason, they chose the constant γ as the value of the constant amplitude. Interestingly, however, constant-modulus algorithms can still function properly even if the underlying data do not possess the constant-modulus property (e.g., data arising from a QAM constellation). This fact was noted earlier by Godard (1980), whose motivation was to perform blind equalization on such disperse constellations. For this reason, the constant γ in Godard (1980) was chosen as some statistical measure of the dispersion in the data (see Probs. IV.16 and IV.18 for further explanation). Compared with RCA, the CMA methods are more reliable. Nevertheless, their implementations are more complex since they tend to require the use of rotators at the output of the equalizers (see Treichler, Larimore, and Harp (1998)).

Since these earlier contributions, much subsequent work has been done on blind algorithms, e.g., by Macchi and Eweda (1984), Benveniste and Goursat (1984), Foschini (1985), Picchi and Prati (1987), and Shalvi and Weinstein (1990). In Picchi and Prati (1987) a "stop-and-go" algorithm was introduced that employs a flag to halt adaptation if the error signal is deemed unreliable; the algorithm is examined in Computer Project III.4. In Yang, Werner, and Dumont (1997), a multi-modulus algorithm (MMA) was introduced that is suitable for two-dimensional modulation schemes (see Prob. III.35); this algorithm was studied in the context of broadband access systems in Werner et al. (1999).

The article by Johnson (1998) provides a survey of the developments in the area of blind algorithms until the late 1990s, including applications to fractionally-spaced equalization (see also Mai and Sayed (2000)). Constant modulus algorithms have also been used for adaptive beamforming (e.g., Keerthi, Mathur, and Shynk (1998)) and for interference cancellation (e.g., Kwon, Un, and Lee (1992)).

Adaptive equalization. In the concluding remarks of Chapter 6, we commented on the early history of channel equalization and how the mean-square error criterion was proposed for this purpose by Widrow (1966), Gersho (1969), and Proakis and Miller (1969). In this chapter, we described how channel equalizers could be designed adaptively, as opposed to the closed-form solution methods of Secs. 5.4 and 6.4. Historically, the LMS algorithm was developed at the right time, in the early 1960s, right when interest in equalization was starting to build up (Lucky (1965)). Soon afterwards, the use of adaptive filters for equalization purposes was studied in greater detail by Gersho (1969), Proakis and Miller (1969), Ungerboeck (1972), and Salz (1973). In these adaptive implementations, an equalizer would have two modes of operations: a training mode (Gersho (1969)) whereby data that are known to the receiver and the transmitter are used to train the equalizer, and a decision-directed mode (Salz (1973)), whereby decisions are used to train the equalizer.

Problems and Computer Projects

PROBLEMS

Problem III.1 (Multiple step-sizes) Refer to expression (8.18) and choose the matrix B as $B = \text{diag}\{b(1), b(2), \ldots, b(M)\}$ with $b(i) > 0$. In this case, the recursion (8.20) is replaced by $w_i = w_{i-1} + \mu B[R_{du} - R_u w_{i-1}]$. This scheme associates one step-size with each entry of the weight vector w_i. Follow the discussion in Sec. 8.2 and derive a necessary and sufficient condition on μ in order to guarantee convergence of w_i to $w^o = R_u^{-1} R_{du}$.

Problem III.2 (Product of infinitely many numbers) Consider a scalar recursion of the form $x(i) = a(i)x(i-1)$ for $i \geq 0$, and assume $a(i) = e^{-1/(i+1)^2}$.

(a) Verify that $0 < a(i) < 1$ for all finite i.

(b) Let $p(i) = \prod_{j=0}^{i} a(j)$. Show that $p(i)$ converges to $e^{-\pi^2/6}$, which is a finite positive number. *Hint:* The series $\sum_{j=1}^{\infty}(1/j^2)$ converges to $\pi^2/6$.

Problem III.3 (Optimal step-size) Refer to expression (9.16) for the optimal step-size. Verify that it is equivalent to the following:

$$\mu^o(i) = \frac{\|\nabla_w J(w_{i-1})\|^2}{[\nabla_w J(w_{i-1})]\ R_u\ [\nabla_w J(w_{i-1})]^*}$$

in terms of the squared Euclidean norm of the gradient vector in the numerator, and the weighted squared Euclidean norm of the same vector in the denominator.

Problem III.4 (Convergent step-size sequence) Consider the steepest-descent algorithm (9.13) with a time-variant step-size. Assume that $\mu(i)$ converges to a positive value, say, $\mu(i) \to \alpha > 0$ as $i \longrightarrow \infty$. Show that if α satisfies $\alpha < 2/\lambda_{\max}$, then w_i converges to w^o.

Problem III.5 (Optimal step-size) Consider the optimal step-size (9.16) in the iteration-dependent case of the steepest-descent algorithm. Show that $1/\lambda_{\max} \leq \mu^o(i) \leq 1/\lambda_{\min}$, where $\lambda_{\max}$ and $\lambda_{\min}$ denote the largest and smallest eigenvalues of R_u. Conclude that $\sum_{i=0}^{\infty} \mu^o(i) = \infty$.

Problem III.6 (Contour curves) Consider the steepest-descent algorithm (9.13) with the optimal time-variant step-size (9.16).

(a) Show that $J(w_i) = J(w_{i-1}) - \mu^o(i)\tilde{w}_{i-1}^* R_u^2 \tilde{w}_{i-1}$.

(b) Assume $M = 2$ (i.e., the size of w^o is 2×1). Sketch the elliptic contours of constant mean-square error and explain how the optimized algorithm moves from one elliptic curve to another.

Problem III.7 (Interfering signals) Refer to Prob. I.14. Suggest a steepest-descent algorithm for estimating $\boldsymbol{x}_1$ from $\boldsymbol{y}$.

Problem III.8 (Prediction problem) A zero-mean stationary random process $\{\boldsymbol{u}(\cdot)\}$ is generated by passing a zero-mean white sequence $\{\boldsymbol{v}(\cdot)\}$ with variance σ_v^2 through a second-order auto-regressive model, namely, $\boldsymbol{u}(i) + \alpha\boldsymbol{u}(i-1) + \beta\boldsymbol{u}(i-2) = \boldsymbol{v}(i),\ i > -\infty$, where α and β are real numbers such that the roots of the characteristic equation $1 + \alpha z^{-1} + \beta z^{-2} = 0$ are strictly inside the unit circle. We wish to design a second-order predictor for the process $\boldsymbol{u}(\cdot)$ of the form $\hat{\boldsymbol{u}}(i) = \begin{bmatrix} \boldsymbol{u}(i-1) & \boldsymbol{u}(i-2) \end{bmatrix} w^o$, for some 2×1 vector w^o.

(a) Verify that $\{\alpha, \beta\}$ must satisfy $|\beta| < 1$ and $|\alpha| < 1 + \beta$.

(b) Define the data vector $\boldsymbol{u} = \begin{bmatrix} \boldsymbol{u}(i-1) & \boldsymbol{u}(i-2) \end{bmatrix}$ and the desired signal $\boldsymbol{d} = \boldsymbol{u}(i)$. Let $R_u = \mathsf{E}\,\boldsymbol{u}^*\boldsymbol{u}$ and $R_{du} = \mathsf{E}\,\boldsymbol{d}\boldsymbol{u}^*$. Show that

$$\left[\begin{array}{c|c} R_u & R_{du} \end{array}\right] = \frac{\sigma_v^2}{(1-\beta)[(1+\beta)^2 - \alpha^2]} \left[\begin{array}{cc|c} 1+\beta & -\alpha & -\alpha \\ -\alpha & 1+\beta & \alpha^2 - \beta^2 - \beta \end{array}\right]$$

Establish that $(1-\beta)[(1+\beta)^2 - \alpha^2] > 0$.

(c) Show that the optimal weight vector w^o is given by $w^o = \text{col}\{-\alpha, -\beta\}$. Could you have guessed this answer more directly without evaluating the product $R_u^{-1} R_{du}$?

(d) Verify that the eigenvalue spread of R_u is $\rho = (\beta + 1 + |\alpha|)/(\beta + 1 - |\alpha|)$. Design a steepest-descent algorithm that determines w^o iteratively. Provide a condition on the step-size μ in terms of $\{\alpha, \beta\}$ in order to guarantee convergence.

(e) Show that the value of the step-size that yields fastest convergence, and the resulting time-constant, are

$$\mu^o = \frac{1-\beta}{1+\beta} \frac{(1+\beta)^2 - \alpha^2}{\sigma_v^2}, \qquad \tau^o = \frac{-1}{2\ln\left(|\alpha|/(\beta+1)\right)}$$

Problem III.9 (Logarithm function) Establish that the logarithm function satisfies the following properties:

(a) For any $0 \le y < 1$, it holds that $\ln(1-y) \le -y$.

(b) There exist $0 < b < 1$ and $a > 0$ such that $-ay \le \ln(1-y)$ for any $y \in [0, b]$.

Problem III.10 (Convergence proof) The purpose of this problem is to establish the statement of Thm. 9.1. Let $\tilde{w}_i = w^o - w_i$ and introduce the eigen-decomposition $R_u = U\Lambda U^*$. Define the transformed vector $x_i = U^* \tilde{w}_i$ and let $x_k(i)$ denote its $k-$th entry.

(a) Assume first that the step-size sequence is divergent and let us establish that w_i converges to w^o. Thus since $\lambda_k > 0$, and using the fact that $\mu(i) \to 0$, conclude that there exists a large enough i_o such that $0 < 1 - \mu(i)\lambda_k \le 1$ for all k and for all $i > i_o$. Use the result of part (a) of Prob. III.9 to conclude that $\sum_{j=i_o+1}^{\infty} \ln(1 - \mu(j)\lambda_k) = -\infty$.

(b) Verify that $x_k(i)$ satisfies the recursion $x_k(i) = (1 - \mu(i)\lambda_k)x_k(i-1)$ and conclude that

$$\lim_{i\to\infty} \ln\left[\frac{|x_k(i)|}{|x_k(i_o)|}\right] = -\infty$$

and, consequently, $x_k(i) \to 0$. *Remark.* The argument in parts (a) and (b) shows that if the step-size sequence is divergent then $\tilde{w}_i \to 0$. We now examine the converse statement.

(c) Let us now assume that w_i converges to w^o and let us establish that the step-size sequence is divergent. Thus assume $x_k(i) \to 0$ and show that it implies $\sum_{j=i_o+1}^{\infty} \ln(1 - \mu(j)\lambda_k) = -\infty$.

(d) Assume that i_o is large enough so that not only $0 < 1 - \mu(j)\lambda_k < 1$ for all k and $j > i_o$, but also $\mu(j)\lambda_k < b$. Use the result of part (b) of Prob. III.9 to conclude that

$$-a\lambda_k \sum_{j=i_o+1}^{\infty} \mu(j) \le \sum_{j=i_o+1}^{\infty} \ln(1 - \mu(j)\lambda_k) = -\infty$$

That is, conclude that the sequence $\{\mu(i)\}$ is divergent.

Remark. The above arguments are patterned along those given in Macchi (1995, pp. 65–67).

Problem III.11 (Regularized Newton's method) Consider the steepest-descent recursion of Remark 2,

$$w_i = w_{i-1} - \mu[\epsilon I + \nabla_w^2 J(w_{i-1})]^{-1} \, [\nabla J(w_{i-1})]^*, \quad w_{-1} = \text{initial guess}$$

for some $\epsilon > 0$. For the quadratic cost function $J(w)$ of (8.8) we have $\nabla_w^2 J(w_{i-1}) = R_u > 0$ and $\nabla_w J(w_{i-1}) = (R_{du} - R_u w_{i-1})^*$.

(a) Show that a necessary and sufficient condition on μ that guarantees convergence of w_i to the minimizing argument of $J(w)$ is $0 < \mu < 2 + 2\epsilon/\lambda_{\max}$.

(b) Find the optimum step-size μ^o at which the convergence rate is maximized.

Problem III.12 (Leaky variant of steepest-descent) Consider the modified optimization problem

$$\min_w \left[J^\alpha(w) \stackrel{\Delta}{=} \alpha\|w\|^2 \; + \; \mathsf{E}\,|\boldsymbol{d} - \boldsymbol{u}w|^2 \right]$$

where α is a positive real number and $J^\alpha(w)$ is the new cost function (it is dependent on α). In the text we studied the case $\alpha = 0$ (see expression (8.7) for $J(w)$). The above modified cost function penalizes the energy (or squared norm) of the vector w, and is therefore useful in situations where we want to avoid solutions with potentially large norms.

(a) Show that the optimal solution is given by $w^\alpha = [R_u + \alpha I]^{-1} R_{du}$. Compute the resulting minimum cost, $J^\alpha(w^\alpha)$, and show that $J^\alpha(w^\alpha) > \sigma_d^2 - R_{ud} R_u^{-1} R_{du} = J_{\min}$, where $J_{\min}$ is the minimum cost associated with $J(w)$ (cf. (8.11)).

(b) Let $\delta w = w^\alpha - w^o$ denote the difference between the new solution w^α and the linear least-mean squares solution w^o of (8.4). Show that $R_{ud}\delta w = J_{\min} - J^\alpha(w^\alpha)$.

(c) Justify the validity of the following steepest-descent method for determining w^α:

$$\boxed{w_i^\alpha = (1 - \mu\alpha) w_{i-1}^\alpha \; + \; \mu[R_{du} - R_u w_{i-1}^\alpha], \quad w_{-1}^\alpha \; = \; \text{initial guess}}$$

Show that w_i^α converges to w^α if, and only if, $0 < \mu < 2/(\lambda_{\max} + \alpha)$. Show also that the optimal step-size for fastest convergence is $\mu^\alpha = 2/(\lambda_{\max} + \lambda_{\min} + 2\alpha)$.

(d) Let μ^o denote the optimal step-size choice for the standard steepest-descent method (with $\alpha = 0$, i.e., cf. (9.5)). Compare μ^α and μ^o.

Remark. We see from the result in part (a) that the inclusion of the term $\alpha\|w\|^2$ in the cost function $J^\alpha(w)$ has the effect of modifying the input covariance matrix from R_u to $R_u + \alpha I$. This can be interpreted as adding a noise vector $\boldsymbol{v}$ to $\boldsymbol{u}$, with the individual entries of $\boldsymbol{v}$ arising from a zero-mean white-noise process with variance α. This process of disturbing the input $\boldsymbol{u}$ with entries from a white-noise sequence is known as *dithering*. Its effect is to provide a mechanism for controlling the size of the solution vector w^α; it also results in a covariance matrix with smaller eigenvalue spread (see Prob. III.13). Its disadvantage is that the optimal solution w^α will be distinct from the desired solution w^o.

Problem III.13 (One effect of dithering) Consider the same setting as in Prob. III.12. Show that the eigenvalue spread of $R_u + \alpha I$ is smaller than the eigenvalue spread of R_u.

Problem III.14 (l_1-norm of complex variables) Let x be a nonzero real-valued variable. Verify that, for $x \neq 0$,

$$\frac{\mathsf{d}|x|}{\mathsf{d}x} = \mathsf{sign}(x) \quad \text{where} \quad \mathsf{sign}(x) \stackrel{\Delta}{=} \begin{cases} +1 & x > 0 \\ -1 & x < 0 \end{cases}$$

[At $x = 0$, we shall define from now on $\mathsf{sign}(0) = 0$.] Now assume x is complex-valued with real and imaginary parts denoted by x_r and x_i, respectively. Define its l_1-norm as follows $|x| = |x_\mathsf{r}| + |x_\mathsf{i}|$. That is, we add the absolute values of its real and imaginary parts. [The result can be interpreted as the l_1-norm of the vector $\text{col}\{x_\mathsf{r}, x_\mathsf{i}\}$.] Show that the complex gradient (cf. Chapter C) of $|x|$ with respect to $x \neq 0$ is given by

$$\frac{\mathsf{d}|x|}{\mathsf{d}x} = \frac{1}{2}\left[\mathsf{sign}(x_\mathsf{r}) - j\,\mathsf{sign}(x_\mathsf{i})\right]$$

Problem III.15 (Sign-error algorithm) Consider two zero-mean random variables $\boldsymbol{d}$ and $\boldsymbol{u}$ where $\boldsymbol{d}$ is scalar-valued and $\boldsymbol{u}$ is a row vector. We are interested in minimizing the expected value of the l_1-norm of the error $\boldsymbol{e} = \boldsymbol{d} - \boldsymbol{u}w$ (cf. Prob. III.14), i.e., $\min_w \mathsf{E}\,|\boldsymbol{e}|$.

(a) Let $\boldsymbol{e} = \boldsymbol{e}_\mathsf{r} + j\boldsymbol{e}_\mathsf{i}$. Show that the complex-gradient of the cost function $J(w) = \mathsf{E}\,|\boldsymbol{e}|$ with respect to w is given by $\nabla_w J(w) = -\mathsf{E}\left(\boldsymbol{u}\;[\mathsf{sign}(\boldsymbol{e}_\mathsf{r}) - j\mathsf{sign}(\boldsymbol{e}_\mathsf{i})]\right)/2$.

(b) Conclude that a steepest-descent method can be obtained via the recursion

$$w_i = w_{i-1} + \frac{\mu}{2}\mathsf{E}\,\boldsymbol{u}^*\left\{\mathsf{sign}[\boldsymbol{e}_\mathsf{r}(i)] + j\mathsf{sign}[\boldsymbol{e}_\mathsf{i}(i)]\right\}$$

for some positive step-size μ and where $\boldsymbol{e}(i) = \boldsymbol{d} - \boldsymbol{u}w_{i-1} = \boldsymbol{e}_\mathsf{r}(i) + j\boldsymbol{e}_\mathsf{i}(i)$.

Remark. A more compact representation can be obtained if we define the sign of a complex number as follows:

$$\mathsf{csgn}(x) \stackrel{\Delta}{=} \mathsf{sign}(x_\mathsf{r}) + j\mathsf{sign}(x_\mathsf{i})$$

Observe that we are writing csgn instead of sign; we reserve the notation sign for the sign of a real number. With this definition, we can rewrite the above steepest-descent recursion as

$$\boxed{w_i = w_{i-1} + \mu\mathsf{E}\,\boldsymbol{u}^*\mathsf{csgn}[\boldsymbol{e}(i)]}$$

in terms of $\boldsymbol{e}(i)$ and for a scaled step-size, which we still denote by μ.

Problem III.16 (Least-mean-fourth (LMF) criterion) Consider zero-mean random variables $\boldsymbol{d}$ and $\boldsymbol{u}$, with $\boldsymbol{u}$ a row vector. An optimal weight vector w^o is to be chosen by minimizing the cost function

$$\min_w \mathsf{E}\,|\boldsymbol{d} - \boldsymbol{u}w|^{2L}$$

for some positive integer L. Observe that we are now minimizing the moment of order $2L$ of the error signal rather than its variance, as was done in Sec. 8.1.

(a) Argue that the cost function $\mathsf{E}\,|\boldsymbol{d} - \boldsymbol{u}w|^{2L}$ is convex in w and that, therefore, it cannot have local minima.

(b) Assume in this part that $\{\boldsymbol{d}, \boldsymbol{u}, w\}$ are real and scalar-valued. Assume further that $\boldsymbol{d}$ and $\boldsymbol{u}$ are jointly Gaussian and that $L = 2$. Verify that the cost function in this case reduces to

$$J(w) = 3\sigma_d^4 - 12\sigma_d^2\sigma_{du}w + 6(\sigma_u^2\sigma_d^2 + 2\sigma_{du}^2)w^2 - 12\sigma_u^2\sigma_{du}w^3 + 3\sigma_u^4 w^4$$

where $\sigma_u^2 = \mathsf{E}\,\boldsymbol{u}^2$, $\sigma_d^2 = \mathsf{E}\,\boldsymbol{d}^2$ and $\sigma_{du} = \mathsf{E}\,\boldsymbol{du}$. Show that $w^o = \sigma_{du}/\sigma_u^2$ is a minimum of $J(w)$. Is it a global minimum? Assume $\sigma_d^2 = 1$, $\sigma_u^2 = 1$ and $\sigma_{du} = 0.7$. Plot $J(w)$.

(c) Let z be a complex-valued variable and consider the function $f(z) = |z|^{2L}$. Verify that $\mathsf{d}f/\mathsf{d}z = Lz^*|z|^{2(L-1)}$.

(d) Let $\boldsymbol{e} = \boldsymbol{d} - \boldsymbol{u}w$ and denote the cost function by $J(w) = \mathsf{E}\,|\boldsymbol{e}|^{2L}$. Use the composition rule of differentiation to verify that $\nabla_w J(w) = -L\;\mathsf{E}\,\boldsymbol{u}\boldsymbol{e}^*|\boldsymbol{e}|^{2(L-1)}$. Conclude that a steepest-descent implementation for finding a minimizing solution of $J(w)$ is given by (note that we blended the constant L into μ):

$$\boxed{w_i = w_{i-1} + \mu\mathsf{E}\,\boldsymbol{u}^*\boldsymbol{e}(i)\;|\boldsymbol{e}(i)|^{2(L-1)}}$$

for some step-size $\mu > 0$ and where $\boldsymbol{e}(i) = \boldsymbol{d} - \boldsymbol{u}w_{i-1}$.

Remark. When $L = 2$, the criterion corresponds to solving a least-mean-fourth (LMF) estimation problem. The recursion in this case would take the form

$$\boxed{w_i = w_{i-1} + \mu\mathsf{E}\,\boldsymbol{u}^*\boldsymbol{e}(i)\;|\boldsymbol{e}(i)|^2}$$

Note also that when all variables are real-valued, the recursion of part (d) reduces to

$$\boxed{w_i = w_{i-1} + \mu\mathsf{E}\,\boldsymbol{u}^\mathsf{T}\boldsymbol{e}^{2L-1}(i)}$$

and in the LMF case it becomes

$$\boxed{w_i = w_{i-1} + \mu\mathsf{E}\,\boldsymbol{u}^\mathsf{T}\boldsymbol{e}^3(i)}$$

Problem III.17 (Least-mean-mixed-norm (LMMN) criterion) Consider zero-mean random variables $\boldsymbol{d}$ and $\boldsymbol{u}$, where $\boldsymbol{u}$ is a row vector. An optimal weight vector w^o is to be chosen so as to minimize the cost function

$$\min_{w} \mathsf{E}\left[\delta|\boldsymbol{e}|^2 + \frac{1}{2}(1-\delta)|\boldsymbol{e}|^4\right]$$

where $\boldsymbol{e} = \boldsymbol{d} - \boldsymbol{u}w$ and $0 \leq \delta \leq 1$. Observe that this cost function reduces to the least-mean-squares and least-mean-fourth criteria at the extreme points $\delta = 1$ and $\delta = 0$, respectively. Other values of δ allow for a tradeoff between both criteria. Verify that

$$\nabla_w J(w) = -\mathsf{E}\boldsymbol{u}\left[\delta\boldsymbol{e}^* + (1-\delta)\boldsymbol{e}^*|\boldsymbol{e}|^2\right]$$

Conclude that a steepest-descent implementation for finding a minimizing solution of $J(w)$ is given by

$$\boxed{w_i = w_{i-1} + \mu\mathsf{E}\boldsymbol{u}^*\boldsymbol{e}(i)\left[\delta + (1-\delta)|\boldsymbol{e}(i)|^2\right]}$$

for some step-size $\mu > 0$ and where $\boldsymbol{e}(i) = \boldsymbol{d} - \boldsymbol{u}w_{i-1}$. *Remark.* For real-data, the above recursion reduces to

$$\boxed{w_i = w_{i-1} + \mu\mathsf{E}\boldsymbol{u}^{\mathsf{T}}\left[\delta\boldsymbol{e}(i) + (1-\delta)\boldsymbol{e}^3(i)\right]}$$

Problem III.18 (Constant-modulus criterion) Let $\boldsymbol{u}$ be a zero-mean row vector and w an unknown column vector that we wish to determine so as to minimize the cost function $J(w) = \mathsf{E}\left(\gamma - |\boldsymbol{u}w|^2\right)^2$, for a given positive number γ.

(a) Let $R_u = \mathsf{E}\boldsymbol{u}^*\boldsymbol{u}$. Show that

$$J(w) = \gamma^2 + w^*[-2\gamma R_u + \mathsf{E}(\boldsymbol{u}^*\boldsymbol{u}\,|\boldsymbol{u}w|^2)]w$$

and

$$\nabla_w J(w) = -2w^*\left(\gamma R_u - \mathsf{E}\left(\boldsymbol{u}^*\boldsymbol{u}|\boldsymbol{u}w|^2\right)\right)$$

Conclude that a steepest-descent algorithm for the minimization of $J(w)$ is given by

$$\boxed{w_i = w_{i-1} + \mu\left(\gamma R_u w_{i-1} - \mathsf{E}\left[\boldsymbol{u}^*\boldsymbol{u}w_{i-1}|\boldsymbol{u}w_{i-1}|^2\right]\right)}$$

for some step-size $\mu > 0$.

(b) Assume that w is two-dimensional, say with entries $w = \text{col}\{\alpha, \beta\}$, and that $\boldsymbol{u}$ is a circular Gaussian random vector (cf. Sec. A.5) with $R_u = \text{diag}\{1, 2\}$.

(b.1) Verify that $\mathsf{E}|\boldsymbol{u}w|^2 = |\alpha|^2 + 2|\beta|^2$ and $\mathsf{E}|\boldsymbol{u}w|^4 = 2\left(|\alpha|^2 + 2|\beta|^2\right)^2$. Conclude that $J(w) = \gamma^2 + 2\left(|\alpha|^2 + 2|\beta|^2\right)^2 - 2\gamma\left(|\alpha|^2 + 2|\beta|^2\right)$. *Remark.* If $\boldsymbol{z}$ is a zero-mean scalar complex-valued Gaussian random variable with variance $\mathsf{E}|\boldsymbol{z}|^2 = \sigma_z^2$, then its fourth-moment is $\mathsf{E}|\boldsymbol{z}|^4 = 2\sigma_z^4$. If $\boldsymbol{z}$ were real-valued instead, its fourth moment would be $\mathsf{E}|\boldsymbol{z}|^4 = 3\sigma_z^4$. Verify these claims.

(b.2) Conclude also that $\nabla_w J(w) = 2(-\gamma + 2|\alpha|^2 + 4|\beta|^2)\begin{bmatrix}\alpha^* & 2\beta^*\end{bmatrix}$. Argue that $J(w)$ has a local maximum at the point $\alpha = \beta = 0$ and global minima at all points lying on the ellipse $|\alpha|^2 + 2|\beta|^2 = \gamma/2$. Argue further that the minimum value of $J(w)$ is equal to $\gamma^2/2$.

(b.3) Conclude that the steepest-descent algorithm of part (a) reduces in this case to the form

$$\begin{bmatrix}\alpha(i)\\ \beta(i)\end{bmatrix} = \begin{bmatrix}\alpha(i-1)\\ \beta(i-1)\end{bmatrix} + \mu(\gamma - 2|\alpha(i-1)|^2 - 4|\beta(i-1)|^2)\begin{bmatrix}\alpha(i-1)\\ 2\beta(i-1)\end{bmatrix}$$

and that the weight estimates are always real-valued if the initial condition is real-valued. This problem is pursued further ahead in a computer project.

Remark. By minimizing the cost function $J(w)$ we are forcing the magnitude of uw to be close to the constant $\sqrt{\gamma}$; hence, the name constant-modulus criterion. Such cost functions are used in blind channel equalization applications — see Computer Project III.4.

Problem III.19 (Reduced-constellation criterion) Let u be a zero-mean row vector and w an unknown column vector that we wish to determine so as to minimize the cost function

$$J(w) = \mathsf{E}\left[\frac{1}{2}|uw|^2 - \gamma|uw|_1\right]$$

for a given constant number γ and in terms of the l_1-norm of uw, as defined in Prob. III.14. Let $R_u = \mathsf{E}\,u^*u$.

(a) Show that a steepest-descent algorithm for the minimization of $J(w)$ is given by

$$\boxed{w_i = w_{i-1} - \mu\left[\, R_u w_{i-1} - \gamma\mathsf{E}\ \left(\mathsf{csgn}(uw_{i-1})u^*\right)\,\right]}$$

for some step-size $\mu > 0$.

(b) Show that minimizing $J'(w) = \mathsf{E}\ |uw - \gamma\cdot\mathsf{csgn}(uw)|^2$ is equivalent to minimizing the cost function above. Conclude that $J(w)$ is attempting to minimize the distance between uw and the four points $\{\pm\gamma \pm j\gamma\}$.

Remark. Compare the above conclusion with the constant-modulus criterion of Prob. III.18, which attempts to minimize the distance between uw and all points on the circle of radius $\sqrt{\gamma}$. Both criteria are useful for blind channel equalization.

Problem III.20 (Multi-modulus criterion) Let u be a zero-mean row vector and w an unknown column vector that we wish to determine so as to minimize the cost function

$$J(w) = \mathsf{E}\left\{\ [(\mathsf{Re}(uw))^2 - \gamma]^2 \ +\ [(\mathsf{Im}(uw))^2 - \gamma]^2\ \right\}$$

for a given constant number γ, and where $\mathsf{Re}(\cdot)$ and $\mathsf{Im}(\cdot)$ denote the real and imaginary parts of their arguments. Let $R_u = \mathsf{E}\,u^*u$. Show that a steepest-descent algorithm for the minimization of $J(w)$ is given by

$$a = \mathsf{Re}(uw_{i-1}),\ \ b = \mathsf{Im}(uw_{i-1}),\ \ e(i) = a[\gamma - a^2] + jb[\gamma - b^2],\ \ w_i = w_{i-1} + \mu\mathsf{E}\,u^*e(i)$$

for some step-size $\mu > 0$. *Remark.* By minimizing the cost function $J(w)$ we are forcing the real and imaginary parts of uw to be close to the values $\pm\sqrt{\gamma}$. In other words, $J(w)$ attempts to minimize the distance between uw and the horizontal and vertical lines located at $\pm\sqrt{\gamma}$.

Problem III.21 (Another constant-modulus criterion) Let u be a zero-mean row vector and w an unknown column vector that we wish to determine so as to minimize the cost function

$$J(w) = \mathsf{E}\left[\frac{1}{2}|uw|^2 - \gamma|uw|\right]$$

for a given constant number γ and in terms of the magnitude of uw. Let $R_u = \mathsf{E}\,u^*u$.

(a) Show that a steepest-descent algorithm for the minimization of $J(w)$ is given by

$$\boxed{w_i = w_{i-1} - \mu\left[\, R_u w_{i-1} - \gamma\mathsf{E}\ \left(\frac{uw}{|uw|}u^*\right)\right]}$$

for some step-size $\mu > 0$.

(b) Show that minimizing $J'(w) = \mathsf{E}\,(\gamma - |uw|)^2$ is equivalent to minimizing the cost function above. Conclude that $J(w)$ is attempting to minimize the distance between $|uw|$ and the circle of radius γ.

Problem III.22 (Constrained mean-square error) Consider the constrained optimization problem described in Prob. II.36, namely,

$$\boxed{\min_{w} \mathsf{E}\,|\boldsymbol{d} - \boldsymbol{u}w|^2 \quad \text{subject to } c^*w = \alpha}$$

where c is a known $M \times 1$ vector and α is a known real scalar. Verify that a steepest-descent recursion for estimating the optimal solution w^o is given by

$$\boxed{w_i = w_{i-1} + \mu\left[\mathrm{I} - \frac{cc^*}{\|c\|^2}\right](R_{du} - R_u w_{i-1})}$$

where $R_u = \mathsf{E}\,\boldsymbol{u}^*\boldsymbol{u} > 0$ and $R_{du} = \mathsf{E}\,\boldsymbol{d}\boldsymbol{u}^*$. *Hint:* Use the extended cost function $J_e(w, \lambda)$ of Prob. II.36 and enforce the condition that the successive weight vectors $\{w_i\}$, including the initial condition, must satisfy the constraint $c^*w_i = \alpha$.

Problem III.23 (Homogeneous difference equation) Consider the homogeneous difference equation $x_i = Ax_{i-1}$ with arbitrary initial vector x_{-1}. Show that x_i tends to the zero vector if, and only if, all the eigenvalues of A have strictly less than unit magnitude. *Hint:* One possibility is to use the Jordan decomposition of the matrix A — see Prob. II.2.

Problem III.24 (Power estimate) Refer to (11.9) and assume that the entries $\{u(j)\}$ arise from observations of a white random sequence $\{\boldsymbol{u}(j)\}$ with variance σ_u^2. In this way, the quantity $p(i)$ can be interpreted as a realization of a random variable $\boldsymbol{p}(i)$ satisfying $\boldsymbol{p}(i) = \beta\boldsymbol{p}(i-1) + (1-\beta)|\boldsymbol{u}(i)|^2$, $\boldsymbol{p}(-1) = 0$. Show that $\mathsf{E}\,\boldsymbol{p}(i) \longrightarrow \sigma_u^2$ as $i \longrightarrow \infty$.

Problem III.25 (Perturbation property of ϵ-NLMS) Follow the derivation in Sec. 10.4 and show that the solution to the optimization problem

$$\min_{w_i} \|w_i - w_{i-1}\|^2 \quad \text{subject to } r(i) = \left(1 - \frac{\mu\|u_i\|^2}{\epsilon + \|u_i\|^2}\right)e(i)$$

is given by the ϵ−NLMS recursion (11.4).

Problem III.26 (Leaky-LMS) The leaky-LMS algorithm is the stochastic gradient version of the steepest-descent method of Prob. III.12. Replace R_{du} by $d(i)u_i^*$, R_u by $u_i^*u_i$, and verify that this leads to the following recursion:

$$\boxed{w_i^\alpha = (1 - \mu\alpha)w_{i-1}^\alpha + \mu u_i^*[d(i) - u_i w_{i-1}^\alpha], \quad i \geq 0}$$

Define the *a posteriori* and *a priori* output errors $r^\alpha(i) = d(i) - u_i w_i^\alpha$ and $e^\alpha(i) = d(i) - u_i w_{i-1}^\alpha$. Verify that $r^\alpha(i) = \left[1 - \alpha\mu - \mu\|u_i\|^2\right]e^\alpha(i) + \alpha\mu d(i)$.

Problem III.27 (Drift problem) Assume the regressors u_i are scalars, say $u(i)$, and given by $u(i) = 1/\sqrt{i+1}$. Let $\mu = 1$ and $d(i) = u(i)w^o + v(i)$ with $w^o = 0$ and $v(i) = c$ for all i.

(a) Verify that the LMS update for the weight estimate $w(i)$ gives

$$w(i) = \frac{c}{i+1}\sum_{k=1}^{i+1}\sqrt{k} \geq \frac{2c}{3}\sqrt{i+1}$$

and, therefore, $w(i) \to \infty$ as $i \to \infty$ (no matter how small c is).

(b) Consider instead the leaky-LMS update of Prob. III.26 for a generic step-size μ,

$$w^\alpha(i) = \left(1 - \mu\alpha - \frac{\mu}{i+1}\right)w^\alpha(i-1) + \frac{\mu c}{\sqrt{i+1}}$$

Show that this recursion results in a bounded sequence $\{w^\alpha(i)\}$ if $0 < \mu < 2/(\alpha+1)$.

Remark. This example shows that the weight estimates computed by LMS can grow slowly to large values, even when the noise is small — see Prob. IV.39 for a more detailed example.

Problem III.28 (Constrained LMS) Refer to the steepest-descent algorithm of Prob. III.22, which pertains to the constrained optimization problem

$$\boxed{\min_{w} \mathsf{E}\,|\boldsymbol{d} - \boldsymbol{u}w|^2 \quad \text{subject to } c^* w = \alpha}$$

where c is a known $M \times 1$ vector and α is a known real scalar. Verify that a stochastic-gradient algorithm for approximating the optimal solution w^o is given by

$$\boxed{w_i = w_{i-1} + \mu \left[\mathrm{I} - \frac{cc^*}{\|c\|^2} \right] u_i^* [d(i) - u_i w_{i-1}]}$$

in terms of realizations $\{d(i), u_i\}$ for $\{\boldsymbol{d}, \boldsymbol{u}\}$, and starting from an initial condition that satisfies $c^* w_{-1} = \alpha$.

Problem III.29 (Sign-error LMS) Refer to the statement of Prob. III.15. Show that the corresponding stochastic-gradient method is given by the following so-called sign-error LMS algorithm,

$$\boxed{w_i = w_{i-1} + \mu u_i^* \mathsf{csgn}[d(i) - u_i w_{i-1}], \quad i \geq 0}$$

where the complex-sign function is as defined in (12.1).

Problem III.30 (LMF algorithm) Refer to Prob. III.16 and assume $L = 2$. Argue that a stochastic gradient implementation is given by

$$\boxed{w_i = w_{i-1} + \mu u_i^* e(i)|e(i)|^2, \qquad e(i) = d(i) - u_i w_{i-1}, \quad i \geq 0}$$

Problem III.31 (LMMN algorithm) Refer to Prob. III.17. Argue that a stochastic-gradient implementation is given by

$$\boxed{w_i = w_{i-1} + \mu u_i^* e(i)[\delta + (1-\delta)|e(i)|^2], \qquad e(i) = d(i) - u_i w_{i-1}, \quad i \geq 0}$$

Problem III.32 (Least mean-phase (LMP) algorithm) Let $\boldsymbol{d}$ and $\boldsymbol{u}$ be zero-mean random variables, with $\boldsymbol{d}$ being a scalar and $\boldsymbol{u}$ a $1 \times M$ vector. Introduce the phase-error cost function $J_{\text{pe}}(\boldsymbol{w}) = \mathsf{E}\,|\text{phase}(\boldsymbol{d}) - \text{phase}(\boldsymbol{u}\boldsymbol{w})|^m = \mathsf{E}\,|\angle \boldsymbol{d} - \angle \boldsymbol{u}\boldsymbol{w}|^m$, where $m = 1, 2$ and $\boldsymbol{w}$ is an unknown weight vector to be estimated. Consider further the squared-error cost function $J_{\text{se}}(\boldsymbol{w}) = \mathsf{E}\,|\boldsymbol{d} - \boldsymbol{u}\boldsymbol{w}|^2$ and let $J(\boldsymbol{w}) = k_1 J_{\text{se}}(\boldsymbol{w}) + k_2 J_{\text{pe}}(\boldsymbol{w})$, where k_1 and k_2 define the contribution of each term to the overall cost function.

(a) Verify that for $m = 2$, a stochastic gradient implementation is given by

$$w_i = w_{i-1} + \mu_1 u_i^* (d(i) - u_i w_{i-1}) + \mu_2 (\angle d(i) - \angle u_i w_{i-1}) \frac{j u_i^*}{(u_i w_{i-1})^*}$$

where μ_1 and μ_2 are step-size parameters.

(b) Likewise, verify that $m = 1$, we obtain instead

$$w_i = w_{i-1} + \mu_1 u_i^* (d(i) - u_i w_{i-1}) + \mu_2 \mathsf{sign}(\angle d(i) - \angle u_i w_{i-1}) \frac{j u_i^*}{(u_i w_{i-1})^*}$$

Remark. In some applications, the squared-error is not the primary parameter affecting the performance of the system. The information bits may be carried over the phase of the transmitted signal, in which case it is useful to consider cost functions that relate to both the error magnitude and the phase error — see Tarighat and Sayed (2004).

Problem III.33 (Constant-modulus algorithm) The constant-modulus algorithm CMA2-2 is a stochastic gradient version of the steepest-descent algorithm developed in Prob. III.18.

(a) Replace the term $\gamma R_u w_{i-1} - \mathsf{E}[\, u^* u w_{i-1} |u w_{i-1}|^2]$ by the instantaneous approximation $\gamma u_i^* u_i w_{i-1} - u_i^* u_i w_{i-1} |u_i w_{i-1}|^2$, and define $z(i) = u_i w_{i-1}$. Verify that this leads to the recursion

$$\boxed{w_i = w_{i-1} + \mu u_i^* z(i) \left[\gamma - |z(i)|^2\right], \quad z(i) = u_i w_{i-1}, \quad i \geq 0}$$

Remark. This recursion is known as CMA2-2. The numbers 2-2 refer to the fact that the cost function in this case (cf. Prob. III.18) is of the form $\mathsf{E}\left(\gamma - |uw|^2\right)^2$, which is a special case of the more general cost function $\mathsf{E}\,(\gamma - |uw|^p)^q$ for the values $p = 2$ and $q = 2$.

(b) Can you guarantee, as in the steepest-descent method of Prob. III.18, that the estimates w_i in CMA2-2 will always be real-valued for any real-valued initial condition w_{-1}? Justify your answer.

Problem III.34 (Reduced-constellation algorithm) The reduced-constellation algorithm (RCA) is a stochastic gradient version of the steepest-descent algorithm developed in Prob. III.19. Replace the expectations in the recursion of part (a) of Prob. III.19 by instantaneous approximations and verify that this substitution leads to the following recursion. Let $z(i) = u_i w_{i-1}$. Then

$$\boxed{w_i = w_{i-1} + \mu u_i^* \left(\gamma \mathsf{csgn}(z(i)) \; - \; z(i)\right)}$$

for some step-size $\mu > 0$.

Remark. Recall from the discussion in Prob. III.19 that RCA attempts to minimize the distance between the output of the filter and four points in the complex plane, namely, $\gamma(\pm 1 \pm j)$. In other words, for multi-level constellations, RCA attempts to minimize the mean-square error between the output of the filter and a reduced number of symbols, which may not belong to the original signal constellation. Compared with CMA, the RCA method is simple to implement but tends to face convergence difficulties (see, e.g., Werner et al. (1999)).

Problem III.35 (Multi-modulus algorithm) The multi-modulus algorithm (MMA) is a stochastic-gradient version of the steepest-descent algorithm developed in Prob. III.20. Replace the expectations in the recursion of Prob. III.20 by instantaneous approximations and verify that this substitution leads to the following algorithm. Let $z(i) = u_i w_{i-1}$. Then

$$\left\{\begin{array}{l} z(i) = u_i w_{i-1} \\ a(i) = \mathsf{Re}[z(i)] \\ b(i) = \mathsf{Im}[z(i)] \\ e(i) = a(i)[\gamma - a^2(i)] \; + \; jb(i)[\gamma - b^2(i)] \\ w_i = w_{i-1} + \mu u_i^* e(i) \end{array}\right.$$

for some step-size $\mu > 0$. How is this recursion different from that of CMA2-2 from Prob. III.33?

Remark. The MMA scheme was proposed by Yang, Werner, and Dumont (1997). Recall from Prob. III.20 that MMA attempts to minimize the distance between the output of the filter and the horizontal and vertical lines located at $\pm\sqrt{\gamma}$. Specifically, RCA attempts to minimize the dispersion between the real and imaginary parts of the filter output around the value of γ.

Problem III.36 (Another constant-modulus algorithm) The constant modulus algorithm CMA1-2 is a stochastic-gradient version of the steepest-descent algorithm developed in Prob. III.21.

(a) Replace the expectations in the recursion of part (a) of Prob. III.21 by instantaneous approximations and verify that this substitution leads to the following recursion. Let $z(i) = u_i w_{i-1}$.

Then

$$w_i = w_{i-1} + \mu u_i^* \left(\gamma \frac{z(i)}{|z(i)|} - z(i) \right)$$

for some step-size $\mu > 0$ (when $z(i) = 0$ we set $w_i = w_{i-1}$).

Remark. This recursion is known as CMA1-2 because the cost function in this case (cf. Prob. III.21) is of the form $\mathsf{E}\,(\gamma - |uw|)^2$, which is a special case of the more general cost function $\mathsf{E}\,(\gamma - |uw|^p)^q$ for the values $p = 1$ and $q = 2$.

(b) Let $w_i = w_{i-1} + \delta w$. Given w_{i-1}, show that the weight vector w_i with the smallest perturbation δw that solves the optimization problem

$$\min_{w_i} |u_i w_i - u_i w_{i-1}|^2 \qquad \text{subject to } |u_i w_i| = \gamma$$

is given by

$$w_i = w_{i-1} + \frac{u_i^*}{\|u_i\|^2} \left(\gamma \frac{z(i)}{|z(i)|} - z(i) \right)$$

Remark. Inserting a step-size μ into the above recursion leads to the normalized CMA algorithm,

$$w_i = w_{i-1} + \mu \frac{u_i^*}{\|u_i\|^2} \left(\gamma \frac{z(i)}{|z(i)|} - z(i) \right)$$

Problem III.37 (Gauss-Newton algorithm) A stochastic-gradient method that is similar in nature to the RLS algorithm of Sec. 14.1 can be obtained in the following manner.

(a) Let $\epsilon(i) = \lambda^{i+1}\epsilon/(i+1)$ and $\mu(i) = \mu$. Use the same approximations as in the RLS case for R_u and $(R_{du} - R_u w_{i-1})$ to verify that the regularized Newton's recursion (14.1) reduces to

$$w_i = w_{i-1} + \mu \Phi_i^{-1} u_i^* [d(i) - u_i w_{i-1}]$$

where

$$\Phi_i \triangleq \left(\frac{\lambda^{i+1}\epsilon}{(i+1)} \mathrm{I} + \frac{1}{i+1} \sum_{j=0}^{i} \lambda^{i-j} u_j^* u_j \right), \quad i \geq 0$$

(b) Show that Φ_i satisfies the recursion $\Phi_i = \lambda[1 - \alpha(i)]\Phi_{i-1} + \alpha(i) u_i^* u_i$ for $i \geq 1$, with initial condition $\Phi_0 = \lambda\epsilon\mathrm{I} + u_0^* u_0$, and where $\alpha(i) = 1/(i+1)$.

(c) Define $P_i = \Phi_i^{-1}$. Show that

$$\begin{aligned} P_i &= \frac{\lambda^{-1}}{[1-\alpha(i)]} \left[P_{i-1} - \frac{\lambda^{-1} P_{i-1} u_i^* u_i P_{i-1}}{\frac{[1-\alpha(i)]}{\alpha(i)} + \lambda^{-1} u_i P_{i-1} u_i^*} \right], \quad i \geq 1 \\ w_i &= w_{i-1} + \mu P_i u_i^* [d(i) - u_i w_{i-1}], \quad i \geq 0 \end{aligned}$$

with initial condition

$$P_0 = \frac{1}{\lambda\epsilon} \left[\mathrm{I} - \frac{u_0^* u_0}{\lambda\epsilon + \|u_0\|^2} \right]$$

(d) Repeat the calculations in Tables 5.7 and 5.8 and estimate the amount of computations that are required per iteration for both cases of real and complex-valued data.

Problem III.38 (Simplified GN algorithm) Consider the same setting as in Prob. III.37. An alternative form of the GN algorithm can be obtained by replacing $\alpha(i)$ by a constant positive number α (usually small, say $0 < \alpha < 0.1$).

(a) Verify that the recursions of part (c) of Prob. III.37 reduce to

$$\begin{aligned} P_i &= \frac{\lambda^{-1}}{1-\alpha} \left[P_{i-1} - \frac{\lambda^{-1} P_{i-1} u_i^* u_i P_{i-1}}{\frac{(1-\alpha)}{\alpha} + \lambda^{-1} u_i P_{i-1} u_i^*} \right], \quad P_{-1} = \epsilon^{-1}\mathrm{I} \\ w_i &= w_{i-1} + \mu P_i u_i^* [d(i) - u_i w_{i-1}], \quad i \geq 0 \end{aligned}$$

(b) Show further that the above recursions could have been derived directly from Newton's recursion (14.1) by using the following sample average approximation for R_u, $\widehat{R}_u = \alpha \sum_{j=0}^{i} [\lambda(1-\alpha)]^{i-j} u_j^* u_j$, along with the choice $\epsilon(i) = \lambda^{(i+1)}(1-\alpha)^{(i+1)}\epsilon$.

Problem III.39 (Sample covariance matrices) Consider the sample covariance matrix $\widehat{R}_u = \frac{1}{i+1}\sum_{j=0}^{i} u_j^* u_j$ used in Sec. 14.1 (with exponential weighting). Let us denote $\widehat{R}_u$ by $\widehat{R}_{u,i}$ to indicate that it is based on data up to time i.

(a) Verify that $\widehat{R}_{u,i}$ satisfies the recursion $\widehat{R}_{u,i} = \frac{i}{i+1}\widehat{R}_{u,i-1} + \frac{1}{i+1}u_i^* u_i$.

(b) Consider instead the sample covariance matrix $\widehat{R}_{u,i} = \alpha \sum_{j=0}^{i}(1-\alpha)^{i-j} u_j^* u_j$ used in Prob. III.38. Verify now that $\widehat{R}_{u,i} = (1-\alpha)\widehat{R}_{u,i-1} + \alpha u_i^* u_i$.

Problem III.40 (Conversion factor) Show that the RLS algorithm can be written in the equivalent form

$$w_i = w_{i-1} + \frac{\lambda^{-1}P_{i-1}\,u_i^*}{1+\lambda^{-1}u_i P_{i-1} u_i^*}\,[d(i) - u_i w_{i-1}]\,, \quad i \geq 0.$$

Show further that $r(i) = \gamma(i)e(i)$ where $\gamma(i) = 1/(1+\lambda^{-1}u_i P_{i-1} u_i^*)$ and $r(i)$ and $e(i)$ denote the *a posteriori* and *a priori* output errors, $r(i) = d(i) - u_i w_i$ and $e(i) = d(i) - u_i w_{i-1}$. Conclude that, for all i, $|r(i)| \leq |e(i)|$. *Remark.* The coefficient $\gamma(i) = 1/(1+\lambda^{-1}u_i P_{i-1} u_i^*)$ is called the *conversion factor* since it converts the *a priori* error to the *a posteriori error*. We shall have more to say about the RLS algorithm, its properties, and its variants in Parts VII (*Least-Squares Methods*) through X (*Lattice Filters*).

Problem III.41 (Bound on the step-size for ϵ−APA) Refer to the discussion in Sec. 13.3 and to the definitions of the error vectors $\{e_i, r_i\}$.

(a) Use the constraint in (13.8) to show that $\|r_i\|^2 < \|e_i\|^2$ if, and only if, the matrix $\mathrm{I} - A$ is positive-definite, where

$$A \triangleq \left(\mathrm{I} - \mu U_i U_i^*(\epsilon\mathrm{I} + U_i U_i^*)^{-1}\right)^* \left(\mathrm{I} - \mu U_i U_i^*(\epsilon\mathrm{I} + U_i U_i^*)^{-1}\right)$$

Show further that $\|r_i\|^2 = \|e_i\|^2$ if, and only if, $e_i = 0$.

(b) Let $U_i U_i^* = V_i \Gamma_i V_i^*$ denote the eigen-decomposition of the $K \times K$ matrix $U_i U_i^*$, where V_i is unitary and $\Gamma_i = \text{diag}\{\gamma_0(i), \gamma_1(i), \ldots, \gamma_{K-1}(i)\}$ contains the corresponding eigenvalues. Show that $\mathrm{I} - A = \mu V_i \Gamma_i'(2\mathrm{I} - \mu\Gamma_i')V_i^*$, where $\Gamma_i' = \Gamma_i(\epsilon\mathrm{I} + \Gamma_i)^{-1}$.

(c) Conclude that $\mathrm{I} - A > 0$ if, and only if,

$$0 < \mu < \min_{0 \leq k \leq K-1} \frac{2(\epsilon + \gamma_k(i))}{\gamma_k(i)}$$

Conclude that $0 < \mu < 2$ guarantees $\|r_i\|^2 \leq \|e_i\|^2$.

Problem III.42 (Least-perturbation property of ϵ-APA) Refer to the optimization problem (13.8).

(a) Introduce the difference $\delta w = w_i - w_{i-1}$. Show that the constraint amounts to the requirement $U_i \delta w = \mu U_i U_i^* (\epsilon\mathrm{I} + U_i U_i^*)^{-1} e_i$.

(b) Verify that the choice $\delta w^o = \mu U_i^*(\epsilon\mathrm{I} + U_i U_i^*)^{-1} e_i$ satisfies the constraint.

(c) Now complete the argument, as in Sec. 10.4, to show that the solution of (13.8) leads to the ϵ-APA recursion (13.5).

Problem III.43 (Gram-Schmidt orthogonalization) Consider three row vectors $\{u_1, u_2, u_3\}$ and define the transformed vectors (also called residuals):

$$\tilde{u}_1 = u_1, \quad \tilde{u}_2 = u_2 - u_2\frac{\tilde{u}_1^*\tilde{u}_1}{\|\tilde{u}_1\|^2}, \quad \tilde{u}_3 = u_3 - u_3\frac{\tilde{u}_1^*\tilde{u}_1}{\|\tilde{u}_1\|^2} - u_3\frac{\tilde{u}_2^*\tilde{u}_2}{\|\tilde{u}_2\|^2}$$

(a) Verify that the residual vectors so obtained are orthogonal to each other, i.e., show that $\tilde{u}_i\tilde{u}_j^* = 0$ for $i \neq j$.

(b) Verify further that $\{u_1, u_2, u_3\}$ and $\{\tilde{u}_1, \tilde{u}_2, \tilde{u}_3\}$ are related via an invertible lower triangular matrix as follows:

$$\begin{bmatrix} \tilde{u}_1 \\ \tilde{u}_2 \\ \tilde{u}_3 \end{bmatrix} = \begin{bmatrix} 1 & & \\ -\dfrac{u_2\tilde{u}_1^*}{\|\tilde{u}_1\|^2} & 1 & \\ -\dfrac{u_3}{\|\tilde{u}_1\|^2}\left(\mathrm{I} - \dfrac{\tilde{u}_2^* u_2}{\|\tilde{u}_2\|^2}\right)\tilde{u}_1^* & -\dfrac{u_3\tilde{u}_2^*}{\|\tilde{u}_2\|^2} & 1 \end{bmatrix} \begin{bmatrix} u_1 \\ u_2 \\ u_3 \end{bmatrix}$$

Problem III.44 (APA with orthogonal correction factors) Consider an APA update of order $K = 3$. i.e., $e_i = d_i - U_i w_{i-1}$ and $w_i = w_{i-1} + \mu U_i^* \left(U_i U_i^*\right)^{-1} e_i$, where

$$U_i = \begin{bmatrix} u_i \\ u_{i-1} \\ u_{i-2} \end{bmatrix}, \qquad d_i = \begin{bmatrix} d(i) \\ d(i-1) \\ d(i-2) \end{bmatrix}$$

In this problem we want to show that APA can be implemented in an equivalent form that involves only orthogonal regression vectors. So assume that, at each iteration i, the regressors $\{u_i, u_{i-1}, u_{i-2}\}$ are first orthogonalized, as described in Prob. III.43, and let $\{\tilde{u}_i, \tilde{u}_{i-1}, \tilde{u}_{i-2}\}$ denote the corresponding residual vectors (with $\tilde{u}_i = u_i$). Let L_i denote the lower-triangular matrix relating the residuals to the regressors, i.e.,

$$\tilde{U}_i \triangleq \begin{bmatrix} \tilde{u}_i \\ \tilde{u}_{i-1} \\ \tilde{u}_{i-2} \end{bmatrix} = L_i U_i$$

(a) Verify that the APA update can be written as $w_i = w_{i-1} + \mu \tilde{U}_i^* \left(\tilde{U}_i \tilde{U}_i^*\right)^{-1} L_i e_i$, where $\tilde{U}_i \tilde{U}_i^*$ is diagonal and given by $\tilde{U}_i \tilde{U}_i^* = \text{diag}\left\{ \|\tilde{u}_i\|^2, \|\tilde{u}_{i-1}\|^2, \|\tilde{u}_{i-2}\|^2 \right\}$.

(b) Show that the entries of $L_i e_i$ are given by $L_i e_i = \text{col}\{e(i), e^{(1)}(i-1), e^{(2)}(i-2)\}$, where

$$e(i) = d(i) - u_i w_{i-1}, \quad e^{(1)}(i-1) = d(i-1) - u_{i-1} w_{i-1}^{(1)}, \quad e^{(2)}(i-2) = d(i-2) - u_{i-2} w_{i-1}^{(2)}$$

and $\{w_{i-1}^{(1)}, w_{i-1}^{(2)}\}$ are intermediate corrections to w_{i-1} defined by

$$w_{i-1}^{(1)} \triangleq w_{i-1} + \frac{\tilde{u}_i^*}{\|\tilde{u}_i\|^2} e(i), \quad w_{i-1}^{(2)} \triangleq w_{i-1}^{(1)} + \frac{\tilde{u}_{i-1}^*}{\|\tilde{u}_{i-1}\|^2} e^{(1)}(i-1)$$

(c) Conclude that a general $K-$th order APA update can be equivalently implemented as follows. For each time instant i, start with $w_{i-1}^{(0)} = w_{i-1}$ and repeat for $k = 0, 1, \ldots, K-1$:

$$\boxed{\begin{aligned} e^{(k)}(i-k) &= d(i-k) - u_{i-k} w_{i-1}^{(k)} \\ w_{i-1}^{(k+1)} &= w_{i-1}^{(k)} + \frac{\tilde{u}_{i-k}^*}{\|\tilde{u}_{i-k}\|^2} e^{(k)}(i-k) \end{aligned}}$$

Then set

$$\boxed{w_i = w_{i-1} + \mu \sum_{k=0}^{K-1} \frac{\tilde{u}_{i-k}^*}{\|\tilde{u}_{i-k}\|^2} e^{(k)}(i-k)}$$

Remark. This form of the algorithm is sometimes referred to as APA or NLMS with orthogonal correction factors (APA-OCF or NLMS-OCF) since the regressors $\{\tilde{u}_i\}$ that are used in the update of the weight vector are orthogonal to each other.

(d) Verify further from part (c) that w_i is given by the convex combination

$$\boxed{w_i = (1-\mu) w_{i-1} + \mu w_{i-1}^{(K)}}$$

Problem III.45 (LMS as a notch filter) Consider an LMS adaptive filter with a regression vector u_i with shift-structure, i.e., its entries are delayed replicas of an input sequence $\{u(i)\}$, as in an FIR implementation,

$$u_i = \left[\begin{array}{cccc} u(i) & u(i-1) & \ldots & u(i-M+1) \end{array}\right]$$

Assume $u(i)$ is sinusoidal, say $u(i) = e^{j\omega_o i}$ for some ω_o. Assume also that the filter is initially at rest. Let w_i denote the coefficients of the LMS filter, which are adapted according to the rule $w_i = w_{i-1} + \mu u_i^*[d(i) - u_i w_{i-1}]$. Although the coefficients of the filter vary with time, due to the adaptation process, it turns out that the input-output mapping is actually time-invariant in this case.

(a) Show that the transfer function from the desired signal $d(i)$ to the error signal $e(i) = d(i) - u_i w_{i-1}$ is given by

$$\frac{E(z)}{D(z)} = \frac{z - e^{j\omega_o}}{z + (\mu M - 1)e^{j\omega_o}}$$

(b) Assume the adaptive filter is trained instead by $\epsilon-$NLMS. Show that the same transfer function becomes

$$\frac{E(z)}{D(z)} = \frac{z - e^{j\omega_o}}{z + \left(\dfrac{\mu M}{\epsilon + M} - 1\right) e^{j\omega_o}}$$

What does this result reduce to when $M \rightarrow \infty$?

(c) Assume the filter is trained using the power normalized $\epsilon-$NLMS algorithm (cf. Alg. 11.2). Show that in the limit, as $i \rightarrow \infty$, the transfer function from $d(i)$ to $e(i)$ is again given by

$$\frac{E(z)}{D(z)} = \frac{z - e^{j\omega_o}}{z + \left(\dfrac{\mu M}{\epsilon + M} - 1\right) e^{j\omega_o}}$$

Remark. The results indicate that, with a sinusoidal excitation, the LMS filter behaves like a notch filter with a notch frequency at ω_o. Applications to adaptive noise cancelling of sinusoidal interferences can be found in Widrow et al. (1976) and Glover (1977).

Problem III.46 (Adaptive line enhancement) Let $\boldsymbol{d}(i) = \boldsymbol{s}(i) + \boldsymbol{v}(i)$ denote a zero-mean random sequence that consists of two components: a signal component, $\boldsymbol{s}(i)$, and a noise component, $\boldsymbol{v}(i)$. Let $r_s(k)$ and $r_v(k)$ denote the auto-correlation sequences of $\{\boldsymbol{s}(i), \boldsymbol{v}(i)\}$, assumed stationary, i.e., $r_s(k) = \mathsf{E}\,\boldsymbol{s}(i)\boldsymbol{s}^*(i-k)$ and $r_v(k) = \mathsf{E}\,\boldsymbol{v}(i)\boldsymbol{v}^*(i-k)$.

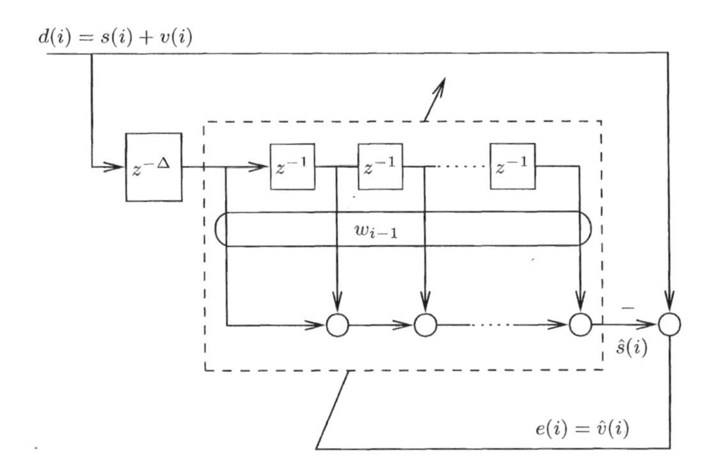

FIGURE III.1 An adaptive structure for line signal enhancement.

Assume that $r_s(k)$ is negligible for $k > \delta_s$, while $r_v(k)$ is negligible for $k > \delta_n$ with $\delta_s \gg \delta_n$. We say that $s(i)$ corresponds to the narrowband component of $d(i)$ while $v(i)$ corresponds to the wideband component of $d(i)$. The adaptive structure of Fig. III.1 is suggested for use in separating the signal $s(i)$ from the noise $v(i)$. As the figure indicates, realizations $d(i)$ are used as the reference sequence, and a delayed replica of these same realizations is used as input to the tapped delay line. The value of the delay Δ is chosen to satisfy $\delta_n < \Delta < \delta_s$, and the filter taps are trained using LMS, for example. It is claimed that the output of the filter, $u_i w_{i-1}$, provides estimates of the signal component, while the error signal, $e(i)$, provides estimates of the noise component. That is, $\hat{v}(i) = e(i)$ and $\hat{s}(i) = u_i w_{i-1}$. Justify the validity of this claim. *Hint:* Refer to Prob. II.13.

Remark. The adaptive structure described in this problem is known as an adaptive line enhancer (ALE); it permits separating a narrowband signal (e.g., a sinusoid) from a wideband noise signal. The ALE was originally developed by Widrow et al. (1975). Its performance was later studied in some detail by Zeidler et al. (1978), Rickard and Zeidler (1979), Treichler (1979), and Zeidler (1990).

COMPUTER PROJECTS

Project III.1 (Constant-modulus criterion) Refer to Prob. III.18, where we introduced the cost function $J(w) = \mathsf{E}\left(\gamma - |uw|^2\right)^2$, for a given positive constant γ. This cost arises in the context of blind equalization where it is used to derive blind adaptive filters — see, e.g., Prob. III.33. In this project, we use $J(w)$ to highlight some of the issues that arise in the design of steepest-descent methods.

(a) Assume first that w is one-dimensional and $\boldsymbol{u}$ is scalar-valued with variance $\sigma_u^2 = \mathsf{E}|\boldsymbol{u}|^2$. Assume further that $\boldsymbol{u}$ and w are real-valued and that $\boldsymbol{u}$ is Gaussian so that its fourth moment is given by $\mathsf{E}\,\boldsymbol{u}^4 = 3\sigma_u^4$. Verify that, under these conditions, the cost function $J(w)$ evaluates to $J(w) = \gamma^2 - 2\gamma\sigma_u^2 w^2 + 3\sigma_u^4 w^4$. Conclude that $J(w)$ has a local maximum at $w = 0$ and two global minima at $w^o = \pm\sqrt{\gamma/3\sigma_u^2}$. Plot $J(w)$ and determine the values of w^o using $\gamma = 1$ and $\sigma_u^2 = 0.5$. Find also the corresponding minimum cost.

(b) Argue that a steepest-descent method for minimizing $J(w)$ can be taken as

$$w(i) = w(i-1) + 4\mu\sigma_u^2 w(i-1)[\gamma - 3\sigma_u^2 w^2(i-1)], \quad w(-1) = \text{initial guess}$$

with scalar estimates $\{w(i)\}$. Does this method converge to a global minimum if the initial weight guess is chosen as $w(-1) = 0$? Why?

(c) Simulate the steepest-descent recursion of part (b) in the following cases and comment on its behavior in each case:

1. $\mu = 0.2$ and $w(-1) = 0.3$ or $w(-1) = -0.3$.
2. $\mu = 0.6$ and $w(-1) = 0.3$ or $w(-1) = -0.3$.
3. $\mu = 1$ and $w(-1) = -0.2$.

In each case, plot the evolution of $w(i)$ as a function of time. Plot also the graph $J[w(i)] \times w(i)$.

(d) Now consider the setting of part (b) in Prob. III.18 where w is two-dimensional.

(d.1) Generate a plot of the contour curves of the cost function for $\gamma = 1$, i.e.,

$$J(w) = \gamma^2 + 2\left(|\alpha|^2 + 2|\beta|^2\right)^2 - 2\gamma\left(|\alpha|^2 + 2|\beta|^2\right)$$

(d.2) Choose $\mu = 0.02$ and apply the resulting algorithm,

$$\begin{bmatrix} \alpha(i) \\ \beta(i) \end{bmatrix} = \begin{bmatrix} \alpha(i-1) \\ \beta(i-1) \end{bmatrix} + \mu\left(\gamma - 2|\alpha(i-1)|^2 - 4|\beta(i-1)|^2\right) \begin{bmatrix} \alpha(i-1) \\ 2\beta(i-1) \end{bmatrix}$$

starting from the initial conditions $w_{-1} = \text{col}\{1, -0.25\}$ and $w_{-1} = \text{col}\{-1, 0.25\}$. Iterate over a period of length $N = 1000$. Plot the trajectory of the weight estimates superimposed on the contour curves of $J(w)$.

Adaptive filters are used in many applications and we cannot attempt to cover all of them in a textbook. Instead, we shall illustrate the use of adaptive filters in selected applications of heightened interest, including channel equalization, channel estimation, and echo cancellation. The computer projects in this section will focus on channel equalization. In later parts of the book, the computer projects will consider applications involving line echo cancellation, channel tracking, channel estimation, acoustic echo cancellation, and active noise control. In addition, in some of the problems, other applications are considered, such as adaptive line enhancement in Prob. III.46.

Project III.2 (Constant-modulus algorithm) In Prob. III.18 we introduced the constant-modulus criterion

$$\min_w \mathsf{E}\left(\gamma - |uw|^2\right)^2$$

and developed a steepest-descent method for it, namely

$$w_i = w_{i-1} + \mu\left(\gamma R_u w_{i-1} - \mathsf{E}\left[u^* u w_{i-1} |u w_{i-1}|^2\right]\right)$$

We reconsidered this method in Prob. III.33 and derived the corresponding stochastic gradient approximation, known as CMA2-2, namely,

$$w_i = w_{i-1} + \mu u_i^* z(i)\left[\gamma - |z(i)|^2\right], \quad z(i) = u_i w_{i-1}$$

In this project we compare the performance of the steepest-descent method and the CMA2-2 recursion. For this purpose, we set $\gamma = 1$ and let u be a 2-dimensional circular Gaussian random vector with covariance matrix $R_u = \text{diag}\{1, 0\}$. We showed in Prob. III.18 that, under these conditions, the steepest-descent method collapses to the form

$$\begin{bmatrix} \alpha(i) \\ \beta(i) \end{bmatrix} = \begin{bmatrix} \alpha(i-1) \\ \beta(i-1) \end{bmatrix} + \mu\left(\gamma - 2|\alpha(i-1)|^2 - 4|\beta(i-1)|^2\right)\begin{bmatrix} \alpha(i-1) \\ 2\beta(i-1) \end{bmatrix}$$

where $w_i = \text{col}\{\alpha(i), \beta(i)\}$. In addition, we showed in Prob. III.18 that the corresponding cost function evaluates to

$$J(w_i) = \gamma^2 + 2\left(|\alpha(i)|^2 + 2|\beta(i)|^2\right)^2 - 2\gamma\left(|\alpha(i)|^2 + 2|\beta(i)|^2\right)$$

(a) Plot the learning curve $J(i) = J(w_{i-1})$ for the steepest-descent method; here w_{i-1} is the weight estimate that results from the steepest-descent recursion. Plot also an ensemble-average learning curve $\widehat{J}(i)$ for the CMA2-2 algorithm that is generated as follows:

$$\widehat{J}(i) = \frac{1}{L}\sum_{j=1}^{L}\left(\gamma - \left|z^{(j)}(i)\right|^2\right)^2, \quad i \geq 0$$

with the data generated from L experiments, say of duration N each. Choose $L = 200$ and $N = 500$. Use $\mu = 0.001$ and $w_{-1} = \text{col}\{-0.8, 0.8\}$. *Remark.* We are assuming complex-valued regressors u, with the two entries of u having variances $\{1, 2\}$. In order to generate such regressors, create four separate *real-valued* zero-mean and independent Gaussian variables $\{a, b, p, q\}$ with variances $\{1/2, 1/2, 1, 1\}$, respectively, and then set $u = [a + jb \;\; p + jq]$.

(b) Plot the contour curves of the cost function $J(w) = \mathsf{E}(\gamma - |uw|^2)^2$ superimposed on the four weight trajectories that are generated by CMA2-2 for the following four choices of initial conditions:

$$w_{-1} \in \left\{\begin{bmatrix} 0.8 & 0.8 \end{bmatrix}, \begin{bmatrix} -0.8 & 0.8 \end{bmatrix}, \begin{bmatrix} 0.8 & -0.8 \end{bmatrix}, \begin{bmatrix} -0.8 & -0.8 \end{bmatrix}\right\}$$

Use $\mu = 0.001$ and $N = 1000$ iterations in each case. Print the final value of the weight vector estimate in each case. Show also the trajectories of the weight estimates that are generated by the steepest-descent method. *Remark.* Although the weight estimates in the steepest-descent method are always real-valued for a real-valued initial condition w_{-1}, the same is not true for CMA2-2. However, the imaginary parts of the successive weight estimates will be small compared to the real parts. For this reason, when plotting the weight trajectories, we shall ignore the imaginary parts.

Project III.3 (Adaptive channel equalization) In Computer Projects II.1 and II.3 we dealt with the design of minimum mean-square error equalizers. In this project we examine the design of *adaptive* equalizers. We consider the same channel of Computer Project II.3, namely,

$$C(z) = 0.5 + 1.2z^{-1} + 1.5z^{-2} - z^{-3}$$

and proceed to design an adaptive linear equalizer for it. The equalizer structure is shown in Fig. III.2. Symbols $\{s(i)\}$ are transmitted through the channel and corrupted by additive complex-valued white noise $\{v(i)\}$. The received signal $\{u(i)\}$ is processed by the FIR equalizer to generate estimates $\{\hat{s}(i-\Delta)\}$, which are fed into a decision device. The equalizer possesses two modes of operation: a training mode during which a delayed replica of the input sequence is used as a reference sequence, and a decision-directed mode during which the output of the decision-device replaces the reference sequence. The input sequence $\{s(i)\}$ is chosen from a quadrature-amplitude modulation (QAM) constellation (e.g., 4-QAM, 16-QAM, 64-QAM, or 256-QAM).

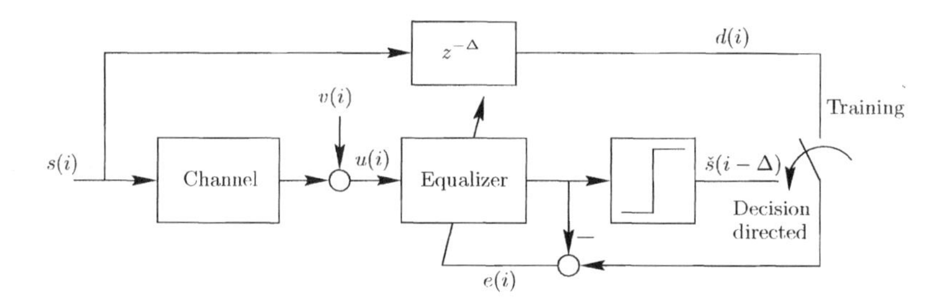

FIGURE III.2 An adaptive linear equalizer operating in two modes: training mode and decision-direction mode.

(a) Write a program that trains the adaptive filter with 500 symbols from a QPSK constellation, followed by decision-directed operation during 5000 symbols from a 16-QAM constellation. Choose the noise variance σ_v^2 in order to enforce an SNR level of 30 dB at the input of the equalizer. Note that symbols chosen from QAM constellations do not have unit variance. For this reason, the noise variance needs to be adjusted properly for different QAM orders in order to enforce the desired SNR level — see Prob. II.16. Choose $\Delta = 15$ and equalizer length $L = 35$. Use $\epsilon-$NLMS to train the equalizer with step-size $\mu = 0.4$ and $\epsilon = 10^{-6}$. Plot the scatter diagrams of $\{s(i), u(i), \hat{s}(i-\Delta)\}$.

(b) For the same setting as part (a), plot and compare the scatter diagrams that would result at the output of the equalizer if training is performed only for 150, 300, and 500 iterations. Repeat the simulations using LMS with $\mu = 0.001$.

(c) Now assume the transmitted data are generated from a 256-QAM constellation rather than a 16-QAM constellation. Plot the scatter diagrams of the output of the equalizer, when trained with ϵ-NLMS using 500 training symbols.

(d) Generate symbol-error-rate (SER) curves versus signal-to-noise ratio (SNR) at the input of the equalizer for $(4, 16, 64, 256)$-QAM data. Let the SNR vary between 5 dB and 30 dB in increments of 1 dB.

(e) Continue with SNR at 30 dB. Design a decision-feedback equalizer with $L = 10$ feedforward taps and $Q = 2$ feedback taps. Use $\Delta = 7$ and plot the resulting scatter diagram of the output of the equalizer. Repeat for $L = 20$, $Q = 2$ and $\Delta = 10$. In both cases, choose the transmitted data from a 64-QAM constellation.

(f) Generate SER curves versus SNR at the input of the DFE for $(4, 16, 64, 256)$-QAM data. Let the SNR vary between 5 dB and 30 dB. Compare the performance of the DFE with that of the linear equalizer of part (d).

(g) Load the file channel, which contains the impulse response sequence of a more challenging channel with spectral nulls. Set the SNR level at the input of the equalizer to 40 dB and select a linear equalizer structure with 55 taps. Set also the delay at $\Delta = 30$. Train the equalizer using $\epsilon-$NLMS for 2000 iterations before switching to decision-directed operation. Plot the resulting scatter diagram of the output of the equalizer. Now train it again using RLS for 100 iterations before switching to decision-directed operation, and plot the resulting scatter diagram. Compare both diagrams.

Project III.4 (Blind adaptive equalization) We consider the same channel used in Computer Project III.3,

$$C(z) = 0.5 + 1.2z^{-1} + 1.5z^{-2} - z^{-3}$$

and proceed to design blind adaptive equalizers for it. The equalizer structure is shown in Fig. III.3. Symbols $\{s(i)\}$ are transmitted through the channel and corrupted by additive complex-valued white noise $\{v(i)\}$. The received signal $\{u(i)\}$ is processed by a linear equalizer, whose outputs $\{z(i)\}$ are fed into a decision device to generate $\{\check{s}(i - \Delta)\}$. These signals are delayed decisions and the value of Δ is determined by the delay that the signals undergo when travelling through the channel and the equalizer. In this project, the equalizer is supposed to operate blindly, i.e., without a reference sequence and therefore without a training mode. Most blind algorithms use the output of the equalizer, $z(i)$, to generate an error signal $e(i)$, which is used to adapt the equalizer coefficients according to the rule

$$w_i = w_{i-1} + \mu u_i^* e(i)$$

where u_i is the regressor at time i. Some blind algorithms use the output of the decision device, $\check{s}(i - \Delta)$, to evaluate $e(i)$ (e.g., the "stop-and-go" variant described in part (d) below).

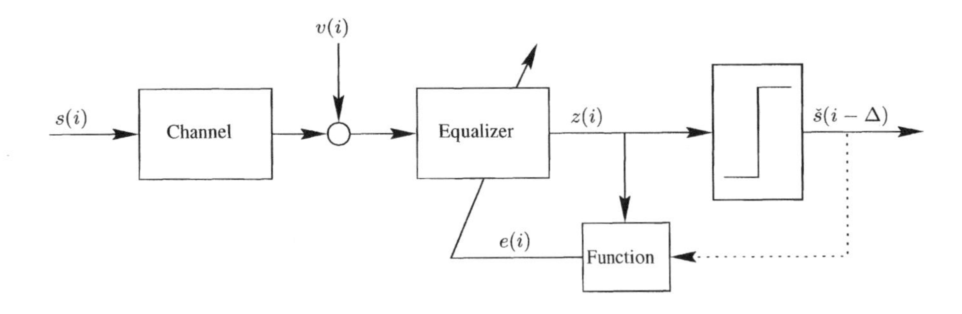

FIGURE III.3 A representation of a general structure for blind adaptive equalization.

(a) Write a program that transmits 10000 QPSK symbols through the channel and trains a 35-tap equalizer using CMA2-2,

$$w_i = w_{i-1} + \mu u_i^* z(i) \left[\gamma - |z(i)|^2\right], \quad z(i) = u_i w_{i-1}$$

Choose the value of γ as $\mathsf{E}\,|s|^4/\mathsf{E}\,|s|^2$, which is in terms of the second and fourth-order moments of the symbol constellation. For QPSK data, $\gamma = 1$. Set the SNR level at the input of the equalizer to 30 dB and use $\mu = 0.001$. Plot the impulse responses of the channel, the equalizer, and the combination channel-equalizer at the end of the adaptation process. How much delay does the signal undergo in travelling through the channel and the equalizer? Plot the scatter diagrams of the transmitted sequence, the received sequence, and the sequence at the output of the equalizer. Ignore the first 2000 transmissions and count the number of erroneous decisions in the remaining decisions (you should take into account the delay introduced by the channel-equalizer system).

(b) Repeat the simulations of part (a) using 16-QAM data, for which $\gamma = 13.2$ (verify this value). Use $\mu = 0.000001$. Run the simulation for 30000 symbols and ignore the first 15000 for error calculation. These numbers are larger than in the QPSK case of part (a), and the step-size is also significantly smaller, since the equalizer converges at a slower pace now.

(c) Repeat the simulations of part (b) using the multi-modulus algorithm (MMA) of Prob. III.35.

(d) Repeat the simulations of part (b) using the following three additional blind adaptive algorithms:

(d.1) CMA1-2 from Prob. III.36, where γ is now chosen as $\gamma = \mathsf{E}\,|s|^2/\mathsf{E}\,|s|$ in terms of the second moment of the symbol constellation divided by the mean of the magnitude of the symbols. For 16-QAM data we find $\gamma = 3.3385$ (verify this value). Use $\mu = 0.0001$ and increase the SNR level at the input of the equalizer to 60 dB. Simulate for 30000 iterations and plot the scatter diagram of the output of the equalizer after ignoring the first 15000 samples.

(d.2) The reduced constellation algorithm (RCA) of Prob. III.34, where γ is now chosen as $\gamma = \mathsf{E}\,|s|^2/\mathsf{E}\,|s|_1$, in terms of the second moment of the symbol constellation divided by the mean of the l_1-norm of the symbols (remember that the l_1 norm of a complex number amounts to adding the absolute values of its real and imaginary parts, as in Prob. III.14). For 16-QAM data we find $\gamma = 2.5$ (verify this value). Use the same step-size and same simulation duration as part (d.1).

(d.3) The "stop-and-go" algorithm is a blind adaptation scheme that employs the decision-directed error,

$$e_d(i) = \check{s}(i-\Delta) - z(i)$$

It also employs a flag to indicate how reliable $e_d(i)$ is. The flag is set by comparing $e_d(i)$ to another error signal, say the one used by the RCA recursion,

$$e_s(i) = \gamma \mathsf{csgn}(z(i)) - z(i), \qquad \gamma = \mathsf{E}\,|s|^2/\mathsf{E}\,|s|_1$$

If the complex signs of $\{e_d(i), e_s(i)\}$ differ, then $e_d(i)$ is assumed unreliable and the flag is set to zero (see Picchi and Prati (1987)). More explicitly, the stop-and-go recursion takes the form:

$$\left\{\begin{array}{l} z(i) = u_i w_{i-1} \\ e_d(i) = \check{s}(i-\Delta) - z(i) \qquad \text{(decision-directed error)} \\ e_s(i) = \gamma \mathsf{csgn}(z(i)) - z(i) \qquad \text{(RCA error)} \\ \\ f(i) = \left\{\begin{array}{ll} 1 & \text{if } \mathsf{csgn}(e_d(i)) = \mathsf{csgn}(e_s(i)) \\ \\ 0 & \text{if } \mathsf{csgn}(e_d(i)) \neq \mathsf{csgn}(e_s(i)) \end{array}\right. \qquad \text{(a flag)} \\ \\ e(i) = f(i) e_d(i) \\ w_i = w_{i-1} + \mu u_i^* e(i) \end{array}\right.$$

Use the same step-size and simulation duration as in part (d.1).

PART **IV**

MEAN-SQUARE PERFORMANCE

CHAPTER 15

Energy Conservation

In Part III (*Stochastic–Gradient Algorithms*) and its problems, we developed stochastic gradient approximations for several steepest-descent methods. The approximations were obtained by replacing exact covariance and cross-correlation quantities by instantaneous estimates. The resulting algorithms operate on actual data realizations and they lead to adaptive filter implementations. However, stochastic approximations introduce gradient noise and, consequently, the performance of adaptive filters will degrade in comparison with the performance of the original steepest-descent methods.

The purpose of this chapter, and of the subsequent chapters in this part (*Mean-Square Performance*) and in Part V (*Transient Performance*), is to describe a unifying framework for the evaluation of the performance of adaptive filters. This objective is rather challenging, especially since adaptive filters are, by design, time-variant, stochastic, and nonlinear systems. Their update recursions not only depend on the reference and regression data in a nonlinear and time-variant manner, but the data they employ are also stochastic in nature. For this reason, the study of the performance of adaptive algorithms is a formidable task, so much so that *exact* performance analyses are rare and limited to special cases. It is customary to introduce simplifying assumptions in order to make the performance analyses more tractable. Fortunately, most assumptions tend to lead to reasonable agreements between theory and practice.

The framework developed in this chapter, and pursued further in the following chapters, relies on energy conservation arguments. While the performance of different adaptive filters tends to be studied separately in the literature, the framework adopted in our presentation applies uniformly across different classes of filters. In particular, the same framework is used for steady-state analysis, tracking analysis, and transient analysis. In other words, energy-based arguments stand out as a common theme that runs throughout our treatment of the performance of adaptive filters.

15.1 PERFORMANCE MEASURE

Before plunging into a detailed study of adaptive filter performance, we need to explain some of the issues that arise in this context, including the need to adopt a common performance measure and the need to model adaptive algorithms in terms of *stochastic* equations. We use the LMS filter as a motivation for our explanations.

Thus, recall that the steepest-descent iteration (8.20), namely,

$$w_i = w_{i-1} + \mu[R_{du} - R_u w_{i-1}] \tag{15.1}$$

was reduced to the LMS recursion (10.10), i.e.,

$$w_i = w_{i-1} + \mu u_i^*[d(i) - u_i w_{i-1}] \tag{15.2}$$

by replacing the second-order moments $R_{du} = \mathsf{E}\,\boldsymbol{d}\boldsymbol{u}^*$ and $R_u = \mathsf{E}\,\boldsymbol{u}^*\boldsymbol{u}$ by the instantaneous approximations

$$R_{du} \approx d(i)u_i^* \quad \text{and} \quad R_u \approx u_i^* u_i \tag{15.3}$$

Other adaptive algorithms were obtained in Chapter 10 by using similar instantaneous approximations. Recall further that we examined the convergence properties of (15.1) in Chapter 8 in some detail. Specifically, we established in Thm. 8.2 that by choosing the step-size μ such that

$$0 < \mu < 2/\lambda_{\max} \tag{15.4}$$

where $\lambda_{\max}$ is the largest eigenvalue of R_u, the successive weight estimates w_i of (15.1) are guaranteed to converge to the solution w^o of the normal equations, i.e., to the vector

$$w^o = R_u^{-1} R_{du} \tag{15.5}$$

that solves the least-mean-squares problem

$$\min_w \mathsf{E}\,|\boldsymbol{d} - \boldsymbol{u}w|^2 \tag{15.6}$$

Correspondingly, the learning curve of the steepest-descent method (15.1), namely,

$$\begin{aligned} J(i) &= \mathsf{E}\,|\boldsymbol{d} - \boldsymbol{u}w_{i-1}|^2 \\ &= \sigma_d^2 - R_{du}^* w_{i-1} - w_{i-1}^* R_{du} + w_{i-1}^* R_u w_{i-1} \end{aligned}$$

is also guaranteed to converge to the minimum cost of (15.6), i.e.,

$$\begin{aligned} J(i) \to J_{\min} &\triangleq \mathsf{E}\,|\boldsymbol{d} - \boldsymbol{u}w^o|^2 \\ &= \sigma_d^2 - R_{ud} R_u^{-1} R_{du} \end{aligned} \tag{15.7}$$

where $\sigma_d^2 = \mathsf{E}\,|\boldsymbol{d}|^2$.

Obviously, the behavior of the weight estimates w_i that are generated by the stochastic-gradient approximation (15.2) is more complex than the behavior of the weight estimates w_i that are generated by the steepest-descent method (15.1). This is because the w_i from (15.2) need not converge to w^o anymore due to gradient noise. It is the purpose of Parts IV (*Mean-Square Performance*) and V (*Transient Performance*) to examine the effect of gradient noise on filter performance, not only for LMS but also for several other adaptive filters.

Stochastic Equations

First, however, in all such performance studies, it is necessary to regard (or treat) the update equation of an adaptive filter as a *stochastic* difference equation rather than a *deterministic* difference equation. What this means is that we need to regard the variables that appear in an update equation of the form (15.2) as *random variables*. Recall our convention in this book that random variables are represented in boldface, while realizations of random variables are represented in normal font.

For this reason, we shall write from now on $\{\boldsymbol{d}(i), \boldsymbol{u}_i\}$, with boldface letters, instead of $\{d(i), u_i\}$. The notation $\boldsymbol{d}(i)$ refers to a zero-mean random variable with variance σ_d^2, while $\boldsymbol{u}_i$ denotes a zero-mean row vector with covariance matrix R_u,

$$\mathsf{E}\,|\boldsymbol{d}(i)|^2 = \sigma_d^2, \quad \mathsf{E}\,\boldsymbol{u}_i^*\boldsymbol{u}_i = R_u, \quad \mathsf{E}\,\boldsymbol{d}(i)\boldsymbol{u}_i^* = R_{du}$$

In the same vein, we shall replace the weight estimates w_i and w_{i-1} in the update equation of an adaptive algorithm by $\boldsymbol{w}_i$ and $\boldsymbol{w}_{i-1}$, respectively, since, by being functions of $\{\boldsymbol{d}(i), \boldsymbol{u}_i\}$, they become random variables as well.

In this way, the stochastic equation that corresponds to the LMS filter (15.2) would be

$$\boldsymbol{w}_i = \boldsymbol{w}_{i-1} + \mu \boldsymbol{u}_i^*[\boldsymbol{d}(i) - \boldsymbol{u}_i \boldsymbol{w}_{i-1}], \qquad \text{(a stochastic equation)}$$

with the initial condition $\boldsymbol{w}_{-1}$ also treated as a random vector. When this equation is implemented as an adaptive algorithm, it would operate on observations $\{d(i), u_i\}$ of the random quantities $\{\boldsymbol{d}(i), \boldsymbol{u}_i\}$, in which case the stochastic equation would be replaced by our earlier *deterministic* description (15.2) for LMS, namely,

$$w_i = w_{i-1} + \mu u_i^*[d(i) - u_i w_{i-1}], \qquad \text{(a deterministic equation)}$$

Similar considerations are valid for the update equations of all other adaptive algorithms. In all of them, we replace the deterministic quantities $\{d(i), u_i, w_i, w_{i-1}\}$ by random quantities $\{\boldsymbol{d}(i), \boldsymbol{u}_i, \boldsymbol{w}_i, \boldsymbol{w}_{i-1}\}$ in order to obtain the corresponding stochastic equations.

Excess Mean-Square Error and Misadjustment

Now in order to compare the performance of adaptive filters, it is customary to adopt a *common* performance measure across filters (even though different filters may have been derived by minimizing different cost functions). The criterion that is most widely used in the literature of adaptive filtering is the *steady-state mean-square error* (MSE) criterion, which is defined as

$$\boxed{\mathsf{MSE} \triangleq \lim_{i\to\infty} \mathsf{E}\,|\boldsymbol{e}(i)|^2} \tag{15.8}$$

where $\boldsymbol{e}(i)$ denotes the *a priori* output estimation error,

$$\boxed{\boldsymbol{e}(i) \triangleq \boldsymbol{d}(i) - \boldsymbol{u}_i \boldsymbol{w}_{i-1}} \tag{15.9}$$

Obviously, if the weight estimator $\boldsymbol{w}_{i-1}$ in (15.9) is replaced by the optimal solution w^o of (15.6), then the value of the MSE would coincide with the minimum cost (15.7), namely,

$$J_{\min} = \sigma_d^2 - R_{ud} R_u^{-1} R_{du}$$

For this reason, it is common to define the *excess-mean-square error* (EMSE) of an adaptive filter as the difference

$$\boxed{\mathsf{EMSE} \triangleq \mathsf{MSE} - J_{\min}} \tag{15.10}$$

It is also common to define a relative measure of performance, called *misadjustment*, as

$$\boxed{\mathcal{M} \triangleq \mathsf{EMSE}/J_{\min}} \tag{15.11}$$

15.2 STATIONARY DATA MODEL

Therefore, given a stochastic difference equation describing an adaptive filter, we are interested in evaluating its EMSE. In order to pursue this objective, and in order to facilitate the ensuing performance analysis, we also need to adopt a model for the data $\{\boldsymbol{d}(i), \boldsymbol{u}_i\}$.

To begin with, recall from the orthogonality principle of linear least-mean-squares estimation (cf. Thm. 4.1) that the solution w^o of (15.6) satisfies the uncorrelatedness property

$$\mathsf{E}\,\boldsymbol{u}_i^*(\boldsymbol{d}(i) - \boldsymbol{u}_i w^o) = 0$$

Let $\boldsymbol{v}(i)$ denote the estimation error (residual), i.e.,

$$\boldsymbol{v}(i) = \boldsymbol{d}(i) - \boldsymbol{u}_i w^o$$

Then we can re-express this result by saying that $\{\boldsymbol{d}(i), \boldsymbol{u}_i\}$ are related via

$$\boldsymbol{d}(i) = \boldsymbol{u}_i w^o + \boldsymbol{v}(i) \tag{15.12}$$

in terms of a signal $\boldsymbol{v}(i)$ that is uncorrelated with $\boldsymbol{u}_i$. The variance of $\boldsymbol{v}(i)$ is obviously equal to the minimum cost J_{min} from (15.7), i.e.,

$$\sigma_v^2 \triangleq \mathsf{E}\,|\boldsymbol{v}(i)|^2 = J_{\text{min}} = \sigma_d^2 - R_{ud}R_u^{-1}R_{du} \tag{15.13}$$

Linear Regression Model

What the above argument shows is that given any random variables $\{\boldsymbol{d}(i), \boldsymbol{u}_i\}$ with second-order moments $\{\sigma_d^2, R_u, R_{du}\}$, we can always assume that $\{\boldsymbol{d}(i), \boldsymbol{u}_i\}$ are related via a linear model of the form (15.12), for some w^o, with the variable $\boldsymbol{v}(i)$ playing the role of a disturbance that is uncorrelated with $\boldsymbol{u}_i$, i.e.,

$$\boxed{\boldsymbol{d}(i) = \boldsymbol{u}_i w^o + \boldsymbol{v}(i), \quad \mathsf{E}\,|\boldsymbol{v}(i)|^2 = \sigma_v^2, \quad \mathsf{E}\,\boldsymbol{v}(i)\boldsymbol{u}_i^* = 0} \tag{15.14}$$

However, in order to make the performance analyses of adaptive filters more tractable, we usually need to adopt the stronger *assumption* that

$$\boxed{\text{The sequence } \{\boldsymbol{v}(i)\} \text{ is i.i.d. and independent of all } \{\boldsymbol{u}_j\}} \tag{15.15}$$

Here, the notation i.i.d. stands for "independent and identically distributed". Condition (15.15) on $\boldsymbol{v}(i)$ is an assumption because, as explained above, the signal $\boldsymbol{v}(i)$ in (15.14) is only uncorrelated with $\boldsymbol{u}_i$; it is not necessarily independent of $\boldsymbol{u}_i$, or of all $\{\boldsymbol{u}_j\}$ for that matter. Still, there are situations when conditions (15.14) and (15.15) hold simultaneously, e.g., in the channel estimation application of Sec. 10.5. In that application, it is usually justified to expect the noise sequence $\{\boldsymbol{v}(i)\}$ to be i.i.d. and independent of all other data, including the regression data.

Given the above, we shall therefore adopt the following data model in our studies of the performance of adaptive filters. We shall assume that the data $\{\boldsymbol{d}(i), \boldsymbol{u}_i\}$ satisfy the following conditions:

(15.16)

- (a) There exists a vector w^o such that $\boldsymbol{d}(i) = \boldsymbol{u}_i w^o + \boldsymbol{v}(i)$.
- (b) The noise sequence $\{\boldsymbol{v}(i)\}$ is i.i.d. with variance $\sigma_v^2 = \mathsf{E}\,|\boldsymbol{v}(i)|^2$.
- (c) The noise sequence $\{\boldsymbol{v}(i)\}$ is independent of $\boldsymbol{u}_j$ for all i, j.
- (d) The initial condition $\boldsymbol{w}_{-1}$ is independent of all $\{\boldsymbol{d}(j), \boldsymbol{u}_j, \boldsymbol{v}(j)\}$.
- (e) The regressor covariance matrix is $R_u = \mathsf{E}\,\boldsymbol{u}_i^*\boldsymbol{u}_i > 0$.
- (f) The random variables $\{\boldsymbol{d}(i), \boldsymbol{v}(i), \boldsymbol{u}_i\}$ have zero means.

We refer to the above model as describing a *stationary* environment, i.e., an environment with constant quantities $\{w^o, R_u, \sigma_v^2\}$. In Chapter 20 we shall modify the model in order to accommodate filter operation in *nonstationary* environments where, for example, w^o will be time-variant — see Sec. 20.2.

Useful Independence Results

An important consequence of the data model (15.16) is that, at any particular time instant i, the noise variable $\boldsymbol{v}(i)$ will be independent of all previous weight estimators $\{\boldsymbol{w}_j, j < i\}$. This fact follows easily from examining the update equation of an adaptive filter. Consider, for instance, the LMS recursion

$$\boldsymbol{w}_i = \boldsymbol{w}_{i-1} + \mu \boldsymbol{u}_i^* [\boldsymbol{d}(i) - \boldsymbol{u}_i \boldsymbol{w}_{i-1}], \quad \boldsymbol{w}_{-1} = \text{initial condition}$$

By iterating the recursion we find that, at any time instant j, the weight estimator $\boldsymbol{w}_j$ can be expressed as a function of $\boldsymbol{w}_{-1}$, the reference signals $\{\boldsymbol{d}(j), \boldsymbol{d}(j-1), \ldots, \boldsymbol{d}(0)\}$, and the regressors $\{\boldsymbol{u}_j, \boldsymbol{u}_{j-1}, \ldots, \boldsymbol{u}_0\}$. We denote this dependency generically as

$$\boldsymbol{w}_j = \mathcal{F}[\, \boldsymbol{w}_{-1};\ \boldsymbol{d}(j), \boldsymbol{d}(j-1), \ldots, \boldsymbol{d}(0);\ \boldsymbol{u}_j, \boldsymbol{u}_{j-1}, \ldots, \boldsymbol{u}_0 \,] \tag{15.17}$$

for some function $\mathcal{F}$. A similar dependency holds for other adaptive schemes.

Now $\boldsymbol{v}(i)$ can be seen to be independent of each one of the terms appearing as an argument of $\mathcal{F}$ in (15.17), so that $\boldsymbol{v}(i)$ will be independent of $\boldsymbol{w}_j$ for all $j < i$. Indeed, the independence of $\boldsymbol{v}(i)$ from $\{\boldsymbol{w}_{-1}, \boldsymbol{u}_j, \ldots, \boldsymbol{u}_0\}$ is obvious by assumption, while its independence from $\{\boldsymbol{d}(j), \ldots, \boldsymbol{d}(0)\}$ can be seen as follows. Consider $\boldsymbol{d}(j)$ for example. Then from $\boldsymbol{d}(j) = \boldsymbol{u}_j w^o + \boldsymbol{v}(j)$ we see that $\boldsymbol{d}(j)$ is a function of $\{\boldsymbol{u}_j, \boldsymbol{v}(j)\}$, both of which are independent of $\boldsymbol{v}(i)$.

Given that $\boldsymbol{v}(i)$ is independent of $\{\boldsymbol{w}_j, j < i\}$, it also follows that $\boldsymbol{v}(i)$ is independent of $\{\tilde{\boldsymbol{w}}_j, j < i\}$, where $\tilde{\boldsymbol{w}}_j$ denotes the weight-error vector:

$$\boxed{\tilde{\boldsymbol{w}}_j \overset{\Delta}{=} w^o - \boldsymbol{w}_j}$$

Moreover, $\boldsymbol{v}(i)$ is also independent of the *a priori* estimation error $\boldsymbol{e}_a(i)$, defined by

$$\boxed{\boldsymbol{e}_a(i) \overset{\Delta}{=} \boldsymbol{u}_i \tilde{\boldsymbol{w}}_{i-1}}$$

This variable measures the difference between $\boldsymbol{u}_i w^o$ and $\boldsymbol{u}_i \boldsymbol{w}_{i-1}$, i.e., it measures how close the estimator $\boldsymbol{u}_i \boldsymbol{w}_{i-1}$ is to the optimal linear estimator of $\boldsymbol{d}(i)$, namely $\hat{\boldsymbol{d}}(i) = \boldsymbol{u}_i w^o$. We summarize the above independence properties in the following statement.

Lemma 15.1 (Useful properties) From the data model (15.16), it follows that $\boldsymbol{v}(i)$ is independent of each of the following:

$$\{\boldsymbol{w}_j \text{ for } j < i\}, \quad \{\tilde{\boldsymbol{w}}_j \text{ for } j < i\}, \quad \text{and} \quad \boldsymbol{e}_a(i) = \boldsymbol{u}_i \tilde{\boldsymbol{w}}_{i-1}$$

Alternative Expression for the EMSE

Using model (15.16), and the independence results of Lemma 15.1, we can determine a more compact expression for the EMSE of an adaptive filter. Recall from (15.8) and (15.10) that, by definition,

$$\text{EMSE} = \lim_{i \to \infty} \mathsf{E}\, |\boldsymbol{e}(i)|^2 - J_{\min} \tag{15.18}$$

where, as we already know from (15.13), $J_{\min} = \sigma_v^2$. Using

$$e(i) = d(i) - u_i w_{i-1}$$

and the linear model (15.16), we find that

$$e(i) = v(i) + u_i(w^o - w_{i-1})$$

That is,

$$\boxed{e(i) = v(i) + e_a(i)} \tag{15.19}$$

Now the independence of $v(i)$ and $e_a(i)$, as stated in Lemma 15.1, gives

$$\mathsf{E}\,|e(i)|^2 \;=\; \mathsf{E}\,|v(i)|^2 \;+\; \mathsf{E}\,|e_a(i)|^2 \;=\; \sigma_v^2 \;+\; \mathsf{E}\,|e_a(i)|^2 \tag{15.20}$$

so that substituting into (15.18) we get

$$\boxed{\mathsf{EMSE} = \lim_{i\to\infty} \mathsf{E}\,|e_a(i)|^2} \tag{15.21}$$

In other words, the EMSE can be computed by evaluating the steady-state mean-square value of the *a priori* estimation error $e_a(i)$. We shall use this alternative representation to evaluate the EMSE of several adaptive algorithms in this chapter. Likewise, from (15.8) and (15.20), we have

$$\boxed{\mathsf{MSE} \;=\; \mathsf{EMSE} \;+\; \sigma_v^2} \tag{15.22}$$

Error Quantities

For ease of reference, we collect in Table 15.1 the definitions of several error measures. Note that we refer to both $\{e(i), e_a(i)\}$ as *a priori* errors since they rely on w_{i-1}, while we refer to both $\{r(i), e_p(i)\}$ as *a posteriori* errors since they rely on the updated weight estimator w_i. Note also that we distinguish between $e(i)$ and $e_a(i)$ by referring to the first one as *output* error. In the sequel, the estimation errors $\{e_a(i), e_p(i)\}$ will play a prominent role.

TABLE 15.1 Definitions of several estimation errors.

Error	Definition	Interpretation
$e(i)$	$d(i) - u_i w_{i-1}$	*a priori* output estimation error
$r(i)$	$d(i) - u_i w_i$	*a posteriori* output estimation error
$\tilde{w}_i$	$w^o - w_i$	weight error vector
$e_a(i)$	$u_i \tilde{w}_{i-1}$	*a priori* estimation error
$e_p(i)$	$u_i \tilde{w}_i$	*a posteriori* estimation error

15.3 ENERGY CONSERVATION RELATION

Our approach to the performance analysis of adaptive filters in Parts IV (*Mean-Square Performance*) and V (*Transient Performance*) is based on an energy conservation relation that holds for general data $\{\boldsymbol{d}(i), \boldsymbol{u}_i\}$ (it does *not* even require the assumptions (15.16)).

In order to motivate this relation, we consider adaptive filter updates of the generic form:

$$\boxed{\boldsymbol{w}_i = \boldsymbol{w}_{i-1} + \mu\, \boldsymbol{u}_i^*\, g[\boldsymbol{e}(i)], \quad \boldsymbol{w}_{-1} = \text{initial condition}} \tag{15.23}$$

where $g[\cdot]$ denotes some function of the *a priori* output error signal,

$$\boldsymbol{e}(i) = \boldsymbol{d}(i) - \boldsymbol{u}_i \boldsymbol{w}_{i-1}$$

Updates of this form are said to correspond to filters with *error nonlinearities*. Several of the adaptive algorithms that we introduced in Chapter 10 are special cases of (15.23) with proper selection of the error function $g[\cdot]$, as shown in Table 15.2. Observe that the listing in the table excludes leaky-LMS, RLS, APA, and CMA algorithms. We shall study these algorithms later by following a similar procedure — see, e.g., Chapter 19 and Probs. IV.7, IV.16–IV.18 and V.29–V.30. We can also study update equations with *data* (as opposed to error) nonlinearities, say, of the form

$$\boldsymbol{w}_i = \boldsymbol{w}_{i-1} + \mu\, g[\boldsymbol{u}_i]\boldsymbol{u}_i^* \boldsymbol{e}(i), \quad \boldsymbol{w}_{-1} = \text{initial condition}$$

for some positive function $g[\cdot]$ of the regression data, $g[\boldsymbol{u}_i] > 0$ (the function $g[\cdot]$ could also be matrix-valued). For example, $\epsilon-$NLMS is a special case of this class of filters by choosing $g[\boldsymbol{u}_i] = 1/(\epsilon + \|\boldsymbol{u}_i\|^2)$. Likewise, LMS is a special case by choosing $g[\boldsymbol{u}_i] = 1$. Most of the results in this chapter, especially those pertaining to the energy-conservation relation of this section, apply to filters with data nonlinearities — see, e.g., Probs. IV.9 and V.22. In the sequel we focus on updates with error nonlinearities.

TABLE 15.2 Examples of error functions for several adaptive algorithms: ϵ is a small positive number, $0 \leq \delta \leq 1$, and $\boldsymbol{p}(i)$ is a positive quantity whose computation will be explained later.

Algorithm	**Error function**
LMS	$g[\boldsymbol{e}(i)] = \boldsymbol{e}(i)$
ϵ-NLMS	$g[\boldsymbol{e}(i)] = \boldsymbol{e}(i)/(\epsilon + \|\boldsymbol{u}_i\|^2)$
ϵ-NLMS with power normalization	$g[\boldsymbol{e}(i)] = \boldsymbol{e}(i)/(\epsilon + \boldsymbol{p}(i))$
LMF	$g[\boldsymbol{e}(i)] = \boldsymbol{e}(i)\|\boldsymbol{e}(i)\|^2$
LMMN	$g[\boldsymbol{e}(i)] = \boldsymbol{e}(i)\left[\delta + (1-\delta)\|\boldsymbol{e}(i)\|^2\right]$
sign-error LMS	$g[\boldsymbol{e}(i)] = \text{csgn}[\boldsymbol{e}(i)]$

The update recursion (15.23) can be rewritten in terms of the weight-error vector

$$\tilde{\boldsymbol{w}}_i = w^o - \boldsymbol{w}_i$$

Subtracting both sides of (15.23) from w^o we get

$$w^o - \boldsymbol{w}_i = w^o - \boldsymbol{w}_{i-1} - \mu\, \boldsymbol{u}_i^*\, g[\boldsymbol{e}(i)], \quad \tilde{\boldsymbol{w}}_{-1} = \text{initial condition}$$

or, equivalently,

$$\boxed{\tilde{\boldsymbol{w}}_i = \tilde{\boldsymbol{w}}_{i-1} - \mu \boldsymbol{u}_i^* g[\boldsymbol{e}(i)]} \tag{15.24}$$

In addition, if we multiply both sides of (15.24) by $\boldsymbol{u}_i$ from the left we find that the *a priori* and *a posteriori* estimation errors $\{\boldsymbol{e}_a(i), \boldsymbol{e}_p(i)\}$ are related via:

$$\boxed{\boldsymbol{e}_p(i) = \boldsymbol{e}_a(i) - \mu \|\boldsymbol{u}_i\|^2 g[\boldsymbol{e}(i)]} \tag{15.25}$$

where $\{\boldsymbol{e}_p(i), \boldsymbol{e}_a(i)\}$ were defined in Table 15.1 as

$$\boxed{\boldsymbol{e}_a(i) = \boldsymbol{u}_i \tilde{\boldsymbol{w}}_{i-1}, \quad \boldsymbol{e}_p(i) = \boldsymbol{u}_i \tilde{\boldsymbol{w}}_i} \tag{15.26}$$

Expressions (15.24)–(15.25) provide an alternative description of the adaptive filter (15.23) in terms of the error quantities $\{\boldsymbol{e}_a(i), \boldsymbol{e}_p(i), \tilde{\boldsymbol{w}}_i, \tilde{\boldsymbol{w}}_{i-1}, \boldsymbol{e}(i)\}$. This description is useful since we are often interested in questions related to the behavior of these errors, such as:

1. Steady-state behavior, which relates to determining the steady-state values of $\mathsf{E}\,\|\tilde{\boldsymbol{w}}_i\|^2$, $\mathsf{E}\,|\boldsymbol{e}_a(i)|^2$, and $\mathsf{E}\,|\boldsymbol{e}(i)|^2$.

2. Stability, which relates to determining the range of values of the step-size μ over which the variances $\mathsf{E}\,|\boldsymbol{e}_a(i)|^2$ and $\mathsf{E}\,\|\tilde{\boldsymbol{w}}_i\|^2$ remain bounded.

3. Transient behavior, which relates to studying the time evolution of the curves $\mathsf{E}\,|\boldsymbol{e}_a(i)|^2$ and $\{\mathsf{E}\,\tilde{\boldsymbol{w}}_i, \mathsf{E}\,\|\tilde{\boldsymbol{w}}_i\|^2\}$.

In order to address questions of this kind, we shall rely on an energy equality that relates the squared norms of the errors $\{\boldsymbol{e}_a(i), \boldsymbol{e}_p(i), \tilde{\boldsymbol{w}}_{i-1}, \tilde{\boldsymbol{w}}_i\}$.

Algebraic Derivation

To derive the energy relation, we first combine (15.24)–(15.25) to eliminate the error nonlinearity $g[\cdot]$ from (15.24), i.e., we solve for $g[\cdot]$ from (15.25) and then substitute into (15.24), as done below. What this initial step means is that the resulting energy relation will hold irrespective of the error nonlinearity. We distinguish between two cases:

1. $\boldsymbol{u}_i = 0$. This is a degenerate situation. In this case, it is obvious from (15.24) and (15.25) that $\tilde{\boldsymbol{w}}_i = \tilde{\boldsymbol{w}}_{i-1}$ and $\boldsymbol{e}_a(i) = \boldsymbol{e}_p(i)$ so that

$$\|\tilde{\boldsymbol{w}}_i\|^2 = \|\tilde{\boldsymbol{w}}_{i-1}\|^2 \quad \text{and} \quad |\boldsymbol{e}_a(i)|^2 = |\boldsymbol{e}_p(i)|^2 \tag{15.27}$$

2. $\boldsymbol{u}_i \neq 0$. In this case, we use (15.25) to solve for $g[\boldsymbol{e}(i)]$,

$$g[\boldsymbol{e}(i)] = \frac{1}{\mu \|\boldsymbol{u}_i\|^2}[\boldsymbol{e}_a(i) - \boldsymbol{e}_p(i)]$$

and substitute into (15.24) to obtain

$$\tilde{\boldsymbol{w}}_i = \tilde{\boldsymbol{w}}_{i-1} - \frac{\boldsymbol{u}_i^*}{\|\boldsymbol{u}_i\|^2}[\boldsymbol{e}_a(i) - \boldsymbol{e}_p(i)] \tag{15.28}$$

This relation involves the four errors $\{\tilde{w}_i, \tilde{w}_{i-1}, e_a(i), e_p(i)\}$; observe that even the step-size μ is cancelled out. Expression (15.28) can be rearranged as

$$\tilde{w}_i + \frac{u_i^*}{\|u_i\|^2} e_a(i) = \tilde{w}_{i-1} + \frac{u_i^*}{\|u_i\|^2} e_p(i) \tag{15.29}$$

On each side of this identity we have a combination of *a priori* and *a posteriori* errors. By evaluating the energies (i.e., the squared Euclidean norms) of both sides we find, after a straightforward calculation, that the following energy equality holds:

$$\|\tilde{w}_i\|^2 + \frac{1}{\|u_i\|^2}|e_a(i)|^2 = \|\tilde{w}_{i-1}\|^2 + \frac{1}{\|u_i\|^2}|e_p(i)|^2 \tag{15.30}$$

Interesting enough, this equality simply amounts to adding the energies of the individual terms of (15.29); the cross-terms cancel out. This is one advantage of working with the energy relation (15.30): irrelevant cross-terms are eliminated so that one does not need to worry later about evaluating their expectations.

The results in both cases of zero and nonzero regression vectors can be combined together by using a common notation. Define $\bar{\mu}(i) = \left(\|u_i\|^2\right)^\dagger$, in terms of the pseudo-inverse operation. Recall that the pseudo-inverse of a nonzero scalar is equal to its inverse value, while the pseudo-inverse of zero is equal to zero. That is,

$$\boxed{\bar{\mu}(i) \triangleq \begin{cases} 1/\|u_i\|^2 & \text{if } u_i \neq 0 \\ 0 & \text{otherwise} \end{cases}} \tag{15.31}$$

Using $\bar{\mu}(i)$, we can combine (15.27) and (15.30) into a single identity as

$$\boxed{\|\tilde{w}_i\|^2 + \bar{\mu}(i)|e_a(i)|^2 = \|\tilde{w}_{i-1}\|^2 + \bar{\mu}(i)|e_p(i)|^2} \tag{15.32}$$

We can alternatively express (15.32) as

$$\|u_i\|^2 \cdot \|\tilde{w}_i\|^2 \ + \ |e_a(i)|^2 = \|u_i\|^2 \cdot \|\tilde{w}_{i-1}\|^2 \ + \ |e_p(i)|^2$$

Theorem 15.1 (Energy conservation relation) For adaptive filters of the form (15.23), and for any data $\{d(i), u_i\}$, it always holds that

$$\|\tilde{w}_i\|^2 + \bar{\mu}(i)|e_a(i)|^2 = \|\tilde{w}_{i-1}\|^2 + \bar{\mu}(i)|e_p(i)|^2$$

where $e_a(i) = u_i\tilde{w}_{i-1}$, $e_p(i) = u_i\tilde{w}_i$, $\tilde{w}_i = w^o - w_i$, and $\bar{\mu}(i)$ is defined as in (15.31).

The important fact to emphasize here is that *no approximations* have been used to establish the energy relation (15.32); it is an exact relation that shows how the energies of the weight-error vectors at two successive time instants are related to the energies of the *a priori* and *a posteriori* estimation errors.

Remark 15.1 (Interpretations of the energy relation) In App. 15.A we provide several interpretations for the energy relation (15.32): one interpretation is geometric in terms of vector projections, a second interpretation is physical and relates to Snell's law for light propagation, and a third interpretation is system-theoretic and relates to feedback concepts. These interpretations are *not* needed for the subsequent derivations in this chapter. Nevertheless, they provide the reader with some insights into the energy relation.

◇

15.4 VARIANCE RELATION

Relation (15.32) has important ramifications in the study of adaptive filters. In this chapter, and the remaining chapters of this part, we shall focus on its significance to the steady-state performance, tracking analysis, and finite-precision analysis of adaptive filters. In Part V (*Transient Performance*) we shall apply it to transient analysis, and in Part XI (*Robust Filters*) we shall examine its significance to robustness analysis. In the course of these discussions, it will become clear that the energy-conservation relation (15.32) provides a unifying framework for the performance analysis of adaptive filters.

With regards to steady-state performance, which is the subject matter of this chapter, it has been common in the literature to study the steady-state performance of an adaptive filter as the limiting behavior of its transient performance (which is concerned with the study of the time evolution of $\mathsf{E}\,\|\tilde{\boldsymbol{w}}_i\|^2$). As we shall see in Part V (*Transient Performance*), transient analysis is a more demanding task to pursue and it tends to require a handful of additional assumptions and restrictions on the data. In this way, steady-state results that are obtained as the limiting behavior of a transient analysis would be governed by the same restrictions on the data. In our treatment, on the other hand, we separate the study of the steady-state performance of an adaptive filter from the study of its transient performance. In so doing, it becomes possible to pursue the steady-state analysis in several instances under weaker assumptions than those required by a full blown transient analysis.

Steady-State Filter Operation

To initiate our steady-state performance studies, we first explain what is meant by an adaptive filter operating in *steady-state*.

Definition 15.1 (Steady-state filter) An adaptive filter will be said to operate in steady-state if it holds that

$$\begin{aligned} \mathsf{E}\,\tilde{\boldsymbol{w}}_i &\longrightarrow s, \quad \text{as } i \to \infty \qquad (15.33) \\ \mathsf{E}\,\tilde{\boldsymbol{w}}_i\tilde{\boldsymbol{w}}_i^* &\longrightarrow C, \quad \text{as } i \to \infty \qquad (15.34) \end{aligned}$$

where s and C are some finite constants (usually $s = 0$).

In other words, the mean and covariance matrix of the weight error vector of a steady-state filter tend to some finite constant values $\{s, C\}$. In particular, it follows that the following condition holds

$$\mathsf{E}\,\|\tilde{\boldsymbol{w}}_i\|^2 = \mathsf{E}\,\|\tilde{\boldsymbol{w}}_{i-1}\|^2 = c, \quad \text{as } i \to \infty \qquad (15.35)$$

where $c = \text{Tr}(C)$. Of course, not every adaptive filter implementation can be guaranteed to reach steady-state operation. For example, as we shall see later in Chapter 24, if the step-size μ in an LMS implementation is not sufficiently small, the filter can diverge with the error signals $e_a(i)$ and $\tilde{\boldsymbol{w}}_i$ growing unbounded. We shall have much more to say in Part V (*Transient Performance*) about conditions on the step-size parameter in order to guarantee filter convergence to steady-state (i.e., stability). The main conclusion of Part V (*Transient Performance*) will be that sufficiently small step-sizes in general guarantee the convergence of adaptive filters to steady-state operation. By assuming in the current chapter that a filter is operating in steady-state, we are in effect attempting to quantify beforehand the performance that can be expected from the filter once it reaches steady-state. Such qualifications are useful at the design stage.

Variance Relation for Steady-State Performance

In order to explain how (15.32) is useful in evaluating the steady-state performance of an adaptive filter, we recall from (15.21) that we are interested in evaluating the steady-state variance of $e_a(i)$. Now taking expectations of both sides of (15.32) we get

$$\boxed{\mathsf{E}\,\|\tilde{w}_i\|^2 + \mathsf{E}\,\bar{\mu}(i)|e_a(i)|^2 = \mathsf{E}\,\|\tilde{w}_{i-1}\|^2 + \mathsf{E}\,\bar{\mu}(i)|e_p(i)|^2} \tag{15.36}$$

where the expectation is with respect to the distributions of the random variables $\{d(i), u_i\}$. Taking the limit of (15.36) as $i \to \infty$ and using the steady-state condition (15.35), we obtain

$$\mathsf{E}\,\bar{\mu}(i)|e_a(i)|^2 = \mathsf{E}\,\bar{\mu}(i)|e_p(i)|^2, \quad \text{as } i \to \infty \tag{15.37}$$

This equality is in terms of $\{e_a(i), e_p(i)\}$. However, from (15.25) we know how $e_p(i)$ is related to $e_a(i)$. Substituting into (15.37) we get

$$\mathsf{E}\,\bar{\mu}(i)|e_a(i)|^2 = \mathsf{E}\,\bar{\mu}(i)\left|e_a(i) - \mu\|u_i\|^2 g[e(i)]\right|^2, \quad \text{as } i \to \infty \tag{15.38}$$

Expanding the term on the right-hand side and simplifying leads to (we are omitting the argument of g for compactness of notation):

$$\begin{aligned} \bar{\mu}(i)\left|e_a(i) - \mu\|u_i\|^2 g\right|^2 &= \bar{\mu}(i)|e_a(i)|^2 + \mu^2\|u_i\|^2|g|^2 - \mu e_a(i)g^* - \mu e_a^*(i)g \\ &= \bar{\mu}(i)|e_a(i)|^2 + \mu^2\|u_i\|^2|g|^2 - 2\mu\mathsf{Re}\left(e_a^*(i)g\right) \end{aligned} \tag{15.39}$$

where in the first equality we used the fact that

$$\bar{\mu}(i)\|u_i\|^4 = \|u_i\|^2 \quad \text{and} \quad \bar{\mu}(i)\|u_i\|^2 e_a(i)g^* = e_a(i)g^*$$

for all u_i (whether $u_i = 0$ or otherwise). Taking expectation of the right-hand side of (15.39) and substituting into (15.38) we obtain

$$\boxed{\mu\mathsf{E}\left(\|u_i\|^2 \cdot |g[e(i)]|^2\right) = 2\mathsf{Re}\left(\mathsf{E}\,e_a^*(i)g[e(i)]\right), \quad \text{as } i \to \infty} \tag{15.40}$$

in terms of the real part of $e_a^*(i)g[e(i)]$. Alternatively, this result can be obtained by starting directly from the weight-error vector recursion (15.24) and equating the squared Euclidean norms of both sides, i.e.,

$$\begin{aligned} \|\tilde{w}_i\|^2 &= \|\tilde{w}_{i-1} - \mu u_i^* g\|^2 \\ &= \|\tilde{w}_{i-1}\|^2 + \mu^2\|u_i\|^2|g|^2 - 2\mu\mathsf{Re}(e_a^*(i)g) \end{aligned}$$

Taking expectations of both sides as $i \to \infty$ gives (15.40). This alternative argument masks the presence of the useful energy conservation relation (15.32). In summary, we find that under expectation and in steady-state, the energy relation (15.32) leads to the following equivalent form.

Theorem 15.2 (Variance relation) For adaptive filters of the form (15.23) and for any data $\{d(i), u_i\}$, assuming filter operation in steady-state, the following relation holds:

$$\mu \mathsf{E}\left(\|u_i\|^2 \cdot |g[e(i)]|^2\right) = 2\mathsf{Re}\left(\mathsf{E}\, e_a^*(i) g[e(i)]\right), \quad \text{as } i \to \infty$$

For real-valued data, this variance relation becomes

$$\mu \mathsf{E}\left(\|u_i\|^2 \cdot g^2[e(i)]\right) = 2\mathsf{E}\, e_a(i) g[e(i)], \quad \text{as } i \to \infty$$

We remark again that the variance relation (15.40) is *exact*, since it holds without any approximations or assumptions (except for the assumption that the filter is operating in steady-state, which is necessary if one is interested in evaluating the steady-state performance of a filter). We refer to (15.40) as a variance relation since it will be our starting point for evaluating the variance $\mathsf{E}\,|e_a(i)|^2$ for different adaptive filters. The results of both Thms. 15.1 and 15.2 do *not* require the analysis model (15.16); they hold for general data $\{d(i), u_i\}$.

Relevance to Mean-Square Performance Analysis

However, when model (15.16) is assumed, then we know from (15.19) that $e(i)$ can be expressed in terms of $e_a(i)$ as

$$e(i) = e_a(i) + v(i)$$

In this way, relation (15.40) becomes an identity involving $e_a(i)$ and, in principle, it could be solved to evaluate the EMSE of an adaptive filter, i.e., to compute $\mathsf{E}\,|e_a(\infty)|^2$. We say "in principle" because, although (15.40) is an exact result, different choices for the error function $g[\cdot]$ can make the solution for $\mathsf{E}\,|e_a(\infty)|^2$ easier for some cases than others. It is at this stage that simplifying assumptions become necessary. We shall illustrate this point for several adaptive filters in the sections that follow.

In order to simplify the notation, we shall employ the symbol ζ to refer to the EMSE of an adaptive filter, i.e.,

$$\zeta = \mathsf{E}\,|e_a(\infty)|^2$$

For example, the EMSE of LMS will be denoted by ζ^{LMS}. Its misadjustment will be denoted by $\mathcal{M}^{\text{LMS}}$. Similar notation will be used for other algorithms. In view of the analysis model (15.16), which enables us to identify $J_{\min}$ as σ_v^2, we obtain from (15.11) that the misadjustment of an adaptive filter is related to its EMSE via

$$\boxed{\mathcal{M} = \text{EMSE}/\sigma_v^2}$$

We limit our derivations in the sequel to determining expressions for the EMSE of several adaptive filters. Expressions for the misadjustment would follow by dividing the result by σ_v^2.

15.A APPENDIX: ENERGY RELATION INTERPRETATIONS

We end this chapter with three interpretations for the energy relation (15.32). The first one is geometric and relates to the projection of vectors onto one another. The second interpretation is physical and relates to Snell's law in optics, and the third interpretation is system-theoretic and relates to feedback concepts. These interpretations provide useful insights into the nature and origin of the energy relation.

Geometric Interpretation

The energy relation (15.32) can be motivated geometrically by observing that every update of the form (15.23) enforces a certain geometric constraint. Specifically, from the weight error recursion (15.24) we have that

$$\tilde{\boldsymbol{w}}_{i-1} - \tilde{\boldsymbol{w}}_i = \mu \boldsymbol{u}_i^* g[\boldsymbol{e}(i)]$$

which shows that the difference $\tilde{\boldsymbol{w}}_{i-1} - \tilde{\boldsymbol{w}}_i$ is parallel to the regression vector $\boldsymbol{u}_i^*$ (we ignore in this argument the degenerate case $\boldsymbol{u}_i = 0$, for which the energy relation trivializes to $\|\tilde{\boldsymbol{w}}_i\|^2 = \|\tilde{\boldsymbol{w}}_{i-1}\|^2$). This property is depicted in Fig. 15.1, which shows the vectors $\{\tilde{\boldsymbol{w}}_{i-1}, \tilde{\boldsymbol{w}}_i, \boldsymbol{u}_i^*\}$ drawn from the origin of the $M-$dimensional space in which they lie. The distances from the vertices of the error vectors $\{\tilde{\boldsymbol{w}}_i, \tilde{\boldsymbol{w}}_{i-1}\}$ to $\boldsymbol{u}_i^*$ should therefore agree. The energy-conservation relation (15.32) is a statement of this fact, as we now elaborate.

First we need to explain the notion of projecting one vector onto another. We shall study such projection problems in great detail in Sec. 29.1 in the context of least-squares problems; at that point much of the discussion below will become self-evident. Here we follow a more elementary exposition.

Let θ denote the *acute* angle between two *real-valued* column vectors $\{x, y\}$. That is, $\theta \in [0, \pi/2]$ and its squared cosine is given by

$$\cos^2(\theta) \triangleq \frac{|y^{\mathsf{T}}x|^2}{\|x\|^2 \cdot \|y\|^2} \tag{15.41}$$

where $\|\cdot\|$ denotes the Euclidean norm of its argument. The ratio $y^{\mathsf{T}}x/(\|x\| \cdot \|y\|)$ always evaluates to a real number that lies within the interval $[-1, 1]$. This fact follows from Cauchy-Schwartz inequality, which states that for any two vectors $\{x, y\}$, it holds that $|y^{\mathsf{T}}x|^2 \leq \|y\|^2 \cdot \|x\|^2$. If we project x onto y (see Fig. 15.2), then the squared norm of this projection is $\|x\|^2\cos^2(\theta)$. Likewise, if we project x onto the orthogonal direction to y, then the squared norm of this projection is $\|x\|^2\sin^2(\theta)$ where

$$\sin^2(\theta) = 1 - \frac{|y^{\mathsf{T}}x|^2}{\|x\|^2 \cdot \|y\|^2}$$

Similar conclusions hold when the vectors $\{x, y\}$ are *complex-valued*. That is, if we project x onto y then the squared norm of this projection would still be given by $\|x\|^2\cos^2(\theta)$ where the term $\cos^2(\theta)$ is now defined as

$$\boxed{\cos^2(\theta) \triangleq \frac{|y^*x|^2}{\|x\|^2 \cdot \|y\|^2}} \tag{15.42}$$

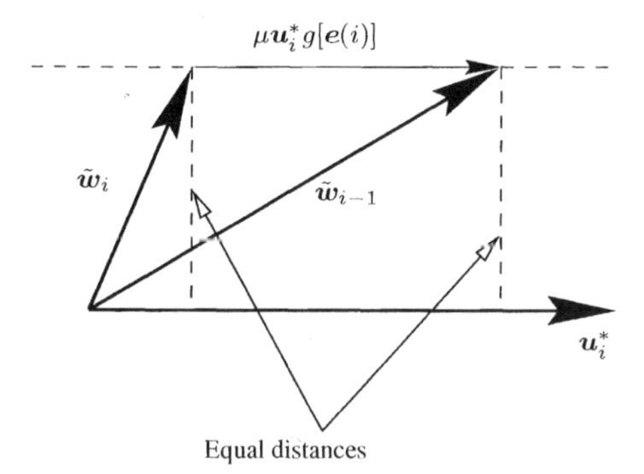

FIGURE 15.1 Geometric interpretation of adaptive updates of the form (15.23). The difference $\tilde{\boldsymbol{w}}_{i-1} - \tilde{\boldsymbol{w}}_i$ remains parallel to $\boldsymbol{u}_i^*$.

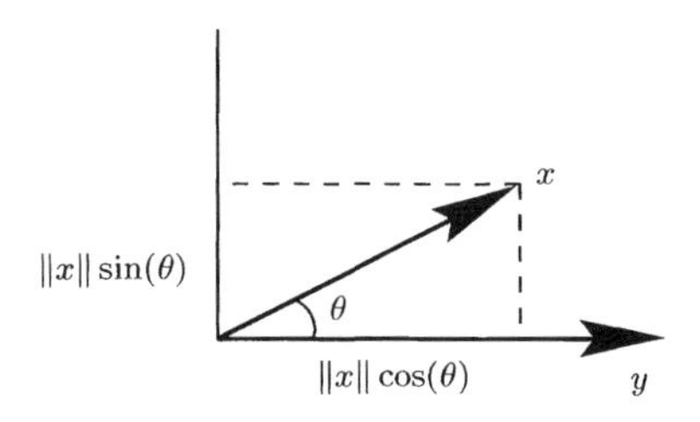

FIGURE 15.2 Projecting x onto y and onto the orthogonal direction to y.

with y^*x replacing $y^\mathsf{T}x$. Likewise, if we project x onto the orthogonal direction to y, then the squared norm of the resulting projection would again be given by $\|x\|^2\sin^2(\theta)$ where now

$$\boxed{\sin^2(\theta) \triangleq 1 - \frac{|y^*x|^2}{\|x\|^2\cdot\|y\|^2}} \tag{15.43}$$

Returning to Fig. 15.2, we therefore find that the squared distance from the endpoint of $\tilde{\boldsymbol{w}}_{i-1}$ to $\boldsymbol{u}_i^*$ is equal to $\|\tilde{\boldsymbol{w}}_{i-1}\|^2\sin^2(\theta_{i-1})$, where (cf. (15.43)):

$$\sin^2(\theta_{i-1}) \triangleq 1 - \frac{|\boldsymbol{u}_i\tilde{\boldsymbol{w}}_{i-1}|^2}{\|\tilde{\boldsymbol{w}}_{i-1}\|^2\cdot\|\boldsymbol{u}_i^*\|^2} = 1 - \frac{|\boldsymbol{e}_a(i)|^2}{\|\tilde{\boldsymbol{w}}_{i-1}\|^2\cdot\|\boldsymbol{u}_i\|^2}$$

and θ_{i-1} is the acute angle between $\tilde{\boldsymbol{w}}_{i-1}$ and $\boldsymbol{u}_i^*$. Similarly, the squared distance from the endpoint of $\tilde{\boldsymbol{w}}_i$ to $\boldsymbol{u}_i^*$ is equal to $\|\tilde{\boldsymbol{w}}_i\|^2\sin^2(\theta_i)$, where

$$\sin^2(\theta_i) \triangleq 1 - \frac{|\boldsymbol{u}_i\tilde{\boldsymbol{w}}_i|^2}{\|\tilde{\boldsymbol{w}}_i\|^2\cdot\|\boldsymbol{u}_i^*\|^2} = 1 - \frac{|\boldsymbol{e}_p(i)|^2}{\|\tilde{\boldsymbol{w}}_i\|^2\cdot\|\boldsymbol{u}_i\|^2}$$

and θ_i is the acute angle between $\tilde{\boldsymbol{w}}_i$ and $\boldsymbol{u}_i^*$. Since the distances from the vertices of $\{\tilde{\boldsymbol{w}}_i, \tilde{\boldsymbol{w}}_{i-1}\}$ to $\boldsymbol{u}_i^*$ should agree, it must hold that

$$\|\tilde{\boldsymbol{w}}_{i-1}\|^2\left[1 - \frac{|\boldsymbol{e}_a(i)|^2}{\|\tilde{\boldsymbol{w}}_{i-1}\|^2\cdot\|\boldsymbol{u}_i\|^2}\right] = \|\tilde{\boldsymbol{w}}_i\|^2\left[1 - \frac{|\boldsymbol{e}_p(i)|^2}{\|\tilde{\boldsymbol{w}}_i\|^2\cdot\|\boldsymbol{u}_i\|^2}\right]$$

This equality is simply the energy relation (15.32).

Physical Interpretation

The geometric argument reveals that the distances from the vertices of $\tilde{\boldsymbol{w}}_{i-1}$ and $\tilde{\boldsymbol{w}}_i$ to $\boldsymbol{u}_i^*$ should coincide and that, therefore,

$$\boxed{\|\tilde{\boldsymbol{w}}_{i-1}\|^2\sin^2(\theta_{i-1}) = \|\tilde{\boldsymbol{w}}_i\|^2\sin^2(\theta_i)} \tag{15.44}$$

where the quantities $\{\sin^2(\theta_{i-1}), \sin^2(\theta_i)\}$ are defined by the ratios

$$\boxed{\sin^2(\theta_{i-1}) \triangleq 1 - \frac{|\boldsymbol{e}_a(i)|^2}{\|\tilde{\boldsymbol{w}}_{i-1}\|^2\cdot\|\boldsymbol{u}_i\|^2} \quad \text{and} \quad \sin^2(\theta_i) \triangleq 1 - \frac{|\boldsymbol{e}_p(i)|^2}{\|\tilde{\boldsymbol{w}}_i\|^2\cdot\|\boldsymbol{u}_i\|^2}} \tag{15.45}$$

Equality (15.44) reminds us of a famous result in optics relating the refraction indices of two mediums with the sines of the incident and refracted rays of light. More specifically, consider two mediums with refraction indices $\{\eta_1, \eta_2\}$. Consider further a ray of light that impinges on the layer separating both mediums at an angle θ_1 relative to the vertical direction (see Fig. 15.3). The refracted ray of light exits the layer from the other side at an angle θ_2, also relative to the vertical direction. Both angles $\{\theta_1, \theta_2\}$ are restricted to the interval $[0, \pi/2]$. It is a well-known result from optics, known as Snell's law, that the quantities $\{\eta_1, \eta_2, \theta_1, \theta_2\}$ satisfy $\eta_1\sin\theta_1 = \eta_2\sin\theta_2$.

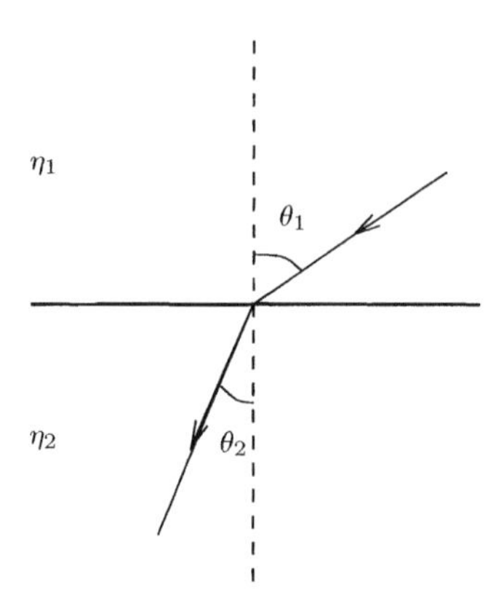

FIGURE 15.3 A ray of light impinging on the layer separating two mediums with different refraction indices.

Now observe that it follows from (15.44) that

$$\boxed{\|\tilde{\boldsymbol{w}}_{i-1}\| \sin(\theta_{i-1}) \;=\; \|\tilde{\boldsymbol{w}}_i\| \sin(\theta_i)} \tag{15.46}$$

in terms of the acute angles $\{\theta_i, \theta_{i-1}\}$ between $\{\tilde{\boldsymbol{w}}_i, \tilde{\boldsymbol{w}}_{i-1}\}$ and $\boldsymbol{u}_i^*$. This suggests that we can associate with the operation of an adaptive filter a fictitious ray travelling from one medium to another. The magnitudes $\|\tilde{\boldsymbol{w}}_{i-1}\|$ and $\|\tilde{\boldsymbol{w}}_i\|$ play the role of refraction indices of the mediums, and the layer separating both mediums is along the direction perpendicular to $\boldsymbol{u}_i^*$. The angles $\{\theta_{i-1}, \theta_i\}$ play the role of the incidence and refraction angles of the ray. Of course, the dynamic nature of the adaptive filter update is such that the values of $\{\|\tilde{\boldsymbol{w}}_{i-1}\|, \|\tilde{\boldsymbol{w}}_i\|, \theta_{i-1}, \theta_i\}$ change with time, as well as the direction of the separation layer.

Alternatively, we can interpret the result (15.46) graphically as shown in Fig. 15.4. An incident vector of norm $\|\tilde{\boldsymbol{w}}_{i-1}\|$ impinges on the separation layer at an angle θ_{i-1} with $\boldsymbol{u}_i^*$, while a refracted vector of norm $\|\tilde{\boldsymbol{w}}_i\|$ leaves $\boldsymbol{u}_i^*$ at an angle θ_i with $\boldsymbol{u}_i^*$. The norms $\{\|\tilde{\boldsymbol{w}}_{i-1}\|, \|\tilde{\boldsymbol{w}}_i\|\}$ and the angles $\{\theta_{i-1}, \theta_i\}$ satisfy (15.46) so that the energy relation amounts to stating that the projections of these vectors along the horizontal direction have equal norms.

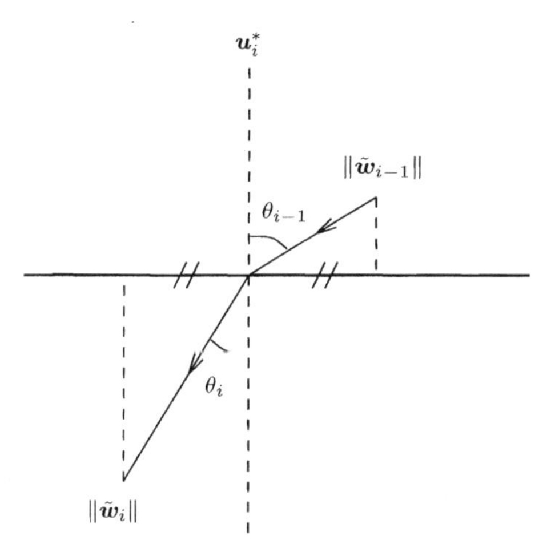

FIGURE 15.4 Snell's law interpretation of the energy-conservation relation (15.32). The norms of the horizontal projections should agree in view of (15.46).

System-Theoretic Interpretation

The energy relation (15.32) also admits a system-theoretic interpretation; it allows us to represent an adaptive filter as the interconnection of a lossless mapping and a feedback path — see Fig. 15.5. The interpretation is useful in studying the robustness properties of adaptive filters; as we shall do later in Part XI (*Robust Filters*). While we do not use this interpretation in this chapter, it is worthwhile presenting it here.

First we remark that whenever two vectors, say, x and y, have identical Euclidean norms, i.e., whenever $\|x\|^2 = \|y\|^2$, there should exist a lossless or energy-preserving mapping between them. In more precise terms, whenever $\|x\|^2 = \|y\|^2$, it can be shown that this is equivalent to the existence of a unitary matrix U that takes one vector to the other, say, $x = Uy$ (so that $\|x\|^2 = y^*U^*Uy = \|y\|^2$). We shall establish this result later in Sec. 33.3. The statement is obviously true for scalars since $|x|^2 = |y|^2$ implies $x = ye^{j\theta}$ for some angle $\theta \in [-\pi, \pi]$. In the vector case, the phase argument $e^{j\theta}$ is replaced by a unitary matrix. Now note that if we introduce the two column vectors

$$\text{col}\left\{\tilde{\boldsymbol{w}}_{i-1}, \sqrt{\bar{\mu}(i)}\, \boldsymbol{e}_p(i)\right\} \quad \text{and} \quad \text{col}\left\{\tilde{\boldsymbol{w}}_i, \sqrt{\bar{\mu}(i)}\, \boldsymbol{e}_a(i)\right\}$$

then the energy relation (15.32) states that these two column-vectors have identical Euclidean norms. Therefore, there should exist an energy-preserving (or lossless or unitary) mapping between them. Actually, it is not hard to determine an expression for this mapping. Using $\boldsymbol{e}_a(i) = \boldsymbol{u}_i\tilde{\boldsymbol{w}}_{i-1}$ and (15.29) we can write

$$\begin{bmatrix} \tilde{\boldsymbol{w}}_i \\ \sqrt{\bar{\mu}(i)}\, \boldsymbol{e}_a(i) \end{bmatrix} = \underbrace{\begin{bmatrix} \mathrm{I} - \bar{\mu}(i)\boldsymbol{u}_i^*\boldsymbol{u}_i & \sqrt{\bar{\mu}(i)}\, \boldsymbol{u}_i^* \\ \sqrt{\bar{\mu}(i)}\, \boldsymbol{u}_i & 0 \end{bmatrix}}_{\triangleq\, \mathcal{T}} \begin{bmatrix} \tilde{\boldsymbol{w}}_{i-1} \\ \sqrt{\bar{\mu}(i)}\, \boldsymbol{e}_p(i) \end{bmatrix} \tag{15.47}$$

where we are denoting the lossless mapping by $\mathcal{T}$ (see Prob. IV.21). Observe that $\mathcal{T}$ is solely dependent on the regression data $\boldsymbol{u}_i$. Furthermore, since, as explained in Sec. 15.2, the data $\{\boldsymbol{d}(i), \boldsymbol{u}_i\}$ can always be assumed to satisfy item (a) of model (15.16) for some noise sequence $\boldsymbol{v}(i)$, namely, $\boldsymbol{d}(i) = \boldsymbol{u}_i w^o + \boldsymbol{v}(i)$, then we can express $\boldsymbol{e}(i)$ as $\boldsymbol{e}(i) = \boldsymbol{e}_a(i) + \boldsymbol{v}(i)$ (cf. (15.19)). In this way, the right-hand side of expression (15.25) is dependent solely on $\{\boldsymbol{e}_a(i), \boldsymbol{v}(i), \boldsymbol{u}_i\}$ so that $\boldsymbol{e}_p(i)$ can be obtained from $\boldsymbol{e}_a(i)$ via a feedback connection as shown in Fig. 15.5.

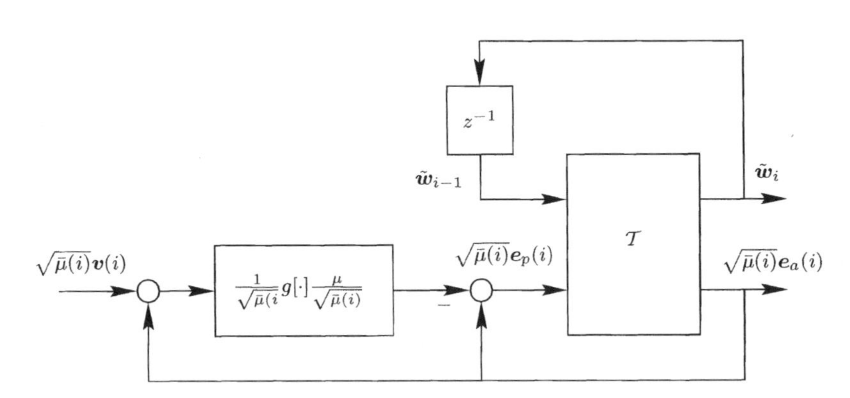

FIGURE 15.5 Every adaptive filter of the form (15.23) can be represented as an interconnection of a lossless feedforward mapping and a feedback loop.

CHAPTER 16

Performance of LMS

We now move on to illustrate the use of the variance relation (15.40) in evaluating the steady-state performance of adaptive filters. We start with the simplest of algorithms, namely, LMS.

16.1 VARIANCE RELATION

Thus, assume that the data $\{d(i), u_i\}$ satisfy model (15.16) and consider the LMS recursion

$$w_i = w_{i-1} + \mu u_i^* e(i) \tag{16.1}$$

for which

$$g[e(i)] = e(i) = e_a(i) + v(i) \tag{16.2}$$

Relation (15.40) then becomes

$$\mu \mathsf{E}\,\|u_i\|^2 |e_a(i) + v(i)|^2 = 2\mathsf{Re}\big(\mathsf{E}\, e_a^*(i)[e_a(i) + v(i)]\big), \quad i \to \infty \tag{16.3}$$

Several terms in this equality get cancelled. We shall carry out the calculations rather slowly in this section for illustration purposes only. Later, when similar calculations are called upon, we shall be less detailed.

To begin with, the expression on the left-hand side of (16.3) expands to

$$\begin{aligned} \mu \mathsf{E}\,\|u_i\|^2 |e_a(i) + v(i)|^2 &= \mu \mathsf{E}\,\|u_i\|^2 [\, |e_a(i)|^2 + |v(i)|^2 + e_a^*(i)v(i) + e_a(i)v^*(i)\,] \\ &= \mu \mathsf{E}\,\|u_i\|^2 |e_a(i)|^2 \;+\; \mu\sigma_v^2 \mathsf{E}\,\|u_i\|^2 \\ &= \mu \mathsf{E}\,\|u_i\|^2 |e_a(i)|^2 \;+\; \mu\sigma_v^2 \mathsf{Tr}(R_u) \end{aligned} \tag{16.4}$$

where we used the fact that $v(i)$ is independent of both u_i and $e_a(i)$ (recall Lemma 15.1), so that the cross-terms involving $\{v(i), e_a(i), u_i\}$ cancel out. We also used the fact that

$$\mathsf{E}\,\|u_i\|^2 = \mathsf{Tr}(R_u) \quad \text{and} \quad \mathsf{E}\,|v(i)|^2 - \sigma_v^2$$

Similarly, the expression on the right-hand side of (16.3) simplifies to $2\mathsf{E}\,|e_a(i)|^2$, which is simply $2\zeta^{\mathsf{LMS}}$ as $i \to \infty$. Therefore, equality (16.3) amounts to

$$\boxed{\zeta^{\mathsf{LMS}} = \frac{\mu}{2}\left[\mathsf{E}\,\|u_i\|^2 |e_a(i)|^2 \;+\; \sigma_v^2 \mathsf{Tr}(R_u)\right], \quad \text{as } i \to \infty} \tag{16.5}$$

This expression has been arrived at without approximations. Still, it requires that we evaluate the steady-state value of the expectation $\mathsf{E}\,\|u_i\|^2 |e_a(i)|^2$ in order to arrive at the EMSE of LMS. Some assumptions will now need to be introduced in order to proceed with the analysis, even for this simplest of algorithms!

We shall examine three scenarios. One scenario assumes sufficiently small step-sizes, while another relies on a useful separation assumption. The third scenario assumes regressors with Gaussian distribution.

16.2 SMALL STEP-SIZES

Expression (16.5) suggests that small step-sizes lead to small $\mathsf{E}\,|e_a(i)|^2$ in steady-state and, consequently, to a high likelihood of small values for $e_a(i)$ itself. So assume μ is small enough so that, in *steady-state*, the contribution of the term $\mathsf{E}\,\|u_i\|^2|e_a(i)|^2$ can be neglected, say,

$$\mathsf{E}\,\|u_i\|^2|e_a(i)|^2 \;\ll\; \sigma_v^2\mathsf{Tr}(R_u)$$

Then, we find from (16.5) that the EMSE can be approximated by

$$\boxed{\zeta^{\mathsf{LMS}} = \frac{\mu\sigma_v^2\mathsf{Tr}(R_u)}{2}} \qquad \text{(for sufficiently small } \mu) \tag{16.6}$$

16.3 SEPARATION PRINCIPLE

If the step-size is not sufficiently small, but still small enough to guarantee filter convergence — as will be discussed in Chapter 24, we can derive an alternative approximation for the EMSE from (16.5); the resulting expression will hold over a wider range of step-sizes. To do so, here and in several other places in this chapter and in subsequent chapters, we shall rely on the following assumption:

$$\boxed{\text{At steady-state, } \|u_i\|^2 \text{ is independent of } e_a(i)} \tag{16.7}$$

We shall refer to this condition as the *separation* assumption or the separation principle. Alternatively, we could assume instead that

$$\boxed{\text{At steady-state, } \|u_i\|^2 \text{ is independent of } e(i)} \tag{16.8}$$

with $e_a(i)$ replaced by $e(i)$. This condition is equivalent to (16.7) since $e(i) = e_a(i)+v(i)$ and $\|u_i\|^2$ is independent of $v(i)$ (as follows from Lemma 15.1).

Of course, assumption (16.7) is only exact in some special cases, e.g., when the successive regressors have constant Euclidean norms, since then $\|u_i\|^2$ becomes a constant; this situation occurs when the entries of u_i arise from a finite alphabet with constant magnitude — see Prob. IV.2. More generally, the assumption is reasonable at *steady-state* since the behavior of $e_a(i)$ in the limit is likely to be less sensitive to the regression (input) data.

Assumption (16.7) allows us to separate the expectation $\mathsf{E}\,\|u_i\|^2|e_a(i)|^2$, which appears in (16.5), into the product of two expectations:

$$\mathsf{E}\,\left(\|u_i\|^2\cdot|e_a(i)|^2\right) \;=\; \left(\mathsf{E}\,\|u_i\|^2\right)\cdot\left(\mathsf{E}\,|e_a(i)|^2\right) \;=\; \mathsf{Tr}(R_u)\zeta^{\mathsf{LMS}}, \quad i\to\infty \tag{16.9}$$

In order to illustrate this approximation, we show in Fig. 16.1 the result of simulating a 20-tap LMS filter over 1000 experiments. The figure shows the ensemble-averaged curves that correspond to the quantities

$$\mathsf{E}\,\left(\|u_i\|^2\cdot|e_a(i)|^2\right) \qquad \text{and} \qquad \left(\mathsf{E}\,\|u_i\|^2\right)\cdot\left(\mathsf{E}\,|e_a(i)|^2\right)$$

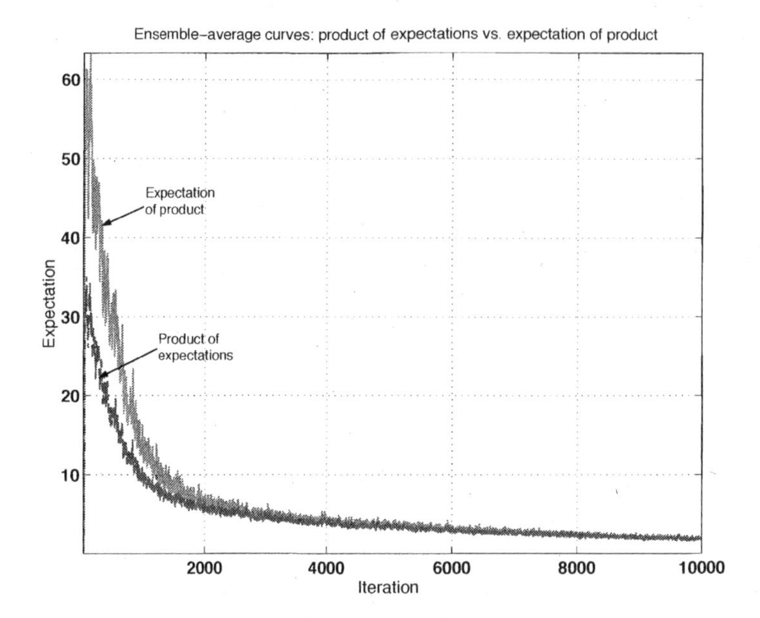

FIGURE 16.1 Ensemble-average curves for the expectation of the product, $\mathsf{E}\left(\|\boldsymbol{u}_i\|^2|\boldsymbol{e}_a(i)|^2\right)$ (*upper curve*), and for the product of expectations, $\left(\mathsf{E}\,\|\boldsymbol{u}_i\|^2\right)\cdot\left(\mathsf{E}\,|\boldsymbol{e}_a(i)|^2\right)$ (*lower curve*), for a 20-tap LMS filter with step-size $\mu = 0.001$. The curves are obtained by averaging over 1000 experiments.

It is seen that both curves tend to each other as time progresses so that it is reasonable to use the separation assumption (16.7) to approximate the expectation of the product as the product of expectations.

Now substituting (16.9) into (16.5) leads to the following expression for the EMSE of LMS:

$$\boxed{\zeta^{\text{LMS}} = \frac{\mu\sigma_v^2\text{Tr}(R_u)}{2-\mu\text{Tr}(R_u)}} \qquad \text{(over a wider range of } \mu) \tag{16.10}$$

This result will be revisited in Chapter 23; see the discussion following the statement of Thm. 23.3.

16.4 WHITE GAUSSIAN INPUT

One particular case for which the term $\mathsf{E}\,\|\boldsymbol{u}_i\|^2|\boldsymbol{e}_a(i)|^2$ that appears in (16.5) can be evaluated in closed-form occurs when $\boldsymbol{u}_i$ has a circular Gaussian distribution with a diagonal covariance matrix, say,

$$R_u = \sigma_u^2\text{I}, \quad \sigma_u^2 > 0 \tag{16.11}$$

That is, when the probability density function of $\boldsymbol{u}_i$ is of the form (cf. Lemma A.1):

$$f_{\boldsymbol{u}}(u) = \frac{1}{\pi^M}\frac{1}{\det R_u}\exp^{\left\{-uR_u^{-1}u^*\right\}} = \frac{1}{(\pi\sigma_u^2)^M}\exp^{\left\{-\|u\|^2/\sigma_u^2\right\}}$$

The diagonal structure of R_u amounts to saying that the entries of $\boldsymbol{u}_i$ are uncorrelated among themselves and that each has variance σ_u^2. The analysis can still be carried out in closed form even without this whiteness assumption; it suffices to require the regressors to be Gaussian. Moreover, R_u does not need to be a scaled multiple of the identity. We treat this more general situation in Sec. 23.1; see Prob. IV.19 for further motivation and the discussion following Thm. 23.3.

In addition to (16.11), we shall assume in this subsection that

$$\boxed{\text{At steady state, } \tilde{\boldsymbol{w}}_{i-1} \text{ is independent of } \boldsymbol{u}_i} \tag{16.12}$$

Conditions (16.11) and (16.12) enable us to evaluate $\mathsf{E}\,\|\boldsymbol{u}_i\|^2|\boldsymbol{e}_a(i)|^2$ explicitly. Before doing so, however, it is worth pointing out that to perform this task, it has been common in the literature to rely not on (16.12) but instead on a set of conditions known collectively as the *independence assumptions*. These assumptions require the data $\{\boldsymbol{d}(i), \boldsymbol{u}_i\}$ to satisfy the following conditions:

(i) The sequence $\{\boldsymbol{d}(i)\}$ is i.i.d.

(ii) The sequence $\{\boldsymbol{u}_i\}$ is also i.i.d.

(iii) Each $\boldsymbol{u}_i$ is independent of previous $\{\boldsymbol{d}_j, j < i\}$.

(iv) Each $\boldsymbol{d}(i)$ is independent of previous $\{\boldsymbol{u}_j, j < i\}$.

(v) The $\boldsymbol{d}(i)$ and $\boldsymbol{u}_i$ are jointly Gaussian.

(vi) In the case of complex-valued data, the $\boldsymbol{d}(i)$ and $\boldsymbol{u}_i$ are individually and jointly circular random variables, i.e., they satisfy $\mathsf{E}\,\boldsymbol{u}_i^\mathsf{T}\boldsymbol{u}_i = 0$, $\mathsf{E}\,\boldsymbol{d}^2(i) = 0$, and $\mathsf{E}\,\boldsymbol{u}_i^\mathsf{T}\boldsymbol{d}(i) = 0$.

The independence assumptions (i)–(vi) are in general restrictive since, in practice, the sequence $\{\boldsymbol{u}_i\}$ is rarely i.i.d. Consider, for example, the case in which the regressors $\{\boldsymbol{u}_i\}$ correspond to state vectors of an FIR implementation, as in the channel estimation application of Sec. 10.5. In this case, two successive regressors share common entries and cannot be statistically independent. Still, when the step-size is sufficiently small, the conclusions that are obtained under the independence assumptions (i)–(vi) tend to be realistic — see App. 24.A. This may explain their widespread use in the adaptive filtering literature. While restrictive, they provide significant simplifications in the derivations and tend to lead to results that match reasonably well with practice for small step-sizes. The key question of course is how *large* the step-size can be in order to validate the conclusions of an independence analysis. There does not seem to exist a clear answer to this inquiry in the literature.

Condition (16.12) is less restrictive than the independence assumptions (i)–(vi). Actually, assumption (16.12) is implied by the independence conditions. To see this, recall from the discussion in Sec. 15.2 that $\tilde{\boldsymbol{w}}_{i-1}$ is a function of the variables $\{\boldsymbol{w}_{-1}; \boldsymbol{d}(i-1), \ldots, \boldsymbol{d}(0); \boldsymbol{u}_{i-1}, \ldots, \boldsymbol{u}_0\}$. Therefore, if the sequence $\boldsymbol{u}_i$ is assumed i.i.d., and if $\boldsymbol{u}_i$ is independent of all previous $\{\boldsymbol{d}(j)\}$ and of $\boldsymbol{w}_{-1}$, then $\boldsymbol{u}_i$ will be independent of $\tilde{\boldsymbol{w}}_{i-1}$ *for all* i. Note, in addition, that condition (16.12) is only requiring the independence of $\{\tilde{\boldsymbol{w}}_{i-1}, \boldsymbol{u}_i\}$ to hold in *steady-state*; which is a considerably weaker assumption than what is implied by the full blown independence assumptions (i)–(vi). Moreover, assumption (16.12) is reasonable for small step-sizes μ. Intuitively, this is because the update term in (15.24) is relatively small for small μ and the statistical dependence of $\tilde{\boldsymbol{w}}_{i-1}$ on $\boldsymbol{u}_i$ becomes weak. Furthermore, in steady-state, the error $\boldsymbol{e}(i)$ is also small, which makes the update term in (15.24) even smaller.

Remark 16.1 (Independence) In our development in this chapter we do *not* adopt the independence assumptions (i)–(vi). Instead, we rely almost exclusively on the separation condition (16.7). It is only in this subsection, for Gaussian regressors, that we also use assumption (16.12) in order to show how to re-derive some known results for Gaussian data from the variance relation (15.40). ◇

So let us return to the term $\mathsf{E}\,\|\boldsymbol{u}_i\|^2|\boldsymbol{e}_a(i)|^2$ in (16.5) and show how it can be evaluated under (16.12), and under the assumption of circular Gaussian regressors. First we show how to express $\mathsf{E}\,\|\boldsymbol{u}_i\|^2|\boldsymbol{e}_a(i)|^2$ in terms of $\mathsf{E}\,|\boldsymbol{e}_a(i)|^2$ (see (16.17) further ahead). Thus, note the following sequence of identities:

$$\begin{aligned}
\mathsf{E}\,\|\boldsymbol{u}_i\|^2|\boldsymbol{e}_a(i)|^2 &= \mathsf{E}\left(\boldsymbol{u}_i\boldsymbol{u}_i^*(\boldsymbol{u}_i\tilde{\boldsymbol{w}}_{i-1}\tilde{\boldsymbol{w}}_{i-1}^*\boldsymbol{u}_i^*)\right)\\
&= \mathsf{E}\ \mathsf{Tr}\left(\boldsymbol{u}_i\boldsymbol{u}_i^*\boldsymbol{u}_i\tilde{\boldsymbol{w}}_{i-1}\tilde{\boldsymbol{w}}_{i-1}^*\boldsymbol{u}_i^*\right)\\
&= \mathsf{E}\ \mathsf{Tr}\left(\boldsymbol{u}_i^*\boldsymbol{u}_i\tilde{\boldsymbol{w}}_{i-1}\tilde{\boldsymbol{w}}_{i-1}^*\boldsymbol{u}_i^*\boldsymbol{u}_i\right)\\
&= \mathsf{Tr}\ \mathsf{E}\left(\boldsymbol{u}_i^*\boldsymbol{u}_i\tilde{\boldsymbol{w}}_{i-1}\tilde{\boldsymbol{w}}_{i-1}^*\boldsymbol{u}_i^*\boldsymbol{u}_i\right)
\end{aligned}\tag{16.13}$$

where in the second equality we used the fact that the trace of a scalar is equal to the scalar itself, and in the third equality we used the property that $\mathsf{Tr}(AB) = \mathsf{Tr}(BA)$ for any matrices A and B of compatible dimensions.

We now evaluate the term $\mathsf{E}\left(\boldsymbol{u}_i^*\boldsymbol{u}_i\tilde{\boldsymbol{w}}_{i-1}\tilde{\boldsymbol{w}}_{i-1}^*\boldsymbol{u}_i^*\boldsymbol{u}_i\right)$, which is a covariance matrix. To do so, we recall the following property of conditional expectations, namely, that for any two random variables $\boldsymbol{x}$ and $\boldsymbol{y}$, it holds that $\mathsf{E}\,\boldsymbol{x} = \mathsf{E}\left(\mathsf{E}\,[\boldsymbol{x}|\boldsymbol{y}]\right)$ — see (1.4). Therefore, in steady-state,

$$\begin{aligned}
\mathsf{E}\left(\boldsymbol{u}_i^*\boldsymbol{u}_i\tilde{\boldsymbol{w}}_{i-1}\tilde{\boldsymbol{w}}_{i-1}^*\boldsymbol{u}_i^*\boldsymbol{u}_i\right) &= \mathsf{E}\left[\mathsf{E}\left(\boldsymbol{u}_i^*\boldsymbol{u}_i\tilde{\boldsymbol{w}}_{i-1}\tilde{\boldsymbol{w}}_{i-1}^*\boldsymbol{u}_i^*\boldsymbol{u}_i \mid \boldsymbol{u}_i\right)\right]\\
&= \mathsf{E}\left[\boldsymbol{u}_i^*\boldsymbol{u}_i\ \mathsf{E}\left(\tilde{\boldsymbol{w}}_{i-1}\tilde{\boldsymbol{w}}_{i-1}^* \mid \boldsymbol{u}_i\right)\boldsymbol{u}_i^*\boldsymbol{u}_i\right]\\
&= \mathsf{E}\left(\boldsymbol{u}_i^*\boldsymbol{u}_i C_{i-1}\boldsymbol{u}_i^*\boldsymbol{u}_i\right)
\end{aligned}\tag{16.14}$$

where in the last step we used assumption (16.12), namely, that $\tilde{\boldsymbol{w}}_{i-1}$ and $\boldsymbol{u}_i$ are independent so that

$$\mathsf{E}\left(\tilde{\boldsymbol{w}}_{i-1}\tilde{\boldsymbol{w}}_{i-1}^* \mid \boldsymbol{u}_i\right) = \mathsf{E}\,\tilde{\boldsymbol{w}}_{i-1}\tilde{\boldsymbol{w}}_{i-1}^* \ \stackrel{\Delta}{=}\ C_{i-1}$$

We are also denoting the covariance matrix of $\tilde{\boldsymbol{w}}_{i-1}$ by C_{i-1}. We do not need to know the value of C_{i-1}, as the argument will demonstrate — see the remark following (16.17). We are then reduced to evaluating the expression $\mathsf{E}\,\boldsymbol{u}_i^*\boldsymbol{u}_i C_{i-1}\boldsymbol{u}_i^*\boldsymbol{u}_i$. Due to the circular Gaussian assumption on $\boldsymbol{u}_i$, this term has the same form as the general term that we evaluated earlier in Lemma A.3 for Gaussian variables, with the identifications

$$\boldsymbol{z} \leftarrow \boldsymbol{u}_i^*, \quad W \leftarrow C_{i-1}, \quad \Lambda \leftarrow \sigma_u^2\mathrm{I}$$

so that we can use the result of that lemma to write

$$\mathsf{E}\,\boldsymbol{u}_i^*\boldsymbol{u}_i C_{i-1}\boldsymbol{u}_i^*\boldsymbol{u}_i = \sigma_u^4\left[\mathsf{Tr}(C_{i-1})\mathrm{I} + C_{i-1}\right]\tag{16.15}$$

Substituting this equality into (16.13) we obtain

$$\mathsf{E}\,\|\boldsymbol{u}_i\|^2|\boldsymbol{e}_a(i)|^2 = \mathsf{Tr}\left[\sigma_u^4\mathsf{Tr}(C_{i-1})\mathrm{I} + \sigma_u^4 C_{i-1}\right] = (M+1)\sigma_u^4\mathsf{Tr}(C_{i-1})\tag{16.16}$$

Now repeating the same argument that led to (16.14), we also find that

$$\begin{aligned}
\mathsf{E}\,|\boldsymbol{e}_a(i)|^2 = \mathsf{E}\left(\boldsymbol{u}_i\tilde{\boldsymbol{w}}_{i-1}\tilde{\boldsymbol{w}}_{i-1}^*\boldsymbol{u}_i^*\right) &= \mathsf{E}\left(\boldsymbol{u}_i C_{i-1}\boldsymbol{u}_i^*\right)\\
&= \mathsf{E}\left(\boldsymbol{u}_i C_{i-1}\boldsymbol{u}_i^*\right)\\
&= \mathsf{Tr}\ \mathsf{E}\left(\boldsymbol{u}_i^*\boldsymbol{u}_i C_{i-1}\right)\\
&= \mathsf{Tr}\ \mathsf{E}\left(R_u C_{i-1}\right)\\
&= \sigma_u^2\mathsf{Tr}\ \mathsf{E}\left(C_{i-1}\right)
\end{aligned}$$

This expression relates $\text{Tr}(C_{i-1})$ to $\mathsf{E}\,|e_a(i)|^2$. Substituting into (16.16) we obtain

$$\mathsf{E}\,\|u_i\|^2|e_a(i)|^2 \;=\; (M+1)\sigma_u^2\mathsf{E}\,|e_a(i)|^2 \tag{16.17}$$

This relation expresses the desired term $\mathsf{E}\,\|u_i\|^2|e_a(i)|^2$ as a scaled multiple of $\mathsf{E}\,|e_a(i)|^2$ alone — observe that C_{i-1} is cancelled out. Using this result in (16.5), we get

$$\boxed{\zeta^{\text{LMS}} = \frac{\mu M\sigma_v^2\sigma_u^2}{2-\mu(M+1)\sigma_u^2} \quad \text{(for complex-valued data)}} \tag{16.18}$$

The above derivation assumes complex-valued data. If the data were real-valued, then the same arguments would still apply with the only exception of Lemma A.3. Instead, we would employ the result of Lemma A.2 and replace (16.15) by

$$\mathsf{E}\,u_i^{\mathsf{T}}u_iC_{i-1}u_i^{\mathsf{T}}u_i \;=\; \sigma_u^4\left[\text{Tr}(C_{i-1})\mathrm{I}+2C_{i-1}\right]$$

with an additional scaling factor of 2 (now $C_{i-1} = \mathsf{E}\,\tilde{w}_{i-1}\tilde{w}_{i-1}^{\mathsf{T}}$). Then (16.16) and (16.17) would become

$$\mathsf{E}\,\|u_i\|^2e_a^2(i) = (M+2)\sigma_u^4\text{Tr}(C_{i-1}) = (M+2)\sigma_u^2\mathsf{E}\,e_a^2(i) \tag{16.19}$$

and the resulting expression for the EMSE is

$$\boxed{\zeta^{\text{LMS}} = \frac{\mu M\sigma_v^2\sigma_u^2}{2-\mu(M+2)\sigma_u^2} \quad \text{(for real-valued data)}} \tag{16.20}$$

16.5 STATEMENT OF RESULTS

We summarize the earlier results for the LMS filter in the following statement. A conclusion that stands out from the expressions in the lemma is that the performance of LMS is dependent on the filter length M, the step-size μ, and the input covariance matrix R_u.

Lemma 16.1 (EMSE of LMS) Consider the LMS recursion (16.1) and assume the data $\{d(i), u_i\}$ satisfy model (15.16). Then its EMSE can be approximated by the following expressions:

1. For sufficiently small step-sizes, it holds that $\zeta^{\text{LMS}} = \mu\sigma_v^2\text{Tr}(R_u)/2$.

2. Under the separation assumption (16.7), it holds that

$$\zeta^{\text{LMS}} = \mu\sigma_v^2\text{Tr}(R_u)/[2-\mu\text{Tr}(R_u)]$$

3. If u_i is Gaussian with $R_u = \sigma_u^2\mathrm{I}$, and under the steady-state assumption (16.12), it holds that

$$\zeta^{\text{LMS}} = \mu M\sigma_v^2\sigma_u^2/[2-\mu(M+\gamma)\sigma_u^2]$$

where $\gamma = 2$ if the data are real-valued and $\gamma = 1$ if the data are complex-valued and u_i circular. Here M is the dimension of u_i.

In all cases, the misadjustment is obtained by dividing the EMSE by σ_v^2.

16.6 SIMULATION RESULTS

Figures 16.2–16.4 show the values of the steady-state MSE of a 10-tap LMS filter for different choices of the step-size and for different signal conditions. The theoretical values are obtained by using the expressions from Lemma 16.1. For each step-size, the experimental value is obtained by running LMS for 4×10^5 iterations and averaging the squared-error curve $\{|e(i)|^2\}$ over 100 experiments in order to generate the ensemble-average curve. The average of the last 5000 entries of the ensemble-average curve is then used as the experimental value for the MSE. The data $\{\boldsymbol{d}(i), \boldsymbol{u}_i\}$ are generated according to model (15.16) using Gaussian noise with variance $\sigma_v^2 = 0.001$.

In Fig. 16.2, the regressors $\{\boldsymbol{u}_i\}$ do not have shift structure (i.e., they do not correspond to regressors that arise from a tapped-delay-line implementation). The regressors are generated as independent realizations of a Gaussian distribution with a covariance matrix R_u whose eigenvalue spread is $\rho = 5$. Observe from the leftmost plot how expression (16.6) leads to a good fit between theory and practice for small step-sizes. On the other hand, as can be seen from the rightmost plot, expression (16.10) provides a better fit over a wider range of step-sizes.

In Fig. 16.3, the regressors $\{\boldsymbol{u}_i\}$ have shift structure and they are generated by feeding correlated data $\{\boldsymbol{u}(i)\}$ into a tapped delay line. The correlated data are obtained by filtering a unit-variance i.i.d. Gaussian random process $\{\boldsymbol{s}(i)\}$ through a first-order auto-regressive model with transfer function

$$\sqrt{1-a^2}/(1-az^{-1})$$

and $a = 0.8$. It is shown in Prob. IV.1 that the auto-correlation sequence of the resulting process $\{\boldsymbol{u}(i)\}$ is

$$r(k) = \mathsf{E}\,\boldsymbol{u}(i)\boldsymbol{u}(i-k) = a^{|k|}$$

for all integer values k. In this way, the covariance matrix R_u of the regressor $\boldsymbol{u}_i$ is a 10×10 Toeplitz matrix with entries $\{a^{|i-j|}, 0 \leq i, j \leq M-1\}$.

In Fig. 16.4, regressors with shift structure are again used but they are now generated by feeding into the tapped delay line a unit-variance *white* (as opposed to correlated) process so that $R_u = \sigma_u^2 \mathrm{I}$ with $\sigma_u^2 = 1$. This situation allows us to verify the third result in Lemma 16.1. It is seen from all these simulations that the expressions of Lemma 16.1 provide reasonable approximations for the EMSE of the LMS filter. In particular, expression

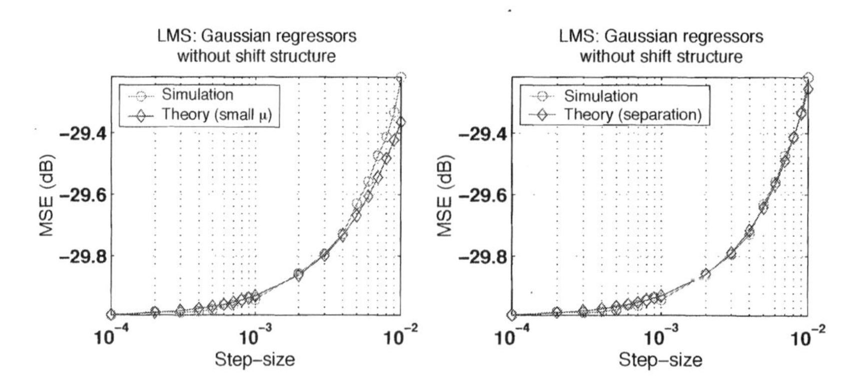

FIGURE 16.2 Theoretical and simulated MSE for a 10-tap LMS filter with $\sigma_v^2 = 0.001$ and Gaussian regressors *without* shift structure. The leftmost plot compares the simulated MSE with expression (16.6), which was derived under the assumption of small step-sizes. The rightmost plot uses expression (16.10), which was derived using the separation assumption (16.7).

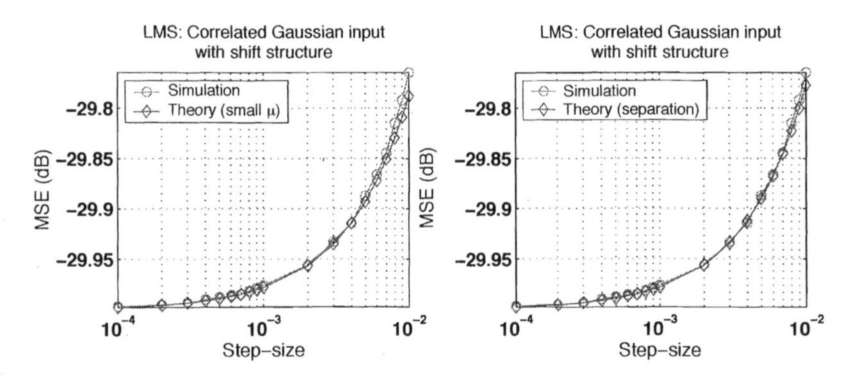

FIGURE 16.3 Theoretical and simulated MSE for a 10-tap LMS filter with $\sigma_v^2 = 0.001$ and regressors *with* shift structure. The regressors are generated by feeding *correlated* data into a tapped delay line. The leftmost plot compares the simulated MSE with expression (16.6), which was derived under the assumption of small step-sizes. The rightmost plot uses expression (16.10), which was derived using the separation assumption (16.7).

(16.10), which was derived under the separation assumption (16.7), provides a good match between theory and practice.

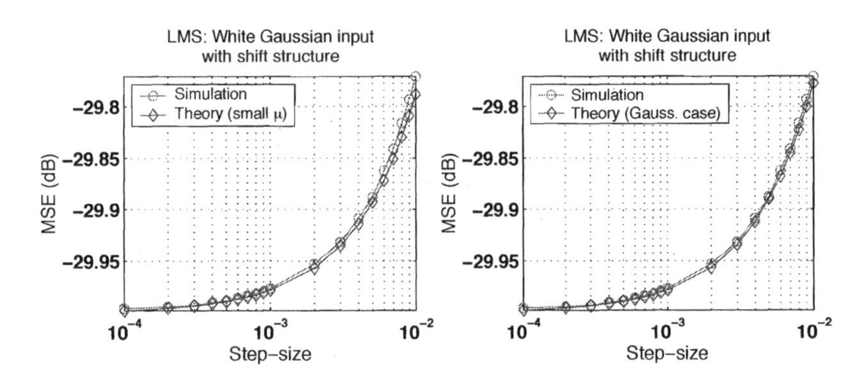

FIGURE 16.4 Theoretical (using (16.20)) and simulated MSE for a 10-tap LMS filter with $\sigma_v^2 = 0.001$ and regressors with shift structure. The regressors are generated by feeding *white* Gaussian input data with unit variance into a tapped delay line.

CHAPTER 17

Performance of NLMS

We now illustrate the use of the variance relation (15.40) in evaluating the steady-state performance of the ϵNLMS algorithm,

$$w_i = w_{i-1} + \frac{\mu}{\epsilon + \|u_i\|^2} u_i^* e(i) \tag{17.1}$$

for which

$$g[e(i)] = \frac{e(i)}{\epsilon + \|u_i\|^2} = \frac{e_a(i) + v(i)}{\epsilon + \|u_i\|^2} \tag{17.2}$$

The data $\{d(i), u_i, v(i)\}$ are assumed to satisfy model (15.16). The variance relation (15.40) in this case becomes

$$\mu \mathsf{E}\left(\frac{\|u_i\|^2 |e_a(i) + v(i)|^2}{(\epsilon + \|u_i\|^2)^2}\right) = 2\mathsf{Re}\left\{\mathsf{E}\, e_a^*(i)\left(\frac{e_a(i) + v(i)}{\epsilon + \|u_i\|^2}\right)\right\}, \quad i \to \infty \tag{17.3}$$

Again, several terms in this equality get cancelled. Since the arguments are similar to what we did for the LMS case in the previous chapter, we shall be brief and only highlight the main steps.

17.1 SEPARATION PRINCIPLE

Note first that by expanding both sides of (17.3) we get

$$\mu \mathsf{E}\left(\frac{\|u_i\|^2 |e_a(i)|^2}{(\epsilon + \|u_i\|^2)^2}\right) + \mu\sigma_v^2 \mathsf{E}\left(\frac{\|u_i\|^2}{(\epsilon + \|u_i\|^2)^2}\right) = 2\mathsf{E}\left(\frac{|e_a(i)|^2}{\epsilon + \|u_i\|^2}\right) \tag{17.4}$$

In order to simplify this equation, we resort to the same separation principle (16.7) that we used in the LMS case, namely, that at steady-state, $\|u_i\|^2$ is independent of $|e_a(i)|^2$. Under this condition, equality (17.4) becomes

$$\mu \mathsf{E}\left(\frac{\|u_i\|^2}{(\epsilon + \|u_i\|^2)^2}\right) \mathsf{E}\,|e_a(i)|^2 + \mu\sigma_v^2 \mathsf{E}\left(\frac{\|u_i\|^2}{(\epsilon + \|u_i\|^2)^2}\right) = 2\mathsf{E}\left(\frac{1}{\epsilon + \|u_i\|^2}\right) \mathsf{E}\,|e_a(i)|^2$$

If we define the quantities (which are solely dependent on the statistics of the regression data):

$$\alpha_u \triangleq \mathsf{E}\left(\frac{\|u_i\|^2}{(\epsilon + \|u_i\|^2)^2}\right), \qquad \eta_u \triangleq \mathsf{E}\left(\frac{1}{\epsilon + \|u_i\|^2}\right) \tag{17.5}$$

then the above equality can be written more compactly as

$$(2\eta_u - \mu\alpha_u)\mathsf{E}\,|e_a(i)|^2 = \mu\sigma_v^2\alpha_u, \quad i \to \infty$$

so that

$$\zeta^{\epsilon-\mathsf{NLMS}} = \frac{\mu\alpha_u\sigma_v^2}{2\eta_u - \mu\alpha_u} \tag{17.6}$$

When the regularization parameter ϵ is sufficiently small, which is usually the case, then its effect can be ignored and the definitions of α_u and η_u coincide, namely, $\alpha_u = \eta_u = \mathsf{E}\,\left(1/\|\boldsymbol{u}_i\|^2\right)$ In this case, expression (17.6) reduces to

$$\boxed{\zeta^{\epsilon-\mathsf{NLMS}} = \frac{\mu\sigma_v^2}{2-\mu}, \quad \text{(when } \epsilon \text{ is small)}} \tag{17.7}$$

which is independent of the regression data.

An alternative expression for the EMSE can be obtained by using the assumption $\epsilon \approx 0$ in order to initially simplify (17.4) into

$$\mu\mathsf{E}\left(\frac{|e_a(i)|^2}{\|\boldsymbol{u}_i\|^2}\right) + \mu\sigma_v^2\mathsf{E}\left(\frac{1}{\|\boldsymbol{u}_i\|^2}\right) = 2\mathsf{E}\left(\frac{|e_a(i)|^2}{\|\boldsymbol{u}_i\|^2}\right)$$

Then we appeal to the following steady-state approximation, instead of the separation assumption (16.7),

$$\mathsf{E}\left(\frac{|e_a(i)|^2}{\|\boldsymbol{u}_i\|^2}\right) \approx \frac{\mathsf{E}\,|e_a(i)|^2}{\mathsf{E}\,\|\boldsymbol{u}_i\|^2} = \frac{\mathsf{E}\,|e_a(i)|^2}{\mathsf{Tr}(R_u)} \quad \text{as } i \longrightarrow \infty \tag{17.8}$$

and use it to find that the filter EMSE can also be approximated by

$$\boxed{\zeta^{\epsilon-\mathsf{NLMS}} = \frac{\mu\sigma_v^2}{2-\mu}\mathsf{Tr}(R_u)\mathsf{E}\left(\frac{1}{\|\boldsymbol{u}_i\|^2}\right), \quad \text{(when } \epsilon \text{ is small)}} \tag{17.9}$$

Lemma 17.1 (EMSE of ϵ-NLMS) Consider the ϵ-NLMS recursion (17.1) and assume the data $\{\boldsymbol{d}(i), \boldsymbol{u}_i\}$ satisfy model (15.16). Then, under the separation assumption (16.7), its EMSE can be approximated by

$$\zeta^{\epsilon-\mathsf{NLMS}} = \frac{\mu\sigma_v^2}{2-\mu} \quad \text{(first approximation)}$$

or, under the steady-state approximation (17.8),

$$\zeta^{\epsilon-\mathsf{NLMS}} = \frac{\mu\sigma_v^2}{2-\mu}\mathsf{Tr}(R_u)\mathsf{E}\left(\frac{1}{\|\boldsymbol{u}_i\|^2}\right) \quad \text{(second approximation)}$$

The misadjustment is obtained by dividing the EMSE by σ_v^2.

The expressions for $\zeta^{\epsilon-\mathsf{NLMS}}$ in the above statement reveal why the performance of $\epsilon-$NLMS is less sensitive to the statistics of the regression data than LMS. Observe, for example, that the expression $\zeta^{\epsilon-\mathsf{NLMS}} = \mu\sigma_v^2/(2-\mu)$ is independent of R_u.

17.2 SIMULATION RESULTS

Figures 17.1 and 17.2 show the values of the steady-state MSE of a 10-tap $\epsilon-$NLMS filter for different choices of the step-size with $\epsilon = 10^{-6}$. The theoretical values are obtained by using the expressions of Lemma 17.1. For each step-size, the experimental value is obtained by running $\epsilon-$NLMS for 1.5×10^5 iterations and averaging the squared-error curve over 100 experiments in order to generate the ensemble-average curve. The average of the last 5000 entries of the ensemble-average curve is used as the experimental value for the MSE. The data $\{\boldsymbol{d}(i), \boldsymbol{u}_i\}$ are generated according to model (15.16) using Gaussian noise with variance $\sigma_v^2 = 0.001$.

In Fig. 17.1, the regressors $\{\boldsymbol{u}_i\}$ do not have shift structure and they are generated as independent realizations of a Gaussian distribution with a covariance matrix R_u whose eigenvalue spread is $\rho = 5$. The range of step-sizes is between 0.01 and 1. Observe from the rightmost plot how expression (17.9) leads, in this case, to a good fit between theory and practice over a wider range of step-sizes, while expression (17.7) provides a good fit for smaller step-sizes as seen from the leftmost plot.

In Fig. 17.2, the regressors $\{\boldsymbol{u}_i\}$ have shift structure and are generated by feeding correlated data $\{\boldsymbol{u}(i)\}$ into a tapped delay line. The correlated data are obtained by passing a unit-variance i.i.d. Gaussian random process $\{\boldsymbol{s}(i)\}$ through a first-order auto-regressive model with transfer function $\sqrt{1-a^2}/(1-az^{-1})$ and $a = 0.8$. In this simulation, it is seen that expression (17.7) results in a better fit between theory and practice when compared with expression (17.9).

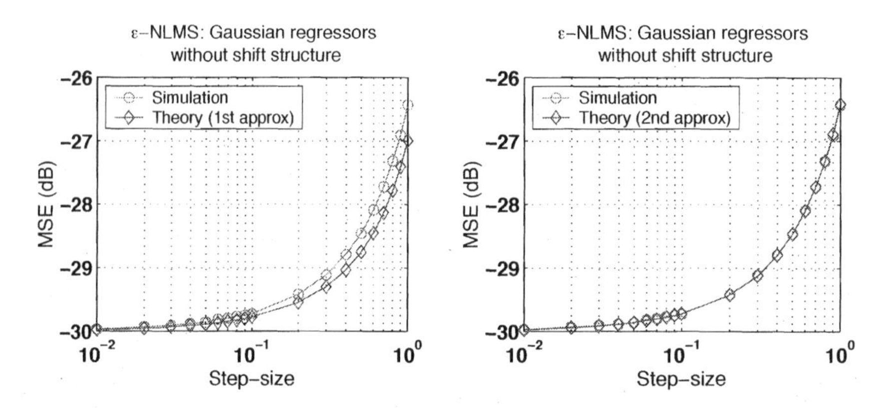

FIGURE 17.1 Theoretical and simulated MSE for a 10-tap $\epsilon-$NLMS filter with $\sigma_v^2 = 0.001$ and Gaussian regressors *without* shift structure. The leftmost plot compares the simulated MSE with the value that results from (17.7), while the rightmost plot uses expression (17.9).

17.A APPENDIX: RELATING NLMS TO LMS

In addition to the treatment given in the body of this chapter, there are alternative ways to study the performance of $\epsilon-$NLMS. One such way is to reduce the ϵ-NLMS recursion (17.1) to an LMS recursion via a suitable change of variables. Thus, introduce the transformed variables:

$$\boxed{\breve{\boldsymbol{u}}_i \triangleq \frac{\boldsymbol{u}_i}{\sqrt{\epsilon + \|\boldsymbol{u}_i\|^2}}, \qquad \breve{\boldsymbol{d}}(i) \triangleq \frac{\boldsymbol{d}(i)}{\sqrt{\epsilon + \|\boldsymbol{u}_i\|^2}}, \qquad \breve{\boldsymbol{v}}(i) \triangleq \frac{\boldsymbol{v}(i)}{\sqrt{\epsilon + \|\boldsymbol{u}_i\|^2}}} \tag{17.10}$$

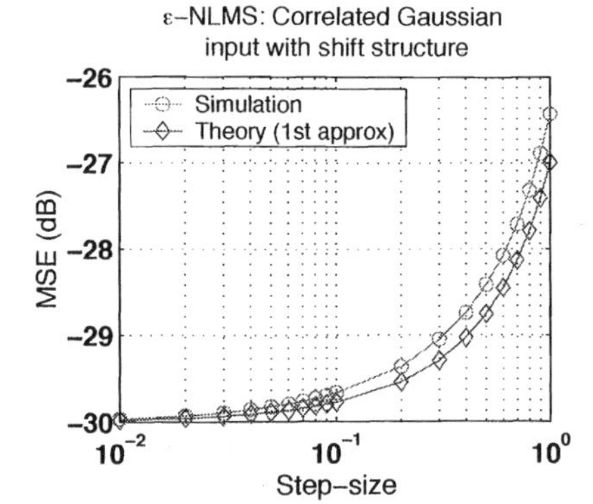

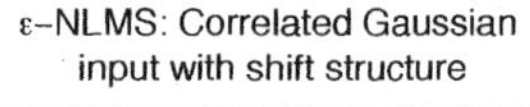

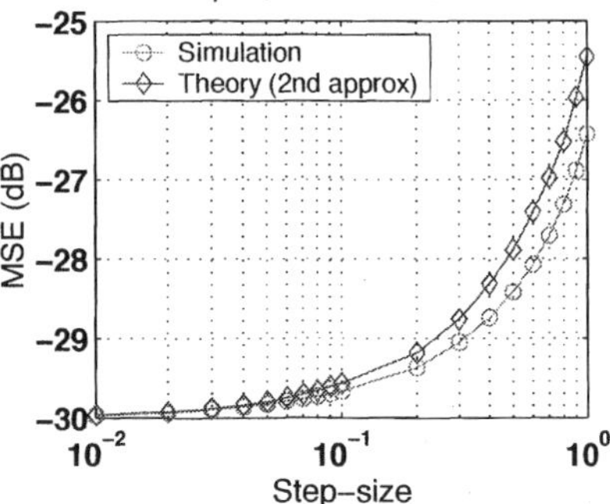

FIGURE 17.2 Theoretical and simulated MSE for a 10-tap $\epsilon-$NLMS filter with $\sigma_v^2 = 0.001$ and Gaussian regressors *with* shift structure. The leftmost plot compares the simulated MSE with the value that results from (17.7), while the rightmost plot uses expression (17.9).

Then the $\epsilon-$NLMS recursion (17.1) becomes

$$\boldsymbol{w}_i = \boldsymbol{w}_{i-1} + \mu\breve{\boldsymbol{u}}_i^*\breve{e}(i), \qquad \text{where} \quad \breve{e}(i) = \breve{\boldsymbol{d}}(i) - \breve{\boldsymbol{u}}_i\boldsymbol{w}_{i-1}$$

which is simply an LMS recursion in the variables $\{\breve{\boldsymbol{d}}(i), \breve{\boldsymbol{u}}_i\}$. Moreover, given model (15.16) for $\{\boldsymbol{d}(i), \boldsymbol{u}_i\}$, it follows that the variables $\{\breve{\boldsymbol{d}}(i), \breve{\boldsymbol{u}}_i\}$ satisfy a similar model with the same w^o. More specifically,

(a) There exists a vector w^o such that $\breve{\boldsymbol{d}}(i) = \breve{\boldsymbol{u}}_i w^o + \breve{\boldsymbol{v}}(i)$.
(b) The noise sequence $\{\breve{\boldsymbol{v}}(i)\}$ is orthogonal with $\breve{\sigma}_v^2 = \mathsf{E}\,|\breve{\boldsymbol{v}}(i)|^2 = \sigma_v^2\mathsf{E}\,\left(1/(\epsilon + \|\boldsymbol{u}_i\|^2)\right)$.
(c) The noise sequence $\breve{\boldsymbol{v}}(i)$ is orthogonal to $\breve{\boldsymbol{u}}_j$ for all $i \neq j$.
(d) The initial condition $\boldsymbol{w}_{-1}$ is independent of all $\{\breve{\boldsymbol{d}}(j), \breve{\boldsymbol{u}}_j, \breve{\boldsymbol{v}}(j)\}$.
(e) The regressors have $\breve{R}_u = \mathsf{E}\,\breve{\boldsymbol{u}}_i^*\breve{\boldsymbol{u}}_i = \mathsf{E}\,\left(\boldsymbol{u}_i^*\boldsymbol{u}_i/(\epsilon + \|\boldsymbol{u}_i\|^2)\right) > 0$.
(f) The random variables $\{\breve{\boldsymbol{v}}(i), \breve{\boldsymbol{u}}_i\}$ are not necessarily zero mean (see Prob. IV.20).

(17.11)

The main differences in relation to model (15.16) are conditions (b), (c), and (f). This is because the sequence $\{\breve{\boldsymbol{v}}(i)\}$ is not i.i.d. any longer but only orthogonal. In other words, it now satisfies $\mathsf{E}\,\breve{\boldsymbol{v}}(i)\breve{\boldsymbol{v}}^*(j) = 0$ for all i, j. Moreover, $\breve{\boldsymbol{v}}(i)$ is not independent of $\breve{\boldsymbol{u}}_i$ since its definition involves $\boldsymbol{u}_i$. Instead, $\breve{\boldsymbol{v}}(i)$ is orthogonal to $\breve{\boldsymbol{u}}_i$, i.e., it satisfies $\mathsf{E}\,\breve{\boldsymbol{u}}_j\breve{\boldsymbol{v}}^*(i) = 0$ for all i, j. Still, the result of Lemma 15.1 will hold with the qualification "independent of" replaced by "orthogonal to", and with the variables $\{\boldsymbol{v}(i), \boldsymbol{e}_a(j)\}$ replaced by $\{\breve{\boldsymbol{v}}(i), \breve{\boldsymbol{e}}_a(j)\}$, where $\breve{\boldsymbol{e}}_a(j) = \breve{\boldsymbol{u}}_j\tilde{\boldsymbol{w}}_{j-1}$. Under these conditions, the derivation that led to (16.5) could be repeated, and it would lead to a similar relation of the form

$$\breve{\zeta} = \mu\left[\mathsf{E}\,\|\breve{\boldsymbol{u}}_i\|^2|\breve{\boldsymbol{e}}_a(i)|^2 \;+\; \breve{\sigma}_v^2\mathsf{Tr}(\breve{R}_u)\right], \quad \text{as } i \longrightarrow \infty$$

This line of reasoning, however, would only allow us to evaluate the transformed variance $\mathsf{E}\,|\breve{\boldsymbol{e}}_a(\infty)|^2$, as opposed to the desired variance $\mathsf{E}\,|\boldsymbol{e}_a(\infty)|^2$. But it follows from the definitions of $\boldsymbol{e}_a(i)$ and $\breve{\boldsymbol{e}}_a(i)$ that

$$\frac{1}{\epsilon + \|\boldsymbol{u}_i\|^2} \cdot |\boldsymbol{e}_a(i)|^2 \;=\; |\breve{\boldsymbol{e}}_a(i)|^2$$

so that, by appealing again to the same kind of approximation (17.8), we can relate $\mathsf{E}\,|\boldsymbol{e}_a(\infty)|^2$ to $\mathsf{E}\,|\breve{\boldsymbol{e}}_a(\infty)|^2$ as follows:

$$\frac{\zeta^{\epsilon-\mathsf{NLMS}}}{\epsilon + \mathsf{Tr}(R_u)} \;=\; \mathsf{E}\,|\breve{\boldsymbol{e}}_a(i)|^2, \quad i \to \infty \tag{17.12}$$

Therefore, using the result of Lemma 16.1 for sufficiently small step-sizes we get

$$\lim_{i\to\infty} \mathsf{E}\,|\check{e}_a(i)|^2 = \frac{\mu\check{\sigma}_v^2}{2}\cdot \mathsf{Tr}(\check{R}_u) = \frac{\mu\check{\sigma}_v^2}{2}\cdot \mathsf{Tr}\left(\mathsf{E}\left(\frac{\boldsymbol{u}_i^*\boldsymbol{u}_i}{\epsilon+\|\boldsymbol{u}_i\|^2}\right)\right)$$

and using (17.12), we obtain

$$\begin{aligned}\zeta^{\epsilon-\mathsf{NLMS}} &= \frac{\mu\check{\sigma}_v^2(\epsilon+\mathsf{Tr}(R_u))}{2}\cdot \mathsf{Tr}\left(\mathsf{E}\left(\frac{\boldsymbol{u}_i^*\boldsymbol{u}_i}{\epsilon+\|\boldsymbol{u}_i\|^2}\right)\right)\\ &= \frac{\mu\sigma_v^2(\epsilon+\mathsf{Tr}(R_u))}{2}\cdot \mathsf{E}\left(\frac{1}{\epsilon+\|\boldsymbol{u}_i\|^2}\right)\cdot \mathsf{Tr}\left(\mathsf{E}\left(\frac{\boldsymbol{u}_i^*\boldsymbol{u}_i}{\epsilon+\|\boldsymbol{u}_i\|^2}\right)\right)\end{aligned}$$

However, when $\epsilon \approx 0$, $\mathsf{Tr}\left(\mathsf{E}\left(\boldsymbol{u}_i^*\boldsymbol{u}_i/(\epsilon+\|\boldsymbol{u}_i\|^2)\right)\right) \approx 1$, in which case the expression for the EMSE becomes

$$\boxed{\zeta^{\epsilon-\mathsf{NLMS}} = \frac{\mu\sigma_v^2}{2}\mathsf{Tr}(R_u)\mathsf{E}\left(\frac{1}{\|\boldsymbol{u}_i\|^2}\right)} \tag{17.13}$$

This result agrees with the second expression in Lemma 17.1 for small μ. We shall revisit the result in Probs. V.18 and V.20 without the above approximations for the expectations, but with an independence condition on the regressors (namely, condition (22.23)).

CHAPTER 18

Performance of Sign-Error LMS

We now illustrate the use of the variance relation (15.40) in evaluating the steady-state performance of the sign-error LMS algorithm,

$$\boldsymbol{w}_i = \boldsymbol{w}_{i-1} + \mu \boldsymbol{u}_i^* \mathsf{csgn}[\boldsymbol{e}(i)] \tag{18.1}$$

for which

$$g[\boldsymbol{e}(i)] = \mathsf{csgn}[\boldsymbol{e}(i)] = \mathsf{csgn}[\boldsymbol{e}_a(i) + \boldsymbol{v}(i)] \tag{18.2}$$

where the data $\{\boldsymbol{d}(i), \boldsymbol{u}_i, \boldsymbol{v}(i)\}$ are assumed to satisfy model (15.16). To proceed, we need to distinguish between real-valued data and complex-valued data. This is because the definition of the sign function is different in both cases. The final expressions for the EMSE, however, will turn out to be identical except for a scaling factor.

18.1 REAL-VALUED DATA

In this case the sign function is defined as

$$\mathsf{csgn}[x] = \mathsf{sign}[x] = \begin{cases} 1 & x > 0 \\ -1 & x < 0 \\ 0 & x = 0 \end{cases}$$

Using the fact that $g^2(x) = 1$ almost everywhere on the real line, the variance relation (15.40) becomes

$$\mu \mathsf{Tr}(R_u) = 2\mathsf{E}\, \boldsymbol{e}_a(i)\mathsf{sign}[\boldsymbol{e}(i)] = 2\mathsf{E}\, \boldsymbol{e}_a(i)\mathsf{sign}[\boldsymbol{e}_a(i) + \boldsymbol{v}(i)], \quad i \to \infty \tag{18.3}$$

In order to arrive at the value of the EMSE we need to evaluate the expectation on the right-hand side. For this purpose, we shall rely on the following assumption:

$$\boxed{\text{The estimation error } \boldsymbol{e}(i) \text{ and the noise } \boldsymbol{v}(i) \text{ are jointly Gaussian}} \tag{18.4}$$

This condition is violated in general. For example, even if we assume that $\boldsymbol{v}(i)$ and $\boldsymbol{u}_i$ in model (15.16) are Gaussian, so that $\boldsymbol{d}(i)$ is Gaussian as well, this assumption alone does not imply that the estimation error

$$\boldsymbol{e}(i) = \boldsymbol{d}(i) - \boldsymbol{u}_i \boldsymbol{w}_{i-1}$$

will be Gaussian. This is because $\boldsymbol{e}(i)$ depends nonlinearly on the data $\{\boldsymbol{u}_j, \boldsymbol{v}(j)\}$ through its dependence on $\boldsymbol{w}_{i-1}$. However, $\boldsymbol{e}(i)$ would be Gaussian *conditioned* on $\boldsymbol{w}_{i-1}$. In other words, if we assume w_{i-1} is constant, then $e(i)$ will be Gaussian. This observation suggests that when the step-size is sufficiently small, so that the weight estimator w_{i-1} varies slowly with time, assumption (18.4) could be reasonable. This argument also suggests that we could replace (18.4) by the following:

$$\boxed{u(i) \text{ and } v(i) \text{ are Gaussian and } \mu \text{ is sufficiently small}} \tag{18.5}$$

To illustrate the Gaussianity assumption on $e(i)$, we plot in Fig. 18.1 a histogram of the distribution of $e(i)$ at the time instant $i = 99000$. The histogram is obtained by simulating a 20-tap sign-error LMS filter over 400 experiments and using $\mu = 5 \times 10^{-6}$. The input to the tapped-delay-line filter is correlated data obtained by passing a unit-variance i.i.d. Gaussian random process through a first-order auto-regressive model with transfer function given by

$$\frac{\sqrt{1-a^2}}{(1-az^{-1})}, \quad a = 0.8$$

The histogram suggests that the Gaussianity assumption on the distribution of $e(i)$ is reasonable.

The motivation for introducing assumption (18.4) is that it allows us to express the expectation

$$\mathsf{E}\, e_a(i)\mathsf{sign}[e_a(i) + v(i)]$$

in terms of

$$\mathsf{E}\, e_a(i)[e_a(i) + v(i)]$$

with the sign function removed. This is achieved by resorting to a special case of a result known as Price's theorem (see Prob. IV.10). The special case we are interested in states

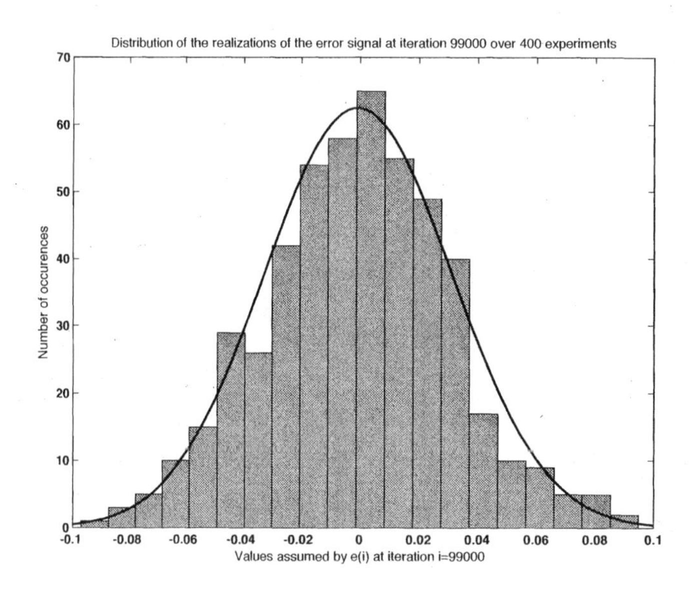

FIGURE 18.1 Histogram of the distribution of $e(i)$ at time instant $i = 99000$. The graph is obtained from 400 runs of a 20-tap sign-error LMS filter with step-size $\mu = 5 \times 10^{-6}$ and correlated input data.

that for any two real-valued zero-mean and jointly Gaussian random variables $\boldsymbol{a}$ and $\boldsymbol{b}$, it holds that

$$\boxed{\mathsf{E}\,\boldsymbol{a}\ \mathsf{sign}(\boldsymbol{b}) \;=\; \sqrt{\frac{2}{\pi}}\,\frac{1}{\sigma_b}\,\mathsf{E}\,\boldsymbol{ab}, \quad \text{where} \quad \sigma_b^2 = \mathsf{E}\,\boldsymbol{b}^2} \tag{18.6}$$

Now, in view of assumption (18.4), we have that $\boldsymbol{e}_a(i)$ and $\boldsymbol{e}(i)$ are jointly Gaussian. Using Price's theorem with the identifications

$$\boldsymbol{a} \leftarrow \boldsymbol{e}_a(i), \quad \boldsymbol{b} \leftarrow \boldsymbol{e}(i), \quad \sigma_b^2 \leftarrow \mathsf{E}\,\boldsymbol{e}_a^2(i) + \sigma_v^2, \quad \mathsf{E}\,\boldsymbol{ab} \leftarrow \mathsf{E}\,\boldsymbol{e}_a^2(i)$$

and substituting into (18.3), we get

$$\mu \mathsf{Tr}(R_u) = 2\sqrt{\frac{2}{\pi}}\,\frac{\mathsf{E}\,\boldsymbol{e}_a^2(i)}{\sqrt{\mathsf{E}\,\boldsymbol{e}_a^2(i) + \sigma_v^2}}, \quad i \to \infty \tag{18.7}$$

Solving for

$$\zeta^{\text{sign-error LMS}} \overset{\Delta}{=} \mathsf{E}\,\boldsymbol{e}_a^2(\infty)$$

we find that

$$\boxed{\zeta^{\text{sign-error LMS}} = \frac{\alpha}{2}\left(\alpha + \sqrt{\alpha^2 + 4\sigma_v^2}\right), \quad \text{where} \quad \alpha = \sqrt{\frac{\pi}{8}}\,\mu \mathsf{Tr}(R_u)} \tag{18.8}$$

18.2 COMPLEX-VALUED DATA

When the data are complex-valued, the definition of the csgn function is

$$\mathsf{csgn}[x] = \mathsf{sign}(x_\mathsf{r}) + j\mathsf{sign}(x_\mathsf{i})$$

where $x = x_\mathsf{r} + jx_\mathsf{i}$ — see Prob. III.15. Using the fact that now $|g(x)|^2 = 2$ almost everywhere in the complex plane, the variance relation (15.40) becomes

$$2\mu \mathsf{Tr}(R_u) = 2\mathsf{Re}\big(\mathsf{E}\,\boldsymbol{e}_a^*(i)\mathsf{csgn}[\boldsymbol{e}_a(i) + \boldsymbol{v}(i)]\big), \quad i \to \infty \tag{18.9}$$

Similarly to assumption (18.4) in the real case, we also need to introduce certain Gaussian assumptions on $\{\boldsymbol{e}(i), \boldsymbol{v}(i)\}$:[6]

1. The real parts of $\boldsymbol{e}(i)$ and $\boldsymbol{v}(i)$ are jointly Gaussian.
2. The imaginary parts of $\boldsymbol{e}(i)$ and $\boldsymbol{v}(i)$ are jointly Gaussian.
3. The real and imaginary parts of $\boldsymbol{e}(i)$ have identical variances.
4. The real parts of $\{\boldsymbol{e}(i), \boldsymbol{v}(i)\}$ are independent of their imaginary parts.

Then if we employ the extension of Price's theorem to complex data (as given by Prob. IV.11), we can rewrite (18.9) as follows:

$$\begin{aligned} 2\mu \mathsf{Tr}(R_u) &= 2\sqrt{\frac{2}{\pi}}\,\frac{\sqrt{2}}{\sqrt{\mathsf{E}\,|\boldsymbol{e}_a(i)|^2 + \sigma_v^2}}\ \mathsf{Re}\big(\mathsf{E}\,\boldsymbol{e}_a^*(i)[\boldsymbol{e}_a(i) + \boldsymbol{v}(i)]\big) \\ &= 2\sqrt{\frac{2}{\pi}}\,\frac{\sqrt{2}\,\mathsf{E}\,|\boldsymbol{e}_a(i)|^2}{\sqrt{\mathsf{E}\,|\boldsymbol{e}_a(i)|^2 + \sigma_v^2}} \end{aligned} \tag{18.10}$$

[6]For item (3), recall that if a complex-valued random variable $\boldsymbol{x}$ is circular, then its real and imaginary parts have equal variances, as can be verified from the defining property $\mathsf{E}\,\boldsymbol{x}^2 = 0$.

which has a form similar to (18.7). Hence, we are led to the following conclusion.

Lemma 18.1 (EMSE of sign-error LMS) Consider the sign-error LMS recursion (18.1) and assume the data $\{\boldsymbol{d}(i), \boldsymbol{u}_i\}$ satisfy model (15.16). Assume further that $\{\boldsymbol{v}(i), \boldsymbol{u}_i\}$ are Gaussian and that the step-size is sufficiently small, as indicated by (18.5). Then its EMSE can be approximated by

$$\zeta^{\text{sign-error LMS}} = \frac{\alpha}{2}\left(\alpha + \sqrt{\alpha^2 + 4\sigma_v^2}\right) \quad \text{where} \quad \alpha = \sqrt{\frac{\gamma\pi}{8}}\;\mu\text{Tr}(R_u)$$

with $\gamma = 1$ for real-valued data and $\gamma = 2$ for complex-valued data. The misadjustment is obtained by dividing the EMSE by σ_v^2.

18.3 SIMULATION RESULTS

Figures 18.2 and 18.3 show the values of the steady-state MSE of a 5-tap sign-error LMS filter for different choices of the step-size and for different signal conditions. The theoretical values are obtained by using the expression from Lemma 18.1. For each step-size, the experimental value is obtained by running the algorithm for 5×10^5 iterations and averaging the squared-error curve over 100 experiments in order to generate the ensemble-average curve. The average of the last 10000 entries of the ensemble-average curve is used as the experimental value for the MSE. The data $\{\boldsymbol{d}(i), \boldsymbol{u}_i\}$ are generated according to model (15.16) using Gaussian noise with variance $\sigma_v^2 = 0.001$.

Figure 18.2 illustrates the situation where the regressors do not have shift structure and they are generated from a Gaussian distribution with a covariance matrix R_u whose eigenvalue spread is $\rho = 5$. In Fig. 18.3, the regressors have shift structure and they are obtained for the top plot by feeding the filter with correlated data obtained by passing a unit-variance i.i.d. Gaussian random process through a first-order auto-regressive model with transfer function $\sqrt{1-a^2}/(1-az^{-1})$ and $a = 0.8$. For the bottom plot of Fig. 18.3, the input data is white with unit variance.

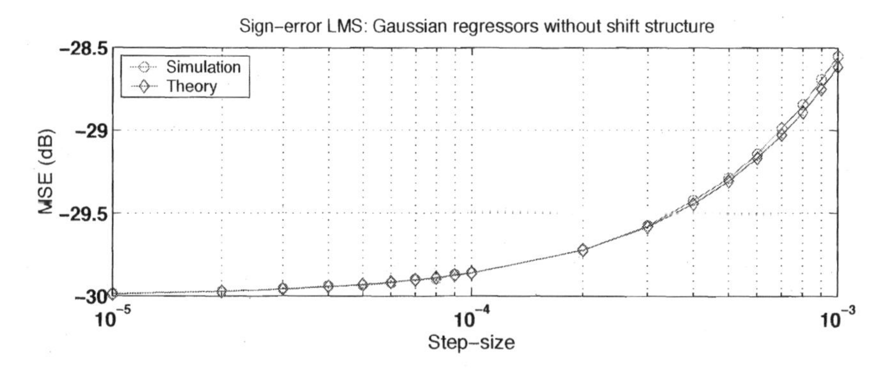

FIGURE 18.2 Theoretical and simulated MSE for a 5-tap sign-error LMS filter with $\sigma_v^2 = 0.001$, and Gaussian regressors *without* shift structure.

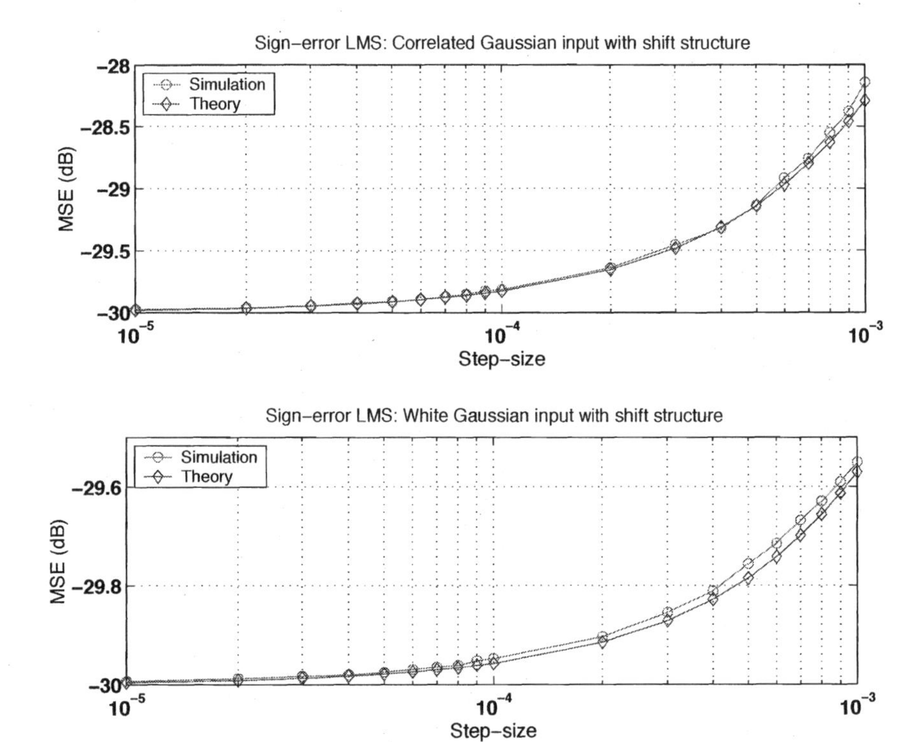

FIGURE 18.3 Theoretical and simulated MSE for a 5-tap sign-error LMS filter with $\sigma_v^2 = 0.001$, and Gaussian regressors *with* shift structure.

CHAPTER 19

Performance of RLS and Other Filters

We now examine the performance of the RLS algorithm and comment on the performance of several other adaptive filters.

19.1 PERFORMANCE OF RLS

We consider the RLS algorithm of Sec. 14.1, namely,

$$\boldsymbol{P}_i = \lambda^{-1}\left[\boldsymbol{P}_{i-1} - \frac{\lambda^{-1}\boldsymbol{P}_{i-1}\boldsymbol{u}_i^*\boldsymbol{u}_i\boldsymbol{P}_{i-1}}{1+\lambda^{-1}\boldsymbol{u}_i\boldsymbol{P}_{i-1}\boldsymbol{u}_i^*}\right] \quad (19.1)$$

$$\boldsymbol{w}_i = \boldsymbol{w}_{i-1} + \boldsymbol{P}_i\,\boldsymbol{u}_i^*[\boldsymbol{d}(i) - \boldsymbol{u}_i\boldsymbol{w}_{i-1}], \quad i \geq 0 \quad (19.2)$$

with initial condition $\boldsymbol{P}_{-1} = \epsilon^{-1}\mathrm{I}$ and $0 \ll \lambda \leq 1$. The scalar ϵ is a small positive number and λ is usually close to one. We are using a boldface letter $\boldsymbol{P}_i$ to indicate that $\boldsymbol{P}_i$ is a random variable due to its dependence on the regressors $\{\boldsymbol{u}_j\}$. Also, recall from (14.3) and (14.5) that

$$\boxed{\boldsymbol{P}_i^{-1} = \lambda^{i+1}\epsilon\mathrm{I} + \sum_{j=0}^{i}\lambda^{i-j}\boldsymbol{u}_j^*\boldsymbol{u}_j} \quad (19.3)$$

which shows that $\boldsymbol{P}_i > 0$ for all finite i.

Compared with the update recursion (15.23), which was seen to be useful in studying the performance of several adaptive filters in the previous sections, we now find that the RLS update differs in a special way; it includes the matrix factor $\boldsymbol{P}_i$, which appears multiplying $\boldsymbol{u}_i^*$ from the left in (19.2). Still, the energy-conservation approach of Sec. 15.3 can be extended to treat this case, and even some other more general cases (see, e.g., Prob. IV.9 as well as Part V (*Transient Performance*) and the problems therein). Since the arguments in the sequel are similar to what we encountered in Sec. 15.3 while deriving the energy relation (15.32), we shall be brief and only highlight the ideas that are particular to RLS.

Energy-Conservation Relation

To begin with, the update recursion (19.2) can be rewritten in terms of the weight error vector $\tilde{\boldsymbol{w}}_i = w^o - \boldsymbol{w}_i$ as

$$\boxed{\tilde{\boldsymbol{w}}_i = \tilde{\boldsymbol{w}}_{i-1} - \boldsymbol{P}_i\boldsymbol{u}_i^*\boldsymbol{e}(i)} \quad (19.4)$$

If we multiply both sides of this recursion by $\boldsymbol{u}_i$ from the left we find that the *a priori* and *a posteriori* estimation errors are related via:

$$\boxed{\boldsymbol{e}_p(i) = \boldsymbol{e}_a(i) - \|\boldsymbol{u}_i\|^2_{\boldsymbol{P}_i}\boldsymbol{e}(i)} \quad (19.5)$$

where

$$\boxed{e_a(i) = \boldsymbol{u}_i\tilde{\boldsymbol{w}}_{i-1}, \quad e_p(i) = \boldsymbol{u}_i\tilde{\boldsymbol{w}}_i} \tag{19.6}$$

and where the notation $\|x\|_W^2$ stands for the squared-weighted Euclidean norm of a vector, namely, for a column vector x, $\|x\|_W^2 = x^*Wx$.

To extend the energy relation to this case, we combine (19.4)–(19.5) and proceed exactly as in Sec. 15.3. For example, when $\boldsymbol{u}_i \neq 0$, we use (19.5) to eliminate $\boldsymbol{e}(i)$ from (19.4). This calculation leads to

$$\tilde{\boldsymbol{w}}_i + \frac{\boldsymbol{P}_i\boldsymbol{u}_i^*}{\|\boldsymbol{u}_i\|_{\boldsymbol{P}_i}^2}\boldsymbol{e}_a(i) = \tilde{\boldsymbol{w}}_{i-1} + \frac{\boldsymbol{P}_i\boldsymbol{u}_i^*}{\|\boldsymbol{u}_i\|_{\boldsymbol{P}_i}^2}\boldsymbol{e}_p(i)$$

By equating the squared weighted norms on both sides of this equality, using $\boldsymbol{P}_i^{-1}$ as a weighting matrix, we arrive at

$$\boxed{\|\tilde{\boldsymbol{w}}_i\|_{\boldsymbol{P}_i^{-1}}^2 + \bar{\mu}(i)|\boldsymbol{e}_a(i)|^2 = \|\tilde{\boldsymbol{w}}_{i-1}\|_{\boldsymbol{P}_i^{-1}}^2 + \bar{\mu}(i)|\boldsymbol{e}_p(i)|^2} \tag{19.7}$$

where

$$\bar{\mu}(i) \triangleq \left(\|\boldsymbol{u}_i\|_{\boldsymbol{P}_i}^2\right)^{\dagger} = \begin{cases} 1/\|\boldsymbol{u}_i\|_{\boldsymbol{P}_i}^2 & \text{if } \boldsymbol{u}_i \neq 0 \\ 0 & \text{otherwise} \end{cases} \tag{19.8}$$

The equality (19.7) extends the energy-conservation relation (15.32) to the RLS context. Observe that the main distinction is the appearance of the weighting matrices $\{\boldsymbol{P}_i^{-1}, \boldsymbol{P}_i\}$: the former is used as a weighting factor for the weight-error vectors while the latter is used as a weighting factor for the regressor.

We are interested in evaluating the performance of RLS in steady-state. To do so, we shall call upon the steady-state condition (cf. Def. 15.1),

$$\mathsf{E}\,\tilde{\boldsymbol{w}}_i\tilde{\boldsymbol{w}}_i^* = \mathsf{E}\,\tilde{\boldsymbol{w}}_{i-1}\tilde{\boldsymbol{w}}_{i-1}^* = C, \qquad \text{as } i \to \infty \tag{19.9}$$

in order to transform (19.7) into a variance relation (as was done in Sec. 15.4). However, the presence of the matrices $\{\boldsymbol{P}_i, \boldsymbol{P}_i^{-1}\}$ in (19.7) makes this step challenging; this is because the matrices $\{\boldsymbol{P}_i, \boldsymbol{P}_i^{-1}\}$ are dependent not only on $\boldsymbol{u}_i$ but also on all prior regressors $\{\boldsymbol{u}_j, j \leq i\}$. For this reason, in order to make the performance analysis of RLS more tractable, whenever necessary, we shall approximate and replace the random variables $\{\boldsymbol{P}_i^{-1}, \boldsymbol{P}_i\}$ in steady-state by their respective mean values. In a sense, this approximation amounts to an ergodicity assumption on the regressors. For an ergodic random process, the time-average coincides with the ensemble average. What this means in the context of RLS is the following. Observe from expression (19.3) that, as $i \to \infty$, $\boldsymbol{P}_i^{-1}$ can be regarded as a weighted sum (or a time-average) of infinitely many terms of the form $\boldsymbol{u}_k^*\boldsymbol{u}_k$. Assuming these terms are realizations of a process $\boldsymbol{u}$ that is ergodic in its second moment, then we can approximate $\boldsymbol{P}_i^{-1}$ by its mean value.

Steady-State Approximation

To begin with, note from (19.3) that

$$\boldsymbol{P}_i^{-1} = \lambda^{i+1}\epsilon\mathrm{I} + \left[\boldsymbol{u}_i^*\boldsymbol{u}_i + \lambda\boldsymbol{u}_{i-1}^*\boldsymbol{u}_{i-1} + \ldots + \lambda^{i-1}\boldsymbol{u}_1^*\boldsymbol{u}_1 + \lambda^i\boldsymbol{u}_0^*\boldsymbol{u}_0\right] \tag{19.10}$$

so that, as $i \to \infty$, and since $\lambda < 1$, the steady-state mean value of $\boldsymbol{P}_i^{-1}$ is given by

$$\boxed{\lim_{i\to\infty} \mathsf{E}\left(\boldsymbol{P}_i^{-1}\right) = \frac{R_u}{1-\lambda} \triangleq P^{-1}} \tag{19.11}$$

We denote the result by P^{-1}. The mean value of $\boldsymbol{P}_i$, on the other hand, is considerably harder to evaluate. So we shall satisfy ourselves with the approximation

$$\mathsf{E}\,\boldsymbol{P}_i \approx \left[\mathsf{E}\,\boldsymbol{P}_i^{-1}\right]^{-1} = (1-\lambda)R_u^{-1} = P, \quad \text{as } i \to \infty \tag{19.12}$$

This is an approximation, of course, because even though $\boldsymbol{P}_i$ and $\boldsymbol{P}_i^{-1}$ are the inverses of one another, it does not hold that their expected values will have the same inverse relation. Consider, for example, a random variable $\boldsymbol{x}$ that is uniformly distributed inside the interval $[1,2]$. Its mean is $3/2$. On the other hand, the random variable $\boldsymbol{y} = 1/\boldsymbol{x}$ has mean value equal to $\ln 2$. Still, approximation (19.12) is generally reasonable for Gaussian regressors. In order to illustrate this fact, we plot in Fig. 19.1 a curve showing how the matrices $\mathsf{E}\,\boldsymbol{P}_i$ and $(1-\lambda)R_u^{-1}$ compare for different forgetting factors. The curve in the figure is generated by running an RLS filter of order 5 over 2000 iterations and averaging the results over 50 experiments. The regressors are chosen as independent realizations of a Gaussian distribution with a covariance matrix R_u whose eigenvalue spread is $\rho = 5$. The steady-state value of $\mathsf{E}\,\boldsymbol{P}_i$ is estimated by averaging the last 200 values of $\boldsymbol{P}_i$ in each run over all experiments. The curve plots the following relative measure of closeness between $\mathsf{E}\,\boldsymbol{P}_i$ and $(1-\lambda)R_u^{-1}$:

$$\kappa \triangleq \frac{\|\mathsf{E}\,\boldsymbol{P}_i - (1-\lambda)R_u^{-1}\|}{\|(1-\lambda)R_u^{-1}\|}$$

in terms of the ratio of the norm of the difference to the norm of $(1-\lambda)R_u^{-1}$. The matrix norm used here is the maximum singular value of the matrix — see Sec. B.6; other matrix norms can be used as well. It is seen from the figure that there is a good match between $\mathsf{E}\,\boldsymbol{P}_i$ and $(1-\lambda)R_u^{-1}$, especially for forgetting factors that are close to one.

For example, for $\lambda = 0.995$, the following steady-state value for $\mathsf{E}\,\boldsymbol{P}_i$ was estimated in this simulation by means of ensemble averaging:

$$\mathsf{E}\,\boldsymbol{P}_i \approx 10^{-3}\cdot\begin{bmatrix} 5.1176 & \times & \times & \times & \times \\ \times & 1.8147 & \times & \times & \times \\ \times & \times & 4.8365 & \times & \times \\ \times & \times & \times & 1.1968 & \times \\ \times & \times & \times & \times & 1.0239 \end{bmatrix}$$

with relatively small off-diagonal elements, while the value of $(1-\lambda)R_u^{-1}$ was

$$(1-\lambda)R_u^{-1} = 10^{-3}\cdot\text{diag}\{5.0, 1.8, 4.7, 1.2, 1.0\}$$

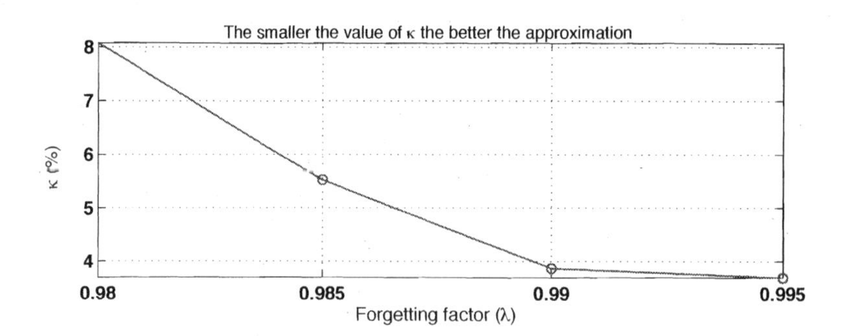

FIGURE 19.1 A plot of the relative difference between $\mathsf{E}\,\boldsymbol{P}_i$ and $(1-\lambda)R_u^{-1}$ for different values of λ and for Gaussian regressors.

Filter Performance

Returning to the energy-conservation relation (19.7), replacing $\boldsymbol{P}_i^{-1}$ by its assumed mean value P^{-1}, we find that $\mathsf{E}\,\|\tilde{\boldsymbol{w}}_i\|^2_{\boldsymbol{P}_i^{-1}}$ in steady-state evaluates to

$$\mathsf{E}\,\|\tilde{\boldsymbol{w}}_i\|^2_{\boldsymbol{P}_i^{-1}} \;\approx\; \mathsf{E}\,\|\tilde{\boldsymbol{w}}_i\|^2_{P^{-1}} \;=\; \mathsf{Tr}(CP^{-1})$$

Likewise, in view of the steady-state assumption (19.9),

$$\mathsf{E}\,\|\tilde{\boldsymbol{w}}_{i-1}\|^2_{\boldsymbol{P}_i^{-1}} \;\approx\; \mathsf{E}\,\|\tilde{\boldsymbol{w}}_{i-1}\|^2_{P^{-1}} \;=\; \mathsf{Tr}(CP^{-1})$$

In other words, we find that in steady-state

$$\boxed{\mathsf{E}\,\|\tilde{\boldsymbol{w}}_i\|^2_{\boldsymbol{P}_i^{-1}} \;=\; \mathsf{E}\,\|\tilde{\boldsymbol{w}}_{i-1}\|^2_{\boldsymbol{P}_i^{-1}}} \tag{19.13}$$

This steady-state condition is the extension of (15.35) to the RLS context.

Now taking expectations of both sides of (19.7), and using (19.13), we arrive at

$$\mathsf{E}\,\bar{\mu}(i)|\boldsymbol{e}_a(i)|^2 = \mathsf{E}\,\bar{\mu}(i)|\boldsymbol{e}_p(i)|^2, \quad \text{as } i \to \infty \tag{19.14}$$

However, from (19.5) we know how $\boldsymbol{e}_p(i)$ is related to $\boldsymbol{e}_a(i)$. Substituting into (19.14) we get

$$\mathsf{E}\,\bar{\mu}(i)|\boldsymbol{e}_a(i)|^2 = \mathsf{E}\,\bar{\mu}(i)\left|\boldsymbol{e}_a(i) - \|\boldsymbol{u}_i\|^2_{\boldsymbol{P}_i}\boldsymbol{e}(i)\right|^2, \quad \text{as } i \to \infty \tag{19.15}$$

which upon expansion and simplification, reduces to

$$\boxed{\mathsf{E}\,\|\boldsymbol{u}_i\|^2_{\boldsymbol{P}_i}|\boldsymbol{e}(i)|^2 = 2\mathsf{Re}\big(\mathsf{E}\,\boldsymbol{e}_a^*(i)\boldsymbol{e}(i)\big), \quad \text{as } i \to \infty} \tag{19.16}$$

This variance relation is the extension of (15.40) to the RLS case. Now, since the data $\{\boldsymbol{d}(i), \boldsymbol{u}_i\}$ satisfy conditions (15.16), we also have that $\boldsymbol{e}(i) = \boldsymbol{e}_a(i) + \boldsymbol{v}(i)$, and (19.16) becomes

$$\boxed{\sigma_v^2\mathsf{E}\,\|\boldsymbol{u}_i\|^2_{\boldsymbol{P}_i} + \mathsf{E}\,\|\boldsymbol{u}_i\|^2_{\boldsymbol{P}_i}|\boldsymbol{e}_a(i)|^2 = 2\mathsf{E}\,|\boldsymbol{e}_a(i)|^2 \;=\; 2\zeta^{\mathsf{RLS}}, \quad i \to \infty} \tag{19.17}$$

To proceed we resort to a separation condition similar to (16.7), namely, we assume that

$$\boxed{\text{At steady-state, } \; \|\boldsymbol{u}_i\|^2_{\boldsymbol{P}_i} \; \text{ is independent of } \; \boldsymbol{e}_a(i)} \tag{19.18}$$

This condition allows us to separate the expectation $\mathsf{E}\,\|\boldsymbol{u}_i\|^2_{\boldsymbol{P}_i}|\boldsymbol{e}_a(i)|^2$ into the product of two expectations, i.e.,

$$\mathsf{E}\,\left(\|\boldsymbol{u}_i\|^2_{\boldsymbol{P}_i}\cdot|\boldsymbol{e}_a(i)|^2\right) = \left(\mathsf{E}\,\|\boldsymbol{u}_i\|^2_{\boldsymbol{P}_i}\right)\cdot\left(\mathsf{E}\,|\boldsymbol{e}_a(i)|^2\right)$$

If we now replace $\boldsymbol{P}_i$ by its assumed mean value, we obtain the approximation

$$\mathsf{E}\,\|\boldsymbol{u}_i\|^2_{\boldsymbol{P}_i} \;\approx\; \mathsf{E}\,\|\boldsymbol{u}_i\|^2_{P} \;=\; \mathsf{Tr}(R_uP) \;=\; (1-\lambda)M \tag{19.19}$$

Substituting into (19.17) we get

$$\boxed{\zeta^{\mathsf{RLS}} = \frac{\sigma_v^2(1-\lambda)M}{2-(1-\lambda)M}} \tag{19.20}$$

Usually, the value of λ is very close to one, so that

$$\boxed{\zeta^{\text{RLS}} \approx \frac{\sigma_v^2(1-\lambda)M}{2}} \tag{19.21}$$

Lemma 19.1 (EMSE of RLS) Consider the RLS recursion (19.2)–(19.1) and assume the data $\{\boldsymbol{d}(i), \boldsymbol{u}_i\}$ satisfy model (15.16). Introduce further the approximations (19.12) and (19.19). Then, under the separation condition (19.18), the EMSE of RLS can be approximated by

$$\zeta^{\text{RLS}} = \frac{\sigma_v^2(1-\lambda)M}{2-(1-\lambda)M} \quad \text{or} \quad \zeta^{\text{RLS}} = \frac{\sigma_v^2(1-\lambda)M}{2}$$

The misadjustment is obtained by dividing the EMSE by σ_v^2.

A conclusion that stands out from the expression in the lemma is that the performance of RLS is independent of the input covariance matrix R_u. [The mean-square performance of RLS is further studied in Probs. IV.15 and V.36.]

Simulation Results

Figure 19.2 shows the values of the steady-state MSE of a 5-tap RLS filter for different choices of the forgetting parameter λ. The theoretical values are obtained by using the expressions from Lemma 19.1. The experimental value is obtained by running the algorithm for 2000 iterations and averaging the squared-error curve over 300 experiments in order to generate the ensemble-average curve. The average of the last 1000 entries of the ensemble-average curve is used as the experimental value for the MSE. The data $\{\boldsymbol{d}(i), \boldsymbol{u}_i\}$ are generated according to model (15.16) using Gaussian noise with variance $\sigma_v^2 = 0.001$.

Two situations are illustrated in the figure. For the top plot, the regressors do not have shift structure and they are generated as independent realizations of a Gaussian distribution with a covariance matrix R_u whose eigenvalue spread is $\rho = 5$. For the bottom plot, the regressors have shift structure and they are obtained by feeding the filter with a correlated input process that is obtained by passing a unit-variance i.i.d. Gaussian random process through a first-order auto-regressive model with transfer function $\sqrt{1-a^2}/(1-az^{-1})$ and $a = 0.8$.

19.2 PERFORMANCE OF OTHER FILTERS

The same line of reasoning used so far in Part IV (*Mean-Square Performance*) is used in the problems, and also in Part V (*Transient Performance*), to study the mean-square performance of several other adaptive filters. For ease of reference, we list here the locations in the text that study some of these filters.

1. **Leaky-LMS**. The leaky-LMS filter is described by the recursion (see Alg. 12.2):

$$\boldsymbol{w}_i = (1-\mu\alpha)\boldsymbol{w}_{i-1} + \mu\boldsymbol{u}_i^*[\boldsymbol{d}(i) - \boldsymbol{u}_i\boldsymbol{w}_{i-1}], \quad i \geq 0$$

which looks similar to the LMS recursion (16.1) except for the presence of the factor α. The mean-square performance of this algorithm is treated in Probs. V.30 and V.32, with Prob. V.30 providing an expression for the EMSE of the filter when $R_u =$

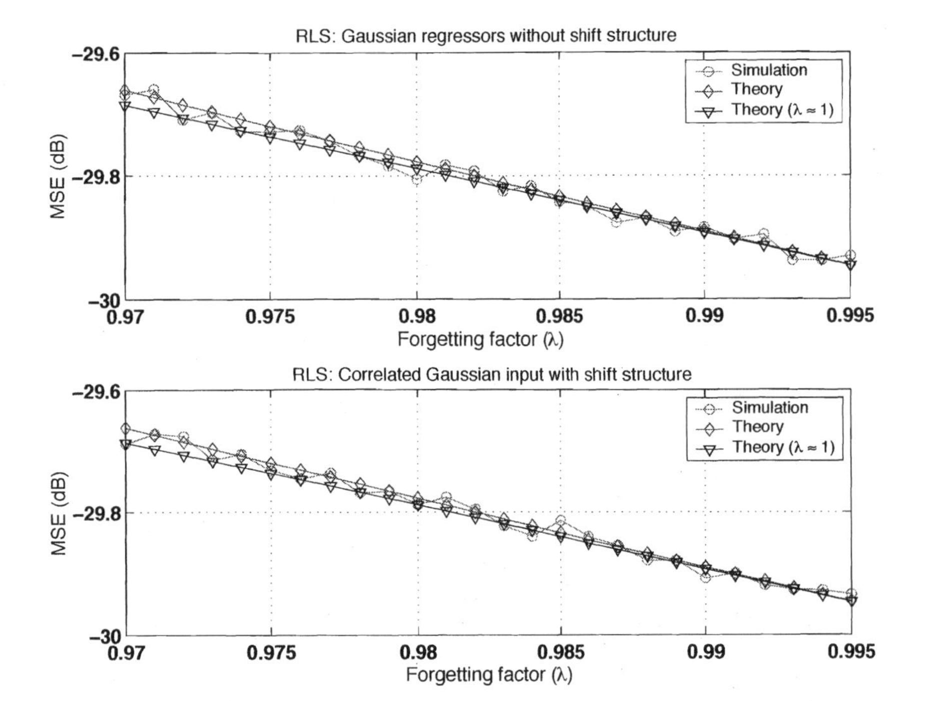

FIGURE 19.2 Theoretical and simulated MSE for a 5-tap RLS filter with $\sigma_v^2 = 0.001$, and Gaussian regressors with and without shift structure.

$\sigma_u^2 \mathrm{I}$ and Prob. V.32 providing an expression for the EMSE of the filter when the regressors are circular Gaussian. In addition, Prob. V.29 studies the mean-square performance of a more general class of leaky filters.

2. **ϵ-NLMS with power normalization.** The algorithm is described by the recursions:

$$\begin{aligned} p(i) &= \beta p(i-1) + (1-\beta)|u(i)|^2, \quad p(-1) = 0 \quad &(19.22) \\ w_i &= w_{i-1} + \frac{\mu}{\epsilon + p(i)} u_i^* e(i) \quad &(19.23) \end{aligned}$$

where β is a positive number in the interval $0 < \beta < 1$. Compared with the update equation for ϵ-NLMS in (17.1), we see that the squared-norm of the regressor, $\|u_i\|^2$, is replaced by the power estimate $p(i)$. The mean-square performance of the algorithm is treated in Prob. IV.5 for Gaussian regressors with shift structure.

3. **LMF and LMMN.** The LMMN algorithm is described by (cf. Alg. 12.4):

$$w_i = w_{i-1} + \mu u_i^* e(i)[\delta + (1-\delta)|e(i)|^2], \quad 0 \leq \delta \leq 1 \qquad (19.24)$$

which employs a linear combination of the $e(i)$ and $e(i)|e(i)|^2$. The least-mean fourth (LMF) algorithm corresponds to the special case $\delta = 0$,

$$w_i = w_{i-1} + \mu u_i^* e(i)|e(i)|^2$$

The mean-square performance of these algorithms is treated in Probs. IV.6.

4. ϵ**–APA**. The ϵ–APA algorithm is described by the recursions (cf. Sec. 13.1):

$$\begin{aligned} \boldsymbol{U}_i &= \begin{bmatrix} \boldsymbol{u}_i \\ \boldsymbol{u}_{i-1} \\ \vdots \\ \boldsymbol{u}_{i-1+K} \end{bmatrix}, \quad \boldsymbol{d}_i = \begin{bmatrix} \boldsymbol{d}(i) \\ \boldsymbol{d}(i-1) \\ \vdots \\ \boldsymbol{d}(i-1+K) \end{bmatrix} & (19.25) \\ \boldsymbol{e}_i &= \boldsymbol{d}_i - \boldsymbol{U}_i \boldsymbol{w}_{i-1} & (19.26) \\ \boldsymbol{w}_i &= \boldsymbol{w}_{i-1} + \mu \boldsymbol{U}_i^* \left(\epsilon \mathrm{I} + \boldsymbol{U}_i \boldsymbol{U}_i^*\right)^{-1} \boldsymbol{e}_i, \quad i \geq 0 & (19.27) \end{aligned}$$

where ϵ is a small positive number and $K \leq M$. The energy conservation approach of Sec. 15.3 is extended to this case and the mean-square performance of the algorithm is studied in Prob. IV.7.

5. **Sign-regressor LMS**. The sign-regressor LMS filter is described by the recursion

$$\boldsymbol{w}_i = \boldsymbol{w}_{i-1} + \mu \cdot \mathsf{csgn}[\boldsymbol{u}_i^*] \cdot [\boldsymbol{d}(i) - \boldsymbol{u}_i \boldsymbol{w}_{i-1}], \quad i \geq 0$$

Compared with the sign-error LMS recursion (18.1), we see that the sign function is now applied to the regressor as opposed to the error signal, $\boldsymbol{e}(i) = \boldsymbol{d}(i) - \boldsymbol{u}_i \boldsymbol{w}_{i-1}$. The mean-square performance of this algorithm is studied in Prob. V.25.

6. **CMA**. The CMA2-2 algorithm is described by the recursion

$$\boldsymbol{w}_i = \boldsymbol{w}_{i-1} + \mu \boldsymbol{u}_i^* \boldsymbol{z}(i)[\gamma - |\boldsymbol{z}(i)|^2], \quad \boldsymbol{z}(i) = \boldsymbol{u}_i \boldsymbol{w}_{i-1}, \quad i \geq 0$$

where γ is a positive scalar (see Alg. 12.6). This algorithm can be viewed as a special case of the more general recursion

$$\boldsymbol{w}_i = \boldsymbol{w}_{i-1} + \mu \boldsymbol{u}_i^* g[\boldsymbol{z}(i)]$$

where $g[\boldsymbol{z}(i)]$ is an arbitrary function of the filter output, $\boldsymbol{z}(i)$. In Chapters 15–18, we focused on adaptive filters of the form (15.23), which used $g[\boldsymbol{e}(i)]$ instead of $g[\boldsymbol{z}(i)]$. Of course, $g[\boldsymbol{e}(i)]$ is a function of $\boldsymbol{z}(i)$ since $\boldsymbol{e}(i) = \boldsymbol{d}(i) - \boldsymbol{z}(i)$. The above formulation with $g[\boldsymbol{z}(i)]$ is more general and it accommodates other filters, such as CMA2-2. The derivations in Secs. 15.3 and 15.4 still apply to this broader formulation with $g[\boldsymbol{z}(i)]$.

The main issue that arises while studying CMA2-2, in contrast to the adaptive filters already studied in this part, is the absence of a reference sequence $\boldsymbol{d}(i)$ and, correspondingly, of an explicit weight vector w^o relative to which we may define the weight-error vector, $\tilde{\boldsymbol{w}}_i = w^o - \boldsymbol{w}_i$. This issue is addressed in Probs. IV.16–IV.17, where expressions for the mean-square performance of CMA2-2 are derived for both real- and complex-valued data. Problem IV.18 extends the results to CMA1-2, which is described by the following recursion (see Alg. 12.5):

$$\boldsymbol{w}_i = \boldsymbol{w}_{i-1} + \mu \boldsymbol{u}_i^* \left[\gamma \frac{\boldsymbol{z}(i)}{|\boldsymbol{z}(i)|} - \boldsymbol{z}(i)\right], \quad \boldsymbol{z}(i) = \boldsymbol{u}_i \boldsymbol{w}_{i-1}, \quad i \geq 0$$

19.3 PERFORMANCE TABLE FOR SMALL STEP-SIZES

We find it useful to list in Table 19.1 the EMSE of several adaptive filters under the assumption of sufficiently small step-sizes. The expressions in the table are obtained as approximations from the results derived in this chapter and they serve as convenient references for comparing algorithms. The table also lists points of reference for each algorithm.

TABLE 19.1 Approximate expressions for the excess mean-square performance of several adaptive filters for sufficiently small step-sizes.

Algorithm	**EMSE**	**Reference**
LMS	$\mu\sigma_v^2\text{Tr}(R_u)/2$	Lemma 16.1
ϵ-NLMS	$\frac{\mu\sigma_v^2}{2-\mu}\text{Tr}(R_u)\text{E}\left(\frac{1}{\lVert\boldsymbol{u}_i\rVert^2}\right)$	Lemma 17.1
ϵ-NLMS with power normalization	$\frac{\mu(1+\beta)M\sigma_v^2}{2\left[\gamma(1-\beta)+2\beta\right]-\mu M(1+\beta)}$	Problem IV.5
sign-error LMS	$\frac{\alpha}{2}\left(\alpha+\sqrt{\alpha^2+4\sigma_v^2}\,\right),\ \alpha=\sqrt{\frac{\gamma\pi}{8}}\,\mu\text{Tr}(R_u)$	Lemma 18.1
LMF	$\mu\xi_v^6\text{Tr}(R_u)/2\sigma_v^2$	Problem IV.6
LMMN	$\mu a\text{Tr}(R_u)/2b$	Problem IV.6
leaky-LMS	$\frac{\mu\sigma_v^2}{2}\text{Tr}[R_u^2(R_u+\alpha\text{I})^{-1}]$	Problem V.32
sign-regressor LMS	$\mu\sigma_v^2M/\left(\sqrt{\frac{8}{\pi\sigma_u^2}}-\mu M\right)$	Problem V.25
ϵ-APA	$\frac{\mu\sigma_v^2}{2-\mu}\text{Tr}(R_u)\text{E}\left(\frac{K}{\lVert\boldsymbol{u}_i\rVert^2}\right)$	Problem IV.7
RLS	$\sigma_v^2(1-\lambda)M/2$	Lemma 19.1
CMA2-2	$\mu\,\frac{\text{E}\,(\gamma^2\lvert\boldsymbol{s}\rvert^2-2\gamma\lvert\boldsymbol{s}\rvert^4+\lvert\boldsymbol{s}\rvert^6)}{2\text{E}\,(2\lvert\boldsymbol{s}\rvert^2-\gamma)}\,\text{Tr}(R_u)$	Problem IV.17
CMA1-2	$\mu\left(\gamma^2+\text{E}\,\lvert\boldsymbol{s}\rvert^2-2\gamma\text{E}\,\lvert\boldsymbol{s}\rvert\right)\text{Tr}(R_u)/2$	Problem IV.18

CHAPTER 20

Nonstationary Environments

A fundamental feature of adaptive filters is their ability to track variations in the underlying signal statistics. This is because by relying on instantaneous data, the statistical properties of the weight vector and error signals are able to react to changes in the input signal properties. The purpose of this chapter, and the next one, is to characterize the tracking ability of adaptive filters for nonstationary environments.

We shall continue to rely on the energy-conservation framework of Chapter 15 and use it to derive expressions for the excess mean-square error of an adaptive filter when the input signal properties vary with time. The presentation will reveal that there are actually minor differences between mean-square analysis and tracking analysis and that, in particular, tracking results can be obtained almost by inspection from the mean-square results of the prior chapters.

20.1 MOTIVATION

In order to motivate our setup for tracking analysis, we start by reviewing the basic linear least-mean-squares estimation problem of Sec. 15.1. Thus, let $\boldsymbol{d}$ and $\boldsymbol{u}$ be zero-mean random variables with second-order moments

$$\mathsf{E}\,|\boldsymbol{d}|^2 = \sigma_d^2, \qquad \mathsf{E}\,\boldsymbol{d}\boldsymbol{u}^* = R_{du}, \qquad \mathsf{E}\,\boldsymbol{u}^*\boldsymbol{u} = R_u > 0$$

The coefficient vector w^o that estimates $\boldsymbol{d}$ from $\boldsymbol{u}$ optimally in the linear least-mean-squares sense, i.e., the vector that solves

$$\min_{w}\ \mathsf{E}\ |\boldsymbol{d} - \boldsymbol{u}w|^2 \tag{20.1}$$

is given by $w^o = R_u^{-1}R_{du}$. The corresponding minimum cost is

$$J_{\min} = \sigma_d^2 \ - \ R_{ud}R_u^{-1}R_{du} = \mathsf{E}\,|\boldsymbol{d} - \boldsymbol{u}w^o|^2 \tag{20.2}$$

In Chapter 10 we developed stochastic-gradient algorithms (i.e., adaptive filters) for approximating w^o. These algorithms rely on data $\{\boldsymbol{d}(i), \boldsymbol{u}_i\}$ with moments $\{\sigma_d^2, R_{du}, R_u\}$ and use update equations of the form, say, for the case of an LMS implementation,

$$\boldsymbol{w}_i = \boldsymbol{w}_{i-1} \ + \ \mu\boldsymbol{u}_i^*\boldsymbol{e}(i), \qquad \boldsymbol{w}_{-1} = \text{ initial condition}$$

or some other update form. In Chapter 15 we evaluated the performance of such filters by measuring the excess mean-square error that is left in steady-state, namely, by computing the difference

$$\mathsf{EMSE} = \lim_{i\to\infty}\ \mathsf{E}\,|\boldsymbol{e}(i)|^2 \ - \ J_{\min} \tag{20.3}$$

where

$$e(i) = d(i) - u_i w_{i-1} \tag{20.4}$$

is the estimation error that results from the adaptive implementation.

Now if the moments $\{\sigma_d^2, R_{du}, R_u\}$ vary with time, say, if

$$\mathsf{E}\,|d(i)|^2 = \sigma_{d,i}^2, \qquad \mathsf{E}\,d(i)u_i^* = R_{du,i}, \qquad \mathsf{E}\,u_i^* u_i = R_{u,i} \tag{20.5}$$

then the optimal weight vector w^o will also vary with time. Specifically, the coefficient vector for estimating $d(i)$ from u_i in the linear least-mean-squares sense will be given by

$$w_i^o = R_{u,i}^{-1} R_{du,i} \tag{20.6}$$

with minimum cost

$$J_{\min}(i) \;=\; \sigma_{d,i}^2 - R_{ud,i} R_{u,i}^{-1} R_{du,i} \;=\; \mathsf{E}\,|d(i) - u_i w_i^o|^2 \tag{20.7}$$

If the $\{R_{du,i}, R_{u,i}\}$ vary slowly with time, then it is justifiable to expect that an adaptive filter will have sufficient time to track the optimal solution w_i^o. If, on the other hand, the moments $\{R_{du,i}, R_{u,i}\}$ vary rapidly with time, then this task becomes challenging (and, at times, impossible).

The purpose of a tracking analysis is to quantify how well an adaptive filter performs under such changing conditions in the signal statistics. In order to make the analysis tractable, it is customary to assume that the statistics of the data vary in a certain manner rather than arbitrarily. For instance, the model that we shall adopt in the next section assumes that R_u and $J_{\min}$ remain fixed, while only $\{\sigma_d^2, R_{du}\}$ may vary with time.

20.2 NONSTATIONARY DATA MODEL

Thus, recall from the discussion in Sec. 15.2 that any given data $\{d(i), u_i\}$ can be assumed to be related via a linear model of the form

$$\boxed{d(i) = u_i w_i^o + v(i)} \tag{20.8}$$

where w_i^o is the coefficient vector (20.6) that estimates $d(i)$ from u_i optimally in the linear least-mean squares sense. Moreover, $v(i)$ is uncorrelated with u_i and has variance $\sigma_v^2(i) = J_{\min}(i)$. However, as was also the case in Sec. 15.2, we shall impose the stronger assumption that

$$\boxed{\{v(i)\} \text{ is i.i.d. with constant variance } \sigma_v^2, \text{ and is independent of all } \{u_j\}} \tag{20.9}$$

This condition on $v(i)$ is an assumption because the signal $v(i)$ in the model (20.8) is only uncorrelated with u_i; it is not necessarily independent of u_i, or of all $\{u_j\}$ for that matter. Moreover, the variance of $v(i)$ is not constant. Still there are situations where conditions (20.8) and (20.9) hold simultaneously, e.g., in the channel estimation application of Sec. 10.5. In that application, it is usually justified to expect the noise sequence $\{v(i)\}$ to be i.i.d. and independent of all other data, including the regression data.

Random-Walk Model

In addition to the model (20.8)–(20.9) for the data $\{d(i), u_i\}$, we shall also adopt a model for the variations in the weight vector w_i^o. It is more convenient to adopt a model for the variation in w_i^o than a model for the variations in the statistics $\{\sigma_d^2, R_{du}\}$. One particular model that is widely used in the adaptive filtering literature is a first-order random-walk model. The model assumes that w_i^o undergoes random variations of the form

$$\boxed{\boldsymbol{w}_i^o = \boldsymbol{w}_{i-1}^o + \boldsymbol{q}_i} \tag{20.10}$$

with $\boldsymbol{q}_i$ denoting some random perturbation that is independent of $\{\boldsymbol{u}_j, \boldsymbol{v}(j)\}$ for all i, j. Observe that we are now using boldface letters for $\{\boldsymbol{w}_i^o, \boldsymbol{w}_{i-1}^o\}$. This is because they become random variables due to the presence of the random quantity $\boldsymbol{q}_i$. The sequence $\{\boldsymbol{q}_i\}$ is assumed to be i.i.d., zero-mean, with covariance matrix

$$\boxed{\mathsf{E}\,\boldsymbol{q}_i\boldsymbol{q}_i^* = Q} \tag{20.11}$$

It is easy to see from (20.10) that

$$\mathsf{E}\,\boldsymbol{w}_i^o = \mathsf{E}\,\boldsymbol{w}_{i-1}^o \tag{20.12}$$

so that the $\{\boldsymbol{w}_i^o\}$ have a constant mean, which we shall denote by w^o,

$$\boxed{\mathsf{E}\,\boldsymbol{w}_i^o \triangleq w^o}$$

The initial condition for model (20.10) is modeled as a random variable $\boldsymbol{w}_{-1}^o$, with mean w^o and independent of all other variables, $\{\boldsymbol{q}_i, \boldsymbol{v}(i), \boldsymbol{u}_i\}$ for all i.

How Appropriate is this Model?

Although widely adopted in the literature, model (20.10) is not necessarily meaningful. For one thing, the covariance matrix of $\boldsymbol{w}_i^o$ grows unbounded. To see this, observe from

$$\boldsymbol{w}_i^o - w^o = \boldsymbol{w}_{i-1}^o - w^o + \boldsymbol{q}_i \tag{20.13}$$

that

$$\mathsf{E}\,(\boldsymbol{w}_i^o - w^o)(\boldsymbol{w}_i^o - w^o)^* = \mathsf{E}\,(\boldsymbol{w}_{i-1}^o - w^o)(\boldsymbol{w}_{i-1}^o - w^o)^* + Q$$

This means that, at each time instant i, a nonnegative-definite matrix Q is added to the covariance matrix of $\boldsymbol{w}_{i-1}^o$ in order to obtain the covariance matrix of $\boldsymbol{w}_i^o$. As a result, the covariance matrix of $\boldsymbol{w}_i^o$ becomes unbounded as time progresses. A more adequate model for tracking analysis would be to replace (20.10), or equivalently (20.13), by

$$(\boldsymbol{w}_i^o - w^o) = \alpha(\boldsymbol{w}_{i-1}^o - w^o) + \boldsymbol{q}_i \tag{20.14}$$

for some scalar $|\alpha| < 1$. In this case, the covariance matrix of $\boldsymbol{w}_i^o$ would tend to a finite steady-state value given by

$$\lim_{i\to\infty} \mathsf{E}\,(\boldsymbol{w}_i^o - w^o)(\boldsymbol{w}_i^o - w^o)^* = Q/(1-|\alpha|^2)$$

Still, it is customary in the literature to assume that the value of α is sufficiently close to one, and to use model (20.10). The main reason for assuming $\alpha \approx 1$ is to simplify the derivations during the tracking analysis. It is for this reason that we shall also proceed with the simple (yet contrived) model (20.10) in the body of the chapter in order to illustrate the key concepts. However, in the problems, and especially Probs. IV.29–IV.33, we extend the analysis to models of the form (20.14), which are rewritten in Prob. IV.29 in the equivalent form

$$\left\{\begin{array}{lll} \boldsymbol{w}_i^o & = & w^o + \boldsymbol{\theta}_i \\ \boldsymbol{\theta}_i & = & \alpha\,\boldsymbol{\theta}_{i-1} + \boldsymbol{q}_i, \quad 0 \le |\alpha| < 1 \end{array}\right. \tag{20.15}$$

Actually, Prob. IV.29 examines a more general model that also incorporates the effect of what we refer to as cyclic nonstationarities. Such nonstationarities arise, for example, as the result of carrier frequency offsets between transmitters and receivers in digital communication systems. In the computer project at the end of this part, we provide an example that shows how models of the form (20.14) arise in applications by studying the problem of tracking a Rayleigh fading channel in a wireless communications environment.

Data Model

To summarize, we shall adopt the following model in our study of the tracking performance of adaptive filters in the body of the chapter. Specifically, we shall assume that the data $\{\boldsymbol{d}(i), \boldsymbol{u}_i\}$ satisfy the following conditions:

(a) There exists a vector $\boldsymbol{w}_i^o$ such that $\boldsymbol{d}(i) = \boldsymbol{u}_i \boldsymbol{w}_i^o + \boldsymbol{v}(i)$.
(b) The weight vector varies according to $\boldsymbol{w}_i^o = \boldsymbol{w}_{i-1}^o + \boldsymbol{q}_i$.
(c) The noise sequence $\{\boldsymbol{v}(i)\}$ is i.i.d. with constant variance $\sigma_v^2 = \mathsf{E}\,|\boldsymbol{v}(i)|^2$.
(d) The noise sequence $\{\boldsymbol{v}(i)\}$ is independent of $\boldsymbol{u}_j$ for all i, j.
(e) The sequence $\boldsymbol{q}_i$ has covariance Q and is independent of $\{\boldsymbol{v}(j), \boldsymbol{u}_j\}$ for all i, j.
(f) The initial conditions $\{\boldsymbol{w}_{-1}, \boldsymbol{w}_{-1}^o\}$ are independent of all $\{\boldsymbol{d}(j), \boldsymbol{u}_j, \boldsymbol{v}(j), \boldsymbol{q}_j\}$.
(g) The regressor covariance matrix is denoted by $R_u = \mathsf{E}\,\boldsymbol{u}_i^* \boldsymbol{u}_i > 0$.
(h) The random variables $\{\boldsymbol{d}(i), \boldsymbol{v}(i), \boldsymbol{u}_i, \boldsymbol{q}_i\}$ are zero mean.
(i) The weight vector $\boldsymbol{w}_i^o$ has constant mean w^o.

(20.16)

We refer to these conditions as describing a *nonstationary* environment. Observe that in this model, the covariance matrix of the regression data is assumed to be constant and equal to R_u; likewise for the (co-)variances of the noise components $\{\boldsymbol{v}(i), \boldsymbol{q}_i\}$. The constancy of σ_v^2 means that, although $\boldsymbol{w}_i^o$ is varying with time, the problem of estimating $\boldsymbol{d}(i)$ from $\boldsymbol{u}_i$ optimally in the linear least-mean-squares sense is such that it has a constant minimum cost for all i,

$$J_{\min} = \sigma_v^2 \tag{20.17}$$

This is because, as explained in the beginning of Sec. 15.2, $\boldsymbol{v}(i)$ plays the role of the estimation error that results from estimating $\boldsymbol{d}(i)$ from $\boldsymbol{u}_i$.

Useful Independence Results

A useful consequence of model (20.16) is that at any particular time instant i, the noise variable $\boldsymbol{v}(i)$ is independent of all previous weight estimators $\{\boldsymbol{w}_j,\ j < i\}$. This fact follows easily from examining the update equation of an adaptive filter. Consider, for instance, the LMS recursion

$$\boldsymbol{w}_i = \boldsymbol{w}_{i-1} + \mu \boldsymbol{u}_i^* [\boldsymbol{d}(i) - \boldsymbol{u}_i \boldsymbol{w}_{i-1}], \quad \boldsymbol{w}_{-1} = \text{initial condition}$$

By iterating this recursion we find that, for any time instant j, the weight estimator $\boldsymbol{w}_j$ is a function of $\boldsymbol{w}_{-1}$, the reference signals $\{\boldsymbol{d}(j), \boldsymbol{d}(j-1), \ldots, \boldsymbol{d}(0)\}$, and the regressors $\{\boldsymbol{u}_j, \boldsymbol{u}_{j-1}, \ldots, \boldsymbol{u}_0\}$. We can represent this dependency generically as

$$\boldsymbol{w}_j = \mathcal{F}\,[\,\boldsymbol{w}_{-1};\ \boldsymbol{d}(j), \ldots, \boldsymbol{d}(0);\ \boldsymbol{u}_j, \ldots, \boldsymbol{u}_0\,] \tag{20.18}$$

for some function $\mathcal{F}$. A similar dependency holds for other adaptive schemes.

Now $\boldsymbol{v}(i)$ is independent of each one of the terms appearing as an argument of $\mathcal{F}$ in (20.18) so that $\boldsymbol{v}(i)$ is independent of $\boldsymbol{w}_j$ for all $j < i$. The independence of $\boldsymbol{v}(i)$ from $\{\boldsymbol{w}_{-1}, \boldsymbol{u}_j, \ldots, \boldsymbol{u}_0\}$ is obvious by assumption, while its independence from $\{\boldsymbol{d}(j), \ldots, \boldsymbol{d}(0)\}$ can be seen as follows. Consider $\boldsymbol{d}(j)$ for example. Then from

$$\boldsymbol{d}(j) = \boldsymbol{u}_j \boldsymbol{w}_j^o + \boldsymbol{v}(j) \;=\; \boldsymbol{u}_j \left(\boldsymbol{w}_{-1}^o + \sum_{k=0}^{j} \boldsymbol{q}_k \right) \;+\; \boldsymbol{v}(j)$$

we see that $\boldsymbol{d}(j)$ is a function of $\{\boldsymbol{u}_j, \boldsymbol{v}(j), \boldsymbol{w}_{-1}^o, \boldsymbol{q}_0, \ldots, \boldsymbol{q}_j\}$, all of which are independent of $\boldsymbol{v}(i)$ for $j < i$. We therefore conclude that $\boldsymbol{v}(i)$ is independent of $\boldsymbol{w}_j$ for all $j < i$.

It is also immediate to verify that $\boldsymbol{v}(i)$ is independent of $\tilde{\boldsymbol{w}}_j$ for all $j < i$, where $\tilde{\boldsymbol{w}}_j$ denotes the weight-error vector that is *now* defined in terms of the time-variant weight vector $\boldsymbol{w}_j^o$,

$$\boxed{\tilde{\boldsymbol{w}}_j \;\stackrel{\Delta}{=}\; \boldsymbol{w}_j^o - \boldsymbol{w}_j}$$

It also follows that $\boldsymbol{v}(i)$ is independent of the *a priori* estimation error $\boldsymbol{e}_a(i)$, which in the nonstationary case is defined as

$$\boxed{\boldsymbol{e}_a(i) \;\stackrel{\Delta}{=}\; \boldsymbol{u}_i \boldsymbol{w}_i^o - \boldsymbol{u}_i \boldsymbol{w}_{i-1}}$$

This definition is consistent with our earlier definition in the stationary case in Sec. 15.2, namely, $\boldsymbol{e}_a(i) = \boldsymbol{u}_i w^o - \boldsymbol{u}_i \boldsymbol{w}_{i-1}$. In both cases, $\boldsymbol{e}_a(i)$ is a measure of the error in estimating the *uncorrupted* part of $\boldsymbol{d}(i)$. However, observe that in the nonstationary case, we cannot write $\boldsymbol{e}_a(i) = \boldsymbol{u}_i \tilde{\boldsymbol{w}}_{i-1}$ since $\tilde{\boldsymbol{w}}_{i-1} = \boldsymbol{w}_{i-1}^o - \boldsymbol{w}_{i-1}$.

We summarize the above independence properties in the following statement.

Lemma 20.1 (Useful properties) From the data model (20.16), it holds that $\boldsymbol{v}(i)$ is independent of each of the following:

$$\{\boldsymbol{w}_j \text{ for } j < i\}, \;\; \{\tilde{\boldsymbol{w}}_j = \boldsymbol{w}_j^o - \boldsymbol{w}_j \text{ for } j < i\}, \;\; \text{and} \;\; \boldsymbol{e}_a(i) = \boldsymbol{u}_i(\boldsymbol{w}_i^o - \boldsymbol{w}_{i-1})$$

Alternative Expression for the EMSE

Using model (20.16), and the result of Lemma 20.1, we can express the filter EMSE in an alternative form. Indeed, using $\boldsymbol{e}(i) = \boldsymbol{d}(i) - \boldsymbol{u}_i \boldsymbol{w}_{i-1}$, and part (a) of model (20.16), we have $\boldsymbol{e}(i) = \boldsymbol{v}(i) + \boldsymbol{u}_i(\boldsymbol{w}_i^o - \boldsymbol{w}_{i-1})$, i.e.,

$$\boxed{\boldsymbol{e}(i) = \boldsymbol{v}(i) + \boldsymbol{e}_a(i)} \tag{20.19}$$

Now since $\boldsymbol{v}(i)$ and $\boldsymbol{e}_a(i)$ are independent, and $\boldsymbol{v}(i)$ has zero mean, it follows from (20.19) that

$$\mathsf{E}\,|\boldsymbol{e}(i)|^2 \;=\; \mathsf{E}\,|\boldsymbol{v}(i)|^2 \;+\; \mathsf{E}\,|\boldsymbol{e}_a(i)|^2$$

The first term on the right-hand side is σ_v^2 which, as explained above in (20.17), coincides with $J_{\min}$ and, hence,

$$\mathsf{E}\,|\boldsymbol{e}(i)|^2 - J_{\min} = \mathsf{E}\,|\boldsymbol{e}_a(i)|^2$$

Substituting this equality into definition (20.3) for the EMSE we arrive at the equivalent characterization

$$\boxed{\text{EMSE} = \lim_{i\to\infty} \mathsf{E}\,|e_a(i)|^2} \tag{20.20}$$

In other words, the EMSE of an adaptive filter operating in a nonstationary environment can be computed by evaluating the steady-state variance of the estimation error $e_a(i)$.

Degree of Nonstationarity

A lower bound on the EMSE can be determined as follows. Using $w_i^o = w_{i-1}^o + q_i$ we have

$$\begin{aligned} e_a(i) &= u_i w_i^o - u_i w_{i-1} \\ &= u_i(w_{i-1}^o + q_i) - u_i w_{i-1} \\ &= u_i(w_{i-1}^o - w_{i-1}) + u_i q_i \end{aligned}$$

so that

$$\begin{aligned} \mathsf{E}\,|e_a(i)|^2 &= \mathsf{E}\,\left|u_i(w_{i-1}^o - w_{i-1}) + u_i q_i\right|^2 \\ &= \mathsf{E}\,\left|u_i(w_{i-1}^o - w_{i-1})\right|^2 + \mathsf{E}\,|u_i q_i|^2 \\ &\geq \mathsf{E}\,|u_i q_i|^2 \\ &= \text{Tr}(R_u Q), \qquad \text{for all } i \end{aligned}$$

where in the second equality we used the fact that q_i is independent of $u_i(w_{i-1}^o - w_{i-1})$ and, hence, their cross-correlation is zero. In the last equality we used the fact that q_i and u_i are independent. We therefore find that the misadjustment of an adaptive filter in a nonstationary environment is lower bounded by

$$\boxed{\mathcal{M} \geq \text{Tr}(R_u Q)/\sigma_v^2}$$

The ratio on the right-hand side, involving $\{R_u, Q, \sigma_v^2\}$, is equal to the square of what is called the *degree of nonstationarity* (DN) of the data,

$$\boxed{\text{DN} \stackrel{\Delta}{=} \sqrt{\text{Tr}(R_u Q)/\sigma_v^2}}$$

If the value of DN is larger than unity then this means that the statistical variations in the optimal weight vector w_i^o are too fast for the filter to be able to track them (and the misadjustment will be large). On the other hand, if DN $\ll 1$, then the adaptive filter would generally be able to track the variations in the weight vector. In this chapter, we are interested in evaluating the tracking performance of adaptive filters in this latter situation, i.e., when tracking is possible.

Error Quantities

For ease of reference, we collect in Table 20.1 the definitions of several of the error measures introduced so far. Observe that, in contrast to the stationary case shown in Table 15.1, the definition of $\tilde{w}_i$ is now with respect to w_i^o, and the definitions of the *a priori* and *a posteriori* estimation errors $\{e_a(i), e_p(i)\}$ are in terms of w_i^o. Comparing with the definitions in the stationary case, observe that $e_a(i)$ and $e_p(i)$ in Table 20.1 are *not* expressed as $u_i\tilde{w}_{i-1}$ and $u_i\tilde{w}_i$, respectively. While this form of expression is valid for $e_p(i)$, it is not valid for $e_a(i)$ in the nonstationary case. The proper interpretation for $\{e_a(i), e_p(i)\}$, for both cases of stationary and nonstationary models, is to regard them as the errors in estimating the uncorrupted part of $d(i)$ by using $\{u_i w_{i-1}, u_i w_i\}$.

TABLE 20.1 Definitions of several estimation errors.

Error	Definition	Interpretation
$e(i)$	$d(i) - u_i w_{i-1}$	*a priori* output estimation error
$r(i)$	$d(i) - u_i w_i$	*a posteriori* output estimation error
$\tilde{w}_i$	$w_i^o - w_i$	weight estimation error
$e_a(i)$	$u_i w_i^o - u_i w_{i-1}$	*a priori* estimation error
$e_p(i)$	$u_i w_i^o - u_i w_i$	*a posteriori* estimation error

As was done earlier in Chapter 15, before moving on to derive expressions for the EMSE of several adaptive algorithms under model (20.16), we first establish a fundamental energy-conservation result that holds for *general* data $\{d(i), u_i\}$ (i.e., it does *not* require the assumptions (20.16)). The derivation that follows is distinct from the arguments we used earlier in Sec. 15.3 only in that it uses the new definition of $\tilde{w}_i$, which is relative to the time-variant coefficient w_i^o. Therefore, we shall be brief.

20.3 ENERGY CONSERVATION RELATION

We again consider adaptive filters whose update equations are of the form

$$\boxed{w_i = w_{i-1} + \mu u_i^* g[e(i)], \quad w_{-1} = \text{initial condition}} \tag{20.21}$$

where $g[\cdot]$ denotes the error function; several examples were listed in Table 15.2.

The update recursion (20.21) can be written in terms of the weight-error vector $\tilde{w}_i = w_i^o - w_i$. Subtracting both sides of (20.21) from w_i^o gives

$$\boxed{w_i^o - w_i = (w_i^o - w_{i-1}) - \mu u_i^* g[e(i)]} \tag{20.22}$$

Multiplying both sides of this equation by u_i from the left we find that the *a priori* and *a posteriori* errors $\{e_p(i), e_a(i)\}$ are related via

$$\boxed{e_p(i) = e_a(i) - \mu \|u_i\|^2 g[e(i)]} \tag{20.23}$$

Equations (20.22) and (20.23) have the same form as equations (15.24) and (15.25) that we derived earlier in Sec. 15.3. Therefore, following the exact same arguments that we presented in that section, we arrive at the following extension of Thm. 15.1.

Theorem 20.1 (Energy-conservation relation) For any adaptive filter of the form (20.21), and for any data $\{d(i), u_i\}$, it holds that

$$\|w_i^o - w_i\|^2 + \bar{\mu}(i)|e_a(i)|^2 = \|w_i^o - w_{i-1}\|^2 + \bar{\mu}(i)|e_p(i)|^2 \qquad (20.24)$$

where $e_a(i) = u_i(w_i^o - w_{i-1})$, $e_p(i) = u_i(w_i^o - w_i)$, and $\bar{\mu}(i)$ is defined as

$$\bar{\mu}(i) \triangleq \left(\|u_i\|^2\right)^\dagger = \begin{cases} 1/\|u_i\|^2 & \text{if } u_i \neq 0 \\ 0 & \text{otherwise} \end{cases}$$

Comparing (20.24) with the energy-conservation relation (15.32) in the stationary case, we see that the only difference pertains to the interpretation of the terms $w_i^o - w_i$ and $w_i^o - w_{i-1}$ that appear on both sides of (20.24). While the first difference can be recognized as $\tilde{w}_i$, just like the term on the left-hand side of (15.32), the second difference is *not* $\tilde{w}_{i-1}$ since, in the nonstationary case, $\tilde{w}_{i-1}$ is defined as $\tilde{w}_{i-1} = w_{i-1}^o - w_{i-1}$ (in terms of w_{i-1}^o and not w_i^o).

20.4 VARIANCE RELATION

In order to explain the relevance of the energy relation (20.24) to the tracking analysis of adaptive filters, we refer to the data model (20.16) and, in particular, to condition (b). The condition states that w_i^o varies according to the random-walk model

$$w_i^o = w_{i-1}^o + q_i \qquad (20.25)$$

where q_i is an i.i.d. sequence with covariance matrix Q and is independent of the initial conditions $\{w_{-1}^o, w_{-1}\}$, of $\{u_j\}$ for all j, and of $\{d(j)\}$ for all $j < i$. This random-walk model, and the assumptions on q_i, are the only conditions that we require from the data model (20.16) for the derivation in this section. [The conditions on $v(i)$, such as its independence of all u_j and q_j, are not needed here.]

Taking expectations of both sides of the energy-conservation relation (20.24) we get

$$\mathsf{E}\,\|\tilde{w}_i\|^2 + \mathsf{E}\,\bar{\mu}(i)|e_a(i)|^2 = \mathsf{E}\,\|w_i^o - w_{i-1}\|^2 + \mathsf{E}\,\bar{\mu}(i)|e_p(i)|^2 \qquad (20.26)$$

since $\tilde{w}_i = w_i^o - w_i$. Moreover, the model (20.25) allows us to relate $\mathsf{E}\,\|w_i^o - w_{i-1}\|^2$ to $\mathsf{E}\,\|\tilde{w}_{i-1}\|^2$ as follows:

$$\begin{aligned} \mathsf{E}\,\|w_i^o - w_{i-1}\|^2 &= \mathsf{E}\,\|w_{i-1}^o + q_i - w_{i-1}\|^2 \\ &= \mathsf{E}\,\|\tilde{w}_{i-1} + q_i\|^2 \\ &= \mathsf{E}\,(\tilde{w}_{i-1} + q_i)^*(\tilde{w}_{i-1} + q_i) \\ &= \mathsf{E}\,\|\tilde{w}_{i-1}\|^2 + \mathsf{E}\,\|q_i\|^2 + \mathsf{E}\,\tilde{w}_{i-1}^* q_i + \mathsf{E}\,q_i^* \tilde{w}_{i-1} \end{aligned} \qquad (20.27)$$

We can be more explicit about the cross terms $\mathsf{E}\,\tilde{w}_{i-1}^* q_i$ and $\mathsf{E}\,q_i^* \tilde{w}_{i-1}$. Indeed, note that

$$\tilde{w}_{i-1} = w_{i-1}^o - w_{i-1} = \left(w_{-1}^o + \sum_{j=0}^{i-1} q_j\right) - w_{i-1}$$

so that

$$\begin{aligned}
\mathsf{E}\,\tilde{\boldsymbol{w}}_{i-1}^{*}\boldsymbol{q}_i &= \mathsf{E}\left(w_{-1}^{o}+\sum_{j=0}^{i-1}\boldsymbol{q}_j\right)^{*}\boldsymbol{q}_i - \mathsf{E}\,\boldsymbol{w}_{i-1}^{*}\boldsymbol{q}_i \\
&= -\mathsf{E}\,\boldsymbol{w}_{i-1}^{*}\boldsymbol{q}_i
\end{aligned}$$

where we used the fact that $\boldsymbol{q}_i$ is independent of all previous $\boldsymbol{q}_j$ and of w_{-1}^{o}. But since $\boldsymbol{w}_{i-1}$ is a function of the variables

$$\{\ \boldsymbol{u}_{i-1},\ldots,\boldsymbol{u}_0,\ \boldsymbol{d}(i-1),\ldots,\boldsymbol{d}(0)\ \}$$

all of which are independent of $\boldsymbol{q}_i$, we conclude that

$$\mathsf{E}\,\tilde{\boldsymbol{w}}_{i-1}^{*}\boldsymbol{q}_i = 0$$

Likewise, $\mathsf{E}\,\boldsymbol{q}_i^{*}\tilde{\boldsymbol{w}}_{i-1} = 0$. Substituting into (20.27) we find that

$$\mathsf{E}\,\|\boldsymbol{w}_i^{o}-\boldsymbol{w}_{i-1}\|^2 = \mathsf{E}\,\|\tilde{\boldsymbol{w}}_{i-1}\|^2 \ + \ \mathsf{Tr}(Q) \tag{20.28}$$

and (20.26) becomes

$$\boxed{\mathsf{E}\,\|\tilde{\boldsymbol{w}}_i\|^2 \ + \ \mathsf{E}\,\bar{\mu}(i)|\boldsymbol{e}_a(i)|^2 = \mathsf{E}\,\|\tilde{\boldsymbol{w}}_{i-1}\|^2 \ + \ \mathsf{E}\,\bar{\mu}(i)|\boldsymbol{e}_p(i)|^2 \ + \ \mathsf{Tr}(Q)} \tag{20.29}$$

Comparing with relation (15.36) in the stationary case, we see that the only difference is the appearance of the additional term $\mathsf{Tr}(Q)$. All other terms are identical!

Steady-State Performance

Now assume that an adaptive filter is operating in steady-state (cf. Def. 15.1), i.e.,

$$\mathsf{E}\,\|\tilde{\boldsymbol{w}}_i\|^2 = \mathsf{E}\,\|\tilde{\boldsymbol{w}}_{i-1}\|^2, \quad \text{as } i\to\infty \tag{20.30}$$

It then follows from (20.29) that

$$\mathsf{E}\,\bar{\mu}(i)|\boldsymbol{e}_a(i)|^2 = \mathsf{Tr}(Q) \ + \ \mathsf{E}\,\bar{\mu}(i)|\boldsymbol{e}_p(i)|^2, \quad \text{as } i\to\infty$$

Using (20.23), we can replace $\boldsymbol{e}_p(i)$ in terms of $\boldsymbol{e}_a(i)$ and get

$$\mathsf{E}\,\bar{\mu}(i)|\boldsymbol{e}_a(i)|^2 = \mathsf{Tr}(Q) \ + \ \mathsf{E}\,\bar{\mu}(i)\left|\boldsymbol{e}_a(i)-\mu\|\boldsymbol{u}_i\|^2 g[\boldsymbol{e}(i)]\right|^2, \quad \text{as } i\to\infty \tag{20.31}$$

This relation can be simplified by expanding both of its sides, and by following the same steps that we carried out for (15.38), thus leading to the following conclusion.

Theorem 20.2 (Variance relation) Consider any adaptive filter of the form (20.21), and assume filter operation in steady-state. Assume further that

$$d(i) = u_i w_i^o + v(i)$$

where w_i^o varies according to the random-walk model $w_i^o = w_{i-1}^o + q_i$ and q_i is a zero-mean i.i.d. sequence with covariance matrix Q. Moreover, q_i is independent of $\{d(j), j < i\}$ and of $\{u_j, w_{-1}^o\}$ for all j. Then it holds that

$$\mu \mathsf{E}\,\|u_i\|^2|g[e(i)]|^2 \;+\; \mu^{-1}\mathsf{Tr}(Q) = 2\mathsf{Re}\big(\mathsf{E}\,e_a^*(i)g[e(i)]\big), \quad \text{as } i \to \infty \quad (20.32)$$

where $e(i) = e_a(i) + v(i)$. For real-valued data, the above relation becomes

$$\mu \mathsf{E}\,\|u_i\|^2 g^2[e(i)] \;+\; \mu^{-1}\mathsf{Tr}(Q) = 2\mathsf{E}\,e_a(i)g[e(i)], \quad \text{as } i \to \infty \qquad (20.33)$$

Alternatively, the variance relation (20.32) can be obtained by starting directly from the weight-error vector recursion (20.22) and equating the squared Euclidean norms of both sides, i.e.,

$$\begin{aligned}\|w_i^o - w_i\|^2 &= \|(w_i^o - w_{i-1}) - \mu u_i^* g\|^2 \\ &= \|w_i^o - w_{i-1}\|^2 + \mu^2\|u_i\|^2|g|^2 - 2\mu\mathsf{Re}(e_a^*(i)g)\end{aligned}$$

Taking expectations of both sides as $i \to \infty$ and using (20.28) gives (20.32). Comparing expression (20.32) with the variance relation of Thm. 15.2 in the stationary case, we see that the only difference is the appearance of the additional term $\mu^{-1}\mathsf{Tr}(Q)$ on the left-hand side of (20.32). All other terms are identical to those in (15.40). This observation shows that obtaining the EMSE of an adaptive filter in a nonstationary environment is a straightforward extension of the calculations carried out in Chapter 15 for obtaining the EMSE of the filter in a stationary environment.

Relevance to Tracking Analysis

Relation (20.32) is an identity that involves $e_a(i)$ and, in principle, it could be used to evaluate the EMSE of an adaptive filter in a nonstationary environment. We say "in principle" because, although the result (20.32) is exact, different choices for the error function $g[\cdot]$ lead to different equations in $\mathsf{E}\,|e_a(\infty)|^2$, some of which are easier to solve than others. It is at this stage that simplifying assumptions become necessary. We shall illustrate this procedure for several adaptive filters in the sections below. As in Chapter 15, we shall employ the symbol ζ to refer to the EMSE of an adaptive filter.

CHAPTER 21

Tracking Performance

The reader will soon realize that the arguments that are used in the sequel in order to arrive at the nonstationary EMSE, starting from the variance relation (20.32), are almost identical to the arguments used before in Chapters 16–19 while studying the stationary EMSE of several adaptive filters. For this reason, the derivations here are brief.

21.1 PERFORMANCE OF LMS

We start with the simplest of algorithms, namely, LMS. Thus, assume that $\{\boldsymbol{d}(i), \boldsymbol{u}_i\}$ satisfy model (20.16) and consider the LMS recursion

$$\boldsymbol{w}_i = \boldsymbol{w}_{i-1} + \mu \boldsymbol{u}_i^* \boldsymbol{e}(i) \tag{21.1}$$

for which

$$g[\boldsymbol{e}(i)] = \boldsymbol{e}(i) = \boldsymbol{e}_a(i) + \boldsymbol{v}(i) \tag{21.2}$$

Relation (20.32) then becomes

$$\mu \mathsf{E}\, \|\boldsymbol{u}_i\|^2 |\boldsymbol{e}_a(i) + \boldsymbol{v}(i)|^2 \; + \; \mu^{-1}\mathsf{Tr}(Q) = 2\mathsf{Re}\big(\mathsf{E}\, \boldsymbol{e}_a^*(i)[\boldsymbol{e}_a(i) + \boldsymbol{v}(i)]\big) \tag{21.3}$$

Except for the term

$$\mu^{-1}\mathsf{Tr}(Q)$$

on the left-hand side, we note that the identity (21.3) has the same form as the identity (16.3) that appeared in our study of the mean-square performance of LMS in Chapter 16. Therefore, performing the same expansions that we did in that section following (16.3), we can readily verify that (21.3) leads to

$$\boxed{\zeta^{\mathsf{LMS}} = \tfrac{1}{2}\left[\, \mu \mathsf{E}\, \|\boldsymbol{u}_i\|^2 |\boldsymbol{e}_a(i)|^2 \; + \; \mu\sigma_v^2 \mathsf{Tr}(R_u) + \mu^{-1}\mathsf{Tr}(Q)\,\right], \quad \text{as } i \to \infty} \tag{21.4}$$

This expression extends the result (16.5) to the nonstationary case. In order to evaluate $\zeta^{\mathsf{LMS}} = \mathsf{E}\, |\boldsymbol{e}_a(\infty)|^2$, we again examine three cases (the similarities with the arguments in Chapter 16 are obvious).

Small Step-Sizes

Assume first that the step-size μ is such that, in *steady-state*, the term

$$\mu \mathsf{E}\, \|\boldsymbol{u}_i\|^2 |\boldsymbol{e}_a(i)|^2$$

can be neglected when compared to the term

$$\mu\sigma_v^2\text{Tr}(R_u) + \mu^{-1}\text{Tr}(Q)$$

This condition occurs for data with a sufficiently small degree of nonstationarity and for step-sizes μ in the vicinity of $\mu_{\text{opt}}^{\text{LMS}}$ in (21.6). Then, expression (21.4) gives

$$\boxed{\zeta^{\text{LMS}} = \frac{\mu\sigma_v^2\text{Tr}(R_u) + \mu^{-1}\text{Tr}(Q)}{2}} \tag{21.5}$$

This result highlights the effect of the step-size on the performance of LMS. The term $\mu\sigma_v^2\text{Tr}(R_u)$ is the same one we encountered earlier in expression (16.6) while studying the EMSE of LMS in stationary environments. The additional term $\mu^{-1}\text{Tr}(Q)$ reflects the effect of the nonstationarity in the weight vector w_i^o on filter performance. Observe in particular that $\text{Tr}(Q)$ appears multiplied by μ^{-1} so that the larger the step-size the smaller the effect of the nonstationarity on the EMSE. This behavior is intuitive since a larger step-size (usually) signifies faster adaptation, in which case LMS will have a better chance at "learning" and at "following" the data statistics. A small step-size, on the other hand, leads to smaller EMSE under stationary conditions, but it may also lead to poor tracking performance.

This discussion suggests that there exists a compromise choice for the step-size, which is obtained by minimizing (21.5) with respect to μ. Setting the derivative of ζ^{LMS} equal to zero gives

$$\boxed{\mu_{\text{opt}}^{\text{LMS}} = \sqrt{\text{Tr}(Q)/\sigma_v^2\text{Tr}(R_u)}} \tag{21.6}$$

Substituting the above optimal value for μ into (21.5) we find that the resulting minimum EMSE is given by

$$\boxed{\zeta_{\min}^{\text{LMS}} = \sqrt{\sigma_v^2\text{Tr}(R_u)\text{Tr}(Q)}} \tag{21.7}$$

Separation Principle

Rather than neglect the effect of the term $\mu\text{E}\,\|\boldsymbol{u}_i\|^2|\boldsymbol{e}_a(i)|^2$ in steady-state, we can call upon the separation assumption (16.7) that we introduced earlier in Sec. 16.3, namely, that

$$\boxed{\text{At steady-state,}\quad \|\boldsymbol{u}_i\|^2 \text{ is independent of } \boldsymbol{e}_a(i)} \tag{21.8}$$

Using this assumption we have

$$\begin{aligned}
\text{E}\left(\|\boldsymbol{u}_i\|^2 \cdot |\boldsymbol{e}_a(i)|^2\right) &= \left(\text{E}\,\|\boldsymbol{u}_i\|^2\right) \cdot \left(\text{E}\,|\boldsymbol{e}_a(i)|^2\right) \\
&= \text{Tr}(R_u)\text{E}\,|\boldsymbol{e}_a(i)|^2
\end{aligned}$$

so that substituting into (21.4) we obtain

$$\boxed{\zeta^{\text{LMS}} = \frac{\mu\sigma_v^2\text{Tr}(R_u) + \mu^{-1}\text{Tr}(Q)}{2 - \mu\text{Tr}(R_u)}} \tag{21.9}$$

Once more, this expression differs from expression (16.10) in the stationary case only by the additional term $\mu^{-1}\text{Tr}(Q)$ that appears in the numerator. Observe further that if μ is such that $2 - \mu\text{Tr}(R_u) \approx 2$, then (21.9) reduces to (21.5).

Differentiating (21.9) with respect to μ leads to the following expression for the optimal step-size (see Prob. IV.22):

$$\boxed{\mu_{\text{opt}}^{\text{LMS}} = \sqrt{\frac{\text{Tr}(Q)}{\sigma_v^2\text{Tr}(R_u)} + \frac{\left(\text{Tr}(Q)\right)^2}{4\sigma_v^4}} - \frac{\text{Tr}(Q)}{2\sigma_v^2}} \tag{21.10}$$

Substituting into (21.9) we can find the minimum value for the EMSE. We forgo this calculation here.

White Gaussian Input

As discussed in Sec. 16.4, one particular case for which the term $\mathsf{E}\,\|\boldsymbol{u}_i\|^2|\boldsymbol{e}_a(i)|^2$ that appears in (21.4) can be evaluated in closed-form occurs when the regressor $\boldsymbol{u}_i$ has a circular Gaussian distribution with a diagonal covariance matrix of the form

$$R_u = \sigma_u^2\text{I}, \quad \sigma_u^2 > 0 \tag{21.11}$$

The diagonal structure of R_u means that the entries of $\boldsymbol{u}_i$ are uncorrelated among themselves and that each has variance σ_u^2. In addition to (21.11), we also assume that (see the justification in Sec. 16.4):

$$\boxed{\text{At steady state, } \tilde{\boldsymbol{w}}_{i-1} \text{ is independent of } \boldsymbol{u}_i} \tag{21.12}$$

Under conditions (21.11)–(21.12), we showed in Sec. 16.4 that

$$\mathsf{E}\,\|\boldsymbol{u}_i\|^2|\boldsymbol{e}_a(i)|^2 = (M+\gamma)\sigma_u^2\mathsf{E}\,|\boldsymbol{e}_a(i)|^2$$

where $\gamma = 2$ for real-valued data and $\gamma = 1$ for complex-valued data — see (16.17) and (16.19). Substituting into (21.4) we obtain

$$\zeta^{\text{LMS}} = \frac{1}{2}\left[\mu(M+\gamma)\sigma_u^2\zeta^{\text{LMS}} + \mu M\sigma_u^2\sigma_v^2 + \mu^{-1}\text{Tr}(Q)\right]$$

so that

$$\boxed{\zeta^{\text{LMS}} = \frac{\mu M\sigma_v^2\sigma_u^2 + \mu^{-1}\text{Tr}(Q)}{2-\mu(M+\gamma)\sigma_u^2}} \tag{21.13}$$

Here again, the only difference from (16.18) is the presence of the additional term $\mu^{-1}\text{Tr}(Q)$ in the numerator. Differentiating (21.13) with respect to μ gives

$$\boxed{\mu_{\text{opt}}^{\text{LMS}} = \sqrt{\frac{\text{Tr}(Q)}{M\sigma_u^2\sigma_v^2} + \frac{(M+\gamma)^2[\text{Tr}(Q)]^2}{4M^2\sigma_v^2}} - \frac{(M+\gamma)\text{Tr}(Q)}{2M\sigma_v^2}} \tag{21.14}$$

We summarize the results for LMS in Lemma 21.1.

Lemma 21.1 (Tracking EMSE of LMS) Consider the LMS algorithm (21.1) and assume $\{\boldsymbol{d}(i), \boldsymbol{u}_i\}$ satisfy the nonstationary model (20.16) with a sufficiently small degree of nonstationarity. Then its EMSE can be approximated by the following expressions:

1. For small step-sizes, it holds that

$$\zeta^{\text{LMS}} = \frac{\mu\sigma_v^2 \text{Tr}(R_u) + \mu^{-1}\text{Tr}(Q)}{2}$$

$$\mu_{\text{opt}}^{\text{LMS}} = \sqrt{\frac{\text{Tr}(Q)}{\sigma_v^2 \text{Tr}(R_u)}} \quad \text{with} \quad \zeta_{\min}^{\text{LMS}} = \sqrt{\sigma_v^2 \text{Tr}(R_u)\text{Tr}(Q)}$$

2. Under the separation assumption (21.8), it holds that

$$\zeta^{\text{LMS}} = \frac{\mu\sigma_v^2 \text{Tr}(R_u) + \mu^{-1}\text{Tr}(Q)}{2 - \mu\text{Tr}(R_u)}$$

$$\mu_{\text{opt}}^{\text{LMS}} = \sqrt{\frac{\text{Tr}(Q)}{\sigma_v^2 \text{Tr}(R_u)} + \frac{\left(\text{Tr}(Q)\right)^2}{4\sigma_v^4}} - \frac{\text{Tr}(Q)}{2\sigma_v^2}$$

3. If $\boldsymbol{u}_i$ is Gaussian with $R_u = \sigma_u^2 \text{I}$, and under assumption (21.12), it holds that ζ^{LMS} and $\mu_{\text{opt}}^{\text{LMS}}$ are given by (21.13) and (21.14), respectively, where $\gamma = 2$ if the data are real-valued and $\gamma = 1$ if the data are complex-valued with $\boldsymbol{u}_i$ circular. Here M is the dimension of $\boldsymbol{u}_i$.

In all cases, the misadjustment is obtained by dividing the EMSE by σ_v^2. Also, substituting the expressions for μ_{opt} into the expressions for EMSE we find the corresponding optimal EMSE.

Remark 21.1 (Auto-regressive model) The results derived so far assume that the weight vector $\boldsymbol{w}_i^o$ varies according to the contrived model (20.10). However, we extend the results in the problems at the end of this part to a more realistic nonstationary model. For example, a specialization of the result of Prob. IV.33 (obtained by setting the variable Ω in that problem to zero) shows the following. If $\boldsymbol{w}_i^o$ varies according to (20.15) (or, equivalently, (20.14)), for some $|\alpha| \leq 1$, then the EMSE of LMS is approximated by

$$\boxed{\zeta^{\text{LMS}} = \frac{\mu\sigma_v^2 \text{Tr}(R_u) + \mu^{-1}\beta}{2}}$$

where the scalar β is defined by

$$\beta = \frac{2}{1-|\alpha|^2}\text{Re}\left\{\text{Tr}\left[(1-\text{Re}(\alpha))\text{I} + (1-\alpha^*)^2 X_\alpha - \mu(1-\alpha)^2 R_u\right]Q\right\}$$

$$X_\alpha = (\text{I} - \mu R_u)\left[\alpha^*\text{I} - (\text{I} - \mu R_u)^{-1}\right]^{-1}$$

Other approximations for ζ^{LMS} also appear in the problem. It can be seen from the expression for β that $\beta \to \text{Tr}(Q)$ as $\alpha \to 1$, in which case we recover the first result stated in Lemma 21.1.

21.2 PERFORMANCE OF NLMS

We now examine the tracking performance of ϵ-NLMS,

$$\boldsymbol{w}_i = \boldsymbol{w}_{i-1} + \frac{\mu}{\epsilon + \|\boldsymbol{u}_i\|^2}\boldsymbol{u}_i^* \boldsymbol{e}(i) \tag{21.15}$$

for which

$$g[\boldsymbol{e}(i)] = \frac{\boldsymbol{e}(i)}{\epsilon + \|\boldsymbol{u}_i\|^2} = \frac{\boldsymbol{e}_a(i) + \boldsymbol{v}(i)}{\epsilon + \|\boldsymbol{u}_i\|^2} \tag{21.16}$$

and the data $\{\boldsymbol{d}(i), \boldsymbol{u}_i, \boldsymbol{v}(i)\}$ are assumed to satisfy model (20.16). The variance relation (20.32) in this case becomes (with $i \to \infty$):

$$\mu \mathsf{E}\left(\frac{\|\boldsymbol{u}_i\|^2 |\boldsymbol{e}_a(i) + \boldsymbol{v}(i)|^2}{(\epsilon + \|\boldsymbol{u}_i\|^2)^2}\right) + \mu^{-1}\mathsf{Tr}(Q) = 2\mathsf{Re}\left\{\mathsf{E}\,\boldsymbol{e}_a^*(i)\left(\frac{\boldsymbol{e}_a(i) + \boldsymbol{v}(i)}{\epsilon + \|\boldsymbol{u}_i\|^2}\right)\right\} \tag{21.17}$$

Separation Principle

Except for the term $\mu^{-1}\mathsf{Tr}(Q)$, the above equality has the same form as the identity (17.3), which appeared in our study of the mean-square performance of ϵ-NLMS in the stationary case in Sec. 17.1. Therefore, by performing the same expansions that we did in that section following (17.3), we can find that the above equality reduces to

$$(2\eta_u - \mu\alpha_u)\mathsf{E}\,|\boldsymbol{e}_a(i)|^2 = \mu\sigma_v^2\alpha_u + \mu^{-1}\mathsf{Tr}(Q), \quad i \to \infty$$

so that

$$\zeta^{\epsilon-\mathsf{NLMS}} = \frac{\mu\alpha_u\sigma_v^2 + \mu^{-1}\mathsf{Tr}(Q)}{2\eta_u - \mu\alpha_u} \tag{21.18}$$

where

$$\alpha_u \triangleq \mathsf{E}\left(\frac{\|\boldsymbol{u}_i\|^2}{(\epsilon + \|\boldsymbol{u}_i\|^2)^2}\right), \qquad \eta_u \triangleq \mathsf{E}\left(\frac{1}{\epsilon + \|\boldsymbol{u}_i\|^2}\right) \tag{21.19}$$

When the regularization parameter ϵ is small, as is usually the case, the values of η_u and α_u coincide, i.e., $\alpha_u = \eta_u = \mathsf{E}\,(1/\|\boldsymbol{u}_i\|^2)$, and we get

$$\boxed{\zeta^{\epsilon-\mathsf{NLMS}} = \frac{1}{2-\mu}\left[\mu\sigma_v^2 + \frac{\mu^{-1}\mathsf{Tr}(Q)}{\mathsf{E}\,(1/\|\boldsymbol{u}_i\|^2)}\right] \quad \text{(when } \epsilon \text{ is small)}} \tag{21.20}$$

An alternative expression for the EMSE of ϵ–NLMS can be obtained by using the assumption $\epsilon \approx 0$ in order to initially simplify (21.17) into

$$\mu\mathsf{E}\left(\frac{|\boldsymbol{e}_a(i)|^2}{\|\boldsymbol{u}_i\|^2}\right) + \mu\sigma_v^2\mathsf{E}\left(\frac{1}{\|\boldsymbol{u}_i\|^2}\right) + \mu^{-1}\mathsf{Tr}(Q) = 2\mathsf{E}\left(\frac{|\boldsymbol{e}_a(i)|^2}{\|\boldsymbol{u}_i\|^2}\right)$$

Then we appeal to the steady-state approximation (17.8), instead of (21.8), to find that

$$\boxed{\zeta^{\epsilon-\mathsf{NLMS}} = \frac{\mathsf{Tr}(R_u)}{2-\mu}\left[\mu\sigma_v^2\mathsf{E}\left(\frac{1}{\|\boldsymbol{u}_i\|^2}\right) + \mu^{-1}\mathsf{Tr}(Q)\right] \quad \text{(when } \epsilon \text{ is small)}} \tag{21.21}$$

In both cases (21.20)–(21.21), approximate expressions for the optimal step-size can be obtained by differentiating $\zeta^{\epsilon-\mathsf{NLMS}}$ with respect to μ — see Prob. IV.23.

Lemma 21.2 (Tracking EMSE of ϵ-NLMS) Consider the ϵ-NLMS recursion (21.15) and assume $\{d(i), u_i\}$ satisfy the nonstationary model (20.16) with a sufficiently small degree of nonstationarity. Then, under the separation assumption (21.8) and for small ϵ, its EMSE can be approximated by

$$\zeta^{\epsilon-\text{NLMS}} = \frac{1}{2-\mu}\left[\mu\sigma_v^2 + \frac{\mu^{-1}\text{Tr}(Q)}{\text{E}\left(1/\|u_i\|^2\right)}\right]$$

or, under the steady-state approximation (17.8),

$$\zeta^{\epsilon-\text{NLMS}} = \frac{\text{Tr}(R_u)}{2-\mu}\left[\mu\sigma_v^2\text{E}\left(\frac{1}{\|u_i\|^2}\right) + \mu^{-1}\text{Tr}(Q)\right]$$

Optimal choices for the step-size parameter, and the resulting minimum EMSE for these approximations, are given in Prob. IV.23. The misadjustment is obtained by dividing the EMSE by σ_v^2.

21.3 PERFORMANCE OF SIGN-ERROR LMS

We now examine the tracking performance of the sign-error LMS algorithm,

$$w_i = w_{i-1} + \mu u_i^* \text{csgn}[e(i)] \tag{21.22}$$

for which

$$g[e(i)] = \text{csgn}[e(i)] = \text{csgn}[e_a(i) + v(i)] \tag{21.23}$$

and the data $\{d(i), u_i, v(i)\}$ are assumed to satisfy model (20.16). As in Chapter 18, we need to distinguish between the case of real-valued data and complex-valued data.

Real-Valued Data

Assume first that all data are real-valued. In this case,

$$\text{csgn}[x] = \text{sign}[x]$$

so that

$$g^2(x) = 1$$

almost everywhere on the real line, and the variance relation (20.32) becomes

$$\mu\text{Tr}(R_u) + \mu^{-1}\text{Tr}(Q) = 2\text{E}\, e_a(i)\text{sign}[e_a(i) + v(i)], \quad i \to \infty \tag{21.24}$$

Again, except for the term $\mu^{-1}\text{Tr}(Q)$, this identity has the same form as the identity (18.3) that appeared in our study of the mean-square performance of sign-error LMS in Chapter 18 for stationary environments. Therefore, performing the same expansions that followed (18.3) and assuming that (cf. (18.4)):

$$\boxed{\text{The estimation error } e(i) \text{ and the noise } v(i) \text{ are jointly Gaussian}} \tag{21.25}$$

we can invoke Price's theorem (18.6) to find that

$$\mu\mathsf{Tr}(R_u) + \mu^{-1}\mathsf{Tr}(Q) = 2\sqrt{\frac{2}{\pi}}\,\frac{\mathsf{E}\,e_a^2(i)}{\sqrt{\mathsf{E}\,e_a^2(i) + \sigma_v^2}}, \quad \text{as } i \to \infty \tag{21.26}$$

Solving for $\mathsf{E}\,e_a^2(\infty)$, we get

$$\boxed{\zeta^{\text{sign-error LMS}} = \frac{\alpha}{2}\left(\alpha + \sqrt{\alpha^2 + 4\sigma_v^2}\right)} \tag{21.27}$$

where

$$\boxed{\alpha \;\stackrel{\Delta}{=}\; \sqrt{\frac{\pi}{8}}\left[\mu\mathsf{Tr}(R_u) + \mu^{-1}\mathsf{Tr}(Q)\right]} \tag{21.28}$$

It can be seen from (21.27) that the EMSE of sign-error LMS increases with the value of α (since the derivative of the EMSE with respect to α is positive for positive α). This shows that the EMSE can be minimized by selecting a step-size that minimizes α. Differentiating the expression for α with respect to μ and setting the derivative equal to zero we get

$$\mu_{\text{opt}}^{\text{sign-error LMS}} = \sqrt{\mathsf{Tr}(Q)/\mathsf{Tr}(R_u)}$$

The resulting minimum EMSE is

$$\zeta_{\min}^{\text{sign-error LMS}} = \frac{\pi}{4}\mathsf{Tr}(R_u)\mathsf{Tr}(Q)\left(1 + \sqrt{1 + \frac{8\sigma_v^2}{\pi\mathsf{Tr}(R_u)\mathsf{Tr}(Q)}}\right)$$

Complex-Valued Data

When the data are complex-valued, the sign function is defined by

$$\mathsf{csgn}[x] \;\stackrel{\Delta}{=}\; \mathsf{sign}(x_{\mathsf{r}}) + j\mathsf{sign}(x_{\mathsf{i}})$$

where $x = x_{\mathsf{r}} + jx_{\mathsf{i}}$. It follows now that

$$|g(x)|^2 = 2$$

almost everywhere in the complex plane, so that the variance relation (20.32) becomes

$$2\mu\mathsf{Tr}(R_u) + \mu^{-1}\mathsf{Tr}(Q) = 2\mathsf{Re}\left(\mathsf{E}\,e_a^*(i)\mathsf{csgn}[e_a(i) + v(i)]\right), \quad i \to \infty \tag{21.29}$$

Similarly to the real case, and as was done in Sec. 18.2, we can now invoke Price's theorem for complex data (from Prob. IV.11) and rewrite (21.29) as follows:

$$2\mu\mathsf{Tr}(R_u) + \mu^{-1}\mathsf{Tr}(Q) = 2\sqrt{\frac{2}{\pi}}\,\frac{\sqrt{2}\,\mathsf{E}\,|e_a(i)|^2}{\sqrt{\mathsf{E}\,|e_a(i)|^2 + \sigma_v^2}}, \quad i \to \infty$$

which has a form similar to (21.26). This equality can now be solved for $\mathsf{E}\,|e_a(\infty)|^2$.

Lemma 21.3 (Tracking EMSE of sign-error LMS) Consider the sign-error LMS recursion (21.22) and assume the data $\{\boldsymbol{d}(i), \boldsymbol{u}_i\}$ satisfy model (20.16) with a sufficiently small degree of nonstationarity. Assume also that $\{\boldsymbol{v}(i), \boldsymbol{u}_i\}$ are Gaussian and the step-size is small. Then its EMSE as a function of the step-size can be approximated by

$$\zeta^{\text{sign-error LMS}} = \frac{\alpha}{2}\left(\alpha + \sqrt{\alpha^2 + 4\sigma_v^2}\right)$$

where

$$\alpha = \sqrt{\frac{\pi}{8\gamma}}\left[\mu\gamma\text{Tr}(R_u) + \mu^{-1}\text{Tr}(Q)\right]$$

with $\gamma = 1$ for real-valued data and $\gamma = 2$ for complex-valued data. The optimal step-size, and the resulting minimum EMSE, are given by

$$\begin{aligned}
\mu_{\text{opt}}^{\text{sign-error LMS}} &= \sqrt{\frac{\text{Tr}(Q)}{\gamma\text{Tr}(R_u)}} \\
\zeta_{\min}^{\text{sign-error LMS}} &= \frac{\pi}{4}\text{Tr}(R_u)\text{Tr}(Q)\left(1 + \sqrt{1 + \frac{8\sigma_v^2}{\pi\text{Tr}(R_u)\text{Tr}(Q)}}\right)
\end{aligned}$$

The misadjustment is obtained by dividing the EMSE by σ_v^2.

21.4 PERFORMANCE OF RLS

We now examine the tracking performance of the RLS algorithm,

$$\boldsymbol{P}_i = \lambda^{-1}\left[\boldsymbol{P}_{i-1} - \frac{\lambda^{-1}\boldsymbol{P}_{i-1}\boldsymbol{u}_i^*\boldsymbol{u}_i\boldsymbol{P}_{i-1}}{1+\lambda^{-1}\boldsymbol{u}_i\boldsymbol{P}_{i-1}\boldsymbol{u}_i^*}\right] \tag{21.30}$$

$$\boldsymbol{w}_i = \boldsymbol{w}_{i-1} + \boldsymbol{P}_i\,\boldsymbol{u}_i^*[\boldsymbol{d}(i) - \boldsymbol{u}_i\boldsymbol{w}_{i-1}], \quad i \geq 0 \tag{21.31}$$

with initial condition $\boldsymbol{P}_{-1} = \epsilon^{-1}\text{I}$ and $0 \ll \lambda \leq 1$. The scalar ϵ is a small positive number and λ is usually close to one. As explained in Chapter 19, we are using a boldface letter for $\boldsymbol{P}_i$ to indicate that $\boldsymbol{P}_i$ is a random variable due to its dependence on the $\{\boldsymbol{u}_j\}$. Also, recall from (19.3) that

$$\boxed{\boldsymbol{P}_i^{-1} = \lambda^{i+1}\epsilon\text{I} + \sum_{j=0}^{i}\lambda^{i-j}\boldsymbol{u}_j^*\boldsymbol{u}_j} \tag{21.32}$$

which shows that $\boldsymbol{P}_i > 0$ for all finite i.

In a manner similar to Chapter 19, the energy-conservation approach of Sec. 20.3 can be extended in a straightforward manner to treat this case. Indeed, it is immediate to verify that the result of Thm. 20.1 becomes

$$\|\boldsymbol{w}_i^o - \boldsymbol{w}_i\|^2_{\boldsymbol{P}_i^{-1}} + \bar{\mu}(i)|\boldsymbol{e}_a(i)|^2 = \|\boldsymbol{w}_i^o - \boldsymbol{w}_{i-1}\|^2_{\boldsymbol{P}_i^{-1}} + \bar{\mu}(i)|\boldsymbol{e}_p(i)|^2 \tag{21.33}$$

where $e_a(i) = u_i(w_i^o - w_{i-1})$, $e_p(i) = u_i(w_i^o - w_i)$, and $\bar{\mu}(i)$ is defined as

$$\bar{\mu}(i) \triangleq \left(\|u_i\|^2_{P_i}\right)^\dagger = \begin{cases} 1/\|u_i\|^2_{P_i} & \text{if } u_i \neq 0 \\ 0 & \text{otherwise} \end{cases} \tag{21.34}$$

Comparing (21.33) with (20.24), we see that the main distinction is the appearance of the weighting matrices $\{P_i^{-1}, P_i\}$: the former is used as a weighting factor for the weight-error vectors while the latter is used as a weighting factor for the regressors. Recall again that the notation $\|x\|^2_W$ stands for $\|x\|^2_W = x^*Wx$, for a column vector x. The same argument that led to (20.29) will then show that (21.33) leads under expectation to

$$\boxed{\mathsf{E}\,\|\tilde{w}_i\|^2_{P_i^{-1}} + \mathsf{E}\,\bar{\mu}(i)|e_a(i)|^2 = \mathsf{E}\,\|\tilde{w}_{i-1}\|^2_{P_i^{-1}} + \mathsf{E}\,\|q_i\|^2_{P_i^{-1}} + \mathsf{E}\,\bar{\mu}(i)|e_p(i)|^2} \tag{21.35}$$

We are interested in evaluating the tracking performance of RLS in steady-state. For this purpose, we shall employ the steady-state condition (cf. Def. 15.1),

$$\mathsf{E}\,\tilde{w}_i\tilde{w}_i^* = \mathsf{E}\,\tilde{w}_{i-1}\tilde{w}_{i-1}^* = C, \quad \text{as } i \to \infty \tag{21.36}$$

in order to simplify the variance relation (21.33) into a form similar to that in Thm. 20.2. However, the presence of the matrices $\{P_i, P_i^{-1}\}$ in (21.35) makes this step challenging; this is because $\{P_i, P_i^{-1}\}$ are dependent not only on u_i but also on all prior regression vectors, $\{u_j, j \leq i\}$. For this reason, as was done in Chapter 19 and in order to make the performance analysis of RLS more tractable, whenever necessary, we shall approximate and replace the random variables $\{P_i^{-1}, P_i\}$ in steady-state by (cf. (19.11)–(19.12)):

$$\boxed{\lim_{i\to\infty} \mathsf{E}\,\left(P_i^{-1}\right) = \frac{R_u}{1-\lambda} \triangleq P^{-1}} \tag{21.37}$$

and

$$\mathsf{E}\,P_i \approx \left[\mathsf{E}\,P_i^{-1}\right]^{-1} = (1-\lambda)R_u^{-1} = P, \quad \text{as } i \to \infty \tag{21.38}$$

These approximations, along with the steady-state condition (21.36), allow us to write (cf. (19.13)):

$$\boxed{\mathsf{E}\,\|\tilde{w}_i\|^2_{P_i^{-1}} = \mathsf{E}\,\|\tilde{w}_{i-1}\|^2_{P_i^{-1}}} \tag{21.39}$$

so that (21.35) reduces to

$$\mathsf{E}\,\bar{\mu}(i)|e_a(i)|^2 = \mathsf{E}\,\|q_i\|^2_{P_i^{-1}} + \mathsf{E}\,\bar{\mu}(i)|e_p(i)|^2, \quad \text{as } i \to \infty \tag{21.40}$$

However, from (19.5) we know how $e_p(i)$ is related to $e_a(i)$. Substituting into (21.40) we get

$$\mathsf{E}\,\bar{\mu}(i)|e_a(i)|^2 = \mathsf{E}\,\|q_i\|^2_{P_i^{-1}} + \mathsf{E}\,\bar{\mu}(i)\left|e_a(i) - \|u_i\|^2_{P_i}e(i)\right|^2, \quad \text{as } i \to \infty \tag{21.41}$$

which upon expansion and simplification reduces to

$$\boxed{\mathsf{E}\,\|u_i\|^2_{P_i}|e(i)|^2 + \mathsf{E}\,\|q_i\|^2_{P_i^{-1}} = 2\text{Re}\left(\mathsf{E}\,e_a^*(i)e(i)\right), \quad \text{as } i \to \infty} \tag{21.42}$$

This variance relation is the extension of (20.32) to the RLS case. Now since the data $\{d(i), u_i\}$ satisfy model (20.16), we also have that $e(i) = e_a(i) + v(i)$ and (21.42) becomes

$$\boxed{\sigma_v^2 \mathsf{E}\,\|u_i\|^2_{P_i} \; + \; \mathsf{E}\,\|u_i\|^2_{P_i} |e_a(i)|^2 \; + \; \mathsf{E}\,\|q_i\|^2_{P_i^{-1}} = 2\mathsf{E}\,|e_a(i)|^2 \; = \; 2\zeta^{\mathsf{RLS}}, \quad i \to \infty} \tag{21.43}$$

To proceed we resort to the separation condition (16.7), namely, we assume that

$$\boxed{\text{At steady-state, } \; \|u_i\|^2_{P_i} \; \text{ is independent of } \; e_a(i)} \tag{21.44}$$

This condition allows us to separate the expectation $\mathsf{E}\,\|u_i\|^2_{P_i}|e_a(i)|^2$ into the product of two expectations as $\mathsf{E}\left(\|u_i\|^2_{P_i} \cdot |e_a(i)|^2\right) = \left(\mathsf{E}\,\|u_i\|^2_{P_i}\right) \cdot \left(\mathsf{E}\,|e_a(i)|^2\right)$. If we now replace P_i by its mean value, we obtain

$$\mathsf{E}\,\|u_i\|^2_{P_i} \; \approx \; \mathsf{E}\,\|u_i\|^2_P \; = \; \mathsf{Tr}(R_u P) \; = \; (1-\lambda)M \tag{21.45}$$

Moreover, since q_i is independent of all regressors, it is also independent of P_i^{-1}, so that

$$\mathsf{E}\,\|q_i\|^2_{P_i^{-1}} \quad = \quad \mathsf{E}\,\|q_i\|^2_{P^{-1}} \; = \; \mathsf{Tr}(QP^{-1}) \; = \; \mathsf{Tr}(QR_u)/(1-\lambda)$$

Substituting into (21.43) we find

$$\boxed{\zeta^{\mathsf{RLS}} = \frac{\sigma_v^2(1-\lambda)M \; + \; \frac{1}{(1-\lambda)}\mathsf{Tr}(QR_u)}{2-(1-\lambda)M}} \tag{21.46}$$

Lemma 21.4 (Tracking EMSE of RLS) Consider the RLS recursion (21.31)-(21.30) and assume $\{d(i), u_i\}$ satisfy the nonstationary model (20.16) with a sufficiently small degree of nonstationarity. Introduce the approximations (21.38) and (21.45). Then, under the separation condition (21.44), the tracking EMSE of RLS can be approximated by

$$\zeta^{\mathsf{RLS}} = \frac{\sigma_v^2(1-\lambda)M \; + \; \frac{1}{(1-\lambda)}\mathsf{Tr}(QR_u)}{2-(1-\lambda)M}$$

The misadjustment is obtained by dividing the EMSE by σ_v^2. Assuming $(1-\lambda)$ is small so that $2-(1-\lambda)M \approx 2$, the optimal choice of λ that results in minimal EMSE and the minimum EMSE are given by

$$\lambda^{\mathsf{RLS}}_{\text{opt}} = 1 - \frac{1}{\sigma_v}\sqrt{\frac{\mathsf{Tr}(QR_u)}{M}} \quad \text{and} \quad \zeta^{\mathsf{RLS}}_{\min} = \sigma_v\sqrt{M\mathsf{Tr}(QR_u)}$$

21.5 COMPARISON OF TRACKING PERFORMANCE

In order to get an appreciation for the tracking behavior of adaptive filters, we compare in this section the performance of three such filters, namely, LMS, LMF, and LMMN. We focus on the case of small step-sizes, since the expressions for the EMSE are simpler in this case. We also assume real-valued data.

First recall from the statement of Lemma 21.1, and from the results of Prob. IV.25 that, for small step-sizes, the minimum achievable EMSE of the aforementioned three algorithms are given by

$$\zeta_{\min}^{\textsf{LMS}} = \sqrt{\sigma_v^2 \textsf{Tr}(R_u)\textsf{Tr}(Q)}, \ \zeta_{\min}^{\textsf{LMF}} = \frac{\sqrt{\xi_v^6 \textsf{Tr}(R_u)\textsf{Tr}(Q)}}{3\sigma_v^2}, \ \zeta_{\min}^{\textsf{LMMN}} = \frac{\sqrt{a\textsf{Tr}(R_u)\textsf{Tr}(Q)}}{b}$$

where

$$\begin{aligned} a &= \delta^2\sigma_v^2 + 2\delta\bar{\delta}\xi_v^4 + \bar{\delta}^2\xi_v^6, \quad b = \delta + 3\bar{\delta}\sigma_v^2 \\ \bar{\delta} &= 1-\delta, \quad \xi_v^4 = \textsf{E}\,|\boldsymbol{v}(i)|^4, \quad \xi_v^6 = \textsf{E}\,|\boldsymbol{v}(i)|^6 \end{aligned}$$

In this section, we use the ratio of the minimum achievable EMSE for each of LMF and LMMN to that of LMS as a performance measure. For LMF, this ratio is equal to

$$\frac{\zeta_{\min}^{\textsf{LMS}}}{\zeta_{\min}^{\textsf{LMF}}} = \frac{3\sigma_v^3}{\sqrt{\xi_v^6}} \tag{21.47}$$

and is seen to depend only on the statistical properties of the noise sequence $\boldsymbol{v}(i)$. For LMMN, the same ratio is given by

$$\frac{\zeta_{\min}^{\textsf{LMS}}}{\zeta_{\min}^{\textsf{LMMN}}} = \frac{\sigma_v b}{\sqrt{a}} \tag{21.48}$$

which is also dependent on the statistical properties of the noise, as well as on the mixing parameter δ. We specialize these results for some noise distributions.

Gaussian noise. Assume that $\boldsymbol{v}(i)$ is Gaussian. Then $\xi_v^4 = 3\sigma_v^4$ and $\xi_v^6 = 15\sigma_v^6$, so that (21.47) becomes

$$\frac{\zeta_{\min}^{\textsf{LMS}}}{\zeta_{\min}^{\textsf{LMF}}} = \sqrt{\frac{3}{5}} \approx -1.1 \text{ dB}$$

where for a value x, we are using $10\log_{10}(x)$ as its dB equivalent. This result indicates that the minimum achievable EMSE of LMS is less than that of LMF by approximately 1.1 dB, for any value of the noise variance σ_v^2.

For the case of LMMN, expression (21.48) yields

$$\frac{\zeta_{\min}^{\textsf{LMS}}}{\zeta_{\min}^{\textsf{LMMN}}} = \frac{\sigma_v b}{\sqrt{\delta^2\sigma_v^2 + 6\delta\bar{\delta}\sigma_v^4 + 15\bar{\delta}^2\sigma_v^6}} \tag{21.49}$$

and Fig. 21.1 shows a plot of this ratio versus δ for various values of σ_v^2. The figure shows that the ratio is always less than unity for all values of δ and σ_v^2, which reflects the superiority of LMS over LMMN for tracking nonstationary systems in Gaussian noise environments.

Uniform noise. Assume now that the noise $\boldsymbol{v}(i)$ is uniformly distributed over an interval $[-\Delta, \Delta]$. Then it is easy to verify that $\sigma_v^2 = \Delta^2/3$, $\xi_v^4 = \Delta^4/5$, and $\xi_v^6 = \Delta^6/7$, so that from (21.47) we get

$$\frac{\zeta_{\min}^{\textsf{LMS}}}{\zeta_{\min}^{\textsf{LMF}}} = \sqrt{\frac{7}{3}} \approx 3.7 \text{ dB}$$

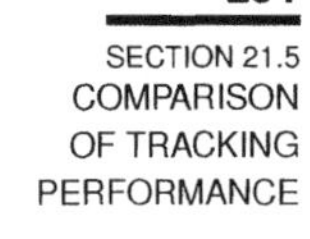

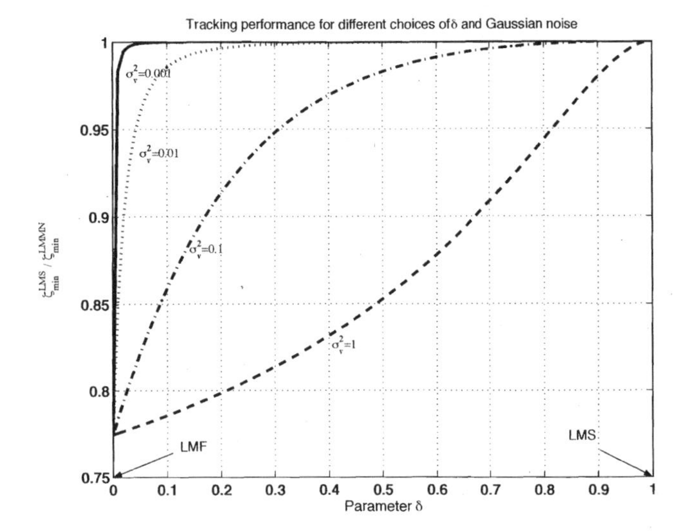

FIGURE 21.1 Comparison of the tracking performance of LMS, LMF, and LMMN for Gaussian noise. The figure plots the ratio (21.49) as a function of δ and σ_v^2.

This result indicates that the minimum achievable EMSE of LMF is now less than that of LMS by approximately 3.7 dB. Figure 21.2 shows a plot of the ratio of the minimum achievable EMSEs for LMS and LMMN versus δ for various values of σ_v^2. The figure shows that this ratio is now always larger than unity for all values of δ and σ_v^2. We can also see that the choice $\delta = 0$ (which corresponds to LMF) results in the best tracking performance. We therefore find that for uniform noise, the LMF algorithm is superior to both LMS and LMMN.

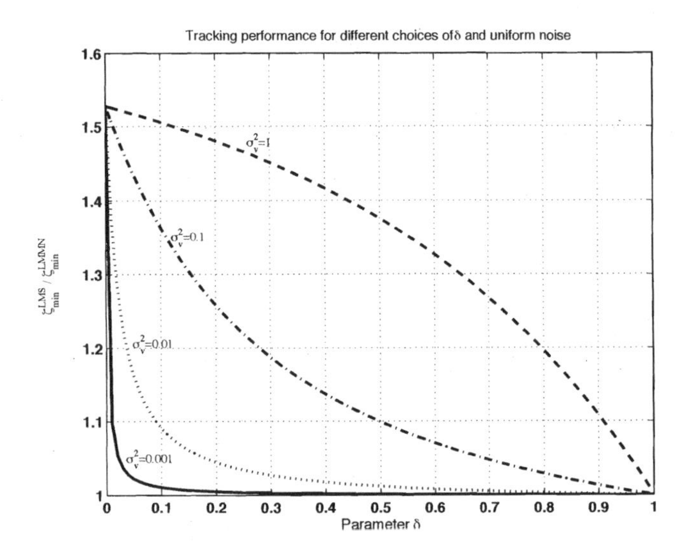

FIGURE 21.2 Comparison of the tracking performance of LMS, LMF, and LMMN for uniform noise. The figure plots the ratio (21.48) as a function of δ and σ_v^2.

Mixed Gaussian and uniform noise. We now consider the case where the noise $v(i)$ is a mixture of Gaussian and uniform distributions (e.g., a mixture of Gaussian system noise and uniformly distributed roundoff errors). In the simulations we generate $v(i)$ as the sum of two Gaussian and uniform random variables with variance ratio 1 to 3. Figure 21.3 shows the ratio of the minimum achievable EMSE of the LMS and LMMN algorithms versus δ for different values of the system noise variance σ_v^2. We see that in this case, the LMMN algorithm has the best tracking performance.

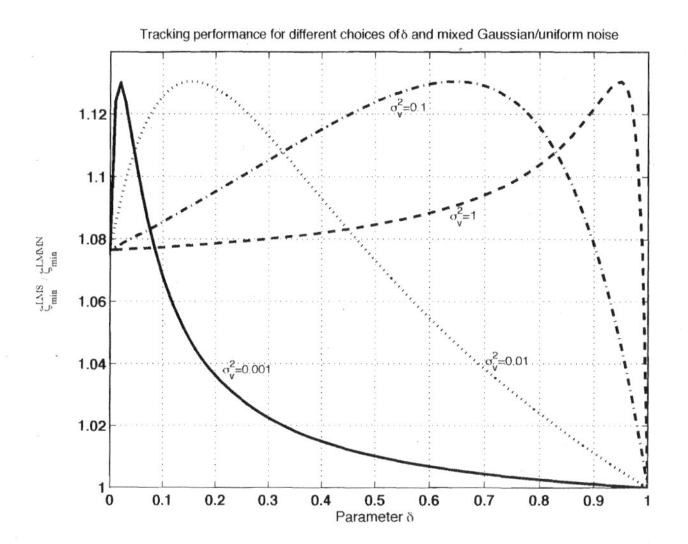

FIGURE 21.3 Comparison of the tracking performance of LMS, LMF, and LMMN for a mixed Gaussian/uniform noise distribution. The figure plots the ratio (21.48) as a function of δ and σ_v^2.

21.6 COMPARING RLS AND LMS

Although the convergence performance of RLS is significantly superior to that of LMS, it does not necessarily follow that the tracking performance of RLS is similarly superior to that of LMS. Actually, there are situations where one algorithm supersedes the other and vice-versa, so much so that a general statement about how their tracking behaviors relate to each other is difficult to make.

To illustrate this fact, recall from Lemma 21.1 that the excess mean-square error of LMS in nonstationary environments, and for sufficiently small step-sizes, can be approximated by

$$\zeta^{\text{LMS}} = \frac{\mu\sigma_v^2\text{Tr}(R_u) + \mu^{-1}\text{Tr}(Q)}{2}$$

with the corresponding optimal choice for the step-size and minimum achievable EMSE given by

$$\mu_{\text{opt}}^{\text{LMS}} = \sqrt{\text{Tr}(Q)/\sigma_v^2\text{Tr}(R_u)}, \qquad \zeta_{\text{min}}^{\text{LMS}} = \sqrt{\sigma_v^2\text{Tr}(R_u)\text{Tr}(Q)}$$

Likewise, from Lemma 21.4 we have that the excess mean-square error of RLS in nonstationary environments, and for forgetting factors that are sufficiently close to one, can be approximated by

$$\zeta^{\text{RLS}} = \frac{\sigma_v^2(1-\lambda)M}{2} + \frac{1}{2(1-\lambda)}\text{Tr}(QR_u)$$

with the corresponding optimal choice for the forgetting factor and minimum achievable EMSE given by

$$\lambda_{\text{opt}}^{\text{RLS}} = 1 - \sqrt{\text{Tr}(QR_u)/\sigma_v^2 M}, \qquad \zeta_{\text{min}}^{\text{RLS}} = \sqrt{\sigma_v^2 M \text{Tr}(QR_u)}$$

It follows that

$$\boxed{\frac{\zeta_{\text{min}}^{\text{RLS}}}{\zeta_{\text{min}}^{\text{LMS}}} = \sqrt{\frac{M\text{Tr}(QR_u)}{\text{Tr}(R_u)\text{Tr}(Q)}}}$$

from which is it seen that the performance of RLS and LMS will be similar whenever R_u or Q is a multiple of the identity matrix. For other choices of $\{Q, R_u\}$, one algorithm may perform better than the other. Three examples for different choices of Q are listed in Table 21.1: (i) $Q = \sigma_q^2 \text{I}$, i.e., the covariance matrix of the random perturbation vector $\boldsymbol{q}_i$ is a multiple of the identity; (ii) Q is a multiple of R_u and (iii) Q is a multiple of R_u^{-1}. It is seen from the results in the table that the performance of LMS is similar to that of RLS in case (i), while LMS is superior in case (ii), and RLS is superior in case (iii). The conclusions in cases (ii) and (iii) follow from the fact that for any $M \times M$ positive-definite matrix R_u, it always holds that (see Prob. IV.28):

$$\boxed{[\text{Tr}(R_u)]^2 \leq M\text{Tr}(R_u^2) \quad \text{and} \quad M^2 \leq \text{Tr}(R_u)\text{Tr}(R_u^{-1})} \qquad (\text{for any } R_u > 0)$$

Of course, these results on the tracking performance of RLS and LMS assume filter operation in environments with a small degree of nonstationarity.

TABLE 21.1 Minimum achievable excess-mean-square error of LMS and RLS (i.e., $\zeta_{\text{min}}^{\text{LMS}}$ and $\zeta_{\text{min}}^{\text{RLS}}$) for three choices of the nonstationary covariance matrix Q in comparison to the regressor covariance matrix R_u. In the table, α^2 is some constant value.

	$Q = \sigma_q^2 \text{I}$	$Q = \alpha^2 R_u$	$Q = \alpha^2 R_u^{-1}$
$\frac{\zeta_{\text{min}}^{\text{RLS}}}{\zeta_{\text{min}}^{\text{LMS}}}$	1	$\frac{\sqrt{M\text{Tr}(R_u^2)}}{\text{Tr}(R_u)} \geq 1$	$\frac{M}{\sqrt{\text{Tr}(R_u)\text{Tr}(R_u^{-1})}} \leq 1$
	similar performance	LMS is superior	RLS is superior

21.7 PERFORMANCE OF OTHER FILTERS

The same line of reasoning used in the previous sections is also used in the problems, and in Part V (*Transient Performance*), to study the tracking performance of other adaptive filters. For ease of reference, we list here the locations in the text that study these filters.

1. **ϵ-NLMS with power normalization**. The algorithm is described by the recursions:

$$\begin{aligned} p(i) &= \beta p(i-1) + (1-\beta)|u(i)|^2, \quad p(-1)=0 && (21.50) \\ w_i &= w_{i-1} + \frac{\mu}{\epsilon + p(i)} u_i^* e(i) && (21.51) \end{aligned}$$

where β is a positive number in the interval $0 < \beta < 1$. The tracking performance of the algorithm is studied in Prob. IV.24 for Gaussian regressors with shift structure by extending the arguments of Sec. 21.2.

2. **LMF and LMMN**. The LMMN algorithm is described by (cf. Alg. 12.4):

$$w_i = w_{i-1} + \mu u_i^* e(i)[\delta + (1-\delta)|e(i)|^2], \quad 0 \le \delta \le 1 \tag{21.52}$$

which employs a linear combination of the $e(i)$ and $e(i)|e(i)|^2$. The least-mean fourth (LMF) algorithm corresponds to the special case $\delta = 0$, i.e., $w_i = w_{i-1} + \mu u_i^* e(i)|e(i)|^2$. The tracking performance of these algorithms is studied in Prob. IV.25.

3. **ϵ–APA**. The ϵ–APA algorithm is described by the recursions (cf. Sec. 13.1):

$$\begin{aligned} U_i &= \begin{bmatrix} u_i \\ u_{i-1} \\ \vdots \\ u_{i-1+K} \end{bmatrix}, \quad d_i = \begin{bmatrix} d(i) \\ d(i-1) \\ \vdots \\ d(i-1+K) \end{bmatrix} && (21.53) \\ e_i &= d_i - U_i w_{i-1} && (21.54) \\ w_i &= w_{i-1} + \mu U_i^* \left(\epsilon \mathrm{I} + U_i U_i^*\right)^{-1} e_i, \quad i \ge 0 && (21.55) \end{aligned}$$

where ϵ is a small positive number and $K \le M$. The energy conservation approach of Sec. 20.3 is extended to this case and the tracking performance of the algorithm is studied in Prob. IV.26.

4. **Leaky-LMS**. The leaky-LMS filter is described by the recursion (see Alg. 12.2):

$$w_i = (1-\mu\alpha)w_{i-1} + \mu u_i^*[d(i) - u_i w_{i-1}], \quad i \ge 0$$

which looks similar to the LMS recursion (21.1) except for the presence of the factor α. The tracking performance of this algorithm is studied in Prob. V.34 for Gaussian regressors.

5. **CMA**. The CMA2-2 algorithm is described by the recursion

$$w_i = w_{i-1} + \mu u_i^* z(i)[\gamma - |z(i)|^2], \quad z(i) = u_i w_{i-1}, \quad i \ge 0$$

where γ is a positive scalar (see Alg. 12.6). The tracking performance of CMA2-2 is studied in Probs IV.34 and IV.35 for both real and complex-valued data. Problem IV.36 extends the results to CMA1-2, which is described by the recursion (see Alg. 12.5):

$$w_i = w_{i-1} + \mu u_i^* \left[\gamma \frac{z(i)}{|z(i)|} - z(i)\right], \quad z(i) = u_i w_{i-1}, \quad i \ge 0$$

21.8 PERFORMANCE TABLE FOR SMALL STEP-SIZES

We find it useful to list in Table 21.2 the tracking EMSE of several adaptive filters under the assumption of sufficiently small step-sizes. The expressions in the table are obtained as approximations from the results derived in the chapter and they serve as convenient references for comparing algorithms. The table also lists points of reference for each algorithm.

TABLE 21.2 Approximate expressions for the excess mean-square performance of adaptive filters in nonstationary environments for small step-sizes.

Algorithm	EMSE	Reference
LMS	$\frac{\mu\sigma_v^2}{2}\text{Tr}(R_u) + \frac{\mu^{-1}}{2}\text{Tr}(Q)$	Lemma 21.1
ϵ-NLMS	$\frac{\mu\sigma_v^2}{2-\mu}\text{Tr}(R_u)\text{E}\left(\frac{1}{\|\boldsymbol{u}_i\|^2}\right) + \frac{\mu^{-1}}{2-\mu}\text{Tr}(R_u)\text{Tr}(Q)$	Lemma 21.2
ϵ-NLMS with power normalization	$\frac{\mu(1+\beta)M\sigma_v^2 + \mu^{-1}\gamma\sigma_u^2(1-\beta)\text{Tr}(Q)}{2\left[\gamma(1-\beta)+2\beta\right] - \mu M(1+\beta)}$	Problem IV.24
sign-error LMS	$\frac{\alpha}{2}\left(\alpha + \sqrt{\alpha^2 + 4\sigma_v^2}\right)$ $\alpha = \sqrt{\frac{\pi}{8\gamma}}\left[\mu\gamma\text{Tr}(R_u) + \mu^{-1}\text{Tr}(Q)\right]$	Lemma 21.3
LMF	$\frac{\mu\xi_v^6}{2\sigma_v^2}\text{Tr}(R_u) + \frac{\mu^{-1}}{2\sigma_v^2}\text{Tr}(Q)$	Problem IV.25
LMMN	$\frac{\mu a}{2b}\text{Tr}(R_u) + \frac{\mu^{-1}}{2b}\text{Tr}(Q)$	Problem IV.25
leaky-LMS	$\frac{\mu\sigma_v^2}{2}\text{Tr}[R_u^2(R_u+\alpha\text{I})^{-1}] + \frac{\mu^{-1}}{2}\text{Tr}[QR_u(R_u+\alpha\text{I})^{-1}]$	Problem V.34
ϵ-APA	$\frac{\mu\sigma_v^2}{2-\mu}\text{Tr}(R_u)\text{E}\left(\frac{K}{\|\boldsymbol{u}_i\|^2}\right) + \frac{\mu^{-1}}{2-\mu}\text{Tr}(R_u)\text{Tr}(Q)$	Problem IV.26
RLS	$\frac{\sigma_v^2(1-\lambda)M + \frac{1}{(1-\lambda)}\text{Tr}(QR_u)}{2-(1-\lambda)M}$	Lemma 21.4
CMA2-2	$\frac{\mu\text{E}\left(\gamma^2\|\boldsymbol{s}\|^2 - 2\gamma\|\boldsymbol{s}\|^4 + \|\boldsymbol{s}\|^6\right)\text{Tr}(R_u) + \mu^{-1}\text{Tr}(Q)}{2\text{E}\left(2\|\boldsymbol{s}\|^2 - \gamma\right)}$	Problem IV.35
CMA1-2	$\frac{\mu}{2}\left(\gamma^2 + \text{E}\,\|\boldsymbol{s}\|^2 - 2\gamma\text{E}\,\|\boldsymbol{s}\|\right)\text{Tr}(R_u) + \frac{\mu^{-1}}{2}\text{Tr}(Q)$	Problem IV.36

Summary and Notes

The chapters in this part describe a procedure for evaluating the mean-square and tracking performance of adaptive filters by relying on energy conservation arguments. Some of the key results in the chapters are reproduced here for ease of reference.

SUMMARY OF MAIN RESULTS

Stationary Environments

1. Consider adaptive filters that are described by stochastic difference equations of the form

$$\boxed{\boldsymbol{w}_i = \boldsymbol{w}_{i-1} + \mu\, \boldsymbol{u}_i^*\, g[\boldsymbol{e}(i)], \quad \boldsymbol{w}_{-1} = \text{initial condition}}$$

where $g[\cdot]$ is some function of $\boldsymbol{e}(i) = \boldsymbol{d}(i) - \boldsymbol{u}_i \boldsymbol{w}_{i-1}$. The mean-square error (MSE) of any such filter is defined as the limiting value of $\mathsf{E}\,|\boldsymbol{e}(i)|^2$,

$$\boxed{\mathsf{MSE} = \lim_{i\to\infty} \mathsf{E}\,|\boldsymbol{e}(i)|^2}$$

2. When the data $\{\boldsymbol{d}(i), \boldsymbol{u}_i\}$ satisfy the stationary model (15.16), then the MSE is also given by

$$\boxed{\mathsf{MSE} = \sigma_v^2 + \lim_{i\to\infty} \mathsf{E}\,|\boldsymbol{e}_a(i)|^2}$$

where $\boldsymbol{e}_a(i) = \boldsymbol{u}_i \tilde{\boldsymbol{w}}_{i-1}$ and $\tilde{\boldsymbol{w}}_i = w^o - \boldsymbol{w}_i$. The limiting variance of the *a priori* error $\boldsymbol{e}_a(i)$ is called the excess mean-square error (EMSE) of the filter,

$$\boxed{\mathsf{EMSE} = \lim_{i\to\infty} \mathsf{E}\,|\boldsymbol{e}_a(i)|^2}$$

3. For any such adaptive filter, and for any data $\{\boldsymbol{d}(i), \boldsymbol{u}_i\}$, the following energy conservation relation holds for all i:

$$\boxed{\|\tilde{\boldsymbol{w}}_i\|^2 + \bar{\mu}(i)|\boldsymbol{e}_a(i)|^2 = \|\tilde{\boldsymbol{w}}_{i-1}\|^2 + \bar{\mu}(i)|\boldsymbol{e}_p(i)|^2}$$

where $\boldsymbol{e}_p(i) = \boldsymbol{u}_i \tilde{\boldsymbol{w}}_i$ and $\bar{\mu}(i)$ is defined as in (15.31).

4. The energy relation is useful in several respects. For the purposes of steady-state analysis, we noted that by taking expectations of both of its sides, and by using the fact that in steady-state $\mathsf{E}\,\|\tilde{\boldsymbol{w}}_i\|^2 = \mathsf{E}\,\|\tilde{\boldsymbol{w}}_{i-1}\|^2$, we arrived at the variance relation:

$$\boxed{\mu \mathsf{E}\,\|\boldsymbol{u}_i\|^2 |g[\boldsymbol{e}(i)]|^2 = 2\mathsf{Re}\left(\mathsf{E}\,\boldsymbol{e}_a^*(i) g[\boldsymbol{e}(i)]\right), \quad \text{as } i \to \infty}$$

For real-valued data, this relation becomes

$$\mu \mathsf{E}\,\|\boldsymbol{u}_i\|^2 g^2[\boldsymbol{e}(i)] = 2\mathsf{E}\,\boldsymbol{e}_a(i) g[\boldsymbol{e}(i)], \quad \text{as } i \to \infty$$

5. The variance relation was used to derive the EMSE of several adaptive filters: LMS, $\epsilon-$NLMS, sign-error LMS, and RLS. For RLS, we showed how to incorporate weighting into the energy conservation relation. In the problems at the end of this part, and also in Part V (*Transient Performance*), we evaluate the EMSE of other adaptive filters, e.g., LMF, LMMN, $\epsilon-$APA, leaky-LMS, sign-regressor LMS, constrained LMS, CMA2-2, and CMA1-2.

Nonstationary Environments

1. When the data $\{d(i), u_i\}$ satisfy the nonstationary model (20.16), then the MSE is still given by

$$\boxed{\text{MSE} = \sigma_v^2 + \lim_{i\to\infty} \mathsf{E}\,|e_a(i)|^2}$$

where now $e_a(i) = u_i(w_i^o - w_{i-1})$. The limiting variance of the *a priori* error $e_a(i)$ is called the excess mean-square error (EMSE) of the filter,

$$\boxed{\text{EMSE} = \lim_{i\to\infty} \mathsf{E}\,|e_a(i)|^2}$$

2. For any such adaptive filter, and for any data $\{d(i), u_i\}$, the following energy conservation relation holds for all i:

$$\boxed{\|w_i^o - w_i\|^2 + \bar{\mu}(i)|e_a(i)|^2 = \|w_i^o - w_{i-1}\|^2 + \bar{\mu}(i)|e_p(i)|^2}$$

where $e_p(i) = u_i\tilde{w}_i$, $\tilde{w}_i = w_i^o - w_i$, and $\bar{\mu}(i)$ is defined as in Thm. 20.1.

3. By taking expectations of both sides of the energy conservation relation, and by using the fact that in steady-state $\mathsf{E}\,\|\tilde{w}_i\|^2 = \mathsf{E}\,\|\tilde{w}_{i-1}\|^2$, we arrived at the following variance relation for nonstationary environments:

$$\boxed{\mu\mathsf{E}\,\|u_i\|^2|g[e(i)]|^2 + \mu^{-1}\text{Tr}(Q) = 2\text{Re}\big(\mathsf{E}\,e_a^*(i)g[e(i)]\big), \quad \text{as } i\to\infty}$$

For real-valued data, this relation becomes

$$\mu\mathsf{E}\,\|u_i\|^2 g^2[e(i)] + \mu^{-1}\text{Tr}(Q) = 2\mathsf{E}\,e_a(i)g[e(i)], \quad \text{as } i\to\infty$$

4. The variance relation was used to derive the EMSE of several adaptive filters in nonstationary environments: LMS, $\epsilon-$NLMS, sign-error LMS, and RLS. For RLS, we showed how to incorporate weighting into the energy conservation relation. In the problems at the end of this part, and also in Part V (*Transient Performance*), we evaluate the EMSE of other adaptive filters, e.g., LMF, LMMN, $\epsilon-$APA, leaky-LMS, CMA2-2, and CMA1-2.

5. In Probs. IV.29–IV.33 we use the variance relation to study the tracking performance of adaptive filters under a more general nonstationary model than (20.16). Specifically, we assume that the data $\{d(i), u_i, w_i^o\}$ are such that

$$d(i) = u_i w_i^o e^{j\Omega i} + v(i), \quad w_i^o = w^o + \theta_i, \quad \theta_i = \alpha\,\theta_{i-1} + q_i, \quad 0 \le |\alpha| < 1$$

That is, we assume w_i^o undergoes random variations around its mean, w^o, through a first-order auto-regressive model with a pole at α. In addition, we incorporate a parameter Ω to model possible frequency offsets between transmitters and receivers.

Energy Conservation

Some of the features of the energy conservation approach used in this part are the following:

a) It permits the evaluation of steady-state and tracking results without requiring a preliminary transient analysis. In other words, it does not obtain steady-state and tracking results as the limiting case of a transient analysis. In this way, steady-state results are not restricted by the same assumptions that are usually required for a successful transient analysis (see Part V).

b) It does not require knowledge of the weight error covariance matrix since it eliminates the effect of the terms $\mathsf{E}\,\|\tilde{\boldsymbol{w}}_i\|^2$ and $\mathsf{E}\,\|\tilde{\boldsymbol{w}}_{i-1}\|^2$.

c) Since it is easier to manipulate variables algebraically than under expectations, the energy conservation approach facilitates the steady-state analysis by eliminating unnecessary cross-terms.

d) The energy conservation relation admits several interpretations: geometric, system-theoretic, and it relates to Snell's law for light propagation — see App. 15.A.

Variance Relation

The variance relation can be solved for different adaptive algorithms, which correspond to different choices of the function $g[\cdot]$, in order to evaluate the corresponding EMSE, i.e., $\mathsf{E}\,|e_a(\infty)|^2$. In the process of this computation, we usually need to rely on some assumptions in order to simplify the evaluation of some expectations, e.g.,

a) A small step-size assumption.

b) A separation principle: at steady-state, $\|\boldsymbol{u}_i\|^2$ is independent of $\boldsymbol{e}_a(i)$.

c) A Gaussian assumption, namely, assuming the regressors $\boldsymbol{u}_i$ are Gaussian.

d) The independence assumptions, as described in Sec. 16.4. We did not rely on these assumptions but used instead the separation principle of part (b) above.

BIBLIOGRAPHIC NOTES

Stationary Environments

Energy-conservation. The energy-conservation relation (15.32) was originally derived by Sayed and Rupp (1995) and subsequently used by the authors in a series of works, including Rupp and Sayed (1996ab,1997,1998,2000) and Sayed and Rupp (1996,1997,1998), in their studies on the robustness and small gain analysis of adaptive filters. These robustness results will be discussed later in Chapter 44. In this part, we focused on showing how the energy-conservation relation can be used to study the steady-state mean-square performance of adaptive filters in a unified manner. The presentation in Chapters 15–19 follows mainly Yousef and Sayed (1999a,2000a,2001), where the variance relation (15.40) was derived from the energy-conservation relation (15.32), as well as Mai and Sayed (2000) and Shin and Sayed (2004). The last two references study constant-modulus and affine projection algorithms.

Performance results. The steady-state performance of LMS and its variants has been studied extensively in the literature, using a variety of approaches:

1. Expression (16.6) is the same result obtained by Jones, Cavin, and Reed (1982) for the performance of LMS for small step-sizes.

2. Expression (16.10) is the same result obtained by Gardner (1984) for LMS.

3. Expression (17.9) is the same result obtained by Slock (1993) for the performance of NLMS assuming a particular model for the regression data (as explained later in Prob. V.20).

4. Expression (18.8) for the performance of sign-error LMS is the same result obtained by Mathews and Cho (1987) by using, in addition to the Gaussian assumption (18.4), the independence assumptions (i)–(vi) of Sec. 16.4. The derivation in Chapter 18 did not use the independence assumptions.

5. The performance of LMMN for small step-sizes from Table 19.1 is the same result obtained by Tanrikulu and Chambers (1996) using averaging analysis (averaging methods are discussed in App. 24.A). The specialization to LMF, by setting $\delta = 0$, is the same result obtained by Walach and Widrow (1984) by using the independence assumptions.

EMSE of adaptive filters. In Chapters 16–19 we derived the EMSE for several particular choices of $g[\cdot]$ in (15.23). In Prob. 6.7 of Sayed (2003), and following the work of Al-Naffouri and Sayed (2001b,2003b), we derive a general expression for the EMSE of adaptive filters corresponding to choices of $g[\cdot]$that are solely functions of $e(\cdot)$. We do so by appealing to a Gaussian assumption on the distribution of $e_a(i)$, and by using this assumption to rewrite the variance relation in an alternative form. More specifically, it is shown in the aforementioned problem that if we introduce the functions:

$$\boxed{h_U \triangleq \mathsf{E}\,|\, g[e(i)]\,|^2} \quad \text{and} \quad \boxed{h_G \triangleq \frac{\mathsf{Re}(\mathsf{E}\, e_a^*(i)g)}{\mathsf{E}\,|e_a(i)|^2}}$$

Then $\{h_U, h_G\}$ are solely functions of $\mathsf{E}\,|e_a(i)|^2$ and, moreover, the variance relation can be rewritten as

$$\boxed{\zeta = \frac{\mu}{2}\mathsf{Tr}(R_u)\frac{h_U(\zeta)}{h_G(\zeta)}}$$

in terms of the EMSE, ζ. Thus, given $g[\cdot]$, we can evaluate the functions h_U and h_G under the assumed conditions, and then proceed to find a *fixed point* of the above nonlinear equation in ζ. Problems 6.8–6.15 of Sayed (2003) illustrate this method of computation for several adaptive filters, and show that its results are consistent with the results obtained in this part.

Price's theorem. This theorem, which is due to Price (1958), plays a useful role in parts of our analysis, especially when the error signal is assumed to be Gaussian and/or the filter is assumed to be of sufficient length. The theorem can be found in Price (1958) and also in Papoulis (1991). It is further studied in Probs. IV.10–IV.12. In the last problem, a special case known as Bussgang's theorem is derived (Bussgang (1952)). In Prob. V.26, the extension of Price's theorem to complex data is considered (cf. McGee (1969) and van den Bos (1996)).

Relating NLMS and LMS. In App. 17.A we relate $\epsilon-$NLMS to LMS via the change of variables (17.10). This transformation allows us to derive results for $\epsilon-$NLMS from those for LMS. Although this change of variables has been used before in the literature, e.g., by Widrow and Lehr (1990) and An, Brown and Harris (1997), these earlier analyses have some limitations: (1) They consider the case $\epsilon = 0$. (2) The conditions for stability are based on results valid for LMS with Gaussian regressors, while it is clear from (17.10) that the transformed regressor $\check{\boldsymbol{u}}_i$ cannot be Gaussian (since it is bounded), and (3) no attention has been given to the fact that $\check{\boldsymbol{u}}_i$ and $\check{\boldsymbol{v}}(i)$ are still orthogonal random variables (in which case, it can be verified that NLMS still computes unbiased estimates). These issues are resolved in App. 17.A.

Convex combination of adaptive filters. Combinations of adaptive filters provide one useful way to improve adaptive filter performance, whereby the outputs of several filters are mixed together to get an overall output of improved quality (see, e.g., Anderson (1985), Niedźwiecki (1992), and Singer and Feder (1999)). Clearly, the issue of how to optimally combine the component filters is a challenging task. In the work by Arenas-Garcia, Figueiras-Vidal, and Sayed (2006), the mean-square performance of a particular convex combination of two transversal filters is studied by using the energy conservation arguments of this part. Performance expressions are derived that indicate that the method is universal with respect to the component filters, i.e., in steady-state, it performs at least as well as the best component filter. The analysis also suggests combination structures with improved tracking performance; see Prob. IV.4 for a special case.

Adaptive networks. Studies on distributed adaptive processing where filters interact with each other over both time and space in the context of adaptive networks appear in Lopes and Sayed (2006,2007a,b, 2008), Sayed and Lopes (2007), and Cattivelli, Lopes, and Sayed (2008). These references employ the same energy conservation arguments of this part to analyze the effect of temporal and spatial interaction on distributed adaptive filters; see also Prob. V.13.

Colored noise and nonlinear effects. Recall that LMS is derived as a stochastic-gradient approximation for solving the normal equations (8.4), which characterize the solution to the linear least-mean-squares estimation problem (8.1). In Chapter 16, we studied how well LMS is able to approximate the optimal solution of the normal equations by relying on the stationary data model (15.16). A special feature of this model is that the noise sequence $\{v(i)\}$, which corresponds to the optimal residual that results from estimating $d(i)$ from u_i, is assumed to be a white noise sequence. Moreover, $v(i)$ is also assumed to be independent of the regression data u_j for all i, j. In the work by Reuter and Zeidler (1999), the authors consider a special example in the context of channel equalization whereby the noise sequence $\{v(i)\}$ is colored. Specifically, $v(i)$ consists of a narrow-band signal embedded in white noise. In addition, the noise and regression sequences $\{v(i), u_i\}$ are highly correlated. Clearly, in this scenario, the analysis that we have given for the MSE performance of LMS in the text should be adjusted accordingly. For instance, in Reuter and Zeidler (1999) it was shown that, under such conditions on $\{v(i), u_i\}$, the mean-square performance of LMS can even be superior to that of the normal-equations solution. One justification for this observation is that LMS processes the data in a nonlinear fashion. In other words, LMS is in effect a nonlinear filter and this property can lead to improved performance in situations with highly correlated data and noise.

Ergodic approximations for RLS. The approximation (19.12) used for RLS is common in the literature — see, for instance, Eleftheriou and Falconer (1986), Haykin (2000, p. 648), and Manolakis, Ingle, and Kogon (2000, p. 557).

Constant-modulus algorithms. The steady-state performance of constant-modulus algorithms is studied in Probs. IV.16–IV.18 using the same energy-conservation arguments that we employed in the body of the chapter. The derivation in these problems is based on the work by Mai and Sayed (2000). Some of the earlier performance results for constant-modulus schemes appear in the works by Chan and Shynk (1990), Bershad and Roy (1990), Shynk et al. (1991), Li and Ding (1996), Zeng and Tong (1997), and Fijalkow, Manlove, and Johnson (1998). The article by Johnson et al. (1998) provides a comprehensive list of additional references on different aspects of CM algorithms. The work by Shynk et al. (1991) gives some of the earliest approximations for the mean-square error of CMA2-2 under the assumption of Gaussian regression vectors. The work by Bershad and Roy (1990) is also an early work on the performance of CMA2-2 albeit for a particular class of input signals that are modeled by Rayleigh fading sinusoids. The work by Zeng and Tong (1997) studies the mean-square-error of the optimal CM receiver but the effects of adaptation and gradient noise are not considered. The work by Fijalkow, Manlove, and Johnson (1998) obtains an expression for the mean-square error of CMA2-2 using Lyapunov stability and averaging analysis arguments. Their MSE expression is the closest to the results in Probs. IV.16 and IV.17.

An application to echo cancellation. There are two types of echoes in communications systems: line echoes and acoustic echoes. Line echoes in voice communications occur over telephone lines due to circuit imperfections and impedance mismatches (see, e.g., Sondhi and Berkley (1980)). Among the many techniques that have been developed over the years to control the echoes (including echo suppressors), adaptive echo cancellation seems to be the most effective way and is the method of choice in modern implementations. It was first reported in Sondhi (1967); actually, in his article, Sondhi recognizes J. L. Kelly Jr. of Bell Laboratories as being the original proposer of using adaptation for echo cancellation purposes (see Kelly and Logan (1970)).

In Computer Project IV.1 we study an adaptive echo canceller implementation. In that project, an adaptive filter is used to estimate the echo path and to subsequently generate a replica of the echo in order to cancel it. It should be mentioned that, in practice, a complete echo canceller implementation would need to perform additional tasks, besides echo cancellation and adaptation, in order to avoid distorting the speech signals. Among these tasks, we may mention the need to identify signaling tones in order to avoid cancelling them, as well as the need to identify double-talk conditions (i.e., situations when both speakers are simultaneously active) in order to freeze adaptation and avoid filter divergence. In addition, the design should account for the effect of finite-precision computations on the performance of the echo canceller. Later, in Computer Project VI.1 we shall study acoustic echo cancellation, as opposed to line echo cancellation.

Energy conservation. The energy-conservation relation (20.24) is the extension to nonstationary environments of relation (15.32), which was originally derived by Sayed and Rupp (1995) in their studies on the robustness and small gain analysis of adaptive filters (see Chapter 44). The extension (20.24), along with its variance relation (20.32), were derived by Yousef and Sayed (1999a,2001,2002) and used therein, as well as in subsequent works by the same authors, to study the tracking performance of adaptive filters in a unified manner. The presentation in this chapter follows mainly Yousef and Sayed (2000a,2001,2002). In the work by Shin and Sayed (2004), the extension of the tracking analysis to affine projection algorithms is presented.

EMSE of adaptive filters. In Chapter 21 we evaluated the tracking EMSE for several particular choices of $g[\cdot]$ in (20.21). In Prob. 7.4 of Sayed (2003) we derive a general expression for the EMSE of adaptive filters corresponding to generic choices of $g[\cdot]$ (with $g[\cdot]$ being solely a function of $e(\cdot)$). We do so by appealing to a Gaussian assumption on the distribution of $e_a(i)$, and by using this assumption to rewrite the variance relation in an alternative form. It is shown in the aforementioned problem that if we introduce the functions:

$$\boxed{h_U \triangleq \mathsf{E}\,|\,g[e(i)]\,|^2} \quad \text{and} \quad \boxed{h_G \triangleq \frac{\mathsf{Re}(\mathsf{E}\,e_a^*(i)g)}{\mathsf{E}\,|e_a(i)|^2}}$$

then $\{h_U, h_G\}$ are solely functions of $\mathsf{E}\,|e_a(i)|^2$ and, moreover, the variance relation leads to the following equation in terms of the desired filter EMSE,

$$\boxed{\zeta = \frac{\mu \mathsf{Tr}(R_u)h_U(\zeta) + \mu^{-1}\mathsf{Tr}(Q)}{2h_G(\zeta)}}$$

This result indicates that the EMSE can be obtained as the *fixed-point* of a nonlinear equation in ζ. In other words, given $g[\cdot]$, we can evaluate the functions h_U and h_G under the assumed conditions, and then proceed to solve the above nonlinear equation for ζ. Problems 7.5–7.8 of Sayed (2003) illustrate this method of computation and show that its results are consistent with the results obtained in this chapter.

Tracking results. Among the earliest works on the tracking performance of adaptive filters are those of Widrow et al. (1976) and Bershad et al. (1980). The former uses a random-walk model for the variations in the optimal weight vector, while the latter assumes deterministic variations in the optimal weight vector. Both works focused on the transient performance of LMS but were not concerned with its steady-state performance. Steady-state results appeared subsequently in Farden (1981), Benveniste and Ruget (1982), Walach and Widrow (1984), Eweda and Macchi (1985), Eleftheriou and Falconer (1986), Marcos and Macchi (1987), and Benveniste (1987). The work by Benveniste and Ruget (1982) was apparently the first to compare the tracking performance of different adaptive algorithms. More recent analysis for other classes of adaptive filters appear in Eweda (1990b,1994,1997), Hajivandi and Gardner (1990), Cho and Mathews (1990), Guo (1994), Bahai and Sarraf (1997), and Rupp (1998b). The tracking results in Lemma 21.3 for sign-error LMS agree with those derived by Cho and Mathews (1990) and Eweda (1990b). The tracking results for LMMN are from Yousef and Sayed (1999c,2001). The comparison results of the tracking performance of RLS and LMS agree with those derived by Eweda (1994).

Random-walk model. It is customary in the literature to use the random-walk model (20.10) — see, e.g., Haykin (2000, p. 644) and Macchi (1995). However, as explained in Sec. 20.2, this model is not necessarily meaningful since the covariance matrix of w_i^o grows unbounded. In Probs. IV.29–IV.33, we show how to study the tracking performance of adaptive filters by relying instead on model (20.14). In the computer project at the end of the chapter, we provide an example that shows how models of the form (20.14) arise in applications.

Random and cyclic nonstationarities. Besides random channel variations, another source of nonstationarity that is common in communication systems is due to mismatches between the transmitter and receiver carrier generators (or clocks). Such mismatches result in periodic system variations, which can be damaging to the performance of adaptive filters, even for small carrier frequency offsets. Examples to this effect appear in Bahai and Sarraf (1997) and Rupp (1998b). The ability of adaptive filtering algorithms to track such periodic system variations has received little attention in the literature, except perhaps for the works by Hajivandi and Gardner (1990), Rupp (1998b), and Yousef and Sayed (2002).

The work by Rupp (1998b) uses a first-order approximation to examine the performance of LMS in the presence of carrier frequency offsets only. In Probs. IV.29–IV.33, we follow the work of Yousef and Sayed (2002) and show how the energy-conservation approach used in the body of the chapter can be applied to study the performance of a variety of adaptive filters in the *joint* cases of random and cyclic nonstationarities.

Tracking Rayleigh fading channels. In Computer Project IV.2 we illustrate the tracking ability of LMS in the context of a Rayleigh fading and multipath channel. Such channel models are widely used in modeling wireless communications environments (see, e.g., Viterbi (1995), Rappaport (1996), and Verdu (1998)). In the project we describe some of the basic concepts that are used in characterizing Rayleigh channels. The pioneering work on the characterization of fading and multipath conditions is due to Price (1954,1956). Other early contributions include Price and Green (1958) and Kailath (1960b,1961).

Finite precision effects. The performance of adaptive filters is affected adversely when they are implemented in finite-precision arithmetic due to roundoff errors. For example, quantization errors may affect the stability of an adaptive filter and ultimately lead to its divergence. They can also degrade the steady-state performance of the filter causing it to attain a higher mean-square error than what is expected from an infinite-precision analysis. The performance degradation tends to be more serious for recursive-least-squares (RLS) algorithms as opposed to least-mean-squares (LMS) algorithms.

Quantization errors propagate in a highly nontrivial manner, and studying their effect on adaptive filter performance requires several assumptions on how roundoff errors arise. Chapter 8 of Sayed (2003) describes a procedure for evaluating the effect of quantization errors by relying on the same energy conservation arguments that we have used so far in our exposition. The analysis there shows that the effect of roundoff errors on filter performance can be analyzed in a manner similar to how we studied the effect of channel nonstationarities on tracking performance in the current chapter. Specifically, the main conclusion from Sayed (2003, Ch. 8) is that, for sufficiently small step-sizes, the approximate EMSE of an adaptive filter in a quantized environment can be obtained from its EMSE in a nonstationary environment by substituting $\{Q, R_u, \sigma_v^2\}$ by:

$$\bar{Q} \approx Q + 2\sigma_c^2\, \mathrm{I}_M, \quad \bar{R}_u = R_u + \sigma_r^2\, \mathrm{I}_M, \quad \sigma_{\bar{v}}^2 \approx \sigma_v^2 \,+\, 3\sigma_r^2$$

where

$$\sigma_r^2 = \frac{\kappa}{12}\frac{L_r^2}{2^{2B_r}} \quad \text{and} \quad \sigma_c^2 = \frac{\kappa}{12}\frac{L_c^2}{2^{2B_c}}$$

It is assumed that the entries of the weight vector, $\boldsymbol{w}_i$, are quantized to B_c bits (assumed large enough) with saturation level L_c and the entries of $\{\boldsymbol{d}(i), \boldsymbol{u}_i, \boldsymbol{u}_i\boldsymbol{w}_{i-1}, \boldsymbol{e}(i)\}$ are quantized to B_r bits with saturation level L_r. Moreover, $\kappa = 1$ for real data and $\kappa = 2$ for complex data. For example, the EMSE for a finite precision implementation of LMS would be given by

$$\zeta^{\text{LMS}} = \frac{\mu\sigma_{\bar{v}}^2\mathrm{Tr}(\bar{R}_u) + \mu^{-1}\mathrm{Tr}(\bar{Q})}{2 - \mu\mathrm{Tr}(\bar{R}_u)}$$

Finite precision results. There have been extensive studies in the literature on the effect of quantization errors on adaptive filter performance. One of the earliest studies of finite-precision effects on LMS performance was performed by Gitlin, Mazo, and Taylor (1973) followed by Weiss and Mitra (1979). Afterwards, Caraiscos and Liu (1984) examined the effect of finite word-length conditions on the steady-state filter performance assuming floating-point arithmetic, while Alexander (1987) examined the effect of finite word-length on the filter transient performance. A discussion of limited precision effects on filter performance also appears in Cioffi (1987) and Sherwood and Bershad (1987).

Further results for LMS, NLMS, and sign-regressor LMS appear in Chang and Willson (1995), Bermudez and Bershad (1996ab), Eweda, Yousef, and El-Ramly (1998), and Eweda, Younis, and El-Ramly (1998). In the works by Bermudez and Bershad (1996ab), the authors develop a model to account for nonlinearities in the quantization process, including the occurrence of underflow, and use the model to study the performance of quantized LMS. The extension of the energy conservation approach to the case of finite precision implementations can be found in Yousef and Sayed (2000b,2003) and Sayed (2003, Ch. 8).

Drift problem. The fact that the LMS filter can produce unbounded weight estimates in some situations is illustrated in Prob. IV.39. This so-called drift problem has been described in several references including, for example, Gitlin, Meadors, and Weinstein (1982), Ioannou and Kokotovic (1984), Cioffi and Werner (1985), Sethares et al. (1986), Cioffi (1987), and Rupp (1995). The reference Ioannou and Kokotovic (1984) provides an analysis in the adaptive control context. The reference Sethares et al. (1986) studies the drift problem in a deterministic infinite-precision setting, while the references Cioffi and Werner (1985) and Cioffi (1987) consider finite-precision effects. They show that even with zero noise, unbounded growth (i.e., drift) of the weight estimates can happen due to finite-precision arithmetic errors. Such unbounded growth of the LMS estimates usually happens if two conditions are satisfied:

1. The regressor covariance matrix, R_u, is singular.
2. The noise or the finite-precision arithmetic errors have nonzero mean (zero mean variables can become non-zero mean due to finite precision errors — see Prob. IV.38.)

One of the earliest propositions to deal with the lack of sufficient excitation (i.e., with a singular R_u) was by Zahm (1973), in which it was suggested to add a small amount of white noise to the regression data (a procedure known as *dithering*). However, the leakage-based solution is nowadays the preferred way to go.

Circular-leaky LMS. As explained in Prob. IV.39, leaky-LMS helps ameliorate the drift problem of LMS. However, this solution comes at the expense of biased weight estimates. In Nascimento and Sayed (1999), a leaky variant (called circular-leaky LMS) is proposed that avoids the drift problem and guarantees unbiased estimates. There are two modifications with respect to leaky LMS in the circular-leaky variant. First, leakage is applied to a single tap at each iteration and, second, leakage is applied only if the tap magnitude exceeds a pre-specified level.

Table IV.1 summarizes the properties of LMS, leaky-LMS, and circular-leaky LMS for comparison purposes. In the complexity column, we list approximate values for the number of multiplications, additions, multiply-and-accumulate (MA), and if-then (IF) commands necessary for each algorithm; assuming real data.

TABLE IV.1 Comparison of three LMS filters.

Algorithm	Drift problem	Biased when $R_u > 0$	Complexity			
			MA	$\times$	$+$	IF
LMS	YES	NO	$2M$	1	0	0
Leaky LMS	NO	YES	$2M$	$M+1$	0	0
circular-leaky LMS	NO	NO	$2M$	3	2	3

Problems and Computer Projects

PROBLEMS

Problem IV.1 (Auto-regressive process) A unit-variance white-random process $s(i)$ is fed into a first-order auto-regressive model with transfer function $\sqrt{1-a^2}/(1-az^{-1})$, where a is real. The output process is denoted by $u(i)$; it is referred to as an auto-regressive process of order 1, written as AR(1). Assume $|a| < 1$. Show that the auto-correlation sequence of $u(i)$ is given by $r(k) = \mathsf{E}\,u(i)u(i-k) = a^{|k|}$, for all integer values k. If $u(i)$ is fed into an adaptive filter of order M, what is the covariance matrix of the resulting regressor u_i?

Problem IV.2 (Finite alphabets) Consider an adaptive filter with a regressor vector u_i that possesses shift-structure, namely, $u_i = \begin{bmatrix} u(i) & u(i-1) & \ldots & u(i-M+1) \end{bmatrix}$. The entries of $u(i)$ are realizations of a binary random variable, i.e., they are ± 1 with probability 1/2. Assume initially that all variables are real-valued.

(a) Show that the EMSE that would result when the filter is trained using LMS is given by $\zeta^{\text{LMS}} = \mu M\sigma_v^2/(2-\mu M)$. Show that this result is exact, i.e., it holds irrespective of any approximations.

(b) Assuming $\epsilon \ll M$, show that a similar conclusion holds when the filter is trained using $\epsilon-$NLMS with $\zeta^{\epsilon-\text{NLMS}} = \mu\sigma_v^2/(2-\mu)$.

(c) Likewise, assuming $\epsilon \ll 1$, show that for power-normalized $\epsilon-$NLMS it holds that $\zeta^{\epsilon-\text{pNLMS}} = \mu M\sigma_v^2/(2-\mu M)$. *Remark.* Recall from the discussion prior to (11.9) that the step-size for ϵ-NLMS with power normalization is in general M times smaller than the step-size for ϵ-NLMS and, therefore, the above expression for the EMSE agrees with that of part (b) if μ is replaced by μ/M.

(d) Show that the EMSE of sign-error LMS would be the same as in Lemma 18.1 with $\text{Tr}(R_u)$ replaced by M, and that this result holds under the same assumptions stated in the lemma.

(e) Show that the EMSE of LMF would be $\zeta^{\text{LMF}} = \mu M\xi_v^6/(6\sigma_v^2 - 15\mu M\xi_v^4)$. Under what approximation does this result hold?

(f) How would the results change if the $u(i)$ arise from a QPSK constellation instead?

Remark. The separation principle (16.7) is also exact in the case of regressors with modulated inputs, e.g., as in Pfann and Steward (1998).

Problem IV.3 (Second-order filter) Consider an adaptive filter with a 2-dimensional regression vector $u_i = [\,u(i)\; u(i-1)]$. The entries $\{u(j)\}$ are independent random variables satisfying

$$u(j) = \begin{cases} a & \text{with probability } p \\ b & \text{with probability } q \end{cases}$$

where a and b are real numbers with $a > b$.

(a) Determine the values of p and q, and conditions on a and b, such that u_i is zero mean.

(b) Find the covariance matrix of u_i.

(c) Find an expression for the EMSE that would result when the filter is trained using LMS.

(d) Repeat part (c) when the filter is trained using $\epsilon-$NLMS.

Problem IV.4 (Combination of adaptive filters) Consider a combination of two adaptive filters as shown in Fig. IV.1. One filter has a 2×1 weight vector $\boldsymbol{w}_{1,i}$ while the other filter has a 3×1 weight vector $\boldsymbol{w}_{2,i}$; both at the same time instant i. The regression sequence for the top filter is denoted by $\{\boldsymbol{u}_{1,i}\}$ and the regression sequence for the bottom filter is denoted by $\{\boldsymbol{u}_{2,i}\}$. Each regression sequence is i.i.d. and independent of the other. The individual entries of $\boldsymbol{u}_{1,i}$ are chosen independently of each other and lie on the unit circle. Likewise for $\boldsymbol{u}_{2,i}$ except that its entries lie on the circle of radius $\sqrt{2}$.

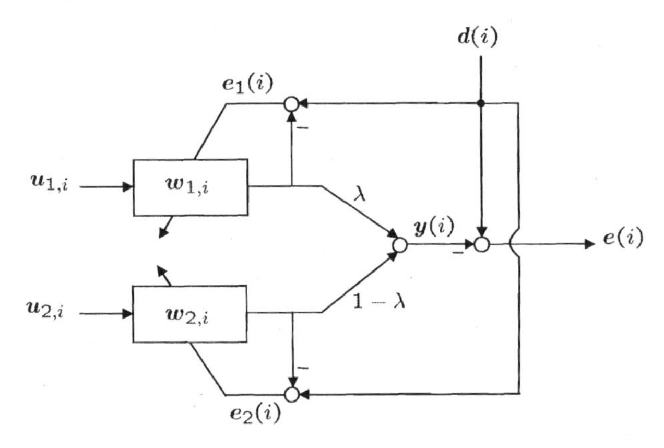

FIGURE IV.1 A combination of two adaptive filters.

Both adaptive filters employ the same reference sequence $\boldsymbol{d}(i)$ to generate their respective output errors:

$$\boldsymbol{e}_1(i) = \boldsymbol{d}(i) - \boldsymbol{u}_{1,i}\boldsymbol{w}_{1,i-1}, \qquad \boldsymbol{e}_2(i) = \boldsymbol{d}(i) - \boldsymbol{u}_{2,i}\boldsymbol{w}_{2,i-1}$$

and the weight vectors are updated according to the following rules:

$$\boldsymbol{w}_{1,i} = \boldsymbol{w}_{1,i-1} + \mu_1 \frac{\boldsymbol{u}_{1,i}^*}{\|\boldsymbol{u}_{1,i}\|^2} \boldsymbol{e}_1(i), \qquad \boldsymbol{w}_{2,i} = \boldsymbol{w}_{2,i-1} + \mu_2 \, \boldsymbol{u}_{2,i}^* \, \boldsymbol{e}_2(i)$$

Let $\{\boldsymbol{y}_1(i), \boldsymbol{y}_2(i)\}$ denote the outputs of the adaptive filters,

$$\boldsymbol{y}_1(i) = \boldsymbol{u}_{1,i}\boldsymbol{w}_{1,i-1}, \qquad \boldsymbol{y}_2(i) = \boldsymbol{u}_{2,i}\boldsymbol{w}_{2,i-1}$$

These outputs are combined by means of a nonnegative scalar $0 \leq \lambda \leq 1$ in a convex manner to generate the output of the combined adaptive structure as follows:

$$\boldsymbol{y}(i) = \lambda \boldsymbol{y}_1(i) + (1 - \lambda)\boldsymbol{y}_2(i)$$

Assume that the data $\{\boldsymbol{d}(i), \boldsymbol{u}_{1,i}, \boldsymbol{u}_{2,i}\}$ satisfy the following conditions:

1. There exists a vector w_1^o such that $\boldsymbol{d}(i) = \boldsymbol{u}_{1,i}w_1^o + \boldsymbol{v}_1(i)$.
2. There exists a vector w_2^o such that $\boldsymbol{d}(i) = \boldsymbol{u}_{2,i}w_2^o + \boldsymbol{v}_2(i)$.
3. The noise sequences $\{\boldsymbol{v}_1(i), \boldsymbol{v}_2(i)\}$ are i.i.d. with variance σ_v^2.
4. The noise sequences $\{\boldsymbol{v}_1(i), \boldsymbol{v}_2(i)\}$ are independent of each other and of $\{\boldsymbol{u}_{1,j}, \boldsymbol{u}_{2,j}\}$ fo all i, j.
5. The initial conditions $\{\boldsymbol{w}_{1,-1}, \boldsymbol{w}_{2,-1}\}$ are independent of all $\{\boldsymbol{d}(j), \boldsymbol{u}_{1,j}, \boldsymbol{u}_{2,j}, \boldsymbol{v}_1(j), \boldsymbol{v}_2(j)\}$.
6. All random variables have zero means.

(a) Find exact expressions for the EMSEs of the two adaptive filters, $\boldsymbol{w}_{1,i}$ and $\boldsymbol{w}_{2,i}$.

(b) Find an expression for the EMSE of the combined adaptive structure in terms of the EMSEs of the individual filters and λ. Determine an optimal value for λ in order to minimize the EMSE of the combined structure? Compare the optimal EMSE to the individual EMSEs.

Remark. For results on a more general covnex combination of adaptive filters, see the work by Arenas-Garcia, Figueiras-Vidal, and Sayed (2006).

Problem IV.5 (EMSE of ϵ-NLMS with power normalization) The purpose of this problem is to extend the derivation of Sec. 17.1 to ϵ-NLMS with power normalization, as described by (19.22)–(19.23).

(a) Repeat the arguments that led to (17.6) to conclude that the EMSE of the above algorithm is still given (17.6) where now $\alpha_u = \mathsf{E}\left(\|\boldsymbol{u}_i\|^2/(\epsilon+\boldsymbol{p}(i))^2\right)$ and $\eta_u = \mathsf{E}\left(1/(\epsilon+\boldsymbol{p}(i))\right)$.

(b) Assume ϵ is small and that the regressor is circular Gaussian with covariance matrix $R_u = \sigma_u^2 \mathrm{I}$. Show that

$$\mathsf{E}\,\|\boldsymbol{u}_i\|^2 = M\sigma_u^2, \qquad \mathsf{E}\;\boldsymbol{p}(i) = \sigma_u^2(1-\beta^{i+1})$$

$$\mathsf{E}\boldsymbol{p}^2(i) = \gamma\sigma_u^4(1-\beta^{2(i+1)})\frac{1-\beta}{1+\beta} + 2\sigma_u^4(1-\beta^{i+1})(1-\beta^i)\frac{\beta}{1+\beta}$$

where $\gamma = 3$ for real data and $\gamma = 2$ for complex data. *Hint.* Recall that the finite sum $a + ar + \ldots + ar^i$, with initial term a and ratio r, evaluates to $a(1-r^{i+1})/(1-r)$. When $i \to \infty$, the sum becomes a geometric series and if $|r| < 1$, it converges to $a/(1-r)$. Recall also that if $\boldsymbol{x}$ is a Gaussian random variable, then $\mathsf{E}\,|\boldsymbol{x}|^4 = 2\sigma_x^4$ when $\boldsymbol{x}$ is complex-valued and circular and $\mathsf{E}\,\boldsymbol{x}^4 = 3\sigma_x^4$ when $\boldsymbol{x}$ is real-valued.

(c) Under the conditions in part (b), use the approximations $\alpha_u \approx \mathsf{E}\,\|\boldsymbol{u}_i\|^2/\mathsf{E}\;\boldsymbol{p}^2(i)$ and $\eta_u \approx 1/\mathsf{E}\;\boldsymbol{p}(i)$ as $i \to \infty$, to justify the expression:

$$\boxed{\zeta^{\epsilon-\mathsf{pNLMS}} = \frac{\mu(1+\beta)M\sigma_v^2}{2\left[\gamma(1-\beta)+2\beta\right] - \mu M(1+\beta)} \qquad \text{(Gaussian regressors)}}$$

for small step-sizes.

Remark. See Sayed (2003, Sec. 6.6.3) for further details.

Problem IV.6 (EMSE of LMF and LMMN) Consider the LMMN algorithm (19.24). The least-mean fourth (LMF) algorithm corresponds to the special case $\delta = 0$. Clearly,

$$g[e] = \delta[\boldsymbol{e}_a + \boldsymbol{v}] \;+\; (1-\delta)\,[\boldsymbol{e}_a + \boldsymbol{v}]\;[\;|\boldsymbol{e}_a|^2 + |\boldsymbol{v}|^2 + \boldsymbol{e}_a^*\boldsymbol{v} + \boldsymbol{e}_a\boldsymbol{v}^*\;]$$

where we are omitting the time index i for compactness of notation. Introduce the symbols $\bar{\delta} = 1-\delta$, $\xi_v^4 = \mathsf{E}\,|\boldsymbol{v}(i)|^4$, and $\xi_v^6 = \mathsf{E}\,|\boldsymbol{v}(i)|^6$, where the scalars $\{\xi_v^4, \xi_v^6\}$ denote the fourth and sixth-order moments of $\boldsymbol{v}(i)$, respectively. The data $\{\boldsymbol{d}(i), \boldsymbol{u}_i, \boldsymbol{v}(i)\}$ are assumed to satisfy model (15.16).

(a) Assume first that all data are real-valued. Using the fact that $\boldsymbol{e}_a(i)$ and $\boldsymbol{v}(i)$ are independent (cf. Lemma 15.1), and ignoring third and higher-order powers of $\boldsymbol{e}_a(i)$, justify the expressions

$$\begin{aligned}
\mathsf{E}\,\boldsymbol{e}_a(i)g[\boldsymbol{e}(i)] &\approx b\mathsf{E}\,\boldsymbol{e}_a^2(i)\\
\mathsf{E}\,\|\boldsymbol{u}_i\|^2 g^2[\boldsymbol{e}(i)] &\approx a\mathsf{Tr}(R_u) + c\mathsf{E}\,\|\boldsymbol{u}_i\|^2\boldsymbol{e}_a^2(i) \;+\; 8\delta\bar{\delta}\left(\mathsf{E}\,\|\boldsymbol{u}_i\|^2\boldsymbol{e}_a(i)\right)\mathsf{E}\,\boldsymbol{v}^3(i)\\
&\quad + 6\bar{\delta}^2\left(\mathsf{E}\,\|\boldsymbol{u}_i\|^2\boldsymbol{e}_a(i)\right)\mathsf{E}\,\boldsymbol{v}^5(i)
\end{aligned}$$

where the constants $\{a, b, c\}$ are defined by

$$a \triangleq \delta^2\sigma_v^2 + 2\delta\bar{\delta}\xi_v^4 + \bar{\delta}^2\xi_v^6, \quad b \triangleq \delta + 3\bar{\delta}\sigma_v^2, \quad c \triangleq \delta^2 + 12\delta\bar{\delta}\sigma_v^2 + 15\bar{\delta}^2\xi_v^4$$

Conclude that the variance relation (15.40) leads to

$$\begin{aligned}
2b\mathsf{E}\,\boldsymbol{e}_a^2(i) &= \mu a\mathsf{Tr}(R_u) + \mu c\mathsf{E}\,\|\boldsymbol{u}_i\|^2\boldsymbol{e}_a^2(i) \;+\; 8\mu\delta\bar{\delta}\left(\mathsf{E}\,\|\boldsymbol{u}_i\|^2\boldsymbol{e}_a(i)\right)\mathsf{E}\,\boldsymbol{v}^3(i)\\
&\quad + 6\mu\bar{\delta}^2\left(\mathsf{E}\,\|\boldsymbol{u}_i\|^2\boldsymbol{e}_a(i)\right)\mathsf{E}\,\boldsymbol{v}^5(i), \quad i \to \infty
\end{aligned}$$

In order to simplify this result, we may consider three cases (as in our study of the mean-square performance of LMS in Chapter 16).

(a.1) Sufficiently small step-sizes. Argue that when μ is sufficiently small we get

$$\boxed{\zeta^{\text{LMMN}} = \frac{\mu a}{2b}\text{Tr}(R_u), \quad \zeta^{\text{LMF}} = \frac{\mu}{2}\left(\frac{\xi_v^6}{3\sigma_v^2}\right)\text{Tr}(R_u)}$$

(a.2) Separation principle. For larger values of μ, use the separation assumption (16.7) and the steady-state assumption (16.12) to conclude that

$$\boxed{\zeta^{\text{LMMN}} = \frac{\mu a \text{Tr}(R_u)}{2b - \mu c \text{Tr}(R_u)}, \quad \zeta^{\text{LMF}} = \frac{\mu \xi_v^6 \text{Tr}(R_u)}{6\sigma_v^2 - 15\mu\xi_v^4 \text{Tr}(R_u)}}$$

(a.3) White Gaussian regressors. Assume that the regression vector $\boldsymbol{u}_i$ is Gaussian with covariance matrix $R_u = \sigma_u^2 \text{I}$, and ignore the terms in $\boldsymbol{v}^3(i)$ and $\boldsymbol{v}^5(i)$. Repeat the argument of Sec. 16 to conclude that

$$\boxed{\zeta^{\text{LMMN}} = \frac{\mu M \sigma_u^2 a}{2b - \mu c(M+2)\sigma_u^2}, \quad \zeta^{\text{LMF}} = \frac{\mu M \sigma_u^2 \xi_v^6}{6\sigma_v^2 - 15\mu(M+2)\sigma_u^2 \xi_v^4}}$$

(b) Assume now that the data are complex-valued. Show that the same results of part (a) still hold with the modifications $b = \delta + 2\bar{\delta}\sigma_v^2$ and $c = \delta^2 + 8\delta\bar{\delta}\sigma_v^2 + 9\bar{\delta}^2\xi_v^4$.

Remark. The argument in this problem follows the approach of Yousef and Sayed (2001) — see also Sayed (2003, Sec. 6.8) for further details.

Problem IV.7 (EMSE of ϵ-APA) The purpose of this problem is to extend the energy-conservation approach of Sec. 15.3 to evaluate the EMSE of the ϵ-APA algorithm (19.25)–(19.27).

(a) Introduce the *a priori* and *a posteriori* error vectors $\boldsymbol{e}_{a,i} = \boldsymbol{U}_i \tilde{\boldsymbol{w}}_{i-1}$ and $\boldsymbol{e}_{p,i} = \boldsymbol{U}_i \tilde{\boldsymbol{w}}_i$, where $\tilde{\boldsymbol{w}}_i = w^o - \boldsymbol{w}_i$. Verify that $\boldsymbol{e}_{p,i} = \boldsymbol{e}_{a,i} - \mu \boldsymbol{U}_i \boldsymbol{U}_i^* \left(\epsilon \text{I} + \boldsymbol{U}_i \boldsymbol{U}_i^*\right)^{-1} \boldsymbol{e}_i$ and establish the relation

$$\boxed{\|\tilde{\boldsymbol{w}}_i\|^2 + \boldsymbol{e}_{a,i}^* \left(\boldsymbol{U}_i \boldsymbol{U}_i^*\right)^{-1} \boldsymbol{e}_{a,i} = \|\tilde{\boldsymbol{w}}_{i-1}\|^2 + \boldsymbol{e}_{p,i}^* \left(\boldsymbol{U}_i \boldsymbol{U}_i^*\right)^{-1} \boldsymbol{e}_{p,i}}$$

This equality extends the energy-conservation relation (15.32) to the ϵ–APA case.

(b) Introduce the matrix quantities $\boldsymbol{A}_i = \left(\epsilon \text{I} + \boldsymbol{U}_i \boldsymbol{U}_i^*\right)^{-1} \boldsymbol{U}_i \boldsymbol{U}_i^* \left(\epsilon \text{I} + \boldsymbol{U}_i \boldsymbol{U}_i^*\right)^{-1}$ and $\boldsymbol{B}_i = \left(\epsilon \text{I} + \boldsymbol{U}_i \boldsymbol{U}_i^*\right)^{-1}$. Show that, in steady-state and under expectation, the energy-conservation relation of part (a) reduces to

$$\boxed{\mu \text{E}\left[\boldsymbol{e}_i^* \boldsymbol{A}_i \boldsymbol{e}_i\right] = \text{E}\left[\boldsymbol{e}_{a,i}^* \boldsymbol{B}_i \boldsymbol{e}_i\right] + \text{E}\left[\boldsymbol{e}_i^* \boldsymbol{B}_i \boldsymbol{e}_{a,i}\right], \quad \text{as } i \to \infty}$$

Ignore the dependency of $\tilde{\boldsymbol{w}}_{i-1}$ on prior noises in steady-state. This variance relation is the extension of (15.40) to the ϵ–APA case.

(c) Consider the following assumptions:

(c.1) The data $\{\boldsymbol{d}(i), \boldsymbol{u}_i\}$ satisfy conditions (15.16).

(c.2) The dependency of $\tilde{\boldsymbol{w}}_{i-1}$ on prior noises is negligible in steady-state.

(c.3) Separation principle. At steady-state, $\boldsymbol{U}_i$ is independent of $\boldsymbol{e}_{a,i}$ and $\text{E}\, \boldsymbol{e}_{a,i} \boldsymbol{e}_{a,i}^* = \left(\text{E}\, |\boldsymbol{e}_a(i)|^2\right) \cdot S$, where $S \approx \text{I}$ for small μ and $S \approx b_0 b_0^{\mathsf{T}}$ for larger μ. Here b_0 denotes the first basis vector, $b_0 = \text{col}\{1, 0, \ldots, 0\}$.

Argue that $\zeta^{\epsilon-\text{APA}} = \mu \sigma_v^2 \text{Tr}(\text{E}\, \boldsymbol{A}_i)/(2\eta_u - \mu \alpha_u)$, where $\alpha_u = \text{Tr}(S \cdot \text{E}[\boldsymbol{A}_i])$ and $\eta_u = \text{Tr}(S \cdot \text{E}[\boldsymbol{B}_i])$.

(d) Two simplifications are possible when the regularization parameter ϵ is sufficiently small.

(d.1) Using $\boldsymbol{A}_i \approx (\boldsymbol{U}_i\boldsymbol{U}_i^*)^{-1}$ and $S \approx \mathrm{I}$, verify that $\zeta^{\epsilon-\mathsf{APA}} \approx \mu\sigma_v^2/(2-\mu)$. On the other hand, using $S \approx b_0 b_0^{\mathsf{T}}$, verify that

$$\boxed{\zeta^{\epsilon-\mathsf{APA}} \approx \frac{\mu\sigma_v^2}{(2-\mu)} \cdot \frac{\mathsf{Tr}(\mathsf{E}\,\boldsymbol{A}_i)}{\mathsf{E}\,\boldsymbol{A}_i(0,0)}}$$

where $\boldsymbol{A}_i(0,0)$ denotes the $(0,0)$ entry of $\boldsymbol{A}_i$.

(d.2) Use the approximations $\mathsf{Tr}(\mathsf{E}\,\boldsymbol{A}_i) \approx \mathsf{E}\left(K/\|\boldsymbol{u}_i\|^2\right)$ and $\mathsf{Tr}(S \cdot \mathsf{E}\,[\boldsymbol{A}_i]) \approx 1/\mathsf{Tr}(R_u)$ to justify

$$\boxed{\zeta^{\epsilon-\mathsf{APA}} = \frac{\mu\sigma_v^2}{(2-\mu)}\,\mathsf{Tr}(R_u)\mathsf{E}\left(\frac{K}{\|\boldsymbol{u}_i\|^2}\right)}$$

Remark. The separation condition in part (c.3) is motivated in Prob. IV.8. The derivation in this problem follows Shin and Sayed (2004) and also Sayed (2003, Sec. 6.10).

Problem IV.8 (APA condition) The purpose of this problem it to motivate the second approximation in condition (c.3) from Prob. IV.7, namely, that at steady-state $\mathsf{E}\,\boldsymbol{e}_{a,i}\boldsymbol{e}_{a,i}^* = \left(\mathsf{E}\,|\boldsymbol{e}_a(i)|^2\right) \cdot S$, where $S \approx \mathrm{I}$ at small μ and $S \approx b_0 b_0^{\mathsf{T}}$ at larger μ, with $b_0 = \mathrm{col}\{1, 0, \ldots, 0\}$.

Consider the *a priori* and *a posteriori* error vectors

$$\boldsymbol{e}_{a,i} = \begin{bmatrix} \boldsymbol{u}_i\tilde{\boldsymbol{w}}_{i-1} \\ \boldsymbol{u}_{i-1}\tilde{\boldsymbol{w}}_{i-1} \\ \vdots \\ \boldsymbol{u}_{i-K+1}\tilde{\boldsymbol{w}}_{i-1} \end{bmatrix}, \quad \boldsymbol{e}_{p,i} = \begin{bmatrix} \boldsymbol{u}_i\tilde{\boldsymbol{w}}_i \\ \boldsymbol{u}_{i-1}\tilde{\boldsymbol{w}}_i \\ \vdots \\ \boldsymbol{u}_{i-K+1}\tilde{\boldsymbol{w}}_i \end{bmatrix}$$

(a) Verify that for small ϵ, $\boldsymbol{e}_{p,i} = \boldsymbol{e}_{a,i} - \mu\boldsymbol{e}_i = (1-\mu)\boldsymbol{e}_{a,i} - \mu\boldsymbol{v}_i$.

(b) Conclude that the variances of the second and third entries of $\boldsymbol{e}_{a,i}$ satisfy

$$\begin{aligned} \mathsf{E}\,|\boldsymbol{u}_{i-1}\tilde{\boldsymbol{w}}_{i-1}|^2 &= (1-\mu)^2\mathsf{E}\,|\boldsymbol{e}_a(i-1)|^2 + \mu^2\sigma_v^2 \\ \mathsf{E}\,|\boldsymbol{u}_{i-2}\tilde{\mathbf{w}}_{i-1}|^2 &= (1-\mu)^4\mathsf{E}\,|\boldsymbol{e}_a(i-2)|^2 + (1-\mu)^2\mu^2\sigma_v^2 + \mu^2\sigma_v^2 \end{aligned}$$

Ignore any dependency between $\tilde{\boldsymbol{w}}_{i-1}$ and $\boldsymbol{v}(i-1)$.

(c) Use similar arguments to approximate the variances of the other entries of $\boldsymbol{e}_{a,i}$. Specifically, use the steady-state condition

$$\mathsf{E}\,|\boldsymbol{e}_a(i)|^2 = \mathsf{E}\,|\boldsymbol{e}_a(i-1)|^2 = \cdots = \mathsf{E}\,|\boldsymbol{e}_a(i-K+1)|^2, \quad \text{as } i \to \infty$$

and neglect the off-diagonal terms of $\mathsf{E}\,\boldsymbol{e}_{a,i}\boldsymbol{e}_{a,i}^*$, to argue that

$$\mathsf{E}\,\boldsymbol{e}_{a,i}\boldsymbol{e}_{a,i}^* \approx \mathsf{E}\,|\boldsymbol{e}_a(i)|^2 D_1 + \mu^2\sigma_v^2 D_2, \quad \text{as } i \to \infty$$

where the diagonal matrices $\{D_1, D_2\}$ are defined by

$$\begin{aligned} D_1 &= \mathrm{diag}\left\{1, (1-\mu)^2, (1-\mu)^4, \ldots, (1-\mu)^{2(K-1)}\right\} \\ D_2 &= \mathrm{diag}\left\{0, 1, 1+(1-\mu)^2, \ldots, 1+\cdots+(1-\mu)^{2K}\right\} \end{aligned}$$

Note that when μ is close to 1 and when the noise variance is relatively small, $D_1 \approx b_0^{\mathsf{T}} b_0$ and $\sigma_v^2 D_2 \approx 0$. Likewise, when μ is relatively small, we get $D_1 \approx \mathrm{I}$.

Problem IV.9 (Filters with data nonlinearities) Consider adaptive filters with update equations of the form

$$\boldsymbol{w}_i = \boldsymbol{w}_{i-1} + \mu\, g[\boldsymbol{u}_i]\boldsymbol{u}_i^*\boldsymbol{e}(i), \quad \boldsymbol{w}_{-1} = \text{initial condition}$$

for some positive scalar-valued function $g[\cdot]$ of the regression data, i.e., $g[\boldsymbol{u}_i] > 0$. Verify that the energy conservation relation of Thm. 15.1 still applies. Verify further that $\{\boldsymbol{e}_p(i), \boldsymbol{e}_a(i), \boldsymbol{e}(i)\}$ are related via $\boldsymbol{e}_p(i) = \boldsymbol{e}_a(i) - \mu\|\boldsymbol{u}_i\|^2 g[\boldsymbol{u}_i]\boldsymbol{e}(i)$.

Problem IV.10 (Price's theorem) Let $\boldsymbol{a}$ and $\boldsymbol{b}$ be scalar real-valued zero-mean jointly Gaussian random variables and denote their correlation by $\rho = \mathsf{E}\,\boldsymbol{ab}$. Price's theorem states that, for any function f of $\{\boldsymbol{a},\boldsymbol{b}\}$ (for which the required derivatives and integrals exist), the following equality holds (Price (1958)):

$$\frac{\partial^n \mathsf{E}\, f(\boldsymbol{a},\boldsymbol{b})}{\partial \rho^n} = \mathsf{E}\left(\frac{\partial^{2n} f(\boldsymbol{a},\boldsymbol{b})}{\partial \boldsymbol{a}^n \partial \boldsymbol{b}^n}\right)$$

in terms of the n-th and $2n-$th order partial derivatives. [In simple terms, Price's theorem allows us to move the expectation on the left-hand side outside of the differentiation operation.]

(a) Choose $n = 1$ and assume $f(\boldsymbol{a},\boldsymbol{b})$ has the form $f(\boldsymbol{a},\boldsymbol{b}) = \boldsymbol{a}g(\boldsymbol{b})$. Verify from Price's theorem that $\partial \mathsf{E}\,\boldsymbol{a}g(\boldsymbol{b})/\partial\rho = \mathsf{E}\,dg/d\boldsymbol{b}$, in terms of the derivative of $g(\cdot)$. Integrate both sides over ρ to establish that $\mathsf{E}\,\boldsymbol{a}g(\boldsymbol{b}) = (\mathsf{E}\,\boldsymbol{ab})\cdot \mathsf{E}\,dg/d\boldsymbol{b}$.

(b) Show further that $\mathsf{E}\,\boldsymbol{b}g(\boldsymbol{b}) = \sigma_b^2 \cdot \mathsf{E}\,dg/d\boldsymbol{b}$ and conclude that the following relation also holds:

$$\boxed{\mathsf{E}\,\boldsymbol{a}g(\boldsymbol{b}) = \frac{\mathsf{E}\,\boldsymbol{ab}}{\sigma_b^2}\cdot \mathsf{E}\,\boldsymbol{b}g(\boldsymbol{b})}$$

(c) Assume $g(\boldsymbol{b}) = \mathsf{sign}(\boldsymbol{b})$. Conclude from part (b) that

$$\boxed{\mathsf{E}\,\boldsymbol{a}\ \mathsf{sign}(\boldsymbol{b}) = \sqrt{\frac{2}{\pi}}\frac{1}{\sigma_b}\,\mathsf{E}\,\boldsymbol{ab}}$$

Problem IV.11 (Price's theorem for complex sign function) Let $\boldsymbol{u}$ and $\boldsymbol{e}$ denote two jointly-Gaussian and complex-valued random variables, where $\boldsymbol{e}$ is scalar-valued and has variance σ_e^2. Let $\boldsymbol{e} = \boldsymbol{e}_\mathsf{r} + j\boldsymbol{e}_\mathsf{i}$, $\boldsymbol{u} = \boldsymbol{u}_\mathsf{r} + j\boldsymbol{u}_\mathsf{i}$, and assume that

1) The real parts of $\boldsymbol{u}$ and $\boldsymbol{e}$ are jointly Gaussian.
2) The imaginary parts of $\boldsymbol{u}$ and $\boldsymbol{e}$ are jointly Gaussian.
3) The real and imaginary parts of $\boldsymbol{e}$ have identical variances.
4) The real parts of $\{\boldsymbol{u},\boldsymbol{e}\}$ are independent of their imaginary parts.

The third condition means that $\mathsf{E}\,\boldsymbol{e}_\mathsf{r}^2 = \mathsf{E}\,\boldsymbol{e}_\mathsf{i}^2$ so that $\sigma_e^2 = 2\sigma_{e_\mathsf{r}}^2$. Using the definition of the sign function for a general complex number from Prob. III.15, namely, $\mathsf{csgn}(\boldsymbol{e}) = \mathsf{sign}(\boldsymbol{e}_\mathsf{r}) + j\mathsf{sign}(\boldsymbol{e}_\mathsf{i})$, verify that $\mathsf{E}\,\mathsf{Re}\{\boldsymbol{u}^*\mathsf{csgn}(\boldsymbol{e})\} = \mathsf{E}\left[\boldsymbol{u}_\mathsf{r}\mathsf{sign}(\boldsymbol{e}_\mathsf{r}) + \boldsymbol{u}_\mathsf{i}\mathsf{sign}(\boldsymbol{e}_\mathsf{i})\right]$. Use Price's theorem for real-valued data from Prob. III.15 to conclude that

$$\boxed{\mathsf{E}\ \mathsf{Re}\{\boldsymbol{u}^*\ \mathsf{csgn}(\boldsymbol{e})\} = \sqrt{\frac{2}{\pi}}\frac{\sqrt{2}}{\sigma_e}\ \mathsf{E}\ \mathsf{Re}\{\boldsymbol{u}^*\boldsymbol{e}\}}$$

Problem IV.12 (Bussgang's theorem) Bussgang's theorem is a special case of Price's theorem. Let $\{\boldsymbol{a},\boldsymbol{b}\}$ be two real zero-mean Gaussian random variables and define the function

$$g(\boldsymbol{b}) \triangleq \int_0^{\boldsymbol{b}} e^{-z^2/\sigma^2}\, dz$$

for some $\sigma > 0$. Bussgang's theorem states that

$$\boxed{\mathsf{E}\,\boldsymbol{a}g(\boldsymbol{b}) = \frac{1}{\sqrt{\frac{\sigma_b^2}{\sigma^2}+1}}\,\mathsf{E}\,\boldsymbol{ab}}$$

The proof of the theorem is as follows. Let $\rho = \mathsf{E}\,\boldsymbol{ab}$. Use Price's general statement from Prob. IV.10 to verify that

$$\frac{\partial \mathsf{E}\,\boldsymbol{a}g(\boldsymbol{b})}{\partial\rho} = \mathsf{E}\left(\frac{\partial^2 \boldsymbol{a}g(\boldsymbol{b})}{\partial\boldsymbol{a}\partial\boldsymbol{b}}\right) = \mathsf{E}\left(e^{-\boldsymbol{b}^2/\sigma_z^2}\right)$$

Integrate both sides of the result of part (a) over ρ to establish Bussgang's theorem (Bussgang (1952)).

Problem IV.13 (Useful identity) Consider two real-valued zero-mean jointly Gaussian random variables $\{\boldsymbol{x}, \boldsymbol{y}\}$ with covariance matrix

$$\mathsf{E}\begin{bmatrix} \boldsymbol{x} \\ \boldsymbol{y} \end{bmatrix}\begin{bmatrix} \boldsymbol{x} & \boldsymbol{y} \end{bmatrix} = \begin{bmatrix} 1 & \rho \\ \rho & 1 \end{bmatrix}$$

That is, $\{\boldsymbol{x}, \boldsymbol{y}\}$ have unit variances and correlation ρ. Define the function

$$g(\boldsymbol{x}, \boldsymbol{y}) = \frac{2}{\pi\sigma^2}\int_0^{\boldsymbol{x}}\int_0^{\boldsymbol{y}} e^{-\alpha^2/2\sigma^2} e^{-\beta^2/2\sigma^2} \mathrm{d}\alpha \mathrm{d}\beta$$

for some $\sigma > 0$.

(a) Verify that $\partial^2 g(\boldsymbol{x}, \boldsymbol{y})/\partial\boldsymbol{x}\partial\boldsymbol{y} = \frac{2}{\pi\sigma^2} e^{-\boldsymbol{x}^2/2\sigma^2} e^{-\boldsymbol{y}^2/2\sigma^2}$, and show that

$$\mathsf{E}\left[\frac{\partial^2 g(\boldsymbol{x}, \boldsymbol{y})}{\partial\boldsymbol{x}\partial\boldsymbol{y}}\right] = \frac{2}{\pi}\frac{1}{\sqrt{(\sigma^2+1)^2 - \rho^2}}$$

(b) Integrate the equality of part (a) over $\rho \in (0, 1)$ and conclude that

$$\int_0^1 \mathsf{E}\left[\frac{\partial^2 g(\boldsymbol{x}, \boldsymbol{y})}{\partial\boldsymbol{x}\partial\boldsymbol{y}}\right] \mathrm{d}\rho = \frac{2}{\pi}\arcsin\left(\frac{1}{1+\sigma^2}\right)$$

(c) Use Price's identity (cf. Prob. IV.10) to conclude that

$$\boxed{\mathsf{E}\, g(\boldsymbol{x}, \boldsymbol{y}) = \frac{2}{\pi}\arcsin\left(\frac{1}{1+\sigma^2}\right)}$$

Problem IV.14 (Performance of constrained LMS) Refer to the constrained LMS filter studied in Prob. III.28, namely,

$$\boxed{\boldsymbol{w}_i = \boldsymbol{w}_{i-1} + \mu P\boldsymbol{u}_i^*[\boldsymbol{d}(i) - \boldsymbol{u}_i\boldsymbol{w}_{i-1}]}$$

where the matrix P is defined by $P = \mathrm{I} - cc^*/\|c\|^2$ for some known $M \times 1$ vector c. Moreover, the initial condition $\boldsymbol{w}_{-1}$ is such that $c^*\boldsymbol{w}_{-1} = \alpha$, where α is a known real scalar. Assume that the data $\{\boldsymbol{d}(i), \boldsymbol{u}_i\}$ satisfy the stationary model (15.16) with the additional requirement that the unknown model w^o is such that $c^*w^o = \alpha$ as well.

(a) Introduce the weighted *a priori* and *a posteriori* errors $\boldsymbol{e}_a^P(i) = \boldsymbol{u}_i P\tilde{\boldsymbol{w}}_{i-1}$ and $\boldsymbol{e}_p^P(i) = \boldsymbol{u}_i P\tilde{\boldsymbol{w}}_i$, where $\tilde{\boldsymbol{w}}_i = w^o - \boldsymbol{w}_i$. Verify that $\boldsymbol{e}_p^P(i) = \boldsymbol{e}_a^P(i) - \mu\|\boldsymbol{u}_i\|_P^2\boldsymbol{e}(i)$, where $\boldsymbol{e}(i) = \boldsymbol{d}(i) - \boldsymbol{u}_i\boldsymbol{w}_{i-1}$ and the notation $\|\boldsymbol{u}_i\|_P^2$ stands for $\boldsymbol{u}_i P\boldsymbol{u}_i^*$.

(b) Follow the arguments of Sec. 15.3 to show that the following energy relation holds:

$$\|\tilde{\boldsymbol{w}}_i\|^2 + \frac{1}{\|\boldsymbol{u}_i\|_P^2}|\boldsymbol{e}_a^P(i)|^2 = \|\tilde{\boldsymbol{w}}_{i-1}\|^2 + \frac{1}{\|\boldsymbol{u}_i\|_P^2}|\boldsymbol{e}_p^P(i)|^2$$

(c) Verify that $\boldsymbol{e}(i) = \boldsymbol{e}_a(i) + \boldsymbol{v}(i)$, where $\boldsymbol{e}_a(i) = \boldsymbol{u}_i\tilde{\boldsymbol{w}}_{i-1}$. Use the steady-state condition $\mathsf{E}\,\|\tilde{\boldsymbol{w}}_i\|^2 = \mathsf{E}\,\|\tilde{\boldsymbol{w}}_{i-1}\|^2$ as $i \to \infty$, to conclude that the above energy relation leads to the following variance relation

$$\mu\mathsf{E}\,\|\boldsymbol{u}_i\|_P^2|\boldsymbol{e}(i)|^2 = 2\mathsf{Re}\big[\mathsf{E}\,\boldsymbol{e}_a^{P*}(i)\boldsymbol{e}_a(i)\big], \quad \text{as } i \longrightarrow \infty$$

Use the fact that $c^*\tilde{\boldsymbol{w}}_{i-1} = 0$ to verify that the variance relation reduces to the following

$$\mu\mathsf{E}\,\|\boldsymbol{u}_i\|_P^2|\boldsymbol{e}_a(i)|^2 + \mu\sigma_v^2\mathsf{Tr}(PR_u) = 2\mathsf{E}\,|\boldsymbol{e}_a(i)|^2, \quad \text{as } i \longrightarrow \infty$$

(d) Argue that for sufficiently small step-sizes, the EMSE of constrained LMS can be approximated by $\zeta^{\text{constrained-LMS}} = \mu\sigma_v^2 \text{Tr}(PR_u)/2$.

(e) Use instead the separation assumption (16.7) to conclude that the EMSE can also be approximated by $\zeta^{\text{constrained-LMS}} = \mu\sigma_v^2 \text{Tr}(PR_u)/(2 - \mu\text{Tr}(PR_u))$.

Problem IV.15 (Performance of RLS by independence) Refer to the RLS algorithm studied in Chapter 19, namely, (19.2)–(19.1) with initial condition $\boldsymbol{w}_{-1} = 0$. Assume further that the data $\{\boldsymbol{d}(i), \boldsymbol{u}_i\}$ satisfy model (15.16).

(a) Show that $\boldsymbol{w}_i$ satisfies the equation $\boldsymbol{w}_i = \boldsymbol{P}_i \boldsymbol{s}_i$, where $\boldsymbol{s}_i = \sum_{j=0}^{i} \lambda^{i-j} \boldsymbol{u}_j^* \boldsymbol{d}(j)$. *Remark.* The equations $\boldsymbol{w}_i = \boldsymbol{P}_i \boldsymbol{s}_i$ will be encountered later in (30.21) when RLS is motivated and derived as the exact recursive solution to a least-squares problem. At that point, we will refer to $\boldsymbol{w}_i = \boldsymbol{P}_i \boldsymbol{s}_i$ as the normal equations.

(b) Use the data model (15.16) to show that

$$\begin{aligned} \lim_{i\to\infty} \boldsymbol{w}_i &= w^o + \lim_{i\to\infty} \left(\boldsymbol{P}_i \sum_{j=0}^{i} \lambda^{i-j} \boldsymbol{u}_j^* \boldsymbol{v}(j) \right) \\ \lim_{i\to\infty} \mathsf{E}\, \tilde{\boldsymbol{w}}_i \tilde{\boldsymbol{w}}_i^* &= \sigma_v^2 \lim_{i\to\infty} \mathsf{E} \left[\boldsymbol{P}_i \left(\sum_{j=0}^{i} \lambda^{2(i-j)} \boldsymbol{u}_j^* \boldsymbol{u}_j \right) \boldsymbol{P}_i \right] \end{aligned}$$

(c) As in (19.11) and (19.12), replace $\boldsymbol{P}_i$ by $(1-\lambda)R_u^{-1}$ and $\left(\sum_{j=0}^{i} \lambda^{2(i-j)} \boldsymbol{u}_j^* \boldsymbol{u}_j\right)$ by $R_u/(1-\lambda^2)$. Conclude that $\lim_{i\to\infty} \mathsf{E}\, \tilde{\boldsymbol{w}}_i \tilde{\boldsymbol{w}}_i^* = \sigma_v^2 (1-\lambda) R_u^{-1}/(1+\lambda)$.

(d) Use the independence condition (16.12) to conclude that the EMSE of RLS is given by $\text{EMSE}^{\text{RLS}} = \sigma_v^2(1-\lambda)M/(1+\lambda)$. Compare this expression with the result of Lemma 19.1.

Problem IV.16 (Performance of CMA2-2 for real-data) In this problem we evaluate the mean-square performance of the CMA2-2 recursion,

$$\boxed{\boldsymbol{w}_i = \boldsymbol{w}_{i-1} + \mu \boldsymbol{u}_i^{\mathsf{T}} \boldsymbol{z}(i)[\gamma - \boldsymbol{z}^2(i)], \quad \boldsymbol{z}(i) = \boldsymbol{u}_i \boldsymbol{w}_{i-1}, \quad i \geq 0}$$

where μ is a positive step-size and γ is a positive scalar (see Alg. 12.6). The main issue that arises while studying CMA2-2, in contrast to the adaptive filters studied in the body of the chapter, is the absence of a reference sequence $\boldsymbol{d}(i)$ and, correspondingly, of an explicit weight vector w^o relative to which we can define the weight error vector, $\tilde{\boldsymbol{w}}_i = w^o - \boldsymbol{w}_i$. This issue can be handled as follows.

CMA2-2 is usually used in the context of channel equalization in communications — see Prob. III.4. Thus let $\{\boldsymbol{s}(i)\}$ denote symbols from a constellation that are transmitted over a communications channel. The received data (i.e., the output of the channel) are denoted by $\boldsymbol{u}(i)$ and are fed into an FIR equalizer that is trained by CMA2-2. The purpose of the equalizer is to reproduce the transmitted data, say to recover $\{\boldsymbol{s}(i-\Delta)\}$ for some delay Δ.

Now assume that the equalization problem is such that there exists a receiver w^o that is able to reproduce the transmitted data $\{\boldsymbol{s}(i)\}$ with some delay Δ, i.e., such that $\boldsymbol{s}(i-\Delta) = \boldsymbol{u}_i w^o$. The task of CMA2-2 then becomes that of attempting to estimate w^o so that the output of the equalizer, $\boldsymbol{z}(i) = \boldsymbol{u}_i \boldsymbol{w}_{i-1}$, would tend to the desired symbol $\boldsymbol{s}(i-\Delta)$. The mean-square performance of CMA2-2 is then measured in terms of how successful it is in achieving this objective, i.e., in terms of the steady-state value of $\mathsf{E}\,|\boldsymbol{s}(i-\Delta) - \boldsymbol{u}_i \boldsymbol{w}_{i-1}|^2$.

In this problem we focus on the case of real-valued data $\{\boldsymbol{s}(i)\}$, e.g., data from a PAM constellation. Problem IV.17 extends the results to complex-valued constellations. We may add that for CMA2-2, one choice for the scalar γ is as $\gamma = \mathsf{E}\,|\boldsymbol{s}(i)|^4/\mathsf{E}\,|\boldsymbol{s}(i)|^2$. The analysis below, however, is for general γ.

(a) Verify that the energy-conservation relation of Thm. 15.1 still holds for CMA2-2, namely, that for any data $\{\boldsymbol{u}_i\}$, $\|\tilde{\boldsymbol{w}}_i\|^2 + \bar{\mu}(i)|\boldsymbol{e}_a(i)|^2 = \|\tilde{\boldsymbol{w}}_{i-1}\|^2 + \bar{\mu}(i)|\boldsymbol{e}_p(i)|^2$, where $\boldsymbol{e}_a(i) = \boldsymbol{u}_i \tilde{\boldsymbol{w}}_{i-1}$, $\boldsymbol{e}_p(i) = \boldsymbol{u}_i \tilde{\boldsymbol{w}}_i$, $\tilde{\boldsymbol{w}}_i = w^o - \boldsymbol{w}_i$, and $\bar{\mu}(i)$ is defined as in (15.31).

(b) Verify also that the variance relation of Thm. 15.2 still holds, namely, for any data $\{u_i\}$, $\mu \mathsf{E}\,\|u_i\|^2 g^2[z(i)] = 2\mathsf{E}\,e_a(i)g[z(i)]$, as $i \to \infty$. Using $g[z] = z[\gamma - z^2]$, show that this relation reduces to

$$\boxed{\mu \mathsf{E}\,\|u_i\|^2 z^2(i)[\gamma - z^2(i)]^2 \;=\; 2\mathsf{E}\,e_a(i)z(i)[\gamma - z^2(i)]}$$

(c) Introduce the following assumptions:

1. The transmitted signal $s(i-\Delta)$ and the estimation error $e_a(i)$ are independent in steady-state, so that $\mathsf{E}\,s(i-\Delta)e_a(i) = 0$ since $s(i-\Delta)$ is assumed zero mean.
2. At steady-state, $\|u_i\|^2$ is independent of $z(i)$.
   This condition extends the separation condition (16.7) to the CMA case.

Using assumptions 1. and 2., and replacing $z(i)$ by $s(i-\Delta) - e_a(i)$, verify that

$$\begin{aligned} \mu \mathsf{E}\,\|u_i\|^2 z^2(i)[\gamma - z^2(i)]^2 &= \mu \mathsf{E}\,(s^2\gamma^2 - 2\gamma s^4 + s^6)\cdot \mathsf{E}\,\|u\|^2 + A \\ 2\mathsf{E}\,e_a(i)z(i)[\gamma - z^2(i)] &= 2\mathsf{E}\,(-e_a^2\gamma + 3s^2 e_a^2 + e_a^4) \end{aligned}$$

where $A = \mu \mathsf{E}\,\left[(\gamma^2 - 12\gamma s^2 + 9s^4)e_a^2\right] \mathsf{E}\,\|u\|^2 + \mathsf{E}\,\left[15 s^2 e_a^4 + e_a^6 - 2\gamma e_a^4\right] \mathsf{E}\,\|u\|^2$. In the above expressions, we dropped the time index from $\{s(i-\Delta), e_a(i)\}$ for compactness of notation. Assume μ and e_a^2 are small enough so as to ignore the terms $\mathsf{E}\,e_a^4$ and A. Conclude that the MSE of CMA2-2 is given by

$$\boxed{\text{MSE}^{\text{CMA2-2}} \;\approx\; \mu\, \frac{\mathsf{E}\,(\gamma^2 s^2 - 2\gamma s^4 + s^6)}{2\mathsf{E}\,(3s^2 - \gamma)}\, \text{Tr}(R_u)}$$

in terms of the second, fourth, and sixth moments of the constellation.

Remark. The results of this problem, and those of Probs. IV.17–IV.18, are based on the work by Mai and Sayed (2000).

Problem IV.17 (Performance of CMA2-2 for complex-data) In this problem we extend the analysis of Prob. IV.16 to complex-valued constellations, in which case CMA2-2 is given by

$$\boxed{w_i = w_{i-1} + \mu u_i^* z(i)[\gamma - |z(i)|^2], \quad z(i) = u_i w_{i-1}, \quad i \geq 0}$$

We now assume, for generality, that there exists an equalizer w^o that reproduces the complex data $\{s(i)\}$ up to some rotation θ, i.e., $u_i w^o = s(i-\Delta)e^{j\theta}$. The purpose of the mean-square performance analysis is to evaluate how well the output of the CMA2-2 implementation, namely, $z(i) = u_i w_{i-1}$, can approximate $s(i-\Delta)e^{j\theta}$.

(a) Verify that the variance relation of Thm. 15.2 still holds, namely, for any data $\{u_i\}$,

$$\mu \mathsf{E}\,\|u_i\|^2 |g[z(i)]|^2 = 2\text{Re}\big(\mathsf{E}\,e_a^*(i)g[z(i)]\big), \quad \text{as } i \to \infty$$

(b) Now introduce the following assumptions:

1. The transmitted signal $s(i-\Delta)$ and the estimation error $e_a(i)$ are independent in steady-state, so that $\mathsf{E}\,s^*(i-\Delta)e_a(i) = 0$ since $s(i-\Delta)$ is assumed zero mean.
2. At steady-state, $\|u_i\|^2$ is independent of $z(i)$.
3. The data $s(i)$ is circular, $\mathsf{E}\,s^2(i) = 0$.
4. The scalar γ satisfies $\mathsf{E}\,(2|s(i)|^2 - \gamma) > 0$.

Repeat the derivation of part (c) of Prob. IV.16 to show that the MSE of CMA2-2 is now given by

$$\boxed{\text{MSE}^{\text{CMA2-2}} \;\approx\; \mu\, \frac{\mathsf{E}\,(\gamma^2|s|^2 - 2\gamma|s|^4 + |s|^6)}{2\mathsf{E}\,(2|s|^2 - \gamma)}\, \text{Tr}(R_u)}$$

in terms of the second, fourth, and sixth moments of the constellation.

(c) Assume the data $\{s(i)\}$ have constant-modulus, $|s(i)| = 1$, and choose $\gamma = 1$. Verify that in this case MSE$= 0$.

Problem IV.18 (Performance of CMA1-2 for complex-data) In this problem we extend the analysis of Probs. IV.16 and IV.17 to CMA1-2, which is described by the following recursion (see Alg. 12.5):

$$\boxed{w_i = w_{i-1} + \mu u_i^* \left[\gamma \frac{z(i)}{|z(i)|} - z(i)\right], \quad z(i) = u_i w_{i-1}, \quad i \geq 0}$$

We also assume, for generality, that there exists an equalizer w^o that reproduces the complex data $\{s(i)\}$ up to some rotation θ and delay Δ, i.e., $u_i w^o = s(i-\Delta)e^{j\theta}$. For CMA1-2, one choice for the scalar γ is $\gamma = \mathsf{E}\,|s(i)|^2/\mathsf{E}\,|s(i)|$. The results below, however, are for general γ. Compared with CMA2-2, the function $g[z]$ is now given by $g[z] = (\gamma z/|z|) - z$. In addition to the assumptions in part (b) of Prob. IV.17, introduce the following:

1. The output $z(i)$ is distributed symmetrically around the transmitted signal $s(i-\Delta)$ in steady-state, so that $\mathsf{E}\,|z(i)| = \mathsf{E}\,|s(i-\Delta)|$.
2. The estimation error $e_a(i)$ is independent of $z(i)/|z(i)|$ in steady-state and $\mathsf{E}\,z(i)/|z(i)| = 0$ so that $\mathsf{E}\,e_a(i)z(i)/|z(i)| = 0$.

(a) Consider again the variance relation $\mu\mathsf{E}\,\|u_i\|^2|g[z(i)]|^2 = 2\mathsf{Re}\big(\mathsf{E}\,e_a^*(i)g[z(i)]\big)$, as $i \to \infty$. Verify that it reduces to

$$\mu\mathsf{E}\,\|u\|^2 \cdot \mathsf{E}\,(\gamma^2 - 2\gamma|z| + |z|^2) \;=\; \mathsf{E}\left(e_a^*\left(\gamma\frac{z}{|z|} - z\right) + e_a\left(\gamma\frac{z}{|z|} - z\right)^*\right)$$

(b) Replacing $\mathsf{E}\,|s| = \mathsf{E}\,|z|$ and $\mathsf{E}\,|z|^2 = \mathsf{E}\,|s|^2 + \mathsf{E}\,|e_a|^2$, and ignoring the term $\mu\mathsf{E}\,|e_a|^2\|u\|^2$ when μ and $e_a(i)$ are small, conclude that the MSE of CMA1-2 is given by

$$\boxed{\mathsf{MSE}^{\mathsf{CMA1-2}} \;=\; \frac{\mu}{2}\left(\gamma^2 + \mathsf{E}\,|s|^2 - 2\gamma\mathsf{E}\,|s|\right) \cdot \mathsf{Tr}(R_u)}$$

Problem IV.19 (Correlated Gaussian regressors) Consider the LMS recursion (16.1) and assume the data $\{d(i), u_i\}$ satisfy model (15.16). Assume further that the steady-state condition (16.12) holds. In this problem we reconsider the performance of LMS for Gaussian regressors, as was done in Sec. 16.4, except that now we do *not* restrict the covariance matrix to be $R_u = \sigma_u^2 \mathrm{I}$. Introduce the eigen-decomposition $R_u = U\Lambda U^*$, where Λ is a diagonal matrix with the eigenvalues of R_u, $\Lambda = \mathrm{diag}\{\lambda_k\}$, and U is a unitary matrix (i.e., it satisfies $UU^* = U^*U = \mathrm{I}$). Define the transformed quantities $\overline{w}_i = U^*\tilde{w}_i$ and $\overline{u}_i = u_i U$.

(a) Argue that $\overline{u}_i$ is circular Gaussian with covariance matrix $\mathsf{E}\,\overline{u}_i^*\overline{u}_i = \Lambda$. Verify also that $e_a(i) = \overline{u}_i\overline{w}_{i-1}$ and $\mathsf{E}\,\|u_i\|^2|e_a(i)|^2 = \mathsf{E}\,\|\overline{u}_i\|^2|e_a(i)|^2$. Let $\overline{C}_{i-1} = \mathsf{E}\,\overline{w}_i\overline{w}_i^*$.

(b) Use the result of Lemma A.3, and the derivation that led to (16.16), to show that

$$\mathsf{E}\,\|\overline{u}_i\|^2|e_a(i)|^2 \;=\; \mathsf{Tr}(\Lambda)\mathsf{Tr}(\overline{C}_{i-1}\Lambda) \;+\; \mathsf{Tr}(\Lambda\overline{C}_{i-1}\Lambda)$$

Likewise, verify that $\mathsf{E}\,|e_a(i)|^2 = \mathsf{Tr}(\Lambda\overline{C}_{i-1})$.

(c) Conclude that, in steady-state, the variance relation (16.5) becomes

$$\zeta^{\mathsf{LMS}} = \frac{\mu}{2}\left[\mathsf{Tr}(\Lambda)\zeta^{\mathsf{LMS}} \;+\; \mathsf{Tr}(\Lambda\overline{C}_\infty\Lambda) \;+\; \sigma_v^2\mathsf{Tr}(\Lambda)\right]$$

Remark 1. Thus observe that in this case, for Gaussian regressors with arbitrary covariance matrix R_u, the factor $\overline{C}_\infty$ does not disappear from the EMSE expression — see Sec. 23.1 and the EMSE expression in Thm. 23.3 for more details on this case.

Remark 2. Observe further that the procedure for finding the filter EMSE through (15.40), as discussed in the body of the chapter, avoids the need to explicitly evaluate $\mathsf{E}\,\|\tilde{w}_i\|^2$ or its steady-state value. This is in contrast to the procedure of this problem, which requires determining the steady-state covariance matrix of $\tilde{\boldsymbol{w}}_i$, before arriving at an expression for the EMSE of the filter.

Problem IV.20 (Regressor transformation) Refer to the discussion in App. 17.A on relating ϵ−NLMS to LMS and consider, in particular, the transformed regressor $\check{\boldsymbol{u}}_i = \boldsymbol{u}_i/\sqrt{\epsilon + \|\boldsymbol{u}_i\|^2}$. Set $\epsilon = 0$ for simplicity in this problem and assume $M = 1$ (i.e., $\boldsymbol{u}_i$ is a scalar). Assume further that the distribution of $\boldsymbol{u}_i$ is not symmetric, say $\boldsymbol{u}_i = 1$ with probability $2/3$ and $\boldsymbol{u}_i = -2$ with probability $1/3$. Verify that $\boldsymbol{u}_i$ is zero-mean while $\check{\boldsymbol{u}}_i$ is not.

Problem IV.21 (Lossless mapping) Refer to equation (15.47) and consider the mapping

$$\mathcal{T} = \begin{bmatrix} \mathrm{I} - \bar{\mu}(i)\boldsymbol{u}_i^*\boldsymbol{u}_i & \sqrt{\bar{\mu}(i)}\ \boldsymbol{u}_i^* \\ \sqrt{\bar{\mu}(i)}\ \boldsymbol{u}_i & 0 \end{bmatrix}$$

Verify that $\mathcal{T}$ is Hermitian and unitary, i.e., $\mathcal{T}^2 = \mathrm{I}$.

Problem IV.22 (Optimal step-size for LMS) Show that the expression for $\mu_{\text{opt}}^{\text{LMS}}$ in (21.10) is the unique minimum of (21.9) over the interval $0 < \mu < 2/\mathrm{Tr}(R_u)$.

Problem IV.23 (Optimal step-sizes for ϵ-NLMS) Refer to the discussion in Sec. 21.2 on the tracking performance of ϵ−NLMS.

(a) Verify by differentiating (21.18) with respect to μ that

$$\mu_{\text{opt}}^{\epsilon-\text{NLMS}} = \sqrt{\frac{\mathrm{Tr}(Q)}{\alpha_u\sigma_v^2} + \frac{(\mathrm{Tr}(Q))^2}{4\eta_u^2\sigma_v^4}} - \frac{\mathrm{Tr}(Q)}{2\eta_u\sigma_v^2}$$

(b) Differentiate instead (21.20) with respect to μ to obtain

$$\mu_{\text{opt}}^{\epsilon-\text{NLMS}} = \frac{1}{2\sigma_v^2}\left(\sqrt{\frac{\mathrm{Tr}(Q)}{\mathsf{E}\left(1/\|\boldsymbol{u}_i\|^2\right)}\left[\frac{\mathrm{Tr}(Q)}{\mathsf{E}\left(1/\|\boldsymbol{u}_i\|^2\right)} + 4\sigma_v^2\right]} - \frac{\mathrm{Tr}(Q)}{\mathsf{E}\left(1/\|\boldsymbol{u}_i\|^2\right)}\right)$$

Verify that this result agrees with the expression of part (a) when η_u and α_u are replaced by $\mathsf{E}\left(1/\|\boldsymbol{u}_i\|^2\right)$. Assume further that the step-size is small enough so that $2 - \mu \approx 2$. Justify the expressions

$$\zeta^{\epsilon-\text{NLMS}} \approx \frac{1}{2}\left(\mu\sigma_v^2 + \frac{\mu^{-1}\mathrm{Tr}(Q)}{\mathsf{E}\left(1/\|\boldsymbol{u}_i\|^2\right)}\right)$$

and

$$\mu_{\text{opt}}^{\epsilon-\text{NLMS}} \approx \sqrt{\frac{\mathrm{Tr}(Q)}{\sigma_v^2\mathsf{E}\left(1/\|\boldsymbol{u}_i\|^2\right)}} \quad \text{with} \quad \zeta_{\min}^{\epsilon-\text{NLMS}} \approx \sqrt{\frac{\sigma_v^2\mathrm{Tr}(Q)}{\mathsf{E}\left(1/\|\boldsymbol{u}_i\|^2\right)}}$$

(c) Differentiate (21.21) with respect to μ and verify that the results of part (b) are still valid, and that the corresponding minimum EMSE for small μ is now given by

$$\zeta_{\min}^{\epsilon-\text{NLMS}} \approx \mathrm{Tr}(R_u)\sqrt{\sigma_v^2\mathrm{Tr}(Q)\mathsf{E}\left(\frac{1}{\|\boldsymbol{u}_i\|^2}\right)}$$

Problem IV.24 (ϵ-NLMS with power normalization) The purpose of this problem is to extend the derivation of Sec. 21.2 to the ϵ-NLMS recursion with power normalization (cf. (21.50)–(21.51)), where the regression vector is assumed to have shift-structure, say

$$\boldsymbol{u}_i = [\ \boldsymbol{u}(i)\ \ \boldsymbol{u}(i-1)\ \boldsymbol{u}(i-2)\ \ldots\ \boldsymbol{u}(i-M+1)\ ]$$

with leading entry $\boldsymbol{u}(i)$. Assume the regression data is circular Gaussian with covariance matrix $R_u = \sigma_u^2\mathrm{I}$. Follow the arguments that led to (21.18), and the simplifications of Sec. 17.1, to conclude that

$$\zeta^{\epsilon-\text{pNLMS}} = \frac{\mu M(1+\beta)\sigma_v^2 + \mu^{-1}\gamma\sigma_u^2(1-\beta)\text{Tr}(Q)}{2\left[\gamma(1-\beta)+2\beta\right] - \mu M(1+\beta)} \qquad \text{(Gaussian regressors)}$$

where $\gamma = 3$ for real data and $\gamma = 2$ for complex data.

Problem IV.25 (Tracking performance of LMMN and LMF) Consider the LMMN algorithm given by (21.52). The least-mean fourth (LMF) algorithm corresponds to the special case $\delta = 0$. Introduce the symbols $\bar{\delta} = 1 - \delta$, $\xi_v^4 = \mathsf{E}\,|\boldsymbol{v}(i)|^4$, and $\xi_v^6 = \mathsf{E}\,|\boldsymbol{v}(i)|^6$, where the scalars $\{\xi_v^4, \xi_v^6\}$ denote the fourth and sixth-order moments of $\boldsymbol{v}(i)$, respectively. The data $\{\boldsymbol{d}(i), \boldsymbol{u}_i, \boldsymbol{v}(i)\}$ are assumed to satisfy model (20.16).

(a) Assume first that all data are real-valued. Repeat the arguments of Prob. IV.6 to arrive at the equality

$$\begin{aligned} 2b\mathsf{E}\,\boldsymbol{e}_a^2(i) &= \mu^{-1}\text{Tr}(Q) + \mu a \text{Tr}(R_u) + \mu c \mathsf{E}\,\|\boldsymbol{u}_i\|^2 \boldsymbol{e}_a^2(i) \\ &\quad + 8\mu\delta\bar{\delta}\left(\mathsf{E}\,\|\boldsymbol{u}_i\|^2 \boldsymbol{e}_a(i)\right)\mathsf{E}\,\boldsymbol{v}^3(i) + 6\mu\bar{\delta}^2\left(\mathsf{E}\,\|\boldsymbol{u}_i\|^2 \boldsymbol{e}_a(i)\right)\mathsf{E}\,\boldsymbol{v}^5(i), \qquad i \to \infty \end{aligned}$$

where $a = \delta^2\sigma_v^2 + 2\delta\bar{\delta}\xi_v^4 + \bar{\delta}^2\xi_v^6$, $b = \delta + 3\bar{\delta}\sigma_v^2$, and $c = \delta^2 + 12\delta\bar{\delta}\sigma_v^2 + 15\bar{\delta}^2\xi_v^4$. In order to simplify this result, we may consider three cases:

(a.1) Sufficiently small step-sizes. Argue that when μ is sufficiently small we get

$$\zeta^{\text{LMMN}} = \frac{\mu a \text{Tr}(R_u) + \mu^{-1}\text{Tr}(Q)}{2b}, \qquad \zeta^{\text{LMF}} = \frac{\mu\xi_v^6 \text{Tr}(R_u) + \mu^{-1}\text{Tr}(Q)}{6\sigma_v^2}$$

Differentiate these expressions with respect to μ to conclude that

$$\mu_{\text{opt}}^{\text{LMMN}} = \sqrt{\frac{\text{Tr}(Q)}{a\text{Tr}(R_u)}}, \qquad \mu_{\text{opt}}^{\text{LMF}} = \sqrt{\frac{\text{Tr}(Q)}{\xi_v^6\text{Tr}(R_u)}}$$

and

$$\zeta_{\min}^{\text{LMMN}} = \frac{\sqrt{a\text{Tr}(Q)\text{Tr}(R_u)}}{b}, \qquad \zeta_{\min}^{\text{LMF}} = \frac{\sqrt{\xi_v^6\text{Tr}(Q)\text{Tr}(R_u)}}{3\sigma_v^2}$$

(a.2) Separation principle. For larger values of μ, use the separation assumption (21.8) and the steady-state assumption (16.12) to conclude that

$$\zeta^{\text{LMMN}} = \frac{\mu a \text{Tr}(R_u) + \mu^{-1}\text{Tr}(Q)}{2b - \mu c \text{Tr}(R_u)}, \qquad \zeta^{\text{LMF}} = \frac{\mu\xi_v^6 \text{Tr}(R_u) + \mu^{-1}\text{Tr}(Q)}{6\sigma_v^2 - 15\mu\xi_v^4\text{Tr}(R_u)}$$

Verify that

$$\mu_{\text{opt}}^{\text{LMMN}} = \sqrt{\text{Tr}(Q)\left(\frac{c^2\text{Tr}(Q)}{4a^2b^2} + \frac{1}{a\text{Tr}(R_u)}\right)} - \frac{c}{2ab}\text{Tr}(Q)$$

(a.3) White Gaussian regressors. Assume that the regression vector $\boldsymbol{u}_i$ is Gaussian with covariance matrix $R_u = \sigma_u^2\mathrm{I}$, and ignore the terms in $\boldsymbol{v}^3(i)$ and $\boldsymbol{v}^5(i)$. Argue that in this case

$$\zeta^{\text{LMMN}} = \frac{\mu M\sigma_u^2 a + \mu^{-1}\text{Tr}(Q)}{2b - \mu c(M+2)\sigma_u^2}, \qquad \zeta^{\text{LMF}} = \frac{\mu M\sigma_u^2\xi_v^6 + \mu^{-1}\text{Tr}(Q)}{6\sigma_v^2 - 15\mu(M+2)\sigma_u^2\xi_v^4}$$

(b) Assume now that the data are complex-valued. Show that the same results of part (a) still hold with the modifications $b = \delta + 2\bar{\delta}\sigma_v^2$ and $c = \delta^2 + 8\delta\bar{\delta}\sigma_v^2 + 9\bar{\delta}^2\xi_v^4$.

Remark. The argument in this problem follows the approach of Yousef and Sayed (2001) — see also Sayed (2003, Sec. 7.8) for further details.

Problem IV.26 (Tracking of ϵ−APA) The purpose of this problem is to extend the energy-conservation approach of Sec. 20.3 to evaluate the EMSE of the ϵ-APA algorithm of Sec. 13.1, namely, (21.53)–(21.55). Assume the data satisfy model (20.16).

(a) Introduce the *a priori* and *a posteriori* error vectors $e_{a,i} = U_i\tilde{w}_{i-1}$ and $e_{p,i} = U_i\tilde{w}_i$, where $\tilde{w}_i = w^o - w_i$. Extend the argument of Prob. IV.7 to show that the result of Thm. 20.1 becomes

$$\boxed{\|w_i^o - w_i\|^2 + e_{a,i}^*(U_iU_i^*)^{-1}e_{a,i} = \|w_i^o - w_{i-1}\|^2 + e_{p,i}^*(U_iU_i^*)^{-1}e_{p,i}}$$

Conclude from the argument that led to (20.29) that the above result reduces under expectation to

$$\boxed{\mathsf{E}\,\|\tilde{w}_i\|^2 + \mathsf{E}e_{a,i}^*(U_iU_i^*)^{-1}e_{a,i} = \mathsf{E}\,\|\tilde{w}_{i-1}\|^2 + \mathsf{E}\,\|q_i\|^2 + \mathsf{E}e_{p,i}^*(U_iU_i^*)^{-1}e_{p,i}}$$

(b) Introduce the matrices

$$\boxed{A_i = (\epsilon\mathrm{I} + U_iU_i^*)^{-1}U_iU_i^*(\epsilon\mathrm{I} + U_iU_i^*)^{-1}, \qquad B_i = (\epsilon\mathrm{I} + U_iU_i^*)^{-1}}$$

Ignore any dependency between v_i and $e_{a,i}$ in steady-state. Argue that, as $i \to \infty$, the result of part (a) leads to the variance relation

$$\boxed{\mu\mathsf{E}\left[e_{a,i}^*A_ie_{a,i}\right] + \mu\mathsf{E}\left[v_i^*A_iv_i\right] + \mathsf{E}\,\|q_i\|^2 = 2\mathsf{E}\left[e_{a,i}^*B_ie_{a,i}\right], \quad \text{as } i \to \infty}$$

(c) Assume that, at steady-state, U_i is independent of $e_{a,i}$ and $\mathsf{E}e_{a,i}e_{a,i}^* = \left(\mathsf{E}\,|e_a(i)|^2\right)\cdot S$, where $S \approx \mathrm{I}$ for small μ and $S \approx b_0b_0^{\mathsf{T}}$ for larger μ. Here b_0 denotes the first basis vector, $b_0 = \text{col}\{1, 0, \ldots, 0\}$. Conclude that for small ϵ, the filter EMSE can be approximated by any of the following expressions:

$$\boxed{\begin{aligned}
\zeta^{\epsilon\text{-APA}} &= \frac{1}{(2-\mu)\mathsf{Tr}(A_i)}\left[\mu\sigma_v^2\mathsf{Tr}(\mathsf{E}\,A_i) + \mu^{-1}\mathsf{Tr}(Q)\right] \\
\zeta^{\epsilon\text{-APA}} &= \frac{1}{(2-\mu)\mathsf{E}\,A_i(0,0)}\left[\mu\sigma_v^2\mathsf{Tr}(\mathsf{E}\,A_i) + \mu^{-1}\mathsf{Tr}(Q)\right] \\
\zeta^{\epsilon\text{-APA}} &= \frac{\mathsf{Tr}(R_u)}{(2-\mu)}\left[\mu\sigma_v^2\mathsf{E}\left(\frac{K}{\|u_i\|^2}\right) + \mu^{-1}\mathsf{Tr}(Q)\right]
\end{aligned}}$$

where $A_i \approx (U_iU_i^*)^{-1}$ and $A_i(0,0)$ denotes the $(0,0)$ entry of A_i.

Problem IV.27 (Tracking of ϵ−APA) Consider the family of affine projection algorithms described by (13.14), for some integers $\{\alpha, D\}$, namely,

$$w_i = w_{i-1-\alpha(K-1)} + \mu U_i^*(\epsilon\mathrm{I} + U_iU_i^*)^{-1}\left[d_i - U_iw_{i-1-\alpha(K-1)}\right]$$

where now $\{U_i, d_i\}$ are replaced by

$$U_i \triangleq \begin{bmatrix} u_i \\ u_{i-D} \\ \vdots \\ u_{i-(K-1)D} \end{bmatrix} \qquad d_i = \begin{bmatrix} d(i) \\ d(i-D) \\ \vdots \\ d(i-(K-1)D) \end{bmatrix}$$

Let $K' = \alpha(K-1)$ and define the *a priori* error vector as $e_{a,i} = U_i(w_i^o - w_{i-1-K'})$.

(a) Repeat the arguments of Prob. IV.26 to to verify that the energy-conservation relation becomes

$$\|w_i^o - w_i\|^2 + e_{a,i}^*(U_iU_i^*)^{-1}e_{a,i} = \|w_i^o - w_{i-1-K'}\|^2 + e_{p,i}^*(U_iU_i^*)^{-1}e_{p,i}$$

which, under expectation, gives

$$\mathsf{E}\,\|\tilde{\boldsymbol{w}}_i\|^2 + \mathsf{E}\,\boldsymbol{e}_{a,i}^*(\boldsymbol{U}_i\boldsymbol{U}_i^*)^{-1}\boldsymbol{e}_{a,i} = \mathsf{E}\,\|\tilde{\boldsymbol{w}}_{i-1-K'}\|^2 + (1+K')\mathsf{Tr}(Q) + \boldsymbol{e}_{p,i}^*(\boldsymbol{U}_i\boldsymbol{U}_i^*)^{-1}\boldsymbol{e}_{p,i}$$

(b) Show that $(2\eta_u - \mu\alpha_u)\mathsf{E}\,|\boldsymbol{e}_a(i)|^2 = \mu\sigma_v^2\mathsf{Tr}(\mathsf{E}\,\boldsymbol{A}_i) + \mu^{-1}(1+K')\mathsf{Tr}(Q)$ as $i \to \infty$, so that the excess mean-square error (EMSE) of the filter is now given by

$$\boxed{\zeta^{\epsilon-\mathsf{APA}} = \frac{\mu\sigma_v^2\mathsf{Tr}(\mathsf{E}\,\boldsymbol{A}_i) + \mu^{-1}(1+K')\mathsf{Tr}(Q)}{(2\eta_u - \mu\alpha_u)}}$$

Problem IV.28 (Trace inequalities) Consider a collection of M positive scalars $\{\lambda_i\}$.

(a) For any two positive numbers a and b, use the fact that $(a-b)^2 \geq 0$ to conclude that $a/b + b/a \geq 2$. Now argue that the term $\left(\sum_{i=1}^M \lambda_i\right) \cdot \left(\sum_{i=1}^M 1/\lambda_i\right)$ consists of M products of λ_i by $1/\lambda_i$ and $M(M-1)/2$ sums of the form $\lambda_i.\frac{1}{\lambda_j} + \lambda_j.\frac{1}{\lambda_i}$, for $i \neq j$. Conclude that $\left(\sum_{i=1}^M \lambda_i\right) \cdot \left(\sum_{i=1}^M 1/\lambda_i\right) \geq M^2$.

(b) Use $\left(\sum_{i=1}^M \lambda_i\right)^2 = \left(\sum_{i=1}^M \lambda_i^2\right) + \left(\sum_{i=1}^M \sum_{j=i+1}^M 2\lambda_i\lambda_j\right)$, and the fact that $2\lambda_i\lambda_j \leq \lambda_i^2 + \lambda_j^2$, to conclude that $\left(\sum_{i=1}^M \lambda_i\right)^2 \leq M\left(\sum_{i=1}^M \lambda_i^2\right)$.

(c) Let R_u be any $M \times M$ Hermitian positive-definite matrix with eigenvalues $\{\lambda_i\}$. Use the results of parts (a) and (b) to establish that $[\mathsf{Tr}(R_u)]^2 \leq M\mathsf{Tr}(R_u^2)$ and $M^2 \leq \mathsf{Tr}(R_u)\mathsf{Tr}(R_u^{-1})$.

Problem IV.29 (Model with frequency offset) Consider data $\{\boldsymbol{d}(i), \boldsymbol{u}_i\}$ that satisfy the linear relation $\boldsymbol{d}(i) = \boldsymbol{u}_i\boldsymbol{w}_i^o e^{j\Omega i} + \boldsymbol{v}(i)$, where $\boldsymbol{v}(i)$ denotes measurement noise and Ω models some constant frequency offset (Ω could be zero as well). Assume further that $\boldsymbol{w}_i^o$ varies according to the auto-regressive model: $\boldsymbol{w}_i^o = w^o + \boldsymbol{\theta}_i$ and $\boldsymbol{\theta}_i = \alpha\boldsymbol{\theta}_{i-1} + \boldsymbol{q}_i$ with $0 \leq |\alpha| < 1$. In other words, $\boldsymbol{w}_i^o$ undergoes random variations around its mean w^o, with the perturbations $\boldsymbol{\theta}_i$ being generated by a first-order auto-regressive model with a pole at α and a random initial condition denoted by $\boldsymbol{\theta}_{-1}$.

The purpose of Probs. IV.29 and IV.30 is to extend the variance analysis of Secs. 20.3 and 20.4 to such more general non-stationary models. Subsequent problems then examine the performance of several adaptive filters under these conditions. So consider adaptive filters of the form (20.21) and define the error quantities:

$$\tilde{\boldsymbol{w}}_i \triangleq \boldsymbol{w}_i^o e^{j\Omega i} - \boldsymbol{w}_i, \quad \boldsymbol{e}_a(i) \triangleq \boldsymbol{u}_i[\boldsymbol{w}_i^o e^{j\Omega i} - \boldsymbol{w}_{i-1}], \quad \boldsymbol{e}_p(i) \triangleq \boldsymbol{u}_i[\boldsymbol{w}_i^o e^{j\Omega i} - \boldsymbol{w}_i]$$

(a) Establish the relations

$$\boxed{\boldsymbol{e}_p(i) = \boldsymbol{e}_a(i) - \mu\|\boldsymbol{u}_i\|^2 g[\boldsymbol{e}(i)], \qquad \tilde{\boldsymbol{w}}_i = \tilde{\boldsymbol{w}}_{i-1} - \mu\boldsymbol{u}_i^* g[\boldsymbol{e}(i)] + \boldsymbol{c}_i e^{j\Omega(i-1)}}$$

where $\boldsymbol{c}_i = w^o(e^{j\Omega} - 1) + \boldsymbol{\theta}_{i-1}(\alpha e^{j\Omega} - 1) + \boldsymbol{q}_i e^{j\Omega}$.

(b) Establish the energy-conservation relation

$$\boxed{\|\tilde{\boldsymbol{w}}_i - \boldsymbol{c}_i e^{j\Omega(i-1)}\|^2 + \bar{\mu}(i)|\boldsymbol{e}_a(i)|^2 = \|\tilde{\boldsymbol{w}}_{i-1}\|^2 + \bar{\mu}(i)|\boldsymbol{e}_p(i)|^2}$$

where $\bar{\mu}(i) = 1/\|\boldsymbol{u}_i\|^2$ if $\boldsymbol{u}_i \neq 0$ and $\bar{\mu}(i) = 0$ otherwise.

(c) Establish also the relations

$$\boxed{\begin{array}{rcl} \boldsymbol{w}_i^o &=& \boldsymbol{w}_{i-1}^o + \boldsymbol{\theta}_i - \boldsymbol{\theta}_{i-1} \\ \boldsymbol{e}(i) &=& \boldsymbol{e}_a(i) + \boldsymbol{v}(i) = \boldsymbol{u}_i\tilde{\boldsymbol{w}}_{i-1} + \boldsymbol{u}_i\left(\boldsymbol{w}_i^o e^{j\Omega} - \boldsymbol{w}_{i-1}^o\right)e^{j\Omega(i-1)} + \boldsymbol{v}(i) \end{array}}$$

(d) Show that the model of this problem encompasses the random-walk model (20.10) as a special case.

Remark. In this problem we adopted a model with a constant frequency offset Ω. The term $e^{j\Omega i}$ in the adopted model could be used, for example, to model Doppler channel variations in a wireless scenario, which result from reflections of the transmitted signal off a remote object moving with constant speed (e.g., Ghosh (1998)). Actually, many digital communication standards use the ability of digital communication receivers to track such Doppler shifts as a performance index for their ability to track time variant channels. A more general model would be to consider a time variant term of the form $e^{j(\Omega i+\phi(i))}$ in order to account for both frequency and phase offsets.

Remark. The results of Probs. IV.29–IV.33 are based on the work by Yousef and Sayed (2002).

Problem IV.30 (Variance relation) Consider the same setting of Prob. IV.29 and assume that $\boldsymbol{q}_i$ is an i.i.d. sequence with covariance matrix Q and independent of the initial conditions $\{\boldsymbol{w}_{-1}, \boldsymbol{\theta}_{-1}\}$, of the regressors $\{\boldsymbol{u}_j\}$ for all j, and of the $\{\boldsymbol{d}(j)\}$ for all $j < i$. Assume further that the filter is operating in steady-state, i.e., $\mathsf{E}\,\|\tilde{\boldsymbol{w}}_i\|^2 = \mathsf{E}\,\|\tilde{\boldsymbol{w}}_{i-1}\|^2$ as $i \to \infty$. Take expectations of both sides of the energy-conservation relation of part (b) in Prob. IV.29 and show that in the limit, as $i \to \infty$, it leads to the following variance relation:

$$\begin{aligned}
2\text{Re}\left(\mathsf{E}\,\boldsymbol{e}_a^*(i)g[\boldsymbol{e}(i)]\right) \;=\;& \mu\mathsf{E}\left(\|\boldsymbol{u}_i\|^2\,|g[\boldsymbol{e}(i)]|^2\right) + \mu^{-1}\text{Tr}(Q) \\
& + \mu^{-1}|1-e^{j\Omega}|^2\|w^o\|^2 + \mu^{-1}|1-\alpha e^{j\Omega}|^2\text{Tr}(\Theta) \\
& - 2\mu^{-1}\text{Re}\left[(1-e^{-j\Omega})w^{o*}\,\mathsf{E}\left((\tilde{\boldsymbol{w}}_{i-1} - \mu\boldsymbol{u}_i^* g[\boldsymbol{e}(i)])e^{-j\Omega(i-1)}\right)\right] \\
& - 2\mu^{-1}\text{Re}\left[(1-\alpha^* e^{-j\Omega})\mathsf{E}\left(\boldsymbol{\theta}_{i-1}^*(\tilde{\boldsymbol{w}}_{i-1} - \mu\boldsymbol{u}_i^* g[\boldsymbol{e}(i)])e^{-j\Omega(i-1)}\right)\right]
\end{aligned}$$

where $\Theta = \lim_{i\to\infty}\mathsf{E}\,\boldsymbol{\theta}_i\boldsymbol{\theta}_i^* = Q/(1-|\alpha|^2)$. *Remark.* Observe that when $\alpha = 1$ and $\Omega = 0$, the last three terms on the right-hand side of the above equality disappear and the relation collapses to the variance relation (20.32).

Problem IV.31 (Mean-weight error by LMS with frequency offset) Here and in Probs. IV.32–IV.33, we use the results of Probs. IV.29–IV.30 to characterize the tracking performance of LMS for the more general nonstationary model of Prob. IV.29. Thus consider the LMS recursion $\boldsymbol{w}_i = \boldsymbol{w}_{i-1} + \mu\boldsymbol{u}_i^*\boldsymbol{e}(i)$ with $\boldsymbol{e}(i) = \boldsymbol{d}(i) - \boldsymbol{u}_i\boldsymbol{w}_{i-1}$, and assume that (21.12) holds, namely,

At steady state, $\tilde{\boldsymbol{w}}_{i-1}$ is independent of $\boldsymbol{u}_i$

Assume further that the data $\{\boldsymbol{d}(i), \boldsymbol{u}_i\}$ are such that:

(a) There exists a vector $\boldsymbol{w}_i^o$ such that $\boldsymbol{d}(i) = \boldsymbol{u}_i\boldsymbol{w}_i^o e^{j\Omega i} + \boldsymbol{v}(i)$.
(b) The weight vector varies according to $\boldsymbol{w}_i^o = w^o + \boldsymbol{\theta}_i$.
(c) The perturbation varies according to $\boldsymbol{\theta}_i = \alpha\boldsymbol{\theta}_{i-1} + \boldsymbol{q}_i$.
(d) The noise sequence $\{\boldsymbol{v}(i)\}$ is i.i.d. with variance $\sigma_v^2 = \mathsf{E}\,|\boldsymbol{v}(i)|^2$.
(e) The noise sequence $\boldsymbol{v}(i)$ is independent of $\boldsymbol{u}_j$ for all i, j.
(f) The sequence $\boldsymbol{q}_i$ has covariance Q and is independent of $\{\boldsymbol{v}(j), \boldsymbol{u}_j\}$ for all i, j.
(g) The initial conditions $\{\boldsymbol{w}_{-1}, \boldsymbol{\theta}_{-1}\}$ are independent of all $\{\boldsymbol{d}(j), \boldsymbol{u}_j, \boldsymbol{v}(j), \boldsymbol{q}_j\}$.
(h) The regressor covariance matrix is denoted by $R_u = \mathsf{E}\,\boldsymbol{u}_i^*\boldsymbol{u}_i > 0$.
(i) The coefficient α satisfies $|\alpha| \leq 1$.

In order to apply the variance relation of Prob. IV.30, we need to evaluate the expectations $\mathsf{E}\,\tilde{\boldsymbol{w}}_{i-1}$ and $\mathsf{E}\,\boldsymbol{\theta}_{i-1}^*\tilde{\boldsymbol{w}}_{i-1}$. The purpose of this problem is to show that, in steady-state, $\mathsf{E}\,\tilde{\boldsymbol{w}}_i$ takes the form

$$\mathsf{E}\,\tilde{\boldsymbol{w}}_i = ve^{j\Omega i}, \qquad i \to \infty$$

for some constant vector v to be determined.

(a) Let $v_i = \mathsf{E}\,\tilde{\boldsymbol{w}}_i$. Show that v_i satisfies $v_i = (\mathrm{I} - \mu R_u)v_{i-1} + (\mathrm{I} - \mu R_u)w^o(e^{j\Omega} - 1)e^{j\Omega(i-1)}$ as $i \to \infty$.

(b) Introduce the eigenvalue decomposition of R_u, say $R_u = U\Lambda U^*$, where U is a unitary matrix and Λ is a diagonal matrix with positive entries $\{\lambda_1, \lambda_2, \ldots, \lambda_M\}$. Let $v_i'(k)$ denote the k–th entry of the transformed vector $v_i' = U^* v_i$. Likewise, let $c'(k)$ denote the $k-$th entry of the transformed vector $c' = U^* w^o(e^{j\Omega} - 1)$. Show that $v_i'(k)$ satisfies the steady-state recursion

$$v_i'(k) = (1 - \mu\lambda_k)v_{i-1}'(k) + (1 - \mu\lambda_k)c'(k)e^{j\Omega(i-1)}, \quad k = 1, \ldots, M, \quad i \to \infty$$

(c) Assume μ satisfies $\mu < 2/\lambda_{\max}$, where $\lambda_{\max}$ denotes the maximum eigenvalue of R_u. Argue from the recursion in part (b) that, in steady-state, $v_i'(k)$ tends to $v_i'(k) = be^{j\Omega i}$, for some constant b.

(d) Conclude that, in steady-state, the vector $v_i = \mathsf{E}\,\tilde{\boldsymbol{w}}_i$ tends to the form $v_i = ve^{j\Omega i}$, where $v = \left[\mathrm{I} - e^{j\Omega}(\mathrm{I} - \mu R_u)^{-1}\right]^{-1} w^o(1 - e^{j\Omega})$.

Problem IV.32 (Cross-correlation by LMS with frequency offset) Consider the setting of Prob. IV.31. We now verify that, in steady-state, the matrix $\mathsf{E}\,\tilde{\boldsymbol{w}}_i\boldsymbol{\theta}_i^*$ takes the form

$$\boxed{\mathsf{E}\,\tilde{\boldsymbol{w}}_i\boldsymbol{\theta}_i^* = We^{j\Omega i}, \qquad i \to \infty}$$

for some constant matrix W to be determined.

(a) Let $W_i = \mathsf{E}\,\tilde{\boldsymbol{w}}_i\boldsymbol{\theta}_i^*$. Show that W_i satisfies $W_i = \alpha^*(\mathrm{I} - \mu R_u)W_{i-1} - (\mathrm{I} - \mu R_u)Ce^{j\Omega(i-1)}$ as $i \to \infty$, where $C = \alpha^*(1 - \alpha e^{j\Omega})\Theta - e^{j\Omega}Q$.

(b) Define the transformed matrices $W_i' = U^* W_i$ and $C' = U^* C$. Use arguments similar to those in parts (c) and (d) of Prob. IV.31 to show that each element of W_i converges to a constant times the exponential sequence $e^{j\Omega i}$.

(c) Conclude that W_i tends to $We^{j\Omega i}$, where $W = \left[\alpha^*\mathrm{I} - e^{j\Omega}(\mathrm{I} - \mu R_u)^{-1}\right]^{-1} C$.

Problem IV.33 (Tracking by LMS with frequency offset) We use the results of Probs. IV.31 and IV.32 to characterize the tracking performance of LMS for the nonstationary model of Prob. IV.29. Use $g[\boldsymbol{e}(i)] = \boldsymbol{e}_a(i) + \boldsymbol{v}(i)$ and substitute the expressions for $\mathsf{E}\,\tilde{\boldsymbol{w}}_i$ and $\mathsf{E}\,\tilde{\boldsymbol{w}}_i\boldsymbol{\theta}_i^*$ into the variance relation of Prob. IV.30. Then proceed to address the following questions.

(a) Assume the step-size is such that the term $\mathsf{E}\,\|\boldsymbol{u}_i\|^2|\boldsymbol{e}_a(i)|^2$ can be neglected. Show that the resulting EMSE is given by

$$\boxed{\zeta^{\mathsf{LMS}} = \frac{\mu}{2}\sigma_v^2\mathsf{Tr}(R_u) + \frac{\mu^{-1}}{2}\beta} \quad \text{(small } \mu\text{)}$$

where

$$\begin{aligned}\beta \;=\;& |1 - e^{j\Omega}|^2 \;\mathsf{Re}\;\mathsf{Tr}\left[\,(\mathrm{I} - 2(X - \mu R_u))W^o\right] \\ &+ |1 - \alpha e^{j\Omega}|^2 \;\mathsf{Re}\;\mathsf{Tr}\left[(\mathrm{I} - 2(\alpha^* X_\alpha + \mu R_u))\Theta\right] \;+\; \mathsf{Re}\;\mathsf{Tr}\left[(\mathrm{I} + 2(e^{j\Omega} - \alpha^*)X_\alpha)Q\right]\end{aligned}$$

with $X = (\mathrm{I} - \mu R_u)\left[\mathrm{I} - e^{j\Omega}(\mathrm{I} - \mu R_u)^{-1}\right]^{-1}$, $X_\alpha = (\mathrm{I} - \mu R_u)\left[\alpha^*\mathrm{I} - e^{j\Omega}(\mathrm{I} - \mu R_u)^{-1}\right]^{-1}$ and $W^o = w^o w^{o*}$.

(b) Use instead the separation principle (21.8) to show that the EMSE evaluates to

$$\boxed{\zeta^{\mathsf{LMS}} = \frac{\mu\sigma_v^2\mathsf{Tr}(R_u) \;+\; \mu^{-1}\beta}{2 - \mu\mathsf{Tr}(R_u)}} \quad \text{(over wider range of } \mu\text{)}$$

(c) Assume Gaussian regressors with $R_u = \sigma_u^2\mathrm{I}$. Show that, in this case, the EMSE evaluates to

$$\boxed{\zeta^{\mathsf{LMS}} = \frac{\mu M\sigma_u^2\sigma_v^2 \;+\; \mu^{-1}\beta}{2 - \mu(M + \gamma)\sigma_u^2}} \quad \text{(Gaussian)}$$

where M is the filter length, $\gamma = 1$ if the $\{\boldsymbol{u}_i\}$ are circular complex-valued and $\gamma = 2$ if the $\{\boldsymbol{u}_i\}$ are real-valued. Moreover, β is as in part (a) with R_u replaced by $\sigma_u^2\mathrm{I}$.

Problem IV.34 (Tracking performance of CMA2-2 for real-data) In this problem we consider the CMA2-2 recursion of Prob. IV.16, namely,

$$\boxed{w_i = w_{i-1} + \mu u_i^{\mathsf{T}} z(i)[\gamma - z^2(i)], \quad z(i) = u_i w_{i-1}, \quad i \geq 0}$$

and study its tracking performance in nonstationary environments. Motivated by the discussion in Prob. IV.16, we assume that there exists an optimal equalizer w_i^o that varies according to the rule $w_i^o = w_{i-1}^o + q_i$, where q_i is i.i.d. with covariance matrix Q. Moreover, q_i is assumed to be independent of all other random variables. The model w_i^o is such that it reproduces the complex data $\{s(i)\}$ up to some delay Δ, i.e., $u_i w_i^o = s(i - \Delta)$. The tracking performance of the algorithm is measured in terms of the steady-state value of $\mathsf{E}\,|s(i-\Delta) - u_i w_{i-1}|^2$.

(a) Verify that the energy-conservation relation of Thm. 20.1 still holds for CMA2-2

(b) Verify also that the variance relation of Thm. 20.2 still holds, namely, for any data $\{u_i\}$, $\mu \mathsf{E}\,\|u_i\|^2 g^2[z(i)] \; + \; \mu^{-1}\mathsf{Tr}(Q) = 2\mathsf{E}\, e_a(i) g[z(i)]$ as $i \to \infty$.

(c) Using $g[z] = z[\gamma - z^2]$, and the assumptions from Prob. IV.16, show that the MSE of CMA2-2 is given by

$$\boxed{\mathsf{MSE}^{\mathsf{CMA2-2}} \approx \frac{\mu \mathsf{E}\,(\gamma^2 s^2 - 2\gamma s^4 + s^6)\; \mathsf{Tr}(R_u) \; + \; \mu^{-1}\mathsf{Tr}(Q)}{2\mathsf{E}\,(3s^2 - \gamma)}}$$

in terms of the second, fourth, and sixth moments of the constellation.

(d) Conclude that, when the denominator below is nonzero, an optimal choice for the step-size that minimizes the MSE is given by

$$\mu_{\text{opt}}^{\mathsf{CMA2-2}} = \sqrt{\frac{\mathsf{Tr}(Q)}{\mathsf{E}\,(\gamma^2 s^2 - 2\gamma s^4 + s^6)\mathsf{Tr}(R_u)}}$$

with the resulting minimum MSE

$$\mathsf{MSE}_{\min}^{\mathsf{CMA2-2}} = \frac{\sqrt{\mathsf{Tr}(Q)\mathsf{E}\,(\gamma^2 s^2 - 2\gamma s^4 + s^6)\mathsf{Tr}(R_u)}}{\mathsf{E}\,(3s^2 - \gamma)}$$

Remark. The results of this problem, and of Probs. IV.35 and IV.36, are based on the work by Yousef and Sayed (1999b,2001).

Problem IV.35 (Tracking performance of CMA2-2 for complex-data) In this problem we extend the results of Prob. IV.34 to complex-valued constellations. Thus consider the CMA2-2 recursion,

$$\boxed{w_i = w_{i-1} + \mu u_i^* z(i)[\gamma - |z(i)|^2], \quad z(i) = u_i w_{i-1}, \quad i \geq 0}$$

and assume there exists an optimal equalizer w_i^o that varies according to the rule $w_i^o = w_{i-1}^o + q_i$, where q_i is i.i.d. with covariance matrix Q. Moreover, q_i is assumed to be independent of all other random variables. The model w_i^o is such that it reproduces the complex data $\{s(i)\}$ up to some rotation θ and delay Δ, i.e., $u_i w_i^o = s(i - \Delta)e^{j\theta}$.

(a) Verify that the variance relation of Thm. 20.2 still holds, namely, for any data $\{u_i\}$,

$$\mu \mathsf{E}\,\|u_i\|^2 |g[z(i)]|^2 \; + \; \mu^{-1}\mathsf{Tr}(Q) \; = \; 2\mathsf{Re}\big(\mathsf{E}\, e_a^*(i) g[z(i)]\big), \quad \text{as } i \to \infty$$

(b) Now introduce the same assumptions as in part (b) of Prob. IV.17 and show that the MSE of CMA2-2 is given by

$$\boxed{\mathsf{MSE}^{\mathsf{CMA2-2}} \approx \frac{\mu \mathsf{E}\,(\gamma^2|s|^2 - 2\gamma|s|^4 + |s|^6)\mathsf{Tr}(R_u) \; + \; \mu^{-1}\mathsf{Tr}(Q)}{2\mathsf{E}\,(2|s|^2 - \gamma)}}$$

in terms of the second, fourth, and sixth moments of the constellation.

(c) Assume the data $\{s(i)\}$ are constant-modulus, $|s(i)| = 1$, and choose $\gamma = 1$. What would the resulting MSE be?

(d) Show that, when the denominator below is nonzero, an optimal choice for the step-size that minimizes the MSE is given by

$$\mu_{\text{opt}}^{\text{CMA2-2}} = \sqrt{\frac{\text{Tr}(Q)}{\text{E}\left(\gamma^2|s|^2 - 2\gamma|s|^4 + |s|^6\right)\text{Tr}(R_u)}}$$

with the resulting minimum MSE

$$\text{MSE}_{\min}^{\text{CMA2-2}} = \frac{\sqrt{\text{Tr}(Q)\text{E}\left(\gamma^2|s|2 - 2\gamma|s|^4 + |s|^6\right)\text{Tr}(R_u)}}{\text{E}\left(2|s|^2 - \gamma\right)}$$

Problem IV.36 (Tracking performance of CMA1-2) Consider the CMA1-2 recursion of Prob. IV.18, namely,

$$\boxed{\boldsymbol{w}_i = \boldsymbol{w}_{i-1} + \mu \boldsymbol{u}_i^*[\gamma \frac{\boldsymbol{z}(i)}{|\boldsymbol{z}(i)|} - \boldsymbol{z}(i)], \quad \boldsymbol{z}(i) = \boldsymbol{u}_i \boldsymbol{w}_{i-1}, \quad i \geq 0}$$

and let us study its tracking performance in nonstationary environments for complex data. We again assume that there exists an optimal equalizer model $\boldsymbol{w}_i^o$ that varies according to the rule $\boldsymbol{w}_i^o = \boldsymbol{w}_{i-1}^o + \boldsymbol{q}_i$, where $\boldsymbol{q}_i$ is i.i.d. with covariance matrix Q. Moreover, $\boldsymbol{q}_i$ is assumed to be independent of all other random variables. The model $\boldsymbol{w}_i^o$ is such that it reproduces the complex data $\{\boldsymbol{s}(i)\}$ up to some rotation θ and delay Δ, i.e., $\boldsymbol{u}_i \boldsymbol{w}_i^o = \boldsymbol{s}(i - \Delta)e^{j\theta}$.

(a) Justify the variance relation $\mu \text{E}\,\|\boldsymbol{u}_i\|^2 |g[\boldsymbol{z}(i)]|^2 + \mu^{-1}\text{Tr}(Q) = 2\text{Re}\big(\text{E}\,\boldsymbol{e}_a^*(i)g[\boldsymbol{z}(i)]\big)$ as $i \to \infty$. Now consider the same assumptions imposed in Prob. IV.18 and show that this relation leads to the following expression for the MSE:

$$\boxed{\text{MSE}^{\text{CMA1-2}} = \frac{\mu}{2}\left(\gamma^2 + \text{E}\,|s|^2 - 2\gamma \text{E}\,|s|\right)\text{Tr}(R_u) + \frac{\mu^{-1}}{2}\text{Tr}(Q)}$$

(b) Conclude that, when the denominator below is nonzero, an optimal choice for the step-size that minimizes the MSE is given by

$$\mu_{\text{opt}}^{\text{CMA1-2}} = \sqrt{\frac{\text{Tr}(Q)}{\text{E}\left(\gamma^2 + |s|^2 - 2\gamma|s|\right)\text{Tr}(R_u)}}$$

and the corresponding minimum MSE is

$$\text{MSE}_{\min}^{\text{CMA1-2}} = \sqrt{\text{Tr}(Q)\text{E}\left(\gamma^2 + |s|^2 - 2\gamma|s|\right)\text{Tr}(R_u)}$$

Problem IV.37 (Feedback mapping) Refer to the discussion in App. 15.A on a system-theoretic interpretation for the energy-conservation relation. How would you modify Fig. 15.5 in order to represent the energy-conservation relation (20.24) for nonstationary models?

Problem IV.38 (Fixed-point arithmetic) Let $b_i \in \{0, 1\}$. Consider a fixed-point representation of the form $\pm 0.b_1 b_2 \ldots b_{B-1}$. This representation is limited to numbers that are less than one in magnitude. Therefore, additions and subtractions may cause overflow if the result lies outside this range, while the multiplication of two fixed-point numbers never causes overflow. Let ϵ denote the machine precision, i.e., the largest absolute difference between a real number a and its fixed point representation, so that $|\text{fx}[a] - a| \leq \epsilon$ for any a. Assume all variables are represented with B bits (including the sign bit) and that rounding is used so that $\epsilon = 2^{-B-1}$.

(a) Consider a random variable $\boldsymbol{a}$ with distribution

$$\boldsymbol{a} = \begin{cases} 0.5 + 2^{-7} & \text{with probability } 0.5 - 2^{-7} \\ -0.5 + 2^{-7} & \text{with probability } 0.5 + 2^{-7} \end{cases}$$

Verify that $\mathsf{E}\,\boldsymbol{a} = 0$.

(b) Assume $\boldsymbol{a}$ is quantized to fixed-point with 7 bits (including the sign bit). Verify that $\mathsf{fx}[\boldsymbol{a}]$ will have the distribution

$$\mathsf{fx}[\boldsymbol{a}] = \begin{cases} 0.5 + 2^{-6} & \text{with probability } 0.5 - 2^{-7} \\ -0.5 & \text{with probability } 0.5 + 2^{-7} \end{cases}$$

and conclude that $\mathsf{E}\,\mathsf{fx}[\boldsymbol{a}] = -2^{-13}$.

Remark. The result of this problem shows that a zero-mean random variable may become non-zero mean after quantization. The small nonzero mean might cause a slow drift of the LMS weight estimate, causing the algorithm to overflow.

Problem IV.39 (Drift problem) All variables in this problem are real-valued. Consider a quantized LMS implementation in a stationary environment with two taps, which according to the discussion in Sayed (2003, Ch. 8) takes the form

$$\boldsymbol{w}_i^q = \boldsymbol{w}_{i-1}^q + \mu \boldsymbol{u}_i^{\mathsf{T}} \left[\boldsymbol{e}_a(i) + \bar{\boldsymbol{v}}(i) \right] - \boldsymbol{p}_i$$

where $\boldsymbol{w}_i^q$ denotes the quantized weight vector, $\boldsymbol{e}_a(i) = \boldsymbol{u}_i \tilde{\boldsymbol{w}}_{i-1}^q$, $\tilde{\boldsymbol{w}}_i^q = w^o - \boldsymbol{w}_i^q$, and $\bar{\boldsymbol{v}}(i)$ and $\boldsymbol{p}_i$ are assumed to be i.i.d. and independent of all other random variables with variances denoted by $\sigma_{\bar{v}}^2$ and R_p, respectively. Assume $\boldsymbol{u}_i = \begin{bmatrix} \boldsymbol{u}(i) & 0 \end{bmatrix}$, where $\boldsymbol{u}(i)$ is and i.i.d. sequence with variance σ_u^2 and independent of all other variables as well. Therefore, in this example, the covariance matrix R_u is singular and given by $R_u = \text{diag}\{\sigma_u^2, 0\}$. Let C_i denote the covariance matrix of $\tilde{\boldsymbol{w}}_i^q$, and partition C_i and $R_p = \mathsf{E}\,\boldsymbol{p}_i \boldsymbol{p}_i^{\mathsf{T}}$ (assumed diagonal) into

$$C_i = \begin{bmatrix} c_1(i) & c_2(i) \\ c_2(i) & c_3(i) \end{bmatrix}, \qquad R_p = \begin{bmatrix} a & 0 \\ 0 & b \end{bmatrix}$$

(a) Show that $c_1(i)$ satisfies $c_1(i) = c_1(i-1)\mathsf{E}\,(1 - \mu \boldsymbol{u}^2)^2 + \mu^2 \sigma_u^2 \sigma_{\bar{v}}^2 + a$.

(b) Conclude further that $c_3(i)$ satisfies $c_3(i) = c_3(i-1) + b$ with a forced term b and that, therefore, it grows unbounded.

Remark. This example shows how the lack of sufficient excitation in the regression data, as determined by the singularity of R_u, may result in weight estimates growing unbounded and overflowing; thus leading to the drift problem of LMS. As explained below in part (c), one way to address the drift problem is to use leakage at the cost of introducing bias — see Probs. V.27–V.32.

(c) Consider now a quantized implementation of leaky-LMS, which according to the discussion in Sayed (2003, Ch. 8) can be modeled as:

$$\boldsymbol{w}_i^q = (1 - \mu\alpha)\boldsymbol{w}_{i-1}^q + \mu \boldsymbol{u}_i^* [\boldsymbol{e}_a(i) + \bar{\boldsymbol{v}}(i)] - \boldsymbol{p}_i$$

Show that $\{c_1(i), c_3(i)\}$ now satisfy the recursions:

$$c_1(i) \;=\; c_1(i-1)\mathsf{E}\,(1 - \mu\alpha - \mu \boldsymbol{u}^2)^2 + \mu^2 \sigma_u^2 \sigma_{\bar{v}}^2 + a, \quad c_3(i) = (1 - \mu\alpha)^2 c_3(i-1) + b$$

Conclude that the drift problem does not occur any longer for sufficiently small step-sizes.

Problem IV.40 (Singular covariance matrix) The result of Prob. IV.39 reveals that a singular covariance matrix can cause LMS to drift in a finite-precision implementation. The case of singular or close-to-singular R_u is not an abstraction. It arises in some applications, e.g., in fractionally-spaced channel equalization (which is discussed in App. 3.B of Sayed (2003)). Figure IV.2 depicts the structure of an adaptive fractionally-spaced equalizer. The received signal is denoted by $\boldsymbol{u}(t)$ and is sampled at the rate of $1/T'$ samples/second. This sampling rate is chosen to be double the symbol rate, $1/T$. The regressor of the equalizer is therefore $\boldsymbol{u}_i = \begin{bmatrix} \boldsymbol{u}(i) & \boldsymbol{u}(i-1) & \ldots & \boldsymbol{u}(i-M+1) \end{bmatrix}$, where $\boldsymbol{u}(i)$ is obtained by sampling $\boldsymbol{u}(t)$ at the rate $1/T' = 2/T$, i.e., $\boldsymbol{u}(i) = \boldsymbol{u}(t)\,|_{t=iT'}$.

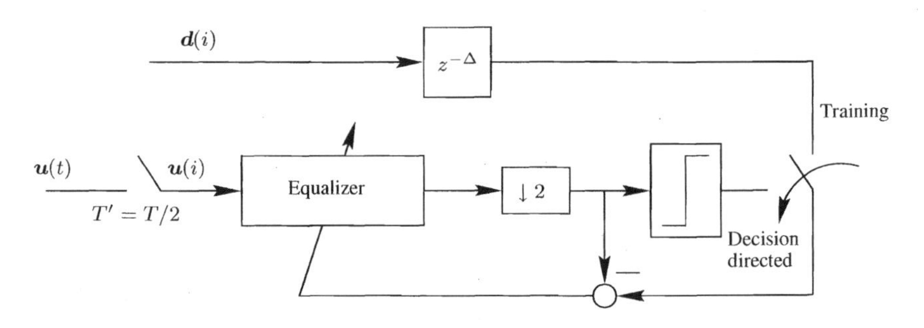

FIGURE IV.2 A structure for adaptive fractionally-spaced equalization with training and decision-directed modes of operation. The update of the weight vector of the equalizer is performed at the symbol rate.

The random process $\boldsymbol{u}(t)$ is assumed to be stationary and its power spectrum is defined by $S_u(j\Omega) = \int_{-\infty}^{\infty} R_u(\tau)e^{-j\Omega\tau}d\tau$, where $j = \sqrt{-1}$ and $R_u(\tau)$ is the auto-correlation function of the random process $\boldsymbol{u}(t)$, i.e., $R_u(\tau) = \mathsf{E}\,\boldsymbol{u}(t)\boldsymbol{u}^*(t-\tau)$. Let $R_u = \mathsf{E}\,\boldsymbol{u}_i\boldsymbol{u}_i^*$ denote the covariance matrix of $\boldsymbol{u}_i$. The purpose of this problem is to show that there are power spectra $S_u(j\Omega)$ for which the matrix R_u can be singular. To do so, we shall show that there exists a nonzero vector x such that $x^*R_ux = 0$.

(a) Let x denote an arbitrary column vector with entries $\{x(1), \ldots, x(M)\}$. Use the inverse Fourier transform to show that

$$x^*R_ux = \sum_{m=1}^{M}\sum_{n=1}^{M} x^*(m)x(n)R_u[(m-n)T'] = \int_{-\infty}^{\infty} S_u(j\Omega)|X(j\Omega)|^2\frac{d\Omega}{2\pi}$$

where $X(j\Omega) \stackrel{\Delta}{=} \sum_{n=1}^{M} x(n)e^{-jn\Omega T'}$. Observe that $X(j\Omega)$ is periodic with period $\Omega_s' = 2\pi/T' = 4\pi/T$. Conclude that R_u is singular if, and only if, a vector x exists such that

$$\int_{-\infty}^{\infty} S_u(j\Omega)|X(j\Omega)|^2\frac{d\Omega}{2\pi} = 0$$

(b) Assume $S_u(j\Omega)$ has bandwidth $\Omega_s = 2\pi/T$ and is given by

$$S_u(j\Omega) = \begin{cases} 1 & 0 \le |\Omega| \le \Omega_s \\ -1 & \Omega_s \le |\Omega| \le \Omega_s' \\ 0 & \text{otherwise} \end{cases}$$

Assume also that x is chosen such that $|X(j\Omega)|^2$ is symmetric around Ω_s. Conclude that the resulting covariance matrix R_u will be singular.

Remark. This problem is a modified version of an example from Gitlin, Meadors, and Weinstein (1982).

COMPUTER PROJECTS

Project IV.1 (Line echo cancellation) In communications over phone lines, a signal travelling from a far-end point to a near-end point is usually reflected at the near-end due to mismatches in circuitry (e.g., hybrid connections). The reflected signal travels back to the far-end point in the form of an echo. As a result, the speaker at the far-end receives, in addition to the desired signal from the near-end speaker, an attenuated replica of his own signal in the form of an echo — see Fig. IV.3.

The echo interferes with the quality of the received signal. A common way to provide better voice quality at both ends is to employ adaptive line echo cancellers (LEC). At the near-end, for example, the signal feeding the LEC is the far-end signal while the reference signal is its reflected version —

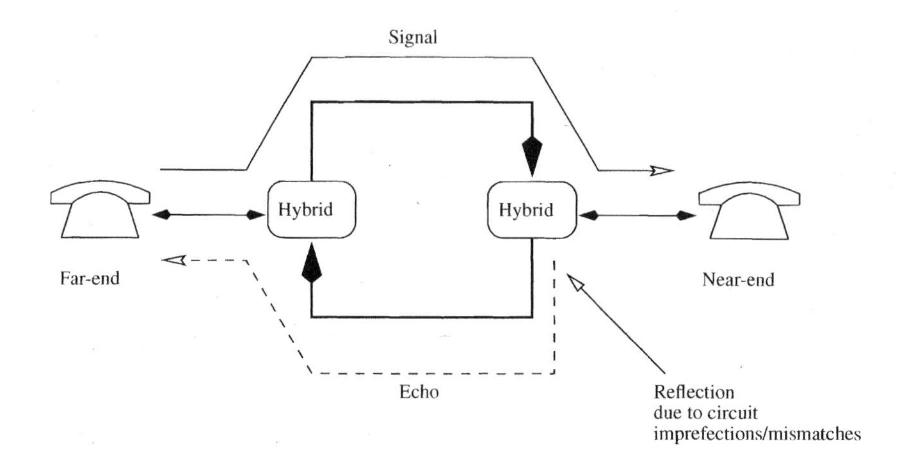

FIGURE IV.3 The signal at the far-end is reflected at the near-end due to circuit mismatches and travels back to the far-end.

see Fig. IV.4. In the figure, the output of the adaptive LEC generates a replica of the echo, and the error signal is therefore a "clean" signal that is transmitted to the far-end. The signals in this project are assumed to be sampled at 8 kHz.

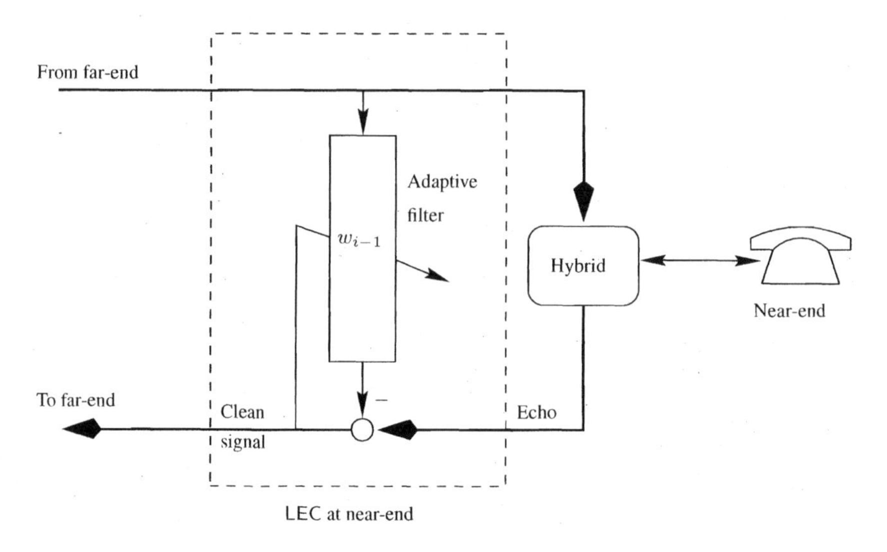

FIGURE IV.4 An adaptive line echo canceller at the near-end.

(a) Load the file path.mat, which contains the impulse response sequence of a typical echo path. Plot the impulse and frequency responses of the echo path.

(b) Load the file css.mat, which contains 5600 samples of a composite source signal; it is a synthetic signal that emulates the properties of speech. Specifically, it contains segments of pause, segments of periodic excitation and segments with white-noise properties. Plot the samples of the CSS data, as well as their spectrum.

(c) Concatenate five such blocks and feed them into the echo path. Plot the resulting echo signal. Estimate the input and output powers in dB using

$$\widehat{P} = 10\log_{10}\left(\frac{1}{N}\sum_{i=1}^{N}|\text{signal}(i)|^2\right)$$

where N denotes the length of the sequence. Evaluate the attenuation in dB that is introduced by the echo path as the signal travels through it; this attenuation is called the echo-return-loss (ERL).

(d) Use 10 blocks of CSS data as far-end signal, and the corresponding output of the echo path as the echo signal. Choose an adaptive line echo canceller with 128 taps. Train the canceller by using as input data the far-end signal, i.e., $u(i) = \text{far\_end}(i)$, and as reference data the echo signal, i.e., $d(i) = \text{echo}(i)$. Use ϵ-NLMS with $\epsilon = 10^{-6}$ and $\mu = 0.25$. Plot the far-end signal, the echo, and the error signal provided by the adaptive filter. Plot also the echo path and its estimate by the adaptive filter at the end of the simulation.

(e) Estimate the steady-state power of the error signal and measure its attenuation in dB relative to the echo signal. Use the last 5600 samples of the signals to estimate their powers. The difference in power is a measure of the attenuation introduced by the LEC and it is called the echo-return-loss-enhancement (ERLE).

(f) Fix the input power at 0 dB and add white Gaussian noise with variance $\sigma_v^2 = 0.0001$ to the echo signal. Train the LEC using 80 blocks of CSS data and measure the steady-state ERLE. Compare the simulated and theoretical ERLEs.

Project IV.2 (Tracking a Rayleigh fading channel) In a wireless communications environment, signals suffer from multiple reflections while travelling from the transmitter to the receiver so that the receiver ends up getting several (almost simultaneous) replicas of the transmitted signal. The reflections are received with different amplitude and phase distortions, and the overall received signal is the combined sum of the reflections. Based on the relative phases of the reflections, the signals may add up constructively or destructively at the receiver. Furthermore, if the transmitter is moving with respect to the receiver, these destructive and constructive interferences will vary with time. This phenomenon is known as channel *fading*.

The impulse response of a single-path (i.e., single-tap) fading channel can be described as

$$\boldsymbol{h}(n) = \gamma\, \boldsymbol{x}(n)\, \delta(n - n_o)$$

where $\{\boldsymbol{x}(n)\}$ is a time-variant complex sequence that models the time-variations in the channel, and n_o is the channel delay. Without loss of generality, the sequence $\{\boldsymbol{x}(n)\}$ is assumed to have unit variance, and the scalar γ is used to model the actual path loss that is introduced by the channel. That is, γ^2 is equal to the power attenuation that a signal will undergo when it travels through the channel.

Several mathematical models can be used to characterize the fading properties of $\{\boldsymbol{x}(n)\}$, and consequently of the channel. A widely used model is known as *Rayleigh fading*. In this case, for each n, the amplitude $|\boldsymbol{x}(n)|$ is assumed to have a Rayleigh distribution (cf. (A.3)), i.e.,

$$\boxed{f_{|\boldsymbol{x}(n)|}(|x(n)|) = |x(n)|\, e^{-|x(n)|^2/2}, \qquad |x(n)| \geq 0}$$

while the phase $\angle \boldsymbol{x}(n)$ is assumed to be uniformly distributed within $[-\pi, \pi]$:

$$\boxed{f(\angle \boldsymbol{x}(n)) = \frac{1}{2\pi}, \qquad -\pi \leq \angle \boldsymbol{x}(n) \leq \pi}$$

In addition, the auto-correlation function of the sequence $\{\boldsymbol{x}(n)\}$, now regarded as a random process, is modeled as a zeroth-order Bessel function of the first kind, namely,

$$\boxed{r(k) \triangleq \mathsf{E}\,\boldsymbol{x}(n)\boldsymbol{x}(n-k) = \mathcal{J}_o\left(2\pi f_D T_s\, k\right), \quad k = \ldots, -1, 0, 1, \ldots}$$

where T_s is the sampling period of the sequence $\{\boldsymbol{x}(n)\}$, f_D is called the maximum Doppler frequency of the Rayleigh fading channel, and the function $\mathcal{J}_o(\cdot)$ is defined by

$$\mathcal{J}_o(y) \stackrel{\Delta}{=} \frac{1}{\pi}\int_0^{\pi} \cos(y \sin\theta) d\theta$$

The Doppler frequency f_D is related to the speed of the mobile user, v, and to the carrier frequency, f_c, as follows:

$$\boxed{f_D = \frac{v f_c}{c}}$$

where c denotes the speed of light, $c = 3 \times 10^8$m/s. This commonly used choice of the auto-correlation function is based on the assumption that all scatterers are uniformly distributed on a circle around the receiver, so that the power spectrum of the channel fading gain $\boldsymbol{x}(t)$, in continuous-time, would have the following well-known U−shaped spectrum,

$$S(f) = \frac{1}{\pi f_D \sqrt{1 - \left(\frac{f}{f_D}\right)^2}}, \qquad |f| \leq f_D$$

Other assumptions on the distribution of the scatterers would lead to different auto-correlation functions.

(a) Assume a carrier frequency of $f_c = 900$MHz. Verify that the Doppler frequency that corresponds to a vehicle moving at the speed of $v = 80$Km/h is $f_D = 66.67$Hz.

(b) Use the program rayleigh.m to generate 2000 samples of a Rayleigh fading coefficient $\boldsymbol{x}(n)$ with Doppler frequency $f_D = 66.67$Hz and sampling period $T_s = 1$ms. Plot its amplitude and phase. Plot also the cumulative distribution function (cdf) of $|\boldsymbol{x}(n)|$ (by definition, for any particular amplitude value κ, the cdf shows the proportion of occurrences smaller than κ). Plot also the amplitude of a Rayleigh fading sequence with the same Doppler frequency but with sampling period $T_s = 1\mu$s. What do you observe?

(c) As explained above, fading is caused by the addition of several reflections that reach the receiver approximately at the same instant. However, in many cases, other reflections might be originated from a far away object such as a mountain or a tall building. These reflections arrive at the receiver with longer delay than the first group of reflections. In such situations, a single-path Rayleigh fading model is not adequate anymore to represent the wireless channel. To model this so-called *multipath* phenomenon, a finite-impulse response model for the channel can be used, say one of the form

$$\boldsymbol{h}(n) = \sum_{k=1}^{L} \gamma_k \boldsymbol{x}_k(n) \delta(n - n_k)$$

where $\{\gamma_k\}$ and $\{\boldsymbol{x}_k(n)\}$ are respectively the path loss and fading sequence of the $k-$th cluster of reflectors, and the $\{n_k\}$ are the cluster delays. The sequences $\{\boldsymbol{x}_k(n)\}$ are modeled as independent Rayleigh fading sequences and the channel is referred to as a *multipath* Rayleigh fading channel.

In this project, we consider a wireless channel with two Rayleigh fading rays; both rays are assumed to fade at the same Doppler frequency of $f_D = 10$Hz. The channel impulse response sequence consists of two zero initial samples (i.e., an initial delay of two samples), followed by a Rayleigh fading ray, followed by another zero sample, and by a second Rayleigh fading ray. In other words, we are assuming a channel length of $M = 5$ taps with only two active Rayleigh fading rays, so that the weight vector that we wish to estimate has the form:

$$\begin{bmatrix} 0 & 0 & \boldsymbol{x}_2(n) & 0 & \boldsymbol{x}_4(n) \end{bmatrix}$$

Train an LMS filter to estimate and track this multipath channel. Assume white input of unit variance is transmitted across the channel and use it to excite the adaptive filter. Assume further that the output of the channel is observed in the presence of white additive Gaussian noise with variance $\sigma_v^2 = 0.001$. Use $\mu = 0.01$ and average the learning curve of LMS over 100 experiments. Plot the learning curve and compute the resulting MSE. Plot also the time evolution of the first ray and its estimate by the LMS filter over a particular experiment. Use the function rayleigh.m to generate time sequences for both channel rays, say of duration 30000 samples, and assume the sampling period is $T_s = 0.8\mu$s.

(d) A first-order approximation for the variation of a Rayleigh fading coefficient $x(n)$ is to assume that $x(n)$ varies according to the auto-regressive model:

$$\boxed{x(n) = r(1)x(n-1) \; + \; \sqrt{1-|r(1)|^2}\,\eta(n)}$$

where $r(1) = \mathcal{J}_o(2\pi f_D T_s)$ and $\eta(n)$ denotes a white noise process with unit-variance. Now since the multipath rays of the channel in part (c) are assumed to fade at the same rate, the above approximation indicates that the variations in the channel weight vector could be approximated as $w_i^o = \alpha w_{i-1}^o + q_i$, where the covariance matrix of q_i is $Q = (1-\alpha^2)\mathrm{I}$ with $\alpha = r(1)$. Of course, the value of α depends on the Doppler frequency. Use this model, and the second expression for the EMSE of LMS from Lemma 21.1, namely,

$$\zeta^{\text{LMS}} = \frac{\mu\sigma_v^2 \text{Tr}(R_u) + \mu^{-1}\text{Tr}(Q)}{2 - \mu\text{Tr}(R_u)}$$

to evaluate the theoretical MSE for part (c); recall that MSE $= \sigma_v^2 +$ EMSE. Compare your answer with the simulated result from part (c).

(e) The expression for the EMSE that is used in part (d), was derived in Lemma 21.1 by using the nonstationary model (20.10), which assumes $\alpha = 1$. Since the model for the Rayleigh fading channel under consideration has

$$\alpha \;=\; r(1) = \mathcal{J}_o(2 \times \pi \times 10 \times 0.8 \times 10^{-6}) \approx 0.99999999936835 \approx 1$$

we are justified to rely on Lemma 21.1. Now recall that in Prob. IV.33 we studied the tracking performance of LMS for a more general nonstationary model, which allows the inclusion of a non-unity α. If we set $\Omega = 0$ in that problem, we find that the EMSE of LMS for a model of the form $w_i^o = \alpha w_{i-1}^o + q_i$ is given by

$$\boxed{\zeta^{\text{LMS}} = \frac{\mu\sigma_v^2 \text{Tr}(R_u) + \mu^{-1}\beta}{2 - \mu\text{Tr}(R_u)}}$$

where

$$\beta = \text{Tr}\left[(\mathrm{I} + 2(1-\alpha)X_\alpha)Q\right], \quad X_\alpha = (\mathrm{I} - \mu R_u)\left[\alpha\mathrm{I} - (\mathrm{I} - \mu R_u)^{-1}\right]^{-1}$$

Compute the theoretical MSE of LMS using this alternative expression and compare it with the simulated and theoretical results of parts (c) and (d).

(f) For the same scenario of parts (c) and (d), vary the Doppler frequency from 10Hz to 25Hz in increments of 5Hz and generate a plot of the MSE as a function of the Doppler frequency. Run the LMS filter for 60000 iterations in each case and average the squared-error curve over 100 experiments. Continue to use $\mu = 0.01$. What do you observe?

(g) Fix the Doppler frequency at 10Hz and run the LMS filter for 20000 iterations for the step-sizes

$$\mu \in \{0.003, 0.005, 0.007, 0.01, 0.03, 0.05, 0.07\}$$

For each case, generate a learning curve by averaging over 50 experiments and compute the mean of the last 200 entries of the resulting curve. Use this value as indicative of the MSE of the algorithm. Generate a plot showing the MSE value versus the step-size. Compare with the theoretical MSE.

PART V

TRANSIENT PERFORMANCE

CHAPTER 22

Weighted Energy Conservation

As is evident by now, adaptive filters are time-variant and nonlinear stochastic systems with inherent learning and tracking abilities. The success of their learning mechanism can be measured in terms of how well they learn the underlying signal statistics given sufficient time (i.e., in terms of their steady-state performance) and in terms of how fast and how stably they adapt to changes in the signal statistics (i.e., in terms of their transient and convergence performance). For this reason, it is customary to study the performance of adaptive filters by examining their transient performance and their steady-state performance. The former is concerned with the stability and convergence rate of an adaptive scheme, while the latter is concerned with the mean-square error that remains in steady-state.

In Part IV (*Mean-Square Performance*) we focused on the steady-state performance of adaptive filters in both stationary and nonstationary environments, as well as on the performance degradation that occurs in finite-precision implementations. In this part we turn our attention to the transient performance of adaptive filters. We continue to rely on the energy conservation arguments of Chapter 15 and show how these same arguments can be used to perform transient analysis in a uniform manner across different algorithms. Compared with the derivations in Chapters15, and for reasons explained below, it will turn out that transient analysis is more conveniently performed by relying on a *weighted* energy-conservation relation, as opposed to the unweighted version that we have employed so far in the book.

We focus in this part on the class of adaptive filters with data normalization for which a more detailed transient analysis is easier to advance. In App. 9.C of Sayed (2003), it is shown how to extend the arguments to the class of adaptive filters with error nonlinearities; the transient analysis for this class of filters is more demanding due to their use of nonlinear error functions.

22.1 DATA MODEL

We rely on the same data model that we adopted in Sec. 15.2 for stationary environments. Thus, let $\boldsymbol{d}(i)$ denote the reference sequence and $\boldsymbol{u}_i$ denote the regressor sequence. We then assume that the data $\{\boldsymbol{d}(i), \boldsymbol{u}_i\}$ satisfy the following conditions:

(a) There exists a vector w^o such that $\boldsymbol{d}(i) = \boldsymbol{u}_i w^o + \boldsymbol{v}(i)$.
(b) The noise sequence $\{\boldsymbol{v}(i)\}$ is i.i.d. with variance $\sigma_v^2 = \mathsf{E}\,|\boldsymbol{v}(i)|^2$.
(c) The noise sequence $\{\boldsymbol{v}(i)\}$ is independent of $\boldsymbol{u}_j$ for all i, j.
(d) The initial condition $\boldsymbol{w}_{-1}$ is independent of all $\{\boldsymbol{d}(j), \boldsymbol{u}_j, \boldsymbol{v}(j)\}$.
(e) The regressor covariance matrix is denoted by $R_u = \mathsf{E}\,\boldsymbol{u}_i^* \boldsymbol{u}_i > 0$.
(f) The random variables $\{\boldsymbol{v}(i), \boldsymbol{u}_i\}$ have zero means.

(22.1)

In the above list, $\boldsymbol{w}_{-1}$ denotes the initial condition for an adaptive filter and it is assumed to be independent of all data. Moreover, in accordance with our convention, the $\boldsymbol{u}_i$ are taken as row vectors. All other vectors are column vectors. The unknown weight vector w^o and the adaptive filter weight vector $\boldsymbol{w}_i$ both have dimensions $M \times 1$. We could have adopted a more general data model than (22.1), such as the nonstationary model of Sec. 20.2. However, it will become clear that the arguments developed in this chapter for the stationary case can be extended to nonstationary models in a straightforward manner (see, e.g., Probs. V.33 and V.34).

22.2 DATA-NORMALIZED ADAPTIVE FILTERS

We shall study filter updates of the form

$$\boxed{\boldsymbol{w}_i = \boldsymbol{w}_{i-1} + \mu \frac{\boldsymbol{u}_i^*}{g[\boldsymbol{u}_i]} \boldsymbol{e}(i), \quad i \geq 0} \tag{22.2}$$

where $\boldsymbol{w}_i$ is an estimate for w^o at iteration i, μ is the step-size,

$$\boxed{\boldsymbol{e}(i) = \boldsymbol{d}(i) - \boldsymbol{u}_i \boldsymbol{w}_{i-1}} \tag{22.3}$$

is the estimation error, and $g[\cdot]$ is some positive-valued function of $\boldsymbol{u}_i$. For example, the choice $g[\boldsymbol{u}_i] = 1$ results in the LMS algorithm, while $g[\boldsymbol{u}_i] = \epsilon + \|\boldsymbol{u}_i\|^2$ results in the $\epsilon -$ NLMS algorithm. One could also study more general data-normalized updates of the form

$$\boldsymbol{w}_i = \boldsymbol{w}_{i-1} + \mu H[\boldsymbol{u}_i] \boldsymbol{u}_i^* \boldsymbol{e}(i), \quad i \geq 0 \tag{22.4}$$

where $H[\cdot]$ is some Hermitian positive-definite matrix-valued (as opposed to scalar-valued) function of $\boldsymbol{u}_i$. Examples of choices for $H[\cdot]$, leading to more general forms of adaptive filters, are treated in the problems at the end of this part (see Probs. V.22, V.23 and V.36).

22.3 WEIGHTED ENERGY CONSERVATION RELATION

Since we shall deal with weighted vector norms on a regular basis in this chapter, we adopt the compact notation $\|x\|_\Sigma^2$ to refer to the weighted squared Euclidean norm of a vector x, i.e.,

$$\boxed{\|x\|_\Sigma^2 \;\stackrel{\Delta}{=}\; x^* \Sigma x}$$

for some Hermitian positive-definite weighting matrix Σ. The choice $\Sigma = \mathrm{I}$ results in the standard squared Euclidean norm of x, i.e.,

$$\|x\|_\mathrm{I}^2 = x^* x = \|x\|^2$$

Although we deal in general with the case $\Sigma > 0$, we shall use the same notation $\|x\|_\Sigma^2$ to denote $x^* \Sigma x$ even when Σ is non-negative definite – see Prob. V.1.

The need to consider weighted norms in the context of a transient analysis of adaptive filters can be motivated as follows. It will be seen that the transient performance of an adaptive filter requires that we study the time evolution of expectations of the form $\mathsf{E}\,\|\tilde{\boldsymbol{w}}_i\|^2$ and $\mathsf{E}\,|\boldsymbol{e}_a(i)|^2$, where the first expectation relates to the weight-error vector,

$$\tilde{\boldsymbol{w}}_i = w^o - \boldsymbol{w}_i$$

while the second expectation relates to the *a priori* estimation error,

$$e_a(i) = \boldsymbol{u}_i \tilde{\boldsymbol{w}}_{i-1}$$

The evaluation of the expectation $\mathsf{E}\,|e_a(i)|^2$ will in turn require that we evaluate a weighted norm of $\tilde{\boldsymbol{w}}_{i-1}$ of the form

$$\mathsf{E}\,\|\tilde{\boldsymbol{w}}_{i-1}\|^2_{R_u}$$

with the particular weighting matrix $\Sigma = R_u$. Now the energy conservation relation that we encountered in Thm. 15.1 involves the squared Euclidean norm of $\tilde{\boldsymbol{w}}_{i-1}$, $\|\tilde{\boldsymbol{w}}_{i-1}\|^2$, and not any weighted version of it. For this reason, we shall first extend our arguments to allow for weighted vector norms. As we shall see, a weighted version of the energy relation can be obtained rather immediately by following arguments similar to those that led to Thm. 15.1.

Thus, let Σ denote any $M \times M$ Hermitian nonnegative-definite matrix (in general, we shall have $\Sigma > 0$). Later we shall see that different choices for Σ are useful to infer different conclusions about the performance of an adaptive filter. Now define the weighted *a priori* and *a posteriori* error signals

$$\boxed{\boldsymbol{e}_a^{\Sigma}(i) \;\overset{\Delta}{=}\; \boldsymbol{u}_i \Sigma \tilde{\boldsymbol{w}}_{i-1}, \qquad \boldsymbol{e}_p^{\Sigma}(i) \;\overset{\Delta}{=}\; \boldsymbol{u}_i \Sigma \tilde{\boldsymbol{w}}_i} \tag{22.5}$$

where

$$\boxed{\tilde{\boldsymbol{w}}_i \;\overset{\Delta}{=}\; w^o - \boldsymbol{w}_i} \tag{22.6}$$

When $\Sigma = \mathrm{I}$, we recover the standard errors

$$\boxed{\boldsymbol{e}_a(i) \;\overset{\Delta}{=}\; \boldsymbol{e}_a^{I}(i) = \boldsymbol{u}_i \tilde{\boldsymbol{w}}_{i-1}, \qquad \boldsymbol{e}_p(i) \;\overset{\Delta}{=}\; \boldsymbol{e}_p^{I}(i) = \boldsymbol{u}_i \tilde{\boldsymbol{w}}_i} \tag{22.7}$$

The energy relation that we seek is one that compares the weighted energies of the error quantities:

$$\{\, \tilde{\boldsymbol{w}}_i, \tilde{\boldsymbol{w}}_{i-1}, \boldsymbol{e}_a^{\Sigma}(i), \boldsymbol{e}_p^{\Sigma}(i) \,\} \tag{22.8}$$

To arrive at the desired relation, we follow the same arguments that we employed before in Sec. 15.3.

First, we rewrite the update recursion (22.2)–(22.3) in terms of the weight-error vector $\tilde{\boldsymbol{w}}_i$. Subtracting both sides of (22.2) from w^o we get

$$\boxed{\tilde{\boldsymbol{w}}_i = \tilde{\boldsymbol{w}}_{i-1} - \mu \frac{\boldsymbol{u}_i^*}{g[\boldsymbol{u}_i]} \boldsymbol{e}(i)} \tag{22.9}$$

If we further multiply both sides of (22.9) by $\boldsymbol{u}_i \Sigma$ from the left we find that the *a priori* and *a posteriori* estimation errors $\{\boldsymbol{e}_p^{\Sigma}(i), \boldsymbol{e}_a^{\Sigma}(i)\}$ are related via

$$\boxed{\boldsymbol{e}_p^{\Sigma}(i) = \boldsymbol{e}_a^{\Sigma}(i) - \mu \frac{\|\boldsymbol{u}_i\|^2_{\Sigma}}{g[\boldsymbol{u}_i]} \boldsymbol{e}(i)} \tag{22.10}$$

We distinguish between two cases:

1. $\|\boldsymbol{u}_i\|^2_{\Sigma} \neq 0$. In this case, we use (22.10) to solve for $\boldsymbol{e}(i)/g[\boldsymbol{u}_i]$,

$$\frac{\boldsymbol{e}(i)}{g[\boldsymbol{u}_i]} = \frac{1}{\mu \|\boldsymbol{u}_i\|^2_{\Sigma}} \left[\boldsymbol{e}_a^{\Sigma}(i) - \boldsymbol{e}_p^{\Sigma}(i)\right]$$

and substitute into (22.9) to get

$$\tilde{w}_i + \frac{u_i^*}{\|u_i\|_\Sigma^2} e_a^\Sigma(i) = \tilde{w}_{i-1} + \frac{u_i^*}{\|u_i\|_\Sigma^2} e_p^\Sigma(i) \tag{22.11}$$

On each side of this identity we have a combination of *a priori* and *a posteriori* errors. By equating the weighted Euclidean norms of both sides of the equation, i.e., by setting

$$\left\|\tilde{w}_i + \frac{u_i^*}{\|u_i\|_\Sigma^2} e_a^\Sigma(i)\right\|_\Sigma^2 = \left\|\tilde{w}_{i-1} + \frac{u_i^*}{\|u_i\|_\Sigma^2} e_p^\Sigma(i)\right\|_\Sigma^2$$

we find, after a straightforward calculation, that the following energy equality holds:

$$\|\tilde{w}_i\|_\Sigma^2 + \frac{1}{\|u_i\|_\Sigma^2}|e_a^\Sigma(i)|^2 = \|\tilde{w}_{i-1}\|_\Sigma^2 + \frac{1}{\|u_i\|_\Sigma^2}|e_p^\Sigma(i)|^2$$

Observe that this equality simply amounts to adding the weighted energies of the individual terms of (22.11); the cross-terms cancel out. Equivalently, we can rewrite the above equality as

$$\boxed{\|\tilde{w}_i\|_\Sigma^2 + \bar{\mu}^\Sigma(i) \cdot |e_a^\Sigma(i)|^2 = \|\tilde{w}_{i-1}\|_\Sigma^2 + \bar{\mu}^\Sigma(i) \cdot |e_p^\Sigma(i)|^2} \tag{22.12}$$

where

$$\bar{\mu}^\Sigma(i) \triangleq \begin{cases} 1/\|u_i\|_\Sigma^2 & \text{if } \|u_i\|_\Sigma^2 \neq 0 \\ 0 & \text{otherwise} \end{cases} \tag{22.13}$$

2. $\|u_i\|_\Sigma^2 = 0$ (since $\Sigma \geq 0$, this implies that $u_i\Sigma = 0$). In this case, it is obvious from (22.10) that $e_a^\Sigma(i) = e_p^\Sigma(i)$ and from (22.9) that $\|\tilde{w}_i\|_\Sigma^2 = \|\tilde{w}_{i-1}\|_\Sigma^2$, so that (22.12) is valid again.

In summary, we arrive at the following statement.

Theorem 22.1 (Weighted energy-conservation relation) For any adaptive filter of the form (22.2), any Hermitian nonnegative-definite matrix Σ, and for any data $\{d(i), u_i\}$, it holds that

$$\|\tilde{w}_i\|_\Sigma^2 + \bar{\mu}^\Sigma(i) \cdot |e_a^\Sigma(i)|^2 = \|\tilde{w}_{i-1}\|_\Sigma^2 + \bar{\mu}^\Sigma(i) \cdot |e_p^\Sigma(i)|^2$$

where $e_a^\Sigma(i) = u_i\Sigma\tilde{w}_{i-1}$, $e_p^\Sigma(i) = u_i\Sigma\tilde{w}_i$, $\tilde{w}_i = w^o - w_i$, and $\bar{\mu}^\Sigma(i)$ is defined by (22.13).

The important fact to emphasize here is that *no approximations* have been used to establish the energy relation (22.12); it is an exact relation that shows how the energies of the weight-error vectors at two successive time instants are related to the energies of the *a priori* and *a posteriori* estimation errors. The special choice $\Sigma = \text{I}$ reduces to the energy relation of Thm. 15.1. In addition, the same geometric, physical, and system-theoretic interpretations that were presented in App. 15.A for the case $\Sigma = \text{I}$ can be extended to the weighted case with little effort — see App. 9.G of Sayed (2003).

We also rewrite below, for later reference, the weight-error recursion (22.9). From the modeling assumption (22.1) we have

$$e(i) = d(i) - u_i w_{i-1} = u_i\tilde{w}_{i-1} + v(i)$$

so that substituting into (22.9) we get the equivalent form

$$\tilde{\boldsymbol{w}}_i = \left(\mathrm{I} - \mu\frac{\boldsymbol{u}_i^*\boldsymbol{u}_i}{g[\boldsymbol{u}_i]}\right)\tilde{\boldsymbol{w}}_{i-1} - \mu\frac{\boldsymbol{u}_i^*}{g[\boldsymbol{u}_i]}\boldsymbol{v}(i) \tag{22.14}$$

Theorem 22.2 (Weight-error recursion) For any adaptive filter of the form (22.2), and for data $\{\boldsymbol{d}(i), \boldsymbol{u}_i\}$ satisfying (22.1), it holds that

$$\tilde{\boldsymbol{w}}_i = \left(\mathrm{I} - \mu\frac{\boldsymbol{u}_i^*\boldsymbol{u}_i}{g[\boldsymbol{u}_i]}\right)\tilde{\boldsymbol{w}}_{i-1} - \mu\frac{\boldsymbol{u}_i^*}{g[\boldsymbol{u}_i]}\boldsymbol{v}(i)$$

where $\tilde{\boldsymbol{w}}_i = w^o - \boldsymbol{w}_i$.

22.4 WEIGHTED VARIANCE RELATION

Relation (22.12) has several useful ramifications in the study of adaptive filters, as was already discussed at some length in Part IV (*Mean-Square Performance*). In this part we focus on its significance to transient analysis. Thus, recall that in Sec. 15.2 we used the energy-conservation relation (15.32) to study the steady-state performance of adaptive filters. In that section, we invoked the steady-state condition

$$\mathsf{E}\,\|\tilde{\boldsymbol{w}}_i\|^2 = \mathsf{E}\,\|\tilde{\boldsymbol{w}}_{i-1}\|^2 \qquad \text{as } i \to \infty$$

in order to cancel the effect of the weight-error vector from both sides of the energy relation, and then used the resulting variance relation (15.40) to evaluate the filter EMSE, i.e., the value of $\mathsf{E}\,|\boldsymbol{e}_a(\infty)|^2$.

In transient analysis, on the other hand, we will be interested in the time evolution of $\mathsf{E}\,\|\tilde{\boldsymbol{w}}_i\|_\Sigma^2$ for some choices of interest for Σ (usually, $\Sigma = \mathrm{I}$ or $\Sigma = R_u$). For this reason, in transient analysis, rather than eliminate the effect of the weight-error vector from (22.12), the contributions of the other error quantities, $\{\boldsymbol{e}_a^\Sigma(i), \boldsymbol{e}_p^\Sigma(i)\}$, will instead be expressed in terms of the weight-error vector itself. In so doing, the energy relation (22.12) will be transformed into a recursion that describes the evolution of $\mathsf{E}\,\|\tilde{\boldsymbol{w}}_i\|_\Sigma^2$ — see, e.g., Eq. (22.21) further ahead.

Variance Relation

Thus, returning to (22.12), and replacing $\boldsymbol{e}_p^\Sigma(i)$ by its equivalent expression (22.10) in terms of $\boldsymbol{e}_a^\Sigma(i)$ and $\boldsymbol{e}(i)$ we get

$$\|\tilde{\boldsymbol{w}}_i\|_\Sigma^2 + \bar{\mu}^\Sigma(i)\cdot\left|\boldsymbol{e}_a^\Sigma(i)\right|^2 = \|\tilde{\boldsymbol{w}}_{i-1}\|_\Sigma^2 + \bar{\mu}^\Sigma(i)\cdot\left|\boldsymbol{e}_a^\Sigma(i) - \frac{\mu\cdot\|\boldsymbol{u}_i\|_\Sigma^2}{g[\boldsymbol{u}_i]}\,\boldsymbol{e}(i)\right|^2$$

or, equivalently, after expanding the rightmost term,

$$\begin{aligned}\|\tilde{\boldsymbol{w}}_i\|_\Sigma^2 &= \|\tilde{\boldsymbol{w}}_{i-1}\|_\Sigma^2 + \frac{\mu^2\|\boldsymbol{u}_i\|_\Sigma^2}{g^2[\boldsymbol{u}_i]}\,|\boldsymbol{e}(i)|^2 \\ &\quad - \frac{\mu}{g[\boldsymbol{u}_i]}\boldsymbol{e}_a^{\Sigma*}(i)\boldsymbol{e}(i) - \frac{\mu}{g[\boldsymbol{u}_i]}\boldsymbol{e}^*(i)\boldsymbol{e}_a^\Sigma(i)\end{aligned} \tag{22.15}$$

Alternatively, this result can be obtained by starting directly from the weight-error vector recursion (22.9) and equating the weighted Euclidean norms of both sides.

Now using the data model (22.1) for $\boldsymbol{d}(i)$, it is clear that $\boldsymbol{e}(i)$ is related to $\boldsymbol{e}_a(i)$ and $\boldsymbol{v}(i)$ as follows:

$$\boldsymbol{e}(i) = \boldsymbol{d}(i) - \boldsymbol{u}_i\boldsymbol{w}_{i-1} = (\boldsymbol{u}_i w^o + \boldsymbol{v}(i)) - \boldsymbol{u}_i\boldsymbol{w}_{i-1} \;=\; \boldsymbol{e}_a(i) + \boldsymbol{v}(i) \tag{22.16}$$

so that substituting into (22.15) we can eliminate $\boldsymbol{e}(i)$ and get

$$\begin{aligned}
\|\tilde{\boldsymbol{w}}_i\|_\Sigma^2 \;=\; & \|\tilde{\boldsymbol{w}}_{i-1}\|_\Sigma^2 \\
& + \frac{\mu^2\|\boldsymbol{u}_i\|_\Sigma^2}{g^2[\boldsymbol{u}_i]}\,|\boldsymbol{e}_a(i)|^2 + \frac{\mu^2\|\boldsymbol{u}_i\|_\Sigma^2}{g^2[\boldsymbol{u}_i]}\,|\boldsymbol{v}(i)|^2 \\
& + \frac{\mu^2\|\boldsymbol{u}_i\|_\Sigma^2}{g^2[\boldsymbol{u}_i]}\,\boldsymbol{v}(i)\boldsymbol{e}_a^*(i) + \frac{\mu^2\|\boldsymbol{u}_i\|_\Sigma^2}{g^2[\boldsymbol{u}_i]}\,\boldsymbol{v}^*(i)\boldsymbol{e}_a(i) \\
& - \frac{\mu}{g[\boldsymbol{u}_i]}\boldsymbol{e}_a^{\Sigma*}(i)\boldsymbol{e}_a(i) - \frac{\mu}{g[\boldsymbol{u}_i]}\boldsymbol{v}(i)\boldsymbol{e}_a^{\Sigma*}(i) \\
& - \frac{\mu}{g[\boldsymbol{u}_i]}\boldsymbol{e}_a^*(i)\boldsymbol{e}_a^\Sigma(i) - \frac{\mu}{g[\boldsymbol{u}_i]}\boldsymbol{v}^*(i)\boldsymbol{e}_a^\Sigma(i)
\end{aligned} \tag{22.17}$$

Most of the factors in this equality disappear under expectation, while other factors can be expressed in terms of $\tilde{\boldsymbol{w}}_{i-1}$. To see this, observe first that

$$\boldsymbol{e}_a^*(i)\boldsymbol{e}_a^\Sigma(i) = \tilde{\boldsymbol{w}}_{i-1}^*\boldsymbol{u}_i^*\boldsymbol{u}_i\Sigma\tilde{\boldsymbol{w}}_{i-1}$$

and

$$\boldsymbol{e}_a^{\Sigma*}(i)\boldsymbol{e}_a(i) = \tilde{\boldsymbol{w}}_{i-1}^*\Sigma\boldsymbol{u}_i^*\boldsymbol{u}_i\tilde{\boldsymbol{w}}_{i-1}$$

so that the first terms on the fourth and fifth lines of (22.17) can be grouped together as a single weighted norm of $\tilde{\boldsymbol{w}}_{i-1}$ as follows:

$$\frac{\mu}{g[\boldsymbol{u}_i]}\boldsymbol{e}_a^{\Sigma*}(i)\boldsymbol{e}_a(i) + \frac{\mu}{g[\boldsymbol{u}_i]}\boldsymbol{e}_a^*(i)\boldsymbol{e}_a^\Sigma(i) \;=\; \frac{\mu}{g[\boldsymbol{u}_i]}\cdot\|\tilde{\boldsymbol{w}}_{i-1}\|^2_{\Sigma\boldsymbol{u}_i^*\boldsymbol{u}_i+\boldsymbol{u}_i^*\boldsymbol{u}_i\Sigma}$$

Likewise,

$$|\boldsymbol{e}_a(i)|^2 = \tilde{\boldsymbol{w}}_{i-1}^*\boldsymbol{u}_i^*\boldsymbol{u}_i\tilde{\boldsymbol{w}}_{i-1} = \|\tilde{\boldsymbol{w}}_{i-1}\|^2_{\boldsymbol{u}_i^*\boldsymbol{u}_i}$$

so that (22.17) becomes

$$\begin{aligned}
\|\tilde{\boldsymbol{w}}_i\|_\Sigma^2 \;=\; & \|\tilde{\boldsymbol{w}}_{i-1}\|_\Sigma^2 \\
& + \frac{\mu^2\|\boldsymbol{u}_i\|_\Sigma^2}{g^2[\boldsymbol{u}_i]}\,\|\tilde{\boldsymbol{w}}_{i-1}\|^2_{\boldsymbol{u}_i^*\boldsymbol{u}_i} \\
& + \frac{\mu^2\|\boldsymbol{u}_i\|_\Sigma^2}{g^2[\boldsymbol{u}_i]}\,|\boldsymbol{v}(i)|^2 \\
& - \frac{\mu}{g[\boldsymbol{u}_i]}\,\|\tilde{\boldsymbol{w}}_{i-1}\|^2_{\Sigma\boldsymbol{u}_i^*\boldsymbol{u}_i+\boldsymbol{u}_i^*\boldsymbol{u}_i\Sigma} \\
& + \frac{\mu^2\|\boldsymbol{u}_i\|_\Sigma^2}{g^2[\boldsymbol{u}_i]}\,\boldsymbol{v}(i)\boldsymbol{e}_a^*(i) + \frac{\mu^2\|\boldsymbol{u}_i\|_\Sigma^2}{g^2[\boldsymbol{u}_i]}\,\boldsymbol{v}^*(i)\boldsymbol{e}_a(i) \\
& - \frac{\mu}{g[\boldsymbol{u}_i]}\boldsymbol{v}(i)\boldsymbol{e}_a^{\Sigma*}(i) - \frac{\mu}{g[\boldsymbol{u}_i]}\boldsymbol{v}^*(i)\boldsymbol{e}_a^\Sigma(i)
\end{aligned} \tag{22.18}$$

Taking expectations of both sides of (22.18) we find

$$\begin{aligned}\mathsf{E}\,\|\tilde{\boldsymbol{w}}_i\|^2_\Sigma \;&=\; \mathsf{E}\,\|\tilde{\boldsymbol{w}}_{i-1}\|^2_\Sigma \;+\; \mathsf{E}\left(\frac{\mu^2\|\boldsymbol{u}_i\|^2_\Sigma}{g^2[\boldsymbol{u}_i]}\,\|\tilde{\boldsymbol{w}}_{i-1}\|^2_{\boldsymbol{u}_i^*\boldsymbol{u}_i}\right)\\ &\quad -\,\mathsf{E}\left(\frac{\mu}{g[\boldsymbol{u}_i]}\,\|\tilde{\boldsymbol{w}}_{i-1}\|^2_{\Sigma\boldsymbol{u}_i^*\boldsymbol{u}_i+\boldsymbol{u}_i^*\boldsymbol{u}_i\Sigma}\right) \;+\; \sigma_v^2\mathsf{E}\left(\frac{\mu^2\|\boldsymbol{u}_i\|^2_\Sigma}{g^2[\boldsymbol{u}_i]}\right)\end{aligned} \tag{22.19}$$

where the expectations of the cross-terms involving $\boldsymbol{v}(i)$ evaluate to zero due to the modeling assumption on $\boldsymbol{v}(i)$ from (22.1), namely, that $\boldsymbol{v}(i)$ is zero-mean, i.i.d., and independent of $\boldsymbol{u}_j$ for all j. Alternatively, we can obtain (22.19) by starting from the weight-error recursion (22.9), equating the weighted norms of both sides, and taking expectations to arrive at (22.19) — see, e.g., the result of Prob. V.27 specialized to $\alpha = 0$.

We can further include some of the multiplicative factors in (22.19) into the weighting matrices and write

$$\begin{aligned}\mathsf{E}\,\|\tilde{\boldsymbol{w}}_i\|^2_\Sigma \;&=\; \mathsf{E}\,\|\tilde{\boldsymbol{w}}_{i-1}\|^2_\Sigma \;+\; \mathsf{E}\left(\|\tilde{\boldsymbol{w}}_{i-1}\|^2_{\frac{\mu^2\|\boldsymbol{u}_i\|^2_\Sigma}{g^2[\boldsymbol{u}_i]}\boldsymbol{u}_i^*\boldsymbol{u}_i}\right)\\ &\quad -\,\mathsf{E}\left(\|\tilde{\boldsymbol{w}}_{i-1}\|^2_{\frac{\mu}{g[\boldsymbol{u}_i]}(\Sigma\boldsymbol{u}_i^*\boldsymbol{u}_i+\boldsymbol{u}_i^*\boldsymbol{u}_i\Sigma)}\right) \;+\; \sigma_v^2\mathsf{E}\left(\frac{\mu^2\|\boldsymbol{u}_i\|^2_\Sigma}{g^2[\boldsymbol{u}_i]}\right)\end{aligned}$$

This equality can be written more compactly as follows. Introduce the random weighting matrix

$$\boxed{\boldsymbol{\Sigma}' \;\triangleq\; \Sigma \;-\; \frac{\mu}{g[\boldsymbol{u}_i]}\Sigma\boldsymbol{u}_i^*\boldsymbol{u}_i \;-\; \frac{\mu}{g[\boldsymbol{u}_i]}\boldsymbol{u}_i^*\boldsymbol{u}_i\Sigma \;+\; \frac{\mu^2\|\boldsymbol{u}_i\|^2_\Sigma}{g^2[\boldsymbol{u}_i]}\boldsymbol{u}_i^*\boldsymbol{u}_i} \tag{22.20}$$

In accordance with our convention, we are using a boldface symbol $\boldsymbol{\Sigma}'$ since $\boldsymbol{\Sigma}'$ is a random quantity (owing to its dependence on the regressor $\boldsymbol{u}_i$). Then

$$\boxed{\mathsf{E}\,\|\tilde{\boldsymbol{w}}_i\|^2_\Sigma = \mathsf{E}\left(\|\tilde{\boldsymbol{w}}_{i-1}\|^2_{\boldsymbol{\Sigma}'}\right) \;+\; \mu^2\sigma_v^2\mathsf{E}\left(\frac{\|\boldsymbol{u}_i\|^2_\Sigma}{g^2[\boldsymbol{u}_i]}\right)} \tag{22.21}$$

In other words, starting from the energy conservation relation (22.12), expanding it, and expressing whatever factors possible in terms of $\tilde{\boldsymbol{w}}_{i-1}$, and then taking expectations to eliminate cross-terms involving the noise variable $\boldsymbol{v}(i)$, we arrive at the variance relation (22.21). This relation shows how the weighted mean-square norm of $\tilde{\boldsymbol{w}}_i$ propagates in time. In particular, observe the important fact that the weighting matrices at the time instants i and $i-1$ are distinct and related via (22.20). Moreover, it is shown in Prob. V.2 that $\boldsymbol{\Sigma}' \geq 0$ when $\Sigma > 0$.

Theorem 22.3 (Weighted variance relation) For adaptive filters of the form (22.2), Hermitian nonnegative-definite matrices Σ, and for data $\{\boldsymbol{d}(i), \boldsymbol{u}_i\}$ satisfying model (22.1), it holds that

$$\begin{aligned}\mathsf{E}\,\|\tilde{\boldsymbol{w}}_i\|^2_\Sigma \;&=\; \mathsf{E}\left(\|\tilde{\boldsymbol{w}}_{i-1}\|^2_{\boldsymbol{\Sigma}'}\right) \;+\; \mu^2\sigma_v^2\mathsf{E}\left(\frac{\|\boldsymbol{u}_i\|^2_\Sigma}{g^2[\boldsymbol{u}_i]}\right)\\ \boldsymbol{\Sigma}' \;&=\; \Sigma \;-\; \mu\Sigma\frac{\boldsymbol{u}_i^*\boldsymbol{u}_i}{g[\boldsymbol{u}_i]} \;-\; \mu\frac{\boldsymbol{u}_i^*\boldsymbol{u}_i}{g[\boldsymbol{u}_i]}\Sigma \;+\; \mu^2\frac{\|\boldsymbol{u}_i\|^2_\Sigma}{g^2[\boldsymbol{u}_i]}\boldsymbol{u}_i^*\boldsymbol{u}_i\end{aligned}$$

Independence Assumption

The recursion of Thm. 22.3 provides a compact characterization of the time-evolution (i.e., dynamics) of the expectation $\mathsf{E}\,\|\tilde{\boldsymbol{w}}_i\|^2_{\Sigma}$. However, more is needed in order for this recursion to permit a tractable transient analysis. This is because the recursion for $\mathsf{E}\,\|\tilde{\boldsymbol{w}}_i\|^2_{\Sigma}$ is hard to propagate as it stands due to the presence of the expectation

$$\mathsf{E}\left(\|\tilde{\boldsymbol{w}}_{i-1}\|^2_{\boldsymbol{\Sigma}'}\right) \;=\; \mathsf{E}\left(\tilde{\boldsymbol{w}}^*_{i-1}\boldsymbol{\Sigma}'\tilde{\boldsymbol{w}}_{i-1}\right) \tag{22.22}$$

This expectation is difficult to evaluate because of the dependence of $\boldsymbol{\Sigma}'$ on $\boldsymbol{u}_i$, and the dependence of $\tilde{\boldsymbol{w}}_{i-1}$ on prior regressors (so that $\tilde{\boldsymbol{w}}_{i-1}$ and $\boldsymbol{\Sigma}'$ are themselves dependent). These dependencies are among the most challenging hurdles in the transient analysis of adaptive filters. One common way to overcome them is to resort to an independence assumption on the regression sequence $\{\boldsymbol{u}_i\}$, namely to assume that

$$\boxed{\text{The sequence } \{\boldsymbol{u}_i\} \text{ is independent and identically distributed}} \tag{22.23}$$

Actually, it is customary in the literature to start the transient analysis of an adaptive filter with the collection of independence assumptions that were described in Sec. 16.4, which included, in addition to the independence condition (22.23), a Gaussian requirement on $\boldsymbol{u}_i$ as well. Our arguments will not require the regressors to be Gaussian.

Although invalid in general, especially for tapped-delay-line implementations, the independence assumption (22.23) is widely used due to the simplifications it introduces into the arguments. Without (22.23), the study of the transient behavior of adaptive filters can become highly challenging, even for simple algorithms. However, there are results in the literature that show that performance results that are obtained using the independence assumption are reasonably close to actual filter performance when the step-size is sufficiently small. We discuss some of these results in App. 24.A. It is for this reason, and in order to simplify the presentation in the chapter, that we shall continue our discussions by using (22.23). Recall, in comparison, that in Part IV (*Mean-Square Performance*) we studied the steady-state and tracking performance of adaptive filters without resorting to the independence assumption. This observation explains why we have opted to order Chapters 15–22 in their present order with Chapter 22 on transient analysis coming last. It is because transient analysis is usually more demanding in terms of conditions and restrictions on the data. In contrast, as was already shown in Part IV (*Mean-Square Performance*), much can be said about the performance of adaptive filters under less restrictive assumptions.

Now note that condition (22.23) guarantees that

$$\boxed{\tilde{\boldsymbol{w}}_{i-1} \text{ is independent of both } \boldsymbol{\Sigma}' \text{ and } \boldsymbol{u}_i} \tag{22.24}$$

This is because $\tilde{\boldsymbol{w}}_{i-1}$ is a function of past regressors and noise, $\{\boldsymbol{u}_j, \boldsymbol{v}(j), j < i\}$ (cf. the explanation after (16.12)), while $\boldsymbol{\Sigma}'$ is a function of $\boldsymbol{u}_i$ alone. Using (22.24) we can then split the expectation $\mathsf{E}\left(\|\tilde{\boldsymbol{w}}_{i-1}\|^2_{\boldsymbol{\Sigma}'}\right)$ into

$$\mathsf{E}\left(\|\tilde{\boldsymbol{w}}_{i-1}\|^2_{\boldsymbol{\Sigma}'}\right) \;=\; \mathsf{E}\left(\|\tilde{\boldsymbol{w}}_{i-1}\|^2_{\mathsf{E}\,[\boldsymbol{\Sigma}']}\right) \tag{22.25}$$

with the weighting matrix $\boldsymbol{\Sigma}'$ replaced by its mean, and which we denote by Σ', i.e.,

$$\boxed{\Sigma' \;\stackrel{\Delta}{=}\; \mathsf{E}\,[\boldsymbol{\Sigma}']}$$

Equality (22.25) follows from the identities

$$\begin{aligned}
\mathsf{E}\left(\|\tilde{\boldsymbol{w}}_{i-1}\|^2_{\boldsymbol{\Sigma}'}\right) &= \mathsf{E}\,\tilde{\boldsymbol{w}}^*_{i-1}\boldsymbol{\Sigma}'\tilde{\boldsymbol{w}}_{i-1} \\
&= \mathsf{E}\left(\mathsf{E}\left[\tilde{\boldsymbol{w}}^*_{i-1}\boldsymbol{\Sigma}'\tilde{\boldsymbol{w}}_{i-1}|\tilde{\boldsymbol{w}}_{i-1}\right]\right) \\
&= \mathsf{E}\,\tilde{\boldsymbol{w}}^*_{i-1}\left[\mathsf{E}\left(\boldsymbol{\Sigma}'|\tilde{\boldsymbol{w}}_{i-1}\right)\right]\tilde{\boldsymbol{w}}_{i-1} \\
&= \mathsf{E}\left(\tilde{\boldsymbol{w}}^*_{i-1}(\mathsf{E}\,\boldsymbol{\Sigma}')\tilde{\boldsymbol{w}}_{i-1}\right) \quad \text{because of (22.24)} \\
&= \mathsf{E}\left(\|\tilde{\boldsymbol{w}}_{i-1}\|^2_{\mathsf{E}\,[\boldsymbol{\Sigma}']}\right)
\end{aligned}$$

Thus, observe that the main value of the independence assumption (22.23) lies in guaranteeing that $\tilde{\boldsymbol{w}}_{i-1}$ is independent of $\boldsymbol{\Sigma}'$, in which case it is possible to use (22.25) and thereby simplify the subsequent derivations. We can see from the expression for $\boldsymbol{\Sigma}'$ in (22.20) that this same conclusion will hold if we replace condition (22.23) by the assumption that $\tilde{\boldsymbol{w}}_{i-1}$ is independent of $\boldsymbol{u}_i^*\boldsymbol{u}_i/g[\boldsymbol{u}_i]$.

In this way, recursion (22.21) is replaced by

$$\boxed{\mathsf{E}\,\|\tilde{\boldsymbol{w}}_i\|^2_{\Sigma} = \mathsf{E}\,\|\tilde{\boldsymbol{w}}_{i-1}\|^2_{\Sigma'} \;+\; \mu^2\sigma_v^2\mathsf{E}\left(\frac{\|\boldsymbol{u}_i\|^2_{\Sigma}}{g^2[\boldsymbol{u}_i]}\right)} \tag{22.26}$$

with two *deterministic* (not random) weighting matrices $\{\Sigma, \Sigma'\}$ and where, by evaluating the expectation of (22.20),

$$\boxed{\Sigma' \triangleq \Sigma \;-\; \mu\Sigma\mathsf{E}\left(\frac{\boldsymbol{u}_i^*\boldsymbol{u}_i}{g[\boldsymbol{u}_i]}\right) \;-\; \mu\mathsf{E}\left(\frac{\boldsymbol{u}_i^*\boldsymbol{u}_i}{g[\boldsymbol{u}_i]}\right)\Sigma \;+\; \mu^2\mathsf{E}\left(\frac{\|\boldsymbol{u}_i\|^2_{\Sigma}}{g^2[\boldsymbol{u}_i]}\boldsymbol{u}_i^*\boldsymbol{u}_i\right)} \tag{22.27}$$

It is further argued in Prob. V.2 that $\Sigma' \geq 0$ when $\Sigma > 0$.

Theorem 22.4 (Weighted variance relation with independence) For adaptive filters of the form (22.2), Hermitian nonnegative-definite matrices Σ, and for data $\{\boldsymbol{d}(i), \boldsymbol{u}_i\}$ satisfying model (22.1) and the independence assumption (22.23), it holds that:

$$\begin{aligned}
\mathsf{E}\,\|\tilde{\boldsymbol{w}}_i\|^2_{\Sigma} &= \mathsf{E}\,\|\tilde{\boldsymbol{w}}_{i-1}\|^2_{\Sigma'} \;+\; \mu^2\sigma_v^2\mathsf{E}\left(\frac{\|\boldsymbol{u}_i\|^2_{\Sigma}}{g^2[\boldsymbol{u}_i]}\right) \\
\Sigma' &= \Sigma \;-\; \mu\Sigma\mathsf{E}\left(\frac{\boldsymbol{u}_i^*\boldsymbol{u}_i}{g[\boldsymbol{u}_i]}\right) \;-\; \mu\mathsf{E}\left(\frac{\boldsymbol{u}_i^*\boldsymbol{u}_i}{g[\boldsymbol{u}_i]}\right)\Sigma \;+\; \mu^2\mathsf{E}\left(\frac{\|\boldsymbol{u}_i\|^2_{\Sigma}}{g^2[\boldsymbol{u}_i]}\boldsymbol{u}_i^*\boldsymbol{u}_i\right)
\end{aligned}$$

Observe that the expression for Σ' is data dependent only; i.e., it depends on $\boldsymbol{u}_i$ alone and does not depend on the weight-error vectors. In this way, the recursion for $\|\tilde{\boldsymbol{w}}_i\|^2_{\Sigma}$ is decoupled from the computation of Σ' in the statement of theorem. Moreover, the expressions for $\mathsf{E}\,\|\tilde{\boldsymbol{w}}_i\|^2_{\Sigma}$ and Σ' show that studying the transient behavior of an adaptive filter requires evaluating the three multivariate moments:

$$\boxed{\mathsf{E}\left(\frac{\|\boldsymbol{u}_i\|^2_{\Sigma}}{g^2[\boldsymbol{u}_i]}\right), \quad \mathsf{E}\left(\frac{\boldsymbol{u}_i^*\boldsymbol{u}_i}{g[\boldsymbol{u}_i]}\right) \quad \text{and} \quad \mathsf{E}\left(\frac{\|\boldsymbol{u}_i\|^2_{\Sigma}}{g^2[\boldsymbol{u}_i]}\boldsymbol{u}_i^*\boldsymbol{u}_i\right)} \tag{22.28}$$

which are solely dependent on $\boldsymbol{u}_i$. Note further that the last moment in the above list appears multiplied by μ^2 in the expression for Σ'. What this means is that sometimes, when the step-size is sufficiently small, this last moment could be ignored in lieu of simplification; see the remark following Thm. 24.1 and also Probs. V.14 and V.38, where this observation is pursued in greater detail.

Finally, taking expectations of both sides of (22.14), using (22.23) and the fact that $\boldsymbol{v}(i)$ is independent of $\boldsymbol{u}_i$, we obtain the following result for the evolution of the mean of the weight-error vector.

Theorem 22.5 (Mean weight error recursion) For any adaptive filter of the form (22.2), and for any data $\{\boldsymbol{d}(i), \boldsymbol{u}_i\}$ satisfying (22.1) and (22.23), it holds that

$$\mathsf{E}\,\tilde{\boldsymbol{w}}_i = \left[\mathrm{I} - \mu\mathsf{E}\left(\frac{\boldsymbol{u}_i^*\boldsymbol{u}_i}{g[\boldsymbol{u}_i]}\right)\right]\mathsf{E}\,\tilde{\boldsymbol{w}}_{i-1} \tag{22.29}$$

Convenient Change of Coordinates

Evaluation of the moments (22.28), and the subsequent analysis, can at times be simplified if we introduce a convenient change of coordinates by appealing to the eigendecomposition of $R_u = \mathsf{E}\,\boldsymbol{u}_i^*\boldsymbol{u}_i$. So let

$$\boxed{R_u = U\Lambda U^*} \tag{22.30}$$

where Λ is diagonal with the eigenvalues of R_u, $\Lambda = \text{diag}\{\lambda_k\}$, and U is unitary (i.e., it satisfies $UU^* = U^*U = \mathrm{I}$). Then define the transformed quantities:

$$\boxed{\overline{\boldsymbol{w}}_i \overset{\Delta}{=} U^*\tilde{\boldsymbol{w}}_i, \qquad \overline{\boldsymbol{u}}_i \overset{\Delta}{=} \boldsymbol{u}_i U, \qquad \overline{\Sigma} \overset{\Delta}{=} U^*\Sigma U} \tag{22.31}$$

Since U is unitary, it is easy to see that

$$\|\tilde{\boldsymbol{w}}_i\|_\Sigma^2 = \|\overline{\boldsymbol{w}}_i\|_{\overline{\Sigma}}^2 \quad \text{and} \quad \|\boldsymbol{u}_i\|_\Sigma^2 = \|\overline{\boldsymbol{u}}_i\|_{\overline{\Sigma}}^2 \tag{22.32}$$

For example,

$$\|\overline{\boldsymbol{w}}_i\|_{\overline{\Sigma}}^2 = \overline{\boldsymbol{w}}_i^*\overline{\Sigma}\overline{\boldsymbol{w}}_i = (\tilde{\boldsymbol{w}}_i^*U)\cdot(U^*\Sigma U)\cdot(U^*\tilde{\boldsymbol{w}}_i) = \tilde{\boldsymbol{w}}_i^*\Sigma\tilde{\boldsymbol{w}}_i = \|\tilde{\boldsymbol{w}}_i\|_\Sigma^2$$

Likewise, for $\|\overline{\boldsymbol{u}}_i\|_{\overline{\Sigma}}^2$. In the special case $\Sigma = \mathrm{I}$, we have $\overline{\Sigma} = \mathrm{I}$, $\|\overline{\boldsymbol{w}}_i\|^2 = \|\tilde{\boldsymbol{w}}_i\|^2$, and $\|\overline{\boldsymbol{u}}_i\|^2 = \|\boldsymbol{u}_i\|^2$.

Now under the change of variables (22.31), the variance relation of Thm. 22.4 retains the same form, namely

$$\boxed{\mathsf{E}\,\|\overline{\boldsymbol{w}}_i\|_{\overline{\Sigma}}^2 = \mathsf{E}\,\|\overline{\boldsymbol{w}}_{i-1}\|_{\overline{\Sigma}'}^2 + \mu^2\sigma_v^2\mathsf{E}\left[\frac{\|\overline{\boldsymbol{u}}_i\|_{\overline{\Sigma}}^2}{g^2[\overline{\boldsymbol{u}}_i]}\right]} \tag{22.33}$$

where

$$\overline{\Sigma}' = U^*\Sigma' U \tag{22.34}$$

The data nonlinearity $g[\cdot]$ is usually invariant under unitary transformations, i.e., $g[\boldsymbol{u}_i] = g[\overline{\boldsymbol{u}}_i]$. This property is obvious for LMS and $\epsilon-$NLMS where $g[\boldsymbol{u}_i] = 1$ and $g[\boldsymbol{u}_i] = \epsilon + \|\boldsymbol{u}_i\|^2$, respectively. However, the invariance property of $g[\cdot]$ is not necessary for our development; if it does not hold, then we would simply continue to work with $g[\boldsymbol{u}_i]$ instead of $g[\overline{\boldsymbol{u}}_i]$.

Continuing, from the equation for Σ' in Thm. 22.4 we find that

$$\boxed{\overline{\Sigma}' = \overline{\Sigma} - \mu\overline{\Sigma}\mathsf{E}\left[\frac{\overline{u}_i^*\overline{u}_i}{g[\overline{u}_i]}\right] - \mu\mathsf{E}\left[\frac{\overline{u}_i^*\overline{u}_i}{g[\overline{u}_i]}\right]\overline{\Sigma} + \mu^2\mathsf{E}\left[\frac{\|\overline{u}_i\|^2_{\overline{\Sigma}}}{g^2[\overline{u}_i]}\overline{u}_i^*\overline{u}_i\right]} \tag{22.35}$$

In the discussion that follows, we shall use either the standard relations (22.26)–(22.27) or their transformed versions (22.33)–(22.35); the transformed versions are particularly useful when the regressors u_i are Gaussian (as we shall see in Sec. 23.1).

Theorem 22.6 (Transformed weighted-variance relation) For any adaptive filter of the form (22.2), any Hermitian nonnegative-definite matrix Σ, and for data $\{d(i), u_i\}$ satisfying model (22.1) and the independence assumption (22.23), it holds that:

$$\begin{aligned} \mathsf{E}\,\|\overline{w}_i\|^2_{\overline{\Sigma}} &= \mathsf{E}\,\|\overline{w}_{i-1}\|^2_{\overline{\Sigma}'} + \mu^2\sigma_v^2\mathsf{E}\left[\frac{\|\overline{u}_i\|^2_{\overline{\Sigma}}}{g^2[\overline{u}_i]}\right] \\ \overline{\Sigma}' &= \overline{\Sigma} - \mu\overline{\Sigma}\mathsf{E}\left[\frac{\overline{u}_i^*\overline{u}_i}{g[\overline{u}_i]}\right] - \mu\mathsf{E}\left[\frac{\overline{u}_i^*\overline{u}_i}{g[\overline{u}_i]}\right]\overline{\Sigma} + \mu^2\mathsf{E}\left[\frac{\|\overline{u}_i\|^2_{\overline{\Sigma}}}{g^2[\overline{u}_i]}\overline{u}_i^*\overline{u}_i\right] \end{aligned}$$

where the transformed variables $\{\overline{w}_i, \overline{u}_i, \overline{\Sigma}, \overline{\Sigma}'\}$ are related to the original variables $\{\tilde{w}_i, u_i, \Sigma, \Sigma'\}$ via (22.31) and (22.34), so that $\mathsf{E}\,\|\overline{w}_i\|^2_{\overline{\Sigma}} = \mathsf{E}\,\|\tilde{w}_i\|^2_{\Sigma}$.

Likewise, the transformed version of the mean-weight error recursion (22.29) is

$$\boxed{\mathsf{E}\,\overline{w}_i = \left[\mathrm{I} - \mu\mathsf{E}\left(\frac{\overline{u}_i^*\overline{u}_i}{g[\overline{u}_i]}\right)\right]\mathsf{E}\,\overline{w}_{i-1}} \tag{22.36}$$

The purpose of the chapters that follow is to show how the above variance and mean relations can be used to characterize the transient performance of data-normalized adaptive filters. In particular, it will be seen that the freedom in selecting the weighting matrix Σ can be used to great advantage in deriving several performance measures.

CHAPTER 23

LMS with Gaussian Regressors

We use the mean and variance relations of the last chapter to study the transient performance of the LMS algorithm,

$$w_i = w_{i-1} + \mu u_i^* e(i) \tag{23.1}$$

for which the data normalization in (22.2) is given by

$$g[u_i] = 1 \tag{23.2}$$

In this case, relations (22.26)–(22.27) and (22.29) become

$$\begin{cases} \mathsf{E}\,\|\tilde{w}_i\|^2_{\Sigma} = \mathsf{E}\,\|\tilde{w}_{i-1}\|^2_{\Sigma'} \;+\; \mu^2\sigma_v^2 \mathsf{E}\,\|u_i\|^2_{\Sigma} \\ \Sigma' = \Sigma - \mu\Sigma\mathsf{E}\,[u_i^* u_i] - \mu\mathsf{E}\,[u_i^* u_i]\,\Sigma \;+\; \mu^2 \mathsf{E}\,\left[\|u_i\|^2_{\Sigma} u_i^* u_i\right] \\ \mathsf{E}\,\tilde{w}_i = [\mathrm{I} - \mu\mathsf{E}\,(u_i^* u_i)]\,\mathsf{E}\,\tilde{w}_{i-1} \end{cases} \tag{23.3}$$

We therefore need to evaluate the three moments:

$$\mathsf{E}\,u_i^* u_i, \qquad \mathsf{E}\,\|u_i\|^2_{\Sigma} \qquad \text{and} \qquad \mathsf{E}\,\|u_i\|^2_{\Sigma} u_i^* u_i \tag{23.4}$$

The first two moments are obvious, and can be evaluated regardless of any assumed distribution for the regression data since

$$\mathsf{E}\,u_i^* u_i \;=\; R_u \qquad \text{(by definition)} \tag{23.5}$$

$$\mathsf{E}\,\|u_i\|^2_{\Sigma} \;=\; \mathsf{E}\,u_i \Sigma u_i^* \;=\; \mathsf{E}\,\mathsf{Tr}(u_i^* u_i \Sigma) \;=\; \mathsf{Tr}(R_u \Sigma) \tag{23.6}$$

The difficulty lies in evaluating the last moment in (23.4). To do so, we shall treat two cases. First we treat the case of Gaussian regressors for which the last moment can be evaluated explicitly. Afterwards, we treat the general case of non-Gaussian regressors.

23.1 MEAN AND VARIANCE RELATIONS

We assume in this chapter that the regressors $\{u_i\}$ arise from a circular Gaussian distribution with covariance matrix R_u (cf. Sec. A.5).We say *circular* because we are treating the general case of complex-valued regressors; otherwise, the circularity assumption is not needed. In the Gaussian case, as we shall explain ahead following (23.11), it is more convenient to work with the transformed versions (22.33)–(22.35) and (22.36) of the variance and mean relations, which for LMS are given by

$$\begin{cases} \mathsf{E}\,\|\overline{\boldsymbol{w}}_i\|^2_{\overline{\Sigma}} = \mathsf{E}\,\|\overline{\boldsymbol{w}}_{i-1}\|^2_{\overline{\Sigma}'} \;+\; \mu^2\sigma_v^2\mathsf{E}\,\|\overline{\boldsymbol{u}}_i\|^2_{\overline{\Sigma}} \\ \overline{\Sigma}' = \overline{\Sigma} - \mu\overline{\Sigma}\mathsf{E}\left[\overline{\boldsymbol{u}}_i^*\overline{\boldsymbol{u}}_i\right] - \mu\mathsf{E}\left[\overline{\boldsymbol{u}}_i^*\overline{\boldsymbol{u}}_i\right]\overline{\Sigma} \;+\; \mu^2\mathsf{E}\left[\|\overline{\boldsymbol{u}}_i\|^2_{\overline{\Sigma}}\overline{\boldsymbol{u}}_i^*\overline{\boldsymbol{u}}_i\right] \\ \mathsf{E}\,\overline{\boldsymbol{w}}_i = \left[\mathrm{I} - \mu\mathsf{E}\left(\overline{\boldsymbol{u}}_i^*\overline{\boldsymbol{u}}_i\right)\right]\mathsf{E}\,\overline{\boldsymbol{w}}_{i-1} \end{cases} \tag{23.7}$$

The moments that we need to evaluate in the transformed domain are

$$\mathsf{E}\,\overline{\boldsymbol{u}}_i^*\overline{\boldsymbol{u}}_i, \qquad \mathsf{E}\,\|\overline{\boldsymbol{u}}_i\|^2_{\overline{\Sigma}} \qquad \text{and} \qquad \mathsf{E}\,\|\overline{\boldsymbol{u}}_i\|^2_{\overline{\Sigma}}\overline{\boldsymbol{u}}_i^*\overline{\boldsymbol{u}}_i \tag{23.8}$$

where the first two are again immediate to compute since

$$\mathsf{E}\;\overline{\boldsymbol{u}}_i^*\overline{\boldsymbol{u}}_i \;=\; \Lambda \tag{23.9}$$

and

$$\mathsf{E}\,\|\overline{\boldsymbol{u}}_i\|^2_{\overline{\Sigma}} = \mathsf{E}\,\overline{\boldsymbol{u}}_i\overline{\Sigma}\overline{\boldsymbol{u}}_i^* \;=\; \mathsf{E}\,\mathsf{Tr}(\overline{\boldsymbol{u}}_i^*\overline{\boldsymbol{u}}_i\overline{\Sigma}) \;=\; \mathsf{Tr}(\Lambda\overline{\Sigma}) \tag{23.10}$$

With regards to the last moment in (23.8), we use the fact that $\overline{\boldsymbol{u}}_i$ is circular Gaussian with a diagonal covariance matrix, and invoke the result of Lemma A.3, to write

$$\begin{aligned} \mathsf{E}\,\|\overline{\boldsymbol{u}}_i\|^2_{\overline{\Sigma}}\overline{\boldsymbol{u}}_i^*\overline{\boldsymbol{u}}_i \;&=\; \mathsf{E}\;\left(\overline{\boldsymbol{u}}_i\overline{\Sigma}\overline{\boldsymbol{u}}_i^*\right)\overline{\boldsymbol{u}}_i^*\overline{\boldsymbol{u}}_i \\ &=\; \mathsf{E}\;\overline{\boldsymbol{u}}_i^*\left(\overline{\boldsymbol{u}}_i\overline{\Sigma}\overline{\boldsymbol{u}}_i^*\right)\overline{\boldsymbol{u}}_i \\ &=\; \Lambda\mathsf{Tr}(\overline{\Sigma}\Lambda) \;+\; \Lambda\overline{\Sigma}\Lambda \end{aligned} \tag{23.11}$$

Recall that the statement of Lemma A.3 requires the variable $\boldsymbol{z}$ to have a diagonal covariance matrix, which explains why we introduced the transformed vector $\overline{\boldsymbol{u}}_i$ and the transformed relations (23.7) in the Gaussian regressor case. Moreover, if the regressors were real-valued rather than complex-valued, then we would invoke Lemma A.2 and use instead

$$\mathsf{E}\,\|\overline{\boldsymbol{u}}_i\|^2_{\overline{\Sigma}}\overline{\boldsymbol{u}}_i^*\overline{\boldsymbol{u}}_i \;=\; \Lambda\mathsf{Tr}(\overline{\Sigma}\Lambda) \;+\; 2\Lambda\overline{\Sigma}\Lambda$$

with an additional factor of 2 compared to (23.11).

Using (23.9)–(23.11), recursions (23.7) become

$$\mathsf{E}\,\|\overline{\boldsymbol{w}}_i\|^2_{\overline{\Sigma}} \;=\; \mathsf{E}\,\|\overline{\boldsymbol{w}}_{i-1}\|^2_{\overline{\Sigma}'} \;+\; \mu^2\sigma_v^2\mathsf{Tr}(\Lambda\overline{\Sigma}) \tag{23.12}$$

$$\overline{\Sigma}' \;=\; \overline{\Sigma} - \mu\overline{\Sigma}\Lambda - \mu\Lambda\overline{\Sigma} \;+\; \mu^2\left[\Lambda\mathsf{Tr}(\overline{\Sigma}\Lambda) \;+\; \Lambda\overline{\Sigma}\Lambda\right] \tag{23.13}$$

$$\mathsf{E}\,\overline{\boldsymbol{w}}_i \;=\; (\mathrm{I} - \mu\Lambda)\,\mathsf{E}\,\overline{\boldsymbol{w}}_{i-1} \tag{23.14}$$

Observe the interesting fact that $\overline{\Sigma}'$ will be diagonal if $\overline{\Sigma}$ is. Now since we are free to choose Σ and, therefore, $\overline{\Sigma}$, we can assume that $\overline{\Sigma}$ is diagonal. Under these conditions, it is possible to rewrite (23.13) in a more compact form in terms of the diagonal entries of $\{\overline{\Sigma}, \overline{\Sigma}'\}$.

Diagonal Notation

To do so, we define the vectors

$$\boxed{\overline{\sigma} \;\triangleq\; \mathrm{diag}\{\overline{\Sigma}\} \qquad \text{and} \qquad \lambda \;\triangleq\; \mathrm{diag}\{\Lambda\}} \tag{23.15}$$

That is, $\{\overline{\sigma}, \lambda\}$ are $M \times 1$ vectors with the diagonal entries of the corresponding matrices; $\overline{\sigma}$ contains the diagonal entries of $\overline{\Sigma}$, while λ contains the diagonal entries of Λ. Actually, in this book, we shall use the notation diag$\{\cdot\}$ in two directions, both of which will be obvious from the context. Writing diag$\{A\}$, for an arbitrary matrix A, extracts the diagonal entries of A into a vector. This is the convention we used in (23.15) to define $\{\overline{\sigma}, \lambda\}$. On the other hand, writing diag$\{a\}$ for a column vector a, results in a diagonal matrix whose entries are obtained from a. Therefore, we shall also write, whenever necessary,

$$\boxed{\overline{\Sigma} = \text{diag}\{\overline{\sigma}\} \qquad \text{and} \qquad \Lambda = \text{diag}\{\lambda\}} \tag{23.16}$$

in order to recover $\{\overline{\Sigma}, \Lambda\}$ from $\{\overline{\sigma}, \lambda\}$.

Linear Vector Relation

Now in terms of the vectors $\{\overline{\sigma}, \lambda\}$, it is easy to see that the matrix relation (23.13) is equivalent to the vector relation

$$\overline{\sigma}' = \left(\mathrm{I} - 2\mu\Lambda + \mu^2\Lambda^2\right)\overline{\sigma} \; + \; \mu^2\left(\lambda^{\mathsf{T}}\overline{\sigma}\right)\lambda$$

which can in turn be written more compactly as

$$\boxed{\overline{\sigma}' = \overline{F}\overline{\sigma}} \tag{23.17}$$

with an $M \times M$ coefficient matrix $\overline{F}$ defined by

$$\boxed{\overline{F} \stackrel{\Delta}{=} \left(\mathrm{I} - 2\mu\Lambda + \mu^2\Lambda^2\right) + \mu^2\lambda\lambda^{\mathsf{T}}} \tag{23.18}$$

Expression (23.17) shows that the relation between the diagonal elements of $\overline{\Sigma}$ and $\overline{\Sigma}'$ is actually *linear*. Moreover, since $\overline{\Sigma}' = \text{diag}\{\overline{\sigma}'\}$, the linear relation (23.17) translates into the matrix relation $\overline{\Sigma}' = \text{diag}\left\{\overline{F}\overline{\sigma}\right\}$.

Variance Relation

We can rewrite recursion (23.12) by using the vectors $\{\overline{\sigma}, \overline{\sigma}'\}$ instead of the matrices $\{\overline{\Sigma}, \overline{\Sigma}'\}$. Using (23.17) and the notation (23.16), recursion (23.12) is equivalent to

$$\mathsf{E}\,\|\overline{\boldsymbol{w}}_i\|^2_{\text{diag}\{\overline{\sigma}\}} \; = \; \mathsf{E}\,\|\overline{\boldsymbol{w}}_{i-1}\|^2_{\text{diag}\{\overline{F}\overline{\sigma}\}} \; + \; \mu^2\sigma_v^2(\lambda^{\mathsf{T}}\overline{\sigma})$$

where, for the last term, we used the fact that $\text{Tr}(\Lambda\overline{\Sigma}) = \lambda^{\mathsf{T}}\overline{\sigma}$. For compactness of notation, we shall drop the diag$\{\cdot\}$ notation from the subscripts and keep the vectors only, so that the above will be rewritten more compactly as

$$\boxed{\mathsf{E}\,\|\overline{\boldsymbol{w}}_i\|^2_{\overline{\sigma}} = \mathsf{E}\,\|\overline{\boldsymbol{w}}_{i-1}\|^2_{\overline{F}\overline{\sigma}} \; + \; \mu^2\sigma_v^2(\lambda^{\mathsf{T}}\overline{\sigma})} \tag{23.19}$$

The vector weighting factors $\{\overline{\sigma}, \overline{F}\overline{\sigma}\}$ in this expression should be understood as compact representations for the actual weighting matrices $\{\text{diag}\{\overline{\sigma}\}, \text{diag}\{\overline{F}\overline{\sigma}\}\}$. In other words, if σ is any column vector, the notation $\|x\|^2_\sigma$ is used to mean

$$\|x\|^2_\sigma \; \stackrel{\Delta}{=} \; \|x\|^2_{\text{diag}\{\sigma\}} \; = \; x^*\Sigma x, \quad \text{where} \quad \Sigma = \text{diag}\{\sigma\}$$

In summary, starting from (23.12)–(23.14), we argued that for Gaussian regressors the weighting matrix $\overline{\Sigma}'$ is diagonal if $\overline{\Sigma}$ is chosen as diagonal, so that (23.12)–(23.13) can be equivalently expressed more compactly as in (23.17)–(23.18) and (23.19), namely,

$$\begin{aligned} \mathsf{E}\,\|\overline{\boldsymbol{w}}_i\|^2_{\overline{\sigma}} &= \mathsf{E}\,\|\overline{\boldsymbol{w}}_{i-1}\|^2_{\overline{F}\overline{\sigma}} + \mu^2\sigma_v^2(\lambda^{\mathsf{T}}\overline{\sigma}) & (23.20)\\ \overline{\sigma}' &= \overline{F}\overline{\sigma} & (23.21)\\ \overline{F} &= (\mathrm{I} - 2\mu\Lambda + \mu^2\Lambda^2) + \mu^2\lambda\lambda^{\mathsf{T}} & (23.22)\\ \mathsf{E}\,\overline{\boldsymbol{w}}_i &= [\mathrm{I} - \mu\Lambda]\,\mathsf{E}\,\overline{\boldsymbol{w}}_{i-1} & (23.23) \end{aligned}$$

Stability and performance analyses are now possible to pursue by using these relations. Recall that in transient analysis we are interested in the time evolution of the expectations $\{\mathsf{E}\,\tilde{\boldsymbol{w}}_i, \mathsf{E}\,\|\tilde{\boldsymbol{w}}_i\|^2\}$ or, equivalently, $\{\mathsf{E}\,\overline{\boldsymbol{w}}_i, \mathsf{E}\,\|\overline{\boldsymbol{w}}_i\|^2\}$ since $\overline{\boldsymbol{w}}_i$ and $\tilde{\boldsymbol{w}}_i$ are related via a unitary matrix as in (22.31). We start with the mean behavior.

23.2 MEAN BEHAVIOR

The behavior of $\mathsf{E}\,\tilde{\boldsymbol{w}}_i$ can be inferred from (23.23). Thus, note that since, by assumption, the initial condition is zero, $\boldsymbol{w}_{-1} = 0$, we get

$$\tilde{\boldsymbol{w}}_{-1} = w^o - \boldsymbol{w}_{-1} = w^o$$

or, equivalently,

$$\overline{\boldsymbol{w}}_{-1} = U^* w^o \stackrel{\Delta}{=} \overline{w}^o$$

The vector w^o is modeled as an unknown constant so that $\mathsf{E}\,\overline{\boldsymbol{w}}_{-1} = \overline{w}^o$. Therefore, iterating (23.23) we find that

$$\mathsf{E}\,\overline{\boldsymbol{w}}_i = (\mathrm{I} - \mu\Lambda)^{i+1}\,\overline{w}^o$$

We can now derive a condition on the step-size in order to guarantee convergence in the mean, i.e., in order to ensure that $\mathsf{E}\,\overline{\boldsymbol{w}}_i \rightarrow 0$ as $i \rightarrow \infty$, which is equivalent to $\mathsf{E}\,\tilde{\boldsymbol{w}}_i \rightarrow 0$. Indeed, since $\mathrm{I} - \mu\Lambda$ is a diagonal matrix, the condition on μ is easily seen to be:

$$|1 - \mu\lambda_k| < 1 \quad \text{for } k = 1, 2, \ldots, M$$

where the $\{\lambda_k\}$ are the entries of Λ (i.e., the eigenvalues of R_u). In other words, μ should satisfy

$$\boxed{\mu < 2/\lambda_{\max}} \tag{23.24}$$

where $\lambda_{\max}$ is the largest eigenvalue of R_u. For such step-sizes, it follows that

$$\boxed{\lim_{i\rightarrow\infty} \mathsf{E}\,\boldsymbol{w}_i = w^o}$$

and we say that the LMS filter is convergent in the mean and, hence, asymptotically unbiased.

23.3 MEAN-SQUARE BEHAVIOR

The study of the mean-square behavior of the filter is more demanding and also more interesting. We start by noting that the desired quantity $\mathsf{E}\,\|\tilde{\boldsymbol{w}}_i\|^2$, which is also equal to $\mathsf{E}\,\|\overline{\boldsymbol{w}}_i\|^2$, can be obtained from the variance recursion (23.20) if $\overline{\Sigma}$ is chosen as $\overline{\Sigma} = \mathrm{I}$ (or, equivalently, $\Sigma = \mathrm{I}$ in view of (22.31)). This corresponds to choosing $\bar{\sigma}$ as the column vector with unit entries, i.e.,

$$\bar{\sigma} = \boxed{\operatorname{col}\{1,1,\ldots,1,1\} \stackrel{\Delta}{=} q} \tag{23.25}$$

which we denote by q. Then (23.20) gives

$$\mathsf{E}\,\|\overline{\boldsymbol{w}}_i\|^2 = \mathsf{E}\,\|\overline{\boldsymbol{w}}_{i-1}\|^2_{\overline{F}q} + \mu^2\sigma_v^2\left(\lambda^{\mathsf{T}} q\right) \tag{23.26}$$

This recursion shows that in order to evaluate $\mathsf{E}\,\|\overline{\boldsymbol{w}}_i\|^2$ we need to know $\mathsf{E}\,\|\overline{\boldsymbol{w}}_{i-1}\|^2_{\overline{F}q}$, with a weighting matrix whose diagonal entries are $\overline{F}q$. Now the quantity $\mathsf{E}\,\|\overline{\boldsymbol{w}}_i\|^2_{\overline{F}q}$ can be inferred from the variance relation (23.20) by writing it for the choice $\bar{\sigma} = \overline{F}q$, namely,

$$\mathsf{E}\,\|\overline{\boldsymbol{w}}_i\|^2_{\overline{F}q} = \mathsf{E}\,\|\overline{\boldsymbol{w}}_{i-1}\|^2_{\overline{F}^2 q} + \mu^2\sigma_v^2\left(\lambda^{\mathsf{T}}\overline{F}q\right) \tag{23.27}$$

We now find from (23.27) that in order to evaluate the term

$$\mathsf{E}\,\|\overline{\boldsymbol{w}}_i\|^2_{\overline{F}q}$$

we need to know

$$\mathsf{E}\,\|\overline{\boldsymbol{w}}_{i-1}\|^2_{\overline{F}^2 q}$$

with a weighting matrix defined by the vector $\overline{F}^2 q$. This term can again be inferred from (23.20) by writing it for the choice $\bar{\sigma} = \overline{F}^2 q$:

$$\mathsf{E}\,\|\overline{\boldsymbol{w}}_i\|^2_{\overline{F}^2 q} = \mathsf{E}\,\|\overline{\boldsymbol{w}}_{i-1}\|^2_{\overline{F}^3 q} + \mu^2\sigma_v^2\left(\lambda^{\mathsf{T}}\overline{F}^2 q\right) \tag{23.28}$$

and a new term with weighting matrix determined by $\overline{F}^3 q$ appears. The natural question is whether this procedure terminates, and whether weighting matrices that correspond to increasing powers of $\overline{F}$ keep coming up? The procedure does terminate. This is because once we write (23.20) for the choice $\bar{\sigma} = \overline{F}^{(M-1)} q$ we get:

$$\mathsf{E}\,\|\overline{\boldsymbol{w}}_i\|^2_{\overline{F}^{(M-1)} q} = \mathsf{E}\,\|\overline{\boldsymbol{w}}_{i-1}\|^2_{\overline{F}^M q} + \mu^2\sigma_v^2\left(\lambda^{\mathsf{T}}\overline{F}^{(M-1)} q\right) \tag{23.29}$$

where the weighting matrix on the right-hand side is now $\overline{F}^M q$. However, we do not need to write (23.20) for the choice $\bar{\sigma} = \overline{F}^M q$ in order to determine $\mathsf{E}\,\|\overline{\boldsymbol{w}}_i\|^2_{\overline{F}^M q}$. This is because this last term can be deduced from the already available weighted factors:

$$\left\{ \mathsf{E}\,\|\overline{\boldsymbol{w}}_i\|^2,\ \mathsf{E}\,\|\overline{\boldsymbol{w}}_i\|^2_{\overline{F}q},\ \mathsf{E}\,\|\overline{\boldsymbol{w}}_i\|^2_{\overline{F}^2 q},\ \ldots,\ \mathsf{E}\,\|\overline{\boldsymbol{w}}_i\|^2_{\overline{F}^{M-1} q} \right\} \tag{23.30}$$

This fact can be seen as follows. With any matrix $\overline{F}$ we associate its characteristic polynomial, defined by

$$p(x) = \det(x\mathrm{I} - \overline{F})$$

It is an $M-$th order polynomial in x,

$$p(x) = x^M + p_{M-1}x^{M-1} + p_{M-2}x^{M-2} + \ldots + p_1 x + p_0$$

with coefficients $\{p_k, p_M = 1\}$ and whose roots coincide with the eigenvalues of $\overline{F}$. Now a famous result in matrix theory, known as the Cayley-Hamilton theorem, states that every matrix satisfies its characteristic equation, i.e., $p(\overline{F}) = 0$. In other words, $\overline{F}$ satisfies

$$\overline{F}^M + p_{M-1}\overline{F}^{M-1} + p_{M-2}\overline{F}^{M-2} + \ldots + p_1\overline{F} + p_0\mathrm{I}_M = 0$$

which means that the $M-$th power of $\overline{F}$ can be expressed as a linear combination of its lower order powers. Using this fact we can write

$$\begin{aligned} \mathsf{E}\,\|\overline{\boldsymbol{w}}_i\|^2_{\overline{F}^M q} &= \mathsf{E}\,\|\overline{\boldsymbol{w}}_i\|^2_{(-p_{M-1}\overline{F}^{M-1} - p_{M-2}\overline{F}^{M-2} - \ldots - p_1\overline{F} - p_0\mathrm{I}_M)q} \\ &= -p_0\mathsf{E}\,\|\overline{\boldsymbol{w}}_i\|^2 \;-\; p_1\mathsf{E}\,\|\overline{\boldsymbol{w}}_i\|^2_{\overline{F}q} \;-\; \ldots \;-\; p_{M-1}\mathsf{E}\,\|\overline{\boldsymbol{w}}_i\|^2_{\overline{F}^{(M-1)}q} \end{aligned}$$

That is,

$$\mathsf{E}\,\|\overline{\boldsymbol{w}}_i\|^2_{\overline{F}^M q} = \sum_{k=0}^{M-1} -p_k\mathsf{E}\,\|\overline{\boldsymbol{w}}_i\|^2_{\overline{F}^k q} \tag{23.31}$$

which expresses $\mathsf{E}\,\|\overline{\boldsymbol{w}}_i\|^2_{\overline{F}^M q}$ as a linear combination of the terms in (23.30), as desired. We can collect the above results into a single self-contained recursion by writing (23.26)–(23.31) as:

$$\underbrace{\begin{bmatrix} \mathsf{E}\,\|\overline{\boldsymbol{w}}_i\|^2 \\ \mathsf{E}\,\|\overline{\boldsymbol{w}}_i\|^2_{\overline{F}q} \\ \mathsf{E}\,\|\overline{\boldsymbol{w}}_i\|^2_{\overline{F}^2 q} \\ \vdots \\ \mathsf{E}\,\|\overline{\boldsymbol{w}}_i\|^2_{\overline{F}^{(M-2)}q} \\ \mathsf{E}\,\|\overline{\boldsymbol{w}}_i\|^2_{\overline{F}^{(M-1)}q} \end{bmatrix}}_{\triangleq\, \mathcal{W}_i} = \underbrace{\begin{bmatrix} 0 & 1 & & & \\ 0 & 0 & 1 & & \\ 0 & 0 & 0 & 1 & \\ \vdots & \vdots & & & \\ 0 & 0 & 0 & & 1 \\ -p_0 & -p_1 & -p_2 & \ldots & -p_{M-1} \end{bmatrix}}_{\triangleq\, \mathcal{F}} \begin{bmatrix} \mathsf{E}\,\|\overline{\boldsymbol{w}}_{i-1}\|^2 \\ \mathsf{E}\,\|\overline{\boldsymbol{w}}_{i-1}\|^2_{\overline{F}q} \\ \mathsf{E}\,\|\overline{\boldsymbol{w}}_{i-1}\|^2_{\overline{F}^2 q} \\ \vdots \\ \mathsf{E}\,\|\overline{\boldsymbol{w}}_{i-1}\|^2_{\overline{F}^{(M-2)}q} \\ \mathsf{E}\,\|\overline{\boldsymbol{w}}_{i-1}\|^2_{\overline{F}^{(M-1)}q} \end{bmatrix}$$

$$+\,\mu^2\sigma_v^2 \underbrace{\begin{bmatrix} \lambda^{\mathsf{T}} q \\ \lambda^{\mathsf{T}}\overline{F}q \\ \lambda^{\mathsf{T}}\overline{F}^2 q \\ \vdots \\ \lambda^{\mathsf{T}}\overline{F}^{M-1} q \end{bmatrix}}_{\triangleq\, \mathcal{Y}}$$

If we introduce the vector and matrix quantities $\{\mathcal{W}_i, \mathcal{F}, \mathcal{Y}\}$ indicated above, then this recursion can be rewritten more compactly as

$$\boxed{\mathcal{W}_i = \mathcal{F}\mathcal{W}_{i-1} + \mu^2\sigma_v^2\mathcal{Y}} \tag{23.32}$$

This is a first-order recursion with a constant coefficient matrix $\mathcal{F}$. In the language of linear system theory, a recursion of the form (23.32) is called a *state-space* recursion with the vector $\mathcal{W}_i$ denoting the state vector. We therefore find that the mean-square behavior of LMS is described by the $M-$dimensional state-space recursion (23.32) with coefficient matrix $\mathcal{F}$. To be more precise, the transient behavior of LMS is described by the combination of both (23.32) and recursion (23.23) for the mean weight-error vector. These two recursions can be grouped together into a single $2M$-dimensional state-space model with a block diagonal coefficient matrix as follows

$$\begin{bmatrix} \mathsf{E}\,\overline{w}_i \\ \mathcal{W}_i \end{bmatrix} = \begin{bmatrix} \mathrm{I}-\mu\Lambda & \\ & \mathcal{F} \end{bmatrix} \begin{bmatrix} \mathsf{E}\,\overline{w}_{i-1} \\ \mathcal{W}_{i-1} \end{bmatrix} + \mu^2\sigma_v^2 \begin{bmatrix} 0 \\ \mathcal{Y} \end{bmatrix}$$

However, since the mean behavior is simpler to study, as we saw before, while the mean-square behavior is considerably richer, and since the combined state-space model is decoupled due to the diagonal block structure of its coefficient matrix, we shall often, for simplicity, refer to the state-space model (23.32) as the model that ultimately determines the transient behavior of an adaptive filter.

The matrix $\mathcal{F}$ in (23.32) has a special structure in this case; it is a companion matrix (namely, a matrix with ones on the upper diagonal and the negatives of the coefficients of $p(x)$ in the last row). A well-known property of such matrices is that their eigenvalues coincide with the roots of $p(x) = 0$, which in turn are the eigenvalues of $\overline{F}$ from (23.22), i.e.,

$$\{\text{ eigenvalues of } \mathcal{F}\ \} = \{\text{ roots of } p(x)\ \} = \{\text{ eigenvalues of } \overline{F}\ \}$$

Recursion (23.32) shows that the transient behavior of LMS is the combined result of the time evolutions of the M variables in (23.30), which are the entries of $\mathcal{W}_i$. For this reason, these variables are called *state* variables; since they determine the state of the filter at any particular time instant.

Lemma 23.1 (Mean-square behavior of complex LMS) Consider the LMS recursion (23.1) and assume the data $\{\boldsymbol{d}(i), \boldsymbol{u}_i\}$ satisfy model (22.1) and the independence assumption (22.23). Assume further that the regressor sequence is circular Gaussian. Then the mean and mean-square behaviors of the filter are characterized by (23.23) and (23.32), namely, by the recursion

$$\begin{bmatrix} \mathsf{E}\,\overline{w}_i \\ \mathcal{W}_i \end{bmatrix} = \begin{bmatrix} \mathrm{I}-\mu\Lambda & \\ & \mathcal{F} \end{bmatrix} \begin{bmatrix} \mathsf{E}\,\overline{w}_{i-1} \\ \mathcal{W}_{i-1} \end{bmatrix} + \mu^2\sigma_v^2 \begin{bmatrix} 0 \\ \mathcal{Y} \end{bmatrix}$$

23.4 MEAN-SQUARE STABILITY

The LMS filter will be said to be mean-square stable if, and only if, the state vector $\mathcal{W}_i$ remains bounded and tends to a steady-state value, regardless of the initial condition $\mathcal{W}_{-1}$. A necessary and sufficient condition for this to hold can be found as follows. For any state-space model of the form

$$x_i = Cx_{i-1} + b$$

a well-known result from linear system theory states that the sequence $\{x_j\}$ remains bounded, regardless of the initial state vector $\{x_{-1}\}$, and that the sequence $\{x_j\}$ will tend to a finite steady-state value as well, if, and only if, all the eigenvalues of C lie inside the open unit disc. Applying this result to (23.32), we conclude that the LMS filter will be mean-square stable if, and only if, all eigenvalues of $\mathcal{F}$ are inside the unit disc or, equivalently, the eigenvalues of $\overline{F}$ from (23.22) should satisfy

$$-1 < \lambda\left(\overline{F}\right) < 1 \tag{23.33}$$

That is, we need $\overline{F}$ to be a stable matrix; here we are writing $\lambda(\overline{F})$ to refer to the eigenvalues (spectrum) of $\overline{F}$.

The lower bound on $\lambda(\overline{F})$ is automatically satisfied because $\overline{F}$ is at least nonnegative-definite (and, hence, its eigenvalues are all nonnegative). This fact can be seen by writing $\overline{F}$ from (23.22) as the sum of two nonnegative-definite matrices:

$$\overline{F} \;=\; (\mathrm{I}-2\mu\Lambda+\mu^2\Lambda^2)+\mu^2\lambda\lambda^{\mathsf{T}} \;=\; (\mathrm{I}-\mu\Lambda)^2+\mu^2\lambda\lambda^{\mathsf{T}} \qquad (23.34)$$

To find a condition on μ for the upper bound on the eigenvalues of $\overline{F}$ to be satisfied, we start by expressing $\overline{F}$ as

$$\overline{F}=\mathrm{I}-\mu A+\mu^2 B \qquad (23.35)$$

where, in this case, the matrices A and B are both positive-definite and given by

$$A \triangleq 2\Lambda, \qquad B \triangleq \Lambda^2+\lambda\lambda^{\mathsf{T}} \qquad (23.36)$$

It is shown in Prob. V.3 that for nonnegative-definite matrices $\overline{F}$ of the form (23.35), its eigenvalues will be upper bounded by one if, and only if, the parameter μ satisfies

$$0<\mu<1/\lambda_{\max}\left(A^{-1}B\right) \qquad (23.37)$$

in terms of the maximum eigenvalue of $A^{-1}B$. It is also shown in that problem that the eigenvalues of $A^{-1}B$ are real and positive.

Let

$$\eta^o=1/\lambda_{\max}(A^{-1}B)$$

The bound (23.37) on μ is simply the smallest positive scalar η that makes the matrix $\mathrm{I}-\eta A^{-1}B$ singular, i.e., it is the smallest η such that

$$\det\left(\mathrm{I}-\eta A^{-1}B\right)=0 \qquad (23.38)$$

Combining (23.37) with (23.24) we find that the condition on μ for the filter to converge in both the mean and mean-square senses is

$$0<\mu<\min\{2/\lambda_{\max},\eta^o\} \qquad (23.39)$$

It turns out that we can be more explicit and show that $\eta^o<2/\lambda_{\max}$ so that the upper bound on μ is ultimately determined by η^o alone. Indeed, using the definitions (23.36) for A and B we have

$$\begin{aligned}\det\left(\mathrm{I}-\eta A^{-1}B\right) &= \det\left(\mathrm{I}-\eta\frac{\Lambda^{-1}}{2}\left[\Lambda^2+\lambda\lambda^{\mathsf{T}}\right]\right)\\ &= \det\left(\mathrm{I}-\frac{\eta}{2}\left[\Lambda+q\lambda^{\mathsf{T}}\right]\right)\end{aligned}$$

We are interested in the smallest value of η that makes $\mathrm{I}-\eta A^{-1}B$ singular. Actually, in view of condition (23.39) on μ, we are only interested in those values of η that lie within the open interval $(0,2/\lambda_{\max})$. If any such η can be found, then $\eta^o<2/\lambda_{\max}$. Since over the interval $\eta\in(0,2/\lambda_{\max})$ the matrix $(\mathrm{I}-\frac{\eta}{2}\Lambda)$ is invertible, we can write

$$\begin{aligned}\det\left(\mathrm{I}-\eta A^{-1}B\right) &= \det\left(\mathrm{I}-\frac{\eta}{2}\Lambda\right)\cdot\det\left(\mathrm{I}-\left[\mathrm{I}-\frac{\eta}{2}\Lambda\right]^{-1}\frac{\eta}{2}q\lambda^{\mathsf{T}}\right)\\ &= \left(1-\lambda^{\mathsf{T}}\left[2\eta^{-1}\mathrm{I}-\Lambda\right]^{-1}q\right)\cdot\det\left(\mathrm{I}-\frac{\eta}{2}\Lambda\right) \qquad (23.40)\end{aligned}$$

where in the last step we used the determinant identity

$$\det(\mathrm{I} - XY) = \det(\mathrm{I} - YX)$$

for any matrices $\{X, Y\}$ of compatible dimensions, so that when X is a column and Y is a row we have

$$\det(\mathrm{I} - xy^{\mathsf{T}}) = 1 - y^{\mathsf{T}}x$$

The values of $\eta \in (0, 2/\lambda_{\max})$ that result in $\det\left(\mathrm{I} - \eta A^{-1}B\right) = 0$ should therefore satisfy

$$\lambda^{\mathsf{T}}\left(2\eta^{-1}\mathrm{I} - \Lambda\right)^{-1} q = 1$$

i.e.,

$$\sum_{k=1}^{M} \frac{\lambda_k \eta}{2 - \lambda_k \eta} = 1 \tag{23.41}$$

Introduce the function

$$f(\eta) \triangleq \sum_{k=1}^{M} \frac{\lambda_k \eta}{2 - \lambda_k \eta} \tag{23.42}$$

and observe that $f(0) = 0$ and $f(\eta)$ is monotonically increasing between 0 and $2/\lambda_{\max}$; this latter claim follows by noting that the derivative of $f(\cdot)$ with respect to η is positive,

$$\frac{df}{d\eta} = \sum_{k=1}^{M} \frac{2\lambda_k}{(2 - \lambda_k \eta)^2} > 0 \quad \text{for } \eta \in [0, 2/\lambda_{\max})$$

Observe further that $f(\eta)$ has a singularity (i.e., a pole) at $\eta = 2/\lambda_{\max}$, so that $f(\eta) \to \infty$ as $\eta \to 2/\lambda_{\max}$ — Fig. 23.1.

Therefore, we conclude that there exists a unique positive η within the interval $(0, 2/\lambda_{\max})$ where the function $f(\eta)$ crosses one, i.e., for which

$$\sum_{k=1}^{M} \frac{\lambda_k \eta^o}{2 - \lambda_k \eta^o} = 1$$

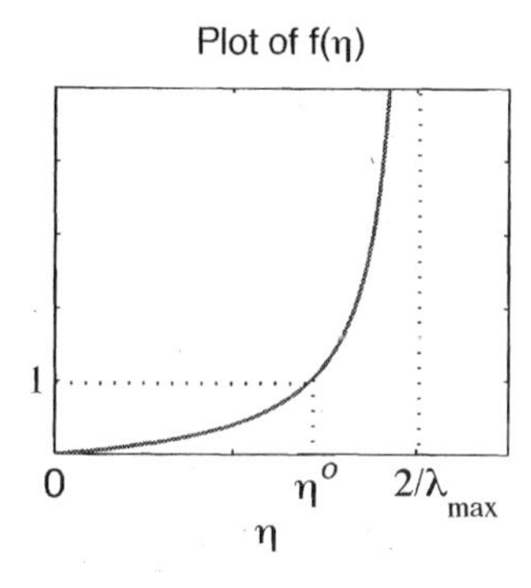

FIGURE 23.1 Behavior of the function $f(\eta)$ defined by (23.42) over the semi-open interval $[0, 2/\lambda_{\max})$.

This value of η is the smallest η that makes $\mathrm{I}-\eta A^{-1}B$ singular and, therefore, it coincides with the desired η^o and is smaller than $2/\lambda_{\max}$. In conclusion, we find that condition (23.39) is equivalent to requiring μ to satisfy

$$\sum_{k=1}^{M} \frac{\lambda_k \mu}{2-\lambda_k \mu} < 1$$

When this happens, the LMS filter will be mean-square stable. In summary, we arrive at the following conclusion.

Theorem 23.1 (Stability of complex LMS) Consider the LMS recursion (23.1) and assume the data $\{\boldsymbol{d}(i), \boldsymbol{u}_i\}$ satisfy model (22.1) and the independence assumption (22.23). Assume further that the regressor sequence is circular Gaussian. Then the LMS filter is mean-square stable (i.e., the state vector $\mathcal{W}_i$ remains bounded and tends to a finite steady-state value) if, and only if, the positive step-size μ satisfies

$$\sum_{k=1}^{M} \frac{\lambda_k \mu}{2-\lambda_k \mu} < 1$$

where the $\{\lambda_k\}$ are the eigenvalues of R_u. The above condition on μ also guarantees convergence in the mean, i.e., $\mathsf{E}\,\boldsymbol{w}_i \rightarrow w^o$.

The arguments leading to the theorem were carried out for complex-valued regressors. If the regressors were real-valued, then, as mentioned after (23.11), this expression would include an additional factor of 2, namely

$$\mathsf{E}\,\|\overline{\boldsymbol{u}}_i\|^2_{\overline{\Sigma}}\overline{\boldsymbol{u}}_i^*\overline{\boldsymbol{u}}_i \;=\; \Lambda \mathsf{Tr}(\overline{\Sigma}\Lambda) \;+\; 2\Lambda\overline{\Sigma}\Lambda$$

If we propagate this factor of 2, then only slight modifications will occur in the derivation. In particular, expressions (23.20)–(23.23) would become

$$\begin{aligned} \mathsf{E}\,\|\overline{\boldsymbol{w}}_i\|^2_{\overline{\sigma}} &= \mathsf{E}\,\|\overline{\boldsymbol{w}}_{i-1}\|^2_{\overline{F}\overline{\sigma}} \;+\; \mu^2\sigma_v^2(\lambda^{\mathsf{T}}\overline{\sigma}) && (23.43)\\ \overline{\sigma}' &= \overline{F}\overline{\sigma} && (23.44)\\ \overline{F} &= (\mathrm{I}-2\mu\Lambda+2\mu^2\Lambda^2)+\mu^2\lambda\lambda^{\mathsf{T}} && (23.45)\\ \mathsf{E}\,\overline{\boldsymbol{w}}_i &= (\mathrm{I}-\mu\Lambda)\mathsf{E}\,\overline{\boldsymbol{w}}_{i-1} && (23.46) \end{aligned}$$

Moreover, the expression for the matrix B in (23.36) would become $B = 2\Lambda^2+\lambda\lambda^{\mathsf{T}}$, and condition (23.41) on η would be replaced by

$$\frac{1}{2}\sum_{k=1}^{M} \frac{\lambda_k \eta}{1-\lambda_k \eta} = 1 \tag{23.47}$$

Theorem 23.2 (Mean-square behavior of real LMS) Consider the LMS recursion (23.1) and assume the data $\{\boldsymbol{d}(i), \boldsymbol{u}_i\}$ satisfy model (22.1) and the independence assumption (22.23). Assume further that the regressor sequence is real-valued Gaussian. Then the LMS filter is mean-square stable (i.e., the state vector $\mathcal{W}_i$ remains bounded and tends to a finite steady-state value) if, and only if, the step-size μ satisfies

$$\frac{1}{2}\sum_{k=1}^{M}\frac{\lambda_k\mu}{1-\lambda_k\mu} < 1$$

where the $\{\lambda_k\}$ are the eigenvalues of R_u. The above condition on μ also guarantees convergence in the mean, i.e., $\mathsf{E}\,\boldsymbol{w}_i \to w^o$. Moreover, the mean and mean-square behaviors of LMS are characterized by recursions (23.23) and (23.32), namely,

$$\begin{bmatrix} \mathsf{E}\,\tilde{\boldsymbol{w}}_i \\ \mathcal{W}_i \end{bmatrix} = \begin{bmatrix} \mathrm{I}-\mu\Lambda & \\ & \mathcal{F} \end{bmatrix}\begin{bmatrix} \mathsf{E}\,\tilde{\boldsymbol{w}}_{i-1} \\ \mathcal{W}_{i-1} \end{bmatrix} + \mu^2\sigma_v^2\begin{bmatrix} 0 \\ \mathcal{Y} \end{bmatrix}$$

(or, more generally, by (23.48) further ahead for other choices of $\bar{\sigma}$ in the mean-square case). The coefficients $\{p_k\}$ that define the matrix $\mathcal{F}$ are now obtained from the characteristic polynomial of the matrix $\overline{F}$ in (23.45).

23.5 STEADY-STATE PERFORMANCE

In the above derivation, we used the variance relation (23.20) to characterize the transient behavior of LMS in terms of the first-order recursion (23.32). We can use the same variance relation to shed further light on the *steady-state* performance of LMS. Of course, we already studied the steady-state performance of LMS in Chapter 16, and derived expressions for its (excess) mean-square error under varied conditions on the step-size and the data. Here we shall use the transient analysis of this section to provide additional insights under the independence assumption (22.23). Clearly, any steady-state results that are obtained by examining the limiting behavior of a transient analysis would be bound by the same assumptions/restrictions that are needed to advance the transient analysis. In contrast, the steady-state analysis that we carried out in Chapter 16 was purposely decoupled from the transient analysis and, in this way, it relied on weaker assumptions than those needed in the current chapter.

In this section, we shall re-examine the excess mean-square error,

$$\mathsf{EMSE} \triangleq \lim_{i\to\infty} \mathsf{E}\,|\boldsymbol{e}_a(i)|^2$$

as well as study the so-called mean-square deviation of the filter, which is defined as

$$\mathsf{MSD} \triangleq \lim_{i\to\infty} \mathsf{E}\,\|\tilde{\boldsymbol{w}}_i\|^2$$

To begin with, let us first note that if model (23.32) is stable, then it will remain stable if q is replaced by any other choice for $\bar{\sigma}$. Indeed, it is straightforward to see from the arguments that led to (23.32) that had we started with any other choice for $\bar{\sigma}$, a similar state-space recursion would have resulted with the same coefficient matrix $\mathcal{F}$, namely

$$\underbrace{\begin{bmatrix} \mathsf{E}\,\|\overline{\boldsymbol{w}}_i\|^2_{\overline{\sigma}} \\ \mathsf{E}\,\|\overline{\boldsymbol{w}}_i\|^2_{\overline{F}\overline{\sigma}} \\ \mathsf{E}\,\|\overline{\boldsymbol{w}}_i\|^2_{\overline{F}^2\overline{\sigma}} \\ \vdots \\ \mathsf{E}\,\|\overline{\boldsymbol{w}}_i\|^2_{\overline{F}^{(M-2)}\overline{\sigma}} \\ \mathsf{E}\,\|\overline{\boldsymbol{w}}_i\|^2_{\overline{F}^{(M-1)}\overline{\sigma}} \end{bmatrix}}_{\triangleq\,\mathcal{W}_i} = \underbrace{\begin{bmatrix} 0 & 1 & & & \\ 0 & 0 & 1 & & \\ 0 & 0 & 0 & 1 & \\ \vdots & & & & \\ 0 & 0 & 0 & & 1 \\ -p_0 & -p_1 & -p_2 & \dots & -p_{M-1} \end{bmatrix}}_{\triangleq\,\mathcal{F}} \begin{bmatrix} \mathsf{E}\,\|\overline{\boldsymbol{w}}_{i-1}\|^2_{\overline{\sigma}} \\ \mathsf{E}\,\|\overline{\boldsymbol{w}}_{i-1}\|^2_{\overline{F}\overline{\sigma}} \\ \mathsf{E}\,\|\overline{\boldsymbol{w}}_{i-1}\|^2_{\overline{F}^2\overline{\sigma}} \\ \vdots \\ \mathsf{E}\,\|\overline{\boldsymbol{w}}_{i-1}\|^2_{\overline{F}^{(M-2)}\overline{\sigma}} \\ \mathsf{E}\,\|\overline{\boldsymbol{w}}_{i-1}\|^2_{\overline{F}^{(M-1)}\overline{\sigma}} \end{bmatrix}$$

$$+\,\mu^2\sigma_v^2 \underbrace{\begin{bmatrix} \lambda^{\mathsf{T}}\overline{\sigma} \\ \lambda^{\mathsf{T}}\overline{F}\overline{\sigma} \\ \lambda^{\mathsf{T}}\overline{F}^2\overline{\sigma} \\ \vdots \\ \lambda^{\mathsf{T}}\overline{F}^{M-1}\overline{\sigma} \end{bmatrix}}_{\triangleq\,\mathcal{Y}}$$

with q replaced by $\overline{\sigma}$. In other words, it would still hold that

$$\boxed{\mathcal{W}_i = \mathcal{F}\mathcal{W}_{i-1} + \mu^2\sigma_v^2\mathcal{Y}} \tag{23.48}$$

where the entries of $\mathcal{W}_i$ are now defined as above for arbitrary $\overline{\sigma}$, while the coefficient matrix $\mathcal{F}$ remains the same as before. Therefore, no matter which $\overline{\sigma}$ we choose, stability of (23.48) is guaranteed by the same condition on $\mathcal{F}$ as in (23.33), namely, its eigenvalues should lie inside the unit disc.

With this issue settled, we can now explain how the freedom in selecting $\overline{\sigma}$ can be useful. To see this, consider the setting of Thm. 23.1 and assume the step-size μ has been chosen to guarantee filter stability. Then recursion (23.20) becomes in steady-state

$$\mathsf{E}\,\|\overline{\boldsymbol{w}}_\infty\|^2_{\overline{\sigma}} \;=\; \mathsf{E}\,\|\overline{\boldsymbol{w}}_\infty\|^2_{\overline{F}\overline{\sigma}} \;+\; \mu^2\sigma_v^2(\lambda^{\mathsf{T}}\overline{\sigma}) \tag{23.49}$$

which is equivalent to

$$\boxed{\mathsf{E}\,\|\overline{\boldsymbol{w}}_\infty\|^2_{(\mathrm{I}-\overline{F})\overline{\sigma}} \;=\; \mu^2\sigma_v^2(\lambda^{\mathsf{T}}\overline{\sigma})} \tag{23.50}$$

This expression can be used to examine the mean-square performance of LMS in an interesting way.

For example, in order to evaluate the MSD of LMS we would need to evaluate the expectation $\mathsf{E}\,\|\overline{\boldsymbol{w}}_\infty\|^2$. Thus, assume that we select the weighting vector $\overline{\sigma}$ in (23.50) as the solution to the linear system of equations $(\mathrm{I}-\overline{F})\overline{\sigma}_{\text{msd}} = q$, i.e., as $\overline{\sigma}_{\text{msd}} = (\mathrm{I}-\overline{F})^{-1}q$. In this way, the weighting vector that appears in (23.50) will become q. Then the left-hand side of (23.50) will coincide with the filter MSD and, therefore, we would be able to conclude that

$$\boxed{\mathsf{MSD} \;=\; \mu^2\sigma_v^2\lambda^{\mathsf{T}}(\mathrm{I}-\overline{F})^{-1}q} \tag{23.51}$$

A more explicit expression for the MSD can be found by evaluating the product $\lambda^{\mathsf{T}}(\mathrm{I}-\overline{F})^{-1}q$. Using expression (23.22) for $\overline{F}$ we have

$$\mathrm{I}-\overline{F} \;=\; 2\mu\Lambda - \mu^2\Lambda^2 - \mu^2\lambda\lambda^{\mathsf{T}} \;\triangleq\; D - \mu^2\lambda\lambda^{\mathsf{T}}$$

where we introduced, for convenience, the matrix

$$D = 2\mu\Lambda - \mu^2\Lambda^2 \tag{23.52}$$

Then, using the matrix inversion formula (5.4),

$$\begin{aligned}
\lambda^{\mathsf{T}}(\mathrm{I}-\overline{F})^{-1}q &= \lambda^{\mathsf{T}}(D-\mu^2\lambda\lambda^{\mathsf{T}})^{-1}q \\
&= \lambda^{\mathsf{T}}\left(D^{-1}+\frac{\mu^2}{1-\mu^2\lambda^{\mathsf{T}}D^{-1}\lambda}D^{-1}\lambda\lambda^{\mathsf{T}}D^{-1}\right)q \\
&= \frac{\lambda^{\mathsf{T}}D^{-1}q}{1-\mu^2\lambda^{\mathsf{T}}D^{-1}\lambda}
\end{aligned} \tag{23.53}$$

Substituting (23.53) into (23.51), we get

$$\boxed{\mathsf{MSD} = \frac{\mu\sigma_v^2\sum_{k=1}^{M}\frac{1}{2-\mu\lambda_k}}{1-\mu\sum_{k=1}^{M}\frac{\lambda_k}{2-\mu\lambda_k}}} \tag{23.54}$$

In a similar vein, we can evaluate the EMSE of LMS. Thus, recall from (22.24) that $\boldsymbol{u}_i$ and $\tilde{\boldsymbol{w}}_{i-1}$ are independent random variables. Then it follows that

$$\begin{aligned}
\mathsf{E}\,|\boldsymbol{e}_a(i)|^2 &= \mathsf{E}\,\tilde{\boldsymbol{w}}_{i-1}^*\boldsymbol{u}_i^*\boldsymbol{u}_i\tilde{\boldsymbol{w}}_{i-1} \\
&= \mathsf{E}\left[\mathsf{E}\left(\tilde{\boldsymbol{w}}_{i-1}^*\boldsymbol{u}_i^*\boldsymbol{u}_i\tilde{\boldsymbol{w}}_{i-1}\,|\,\tilde{\boldsymbol{w}}_{i-1}\right)\right] \\
&= \mathsf{E}\left[\tilde{\boldsymbol{w}}_{i-1}^*\,\mathsf{E}\left(\boldsymbol{u}_i^*\boldsymbol{u}_i\,|\,\tilde{\boldsymbol{w}}_{i-1}\right)\tilde{\boldsymbol{w}}_{i-1}\right] \\
&= \mathsf{E}\left(\tilde{\boldsymbol{w}}_{i-1}^*R_u\tilde{\boldsymbol{w}}_{i-1}\right) \\
&= \|\tilde{\boldsymbol{w}}_{i-1}\|_{R_u}^2 \\
&= \|\overline{\boldsymbol{w}}_{i-1}\|_{\lambda}^2
\end{aligned} \tag{23.55}$$

In other words, in order to determine the EMSE we need to evaluate $\mathsf{E}\,\|\overline{\boldsymbol{w}}_\infty\|_\lambda^2$, with weighting factor $\lambda = \text{diag}\{\Lambda\}$. Therefore, assume that we now select $\overline{\sigma}$ in (23.50) as the solution to the linear system of equations $(\mathrm{I}-\overline{F})\overline{\sigma}_{\text{emse}} = \lambda$, i.e., as $\overline{\sigma}_{\text{emse}} = (\mathrm{I}-\overline{F})^{-1}\lambda$. In this way, the weighting quantity that appears in (23.50) will become λ. Then the left-hand side of (23.50) will coincide with the filter EMSE and we would get

$$\boxed{\mathsf{EMSE} = \mu^2\sigma_v^2\lambda^{\mathsf{T}}(\mathrm{I}-\overline{F})^{-1}\lambda} \tag{23.56}$$

Again a more explicit expression for the EMSE can be found by evaluating the product $\lambda^{\mathsf{T}}(\mathrm{I}-\overline{F})^{-1}\lambda$ in much the same way as we did for the MSD above, leading to

$$\lambda^{\mathsf{T}}(\mathrm{I}-\overline{F})^{-1}\lambda = \frac{\lambda^{\mathsf{T}}D^{-1}\lambda}{1-\mu^2\lambda^{\mathsf{T}}D^{-1}\lambda}$$

so that

$$\boxed{\mathsf{EMSE} = \frac{\mu\sigma_v^2\sum_{k=1}^{M}\frac{\lambda_k}{2-\mu\lambda_k}}{1-\mu\sum_{k=1}^{M}\frac{\lambda_k}{2-\mu\lambda_k}}}$$

The above derivations assume complex-valued regressors. If the regressors were instead real-valued, then only slight modifications will occur. In particular, we would need to use expression (23.45) for $\overline{F}$, in which case the matrix D in (23.52) would be replaced by $D = 2\mu\Lambda - 2\mu^2\Lambda^2$. Repeating the derivation for the MSD and EMSE leads to the result summarized below.

Theorem 23.3 (MSD and EMSE of LMS) Consider the LMS recursion (23.1) and assume the data $\{\boldsymbol{d}(i), \boldsymbol{u}_i\}$ satisfy model (22.1) and the independence assumption (22.23). Then the MSD and EMSE are given by

$$\text{EMSE} = \frac{\mu\sigma_v^2 \sum_{k=1}^{M} \frac{\lambda_k}{2 - s\mu\lambda_k}}{1 - \mu \sum_{k=1}^{M} \frac{\lambda_k}{2 - s\mu\lambda_k}} \qquad \text{MSD} = \frac{\mu\sigma_v^2 \sum_{k=1}^{M} \frac{1}{2 - s\mu\lambda_k}}{1 - \mu \sum_{k=1}^{M} \frac{\lambda_k}{2 - s\mu\lambda_k}}$$

where $s = 1$ if the regressors are circular Gaussian and $s = 2$ if the regressors are real-valued Gaussian. Moreover, the $\{\lambda_k\}$ are the eigenvalues of R_u.

23.6 SMALL STEP-SIZE APPROXIMATIONS

If the step-size is small enough in the sense that $\mu\lambda_k \ll 1$, then the above expressions for the EMSE and MSD simplify to

$$\text{EMSE} = \frac{\mu\sigma_v^2 \text{Tr}(R_u)}{2 - \mu\text{Tr}(R_u)} \quad \text{and} \quad \text{MSD} = \frac{\mu\sigma_v^2 M}{2 - \mu\text{Tr}(R_u)} \tag{23.57}$$

For even smaller step-sizes, we can further approximate the denominators by 2 so that

$$\text{EMSE} = \mu\sigma_v^2 \text{Tr}(R_u)/2 \quad \text{and} \quad \text{MSD} = \mu\sigma_v^2 M/2 \tag{23.58}$$

If, on the other hand, the covariance matrix R_u is a scaled multiple of the identity, say, $R_u = \sigma_u^2 \text{I}$, then the expressions for EMSE and MSD from the theorem reduce to

$$\text{EMSE} = \frac{\mu M \sigma_v^2 \sigma_u^2}{2 - \mu(M+s)\sigma_u^2} \quad \text{and} \quad \text{MSD} = \frac{\mu M \sigma_v^2}{2 - \mu(M+s)\sigma_u^2} \tag{23.59}$$

These expressions for the EMSE coincide with the ones derived earlier in Lemma 16.1. It is reassuring to see how the arguments of Chapter 16 and the arguments of the current chapter complement each other. Keep in mind though that the derivation here relies on the independence condition (22.23), while the derivation of Chapter 16 did not require (22.23). Observe further how the expressions for the filter EMSE and MSD were obtained by simply choosing convenient values for the free parameter $\overline{\sigma}$.

23.A APPENDIX: CONVERGENCE TIME

Besides stability and steady-state performance, the transient analysis developed so far also provides information about other aspects pertaining to the operation of adaptive filters. In this appendix we illustrate how the convergence time of an adaptive filter can be estimated from the results on its transient performance. We consider the example of an LMS filter with white regression data. The case of correlated data, as well as more general data-normalized filters, are treated in App. 9.D of Sayed (2003).

LMS with White-Input Gaussian Data

Consider the LMS recursion

$$\boldsymbol{w}_i = \boldsymbol{w}_{i-1} + \mu \boldsymbol{u}_i^* \boldsymbol{e}(i), \qquad \boldsymbol{e}(i) = \boldsymbol{d}(i) - \boldsymbol{u}_i \boldsymbol{w}_{i-1}, \qquad w_{-1} = 0$$

and assume the data $\{\boldsymbol{d}(i), \boldsymbol{u}_i\}$ satisfy model (22.1) and the independence assumption (22.23). Assume further that the individual entries of $\boldsymbol{u}_i$ are zero-mean, Gaussian, and i.i.d. with variance σ_u^2 and fourth moment $\gamma\sigma_u^4$, where $\gamma = 2$ for complex-valued data and $\gamma = 3$ for real-valued data. It then follows from Thm. 23.3 (see Prob. V.21) that the mean-square deviation of the filter evolves with time according to the following difference equation:

$$\boxed{\mathsf{E}\,\|\tilde{\boldsymbol{w}}_i\|^2 = [1 - 2\mu\sigma_u^2 + \mu^2\sigma_u^4(M + \gamma - 1)]\;\mathsf{E}\,\|\tilde{\boldsymbol{w}}_{i-1}\|^2 \;+\; \mu^2\sigma_v^2\sigma_u^2 M} \tag{23.60}$$

Now since, in this case,

$$\mathsf{E}\,|\boldsymbol{e}_a(i)|^2 = \sigma_u^2 \mathsf{E}\,\|\tilde{\boldsymbol{w}}_{i-1}\|^2$$

and since also

$$\mathsf{E}\,|\boldsymbol{e}(i)|^2 = \sigma_v^2 + \mathsf{E}\,|\boldsymbol{e}_a(i)|^2$$

some simple algebra will show that the learning curve of the filter evolves according to the recursion:

$$\mathsf{E}\,|\boldsymbol{e}(i)|^2 = [1 - 2\mu\sigma_u^2 + \mu^2\sigma_u^4(M + \gamma - 1)]\;\mathsf{E}\,|\boldsymbol{e}(i-1)|^2 \;+\; \mu\sigma_v^2\sigma_u^2[2 - \mu(\gamma - 1)\sigma_u^2]$$

which we rewrite more compactly as

$$\boxed{\mathsf{E}\,|\boldsymbol{e}(i)|^2 = \alpha \mathsf{E}\,|\boldsymbol{e}(i-1)|^2 + \beta} \tag{23.61}$$

with the constants $\{\alpha, \beta\}$ defined by

$$\boxed{\alpha \triangleq 1 - 2\mu\sigma_u^2 + \mu^2\sigma_u^4(M + \gamma - 1), \qquad \beta \triangleq \mu\sigma_v^2\sigma_u^2[2 - \mu(\gamma - 1)\sigma_u^2]}$$

Mean-square stability of the filter requires $|\alpha| < 1$. Actually, it is easy to see that $\alpha > 0$ in this example since α can be written as the sum of two positive terms,

$$\alpha = (1 - \mu\sigma_u^2)^2 \;+\; \mu^2\sigma_u^4(M + \gamma - 2)$$

so that α must satisfy $0 < \alpha < 1$ for stability. It then follows from (23.61) that the steady-state mean-square error (MSE) of the filter is given by

$$\begin{aligned} \text{MSE} &= \mathsf{E}\,|\boldsymbol{e}(\infty)|^2 \\ &= \frac{\beta}{1 - \alpha} \\ &= \frac{\sigma_v^2[2 - \mu(\gamma - 1)\sigma_u^2]}{2 - \mu\sigma_u^2(M + \gamma - 1)} \end{aligned} \tag{23.62}$$

To proceed, we shall characterize the convergence time of an adaptive filter as follows.

Definition 23.1 (Convergence time) The convergence time of an adaptive filter is the number of iterations, $\mathcal{K}$, that is needed for its mean-square error to reach $(1+\epsilon)$ times its steady-state value, for some given $\epsilon > 0$. That is, it is the time $\mathcal{K}$ at which

$$\mathsf{E}\,|\boldsymbol{e}(\mathcal{K})|^2 = (1 + \epsilon)\mathsf{E}\,|\boldsymbol{e}(\infty)|^2$$

For example, choosing $\epsilon = 0.1$ amounts to requiring the mean-square error of the filter to be within 10% of its steady-state value. To apply the definition to LMS, we first rewrite expression (23.61) more conveniently as

$$\left(\mathsf{E}\,|e(i)|^2 - \frac{\beta}{1-\alpha}\right) = \alpha\left(\mathsf{E}\,|e(i-1)|^2 - \frac{\beta}{1-\alpha}\right)$$

i.e., we center the mean-square error around its steady-state value. Iterating we obtain

$$\left(\mathsf{E}\,|e(i)|^2 - \mathsf{MSE}\right) = \alpha^i\left(\mathsf{E}\,|e(0)|^2 - \mathsf{MSE}\right) \tag{23.63}$$

Using

$$e(0) = e_a(0) + v(0) = u_0 w^o + v(0)$$

(since $\tilde{w}_{-1} = w^o$), we can express $\mathsf{E}\,|e(0)|^2$ in terms of the SNR at the output of the model filter w^o. Specifically, we have

$$\mathsf{E}\,|e(0)|^2 = \sigma_v^2 + \sigma_u^2\|w^o\|^2 = \sigma_v^2\left(1 + \frac{\sigma_u^2\|w^o\|^2}{\sigma_v^2}\right) = \sigma_v^2\,(1 + \mathsf{SNR})$$

where the ratio

$$\mathsf{SNR} = \sigma_u^2\|w^o\|^2/\sigma_v^2$$

denotes the output SNR. Substituting into (23.63) gives

$$\left(\mathsf{E}\,|e(i)|^2 - \mathsf{MSE}\right) = \alpha^i\left(\sigma_v^2(1+\mathsf{SNR}) - \mathsf{MSE}\right)$$

Setting

$$i = \mathcal{K}, \quad \mathsf{E}\,|e(\mathcal{K})|^2 = (1+\epsilon)\mathsf{MSE}$$

and solving for $\mathcal{K}$, we arrive at the expression

$$\mathcal{K}\ln(\alpha) = \ln\left[\frac{\epsilon\mathsf{MSE}}{\sigma_v^2(1+\mathsf{SNR}) - \mathsf{MSE}}\right]$$

We can rework this result into an equivalent form in terms of the filter misadjustment as follows. Recall that

$$\mathsf{MSE} = \sigma_v^2 + \mathsf{EMSE}$$

with EMSE denoting the filter *excess* mean-square error (i.e., $\mathsf{EMSE} = \mathsf{E}\,|e_a(\infty)|^2$). Recall further that the filter misadjustment is $\mathcal{M} = \mathsf{EMSE}/\sigma_v^2$. Substituting into the expression for $\mathcal{K}$ we get

$$\boxed{\mathcal{K} = \frac{\ln\left[\dfrac{\epsilon(1+\mathcal{M})}{\mathsf{SNR} - \mathcal{M}}\right]}{\ln(\alpha)}} \quad \text{(iterations)} \tag{23.64}$$

Moreover, from (23.62),

$$\mathsf{EMSE} = \frac{\mu\sigma_v^2\sigma_u^2 M}{2 - \mu\sigma_u^2(M + \gamma - 1)}$$

so that

$$\boxed{\mathcal{M} = \frac{\mu\sigma_u^2 M}{2 - \mu\sigma_u^2(M+\gamma-1)}}$$

The resulting expression (23.64) for the convergence time $\mathcal{K}$ is seen to be dependent on the step-size, the filter length, and the output SNR.

Figure 23.2 plots the theoretical and simulated values of the convergence time $\mathcal{K}$ (for $\epsilon = 0.1$) and the theoretical values for the filter misadjustment $\mathcal{M}$, against the step-size μ, for an LMS implementation with 10 taps, unit-variance Gaussian input, and output SNR at 30 dB. The values obtained for $\mathcal{K}$ are listed in Table 23.1. The simulated values were obtained by averaging over 100 experiments and by running each experiment over 500000 iterations. It is seen from the top plot in Fig. 23.2 that the convergence time decreases as μ increases, which is an expected behavior. The plot uses logarithmic scales for both the vertical and horizontal axes, in which case the dependency of $\ln \mathcal{K}$ on $\ln \mu$ is seen to be linear. It is further seen from the bottom plot in the figure that the filter misadjustment increases with μ. This behavior conflicts with the one exhibited by the convergence time. Often in practice, the designer will be faced with the necessity to compromise on the choice of the step-size: smaller values for better misadjustment vs. larger values for shorter convergence time.

TABLE 23.1 Simulated and experimental convergence time (in terms of number of iterations) for an LMS implementation with 10 taps and using white Gaussian input.

μ	0.00001	0.00005	0.0001	0.0005	0.001	0.005	0.01
$\mathcal{K}_{\rm thy}$	460537	92123	46072	9230	4625	942	482
$\mathcal{K}_{\rm sim}$	479501	92501	46501	9501	4501	501	501

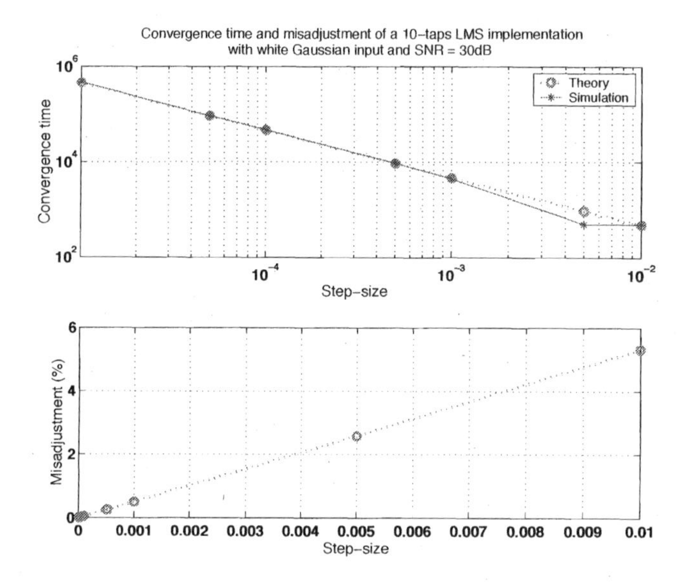

FIGURE 23.2 Plots of the convergence time (*top*) and misadjustment (*bottom*) for a 10-tap LMS implementation with output SNR set at 20 dB.

CHAPTER 24

LMS with non-Gaussian Regressors

We now drop the Gaussian assumption on the regressors and show how the variance relation of Thm. 22.3 can be used to study the performance of LMS in this case as well. Although we are using the words *non-Gaussian regressors* in the title of this chapter, the results herein include the Gaussian case as a special case as well.

24.1 MEAN AND VARIANCE RELATIONS

Thus, refer again to the transformed recursion (23.7), which characterizes the transient performance of LMS. When the regressors $\boldsymbol{u}_i$ were Gaussian, we were able to evaluate the three moments below (see (23.9)–(23.11)):

$$\mathsf{E}\,\overline{\boldsymbol{u}}_i^*\overline{\boldsymbol{u}}_i, \quad \mathsf{E}\,\|\overline{\boldsymbol{u}}_i\|^2_{\overline{\Sigma}} \quad \text{and} \quad \mathsf{E}\,\|\overline{\boldsymbol{u}}_i\|^2_{\overline{\Sigma}}\overline{\boldsymbol{u}}_i^*\overline{\boldsymbol{u}}_i \tag{24.1}$$

In particular, we found that $\mathsf{E}\,\overline{\boldsymbol{u}}_i^*\overline{\boldsymbol{u}}_i$ and $\mathsf{E}\,\|\overline{\boldsymbol{u}}_i\|_{\overline{\Sigma}}\overline{\boldsymbol{u}}_i^*\overline{\boldsymbol{u}}_i$ were simultaneously diagonal and that the weighting matrices $\{\overline{\Sigma},\overline{\Sigma}'\}$ themselves could be made diagonal as well — see (23.13).

However, when the regressors $\boldsymbol{u}_i$ are non-Gaussian, it is generally not possible to express the last moment in (24.1) in closed-form any longer (as we did in (23.11); see though Prob. V.11). In addition, and more importantly perhaps, the moments $\mathsf{E}\,\overline{\boldsymbol{u}}_i^*\overline{\boldsymbol{u}}_i$ and $\mathsf{E}\,\|\overline{\boldsymbol{u}}_i\|_{\overline{\Sigma}}\overline{\boldsymbol{u}}_i^*\overline{\boldsymbol{u}}_i$ need not be simultaneously diagonal anymore. In this way, the weighting matrix $\overline{\Sigma}'$ need not be diagonal even if $\overline{\Sigma}$ is.

Nevertheless, the transient analysis of LMS can still be pursued in much the same way as we did in the Gaussian case if we replace the diag$\{\cdot\}$ notation in (23.15) by an alternative vec$\{\cdot\}$ notation. Before doing so, we remark that since the weighting matrices $\{\overline{\Sigma}',\overline{\Sigma}\}$ are not necessarily diagonal anymore, we shall pursue our analysis by working with the original variance and mean relations (23.3), instead of the transformed variance and mean relations (23.7), namely, we shall now work with

$$\left\{\begin{array}{rcl} \mathsf{E}\,\|\tilde{\boldsymbol{w}}_i\|^2_{\Sigma} & = & \mathsf{E}\,\|\tilde{\boldsymbol{w}}_{i-1}\|^2_{\Sigma'} + \mu^2\sigma_v^2\mathsf{E}\,\|\boldsymbol{u}_i\|^2_{\Sigma} \\ \Sigma' & = & \Sigma - \mu\Sigma\mathsf{E}\left[\boldsymbol{u}_i^*\boldsymbol{u}_i\right] - \mu\mathsf{E}\left[\boldsymbol{u}_i^*\boldsymbol{u}_i\right]\Sigma + \mu^2\mathsf{E}\left[\|\boldsymbol{u}_i\|^2_{\Sigma}\boldsymbol{u}_i^*\boldsymbol{u}_i\right] \\ \mathsf{E}\,\tilde{\boldsymbol{w}}_i & = & (\mathrm{I}-\mu R_u)\,\mathsf{E}\,\tilde{\boldsymbol{w}}_{i-1} \end{array}\right. \tag{24.2}$$

Vector Notation

The diag$\{\cdot\}$ notation allowed us to replace an $M\times M$ matrix by an $M\times 1$ column vector whose entries were the diagonal entries of the matrix. More generally, we shall use the vec$\{\cdot\}$ notation to replace an $M\times M$ arbitrary matrix by an $M^2\times 1$ column vector whose entries are formed by stacking the successive columns of the matrix on top of each other.

We shall therefore write

$$\boxed{\sigma = \text{vec}(\Sigma) \qquad (\sigma \text{ is now } M^2 \times 1)} \tag{24.3}$$

with σ denoting the vectorized version of Σ. Likewise, we shall write r to denote the vectorized version of R_u,

$$\boxed{r = \text{vec}(R_u) \qquad (r \text{ is } M^2 \times 1)} \tag{24.4}$$

and r' to denote to the vectorized version of R_u^{T},

$$\boxed{r' = \text{vec}(R_u^{\mathsf{T}})} \tag{24.5}$$

When the regressors are real-valued, so that $R_u = R_u^{\mathsf{T}}$, the vectors $\{r, r'\}$ will coincide. However, when the regressors are complex-valued, we need to distinguish between r and r'.

In addition, we shall use the notation $\text{vec}^{-1}\{\cdot\}$ to recover a matrix from its vec representation. Thus, writing $\text{vec}^{-1}\{a\}$ for an $M^2 \times 1$ column vector a, results in an $M \times M$ matrix whose entries are obtained by unstacking the elements of a. This choice of notation is in contrast to the $\text{diag}\{\cdot\}$ operation, which is generally accepted as a two-directional operation: it maps diagonal matrices to vectors and vectors into diagonal matrices. Therefore, we shall write

$$\boxed{\Sigma = \text{vec}^{-1}\{\sigma\} \qquad \text{and} \qquad R_u = \text{vec}^{-1}\{r\}} \tag{24.6}$$

to recover $\{\Sigma, R_u\}$ from $\{\sigma, r\}$.

Kronecker Products

The $\text{vec}\{\cdot\}$ notation is most convenient when working with Kronecker products (see Sec. B.7). The Kronecker product of two matrices A and B, say, of dimensions $m_a \times n_a$ and $m_b \times n_b$, respectively, is denoted by $A \otimes B$ and is defined as the $m_a m_b \times n_a n_b$ matrix

$$A \otimes B \triangleq \begin{bmatrix} a_{1,1}B & a_{1,2}B & \ldots & a_{1,n_a}B \\ a_{2,1}B & a_{2,2}B & \ldots & a_{2,n_a}B \\ \vdots & \vdots & & \vdots \\ a_{m_a,1}B & a_{m_a,2}B & \ldots & a_{m_a,n_a}B \end{bmatrix}$$

This operation has several useful properties (see Lemma B.8), but the one that most interests us here is the following. For any matrices $\{P, \Sigma, Q\}$ of compatible dimensions, it holds that

$$\boxed{\text{vec}(P\Sigma Q) = (Q^{\mathsf{T}} \otimes P)\text{vec}(\Sigma)} \tag{24.7}$$

This property tells us how the vec of the product of three matrices is related to the vec of the center matrix.

Linear Vector Relation

With the above notations, we can now verify that expression (24.2) for Σ' in terms of Σ still amounts to a linear relation between the corresponding vectors $\{\sigma, \sigma'\}$; just like it was the case for $\{\overline{\Sigma}, \overline{\Sigma}'\}$ in (23.17). Indeed, applying (24.7) to some of the terms in the expression for Σ' in (24.2) we find that

$$\begin{aligned}
\text{vec}(\Sigma \mathsf{E}\,[\boldsymbol{u}_i^*\boldsymbol{u}_i]) &= ([\mathsf{E}\,\boldsymbol{u}_i^*\boldsymbol{u}_i]^\mathsf{T} \otimes \mathrm{I}_M)\sigma \;=\; (R_u^\mathsf{T} \otimes \mathrm{I}_M)\sigma \\
\text{vec}(\mathsf{E}\,[\boldsymbol{u}_i^*\boldsymbol{u}_i]\,\Sigma) &= (\mathrm{I}_M \otimes [\mathsf{E}\,\boldsymbol{u}_i^*\boldsymbol{u}_i])\sigma \;=\; (\mathrm{I}_M \otimes R_u)\sigma \\
\text{vec}(\mathsf{E}\,\|\boldsymbol{u}_i\|_\Sigma^2 \boldsymbol{u}_i^*\boldsymbol{u}_i) &= \text{vec}(\mathsf{E}\,[\boldsymbol{u}_i^*\boldsymbol{u}_i\Sigma\boldsymbol{u}_i^*\boldsymbol{u}_i]) \;=\; \mathsf{E}\left([\boldsymbol{u}_i^*\boldsymbol{u}_i]^\mathsf{T} \otimes [\boldsymbol{u}_i^*\boldsymbol{u}_i]\right)\sigma
\end{aligned}$$

Taking the vec of both sides of (24.2), and using the above equalities, we find that the weighting vectors $\{\sigma, \sigma'\}$ satisfy a relation similar to (23.21), albeit one that is M^2-dimensional,

$$\boxed{\sigma' = F\sigma} \tag{24.8}$$

where F is $M^2 \times M^2$ and given by

$$\boxed{F \triangleq \mathrm{I}_{M^2} \;-\; \mu(\mathrm{I}_M \otimes R_u) \;-\; \mu(R_u^\mathsf{T} \otimes \mathrm{I}_M) \;+\; \mu^2 \mathsf{E}\left([\boldsymbol{u}_i^*\boldsymbol{u}_i]^\mathsf{T} \otimes [\boldsymbol{u}_i^*\boldsymbol{u}_i]\right)} \tag{24.9}$$

or, more compactly, in factored form:

$$F \triangleq \mathsf{E}\left[(\mathrm{I}_M \;-\; \mu\boldsymbol{u}_i^*\boldsymbol{u}_i)^\mathsf{T} \otimes (\mathrm{I}_M \;-\; \mu\boldsymbol{u}_i^*\boldsymbol{u}_i)\right] \tag{24.10}$$

Variance Relation

We can further rewrite recursion (24.2) for $\mathsf{E}\,\|\tilde{\boldsymbol{w}}_i\|_\Sigma^2$ by using the weighting vectors $\{\sigma, \sigma'\}$ instead of the matrices $\{\Sigma, \Sigma'\}$. Using (24.8) and the notation (24.6), recursion (24.2) becomes

$$\mathsf{E}\,\|\tilde{\boldsymbol{w}}_i\|^2_{\text{vec}^{-1}\{\sigma\}} \;=\; \mathsf{E}\,\|\tilde{\boldsymbol{w}}_{i-1}\|^2_{\text{vec}^{-1}\{F\sigma\}} \;+\; \mu^2\sigma_v^2(r'^\mathsf{T}\sigma)$$

where, for the last term, we used the fact that $\mathsf{E}\,\|\boldsymbol{u}_i\|_\Sigma^2 \;=\; \mathsf{Tr}(R_u\Sigma) \;=\; r'^\mathsf{T}\sigma$. For compactness of notation, we shall drop the $\text{vec}^{-1}\{\cdot\}$ notation from the subscripts and keep the vectors, so that the above can be rewritten more compactly as

$$\boxed{\mathsf{E}\,\|\tilde{\boldsymbol{w}}_i\|_\sigma^2 = \mathsf{E}\,\|\tilde{\boldsymbol{w}}_{i-1}\|_{F\sigma}^2 \;+\; \mu^2\sigma_v^2(r'^\mathsf{T}\sigma)} \tag{24.11}$$

The vector weighting factors $\{\sigma, F\sigma\}$ in this expression should be understood as compact representations for the actual weighting matrices $\{\text{vec}^{-1}\{\sigma\}, \text{vec}^{-1}\{F\sigma\}\}$. In other words, if σ is any column vector, the compact notation $\|x\|_\sigma^2$ denotes

$$\|x\|_\sigma^2 \;\triangleq\; \|x\|^2_{\text{vec}^{-1}\{\sigma\}} \;=\; x^*\Sigma x, \quad \text{where} \quad \Sigma = \text{vec}^{-1}\{\sigma\}$$

In summary, starting from (24.2), we argued that the weighting matrices $\{\Sigma, \Sigma'\}$ can be vectorized, so that (24.2) can be equivalently expressed more compactly as in (24.8)–(24.10) and (24.11), namely,

$$\begin{aligned}
\mathsf{E}\,\|\tilde{\boldsymbol{w}}_i\|_\sigma^2 &= \mathsf{E}\,\|\tilde{\boldsymbol{w}}_{i-1}\|_{F\sigma}^2 \;+\; \mu^2\sigma_v^2(r'^\mathsf{T}\sigma) && (24.12) \\
\sigma' &= F\sigma && (24.13) \\
F &= \mathrm{I}_{M^2} - \mu(\mathrm{I}_M \otimes R_u) - \mu(R_u^\mathsf{T} \otimes \mathrm{I}_M) + \mu^2\mathsf{E}\left([\boldsymbol{u}_i^*\boldsymbol{u}_i]^\mathsf{T} \otimes [\boldsymbol{u}_i^*\boldsymbol{u}_i]\right) && (24.14) \\
\mathsf{E}\,\tilde{\boldsymbol{w}}_i &= (\mathrm{I} - \mu R_u)\,\mathsf{E}\,\tilde{\boldsymbol{w}}_{i-1} && (24.15)
\end{aligned}$$

24.2 MEAN-SQUARE STABILITY AND PERFORMANCE

Although the coefficient matrix in the recursion for $\mathsf{E}\,\tilde{w}_i$ is $(\mathrm{I} - \mu R_u)$, this recursion is equivalent under the unitary transformation (22.31) to (23.23), so that the same condition on μ from (23.24) guarantees convergence in the mean, i.e.,

$$\mu < 2/\lambda_{\max} \tag{24.16}$$

guarantees $\mathsf{E}\,w_i \rightarrow w^o$.

In addition, the mean-square behavior of LMS in the non-Gaussian case is characterized by Eqs. (24.12)–(24.15) in a manner that is similar to equations (23.20)–(23.23) in the Gaussian case. The main difference is the dimension of the variables involved. In the non-Gaussian case, the vector σ is $M^2 \times 1$ and the matrix F is $M^2 \times M^2$. Apart from this difference, all other arguments that were employed before regarding mean-square stability, excess mean-square error, and mean-square deviation extend almost literally to the present case. In so doing, we will be able to conclude that the mean-square behavior of LMS is now characterized by the following M^2-dimensional state-space recursion (see Prob. V.7):

$$\boxed{\mathcal{W}_i = \mathcal{F}\mathcal{W}_{i-1} + \mu^2\sigma_v^2\mathcal{Y}} \tag{24.17}$$

where $\mathcal{F}$ is the companion matrix

$$\mathcal{F} = \begin{bmatrix} 0 & 1 & & & \\ 0 & 0 & 1 & & \\ 0 & 0 & 0 & 1 & \\ \vdots & & & & \\ 0 & 0 & 0 & & 1 \\ -p_0 & -p_1 & -p_2 & \dots & -p_{M^2-1} \end{bmatrix} \qquad (M^2 \times M^2) \tag{24.18}$$

with

$$p(x) \stackrel{\Delta}{=} \det(x\mathrm{I} - F) = x^{M^2} + \sum_{k=0}^{M^2-1} p_k x^k$$

denoting the characteristic polynomial of F in (24.14). Also, $\mathcal{W}_i$ and $\mathcal{Y}$ are now the $M^2 \times 1$ vectors

$$\mathcal{W}_i \stackrel{\Delta}{=} \begin{bmatrix} \mathsf{E}\,\|\tilde{w}_i\|^2_{\sigma} \\ \mathsf{E}\,\|\tilde{w}_i\|^2_{F\sigma} \\ \mathsf{E}\,\|\tilde{w}_i\|^2_{F^2\sigma} \\ \vdots \\ \mathsf{E}\,\|\tilde{w}_i\|^2_{F^{(M^2-1)}\sigma} \end{bmatrix}, \qquad \mathcal{Y} = \begin{bmatrix} r'^{\mathsf{T}}\sigma \\ r'^{\mathsf{T}}F\sigma \\ r'^{\mathsf{T}}F^2\sigma \\ \vdots \\ r'^{\mathsf{T}}F^{M^2-1}\sigma \end{bmatrix} \tag{24.19}$$

for any σ of interest, e.g., most commonly, $\sigma = q$ or $\sigma = r$.

We still need to specify the condition on μ for mean-square stability. To do so, we start by expressing the matrix F in (24.14) in the form

$$F = \mathrm{I}_{M^2} - \mu A + \mu^2 B \tag{24.20}$$

with Hermitian matrices $\{A, B\}$ given by

$$A = (\mathrm{I}_M \otimes R_u) + (R_u^{\mathsf{T}} \otimes \mathrm{I}_M), \qquad B = \mathsf{E}\left([u_i^* u_i]^{\mathsf{T}} \otimes [u_i^* u_i]\right) \tag{24.21}$$

Actually, A is positive-definite and B is nonnegative-definite. We shall assume that the distribution of the regression sequence is such that B is a finite matrix. For mean-square stability we need to find conditions on μ in order to guarantee $-1 < \lambda(F) < 1$. However, contrary to the Gaussian case in (23.34), the matrix F is no longer guaranteed to be nonnegative-definite in general — see Prob. V.4 for an example. In this way, the result of Prob. V.3, which was used in the Gaussian case, is not immediately applicable since it assumed $F \geq 0$. The condition given in that problem, namely, $\mu < 1/\lambda_{\max}(A^{-1}B)$ guarantees $\lambda(F) < 1$. A second condition on μ is needed to enforce $-1 < \lambda(F)$. It is shown in App. 25.A that the matrix F will be stable for values of μ in the range:

$$0 < \mu < \min\left\{\frac{1}{\lambda_{\max}(A^{-1}B)}, \frac{1}{\max\{\lambda(H) \in \mathbb{R}^+\}}\right\} \tag{24.22}$$

where the second condition is in terms of the largest positive real eigenvalue of the following block matrix,

$$\boxed{H \triangleq \begin{bmatrix} A/2 & -B/2 \\ \mathrm{I}_{M^2} & 0 \end{bmatrix}} \quad (2M \times 2M) \tag{24.23}$$

when it exists. If H does not have any real positive eigenvalue, then the corresponding condition is removed from (24.22) and we only require $\mu < 1/\lambda_{\max}(A^{-1}B)$. Conditions (24.16) and (24.22) can be grouped together into a single condition as follows:

$$\boxed{0 < \mu < \min\left\{\frac{2}{\lambda_{\max}}, \frac{1}{\lambda_{\max}(A^{-1}B)}, \frac{1}{\max\{\lambda(H) \in \mathbb{R}^+\}}\right\}} \tag{24.24}$$

Theorem 24.1 (Stability of LMS for non-Gaussian regressors) Assume the data $\{\boldsymbol{d}(i), \boldsymbol{u}_i\}$ satisfy model (22.1) and the independence assumption (22.23). The regressors need not be Gaussian. Then the LMS filter (23.1) is convergent in the mean and is mean-square stable if the step-size μ is chosen to satisfy (24.24), where the matrices A and B are defined by (24.21) and B is finite. Moreover, the transient behavior of LMS is characterized by the M^2-dimensional state-space recursion (24.17)–(24.19), and the mean-square deviation and excess mean-square error are given by

$$\text{MSD} = \mu^2\sigma_v^2 r'^{\mathsf{T}}(\mathrm{I} - F)^{-1}q, \qquad \text{EMSE} = \mu^2\sigma_v^2 r'^{\mathsf{T}}(\mathrm{I} - F)^{-1}r$$

where $r = \text{vec}(R_u)$, $r' = \text{vec}(R_u^{\mathsf{T}})$, and $q = \text{vec}(\mathrm{I})$.

The expressions for the MSD and EMSE in the statement of the theorem are derived in a manner similar to (23.51) and (23.56). They can be rewritten as

$$\text{MSD} = \mu^2\sigma_v^2 \mathsf{E}\,\|\boldsymbol{u}_i\|^2_{(\mathrm{I}-F)^{-1}q} \quad \text{and} \quad \text{EMSE} = \mu^2\sigma_v^2 \mathsf{E}\,\|\boldsymbol{u}_i\|^2_{(\mathrm{I}-F)^{-1}r}$$

or, equivalently, as

$$\boxed{\text{MSD} = \mu^2\sigma_v^2 \text{Tr}(R_u\Sigma_{\text{msd}}) \quad \text{and} \quad \text{EMSE} = \mu^2\sigma_v^2 \text{Tr}(R_u\Sigma_{\text{emse}})}$$

where $\{\Sigma_{\text{msd}}, \Sigma_{\text{emse}}\}$ are the weighting matrices that correspond to the vectors $\sigma_{\text{msd}} = (\mathrm{I} - F)^{-1}q$ and $\sigma_{\text{emse}} = (\mathrm{I} - F)^{-1}r$. That is,

$$\Sigma_{\text{msd}} = \text{vec}^{-1}(\sigma_{\text{msd}}) \quad \text{and} \quad \Sigma_{\text{emse}} = \text{vec}^{-1}(\sigma_{\text{emse}})$$

As a side remark, it is immediate to verify, using $(\mathrm{I}-F)=\mu A-\mu^2 B$ and the expressions for $\{A,B\}$ from (24.21), that the $\{\Sigma_{\text{msd}},\Sigma_{\text{emse}}\}$ so defined satisfy the equations

$$\begin{aligned}\mu(R_u\Sigma_{\text{msd}}+\Sigma_{\text{msd}}R_u)-\mu^2\mathsf{E}\,(\boldsymbol{u}_i^*\boldsymbol{u}_i\Sigma_{\text{msd}}\boldsymbol{u}_i^*\boldsymbol{u}_i) &= \mathrm{I}\\ \mu(R_u\Sigma_{\text{emse}}+\Sigma_{\text{emse}}R_u)-\mu^2\mathsf{E}\,(\boldsymbol{u}_i^*\boldsymbol{u}_i\Sigma_{\text{emse}}\boldsymbol{u}_i^*\boldsymbol{u}_i) &= R_u\end{aligned}$$

24.3 SMALL STEP-SIZE APPROXIMATIONS

While the matrix A in (24.21) is readily available, the difficulty in the non-Gaussian case lies in evaluating the matrix B (which involves fourth-order moments). Only in special cases we may be able to evaluate B in closed form (see, e.g., Prob. V.15). In general, for arbitrary distributions of $\boldsymbol{u}_i$, it may not be possible to evaluate B. Still, it can be proved that as long as B is a finite matrix (i.e., as long as the distribution of the regressors is such that the corresponding matrix B of fourth-order moments is finite), then there always exists a small enough μ that satisfies condition (24.24). In other words, under the data conditions (22.1) and (22.23), the LMS filter can be guaranteed to be mean-square stable for sufficiently small step-sizes, regardless of the distribution of the regressors. This fact is established in Prob. V.38.

Moreover, observe that the expressions for the EMSE and MSD that are given in the theorem need not be easy to evaluate in general since they are defined in terms of F, which requires knowledge of B. However, since the fourth-order moment $\mathsf{E}\,\|\boldsymbol{u}_i\|_\Sigma^2\boldsymbol{u}_i^*\boldsymbol{u}_i$, which is part of expression (24.2) for Σ', appears multiplied by μ^2 in (24.2), its effect can be ignored if the step-size is sufficiently small. In this way, the variance relation (24.2) would reduce to

$$\left\{\begin{aligned}\mathsf{E}\,\|\tilde{\boldsymbol{w}}_i\|_\Sigma^2 &= \mathsf{E}\,\|\tilde{\boldsymbol{w}}_{i-1}\|_{\Sigma'}^2+\mu^2\sigma_v^2\mathsf{E}\,\|\boldsymbol{u}_i\|_\Sigma^2\\ \Sigma' &= \Sigma-\mu\Sigma\mathsf{E}\,[\boldsymbol{u}_i^*\boldsymbol{u}_i]-\mu\mathsf{E}\,[\boldsymbol{u}_i^*\boldsymbol{u}_i]\,\Sigma\end{aligned}\right. \tag{24.25}$$

This relation can then be used to derive simplified expressions for the EMSE and MSD of LMS for Gaussian and non-Gaussian regressors, as done in Prob. V.14 where it is shown that

$$\boxed{\mathsf{EMSE}\approx\mu\sigma_v^2\mathsf{Tr}(R_u)/2\quad\text{and}\quad\mathsf{MSD}\approx\mu\sigma_v^2M/2\quad(\text{small }\mu)}$$

This expression for the EMSE is the same one we derived earlier in Lemma 16.1.

Lemma 24.1 (Performance of LMS) Consider the LMS recursion (23.1) and assume the data $\{\boldsymbol{d}(i),\boldsymbol{u}_i\}$ satisfy model (22.1) and the independence assumption (22.23). Assume further that the matrix B from (24.21) is bounded. Then there always exists a small enough μ such that the LMS filter is mean-square stable. Moreover, it holds that for such sufficiently small step-sizes:

$$\mathsf{EMSE}\approx\mu\sigma_v^2\mathsf{Tr}(R_u)/2\quad\text{and}\quad\mathsf{MSD}\approx\mu\sigma_v^2M/2$$

Learning Curve

Finally, since $\mathsf{E}\,|\boldsymbol{e}_a(i)|^2=\mathsf{E}\,\|\tilde{\boldsymbol{w}}_{i-1}\|_{R_u}^2$, we find that the evolution of $\mathsf{E}\,|\boldsymbol{e}_a(i)|^2$ is described by the top entry of the state vector $\mathcal{W}_i$ in (24.17)–(24.19) with σ chosen as $\sigma=r$. In Probs. V.9 and V.10, we derive a recursion for $\mathsf{E}\,|\boldsymbol{e}_a(i)|^2$. Specifically, we show that

$$\mathsf{E}\,|\boldsymbol{e}_a(i)|^2=\mathsf{E}\,|\boldsymbol{e}_a(i-1)|^2+w^{o*}(A_{i-1}-A_{i-2})w^o+\mu^2\sigma_v^2\mathsf{Tr}(R_uA_{i-2}),\quad i\geq 1$$

with initial condition $\mathsf{E}\,|e_a(0)|^2 = w^{o*}R_u w^o$ (assuming $w_{-1} = 0$ so that $\tilde{w}_{-1} = w^o$), and where the matrix A_i is computed via $A_i = \text{vec}^{-1}(a_i)$, where $a_i = Fa_{i-1}$ with initial conditions $a_{-1} = r$ and $A_{-1} = R_u$. The learning curve of the filter is then given by $\mathsf{E}\,|e(i)|^2 = \sigma_v^2 + \mathsf{E}\,|e_a(i)|^2$.

24.A APPENDIX: AVERAGING ANALYSIS

The independence assumption (22.23) is a widely used condition in the transient analysis of adaptive filters. Although not valid in general, because of the tapped-delay-line structure of the regression data in most filter implementations, its value lies in the simplifications it introduces into the analysis. Without the independence assumption, the transient, convergence, and stability analyses of adaptive filters can become highly challenging, even for the simplest of algorithms. Fortunately, there are results in the literature that show, in one way or another, that performance results that are obtained using the independence assumption are reasonably close to actual filter performance when the step-size is sufficiently small. The purpose of this appendix is threefold: (i) to describe some of these results, (ii) to give an overview of averaging analysis, which is a useful method of filter analysis without the independence assumption (albeit one that assumes infinitesimally small step-sizes), and (iii) to combine energy-conservation arguments with averaging analysis in order to show that the results obtained from an averaging analysis are essentially identical to those obtained from an independence analysis.

N-Dependent Regressors

To begin with, one of the earliest results on the accuracy of the independence analysis assumes tapped-delay-line regressors of the form $u_i = [u(i)\ u(i-1)\ \ldots\ u(i-M+1)]$, with an input sequence $\{u(i)\}$ that is generated by passing an i.i.d. *binary* sequence $\{s(k)\}$ through an FIR filter of length L. Specifically,

$$u(i) = \sum_{k=0}^{L-1} h(k)s(i-k) \tag{24.26}$$

where $\{h(k)\}$ denotes the impulse response sequence of the filter. This model is adequate, for example, for equalization applications with BPSK modulation, in which case $\{h(k)\}$ refers to the channel impulse response. A useful property of the model (24.26) with i.i.d. $\{s(k)\}$ is that two regressors $\{u_i, u_j\}$ are truly independent if the time difference $|i-j|$ exceeds the sum $M+L$.

Using model (24.26), it has been shown by Mazo (1979), in the case of LMS, that results obtained from the independence theory are reasonable approximations for the actual filter performance for small step-sizes. The main conclusion is the following (compare the expression for the EMSE in the statement of the theorem below with the one given by (23.58)).

Theorem 24.2 (Binary model) Assume the input sequence $\{u(i)\}$ is generated as in (24.26) with an i.i.d. binary sequence $\{s(k)\}$. Assume further that the reference sequence $\{d(i)\}$ is modeled as in (22.1) with $R_u > 0$. Then it holds that the actual EMSE of LMS is given by

$$\text{EMSE} = \lim_{i\to\infty} \mathsf{E}\,|e_a(i)|^2 \ = \ \mu\sigma_v^2\text{Tr}(R_u)/2 \ + \ O(\mu^2)$$

There exists a generalization of the above result that allows for input sequences $\{s(k)\}$ that are not restricted to a finite alphabet. Instead, $\{s(k)\}$ is allowed to be an arbitrary i.i.d. sequence — compare again the result below with (23.58) and (23.23).

Theorem 24.3 (i.i.d. model) The same conclusion of Thm. 24.2 holds for arbitrary i.i.d. sequences $\{s(k)\}$. Moreover, $\mathsf{E}\,\tilde{w}_i$ converges to its limiting value approximately as a linear system with modes at $\{1-\mu\lambda_k\}$, where the λ_k are the eigenvalues of R_u.

Theorem 24.3 is due to Jones, Cavin, and Reed (1982), and it can be established under the weaker condition of a noise sequence $\{v(i)\}$ that is not required to be i.i.d. or even independent of the regressors $\{u_k\}$. However, for our purposes here, it is sufficient to state the result, as above, for noise sequences $\{v(i)\}$ that satisfy model (22.1).

The arguments that establish Thms. 24.2–24.3 assume that the step-size is sufficiently small to guarantee filter convergence, and subsequently they establish that the resulting steady-state performance can be predicted reasonably well by independence theory. The next result by Macchi and Eweda (1983) goes a step further and shows that a sufficiently small step-size does exist to guarantee filter convergence even when the regressors are not independent. The result relaxes assumption (24.26) and assumes that the extended sequence $\{d(i), u_i\}$ is such that

$$\{\ldots,(d(i-1),u_{i-1}),(d(i),u_i)\} \quad \text{and} \quad \{(d(i+j),u_{i+j}),(d(i+j+1),u_{i+j+1}),\ldots\}$$

are mutually independent whenever $j > N$ for some integer $N > M/12$. Sequences satisfying this property are said to be N*-dependent*.

Theorem 24.4 (N-dependent regressors) Assume the sequence $\{(d(i), u_i)\}$ is $N-$dependent for some integer $N > M/12$, and assume further that the moments of $\{d(i), u_i\}$ are bounded in the following manner:

$$\mathsf{E}\left(\|u_i\|^{2nN}\right) < \infty \quad \text{for all } n \leq 12 \quad \text{and} \quad \mathsf{E}\left(|d(i)|^{4+\frac{4}{6N-1}}\right) < \infty$$

with $R_u > 0$. Then there exists a pair $(\mu_0,\ \beta)$ of positive real numbers such that the weight-error vector $\tilde{w}_i$ of LMS satisfies

$$\limsup_{i\to\infty} \mathsf{E}\,\|\tilde{w}_i\|^2 \leq \beta\mu \qquad \text{for all step-sizes} \quad \mu \leq \mu_0$$

Averaging Analysis

Theorems 24.2–24.4, and the associated arguments and derivations, are specific to LMS. Similar results for a wider class of adaptive filters can be obtained by appealing to a method of analysis known as averaging analysis. In this framework, the regressor sequence $\{u_i\}$ is not required to be N-dependent anymore; instead it is required to satisfy a certain mixing condition, namely, that the correlation between u_j and u_i "dies out" as the time difference $|i-j|$ increases. The regressor sequence $\{u_i\}$ is also required to be bounded, i.e., there should exist $\beta < \infty$ such that

$$\sup_{i>0} \|u_i\|^2 \leq \beta < 2/\mu \qquad \text{with probability 1}$$

Note that this boundedness condition is not satisfied for some important input distributions, e.g., LMS with Gaussian distributed regressors. Averaging theory further requires the step-size to be vanishingly small. Nevertheless, and unlike the previous results, the theory applies to a larger class of adaptive algorithms, with little modification in the basic theorems.

To describe the mixing condition that is imposed on the regressors $\{u_i\}$ we need the following definition (Durrett (1996)).

Definition 24.1 (Uniform-mixing processes) A random process $\boldsymbol{\xi}_i$ is said to be a uniform- (or ϕ-) mixing process if there exists a sequence $\phi(n)$ satisfying $\lim_{n\to\infty}\phi(n) = 0$ such that

$$\sup_{a,b}\left|\text{Prob}(\boldsymbol{\xi}_{i+n} = a|\boldsymbol{\xi}_i = b) - \text{Prob}(\boldsymbol{\xi}_{i+n} = a)\right| \leq \phi(n) \quad \text{for all } i, n$$

That is, the variables $\boldsymbol{\xi}_{i+n}$ and $\boldsymbol{\xi}_i$ become essentially independent as the time difference $|n|$ grows; the notation $\text{Prob}(\boldsymbol{\xi} = a)$ denotes the probability of the event $\boldsymbol{\xi} = a$.

Examples of ϕ-mixing processes are N-dependent processes, processes generated from bounded white-noise filtered through a stable finite-dimensional linear filter, as well as purely nondeterministic processes. Recall that a process $\{\boldsymbol{\xi}_i\}$ is said to be purely nondeterministic if, for any i, $\boldsymbol{\xi}_i$ cannot be perfectly predicted from the previous elements $\{\boldsymbol{\xi}_{i-1}, \boldsymbol{\xi}_{i-2}, \ldots, \boldsymbol{\xi}_{i-k}\}$ for any k. In other words, if $\hat{\boldsymbol{\xi}}_i$ is an estimator for $\boldsymbol{\xi}_i$ given the past values$\{\boldsymbol{\xi}_{i-1}, \boldsymbol{\xi}_{i-2}, \ldots, \boldsymbol{\xi}_{i-k}\}$, then there is a constant $\sigma^2 > 0$ such that $\text{var}(\hat{\boldsymbol{\xi}}_i - \boldsymbol{\xi}_i) \geq \sigma^2 > 0$ even when $i \to \infty$. Here, $\text{var}(\cdot)$ denotes the variance of its argument.

Now consider a general adaptive filter weight-error update of the form

$$\boxed{\tilde{\boldsymbol{w}}_i = \tilde{\boldsymbol{w}}_{i-1} + \mu \boldsymbol{f}(i, \tilde{\boldsymbol{w}}_{i-1}) \quad \text{with initial condition } \tilde{\boldsymbol{w}}_{-1}} \tag{24.27}$$

The function $\boldsymbol{f}$ is stochastic, i.e., for every i and $\tilde{\boldsymbol{w}}_{i-1}$, $\boldsymbol{f}(i, \tilde{\boldsymbol{w}}_{i-1})$ is a random vector. For example, in the LMS case we have

$$\tilde{\boldsymbol{w}}_i = \tilde{\boldsymbol{w}}_{i-1} - \mu \boldsymbol{u}_i^*[\boldsymbol{u}_i\tilde{\boldsymbol{w}}_{i-1} + \boldsymbol{v}(i)]$$

so that

$$\boldsymbol{f}(i, \tilde{\boldsymbol{w}}_{i-1}) = -\boldsymbol{u}_i^*\boldsymbol{u}_i\tilde{\boldsymbol{w}}_{i-1} - \boldsymbol{u}_i^*\boldsymbol{v}(i)$$

Define the *averaged function* f_{av},

$$\boxed{f_{av}(i, \tilde{\boldsymbol{w}}_{i-1}) \triangleq \mathsf{E}\,\boldsymbol{f}(i, \tilde{\boldsymbol{w}}_{i-1})} \tag{24.28}$$

where $\tilde{\boldsymbol{w}}_{i-1}$ is considered *constant* for the computation of the expected value. Again, for LMS, the averaged function is given by

$$f_{av,\text{LMS}}(i, \tilde{\boldsymbol{w}}_{i-1}) = \mathsf{E}\left(-\boldsymbol{u}_i^*\boldsymbol{u}_i\tilde{\boldsymbol{w}}_{i-1} - \boldsymbol{u}_i^*\boldsymbol{v}(i)\right) = -R_u\tilde{\boldsymbol{w}}_{i-1}$$

where we are assuming that the data $\{\boldsymbol{d}(i), \boldsymbol{u}_i\}$ satisfy model (22.1). Define further the *averaged system*:

$$\boxed{\tilde{\boldsymbol{w}}_i^{av} = \tilde{\boldsymbol{w}}_{i-1}^{av} + \mu f_{av}(i, \tilde{\boldsymbol{w}}_{i-1}^{av}), \qquad \tilde{\boldsymbol{w}}_{-1}^{av} = \tilde{\boldsymbol{w}}_{-1}} \tag{24.29}$$

where the stochastic function $\boldsymbol{f}(\cdot,\cdot)$ in (24.27) is replaced by its averaged value and, accordingly, the corresponding weight-error vectors are denoted by $\tilde{\boldsymbol{w}}_i^{av}$. Again, for LMS, for example, we have

$$\tilde{\boldsymbol{w}}_i^{av,\text{LMS}} = (\mathrm{I} - \mu R_u)\tilde{\boldsymbol{w}}_{i-1}^{av,\text{LMS}}$$

It is clear from this simple example that an averaged system is not useful to estimate the steady-state performance of an adaptive filter (for instance, the noise information is lost). To do so, it is helpful to consider instead a *partially averaged system*, defined as

$$\boxed{\tilde{\boldsymbol{w}}_i^{pav} = [\mathrm{I} + \mu\nabla_{\tilde{w}} f_{av}(0)]\,\tilde{\boldsymbol{w}}_{i-1}^{pav} + \mu[\boldsymbol{f}(i,0) - f_{av}(i,0)]} \tag{24.30}$$

where $\nabla_{\tilde{w}} f_{av}(0)$ denotes the value of the gradient of $f_{av}(i, \tilde{w})$ with respect to $\tilde{w}$ evaluated at the origin. Using the LMS algorithm again as an example, we have

$$\nabla_{\tilde{w}} f_{av}(0) = -R_u \quad \text{and} \quad \boldsymbol{f}(i,0) - f_{av}(i,0) = -\boldsymbol{u}_i^*\boldsymbol{v}(i)$$

so that

$$\tilde{\boldsymbol{w}}_i^{pav,\text{LMS}} = (\text{I} - \mu R_u)\,\tilde{\boldsymbol{w}}_{i-1}^{pav,\text{LMS}} - \mu \boldsymbol{u}_i^* \boldsymbol{v}(i)$$

The following result from Solo and Kong (1995, Chapter 9) now states that if the step-size μ is sufficiently small, and for fairly general adaptive schemes, the original weight-error vector $\tilde{\boldsymbol{w}}_i$ of (24.27) remains close to the partially averaged weight-error vector $\tilde{\boldsymbol{w}}_i^{pav}$ of (24.30), and that the steady-state covariance matrix of $\tilde{\boldsymbol{w}}_i$ is close to that of $\tilde{\boldsymbol{w}}_i^{pav}$ as well.

Theorem 24.5 (Averaging result) Consider a weight-error recursion of the form (24.27) and its averaged forms (24.29) and (24.30), with regressor sequence $\{\boldsymbol{u}_i\}$ assumed to be uniform mixing. Assume further that the following conditions hold:

1. The origin, 0, is an exponentially-stable equilibrium point of the averaged system (24.29) with decay rate $O(\mu)$.
2. The gradient vector $\nabla_{\tilde{w}} f_{av}(i, \tilde{w})$ exists and is continuous at the origin.
3. There exists a positive constant c such that, for any vectors $\boldsymbol{a}$ and $\boldsymbol{b}$, the following so-called Lipschitz condition holds: $\|\nabla_{\tilde{w}} \boldsymbol{f}(i, \boldsymbol{a}) - \nabla_{\tilde{w}} \boldsymbol{f}(i, \boldsymbol{b})\| \leq c\|\boldsymbol{a} - \boldsymbol{b}\|$.

Under these conditions, $\tilde{\boldsymbol{w}}_i$ and $\tilde{\boldsymbol{w}}_i^{pav}$ obtained from (24.27) and (24.30) satisfy

$$\lim_{\mu\to 0} \sup_{i\geq 0} \mathsf{E}\left\|\tilde{\boldsymbol{w}}_i - \tilde{\boldsymbol{w}}_i^{pav}\right\| = 0, \qquad \lim_{\mu\to 0}\lim_{i\to\infty} \mathsf{E}\left\|\tilde{\boldsymbol{w}}_i - \tilde{\boldsymbol{w}}_i^{pav}\right\|^2 = 0$$

$$\lim_{\mu\to 0}\lim_{i\to\infty} \left(\frac{1}{\mu}\mathsf{E}\,\tilde{\boldsymbol{w}}_i\tilde{\boldsymbol{w}}_i^*\right) = \lim_{\mu\to 0}\lim_{i\to\infty} \left(\frac{1}{\mu}\mathsf{E}\,\tilde{\boldsymbol{w}}_i^{pav}\tilde{\boldsymbol{w}}_i^{pav,*}\right)$$

In view of the above statement, and for sufficiently small step-sizes, we may evaluate the performance of an adaptive filter by studying the performance of its partially-averaged system (24.30). For example, for nondeterministic regressors $\{\boldsymbol{u}_i\}$, and assuming the noise sequence $\{\boldsymbol{v}(i)\}$ satisfies model (22.1), the steady-state EMSE performance of LMS can again be deduced to be

$$\text{EMSE} = \lim_{i\to\infty} \mathsf{E}\,|\boldsymbol{e}_a(i)|^2 = \mu\sigma_v^2 \text{Tr}(R_u)/2 + O(\mu^2)$$

as we verify next in greater detail.

Energy Conservation and Averaging Analysis

We can combine energy-conservation arguments with averaging analysis in order to evaluate the performance of adaptive filters. For brevity, we shall focus only on LMS and NLMS for which the functions[7] $\boldsymbol{f}(i, \tilde{\boldsymbol{w}}_{i-1})$, $\nabla_{\tilde{w}} f_{av}(0)$, and $(\boldsymbol{f}(i,0) - f_{av}(i,0))$ are listed in Table 24.1, where we are also introducing the stochastic function $\boldsymbol{f}_{\triangle}(i,0)$ defined via the identity

$$\boxed{\boldsymbol{f}(i,0) - f_{av}(i,0) = \boldsymbol{u}_i^* \boldsymbol{f}_{\triangle}(i,0)}$$

With this definition, the partially-averaged system (24.30) becomes

$$\boxed{\tilde{\boldsymbol{w}}_i^{pav} = [\text{I} + \mu \nabla_{\tilde{w}} f_{av}(0)]\tilde{\boldsymbol{w}}_{i-1}^{pav} + \mu \boldsymbol{u}_i^* \boldsymbol{f}_{\triangle}(i,0)} \tag{24.31}$$

We are now in a position to study the steady-state and transient behavior of such systems.

Steady-state analysis. Assume the data $\{\boldsymbol{d}(i), \boldsymbol{u}_i\}$ satisfy model (22.1) and let $\boldsymbol{e}(i) = \boldsymbol{d}(i) - \boldsymbol{u}_i \boldsymbol{w}_{i-1}$ denote the output error at time i. In steady-state analysis, we are interested in evaluating

[7]For a discussion on a larger class of adaptive filters, including LMF, LMMN, and sign-error LMS, see Shin, Sayed, and Song (2004b).

TABLE 24.1 Listing of functions useful in the averaging analysis of LMS and NLMS.

Algorithm	$f(i,\tilde{w}_{i-1})$	$\nabla_{\tilde{w}} f_{av}(0)$	$f(i,0)-f_{av}(i,0)$	$f_{\triangle}(i,0)$
LMS	$-u_i^*(u_i\tilde{w}_{i-1}+v(i))$	$-R_u$	$-u_i^*v(i)$	$-v(i)$
NLMS	$-\dfrac{u_i^*(u_i\tilde{w}_{i-1}+v(i))}{\lVert u_i\rVert^2}$	$-\mathsf{E}\left[\dfrac{u_i^*u_i}{\lVert u_i\rVert^2}\right]$	$-\dfrac{u_i^*v(i)}{\lVert u_i\rVert^2}$	$-\dfrac{v(i)}{\lVert u_i\rVert^2}$

the steady-state variance of $e(i)$, which translates into the filter mean-square error, i.e., MSE $=\mathsf{E}\,|e(\infty)|^2$. To evaluate the MSE, we shall rely on the same arguments we used in this part and also in Part IV (*Mean-Square Performance*). Let Σ denote any $M\times M$ Hermitian positive-definite matrix, and introduce the weighted *a priori* and *a posteriori* errors

$$e_a^{\Sigma}(i) \;\triangleq\; u_i\Sigma[\mathrm{I}+\mu\nabla_{\tilde{w}}f_{av}(0)]\tilde{w}_{i-1}^{pav}, \qquad e_p^{\Sigma}(i) \;\triangleq\; u_i\Sigma\tilde{w}_i^{pav} \tag{24.32}$$

If we multiply both sides of (24.31) by $u_i\Sigma$ from the left we find that

$$e_p^{\Sigma}(i) = e_a^{\Sigma}(i) + \mu u_i\Sigma u_i^* f_{\triangle}(i,0) \tag{24.33}$$

Solving for $f_{\triangle}(i,0)$ and substituting into (24.31), we arrive at

$$\tilde{w}_i^{pav} + \frac{u_i^*}{\lVert u_i\rVert_{\Sigma}^2}e_a^{\Sigma}(i) = [\mathrm{I}+\mu\nabla_{\tilde{w}}f_{av}(0)]\tilde{w}_{i-1}^{pav} + \frac{u_i^*}{\lVert u_i\rVert_{\Sigma}^2}e_p^{\Sigma}(i)$$

By equating the weighted norms on both sides of this equation, we conclude that the following energy equality should hold:

$$\lVert\tilde{w}_i^{pav}\rVert_{\Sigma}^2 + \bar{\mu}^{\Sigma}(i)\cdot|e_a^{\Sigma}(i)|^2 \;=\; \lVert\tilde{w}_{i-1}^{pav}\rVert_{\Sigma'}^2 + \bar{\mu}^{\Sigma}(i)\cdot|e_p^{\Sigma}(i) \tag{24.34}$$

where

$$\Sigma' \;\triangleq\; [\mathrm{I}+\mu\nabla_{\tilde{w}}f_{av}(0)]^*\Sigma[\mathrm{I}+\mu\nabla_{\tilde{w}}f_{av}(0)]$$

and $\bar{\mu}^{\Sigma}(i) = (\lVert u_i\rVert_{\Sigma}^2)^{\dagger}$. This result is the extension of the energy relation of Thm. 22.1 to the partially-averaged system (24.31).

Replacing $e_p^{\Sigma}(i)$ in (24.34) by its equivalent expression (24.33), taking expectations of both sides, and using $\mathsf{E}\,f_{\triangle}(i,0)=0$, we get

$$\mathsf{E}\,\lVert\tilde{w}_i^{pav}\rVert_{\Sigma}^2 = \mathsf{E}\,\lVert\tilde{w}_{i-1}^{pav}\rVert_{\Sigma'}^2 + \mu^2\mathsf{E}[\,|f_{\triangle}(i,0)|^2\cdot\lVert u_i\rVert_{\Sigma}^2\,] \tag{24.35}$$

For LMS, we have

$$\mathsf{E}[\,|f_{\triangle}(i,0)|^2\cdot\lVert u_i\rVert_{\Sigma}^2\,] \;=\; \mathsf{E}\,|f_{\triangle}(i,0)|^2\cdot\mathsf{E}\,\lVert u_i\rVert_{\Sigma}^2 \;=\; \sigma_v^2\mathsf{E}\,\lVert u_i\rVert_{\Sigma}^2$$

while for NLMS, we shall use the approximation

$$\mathsf{E}[\,|f_{\triangle}(i,0)|^2\cdot\lVert u_i\rVert_{\Sigma}^2\,] = \sigma_v^2\mathsf{E}\left(\frac{\lVert u_i\rVert_{\Sigma}^2}{\lVert u_i\rVert^4}\right) \;\approx\; \frac{\sigma_v^2}{[\mathrm{Tr}(R_u)]^2}\cdot\mathsf{E}\,\lVert u_i\rVert_{\Sigma}^2$$

In this way, the variance relation (24.35) will be replaced by

$$\mathsf{E}\,\lVert\tilde{w}_i^{pav}\rVert_{\Sigma}^2 = \mathsf{E}\,\lVert\tilde{w}_{i-1}^{pav}\rVert_{\Sigma'}^2 + \mu^2\gamma\mathsf{E}\,\lVert u_i\rVert_{\Sigma}^2 \tag{24.36}$$

where the positive scalar γ is defined by

$$\boxed{\gamma = \sigma_v^2 \quad (\text{for LMS}), \qquad \gamma = \frac{\sigma_v^2}{[\text{Tr}(R_u)]^2} \quad (\text{for NLMS})} \tag{24.37}$$

For other adaptive filters, we would proceed with (24.35).

The subsequent analysis can be simplified by appealing to a change of coordinates. Thus, let $R_u = U\Lambda U^*$ denote the eigen-decomposition of R_u, with U unitary, and introduce the transformed quantities:

$$\overline{\boldsymbol{w}}_i^{pav} \triangleq U^*\tilde{\boldsymbol{w}}_i^{pav}, \quad \nabla_{\tilde{w}}\overline{f}_{av}(0) \triangleq U^*\nabla_{\tilde{w}}f_{av}(0)U, \quad \overline{\boldsymbol{u}}_i \triangleq \boldsymbol{u}_iU, \quad \overline{\Sigma} \triangleq U^*\Sigma U$$

Then the variance relation (24.36) retains a similar form:

$$\boxed{\mathsf{E}\,\|\overline{\boldsymbol{w}}_i^{pav}\|^2_{\overline{\Sigma}} = \mathsf{E}\,\|\overline{\boldsymbol{w}}_{i-1}^{pav}\|^2_{\overline{\Sigma}'} + \mu^2\gamma\mathsf{E}\,\|\overline{\boldsymbol{u}}_i\|^2_{\overline{\Sigma}}} \tag{24.38}$$

with

$$\boxed{\overline{\Sigma}' = \overline{\Sigma} + 2\mu\overline{\Sigma}\nabla_{\tilde{w}}\overline{f}_{av}(0) + \mu^2\left[\nabla_{\tilde{w}}\overline{f}_{av}(0)\right]^*\overline{\Sigma}\nabla_{\tilde{w}}\overline{f}_{av}(0)} \tag{24.39}$$

Relation (24.38) can now be used to deduce an expression for the filter MSE as follows. Since, by Thm. 24.5, $\tilde{\boldsymbol{w}}_i$ and $\tilde{\boldsymbol{w}}_i^{pav}$ stay close to each other for small step-sizes, we assume $\boldsymbol{u}_i\tilde{\boldsymbol{w}}_{i-1} \approx \boldsymbol{u}_i\tilde{\boldsymbol{w}}_{i-1}^{pav}$, so that the filter excess-mean-square error, which is defined by

$$\text{EMSE} = \lim_{i\to\infty} \mathsf{E}\,|\boldsymbol{u}_i\tilde{\boldsymbol{w}}_{i-1}|^2$$

can be approximated via

$$\text{EMSE} \approx \lim_{i\to\infty} \mathsf{E}\,|\boldsymbol{u}_i\tilde{\boldsymbol{w}}_{i-1}^{pav}|^2 \;=\; \lim_{i\to\infty} \mathsf{E}\,\|\tilde{\boldsymbol{w}}_{i-1}^{pav}\|^2_{R_u} \;=\; \lim_{i\to\infty} \mathsf{E}\,\|\overline{\boldsymbol{w}}_{i-1}^{pav}\|^2_{\Lambda}$$

where the second equality assumes, in view of (24.31), that $\boldsymbol{u}_i$ and $\tilde{\boldsymbol{w}}_{i-1}^{pav}$ are essentially independent for infinitesimally small step-sizes. Taking the limit of (24.38) as $i \to \infty$, and using the steady-state condition $\mathsf{E}\,\|\overline{\boldsymbol{w}}_i^{pav}\|^2_{\overline{\Sigma}} = \mathsf{E}\,\|\overline{\boldsymbol{w}}_{i-1}^{pav}\|^2_{\overline{\Sigma}}$, we obtain

$$\boxed{\mathsf{E}\,\|\overline{\boldsymbol{w}}_{i-1}^{pav}\|^2_{-2\overline{\Sigma}\nabla_{\tilde{w}}\overline{f}_{av}(0)-\mu[\nabla_{\tilde{w}}\overline{f}_{av}(0)]^*\overline{\Sigma}\nabla_{\tilde{w}}\overline{f}_{av}(0)} = \mu\gamma\text{Tr}(\Lambda\overline{\Sigma})} \tag{24.40}$$

In order to evaluate the EMSE, we need to select $\overline{\Sigma}$ such that the weighting matrix on the left-hand side becomes

$$-2\overline{\Sigma}\nabla_{\tilde{w}}\overline{f}_{av}(0) - \mu[\nabla_{\tilde{w}}\overline{f}_{av}(0)]^*\overline{\Sigma}\nabla_{\tilde{w}}\overline{f}_{av}(0) = \Lambda \tag{24.41}$$

LMS. Using $\nabla_{\tilde{w}}\overline{f}_{av}(0) = -\Lambda$ and $\gamma = \sigma_v^2$, we have that $\overline{\Sigma}$ should be selected to satisfy $2\overline{\Sigma}\Lambda - \mu\Lambda\overline{\Sigma}\Lambda = \Lambda$, that is,

$$\overline{\Sigma} = (2\text{I} - \mu\Lambda)^{-1}$$

Then the left-hand-side of (24.40) becomes the filter EMSE and (24.40) leads to

$$\zeta^{\text{LMS}} = \mu\sigma_v^2\text{Tr}(\Lambda(2\text{I} - \mu\Lambda)^{-1}) = \mu\sigma_v^2\sum_{i=0}^{M-1}\frac{\lambda_i}{2-\mu\lambda_i}$$

For small step-sizes, this expression reduces to $\zeta^{\text{LMS}} = \mu\sigma_v^2\text{Tr}(R_u)/2$, which is the same expression we derived for small step-sizes in Sec. 16.

NLMS. Using $\nabla_{\tilde{w}}\overline{f}_{av}(0) = -\Lambda/\text{Tr}(R_u)$ and $\gamma = \sigma_v^2/[\text{Tr}(R_u)]^2$, then $\overline{\Sigma}$ should be chosen to satisfy

$$2\overline{\Sigma}\frac{\Lambda}{\text{Tr}(R_u)} - \mu\frac{\Lambda\overline{\Sigma}\Lambda}{[\text{Tr}(R_u)]^2} = \Lambda$$

that is,

$$\overline{\Sigma} = (2\text{Tr}(R_u)\text{I} - \mu\Lambda)^{-1}[\text{Tr}(R_u)]^2$$

Then the left-hand-side of (24.40) becomes the filter EMSE and it leads to

$$\zeta^{\text{NLMS}} = \mu\sigma_v^2 \text{Tr}(\Lambda[2\text{Tr}(R_u)\text{I} - \mu\Lambda]^{-1}) = \sum_{i=0}^{M-1} \frac{\mu\sigma_v^2\lambda_i}{2\text{Tr}(R_u) - \mu\lambda_i}$$

For small step-sizes, this expression reduces to $\zeta^{\text{NLMS}} = \mu\sigma_v^2/2$, which is the same expression we derived for small step-sizes in Sec. 17.1.

Transient analysis. The same variance relation (24.38) can be used to characterize the transient behavior of the filters, in addition to the following recursion for the mean weight-error vector (which follows by taking expectations of both sides of (24.31)):

$$\boxed{\mathsf{E}\,\tilde{\boldsymbol{w}}_i^{pav} = [\text{I} + \mu\nabla_{\tilde{w}}\overline{f}_{av}(0)]\mathsf{E}\,\tilde{\boldsymbol{w}}_{i-1}^{pav}} \tag{24.42}$$

Observe that $\overline{\Sigma}'$ in (24.39) will be diagonal if $\overline{\Sigma}$ is. In this way, rather than propagate the diagonal weighting matrices, it is more convenient to rewrite the recursion for $\overline{\Sigma}'$ in terms of its diagonal entries as

$$\overline{\sigma}' = \overline{F}\overline{\sigma} \tag{24.43}$$

where $\sigma = \text{diag}\{\Sigma\}$, $\overline{\sigma} = \text{diag}\{\overline{\Sigma}\}$, and $\lambda = \text{diag}\{\Lambda\}$, while the $M \times M$ matrix $\overline{F}$ is defined by

$$\overline{F} \triangleq [\text{I} + \mu\nabla_{\tilde{w}}\overline{f}_{av}(0)]^2 = \begin{cases} = (\text{I} - \mu\Lambda)^2 & \text{for LMS} \\ \approx -(\text{I} - \mu\Lambda/\text{Tr}(\Lambda))^2 & \text{for NLMS} \end{cases} \tag{24.44}$$

In this way, the recursion for $\mathsf{E}\,\|\overline{\boldsymbol{w}}_i^{pav}\|_{\overline{\Sigma}}^2$ from (24.38) can be rewritten as

$$\boxed{\mathsf{E}\,\|\overline{\boldsymbol{w}}_i^{pav}\|_{\overline{\sigma}}^2 = \mathsf{E}\,\|\overline{\boldsymbol{w}}_{i-1}^{pav}\|_{\overline{F}\overline{\sigma}}^2 + \mu^2\gamma\lambda^{\mathsf{T}}\overline{\sigma}} \tag{24.45}$$

This result extends recursion (25.14) to the partially-averaged system (24.30), except that now the matrix $\overline{F}$ is $M \times M$. We are therefore led to the statement of Thm. 24.6 below, which is justified by the same arguments used in Sec. 25.2 further ahead.

Note that the learning curve of either filter, for infinitesimally small step-sizes, can be estimated by evaluating $\mathsf{E}\,|\boldsymbol{u}_i\tilde{\boldsymbol{w}}_{i-1}|^2 = \mathsf{E}\,\|\overline{\boldsymbol{w}}_{i-1}^{pav}\|_{R_u}^2 = \mathsf{E}\,\|\overline{\boldsymbol{w}}_{i-1}^{pav}\|_{\lambda}^2$, which can be obtained from the top entry of the state vector $\mathcal{W}_i$ for the choice $\overline{\sigma} = \lambda$. It can be verified by experimentation that the learning curve generated in this manner, by means of averaging analysis, agrees well with the learning curve generated from the independence analysis in the body of the chapter, and that both theoretical curves match well with experimental results —see Fig. 24.1.

Mean-square deviation. We can also use the variance relation (24.45) to shed further light on the mean-square performance of the filters. In particular, for each filter, we can re-examine its EMSE as well as its mean-square deviation (MSD), which is defined as

$$\text{MSD} = \lim_{i\to\infty} \mathsf{E}\,\|\tilde{\boldsymbol{w}}_i\|^2$$

Indeed, assuming the step-size μ is chosen to guarantee filter stability, recursion (24.45) becomes in steady-state

$$\mathsf{E}\,\|\overline{\boldsymbol{w}}_\infty^{pav}\|_{(\text{I}-\overline{F})\overline{\sigma}}^2 = \mu^2\gamma(\lambda^{\mathsf{T}}\overline{\sigma}) \tag{24.46}$$

If we now select $\overline{\sigma}$ as the solution to the linear system of equations $(\text{I} - \overline{F})\overline{\sigma} = \text{vec}\{\text{I}\}$, then the weighting quantity that appears in (24.46) reduces to the vector of unit entries. In this way, the left-hand side of (24.46) becomes the filter MSD and (24.46) leads to

$$\boxed{\text{MSD} = \mu^2\gamma\lambda^{\mathsf{T}}(\text{I} - \overline{F})^{-1}\text{vec}\{\text{I}\}} \tag{24.47}$$

Theorem 24.6 (Transient behavior by averaging analysis) Assume $\{\boldsymbol{d}(i), \boldsymbol{u}_i\}$ satisfy model (22.1) and consider the LMS and NLMS algorithms. Assume further that the conditions of Thm. 24.5 hold. Then the transient behavior of either filter is described by the $M-$dimensional model $\mathcal{W}_i = \mathcal{F}\mathcal{W}_{i-1} + \mu^2\gamma\mathcal{Y}$, where γ is defined by (24.37) and

$$\mathcal{F} = \begin{bmatrix} 0 & 1 & & & \\ 0 & 0 & 1 & & \\ 0 & 0 & 0 & 1 & \\ \vdots & & & & \\ 0 & 0 & 0 & & 1 \\ -p_0 & -p_1 & -p_2 & \dots & -p_{M-1} \end{bmatrix} \quad (M \times M)$$

with $p(x) \stackrel{\Delta}{=} \det(x\mathrm{I} - \overline{F}) = x^M + \sum_{k=0}^{M-1} p_k x^k$ denoting the characteristic polynomial of $\overline{F}$ in (24.44). Also,

$$\mathcal{W}_i = \begin{bmatrix} E\|\overline{\boldsymbol{w}}_i^{pav}\|^2_{\overline{\sigma}} \\ E\|\overline{\boldsymbol{w}}_i^{pav}\|^2_{\overline{F}\overline{\sigma}} \\ \vdots \\ E\|\overline{\boldsymbol{w}}_i^{pav}\|^2_{\overline{F}^{M-2}\overline{\sigma}} \\ E\|\overline{\boldsymbol{w}}_i^{pav}\|^2_{\overline{F}^{M-1}\overline{\sigma}} \end{bmatrix}, \quad \mathcal{Y} = \begin{bmatrix} \lambda^{\mathsf{T}}\overline{\sigma} \\ \lambda^{\mathsf{T}}\overline{F}\overline{\sigma} \\ \vdots \\ \lambda^{\mathsf{T}}\overline{F}^{M-2}\overline{\sigma} \\ \lambda^{\mathsf{T}}\overline{F}^{M-1}\overline{\sigma} \end{bmatrix}$$

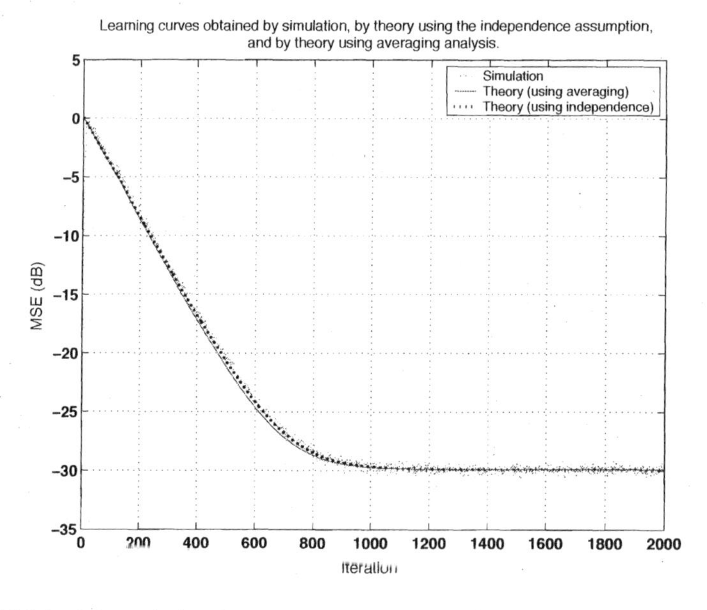

FIGURE 24.1 Theoretical and simulated learning curves for a $10-$tap LMS implementation using $\mu = 0.005$ and Gaussian input with SNR set at 30 dB. The theoretical curves are evaluated with and without the independence assumption (22.23): in one case we use the result of Thm. 24.6, which relies on averaging, and in the other case we rely on the results of Chapter 24 under independence.

CHAPTER 25

Data-Normalized Filters

We extend the transient analysis of the earlier chapters to more general filter recursions, starting with the normalized LMS algorithm.

25.1 NLMS FILTER

We thus consider the ϵ-NLMS recursion

$$\boldsymbol{w}_i = \boldsymbol{w}_{i-1} + \mu \frac{\boldsymbol{u}_i^*}{\epsilon + \|\boldsymbol{u}_i\|^2} \boldsymbol{e}(i) \tag{25.1}$$

for which the data normalization in (22.2) is given by

$$g[\boldsymbol{u}_i] = \epsilon + \|\boldsymbol{u}_i\|^2 \tag{25.2}$$

In this case, relations (22.26)–(22.27) and (22.29) become

$$\begin{cases} \mathsf{E}\,\|\tilde{\boldsymbol{w}}_i\|^2_{\Sigma} = \mathsf{E}\,\|\tilde{\boldsymbol{w}}_{i-1}\|^2_{\Sigma'} \ + \ \mu^2\sigma_v^2 \mathsf{E}\left[\dfrac{\|\boldsymbol{u}_i\|^2_{\Sigma}}{(\epsilon + \|\boldsymbol{u}_i\|^2)^2}\right] \\ \Sigma' = \Sigma - \mu\Sigma\mathsf{E}\left[\dfrac{\boldsymbol{u}_i^*\boldsymbol{u}_i}{\epsilon + \|\boldsymbol{u}_i\|^2}\right] - \mu\mathsf{E}\left[\dfrac{\boldsymbol{u}_i^*\boldsymbol{u}_i}{\epsilon + \|\boldsymbol{u}_i\|^2}\right]\Sigma \ + \ \mu^2\mathsf{E}\left[\dfrac{\|\boldsymbol{u}_i\|^2_{\Sigma}}{(\epsilon + \|\boldsymbol{u}_i\|^2)^2}\boldsymbol{u}_i^*\boldsymbol{u}_i\right] \\ \mathsf{E}\,\tilde{\boldsymbol{w}}_i = \left[\mathrm{I} - \mu\mathsf{E}\left(\dfrac{\boldsymbol{u}_i^*\boldsymbol{u}_i}{\epsilon + \|\boldsymbol{u}_i\|^2}\right)\right]\mathsf{E}\,\tilde{\boldsymbol{w}}_{i-1} \end{cases}$$

and we see that we need to evaluate the moments

$$\mathsf{E}\left[\frac{\|\boldsymbol{u}_i\|^2_{\Sigma}}{(\epsilon + \|\boldsymbol{u}_i\|^2)^2}\right], \quad \mathsf{E}\left[\frac{\boldsymbol{u}_i^*\boldsymbol{u}_i}{\epsilon + \|\boldsymbol{u}_i\|^2}\right], \quad \mathsf{E}\left[\frac{\|\boldsymbol{u}_i\|^2_{\Sigma}}{(\epsilon + \|\boldsymbol{u}_i\|^2)^2}\boldsymbol{u}_i^*\boldsymbol{u}_i\right] \tag{25.3}$$

Unfortunately, closed form expressions for these moments are not available in general, even for Gaussian regressors. Still, we will be able to show that the filter is convergent in the mean and is also mean-square stable for step-sizes satisfying $\mu < 2$, and regardless of the input distribution (Gaussian or otherwise) — see App. 25.B. We therefore treat the general case directly. Since the arguments are similar to those in Chapter 24 for LMS, we shall be brief.

Thus, introduce the $M^2 \times 1$ vectors

$$\sigma \triangleq \text{vec}(\Sigma), \qquad r \triangleq \text{vec}(R_u)$$

as well as the $M^2 \times M^2$ matrices

$$A \triangleq \left[\mathsf{E}\left(\frac{\boldsymbol{u}_i^*\boldsymbol{u}_i}{\epsilon+\|\boldsymbol{u}_i\|^2}\right)^{\mathsf{T}} \otimes \mathrm{I}\right] + \left[\mathrm{I}\otimes \mathsf{E}\left(\frac{\boldsymbol{u}_i^*\boldsymbol{u}_i}{\epsilon+\|\boldsymbol{u}_i\|^2}\right)\right] \tag{25.4}$$

$$B \triangleq \mathsf{E}\left[\left(\frac{\boldsymbol{u}_i^*\boldsymbol{u}_i}{\epsilon+\|\boldsymbol{u}_i\|^2}\right)^{\mathsf{T}} \otimes \left(\frac{\boldsymbol{u}_i^*\boldsymbol{u}_i}{\epsilon+\|\boldsymbol{u}_i\|^2}\right)\right] \tag{25.5}$$

and the $M \times M$ matrix

$$P \triangleq \mathsf{E}\left[\frac{\boldsymbol{u}_i^*\boldsymbol{u}_i}{\epsilon+\|\boldsymbol{u}_i\|^2}\right] \tag{25.6}$$

The matrix A is positive-definite, while B is nonnegative-definite — see Prob. V.6. Applying the vec notation to both sides of the expression for Σ' in (25.3) we find that it reduces to

$$\sigma' = F\sigma \tag{25.7}$$

where F is $M^2 \times M^2$ and given by

$$F \triangleq \mathrm{I} - \mu A + \mu^2 B \tag{25.8}$$

Moreover, the recursion for $\mathsf{E}\,\tilde{\boldsymbol{w}}_i$ can be written as

$$\mathsf{E}\,\tilde{\boldsymbol{w}}_i = (\mathrm{I} - \mu P)\,\mathsf{E}\,\tilde{\boldsymbol{w}}_{i-1} \tag{25.9}$$

The same arguments that were used in Sec. 24.1 will show that the mean-square behavior of ϵ–NLMS is characterized by the following M^2-dimensional state-space model:

$$\boxed{\mathcal{W}_i = \mathcal{F}\mathcal{W}_{i-1} + \mu^2\sigma_v^2\mathcal{Y}} \tag{25.10}$$

where $\mathcal{F}$ is the companion matrix

$$\mathcal{F} = \begin{bmatrix} 0 & 1 & & & \\ 0 & 0 & 1 & & \\ 0 & 0 & 0 & 1 & \\ \vdots & & & & \\ 0 & 0 & 0 & & 1 \\ -p_0 & -p_1 & -p_2 & \cdots & -p_{M^2-1} \end{bmatrix} \quad (M^2 \times M^2)$$

with

$$p(x) \triangleq \det(x\mathrm{I} - F) = x^{M^2} + \sum_{k=0}^{M^2-1} p_k x^k$$

denoting the characteristic polynomial of F in (25.8). $\mathcal{W}_i$ is the $M^2 \times 1$ state vector

$$\mathcal{W}_i \triangleq \begin{bmatrix} \mathsf{E}\,\|\tilde{\boldsymbol{w}}_i\|^2_{\sigma} \\ \mathsf{E}\,\|\tilde{\boldsymbol{w}}_i\|^2_{F\sigma} \\ \mathsf{E}\,\|\tilde{\boldsymbol{w}}_i\|^2_{F^2\sigma} \\ \vdots \\ \mathsf{E}\,\|\tilde{\boldsymbol{w}}_i\|^2_{F^{(M^2-1)}\sigma} \end{bmatrix} \tag{25.11}$$

and the k-th entry of $\mathcal{Y}$ is given by

$$[\mathsf{E}\,\mathcal{Y}]_k = \mathsf{E}\left[\frac{\|\boldsymbol{u}_i\|^2_{F^k\sigma}}{(\epsilon+\|\boldsymbol{u}_i\|^2)^2}\right], \quad k=0,1,\ldots,M^2-1 \tag{25.12}$$

The definitions of $\{\mathcal{W}_i,\mathcal{Y}\}$ are in terms of any σ of interest, e.g., most commonly, $\sigma=q$ or $\sigma=r$. It is shown in App. 25.B that any $\mu<2$ is a sufficient condition for the stability of (25.10).

Theorem 25.1 (Stability of ϵ-NLMS) Consider the $\epsilon-$NLMS recursion (25.1) and assume the data $\{\boldsymbol{d}(i),\boldsymbol{u}_i\}$ satisfy model (22.1) and the independence assumption (22.23). The regressors need not be Gaussian. Then the filter is convergent in the mean and is also mean-square stable for any $\mu<2$. Moreover, the transient behavior of the filter is characterized by the state-space recursion (25.10)–(25.12), and the mean-square deviation and the excess mean-square error are given by

$$\mathsf{MSD} = \mu^2\sigma_v^2\mathsf{E}\left[\frac{\|\boldsymbol{u}_i\|^2_{(\mathrm{I}-F)^{-1}q}}{(\epsilon+\|\boldsymbol{u}_i\|^2)^2}\right], \qquad \mathsf{EMSE} = \mu^2\sigma_v^2\mathsf{E}\left[\frac{\|\boldsymbol{u}_i\|^2_{(\mathrm{I}-F)^{-1}r}}{(\epsilon+\|\boldsymbol{u}_i\|^2)^2}\right]$$

where $r=\text{vec}(R_u)$ and $q=\text{vec}(\mathrm{I})$.

The expressions for the MSD and EMSE in the statement of the theorem are derived in a manner similar to (23.51) and (23.56). They can be rewritten as

$$\boxed{\mathsf{MSD} = \mu^2\sigma_v^2\mathsf{Tr}(S\Sigma_{\text{msd}}) \quad \text{and} \quad \mathsf{EMSE} = \mu^2\sigma_v^2\mathsf{Tr}(S\Sigma_{\text{emse}})}$$

where

$$S \triangleq \mathsf{E}\left(\frac{\boldsymbol{u}_i^*\boldsymbol{u}_i}{(\epsilon+\|\boldsymbol{u}_i\|^2)^2}\right)$$

and the weighting matrices $\{\Sigma_{\text{msd}},\Sigma_{\text{emse}}\}$ correspond to the vectors $\sigma_{\text{msd}}=(\mathrm{I}-F)^{-1}q$ and $\sigma_{\text{emse}}=(\mathrm{I}-F)^{-1}r$. That is,

$$\Sigma_{\text{msd}}=\text{vec}^{-1}(\sigma_{\text{msd}}) \quad \text{and} \quad \Sigma_{\text{emse}}=\text{vec}^{-1}(\sigma_{\text{emse}})$$

Learning Curve

Observe that since $\mathsf{E}\,|\boldsymbol{e}_a(i)|^2 = \mathsf{E}\,\|\tilde{\boldsymbol{w}}_{i-1}\|^2_{R_u}$, we again find that the time evolution of $\mathsf{E}\,|\boldsymbol{e}_a(i)|^2$ is described by the top entry of the state vector $\mathcal{W}_i$ in (25.10)–(25.12) with σ chosen as $\sigma=r$. The learning curve of the filter will be $\mathsf{E}\,|\boldsymbol{e}(i)|^2=\sigma_v^2+\mathsf{E}\,|\boldsymbol{e}_a(i)|^2$.

Small Step-Size Approximations

Several approximations for the EMSE and MSD expressions that appear in the above theorem are derived in Probs. V.18–V.20. The ultimate conclusion from these problems is that for small enough μ and ϵ, we get

$$\boxed{\mathsf{EMSE} = \frac{\mu\sigma_v^2}{2-\mu}\mathsf{E}\left(\frac{1}{\|\boldsymbol{u}_i\|^2}\right)\mathsf{Tr}(R_u) \quad \text{and} \quad \mathsf{MSD} = \frac{\mu\sigma_v^2}{2-\mu}\mathsf{E}\left(\frac{1}{\|\boldsymbol{u}_i\|^2}\right)} \tag{25.13}$$

The expression for the EMSE is the same we derived earlier in Lemma 17.1.

Gaussian Regressors

If the regressors happen to be Gaussian, then it can be shown that the M^2-dimensional state-space model (25.10)–(25.12) reduces to an $M-$dimensional model — this assertion is proved in Probs. V.16 and V.17.

25.2 DATA-NORMALIZED FILTERS

The arguments that were employed in the last two sections for LMS and $\epsilon-$NLMS are general enough and can be applied to adaptive filters with generic data nonlinearities of the form (22.2)–(22.3). To see this, consider again the variance and mean relations (22.26)–(22.27) and (22.29), which are reproduced below:

$$\left\{\begin{array}{rcl}\mathsf{E}\,\|\tilde{\boldsymbol{w}}_i\|^2_\Sigma & = & \mathsf{E}\,\|\tilde{\boldsymbol{w}}_{i-1}\|^2_{\Sigma'} + \mu^2\sigma_v^2\mathsf{E}\left[\dfrac{\|\boldsymbol{u}_i\|^2_\Sigma}{g^2[\boldsymbol{u}_i]}\right] \\ \Sigma' & = & \Sigma - \mu\Sigma\mathsf{E}\left[\dfrac{\boldsymbol{u}_i^*\boldsymbol{u}_i}{g[\boldsymbol{u}_i]}\right] - \mu\mathsf{E}\left[\dfrac{\boldsymbol{u}_i^*\boldsymbol{u}_i}{g[\boldsymbol{u}_i]}\right]\Sigma + \mu^2\mathsf{E}\left[\dfrac{\|\boldsymbol{u}_i\|^2_\Sigma}{g^2[\boldsymbol{u}_i]}\boldsymbol{u}_i^*\boldsymbol{u}_i\right] \\ \mathsf{E}\,\tilde{\boldsymbol{w}}_i & = & \left(\mathrm{I} - \mu\mathsf{E}\left[\dfrac{\boldsymbol{u}_i^*\boldsymbol{u}_i}{g[\boldsymbol{u}_i]}\right]\right)\mathsf{E}\,\tilde{\boldsymbol{w}}_{i-1}\end{array}\right. \tag{25.14}$$

If we now introduce the $M^2 \times M^2$ matrices

$$A \triangleq \left(\mathsf{E}\left[\frac{\boldsymbol{u}_i^*\boldsymbol{u}_i}{g[\boldsymbol{u}_i]}\right]^{\mathsf{T}} \otimes \mathrm{I}\right) + \left(\mathrm{I} \otimes \mathsf{E}\left[\frac{\boldsymbol{u}_i^*\boldsymbol{u}_i}{g[\boldsymbol{u}_i]}\right]\right) \tag{25.15}$$

$$B \triangleq \mathsf{E}\left(\left[\frac{\boldsymbol{u}_i^*\boldsymbol{u}_i}{g[\boldsymbol{u}_i]}\right]^{\mathsf{T}} \otimes \left[\frac{\boldsymbol{u}_i^*\boldsymbol{u}_i}{g[\boldsymbol{u}_i]}\right]\right) \tag{25.16}$$

and the $M \times M$ matrix

$$P \triangleq \mathsf{E}\left[\frac{\boldsymbol{u}_i^*\boldsymbol{u}_i}{g[\boldsymbol{u}_i]}\right] \tag{25.17}$$

then

$$\mathsf{E}\,\tilde{\boldsymbol{w}}_i = (\mathrm{I} - \mu P)\,\mathsf{E}\,\tilde{\boldsymbol{w}}_{i-1}$$

and the expression for Σ' can be written in terms of the linear vector relation

$$\sigma' = F\sigma \tag{25.18}$$

where F is $M^2 \times M^2$ and given by

$$F \triangleq \mathrm{I} - \mu A + \mu^2 B \tag{25.19}$$

Let

$$H = \begin{bmatrix} A/2 & -B/2 \\ \mathrm{I}_{M^2} & 0 \end{bmatrix} \quad (2M^2 \times 2M^2) \tag{25.20}$$

Then the same arguments that were used in Chapter 24 will lead to the statement of Thm. 25.2 listed further ahead. The expressions for the MSD and EMSE in the statement of the theorem are derived in a manner similar to (23.51) and (23.56). They can be rewritten as

$$\boxed{\mathsf{MSD} = \mu^2\sigma_v^2\mathsf{Tr}(S\Sigma_{\text{msd}}) \quad \text{and} \quad \mathsf{EMSE} = \mu^2\sigma_v^2\mathsf{Tr}(S\Sigma_{\text{emse}})}$$

where

$$S = \mathsf{E}\left(\boldsymbol{u}_i^*\boldsymbol{u}_i/g^2[\boldsymbol{u}_i]\right)$$

and the weighting matrices $\{\Sigma_{\text{msd}}, \Sigma_{\text{emse}}\}$ correspond to the vectors $\sigma_{\text{msd}} = (\text{I} - F)^{-1}q$ and $\sigma_{\text{emse}} = (\text{I} - F)^{-1}r$. That is,

$$\Sigma_{\text{msd}} = \text{vec}^{-1}(\sigma_{\text{msd}}) \quad \text{and} \quad \Sigma_{\text{emse}} = \text{vec}^{-1}(\sigma_{\text{emse}})$$

Theorem 25.2 (Stability of data-normalized filters) Consider data normalized adaptive filters of the form (22.2)–(22.3), and assume the data $\{\boldsymbol{d}(i), \boldsymbol{u}_i\}$ satisfy model (22.1) and the independence assumption (22.23). Then the filter is convergent in the mean and is mean-square stable for step-sizes satisfying

$$0 < \mu < \min\left\{2/\lambda_{\max}(P),\ 1/\lambda_{\max}(A^{-1}B),\ 1/\max\left\{\lambda(H) \in \mathbb{R}^+\right\}\right\}$$

where the matrices $\{A, B, P, H\}$ are defined by (25.15)–(25.17) and (25.20) and B is assumed finite. Moreover, the transient behavior of the filter is characterized by the M^2-dimensional state-space recursion $\mathcal{W}_i = \mathcal{F}\mathcal{W}_{i-1} + \mu^2\sigma_v^2\mathcal{Y}$, where $\mathcal{F}$ is the companion matrix

$$\mathcal{F} = \begin{bmatrix} 0 & 1 & & & \\ 0 & 0 & 1 & & \\ 0 & 0 & 0 & 1 & \\ \vdots & & & & \\ 0 & 0 & 0 & & 1 \\ -p_0 & -p_1 & -p_2 & \ldots & -p_{M^2-1} \end{bmatrix} \quad (M^2 \times M^2)$$

with

$$p(x) \triangleq \det(x\text{I} - F) = x^{M^2} + \sum_{k=0}^{M^2-1} p_k x^k$$

denoting the characteristic polynomial of F in (25.19). Also,

$$\mathcal{W}_i \triangleq \begin{bmatrix} \mathsf{E}\,\|\tilde{\boldsymbol{w}}_i\|^2_{\sigma} \\ \mathsf{E}\,\|\tilde{\boldsymbol{w}}_i\|^2_{F\sigma} \\ \mathsf{E}\,\|\tilde{\boldsymbol{w}}_i\|^2_{F^2\sigma} \\ \vdots \\ \mathsf{E}\,\|\tilde{\boldsymbol{w}}_i\|^2_{F^{(M^2-1)}\sigma} \end{bmatrix}, \quad [\mathcal{Y}]_k = \mathsf{E}\left[\frac{\|\boldsymbol{u}_i\|^2_{F^k\sigma}}{g^2[\boldsymbol{u}_i]}\right], \quad k = 0, \ldots, M^2-1$$

for any σ of interest, e.g., $\sigma = q$ or $\sigma = r$. In addition, the mean-square deviation and the excess mean-square error are given by

$$\text{MSD} = \mu^2\sigma_v^2\mathsf{E}\left[\frac{\|\boldsymbol{u}_i\|^2_{(\text{I}-F)^{-1}q}}{g^2[\boldsymbol{u}_i]}\right], \qquad \text{EMSE} = \mu^2\sigma_v^2\mathsf{E}\left[\frac{\|\boldsymbol{u}_i\|^2_{(\text{I}-F)^{-1}r}}{g^2[\boldsymbol{u}_i]}\right]$$

where $r = \text{vec}(R_u)$ and $q = \text{vec}(\text{I})$.

Learning Curve

As before, since $\mathsf{E}\,|e_a(i)|^2 = \mathsf{E}\,\|\tilde{w}_{i-1}\|^2_{R_u}$, we find that the time evolution of $\mathsf{E}\,|e_a(i)|^2$ is described by the top entry of the state vector $\mathcal{W}_i$ in with σ chosen as $\sigma = r$. The learning curve of the filter will be $\mathsf{E}\,|e(i)|^2 = \sigma_v^2 + \mathsf{E}\,|e_a(i)|^2$.

Small Step-Size Approximations

In Prob. V.39 it is shown that under a boundedness requirement on the matrix B of fourth moments, data-normalized adaptive filters can be guaranteed to be mean-square stable for sufficiently small step-sizes. That is, there always exists a small-enough step-size that lies within the stability range described in Thm. 25.2.

Now observe that the performance results of Thm. 25.2 are in terms of the moment matrices $\{A, B, P\}$. These moments are generally hard to evaluate for arbitrary input distributions and data nonlinearities $g[\cdot]$. However, some simplifications occur when the step-size is sufficiently small. This is because, in this case, we may ignore the quadratic term in μ that appears in the expression for Σ' in (25.14), and thereby approximate the variance and mean relations by

$$\boxed{\begin{array}{l} \mathsf{E}\,\|\tilde{w}_i\|^2_{\Sigma} = \mathsf{E}\,\|\tilde{w}_{i-1}\|^2_{\Sigma'} + \mu^2\sigma_v^2\mathsf{E}\left(\dfrac{\|u_i\|^2_{\Sigma}}{g^2[u_i]}\right) \\ \Sigma' = \Sigma - \mu\Sigma P - \mu P\Sigma \\ \mathsf{E}\,\tilde{w}_i = (\mathrm{I} - \mu P)\,\mathsf{E}\,\tilde{w}_{i-1} \end{array}} \tag{25.21}$$

where P is as in (25.17). Using the weighting vector notation, we can write

$$\mathsf{E}\,\|\tilde{w}_i\|^2_{\sigma} = \mathsf{E}\,\|\tilde{w}_{i-1}\|^2_{F\sigma} + \mu^2\sigma_v^2\mathsf{E}\left(\frac{\|u_i\|^2_{\sigma}}{g^2[u_i]}\right) \tag{25.22}$$

$$F = I - \mu A \tag{25.23}$$

where now

$$A = (P^{\mathsf{T}} \otimes \mathrm{I}) + (\mathrm{I} \otimes P)$$

The variance relation (25.22) would then lead to the following approximate expressions for the filter EMSE and MSD:

$$\mathsf{EMSE} = \mu^2\sigma_v^2\mathsf{Tr}(S\Sigma_{\text{emse}}) \quad \text{and} \quad \mathsf{MSD} = \mu^2\sigma_v^2\mathsf{Tr}(S\Sigma_{\text{msd}})$$

where

$$S = \mathsf{E}\,\left(u_i^* u_i / g^2[u_i]\right)$$

and the weighting matrices $\{\Sigma_{\text{emse}}, \Sigma_{\text{msd}}\}$ correspond to the vectors $\sigma_{\text{emse}} = A^{-1}\text{vec}(R_u)/\mu$ and $\sigma_{\text{msd}} = A^{-1}\text{vec}(\mathrm{I})/\mu$. That is, $\{\Sigma_{\text{emse}}, \Sigma_{\text{msd}}\}$ are the unique solutions of the Lyapunov equations

$$\mu P\Sigma_{\text{msd}} + \mu\Sigma_{\text{msd}}P = \mathrm{I} \quad \text{and} \quad \mu P\Sigma_{\text{emse}} + \mu\Sigma_{\text{emse}}P = R_u$$

It is easy to verify that $\Sigma_{\text{msd}} = \mu^{-1}P^{-1}/2$ so that the performance expressions can be rewritten as

$$\boxed{\mathsf{EMSE} = \mu^2\sigma_v^2\mathsf{Tr}(S\Sigma_{\text{emse}}), \qquad \mathsf{MSD} = \mu\sigma_v^2\mathsf{Tr}(SP^{-1})/2}$$

Remark 25.1 (Filters with error nonlinearities) There is more to say about the transient performance of adaptive filters, especially for filters with error nonlinearities in their update equations. This is a more challenging class of filters to study and their performance is examined in App. 9.C of Sayed (2003) by using the same energy-conservation arguments of this part. The derivation used in that appendix to study adaptive filters with error nonlinearities can also be used to provide an alternative simplified transient analysis for data-normalized filters. The derivation is based on a long filter assumption in order to justify a Gaussian condition on the distribution of the *a priori* error signal. Among other results, it is shown in App. 9.C of Sayed (2003) that the transient behavior of data-normalized filters can be approximated by an M-dimensional linear time-invariant state-space model even for non-Gaussian regressors. Appendix 9.E of the same reference further examines the learning abilities of adaptive filters and shows, among other interesting results, that the learning behavior of LMS cannot be fully captured by relying solely on mean-square analysis!

◇

25.A APPENDIX: STABILITY BOUND

Consider a matrix F of the form $F = \mathrm{I} - \mu A + \mu^2 B$ with $A > 0$, $B \geq 0$, and $\mu > 0$. Matrices of this form arise frequently in the study of the mean-square stability of adaptive filters (see, e.g., (25.19)). The purpose of this section is to find conditions on μ in terms of $\{A, B\}$ in order to guarantee that all eigenvalues of F are strictly inside the unit circle, i.e., so that $-1 < \lambda(F) < 1$.

To begin with, in order to guarantee $\lambda(F) < 1$, the step-size μ should be such that (cf. the Rayleigh-Ritz characterization of eigenvalues from Sec. B.1):

$$\max_{\|x\|=1} x^*(\mathrm{I} - \mu A + \mu^2 B)x < 1$$

or, equivalently, $A - \mu B > 0$. The argument in parts (b) and (c) of Prob. V.3 then show that this condition holds if, and only if,

$$\mu < 1/\lambda_{\max}(A^{-1}B) \tag{25.24}$$

Moreover, in order to guarantee $\lambda(F) > -1$, the step-size μ should be such that

$$\min_{\|x\|=1} x^*(\mathrm{I} - \mu A + \mu^2 B)x > -1$$

or, equivalently, $G(\mu) \triangleq 2\mathrm{I} - \mu A + \mu^2 B > 0$. When $\mu = 0$, the eigenvalues of G are all positive and equal to 2. As μ increases, the eigenvalues of G vary continuously with μ. Indeed, the eigenvalues of $G(\mu)$ are the roots of $\det[\lambda\mathrm{I} - G(\mu)] = 0$. This is a polynomial equation in λ and its coefficients are functions of μ. A fundamental result in function theory and matrix analysis states that the zeros of a polynomial depend continuously on its coefficients and, consequently, the eigenvalues of $G(\mu)$ vary continuously with μ. This means that $G(\mu)$ will first become singular before becoming indefinite. For this reason, there is an upper bound on μ, say, $\mu_{\max}$, such that $G(\mu) > 0$ for all $\mu < \mu_{\max}$. This bound on μ is equal to the smallest value of μ that makes $G(\mu)$ singular, i.e., for which $\det[G(\mu)] = 0$. Now note that the determinant of $G(\mu)$ is equal to the determinant of the block matrix

$$K(\mu) \triangleq \begin{bmatrix} 2\mathrm{I} - \mu A & \mu B \\ -\mu\mathrm{I} & \mathrm{I} \end{bmatrix}$$

since

$$\det\left(\begin{bmatrix} X & W \\ Y & Z \end{bmatrix}\right) = \det(Z)\det(X - WZ^{-1}Y)$$

whenever Z is invertible. Moreover, since we can write

$$K(\mu) = \begin{bmatrix} 2\mathrm{I} & 0 \\ 0 & \mathrm{I} \end{bmatrix} - \mu\begin{bmatrix} A & -B \\ \mathrm{I} & 0 \end{bmatrix} = \begin{bmatrix} 2\mathrm{I} & 0 \\ 0 & \mathrm{I} \end{bmatrix}\left(\begin{bmatrix} \mathrm{I} & 0 \\ 0 & \mathrm{I} \end{bmatrix} - \mu\begin{bmatrix} A/2 & -B/2 \\ \mathrm{I} & 0 \end{bmatrix}\right)$$

we find that the condition $\det[K(\mu)] = 0$ is equivalent to $\det(\mathrm{I} - \mu H) = 0$, where

$$H \triangleq \begin{bmatrix} A/2 & -B/2 \\ \mathrm{I} & 0 \end{bmatrix}$$

In this way, the smallest positive μ that results in $\det[K(\mu)] = 0$ is equal to

$$\frac{1}{\max\left\{\lambda(H) \in \mathbb{R}^+\right\}} \tag{25.25}$$

in terms of the largest positive real eigenvalue of H when it exists.

The results (25.24)–(25.25) can be grouped together to yield the condition

$$\boxed{0 < \mu < \min\left\{\frac{1}{\lambda_{\max}(A^{-1}B)}, \frac{1}{\max\left\{\lambda(H) \in \mathbb{R}^+\right\}}\right\}} \tag{25.26}$$

If H does not have any real positive eigenvalue, then the corresponding condition is removed and we only require $\mu < 1/\lambda_{\max}(A^{-1}B)$. The result (25.26) is valid for general $A > 0$ and $B \geq 0$. The above derivation does not exploit any particular structure in the matrices A and B defined by (25.16).

25.B APPENDIX: STABILITY OF NLMS

The purpose of this appendix is to show that for $\epsilon-$NLMS, any $\mu < 2$ is sufficient to guarantee mean-square stability. Thus, refer again to the discussion in Sec. 25.1 and to the definitions of the matrices $\{A, B, P, F\}$ in (25.4)–(25.8). We already know from the result in App. 25.A that stability in the mean and mean-square senses is guaranteed for step-sizes in the range

$$0 < \mu < \min\left\{\frac{2}{\lambda_{\max}(P)}, \frac{1}{\lambda_{\max}(A^{-1}B)}, \frac{1}{\max\left\{\lambda(H) \in \mathbb{R}^+\right\}}\right\}$$

where the third condition is in terms of the largest positive real eigenvalue of the block matrix,

$$H \triangleq \begin{bmatrix} A/2 & -B/2 \\ \mathrm{I}_{M^2} & 0 \end{bmatrix}$$

The first condition on μ, namely, $\mu < 2/\lambda_{\max}(P)$, guarantees convergence in the mean. The second condition on μ, namely, $\mu < 1/\lambda_{\max}(A^{-1}B)$, guarantees $\lambda(F) < 1$. The last condition, $\mu < 1/\max\left\{\lambda(H) \in \mathbb{R}^+\right\}$, enforces $\lambda(F) > -1$. The point now is that these conditions on μ are met by any $\mu < 2$ (i.e., F is stable for any $\mu < 2$). This is because there are some important relations between the matrices $\{A, B, P\}$ in the $\epsilon-$NLMS case. To see this, observe first that the term

$$\boldsymbol{u}_i^*\boldsymbol{u}_i/(\epsilon + \|\boldsymbol{u}_i\|^2) \tag{25.27}$$

which appears in the expression (25.6) for P is generally a rank-one matrix (unless $\boldsymbol{u}_i = 0$); it has $M-1$ zero eigenvalues and one possibly nonzero eigenvalue that is equal to[8] $\|\boldsymbol{u}_i\|^2/(\epsilon + \|\boldsymbol{u}_i\|^2)$. This eigenvalue is less than unity so that

$$\lambda_{\max}\left(\frac{\boldsymbol{u}_i^*\boldsymbol{u}_i}{\epsilon + \|\boldsymbol{u}_i\|^2}\right) \leq 1 \tag{25.28}$$

Now recalling the following Rayleigh-Ritz characterization of the maximum eigenvalue of any Hermitian matrix R (from Sec. B.1):

$$\lambda_{\max}(R) = \max_{\|x\|=1} x^*Rx \tag{25.29}$$

[8]Every rank-one matrix of the form xx^*, where x is a column vector of size M, has $M-1$ zero eigenvalues and one nonzero eigenvalue that is equal to $\|x\|^2$.

we conclude from (25.28) that

$$\max_{\|x\|=1}\left(x^*\frac{u_i^*u_i}{\epsilon+\|u_i\|^2}x\right)\le 1$$

Applying the same characterization (25.29) to the matrix P in (25.6), and using the above inequality, we find that

$$\begin{aligned}\lambda_{\max}(P) &\triangleq \max_{\|x\|=1} x^*Px = \max_{\|x\|=1} x^*\mathsf{E}\left[\frac{u_i^*u_i}{\epsilon+\|u_i\|^2}\right]x = \max_{\|x\|=1}\mathsf{E}\left(x^*\frac{u_i^*u_i}{\epsilon+\|u_i\|^2}x\right)\\ &\le 1\end{aligned} \tag{25.30}$$

In other words, the maximum eigenvalue of P is bounded by one, so that the condition $\mu < 2/\lambda_{\max}(P)$ can be met by any $\mu < 2$.

Let us now examine the condition $\mu < 1/\lambda_{\max}(A^{-1}B)$. Using again the fact that the matrix in (25.27) has rank one, it is shown in Prob. V.8 that

$$2\left[\left(\frac{u_i^*u_i}{\epsilon+\|u_i\|^2}\right)^{\mathsf{T}}\otimes\left(\frac{u_i^*u_i}{\epsilon+\|u_i\|^2}\right)\right] \le \left[\left(\frac{u_i^*u_i}{\epsilon+\|u_i\|^2}\right)^{\mathsf{T}}\otimes \mathrm{I}\right] + \left[\mathrm{I}\otimes\left(\frac{u_i^*u_i}{\epsilon+\|u_i\|^2}\right)\right] \tag{25.31}$$

Taking expectations of both sides, and using the definitions (25.4)–(25.5) for A and B, we conclude that $2B - A \le 0$ so that the condition $\mu < 1/\lambda_{\max}(A^{-1}B)$ can be met by any μ satisfying $\mu < 2$.

What about the third condition on μ in terms of the positive eigenvalues of the matrix H? It turns out that $\mu < 2$ is also sufficient since it already guarantees mean-square convergence of the filter, as can be seen from the following argument. Choosing $\Sigma = \mathrm{I}$ in the variance relation (25.3) we get

$$\left\{\begin{array}{l}\mathsf{E}\,\|\tilde{w}_i\|^2 = \mathsf{E}\,\|\tilde{w}_{i-1}\|^2_{\Sigma'} + \mu^2\sigma_v^2\mathsf{E}\left[\dfrac{\|u_i\|^2}{(\epsilon+\|u_i\|^2)^2}\right]\\ \Sigma' = \mathrm{I} - 2\mu P + \mu^2 S\end{array}\right.$$

where

$$S \triangleq \mathsf{E}\left[\frac{\|u_i\|^2}{(\epsilon+\|u_i\|^2)^2}u_i^*u_i\right]$$

Obviously, $S \le P$ so that $\Sigma' \le \mathrm{I} - 2\mu P + \mu^2 P$ and, hence,

$$\begin{aligned}\mathsf{E}\,\|\tilde{w}_i\|^2 &\le \mathsf{E}\,\|\tilde{w}_{i-1}\|^2_{\mathrm{I}-2\mu P+\mu^2 P} + \mu^2\sigma_v^2\mathsf{E}\left[\frac{\|u_i\|^2}{(\epsilon+\|u_i\|^2)^2}\right]\\ &= \mathsf{E}\left[\tilde{w}_{i-1}^*(\mathrm{I}-2\mu P+\mu^2 P)\tilde{w}_{i-1}\right] + \mu^2\sigma_v^2\mathsf{E}\left[\frac{\|u_i\|^2}{(\epsilon+\|u_i\|^2)^2}\right]\end{aligned}$$

Now from the result of part (a) of Prob. V.6 we know that $R_u > 0$ implies $P > 0$. We also know from (25.30) that $\lambda_{\max}(P) < 1$. Therefore, all the eigenvalues of P are positive and lie inside the open interval $(0, 1)$. Moreover, over the interval $0 < \mu < 2$, the following quadratic function of μ,

$$k(\mu) \triangleq 1 - 2\mu\lambda + \mu^2\lambda$$

assumes values between 1 and $1 - \lambda$ for each of the eigenvalues λ of P. Therefore, it holds that

$$\mathrm{I} - 2\mu P + \mu^2 P \le [1 - 2\mu\lambda_{\min}(P) + \mu^2\lambda_{\min}(P)]\mathrm{I}$$

from which we conclude that

$$\mathsf{E}\,\|\tilde{w}_i\|^2 \le \alpha\mathsf{E}\,\|\tilde{w}_{i-1}\|^2 + \mu^2\sigma_v^2\mathsf{E}\left[\frac{\|u_i\|^2}{(\epsilon+\|u_i\|^2)^2}\right]$$

where the scalar coefficient $\alpha = 1 - 2\mu\lambda_{\min}(P) + \mu^2\lambda_{\min}(P)$ is positive and strictly less than one for $0 < \mu < 2$. It then follows that $\mathsf{E}\,\|\tilde{w}_i\|^2$ remains bounded for all i.

Summary and Notes

The chapters in this part describe a procedure for evaluating the transient behavior of adaptive filters by relying on energy conservation arguments.

SUMMARY OF MAIN RESULTS

1. Consider adaptive filters that are described by data-normalized stochastic difference equations of the form

$$\boxed{\boldsymbol{w}_i = \boldsymbol{w}_{i-1} + \mu\,\frac{\boldsymbol{u}_i^*}{g(\boldsymbol{u}_i)}\boldsymbol{e}(i), \quad \boldsymbol{e}(i) = \boldsymbol{d}(i) - \boldsymbol{u}_i\boldsymbol{w}_{i-1}}$$

where $g[\cdot]$ is some positive-valued function of $\boldsymbol{u}_i$. The mean-square error (MSE) of any such filter is defined as the limiting value of $\mathsf{E}\,|\boldsymbol{e}(i)|^2$,

$$\boxed{\mathsf{MSE} \;=\; \lim_{i\to\infty}\,\mathsf{E}\,|\boldsymbol{e}(i)|^2}$$

Likewise, the mean-square deviation (MSD) of the filter is defined as the limiting value of $\mathsf{E}\,\|\tilde{\boldsymbol{w}}_i\|^2$,

$$\boxed{\mathsf{MSD} \;=\; \lim_{i\to\infty}\,\mathsf{E}\,\|\tilde{\boldsymbol{w}}_i\|^2}$$

where $\tilde{\boldsymbol{w}}_i = w^o - \boldsymbol{w}_i$ and w^o is the unknown vector in the data model (22.1).

2. For any such adaptive filter, and for any data $\{\boldsymbol{d}(i), \boldsymbol{u}_i\}$ and Hermitian nonnegative-definite weighting matrix Σ, the following weighted energy-conservation relation holds for all i:

$$\boxed{\|\boldsymbol{u}_i\|^2_\Sigma \cdot \|\tilde{\boldsymbol{w}}_i\|^2_\Sigma \;+\; |\boldsymbol{e}_a^\Sigma(i)|^2 = \|\boldsymbol{u}_i\|^2_\Sigma \cdot \|\tilde{\boldsymbol{w}}_{i-1}\|^2_\Sigma \;+\; |\boldsymbol{e}_p^\Sigma(i)|^2}$$

where $\boldsymbol{e}_p^\Sigma(i) = \boldsymbol{u}_i\Sigma\tilde{\boldsymbol{w}}_i$ and $\boldsymbol{e}_a^\Sigma(i) = \boldsymbol{u}_i\Sigma\tilde{\boldsymbol{w}}_{i-1}$.

3. The energy relation is useful in several respects. For example, by taking expectations of both of its sides, expressing $\boldsymbol{e}_a^\Sigma(i)$ and $\boldsymbol{e}_p^\Sigma(i)$ in terms of $\tilde{\boldsymbol{w}}_i$, and by assuming the regressors $\{\boldsymbol{u}_i\}$ to be independent and identically distributed, we arrive at the following weighted variance relation:

$$\boxed{\begin{aligned}\mathsf{E}\,\|\tilde{\boldsymbol{w}}_i\|^2_\Sigma &= \mathsf{E}\,\|\tilde{\boldsymbol{w}}_{i-1}\|^2_{\Sigma'} \;+\; \mu^2\sigma_v^2\mathsf{E}\left(\frac{\|\boldsymbol{u}_i\|^2_\Sigma}{g^2[\boldsymbol{u}_i]}\right)\\ \Sigma' &= \Sigma \;-\; \mu\Sigma\mathsf{E}\left(\frac{\boldsymbol{u}_i^*\boldsymbol{u}_i}{g[\boldsymbol{u}_i]}\right) \;-\; \mu\mathsf{E}\left(\frac{\boldsymbol{u}_i^*\boldsymbol{u}_i}{g[\boldsymbol{u}_i]}\right)\Sigma \;+\; \mu^2\mathsf{E}\left(\frac{\|\boldsymbol{u}_i\|^2_\Sigma}{g^2[\boldsymbol{u}_i]}\boldsymbol{u}_i^*\boldsymbol{u}_i\right)\end{aligned}}$$

with a weighting matrix Σ' for the time instant $i-1$. The mean of the weight-error vector also satisfies the recursion

$$\boxed{\mathsf{E}\,\tilde{\boldsymbol{w}}_i = \left[\mathrm{I} - \mu\mathsf{E}\left(\frac{\boldsymbol{u}_i^*\boldsymbol{u}_i}{g[\boldsymbol{u}_i]}\right)\right]\mathsf{E}\,\tilde{\boldsymbol{w}}_{i-1}}$$

4. The expression for Σ', and the recursion for $\mathsf{E}\,\tilde{w}_i$, show that studying the transient behavior of an adaptive filter requires that we evaluate the three moments:

$$\boxed{\mathsf{E}\left(\frac{\|u_i\|^2_\Sigma}{g^2[u_i]}\right), \quad \mathsf{E}\left(\frac{u_i^* u_i}{g[u_i]}\right) \quad \text{and} \quad \mathsf{E}\left(\frac{\|u_i\|^2_\Sigma}{g^2[u_i]} u_i^* u_i\right)}$$

These moments, and especially the third one, are in general hard to evaluate for generic functions $g(\cdot)$. Observe, however, that the third moment appears multiplied by μ^2 in the expression for Σ'. Therefore, if desired, its effect can be neglected for small-step sizes. This line of reasoning is pursued in some of the problems at the end of this part.

5. In Sec. 25.2 we defined the $M^2 \times M^2$ matrices $F \triangleq \mathrm{I} - \mu A + \mu^2 B$,

$$A \triangleq \left(\mathsf{E}\left[\frac{u_i^* u_i}{g[u_i]}\right]^{\mathsf{T}} \otimes \mathrm{I}\right) + \left(\mathrm{I} \otimes \mathsf{E}\left[\frac{u_i^* u_i}{g[u_i]}\right]\right), \quad B \triangleq \mathsf{E}\left(\left[\frac{u_i^* u_i}{g[u_i]}\right]^{\mathsf{T}} \otimes \left[\frac{u_i^* u_i}{g[u_i]}\right]\right)$$

as well as the $M \times M$ and $2M^2 \times 2M^2$ matrices

$$P \triangleq \mathsf{E}\left[\frac{u_i^* u_i}{g[u_i]}\right] \quad \text{and} \quad H \triangleq \begin{bmatrix} A/2 & -B/2 \\ \mathrm{I}_{M^2} & 0 \end{bmatrix}$$

Then we showed that the transient behavior of data-normalized adaptive filters is characterized by the equations:

$$\boxed{\begin{array}{rcll} \mathsf{E}\,\tilde{w}_i & = & (\mathrm{I} - \mu P)\,\mathsf{E}\,\tilde{w}_{i-1} & \text{(mean behavior)} \\ \mathcal{W}_i & = & \mathcal{F}\mathcal{W}_{i-1} + \mu^2\sigma_v^2 \mathcal{Y} & \text{(mean-square behavior)} \end{array}}$$

where $\mathcal{F}$ is the companion matrix

$$\mathcal{F} = \begin{bmatrix} 0 & 1 & & & \\ 0 & 0 & 1 & & \\ 0 & 0 & 0 & 1 & \\ \vdots & & & & \\ 0 & 0 & 0 & & 1 \\ -p_0 & -p_1 & -p_2 & \dots & -p_{M^2-1} \end{bmatrix} \quad (M^2 \times M^2)$$

with

$$p(x) \triangleq \det(x\mathrm{I} - F) = x^{M^2} + \sum_{k=0}^{M^2-1} p_k x^k$$

denoting the characteristic polynomial of F. Also, $\mathcal{W}_i$ is the $M^2 \times 1$ state vector

$$\mathcal{W}_i \triangleq \text{col}\left\{\mathsf{E}\,\|\tilde{w}_i\|^2, \mathsf{E}\,\|\tilde{w}_i\|^2_{Fq}, \mathsf{E}\,\|\tilde{w}_i\|^2_{F^2q}, \dots, \mathsf{E}\,\|\tilde{w}_i\|^2_{F^{(M^2-1)}q}\right\}$$

$q = \text{vec}(\mathrm{I})$, and the k-th entry of $\mathcal{Y}$ is given by

$$[\mathcal{Y}]_k = \mathsf{E}\left[\frac{\|u_i\|^2_{F^k q}}{g^2[u_i]}\right], \quad k = 0, 1, \dots, M^2 - 1$$

6. We further showed that a data-normalized filter is stable in the mean and mean-square senses for step-sizes μ satisfying

$$0 < \mu < \min\left\{\frac{2}{\lambda_{\max}(P)}, \frac{1}{\lambda_{\max}(A^{-1}B)}, \frac{1}{\max\left\{\lambda(H) \in \mathbb{R}^+\right\}}\right\}$$

The corresponding mean-square deviation and excess mean-square error are given by

$$\boxed{\mathsf{MSD} \;=\; \mu^2\sigma_v^2\mathsf{Tr}(S\Sigma_{\rm msd}) \quad \text{and} \quad \mathsf{EMSE} \;=\; \mu^2\sigma_v^2\mathsf{Tr}(S\Sigma_{\rm emse})}$$

where

$$S \;\stackrel{\Delta}{=}\; \mathsf{E}\left(\frac{\boldsymbol{u}_i^*\boldsymbol{u}_i}{g^2[\boldsymbol{u}_i]}\right)$$

and $\{\Sigma_{\rm msd}, \Sigma_{\rm emse}\}$ are the weighting matrices that correspond to the vectors

$$\sigma_{\rm msd} = (\mathrm{I}-F)^{-1}\text{vec}(\mathrm{I}) \quad \text{and} \quad \sigma_{\rm emse} = (\mathrm{I}-F)^{-1}\text{vec}(R_u)$$

That is, $\Sigma_{\rm msd} = \text{vec}^{-1}(\sigma_{\rm msd})$ and $\Sigma_{\rm emse} = \text{vec}^{-1}(\sigma_{\rm emse})$. All these results hold irrespective of the distribution of $\boldsymbol{u}_i$ (e.g., $\boldsymbol{u}_i$ is not required to be Gaussian).

7. In general, it hard to evaluate the matrices $\{A, B, P\}$ in closed-form for arbitrary regressor distributions and, therefore, it is hard to evaluate the upper bound on μ for mean-square stability. However, it is shown in Prob. V.39 that if the distribution of $\boldsymbol{u}_i$ is such that the matrix B is finite, then there always exist small enough step-sizes that guarantee mean-square stability of data-normalized adaptive filters.

8. The above results are specialized in the chapters to some well-known algorithms, such as LMS and ϵ-NLMS.

   a) For LMS, we have $g(\boldsymbol{u}_i) = 1$. When the $\{\boldsymbol{u}_i\}$ are Gaussian, it is shown in Thms. 23.1 and 23.2 that LMS is mean-square stable for step-sizes μ satisfying

   $$\sum_{k=1}^{M}\frac{\lambda_k\mu}{2-s\lambda_k\mu} < 1$$

   with

   $$\mathsf{EMSE} = \frac{\mu\sigma_v^2\sum_{k=1}^{M}\frac{\lambda_k}{2-s\lambda_k\mu}}{1-\mu\sum_{k=1}^{M}\frac{\lambda_k}{2-s\lambda_k\mu}}, \qquad \mathsf{MSD} = \frac{\mu\sigma_v^2\sum_{k=1}^{M}\frac{1}{2-s\lambda_k\mu}}{1-\mu\sum_{k=1}^{M}\frac{\lambda_k}{2-s\lambda_k\mu}}$$

   where $s = 1$ if the regressors are complex-valued and $s = 2$ if the regressors are real-valued.

   b) For $\epsilon-$NLMS, we have $g(\boldsymbol{u}_i) = \epsilon + \|\boldsymbol{u}_i\|^2$. We state in Sec. 25.1 (and prove in App. 25.B) that the filter is mean-square stable for any $\mu < 2$, regardless of the distribution of $\boldsymbol{u}_i$.

   c) For Gaussian regressors, we also argue in Secs. 23.1 and 25.1 that the M^2-dimensional state-space model for $\mathcal{W}_i$ reduces to an $M-$dimensional state-space model for both LMS and $\epsilon-$NLMS.

9. For adaptive filters with error nonlinearities,

$$\boxed{\boldsymbol{w}_i = \boldsymbol{w}_{i-1} + \mu\boldsymbol{u}_i^* g[\boldsymbol{e}(i)], \qquad \boldsymbol{e}(i) = \boldsymbol{d}(i) - \boldsymbol{u}_i\boldsymbol{w}_{i-1}}$$

it is shown in App. 9.C of Sayed (2003) that the variance relation takes the form

$$\boxed{\mathsf{E}\,\|\tilde{\boldsymbol{w}}_i\|_\Sigma^2 = \mathsf{E}\,\|\tilde{\boldsymbol{w}}_{i-1}\|_\Sigma^2 \;+\; \mu^2 h_U\,\mathsf{Tr}(R_u\Sigma) \;-\; 2\mu\mathsf{Re}\big(h_g\mathsf{E}\,\|\tilde{\boldsymbol{w}}_{i-1}\|_{\Sigma R_u}^2\big)}$$

where the functions $\{h_U, h_g\}$ are defined by

$$\boxed{h_U \;\stackrel{\Delta}{=}\; \mathsf{E}\,|g[\boldsymbol{e}(i)]|^2, \qquad h_g \;\stackrel{\Delta}{=}\; \frac{\mathsf{E}\boldsymbol{e}_a^*(i)\cdot g[\,\boldsymbol{e}(i)\,]}{\mathsf{E}\,|\boldsymbol{e}_a(i)|^2}}$$

This variance relation is used in Sayed (2003) to characterize the transient behavior of several adaptive filters with error nonlinearities. It turns out that the transient performance of such filters is more challenging to study due to the nonlinear nature of the resulting state-space model.

BIBLIOGRAPHIC NOTES

Energy conservation. The energy-conservation relation (22.12) with weighting is an extension of the unweighted relation (15.32), which was derived by Sayed and Rupp (1995) in their studies on the robustness and small gain analysis of adaptive filters (see Chapter 44). The extension (22.12), along with its variance relations (22.21) and (22.26)–(22.27), were derived by Al-Naffouri and Sayed (2001a) and used therein, as well as in subsequent works by the same authors to study the transient performance of adaptive filters in a unified manner. The presentation in this part follows mainly Al-Naffouri and Sayed (2001ab,2003ab), Sayed and Al-Naffouri (2001), and Sayed, Al-Naffouri and Nascimento (2003). The extension of the transient analysis to affine projection algorithms is presented in Shin and Sayed (2004) and in Prob. V.37.

Real vs. complex data. Performance analyses in the literature tend to be carried out separately not only for different algorithms, but also for real-valued data as opposed to complex-valued data. In our treatment in this book, we have chosen to maintain a uniform treatment and notation for both cases of real and complex data.

Performance results. Some of the early works on the transient performance of LMS, for both stationary and nonstationary environments, is the article by Widrow et al. (1976). Additional early works on LMS include Jones, Cavin, and Reed (1982) and Feuer and Weinstein (1985). This latter reference derived the necessary and sufficient condition for the mean-square stability of LMS for Gaussian uncorrelated regression data using the independence assumptions (cf. Thms. 23.1 and 23.2), as well as the expressions for the filter EMSE and MSD performance (cf. Thm. 23.3). A related analysis, using a different approach than Feuer and Weinstein (1985), has also been advanced by Horowitz and Senne (1981). The transient behavior of NLMS algorithm, on the other hand, has been studied in a series of works including Bershad (1986b,1987), Tarrab and Feuer (1988), Rupp (1993), and Slock (1993). The transient behavior of sign-error LMS and sign-regressor LMS has been studied by Bershad (1986a,1988), Mathews and Cho (1987), and Eweda (1990a). Studies that involve matrix data nonlinearities, or other forms of data nonlinearities, can be found in Bershad (1986a), Mikhael (1986), Harris, Chabries, and Bishop (1986), and Douglas and Meng (1994b). In all these earlier works, it is often assumed that the regression data is white (i.e., uncorrelated) and/or Gaussian. It is also generally observed that most works study individual algorithms separately.

Good treatments of these earlier developments, but mainly for LMS, appear in the textbooks by Widrow and Stearns (1985), Bellanger (1987), Solo and Kong (1995), Macchi (1995), Haykin (1996), Farhang-Boroujeny (1999), Manolakis, Ingle, and Kogon (2000), Treichler, Johnson, and Larimore (2001), and Diniz (2002). Macchi's (1995) book provides one of the most thorough treatments of the convergence behavior of LMS for both independent and N-dependent regressors, while Solo and Kong's (1995) book relies on averaging theory and small step-size assumptions.

LMS and independence assumptions. The independence assumptions (22.23) (actually, the more detailed assumptions (i)–(vi) of Sec. 16.4) have been applied to the study of LMS since the late 1960's by Widrow et al. (1967,1975) for the case of Gaussian variables. The motivation for their use was mainly to obtain a tractable mathematical framework. However, it was not until Feuer and Weinstein (1985) that a precise study of LMS with independent Gaussian regressors was developed, and an improved derivation was given later by Foley and Boland (1988). An analysis for non-Gaussian variables was given by Hsia (1983), but the results were not as detailed as in the Gaussian case.

NLMS and independence assumptions. A precise analysis of the NLMS algorithm with Gaussian independent regressors was given by Tarrab and Feuer (1988). A simplified data model for performance analysis was later developed by Slock (1993).

Leaky-LMS and independence assumptions. In Probs. V.28–V.32 we follow Sayed and Al-Naffouri (2001) and perform a transient analysis of the leaky-LMS filter for *both* cases of Gaussian and non-Gaussian regressors. We derive necessary and sufficient conditions for its mean-square stability under the independence assumptions. Earlier transient analysis results for leaky-LMS were derived by Mayyas and Aboulnasr (1997) for Gaussian regressors; however, the stability conditions in this latter reference were only sufficient.

Affine projection algorithms. The transient behavior of affine projection algorithms is not as widely studied as that of LMS or NLMS. The available results have progressed more for some variations than others, and most analyses assume particular models for the regression data. For example, in Apolinário, Campos, and Diniz (2000) convergence analyses in the mean and in the mean-square senses are presented for the binormalized data-reusing LMS (BNDR-LMS) algorithm using a particular model for the input signal. Likewise, the convergence results in Sankaran and Beex (2000) focus on APA with orthogonal correction factors (APA-OCF) and rely on a special model for the input signal vector. In Bershad, Linebarger and McLaughlin (2001), the theoretical results of Rupp (1998a) are extended to the evaluation of learning curves assuming a Gaussian autoregressive input model. In Shin and Sayed (2004), a treatment of the transient performance of a family of affine projection algorithms is provided by relying on the energy-conservation arguments of this chapter. Prob. V.37 specializes the transient analysis of this work to ϵ-APA.

How valid are the independence assumptions? The independence assumptions (22.23) (actually, the more detailed assumptions (i)–(vi) of Sec. 16.4) have been applied to the study of LMS since the late 1960's. Although unrealistic for tapped-delay-line implementations, they tend to give good results that match real filter performance for small step-sizes. Several studies in the literature have struggled with this observation in an attempt to explain it. In App. 24.A we describe some of these results. Basically, it is shown in the appendix that if $\mu \approx 0$, the conclusions that are obtained using the independence assumptions are good approximations for the real performance of LMS.

One of the earliest results in this regard is that of Mazo (1979), which established Thm. 24.2. The arguments in this reference assume that the regression data are generated by feeding a binary i.i.d. sequence, say $\{s(i)\}$, through an FIR filter. Mazo's result then ascertains that for tapped-delay-line implementations with such regression data, the conclusions of an independence analysis match the exact filter performance for sufficiently small step-sizes. This result was later strengthened by Jones, Cavin, and Reed (1982) to the case in which $\{s(i)\}$ was not restricted to being binary but could be any i.i.d. sequence. They established Thm. 24.3. A similar conclusion appears in Macchi (1995, p. 114), whereby it is shown that results under slow adaptation are essentially equivalent to those obtained under independence.

The above studies assumed (or required) filter convergence. The result in Macchi and Eweda (1983) closed the gap by showing that there does exist a small enough step-size that guarantees filter convergence even when the regressors are not independent; the authors assumed only that the regressors are N-dependent and established Thm. 24.4 — see also the book by Macchi (1995).

Averaging analysis. The results described in the previous paragraph on the practicality of the independence assumptions for slowly adapting filters are specific to LMS. Another approach for studying the performance of LMS without the independence assumptions, as well as for studying the performance of a larger class of adaptive filters, is to rely on averaging analysis. This approach is reviewed in App. 24.A, where it is seen that it requires the assumption of infinitesimal step-sizes. A key result in this framework is Thm. 24.5 (see Solo and Kong (1995, Chapter 9)), which essentially states that for small step-sizes, the performance of an adaptive filter can be determined by studying the performance of a so-called *averaged* filter; the averaged filter is such that its analysis may not require the independence conditions but still requires $\mu \approx 0$. Examples of studies that are motivated by averaging arguments can be found in Butterweck (1995,2001) and Treichler, Johnson, and Larimore (2001, p. 95). For more detailed treatments of averaging analysis see the books by Kushner (1984), Benveniste, Métivier, and Priouret (1987), Solo and Kong (1995), and Kushner and Yin (1997).

ODE method. Another contribution to the analysis of adaptive filters without the independence assumptions is that of Ljung (1977ab), who developed an approach known as the ordinary-differential-equation (ODE) method. The idea is to reduce the adaptive filter difference equation to a differential equation, and to study the convergence and stability properties of the resulting continuous-time system by using, for example, Lyapunov stability methods. The convenience of this approach relies on the fact that the stability of differential equations is a well-studied subject (see, e.g., Bellman (1953)), and many results from this field could therefore be used in the study of adaptive filters. However, just like the averaging method of App. 24.A, the ODE approach is only applicable to infinitesimal step-sizes ($\mu \approx 0$). Moreover, in his original work, Ljung (1977ab) only studied the case of time-variant step-sizes that tend to zero; a condition that hinders the tracking ability of an adaptive filter in steady-state. Extensions of the ODE method to deal with constant step-sizes were later developed by Kushner (1984) and Benveniste, Métivier, and Priouret (1987).

Briefly, the basic idea of the ODE method as applied to LMS is the following. Starting from the LMS update $\boldsymbol{w}_i = \boldsymbol{w}_{i-1} + \mu \boldsymbol{u}_i^* \boldsymbol{e}(i)$, with $\boldsymbol{e}(i) = \boldsymbol{d}(i) - \boldsymbol{u}_i \boldsymbol{w}_{i-1}$, and iterating it over L iterations we obtain

$$\boldsymbol{w}_{i+L} = \boldsymbol{w}_{i-1} + \mu \sum_{k=i}^{i+L} \boldsymbol{u}_k^* \boldsymbol{e}(k)$$

Assuming stationary and ergodic processes, and a sufficiently small-step size (so that the weight-vectors do not vary appreciably over L iterations), the following approximation becomes possible:

$$\frac{1}{L} \sum_{k=i}^{i+L} \boldsymbol{u}_k^* \boldsymbol{e}(k) \approx \mathsf{E}\, \boldsymbol{u}_i^* \boldsymbol{e}(i) \;=\; R_{du} - R_u \boldsymbol{w}_{i-1}$$

where the expectation is computed conditioned on $\boldsymbol{w}_{i-1}$ (i.e., assuming $\boldsymbol{w}_{i-1}$ is fixed). Then

$$\frac{\boldsymbol{w}_{i+L} - \boldsymbol{w}_{i-1}}{\mu L} \approx [R_{du} - R_u \boldsymbol{w}_{i-1}]$$

which, for small μ, suggests a differential equation in the weight vector of the form:

$$\frac{\mathsf{d}\boldsymbol{w}(t)}{\mathsf{d}t} \;=\; R_{du} - R_u \boldsymbol{w}(t)$$

The convergence properties of this equation can now be studied using classical results from the theory of differential equations. For the above example of LMS, this step is trivial since all eigenvalues of $-R_u$ lie in the open left-half plane and, therefore, the resulting dynamic behavior is stable. For more general adaptive filters, the corresponding differential equations will be more involved.

Which method of analysis is more accurate? There is a lively discussion in the literature on which method is more appropriate for the analysis of adaptive filters. Those who argue against the use of the independence assumptions state, and rightfully so, that these assumptions do not hold for tapped-delay-line implementations. On the other hand, analyses that are based on averaging and ODE methods are only valid for infinitesimal step-sizes. When you judge this fact against the observation that the independence conditions give good results for small step-sizes, then we are back to square one! Thus it seems that if one is interested in studying the performance of adaptive filters for sufficiently small step-sizes, then any of these methods should be fine — see the discussion in App. 24.A. It is truly remarkable that the independence assumptions, being theoretically invalid, still give good results for small step-sizes. There does not seem to exist a complete satisfactory answer to this mystery to this date (except for the early investigations mentioned before by Mazo (1979), Jones, Cavin, and Reed (1982), and Macchi (1995, p. 114)).

Exact performance analysis. The next natural step in the analysis of adaptive filters is to inquire whether it is possible to study their performance without having to assume the independence conditions or even small step-sizes. In this context, the following questions would become relevant and they remain largely unanswered in the literature:

1. How large can the step-size be so that the independence-based approximations are still reasonable? Also, for a given value of the step-size, what is the order of magnitude of the error incurred by using these approximations?

2. What is the real performance of an adaptive filter when the step-size is not small and the independence assumptions are not used?

3. How large the step-size can get without compromising filter (mean-square) stability?
4. What is the step-size that gives the fastest convergence rate?

There are very few results in the literature that predict or confirm the behavior/stability of even LMS under these conditions.

One original work that attempted to study the performance of LMS without independence and slow adaptation assumptions was that of Florian and Feuer (1986). However, as is clear from the title of their article, the arguments were limited to filters with only two-taps (and there is a good reason for this — see below). Their idea was to show that studying the performance of LMS in essence amounts to studying the stability of a state-space model of the form

$$x_{k+1} = \Phi(\mu)x_k + z$$

where z is some constant vector, x_k is a state vector, and Φ is some constant matrix that depends on the step-size. With this model, the largest step-size ($\mu_{\max}$) that guarantees stable performance of LMS is the largest μ for which $\Phi(\mu)$ is still a stable matrix, i.e.,

$$\mu_{\max} \triangleq \sup\{\mu \text{ such that } \rho(\Phi(\mu)) < 1\}$$

where $\rho(\Phi)$ denotes the spectral radius of Φ, i.e., $\rho(\Phi) = \max_i |\lambda_i(\Phi)|$.

Unfortunately, determining $\mu_{\max}$ is a highly nontrivial task for two main reasons. First, the eigenvalues of the matrix Φ depend nonlinearly on μ and, secondly, the dimension of Φ grows extremely fast with the filter length (e.g., for a filter of order $M = 6$ the matrix has dimensions $28,181 \times 28,181$). It is computationally infeasible to work directly with Φ; the approach is feasible only for relatively small filter lengths. It was for this reason that Florian and Feuer (1986) considered only the case $M = 2$ (i.e., a filter with two taps), and later Douglas and Pan (1995) used the same method for orders up to $M = 6$ coupled with a numerical procedure (namely, the power method for sparse matrices) to evaluate the eigenvalues of Φ.

For longer filters, Nascimento and Sayed (1998) developed an alternative procedure for estimating a lower bound for $\mu_{\max}$ that does not work directly with the matrix Φ. Their approach was based on the observation that Φ, although of large dimensions, is both *sparse* and *structured*. These two properties can be exploited to derive a *computable* bound on the step-size for stable performance.

Another notable existence result was given by Solo (1997). This work provides a bound for the step-size that guarantees almost-sure stability of LMS. Unfortunately, however, the bound is not computable and, as explained in App. 9.E of Sayed (2003), almost-sure stability does not necessarily imply mean-square stability or even reasonable performance.

Adaptive filters with error nonlinearities. In this part we focused on adaptive filters with data nonlinearities. Appendix 9.C of Sayed (2003), on the other hand, considers adaptive filters with error nonlinearities. This class of filters is among the most difficult to analyze. For this reason, it has been common in the literature to resort to different techniques and assumptions with the intent of allowing tractable analyses. Some of the most common approaches in this regard are the following.

1. *The use of linearization* as, for example, in Duttweiler (1982), Walach and Widrow (1984), Gibson and Gray (1988), Sethares (1992), and Douglas and Meng (1994a). In this method of analysis, the error nonlinearity is linearized around an operating point and higher-order terms are discarded. Analyses that are based on this technique fail to accurately describe the adaptive filter performance at the early stages of adaptation where the error usually assumes larger values.
2. *Restricting the class of error nonlinearities* as, e.g., in Claasen and Mecklenbräuker (1981), Mathew and Cho (1987), Bershad (1988), Tanrikulu and Chambers (1996), and Chambers, Tanrikulu, and Constantinides (1994). In this case, the analysis is restricted to particular classes of algorithms such as the sign-error LMS algorithm, the least-mean mixed-norm (LMMN) algorithm, the least-mean fourth (LMF) algorithm, and error saturation nonlinearities. By limiting the study to a specific nonlinearity or to a class of nonlinearities, it is possible to avoid linearization and the results become more accurate.
3. *Imposing statistical assumptions on the distribution of the error signals* as, e.g., in Koike (1995) where the elements of the weight-error vector were assumed to be jointly Gaussian

while studying the sign-error LMS algorithm. This particular assumption was shown to be valid asymptotically in Sharma, Sethares, and Bucklew (1996). More accurate is the assumption that the residual error is Gaussian as in Duttweiler (1982) and Bershad and Bonnet (1990), or that its conditional value is Gaussian as in Mathews and Cho (1987) and Bershad (1988). Although it has been argued in Masry and Bullo (1995) that the Gaussian assumption on $e_a(i)$ for the sign-error LMS algorithm does not exactly hold under the independence assumption, still the Gaussianity assumption on $e_a(i)$ is reasonable for long filters by central limit arguments.

4. *Restricting the class of input data*, such as assuming white and/or Gaussian regression data as done, for example, in Claasen and Mecklenbräuker (1981), Duttweiler (1982), Gardner (1984), Gibson and Gray (1988), Rupp (1993), and Douglas and Meng (1994a).

5. *Using the independence assumptions*, whereby the successive regressors are assumed to be independent of each other as in Mazo (1979). Despite being unrealistic, the independence assumptions are among the most heavily used assumptions in adaptive filtering analysis and they tend to lead to results that agree with practice for small step-sizes.

6. *Assuming Gaussian white noise*. Although Gaussianity is not as common as the whiteness assumption, the whiteness condition on the noise allows for more tractable analyses.

Learning curves. Ensemble-average learning curves are commonly used to analyze and illustrate the performance of adaptive filters. They are obtained by averaging several squared-error curves over repeated experiments. Such averaged curves have been used extensively in the literature to extract information about the rate of convergence of an adaptive filter, its steady-state performance, or choices of step-sizes for faster or slower convergence. For *infinitesimal* step-sizes, or under the independence assumptions (22.23), it is known that information extracted from such ensemble-average learning curves provide reasonably accurate information about the real performance of an adaptive filter (see, e.g., the works by Widrow et al. (1976), Macchi and Eweda (1983), Bershad (1986b), Macchi (1995), Solo and Kong (1995), and Kushner and Vázquez-Abad (1996)).

Appendix 9.E of Sayed (2003), however, examines the performance of adaptive schemes for *larger* step-sizes. By larger step-sizes it is not meant step-sizes that are necessarily large, but rather step-sizes that are *not* infinitesimally small. The presentation in the appendix follows the work by Nascimento and Sayed (2000). In the process of comparing results obtained from ensemble-average learning curves with results predicted by an exact theoretical analysis, the authors observed some interesting differences between theory and practice. As shown in their article and in the appendix (see also Sayed and Nascimento (2003)), the differences in behavior can be explained by examining jointly the mean-square and almost-sure convergence of an adaptive filter. Both forms of convergence tend to agree for sufficiently small step-sizes, but they behave differently at larger step-sizes; thus leading to the occurrence of interesting phenomena in the learning behavior of an adaptive filter.

Almost-sure stability of filters. Prior to Nascimento and Sayed (2000), there have been several works in the literature on the almost-sure stability of LMS, e.g., by Bitmead and Anderson (1981), Bitmead, Anderson, and Ng (1986), and Solo (1997), or even more generally on the almost-sure stability of continuous-time systems, e.g., by Kozin (1969) and Parthasarathy and Evan-Iwanowski (1978). These earlier works focused mainly on infinitesimal step-sizes for which the distinctions between mean-square stability and almost-sure stability do not manifest themselves. The earlier works were also not interested in how the mean-square and almost-sure convergence performance of the filters differed for larger step-sizes. For example, Bitmead, Anderson, and Ng (1986) only compared both notions of stability for $\mu \approx 0$, when they in fact agree. Moreover, some of these earlier investigations may suggest that almost-sure stability implies reasonable algorithm performance (Solo (1997)). The material in App. 9.E of Sayed (2003) shows that this is not necessarily the case; an almost-sure stable filter might have poor performance when it is not also mean-square stable. This is because there is a small time interval during which the ensemble-average learning curve tends to stay reasonably close to the (mean-square) theoretical learning curve. In this way, an almost-sure stable, but mean-square unstable, algorithm would likely have its error diverging to a large value before starting to converge.

Problems and Computer Projects

PROBLEMS

Problem V.1 (Weighted norms) Introduce the notation $\|x\|_{\Sigma}^2 = x^*\Sigma x$, where Σ is a Hermitian and non-negative definite matrix.

(a) Show that $\|x\|_{\Sigma} = \sqrt{x^*\Sigma x}$ is a valid vector norm when Σ is positive-definite, i.e., verify that it satisfies all the properties of vector norms as described in Chapter B.

(b) When Σ is singular, which properties of vector norms are violated?

Problem V.2 (Positive-definiteness of the weighting matrix) Refer to expression (22.20) for Σ' and assume $\Sigma > 0$.

(a) Verify that

$$\boldsymbol{\Sigma}' = \left(\mathrm{I} - \mu\frac{\boldsymbol{u}_i^*\boldsymbol{u}_i}{g[\boldsymbol{u}_i]}\right)\Sigma\left(\mathrm{I} - \mu\frac{\boldsymbol{u}_i^*\boldsymbol{u}_i}{g[\boldsymbol{u}_i]}\right)$$

Conclude that $\boldsymbol{\Sigma}' \geq 0$.

(b) Now let $\Sigma' = \mathsf{E}\,\boldsymbol{\Sigma}'$. Clearly, $\Sigma' \geq 0$ in view of part (a). We wish to show that it is actually positive-definite. So assume, to the contrary, that there exists a nonzero constant vector x such that $\Sigma' x = 0$. Define

$$\boldsymbol{y} \triangleq \left(\mathrm{I} - \mu\frac{\boldsymbol{u}_i^*\boldsymbol{u}_i}{g[\boldsymbol{u}_i]}\right)x$$

Show from $\Sigma' x = 0$ that $\boldsymbol{y} = 0$ almost surely. Argue that this is only possible if the regressor $\boldsymbol{u}_i^*$ is non-random (i.e., a constant vector).

Problem V.3 (Special matrix) Consider a nonnegative-definite matrix F of the form $F = \mathrm{I} - \mu A + \mu^2 B$, for some positive-definite matrix A, nonnegative-definite matrix B (assumed nonzero), and a positive scalar μ. We would like to determine a bound on μ such that the eigenvalues of F are less than unity.

(a) Argue that the required condition on the eigenvalues of F is equivalent to $A - \mu B > 0$.

(b) Let $A = U_A \Lambda_A U_A^*$ denote the eigen-decomposition of A, where U_A is unitary and Λ_A is diagonal with positive entries. Argue that the matrices $A - \mu B$ and $\mathrm{I} - \mu\Lambda_A^{-1/2} U_A^* B U_A \Lambda_A^{-1/2}$ are congruent, where $\Lambda_A^{1/2}$ denotes a diagonal matrix with the positive square-roots of the eigenvalues of A. Conclude that the condition on μ is

$$\mu < \frac{1}{\lambda_{\max}(\Lambda_A^{-1/2} U_A^* B U_A \Lambda_A^{-1/2})}$$

(c) Argue that the matrices $\Lambda_A^{-1/2} U_A^* B U_A \Lambda_A^{-1/2}$ and $A^{-1}B$ are similar. Conclude that all the eigenvalues of $A^{-1}B$ are nonnegative, and that an equivalent characterization for the condition on μ is $\mu < 1/\lambda_{\max}(A^{-1}B)$. Conclude further that if B is positive-definite then the eigenvalues of $A^{-1}B$ are positive and real.

(d) Show further that the upper bound $1/\lambda_{\max}(A^{-1}B)$ is equal to the smallest η that results in a singular matrix $A - \eta B$.

Problem V.4 (Indefinite coefficient matrix) Consider the matrix F defined by (24.10) and assume all data are real-valued. Assume also $M = 2$ and generate the regressors as follows:

$$\boldsymbol{u}_i = \begin{cases} \begin{bmatrix} 1 & 0 \end{bmatrix} & \text{with probability } 1/2 \\ \begin{bmatrix} 0 & 1/\sqrt{2} \end{bmatrix} & \text{with probability } 1/2 \end{cases}$$

(a) Show that in this case, $F = \text{diag}\{1 - \mu + \mu^2/2,\ 1 - 3\mu/4,\ 1 - 3\mu/4,\ 1 - \mu/2 + \mu^2/8\}$.

(b) Verify that F can become indefinite. Do the eigenvalues of F cross first $+1$ or -1 as μ increases?

Problem V.5 (MSE of LMS) Consider the setting of Thm. 23.3, and let c denote the constant

$$c \triangleq \sum_{k=1}^{M} \frac{\lambda_k}{2 - s\lambda_k \mu}$$

Verify that the mean-square error (MSE) of LMS, which is the steady-state variance of $\boldsymbol{e}(i)$, is given by $\text{MSE} = \sigma_v^2/(1 - \mu c)$.

Problem V.6 (A positive-definite covariance matrix) Consider regressors $\{\boldsymbol{u}_i\}$ with a positive-definite covariance matrix R_u. Define the modified regressor $\check{\boldsymbol{u}}_i = \boldsymbol{u}_i/\sqrt{\epsilon + \|\boldsymbol{u}_i\|^2}$, and let $\check{R}_u$ denote its covariance matrix.

(a) Prove that $\check{R}_u$ is positive-definite.

(b) Use the properties of Kronecker products from Lemma B.8 to show that for any two $n \times n$ matrices $\{D, E\}$, the eigenvalues of $(\text{I} \otimes D) + (E \otimes \text{I})$ are all the n^2 combinations $\{\lambda_k(D) + \lambda_j(E)\}$ for all $1 \leq j, k \leq n$.

(c) Now consider the matrix A defined by (25.4). Show that $A > 0$.

(d) Likewise, show that the matrix B in (25.5) is nonnegative definite, $B \geq 0$.

Problem V.7 (Transient behavior of LMS) Start from (24.11) and show that

$$\boxed{\mathsf{E}\,\|\tilde{\boldsymbol{w}}_i\|^2 = \mathsf{E}\,\|\tilde{\boldsymbol{w}}_{i-1}\|^2 + \mu^2\sigma_v^2\left(r'^{\mathsf{T}} F^i q\right) - \|w^o\|^2_{F^i(\text{I}-F)q}}$$

where $\{F, r, q\}$ are as defined in Sec. 24.1.

Problem V.8 (Kronecker matrix inequality) Assume Z is a rank-one $M \times M$ matrix. Denote its eigenvalues by $\{\lambda, 0, 0, \ldots, 0\}$ and assume $0 < \lambda < 1$. Introduce the eigen-decomposition $Z = V\Delta V^*$ with V unitary and $\Delta = \text{diag}\{\lambda, 0, \ldots, 0\}$.

(a) Use property (i) from Lemma B.8 to verify that

$$Z^{\mathsf{T}} \otimes Z \;=\; (V^{*\mathsf{T}} \otimes V)(\Delta \otimes \Delta)(V^{\mathsf{T}} \otimes V^*), \quad Z^{\mathsf{T}} \otimes \text{I} = (V^{*\mathsf{T}} \otimes V)(\Delta \otimes \text{I})(V^{\mathsf{T}} \otimes V^*)$$

(b) Let $C = (V^{\mathsf{T}} \otimes V^*)$ and define $y = Cx$ for any nonzero vector x. Denote the entries of y by $\{y(i), i = 1, \ldots, M^2\}$. Verify that

$$x^*(Z^{\mathsf{T}} \otimes Z)x = \lambda^2|y(1)|^2, \qquad x^*(Z^{\mathsf{T}} \otimes \text{I})x = \sum_{i=1}^{M} \lambda^2|y(i)|^2$$

and conclude that $(Z^{\mathsf{T}} \otimes Z) \leq (Z^{\mathsf{T}} \otimes \text{I})$. Likewise, show that $(Z^{\mathsf{T}} \otimes Z) \leq (\text{I} \otimes Z)$. Conclude that $2(Z^{\mathsf{T}} \otimes Z) \leq (Z^{\mathsf{T}} \otimes \text{I}) + (\text{I} \otimes Z)$.

(c) Inequality (25.31) is obviously true when $\boldsymbol{u}_i = 0$. When $\boldsymbol{u}_i \neq 0$, choose $Z = \boldsymbol{u}_i^* \boldsymbol{u}_i / (\epsilon + \|\boldsymbol{u}_i\|^2)$ and use the result of part (b) to establish the validity of (25.31).

Problem V.9 (Learning curves) Consider the general setting of Thm. 25.2, which deals with data-normalized adaptive filters. The purpose of this problem is to use the variance relation (25.14) to characterize not only the time evolution of $\mathsf{E}\,\|\tilde{\boldsymbol{w}}_i\|^2$, as was done in the statement of theorem, but also the time evolution of $\mathsf{E}\,|\boldsymbol{e}_a(i)|^2$, which relates to the learning curve of the filter.

Recall that the learning curve of an adaptive filter is defined by $\mathsf{E}\,|\boldsymbol{e}(i)|^2$, which in view of (22.16) is given by $\mathsf{E}\,|\boldsymbol{e}(i)|^2 = \mathsf{E}\,|\boldsymbol{e}_a(i)|^2 + \sigma_v^2$, so that the learning curve can be equivalently characterized by studying the evolution of $\mathsf{E}\,|\boldsymbol{e}_a(i)|^2$. Under the conditions on the data stated in Thm. 25.2, we know from (23.55) that $\mathsf{E}\,|\boldsymbol{e}_a(i)|^2 = \mathsf{E}\,\|\tilde{\boldsymbol{w}}_{i-1}\|^2_{R_u}$. This means that the learning curve can be determined by studying the evolution of the above weighted norm of $\tilde{\boldsymbol{w}}_{i-1}$. This evolution can be deduced from the variance relation (25.14), namely,

$$\mathsf{E}\,\|\tilde{\boldsymbol{w}}_i\|^2_\sigma = \mathsf{E}\,\|\tilde{\boldsymbol{w}}_{i-1}\|^2_{\sigma'} + \mu^2\sigma_v^2 \mathsf{E}\left[\frac{\|\boldsymbol{u}_i\|^2_\sigma}{g^2[\boldsymbol{u}_i]}\right]$$

by properly choosing the weighting factor σ.

(a) Choose $\sigma = \text{vec}\{R_u\} = r$ (i.e., choose the weighting matrix as $\Sigma = R_u$). Use the weighting matrix recursion (25.18), and iterate the above variance relation, to deduce that

$$\mathsf{E}\,\|\tilde{\boldsymbol{w}}_i\|^2_r = \mathsf{E}\,\|\tilde{\boldsymbol{w}}_{-1}\|^2_{F^{i+1}r} + \mu^2\sigma_v^2 \mathsf{E}\left[\frac{\|\boldsymbol{u}_i\|^2_{(I+F+F^2+\ldots+F^i)r}}{g^2[\boldsymbol{u}_i]}\right]$$

where $\tilde{\boldsymbol{w}}_{-1} = w^o$, assuming $\boldsymbol{w}_{-1} = 0$.

(b) Verify that the expression for $\mathsf{E}\,\|\tilde{\boldsymbol{w}}_i\|^2_r$ from part (a) can be rewritten as $\mathsf{E}\,\|\tilde{\boldsymbol{w}}_i\|^2_r = \|w^o\|^2_{a_i} + \mu^2\sigma_v^2 b(i)$, where the vector a_i and the scalar $b(i)$ satisfy the recursions

$$a_i = Fa_{i-1}, \quad b(i) = b(i-1) + \mathsf{E}\left[\frac{\|\boldsymbol{u}_i\|^2_{a_{i-1}}}{g^2[\boldsymbol{u}_i]}\right], \quad a_{-1} = r, \quad b(-1) = 0$$

(c) Conclude that the following recursion holds for $\mathsf{E}\,|\boldsymbol{e}_a(i)|^2$,

$$\boxed{\mathsf{E}\,|\boldsymbol{e}_a(i)|^2 = \mathsf{E}\,|\boldsymbol{e}_a(i-1)|^2 - \|w^o\|^2_{F^{i-1}(I-F)r} + \mu^2\sigma_v^2 \mathsf{E}\left[\frac{\|\boldsymbol{u}_i\|^2_{F^{i-1}r}}{g^2[\boldsymbol{u}_i]}\right]}$$

(d) Use the recursion of part (c) to re-derive the expression for EMSE of Thm. 25.2.

(e) Specialize the result of part (c) to the LMS algorithm assuming circular Gaussian regressors, and re-derive the expression for the EMSE of LMS from Thm. 23.3.

Problem V.10 (Learning curve of LMS) Assume circular Gaussian regressors $\{\boldsymbol{u}_i\}$. Start from (23.19) and repeat the argument of Prob. V.9. Use the resulting learning curve to re-establish the expression for the EMSE of LMS from Thm. 23.3.

Problem V.11 (Non-Gaussian data) When the regression data are not Gaussian, it may still be possible to evaluate the last moment in (24.1) explicitly, in which case one can rely on the techniques of Chapter 23 as opposed to Chapter 24. Thus consider an adaptive filter with a 2-dimensional regression vector $\boldsymbol{u}_i = [\ \boldsymbol{u}_i(1)\ \ \boldsymbol{u}_i(2)]$. At every time instant i, the entries $\{\boldsymbol{u}_i(1), \boldsymbol{u}_i(2)\}$ are chosen independently of each other as follows:

$$\boldsymbol{u}_i(1) = \begin{cases} 2a & \text{with probability } 1/3 \\ -a & \text{with probability } 2/3 \end{cases} \qquad \boldsymbol{u}_i(2) = \begin{cases} -b & \text{with probability } 3/4 \\ 3b & \text{with probability } 1/4 \end{cases}$$

where a and b are positive real numbers. Moreover, the successive regression vectors $\{\boldsymbol{u}_i\}$ are independent of each other.

(a) Find the variances of $\boldsymbol{u}_i(1)$ and $\boldsymbol{u}_i(2)$ and the covariance matrix of $\boldsymbol{u}_i$.

(b) Characterize the learning curve of the LMS filter.

(c) Find exact expressions for the EMSE and MSD that would result when the filter is trained using LMS.

(d) Find conditions on the step-size to ensure stability in the mean and mean-square senses.

(e) How would the results of parts (c) and (d) change if $\boldsymbol{u}_i(1)$ and $\boldsymbol{u}_i(2)$ were replaced by independent zero-mean Gaussian random variables with the same variances?

Problem V.12 (Combination of adaptive filters) Refer to Prob. IV.4.

(a) Find exact conditions on the step-sizes μ_1 and μ_2 in order to guarantee the mean-square convergence of the combined adaptive structure.

(b) Find exact expressions for the MSDs of the two adaptive filters, $\boldsymbol{w}_{1,i}$ and $\boldsymbol{w}_{2,i}$.

(c) Find an expression describing the time evolution of the learning curve $\mathsf{E}\,|\boldsymbol{e}(i)|^2$, where $\boldsymbol{e}(i) = \boldsymbol{d}(i) - \boldsymbol{y}(i)$.

Problem V.13 (Diffusion strategy) Consider two nodes at spatial locations 1 and 2. At each time instant i, each node $k = 1, 2$ has access to data $\{\boldsymbol{d}_i^{(k)}, \boldsymbol{u}_i^{(k)}\}$ satisfying the standard data model

$$\boldsymbol{d}_i^{(k)} = \boldsymbol{u}_i^{(k)} w^o + \boldsymbol{v}_i^{(k)}, \quad k = 1, 2$$

for the same unknown vector w^o. The zero-mean regression sequences $\{\boldsymbol{u}_i^{(k)}\}$ are independent over both time and space (i.e., over i and k) with covariance matrices $\{R_u^{(k)}\}$. Likewise, the noise sequences $\{\boldsymbol{v}_i^{(k)}\}$ are both temporally and spatially white with variances $\{\sigma_v^{(k)}\}$. As indicated in Fig. V.1, each node k runs an LMS type filter that uses both temporal and spatial data as follows:

spatial information :

$$\begin{aligned} \boldsymbol{\phi}_i^{(1)} &= \alpha \boldsymbol{w}_{i-1}^{(1)} + (1-\alpha)\boldsymbol{w}_{i-1}^{(2)} \\ \boldsymbol{\phi}_i^{(2)} &= \beta \boldsymbol{w}_{i-1}^{(2)} + (1-\beta)\boldsymbol{w}_{i-1}^{(1)} \end{aligned}$$

temporal information :

$$\begin{aligned} \boldsymbol{w}_i^{(1)} &= \boldsymbol{\phi}_i^{(1)} + \mu \boldsymbol{u}_i^{(1)} \left[\boldsymbol{d}_i^{(1)} - \boldsymbol{u}_i^{*(1)} \boldsymbol{\phi}_i^{(1)}\right] \\ \boldsymbol{w}_i^{(2)} &= \boldsymbol{\phi}_i^{(2)} + \mu \boldsymbol{u}_i^{(2)} \left[\boldsymbol{d}_i^{(2)} - \boldsymbol{u}_i^{*(2)} \boldsymbol{\phi}_i^{(2)}\right] \end{aligned}$$

That is, each node first combines the existing weight estimates at both nodes to generate the intermediate estimates $\{\boldsymbol{\phi}_i^{(1)}, \boldsymbol{\phi}_i^{(2)}\}$. The combination uses coefficients $\{0 \le \alpha, \beta \le 1\}$. Subsequently, the intermediate vectors are updated to $\{\boldsymbol{w}_i^{(1)}, \boldsymbol{w}_i^{(2)}\}$, and the process continues.

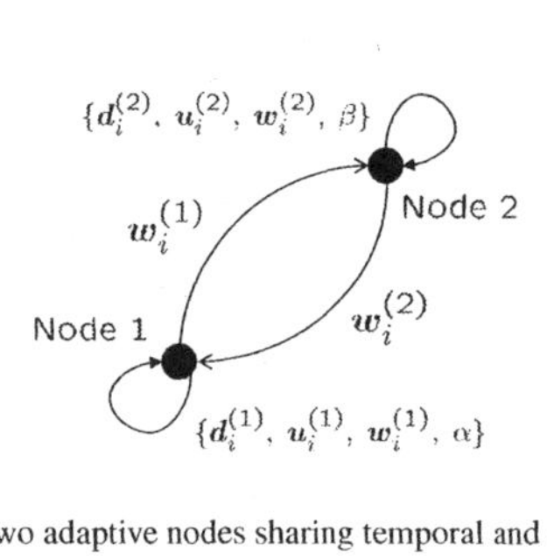

FIGURE V.1 Two adaptive nodes sharing temporal and spatial information.

(a) Find a condition on the step-size μ such that the mean behavior of the weight-error vectors of both adaptive filters is stable.

(b) How would the condition on μ change if $\alpha = 1$ and $\beta = 1$, i.e., if the nodes do not share spatial information?

(c) Assume $\alpha = \beta = 1/2$. Argue that for the same step-size, the means of the weight error vectors will generally converge faster to zero than when there is no spatial cooperation among the nodes (as in part (b)).

Remark. This example is a special case of more general distributed adaptive filters derived by Lopes Sayed (2006,2007a,b,2008), Sayed and Lopes (2007), and Cattivelli, Lopes, and Sayed (2008). The distributed schemes incorporate various forms of spatial cooperation among nodes. The term diffusion that appears in the title of this problem refers generally to the manner by which information is shared or diffused among the nodes in a neighborhood.

Problem V.14 (Performance of LMS with small step-size) Consider the discussion following Thm. 24.1, namely, that for a sufficiently small step-size, the variance relation (24.2) can be approximated by

$$\left\{\begin{array}{lcl} \mathsf{E}\,\|\tilde{w}_i\|^2_{\Sigma} & = & \mathsf{E}\,\|\tilde{w}_{i-1}\|^2_{\Sigma'} + \mu^2\sigma_v^2\mathsf{E}\,\|u_i\|^2_{\Sigma} \\ \Sigma' & = & \Sigma - \mu\Sigma\mathsf{E}\,[u_i^* u_i] - \mu\mathsf{E}\,[u_i^* u_i]\,\Sigma \end{array}\right.$$

Introduce the eigen-decomposition $R_u = U\Lambda U^*$, where Λ is diagonal with the eigenvalues of R_u and U is a unitary matrix. Introduce further the transformed quantities (22.31).

(a) Verify that the above variance relation becomes

$$\left\{\begin{array}{lcl} \mathsf{E}\,\|\overline{w}_i\|^2_{\overline{\Sigma}} & = & \mathsf{E}\,\|\overline{w}_{i-1}\|^2_{\overline{\Sigma}'} + \mu^2\sigma_v^2\mathsf{E}\,\|\overline{u}_i\|^2_{\overline{\Sigma}} \\ \overline{\Sigma}' & = & \overline{\Sigma} - \mu\overline{\Sigma}\Lambda - \mu\Lambda\overline{\Sigma} \end{array}\right.$$

Conclude that $\overline{\Sigma}'$ is diagonal if $\overline{\Sigma}$ is.

(b) Let $\overline{\sigma} = \text{diag}\{\overline{\Sigma}\}$ and $\lambda = \text{diag}\{\Lambda\}$. Verify that the expression for $\overline{\Sigma}'$ reduces to $\overline{\sigma}' = \overline{F}\overline{\sigma}$, where $\overline{F} = \mathrm{I} - 2\mu\Lambda$.

(c) Follow the arguments that led to Thm. 23.3 and show that $\mathsf{EMSE} = \mu\sigma_v^2\mathsf{Tr}(R_u)/2$ and $\mathsf{MSD} = \mu\sigma_v^2 M/2$. Compare the expression for EMSE with that in Lemma 16.1.

Problem V.15 (Numerical example) Consider the setting of Thm. 24.1 and assume all variables are real-valued. Set $M = 2$, i.e., the regressors are two-dimensional, say $u_i = \left[\begin{array}{cc} u_1(i) & u_2(i) \end{array}\right]$, and assume the entries $\{u_k(i)\}$ are i.i.d. and uniformly distributed between -1 and 1.

(a) Verify that $R_u = \mathrm{I}_2/3$, $(\mathrm{I}_2 \otimes R_u) = (R_u \otimes \mathrm{I}_2) = \mathrm{I}_4/3$, and

$$\mathsf{E}\,(u_i^{\mathsf{T}} u_i \otimes u_i^{\mathsf{T}} u_i) = \left[\begin{array}{cccc} 1/5 & 0 & 0 & 1/9 \\ 0 & 1/9 & 1/9 & 0 \\ 0 & 1/9 & 1/9 & 0 \\ 1/9 & 0 & 0 & 1/5 \end{array}\right]$$

(b) Evaluate $\mathsf{E}\,\|\tilde{w}_\infty\|^2$ for the choices $\mu = 0.01$ and $\mu = 0.5$. For what range of step-sizes is the filter mean-square stable?

Problem V.16 (Diagonal moments) Consider row vectors $\overline{u}_i$ that are zero-mean circular Gaussian with a diagonal covariance matrix Λ. Define the matrices

$$A \triangleq \mathsf{E}\left[\frac{\overline{u}_i^*\overline{u}_i}{\epsilon + \|\overline{u}_i\|^2}\right] + \mathsf{E}\left[\frac{\overline{u}_i^*\overline{u}_i}{\epsilon + \|\overline{u}_i\|^2}\right]^{\mathsf{T}}, \quad B' \triangleq \mathsf{E}\left[\frac{\|\overline{u}_i\|^2_{\overline{\Sigma}}}{(\epsilon + \|\overline{u}_i\|^2)^2}\overline{u}_i^*\overline{u}_i\right]$$

(a) Verify that the $(j,k)-$th element of A is given by

$$A_{j,k} = \mathsf{E}\left[\frac{\overline{u}_i^*(j)\overline{u}_i(k)}{\epsilon + \|\overline{u}_i\|^2}\right] + \mathsf{E}\left[\frac{\overline{u}_i^*(k)\overline{u}_i(j)}{\epsilon + \|\overline{u}_i\|^2}\right]^{\mathsf{T}}$$

where $\overline{u}_i(j)$ denotes the $j-$th entry of $\overline{u}_i$. For $j \neq k$, argue that the first term in the above expression for $A_{j,k}$ is an odd function of $\overline{u}_i(k)$, while the second term is an odd function of $\overline{u}_i(j)$. Conclude that $A_{j,k} = 0$ for $j \neq k$.

(b) Assume $\overline{\Sigma}$ is diagonal. Use a similar argument to establish that B' is diagonal. Verify that the k-th diagonal element of B' can be expressed as

$$B'_{k,k} \;=\; \mathsf{E}\left[\frac{\|\overline{u}_i\|^2_{\overline{\Sigma}}}{(\epsilon+\|\overline{u}_i\|^2)^2}|\overline{u}_i(k)|^2\right] = \mathsf{E}\left[\frac{|\overline{u}_i(k)|^2}{(\epsilon+\|\overline{u}_i\|^2)^2}\;[\overline{u}_i \circ (\overline{u}_i^*)^{\mathsf{T}}]\right]\overline{\sigma}$$

where the notation $\circ$ stands for Hadamard (or elementwise) product of two vectors.

(c) Let b' denote a vector with the diagonal entries of B', i.e., $b' = \text{diag}\{B'\}$. Verify that b' can be expressed as $b' = B\overline{\sigma}$ where B is the positive-definite matrix

$$B = \mathsf{E}\left[\frac{[\overline{u}_i^* \circ \overline{u}_i^{\mathsf{T}}]\;[\overline{u}_i \circ (\overline{u}_i^*)^{\mathsf{T}}]}{(\epsilon+\|\overline{u}_i\|^2)^2}\right] \;=\; \mathsf{E}\left[\frac{[\overline{u}_i^*\overline{u}_i]^{\mathsf{T}} \circ [\overline{u}_i^*\overline{u}_i]}{(\epsilon+\|\overline{u}_i\|^2)^2}\right]$$

and $\overline{\sigma}$ is the vector containing the diagonal entries of $\overline{\Sigma}$.

Problem V.17 (Gaussian regressors and ϵ-NLMS) Refer to the discussion in Sec. 25.1 on the transient performance of ϵ-NLMS and, in particular, to Thm. 25.1. The purpose of this problem is to show that the M^2-dimensional problem stated in the theorem reduces to an $M-$dimensional model when the regressors are Gaussian.

Thus assume the $\{u_i\}$ are circular Gaussian so that the transformed regressors $\overline{u}_i$ defined by (22.31) are also circular Gaussian. Since the covariance matrix of $\overline{u}_i$ is diagonal, and equal to Λ, the individual entries of $\overline{u}_i$ will be independent of each other. With Gaussian regressors, it is convenient to work with the transformed versions (22.33)–(22.35) and (22.36) of the variance and mean relations, which in the ϵ-NLMS case are given by

$$\begin{cases} \mathsf{E}\,\|\overline{w}_i\|^2_{\overline{\Sigma}} = \mathsf{E}\,\|\overline{w}_{i-1}\|^2_{\overline{\Sigma}'} \;+\; \mu^2\sigma_v^2\mathsf{E}\left[\dfrac{\|\overline{u}_i\|^2_{\overline{\Sigma}}}{(\epsilon+\|\overline{u}_i\|^2)^2}\right] \\ \overline{\Sigma}' = \overline{\Sigma} - \mu\overline{\Sigma}\mathsf{E}\left[\dfrac{\overline{u}_i^*\overline{u}_i}{\epsilon+\|\overline{u}_i\|^2}\right] - \mu\mathsf{E}\left[\dfrac{\overline{u}_i^*\overline{u}_i}{\epsilon+\|\overline{u}_i\|^2}\right]\overline{\Sigma} \;+\; \mu^2\mathsf{E}\left[\dfrac{\|\overline{u}_i\|^2_{\overline{\Sigma}}}{(\epsilon+\|\overline{u}_i\|^2)^2}\overline{u}_i^*\overline{u}_i\right] \\ \mathsf{E}\,\overline{w}_i = \left[\mathrm{I} - \mu\mathsf{E}\left(\dfrac{\overline{u}_i^*\overline{u}_i}{\epsilon+\|\overline{u}_i\|^2}\right)\right]\mathsf{E}\,\overline{w}_{i-1} \end{cases}$$

The moments that need to be evaluated are therefore

$$\mathsf{E}\left[\frac{\|\overline{u}_i\|^2_{\overline{\Sigma}}}{(\epsilon+\|\overline{u}_i\|^2)^2}\right],\quad \mathsf{E}\left[\frac{\overline{u}_i^*\overline{u}_i}{\epsilon+\|\overline{u}_i\|^2}\right],\quad \mathsf{E}\left[\frac{\|\overline{u}_i\|^2_{\overline{\Sigma}}}{(\epsilon+\|\overline{u}_i\|^2)^2}\overline{u}_i^*\overline{u}_i\right]$$

(a) Use the results of Prob. V.16 to conclude that $\overline{\Sigma}'$ will be diagonal when $\overline{\Sigma}$ is.

(b) Verify that the expression for $\overline{\Sigma}'$ can be rewritten as $\overline{\sigma}' = \overline{F}\overline{\sigma}$, where the $M \times M$ matrix $\overline{F}$ is given by $\overline{F} = \mathrm{I} - \mu A + \mu^2 B$ with

$$A \;\triangleq\; \mathsf{E}\left[\frac{\overline{u}_i^*\overline{u}_i}{\epsilon+\|\overline{u}_i\|^2}\right] + \mathsf{E}\left[\frac{\overline{u}_i^*\overline{u}_i}{\epsilon+\|\overline{u}_i\|^2}\right]^{\mathsf{T}} \;=\; P + P^{\mathsf{T}}$$

$$B \;\triangleq\; \mathsf{E}\left[\frac{[\overline{u}_i^*\overline{u}_i]^{\mathsf{T}} \circ [\overline{u}_i^*\overline{u}_i]}{(\epsilon+\|\overline{u}_i\|^2)^2}\right],\qquad P \triangleq \mathsf{E}\left[\frac{\overline{u}_i^*\overline{u}_i}{\epsilon+\|\overline{u}_i\|^2}\right]$$

where the notation $\circ$ stands for Hadamard (or elementwise) product of two vectors.

(c) Conclude also that the recursion for $\mathsf{E}\,\overline{w}_i$ can be written as $\mathsf{E}\,\overline{w}_i = (\mathrm{I} - \mu P)\,\mathsf{E}\,\overline{w}_{i-1}$, and that the transient behavior of $\epsilon-$NLMS is characterized by the M-dimensional state-space equation $\mathcal{W}_i = \mathcal{F}\mathcal{W}_{i-1} + \mu^2\sigma_v^2\mathcal{Y}$, where

$$\mathcal{F} = \begin{bmatrix} 0 & 1 & & & \\ 0 & 0 & 1 & & \\ 0 & 0 & 0 & 1 & \\ \vdots & & & & \\ 0 & 0 & 0 & & 1 \\ -p_0 & -p_1 & -p_2 & \ldots & -p_{M-1} \end{bmatrix}_{M\times M} \qquad \mathcal{W}_i = \begin{bmatrix} \mathsf{E}\,\|\overline{\boldsymbol{w}}_i\|^2_{\overline{\sigma}} \\ \mathsf{E}\,\|\overline{\boldsymbol{w}}_i\|^2_{\overline{F}\overline{\sigma}} \\ \mathsf{E}\,\|\overline{\boldsymbol{w}}_i\|^2_{\overline{F}^2\overline{\sigma}} \\ \vdots \\ \mathsf{E}\,\|\overline{\boldsymbol{w}}_i\|^2_{\overline{F}^{(M-1)}\overline{\sigma}} \end{bmatrix}$$

with

$$p(x) \triangleq \det(x\mathrm{I} - \overline{F}) = x^M + \sum_{k=0}^{M-1} p_k x^k$$

denoting the characteristic polynomial of the matrix $\overline{F}$, and where $\overline{\sigma}$ is an $M-$dimensional vector and the k-th entry of $\mathcal{Y}$ is

$$[\mathsf{E}\,\mathcal{Y}]_k = \left[\frac{\|\overline{\boldsymbol{u}}_i\|^2_{\overline{F}^k\overline{\sigma}}}{(\epsilon + \|\overline{\boldsymbol{u}}_i\|^2)^2}\right], \qquad k = 0, \ldots, M-1$$

Remark. These definitions of $\{\mathcal{W}_i, \mathcal{Y}\}$ are in terms of any $\overline{\sigma}$ of interest, e.g., most commonly, $\overline{\sigma} = q$ or $\overline{\sigma} = \lambda$, where $\lambda = \text{diag}(\Lambda)$.

(d) Verify that the mean-square deviation and the excess mean-square error of the algorithm are given by

$$\mathsf{MSD} = \mu^2\sigma_v^2\mathsf{E}\left[\frac{\|\overline{\boldsymbol{u}}_i\|^2_{(\mathrm{I}-\overline{F})^{-1}q}}{(\epsilon + \|\overline{\boldsymbol{u}}_i\|^2)^2}\right], \qquad \mathsf{EMSE} = \mu^2\sigma_v^2\mathsf{E}\left[\frac{\|\overline{\boldsymbol{u}}_i\|^2_{(\mathrm{I}-\overline{F})^{-1}\lambda}}{(\epsilon + \|\overline{\boldsymbol{u}}_i\|^2)^2}\right]$$

(e) Verify that the expressions of part (d) can be rewritten as

$$\boxed{\mathsf{MSD} = \mu^2\sigma_v^2\mathsf{Tr}(\overline{S}\,\overline{\Sigma}_{\text{msd}}) \quad \text{and} \quad \mathsf{EMSE} = \mu^2\sigma_v^2\mathsf{Tr}(\overline{S}\,\overline{\Sigma}_{\text{emse}})}$$

where $\overline{S} = \mathsf{E}\left(\overline{\boldsymbol{u}}_i^*\overline{\boldsymbol{u}}_i/(\epsilon + \|\overline{\boldsymbol{u}}_i\|^2)^2\right)$ and $\{\overline{\Sigma}_{\text{msd}}, \overline{\Sigma}_{\text{emse}}\}$ are the weighting matrices that correspond to the vectors $\overline{\sigma}_{\text{msd}} = (\mathrm{I} - \overline{F})^{-1}q$ and $\overline{\sigma}_{\text{emse}} = (\mathrm{I} - \overline{F})^{-1}\lambda$. That is, $\overline{\Sigma}_{\text{msd}} = \text{vec}^{-1}(\overline{\sigma}_{\text{msd}})$ and $\overline{\Sigma}_{\text{emse}} = \text{vec}^{-1}(\overline{\sigma}_{\text{emse}})$.

Problem V.18 (Performance of ϵ-NLMS with small step-size) In App. 17.A we argued that an ϵ-NLMS recursion can be reduced to an equivalent LMS recursion by using the transformations (17.10), in which case the $\epsilon-$NLMS update becomes

$$\boldsymbol{w}_i = \boldsymbol{w}_{i-1} + \mu\check{\boldsymbol{u}}_i^*\check{\boldsymbol{e}}(i), \qquad \text{where } \check{\boldsymbol{e}}(i) = \check{\boldsymbol{d}}(i) - \check{\boldsymbol{u}}_i\boldsymbol{w}_{i-1}$$

Moreover, the variables $\{\boldsymbol{d}(i), \boldsymbol{u}_i, \check{\boldsymbol{d}}(i), \check{\boldsymbol{u}}_i\}$ satisfy models (15.16) and (17.11). The variance of $\check{\boldsymbol{u}}_i$ is denoted by $\check{R}_u = \mathsf{E}\,\check{\boldsymbol{u}}_i^*\check{\boldsymbol{u}}_i = \mathsf{E}\left(\boldsymbol{u}_i^*\boldsymbol{u}_i/(\epsilon + \|\boldsymbol{u}_i\|^2)\right)$. Assume the step-size is sufficiently small and consider again the reasoning in Prob. V.14.

(a) Using the above LMS update, argue that a simplified variance relation in this case is given by (see Prob. V.19 for another approximation):

$$\begin{cases} \mathsf{E}\,\|\tilde{\boldsymbol{w}}_i\|^2_\Sigma = \mathsf{E}\,\|\tilde{\boldsymbol{w}}_{i-1}\|^2_{\Sigma'} + \mu^2\check{\sigma}_v^2\mathsf{E}\,\|\check{\boldsymbol{u}}_i\|^2_\Sigma \\ \Sigma' = \Sigma - \mu\Sigma\mathsf{E}\,[\check{\boldsymbol{u}}_i^*\check{\boldsymbol{u}}_i] - \mu\mathsf{E}\,[\check{\boldsymbol{u}}_i^*\check{\boldsymbol{u}}_i]\,\Sigma \end{cases}$$

(b) Introduce the eigen-decomposition $\check{R}_u = \check{U}\check{\Lambda}\check{U}^*$, where $\check{\Lambda}$ is diagonal with the eigenvalues of $\check{R}_u$ and $\check{U}$ is unitary. Introduce further the transformed quantities $\overline{\boldsymbol{w}}_i = \check{U}^*\tilde{\boldsymbol{w}}_i$, $\overline{\boldsymbol{u}}_i = \check{\boldsymbol{u}}_i\check{U}$, and $\overline{\Sigma} = \check{U}^*\Sigma\check{U}$. Verify that the above variance relation becomes

$$\begin{cases} \mathsf{E}\,\|\overline{\boldsymbol{w}}_i\|^2_{\overline{\Sigma}} = \mathsf{E}\,\|\overline{\boldsymbol{w}}_{i-1}\|^2_{\overline{\Sigma}'} + \mu^2\check{\sigma}_v^2\mathsf{E}\,\|\overline{\boldsymbol{u}}_i\|^2_{\overline{\Sigma}} \\ \overline{\Sigma}' = \overline{\Sigma} - \mu\overline{\Sigma}\check{\Lambda} - \mu\check{\Lambda}\overline{\Sigma} \end{cases}$$

Conclude that $\overline{\Sigma}'$ is diagonal if $\overline{\Sigma}$ is.

(c) Let $\overline{\sigma} = \text{diag}\{\overline{\Sigma}\}$ and $\check{\lambda} = \text{diag}\{\check{\Lambda}\}$. Verify that the expression for $\overline{\Sigma}'$ reduces to $\overline{\sigma}' = \overline{F}\overline{\sigma}$, where $\overline{F} = \text{I} - 2\mu\check{\Lambda}$.

(d) Follow the arguments that led to Thm. 23.3 to verify that $\text{MSD} = \mu\check{\sigma}_v^2 M/2$, where $\check{\sigma}_v^2 = \text{E}\,|\check{v}(i)|^2 = \sigma_v^2 \text{E}\,\left(1/(\epsilon + \|u_i\|^2)\right)$.

Problem V.19 (Another approximation for ϵ-NLMS) Consider the same setting of Prob. V.18 and continue to assume a sufficiently small step-size.

(a) Use (25.3) to justify the following simplified variance relation:

$$\left\{\begin{array}{rcl} \text{E}\,\|\tilde{w}_i\|^2_{\Sigma} & = & \text{E}\,\|\tilde{w}_{i-1}\|^2_{\Sigma'} + \mu^2\sigma_v^2 \text{E}\left[\dfrac{\|\check{u}_i\|^2_{\Sigma}}{\epsilon + \|u_i\|^2}\right] \\ \Sigma' & = & \Sigma - \mu\Sigma\text{E}\,[\check{u}_i^*\check{u}_i] - \mu\text{E}\,[\check{u}_i^*\check{u}_i]\,\Sigma \end{array}\right.$$

(b) Verify that the above variance relation can be rewritten as

$$\left\{\begin{array}{rcl} \text{E}\,\|\overline{w}_i\|^2_{\overline{\Sigma}} & = & \text{E}\,\|\overline{w}_{i-1}\|^2_{\overline{\Sigma}'} + \mu^2\sigma_v^2 \text{E}\left[\dfrac{\|\check{u}_i\|^2_{\check{U}\overline{\Sigma}\check{U}^*}}{\epsilon + \|u_i\|^2}\right] \\ \overline{\Sigma}' & = & \overline{\Sigma} - \mu\overline{\Sigma}\check{\Lambda} - \mu\check{\Lambda}\overline{\Sigma} \end{array}\right.$$

Conclude that $\overline{\Sigma}'$ is diagonal if $\overline{\Sigma}$ is.

(c) Define the covariance matrix

$$\widehat{R}_u \triangleq \text{E}\left(\frac{\check{u}_i^*\check{u}_i}{\epsilon + \|u_i\|^2}\right) = \text{E}\left(\frac{u_i^* u_i}{(\epsilon + \|u_i\|^2)^2}\right)$$

and justify the following expressions for the MSD and EMSE:

$$\boxed{\text{MSD} = \mu\sigma_v^2 \text{Tr}(\check{R}_u^{-1}\widehat{R}_u)/2, \qquad \text{EMSE} = \mu\sigma_v^2 \text{Tr}(\widehat{R}_u \Sigma_{\text{emse}})}$$

where Σ_{emse} is the unique solution to the Lyapunov equation $\check{R}_u\Sigma_{\text{emse}} + \Sigma_{\text{emse}}\check{R}_u = R_u$.

Problem V.20 (Alternative independence analysis of NLMS) There is an alternative approximation method to simplify the analysis of NLMS. Unlike the results we discussed in Secs. 17.1 and 25.1, this method requires $\epsilon = 0$. Thus consider the NLMS recursion

$$w_i = w_{i-1} + \mu\frac{u_i^*}{\|u_i\|^2}e(i), \qquad e(i) = d(i) - u_i w_{i-1}$$

where the data $\{d(i), u_i\}$ are still assumed to satisfy model (22.1) and condition (22.23). In addition, there is one more assumption on the distribution of the regressors. Let $R_u = U\Lambda U^*$ denote the eigen-decomposition of R_u. The columns of U are the eigenvectors of R_u and they will be denoted by $\{q_i\}$; they all have unit norm. Assume that the regressors u_i^* are modeled as $u_i^* = r(i)s(i)h_i$, where the random variables $\{r(i), s(i), h_i\}$ are independent, $r(i)$ has the same probability distribution as $\|u_i\|$, $s(i) = \pm 1$ with probability $1/2$, and $\text{Prob}(h_i = q_j) = \lambda_j/\text{Tr}(R_u)$. The idea of this construction is to assume a regression distribution that simplifies the analysis; it is chosen such that the first and second-order moments of the u_i so generated coincide with those of the original u_i. Since for small step-sizes, the filter performance depends on these lower-order moments, the approximation tends to be reasonable.

(a) Show that

$$\text{E}\,h_i h_i^* = R_u/\text{Tr}(R_u), \quad \text{E}\,h_i h_i^* h_i h_i^* = R_u/\text{Tr}(R_u)$$

and

$$\text{E}\,h_i^* R_u^{-1} h_i = M/\text{Tr}(R_u), \quad \text{E}\,h_i h_i^* R_u^{-1} h_i h_i^* = \text{I}/\text{Tr}(R_u)$$

(b) Refer to the variance relation (25.3), with $\epsilon = 0$. Using the assumed model for $\boldsymbol{u}_i$, verify that these relations reduce to

$$\begin{cases} \mathsf{E}\,\|\tilde{\boldsymbol{w}}_i\|^2_{\Sigma} &= \mathsf{E}\,\|\tilde{\boldsymbol{w}}_{i-1}\|^2_{\Sigma'} + \mu^2\sigma_v^2\mathsf{E}\,\|\boldsymbol{h}_i\|^2_{\Sigma}\mathsf{E}\left(\dfrac{1}{\boldsymbol{r}^2(i)}\right) \\ \Sigma' &= \Sigma - \mu\Sigma(\mathsf{E}\,\boldsymbol{h}_i\boldsymbol{h}_i^*) - \mu(\mathsf{E}\,\boldsymbol{h}_i\boldsymbol{h}_i^*)\Sigma + \mu^2\mathsf{E}\,\boldsymbol{h}_i\boldsymbol{h}_i^*\Sigma\boldsymbol{h}_i\boldsymbol{h}_i^* \end{cases}$$

(c) Assume first $\Sigma = \mathrm{I}$. Show that $\Sigma' = \mathrm{I} - \mu(2-\mu)\frac{R_u}{\mathsf{Tr}(R_u)}$. Assume now $\Sigma = \frac{\mathsf{Tr}(R_u)}{\mu(2-\mu)}R_u^{-1}$, show that

$$\Sigma' = \frac{\mathsf{Tr}(R_u)}{\mu(2-\mu)}R_u^{-1} - \mathrm{I}$$

(i.e., verify that $\Sigma - \Sigma' = \mathrm{I}$).

(d) Take the limit of both sides of the recursion for $\mathsf{E}\,\|\tilde{\boldsymbol{w}}_i\|^2_{\Sigma}$ to conclude that

$$\mathsf{E}\,\|\tilde{\boldsymbol{w}}_\infty\|^2_{\Sigma-\Sigma'} = \mu^2\sigma_v^2\mathsf{E}\,\|\boldsymbol{h}_i\|^2_{\Sigma}\mathsf{E}\left(\frac{1}{\boldsymbol{r}^2(i)}\right)$$

Use the choice $\Sigma = \dfrac{\mathsf{Tr}(R_u)}{\mu(2-\mu)}R_u^{-1}$ to conclude that the MSD $= \mathsf{E}\,\|\tilde{\boldsymbol{w}}_\infty\|^2$ is given by

$$\boxed{\mathsf{MSD} = \frac{\mu\sigma_v^2 M}{2-\mu}\mathsf{E}\left(\frac{1}{\|\boldsymbol{u}_i\|^2}\right)}$$

(e) Choose now $\Sigma = \frac{\mathsf{Tr}(R_u)}{\mu(2-\mu)}\mathrm{I}$, and verify that $\Sigma - \Sigma' = R_u$. With this choice for Σ, conclude that

$$\mathsf{E}\,\|\tilde{\boldsymbol{w}}_\infty\|^2_{R_u} = \mu^2\sigma_v^2\frac{\mathsf{Tr}(R_u)}{\mu(2-\mu)}\mathsf{E}\,\|\boldsymbol{h}_i\|^2\mathsf{E}\left(\frac{1}{\boldsymbol{r}^2(i)}\right)$$

so that the EMSE $= \mathsf{E}\,\|\tilde{\boldsymbol{w}}_\infty\|^2_{R_u}$ is given by

$$\boxed{\mathsf{EMSE} = \frac{\mu\sigma_v^2}{2-\mu}\mathsf{Tr}(R_u)\mathsf{E}\left(\frac{1}{\|\boldsymbol{u}_i\|^2}\right)}$$

Compare with Lemma 17.1.

Remark. The model used in this problem for $\boldsymbol{u}_i$ has been suggested by Slock (1993), and the resulting performance results agree with those derived by him.

Problem V.21 (LMS with white input) Consider the LMS recursion (23.1) and assume the data $\{\boldsymbol{d}(i), \boldsymbol{u}_i\}$ satisfy (22.1) and the independence assumption (22.23). Assume further that the individual entries of the regressor $\boldsymbol{u}_i$ are zero-mean i.i.d. with variance σ_u^2 and fourth moment ξ_u^4. In this way, the covariance matrix is diagonal, $R_u = \sigma_u^2\mathrm{I}$.

(a) Assume Σ is diagonal and let $\sigma = \mathsf{diag}\{\Sigma\}$. Verify that the expectation $\mathsf{E}\,[\|\boldsymbol{u}_i\|^2_{\Sigma}\boldsymbol{u}_i^*\boldsymbol{u}_i]$ is diagonal with entries given by $B\sigma$, where $B = \sigma_u^4 qq^{\mathsf{T}} + (\xi_u^4 - \sigma_u^4)\mathrm{I}$ and q is the column vector with unit entries.

(b) Refer to the variance relation of Thm. 22.4, specialized to LMS, and verify that Σ' will be diagonal if Σ is. In particular, set $\Sigma = \mathrm{I}$ and define $\kappa^2 = \xi_u^4 + (M-1)\sigma_u^4$. Verify that this leads to $\Sigma' = (1 - 2\mu\sigma_u^2 + \mu^2\kappa^2)\mathrm{I}$.

(c) Conclude that the transient behavior of LMS is now described by the one-dimensional difference equation $\mathsf{E}\,\|\tilde{\boldsymbol{w}}_i\|^2 = (1 - 2\mu\sigma_u^2 + \mu^2\kappa^2)\mathsf{E}\,\|\tilde{\boldsymbol{w}}_{i-1}\|^2 + \mu^2\sigma_v^2\sigma_u^2 M$.

(d) Show that the MSD and EMSE of LMS under these conditions are given by

$$\mathsf{MSD} = \frac{\mu\sigma_v^2\sigma_u^2 M}{2\sigma_u^2 - \mu\kappa^2} \quad \text{and} \quad \mathsf{EMSE} = \sigma_u^2\cdot\mathsf{MSD}$$

(e) Assume the individual entries of $\boldsymbol{u}_i$ have a Gaussian distribution, so that $\xi_u^4 = 2\sigma_u^4$ (for complex-valued random variables; $\xi_u^4 = 3\sigma_u^4$ for real-valued random variables). Verify that in this case

$$\mathsf{E}\,\|\tilde{w}_i\|^2 = (1 - 2\mu\sigma_u^2 + \mu^2\sigma_u^4(M+1))\mathsf{E}\,\|\tilde{w}_{i-1}\|^2 \; + \; \mu^2\sigma_v^2\sigma_u^2 M$$

and

$$\text{MSD} = \frac{\mu\sigma_v^2 M}{2 - \mu\sigma_u^2(M+1)}, \qquad \text{EMSE} = \frac{\mu\sigma_v^2\sigma_u^2 M}{2 - \mu\sigma_u^2(M+1)}$$

(f) Show that the expressions of part (e) agree with what you would obtain from the result of Thm. 23.3.

Problem V.22 (Matrix data-nonlinearities) Consider adaptive filter updates of the form

$$\begin{aligned} w_i &= w_{i-1} + \mu H[u_i]u_i^* e(i), \quad i \geq 0 \\ e(i) &= d(i) - u_i w_{i-1} \end{aligned}$$

where $H[\cdot]$ is some positive-definite Hermitian matrix-valued (as opposed to scalar-valued) function of u_i. Assume further that the data $\{d(i), u_i\}$ satisfy the modeling assumptions (22.1).

For any Hermitian positive-definite matrix Σ, define the *a priori* and *a posterior* estimation errors $e_a^{H\Sigma}(i) = u_i H[u_i]\Sigma\tilde{w}_{i-1}$ and $e_p^{H\Sigma}(i) = u_i H[u_i]\Sigma\tilde{w}_i$, where $\tilde{w}_i = w^o - w_i$.

(a) Repeat the derivation of Sec. 22.3 to establish the following energy conservation relation:

$$\|u_i\|^2_{H\Sigma H} \cdot \|\tilde{w}_i\|^2_\Sigma \; + \; |e_a^{H\Sigma}(i)|^2 = \|u_i\|^2_{H\Sigma H} \cdot \|\tilde{w}_{i-1}\|^2_\Sigma \; + \; |e_p^{H\Sigma}(i)|^2$$

where $\|u_i\|^2_{H\Sigma H} = u_i(H[u_i]\Sigma H[u_i])u_i^*$.

(b) Assume further that the regressors satisfy the independence assumption (22.23). Repeat the derivation of Sec. 22.4 in order to extend the result of Thm. 22.4 to the present context, i.e., establish the validity of the following variance relation:

$$\begin{aligned} &\mathsf{E}\,\|\tilde{w}_i\|^2_\Sigma = \mathsf{E}\,\|\tilde{w}_{i-1}\|^2_{\Sigma'} \; + \; \mu^2\sigma_v^2\mathsf{E}\,\|u_i\|^2_{H\Sigma H} \\ &\Sigma' = \Sigma \; - \; \mu\Sigma\mathsf{E}\,(H[u_i]u_i^*u_i) \; - \; \mu\mathsf{E}\,(u_i^*u_iH[u_i])\Sigma \; + \; \mu^2\mathsf{E}\,\left(\|u_i\|^2_{H\Sigma H}u_i^*u_i\right) \end{aligned}$$

(c) Verify also that $\tilde{w}_i = (\mathrm{I} - \mu H[u_i]u_i^*u_i)\,\tilde{w}_{i-1} - \mu H[u_i]u_i^* v(i)$.

Problem V.23 (Filters with data nonlinearities) Verify that the algorithms listed below can be regarded as special cases of the filter update with matrix data nonlinearities of Prob. V.22 for different choices of $H[\cdot]$. In each case, identify the matrix H.

(a) LMS with multiple step-sizes: $w_i = w_{i-1} + \text{diag}\{\mu_1, \mu_2, \ldots, \mu_M\}u_i^*[d(i) - u_i w_{i-1}]$.

(b) NLMS with individual power estimates:

$$w_i = w_{i-1} + \mu\text{diag}\left\{1/p_1(i), 1/p_2(i), \ldots, 1/p_M(i)\right\} u_i^*\left[d(i) - u_i w_{i-1}\right]$$

where each $p_k(i)$ is updated as follows:

$$p_k(i) = \beta p_k(i-1) + (1-\beta)|u_i(k)|^2, \quad 0 \ll \beta < 1, \;\; p_k(-1) = \epsilon$$

and $u_i(k)$ denotes the $k-$th entry of u_i.

Problem V.24 (Stability of data-normalized filters) Refer to the discussion on data normalized filters in Sec. 25.2 and consider the variance relation (25.14). Let

$$P \triangleq \mathsf{E}\left[\frac{u_i^*u_i}{g[u_i]}\right], \qquad S \triangleq \mathsf{E}\left[\frac{u_i^*u_i}{g^2[u_i]}\right], \qquad X \triangleq \mathsf{E}\left[\frac{\|u_i\|^2}{g^2[u_i]}u_i^*u_i\right]$$

and choose $\Sigma = \mathrm{I}$.

(a) Verify that the relations (25.14) can be written as

$$\mathsf{E}\,\|\tilde{\boldsymbol{w}}_i\|^2 = \mathsf{E}\,\|\tilde{\boldsymbol{w}}_{i-1}\|^2_{\Sigma'} + \mu^2\sigma_v^2\mathsf{Tr}(S),\ \ \Sigma' = \mathrm{I} - 2\mu P + \mu^2 X,\ \ \mathsf{E}\,\tilde{\boldsymbol{w}}_i = (\mathrm{I}-\mu P)\,\mathsf{E}\,\tilde{\boldsymbol{w}}_{i-1}$$

(b) Assume the moments of $\boldsymbol{u}_i$ and the function $g[\cdot]$ are such that $X = cP$ for some constant $c > 0$. This situation arises, for example, while studying the sign-regressor LMS algorithm (see Prob. V.25). Show that $\mathsf{E}\,\|\tilde{\boldsymbol{w}}_i\|^2$ converges if, and only if, $\mu < 2/c$.

Remark. The scenario studied in this problem arises in some studies of LMS itself (see, e.g., Macchi (1995, pp. 161–163)). See also App. 9.C of Sayed (2003).

Problem V.25 (Sign-regressor LMS algorithm) Assume all data are real-valued, and consider an adaptive filter update of the form:

$$\boxed{\boldsymbol{w}_i = \boldsymbol{w}_{i-1} + \mu\mathsf{sign}[\boldsymbol{u}_i]^{\mathsf{T}}[\boldsymbol{d}(i) - \boldsymbol{u}_i\boldsymbol{w}_{i-1}]}$$

where $\mathsf{sign}[\boldsymbol{u}_i]$ is a column vector with the signs of the entries of $\boldsymbol{u}_i$.

(a) Argue that this algorithm can be regarded as a special case of the general update form of Prob. V.22 for some matrix data nonlinearity that is implicity defined by the identity $\mathsf{sign}[\boldsymbol{u}_i]^{\mathsf{T}} = H[\boldsymbol{u}_i]\boldsymbol{u}_i^{\mathsf{T}}$.

(b) Verify that the variance relation of part (b) of Prob. V.22 becomes

$$\begin{cases} \mathsf{E}\,\|\tilde{\boldsymbol{w}}_i\|^2_{\Sigma} = \mathsf{E}\,\|\tilde{\boldsymbol{w}}_{i-1}\|^2_{\Sigma'} + \mu^2\sigma_v^2\mathsf{E}\left[\,\|\mathsf{sign}[\boldsymbol{u}_i]\|^2_{\Sigma}\,\right] \\ \Sigma' = \Sigma - \mu\Sigma\mathsf{E}\left(\mathsf{sign}[\boldsymbol{u}_i]^{\mathsf{T}}\boldsymbol{u}_i\right) - \mu\mathsf{E}\left(\boldsymbol{u}_i^{\mathsf{T}}\mathsf{sign}[\boldsymbol{u}_i]\right)\Sigma + \mu^2\mathsf{E}\left[\,\|\mathsf{sign}[\boldsymbol{u}_i]\|^2_{\Sigma}\boldsymbol{u}_i^{\mathsf{T}}\boldsymbol{u}_i\,\right] \end{cases}$$

(c) Assume that the individual entries of $\boldsymbol{u}_i$ have variance σ_u^2, and that $\boldsymbol{u}_i$ has a Gaussian distribution. Use Price's theorem (cf. (18.6)) to verify that

$$\mathsf{E}\left[\mathsf{sign}[\boldsymbol{u}_i]^{\mathsf{T}}\boldsymbol{u}_i\right] = \sqrt{\frac{2}{\pi\sigma_u^2}}R_u$$

Conclude that the weighting matrix of part (b) reduces to

$$\Sigma' = \Sigma - \mu\sqrt{\frac{2}{\pi\sigma_u^2}}\Sigma R_u - \mu\sqrt{\frac{2}{\pi\sigma_u^2}}R_u\Sigma + \mu^2\mathsf{E}\left[\,\|\mathsf{sign}[\boldsymbol{u}_i]\|^2_{\Sigma}\boldsymbol{u}_i^{\mathsf{T}}\boldsymbol{u}_i\,\right]$$

Choose $\Sigma = \mathrm{I}$ and verify that $\Sigma' = \mathrm{I} + \mu\left(\mu M - 2\sqrt{\frac{2}{\pi\sigma_u^2}}\right)R_u$.

(d) Use the result of Prob. V.24 to show that $\mathsf{E}\,\|\tilde{\boldsymbol{w}}_i\|^2$ converges if, and only if,

$$\boxed{0 < \mu < \frac{1}{M}\sqrt{\frac{8}{\pi\sigma_u^2}}}$$

Conclude further that

$$\boxed{\mathsf{FMSE} = \frac{\mu\sigma_v^2 M}{\sqrt{\frac{8}{\pi\sigma_u^2}} - \mu M}}$$

Remark. The conditions on μ for mean-square stability and the expression for the filter EMSE coincide with those derived by Eweda (1990a). It should be noted though that replacement of the regressor by its sign limits the range of search directions that are followed by the adaptive algorithm and, therefore, performance degradation in terms of convergence speed (and even possibly divergence) can occur relative to a plain LMS implementation. Actually, it is possible to determine examples of input sequences that can cause sign-regressor LMS to diverge while LMS still converges. For more discussion on these issues, see Sethares et al. (1988) and Sethares and Johnson (1989).

Problem V.26 (Price's theorem for complex variables) Let $\boldsymbol{a}$ and $\boldsymbol{b}$ be scalar complex-valued zero-mean jointly Gaussian and circular random variables. Let $\rho = \mathsf{E}\,\boldsymbol{a}\boldsymbol{b}$. The complex form of

Price's theorem states that for any function f of $\{\boldsymbol{a}, \boldsymbol{b}\}$ (for which the required derivatives and integrals exist), it holds that

$$\frac{\partial \mathsf{E}\, f(\boldsymbol{a},\boldsymbol{b})}{\partial \rho} = \mathsf{E}\left(\frac{\partial^2 f(\boldsymbol{a},\boldsymbol{b})}{\partial \boldsymbol{a}\partial \boldsymbol{b}}\right)$$

where differentiation with respect to a complex variable is defined as in Chapter C.

(a) Assume the function $f(\boldsymbol{a},\boldsymbol{b})$ has the form $f(\boldsymbol{a},\boldsymbol{b}) = \boldsymbol{a}g(\boldsymbol{b})$. Verify from Price's theorem that $\mathsf{E}\,\boldsymbol{a}g(\boldsymbol{b}) = (\mathsf{E}\,\boldsymbol{a}\boldsymbol{b}) \cdot \mathsf{E}\,(\mathrm{d}g(\boldsymbol{b})/\mathrm{d}\boldsymbol{b})$.

(b) Show further that $\mathsf{E}\,\boldsymbol{b}^* g(\boldsymbol{b}) = \mathsf{E}\,|\boldsymbol{b}|^2 \cdot \mathsf{E}\,\mathrm{d}g(\boldsymbol{b})/\mathrm{d}\boldsymbol{b}$ and conclude that the following relation also holds:

$$\boxed{\mathsf{E}\,\boldsymbol{a}g(\boldsymbol{b}) = \frac{\mathsf{E}\,\boldsymbol{a}\boldsymbol{b}}{\mathsf{E}\,|\boldsymbol{b}|^2} \cdot \mathsf{E}\,\boldsymbol{b}^* g(\boldsymbol{b})}$$

Remark. For more details on Price's theorem for complex data, see McGee (1969) and van den Bos (1996).

Problem V.27 (Leaky variance relation) In this problem we extend the energy-conservation and variance relations of Thms. 22.1, 22.3 and 22.4 to leaky adaptive updates of the form

$$\boxed{\begin{aligned} \boldsymbol{w}_i &= (1-\mu\alpha)\boldsymbol{w}_{i-1} + \mu\frac{\boldsymbol{u}_i^*}{g[\boldsymbol{u}_i]}\boldsymbol{e}(i), \quad i \geq 0 \\ \boldsymbol{e}(i) &= \boldsymbol{d}(i) - \boldsymbol{u}_i\boldsymbol{w}_{i-1} \end{aligned}}$$

where α is a positive scalar and $g[\cdot]$ is some positive-valued function of $\boldsymbol{u}_i$. For example, the choice $g[\boldsymbol{u}_i] = 1$ results in the leaky-LMS filter. The data $\{\boldsymbol{d}(i), \boldsymbol{u}_i\}$ are still assumed to satisfy model (22.1).

(a) Verify that the weight-error vector, $\tilde{\boldsymbol{w}}_i = w^o - \boldsymbol{w}_i$, satisfies the recursion

$$\tilde{\boldsymbol{w}}_i = (1-\alpha\mu)\tilde{\boldsymbol{w}}_{i-1} - \mu\frac{\boldsymbol{u}_i^*}{g[\boldsymbol{u}_i]}\boldsymbol{e}(i) + \alpha\mu w^o$$

(b) Equating the weighted norms of both sides of the above equality, for some arbitrary Hermitian positive-definite weighting matrix Σ, verify that

$$\begin{aligned} \|\tilde{\boldsymbol{w}}_i\|_\Sigma^2 &= \left\|(1-\alpha\mu)\tilde{\boldsymbol{w}}_{i-1} - \mu\frac{\boldsymbol{u}_i^*}{g[\boldsymbol{u}_i]}\boldsymbol{e}_a(i) + \mu\alpha w^o\right\|_\Sigma^2 + \mu^2\frac{\|\boldsymbol{u}_i\|_\Sigma^2}{g^2[\boldsymbol{u}_i]}|\boldsymbol{v}(i)|^2 \\ &\quad - \frac{2\mu}{g[\boldsymbol{u}_i]}\mathsf{Re}\left\{\boldsymbol{u}_i\boldsymbol{v}^*(i)\left[(1-\alpha\mu)\tilde{\boldsymbol{w}}_{i-1} - \mu\frac{\boldsymbol{u}_i^*}{g[\boldsymbol{u}_i]}\boldsymbol{e}_a(i) + \mu\alpha w^o\right]\right\} \end{aligned}$$

(c) Use the condition on $\boldsymbol{v}(i)$ from the data model (22.1), and take expectations of both sides of the equality of part (b), to establish the variance relation:

$$\boxed{\begin{aligned} \mathsf{E}\,\|\tilde{\boldsymbol{w}}_i\|_\Sigma^2 &= \mathsf{E}\,\|\tilde{\boldsymbol{w}}_{i-1}\|_{\boldsymbol{\Sigma}'}^2 + \alpha^2\mu^2\|w^o\|_\Sigma^2 + \mu^2\sigma_v^2\mathsf{E}\left(\frac{\|\boldsymbol{u}_i\|_\Sigma^2}{g^2[\boldsymbol{u}_i]}\right) \\ &\quad + 2\alpha\mu\mathsf{Re}\left\{w^{o*}\Sigma\mathsf{E}\left[(\mathrm{I}-\mu\boldsymbol{U}_i)\tilde{\boldsymbol{w}}_{i-1}\right]\right\} \end{aligned}}$$

where

$$\boldsymbol{\Sigma}' = (1-\alpha\mu)^2\Sigma - \mu(1-\alpha\mu)\Sigma\frac{\boldsymbol{u}_i^*\boldsymbol{u}_i}{g[\boldsymbol{u}_i]} - \mu(1-\alpha\mu)\frac{\boldsymbol{u}_i^*\boldsymbol{u}_i}{g[\boldsymbol{u}_i]}\Sigma + \mu^2\boldsymbol{u}_i^*\frac{\|\boldsymbol{u}_i\|_\Sigma^2}{g^2[\boldsymbol{u}_i]}\boldsymbol{u}_i$$

and

$$\boxed{\boldsymbol{U}_i \triangleq \alpha\mathrm{I} + \frac{\boldsymbol{u}_i^*\boldsymbol{u}_i}{g[\boldsymbol{u}_i]}}$$

Verify further that the recursion for the weighting matrix can be rewritten more compactly as

$$\boxed{\boldsymbol{\Sigma}' = \Sigma - \mu\boldsymbol{U}_i\Sigma - \mu\Sigma\boldsymbol{U}_i + \mu^2\boldsymbol{U}_i\Sigma\boldsymbol{U}_i}$$

The above variance relation is the extension of Thm. 22.3 to leaky data-normalized filters.

(d) Now assume that the regressors $\{\boldsymbol{u}_i\}$ satisfy the independence assumption (22.23). Conclude that the variance relation of part (c) reduces to the following:

$$\boxed{\begin{array}{rcl} \mathsf{E}\,\|\tilde{\boldsymbol{w}}_i\|^2_{\Sigma} & = & \mathsf{E}\,\|\tilde{\boldsymbol{w}}_{i-1}\|^2_{\Sigma'} \;+\; \alpha^2\mu^2\|w^o\|^2_{\Sigma} \;+\; \mu^2\sigma_v^2\mathsf{E}\left(\dfrac{\|\boldsymbol{u}_i\|^2_{\Sigma}}{g^2[\boldsymbol{u}_i]}\right) \\ & & +\; 2\alpha\mu\mathsf{Re}\left\{w^{o*}\Sigma J \mathsf{E}\,\tilde{\boldsymbol{w}}_{i-1}\right\} \\ \mathsf{E}\,\tilde{\boldsymbol{w}}_i & = & J\mathsf{E}\,\tilde{\boldsymbol{w}}_{i-1} \;+\; \alpha\mu w^o \\ \Sigma' & = & \Sigma - \mu(\mathsf{E}\,\boldsymbol{U}_i)\Sigma - \mu\Sigma(\mathsf{E}\,\boldsymbol{U}_i) \;+\; \mu^2\mathsf{E}\left(\boldsymbol{U}_i\Sigma\boldsymbol{U}_i\right) \end{array}}$$

where $J = \mathrm{I} - \mu\mathsf{E}\,(\boldsymbol{U}_i)$ and $\Sigma' = \mathsf{E}\,\boldsymbol{\Sigma}'$. These recursions extend the statements of Thms. 22.4 and 22.5 to leaky data-normalized filters.

Remark. The results of Probs. V.27–V.32 are based on the work by Sayed and Al-Naffouri (2001).

Problem V.28 (Transient performance of leaky filters) Consider the setting of Prob. V.27. We now follow the discussion of Sec. 24.1 in order to derive a state-space model that characterizes the transient behavior of leaky adaptive filters, in the general case of regressors that are not necessarily Gaussian. We do so by appealing to the vec$\{\cdot\}$ notation introduced in that section. Thus let $\sigma = \text{vec}\{\Sigma\}$, i.e., σ is an $M^2 \times 1$ column vector that is obtained by stacking the columns of Σ on top of each other.

(a) Verify that the expression for Σ' from part (d) of Prob. V.27 reduces to the linear relation

$$\boxed{\sigma' = F\sigma, \qquad F \triangleq \mathsf{E}\left\{(\mathrm{I}_M - \mu\boldsymbol{U}_i)^{\mathsf{T}} \otimes (\mathrm{I}_M - \mu\boldsymbol{U}_i)\right\}}$$

with an $M^2 \times M^2$ matrix F.

(b) Show that the recursion for the mean weight-error vector, $\mathsf{E}\,\tilde{\boldsymbol{w}}_i$, is stable if, and only if, the step-size μ satisfies

$$0 < \mu < \frac{2}{\alpha \;+\; \lambda_{\max}\left(\mathsf{E}\left[\frac{\boldsymbol{u}_i^*\boldsymbol{u}_i}{g[\boldsymbol{u}_i]}\right]\right)}$$

Assuming $\boldsymbol{w}_{-1} = 0$ and, therefore, $\mathsf{E}\,\tilde{\boldsymbol{w}}_{-1} = w^o$, iterate the recursion for $\mathsf{E}\,\tilde{\boldsymbol{w}}_i$ and show that $\mathsf{E}\,\tilde{\boldsymbol{w}}_{i-1} = C_i w^o$ for $i \geq 0$, where $C_i = J^i + \alpha\mu(I + J + \ldots + J^{i-1})$. Conclude that $2\alpha\mu\mathsf{Re}\left\{w^{o*}\Sigma J\mathsf{E}\,\tilde{\boldsymbol{w}}_{i-1}\right\} \;=\; \alpha\mu\|w^o\|^2_{\Sigma J C_i + C_i^* J\Sigma}$.

(c) Show that the transient behavior of leaky data-normalized adaptive filters is characterized by the following state-space model $\mathcal{W}_i = \mathcal{F}\mathcal{W}_{i-1} + \mu\mathcal{Y}_i$, where

$$\mathcal{W}_i \triangleq \begin{bmatrix} \mathsf{E}\,\|\tilde{\boldsymbol{w}}_i\|^2_{\sigma} \\ \mathsf{E}\,\|\tilde{\boldsymbol{w}}_i\|^2_{F\cdot\sigma} \\ \mathsf{E}\,\|\tilde{\boldsymbol{w}}_i\|^2_{F^2\cdot\sigma} \\ \vdots \\ \mathsf{E}\,\|\tilde{\boldsymbol{w}}_i\|^2_{F^{M^2-1}\cdot\sigma} \end{bmatrix}, \quad \mathcal{F} = \begin{bmatrix} 0 & 1 & & & & \\ 0 & 0 & 1 & & & \\ 0 & 0 & 0 & 1 & & \\ \vdots & & & & \ddots & \\ 0 & 0 & 0 & & & 1 \\ -p_0 & -p_1 & -p_2 & \ldots & & -p_{M^2-1} \end{bmatrix}_{M^2\times M^2}$$

with

$$p(x) \triangleq \det(x\mathrm{I} - F) = x^{M^2} \;+\; \sum_{k=0}^{M^2-1} p_k x^k$$

denoting the characteristic polynomial of F. Moreover,

$$\mathcal{Y}_i = \mu\sigma_v^2 \begin{bmatrix} \mathsf{E}\left(\frac{\|\boldsymbol{u}_i\|_\sigma^2}{g^2[\boldsymbol{u}_i]}\right) \\ \mathsf{E}\left(\frac{\|\boldsymbol{u}_i\|_{F\sigma}^2}{g^2[\boldsymbol{u}_i]}\right) \\ \vdots \\ \mathsf{E}\left(\frac{\|\boldsymbol{u}_i\|_{F^{M^2-1}\sigma}^2}{g^2[\boldsymbol{u}_i]}\right) \end{bmatrix} + \alpha \begin{bmatrix} \|w^o\|^2_{(\alpha\mu\mathrm{I}+S_i)\cdot\sigma} \\ \|w^o\|^2_{(\alpha\mu\mathrm{I}+S_i)F\sigma} \\ \|w^o\|^2_{(\alpha\mu\mathrm{I}+S_i)F^2\sigma} \\ \vdots \\ \|w^o\|^2_{(\alpha\mu\mathrm{I}+S_i)F^{M^2-1}\sigma} \end{bmatrix}$$

where S_i is the $M^2 \times M^2$ matrix $S_i = ((JC_i)^\mathsf{T} \otimes \mathrm{I}_M) + (\mathrm{I}_M \otimes C_i J)$.

(d) Verify that F can be expressed as $F = \mathrm{I} - \mu A + \mu^2 B$, where $B = \mathsf{E}(\boldsymbol{U}_i^\mathsf{T} \otimes \boldsymbol{U}_i)$ and $A = (\mathsf{E}(\boldsymbol{U}_i)^\mathsf{T} \otimes \mathrm{I}) + (\mathrm{I} \otimes \mathsf{E}(\boldsymbol{U}_i))$. Show that the adaptive filter is mean-square stable for step-sizes in the range

$$0 < \mu < \min\left\{ \frac{2}{\alpha + \lambda_{\max}\left(\mathsf{E}\left[\frac{\boldsymbol{u}_i^*\boldsymbol{u}_i}{g[\boldsymbol{u}_i]}\right]\right)},\ \frac{1}{\lambda_{\max}(A^{-1}B)},\ \frac{1}{\max\{\lambda(H) \in \mathbb{R}^+\}} \right\}$$

where

$$H = \begin{bmatrix} A/2 & -B/2 \\ \mathrm{I}_{M^2} & 0 \end{bmatrix}$$

(e) How do the results of parts (a)–(d) simplify when the data are real-valued?

Problem V.29 (Mean-square performance of leaky adaptive filters). Consider the setting of Probs. V.27 and V.28. We follow the discussion that led to the MSD and EMSE expressions in Thm. 24.1 to derive similar expressions for the mean-square performance of leaky data-normalized adaptive filters.

(a) Using the fact that in steady-state,

$$\lim_{i\to\infty} \mathsf{E}\,\|\tilde{\boldsymbol{w}}_i\|_\sigma^2 = \lim_{i\to\infty} \mathsf{E}\,\|\tilde{\boldsymbol{w}}_{i-1}\|_\sigma^2, \qquad \lim_{i\to\infty} \mathsf{E}\,\tilde{\boldsymbol{w}}_i = \lim_{i\to\infty} \mathsf{E}\,\tilde{\boldsymbol{w}}_{i-1}$$

conclude that

$$\lim_{i\to\infty} \mathsf{E}\,\tilde{\boldsymbol{w}}_i = \alpha\mu(\mathrm{I}-J)^{-1}w^o, \quad \lim_{i\to\infty} \|\mathsf{E}\,\tilde{\boldsymbol{w}}_i\|^2_{(\mathrm{I}-F)\sigma} = \mu^2\sigma_v^2\mathsf{E}\left(\frac{\|\boldsymbol{u}_i\|_\sigma^2}{g^2[\boldsymbol{u}_i]}\right) + \alpha^2\mu^2\|w^o\|^2_{T\sigma}$$

where T is $M^2 \times M^2$ and given by $T = \mathrm{I} + ((\mathrm{I}-J^\mathsf{T})^{-1}J^\mathsf{T} \otimes \mathrm{I}) + (\mathrm{I} \otimes (\mathrm{I}-J)^{-1}J)$.

(b) Use part (a) to establish the following expressions for the MSD and EMSE of leaky data-normalized adaptive filters:

$$\begin{aligned} \mathsf{MSD} &= \mu^2\sigma_v^2\mathsf{E}\left(\frac{\|\boldsymbol{u}_i\|^2_{(\mathrm{I}-F)^{-1}q}}{g^2[\boldsymbol{u}_i]}\right) + \alpha^2\mu^2\|w^o\|^2_{T(\mathrm{I}-F)^{-1}q} \\ \mathsf{EMSE} &= \mu^2\sigma_v^2\mathsf{E}\left(\frac{\|\boldsymbol{u}_i\|^2_{(\mathrm{I}-F)^{-1}r}}{g^2[\boldsymbol{u}_i]}\right) + \alpha^2\mu^2\|w^o\|^2_{T(\mathrm{I}-F)^{-1}r} \end{aligned}$$

where q is the $M^2 \times 1$ vector $q = \text{vec}(\mathrm{I})$ and $r = \text{vec}\{R_u\}$ is also $M^2 \times 1$.

(c) Verify that the expressions for the MSD and EMSE from part (b) can be rewritten as

$$\begin{aligned} \mathsf{MSD} &= \mu^2\sigma_v^2\mathsf{Tr}(S\Sigma_{\text{msd}}) + \alpha^2\mu^2\|w^o\|^2_{T\Sigma_{\text{msd}}} \\ \mathsf{EMSE} &= \mu^2\sigma_v^2\mathsf{Tr}(S\Sigma_{\text{emse}}) + \alpha^2\mu^2\|w^o\|^2_{T\Sigma_{\text{emse}}} \end{aligned}$$

where $S = \mathsf{E}\left(\boldsymbol{u}_i^*\boldsymbol{u}_i/g^2[\boldsymbol{u}_i]\right)$ and $\{\Sigma_{\text{msd}}, \Sigma_{\text{emse}}\}$ are the weighting matrices that correspond to the vectors $\sigma_{\text{msd}} = (\mathrm{I}-F)^{-1}q$ and $\sigma_{\text{emse}} = (\mathrm{I}-F)^{-1}r$. That is, $\Sigma_{\text{msd}} = \text{vec}^{-1}(\sigma_{\text{msd}})$ and $\Sigma_{\text{emse}} = \text{vec}^{-1}(\sigma_{\text{emse}})$.

Problem V.30 (Leaky-LMS with white input) Consider the settings of Probs. V.27–V.29 specialized to the case $g[\boldsymbol{u}_i] = 1$, which corresponds to the leaky-LMS update $\boldsymbol{w}_i = (1-\alpha\mu)\boldsymbol{w}_{i-1} + \mu\boldsymbol{u}_i^*\boldsymbol{e}(i)$. In this problem we assume further that the entries of $\boldsymbol{u}_i$ are i.i.d. with variance σ_u^2 and fourth moment ξ_u^4, so that $R_u = \sigma_u^2\mathrm{I}$. Refer to the variance relation in part (d) of Prob. V.27 and set $\Sigma = \mathrm{I}$.

(a) Let $\kappa^2 = \alpha^2 + 2\alpha\sigma_u^2 + \xi_u^4 + (M-1)\sigma_u^4$. Show that $\Sigma' = \left(1 - 2\mu(\alpha+\sigma_u^2) + \mu^2\kappa^2\right)\mathrm{I}$, and conclude that

$$\boxed{\begin{aligned} \mathsf{E}\,\|\tilde{\boldsymbol{w}}_i\|^2 &= \left(1 - 2\mu(\alpha+\sigma_u^2) + \mu^2\kappa^2\right)\mathsf{E}\,\|\tilde{\boldsymbol{w}}_{i-1}\|^2 \;+\; \alpha^2\mu^2\|w^o\|^2 \;+ \\ &\quad\; \mu^2\sigma_v^2\sigma_u^2 M \;+\; 2\alpha\mu[1-\mu(\alpha+\sigma_u^2)]\mathsf{Re}\,\{w^{o*}\mathsf{E}\,\tilde{\boldsymbol{w}}_{i-1}\} \\ \mathsf{E}\,\tilde{\boldsymbol{w}}_i &= (1-\mu(\alpha+\sigma_u^2))\mathsf{E}\,\tilde{\boldsymbol{w}}_{i-1} \;+\; \alpha\mu w^o \end{aligned}}$$

(b) Argue that in steady-state $\lim_{i\to\infty}\mathsf{E}\,\tilde{\boldsymbol{w}}_i = \frac{\alpha}{\alpha+\sigma_u^2}\,w^o$ and, therefore,

$$\boxed{\begin{aligned} \mathsf{MSD} &= \frac{\mu\sigma_v^2\sigma_u^2 M}{2(\alpha+\sigma_u^2) - \mu^2\kappa^2} \;+\; \frac{\mu\alpha^2(2-\mu(\alpha+\sigma_u^2))}{[2(\alpha+\sigma_u^2) - \mu^2\kappa^2][\alpha+\sigma_u^2]}\,\|w^o\|^2 \\ \mathsf{EMSE} &= \sigma_u^2\cdot\mathsf{MSD} \end{aligned}}$$

(c) Assume the individual entries of $\boldsymbol{u}_i$ have a Gaussian distribution, so that $\xi_u^4 = 2\sigma_u^4$ (for complex-valued random variables; $\xi_u^4 = 3\sigma_u^4$ for real-valued random variables). Evaluate the MSD and EMSE in this case.

Problem V.31 (Leaky-LMS with Gaussian regressors) Consider the setting of Probs. V.27–V.29 specialized to the case $g[\boldsymbol{u}_i] = 1$, which corresponds to the leaky-LMS update

$$\boldsymbol{w}_i = (1-\alpha\mu)\boldsymbol{w}_{i-1} \;+\; \mu\boldsymbol{u}_i^*\boldsymbol{e}(i)$$

In this problem we assume further that the regressors $\boldsymbol{u}_i$ are circular Gaussian, and follow the arguments of Sec. 23.1 in order to evaluate the performance of leaky-LMS. Thus introduce the eigen-decomposition $R_u = U\Lambda U^*$, where Λ is diagonal with the eigenvalues of R_u and U is unitary. Define further the transformed quantities:

$$\begin{array}{llll} \overline{\boldsymbol{w}}_i \triangleq U^*\tilde{\boldsymbol{w}}_i, & \overline{w}^o \triangleq U^*w^o, & \overline{\boldsymbol{u}}_i \triangleq \boldsymbol{u}_iU, & \overline{\Sigma} \triangleq U^*\Sigma U \\ \overline{\boldsymbol{U}}_i \triangleq U^*\boldsymbol{U}_iU, & \overline{J} \triangleq U^*JU, & \overline{C}_i \triangleq U^*C_iU & \end{array}$$

as well as $\Lambda^\alpha \triangleq \Lambda + \alpha\mathrm{I}$ with diagonal entries $\{\lambda_j^\alpha = \lambda_j + \alpha\}$. The quantities $\{\boldsymbol{U}_i, J, C_i\}$ are defined as in Probs. V.27–V.28.

(a) Verify that $\mathsf{E}\,(\overline{\boldsymbol{U}}_i\overline{\Sigma}\overline{\boldsymbol{U}}_i) = \Lambda^\alpha\overline{\Sigma}\Lambda^\alpha + \Lambda\mathsf{Tr}(\overline{\Sigma}\Lambda)$. Now refer to the recursion for Σ' from part (d) of Prob. V.27 and show that it leads to

$$\overline{\Sigma}' = \overline{\Sigma} - \mu\Lambda^\alpha\overline{\Sigma} - \mu\overline{\Sigma}\Lambda^\alpha \;+\; \mu^2\left[\Lambda^\alpha\overline{\Sigma}\Lambda^\alpha \;+\; \Lambda\mathsf{Tr}(\overline{\Sigma}\Lambda)\right]$$

Verify that $\overline{\Sigma}'$ will be diagonal if $\overline{\Sigma}$ is.

(b) Let $\overline{\sigma} = \mathrm{diag}\{\overline{\Sigma}\}$ denote the $M\times 1$ vector with the diagonal entries of $\overline{\Sigma}$. Likewise, let $\lambda = \mathrm{diag}\{\Lambda\}$ and $\lambda^\alpha = \mathrm{diag}\{\Lambda^\alpha\}$. Show that the above relation for $\overline{\Sigma}'$ leads to $\overline{\sigma}' = \overline{F}\overline{\sigma}$, where $\overline{F}$ now denotes the $M\times M$ matrix (compare with the expression for $\overline{F}$ in (23.22) for the non-leaky LMS case):

$$\boxed{\overline{F} \triangleq \left(\mathrm{I} - 2\mu\Lambda^\alpha + \mu^2(\Lambda^\alpha)^2\right) + \mu^2\lambda\lambda^{\mathsf{T}}}$$

(c) Verify that the matrices $\overline{J}$ and $\overline{C}_i$ are diagonal. In particular,

$$\boxed{\overline{J} = \mathsf{E}\,(\mathrm{I} - \mu\overline{\boldsymbol{U}}_i) = \mathrm{I} - \mu\Lambda^\alpha}$$

Conclude that the variance relation from part (d) of Prob. V.27 leads to

$$\begin{aligned} \mathsf{E}\,\|\overline{w}_i\|^2_{\overline{\sigma}} &= \mathsf{E}\,\|\overline{w}_{i-1}\|^2_{\overline{\sigma}'} + \mu^2\sigma_v^2(\lambda^{\mathsf{T}}\overline{\sigma}) + \alpha\mu\|\overline{w}^o\|^2_{(\alpha\mu\mathrm{I}+2\overline{J}\overline{C}_i)\overline{\sigma}} \\ \mathsf{E}\,\overline{w}_i &= \overline{J}\mathsf{E}\,\overline{w}_{i-1} + \alpha\mu\overline{w}^o \end{aligned}$$

(d) Show that the transient behavior of leaky-LMS is characterized by the following state-space model of dimension M:

$$\mathcal{W}_i = \mathcal{F}\mathcal{W}_{i-1} + \mu\mathcal{Y}_i$$

where

$$\mathcal{W}_i \triangleq \begin{bmatrix} \mathsf{E}\,\|\overline{w}_i\|^2_{\overline{\sigma}} \\ \mathsf{E}\,\|\overline{w}_i\|^2_{\overline{F}\overline{\sigma}} \\ \mathsf{E}\,\|\overline{w}_i\|^2_{\overline{F}^2\overline{\sigma}} \\ \vdots \\ \mathsf{E}\,\|\overline{w}_i\|^2_{\overline{F}^{M-1}\overline{\sigma}} \end{bmatrix}, \quad \mathcal{F} = \begin{bmatrix} 0 & 1 & & & \\ 0 & 0 & 1 & & \\ 0 & 0 & 0 & 1 & \\ \vdots & & & & \\ 0 & 0 & 0 & & 1 \\ -p_0 & -p_1 & -p_2 & \ldots & -p_{M-1} \end{bmatrix}_{M\times M}$$

with

$$p(x) \triangleq \det(x\mathrm{I} - \overline{F}) = x^M + \sum_{k=0}^{M-1} p_k x^k$$

denoting the characteristic polynomial of $\overline{F}$. Moreover,

$$\mathcal{Y}_i = \mu\sigma_v^2 \begin{bmatrix} \lambda^{\mathsf{T}}\overline{\sigma} \\ \lambda^{\mathsf{T}}\overline{F}\overline{\sigma} \\ \lambda^{\mathsf{T}}\overline{F}^2\overline{\sigma} \\ \vdots \\ \lambda^{\mathsf{T}}\overline{F}^{M-1}\overline{\sigma} \end{bmatrix} + \alpha \begin{bmatrix} \|w^o\|^2_{(\alpha\mu\mathrm{I}+2\overline{J}\overline{C}_i)\overline{\sigma}} \\ \|w^o\|^2_{(\alpha\mu\mathrm{I}+2\overline{J}\overline{C}_i)\overline{F}\overline{\sigma}} \\ \|w^o\|^2_{(\alpha\mu\mathrm{I}+2\overline{J}\overline{C}_i)\overline{F}^2\overline{\sigma}} \\ \vdots \\ \|w^o\|^2_{(\alpha\mu\mathrm{I}+2\overline{J}\overline{C}_i)\overline{F}^{M-1}\overline{\sigma}} \end{bmatrix}$$

(e) Express $\overline{F}$ as $\overline{F} = \mathrm{I} - \mu A + \mu^2 B$, where $A = 2\Lambda^\alpha$ and $B = \mu^2(\Lambda^\alpha)^2 + \lambda\lambda^{\mathsf{T}}$. Follow the arguments that led to (23.41) to conclude that the leaky-LMS filter is mean-square stable if, and only if, the step-size μ is chosen such that

$$\sum_{k=1}^{M} \frac{\mu\lambda_k^2}{\lambda_k^\alpha[2 - \mu\lambda_k^\alpha]} < 1$$

Problem V.32 (Performance of leaky-LMS with Gaussian regressors) Consider the same setting as Prob. V.31. Here we wish to follow the discussion that led to the results of Thm. 23.3 in order to determine expressions for the MSD and EMSE of leaky-LMS under Gaussian regressors.

(a) Use the variance relation of part (c) of Prob. V.31 to verify that, in steady-state,

$$\mathsf{E}\,\|\overline{w}_\infty\|^2 = \mu^2\sigma_v^2(\lambda^{\mathsf{T}}(\mathrm{I}-\overline{F})^{-1}q) + \alpha^2\mu^2\|\overline{w}^o\|^2_{(\mathrm{I}+2\overline{J}(\mathrm{I}-\overline{J})^{-1})(\mathrm{I}-\overline{F})^{-1}q}$$

where q is the $M-$dimensional column vector with unit entries.

(b) Show that

$$\lambda^{\mathsf{T}}(\mathrm{I}-\overline{F})^{-1}q = \frac{\lambda^{\mathsf{T}}D^{-1}q}{1-\mu^2\lambda^{\mathsf{T}}D^{-1}\lambda}$$

where $D = 2\mu\Lambda^{\alpha} - \mu^2(\Lambda^{\alpha})^2$. Conclude that

$$\boxed{\mathsf{MSD} \;=\; \mu\frac{\sum_{k=1}^{M}\left(\frac{\sigma_v^2\lambda_k \;+\; \alpha^2\left(\frac{2}{\mu\lambda_k^{\alpha}}-1\right)|\overline{w}^o(k)|^2}{\lambda_k^{\alpha}[2-\mu\lambda_k^{\alpha}]}\right)}{1-\mu\sum_{k=1}^{M}\frac{\lambda_k^2}{\lambda_k^{\alpha}[2-\mu\lambda_k^{\alpha}]}}}$$

where $\overline{w}^o(k)$ denotes the $k-$th entry of $\overline{w}^o$.

(c) In a similar vein, verify that

$$\mathsf{EMSE} \;=\; \mu^2\sigma_v^2\lambda^{\mathsf{T}}(\mathrm{I}-\overline{F})^{-1}\lambda \;+\; \alpha^2\mu^2\|\overline{w}^o\|^2_{(\mathrm{I}+2\overline{J}(\mathrm{I}-\overline{J})^{-1})(\mathrm{I}-\overline{F})^{-1}\lambda}$$

and simplify the expression to

$$\boxed{\mathsf{EMSE} \;=\; \mu\frac{\sum_{k=1}^{M}\left(\frac{\sigma_v^2\lambda_k^2 \;+\; \alpha^2\lambda_k\left(\frac{2}{\mu\lambda_k^{\alpha}}-1\right)|\overline{w}^o(k)|^2}{\lambda_k^{\alpha}[2-\mu\lambda_k^{\alpha}]}\right)}{1-\mu\sum_{k=1}^{M}\frac{\lambda_k^2}{\lambda_k^{\alpha}[2-\mu\lambda_k^{\alpha}]}}}$$

Problem V.33 (Tracking performance of LMS) The energy conservation argument can also be used to study the transient performance of adaptive filters in nonstationary environments. We illustrate this fact by considering the LMS algorithm

$$\boldsymbol{w}_i = \boldsymbol{w}_{i-1} \;+\; \mu\boldsymbol{u}_i^*[\boldsymbol{d}(i) - \boldsymbol{u}_i\boldsymbol{w}_{i-1}]$$

where the data $\{\boldsymbol{d}(i), \boldsymbol{u}_i\}$ are now assumed to satisfy the nonstationary model (20.16) instead of the stationary model (22.1), namely

(1) There exists a vector $\boldsymbol{w}_i^o$ such that $\boldsymbol{d}(i) = \boldsymbol{u}_i\boldsymbol{w}_i^o \;+\; \boldsymbol{v}(i)$.
(2) The weight vector varies according to $\boldsymbol{w}_i^o = \boldsymbol{w}_{i-1}^o \;+\; \boldsymbol{q}_i$.
(3) The noise sequence $\{\boldsymbol{v}(i)\}$ is i.i.d. with variance $\sigma_v^2 = \mathsf{E}\,|\boldsymbol{v}(i)|^2$.
(4) The noise sequence $\boldsymbol{v}(i)$ is independent of $\boldsymbol{u}_j$ for all i, j.
(5) The sequence $\boldsymbol{q}_i$ has covariance Q and is independent of $\{\boldsymbol{v}(j), \boldsymbol{u}_j\}$ for all i, j.
(6) The initial conditions $\{\boldsymbol{w}_{-1}, \boldsymbol{w}_{-1}^o\}$ are independent of all $\{\boldsymbol{d}(j), \boldsymbol{u}_j, \boldsymbol{v}(j), \boldsymbol{q}_j\}$.
(7) The regressor covariance matrix is $R_u = \mathsf{E}\,\boldsymbol{u}_i^*\boldsymbol{u}_i > 0$.
(8) The variables $\{\boldsymbol{v}(i), \boldsymbol{u}_i, \boldsymbol{q}_i\}$ have zero means.

Assume in addition that the regressors $\{\boldsymbol{u}_i\}$ satisfy the independence condition (22.23). The arguments we employed in Sec. 22.3 and Chapter 24 can be extended in a rather straightforward manner.

(a) Establish the following variance relation (which extends the result of Thm. 22.4 in the LMS case to nonstationary models):

$$\begin{aligned}\mathsf{E}\,\|\tilde{\boldsymbol{w}}_i\|^2_{\Sigma} &= \mathsf{E}\,\|\tilde{\boldsymbol{w}}_{i-1}\|^2_{\Sigma'} \;+\; \mu^2\sigma_v^2\mathsf{Tr}(R_u\Sigma) \;+\; \mathsf{Tr}(Q\Sigma)\\ \Sigma' &= \Sigma \;-\; \mu\Sigma\mathsf{E}\,\boldsymbol{u}_i^*\boldsymbol{u}_i \;-\; \mu\mathsf{E}\,\boldsymbol{u}_i^*\boldsymbol{u}_i\Sigma \;+\; \mu^2\mathsf{E}\,\|\boldsymbol{u}_i\|^2_{\Sigma}\boldsymbol{u}_i^*\boldsymbol{u}_i\end{aligned}$$

We thus see that an additional driving term appears in the first recursion, namely, $\mathsf{Tr}(Q\Sigma) = \mathsf{E}\,\|\boldsymbol{q}_i\|^2_{\Sigma}$.

(b) Assume further that the regressors $\{\boldsymbol{u}_i\}$ are circular Gaussian. Argue that the same condition on the step-size μ from Thm. 23.2 is necessary and sufficient for mean-square stability of the filter.

(c) Define $\overline{q}_i = U^* q_i$, whose covariance matrix is $\overline{Q} = U^* Q U$. Extend the result of Thm. 23.3 to this case and show that the MSD and EMSE are given by

$$\text{EMSE} = \frac{\mu\sigma_v^2 \sum_{k=1}^{M} \frac{\lambda_k}{2-\mu\lambda_k}}{1-\mu\sum_{k=1}^{M}\frac{\lambda_k}{2-\mu\lambda_k}} + \mu^{-1}\frac{\text{Tr}\left(\overline{Q}\text{diag}\left\{\frac{1}{2-\mu\lambda_k}\right\}\right)}{1-\mu\sum_{k=1}^{M}\frac{\lambda_k}{2-\mu\lambda_k}}$$

$$\text{MSD} = \frac{\mu\sigma_v^2 \sum_{k=1}^{M} \frac{1}{2-\mu\lambda_k}}{1-\mu\sum_{k=1}^{M}\frac{\lambda_k}{2-\mu\lambda_k}} + \mu^{-1}\frac{\text{Tr}\left(\overline{Q}\text{diag}\left\{\frac{1}{2\lambda_k-\mu\lambda_k^2}\right\}\right)}{1-\mu\sum_{k=1}^{M}\frac{\lambda_k}{2-\mu\lambda_k}}$$

(d) Assume $Q = \sigma_q^2 \mathrm{I}$. Verify that the expressions of part (c) simplify to

$$\text{EMSE} = \frac{\sum_{k=1}^{M}\frac{\mu^{-1}\sigma_q^2+\mu\sigma_v^2\lambda_k}{2-\mu\lambda_k}}{1-\mu\sum_{k=1}^{M}\frac{\lambda_k}{2-\mu\lambda_k}} \qquad \text{MSD} = \frac{\sum_{k=1}^{M}\frac{\mu^{-1}\sigma_q^2+\mu\sigma_v^2\lambda_k}{2\lambda_k-\mu\lambda_k^2}}{1-\mu\sum_{k=1}^{M}\frac{\lambda_k}{2-\mu\lambda_k}}$$

Problem V.34 (Tracking performance of leaky-LMS) The purpose of this problem is to extend the results of Prob. V.33 to leaky-LMS,

$$w_i = (1-\alpha\mu)w_{i-1} + \mu u_i^* e(i)$$

Thus assume that the data $\{d(i), u_i\}$ satisfy the nonstationary model of Prob. V.33 with the exception of item (2), which is replaced by $w_i^o = w^o + q_i$ for some unknown vector w^o. Assume in addition that the regressors u_i satisfy the independence condition (22.23).

(a) Verify that the variance relation of part (d) of Prob. V.27 should be modified as follows (using $g[u_i] = 1$):

$$\begin{aligned}
\mathsf{E}\,\|\tilde{w}_i\|_\Sigma^2 &= \mathsf{E}\,\|\tilde{w}_{i-1}\|_{\Sigma'}^2 + \alpha^2\mu^2\|w^o\|_\Sigma^2 + \mu^2\sigma_v^2\mathsf{E}\,\|u_i\|_\Sigma^2 + \text{Tr}(Q\Sigma) \\
&\quad 2\alpha\mu\text{Re}\left\{w^{o*}\Sigma J\mathsf{E}\,\tilde{w}_{i-1}\right\} \\
\mathsf{E}\,\tilde{w}_i &= J\mathsf{E}\,\tilde{w}_{i-1} + \alpha\mu w^o \\
\Sigma' &= \Sigma - \mu(\mathsf{E}\,U_i)\Sigma - \mu\Sigma(\mathsf{E}\,U_i) + \mu^2\mathsf{E}\,(U_i\Sigma U_i)
\end{aligned}$$

That is, an additional term $\text{Tr}(Q\Sigma)$ appears in the recursion for $\mathsf{E}\,\|\tilde{w}_i\|_\Sigma^2$.

(b) Assume further that the regressors u_i are circular Gaussian and consider the transformations of Prob. V.31. Repeat the arguments of Prob. V.32 to conclude that

$$\text{MSD} = \mu\frac{\sum_{k=1}^{M}\left(\frac{\sigma_v^2\lambda_k + \alpha^2\left(\frac{2}{\mu\lambda_k^\alpha}-1\right)|\overline{w}^o(k)|^2}{\lambda_k^\alpha[2-\mu\lambda_k^\alpha]}\right)}{1-\mu\sum_{k=1}^{M}\frac{\lambda_k^2}{\lambda_k^\alpha[2-\mu\lambda_k^\alpha]}} + \frac{\text{Tr}\left(\overline{Q}\text{diag}\left\{\frac{1}{\mu\lambda_k^\alpha[2-\mu\lambda_k^\alpha]}\right\}\right)}{1-\mu\sum_{k=1}^{M}\frac{\lambda_k^2}{\lambda_k^\alpha[2-\mu\lambda_k^\alpha]}}$$

where $\overline{Q} = U^* Q U$ and $w^o(k)$ denotes the $k-$th entry of $\overline{w}^o$. Conclude also that

$$\text{EMSE} = \mu\frac{\sum_{k=1}^{M}\left(\frac{\sigma_v^2\lambda_k^2 + \alpha^2\lambda_k\left(\frac{2}{\mu\lambda_k^\alpha}-1\right)|\overline{w}^o(k)|^2}{\lambda_k^\alpha[2-\mu\lambda_k^\alpha]}\right)}{1-\mu\sum_{k=1}^{M}\frac{\lambda_k^2}{\lambda_k^\alpha[2-\mu\lambda_k^\alpha]}} + \frac{\text{Tr}\left(\overline{Q}\text{diag}\left\{\frac{\lambda_k}{\mu\lambda_k^\alpha[2-\mu\lambda_k^\alpha]}\right\}\right)}{1-\mu\sum_{k=1}^{M}\frac{\lambda_k^2}{\lambda_k^\alpha[2-\mu\lambda_k^\alpha]}}$$

(c) Assume $Q = \sigma_q^2 \mathrm{I}$. Verify that the expressions of part (b) simplify to

$$\mathsf{MSD} = \frac{\sum_{k=1}^{M}\left(\frac{\mu^{-1}\sigma_q^2 + \mu\sigma_v^2\lambda_k + \mu\alpha^2\left(\frac{2}{\mu\lambda_k^\alpha} - 1\right)|\overline{w}^o(k)|^2}{\lambda_k^\alpha[2 - \mu\lambda_k^\alpha]}\right)}{1 - \mu\sum_{k=1}^{M}\frac{\lambda_k^2}{\lambda_k^\alpha[2-\mu\lambda_k^\alpha]}}$$

and

$$\mathsf{EMSE} = \frac{\sum_{k=1}^{M}\left(\frac{\mu^{-1}\sigma_q^2 + \mu\sigma_v^2\lambda_k^2 + \mu\alpha^2\lambda_k\left(\frac{2}{\mu\lambda_k^\alpha} - 1\right)|\overline{w}^o(k)|^2}{\lambda_k^\alpha[2 - \mu\lambda_k^\alpha]}\right)}{1 - \mu\sum_{k=1}^{M}\frac{\lambda_k^2}{\lambda_k^\alpha[2-\mu\lambda_k^\alpha]}}$$

Problem V.35 (Mean performance of RLS) Consider the RLS algorithm

$$\begin{aligned} \boldsymbol{P}_i &= \lambda^{-1}\left[\boldsymbol{P}_{i-1} - \frac{\lambda^{-1}\boldsymbol{P}_{i-1}\boldsymbol{u}_i^*\boldsymbol{u}_i\boldsymbol{P}_{i-1}}{1+\lambda^{-1}\boldsymbol{u}_i\boldsymbol{P}_{i-1}\boldsymbol{u}_i^*}\right] \\ \boldsymbol{w}_i &= \boldsymbol{w}_{i-1} + \boldsymbol{P}_i\,\boldsymbol{u}_i^*[\boldsymbol{d}(i) - \boldsymbol{u}_i\boldsymbol{w}_{i-1}], \quad i \geq 0 \end{aligned}$$

and recall from the discussion in Sec. 14.1, and especially from recursion (14.4), that

$$\boldsymbol{P}_i^{-1} = \lambda\boldsymbol{P}_{i-1}^{-1} + \boldsymbol{u}_i^*\boldsymbol{u}_i, \qquad \boldsymbol{P}_{-1}^{-1} = \epsilon\mathrm{I}$$

(a) Let $\tilde{\boldsymbol{w}}_i = w^o - \boldsymbol{w}_i$ where we are assuming the data $\{\boldsymbol{d}(i), \boldsymbol{u}_i\}$ satisfy model (22.1). Show that

$$\boxed{\boldsymbol{P}_i^{-1}\tilde{\boldsymbol{w}}_i = \lambda\boldsymbol{P}_{i-1}^{-1}\tilde{\boldsymbol{w}}_{i-1} + \boldsymbol{u}_i^*\boldsymbol{v}(i)}$$

and, conclude that, (assuming $\boldsymbol{w}_{-1} = 0$)

$$\boldsymbol{P}_i^{-1}\tilde{\boldsymbol{w}}_i = \lambda^{i+1}\epsilon w^o + \sum_{j=0}^{i}\lambda^{i-j}\boldsymbol{u}_j^*\boldsymbol{v}(j)$$

(b) Verify that

$$\boxed{\mathsf{E}\,\tilde{\boldsymbol{w}}_i = \epsilon\lambda^{i+1}(\mathsf{E}\,\boldsymbol{P}_i)w^o}$$

Assuming $0 < \lambda < 1$ and $\mathsf{E}\,\boldsymbol{P}_i > 0$ as $i \to \infty$, conclude that $\mathsf{E}\,\tilde{\boldsymbol{w}}_i \to 0$ as $i \to \infty$. In other words, conclude that the exponentially-weighted RLS algorithm is asymptotically unbiased.

Problem V.36 (Mean-square performance of RLS) The class of adaptive filters with matrix data nonlinearities (as described in Prob. V.22) can be generalized to include the RLS algorithm as well:

$$\begin{aligned} \boldsymbol{P}_i &= \lambda^{-1}\left[\boldsymbol{P}_{i-1} - \frac{\lambda^{-1}\boldsymbol{P}_{i-1}\boldsymbol{u}_i^*\boldsymbol{u}_i\boldsymbol{P}_{i-1}}{1+\lambda^{-1}\boldsymbol{u}_i\boldsymbol{P}_{i-1}\boldsymbol{u}_i^*}\right] \\ \boldsymbol{w}_i &= \boldsymbol{w}_{i-1} + \boldsymbol{P}_i\,\boldsymbol{u}_i^*[\boldsymbol{d}(i) - \boldsymbol{u}_i\boldsymbol{w}_{i-1}], \quad i \geq 0 \end{aligned}$$

Thus assume that the data $\{\boldsymbol{d}(i), \boldsymbol{u}_i\}$ satisfy (22.1). Assume further that the regressors satisfy the independence assumption (22.23).

In Prob. V.22 we studied the following class of algorithms:

$$\boxed{\begin{aligned} &\boldsymbol{w}_i = \boldsymbol{w}_{i-1} + \mu H[\boldsymbol{u}_i]\boldsymbol{u}_i^*\boldsymbol{e}(i), \quad i \geq 0 \\ &\boldsymbol{e}(i) = \boldsymbol{d}(i) - \boldsymbol{u}_i\boldsymbol{w}_{i-1} \end{aligned}}$$

where $H[\cdot]$ is a Hermitian positive-definite matrix of $\boldsymbol{u}_i$. If we set $\mu = 1$ and allow $H[\cdot]$ to be dependent not only on $\boldsymbol{u}_i$ but also on all prior regressors, then the RLS update becomes a special case (since $\boldsymbol{P}_i$ is determined by the regressors $\{\boldsymbol{u}_k, k \leq i\}$ and we can take $H[\cdot] = \boldsymbol{P}_i$).

(a) Use arguments similar to Prob. V.22 to conclude that the following relation holds for RLS:

$$\begin{aligned} \mathsf{E}\,\|\tilde{\boldsymbol{w}}_i\|^2_{\Sigma} &= \mathsf{E}\,\left(\|\tilde{\boldsymbol{w}}_{i-1}\|^2_{\Sigma'}\right) + \sigma_v^2 \mathsf{E}\,\|\boldsymbol{u}_i\|^2_{\boldsymbol{P}_i\Sigma\boldsymbol{P}_i} \\ \Sigma' &= \Sigma - \Sigma\boldsymbol{P}_i\boldsymbol{u}_i^*\boldsymbol{u}_i - \boldsymbol{u}_i^*\boldsymbol{u}_i\boldsymbol{P}_i\Sigma + \|\boldsymbol{u}_i\|^2_{\boldsymbol{P}_i\Sigma\boldsymbol{P}_i}\boldsymbol{u}_i^*\boldsymbol{u}_i \end{aligned}$$

(b) In Chapter 19, while studying the mean-square performance of RLS, we argued that the dependence of $\boldsymbol{P}_i$ on present and past regressors makes the analysis challenging. For this reason, whenever necessary, we replaced the random variables $\{\boldsymbol{P}_i^{-1}, \boldsymbol{P}_i\}$ by (cf. (19.11) and (19.12)):

$$\begin{aligned} P^{-1} &\triangleq \lim_{i\to\infty} \mathsf{E}\,\left(\boldsymbol{P}_i^{-1}\right) = \frac{R_u}{1-\lambda} \\ \mathsf{E}\,\boldsymbol{P}_i &\approx P = (1-\lambda)R_u^{-1} \end{aligned}$$

Use these substitutions to verify that the relation of part (a) becomes

$$\begin{aligned} \mathsf{E}\,\|\tilde{\boldsymbol{w}}_i\|^2_{\Sigma} &= \mathsf{E}\,\|\tilde{\boldsymbol{w}}_{i-1}\|^2_{\Sigma'} + \sigma_v^2(1-\lambda)^2 \mathsf{E}\,\|\boldsymbol{u}_i\|^2_{R_u^{-1}\Sigma R_u^{-1}} \\ \Sigma' &= \Sigma - 2(1-\lambda)\Sigma + (1-\lambda)^2 \mathsf{E}\,\left[\|\boldsymbol{u}_i\|^2_{R_u^{-1}\Sigma R_u^{-1}}\boldsymbol{u}_i^*\boldsymbol{u}_i\right] \end{aligned}$$

(c) Assume now that the regressors $\{\boldsymbol{u}_i\}$ are circular Gaussian. As in Sec. 23.1, introduce the eigen-decomposition $R_u = U\Lambda U^*$, and define the transformed variables $\overline{\boldsymbol{w}}_i = U^*\tilde{\boldsymbol{w}}_i$, $\overline{\boldsymbol{u}}_i = \boldsymbol{u}_i U$, and $\overline{\Sigma} \triangleq U^*\Sigma U$. Verify that

$$\mathsf{E}\,\left[\|\overline{\boldsymbol{u}}_i\|^2_{\Lambda^{-1}\overline{\Sigma}\Lambda^{-1}}\overline{\boldsymbol{u}}_i^*\overline{\boldsymbol{u}}_i\right] = \Lambda\mathsf{Tr}(\Lambda^{-1}\overline{\Sigma}) + \overline{\Sigma}$$

and conclude that the variance relation of part (b) becomes

$$\begin{aligned} &\mathsf{E}\,\|\overline{\boldsymbol{w}}_i\|^2_{\overline{\Sigma}} = \mathsf{E}\,\|\overline{\boldsymbol{w}}_{i-1}\|^2_{\overline{\Sigma}'} + \sigma_v^2(1-\lambda)^2 \mathsf{E}\,\|\overline{\boldsymbol{u}}_i\|^2_{\Lambda^{-1}\overline{\Sigma}\Lambda^{-1}} \\ &\overline{\Sigma}' = \lambda^2\overline{\Sigma} + (1-\lambda)^2\Lambda\mathsf{Tr}(\Lambda^{-1}\overline{\Sigma}) \end{aligned}$$

Conclude that $\overline{\Sigma}'$ will be diagonal if $\overline{\Sigma}$ is diagonal.

(d) Define the $M-$dimensional vectors $b = \text{diag}\{\Lambda\}$, $a = \text{diag}\{\Lambda^{-1}\}$, and $\overline{\sigma} = \text{diag}\{\overline{\Sigma}\}$. Observe that, contrary to the notation in the previous problems and in the chapter, we are denoting the vector corresponding to Λ by b in order to avoid confusion with the forgetting factor λ for RLS. Verify that the recursion for $\overline{\Sigma}'$ reduces to

$$\overline{\sigma}' = \overline{F}\overline{\sigma}, \qquad \overline{F} \triangleq \lambda^2 \mathrm{I} + (1-\lambda)^2 ba^{\mathsf{T}}$$

(e) Let q denote the column vector with unit entries. Show that

$$\text{diag}\left((\mathrm{I}-\overline{F})^{-1}q\right) = \frac{1}{1-\lambda^2}\left[\mathrm{I} + \frac{\mathsf{Tr}(\Lambda^{-1})\Lambda}{\frac{1+\lambda}{1-\lambda} - M}\right]$$

Now choosing $\overline{\Sigma} = \mathrm{I}$ and considering the steady-state value of the recursion for $\|\overline{\boldsymbol{w}}_i\|^2_{\overline{\Sigma}}$, conclude that

$$\mathsf{MSD} = \frac{\sigma_v^2}{\frac{1+\lambda}{1-\lambda} - M} \cdot \sum_{k=1}^{M}\left(\frac{1}{\lambda_k}\right)$$

Show also that

$$\mathsf{EMSE} = \frac{\sigma_v^2 M}{\frac{1+\lambda}{1-\lambda} - M}$$

Compare with the EMSE expression from Lemma 19.1.

(f) How do the EMSE and MSD expressions of part (e) simplify when $\lambda = 1$?

Problem V.37 (Transient performance of $\epsilon-$APA) In this problem we study the transient performance of the $\epsilon-$APA algorithm (21.53)–(21.55) with $K \leq M$. Thus introduce the weight-error vector, and the weighted *a priori* and *a posteriori* error vectors, $\tilde{\boldsymbol{w}}_i = w^o - \boldsymbol{w}_i$, $\boldsymbol{e}_{a,i}^{\Sigma} = \boldsymbol{U}_i\Sigma\tilde{\boldsymbol{w}}_{i-1}$ and $\boldsymbol{e}_{p,i}^{\Sigma} = \boldsymbol{U}_i\Sigma\tilde{\boldsymbol{w}}_i$, where Σ is any Hermitian positive-definite weighting matrix.

(a) Verify that $\tilde{\boldsymbol{w}}_i$ satisfies the recursion

$$\tilde{\boldsymbol{w}}_i = \left(\mathrm{I} - \mu\boldsymbol{U}_i^*(\epsilon\mathrm{I} + \boldsymbol{U}_i\boldsymbol{U}_i^*)^{-1}\boldsymbol{U}_i\right)\tilde{\boldsymbol{w}}_{i-1} - \mu\boldsymbol{U}_i^*(\epsilon\mathrm{I} + \boldsymbol{U}_i\boldsymbol{U}_i^*)^{-1}\boldsymbol{v}_i$$

where $\boldsymbol{v}_i = \mathrm{col}\{\boldsymbol{v}(i), \boldsymbol{v}(i-1), \ldots, \boldsymbol{v}(i-K+1)\}$.

(b) Follow arguments similar to those in Sec. 22.3 to establish the following weighted energy-conservation relation:

$$\boxed{\|\tilde{\boldsymbol{w}}_i\|_{\Sigma}^2 + \boldsymbol{e}_{a,i}^{\Sigma*}\left(\boldsymbol{U}_i\Sigma\boldsymbol{U}_i^*\right)^{-1}\boldsymbol{e}_{a,i}^{\Sigma} = \|\tilde{\boldsymbol{w}}_{i-1}\|_{\Sigma}^2 + \boldsymbol{e}_{p,i}^{\Sigma*}\left(\boldsymbol{U}_i\Sigma\boldsymbol{U}_i^*\right)^{-1}\boldsymbol{e}_{p,i}^{\Sigma}}$$

This relation is the extension of Thm. 22.1 to the context of affine projection algorithms.

(c) Follow further the arguments of Sec. 22.4, and ignore dependencies between $\boldsymbol{v}_i$ and $\{\boldsymbol{e}_{a,i}, \boldsymbol{e}_{a,i}^{\Sigma}\}$, to establish the following variance relation, which extends the result of Thm. 22.3:

$$\mathsf{E}\,\|\tilde{\boldsymbol{w}}_i\|_{\Sigma}^2 = \mathsf{E}\left(\|\tilde{\boldsymbol{w}}_{i-1}\|_{\boldsymbol{\Sigma}'}^2\right) + \mu^2\mathsf{E}\,\boldsymbol{v}_i^*\boldsymbol{A}_i^{\Sigma}\boldsymbol{v}_i$$

where $\boldsymbol{\Sigma}'$ is the random matrix

$$\boldsymbol{\Sigma}' = \Sigma - \mu\Sigma\boldsymbol{P}_i - \mu\boldsymbol{P}_i\Sigma + \mu^2\left(\boldsymbol{U}_i^*\boldsymbol{A}_i^{\Sigma}\boldsymbol{U}_i\right)$$

and $\{\boldsymbol{P}_i, \boldsymbol{A}_i^{\Sigma}\}$ are defined by

$$\boldsymbol{P}_i \;\triangleq\; \boldsymbol{U}_i^*(\epsilon\mathrm{I} + \boldsymbol{U}_i\boldsymbol{U}_i^*)^{-1}\boldsymbol{U}_i, \quad \boldsymbol{A}_i^{\Sigma} \triangleq (\epsilon\mathrm{I} + \boldsymbol{U}_i\boldsymbol{U}_i^*)^{-1}\boldsymbol{U}_i\Sigma\boldsymbol{U}_i^*(\epsilon\mathrm{I} + \boldsymbol{U}_i\boldsymbol{U}_i^*)^{-1}$$

(d) If we were to extend the independence assumption (22.23) to the current context, we would require the matrix sequence $\{\boldsymbol{U}_i\}$ to be i.i.d. This assumption would then guarantee that $\tilde{\boldsymbol{w}}_{i-1}$ is independent of $\boldsymbol{\Sigma}'$. However, requiring the sequence $\{\boldsymbol{U}_i\}$ to be i.i.d. is a strong condition (actually more so than the usual condition (22.23) since each $\boldsymbol{U}_i$ consists of successive regressors). However, it can be seen from the expression for $\boldsymbol{\Sigma}'$ that it is sufficient for our purposes to require $\tilde{\boldsymbol{w}}_{i-1}$ to be independent of $\boldsymbol{P}_i$; this can be a weaker condition and more likely to hold. As an illustration, consider the special case in which $\boldsymbol{U}_i$ is square and invertible with $\epsilon \approx 0$. In this case, $\boldsymbol{P}_i = \mathrm{I}$ and is independent of $\tilde{\boldsymbol{w}}_{i-1}$. Use this assumption to verify that the variance relation of part (c) becomes

$$\boxed{\begin{array}{l} \mathsf{E}\,\|\tilde{\boldsymbol{w}}_i\|_{\Sigma}^2 = \mathsf{E}\,\|\tilde{\boldsymbol{w}}_{i-1}\|_{\Sigma'}^2 + \mu^2\mathsf{E}\,\boldsymbol{v}_i^*\boldsymbol{A}_i^{\Sigma}\boldsymbol{v}_i \\ \\ \Sigma' = \Sigma - \mu\Sigma(\mathsf{E}\,\boldsymbol{P}_i) - \mu(\mathsf{E}\,\boldsymbol{P}_i)\Sigma + \mu^2\mathsf{E}\left[\boldsymbol{U}_i^*\boldsymbol{A}_i^{\Sigma}\boldsymbol{U}_i\right] \end{array}}$$

where Σ' is now a deterministic matrix. The above variance relation extends the result of Thm. 22.4 to $\epsilon-$APA. Show further that the mean of the weight-error vector satisfies

$$\boxed{\mathsf{E}\,\tilde{\boldsymbol{w}}_i = \left[\mathrm{I} - \mu\mathsf{E}\,(\boldsymbol{P}_i)\right]\mathsf{E}\,\tilde{\boldsymbol{w}}_{i-1}}$$

(e) As in Sec. 24.1, introduce the vec$(\cdot)$ notation, $\sigma = \mathrm{vec}(\Sigma)$. Argue that the variance relation of part (d) can be rewritten in the form

$$\boxed{\mathsf{E}\,\|\tilde{\boldsymbol{w}}_i\|_{\sigma}^2 = \mathsf{E}\,\|\tilde{\boldsymbol{w}}_{i-1}\|_{F\sigma}^2 + \mu^2\sigma_v^2(\gamma^{\mathsf{T}}\sigma)}$$

where $\gamma = \text{vec}\{\mathsf{E}\,\boldsymbol{U}_i^*(\epsilon \mathrm{I} + \boldsymbol{U}_i\boldsymbol{U}_i^*)^{-2}\boldsymbol{U}_i\}$ and

$$\boxed{F \triangleq \mathrm{I} - \mu(\mathsf{E}\,[\boldsymbol{P}_i^{\mathsf{T}}] \otimes \mathrm{I}) - \mu(\mathrm{I} \otimes \mathsf{E}\,[\boldsymbol{P}_i]) + \mu^2 \mathsf{E}\,[\boldsymbol{P}_i^{\mathsf{T}} \otimes \boldsymbol{P}_i]}$$

(f) Conclude that the transient performance of ϵ−APA is described by the M^2-dimensional state-space recursion $\mathcal{W}_i = \mathcal{F}\mathcal{W}_{i-1} + \mu^2\sigma_v^2\mathcal{Y}$ where

$$\mathcal{F} = \begin{bmatrix} 0 & 1 & & & \\ 0 & 0 & 1 & & \\ 0 & 0 & 0 & 1 & \\ \vdots & & & & \\ 0 & 0 & 0 & & 1 \\ -p_0 & -p_1 & -p_2 & \ldots & -p_{M^2-1} \end{bmatrix}_{M^2 \times M^2} \qquad \mathcal{W}_i = \begin{bmatrix} \mathsf{E}\,\|\tilde{\boldsymbol{w}}_i\|_{\sigma}^2 \\ \mathsf{E}\,\|\tilde{\boldsymbol{w}}_i\|_{F\sigma}^2 \\ \mathsf{E}\,\|\tilde{\boldsymbol{w}}_i\|_{F^2\sigma}^2 \\ \vdots \\ \mathsf{E}\,\|\tilde{\boldsymbol{w}}_i\|_{F^{(M^2-1)}\sigma}^2 \end{bmatrix}$$

with

$$p(x) \triangleq \det(x\mathrm{I} - F) = x^{M^2} + \sum_{k=0}^{M^2-1} p_k x^k$$

denoting the characteristic polynomial of F and $[\mathcal{Y}]_k = \gamma^{\mathsf{T}} F^k \sigma$ for $k = 0, \ldots, M^2 - 1$.

(g) Conclude further that ϵ−APA is stable in the mean and mean-square senses for step-sizes in the range

$$0 < \mu < \min\left\{\frac{2}{\lambda_{\max}(\mathsf{E}\,\boldsymbol{P}_i)}, \frac{1}{\lambda_{\max}(A^{-1}B)}, \frac{1}{\max\{\lambda(H) \in \mathbb{R}^+\}}\right\}$$

where

$$A = (\mathsf{E}\,[\boldsymbol{P}_i] \otimes \mathrm{I}) + (\mathrm{I} \otimes \mathsf{E}\,[\boldsymbol{P}_i]), \quad B = \mathsf{E}\,[\boldsymbol{P}_i \otimes \boldsymbol{P}_i], \quad H = \begin{bmatrix} A/2 & -B/2 \\ \mathrm{I}_{M^2} & 0 \end{bmatrix}$$

(h) Deduce that the MSD and EMSE of ϵ−APA are given by

$$\text{MSD} = \mu^2\sigma_v^2\gamma^{\mathsf{T}}(\mathrm{I} - F)^{-1}\text{vec}(\mathrm{I}) \quad \text{and} \quad \text{EMSE} = \mu^2\sigma_v^2\gamma^{\mathsf{T}}(\mathrm{I} - F)^{-1}\text{vec}(R_u)$$

Remark. The results of this problem are based on the work by Shin and Sayed (2003a,b).

Problem V.38 (Stable LMS with small step-size) Consider the setting of Thm. 24.1 where $F = \mathrm{I}_{M^2} - \mu A + \mu^2 B$, with

$$A = (\mathrm{I}_M \otimes [\mathsf{E}\,\boldsymbol{u}_i^*\boldsymbol{u}_i]) + ([\mathsf{E}\,\boldsymbol{u}_i^*\boldsymbol{u}_i]^{\mathsf{T}} \otimes \mathrm{I}_M), \qquad B = \mathsf{E}\left([\boldsymbol{u}_i^*\boldsymbol{u}_i]^{\mathsf{T}} \otimes [\boldsymbol{u}_i^*\boldsymbol{u}_i]\right)$$

This theorem relates to the performance of LMS for non-Gaussian regressors. As explained in the arguments that led to the theorem, the key difficulty lies in evaluating the last moment in (23.6), which translates into a difficulty in evaluating the matrix B itself. The purpose of this problem is to show that, although we may not be able to evaluate B, if the distribution of $\boldsymbol{u}_i$ is such that B is finite, then we can guarantee the existence of a small enough step-size for which F (and, hence, the filter) is stable. To establish the result we use the property of Kronecker products stated in Prob. V.6, as well as the characterization (25.29) for the maximum eigenvalue of Hermitian matrices.

(a) Verify that the eigenvalues of $\mathrm{I}_{M^2} - \mu A$ are given by $1 - \mu(\lambda_k + \lambda_j)$ for all $1 \leq j, k \leq M$.

(b) Assume that B is finite. We mean that B is finite relative to some matrix norm. One such norm could be the Frobenius norm of B, which is defined as the Euclidean norm of $\text{vec}\{B\}$. Argue that the maximum eigenvalue of F satisfies $\lambda_{\max}(F) \leq 1 - 2\mu\lambda_{\min} + \mu^2\beta$, for some finite positive scalar β. Conclude that there exists a small enough μ such that F is stable (and, hence, the filter is mean-square stable).

Problem V.39 (Stable data-normalized filters) Consider the setting of Thm. 25.2 where $F = \mathrm{I}_{M^2} - \mu A + \mu^2 B$, with

$$A = \left(\left[\mathsf{E}\, \frac{u_i^* u_i}{g[u_i]} \right]^{\mathsf{T}} \otimes \mathrm{I} \right) + \left(\mathrm{I} \otimes \left[\mathsf{E}\, \frac{u_i^* u_i}{g[u_i]} \right] \right), \quad B = \mathsf{E} \left(\left[\frac{u_i^* u_i}{g[u_i]} \right]^{\mathsf{T}} \otimes \left[\frac{u_i^* u_i}{g[u_i]} \right] \right)$$

Let $P = \mathsf{E}\, u_i^* u_i / g[u_i]$. The matrix P is clearly non-negative definite. Assume $P > 0$ so that its minimum eigenvalue is positive. Assume further that B is finite. Repeat the argument of Prob. V.38 to show that there exists a finite positive scalar β such that $\lambda(F) < 1$ for any $\mu < 2\lambda_{\min}(P)/\beta$.

COMPUTER PROJECT

Project V.1 (Transient behavior of LMS) In this project we examine the transient behavior of LMS and verify some of the results derived in the chapter for both cases of Gaussian and non-Gaussian data.

Thus consider a real-valued regression sequence $\{u_i\}$ with covariance matrix R_u whose eigenvalue spread we set at $\rho = 5$. Let the noise variance be $\sigma_v^2 = 0.001$ and fix the filter order at $M = 5$.

(a) Generate a covariance matrix R_u whose eigenvalue spread is 5. This can be achieved, for example, by choosing R_u to be diagonal with smallest eigenvalue at 1 and largest eigenvalue at 5. Generate independent and identically distributed regression vectors $\{u_i\}$ from a Gaussian distribution with covariance matrix R_u. For each time instant i, generate also the reference signal as follows:

$$d(i) = u_i w^o + v(i)$$

where $v(i)$ is white Gaussian noise with variance 0.001, and w^o is some arbitrary weight vector that we wish to estimate. Adjust the norm of w^o to unity.

(b) Fix initially the step-size at $\mu = 0.01$ and train LMS for 10000 iterations. Average the squared-weight-error curve, $\|\tilde{w}_i\|^2$, over 30 experiments and generate an ensemble-average curve for $\mathsf{E}\,\|\tilde{w}_i\|^2$. Use recursion (23.32) to generate the theoretical curve for $\mathsf{E}\,\|\tilde{w}_i\|^2$. Observe that since in this project we are choosing R_u to be diagonal and, hence, $R_u = \Lambda$, it holds that $\overline{w}_i = \tilde{w}_i$. Compare the simulated and theoretical curves. Use the simulated curve to estimate the MSD. Compare this value with the one predicted by theory through expression (23.54).

(c) Repeat the simulation of part (b) and generate a curve of the MSD performance as a function of the step-size. Simulate for the values

$$\mu \in \{\, 0.03,\ 0.05,\ 0.07,\ 0.08,\ 0.09,\ 0.095,\ 0.1,\ 0.125\}$$

At what value of the step-size does the filter become unstable? Now recall that according to Thm. 23.2, the LMS filter remains mean-square stable for all step-sizes that satisfy the condition

$$f(\mu) \;\triangleq\; \frac{1}{2} \sum_{k=1}^{M} \frac{\lambda_k \mu}{1 - \mu\lambda_k} < 1$$

Plot the function $f(\mu)$. At what value of μ does $f(\mu)$ exceed one? Compare this theoretical value with the simulated value.

(d) Repeat the simulation of part (b) except that now the regression vectors $\{u_i\}$ are selected as independent and identically distributed realizations of a *uniform* distribution with covariance matrix R_u. More specifically, each u_i is generated as follows. Select first random vectors s_i, of the same size as u_i, but with i.i.d. entries that are uniformly distributed within the interval $[-1, 1]$. In this way, each entry of s_i has variance $1/12$. Then set

$$u_i \;=\; \sqrt{12} \cdot s_i \cdot R_u^{1/2}$$

where, since R_u is diagonal, $R_u^{1/2}$ is the diagonal matrix with the positive square-roots of the entries of R_u. Verify analytically that the vectors $\boldsymbol{u}_i$ generated in this manner have covariance matrix R_u.

Generate also a uniform noise sequence $\boldsymbol{v}(i)$ with variance $\sigma_v^2 = 0.001$. Use the data $\{\boldsymbol{d}(i), \boldsymbol{u}_i\}$ so obtained to simulate the operation of LMS with step-size $\mu = 0.05$. Use the recursion in Prob. V.7 to generate the theoretical curve for $\mathsf{E}\,\|\tilde{\boldsymbol{w}}_i\|^2$. Compare the simulated and theoretical curves. Construct also a curve for the MSD performance of the filter for the same range of step-sizes as in part (b). At what value of the step-size does the filter become unstable? Now recall that according to Thm. 24.1, the LMS filter remains mean-square stable for step-sizes that satisfy condition (24.24), namely,

$$0 < \mu < \min\left\{\frac{2}{\lambda_{\max}}, \frac{1}{\lambda_{\max}(A^{-1}B)} \frac{1}{\max\left\{\lambda(H) \in \mathbb{R}^{+}\right\}}\right\}$$

where $\{A, B\}$ are given by (24.21); for real-valued data, these matrices reduce to

$$A = (\mathrm{I}_M \otimes R_u) + (R_u \otimes \mathrm{I}_M), \qquad B = \mathsf{E}\left(\left[\boldsymbol{u}_i^{\mathsf{T}}\boldsymbol{u}_i\right] \otimes \left[\boldsymbol{u}_i^{\mathsf{T}}\boldsymbol{u}_i\right]\right)$$

and

$$H = \begin{bmatrix} A/2 & -B/2 \\ \mathrm{I}_{M^2} & 0 \end{bmatrix}$$

Estimate the matrices A and B via ensemble-averaging and evaluate the upper bound on μ for mean-square stability. Compare this result with the one obtained from the simulated MSD curve.

PART VI

BLOCK ADAPTIVE FILTERS

CHAPTER 26

Transform Domain Adaptive Filters

The convergence performance of LMS-type filters is highly dependent on the correlation of the input data and, in particular, on the eigenvalue spread of the covariance matrix of the regression data. In addition, the computational complexity of this class of filters is proportional to the filter length and, therefore, it can become prohibitive for long tapped delay lines. The purpose of this part is to describe three other classes of adaptive filters that address the two concerns of complexity and convergence, namely, transform-domain adaptive filters, block adaptive filters, and subband adaptive filters.

Transform-domain filters exploit the de-correlation properties of some well-known signal transforms, such as the discrete Fourier transform (DFT) and the discrete cosine transform (DCT), in order to pre-whiten the input data and speed up filter convergence. The resulting improvement in performance is usually a function of the data correlation and, therefore, the degree of success in achieving the desired objective varies from one signal correlation to another. The computational cost continues to be $O(M)$ operations per sample for a filter of length M.

Block adaptive filters, on the other hand, reduce the computational cost by a factor $\alpha > 1$, while at the same time improving the convergence speed. This is achieved by processing the data on a block-by-block basis, as opposed to a sample-by-sample basis, and by exploiting the fact that many signal transforms admit efficient implementations. However, the reduction in cost and the improvement in convergence speed come at a cost. Block implementations tend to suffer from a delay problem in the signal path, and this delay results from the need to collect blocks of data before processing.

The class of subband adaptive filters is related to the class of block adaptive filters, except that it attempts to achieve better pre-whitening (or band partitioning) of the data via selection of what are called prototype filters for their analysis and synthesis filter banks. While subband filters also succeed in reducing the computational cost by a factor $\alpha > 1$, their convergence and mean-square performance can be less than that of block filters. This is because the design of the analysis and synthesis filter banks is usually decoupled from the adaptive filter design and, in this process, performance degradation can occur.

In this part, we study in some detail these classes of algorithms and comment on some of their additional properties. We start with transform-domain filters.

26.1 TRANSFORM-DOMAIN FILTERS

Recall from the discussions in App. 23.A, Chapter 24, and also Prob. V.14, that the performance of LMS is sensitive to the correlation of the input sequence $\{\boldsymbol{u}(i)\}$ and, more specifically, to the eigenvalues and eigenvalue spread of the covariance matrix

$$R_u = \mathsf{E}\,\boldsymbol{u}_i^*\boldsymbol{u}_i$$

where

$$u_i = \begin{bmatrix} u(i) & u(i-1) & \ldots & u(i-M+1) \end{bmatrix}$$

denotes the regression vector. The smaller eigenvalues of R_u contribute to slower convergence and the larger eigenvalues limit the range of allowed step-sizes and, thereby, limit the learning abilities of the filter. Best convergence and learning performance would result when all the eigenvalues of R_u are equal. This situation requires the input data to be white so that R_u would have the form $R_u = \sigma_u^2 \mathrm{I}$ for some variance σ_u^2.

However, in general, the input data is colored and the eigenvalues of R_u can vary significantly in value from the smallest to the largest. One strategy to improve filter performance is to attempt to pre-whiten the data prior to adaptation. If the auto-correlation of the input sequence is known, then we could use this information to construct a filter that pre-whitens the data, as we shall explain shortly. However, since the statistics of the input data are seldom known beforehand, the design of such pre-whitening filters is generally not possible. Still, there are ways to achieve this objective in an approximate manner. One such way is to transform the regressor prior to adaptation by some pre-selected unitary transformation, such as the discrete Fourier transform (DFT) or the discrete cosine transform (DCT). These transforms have useful de-correlation properties that help improve the convergence performance of LMS for correlated input data.

Pre-Whitening Filters

We first explain how to pre-whiten the data when the second-order statistics of the input sequence $\{u(i)\}$ is known. Thus, assume that $\{u(i)\}$ is zero-mean and wide-sense stationary, with known auto-correlation function

$$r(k) = \mathsf{E}\, u(i)u^*(i-k), \qquad \text{for} \quad k = 0, \pm 1, \pm 2, \ldots$$

In order to determine the pre-whitening filter from knowledge of $\{r(k)\}$, we need to explain briefly the useful notions of power spectrum and spectral factorization.

Spectral Factorization

The $z-$spectrum of a wide-sense stationary random process $\{u(i)\}$ is denoted by $S_u(z)$ and is defined as the $z-$transform of $\{r(k)\}$,

$$\boxed{S_u(z) \stackrel{\Delta}{=} \sum_{k=-\infty}^{\infty} r(k)z^{-k}} \tag{26.1}$$

Of course, this definition makes sense only for those values of z for which the series converges. For our purposes, it suffices to assume that $\{r(k)\}$ is exponentially bounded, i.e.,

$$|r(k)| \leq \beta a^{|k|} \tag{26.2}$$

for some $\beta > 0$ and $0 < a < 1$. In this case the series (26.1) is absolutely convergent for all values of z in the annulus $a < |z| < a^{-1}$, i.e., it satisfies

$$\sum_{k=-\infty}^{\infty} |r(k)| \cdot |z^{-k}| < \infty \qquad \text{for all} \quad a < |z| < 1/a$$

We then say that the interval $a < |z| < a^{-1}$ defines the region of convergence (ROC) of $S_u(z)$. Since this ROC includes the unit circle, we establish that $S_u(z)$ cannot have poles on the unit circle. Evaluating $S_u(z)$ on the unit circle then leads to what is called the power spectrum (or the power-spectral-density function) of the random process $\{\boldsymbol{u}(i)\}$:

$$\boxed{S_u(e^{j\omega}) \triangleq \sum_{k=-\infty}^{\infty} r(k)e^{-j\omega k}} \tag{26.3}$$

The power spectrum has two important properties:

1. Hermitian symmetry, i.e., $S_u(e^{j\omega}) = [S_u(e^{j\omega})]^*$ and $S_u(e^{j\omega})$ is therefore real.

2. Nonnegativity on the unit circle, i.e., $S_u(e^{j\omega}) \geq 0$ for $0 \leq \omega \leq 2\pi$.

The first property is easily checked from the definition of $S_u(e^{j\omega})$ since

$$r(k) = r^*(-k)$$

Actually, and more generally, the $z-$spectrum satisfies the para-Hermitian symmetry property

$$S_u(z) = [S_u(1/z^*)]^* \tag{26.4}$$

That is, if we replace z by $1/z^*$ and conjugate the result, then we recover $S_u(z)$ again. The second claim regarding the nonnegativity of $S_u(e^{j\omega})$ on the unit circle is more demanding to establish; it is proven in Prob. VI.2 under assumption (26.2).

To continue, we shall assume that $S_u(z)$ is a proper *rational* function and that it does not have zeros on the unit-circle so that

$$S_u(e^{j\omega}) > 0 \quad \text{for all} \quad -\pi \leq \omega \leq \pi \tag{26.5}$$

Then using the para-Hermitian symmetry property (26.4), it is easy to see that for every pole (or zero) at a point ξ, there must exist a pole (respectively a zero) at $1/\xi^*$. In addition, it follows from the fact that $S_u(z)$ does not have poles and zeros on the unit circle that any such rational function $S_u(z)$ can be factored as

$$S_u(z) = \sigma_u^2 \cdot \left(\frac{\prod_{l=1}^{m}(z - z_l)(z^{-1} - z_l^*)}{\prod_{l=1}^{n}(z - p_l)(z^{-1} - p_l^*)} \right) \tag{26.6}$$

for some positive scalar σ_u^2, and for poles and zeros satisfying $|z_l| < 1$ and $|p_l| < 1$. The spectral factorization of $S_u(z)$ is now defined as a factorization of the form

$$S_u(z) = \sigma_u^2 A(z)\,[A(1/z^*)]^* \tag{26.7}$$

where $\{\sigma_u^2, A(z)\}$ satisfy the following conditions:

1. σ_u^2 is a positive scalar.

2. $A(z)$ is normalized to unity at infinity, i.e., $A(\infty) = 1$.

3. $A(z)$ is a rational minimum-phase function (i.e., its poles and zeros are inside the unit circle).

The normalization $A(\infty) = 1$ makes the choice of $A(z)$ unique since otherwise infinitely many choices for $\{\sigma_u^2, A(z)\}$ would exist. In order to determine $A(z)$ from (26.6) we just

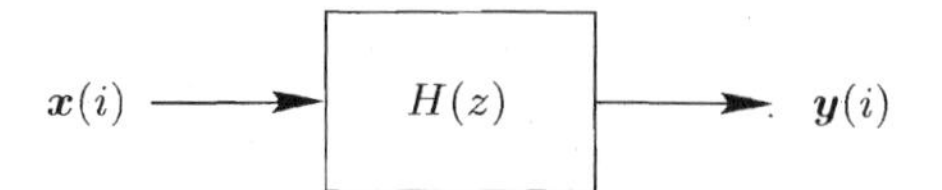

FIGURE 26.1 Filtering of a wide-sense stationary random process $\{x(i)\}$ by a stable linear system $H(z)$.

have to extract the poles and zeros that are inside the unit circle. However, in order to meet the normalization condition $A(\infty) = 1$, we take

$$\boxed{A(z) = z^{n-m} \cdot \left(\frac{\prod_{l=1}^{m} (z - z_l)}{\prod_{l=1}^{n} (z - p_l)} \right) = \frac{\prod_{l=1}^{m} (1 - z_l z^{-1})}{\prod_{l=1}^{n} (1 - p_l z^{-1})}} \tag{26.8}$$

The spectral factor $A(z)$ so defined has a useful interpretation. To see this, we first recall the following result. Let $\{x(i)\}$ be a wide-sense stationary random process with $z-$spectrum $S_x(z)$, and assume it is fed into a stable linear system with transfer function $H(z)$, as shown in Fig. 26.1. Then the output process $\{y(i)\}$ is also wide-sense stationary and it can be verified, by direct calculation, that its $z-$spectrum is related to $\{S_x(z), H(z)\}$ via

$$S_y(z) = H(z) S_x(z) \left[H(1/z^*) \right]^*$$

Therefore, returning to $A(z)$, if we feed the process $\{u(i)\}$ through the filter $1/[\sigma_u A(z)]$, as shown in Fig. 26.2, then the $z-$spectrum of the output process, denoted by $\{\bar{u}(\cdot)\}$, will be

$$S_{\bar{u}}(z) = \frac{1}{\sigma_u A(z)} S_u(z) \frac{1}{\sigma_u A^*(z^{-*})} = \frac{1}{\sigma_u A(z)} \left[\sigma_u^2 A(z) A^*(z^{-*}) \right] \frac{1}{\sigma_u A^*(z^{-*})} = 1$$

In other words, the process $\{\bar{u}(\cdot)\}$ becomes white with unit-variance and, by inverse $z-$transformation, its auto-correlation sequence will be $\bar{r}(k) = \mathsf{E}\,\bar{u}(i)\bar{u}^*(i-k) = \delta(k)$.

LMS *Adaptation*

Consider now an LMS filter that is adapted by using $\{\bar{u}(i)\}$ as regression data, instead of $\{u(i)\}$, as shown in Fig. 26.3, with the reference sequence $d(i)$ also filtered by $1/(\sigma_u A(z))$, i.e.,

$$\boxed{\bar{w}_i = \bar{w}_{i-1} + \mu \bar{u}_i^* \bar{e}(i), \quad \bar{e}(i) = \bar{d}(i) - \bar{u}_i \bar{w}_{i-1}, \quad \bar{w}_{-1} = 0} \tag{26.9}$$

where

$$\bar{u}_i = [\bar{u}(i) \ \bar{u}(i-1) \ \ldots \ \bar{u}(i-M+1)]$$

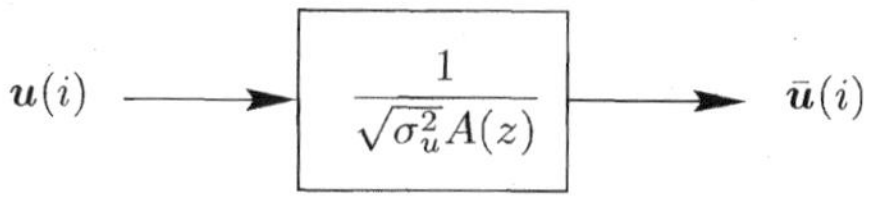

$$u(i) \longrightarrow \boxed{\frac{1}{\sqrt{\sigma_u^2} A(z)}} \longrightarrow \bar{u}(i)$$

FIGURE 26.2 Pre-whitening of $\{u(i)\}$ by using the inverse of the spectral factor of $S_u(z)$.

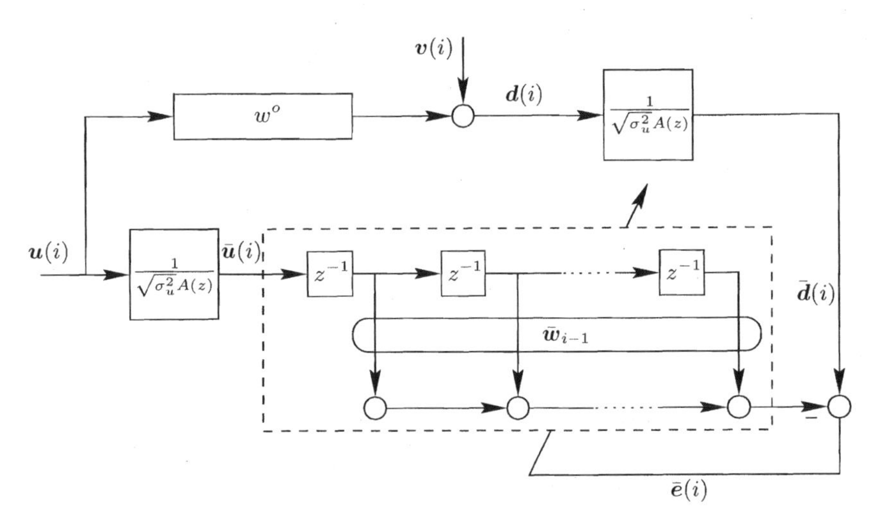

FIGURE 26.3 An adaptive filter implementation with a pre-whitening filter.

denotes the regression data and $\bar{w}_i$ denotes the resulting weight vector. In Fig. 26.3, it is assumed that the data $\{d(i), u_i\}$ satisfy the linear regression model $d(i) = u_i w^o + v(i)$, for some unknown w^o. The covariance matrix of the transformed regressor is seen to be $R_{\bar{u}} = \mathsf{E}\,\bar{u}_i^*\bar{u}_i = \mathrm{I}$, with an eigenvalue spread of unity. In this way, the convergence performance of the filter will be improved relative to an LMS implementation that relies solely on $\{d(i), u(i)\}$. We illustrate this fact later in Fig. 26.5.

As an example, consider a random process $\{u(i)\}$ with an exponential auto-correlation sequence of the form

$$r(k) = a^{|k|}, \qquad k = \ldots, -1, 0, 1, \ldots \tag{26.10}$$

with $|a| < 1$. Its $z-$spectrum is given by

$$S_u(z) = \sum_{k=-\infty}^{\infty} a^{|k|} z^{-k} = \frac{1 - |a|^2}{(1 - az^{-1})(1 - a^* z)}$$

and the corresponding spectral factor is therefore $A(z) = 1/(1 - az^{-1})$ with $\sigma_u^2 = 1 - |a|^2$. This result shows that a random process with exponential auto-correlation function of the form (26.10) can be whitened by passing it through the first-order FIR filter $(1 - az^{-1})/(\sqrt{1 - |a|^2}$.

Unitary Transformations

Since the statistics of the input data are rarely known in advance, and since the data itself may not even be stationary, the design of a pre-whitening filter $1/A(z)$ is usually not possible in the manner explained above. Still, there are ways to approximately pre-whiten the data, with varied degrees of success depending on the nature of the data. One such way is to transform the regressors prior to adaptation by some pre-selected unitary transformation, such as the DFT or the DCT, as we now explain.

Consider the standard LMS implementation

$$w_i = w_{i-1} + \mu u_i^*[d(i) - u_i w_{i-1}] \tag{26.11}$$

with regressor $\boldsymbol{u}_i$ and associated covariance matrix $R_u = \mathsf{E}\,\boldsymbol{u}_i^*\boldsymbol{u}_i$. Let T denote an arbitrary unitary matrix of size $M \times M$, i.e., $TT^* = T^*T = \mathrm{I}$. For example, T could be chosen in terms of the discrete Fourier transform matrix (DFT),

$$[F]_{km} \triangleq \frac{1}{\sqrt{M}} e^{-\frac{j2\pi mk}{M}} \qquad k,m = 0,1,\ldots,M-1 \quad \text{(DFT)} \tag{26.12}$$

or the discrete cosine transform matrix (DCT),

$$[C]_{km} \triangleq \alpha(k)\cos\left(\frac{k(2m+1)\pi}{2M}\right), \qquad k,m = 0,1,\ldots,M-1 \quad \text{(DCT)} \tag{26.13}$$

where

$$\alpha(0) = 1/\sqrt{M} \quad \text{and} \quad \alpha(k) = \sqrt{2/M}, \quad \text{for } k \neq 0$$

Also, k indicates the row index and m the column index. The scaling factor $1/\sqrt{M}$ in the expression (26.12) for the DFT matrix F is added here in order to result in a unitary transformation since then F satisfies $FF^* = F^*F = \mathrm{I}$. Moreover, there are other variants of the discrete cosine transform; the one we choose is widely used. There are also discrete sine transforms (DST). The arguments in this section apply to other unitary transforms as well. Observe that F is symmetric ($F^\mathsf{T} = F$) while C is real but nonsymmetric; yet both are unitary. We usually choose $T = F$ or $T = C^\mathsf{T}$, or some other unitary matrix. Once T is selected, we then define the transformed regressor

$$\boxed{\bar{\boldsymbol{u}}_i = \boldsymbol{u}_i T}$$

whose covariance matrix is related to R_u via

$$R_{\bar{u}} = \mathsf{E}\,\bar{\boldsymbol{u}}_i^*\bar{\boldsymbol{u}}_i = T^*\left(\mathsf{E}\,\boldsymbol{u}_i^*\boldsymbol{u}_i\right)T$$

that is,

$$\boxed{R_{\bar{u}} = T^* R_u T} \tag{26.14}$$

If we further let

$$\bar{\boldsymbol{w}}_i = T^*\boldsymbol{w}_i$$

and if we multiply both sides of (26.11) by T^* from the left, then (26.11) can be rewritten in terms of the transformed weight vector as (see Fig. 26.4):

$$\bar{\boldsymbol{w}}_i = \bar{\boldsymbol{w}}_{i-1} + \mu\bar{\boldsymbol{u}}_i^*[\boldsymbol{d}(i) - \bar{\boldsymbol{u}}_i\bar{\boldsymbol{w}}_{i-1}], \qquad \bar{\boldsymbol{w}}_{-1} = 0 \tag{26.15}$$

since $\boldsymbol{u}_i\boldsymbol{w}_{i-1} = \bar{\boldsymbol{u}}_i\bar{\boldsymbol{w}}_{i-1}$ in view of the unitarity of T. Observe that the reference sequence $\{\boldsymbol{d}(i)\}$ remains unchanged. Recursion (26.15) is a standard LMS implementation with regression data $\bar{\boldsymbol{u}}_i$ instead of $\boldsymbol{u}_i$. However, since T is unitary and therefore $T^* = T^{-1}$, the relation between $\{R_u, R_{\bar{u}}\}$ has the form of a similarity transformation and such transformations are known to preserve eigenvalues.[9] This means that $\{R_u, R_{\bar{u}}\}$ will have the same eigenvalues and the same eigenvalue spread. Consequently, the implementation (26.15) will face similar convergence limitations as the implementation (26.11). More needs to be done in order to achieve improvement in performance.

[9]Two square matrices A and B are said to be similar if they are related via $B = T^{-1}AT$ for some invertible matrix T. Similarity transformations preserve eigenvalues and, hence, both A and B will have the same eigenvalues — see Prob. VI.1.

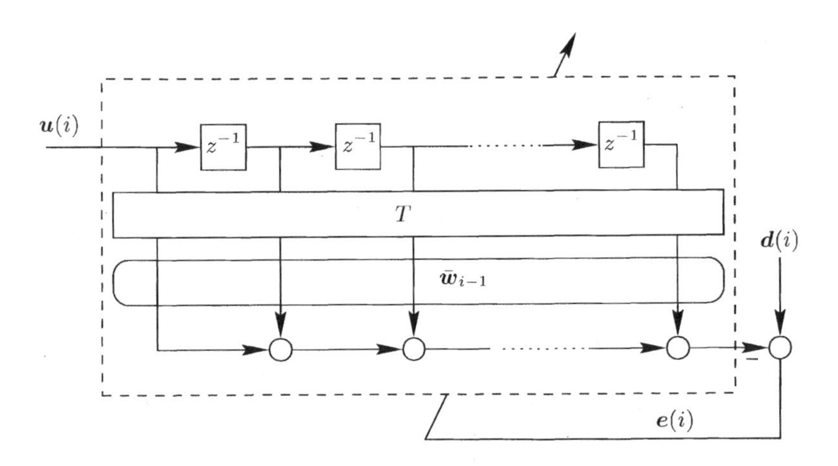

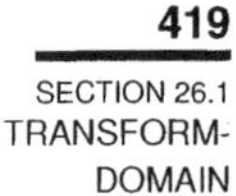

FIGURE 26.4 Transform-domain adaptive filter implementation, where T is generally a unitary transformation.

So assume that we replace the LMS update (26.15) by

$$\bar{w}_i = \bar{w}_{i-1} \;+\; \mu D^{-1}\bar{u}_i^*[d(i) - \bar{u}_i\bar{w}_{i-1}], \qquad \bar{w}_{-1} = 0 \tag{26.16}$$

where, in addition to using the transformed regressor $\bar{u}_i$, we also employ a diagonal normalization matrix D with positive entries to be determined. We can regard (26.16) as an LMS update with a matrix step-size that is equal to μD^{-1}. Let $D^{1/2}$ denote the diagonal matrix whose entries are the positive square-roots of the entries of D. If we multiply both sides of (26.16) by $D^{1/2}$ we find that it reduces to

$$w_i' = w_{i-1}' \;+\; \mu u_i'^*[d(i) - u_i' w_{i-1}'], \qquad w_{-1}' = 0 \tag{26.17}$$

where

$$w_i' \triangleq D^{1/2}\bar{w}_i, \qquad u_i' \triangleq \bar{u}_i D^{-1/2} = u_i T D^{-1/2}$$

Recursion (26.17) shows that the transform-domain algorithm (26.16) is equivalent to an LMS implementation with regression data $\{u_i'\}$. Consequently, the performance of (26.16) will be similar to that of an LMS filter with regression covariance matrix

$$R_{u'} = \mathsf{E}\, u_i'^* u_i' \;=\; D^{-1/2}T^* R_u T D^{-1/2} \tag{26.18}$$

The point to note now is that the relation between $R_{u'}$ and R_u is not a similarity transformation any longer and, therefore, their eigenvalues and eigenvalue spreads are generally different. An ideal choice for D would be to try to force $R_{u'}$ in (26.18) to become the identity matrix (or a multiple of the identity).

Let $R_u = U\Lambda U^*$ denote the eigen-decomposition of R_u, and assume we choose T and D as

$$T = U \qquad \text{and} \qquad D = \Lambda \tag{26.19}$$

Then $R_{\bar{u}} = \Lambda$ and $R_{u'} = \mathrm{I}$, i.e., these choices of T and D de-correlate the entries of $\bar{u}_i$ and the variances of the individual entries of $\bar{u}_i$ are the $\{\lambda_k\}$, i.e., the eigenvalues of R_u. This choice of T is known as the Karhunen Loève transform (KLT). However, using the KLT is not practical since it requires knowledge of R_u (in order to construct T and D) and this information is generally lacking in adaptive implementations.

The alternative would be to replace T by a unitary transformation that does not require prior knowledge of R_u. We often choose T as the DFT or DCT matrices, $T = F$ or $T = C^{\mathsf{T}}$ from (26.12)–(26.13). Of course, applying the DFT or the DCT to $\boldsymbol{u}_i$ does not generally result in a transformed regressor $\bar{\boldsymbol{u}}_i$ with a diagonal covariance matrix $R_{\bar{u}}$.The degree of success in transforming R_u to close-to-diagonal by F or C^{T} is dependent on R_u itself; the transformations can be more successful in some cases and less successful in others. In general, these transformations tend to result in covariance matrices $R_{\bar{u}}$ that are reasonably close to diagonal, i.e.,

$$R_{\bar{u}} = T^* R_u T \approx \text{diagonal}$$

In this way, the entries of the resulting regressor $\bar{\boldsymbol{u}}_i$ will be uncorrelated to a reasonable extent, and their variances could be estimated recursively as follows:

$$\boldsymbol{\lambda}_k(i) = \beta\boldsymbol{\lambda}_k(i-1) + (1-\beta)|\bar{\boldsymbol{u}}_i(k)|^2, \quad k = 0, 1, \ldots, M-1$$

using $0 \ll \beta < 1$ (e.g., $\beta = 0.9$), and where $\bar{\boldsymbol{u}}_i(k)$ denotes the $k-$th entry of

$$\bar{\boldsymbol{u}}_i = [\bar{\boldsymbol{u}}_i(0)\ \bar{\boldsymbol{u}}_i(1)\ \ldots\ \bar{\boldsymbol{u}}_i(M-1)]$$

The estimators $\{\boldsymbol{\lambda}_k(i)\}$ can in turn be used to construct an estimator for the step-size matrix D for iteration i as $\boldsymbol{D}_i = \text{diag}\{\boldsymbol{\lambda}_k(i)\}$. Now if we write down (26.16) for each entry of the weight vector, using the just constructed $\boldsymbol{D}_i$, we get

$$\begin{aligned}
\boldsymbol{e}(i) &= \boldsymbol{d}(i) - \bar{\boldsymbol{u}}_i\bar{\boldsymbol{w}}_{i-1} \\
\bar{\boldsymbol{w}}_k(i) &= \bar{\boldsymbol{w}}_k(i-1) + \frac{\mu}{\boldsymbol{\lambda}_k(i)}\bar{\boldsymbol{u}}^*(i-k)\boldsymbol{e}(i), \quad k = 0, 1, \ldots, M-1
\end{aligned}$$

which can be seen to be an NLMS-type update with power normalization. In summary, we arrive at the following statement; the result is stated in terms of realizations $\{d(i), u_i, \lambda_k(i)\}$ for the random variables $\{\boldsymbol{d}(i), \boldsymbol{u}_i, \boldsymbol{\lambda}_k(i)\}$.

Algorithm 26.1 (General transform-domain LMS) Consider a zero-mean random variable $\boldsymbol{d}$ with realizations $\{d(0), d(1), \ldots\}$, and a zero-mean random row vector $\boldsymbol{u}$ with realizations $\{u_0, u_1, \ldots\}$. The weight vector w^o that solves

$$\min_w \mathsf{E}\,|\boldsymbol{d} - \boldsymbol{u}w|^2$$

can be approximated iteratively via $w_i = T\bar{w}_i$, where T is some pre-selected unitary transformation and $\bar{w}_i$ is updated as follows. Start with $\lambda_k(-1) = \epsilon$ (a small positive number) and $\bar{w}_{-1} = 0$, and repeat for $i \geq 0$:

$$\begin{aligned}
\bar{u}_i &= u_i T \triangleq [\ \bar{u}_i(0)\ \ \bar{u}_i(1)\ \ \ldots\ \ \bar{u}_i(M-1)\ ] \\
\bar{u}_i(k) &= k\text{-th entry of } \bar{u}_i \\
\lambda_k(i) &= \beta\lambda_k(i-1) + (1-\beta)|\bar{u}_i(k)|^2, \quad k = 0, 1, \ldots, M-1 \\
D_i &= \text{diag}\{\lambda_k(i)\} \\
e(i) &= d(i) - \bar{u}_i\bar{w}_{i-1} \\
\bar{w}_i &= \bar{w}_{i-1} + \mu D_i^{-1}\bar{u}_i^* e(i)
\end{aligned}$$

where μ is a positive step-size (usually small) and $0 \ll \beta < 1$.

Example 26.1 (Decorrelation properties)

In order to illustrate the de-correlation properties of the DFT and DCT transforms, consider the same random process $\{\boldsymbol{u}(i)\}$ as before with the exponential auto-correlation sequence (26.10). Choose $a = 0.95$ and $M = 4$. Then the covariance matrix R_u is given by

$$R_u = \begin{bmatrix} 1.000000 & 0.950000 & 0.902500 & 0.857375 \\ 0.950000 & 1.000000 & 0.950000 & 0.902500 \\ 0.902500 & 0.950000 & 1.000000 & 0.950000 \\ 0.857375 & 0.902500 & 0.950000 & 1.000000 \end{bmatrix}$$

Choosing $T = F$ and applying it to R_u as indicated by (26.14), we obtain, after rounding to five decimal places,

$$R_{\bar{u}} \quad = \quad F^* R_u F = \begin{bmatrix} \boxed{3.75619} & -0.02316(1+j) & 0 & -0.02316(1-j) \\ -0.02316(1-j) & \boxed{0.09750} & 0.02316(1+j) & j0.04631 \\ 0 & 0.02316(1-j) & \boxed{0.04881} & 0.02316(1+j) \\ -0.02316(1+j) & -j0.04631 & 0.02316(1-j) & \boxed{0.09750} \end{bmatrix}$$

while choosing $T = C^{\mathsf{T}}$ and applying it to R_u leads to

$$R_{\bar{u}} = C R_u C^{\mathsf{T}} = \begin{bmatrix} \boxed{3.75619} & 0 & -0.04631 & 0 \\ 0 & \boxed{0.16265} & 0 & -0.00084 \\ -0.04631 & 0 & \boxed{0.05119} & 0 \\ 0 & -0.00084 & 0 & \boxed{0.02998} \end{bmatrix}$$

We see that the resulting matrices $R_{\bar{u}}$ are diagonally dominant and that the DCT transformation is more successful in transforming R_u to a close-to-diagonal matrix. Further comparisons between both transforms in the context of LMS adaptation are given later in Fig. 26.5.

◇

26.2 DFT-DOMAIN LMS

The computational complexity of the transform-domain LMS filter of Alg. 26.1 depends on the selection of T and on the manner by which this transformation is implemented. Assume, for instance, that T is chosen as the DFT matrix (26.12). Even if the calculation $\bar{u}_i = u_i T$ is performed using the fast Fourier transform (FFT), this step would require $O(M \log_2 M)$ operations per iteration. This cost is higher than the usual $O(M)$ figure that is required by the standard LMS implementation (26.11).

Still, the transformed filter can be implemented at a cost of $O(M)$ operations per iteration by exploiting the fact that two successive regressors $\{u_{i-1}, u_i\}$ share most of their entries. Thus, consider the regressors $\{u_{i-1}, u_i\}$:

$$\begin{aligned} u_{i-1} &= \Big[\boxed{u(i-1) \quad u(i-2) \quad \ldots \quad u(i-M+1)} \quad u(i-M) \Big] \\ u_i &= \Big[u(i) \quad \boxed{u(i-1) \quad u(i-2) \quad \ldots \quad u(i-M+1)} \Big] \end{aligned}$$

and their transformed versions

$$\bar{u}_i = u_i T \quad \text{and} \quad \bar{u}_{i-1} = u_{i-1} T$$

We can evaluate $\bar{u}_i$ directly from $\bar{u}_{i-1}$ and from the entries $\{u(i), u(i-M)\}$ as follows. The k-th elements of $\bar{u}_i$ and $\bar{u}_{i-1}$ are equal to the inner products between u_i and $\boldsymbol{u}_{i-1}$ and the $k-$th column of F, respectively, and are therefore given by

$$\begin{aligned}\bar{u}_i(k) &= \frac{1}{\sqrt{M}}\sum_{m=0}^{M-1} u(i-m)e^{-\frac{j2\pi mk}{M}}\\ \bar{u}_{i-1}(k) &= \frac{1}{\sqrt{M}}\sum_{m=0}^{M-1} u(i-1-m)e^{-\frac{j2\pi mk}{M}}\end{aligned}$$

Now note that, using the change of variables $n = m+1$,

$$\begin{aligned}\bar{u}_{i-1}(k) &= \frac{1}{\sqrt{M}}\sum_{n=1}^{M} u(i-n)e^{-\frac{j2\pi(n-1)k}{M}} = \frac{1}{\sqrt{M}}e^{\frac{j2\pi k}{M}}\left(\sum_{n=1}^{M} u(i-n)e^{-\frac{j2\pi nk}{M}}\right)\\ &= \frac{1}{\sqrt{M}}e^{\frac{j2\pi k}{M}}\left(\sum_{n=1}^{M-1} u(i-n)e^{-\frac{j2\pi nk}{M}}\right) + \frac{1}{\sqrt{M}}e^{\frac{j2\pi k}{M}}u(i-M)\\ &= e^{\frac{j2\pi k}{M}}\left[\bar{u}_i(k) - \frac{1}{\sqrt{M}}u(i)\right] + \frac{1}{\sqrt{M}}e^{\frac{j2\pi k}{M}}u(i-M)\end{aligned}$$

so that

$$\bar{u}_i(k) = e^{-\frac{j2\pi k}{M}}\bar{u}_{i-1}(k) + [u(i) - u(i-M)]/\sqrt{M}$$

Collecting this result for $k = 0, 1, \ldots, M-1$ we arrive at the relation $\{\bar{u}_i, \bar{u}_{i-1}\}$ as

$$\boxed{\bar{u}_i = \bar{u}_{i-1}S + \frac{1}{\sqrt{M}}\{u(i) - u(i-M)\}\begin{bmatrix} 1 & 1 & \ldots & 1\end{bmatrix}} \qquad (26.20)$$

where S is the diagonal matrix defined in the statement below. Expression (26.20) allows us to evaluate the transformed regressors iteratively at the cost of $O(M)$ operations per iteration.

Algorithm 26.2 (DFT-domain LMS) Consider the setting of Alg. 26.1. The weight vector w^o can be approximated iteratively via $w_i = F\bar{w}_i$, where F is the unitary DFT matrix (26.12) and $\bar{w}_i$ is updated as follows. Define the $M \times M$ diagonal matrix

$$S \triangleq \text{diag}\left\{1, e^{\frac{-j2\pi}{M}}, e^{\frac{-j4\pi k}{M}}, \ldots, e^{\frac{-j2\pi(M-1)}{M}}\right\}$$

Then start with $\lambda_k(-1) = \epsilon$ (a small positive number), $\bar{w}_{-1} = 0$, $\bar{u}_{-1} = 0$, and repeat for $i \geq 0$:

$$\begin{aligned}\bar{u}_i &= \bar{u}_{i-1}S + \frac{1}{\sqrt{M}}\{u(i) - u(i-M)\}\begin{bmatrix} 1 & 1 & \ldots & 1\end{bmatrix}\\ \bar{u}_i(k) &= k\text{-th entry of } \bar{u}_i\\ \lambda_k(i) &= \beta\lambda_k(i-1) + (1-\beta)|\bar{u}_i(k)|^2, \quad k = 0, 1, \ldots, M-1\\ D_i &= \text{diag}\{\lambda_k(i)\}\\ e(i) &= d(i) - \bar{u}_i\bar{w}_{i-1}\\ \bar{w}_i &= \bar{w}_{i-1} + \mu D_i^{-1}\bar{u}_i^* e(i)\end{aligned}$$

where μ is a positive step-size (usually small) and $0 \ll \beta < 1$. The computational cost of this algorithm is $O(M)$ operations per iteration.

26.3 DCT-DOMAIN LMS

A similar derivation can be carried out when the transformation T is chosen as $T = C^{\mathsf{T}}$, where C is the DCT matrix (26.13). Let

$$\bar{u}_i = u_i T, \qquad \bar{u}_{i-1} = u_{i-1}T, \qquad \bar{u}_{i-2} = u_{i-2}T$$

In contrast to the DFT case, it turns out that there is now a relation between three successive transformed regressors. It is shown in App. 26.A that the following relation holds:

$$\boxed{\bar{u}_i = \bar{u}_{i-1}S \;-\; \bar{u}_{i-2} + \begin{bmatrix} \phi(0) & \phi(1) & \ldots & \phi(M-1) \end{bmatrix}} \tag{26.21}$$

where S is a diagonal matrix and the $\{\phi(k)\}$ are scalars defined in the following statement.

Algorithm 26.3 (DCT-domain LMS) Consider the setting of Alg. 26.1. The weight vector w^o can be approximated iteratively via $w_i = C^{\mathsf{T}}\bar{w}_i$, where C is the unitary DCT matrix (26.13) and $\bar{w}_i$ is updated as follows. Define the $M \times M$ diagonal matrix

$$S = \text{diag}\left\{2\cos\left(k\pi/M\right),\; k = 0, 1, \ldots, M-1\right\}$$

Then start with $\lambda_k(-1) = \epsilon$ (a small positive number), $\bar{w}_{-1} = 0$, $\bar{u}_{-1} = 0$, and repeat for $i \geq 0$:

$$\begin{aligned}
a(k) &= [u(i) - u(i-1)]\cos\left(\frac{k\pi}{2M}\right), \quad k = 0, 1, \ldots, M-1 \\
b(k) &= (-1)^k\,[u(i-M) - u(i-M-1)]\cos\left(\frac{k\pi}{2M}\right), \quad k = 0, \ldots, M-1 \\
\phi(k) &= \alpha(k)\,[a(k) - b(k)], \quad k = 0, 1, \ldots, M-1 \\
\bar{u}_i &= \bar{u}_{i-1}S \;-\; \bar{u}_{i-2} \;+\; \begin{bmatrix} \phi(0) & \phi(1) & \ldots & \phi(M-1) \end{bmatrix} \\
\bar{u}_i(k) &= k\text{-th entry of } \bar{u}_i \\
\lambda_k(i) &= \beta\lambda_k(i-1) \;+\; (1-\beta)|\bar{u}_i(k)|^2, \qquad k = 0, 1, \ldots, M-1 \\
D_i &= \text{diag}\{\lambda_k(i)\} \\
e(i) &= d(i) - \bar{u}_i\bar{w}_{i-1} \\
\bar{w}_i &= \bar{w}_{i-1} \;+\; \mu D_i^{-1}\bar{u}_i^* e(i)
\end{aligned}$$

where μ is a positive step-size (usually small) and $0 \ll \beta < 1$. The computational cost of this algorithm is $O(M)$ operations per iteration.

Example 26.2 (Performance comparison)

Figure 26.5 compares the performance of four LMS implementations for a first-order auto-regressive process $\boldsymbol{u}(i)$ with a in (26.10) chosen as 0.95. The filter order is set to $M = 8$ and the ensemble-average learning curves are generated by averaging over 300 experiments. The step-size is set to $\mu = 0.01$ and the noise variance at -40 dB. It is seen from the figure that DCT-LMS and DFT-LMS exhibit faster convergence than a standard LMS implementation. Moreover, in the figure, the learning curve for LMS with a pre-whitening filter is evaluated by transforming the error signal $\bar{e}(i)$ of Fig. 26.3 to the $e(i)$ domain by filtering $\bar{e}(i)$ through $\sqrt{\sigma_u^2}A(z)$.

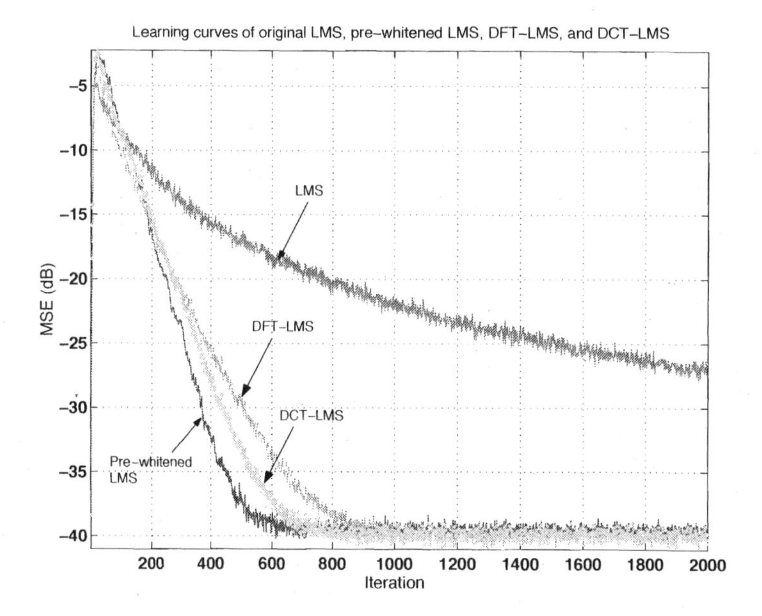

FIGURE 26.5 A comparison of the learning curves of plain LMS, DFT-domain LMS, DCT-domain LMS, and LMS with pre-whitening for a first-order auto-regressive input process.

◇

26.A APPENDIX: DCT-TRANSFORMED REGRESSORS

To arrive at (26.21), we start by noting that the k-th element of $\bar{u}_i$ is equal to the inner product between u_i and the k–th column of C^{T} and is therefore given by

$$\begin{aligned}\bar{u}_i(k) &= \alpha(k)\sum_{m=0}^{M-1} u(i-m)\cos\left[\frac{k(2m+1)\pi}{2M}\right]\\ &= \alpha(k)\left[u(i)\cos\left(\frac{k\pi}{2M}\right) + A - \alpha(k)u(i-M)\cos\left(\frac{k(2M+1)\pi}{2M}\right)\right]\end{aligned}$$

where we defined

$$A \triangleq \frac{1}{2}\sum_{n=0}^{M-1} u(i-n-1)\left(e^{\frac{jk(2n+3)\pi}{2M}} + e^{\frac{-jk(2n+3)\pi}{2M}}\right)$$

and where we employed a change of variables from m to $n = m-1$. Now note that

$$A = \frac{1}{2}e^{\frac{j2k\pi}{2M}}\sum_{n=0}^{M-1} u(i-n-1)e^{\frac{jk(2n+1)\pi}{2M}} + \frac{1}{2}u(i-n-1)e^{-\frac{j2k\pi}{2M}}\sum_{n=0}^{M-1} e^{\frac{-jk(2n+1)\pi}{2M}}$$

so that

$$\begin{aligned}A &= 2\cos\left(\frac{2k\pi}{2M}\right)\sum_{n=0}^{M-1}u(i-n-1)\cos\left(\frac{k(2n+1)\pi}{2M}\right)\\ &\quad -\frac{1}{2}e^{\frac{-j2k\pi}{2M}}\sum_{n=0}^{M-1}u(i-n-1)e^{\frac{jk(2n+1)\pi}{2M}} - \frac{1}{2}e^{\frac{j2k\pi}{2M}}\sum_{n=0}^{M-1}u(i-n-1)e^{\frac{-jk(2n+1)\pi}{2M}}\end{aligned}$$

which gives

$$\begin{aligned}A &= \frac{2}{\alpha(k)}\cos\left(\frac{k\pi}{M}\right)\bar{u}_{i-1}(k) - \frac{1}{2}\sum_{n=0}^{M-1}u(i-n-1)\left[e^{\frac{jk(2n-1)\pi}{2M}} + e^{\frac{-jk(2n-1)\pi}{2M}}\right]\\ &= \frac{2}{\alpha(k)}\cos\left(\frac{k\pi}{M}\right)\bar{u}_{i-1}(k) - \frac{1}{2}\sum_{m=-1}^{M-2}u(i-m-2)\left[e^{\frac{jk(2m+1)\pi}{2M}} + e^{\frac{-jk(2m+1)\pi}{2M}}\right]\\ &= \frac{2}{\alpha(k)}\cos\left(\frac{k\pi}{M}\right)\bar{u}_{i-1}(k) - \sum_{m=-1}^{M-2}u(i-m-2)\cos\left(\frac{k(2m+1)\pi}{2M}\right)\\ &= \frac{2}{\alpha(k)}\cos\left(\frac{k\pi}{M}\right)\bar{u}_{i-1}(k)\\ &\quad - \left[\frac{1}{\alpha(k)}\bar{u}_{i-2}(k) + u(i-1)\cos\left(\frac{k\pi}{2M}\right) - u(i-M-1)\cos\left(\frac{k(2M-1)\pi}{2M}\right)\right]\end{aligned}$$

In summary, we arrive at the relation

$$\bar{u}_i(k) = 2\cos\left(\frac{k\pi}{M}\right)\bar{u}_{i-1}(k) - \bar{u}_{i-2}(k) + \phi(k)$$

where

$$\begin{aligned}\phi(k) &\triangleq \alpha(k)\left[a(k) - b(k)\right]\\ a(k) &\triangleq [u(i) - u(i-1)]\cos\left(\frac{k\pi}{2M}\right)\\ b(k) &\triangleq (-1)^k\left[u(i-M) - u(i-M-1)\right]\cos\left(\frac{k\pi}{2M}\right)\end{aligned}$$

and where we used the fact that

$$\cos\left(\frac{k(2M+1)\pi}{2M}\right) = \cos\left(\frac{k(2M-1)\pi}{2M}\right) = (-1)^k\cos\left(\frac{k\pi}{2M}\right)$$

Collecting this result for $k = 0, 1, \ldots, M-1$ we arrive at the desired relation (26.21).

CHAPTER 27

Efficient Block Convolution

The transform-domain adaptive filters of the previous chapter were motivated by the desire to improve the convergence performance of LMS by exploiting the de-correlation properties of unitary transforms such as the DFT and the DCT. It turns out that these same transforms, and other similar ones, are also useful in reducing the computational cost per iteration of LMS below the $O(M)$ figure. This cost reduction can be achieved by processing the data on a *block-by-block* basis rather than on a sample-by-sample basis.

27.1 MOTIVATION

As motivation, consider the setting of Fig. 27.1, which shows an FIR channel of length M; assumed long. The channel is excited by a zero-mean random sequence $\{\boldsymbol{u}(i)\}$ and its output is another zero-mean random sequence $\{\boldsymbol{y}(i)\}$. At any particular time instant i, the state of the channel is captured by the regression vector

$$\boldsymbol{u}_i = [\boldsymbol{u}(i)\ \boldsymbol{u}(i-1)\ \boldsymbol{u}(i-2)\ \ldots\ \boldsymbol{u}(i-M+1)]$$

and its output is measured in the presence of noise,

$$\boldsymbol{d}(i) = \boldsymbol{u}_i g + \boldsymbol{v}(i) \tag{27.1}$$

where the column vector g represents the channel impulse response, and $\boldsymbol{v}(i)$ is a zero-mean noise sequence uncorrelated with $\boldsymbol{u}_i$. Let $G(z)$ denote the transfer function associated with g, i.e.,

$$G(z) \stackrel{\Delta}{=} g(0) + g(1)z^{-1} + g(2)z^{-2} + \ldots + g(M-1)z^{-M+1} = \sum_{k=0}^{M-1} g(k)z^{-k} \tag{27.2}$$

where the $\{g(k)\}$ are the individual samples of g. Let also $\{u_i, d(i), y(i)\}$ denote observed values for the random variables $\{\boldsymbol{u}_i, \boldsymbol{d}(i), \boldsymbol{y}(i)\}$.

An LMS adaptive implementation for estimating g is depicted in Fig. 27.2, with adaptation equations given by

$$\hat{d}(i) = u_i w_{i-1}, \quad e(i) = d(i) - \hat{d}(i), \quad w_i = w_{i-1} + \mu u_i^* e(i) \tag{27.3}$$

This implementation requires $O(M)$ operations per iteration and when M is large, the cost can be prohibitive. For example, in acoustic echo cancellation, a few thousand taps may be needed to adequately model the echo path. In such situations, we must seek more efficient adaptive implementations.

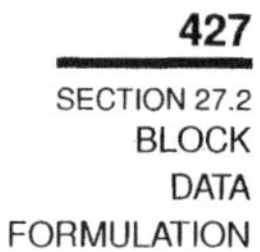

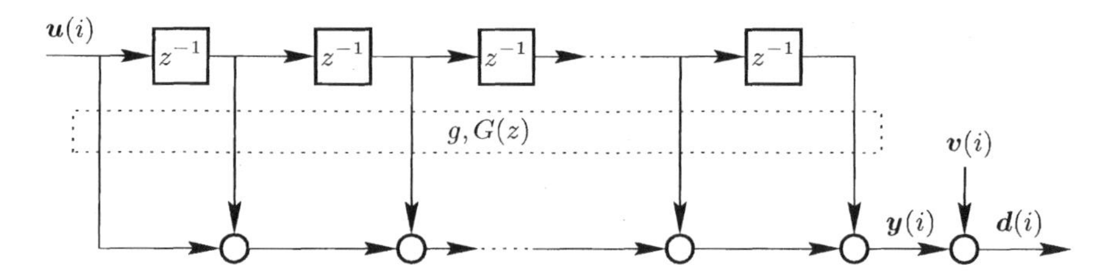

FIGURE 27.1 Noisy measurements of the output of a long FIR channel with an unknown impulse response vector g and transfer function $G(z)$.

As we shall see, block adaptive filters are well suited for such scenarios. While these filters are essentially equivalent to the LMS procedure (27.3), they nevertheless evaluate the error sequence $\{e(i)\}$ and the estimates $\{\hat{d}(i)\}$ in a more efficient manner. They do so by working with transformed regressors *and* by processing the data on a block-by-block basis. In this way, they end up reducing the computational cost by a factor $\alpha > 1$ compared with a plain LMS implementation. Besides computational efficiency, block adaptive filters also exhibit better convergence performance than LMS. This is because the eigenvalue spread of the covariance matrix of the transformed regression data is usually reduced relative to that of the original regression data as a result of a band partitioning property; this property will be explained in greater detail in Sec. 28.2.

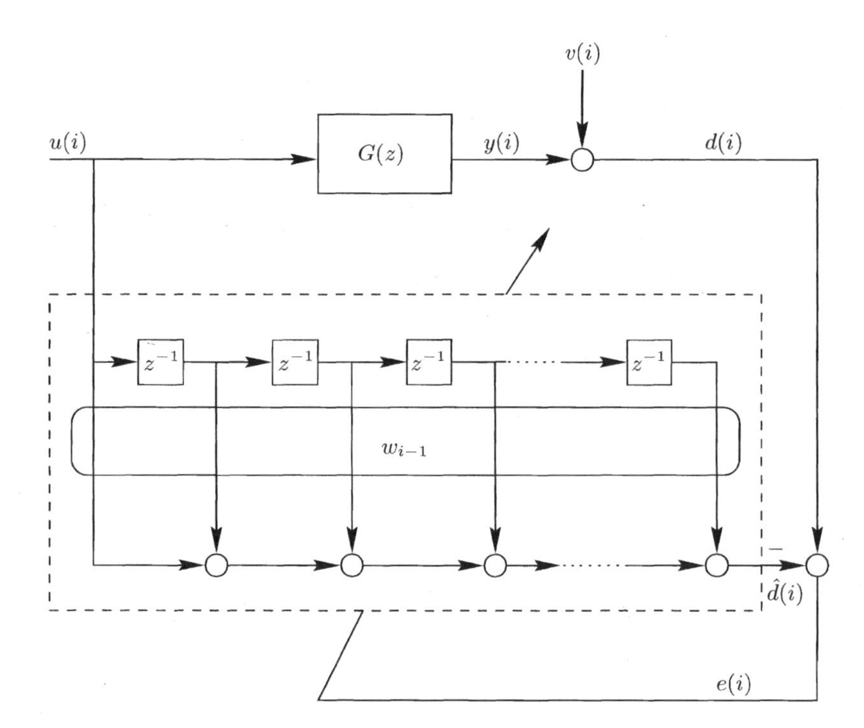

FIGURE 27.2 A structure for adaptive channel estimation.

27.2 BLOCK DATA FORMULATION

The first step toward developing efficient block adaptive filters is to explain how a long FIR filter, such as the one shown in Fig. 27.1 and *independent* of the adaptive context, can be implemented in an equivalent form that operates on blocks of data rather than on one sample at a time. We find it convenient to motivate the block implementations by working with the $z-$transform notation, while traditional derivations of block adaptive filters tend to be carried out in the time-domain. The $z-$domain arguments allows one to exploit the block structure to great extent and to motivate families of block adaptive filters that use other orthogonal transformations — see, e.g., Apps. 10.D and 10.E of Sayed (2003). The $z-$domain arguments also bring forth connections between block adaptive filters and another class of filters known as subband adaptive filters (see Sec. 28.2).

Consider a long impulse response sequence g and its transfer function $G(z)$, as in (27.2). We refer to $G(z)$ as the *fullband* filter. Let also $\{Y(z), U(z)\}$ denote the $z-$transforms of its causal input and output sequences $\{y(i), u(i)\}$,

$$Y(z) \stackrel{\Delta}{=} \sum_{i=0}^{\infty} y(i)z^{-i} \qquad\qquad U(z) \stackrel{\Delta}{=} \sum_{i=0}^{\infty} u(i)z^{-i}$$

Due to causality, these transforms are assumed to exist outside some circular domains, say, $|z| > r_y$ and $|z| > r_u$, respectively, for some positive scalars $\{r_y, r_u\}$. Then the input/output filter relation $y(i) = u_i g$ translates into

$$\boxed{Y(z) = G(z)U(z)} \tag{27.4}$$

which is depicted in Fig. 27.3. In this implementation, scalar entries $u(i)$ are fed into the channel $G(z)$ and scalar outputs $y(i)$ are obtained as a result.

Now we shall derive an alternative implementation that processes several input samples simultaneously and generates the corresponding output samples also simultaneously. In this implementation, data will be processed in a block manner, say, in blocks of size B each. To see how this can be done, we start by defining column vectors of length B as follows:

$$u_{B,n} \stackrel{\Delta}{=} \begin{bmatrix} u(nB) \\ u(nB-1) \\ \vdots \\ u((n-1)B+1) \end{bmatrix}, \qquad y_{B,n} \stackrel{\Delta}{=} \begin{bmatrix} y(nB) \\ y(nB-1) \\ \vdots \\ y((n-1)B+1) \end{bmatrix} \tag{27.5}$$

where the integer $n = 0, 1, \ldots$ is used as a block *index*. For example, assuming the block size is $B = 3$, the block vectors $\{u_{B,n}, y_{B,n}\}$ will result from partitioning the streams of

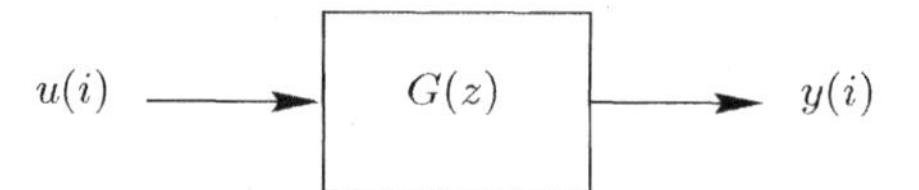

FIGURE 27.3 Fullband processing whereby signals are processed on a sample-by-sample basis.

data $\{u(i), y(i)\}$ into successive blocks of size B each:

$$\underbrace{0 \quad 0 \quad u(0)}_{u_{3,0}} \quad \underbrace{u(1) \quad u(2) \quad u(3)}_{u_{3,1}} \quad \underbrace{u(4) \quad u(5) \quad u(6)}_{u_{3,2}} \quad \ldots$$

$$\underbrace{0 \quad 0 \quad y(0)}_{y_{3,0}} \quad \underbrace{y(1) \quad y(2) \quad y(3)}_{y_{3,1}} \quad \underbrace{y(4) \quad y(5) \quad y(6)}_{y_{3,2}} \quad \ldots$$

We shall assume, without loss of generality, that the filter length M and the block size B are such that the ratio M/B is an integer. Usually, both M and B are powers of 2, e.g., $M = 1024$ and $B = 32$ or some other values. Let $\{\mathcal{U}_B(z), \mathcal{Y}_B(z)\}$ denote the $z-$transforms of the so-defined block causal sequences $\{u_{B,n}, y_{B,n}\}$, i.e.,

$$\mathcal{Y}_B(z) \triangleq \sum_{n=0}^{\infty} y_{B,n} z^{-n} \qquad\qquad \mathcal{U}_B(z) \triangleq \sum_{n=0}^{\infty} u_{B,n} z^{-n}$$

Note that we are using calligraphic letters to refer to *vector* or *matrix* functions of z. The transforms $\{\mathcal{U}_B(z), \mathcal{Y}_B(z)\}$ are $B \times 1$ vector functions of z. Just like (27.4), there is also a relation between $\{\mathcal{Y}_B(z), \mathcal{U}_B(z)\}$. Indeed, let

$$\{P_k(z), \;\; k = 0, 1, \ldots, B-1\}$$

denote the so-called polyphase components of the fullband filter $G(z)$: these are M/B-long FIR filters defined as follows:

$$\left\{\begin{array}{lcl} P_0(z) & = & g(0) + g(B)z^{-1} + g(2B)z^{-2} + \cdots \\ P_1(z) & = & g(1) + g(B+1)z^{-1} + g(2B+1)z^{-2} + \cdots \\ P_2(z) & = & g(2) + g(B+2)z^{-1} + g(2B+2)z^{-2} + \cdots \\ & \vdots & \\ P_{B-1}(z) & = & g(B-1) + g(2B-1)z^{-1} + g(3B-1)z^{-2} + \cdots \end{array}\right. \tag{27.6}$$

That is, the first B coefficients of $G(z)$ are the leading coefficients of the $\{P_k(z)\}$; the next B coefficients of $G(z)$ are the second coefficients of the $\{P_k(z)\}$, and so on. For example, assume $B = 3$ and $M = 12$. Then $G(z)$ will have 3 polyphase components that are given by

$$\left\{\begin{array}{lcl} P_0(z) & = & g(0) + g(3)z^{-1} + g(6)z^{-2} + g(9)z^{-3} \\ P_1(z) & = & g(1) + g(4)z^{-1} + g(7)z^{-2} + g(10)z^{-3} \\ P_2(z) & = & g(2) + g(5)z^{-1} + g(8)z^{-2} + g(11)z^{-3} \end{array}\right.$$

In this way, as indicated in the diagram below, every third coefficient of $G(z)$ is copied into the relevant polyphase component:

$g(0)$	$g(1)$	$g(2)$	$g(3)$	$g(4)$	$g(5)$	$g(6)$	$g(7)$	$g(8)$	$\ldots$
P_0	P_1	P_2	P_0	P_1	P_2	P_0	P_1	P_2	

To proceed, we collect the $\{P_k(z)\}$ into a $B \times B$ matrix $\mathcal{G}(z)$ as follows, e.g., for $B = 3$,

$$\mathcal{G}(z) = \left[\begin{array}{ccc} P_0(z) & P_1(z) & P_2(z) \\ z^{-1}P_2(z) & P_0(z) & P_1(z) \\ z^{-1}P_1(z) & z^{-1}P_2(z) & P_0(z) \end{array}\right] \tag{27.7}$$

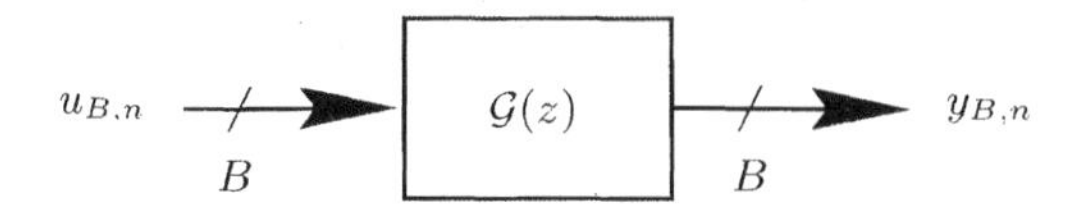

FIGURE 27.4 Block processing whereby signals are processed on a block-by-block basis.

The first row of $\mathcal{G}(z)$ contains all the polyphase components $\{P_k(z)\}$. Moreover, $\mathcal{G}(z)$ has a pseudo-circulant structure. A pseudo-circulant matrix function is essentially a circulant matrix function with the exception that all entries below the main diagonal are further multiplied by z^{-1}. Recall that a circulant matrix is a Toeplitz matrix (i.e., one with identical entries along the diagonals) with the additional property that its first row is circularly shifted to the right, one shift at a time, in order to form the other rows.

With the functions $\{\mathcal{Y}_B(z), \mathcal{U}_B(z), \mathcal{G}(z)\}$ defined in this manner, some straightforward algebra will show that expression (27.4) leads to the following block relation (recall Prob. II.33):

$$\boxed{\mathcal{Y}_B(z) = \mathcal{G}(z)\mathcal{U}_B(z)} \tag{27.8}$$

The result is depicted in Fig. 27.4. In this implementation, block entries $u_{B,n}$ are fed into the matrix filter $\mathcal{G}(z)$ and block outputs $y_{B,n}$ are obtained as a result. Comparing the implementations of Figs. 27.3 and 27.4, we see that the former relies on the transfer function $G(z)$ while the latter relies on the matrix transfer function $\mathcal{G}(z)$.

Although we have arrived at a block implementation scheme for $G(z)$, this solution is still inefficient, and more needs to be done in order to transform it into a truly efficient block processing scheme. Specifically, note that the polyphase components $P_k(z)$ in (27.7) are polynomials in z^{-1}, and each one of them has in general degree $M/B - 1$. It follows that $\mathcal{G}(z)$ is a matrix polynomial function with highest degree equal to M/B. This means that we can express $\mathcal{G}(z)$ in the form

$$\mathcal{G}(z) = G_0 + G_1 z^{-1} + G_2 z^{-2} + \ldots + G_{M/B} z^{-M/B} \tag{27.9}$$

for some $B \times B$ coefficients $\{G_k\}$. Then we could envision a block FIR structure for implementing $\mathcal{G}(z)$ in a manner similar to Fig. 27.1, with block coefficients $\{G_k\}$ and block input and output vectors $\{u_{B,n}, y_{B,n}\}$. However, such a structure would be inefficient for two reasons. First, the $M/B + 1$ block coefficients $\{G_k\}$ in (27.9) amount to a total of $B(M + B)$ scalar coefficients, which is essentially B times larger than the $(M + 1)$ coefficients $\{g(k)\}$ we started with for the original fullband filter implementation of Fig. 27.1. Second, the coefficients $\{G_k\}$ themselves are highly structured, as can be seen from (27.7), and we should be able to exploit this structure to our advantage.

Our purpose is therefore to show how to devise an *efficient* structure for implementing $\mathcal{G}(z)$ and, hence, for carrying out the original convolution of Fig. 27.1 in a block manner. Once this is done, we shall then use the resulting block structure as the launching pad for developing block *adaptive* implementations instead of the tapped-delay-line implementation of Fig. 27.2. These block adaptive solutions will be efficient in that they will require less computations than the LMS implementation (27.3) itself.

As a prelude to our arguments, we start by noting that the matrix $\mathcal{G}(z)$ in (27.7) can be factored as

$$\boxed{\mathcal{G}(z) = \mathcal{P}(z)\mathcal{Q}(z)} \tag{27.10}$$

where $\mathcal{P}(z)$ is a $B \times (2B-1)$ matrix function with Toeplitz structure, e.g., for $B = 3$,

$$\mathcal{P}(z) = \begin{bmatrix} P_0(z) & P_1(z) & P_2(z) & 0 & 0 \\ 0 & P_0(z) & P_1(z) & P_2(z) & 0 \\ 0 & 0 & P_0(z) & P_1(z) & P_2(z) \end{bmatrix}; \quad B \times (2B-1) \tag{27.11}$$

and $\mathcal{Q}(z)$ is a $(2B-1) \times B$ matrix with a leading identity block and a lower block with unit delays, say, for $B = 3$ again,

$$\mathcal{Q}(z) = \begin{bmatrix} 1 & 0 & 0 \\ 0 & 1 & 0 \\ 0 & 0 & 1 \\ \hline z^{-1} & 0 & 0 \\ 0 & z^{-1} & 0 \end{bmatrix}; \quad (2B-1) \times B \tag{27.12}$$

27.3 BLOCK CONVOLUTION

The initial step in our argument is to show how to use the discrete Fourier transform (DFT) in order to arrive at an efficient implementation of the block processing scheme of Fig. 27.4. For this purpose, we first remark that since it is usually desirable to work with sequences whose lengths are powers of 2 when dealing with the DFT, it is convenient to redefine the matrices $\mathcal{P}(z)$ and $\mathcal{Q}(z)$ in (27.11) and (27.12) as

$$\bar{\mathcal{P}}(z) = \begin{bmatrix} P_0(z) & P_1(z) & P_2(z) & 0 & 0 & 0 \\ 0 & P_0(z) & P_1(z) & P_2(z) & 0 & 0 \\ 0 & 0 & P_0(z) & P_1(z) & P_2(z) & 0 \end{bmatrix}; \quad (B \times 2B) \tag{27.13}$$

and

$$\bar{\mathcal{Q}}(z) = \begin{bmatrix} 1 & 0 & 0 \\ 0 & 1 & 0 \\ 0 & 0 & 1 \\ \hline z^{-1} & 0 & 0 \\ 0 & z^{-1} & 0 \\ 0 & 0 & z^{-1} \end{bmatrix}; \quad (2B \times B) \tag{27.14}$$

with an additional zero column added to $\mathcal{P}(z)$ and an additional row added to $\mathcal{Q}(z)$. Of course, the product $\bar{\mathcal{P}}(z)\bar{\mathcal{Q}}(z)$ remains equal to $\mathcal{G}(z)$. Now, however, $\bar{\mathcal{P}}(z)$ is $B \times 2B$ and $\bar{\mathcal{Q}}(z)$ is $2B \times B$, and the factor $2B$ will be a power of 2 whenever B is, which is generally the case.

Transfer Function Formulation

With the matrices $\{\bar{\mathcal{P}}(z), \bar{\mathcal{Q}}(z)\}$ so defined, we embed $\bar{\mathcal{P}}(z)$ into a $2B \times 2B$ circulant matrix $\mathcal{C}(z)$, say, for $B = 3$,

$$\mathcal{C}(z) = \begin{bmatrix} P_0(z) & P_1(z) & P_2(z) & 0 & 0 & 0 \\ 0 & P_0(z) & P_1(z) & P_2(z) & 0 & 0 \\ 0 & 0 & P_0(z) & P_1(z) & P_2(z) & 0 \\ \hline 0 & 0 & 0 & P_0(z) & P_1(z) & P_2(z) \\ P_2(z) & 0 & 0 & 0 & P_0(z) & P_1(z) \\ P_1(z) & P_2(z) & 0 & 0 & 0 & P_0(z) \end{bmatrix} \tag{27.15}$$

Note that $\bar{\mathcal{P}}(z)$ can be recovered from the top B rows of $\mathcal{C}(z)$ via

$$\boxed{\bar{\mathcal{P}}(z) = [\mathrm{I}_B \ \ 0_{B\times B}]\mathcal{C}(z)} \tag{27.16}$$

where I_B is the $B \times B$ identity matrix and $0_{B\times B}$ is the $B \times B$ null matrix. The main reason for embedding $\bar{\mathcal{P}}(z)$ into $\mathcal{C}(z)$ is the following result. Let

$$\boxed{[F]_{km} \stackrel{\Delta}{=} e^{-\frac{j2\pi mk}{2B}} \qquad k,m = 0,1,\ldots,2B-1} \tag{27.17}$$

denote the entries of the DFT matrix of size $2B \times 2B$. In contrast to (26.12), from now on we shall use F to refer to the standard DFT matrix without the scaling factor $1/\sqrt{M}$; the scaling was used earlier in (26.12) while studying transform-domain adaptive filters in order to enforce a unitary F. Now it is a well-known result that any circulant matrix, such as $\mathcal{C}(z)$ above, can be diagonalized by the DFT matrix — see Prob. VI.6. More specifically, it holds that

$$\boxed{\mathcal{C}(z) = F^*\mathcal{L}(z)F} \tag{27.18}$$

for some $2B \times 2B$ diagonal matrix function $\mathcal{L}(z)$ with entries

$$\mathcal{L}(z) = \text{diag}\{L_0(z), L_1(z), \ldots, L_{2B-1}(z)\} \tag{27.19}$$

Each $L_k(z)$ is an FIR transfer function with M/B coefficients — see Prob. VI.8. Using the fact that $F = F^{\mathsf{T}}$, it is easy to verify by transposing (27.18) that the entries of the first row of $\mathcal{C}(z)$ are related to the diagonal entries of $\mathcal{L}(z)$ via, e.g., for $B = 3$ — see Prob. VI.9:

$$\begin{bmatrix} P_0(z) \\ P_1(z) \\ P_2(z) \\ 0 \\ 0 \\ 0 \end{bmatrix} = F \begin{bmatrix} L_0(z) \\ L_1(z) \\ L_2(z) \\ L_3(z) \\ L_4(z) \\ L_5(z) \end{bmatrix} \tag{27.20}$$

This important relation tells us how to map the polyphase components $\{P_k(z)\}$ into the diagonal components $\{L_k(z)\}$ and vice-versa. It should be clear though that although any diagonal matrix $\mathcal{L}(z)$ in (27.18) will always result in a circulant matrix $\mathcal{C}(z)$, it does not hold that any such $\mathcal{L}(z)$ will result in a circulant matrix $\mathcal{C}(z)$ that has the special form (27.15). This is because the transformation (27.20) requires the $\{L_k(z)\}$ to be such that the last B entries of the transformed vector are zero.

Combining (27.16) and (27.18) we can express the factorization

$$\mathcal{G}(z) = \bar{\mathcal{P}}(z)\bar{\mathcal{Q}}(z)$$

as

$$\boxed{\mathcal{G}(z) = \underbrace{[\mathrm{I}_B \ \ 0_{B\times B}]\mathcal{C}(z)}_{\bar{\mathcal{P}}(z)}\bar{\mathcal{Q}}(z) = [\mathrm{I}_B \ \ 0_{B\times B}]\underbrace{F^*\mathcal{L}(z)F}_{\mathcal{C}(z)}\bar{\mathcal{Q}}(z)} \tag{27.21}$$

This result shows that the transfer matrix function $\mathcal{G}(z)$ in Fig. 27.4 can be obtained as the top B rows of the transfer matrix function $F^*\mathcal{L}(z)F\bar{\mathcal{Q}}(z)$. In other words, the mapping from $u_{B,n}$ to $y_{B,n}$ in Fig. 27.4 can be alternatively implemented as shown in Fig. 27.5

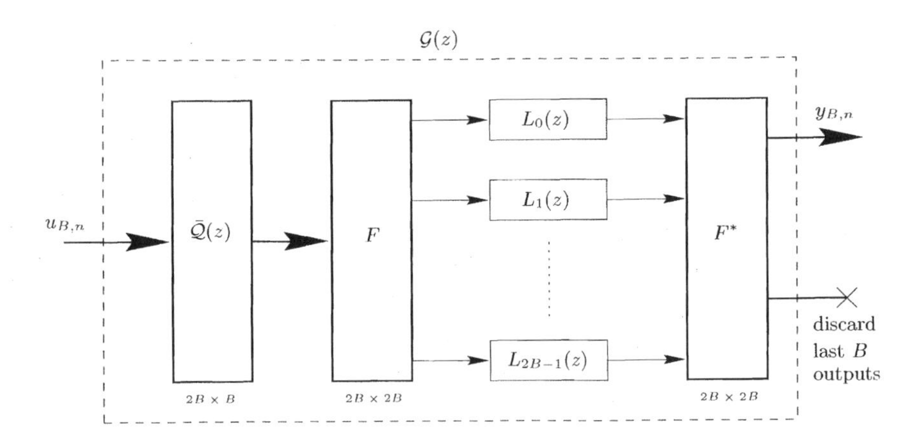

FIGURE 27.5 Equivalent implementation of the mapping in Fig. 27.4 for block convolution in terms of the DFT matrix.

in terms of $\{F, L_k(z), \bar{Q}(z)\}$. We shall refer to the $\{L_k(z)\}$ as the *subband filters*. The reason for the terminology will become clear later in Sec. 28.2 where we explain that block adaptive filters essentially partition the signal bandwidth into several frequency subbands and then process the signals within these subbands.

Time-Domain Formulation

In the time-domain, the implementation of Fig. 27.5 amounts to performing the following operations. First, applying the $2B \times B$ matrix function $\bar{Q}(z)$ to $u_{B,n}$ results in the output vector

$$\begin{bmatrix} u_{B,n} \\ u_{B,n-1} \end{bmatrix} \tag{27.22}$$

which, for example, for block size $B = 3$ and at $n = 2$, has the form

$$\begin{bmatrix} u(6) \\ u(5) \\ u(4) \\ \hline u(3) \\ u(2) \\ u(1) \end{bmatrix}$$

For this reason, we shall say that the effect of $\bar{Q}(z)$ is to perform a serial-to-parallel (S/P) conversion of the data. Another way to implement (or describe) this S/P conversion process is as follows. Let $\downarrow B$ denote a decimator (more specifically, a downsampler) of order B, denoted by

$$u(i) \longrightarrow \boxed{\downarrow B} \longrightarrow z(n)$$

where the output $z(n)$ is related to the input $u(i)$ as follows

$$z(n) = u(nB), \qquad n = 0, 1, 2, \ldots$$

In other words, the sequence at the output of the decimator is obtained by keeping the samples of the input sequence that occur at multiples of B, and ignoring the other samples.

In this way, the output of the decimator is a lower-rate representation of its input sequence; the original rate is scaled down by a factor of B. Observe that we are using the time index n for the lower-rate signal and the time index i for the higher-rate signal. Using the decimation-block representation, we can construct the block data vector (27.22) from the input sequence $\{u(i)\}$ as shown in Fig. 27.6; after every B input samples, a vector $\text{col}\{u_{B,n}, u_{B,n-1}\}$ is formed as in (27.22).

Once this is done, the vector (27.22) is processed by the DFT matrix F, resulting in a $2B \times 1$ transformed vector, whose entries we denote by

$$u'_{2B,n} \triangleq \begin{bmatrix} u'_0(n) \\ u'_1(n) \\ \vdots \\ u'_{2B-1}(n) \end{bmatrix} \triangleq F \begin{bmatrix} u_{B,n} \\ u_{B,n-1} \end{bmatrix}$$

i.e., we shall use primes from now on to denote transformed data. The entries of $\{u'_j(n)\}$ are then fed into the subband filters $\{L_k(z)\}$ and the resulting outputs are processed by F^*. The top B outputs of F^* are the desired output vector $y_{B,n}$.

In a manner similar to constructing $u_{B,n}$ from the $\{u(i)\}$ by means of decimators, we can conversely recover the output $\{y(i)\}$ from $y_{B,n}$ by means of *interpolators* or a parallel-to-serial (P/S) conversion. While the description that follows is not necessary for our future discussions, it is nevertheless a useful precursor to our treatment of subband adaptive filters later in Sec. 28.2.

Let $\uparrow B$ denote an interpolator (i.e., an upsampler) of order B, denoted by

$$z(n) \longrightarrow \boxed{\uparrow B} \longrightarrow y(i)$$

where the output $y(i)$ is related to the input $z(n)$ as follows:

$$y(i) = \begin{cases} z\left(\frac{i}{B}\right) & \text{if } i/B \text{ is an integer} \\ 0 & \text{otherwise} \end{cases}$$

In other words, the sequence at the output of the interpolator is obtained by adding $(B-1)$ zeros between the samples of the input sequence. In this way, the output of the interpolator is at the same original rate as $\{u(i)\}$.

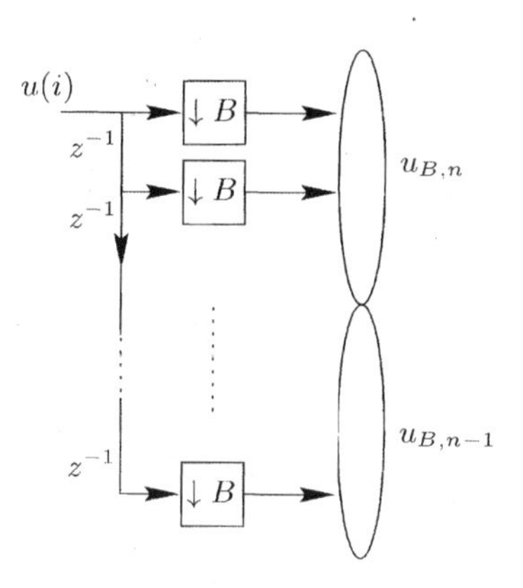

FIGURE 27.6 Formation of the block data vector $\text{col}\{u_{B,n}, u_{B,n-1}\}$ by means of a delay line with $2B$ decimators. This decimation-based structure is equivalent to processing by $\bar{Q}(z)$ in Fig. 27.5.

Using the interpolator-block representation, we can recover the $\{y(i)\}$ from $\{y_{B,n}\}$ as shown in Fig. 27.7; the structure in the figure performs parallel-to-serial data conversion. In order to understand how the structure functions, assume $B = 3$ so that for the first three initial block iterations:

$$y_{B,2} = \begin{bmatrix} y(6) \\ y(5) \\ y(4) \end{bmatrix}, \quad y_{B,1} = \begin{bmatrix} y(3) \\ y(2) \\ y(1) \end{bmatrix}, \quad y_{B,0} = \begin{bmatrix} y(0) \\ 0 \\ 0 \end{bmatrix}$$

Clearly, the block at $n = 2$ becomes available at time instant $i = 8$. The signals at the output of each of the interpolators are then

$$y(6)\ y(3)\ y(0) \longrightarrow \boxed{\uparrow 3} \longrightarrow y(6)\ 0\ 0\ y(3)\ 0\ 0\ y(0)$$

$$y(5)\ y(2)\ 0 \longrightarrow \boxed{\uparrow 3} \longrightarrow y(5)\ 0\ 0\ y(2)\ 0\ 0\ 0$$

$$y(4)\ y(1)\ 0 \longrightarrow \boxed{\uparrow 3} \longrightarrow y(4)\ 0\ 0\ y(1)\ 0\ 0\ 0$$

Therefore, when $n = 2$, which corresponds to time instant $i = 8$, the first signal appearing at the output of Fig. 27.7 would be $y(4) = y(i - B + 1)$, followed by $y(5)$ and $y(6)$. We thus see that a delay of $(B - 1)$ samples is introduced in the signal path.

With these remarks we conclude that the structure of Fig. 27.5 amounts in the time-domain to the operations shown in Fig. 27.8, with the outputs of the $\{L_k(z)\}$ filters denoted by

$$\{y'_k(n), k = 0, 1 \ldots, 2B - 1\}$$

We therefore find that the fullband FIR implementation of Fig. 27.1 can be equivalently implemented as a bank of $2B$ FIR filters, $\{L_k(z)\}$, of length M/B each and which operate at $1/B$ the original data rate. In addition, the input signals to these shorter filters are obtained after processing a block of $2B$ input data, shown in (27.22), by the DFT matrix. Observe that while the implementation of Fig. 27.1 relies on the M coefficients of $G(z)$, the one in Fig. 27.8 relies on a total of $2B \times (M/B) = 2M$ coefficients of the filters $\{L_k(z)\}$. So it may seem, at first sight, that any subsequent adaptive implementation that estimates the $2M$ coefficients of $\{L_k(z)\}$ would be costlier than the one that estimates the M coefficients of $G(z)$ (as in Fig. 27.2). However, as we shall see, this is not the case

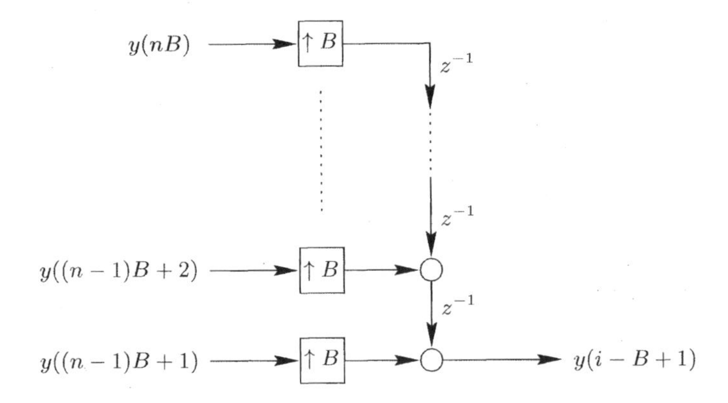

FIGURE 27.7 Reconstruction of the sequence $\{y(i)\}$ from the entries of $y_{B,n}$ by means of B interpolators. Observe that the output sequence is $y(i - B + 1)$ with a delay of $B - 1$ samples.

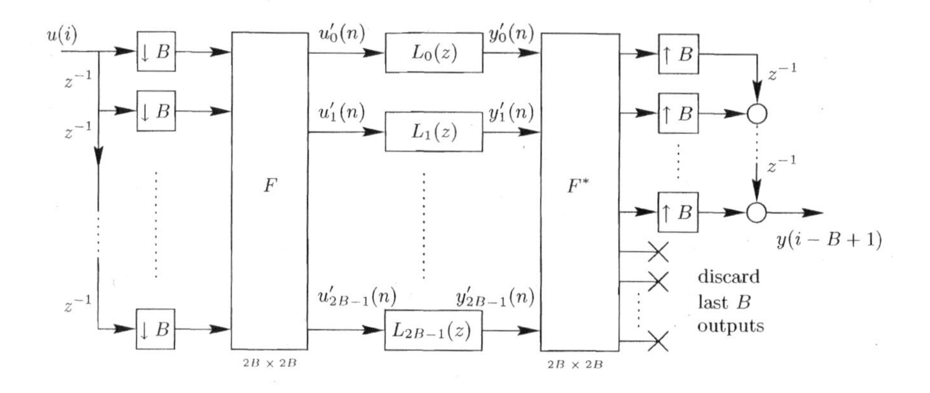

FIGURE 27.8 Time-domain equivalent of the block implementation of Fig. 27.5, using $2B$ decimators at the input and B interpolators at the output. A delay of $(B-1)$ samples is introduced in the signal path.

estimates the $2M$ coefficients of $\{L_k(z)\}$ would be costlier than the one that estimates the M coefficients of $G(z)$ (as in Fig. 27.2). However, as we shall see, this is not the case because the filters $L_k(z)$ operate at a sampling rate that is B times slower than $G(z)$ (and, therefore, their coefficients will only need to be adapted at this slower rate). This difference in the rate of operation leads to significant savings in computation.

In summary, so far, we have described two ways for implementing a transfer function $G(z)$. The first one is the fullband implementation of Fig. 27.3 whereby the output of $G(z)$ is evaluated one sample at a time by convolving the sequence $\{u(i)\}$ with the impulse response sequence of $G(z)$, i.e., via the inner product $y(i) = u_i g$, where u_i is the M–dimensional regressor

$$u_i = [u(i)\ u(i-1)\ \ldots\ u(i-M+1)]$$

The inconvenience of this procedure is that it requires convolution with the long filter $G(z)$.

The second implementation is the one shown in Fig. 27.8, which relies on a bank of $2B$ filters $\{L_k(z)\}$ applied to the transformed data $\{u'_k(n)\}$. This structure operates in a block manner, with the processing by the input DFT transformation performed after each block of B input data is collected, as indicated by Fig. 27.6. This implementation is more efficient than that of Fig. 27.1, as we are going to show in the next subsection. Nevertheless it suffers from a delay problem since it introduces a delay of $(B-1)$ samples in the signal path. In Prob. VI.14 we show how to modify the frequency-domain implementation of Fig. 27.8 in order to remove the delay problem. Basically, a direct convolution path of length B is added to Fig. 27.8.

Computational Complexity

Let us now compare the computational complexity of the implementations of Figs. 27.1 and 27.8. To do so, we shall only focus on the required number of multiplications per input sample. We shall also assume that a DFT of size K requires approximately $\frac{K}{2}\log_2(K)$ complex multiplications when evaluated by means of the fast Fourier transform (FFT).

In the direct convolution implementation of Fig. 27.1 we need to compute an inner product of order M, which translates into M complex multiplications per input sample. In the DFT-based implementation of Fig. 27.8, on the other hand, the following steps are necessary for *each block* of data of size B:

1. A transform of size $2B$ to compute the product $F \cdot \text{col}\{u_{B,n}, u_{B,n-1}\}$. This step requires $B\log_2(2B)$ complex multiplications.

2. Filtering by $2B$ filters $\{L_k(z)\}$ of order M/B each. Each filtering operation requires an inner product of size M/B and, therefore, M/B complex multiplications. In total, this step requires $2M$ complex multiplications.

3. A second transform of size $2B$ to generate the output signals. This step requires $B\log_2(2B)$ complex multiplications.

Steps 1–3 add up to $2M + 2B\log_2 2B$ complex multiplications for each *block* of input data of size B.

In addition, the filters $\{L_k(z)\}$ themselves need to be determined from $G(z)$. From (27.20) we see that the coefficients of $\{L_k(z)\}$ are found by applying F^* to a matrix with the polyphase components of $G(z)$. Specifically, using

$$F^*F = 2B \cdot \mathrm{I}_{2B}$$

we have that

$$\begin{bmatrix} L_0(z) \\ L_1(z) \\ \\ \vdots \\ L_{2B-1}(z) \end{bmatrix} = \frac{1}{2B} F^* \begin{bmatrix} P_0(z) \\ \vdots \\ P_{B-1}(z) \\ 0_{B\times 1} \end{bmatrix} \tag{27.23}$$

e.g., for $B = 3$ and $M = 12$,

$$\underbrace{\begin{bmatrix} l_{00} & l_{01} & l_{02} & l_{03} \\ l_{10} & l_{11} & l_{12} & l_{13} \\ l_{20} & l_{21} & l_{22} & l_{23} \\ l_{30} & l_{31} & l_{32} & l_{33} \\ l_{40} & l_{41} & l_{42} & l_{43} \\ l_{50} & l_{51} & l_{52} & l_{53} \end{bmatrix}}_{2B\times M/B} = \frac{1}{6}F^* \underbrace{\begin{bmatrix} g(0) & g(3) & g(6) & g(9) \\ g(1) & g(4) & g(7) & g(10) \\ g(2) & g(5) & g(8) & g(11) \\ 0 & 0 & 0 & 0 \\ 0 & 0 & 0 & 0 \\ 0 & 0 & 0 & 0 \end{bmatrix}}_{2B\times M/B} \tag{27.24}$$

This step therefore requires computing M/B DFT's of size $2B$ each, which amounts to a cost of $M\log_2(2B)$ complex multiplications. If we normalize by the size of the block, we get a cost on the order of $\frac{M}{B}\log_2(2B)$ complex multiplications per *input* sample. For a fixed $G(z)$, this cost corresponds to a one-time overhead since the computation of the $\{L_k(z)\}$ is done once and used thereafter for all input samples. However, as we shall see, for adaptive implementations, the calculation of the $\{L_k(z)\}$ needs to be repeated once for every block of data.

We therefore find that the cost associated with the implementation of Fig. 27.8 is approximately

$$\boxed{\tfrac{2M}{B} + \left(\tfrac{M}{B} + 2\right)\log_2(2B) \quad \text{complex multiplications per input } \textit{sample}}$$

The main conclusion is that while the direct convolution method of Fig. 27.1 requires $O(M)$ operations per sample, the frequency-domain implementation of Fig. 27.8 requires

$O(2M/B)$ operations per sample; the reduction in complexity is determined by the block size B. However, larger values for B result in longer delays in the signal path.

Algorithm 27.1 (Block convolution via DFT) Consider an input/output mapping as shown in Fig. 27.1 for some transfer function $G(z)$ with M taps. The direct convolution operation can be implemented more efficiently as shown in Figs. 27.5 and 27.8 Specifically, choose a block size B; usually M and B are powers of 2 and M/B is an integer.

Bank of filters. Determine the polyphase components of $G(z)$ of order M/B, as defined by (27.6), and use (27.23)–(27.24) to determine the $2B$ filters $\{L_k(z)\}$ of size M/B each.

Block filtering. Start with $u_{B,-1} = 0$ and repeat for $n \geq 0$:

1. Construct the block vectors:

$$u_{B,n} = \text{col}\{u(nB), \ldots, u((n-1)B+1)\}, \quad u_{2B,n} = \text{col}\{u_{B,n}, u_{B,n-1}\}$$

2. Perform the DFT transformation:

$$u'_{2B,n} = Fu_{2B,n} \triangleq \text{col}\{u'_k(n),\ k = 0, 1, \ldots, 2B-1\}$$

and filter the entries $\{u'_k(n)\}$ through the subband filters $\{L_k(z)\}$. Let $y'_{2B,n} = \text{col}\{y'_k(n), k = 0, 1, \ldots, 2B-1\}$ be a vector of size $2B \times 1$ that collects the outputs of these filters at iteration n.

3. Perform the DFT transformation:

$$\begin{bmatrix} y_{B,n} \\ \times \end{bmatrix} = F^* y'_{2B,n}$$

where $\times$ denotes B entries to be ignored, and $y_{B,n} = \text{col}\{y(nB), y(nB-1), \ldots, y((n-1)B+1)\}$.

Example 27.1 (Cost comparison)

In order to compare the computational costs of the direct convolution method of Fig. 27.3 and the frequency-domain implementation of Fig. 27.8, we plot in Fig. 27.9 two curves that show how the ratio

$$\frac{\text{frequency-domain cost}}{\text{direct convolution cost}} = \frac{\frac{2M}{B} + \left(\frac{M}{B} + 2\right)\log_2(2B)}{M} \tag{27.25}$$

varies as a function of M and B. In the top plot of the figure we fix the fullband length at $M = 1024$ taps and vary the block size B in powers of 2 between 2 and 1024. We see that the ratio drops below one (and, hence, the frequency-domain implementation becomes more efficient) for values of

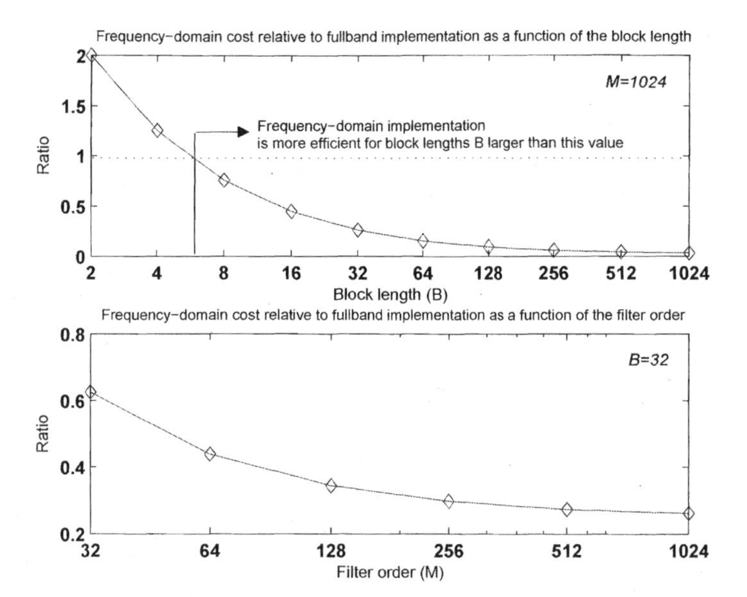

FIGURE 27.9 Plots of the ratio (27.25) for fixed M and varying B (*top curve*) and fixed B and varying M (*bottom curve*). The ratio compares the cost of the fullband and frequency-domain implementations of Figs. 27.3 and 27.8. When the ratio drops below one, the frequency-domain implementation becomes more efficient than the fullband implementation.

B larger than or equal to 4. In the bottom plot we fix the block length at $B = 32$ and vary the fullband filter length M in powers of 2 between 32 and 1024.

◇

Remark 27.1 (Overlap-save and overlap-add structures) The implementation of Fig. 27.8 and Alg. 27.1 is referred to as an overlap-save block-convolution procedure. The term *overlap-save* is used to indicate that successive input blocks $\{u_{2B,n}\}$ are overlapped prior to the DFT operation and only half of each block is saved. Observe, for example, that

$$u_{2B,n} = \begin{bmatrix} \boxed{u_{B,n}} \\ u_{B,n-1} \end{bmatrix}, \qquad u_{2B,n+1} = \begin{bmatrix} u_{B,n+1} \\ \boxed{u_{B,n}} \end{bmatrix}$$

which shows that $u_{2B,n}$ and $u_{2B,n+1}$ share a common sub-block. There is an alternative *overlap-add* procedure for block convolution. In this case, the successive input blocks $\{u_{2B,n}\}$ do not share a common sub-block. This structure, and the corresponding block adaptive filters, are discussed in App. 28.B.

CHAPTER 28

Block and Subband Adaptive Filters

Now that we know how to implement a long FIR filter (i.e., a long fullband convolution) efficiently by means of block processing, we can proceed to show how this result can be used to develop efficient block *adaptive* filters.

28.1 DFT BLOCK ADAPTIVE FILTERS

Thus, refer to the block implementation of Fig. 27.8. In a manner similar to (27.5), we define the vectors

$$d_{B,n} \triangleq \begin{bmatrix} d(nB) \\ d(nB-1) \\ \vdots \\ d((n-1)B+1) \end{bmatrix}, \quad v_{B,n} \triangleq \begin{bmatrix} v(nB) \\ v(nB-1) \\ \vdots \\ v((n-1)B+1) \end{bmatrix}$$

We further let l_k denote the $M/B \times 1$ weight vector associated with $L_k(z)$, and introduce the corresponding $1 \times \frac{M}{B}$ regression vector

$$u'_{k,n} = \begin{bmatrix} u'_k(n) & u'_k(n-1) & \ldots & u'_k(n-\frac{M}{B}+1) \end{bmatrix}$$

Then the output of each $L_k(z)$ is given by the inner product $u'_{k,n}l_k$. With this notation, we can express $d_{B,n}$ in terms of the subband filter coefficients $\{l_k\}$ as follows. Collect the regressors $\{u'_{k,n}\}$ into a block diagonal matrix $\mathcal{U}'_n$,

$$\mathcal{U}'_n \triangleq \begin{bmatrix} u'_{0,n} & & & \\ & u'_{1,n} & & \\ & & \ddots & \\ & & & u'_{2B-1,n} \end{bmatrix}; \quad (2B \times 2M) \tag{28.1}$$

and the corresponding filter coefficients into a $2M \times 1$ vector

$$l \triangleq \begin{bmatrix} l_0 \\ l_1 \\ \vdots \\ l_{2B-1} \end{bmatrix}; \quad (2M \times 1) \tag{28.2}$$

Then the vector $y'_{2B,n}$ at the output of the bank of filters $\{L_k(z)\}$ in Fig. 27.8 is equal to the product $U'_n l$, so that the output $y_{B,n}$ is given by

$$y_{B,n} = \begin{bmatrix} \mathrm{I}_B & 0_{B\times B} \end{bmatrix} F^*\mathcal{U}'_n l$$

and, consequently, since $d_{B,n} = y_{B,n} + v_{B,n}$, we have

$$\boxed{d_{B,n} = \begin{bmatrix} \mathrm{I}_B & 0_{B\times B} \end{bmatrix} F^*\mathcal{U}'_n l + v_{B,n}} \tag{28.3}$$

We can rewrite this relation in terms of *stochastic* data as

$$\boldsymbol{d}_{B,n} = \begin{bmatrix} \mathrm{I}_B & 0_{B\times B} \end{bmatrix} F^*\boldsymbol{\mathcal{U}}'_n l + \boldsymbol{v}_{B,n} \tag{28.4}$$

with the boldface symbols $\{\boldsymbol{d}_{B,n}, \boldsymbol{\mathcal{U}}'_n, \boldsymbol{v}_{B,n}\}$ denoting random variables whose realizations are $\{d_{B,n}, \mathcal{U}'_n, v_{B,n}\}$. We can now motivate an adaptive implementation for estimating the channel g (or, correspondingly, its subband filters $\{l_k\}$) by posing the problem of estimating l from $\boldsymbol{d}_{B,n}$ in (28.4) in the linear least-mean-squares sense, namely, by solving a problem of the form

$$\min_{\overline{w}} \mathsf{E} \left| \boldsymbol{d}_{B,n} - \begin{bmatrix} \mathrm{I}_B & 0_{B\times B} \end{bmatrix} F^*\boldsymbol{\mathcal{U}}'_n \overline{w} \right|^2 \tag{28.5}$$

where $\overline{w}$ is $2M \times 1$. In this problem, the matrix $\begin{bmatrix} \mathrm{I}_B & 0_{B\times B} \end{bmatrix} F^*\boldsymbol{\mathcal{U}}'_n$ plays the role of the regression data. An LMS procedure for estimating l would then operate on the realizations $\{d_{B,n}, \mathcal{U}'_n\}$ and take the form

$$\begin{aligned} \hat{d}_{B,n} &= \begin{bmatrix} \mathrm{I}_B & 0_{B\times B} \end{bmatrix} F^*\mathcal{U}'_n \overline{w}_{n-1} && (28.6) \\ e_{B,n} &= d_{B,n} - \hat{d}_{B,n} && (28.7) \\ \overline{w}_n &= \overline{w}_{n-1} + \mu \mathcal{U}'^*_n F \begin{bmatrix} \mathrm{I}_B \\ 0_{B\times B} \end{bmatrix} e_{B,n} && (28.8) \end{aligned}$$

Equations (28.6)–(28.8) are equivalent to

$$\overline{w}_n = \overline{w}_{n-1} + \mu \mathcal{U}'^*_n e'_{2B,n}, \quad \overline{w}_{-1} = 0 \tag{28.9}$$

where we are defining the $2B \times 1$ transformed error vector

$$\boxed{e'_{2B,n} \triangleq F \begin{bmatrix} \mathrm{I}_B \\ 0_{B\times B} \end{bmatrix} e_{B,n}} \quad (2B \times 1) \tag{28.10}$$

From the $2M \times 1$ vector $\overline{w}_n$ we can recover estimates for the coefficients of $\{L_k(z)\}$; they are simply stacked on top of each other in $\overline{w}_n$. Specifically, from (28.9), the update for the estimate of the $k-$th weight vector l_k is given by (in terms of the $k-$th entry of $e'_{2B,n}$ and the $k-$th regression vector $u'_{n,k}$):

$$l_{k,n} = l_{k,n-1} + \mu u'^*_{k,n} e'_k(n), \qquad l_{k,-1} = 0, \qquad k = 0, 1, \ldots, 2B-1 \tag{28.11}$$

Unconstrained Filter Implementation

Usually, an NLMS-type update with power normalization is employed instead of LMS so that (28.11) is replaced by

$$\boxed{l_{k,n} = l_{k,n-1} + \frac{\mu}{\lambda_k(n)} u'^*_{k,n} e'_k(n), \qquad l_{k,-1} = 0, \qquad k = 0, 1, \ldots, 2B-1} \tag{28.12}$$

with each $\lambda_k(n)$ evaluated via

$$\boxed{\lambda_k(n) = \beta\lambda_k(n-1) + (1-\beta)|u'_k(n)|^2, \qquad 0 \ll \beta < 1}$$

with initial condition $\lambda_k(-1) = \epsilon$ (a small positive number) and where $0 \ll \beta < 1$ (e.g., $\beta = 0.9$). The use of power normalization helps improve the convergence performance of the algorithm. This is because, as we are going to explain later in Sec. 28.2, the spectrum of the sequence $\{u'_k(n)\}$ at the input of each subband filter is approximately flat. Then the $\{\lambda_k(n)\}$ provide estimates for the input powers across the subband filters and normalization by them helps normalize the signal powers across the subbands.

In summary, we arrive at the statement below. Figure 28.1 shows a block diagram representation of the algorithm. This implementation introduces a delay of $B-1$ samples in the evaluation of the sequences $\{\hat{d}(i), e(i)\}$ due to the parallelization of $\{d(i), u(i)\}$, as explained in Fig. 27.8. Later, in App. 28.A we comment on how Fig. 28.1 can be modified to ameliorate this delay problem (see Figs. 28.9 and 28.10).

Algorithm 28.1 (Unconstrained DFT block filter) Consider the adaptive channel estimation scheme shown in Fig. 27.2, where w_{i-1} is $M \times 1$ and the regressor u_i is $1 \times M$. For relatively long channels (i.e., large M), a more efficient adaptive procedure for generating the estimates $\hat{d}(i)$, and the corresponding error sequence $\{e(i)\}$, is the following. Choose a block size B; usually M and B are powers of 2 and M/B is an integer. Select $0 \ll \beta < 1$, set $l_{k,-1} = 0$, $\lambda_k(-1) = \epsilon$ (a small positive number), and repeat for $n \geq 0$:

$$\begin{aligned}
u_{B,n} &= \text{col}\{u(nB), u(nB-1), \ldots, u((n-1)B+1)\} \\
u'_{2B,n} &= F\begin{bmatrix} u_{B,n} \\ u_{B,n-1} \end{bmatrix} \stackrel{\Delta}{=} \text{col}\{u'_k(n),\ k = 0,1,\ldots,2B-1\} \\
\lambda_k(n) &= \beta\lambda_k(n-1) + (1-\beta)|u'_k(n)|^2, \quad k = 0,1,\ldots,2B-1 \\
u'_{k,n} &= \begin{bmatrix} u'_k(n) & \ldots & u'_k(n-\frac{M}{B}+1) \end{bmatrix}, \quad k = 0,1,\ldots,2B-1 \\
y'_k(n) &= u'_{k,n} l_{k,n-1}, \quad k = 0,1,\ldots,2B-1 \\
\hat{d}_{B,n} &= \begin{bmatrix} \mathrm{I}_B & 0_{B\times B} \end{bmatrix} F^* \text{col}\left\{y'_0(n), \ldots, y'_{2B-1}(n)\right\} \\
e_{B,n} &= d_{B,n} - \hat{d}_{B,n} \\
e'_{2B,n} &= F\begin{bmatrix} \mathrm{I}_B \\ 0_{B\times B} \end{bmatrix} e_{B,n} \stackrel{\Delta}{=} \text{col}\{e'_k(n),\ k = 0,1,\ldots,2B-1\} \\
l_{k,n} &= l_{k,n-1} + \frac{\mu}{\lambda_k(n)} u'^*_{k,n} e'_k(n), \quad k = 0,1,\ldots,2B-1
\end{aligned}$$

At each block n, the entries of $\{\hat{d}_{B,n}, e_{B,n}\}$ correspond to

$$\begin{aligned}
\hat{d}_{B,n} &= \text{col}\{\hat{d}(nB), \hat{d}(nB-1), \ldots, \hat{d}((n-1)B+1)\} \\
e_{B,n} &= \text{col}\{e(nB), e(nB-1), \ldots, e((n-1)B+1)\}
\end{aligned}$$

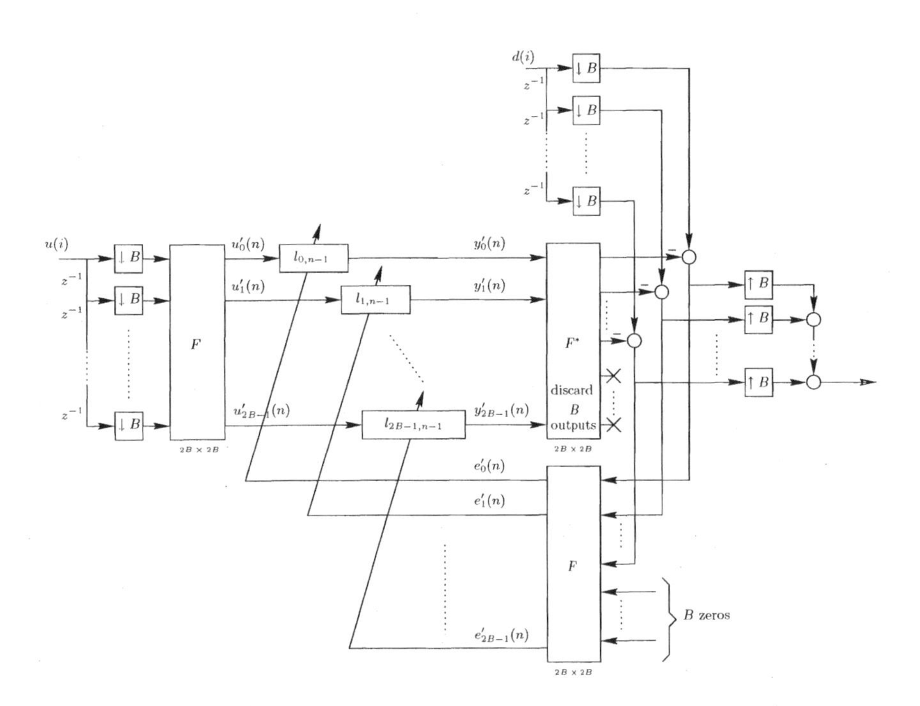

FIGURE 28.1 The unconstrained DFT-based block adaptive filter of Alg. 28.1. The input signal $u(i)$ is processed by a bank of $2B$ decimators, while the reference signal $d(i)$ is processed by a bank of B decimators. The error sequence $\{e(i)\}$ is generated, with a delay of $(B-1)$ samples, by using a bank of B interpolators.

Alternatively, a standard $\epsilon-$NLMS recursion could be employed instead of (28.12), say,

$$l_{k,n} = l_{k,n-1} + \frac{\mu}{\epsilon + \|u'_{k,n}\|^2} u'^{*}_{k,n} e'_k(n), \qquad l_{k,-1} = 0 \tag{28.13}$$

with the step-size in (28.12) M/B times smaller than the step-size in (28.13), as explained following (11.10) in our discussions on ϵ-NLMS with power normalization (recall that M/B is the length of each of the subband filters). The computer project at the end of the chapter compares the performance of (28.12) and (28.13).

It is useful to note that the sequences $\{\hat{d}(i), e(i)\}$ that are generated by Alg. 28.1 do not match exactly the sequences $\{\hat{d}(i), e(i)\}$ of the fullband adaptive implementation of Fig. 27.2. There are at least two reasons for this difference:

a) First, even if we use NLMS with power normalization instead of LMS in (27.3), the normalization that is performed by Alg. 28.1 is carried out in the frequency-domain and across the bank of subband filters, with a separate normalization for each filter.

b) A more significant difference is the fact that, in general, the successive filter estimates $\{l_{k,n}\}$ that are computed by Alg. 28.1 do not necessarily satisfy the constraint

(27.20); i.e., in the time-domain, the transformation

$$F\begin{bmatrix} l_{0,n}^{\mathsf{T}} \\ l_{1,n}^{\mathsf{T}} \\ \vdots \\ l_{2B-1,n}^{\mathsf{T}} \end{bmatrix} \tag{28.14}$$

need not result in a matrix whose last B rows are zero, as required by (27.20) — see also (27.24). In this way, the resulting estimates for the polyphase components $\{P_k(z)\}$ will not correspond to a circulant matrix function of the form (27.15). It is for this reason that the implementation of Alg. 28.1 is referred to as *unconstrained*; the qualification means that the successive weight estimates $\{l_{k,n}\}$ are not being constrained so as to guarantee the structural requirement (27.20).

Constrained Filter Implementations

One way to enforce the constraint (27.20) into recursion (28.12) is as follows. After each iteration n, we multiply (28.14) by $(\mathrm{I}\oplus 0_{B\times B})$ from the left, i.e., we perform the operation

$$\begin{bmatrix} \mathsf{X} \\ 0_{B\times B} \end{bmatrix} \longleftarrow \begin{bmatrix} \mathrm{I}_B & \\ & 0_{B\times B} \end{bmatrix} F\begin{bmatrix} l_{0,n}^{\mathsf{T}} \\ l_{1,n}^{\mathsf{T}} \\ \vdots \\ l_{2B-1,n}^{\mathsf{T}} \end{bmatrix}$$

which amounts to zeroing out the lower B components. Then the rows of X will be estimates for the coefficients of the polyphase components $\{P_k(z)\}$. More explicitly,

$$\begin{bmatrix} p_{0,n}^{\mathsf{T}} \\ p_{1,n}^{\mathsf{T}} \\ \vdots \\ p_{B-1,n}^{\mathsf{T}} \\ \hline 0_{B\times B} \end{bmatrix} = \begin{bmatrix} \mathrm{I}_B & \\ & 0_{B\times B} \end{bmatrix} F\begin{bmatrix} l_{0,n}^{\mathsf{T}} \\ l_{1,n}^{\mathsf{T}} \\ \vdots \\ l_{2B-1,n}^{\mathsf{T}} \end{bmatrix} \tag{28.15}$$

where $p_{k,n}$ denotes a column vector with the estimates for the coefficients of the $k-$th polyphase component $P_k(z)$. With the estimates $\{p_{k,n}\}$ so defined, we can return to the frequency domain by multiplying the result by F^*, as required by (27.24), in order to enforce the desired constraint on the subband filters. We shall denote the resulting subband filters by $\{l_{k,n}^c\}$, with a superscript c used to indicate that they are obtained from the $\{l_{k,n}\}$ by enforcing the constraint (27.20):

$$\begin{bmatrix} l_{0,n}^{c\mathsf{T}} \\ l_{1,n}^{c\mathsf{T}} \\ \vdots \\ l_{2B-1,n}^{c\mathsf{T}} \end{bmatrix} \triangleq \frac{1}{2B}F^*\begin{bmatrix} p_{0,n}^{\mathsf{T}} \\ p_{1,n}^{\mathsf{T}} \\ \vdots \\ p_{B-1,n}^{\mathsf{T}} \\ \hline 0_{B\times B} \end{bmatrix} \tag{28.16}$$

Thus, note that the estimates $\{l_{k,n}^c\}$ are now such that the product

$$F\begin{bmatrix} l_{0,n}^{c\mathsf{T}} \\ l_{1,n}^{c\mathsf{T}} \\ \vdots \\ l_{2B-1,n}^{c\mathsf{T}} \end{bmatrix}$$

satisfies the constraint (27.20), with the required zeros on the left-hand side.

In summary, we arrive at the statement of Alg. 28.2 below. The main difference in relation to the unconstrained implementation of Alg. 28.1 occurs in the evaluation of the constrained weight vectors $\{l^c_{k,n}\}$ and their use in computing the filter outputs $\{y'_k(n)\}$. All other recursions remain unchanged. Figure 28.2 shows a block diagram representation of the algorithm; the errors $\{e'_k(n)\}$ are used to adapt the unconstrained vectors $\{l_{k,n}\}$, which are subsequently corrected to $\{l^c_{k,n}\}$. In the figure, we only show the constrained weight vectors $\{l^c_{k,n}\}$. The filters of Algs. 28.1 and 28.2 are usually referred to as *multi-delay filters* (or MDF) with the qualification "multi-delay" used to mean that each of the $2B$ subband filters $\{L_k(z)\}$ has multiple coefficients. Of course, a special case would be to choose $B = M$ (i.e., to choose the block size equal to the fullband filter length), in which case each subband filter will consist of a single coefficient.

Algorithm 28.2 (Constrained DFT block filter) Consider the adaptive channel estimation scheme shown in Fig. 27.2, where w_{i-1} is $M \times 1$ and the regressor u_i is $1 \times M$. For relatively long channels (i.e., large M), a more efficient adaptive procedure for generating the estimates $\hat{d}(i)$, and the corresponding error sequence $\{e(i)\}$, is the following. Choose a block size B; usually M and B are powers of 2 and M/B is an integer. Select $0 \ll \beta < 1$, set $l_{k,-1} = l^c_{k,-1} = 0$, $\lambda_k(-1) = 0$, and repeat for $n \geq 0$:

$$\begin{aligned}
u_{B,n} &= \text{col}\{u(nB), u(nB-1), \ldots, u((n-1)B+1)\} \\
u'_{2B,n} &= F\begin{bmatrix} u_{B,n} \\ u_{B,n-1} \end{bmatrix} \triangleq \text{col}\{u'_k(n),\ k = 0,1,\ldots,2B-1\}
\end{aligned}$$

$$\begin{aligned}
\lambda_k(n) &= \beta\lambda_k(n-1) + (1-\beta)|u'_k(n)|^2, \quad k = 0,1,\ldots,2B-1 \\
u'_{k,n} &= \begin{bmatrix} u'_k(n) & \ldots & u'_k(n-\frac{M}{B}+1) \end{bmatrix}, \quad k = 0,1,\ldots,2B-1 \\
y'_k(n) &= u'_{k,n} l^c_{k,n-1}, \quad k = 0,1,\ldots,2B-1
\end{aligned}$$

$$\begin{aligned}
\hat{d}_{B,n} &= \begin{bmatrix} \mathrm{I}_B & 0_{B\times B} \end{bmatrix} F^* \text{col}\left\{y'_0(n),\ldots,y'_{2B-1}(n)\right\} \\
e_{B,n} &= d_{B,n} - \hat{d}_{B,n} \\
e'_{2B,n} &= F\begin{bmatrix} \mathrm{I}_B \\ 0_{B\times B} \end{bmatrix} e_{B,n} \triangleq \text{col}\{e'_k(n),\ k = 0,1,\ldots,2B-1\}
\end{aligned}$$

$$l_{k,n} = l_{k,n-1} + \frac{\mu}{\lambda_k(n)} u'^*_{k,n} e'_k(n), \quad k = 0,1,\ldots,2B-1$$

$$\begin{bmatrix} l^{c\mathsf{T}}_{0,n} \\ l^{c\mathsf{T}}_{1,n} \\ \vdots \\ l^{c\mathsf{T}}_{2B-1,n} \end{bmatrix} = \frac{1}{2B} F^* \begin{bmatrix} \mathrm{I}_B & \\ & 0_{B\times B} \end{bmatrix} F \begin{bmatrix} l^{\mathsf{T}}_{0,n} \\ l^{\mathsf{T}}_{1,n} \\ \vdots \\ l^{\mathsf{T}}_{2B-1,n} \end{bmatrix}$$

At each block n, the entries of $\{\hat{d}_{B,n}, e_{B,n}\}$ correspond to

$$\begin{aligned}
\hat{d}_{B,n} &= \text{col}\{\hat{d}(nB), \hat{d}(nB-1), \ldots, \hat{d}((n-1)B+1)\} \\
e_{B,n} &= \text{col}\{e(nB), e(nB-1), \ldots, e((n-1)B+1)\}
\end{aligned}$$

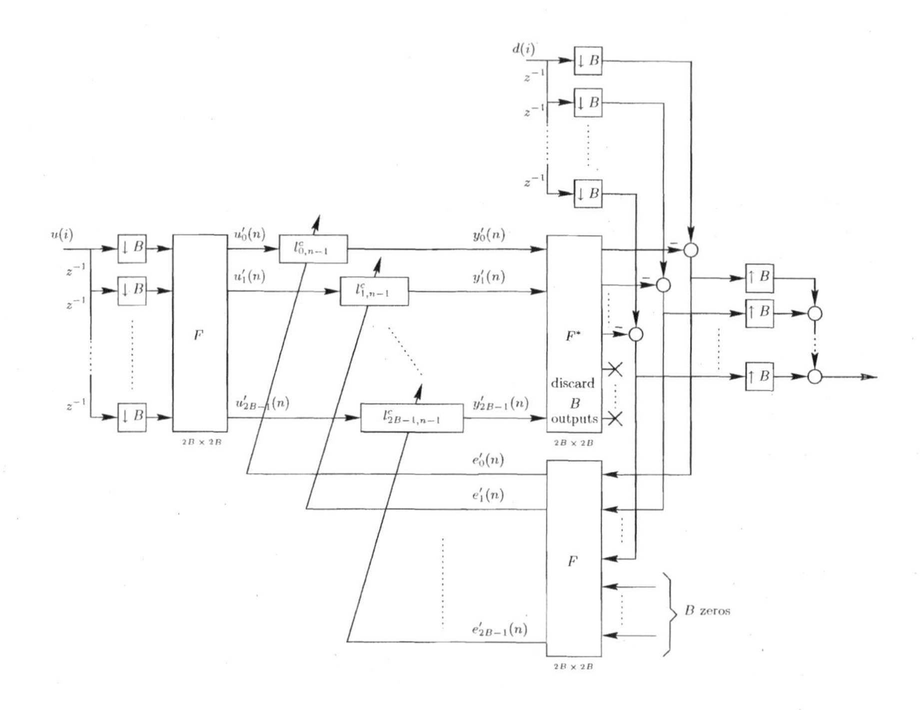

FIGURE 28.2 An implementation of the constrained DFT block adaptive filter of Alg. 28.2. The input signal $u(i)$ is processed by a bank of $2B$ decimators, while the reference signal $d(i)$ is processed by a bank of B decimators. The error sequence $\{e(i)\}$ is generated, with a delay of $(B-1)$ samples, by using a bank of B interpolators. Moreover, the errors $\{e'_k(n)\}$ are used to adapt the unconstrained vectors $\{l_{k,n}\}$, which are subsequently corrected to $\{l^c_{k,n}\}$. In the figure, we only show the constrained weight vectors $\{l^c_{k,n}\}$.

Computational Complexity

The overall computational requirements of the block adaptive filters of Algs. 28.1–28.2, and of their other constrained and overlap-add versions in Apps. 28.A and 28.B, are comparable (i.e., the algorithms require essentially the same order of computations). For this reason, it is enough to study the complexity of just one of the algorithms. We choose the multi-delay filter of Alg. 28.2 or Fig. 28.2. The overall complexity of the algorithm can be divided into four parts:

1. **Subband decomposition** of the input and error signals $\{u(i), e(i)\}$ in order to form the transformed vectors $\{u'_{2B,n}, e'_{2B,n}\}$. This step requires two DFTs of size $2B$ each for each block of B input samples. Assuming that the cost of a $K-$size DFT is $\frac{K}{2}\log_2 K$ complex operations, we find that the total cost is

$$\frac{1}{B}\cdot 2\cdot(B\log_2(2B)) \;=\; 2\log_2 2B \quad \text{operations per input sample}$$

2. **Updating of the subband filters**. This step requires that we update $2B$ filters of length M/B each by using NLMS with power normalization. The updates are performed once every B input samples. Now since the update of a K-long NLMS filter

requires approximately $2K$ complex operations, we find that the total cost is

$$\frac{1}{B} \cdot 2B \cdot \frac{2M}{B} = \frac{4M}{B} \quad \text{operations per input sample}$$

3. **Enforcement of the constraint (27.20)**. This requires M/B transforms of size $2B$ for each block of B input samples. The complexity of this part is similar to the first one, except that here we need to compute M/B transforms rather than only 2, leading to

$$\frac{1}{B} \cdot \frac{M}{B} \cdot (B \log_2 2B) = \frac{M}{B} \log_2(2B) \quad \text{operations per input sample}$$

   The computational burden of this part can be reduced if we apply the constraint less often than every B samples, say, every pB samples where p is an integer larger than one.

4. **Inverse transformation**. This step requires one DFT of size $2B$ to map the signals $\{y'_k(n)\}$ into the time-domain signals $\{y_k(n)\}$. The cost would be

$$B \log_2(2B) \quad \text{operations per input sample}$$

In summary, we find that the computational complexity per input sample of the multi-delay filter of Alg. 28.2, and similarly of the other block adaptive filters, is on the order of

$$\boxed{\tfrac{4M}{B} + \left(\tfrac{M}{B} + 3\right) \log_2(2B) \quad \text{operations per input sample}}$$

The main conclusion is that the cost is $O(M/B)$ operations per input sample.

Remark 28.1 (Other block adaptive filters) The DFT block adaptive filters described so far, and also in Apps. 28.A and 28.B, are derived by embedding the matrix $\bar{\mathcal{P}}(z)$ of (27.13) into the circulant matrix $\mathcal{C}(z)$ in (27.15). This latter matrix was then diagonalized by the DFT, as shown by (27.18). Now we could have embedded $\bar{\mathcal{P}}(z)$ [or $\mathcal{P}(z)$ from (27.11)] into other matrices that are not necessarily circulant, but which can still be diagonalized by some other transforms, say, by the trigonometric transforms. In this way, we would be able to derive other block adaptive structures. This point of view is pursued in Apps. 10.D and 10.E of Sayed (2003), where it is shown how to derive block adaptive filters that are based on the discrete-cosine and discrete Hartley transforms (DCT and DHT).

28.2 SUBBAND ADAPTIVE FILTERS

Besides transform-domain and block adaptive filters, there is another class of adaptive algorithms that achieves computational savings and improvement in performance over a conventional LMS implementation. This third class of algorithms, known as *subband adaptive filters*, has a close relation to DFT-based block adaptive filters and we shall exploit this connection to motivate subband adaptive filters.

Define the filters

$$H_k(z) \triangleq \sum_{n=0}^{2B-1} z^{-n} e^{\frac{j2\pi kn}{2B}}, \qquad k = 0, 1, \ldots, 2B-1$$

In particular, $H_0(z)$ is the moving-average (low-pass) filter

$$H_0(z) = 1 + z^{-1} + \ldots + z^{-(2B-1)}$$

with a rectangular window as its impulse response, while the remaining filters are related to $H_0(z)$ via

$$H_k(z) = H_0\left(ze^{-j2\pi k/2B}\right)$$

or, in terms of their frequency responses,

$$H_k(e^{j\omega}) = H_0\left(e^{j\left(\omega - \frac{\pi k}{B}\right)}\right), \qquad k = 0, 1, \ldots, 2B-1$$

The filter $H_0(z)$ is called the *prototype* filter since the other filters $\{H_k(z)\}$ are generated from it. Its frequency response is given by

$$H_0(e^{j\omega}) = \begin{cases} 2B, & \omega = 0 \\ e^{-j\omega\frac{2B-1}{2}} \cdot \dfrac{\sin(\omega B)}{\sin\frac{\omega}{2}}, & \text{otherwise} \end{cases}$$

and is shown in Fig. 28.3 when $B = 8$. Its first zero crossing occurs at $\omega_0 = \pi/B$, which is effectively half the width of its main lobe. We thus say that its approximate bandwidth is $2\pi/B$ radians/sample. The attenuation of the first side lobe is approximately 13 dB relative to the main lobe. The frequency responses of all other filters $\{H_k(z)\}$ are obtained from $H_0(e^{j\omega})$ by shifting the latter to the frequencies $\{\pi k/B\}$.

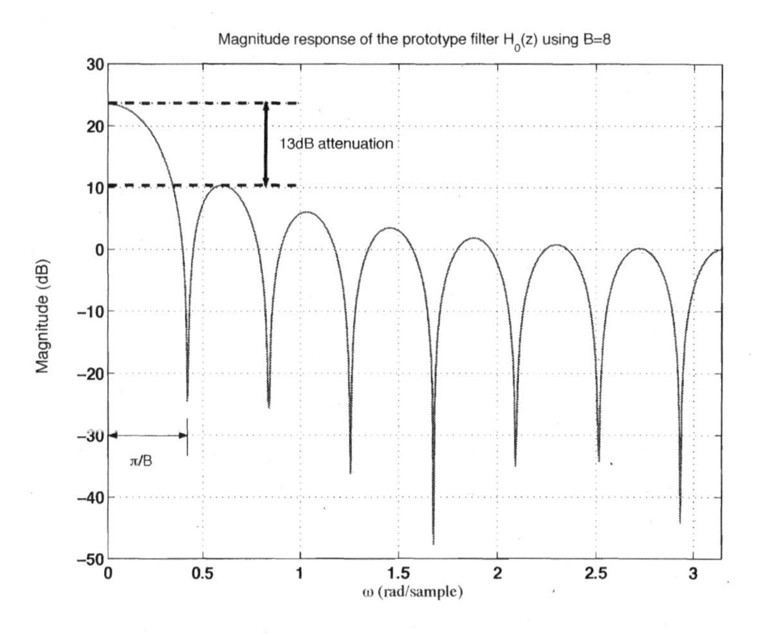

FIGURE 28.3 Magnitude frequency response of the prototype filter $H_0(z)$ for $B = 8$. The width of the main lobe is $2\pi/B = \pi/4$ rad/sample.

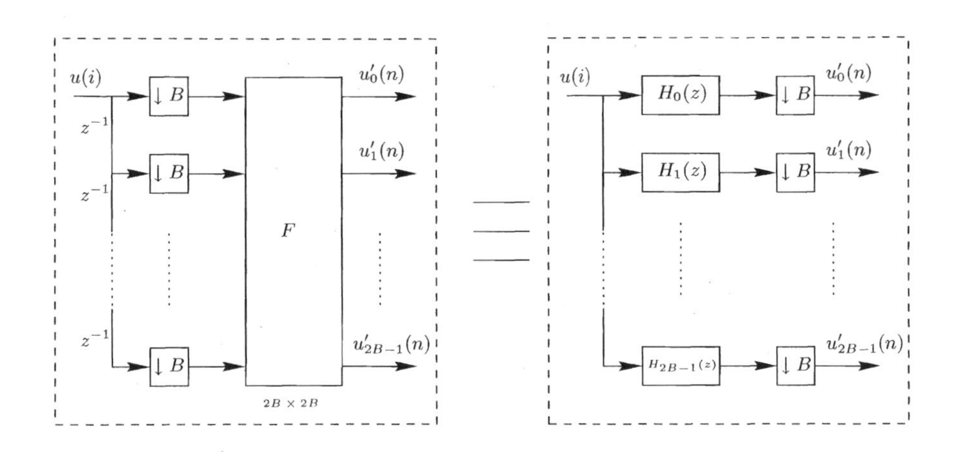

FIGURE 28.4 Two equivalent representations in terms of the DFT matrix and a set of DFT-modulated bandpass filters $\{H_k(z)\}$.

Now using the expression for the DFT matrix from (27.17), it is easy to see from the definition of the $\{H_k(z)\}$ that they are obtained as follows:

$$\begin{bmatrix} H_0(z) \\ H_1(z) \\ \vdots \\ H_{2B-1}(z) \end{bmatrix} = F \begin{bmatrix} 1 \\ z^{-1} \\ \vdots \\ z^{-(2B-1)} \end{bmatrix}$$

This result shows that the DFT transformations from $\{u(i)\}$ to the $\{u'_k(n)\}$ on the left-hand side of Fig. 28.4 can be redrawn equivalently as shown on the right-hand side of the same figure in terms of the DFT-modulated bandpass filters $\{H_k(z)\}$. In other words, the step on the left involving decimation followed by DFT can be interpreted as attempting to split the bandwidth of the original input signal $\{u(i)\}$ into a set of $2B$ partially exclusive and equally wide bands by filtering through the $\{H_k(z)\}$. It is for this reason that the bank of filters $\{H_k(z)\}$ is called the *uniform* DFT *analysis filter bank*. If we examine the unconstrained and constrained DFT-based adaptive structures of Figs. 28.1 and 28.2, we see that they both employ the DFT analysis filter bank at their inputs. Of course, since the bandwidths of the filters $\{H_k(z)\}$ overlap, and since the side lobe attenuation in each filter is not high enough relative to the main lobe, the bands of the resulting signals $\{u'_k(n)\}$ are not necessarily mutually exclusive or well separated.

Analysis Filter Bank

More generally, we may consider employing different choices for the prototype filter $H_0(z)$, and also even change the number of analysis filters used as well as the value of the decimation factor B. These choices would be guided by the desire to arrive at filter structures that result in better band partitioning.

So assume we select some prototype filter $H(z)$ and use it to define an analysis filter bank with K filters as follows:

$$H_k(z) = H\left(ze^{-j2\pi k/K}\right), \qquad k = 0, 1, \ldots, K-1 \tag{28.17}$$

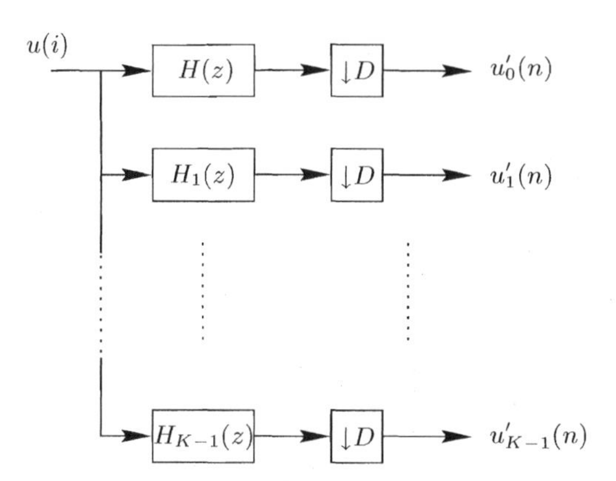

FIGURE 28.5 A uniform DFT-based analysis filter bank.

i.e., with frequency responses

$$H_k(e^{j\omega}) = H\left(e^{j(\omega-\omega_k)}\right), \qquad \omega_k = 2\pi k/K, \qquad k = 0, 1, \ldots, K-1$$

In this case, the interval $[0, 2\pi]$ is divided into K equally spaced subbands of width $2\pi/K$ each. Now since the signals at the outputs of the $\{H_k(z)\}$ have bandwidths that are smaller than the bandwidth of the fullband signal $\{u(i)\}$, the outputs of $\{H_k(z)\}$ can be decimated (i.e., down-sampled to a lower rate), say, by some factor D, prior to further processing. The resulting structure is shown in Fig. 28.5. It is still called a uniform DFT-based analysis filter bank; the difference now, compared with Fig. 28.4, is that the number of analysis filters, K, and the decimation factor, D, are not necessarily related as before. Moreover, the choice of $H(z)$ is also left to the designer. The analysis filter bank $\{H(z), H_1(z), \ldots, H_{K-1}(z)\}$ is usually chosen such that the bandwidth of the input signal $u(i)$ is divided uniformly into K bands of equal bandwidths. In this way, the bandwidth of the output of each of the analysis filters is K times smaller than the bandwidth of the original signal $u(i)$ and, therefore, we can decimate the outputs of the analysis filters by some factor $D \leq K$.

The special choice $D = K$ results in what is called a *maximally* or critically decimated analysis filter bank. This is because, in this case, the total number of samples after decimation is the same as the total number of input samples. Thus, observe that N input samples $\{u(i)\}$ lead to N samples at the output of each analysis filter and, therefore, to a total of KN samples at the output of these filters. After decimation by a factor $D = K$, we are left again with N samples. Choices of D such that $D < K$, on the other hand, lead to what are called *oversampled* or noncritically sampled analysis filter banks. In these cases, N input samples generate more than N total samples at the output of the decimators. If $D > K$, then aliasing occurs in the subbands.

Synthesis Filter Bank

The step of recovering the input signal $\{u(i)\}$ from its subband components $\{u'_k(n)\}$ in Fig. 28.5 is known as *synthesis*. To do so, the signals $\{u'_k(n)\}$ are first upsampled by appending $D-1$ zeros after each sample, i.e., the $\{u'_k(n)\}$ are transformed into $\{s'_k(i)\}$ as follows:

$$u'_k(n) \longrightarrow \boxed{\uparrow D} \longrightarrow s'_k(i)$$

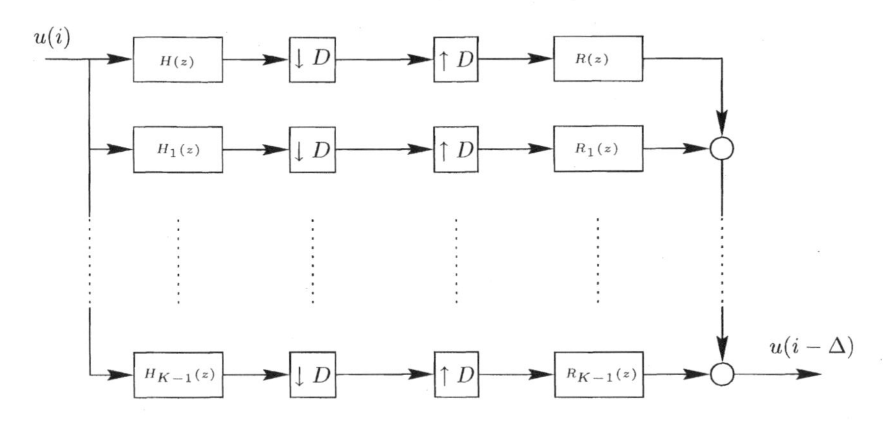

FIGURE 28.6 A cascade of analysis and synthesis filter banks.

where $s'_k(i)$ is related to $u'_k(n)$ via

$$s'_k(i) = \begin{cases} u'_k\left(\frac{i}{D}\right) & \text{if } i/D \text{ is an integer} \\ 0 & \text{otherwise} \end{cases}$$

Now since, for any interpolator, the spectrum of its output signal consists of D repetitions of the spectrum of its input signal (see Prob. VI.17), we need to filter $\{s'_k(i)\}$ by some lowpass filter in order to remove the repetitions in the spectrum.

Thus, select some lowpass prototype filter $R(z)$ and use it to define a synthesis filter bank with K filters as follows:

$$R_k(z) = R\big(ze^{-j2\pi k/K}\big), \qquad k = 0, 1, \ldots, K-1$$

i.e., with frequency responses

$$R_k(e^{j\omega}) = R\big(e^{j(\omega-\omega_k)}\big), \qquad \omega_k = 2\pi k/K, \qquad k = 0, 1, \ldots, K-1$$

Considerable investigations in the literature have gone into the design of the prototype filters $\{H(z), R(z)\}$, and consequently into the design of analysis and synthesis filter banks, in order to reduce the distortion and delay between the input and output nodes (such as the design of perfect reconstruction filter banks). Such studies are well-documented in textbooks on multi-rate discrete-time signal processing (see, e.g., Vaidyanathan (1993)). When the subband signals are modified (due to overlaps in their spectra), cross-filters may be needed in order to reduce signal distortion (see, e.g., Gilloire and Vetterli (1992)).

Structures for Subband Filtering

Figures. 28.7 and 28.8 show two basic configurations for adaptive subband filtering, commonly referred to as *open-loop* and *closed-loop* configurations.

In Fig. 28.7, the input sequence $\{u(i)\}$ and the reference sequence $\{d(i)\}$ are processed by the same analysis filter bank to produce subband signals $\{u'_k(n), d'_k(n)\}$. The subband signals are then used to train adaptive filters $\{l_{k,n-1}\}$ by using the subband error sequences

$$e'_k(n) = d'_k(n) - y'_k(n)$$

where the $\{y'_k(n)\}$ are the outputs of the subband adaptive filters. These outputs are in turn applied to a synthesis filter bank to generate the signal $\hat{d}(i-\Delta)$, for some delay Δ. The reason for the presence of Δ is because filter banks introduce delay in the signal path. One major drawback of this open-loop structure for adaptive filtering is that it usually results in higher mean-square error performance since the algorithm is attempting to minimize the variance of the subband errors $\{e'_k(n)\}$, and not the variance of the fullband errors $\{e(i) = d(i) - \hat{d}(i)\}$.

In Fig. 28.8, on the other hand, the error signal $e(i)$ is evaluated in fullband and then converted to subband by means of the analysis filter bank. In other words, the reference sequence $d(i)$ is now kept in fullband. Although this scheme tends to show better mean-square performance than the open-loop scheme due to the fullband error feedback mechanism, this improvement comes at the expense of a delay in the error feedback path, which degrades the convergence performance. In particular, observe that there is the delay introduced by the two filter banks between the outputs of the subband adaptive filters and the subband errors.

Another problem with both the open- and closed-loop subband structures is the lack of optimality in their construction. As we have seen throughout the discussions so far in the book, adaptive structures are usually derived as approximations to some optimality criterion, such as the mean-square error criterion. However, this is not generally the case with the structures of Figs. 28.7 and 28.8. This is because they are motivated mostly by the desire to partition the input signal bandwidth into smaller "separate" bands over which the

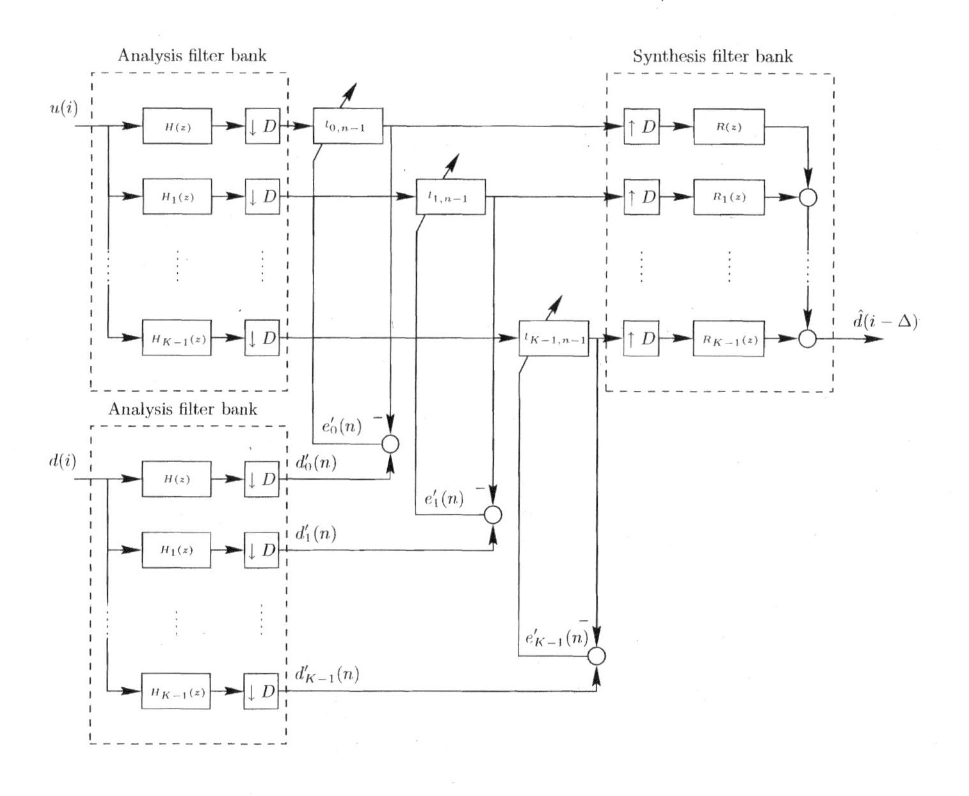

FIGURE 28.7 Open-loop structure for subband adaptive filtering.

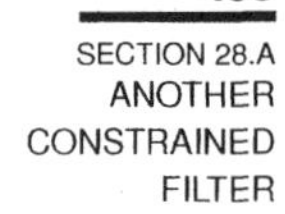

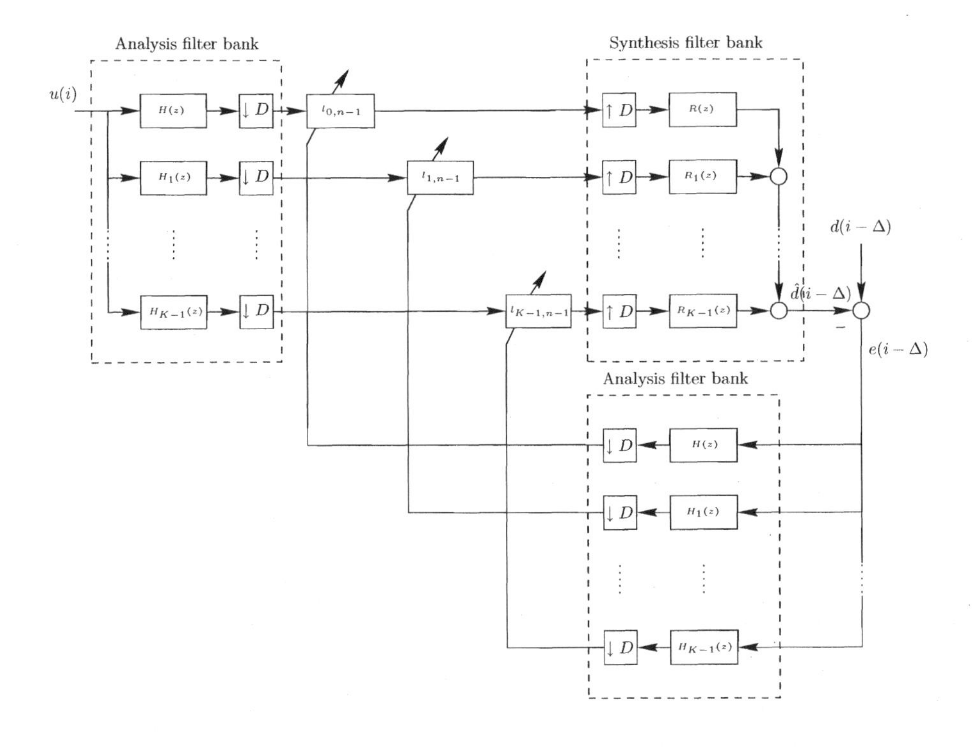

FIGURE 28.8 Closed-loop structure for subband adaptive filtering.

spectra are close to flat. There is no guarantee that these implementations will ultimately provide the error and estimate sequences $\{e(i), \hat{d}(i)\}$ that would result from an optimal or close-to-optimal mean-square-error formulation; contrast this situation, for example, with the DFT-based schemes of Algs. 28.2 and 28.3, which followed from the solution to (28.5). In addition, no constraints on the subband weight vectors are usually applied during adaptation of the open- and closed-loop structures.

28.A APPENDIX: ANOTHER CONSTRAINED FILTER

There is an alternative to the constrained filter of Alg. 28.2. Observe that the error vector $e_{B,n}$ in the algorithm is computed using the frequency-domain data $\{y'_k(n)\}$. This error vector can be evaluated in another manner by convolving the input sequence $\{u(i)\}$ directly with an estimate of the fullband filter $G(z)$ itself. More specifically, from (28.15) we have that

$$\begin{bmatrix} p_{0,n-1}^{\mathsf{T}} \\ p_{1,n-1}^{\mathsf{T}} \\ p_{2,n-1}^{\mathsf{T}} \\ \vdots \\ p_{B-1,n-1}^{\mathsf{T}} \end{bmatrix} = \begin{bmatrix} \mathrm{I}_B & 0_{B\times B} \end{bmatrix} F \begin{bmatrix} l_{0,n-1}^{\mathsf{T}} \\ l_{1,n-1}^{\mathsf{T}} \\ l_{2,n-1}^{\mathsf{T}} \\ \vdots \\ l_{2B-1,n-1}^{\mathsf{T}} \end{bmatrix} \tag{28.18}$$

This expression provides estimates for the polyphase components of $G(z)$ at block iteration $n-1$. Using the relation between $G(z)$ and its polyphase components $\{P_k(z)\}$ from (27.6), the above

$\{p_{k,n-1}\}$ can be used to estimate the impulse response sequence g itself. If we denote the estimate of g at block iteration $n-1$ by w_{n-1}, then

$$w_{n-1} = \text{vec}\left(\begin{bmatrix} p_{0,n-1}^{\mathsf{T}} \\ p_{1,n-1}^{\mathsf{T}} \\ p_{2,n-1}^{\mathsf{T}} \\ \vdots \\ p_{B-1,n-1}^{\mathsf{T}} \end{bmatrix}\right) \tag{28.19}$$

In other words, we simply stack the columns of (28.18) on top of each other. Then, at block iteration n, we can generate the error signals as follows:

$$\begin{aligned} \hat{d}(nB-j) &= u_{nB-j}w_{n-1} \quad \text{(convolution in fullband)} \\ e(nB-j) &= d(nB-j) - \hat{d}(nB-j), \quad j = 0, 1, \ldots, B-1 \end{aligned} \tag{28.20}$$

where

$$u_{nB-j} = \begin{bmatrix} u(nB-j) & u(nB-j-1) & \ldots & u(nB-j-M+1) \end{bmatrix}$$

is the regressor of the fullband filter at time $nB-j$. The sequence $\{e(i)\}$ so generated is then decimated by a factor of B to construct $e_{B,n}$. The resulting filter structure is shown in Fig. 28.9. Compared with Fig. 28.2, we see that only two DFT transformations are used, in addition to a fullband convolution. The constraint (27.20) is enforced in the process of mapping the subband filters $\{l_{k,n-1}\}$ into the fullband filter w_{n-1} according to (28.19). We therefore find that the constrained filter of Alg. 28.2 can also be implemented as Alg. 28.3.

We still need to discuss how to evaluate the inner product $\hat{d}(nB-j) = u_{nB-j}w_{n-1}$ in an efficient manner. Otherwise, since the vectors involved are $M-$dimensional, the resulting filter implementation ends up being inefficient and we defeat our original objective of reduced computational cost. Now we already know from Sec. 27.2 how to evaluate inner products (or convolutions) efficiently by using Alg. 27.1. However, two issues arise here that were not relevant before:

a) First, if we employ the structure of Fig. 27.8 to evaluate the inner product $\hat{d}(nB-j) = u_{nB-j}w_{n-1}$, with w_{n-1} playing the role of g, then we would end up with two frequency-domain structures: one is used for efficient convolution while the other is used for block filter adaptation, as in Fig. 28.10. Obviously, there is no need for the block sizes in both structures to be identical. For instance, the block size for adaptation could be B and the one for convolution could be some other value R, with the constraint that B is a multiple of R, e.g., $B = mR$ for some integer m. The structure of Fig. 28.10 then functions as follows:

   1. Starting from the subband weights $\{l_{k,n-1}\}$, an estimate for the fullband filter is computed, i.e., w_{n-1}. The polyphase components of order M/R of w_{n-1} are determined and used via (27.24) to find the corresponding bank of $2R$ subband filters for the convolution structure, which we denote by $\{\bar{L}_{k,n-1}(z)\}$.

   2. At block iteration n, by the time the adaptive structure finishes forming the $n-$th block of data, the convolution structure would have operated on m sub-blocks of size R each and generated the outputs $\{\hat{d}(nB-j), j = 0, \ldots, B-1\}$ (which are the entries of the block vector $\hat{d}_{B,n}$). These signals are used by the adaptive structure to compute $e_{B,n}$ and to update the subband weights to $\{l_{k,n}\}$.

b) The polyphase components $\{\bar{L}_{k,n-1}(z)\}$ of the convolution structure are updated after every block iteration, i.e., after every new estimate w_n is generated. The states (i.e., initial conditions) of the subband filters within the convolution structure should of course propagate from one iteration to another.

One advantage of the constrained filter of Figs. 28.9 and 28.10 over the constrained filter of Fig. 28.2 is that the delay problem associated with the convolution structure of Fig. 28.10 can be addressed by using the modifications suggested in Prob. VI.14.

Algorithm 28.3 (Another constrained DFT block filter) Consider the adaptive channel estimation scheme shown in Fig. 27.2, where w_{i-1} is $M \times 1$ and the regressor u_i is $1 \times M$. For relatively long channels (i.e., large M), a more efficient adaptive procedure for generating the estimates $\hat{d}(i)$, and the corresponding error sequence $\{e(i)\}$, is the following. Choose a block size B; usually M and B are powers of 2 and M/B is an integer. Select $0 \ll \beta < 1$, set $l_{k,-1} = l^c_{k,-1} = 0$, $w_{-1} = 0$, $u_{-1} = 0$, $u_{B,-1} = 0$, $\lambda_k(-1) = \epsilon$ (a small positive number), and repeat for $n \geq 0$:

Filtering in fullband. Generate the entries of $e_{B,n}$ by computing for $j = 0, 1, \ldots, B-1$:

$$\begin{aligned}
u_{nB-j} &= \begin{bmatrix} u(nB-j) & u(nB-j-1) & \ldots & u(nB-j-M+1) \end{bmatrix} \\
\hat{d}(nB-j) &= u_{nB-j} w_{n-1} \\
e(nB-j) &= d(nB-j) - \hat{d}(nB-j)
\end{aligned}$$

Block input and error signals. Construct the block vectors:

$$\begin{aligned}
u_{B,n} &= \text{col}\{u(nB), \ldots, u((n-1)B+1)\} \\
u_{2B,n} &= \text{col}\{u_{B,n}, u_{B,n-1}\} \\
e_{B,n} &= \text{col}\{e(nB), \ldots, e((n-1)B+1)\}
\end{aligned}$$

Frequency transformations. Perform the DFT transformations:

$$u'_{2B,n} = F u_{2B,n} \qquad e'_{2B,n} = F \begin{bmatrix} \mathrm{I}_B \\ 0_{B\times B} \end{bmatrix} e_{B,n}$$

and let $\{e'_k(n), u'_k(n)\}$ denote the k-th entries of $\{e'_{2B,n}, u'_{2B,n}\}$. Let also

$$u'_{k,n} = \begin{bmatrix} u'_k(n) & u'_k(n-1) & \ldots & u'_k(n - \frac{M}{B} + 1) \end{bmatrix}$$

Adaptation. For each subband filter $k = 0, \ldots, 2B-1$, adapt its coefficients:

$$\begin{aligned}
\lambda_k(n) &= \beta \lambda_k(n-1) + (1-\beta)|u'_k(n)|^2 \\
l_{k,n} &= l_{k,n-1} + \frac{\mu}{\lambda_k(n)} u'^*_{k,n} e'_k(n)
\end{aligned}$$

Subband/fullband mapping. Set

$$w_n = \text{vec}\left([\mathrm{I}_B \;\; 0_{B\times B}] F \begin{bmatrix} l^{\mathsf{T}}_{0,n} \\ \vdots \\ l^{\mathsf{T}}_{2B-1,n} \end{bmatrix} \right)$$

where the operation $\text{vec}(\cdot)$ stacks the columns of its argument on top of each other.

28.B APPENDIX: OVERLAP-ADD BLOCK ADAPTIVE FILTERS

The filters of Algs. 28.1, 28.2, and 28.3 are referred to as *overlap-save* filters. As explained in Remark 1 in the text, the term *overlap-save* is used to indicate that successive input blocks $\{u_{2B,n}\}$ are overlapped prior to the DFT operation and only half of each block is saved. In this appendix,

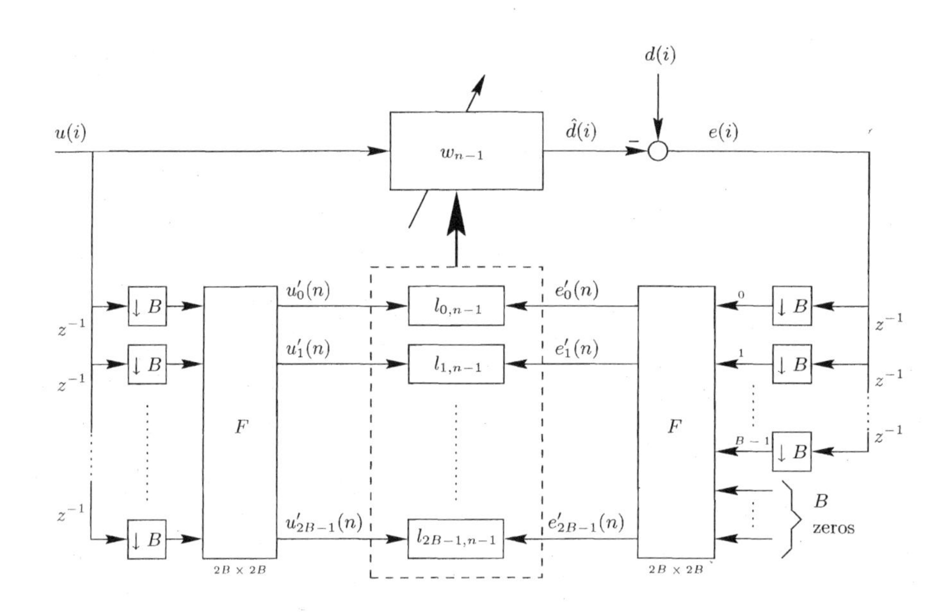

FIGURE 28.9 An implementation of the constrained DFT block adaptive filter of Alg. 28.3. The input sequence $\{u(i)\}$ is processed by a bank of $2B$ decimators, and is also convolved in fullband with w_{n-1} in order to generate $\hat{d}(i)$. The error sequence $\{e(i)\}$ is processed by a band of B decimators and used to generate the transformed errors $\{e'_k(n)\}$ for subband adaptation. The mapping from the $\{l_{k,n-1}\}$ to w_{n-1} is done according to (28.19).

we derive alternative *overlap-add* structures for DFT-based block adaptive filtering. In this case, the successive input blocks $\{u_{2B,n}\}$ do not share a common sub-block. These structures can be motivated as follows. Rather than factor $\mathcal{G}(z)$ as in (27.10) with $\{\bar{\mathcal{P}}(z), \bar{\mathcal{Q}}(z)\}$ given by (27.13)–(27.14), we can equivalently factor it as

$$\mathcal{G}(z) = \bar{\mathcal{Q}}_1(z)\bar{\mathcal{P}}_1(z)$$

where now, e.g., for $B = 3$ again,

$$\bar{\mathcal{P}}_1(z) = \begin{bmatrix} 0 & 0 & 0 \\ P_2(z) & 0 & 0 \\ P_1(z) & P_2(z) & 0 \\ P_0(z) & P_1(z) & P_2(z) \\ 0 & P_0(z) & P_1(z) \\ 0 & 0 & P_0(z) \end{bmatrix}, \quad \bar{\mathcal{Q}}_1(z) = \begin{bmatrix} z^{-1} & 0 & 0 & 1 & 0 & 0 \\ 0 & z^{-1} & 0 & 0 & 1 & 0 \\ 0 & 0 & z^{-1} & 0 & 0 & 1 \end{bmatrix} \tag{28.21}$$

Following the derivation in Sec. 27.2, we can embed $\bar{\mathcal{P}}_1(z)$ into the *same* circulant matrix $\mathcal{C}(z)$ in (27.15), and $\bar{\mathcal{P}}_1(z)$ can be recovered from the last columns of $\mathcal{C}(z)$ as

$$\bar{\mathcal{P}}_1(z) = \mathcal{C}(z) \begin{bmatrix} 0_{B\times B} \\ \mathrm{I}_B \end{bmatrix}$$

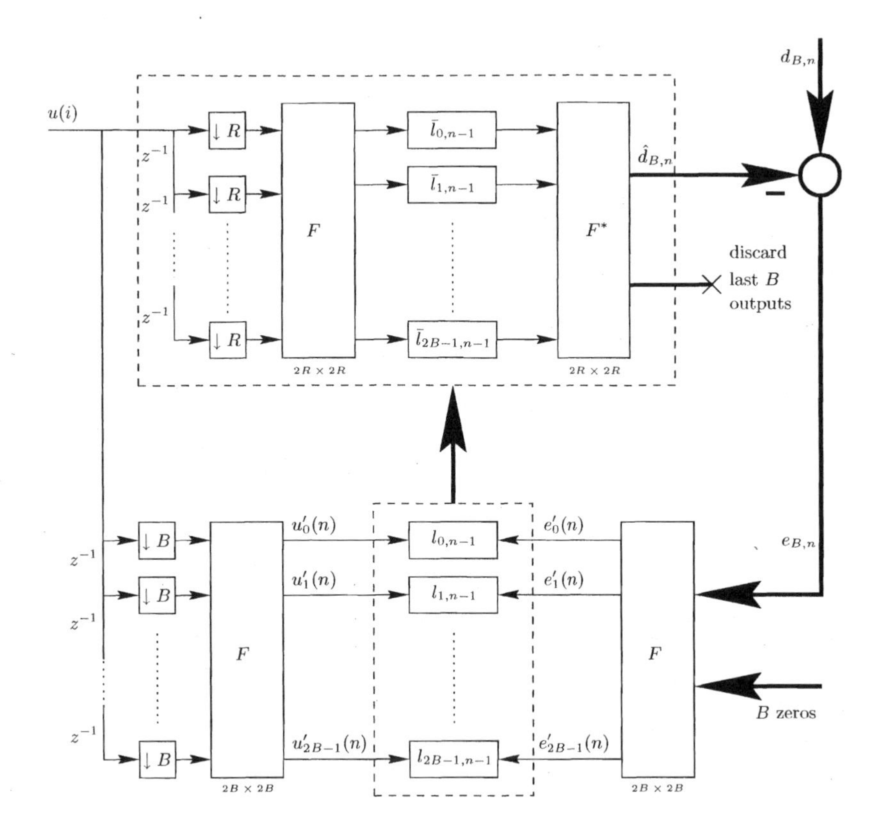

FIGURE 28.10 A second implementation of the constrained DFT block adaptive filter of Alg. 28.3 with frequency-domain structures for both adaptation and convolution. The input signal $u(i)$ is processed by two banks of $2R$ and $2B$ decimators for convolution and adaptation, respectively. The $\{\bar{l}_{k,n-1}\}$ are the subband filters that correspond to the fullband weight vector w_{n-1}, which in turn is obtained by mapping the subband filters $\{l_{k,n-1}\}$ according to the constraint (28.19).

Then (27.21) becomes

$$\boxed{\mathcal{G}(z) \;=\; \underbrace{\bar{\mathcal{Q}}_1(z)\,\mathcal{C}(z)\begin{bmatrix} 0_{B\times B} \\ \mathrm{I}_B \end{bmatrix}}_{\bar{\mathcal{P}}_1(z)} \;=\; \bar{\mathcal{Q}}_1(z)\,\underbrace{F^*\mathcal{L}(z)F}_{\mathcal{C}(z)}\begin{bmatrix} 0_{B\times B} \\ \mathrm{I}_B \end{bmatrix}} \tag{28.22}$$

This result shows that the mapping from $u_{B,n}$ to $y_{B,n}$ in Fig. 27.4 can be alternatively implemented as shown in Fig. 28.11 — compare with Fig. 28.5. Observe that the outputs of the transformation by F^* are combined as shown in the figure to yield $y_{B,n}$. The delays at the output of Fig. 28.11 can be moved prior to the subband filters $\{L_k(z)\}$. To see this, we first express $\bar{\mathcal{Q}}_1(z)$ as

$$\bar{\mathcal{Q}}_1(z) = \begin{bmatrix} 0_{B\times B} & \mathrm{I}_B \end{bmatrix} + z^{-1}\begin{bmatrix} \mathrm{I}_B & 0_{B\times B} \end{bmatrix}$$

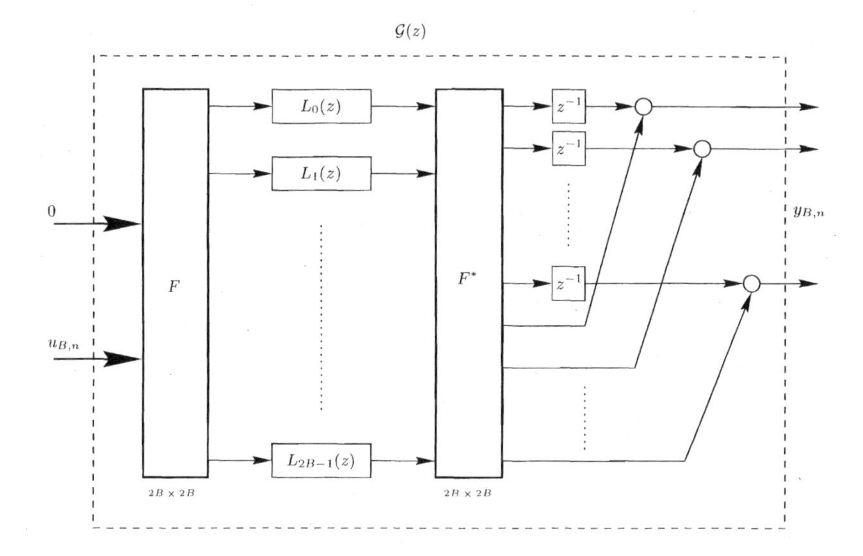

FIGURE 28.11 An alternative implementation of the mapping in Fig. 27.4 for block estimation; compare with Fig. 27.5.

so that expression (28.22) for $\mathcal{G}(z)$ becomes

$$\mathcal{G}(z) = \begin{bmatrix} 0_{B\times B} & \mathrm{I}_B \end{bmatrix} F^*\mathcal{L}(z)F \begin{bmatrix} 0_{B\times B} \\ \mathrm{I}_B \end{bmatrix} + z^{-1}\begin{bmatrix} \mathrm{I}_B & 0_{B\times B} \end{bmatrix} F^*\mathcal{L}(z)F \begin{bmatrix} 0_{B\times B} \\ \mathrm{I}_B \end{bmatrix} \tag{28.23}$$

Now it can be verified that (see Prob. VI.11):

$$F\begin{bmatrix} \mathrm{I}_B \\ 0_{B\times B} \end{bmatrix} = JF\begin{bmatrix} 0_{B\times B} \\ \mathrm{I}_B \end{bmatrix} \tag{28.24}$$

where $J = \text{diag}\{1, -1, 1, -1, \ldots, 1, -1\}$ is the $2B \times 2B$ diagonal matrix with alternating $\pm 1's$. Substituting into (28.23) we find that $\mathcal{G}(z)$ can be rewritten more compactly as

$$\boxed{\mathcal{G}(z) = \begin{bmatrix} 0_{B\times B} & \mathrm{I}_B \end{bmatrix} F^*\mathcal{L}(z)(\mathrm{I} + z^{-1}J)F \begin{bmatrix} 0_{B\times B} \\ \mathrm{I}_B \end{bmatrix}} \tag{28.25}$$

This result shows that the implementation of Fig. 28.11 can be redrawn as shown in Fig. 28.12, where the signals $\{s'_k(n)\}$ are generated via

$$s'_k(n) = u'_k(n) + (-1)^k u'_k(n-1), \quad k = 0, 1, \ldots, 2B - 1 \tag{28.26}$$

In summary, we find that the efficient block convolution of Alg. 27.1 can also be attained as Alg. 28.4.

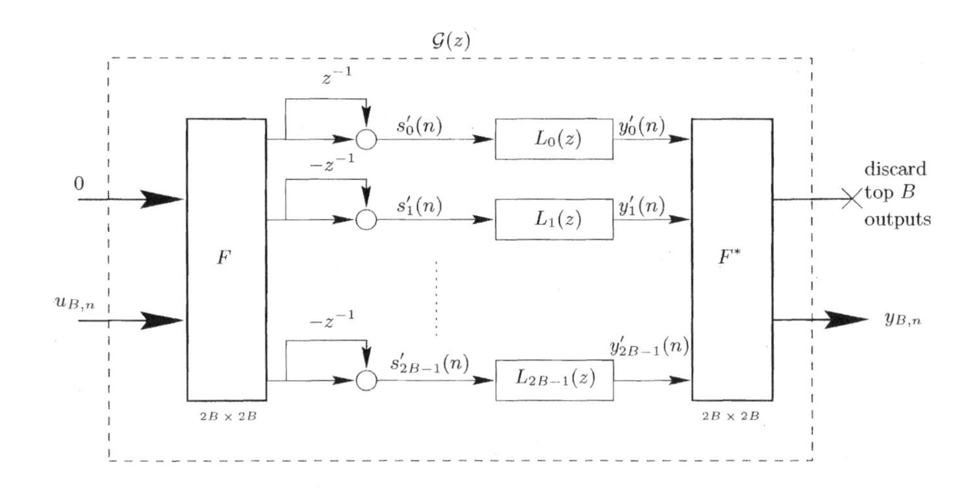

FIGURE 28.12 The delays at the output of Fig. 28.11 are moved prior to the subband filters.

Algorithm 28.4 (Block overlap-add convolution via DFT) Consider the mapping shown in Fig. 27.3 for some transfer function $G(z)$ with M taps. The direct convolution operation can be implemented more efficiently as shown in Fig. 28.12. Specifically, choose a block size B; usually M and B are powers of 2 and M/B is an integer.

Bank of filters. Determine the polyphase components of $G(z)$ of order M/B, as defined by (27.6), and use (27.23)–(27.24) to determine the $2B$ filters $\{L_k(z)\}$ of size M/B each.

Block filtering. Start with $u_{B,-1} = 0$ and repeat for $n \geq 0$:

1. Construct the block vectors:

$$u_{B,n} = \text{col}\{u(nB), \ldots, u((n-1)B+1)\}, \quad u_{2B,n} = \text{col}\{0, u_{B,n}\}$$

2. Perform the DFT transformation $u'_{2B,n} = Fu_{2B,n}$ and let

$$s'_{2B,n} = u'_{2B,n} + Ju'_{2B,n-1} \triangleq \text{col}\{s'_k(n),\ k = 0, 1, \ldots, 2B-1\}$$

Filter the entries $\{s'_k(n)\}$ through the subband filters $\{L_k(z)\}$. Let $y'_{2B,n} = \text{col}\{y'_k(n),\ k = 0, 1, \ldots, 2B-1\}$ denote the outputs of these filters at iteration n.

3. Perform the DFT transformation:

$$\begin{bmatrix} \times \\ y_{B,n} \end{bmatrix} = F^* y'_{2B,n}$$

where $\times$ denotes B entries to be ignored, and $y_{B,n} = \text{col}\{y(nB), y(nB-1), \ldots, y((n-1)B+1)\}$.

We can now proceed as in Sec. 28.1 and derive adaptive implementations as well. Since the arguments are similar to what we did before, we shall be brief and only highlight the differences. Introduce the $1 \times M/B$ regression vector

$$s'_{k,n} = \begin{bmatrix} s'_k(n) & s'_k(n-1) & \dots & s'_k(n-M/B+1) \end{bmatrix}$$

where the $\{s'_k(n)\}$ denote the inputs to the subband filters, as defined by (28.26). Collect further the regressors $\{s'_{k,n}\}$ into a block diagonal matrix $\mathcal{S}'_n$,

$$\mathcal{S}'_n \triangleq \text{diag}\{s'_{0,n}, s'_{1,n}, \dots, s'_{2B-1,n}\}; \qquad (2B \times 2M) \tag{28.27}$$

Then the same argument that led to (28.3) gives

$$\boxed{d_{B,n} = \begin{bmatrix} 0_{B\times B} & \text{I}_B \end{bmatrix} F^* \mathcal{S}'_n l + v_{B,n}} \tag{28.28}$$

Comparing with (28.3) we see that the main distinction is in the use of $\mathcal{S}'_n$ in place of $\mathcal{U}'_n$. In addition, $[0_{B\times B} \;\; \text{I}_B]$ replaces $[\text{I}_B \;\; 0_{B\times B}]$. Therefore, the results of Sec. 28.1 and also App. 28.A will extend with minor modifications. In particular, the unconstrained filter of Alg. 28.1 becomes Alg. 28.5. Figure 28.13 shows a block diagram representation of Alg. 28.5. In a similar manner, the constrained filters of Algs. 28.2 and 28.3 become Algs. 28.6 and 28.7, respectively, which are illustrated in Figs. 28.14 and 28.15.

Algorithm 28.5 (Unconstrained overlap-add DFT block filter) Refer to the setting of Alg. 28.1 and let $J = \text{diag}\{1, -1, 1, -1, \dots, 1, -1\}$ be a $2B \times 2B$ matrix with alternating $\pm 1's$. Repeat for $n \geq 0$:

$$\begin{aligned}
u_{B,n} &= \text{col}\{u(nB), \dots, u((n-1)B+1)\} \\
u'_{2B,n} &= F\begin{bmatrix} 0_{B\times 1} \\ u_{B,n} \end{bmatrix} \\
s'_{2B,n} &= u'_{2B,n} + Ju'_{2B,n-1} \triangleq \text{col}\{s'_k(n),\ k = 0, 1, \dots, 2B-1\} \\
\lambda_k(n) &= \beta\lambda_k(n-1) + (1-\beta)|s'_k(n)|^2, \quad k = 0, 1, \dots, 2B-1 \\
s'_{k,n} &= \begin{bmatrix} s'_k(n) & \dots & s'_k(n - \frac{M}{B} + 1) \end{bmatrix}, \quad k = 0, 1, \dots, 2B-1 \\
y'_k(n) &= s'_{k,n} l_{k,n-1}, \quad k = 0, 1, \dots, 2B-1 \\
\hat{d}_{B,n} &= \begin{bmatrix} 0_{B\times B} & \text{I}_B \end{bmatrix} F^* \text{col}\{y'_0(n), \dots, y'_{2B-1}(n)\} \\
e_{B,n} &= d_{B,n} - \hat{d}_{B,n} \\
e'_{2B,n} &= F\begin{bmatrix} 0_{B\times B} \\ \text{I}_B \end{bmatrix} e_{B,n} \triangleq \text{col}\{e'_k(n),\ k = 0, 1, \dots, 2B-1\} \\
l_{k,n} &= l_{k,n-1} + \frac{\mu}{\lambda_k(n)} u'^*_{k,n} e'_k(n), \quad k = 0, 1, \dots, 2B-1
\end{aligned}$$

At each block n, the entries of $\{\hat{d}_{B,n}, e_{B,n}\}$ correspond to

$$\begin{aligned}
\hat{d}_{B,n} &= \text{col}\{\hat{d}(nB), \hat{d}(nB-1), \dots, \hat{d}((n-1)B+1)\} \\
e_{B,n} &= \text{col}\{e(nB), e(nB-1), \dots, e((n-1)B+1)\}
\end{aligned}$$

Algorithm 28.6 (Constrained overlap-add DFT block filter I) Consider the same setting of Alg. 28.2, and repeat for $n \geq 0$:

$$\begin{aligned}
u_{B,n} &= \text{col}\{u(nB), u(nB-1), \ldots, u((n-1)B+1)\} \\
u'_{2B,n} &= F\begin{bmatrix} 0_{B\times 1} \\ u_{B,n} \end{bmatrix} \\
s'_{2B,n} &= u'_{2B,n} + Ju'_{2B,n-1} \\
&\triangleq \text{col}\{s'_k(n),\ k=0,1,\ldots,2B-1\} \\
\lambda_k(n) &= \beta\lambda_k(n-1) + (1-\beta)|s'_k(n)|^2, \quad k=0,1,\ldots,2B-1 \\
s'_{k,n} &= \begin{bmatrix} s'_k(n) & \ldots & s'_k(n-\frac{M}{B}+1) \end{bmatrix}, \quad k=0,1,\ldots,2B-1 \\
y'_k(n) &= s'_{k,n} l^{c}_{k,n-1}, \quad k=0,1,\ldots,2B-1 \\
\hat{d}_{B,n} &= \begin{bmatrix} 0_{B\times B} & \text{I}_B \end{bmatrix} F^* \text{col}\left\{y'_0(n),\ldots,y'_{2B-1}(n)\right\} \\
e_{B,n} &= d_{B,n} - \hat{d}_{B,n} \\
e'_{2B,n} &= F\begin{bmatrix} 0_{B\times B} \\ \text{I}_B \end{bmatrix} e_{B,n} \\
&\triangleq \text{col}\{e'_k(n),\ k=0,1,\ldots,2B-1\} \\
l_{k,n} &= l_{k,n-1} + \frac{\mu}{\lambda_k(n)} u'^{*}_{k,n} e'_k(n), \quad k=0,1,\ldots,2B-1
\end{aligned}$$

$$\begin{bmatrix} l^{c\mathsf{T}}_{0,n} \\ l^{c\mathsf{T}}_{1,n} \\ \vdots \\ l^{c\mathsf{T}}_{2B-1,n} \end{bmatrix} = \frac{1}{2B} F^* \begin{bmatrix} \text{I}_B & \\ & 0_{B\times B} \end{bmatrix} F \begin{bmatrix} l^{\mathsf{T}}_{0,n} \\ l^{\mathsf{T}}_{1,n} \\ \vdots \\ l^{\mathsf{T}}_{2B-1,n} \end{bmatrix}$$

At each block n, the entries of $\{\hat{d}_{B,n}, e_{B,n}\}$ correspond to

$$\begin{aligned}
\hat{d}_{B,n} &= \text{col}\{\hat{d}(nB), \hat{d}(nB-1), \ldots, \hat{d}((n-1)B+1)\} \\
e_{B,n} &= \text{col}\{e(nB), e(nB-1), \ldots, e((n-1)B+1)\}
\end{aligned}$$

Algorithm 28.7 (Constrained overlap-add DFT block filter II) Consider again the setting of Alg. 28.3, and repeat for $n \geq 0$:

Filtering in fullband. For $j = 0 : B-1$, generate the entries of the vector $e_{B,n}$ by computing:

$$\begin{aligned}
u_{nB-j} &= \begin{bmatrix} u(nB-j) & u(nB-j-1) & \ldots & u(nB-j-M+1) \end{bmatrix} \\
\hat{d}(nB-j) &= u_{nB-j}w_{n-1} \\
e(nB-j) &= d(nB-j) - \hat{d}(nB-j)
\end{aligned}$$

Block input and error signals. Construct the block vectors

$$\begin{aligned}
u_{B,n} &= \text{col}\{u(nB), \ldots, u((n-1)B+1)\} \\
u_{2B,n} &= \text{col}\{0, u_{B,n}\} \\
e_{B,n} &= \text{col}\{e(nB), \ldots, e((n-1)B+1)\}
\end{aligned}$$

Frequency transformations. Perform the DFT transformations:

$$\begin{aligned}
u'_{2B,n} &= Fu_{2B,n} \\
s'_{2B,n} &= u'_{2B,n} + Ju'_{2B,n-1} \\
e'_{2B,n} &= F\begin{bmatrix} 0_{B\times B} \\ \text{I}_B \end{bmatrix} e_{B,n}
\end{aligned}$$

and let $\{e'_k(n), s'_k(n)\}$ denote the k-th entries of $\{e'_{2B,n}, s'_{2B,n}\}$. Let also

$$s'_{k,n} = \begin{bmatrix} s'_k(n) & s'_k(n-1) & \ldots & s'_k(n-\frac{M}{B}+1) \end{bmatrix}$$

Adaptation. For each subband filter $k = 0, \ldots, 2B-1$, adapt its coefficients:

$$\begin{aligned}
\lambda_k(n) &= \beta\lambda_k(n-1) + (1-\beta)|s'_k(n)|^2 \\
l_{k,n} &= l_{k,n-1} + \frac{\mu}{\lambda_k(n)} s'^*_{k,n} e'_k(n)
\end{aligned}$$

Subband/fullband mapping. Set

$$w_n = \text{vec}\left([\text{I}_B \ \ 0_{B\times B}]F \begin{bmatrix} l^{\mathsf{T}}_{0,n} \\ \vdots \\ l^{\mathsf{T}}_{2B-1,n} \end{bmatrix} \right)$$

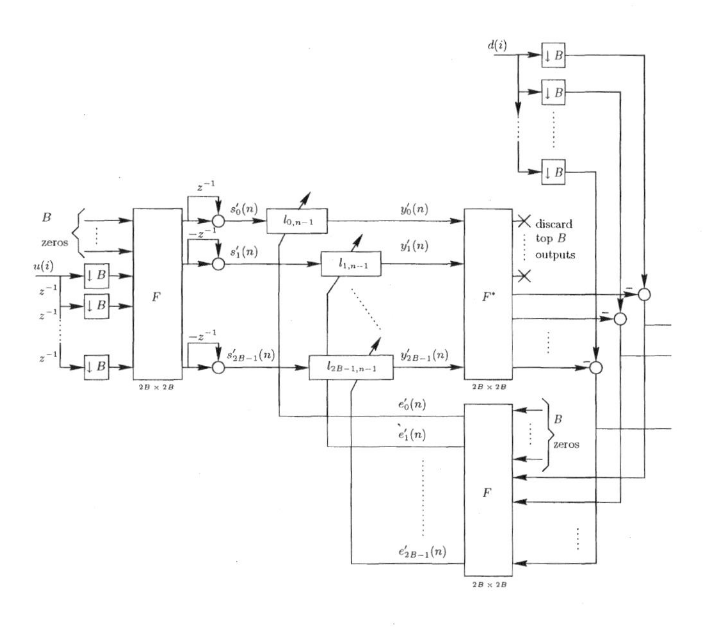

FIGURE 28.13 Block diagram representation of the unconstrained overlap-add DFT block adaptive filter of Alg. 28.5.

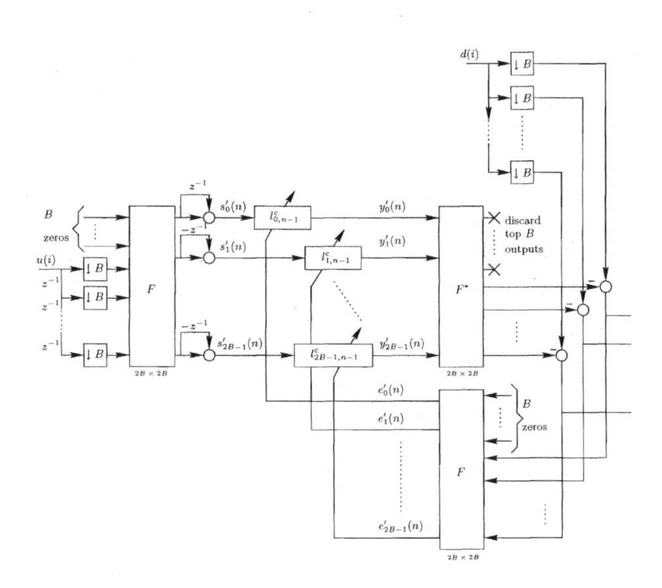

FIGURE 28.14 Block diagram representation of the constrained overlap-add DFT block adaptive filter of Alg. 28.6.

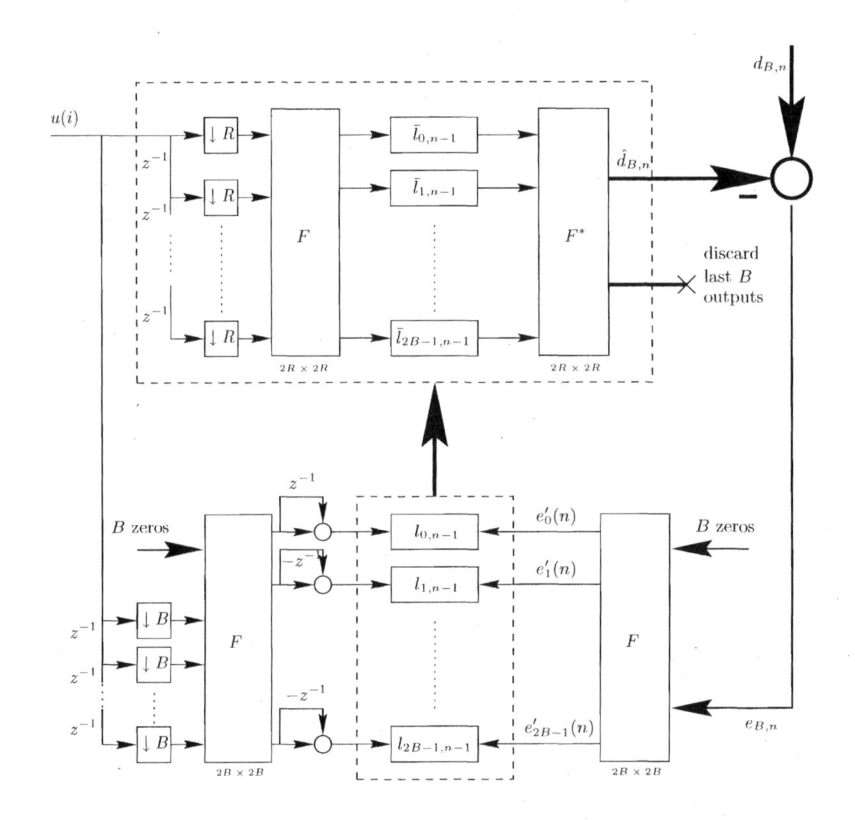

FIGURE 28.15 The constrained overlap-add DFT block adaptive filter of Alg. 28.7 with frequency-domain structures for both adaptation and convolution.

Summary and Notes

The chapters in this part describe several adaptive structures that are meant to reduce the computational requirements of LMS and improve its convergence speed.

SUMMARY OF MAIN RESULTS

1. The text describes three classes of efficient adaptive structures: transform-domain filters, block adaptive filters, and subband adaptive filters.
2. Transform-domain filters exploit the de-correlation properties of some unitary signal transforms, such as the discrete Fourier transform (DFT) and the discrete cosine transform (DCT), in order to pre-whiten the input data and speed up filter convergence. Designing a transform-domain filter usually involves three steps:
   - (i) Replacing the regression vector u_i by a transformed version $\bar{u}_i = u_i T$, where T is a unitary transformation.
   - (ii) Employing a diagonal step-size matrix that consists of estimates of the inverse powers in the individual entries of $\bar{u}_i$.
   - (iii) Making sure that the elements of the transformed regressor can be evaluated efficiently in $O(M)$ operations.

   Steps (i)–(iii) only help improve the convergence speed of LMS; they do not result in a reduction of computational complexity. Two types of transformed filters were discussed in the text: DFT-LMS and DCT-LMS.
3. Block adaptive filters reduce the computational cost of LMS by some factor $\alpha > 1$, while at the same time improving the convergence speed. This is achieved by processing the data on a block-by-block basis and by exploiting the fact that many signal transforms admit efficient implementations. Designing a block adaptive filter generally involves three steps:
   - (i) Choosing a signal transform, such as the DFT, and choosing a block size and an order for the subband filters. The number of subband filters is related to the block size and to the type of signal transform used.
   - (ii) Collecting blocks of input data and transforming them by the chosen signal transform. The manner by which the data blocks are processed depends on the transform. The transformed data then play the role of regression data to the subband filters.
   - (iii) There are mainly two ways to obtain the error signals for training the subband filters:
     - (iii.1) In one implementation, the estimates $\{\hat{d}(i)\}$ are generated by signal transforming the outputs of the subband filters.
     - (iii.2) In another implementation, the estimates $\{\hat{d}(i)\}$ are generated through fullband convolution.

     The estimates $\{\hat{d}(i)\}$ are subtracted from the original reference sequence $\{d(i)\}$ in order to generate the time-domain error sequence $\{e(i)\}$, which is then transformed to the frequency domain and used to train the subband filters.

4. Two kinds of block adaptive implementations are described: unconstrained and constrained with the latter having superior performance. In addition, for the constrained case, two implementations are possible with $\{\hat{d}(i)\}$ generated either through convolution in fullband or through transforming the outputs of the subband filters.

5. The derivation of block adaptive filters is carried out by using a convenient embedding, whereby a certain transfer matrix function is embedded into a larger matrix with special structure. The structure is such that it can be diagonalized by the DFT. Other embeddings are possible that result in structures that are diagonalizable by other common transforms such as the DCT and DHT. The DFT-based block adaptive filters are derived in the text and in Apps. 28.A and 28.B. On the other hand, DCT and DHT-based structures can be found in Apps. 10.D and 10.E of Sayed (2003).

6. Subband adaptive filters are related to the DFT-based block adaptive filters, except that they attempt to achieve better band partitioning of the data via selection of more suitable prototype filters for the analysis and synthesis banks. There are two basic configurations for subband filtering: open-loop and closed-loop. In closed-loop, the error sequence is generated in the fullband time-domain and then processed by the analysis filter bank. In open-loop, subband error signals are used to train the subband filters.

BIBLIOGRAPHIC NOTES

Transform-domain filters. The idea of a transform-domain adaptive filter was first proposed by Narayan and Peterson (1981) and Narayan, Peterson and Narasimha (1983). Another early work is by Marshall, Jenkins, and Murphy (1989), which provides a geometric interpretation of the effect of step-normalization on the performance of transform-domain LMS. The DCT-domain LMS filter of Sec. 26.3 was proposed by Beaufays and Widrow (1994) and Beaufays (1995); the form we presented in Alg. 26.3 is slightly different.

Terminology. The block filters we derived in Sec. 28.1 by using the DFT transform are generally referred to in the literature as "block adaptive filters" or "block LMS filters". We have opted to use a more specific terminology and attach the "DFT" qualification to the name. Thus we refer to them as "DFT-based block adaptive filters", "DFT block adaptive filters" or simply "DFT block LMS filters". We do so in order to distinguish these forms from other block adaptive filters proposed by Merched and Sayed (2000a), and are described in Apps. 10.D and 10.E of Sayed (2003). These alternative forms are based on other signal transforms, such as the discrete-cosine and discrete Hartley transforms. Thus one can refer to DFT-, DCT-, and DHT-block adaptive filters.

Block adaptive filters. In our treatment, we study directly block adaptive filters with multiple taps in the subbands (i.e., multi-delay filters). This is not how these filters were historically developed. The earliest block filters were derived in the early 1980s with a *single* tap in the subbands by Ferrara (1980) and Clark, Mitra, and Parker (1981). The multi-delay filters appeared later and were derived independently by Asharif et al. (1986a,1986b), Soo and Pang (1987,1990), and Sommen (1989), under different names (specifically, frequency bin adaptive filter, multi-delay frequency block LMS, and partitioned frequency block LMS, respectively). Other early works on block filters include those by Xu and Grenier (1989), Petraglia and Mitra (1991,1993), Shynk (1992), and Borallo and Otero (1992). All these works were specific to the DFT domain.

Embeddings. The presentation in Chapters 27 and 28 follows Merched and Sayed (2000a), where DCT- and DHT-block adaptive filters are also derived. The arguments in these chapters, even for the DFT case, are different from the original derivations of multi-delay filters. Specifically, the derivation in Sec. 28.1 relies on embedding certain transfer matrices into larger *structured* block matrices that can be diagonalized by the DFT matrix and also by other trigonometric transforms (using, e.g., results from Bini and Favati (1993) and Heinig and Rost (1998) on structured matrix computations — see also Kailath and Sayed (1995)). One useful consequence of the embedding approach is that it can handle extensions of the block adaptive structure to other transform domains such as the DCT and the DHT (see Apps. 10.D and 10.E of Sayed (2003)). Similar embedding techniques in the DFT domain were used before by Lin and Mitra (1996) in the design of efficient block implementations for digital filters. It should be noted that the DCT- and DHT-based schemes require only real arithmetic and, since efficient algorithms exist for computing the DCT and the DHT (see, e.g., Vetterli and Nussbaumer (1984) and Ersoy (1997)), these schemes also lead to efficient adaptive structures.

Power normalization. The use of NLMS, or of a power-normalized version of it as in Sec. 28.1, in order to train the subband filters, can improve the rate of convergence of block filters as indicated by the analyses in Sommen et al. (1987) and Lee and Un (1989).

Constrained vs. unconstrained implementations. In Mansoni and Gray (1982), it was shown that the unconstrained DFT-based block adaptive filter can still work well if the filter length M is sufficiently large. Still, unconstrained implementations are slow to converge (especially for $M/B > 1$) and they do not converge to the optimal solution. On the other hand, it has been verified experimentally that the constraint does not need to be applied at every iteration.

Subband adaptive filters. The concept of subband adaptive filtering was introduced by Furukawa (1984) and Kellermann (1984,1985). Other early works in this area include Yasukawa and Shimada (1987,1993), Chen et al. (1988), Gilloire and Vetterli (1988,1992), and Petraglia and Mitra (1993). Some further results on subband structures can be found, e.g., in Morgan and Thi (1995), Pradhan and Reddy (1999), Farhang-Boroujeny (1999), Petraglia, Alves, and Diniz (2000), and the references therein. The work by Morgan and Thi (1995) provides a useful approach to handling the delay problem that plagues subband architectures — see Prob. VI.14. The book by Farhang-Boroujney (1999) provides a good treatment of subband adaptive filters with emphasis on design issues.

Block and subband filters. When the open- and closed-loop subband architectures of Sec. 28.2 were proposed in the literature, the DFT-block adaptive filters of Sec. 28.1 did not even exist. So the derivations of these subband structures were not originally motivated in the same manner we discussed in Sec. 28.2, by showing how block and subband adaptive filters relate to each other. In fact, one can verify that even the technical language used in the derivation of the DFT-block adaptive structure, e.g., in Soo and Pang (1987,1990), has little resemblance with the language of modern multi-rate systems and filter banks (e.g., as in Vaidyanathan (1993)).

An application to acoustic echo cancellation. As mentioned in the concluding remarks of Part IV (*Mean-Square Performance*), there are two types of echoes in communications systems: line echoes and acoustic echoes. In Computer Project IV.1 we studied line echo cancellation and in Computer Project VI.1 we shall study acoustic echo cancellation using block adaptive filters. Acoustic echoes occur in hands-free telephony and teleconferencing and they result from the reflection of sound waves inside an enclosure, and from the acoustic coupling between the microphone and the loudspeaker (see, e.g., Benesty et al. (2001)). Compared with line echo problems, the duration of the echo path tends to be longer for acoustic echoes. This is one reason why block and subband adaptive filters are popular choices for acoustic echo cancellation applications.

Problems and Computer Projects

PROBLEMS

Problem VI.1 (Similarity transformations) Let A and B be two similar matrices, i.e., $B = T^{-1}AT$ for some square invertible matrix T. Show that A and B have the same eigenvalues.

Problem VI.2 (Nonnegativity of power spectra) Let $r(k) = \mathsf{E}\,\boldsymbol{u}(i)\boldsymbol{u}^*(i-k)$ and assume this auto-correlation sequence is exponentially bounded as in (26.2). Introduce the corresponding power spectrum (26.3). Let R_M be the Hermitian Toeplitz covariance matrix whose first column is $\text{col}\{r(0),\ldots,r(M-1)\}$. Pick any finite scalar a and define $b = \text{col}\{a, ae^{-j\omega},\ldots,ae^{-j(M-1)\omega}\}$, where $j=\sqrt{-1}$. Show that

$$0 \leq \frac{1}{M}b^*R_Mb = a^*\left[\sum_{k=-(M-1)}^{M-1} r(k)e^{-j\omega k}\right]a \;-\; \frac{1}{M}\sum_{k=-(M-1)}^{M-1} |k|\cdot|a|^2 r(k)e^{-j\omega k}$$

Now take the limit as $M\to\infty$ to conclude that $S_u(e^{j\omega})\geq 0$.

Problem VI.3 (Diagonal weighting matrix) Consider the stochastic version of the general transform domain LMS recursion stated in Alg. 26.1, namely,

$$\begin{aligned}
\bar{\boldsymbol{u}}_i &= \boldsymbol{u}_iT, \quad \bar{\boldsymbol{u}}_i(k) = k\text{-th entry of } \bar{\boldsymbol{u}}_i \\
\boldsymbol{\lambda}_k(i) &= \beta\boldsymbol{\lambda}_k(i-1) \;+\; (1-\beta)|\bar{\boldsymbol{u}}_i(k)|^2, \qquad k=0,1,\ldots,M-1, \quad \boldsymbol{D}_i = \text{diag}\{\boldsymbol{\lambda}_k(i)\} \\
\bar{\boldsymbol{w}}_i &= \bar{\boldsymbol{w}}_{i-1} \;+\; \mu\boldsymbol{D}_i^{-1}\bar{\boldsymbol{u}}_i^*[\boldsymbol{d}(i) - \bar{\boldsymbol{u}}_i\bar{\boldsymbol{w}}_{i-1}]
\end{aligned}$$

Let $[R_{\bar{u}}]_{k,k}$ denote the $k-$th diagonal entry of the covariance matrix $R_{\bar{u}} = \mathsf{E}\,\bar{\boldsymbol{u}}_i^*\bar{\boldsymbol{u}}_i$. Argue that,

$$\mathsf{E}\,\boldsymbol{D}_i \;\longrightarrow\; \text{diag}\left\{\,[R_{\bar{u}}]_{0,0},\; [R_{\bar{u}}]_{1,1},\;\ldots,[R_{\bar{u}}]_{M-1,M-1}\right\} \;\stackrel{\Delta}{=}\; \text{diagonal}\{R_{\bar{u}}\}, \quad \text{as} \quad i\to\infty$$

In other words, $\mathsf{E}\,\boldsymbol{D}_i$ tends to a diagonal matrix that agrees with the diagonal of $R_{\bar{u}}$.

Problem VI.4 (Mean-square performance) Consider the same setting of Prob. VI.3 and assume the data $\{\boldsymbol{d}(i),\boldsymbol{u}_i\}$ satisfy model (15.16). Follow the arguments used in Sec. 19.1 while evaluating the performance of RLS to justify the following approximation for the excess mean-square error of the general transform-domain LMS algorithm: $\zeta^{\text{transform LMS}} = \mu\sigma_v^2M/(2-\mu M)$.

Problem VI.5 (Tracking performance) Consider the same setting of Prob. VI.3 and assume the data $\{\boldsymbol{d}(i),\boldsymbol{u}_i\}$ satisfythe nonstationary model (20.16) with a sufficiently small degree of nonstationarity. Follow the arguments used in Sec. 21.4 while evaluating the performance of RLS to justify the following approximation for the excess mean-square error of the general transform-domain LMS algorithm:

$$\zeta^{\text{transform LMS}} = \frac{\mu\sigma_v^2M + \mu^{-1}\text{Tr}\left(\bar{Q}\cdot\text{diagonal}\{R_{\bar{u}}\}\right)}{2-\mu M}$$

where $\bar{Q} = T^*QT$.

Problem VI.6 (Circulant matrices) Consider a 4×4 circulant matrix C and the 4×4 DFT matrix F, namely,

$$C = \begin{bmatrix} c_0 & c_1 & c_2 & c_3 \\ c_3 & c_0 & c_1 & c_2 \\ c_2 & c_3 & c_0 & c_1 \\ c_1 & c_2 & c_3 & c_0 \end{bmatrix}, \quad [F]_{mk} \stackrel{\Delta}{=} e^{-\frac{j2\pi mk}{4}}, \quad m,k = 0,1,2,3$$

(a) Show that the columns of F^* are eigenvectors of C. What are the corresponding eigenvalues? Conclude that C is diagonalized by the DFT matrix.

(b) Replace C by a circulant matrix *function*, $\mathcal{C}(z)$. Use the result of part (a) to argue that $\mathcal{C}(z)$ is also diagonalized by the DFT matrix.

Problem VI.7 (Diagonalization of circulant matrices) Let $C = F^*DF$ denote the diagonalization of an $M \times M$ circulant matrix C by means of the DFT matrix F. Collect the entries of the diagonal matrix D into a column vector $d = \text{diag}\{D\}$. Verify that d is proportional to the DFT of the first column of C, specifically, $d = FCe_1/M$, where e_1 is the first basis vector.

Problem VI.8 (Order of subband filters) Verify that the DFT matrix (27.17) satisfies $FF^* = 2B \cdot \mathrm{I}_{2B}$, and use (27.18) to write $\mathcal{L}(z) = \frac{1}{4B^2} \cdot F\mathcal{C}(z)F^*$. Conclude that each $L_k(z)$ is a linear combination of the polyphase components $\{P_k(z)\}$. Conclude further that each $L_k(z)$ is a polynomial in z^{-1} with M/B coefficients.

Problem VI.9 (Constraint condition) Use $F = F^{\mathsf{T}}$ and (27.18) to write $\mathcal{C}^{\mathsf{T}}(z) = F\mathcal{L}(z)F^*$. Multiply this equality by $\text{col}\{1, 0, \ldots, 0\}$ from both sides to conclude that (27.20) holds.

Problem VI.10 (Relating fullband and subband filters) Refer to the discussion in Sec. 27.3 and to the relation between the polyphase components $\{P_k(z)\}$ and the subband filters $\{L_k(z)\}$. Assume $B = M$ so that the filters $\{P_k(z), L_k(z)\}$ are all constants, namely, $P_k(z) = g(k)$ and $L_k(z) = l(k)$. Here, the $\{g(k), k = 0, 1, \ldots, M-1\}$ denote the impulse response coefficients of the wideband filter $G(z)$ in (27.2). Moreover, the $\{l(k), k = 0, 1, \ldots, 2M-1\}$ denote the coefficients of the subband filters $\{L_k(z)\}$. Follow the argument that led to (27.20) to show that the fullband coefficients $\{g(k)\}$ and the subband coefficients $\{l(k)\}$ satisfy the following relations:

$$\begin{bmatrix} l(0) \\ l(1) \\ \vdots \\ l(2M-1) \end{bmatrix} = \frac{1}{2M} F^* \begin{bmatrix} g(0) \\ \vdots \\ g(M-1) \\ \hline 0_{M \times 1} \end{bmatrix} = \frac{1}{2M} F \begin{bmatrix} g(0) \\ \hline 0_{M \times 1} \\ \hline g(M-1) \\ \vdots \\ g(1) \end{bmatrix}$$

Problem VI.11 (Property of the DFT) Use expression (27.17) to establish (28.24).

Problem VI.12 (Block LMS) Recall from the discussions in Sec. 27.1 that our motivation is to estimate the weight vector g in (27.1) in an efficient manner. The block recursions (28.6)–(28.8) provide one efficient DFT-based solution. Assume $B = M$, in which case the subband filters $\{l_{k,n-1}\}$ in Fig. 28.1 have a single coefficient each. Let w_n denote the $M \times 1$ estimate of the weight vector g at block iteration n. In view of the result of Prob. VI.10, the entries of the $M \times 1$ time-domain estimate w_n and the entries of the $2M \times 1$ frequency-domain estimate $\overline{w}_n$ can be assumed to be related to each other as follows:

$$\begin{bmatrix} \overline{w}_n(0) \\ \overline{w}_n(1) \\ \vdots \\ \overline{w}_n(2M-1) \end{bmatrix} = \frac{1}{2M} F^* \begin{bmatrix} w_n(0) \\ \vdots \\ w_n(M-1) \\ \hline 0_{M \times 1} \end{bmatrix} = \frac{1}{2M} F \begin{bmatrix} w_n(0) \\ \hline 0_{M \times 1} \\ \hline w_n(M-1) \\ \vdots \\ w_n(1) \end{bmatrix}$$

Use this relation to establish that the DFT-based block recursions (28.6)–(28.8) are equivalent to the following (time-domain) block LMS recursion:

$$\begin{aligned} w_n &= w_{n-1} + \mu \sum_{i=0}^{M-1} u_{nM-i}^* e(nM-i) \\ e(nM-i) &= d(nM-i) - u_{nM-i} w_{n-1} \\ u_{nM-i} &= \left[\, u(nM-i) \quad u(nM-i-1) \quad \ldots, u((n-1)M-i+1) \,\right] \quad (1 \times M) \end{aligned}$$

Remark. Typically, in a block LMS implementation, the gradient vector used to update the block weight estimate w_{n-1} to w_n is an average over several instantaneous approximations, say

$$w_n = w_{n-1} + \mu \sum_{i=0}^{B-1} u_{nB-i}^* e(nB-i)$$

for some $B \leq M$, where B is referred to as the block length and u_{nB-i} is still $1 \times M$. Note that all error signals $\{e(nB-i)\}$ over a block are computed using the same weight estimate w_{n-1}. Moreover, while this problem considered the special case $B = M$, the discussion in Chapter 28 studies DFT-based algorithms for the more general case $B < M$ when the subband filters $\{l_{k,n-1}\}$ in Fig. 28.1 have M/B coefficients each.

Problem VI.13 (Mean-square performance of block LMS) Consider a block LMS stochastic recursion of the form

$$\boldsymbol{w}_n = \boldsymbol{w}_{n-1} + \mu \sum_{i=0}^{B-1} \boldsymbol{u}_{nB-i}^* [\boldsymbol{d}(nB-i) - \boldsymbol{u}_{nB-i} \boldsymbol{w}_{n-1}]$$

with block length $B \leq M$, and where $\boldsymbol{w}_n$ is $M \times 1$ and $\boldsymbol{u}_{nB-i}$ is $1 \times M$. Moreover, n is the block index, so that each update is performed after the collection of B data points. Assume the data $\{\boldsymbol{d}(i), \boldsymbol{u}_i\}$ satisfy model (15.16).

(a) Verify that the filter recursion can be expressed in the form

$$\boldsymbol{w}_n = \boldsymbol{w}_{n-1} + \mu \boldsymbol{U}_n^* \boldsymbol{e}_{B,n}$$

where

$$\boldsymbol{U}_n = \begin{bmatrix} \boldsymbol{u}_{nB} \\ \boldsymbol{u}_{nB-1} \\ \vdots \\ \boldsymbol{u}_{(n-1)B+1} \end{bmatrix} \; (B \times M), \quad \boldsymbol{e}_{B,n} = \begin{bmatrix} \boldsymbol{e}(nB) \\ e(nB-1) \\ \vdots \\ \boldsymbol{e}((n-1)B+1) \end{bmatrix} \triangleq \boldsymbol{d}_{B,n} - \boldsymbol{U}_n w_{n-1}$$

(b) Let $\tilde{\boldsymbol{w}}_n = w^o - \boldsymbol{w}_n$. Introduce the *a posteriori* and *a priori* error vectors $\boldsymbol{e}_{p,n} = \boldsymbol{U}_n \tilde{\boldsymbol{w}}_n$ and $\boldsymbol{e}_{a,n} = \boldsymbol{U}_n \tilde{\boldsymbol{w}}_{n-1}$, respectively. Assuming $\boldsymbol{U}_n$ is full rank, establish the energy conservation relation:

$$\|\tilde{\boldsymbol{w}}_n\|^2 + \boldsymbol{e}_{a,n}^* (\boldsymbol{U}_n \boldsymbol{U}_n^*)^{-1} \boldsymbol{e}_{a,n} = \|\tilde{\boldsymbol{w}}_{n-1}\|^2 + \boldsymbol{e}_{p,n}^* (\boldsymbol{U}_n \boldsymbol{U}_n^*)^{-1} \boldsymbol{e}_{p,n}$$

(c) Argue that in steady-state, and under expectation, the relation of part (b) leads to

$$2\mathsf{E}\,\|\boldsymbol{e}_{a,n}\|^2 = \mu \sigma_v^2 B \,\mathsf{Tr}(R_u) \; + \; \mu \mathsf{Tr}\left[\mathsf{E}\left(\boldsymbol{e}_{a,n} \boldsymbol{e}_{a,n}^* \boldsymbol{U}_n \boldsymbol{U}_n^*\right)\right], \quad n \to \infty$$

(d) Assume that, in steady-state, $\mathsf{E}\,\boldsymbol{e}_{a,n} \boldsymbol{e}_{a,n}^* \approx \left(\mathsf{E}\,|\boldsymbol{e}_a(\infty)|^2\right) \cdot \mathrm{I}$ for small μ, and $\boldsymbol{U}_n$ is independent of $\boldsymbol{e}_{a,n}$ (separation principle). Conclude from part (c) that the EMSE of the block LMS filter is approximated by

$$\zeta^{\text{block LMS}} \approx \frac{\mu \sigma_v^2 \mathsf{Tr}(R_u)}{2 - \mu \mathsf{Tr}(R_u)}$$

Problem VI.14 (Delayless implementation) There is an alternative implementation that ameliorates the delay problem associated with the frequency-domain structure of Fig. 27.8. Consider the structure shown in Fig. VI.1 (for the special case $B = 3$). The input sequence $\{u(i)\}$ is convolved directly with the first $B-1$ coefficients of $G(z)$, i.e., with the transfer function $\{g(0) + g(1)z^{-1} + \ldots + g_{B-2}z^{-(B-2)}\}$, in order to generate an intermediate sequence $\{x(i)\}$. The remaining part of $G(z)$, namely

$$G_B(z) \triangleq G(z) - \sum_{k=0}^{B-2} g(k)z^{-k}$$

is implemented block-wise as in Fig. 27.8 by using instead the subband filters that correspond to $G_B(z)$; this structure is depicted in Fig. VI.1 in terms of serial-to-parallel and parallel-to-serial conversion blocks (representing the banks of decimators and interpolators of Fig. 27.8). The B-dimensional vector at the input of the parallel-to-serial converter is denoted by z_i. Assume $B = 3$.

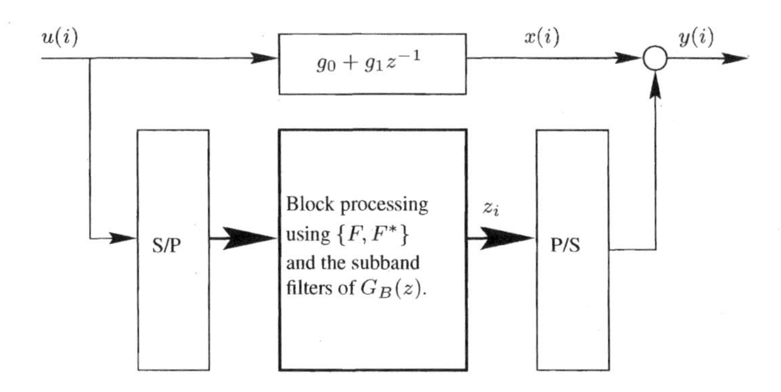

FIGURE VI.1 A delayless alternative to the block implementation of Fig. 27.8. The S/P block employs $2B$ decimators at the downsampling rate of B each, and the P/S block employs B interpolators at the upsampling rate of B each. The block processing relies on $2B \times 2B$ DFT matrices $\{F, F^*\}$ and on $2B$ subband filters corresponding to $G_B(z)$ and, therefore, of order $(M - B + 1)/B$ each.

(a) Verify that, at any particular instant i,

$$x(i) = g(0)u(i) + g(1)u(i-1), \qquad z_i = \begin{bmatrix} g(2)u(i) + g(3)u(i-1) + \ldots \\ g(2)u(i-1) + g(3)u(i-2) + \ldots \\ g(2)u(i-2) + g(3)u(i-3) + \ldots \end{bmatrix}$$

(b) Adding the output of the direct convolution path to the output of the parallel-to-serial converter, verify that

$$\begin{bmatrix} y(i+2) \\ y(i+1) \\ y(i) \end{bmatrix} = \begin{bmatrix} x(i+2) \\ x(i+1) \\ x(i) \end{bmatrix} + z_i$$

$$= \begin{bmatrix} g(0)u(i+2) + g(1)u(i+1) + g(2)u(i) + g(3)u(i-1) + \ldots \\ g(0)u(i+1) + g(1)u(i) + g(2)u(i-1) + g(3)u(i-2) + \ldots \\ g(0)u(i) + g(1)u(i-1) + g(2)u(i-2) + g(3)u(i-3) + \ldots \end{bmatrix}$$

Remark. In other words, $y(i)$ is generated at the output of Fig. VI.1 in response to $u(i)$ and so forth. In this way, the delay problem of Fig. 27.8 is removed. For further details on the results of this problem, see Morgan and Thi (1995).

Problem VI.15 (Unconstrained block filter) Refer to recursion (28.12) for the estimation of the subband adaptive filters in an unconstrained implementation. Define

$$\mathcal{W}_n = \begin{bmatrix} l_{0,n}^{\mathsf{T}} \\ l_{1,n}^{\mathsf{T}} \\ \vdots \\ l_{2B-1,n}^{\mathsf{T}} \end{bmatrix}, \qquad \mathcal{U}_n = \begin{bmatrix} \left(u_{0,n}^{\prime *}\right)^{\mathsf{T}} \\ \left(u_{1,n}^{\prime *}\right)^{\mathsf{T}} \\ \vdots \\ \left(u_{2B-1,n}^{\prime *}\right)^{\mathsf{T}} \end{bmatrix}$$

as well as

$$\mathcal{E}_n = \text{diag}\{e_0'(n), e_1'(n), \ldots, e_{2B-1}'(n)\} \quad \text{and} \quad \Lambda_n = \text{diag}\{\lambda_0(n), \lambda_1(n), \ldots, \lambda_{2B-1}(n)\}$$

where $\{\mathcal{W}_n, \mathcal{U}_n\}$ are $2B \times M/B$ and $\{\mathcal{E}_n, \Lambda_n\}$ are $2B \times 2B$. Verify that the $2B$ recursions (28.12) can be grouped together into a single matrix recursion as follows:

$$\boxed{\begin{array}{l} e_{B,n} = d_{B,n} - \begin{bmatrix} \mathrm{I}_B & 0_{B\times B} \end{bmatrix} F^*\text{diag}\left(\mathcal{W}_{n-1}\mathcal{U}_n^*\right) \\ e_{2B,n}' = F \begin{bmatrix} \mathrm{I}_B \\ 0_{B\times B} \end{bmatrix} e_{B,n} \\ \mathcal{W}_n = \mathcal{W}_{n-1} + \mu\Lambda_n^{-1}\mathcal{E}_n\mathcal{U}_n, \qquad \mathcal{W}_{-1} = 0 \end{array}}$$

where the notation diag$\{A\}$, for a matrix A, is a vector with the diagonal elements of A.

Problem VI.16 (Constrained block filter) Refer to definition (28.16) for the constrained weight-vector estimates, and define the weight matrix

$$\mathcal{W}_n^c = \begin{bmatrix} l_{0,n}^{c\mathsf{T}} \\ l_{1,n}^{c\mathsf{T}} \\ \vdots \\ l_{2B-1,n}^{c\mathsf{T}} \end{bmatrix}$$

Use the result of Prob. VI.15 to show that the constrained estimates $\{l_{k,n}^c\}$ satisfy the matrix recursion:

$$\boxed{\begin{array}{l} e_{B,n} = d_{B,n} - \begin{bmatrix} \mathrm{I}_B & 0_{B\times B} \end{bmatrix} F^*\text{diag}\left(\mathcal{W}_{n-1}^c\mathcal{U}_n^*\right) \\ e_{2B,n}' = F \begin{bmatrix} \mathrm{I}_B \\ 0_{B\times B} \end{bmatrix} e_{B,n} \\ \mathcal{W}_n^c = \mathcal{W}_{n-1}^c + \dfrac{\mu}{2B} F^* \begin{bmatrix} \mathrm{I}_B & \\ & 0_{B\times B} \end{bmatrix} F\Lambda_n^{-1}\mathcal{E}_n\mathcal{U}_n, \quad \mathcal{W}_{-1}^c = 0 \end{array}}$$

Clearly, the factor $1/2B$ can be incorporated into μ. Verify further that $\{\mathcal{W}_n, \mathcal{W}_n^c\}$ are related via

$$\mathcal{W}_n^c = \frac{1}{2B} F^* \begin{bmatrix} \mathrm{I}_B & \\ & 0_{B\times B} \end{bmatrix} F\mathcal{W}_n$$

Problem VI.17 (Decimators and interpolators) Consider the interpolator shown below:

$$x(n) \longrightarrow \boxed{\uparrow B} \longrightarrow y(i)$$

Show that the $z-$transforms of the sequences $\{x(n), y(i)\}$ are related via $Y(z) = X(z^B)$. Conclude that $Y(e^{j\omega}) = X(e^{j\omega B})$ and that, therefore, the frequency spectrum of $\{y(i)\}$ is obtained by compressing the frequency spectrum of $\{x(n)\}$ by a factor B and repeating it periodically over $[0, 2\pi]$. Conclude further that the spectrum of the interpolated signal $\{y(i)\}$ is periodic with period $2\pi/B$.

Likewise, consider the decimator shown below

$$x(i) \longrightarrow \boxed{\downarrow B} \longrightarrow y(n)$$

Show that the $z-$transforms of the sequences $\{x(i), y(n)\}$ are now related as follows:

$$Y(z) = \frac{1}{B}\sum_{k=0}^{B-1} X\left(z^{1/B}e^{-j2\pi k/B}\right)$$

so that

$$Y(e^{j\omega}) = \frac{1}{B}\sum_{k=0}^{B-1} X\left(e^{j\frac{\omega-2\pi k}{B}}\right)$$

In other words, conclude that the spectrum of the decimated signal $\{y(n)\}$ is obtained by expanding the spectrum of $x(i)$ by a factor B and scaling its amplitude by B as well. Conclude that in order to avoid aliasing after decimation, the spectrum of the original signal $\{x(i)\}$ should be limited to the interval $[-\pi/B, \pi/B]$. Figure VI.2 illustrates the spectra of the original, decimated, and interpolated signals for the case $B = 2$.

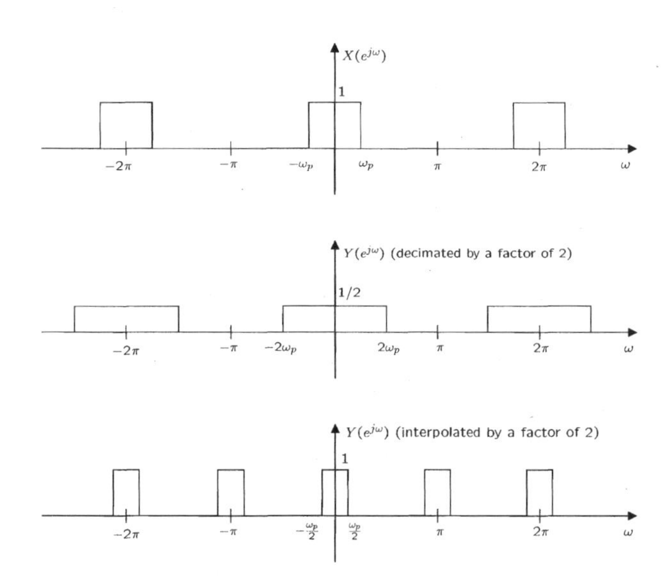

FIGURE VI.2 Spectra of the original (*top*), decimated (*middle*), and interpolated (*bottom*) signals using $B = 2$.

COMPUTER PROJECT

Project VI.1 (Acoustic echo cancellation) Acoustic echo cancellation is a common problem in hands-free telephony, where a person moves freely in a room while talking and listening to a remote speaker. With a loudspeaker and a microphone installed in the same room, as indicated by Fig. VI.3, undesired echoes over the walls interfere with the signal from the local speaker at the microphone. The task of an acoustic echo canceller is to estimate the echo path, reconstruct the echoes, and subtract them from the microphone signal; thus leaving only the signal by the local speaker. Acoustic echo cancellation shares many commonalities with line echo cancellation, as studied in Computer Project IV.1. One main difference is that the length of the acoustic echo path tends to be considerably longer than the length of the line echo path. For this reason, block and subband adaptive filters are useful for applications involving acoustic echo cancellation.

(a) Load the file room.mat, which contains 1024 samples of a measured impulse response sequence of an echo path in a room. Plot the impulse and frequency responses of the echo path.

(b) Load the file css.mat, which contains 5600 samples of a composite source signal. As explained in Computer Project IV.1, this is a synthetic signal that emulates the properties of speech. Concatenate 20 such blocks to form a loudspeaker signal and feed it into the echo path. Train an acoustic echo canceller with 1024 taps using ϵ-NLMS with $\epsilon = 10^{-6}$ and step-sizes $\mu \in \{0.1, 0.25, 0.5, 0.75, 1, 1.25, 1.5, 1.75, 1.9\}$. Use the echo as a reference sequence and the loudspeaker signal as the input to the adaptive filter. Plot the loudspeaker signal, the echo at the microphone, and the error signals that remain after adaptation for both $\mu = 0.5$ and $\mu = 0.1$. For each step-size, compute the relative filter mismatch defined as $\mathcal{S} = \|w_N - \text{room}\|^2/\|\text{room}\|^2$, where w_N is the filter estimate at the end of adaptation. Plot $\mathcal{S}$ as a function of the step-size. Which step-size results in smallest mismatch?

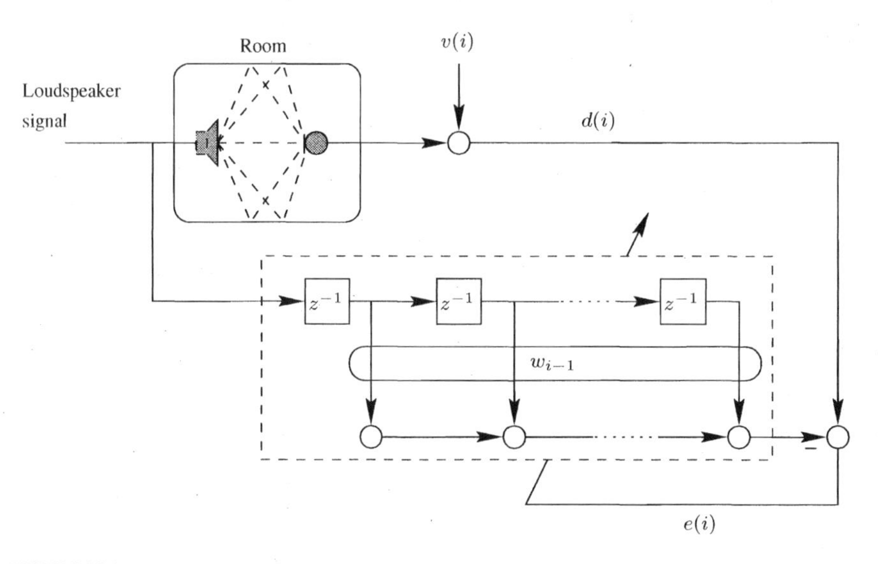

FIGURE VI.3 Adaptive acoustic echo reflections in a room containing a loudspeaker and a microphone.

(c) Limit the length of the loudspeaker signal to $N = 65536$ and choose as block length $B = 32$. Train the constrained DFT-block adaptive filter of Alg. 28.7 with 64 subband filters using both NLMS with power normalization as in (28.12) and standard NLMS as in (28.13). In the latter case set the step-size at $\mu = 0.1$ and in the former case set the step-size at $\mu = 0.1/32$, where 32 is the length of each subband filter. Plot the loudspeaker signal, the resulting echo at the microphone, and the error signals that remain after adaptation by both algorithms. Plot also the resulting mismatches.

(d) Load the file speech.mat and treat it as the signal from the loudspeaker. Adjust its length to a multiple of the block length $B = 32$ and repeat it 20 times for a longer simulation period. Train the acoustic echo canceller using the constrained DFT-block adaptive implementation of Alg. 28.7 with $\mu = 0.03/32$ and listen to the loudspeaker, echo, and error (clean) signals. Also plot the signals.

(e) Using CSS data again, generate a learning curve for ϵ-NLMS using $\mu = 0.5$ and another one for the constrained DFT-block adaptive implementation of Alg. 28.7 using $B = 32$ and $\mu = 0.1/32$. Use as loudspeaker data a first-order auto-regressive process generated by filtering white Gaussian noise through $1/(1 - 0.95z^{-1})$. Normalize the variance of the loudspeaker signal to unity. Generate the curves by averaging over 50 experiments and by smoothing the resulting ensemble-average curves using a sliding window of width 10. Set the noise power at 30 dB below the input signal power.

PART **VII**

LEAST-SQUARES METHODS

CHAPTER 29

Least-Squares Criterion

The earlier parts of this book dealt extensively with the problem of linear *least-mean-squares* estimation, whereby one random variable is estimated from observations of another correlated random variable. For example, in Sec. 8.1 we studied the problem of estimating a zero-mean random variable $\boldsymbol{d}$ from a zero-mean random row vector $\boldsymbol{u}$, by seeking the optimal column vector w that solves

$$\min_w \mathsf{E}\,|\boldsymbol{d} - \boldsymbol{u}w|^2 \tag{29.1}$$

The optimal estimator was found to be

$$\hat{\boldsymbol{d}} = \boldsymbol{u}w^o \qquad \text{where} \qquad w^o = R_u^{-1}R_{du} \tag{29.2}$$

in terms of the second-order moments of $\{\boldsymbol{d}, \boldsymbol{u}\}$, namely,

$$R_u = \mathsf{E}\,\boldsymbol{u}^*\boldsymbol{u} > 0 \quad \text{and} \quad R_{du} = \mathsf{E}\,\boldsymbol{d}\boldsymbol{u}^*$$

The resulting minimum mean-square error was further seen to be given by

$$\text{m.m.s.e.} = \sigma_d^2 - R_{ud}R_u^{-1}R_{du} \;=\; \sigma_d^2 - \sigma_{\hat{d}}^2$$

where $\sigma_d^2 = \mathsf{E}\,|\boldsymbol{d}|^2$ and $\sigma_{\hat{d}}^2 = \mathsf{E}\,|\hat{\boldsymbol{d}}|^2$.

We proceeded in Chapter 8 to devise steepest-descent schemes for evaluating w^o iteratively, and in Chapter 10 we showed how stochastic gradient algorithms can be used to approximate w^o, also iteratively. These latter algorithms were aimed at situations where access to the moments $\{R_{du}, R_u\}$ was not readily available, but only access to realizations $\{d(i), u_i\}$ of the random variables $\{\boldsymbol{d}, \boldsymbol{u}\}$. One such stochastic gradient method was the LMS algorithm (cf. Sec. 10.2):

$$w_i = w_{i-1} + \mu u_i^*[d(i) - u_i w_{i-1}], \qquad w_{-1} = 0$$

Another stochastic gradient method was the exponentially-weighted RLS algorithm described in Sec. 14.1:

$$\begin{aligned} P_i &= \lambda^{-1}\left[P_{i-1} - \frac{\lambda^{-1}P_{i-1}u_i^*u_iP_{i-1}}{1+\lambda^{-1}u_iP_{i-1}u_i^*}\right], \qquad P_{-1} = \epsilon^{-1}\mathrm{I} \\ w_i &= w_{i-1} + P_i\, u_i^*[d(i) - u_i w_{i-1}], \qquad w_{-1} = 0 \end{aligned}$$

for some $0 \ll \lambda \leq 1$. Both LMS and RLS were motivated and derived in Chapter 10 by appealing to instantaneous-gradient approximations.

The purpose of Chapters 29–43 is to study the recursive least-squares algorithm in greater detail. Rather than motivate it as a stochastic gradient *approximation* to a steepest-descent method, as was done in Sec. 14.1, the discussion in these chapters will bring forth deeper insights into the nature of the RLS algorithm. In particular, it will be seen in Chapter 30 that RLS is an *optimal* (as opposed to *approximate*) solution to a well-defined optimization problem. In addition, the discussion will reveal that RLS is very rich in structure, so much so that many equivalent variants exist. While all these variants are mathematically equivalent, they vary among themselves in computational complexity, performance under finite-precision conditions, and even in modularity and ease of implementation.

In this chapter, we start by studying the famed least-squares criterion in its own right. The results obtained here will then be used in Chapter 30 to establish that RLS is in effect a recursive solution to this criterion. In our presentation, we have chosen a treatment that is rich in both geometric and linear algebraic arguments. In so doing, it is hoped that the reader will have an opportunity to appreciate the elegance and beauty of the least-squares theory.

29.1 LEAST-SQUARES PROBLEM

Assume we have available N realizations of the random variables $\boldsymbol{d}$ and $\boldsymbol{u}$, say,

$$\{d(0), d(1), \ldots, d(N-1)\} \quad \text{and} \quad \{u_0, u_1, \ldots, u_{N-1}\}$$

respectively, where the $\{d(i)\}$ are scalars and the $\{u_i\}$ are $1 \times M$. Given the $\{d(i), u_i\}$, and assuming ergodicity,[10] we can approximate the mean-square-error cost in (29.1) by its sample average as

$$\mathsf{E}\,|\mathbf{d} - \boldsymbol{u}w|^2 \;\approx\; \frac{1}{N}\sum_{i=0}^{N-1} |d(i) - u_i w|^2 \tag{29.3}$$

In this way, the optimization problem (29.1) can be replaced by the related problem:

$$\min_{w} \left(\sum_{i=0}^{N-1} |d(i) - u_i w|^2 \right) \tag{29.4}$$

where we have removed the scaling factor $1/N$.

Vector Formulation

The cost function (29.4) can be reformulated in vector notation as follows. We collect the observations $\{d(i)\}$ into an $N \times 1$ vector y and the row vectors $\{u_i\}$ into an $N \times M$ data matrix H:

$$y \;\triangleq\; \begin{bmatrix} d(0) \\ d(1) \\ d(2) \\ \vdots \\ d(N-1) \end{bmatrix}, \qquad H \;\triangleq\; \begin{bmatrix} u_0 \\ u_1 \\ u_2 \\ \vdots \\ u_{N-1} \end{bmatrix}$$

Then (29.4) can be rewritten as

$$\min_{w} \; \|y - Hw\|^2 \tag{29.5}$$

[10] As explained before in Chapter 19, an ergodic random process is one for which time-averages coincide with ensemble averages.

where the notation $\|\cdot\|^2$ denotes the squared Euclidean norm of its argument, namely, $\|a\|^2 = a^*a$ for any column vector a. Problem (29.5) is known as the standard least-squares problem.

Of course, we could have posed problem (29.5) directly, without going through (29.1) and its approximation by (29.3). Here we have opted to motivate the least-squares criterion by relating it to the mean-square-error criterion (29.1), which we studied extensively in Chapters 3–10.

Definition 29.1 (Least-squares problem) Given an $N \times 1$ vector y and an $N \times M$ data matrix H, the least-squares problem seeks an $M \times 1$ vector w that solves $\min_w \|y - Hw\|^2$.

Two cases can occur depending on the relation between the dimensions $\{N, M\}$:

1. **Over-determined least-squares** ($N \geq M$): In this case, the data matrix H has at least as many rows as columns, so that the number of measurements (i.e., the number of entries in y) is at least equal to the number of unknowns (i.e., the number of entries in w). This situation corresponds to an *over-determined* least-squares problem and, as we shall see, (29.5) will either have a unique solution or an infinite number of solutions.

2. **Under-determined least-squares** ($N < M$): In this case, the data matrix H has fewer rows than columns, so that the number of measurements is less than the number of unknowns. This situation corresponds to an *under-determined* least-squares problem for which (29.5) will have an infinite number of solutions.

The purpose of the discussion that follows is to show that all solutions $\widehat{w}$ to the least-squares problem (29.5) are characterized as solutions to the linear system of equations

$$\boxed{H^*H\widehat{w} = H^*y}$$

which are known as the *normal equations*. In addition, the discussion will clarify under what conditions a unique $\widehat{w}$ exists, as opposed to infinitely many, and it will highlight an important orthogonality property of least-squares solutions. In our presentation, we shall use both geometric and algebraic derivations to establish these facts. We start with the geometric argument and later show how to arrive at the same conclusions by means of algebraic arguments.

29.2 GEOMETRIC ARGUMENT

Our objective is to characterize all solutions of (29.5). We thus note first that, for any w, the vector Hw lies in the column span (or range space) of the data matrix H, written as $Hw \in \mathcal{R}(H)$. Therefore, the least-squares criterion (29.5) is such that it seeks a column vector in the range space of H that is closest to y in the Euclidean norm sense. Specifically, the least-squares problem seeks a $\widehat{w}$ such that $H\widehat{w}$ is closest to y.

Now we know from Euclidean geometry that the closest vector to y within $\mathcal{R}(H)$ is such that the residual vector, $y - H\widehat{w}$, is orthogonal to all vectors in $\mathcal{R}(H)$ as illustrated in Fig. 29.1. For readers not familiar with the geometry of vectors in Euclidean space,

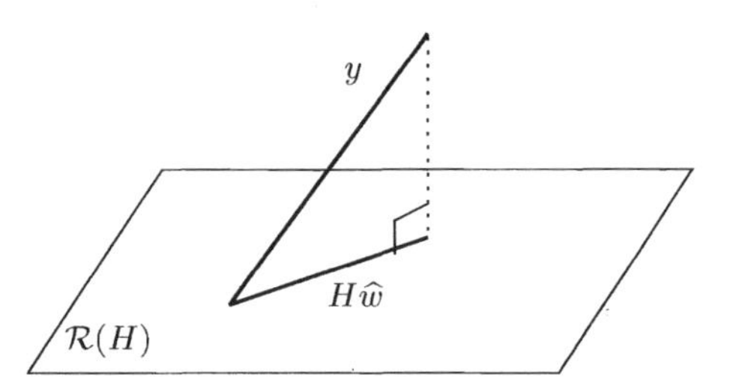

FIGURE 29.1 A least-squares solution is obtained when $y - H\widehat{w}$ is orthogonal to $\mathcal{R}(H)$.

the algebraic derivations in Sec. 29.3 will arrive at the same conclusion and will therefore provide a justification for this claim.

Therefore, it must hold that any candidate solution $\widehat{w}$ should result in a residual vector, $y - H\widehat{w}$, that is orthogonal to Hp, for any vector p or, equivalently, $p^*H^*(y - H\widehat{w}) = 0$. Clearly, the only vector that is orthogonal to any vector p is the zero vector, so that we must have

$$H^*(y - H\widehat{w}) = 0 \tag{29.6}$$

and we conclude that any solution $\widehat{w}$ of the least-squares problem (29.5) must satisfy the so-called normal equations:

$$\boxed{H^*H\widehat{w} = H^*y} \tag{29.7}$$

These equations are always consistent, i.e., a solution $\widehat{w}$ always exists. This is because, as was shown earlier in Sec. B.2, the matrices H^*H and H^* have the same column span, i.e., $\mathcal{R}(H^*) = \mathcal{R}(H^*H)$.

For any solution $\widehat{w}$ of (29.7), we denote the resulting closest vector to y by $\widehat{y} = H\widehat{w}$ and we refer to it as the *projection* of y onto $\mathcal{R}(H)$:

$$\boxed{\widehat{y} = H\widehat{w} \stackrel{\Delta}{=} \text{projection of } y \text{ onto } \mathcal{R}(H)} \tag{29.8}$$

We show in Thm. 29.2 further ahead that even when the normal equations (29.7) have a multitude of solutions $\widehat{w}$, all of them will lead to the *same* value for $\widehat{y} = H\widehat{w}$. After all, from a geometric point of view, projecting y onto $\mathcal{R}(H)$ results in a unique projection $\widehat{y}$. What the many $\widehat{w}'s$ amount to, when they exist, are equivalent representations for this unique $\widehat{y}$ in terms of the columns of H.

We shall denote the corresponding residual vector by

$$\boxed{\widetilde{y} \stackrel{\Delta}{=} y - H\widehat{w}}$$

so that the orthogonality condition (29.6) can be rewritten as

$$\boxed{H^*\widetilde{y} = 0} \qquad \text{(orthogonality condition)} \tag{29.9}$$

We shall often express this orthogonality condition more succinctly as $\widetilde{y} \perp \mathcal{R}(H)$, where the $\perp$ notation is used to mean that $\widetilde{y}$ is orthogonal to any vector in the range space (column span) of H. In particular, since, by construction, $\widehat{y} \in \mathcal{R}(H)$, it also holds that

$$\widetilde{y} \perp \widehat{y} \quad \text{or} \quad \widehat{y}^*\widetilde{y} = 0$$

Let ξ denote the minimum cost of (29.5). It can be evaluated as follows:

$$\begin{aligned}
\xi &= \|y - H\widehat{w}\|^2 \\
&= (y - H\widehat{w})^*(y - H\widehat{w}) \\
&= y^*(y - H\widehat{w}), \quad \text{since } \widehat{w}^* H^* \widetilde{y} = 0 \text{ by } (29.9) \\
&= y^*y - y^*H\widehat{w} \\
&= y^*y - \widehat{w}^*H^*H\widehat{w}, \quad \text{since } y^*H = \widehat{w}^*H^*H \text{ by } (29.7) \\
&= y^*y - \widehat{y}^*\widehat{y}
\end{aligned}$$

That is, we obtain the following equivalent representations for the minimum cost:

$$\boxed{\xi = \|y\|^2 - \|\widehat{y}\|^2 = y^*\widetilde{y}} \tag{29.10}$$

In summary, we arrive at the following statement.

Theorem 29.1 (The normal equations) A vector $\widehat{w}$ solves the least-squares problem (29.5) if, and only if, it satisfies the normal equations

$$H^*H\widehat{w} = H^*y$$

or, equivalently, if and only if, it satisfies the orthogonality condition

$$y - H\widehat{w} \perp \mathcal{R}(H)$$

The normal equations are always consistent, i.e., a solution $\widehat{w}$ always exists and the resulting minimum cost is given by either expression:

$$\xi = \|y\|^2 - \|\widehat{y}\|^2 = y^*\widetilde{y}$$

where $\widehat{y} = H\widehat{w}$ is the projection of y onto $\mathcal{R}(H)$ and $\widetilde{y} = y - \widehat{y}$ is the residual vector.

29.3 ALGEBRAIC ARGUMENTS

Before summarizing the above discussion, and before enumerating additional properties of the least-squares solution, we proceed to re-derive the above results by means of two independent algebraic arguments.

Differentiation Argument

Let $J(w)$ denote the cost function in (29.5), i.e.,

$$J(w) \triangleq \|y - Hw\|^2 = \|y\|^2 - y^*Hw - w^*H^*y + w^*H^*Hw \tag{29.11}$$

Differentiating $J(w)$ with respect to w we find that its gradient vector evaluates to zero at all $\widehat{w}$ that satisfy

$$-y^*H + \widehat{w}^*H^*H = 0$$

which are again the normal equations (29.7). The solution(s) $\widehat{w}$ so obtained correspond to minima of $J(w)$ since its Hessian matrix is nonnegative-definite, i.e.,

$$\nabla_w^2[J(w)] = H^*H \geq 0$$

Completion-of-Squares Argument[12]

Starting from the least-squares criterion (29.5), we can alternatively resort to a completion-of-squares argument in order to characterize its solution(s). We used this technique on a handful of occasions before (e.g., in Sec. 3.3).

Thus, note that $J(w)$ in (29.11) is quadratic in w and it can be written as

$$J(w) = \begin{bmatrix} y^* & w^* \end{bmatrix} \begin{bmatrix} \mathrm{I}_N & -H \\ -H^* & H^*H \end{bmatrix} \begin{bmatrix} y \\ w \end{bmatrix} \tag{29.12}$$

Now the central matrix in (29.12) can be factored into a product of block upper-diagonal-lower triangular matrices as follows (recall (3.12)):

$$\begin{bmatrix} \mathrm{I}_N & -H \\ -H^* & H^*H \end{bmatrix} = \begin{bmatrix} \mathrm{I}_N & D \\ 0 & \mathrm{I}_M \end{bmatrix} \begin{bmatrix} \mathrm{I}_N + DH^* & 0 \\ 0 & H^*H \end{bmatrix} \begin{bmatrix} \mathrm{I}_N & 0 \\ D^* & \mathrm{I}_M \end{bmatrix} \tag{29.13}$$

where D is chosen as any $N \times M$ matrix satisfying the linear equations

$$D(H^*H) = -H \tag{29.14}$$

For compactness of notation, we are not indicating explicitly the dimensions of the block zero entries in (29.13); these dimensions can be inferred from the dimensions of $\{D, H\}$ and the identity matrices. The equations (29.14) are consistent since the matrices H^*H and H^* have the same column spans and, therefore, by writing (29.14) as $H^*HD^* = -H^*$, we conclude that a solution D always exists.

The triangular factorization (29.13), in this general form, i.e., in terms of a matrix D that solves (29.14), allows us to accommodate the case of singular H^*H. Of course, if H^*H were invertible, then $D = -H(H^*H)^{-1}$, and we could replace the factorization (29.13) by the more explicit (and familiar) form:

$$\begin{bmatrix} \mathrm{I}_N & -H \\ -H^* & H^*H \end{bmatrix}$$

$$= \begin{bmatrix} \mathrm{I}_N & -H(H^*H)^{-1} \\ 0 & \mathrm{I}_M \end{bmatrix} \begin{bmatrix} \mathrm{I}_N - H(H^*H)^{-1}H^* & 0 \\ 0 & H^*H \end{bmatrix} \begin{bmatrix} \mathrm{I}_N & 0 \\ -(H^*H)^{-1}H^* & \mathrm{I}_M \end{bmatrix}$$

However, we shall continue to use (29.13) in order to accommodate the general case of possibly singular H^*H. Substituting (29.13) into (29.12) we find that

$$J(w) = y^*(\mathrm{I} + DH^*)y \; + \; (w + D^*y)^*H^*H(w + D^*y) \tag{29.15}$$

where only the second term depends on the unknown w. This second term is always nonnegative since

$$(w + D^*y)^*H^*H(w + D^*y) \; = \; \|H(w + D^*y)\|^2 \geq 0 \quad \text{for any } w$$

[12]This section can be skipped on a first reading.

Therefore,

$$J(w) \geq y^*(\mathrm{I} + DH^*)y, \qquad \text{for any } w$$

and the minimum of $J(w)$ is attained when w is chosen to annihilate the second term of (29.15), i.e., when it is chosen as any $\widehat{w}$ satisfying $H(\widehat{w} + D^*y) = 0$. In this way, we arrive at the conclusion that all solutions to (29.5) are vectors $\widehat{w}$ that satisfy

$$\boxed{H\widehat{w} = -HD^*y, \quad \text{where } D \text{ is any solution to } DH^*H = -H} \tag{29.16}$$

This description is equivalent to the normal equations (29.7). To see this, we multiply (29.16) by H^* from the left, and use the definition of D from (29.14), to conclude that $\widehat{w}$ satisfies the normal equations

$$H^*H\widehat{w} = H^*y \tag{29.17}$$

Conversely, if $\widehat{w}$ satisfies (29.17), then $H^*H\widehat{w} = -H^*HD^*y$ or, equivalently,

$$H^*H(\widehat{w} + D^*y) = 0$$

This means that

$$\widehat{w} + D^*y \in \mathcal{N}(H^*H)$$

But since we already know from Sec. B.2 that the matrices H^*H and H have the same nullspace, we conclude that

$$\widehat{w} + D^*y \in \mathcal{N}(H)$$

That is, $H(\widehat{w} + D^*y) = 0$, which agrees with (29.16).

Note that the equations (29.17) re-confirm our earlier remark in (29.6) that the residual vector, $y - H\widehat{w}$, satisfies the orthogonality condition

$$H^*(y - H\widehat{w}) = 0 \tag{29.18}$$

Moreover, the resulting minimum value of the cost function will be

$$\begin{aligned} J(\widehat{w}) &= y^*(\mathrm{I} + DH^*)y \\ &= y^*y - \widehat{y}^*y, \quad \text{since } \widehat{y} = H\widehat{w} = -HD^*y \text{ by } (29.16) \\ &= y^*y - \widehat{y}^*(\widehat{y} + \widetilde{y}) \\ &= y^*y - \widehat{y}^*\widehat{y}, \quad \text{since } \widehat{y} \in \mathcal{R}(H) \text{ and } \widehat{y}^*\widetilde{y} = 0 \text{ by } (29.18) \\ &= \|y\|^2 - \|\widehat{y}\|^2 \end{aligned}$$

as in (29.10).

29.4 PROPERTIES OF LEAST-SQUARES SOLUTION

Actually, we can say more about the solution of the least-squares problem. The following statement enumerates several additional properties. In particular, it specifies when a unique solution $\widehat{w}$ exists and, perhaps more interestingly, it states that even when many solutions $\widehat{w}$ exist, the projection $\widehat{y} = H\widehat{w}$ is always unique!

Theorem 29.2 (Properties of solutions) The solution of the least-squares problem (29.5) has the following properties:

1. The solution $\widehat{w}$ is unique if, and only if, the data matrix H has full column rank (i.e., all its columns are linearly independent, which necessarily requires $N \geq M$). In this case, $\widehat{w}$ is given by

$$\widehat{w} = (H^*H)^{-1}H^*y$$

This situation occurs only for over-determined least-squares problems.

2. When H^*H is singular, then infinitely many solutions $\widehat{w}$ exist and any two solutions differ by a vector in the nullspace of H, i.e., if $\widehat{w}_1$ and $\widehat{w}_2$ are any two solutions, then $\widehat{w}_1 - \widehat{w}_2 \in \mathcal{N}(H)$. This situation can occur for both over- and under-determined least-squares problems.

3. When many solutions $\widehat{w}$ exist, regardless of which one we pick, the resulting projection vector $\widehat{y} = H\widehat{w}$ is the same and the resulting minimum cost is also the same and given by $\xi = \|y\|^2 - \|\widehat{y}\|^2$.

4. When many solutions $\widehat{w}$ exist, the one that has the smallest Euclidean norm, namely, the one that solves

$$\min_{\widehat{w}} \|\widehat{w}\|^2 \quad \text{subject to} \quad H^*H\widehat{w} = H^*y$$

is given by $\widehat{w} = H^\dagger y$, where $H^\dagger$ denotes the pseudo-inverse of H.

Proof: We proceed in steps.

1. The normal equations $H^*H\widehat{w} = H^*y$ have a unique solution if, and only if, H^*H is invertible. This condition cannot happen when $N < M$ since the product H^*H will be rank deficient. Therefore, we must have $N \geq M$. Moreover, recall that we proved in Sec. B.2 that for any matrix H having at least as many rows as columns, it holds that H^*H is invertible if, and only if, H has full rank. These facts establish the first statement in the theorem.

2. Let r be any nonzero vector in the nullspace of H^*H, i.e., $H^*Hr = 0$. If $\widehat{w}$ solves the normal equations, i.e., if $H^*H\widehat{w} = H^*y$, then so does $\widehat{w}+r$ since $H^*H(\widehat{w}+r) = H^*y$. Therefore, infinitely many solutions to the normal equations exist in this case. Moreover, if $\widehat{w}_1$ and $\widehat{w}_2$ are any two solutions, say, $H^*H\widehat{w}_1 = H^*y$ and $H^*H\widehat{w}_2 = H^*y$ then $H^*H(\widehat{w}_1 - \widehat{w}_2) = 0$. That is, $\widehat{w}_1 - \widehat{w}_2 \in \mathcal{N}(H^*H)$. However, we proved in Sec. B.2 that, for any matrix H, the matrices H^*H and H have the same nullspace and, hence, $H(\widehat{w}_1 - \widehat{w}_2) = 0$. These facts establish the second statement in the theorem.

3. Let $\widehat{w}_1$ and $\widehat{w}_2$ denote any two solutions when multiple solutions exist and let $\widehat{y}_1 = H\widehat{w}_1$ and $\widehat{y}_2 = H\widehat{w}_2$ denote the corresponding projections. Then $\widehat{y}_1 - \widehat{y}_2 = H(\widehat{w}_1 - \widehat{w}_2) = 0$ since, by the second property, $\widehat{w}_1 - \widehat{w}_2 \in \mathcal{N}(H)$. Therefore, $\widehat{y}_1 = \widehat{y}_2$, which establishes the third statement in the theorem.

4. We first remark that, for a general matrix H, the pseudo-inverse is defined in Sec. B.6, where the fourth statement in the theorem is also proven (see Lemma B.7). Here we note that when H has full rank, its pseudo-inverse is given by the following expressions:

$$H^\dagger = \begin{cases} (H^*H)^{-1}H^* & \text{when } N > M \text{ (a “tall” matrix)} \\ H^*(HH^*)^{-1} & \text{when } N < M \text{ (a “fat” matrix)} \\ H^{-1} & \text{when } N = M \text{ (a square matrix)} \end{cases}$$

When H is rank-deficient, it is more convenient to define its pseudo-inverse in terms of its singular value decomposition, as explained in Sec. B.6. [See also Prob. VII.6 for a proof, from first principles, of the fourth statement of the theorem in the under-determined case.]

29.5 PROJECTION MATRICES

We restrict ourselves in this section to the case of over-determined least-squares problems with a full-rank data matrix H (and, hence, $N \geq M$). In this case, the coefficient matrix H^*H is invertible (actually positive-definite) and the least-squares problem (29.5) will have a unique solution that is given by

$$\widehat{w} = (H^*H)^{-1}H^*y$$

with the corresponding projection vector

$$\widehat{y} = H\widehat{w} = H(H^*H)^{-1}H^*y$$

The matrix multiplying y in the above expression is called the *projection* matrix and we denote it by

$$\boxed{\mathcal{P}_H \stackrel{\Delta}{=} H(H^*H)^{-1}H^*, \quad \text{when } H \text{ has full column rank}} \tag{29.19}$$

The designation *projection matrix* stems from the fact that multiplying y by $\mathcal{P}_H$ amounts to projecting it onto the column span of H. Such projection matrices play a prominent role in least-squares theory and they have many useful properties. For example, projection matrices are Hermitian and also idempotent, i.e., they satisfy

$$\boxed{\mathcal{P}_H^* = \mathcal{P}_H, \qquad \mathcal{P}_H^2 = \mathcal{P}_H} \tag{29.20}$$

Note further that the residual vector,

$$\widetilde{y} = y - H\widehat{w}$$

is given by

$$\widetilde{y} \ = \ y - \mathcal{P}_H y \ = \ (\mathrm{I} - \mathcal{P}_H)y \ = \ \mathcal{P}_H^{\perp} y$$

so that the matrix

$$\boxed{\mathcal{P}_H^{\perp} \stackrel{\Delta}{=} \mathrm{I} - \mathcal{P}_H}$$

is called the projection matrix onto the orthogonal complement space of H. It is also easy to see that the minimum cost of the least-squares problem (29.5) can be expressed in terms of $\mathcal{P}_H^{\perp}$ as follows:

$$\begin{aligned} \xi &= y^*y - \widehat{y}^*\widehat{y} \\ &= y^*y - y^*\mathcal{P}_H^*\mathcal{P}_H y \\ &= y^*y - y^*\mathcal{P}_H y, \qquad \text{since } \mathcal{P}_H^*\mathcal{P}_H = \mathcal{P}_H^2 = \mathcal{P}_H \\ &= y^*\mathcal{P}_H^{\perp} y \end{aligned}$$

Lemma 29.1 (Unique solution) When the matrix H has full-column rank (and, hence, $N \geq M$), the least-squares problem (29.5) will have a unique solution that is given by $\widehat{w} = (H^*H)^{-1}H^*y$. Moreover, the projection of y onto $\mathcal{R}(H)$, and the corresponding residual vector, are given by $\widehat{y} = \mathcal{P}_H y$ and $\widetilde{y} = \mathcal{P}_H^{\perp} y$ so that y can be decomposed as

$$y \;=\; \widehat{y} + \widetilde{y} \;=\; \mathcal{P}_H y \;+\; \mathcal{P}_H^{\perp} y$$

with $\|y\|^2 = \|\widehat{y}\|^2 + \|\widetilde{y}\|^2$. The resulting minimum cost is $\xi = y^*\mathcal{P}_H^{\perp} y$.

29.6 WEIGHTED LEAST-SQUARES

It is often the case that weighting is incorporated into the cost function of the least-squares problem, so that (29.5) is replaced by

$$\boxed{\min_w \; (y - Hw)^* W (y - Hw)} \qquad W > 0 \tag{29.21}$$

where W is a Hermitian positive-definite matrix. For example, when W is diagonal, its elements assign different weights to the entries of the error vector $y - Hw$.

We shall often rewrite the cost function in (29.21) more compactly as

$$\boxed{\min_w \; \|y - Hw\|_W^2} \tag{29.22}$$

where, for any column vector x, the notation $\|x\|_W^2$ refers to the weighted Euclidean norm of x, i.e., $\|x\|_W^2 = x^*Wx$.

One way to solve (29.22) is to show that it reduces to the standard form (29.5). To see this, we introduce the eigen-decomposition

$$W = V\Delta V^*$$

where Δ is diagonal with positive entries and V is unitary, i.e., it satisfies $VV^* = V^*V = I$. Let $\Delta^{1/2}$ denote the diagonal matrix whose entries are equal to the positive square-roots of the entries of Δ, and define the change of variables:

$$a \stackrel{\Delta}{=} \Delta^{1/2}V^*y, \qquad A \stackrel{\Delta}{=} \Delta^{1/2}V^*H \tag{29.23}$$

Observe that since both $\Delta^{1/2}$ and V are invertible, it follows that A has full column rank if, and only if, H has full column rank.

Using the variables $\{a, A\}$ so defined, we can rewrite the weighted problem (29.21) in the equivalent form

$$\min_w \; \|a - Aw\|^2 \tag{29.24}$$

which is of the same form as the standard (unweighted) least-squares problem (29.5). Therefore, we can extend all the results we obtained for (29.5) to the weighted version (29.21) by working with (29.24) instead. In particular, we readily conclude from (29.24) that any solution $\widehat{w}$ to (29.21) should satisfy the orthogonality condition $A^*(a - A\widehat{w}) = 0$

(cf. (29.6)), which, upon using the definitions (29.23) for $\{a, A\}$, can be rewritten in terms of the original data $\{y, H\}$ as

$$H^*V\Delta^{1/2}\left(\Delta^{1/2}V^*y - \Delta^{1/2}V^*H\widehat{w}\right) = 0$$

or, equivalently,

$$\boxed{H^*W(y - H\widehat{w}) = 0} \tag{29.25}$$

Comparing with the orthogonality condition (29.6) in the unweighted case, we see that the only difference is the presence of the weighting matrix W. This conclusion suggests that we can extend to the weighted least-squares setting the same geometric properties of the standard least-squares setting if we simply employ the concept of *weighted* inner products. Specifically, for any two column vectors $\{c, d\}$, we can define their weighted inner product as $\langle c, d\rangle_W = c^*Wd$, and then say that c and d are orthogonal whenever their weighted inner product is zero. Using this definition, we can interpret (29.25) to mean that the residual vector, $y - H\widehat{w}$, is orthogonal to the column span of H in a weighted sense, i.e.,

$$\langle q, y - H\widehat{w}\rangle_W = 0 \quad \text{for any } q \in \mathcal{R}(H)$$

We further conclude from (29.25) that the normal equations (29.7) are now replaced by

$$\boxed{H^*WH\widehat{w} = H^*Wy} \tag{29.26}$$

Proceeding in this manner, and applying the results of the previous sections (especially Thms. 29.1 and 29.2) to (29.24), we arrive at the following statement — see Prob. VII.9.

Theorem 29.3 (Weighted least-squares) A vector $\widehat{w}$ is a solution of the weighted least-squares problem (29.21) if, and only if, it satisfies the normal equations $H^*WH\widehat{w} = H^*Wy$. Moreover, the following properties hold:

1. These normal equations are always consistent, i.e., a solution $\widehat{w}$ always exists.

2. The solution $\widehat{w}$ is unique if, and only if, the data matrix H has full column rank, which necessarily requires $N \geq M$. In this case, $\widehat{w}$ is given by $\widehat{w} = (H^*WH)^{-1}H^*Wy$.

3. When H^*WH is singular, which is equivalent to H^*H being singular, then many solutions $\widehat{w}$ exist and any two solutions differ by a vector in the nullspace of H, i.e., if $\widehat{w}_1$ and $\widehat{w}_2$ are any two solutions, then $H(\widehat{w}_1 - \widehat{w}_2) = 0$.

4. When many solutions $\widehat{w}$ exist, regardless of which one we pick, the resulting projection vector $\widehat{y} = H\widehat{w}$ is the same and the resulting minimum cost is also the same and given by either expression: $\xi = \|y\|_W^2 - \|\widehat{y}\|_W^2 = y^*W\widetilde{y}$, where $\widetilde{y} = H\widehat{w}$.

5. When many solutions $\widehat{w}$ exist, the one that has the smallest Euclidean norm, namely, the one that solves

$$\min_{\widehat{w}} \|\widehat{w}\|^2 \text{ subject to } H^*WH\widehat{w} = H^*Wy$$

is given by $\widehat{w} = A^\dagger a$, where $A = \Delta^{1/2}V^*H$ and $a = \Delta^{1/2}V^*y$.

Note that in the special case of over-determined least-squares problems with full-rank data matrices H (and, hence, $N \geq M$), problem (29.21) will have a unique solution that is given by

$$\widehat{w} = (H^*WH)^{-1}H^*Wy$$

with the corresponding projection vector

$$\widehat{y} = H\widehat{w} = H(H^*WH)^{-1}H^*Wy$$

We shall continue to refer to the matrix multiplying y in the above expression as a *projection* matrix,

$$\boxed{\mathcal{P}_H \stackrel{\Delta}{=} H(H^*WH)^{-1}H^*W, \quad \text{when } H \text{ has full column rank}} \tag{29.27}$$

and write $\widehat{y} = \mathcal{P}_H y$. In contrast to the unweighted case (29.20), the matrix $\mathcal{P}_H$ now satisfies the properties:

$$\boxed{\mathcal{P}_H^* W = W\mathcal{P}_H, \qquad \mathcal{P}_H^2 = \mathcal{P}_H, \quad \mathcal{P}_H^* W \mathcal{P}_H = W\mathcal{P}_H}$$

and the minimum cost of (29.21) can be expressed in terms of $\mathcal{P}_H^\perp$ as follows:

$$\xi \;=\; y^*Wy - \widehat{y}^*W\widehat{y} \;=\; y^*Wy - y^*W\mathcal{P}_H y \;=\; y^*W\mathcal{P}_H^\perp y$$

where $\mathcal{P}_H^\perp = \mathrm{I} - \mathcal{P}_H$.

29.7 REGULARIZED LEAST-SQUARES

A second variation of the standard least-squares problem (29.5) is regularized least-squares. In this formulation, we seek a vector $\widehat{w}$ that solves

$$\boxed{\min_w \left[\, (w - \bar{w})^*\Pi(w - \bar{w}) \;+\; \|y - Hw\|^2 \,\right]} \tag{29.28}$$

where, compared with (29.5), we are now incorporating the so-called regularization term $\|w - \bar{w}\|_\Pi^2$. Here, Π is a positive-definite matrix, usually a multiple of the identity, and $\bar{w}$ is a given column vector, usually $\bar{w} = 0$.

One motivation for using regularization is that it allows us to incorporate some *a priori* information about the solution into the problem statement. Assume, for instance, that we set $\Pi = \delta I$ and choose δ as a large positive number. Then, the first term in the cost function (29.28) becomes dominant and it is not hard to imagine that the cost will be minimized by a vector $\widehat{w}$ that is close to $\bar{w}$ in order to offset the dominant effect of this first term. For this reason, we say that a "large" Π reflects high confidence that $\bar{w}$ is a good guess for the solution $\widehat{w}$. On the other hand, a "small" Π indicates a high degree of uncertainty in the initial guess $\bar{w}$.

Another reason for regularization is that it relieves problems associated with rank deficiency in the data matrix H. To clarify this issue, we first need to solve (29.28). The solution can be obtained in many ways (including, e.g., plain differentiation of the cost function with respect to w as was done in Sec. 29.3). We choose instead to solve (29.28) by showing again how it can be reduced to the solution of a standard least-squares problem of the form (29.5). This line of argument helps clarify the role of the orthogonality condition in regularized least-squares.

Thus, introduce the change of variables $z = w - \bar{w}$ and $b = y - H\bar{w}$, so that the regularized cost function in (29.28) becomes

$$\min_z \left[z^* \Pi z + \|b - Hz\|^2 \right] \tag{29.29}$$

Introduce further the eigen-decomposition of Π, say, $\Pi = U\Lambda U^*$ where U is unitary and Λ has positive diagonal entries. Let $\Lambda^{1/2}$ denote the diagonal matrix whose entries are the positive square-roots of the entries of Λ. Then we can rewrite the cost in (29.29) in the equivalent form

$$\min_z \left\| \begin{bmatrix} 0 \\ b \end{bmatrix} - \begin{bmatrix} \Lambda^{1/2}U^* \\ H \end{bmatrix} z \right\|^2 \tag{29.30}$$

This problem is now of the same form as the standard least-squares problem (29.5), where the roles of the vector y and the data matrix H are played by

$$\begin{bmatrix} 0 \\ b \end{bmatrix} \quad \text{and} \quad \begin{bmatrix} \Lambda^{1/2}U^* \\ H \end{bmatrix}$$

respectively. The orthogonality condition (29.6) of least-squares solutions then shows that any solution $\widehat{z}$ must satisfy

$$\begin{bmatrix} \Lambda^{1/2}U^* \\ H \end{bmatrix}^* \left(\begin{bmatrix} 0 \\ b \end{bmatrix} - \begin{bmatrix} \Lambda^{1/2}U^* \\ H \end{bmatrix} \widehat{z} \right) = 0 \tag{29.31}$$

which, upon using $\widehat{z} = \widehat{w} - \bar{w}$, leads to the linear system of equations:

$$\boxed{[\Pi + H^*H]\,(\widehat{w} - \bar{w}) = H^*(y - H\bar{w})} \tag{29.32}$$

These equations characterize the solution of the regularized least-squares problem (29.28). When $\bar{w} = 0$, the equations reduce to

$$[\Pi + H^*H]\,\widehat{w} = H^*y \tag{29.33}$$

Compared with the normal equations (29.7) for the standard least-squares problem (29.5), we see that the presence of the positive-definite matrix Π in (29.33) guarantees an invertible coefficient matrix since $\Pi + H^*H > 0$, regardless of whether H is rank-deficient or not and regardless of how the row and column dimensions of H compare to each other. For this reason, the solution $\widehat{w}$ of the regularized least-squares problem, as given by (29.32), exists and is always unique.

Using the expression given in Thm. 29.1 for the minimum cost of a standard least-squares problem, we can similarly evaluate the minimum cost of the regularized least-squares problem (29.30) as

$$\begin{aligned} \xi &= \begin{bmatrix} 0 \\ b \end{bmatrix}^* \left(\begin{bmatrix} 0 \\ b \end{bmatrix} - \begin{bmatrix} \Lambda^{1/2}U^* \\ H \end{bmatrix} \widehat{z} \right) \\ &= b^*(b - H\widehat{z}) \\ &= (y - H\bar{w})^*[(y - H\bar{w}) - H(\widehat{w} - \bar{w})] \\ &= (y - H\bar{w})^*(y - H\widehat{w}) \\ &= (y - H\bar{w})^*\widetilde{y} \end{aligned} \tag{29.34}$$

where

$$\widetilde{y} = y - \widehat{y} = y - H\widehat{w}$$

An equivalent expression for ξ can be obtained as follows:

$$\begin{aligned}\xi &= (y-H\bar{w})^*\left[(y-H\bar{w}) \;-\; H(\widehat{w}-\bar{w})\right] \\ &= (y-H\bar{w})^*\left[\mathrm{I}-H[\Pi+H^*H]^{-1}H^*\right](y-H\bar{w}) \\ &= (y-H\bar{w})^*\left[\mathrm{I}+H\Pi^{-1}H^*\right]^{-1}(y-H\bar{w}) \end{aligned} \tag{29.35}$$

where we used the matrix inversion lemma (5.4) in the last step. Note that expression (29.34) does not include Π explicitly.

Finally, observe that the orthogonality condition (29.31) can be written as

$$\begin{bmatrix} U\Lambda^{1/2} & H^* \end{bmatrix}\begin{bmatrix} -\Lambda^{1/2}U^*\widehat{z} \\ \widetilde{b} \end{bmatrix} = 0$$

where

$$\widetilde{b} = b - H\widehat{z} = (y-H\bar{w}) - H(\widehat{w}-\bar{w}) = y - \widehat{y} = \widetilde{y}$$

Recalling that $\Pi = U\Lambda U^*$ and $\widehat{z} = \widehat{w} - \bar{w}$, the above orthogonality condition can be rewritten more compactly as

$$\boxed{H^*\widetilde{y} = \Pi(\widehat{w}-\bar{w})} \quad \text{(orthogonality condition)} \tag{29.36}$$

In the absence of regularization, i.e., when $\Pi = 0$, the above result reduces to the standard orthogonality condition (29.9), namely, it becomes $H^*\widetilde{y} = 0$.

Theorem 29.4 (Regularized least-squares) The solution of the regularized least-squares problem (29.28) is always unique and given by

$$\widehat{w} = \bar{w} + \left[\Pi + H^*H\right]^{-1} H^*(y - H\bar{w})$$

The resulting minimum cost is given by either expression:

$$\xi \;=\; (y-H\bar{w})^*\widetilde{y} \;=\; (y-H\bar{w})^*\left[\mathrm{I}+H\Pi^{-1}H^*\right]^{-1}(y-H\bar{w})$$

where $\widetilde{y} = y - \widehat{y}$ and $\widehat{y} = H\widehat{w}$. Moreover, $\widehat{w}$ satisfies the orthogonality condition $H^*\widetilde{y} = \Pi(\widehat{w}-\bar{w})$.

29.8 WEIGHTED REGULARIZED LEAST-SQUARES

We can combine the formulations of Secs. 29.6 and 29.7 and introduce a weighted regularized least-squares problem. The weighted version of (29.28) would have the form

$$\boxed{\min_{w} \;\left[\, (w-\bar{w})^*\Pi(w-\bar{w}) \;+\; (y-Hw)^*W(y-Hw) \,\right]} \tag{29.37}$$

where, as before, W is positive-definite. Actually, with $\Pi > 0$, the weighting matrix W can be allowed to be nonnegative-definite. It is easy to verify that all the expressions in Thm. 29.5 further ahead that do not involve an inverse of W will still hold.

Again, the solution of (29.37) can be obtained in many ways (including plain differentiation with respect to w). Here, as before, we choose to solve (29.37) by showing how it reduces to the standard least-squares problem (29.5). For this purpose, we resort one more time to the eigen-decomposition $W = V\Delta V^*$, and define the normalized quantities

$$a = \Delta^{1/2}V^*y \quad \text{and} \quad A = \Delta^{1/2}V^*H$$

Then the weighted regularized problem (29.37) becomes

$$\min_{w} \left[(w - \bar{w})^*\Pi(w - \bar{w}) + \|a - Aw\|^2 \right] \tag{29.38}$$

which is of the same form as the unweighted regularized least-squares problem (29.28). We can therefore invoke Thm. 29.4, and the definitions of $\{a, A\}$ above, to arrive at the following statement, where the orthogonality condition (29.36) is now replaced by

$$\boxed{H^*W\widetilde{y} = \Pi(\widehat{w} - \bar{w})} \quad \text{(orthogonality condition)} \tag{29.39}$$

Theorem 29.5 (Weighted regularized least-squares) The solution of the weighted regularized least-squares problem (29.37) is always unique and given by

$$\widehat{w} = \bar{w} + \left[\Pi + H^*WH\right]^{-1} H^*W(y - H\bar{w})$$

and the resulting minimum cost is given by

$$\xi \;=\; (y - H\bar{w})^*W\widetilde{y} \;=\; (y - H\bar{w})^* \left[W^{-1} + H\Pi^{-1}H^*\right]^{-1} (y - H\bar{w})$$

where $\widetilde{y} = y - \widehat{y}$ and $\widehat{y} = H\widehat{w}$. Moreover, $\widehat{w}$ satisfies the orthogonality condition $H^*W\widetilde{y} = \Pi(\widehat{w} - \bar{w})$.

For ease of reference, we summarize in Tables 29.1 through 29.3, several of the properties of the least-squares problems studied in this chapter.

TABLE 29.1 Normal equations associated with several least-squares problems.

Problem	**Cost function**	**Normal equations**
Standard least-squares	$\min_{w} \|y - Hw\|^2$	$H^*H\widehat{w} = H^*y$
Weighted least-squares	$\min_{w} \|y - Hw\|_W^2,\ W > 0$	$H^*WH\widehat{w} = H^*Wy$
Regularized least-squares	$\min_{w} \|w - \bar{w}\|_\Pi^2 + \|y - Hw\|^2$ $\Pi > 0$	$(\Pi + H^*H)(\widehat{w} - \bar{w}) = H^*(y - H\bar{w})$
Weighted regularized least-squares	$\min_{w} \|w - \bar{w}\|_\Pi^2 + \|y - Hw\|_W^2$ $\Pi > 0,\ W \geq 0$	$(\Pi + H^*WH)(\widehat{w} - \bar{w}) = H^*W(y - H\bar{w})$

TABLE 29.2 Orthogonality conditions associated with several least-squares problems. In the statements below, $\widetilde{y} = y - \widehat{y}$ where $\widehat{y} = H\widehat{w}$.

Problem	Cost function	Orthogonality condition
Standard least-squares	$\min_w \|y - Hw\|^2$	$H^*\widetilde{y} = 0$
Weighted least-squares	$\min_w \|y - Hw\|_W^2,\ W > 0$	$H^*W\widetilde{y} = 0$
Regularized least-squares	$\min_w \|w - \bar{w}\|_\Pi^2 + \|y - Hw\|^2$ $\Pi > 0$	$H^*\widetilde{y} = \Pi(\widehat{w} - \bar{w})$
Weighted regularized least-squares	$\min_w \|w - \bar{w}\|_\Pi^2 + \|y - Hw\|_W^2$ $\Pi > 0,\ W \geq 0$	$H^*W\widetilde{y} = \Pi(\widehat{w} - \bar{w})$

TABLE 29.3 Minimum costs associated with several least-squares problems. In the statements below, $\widetilde{y} = y - \widehat{y}$ where $\widehat{y} = H\widehat{w}$.

Problem	Cost function	Minimum cost
Standard least-squares	$\min_w \|y - Hw\|^2$	$y^*\widetilde{y}$
Weighted least-squares	$\min_w \|y - Hw\|_W^2,\ W > 0$	$y^*W\widetilde{y}$
Regularized least-squares	$\min_w \|w - \bar{w}\|_\Pi^2 + \|y - Hw\|^2$ $\Pi > 0$	$(y - H\bar{w})^*\widetilde{y}$
Weighted regularized least-squares	$\min_w \|w - \bar{w}\|_\Pi^2 + \|y - Hw\|_W^2$ $\Pi > 0,\ W \geq 0$	$(y - H\bar{w})^*W\widetilde{y}$

CHAPTER 30

Recursive Least-Squares

Now that we have studied in some detail the least-squares problem, we proceed to derive an algorithm for updating its solution. The resulting recursion will be referred to as the Recursive Least-Squares (RLS) algorithm and it will form the basis for most of our discussions in the future chapters.

30.1 MOTIVATION

Given an $N \times 1$ measurement vector y, an $N \times M$ data matrix H and an $M \times M$ positive-definite matrix Π, we saw in Sec. 29.7 that the $M \times 1$ solution to the following regularized least-squares problem:

$$\min_{w} \left[w^*\Pi w + \|y - Hw\|^2 \right] \tag{30.1}$$

is given by

$$\widehat{w} = (\Pi + H^*H)^{-1}H^*y \tag{30.2}$$

where, in comparison with (29.28), we are assuming $\bar{w} = 0$ for simplicity of presentation. The arguments would apply equally well to the case $\bar{w} \neq 0$ — see the remark after Lemma 30.1.

We denote the individual entries of y by $\{d(i)\}$, and the individual rows of H by $\{u_i\}$, say,

$$y = \begin{bmatrix} d(0) \\ d(1) \\ d(2) \\ \vdots \\ d(N-1) \end{bmatrix}, \quad H = \begin{bmatrix} u_0 \\ u_1 \\ u_2 \\ \vdots \\ u_{N-1} \end{bmatrix}$$

so that the solution $\widehat{w}$ in (30.2) is determined by data $\{d(i), u_i\}$ defined up to time $N-1$. In order to indicate this fact explicitly, we shall write w_{N-1} instead of $\widehat{w}$ from now on, with a time subscript $(N-1)$. We shall also write y_{N-1} and H_{N-1} instead of y and H since these quantities are defined in terms of data up to time $N-1$ as well. With this notation, we replace problem (30.1) by

$$\min_{w} \left[w^*\Pi w + \|y_{N-1} - H_{N-1}w\|^2 \right] \tag{30.3}$$

and its solution (30.2) by

$$\boxed{w_{N-1} = (\Pi + H_{N-1}^*H_{N-1})^{-1}H_{N-1}^*y_{N-1}} \tag{30.4}$$

In recursive least-squares, we deal with the issue of an increasing N, and, hence of an increasing amount of data. If, for example, one more row is added to H_{N-1} and one more entry is added to y_{N-1}, leading to

$$y_N = \begin{bmatrix} y_{N-1} \\ d(N) \end{bmatrix}, \qquad H_N = \begin{bmatrix} H_{N-1} \\ u_N \end{bmatrix} \tag{30.5}$$

then the solution w_N of the time-updated least-squares problem

$$\min_w \left[w^*\Pi w + \|y_N - H_N w\|^2 \right] \tag{30.6}$$

would become

$$\boxed{w_N = (\Pi + H_N^* H_N)^{-1} H_N^* y_N} \tag{30.7}$$

Going from (30.4) to (30.7) is referred to as a *time-update* step since it amounts to employing new data $\{d(N), u_N\}$ in addition to all previous data $\{d(j), u_j, 0 \leq j \leq N-1\}$.

Now, performing time-updates by relying on expressions (30.4) and (30.7) is costly both computationally and memory-wise. This is because they require that we invert the $M \times M$ coefficient matrices

$$(\Pi + H_N^* H_N) \quad \text{and} \quad (\Pi + H_{N-1}^* H_{N-1})$$

at the respective time instants. In addition, (30.7) requires that we store in memory the entries of $\{H_{N-1}, y_{N-1}\}$ so that $\{H_N, y_N\}$ could be formed when the new data $\{u_N, d(N)\}$ become available. These two requirements of matrix inversion (requiring $O(M^3)$ operations) and increasing storage capacity can be alleviated by seeking an update method that would compute w_N solely from knowledge of the new data $\{d(N), u_N\}$ and from the previous solution w_{N-1}. Such a method is possible since, as we see from (30.5), the quantities $\{H_{N-1}, y_{N-1}\}$ and $\{H_N, y_N\}$ differ only by the new data $\{u_N, d(N)\}$; all other entries are identical. Exploiting this observation is the basis for deriving the recursive least-squares algorithm.

30.2 RLS ALGORITHM

Introduce the matrices

$$\boxed{P_N \triangleq (\Pi + H_N^* H_N)^{-1}, \qquad P_{N-1} \triangleq (\Pi + H_{N-1}^* H_{N-1})^{-1}} \tag{30.8}$$

with initial condition $P_{-1} = \Pi^{-1}$. Then (30.4) and (30.7) can be written more compactly as

$$\boxed{w_{N-1} = P_{N-1} H_{N-1}^* y_{N-1}, \qquad w_N = P_N H_N^* y_N} \tag{30.9}$$

The time-update relation (30.5) between $\{y_N, H_N\}$ and $\{y_{N-1}, H_{N-1}\}$ can be used to relate P_N to P_{N-1} and w_N to w_{N-1}.

Derivation

To begin with, note that

$$\begin{aligned} P_N^{-1} &= \Pi + H_N^* H_N \\ &= \Pi + H_{N-1}^* H_{N-1} + u_N^* u_N \end{aligned}$$

so that

$$\boxed{P_N^{-1} = P_{N-1}^{-1} + u_N^* u_N} \tag{30.10}$$

Then, by using the matrix inversion identity

$$(A + BCD)^{-1} = A^{-1} - A^{-1}B(C^{-1} + DA^{-1}B)^{-1}DA^{-1} \tag{30.11}$$

with the identifications

$$A \leftarrow P_{N-1}^{-1}, \quad B \leftarrow u_N^*, \quad C \leftarrow 1, \quad D \leftarrow u_N$$

we obtain a recursive formula for updating P_N directly rather than its inverse,

$$\boxed{P_N = P_{N-1} - \frac{P_{N-1}u_N^* u_N P_{N-1}}{1 + u_N P_{N-1} u_N^*}, \qquad P_{-1} = \Pi^{-1}} \tag{30.12}$$

This recursion for P_N also gives one for updating the least-squares solution w_N itself. Using expression (30.9) for w_N, and substituting the above recursion for P_N, we find

$$\begin{aligned} w_N &= P_N \left[H_{N-1}^* y_{N-1} + u_N^* d(N) \right] \\ &= \left(P_{N-1} - \frac{P_{N-1}u_N^* u_N P_{N-1}}{1 + u_N P_{N-1} u_N^*} \right) \left[H_{N-1}^* y_{N-1} + u_N^* d(N) \right] \\ &= \underbrace{P_{N-1} H_{N-1}^* y_{N-1}}_{=w_{N-1}} - \frac{P_{N-1}u_N^*}{1 + u_N P_{N-1} u_N^*} u_N \underbrace{P_{N-1} H_{N-1}^* y_{N-1}}_{=w_{N-1}} \\ &\quad + P_{N-1}u_N^* \left(1 - \frac{u_N P_{N-1} u_N^*}{1 + u_N P_{N-1} u_N^*} \right) d(N) \end{aligned}$$

That is,

$$\boxed{w_N = w_{N-1} + \frac{P_{N-1}u_N^*}{1 + u_N P_{N-1} u_N^*}[d(N) - u_N w_{N-1}], \qquad w_{-1} = 0} \tag{30.13}$$

We summarize the time-updates for $\{P_N, w_N\}$ in Alg. 30.1 below, where we also introduce two important quantities. One is called the conversion factor, for reasons to be explained shortly in Sec. 30.4, and is defined by

$$\boxed{\gamma(N) \overset{\Delta}{=} 1/(1 + u_N P_{N-1} u_N^*)} \tag{30.14}$$

whereas the other is called the gain vector and is defined by

$$\boxed{g_N \overset{\Delta}{=} P_{N-1} u_N^* \gamma(N)} \tag{30.15}$$

Some straightforward algebra, using recursion (30.12) for P_N, shows that $\{g_N, \gamma(N)\}$ are also given by

$$\boxed{\gamma(N) = 1 - u_N P_N u_N^*} \tag{30.16}$$

and

$$\boxed{g_N = P_N u_N^*} \tag{30.17}$$

These alternative expressions for $\{g_N, \gamma(N)\}$ are in terms of P_N while the first two expressions (30.14)–(30.15) are in terms of P_{N-1}. To justify (30.17)–(30.16) simply note the following. Multiplying recursion (30.12) for P_N by u_N^* from the right we get

$$\begin{aligned} P_N u_N^* &= P_{N-1}u_N^* - \frac{P_{N-1}u_N^* u_N P_{N-1}u_N^*}{1+u_N P_{N-1}u_N^*} \\ &= \frac{P_{N-1}u_N^*}{1+u_N P_{N-1}u_N^*} \quad = \quad g_N \end{aligned}$$

By further multiplying the above identity by u_N from the left we get

$$u_N P_N u_N^* = \frac{u_N P_{N-1}u_N^*}{1+u_N P_{N-1}u_N^*}$$

so that, by subtracting 1 from both sides, we obtain (30.16).

Algorithm 30.1 (RLS algorithm) Given $\Pi > 0$, the solution w_N that minimizes the cost

$$w^*\Pi w \ + \ \|y_N - H_N w\|^2$$

can be computed recursively as follows. Start with $w_{-1} = 0$ and $P_{-1} = \Pi^{-1}$ and iterate for $i \geq 0$:

$$\begin{aligned} \gamma(i) &= 1/(1+u_i P_{i-1}u_i^*) \\ g_i &= P_{i-1}u_i^*\gamma(i) \\ w_i &= w_{i-1} + g_i[d(i) - u_i w_{i-1}] \\ P_i &= P_{i-1} - g_i g_i^*/\gamma(i) \end{aligned}$$

At each iteration, it holds that w_i minimizes $w^*\Pi w \ + \ \|y_i - H_i w\|^2$, where $y_i = \text{col}\,\{d(0), d(1), \ldots, d(i)\}$ and the rows of H_i are $\{u_0, u_1, \ldots, u_i\}$. Moreover, $P_i = (\Pi + H_i^* H_i)^{-1}$.

30.3 REGULARIZATION

Observe that regularization is necessary (i.e., the use of $\Pi > 0$), since it guarantees the existence of $P_N = (\Pi + H_N^* H_N)^{-1}$. In the absence of regularization, i.e., when $\Pi = 0$, the above inverse need not exist especially during the initial update stages when H_N has fewer rows than columns or even at later stages if H_N becomes rank deficient. In these situations, the RLS recursions of Alg. 30.1 would not be applicable, not only because the matrix P_N is not defined but also because of the non-practical initialization $P_{-1} = \infty\mathrm{I}$.

In Sec. 35.2 we shall describe an alternative implementation of RLS that addresses these difficulties; it can be used even in the absence of regularization and has better numerical properties than RLS itself. In preparation for the derivation of this alternative algorithm later in Sec. 35.2, we indicate here some of the RLS equations that will be needed in that section.

Thus, note that expression (30.7) for w_N is simply a rewriting of the normal equations

$$(\Pi + H_N^* H_N)w_N = H_N^* y_N \tag{30.18}$$

which characterize the solution of the regularized least-squares problem (30.6). The form (30.18) is more general than (30.7) since it holds even when $(\Pi + H_N^* H_N)$ is singular (e.g., when $\Pi = 0$ and H_N is rank deficient) — recall (29.33) and the discussion thereafter.

Now if we define the $M \times M$ matrix Φ_N and the $M \times 1$ vector s_N,

$$\Phi_N \triangleq \Pi + H_N^* H_N, \qquad s_N \triangleq H_N^* y_N$$

then, just like the recursion (30.10) for P_N^{-1}, we also obtain a recursion for Φ_N:

$$\boxed{\Phi_N = \Phi_{N-1} + u_N^* u_N, \qquad \Phi_{-1} = \Pi} \tag{30.19}$$

and using (30.5), it is immediate to see that s_N satisfies the time-update recursion:

$$\boxed{s_N = s_{N-1} + u_N^* d(N), \qquad s_{-1} = 0} \tag{30.20}$$

With $\{\Phi_N, s_N\}$ so defined, the normal equations (30.18) can be rewritten as

$$\boxed{\Phi_N w_N = s_N} \tag{30.21}$$

The point is that this description is valid even when $\Pi = 0$ since, as we already know from Thm. 29.1, the corresponding normal equations

$$\underbrace{H_N^* H_N}_{\Phi_N} w_N = \underbrace{H_N^* y_N}_{s_N}$$

are consistent so that a solution w_N can always be found (e.g., one could choose the minimum-norm solution when multiple solutions exist). The equations (30.19)–(30.21) will form the basis for the aforementioned alternative implementation of RLS in Sec. 35.2; one that has better properties in finite-precision arithmetic and can be used even if $\Pi = 0$.

30.4 CONVERSION FACTOR

We can also derive a time-update relation for the minimum costs associated with problems (30.3) and (30.6). To do so, we first associate with RLS two error quantities: the *a priori* output error, denoted by $e(N)$,

$$\boxed{e(N) \triangleq d(N) - u_N w_{N-1}} \tag{30.22}$$

and the *a posteriori* output error, denoted by $r(N)$,

$$\boxed{r(N) \triangleq d(N) - u_N w_N} \tag{30.23}$$

It then turns out that the factor $\gamma(N)$ defined by (30.14) serves a useful purpose: it maps $e(N)$ to $r(N)$, i.e., it converts the *a priori* error into its *a posteriori* version. To see this, we replace w_N in the definition of $r(N)$ by its RLS update from Alg. 30.1 to obtain

$$\begin{aligned} r(N) &= d(N) - u_N(\, w_{N-1} + g_N[d(N) - u_N w_{N-1}]\,) \\ &= d(N) - u_N w_{N-1} - u_N g_N e(N) \\ &= e(N) - u_N g_N e(N) \\ &= (1 - u_N g_N) e(N) \end{aligned}$$

Using (30.16) and (30.17) we get

$$\boxed{r(N) = \gamma(N)e(N)}$$

as desired. Observe further from the definition of $\gamma(N)$ in (30.14) that

$$\boxed{0 < \gamma(N) \leq 1} \tag{30.24}$$

so that it always holds that

$$|r(N)| \leq |e(N)|$$

30.5 TIME-UPDATE OF THE MINIMUM COST

Now let $\xi(N-1)$ denote the minimum cost of the regularized least-squares problem (30.3). Likewise, let $\xi(N)$ denote the minimum cost of the time-updated problem (30.6). From Thm. 29.4 we know that $\xi(N-1)$ and $\xi(N)$ are given by

$$\xi(N) \;=\; y_N^*[y_N - H_N w_N], \qquad \xi(N-1) = y_{N-1}^*[y_{N-1} - H_{N-1}w_{N-1}]$$

Using the partitioning (30.5) for $\{y_N, H_N\}$, as well as the RLS update

$$w_N = w_{N-1} + g_N e(N), \qquad e(N) = d(N) - u_N w_{N-1}$$

we can arrive at a relation between $\xi(N)$ and $\xi(N-1)$ by means of the following sequence of calculations:

$$\begin{aligned}
\xi(N) &= \begin{bmatrix} y_{N-1}^* & d^*(N) \end{bmatrix} \begin{bmatrix} y_{N-1} - H_{N-1}[w_{N-1} + g_N e(N)] \\ d(N) - u_N[w_{N-1} + g_N e(N)] \end{bmatrix} \\
&= \underbrace{y_{N-1}^*(y_{N-1} - H_{N-1}w_{N-1})}_{=\xi(N-1)} - y_{N-1}^* H_{N-1} g_N e(N) + d^*(N)e(N)[1 - u_N g_N] \\
&= \xi(N-1) + e(N)\left[d^*(N)\underbrace{(1 - u_N g_N)}_{=\gamma(N)} - y_{N-1}^* H_{N-1} g_N\right] \\
&= \xi(N-1) + e(N)\gamma(N)\left[d^*(N) - \underbrace{y_{N-1}^* H_{N-1} P_{N-1}}_{=w_{N-1}^*} u_N^*\right] \\
&= \xi(N-1) + |e(N)|^2\gamma(N)
\end{aligned}$$

Moreover, since $r(N) = \gamma(N)e(N)$, it holds that

$$|e(N)|^2\gamma(N) \;=\; e(N)r^*(N) \;=\; e^*(N)r(N) \;=\; |r(N)|^2/\gamma(N)$$

In summary, we arrive at the following conclusion.

> **Lemma 30.1 (Estimation errors)** Consider the same setting of Alg. 30.1. At each iteration i, the *a priori* and *a posteriori* estimation errors defined by
>
> $$e(i) = d(i) - u_i w_{i-1} \quad \text{and} \quad r(i) = d(i) - u_i w_i$$
>
> are related via the conversion factor $\gamma(i)$ as $r(i) = \gamma(i)e(i)$. Moreover, the minimum costs of the successive regularized least-squares problems satisfy any of the following time-update relations with initial condition $\xi(-1) = 0$:
>
> $$\begin{aligned} \xi(i) &= \xi(i-1) + e(i)r^*(i) \\ &= \xi(i-1) + \gamma(i)|e(i)|^2 \\ &= \xi(i-1) + |r(i)|^2/\gamma(i) \end{aligned}$$

One final remark is in place. Had we considered a regularized cost function of the form

$$\min_{w} \; \left[(w - \bar{w})^*\Pi(w - \bar{w}) \; + \; \|y_N - H_N w\|^2\right]$$

with a nonzero $\bar{w}$, then the only modification to the RLS algorithm would be in the value of its initial condition, w_{-1}; it becomes $w_{-1} = \bar{w}$. Moreover, the time-update for $\xi(N)$ will still hold. To see this, we only need to repeat the above derivation for $\xi(N)$ starting from the expression (cf. Table 29.3) — see Prob. VII.29:

$$\xi(N) = [y_N - H_N\bar{w}]^*[y_N - H_N w_N]$$

30.6 EXPONENTIALLY-WEIGHTED RLS ALGORITHM

It is more common in adaptive filtering to employ a *weighted* regularized least-squares cost function, as opposed to the unweighted cost in (30.6). More specifically, a diagonal weighting matrix is used whose purpose is to give more weight to recent data and less weight to data from the remote past.

Let λ be a positive scalar, usually very close to one (e.g., $\lambda = 0.998$ or some similar value), say, $0 \ll \lambda \leq 1$, and introduce the diagonal matrix

$$\Lambda_N \stackrel{\Delta}{=} \text{diag}\{\lambda^N, \lambda^{N-1}, \ldots, \lambda, 1\} \tag{30.25}$$

Then replace (30.6) by

$$\min_{w} \left[\lambda^{(N+1)} w^*\Pi w \; + \; (y_N - H_N w)^*\Lambda_N(y_N - H_N w) \right] \tag{30.26}$$

or, more explicitly, by

$$\boxed{\min_{w} \left[\lambda^{(N+1)} w^*\Pi w \; + \; \sum_{j=0}^{N} \lambda^{N-j}|d(j) - u_j w|^2 \right]} \tag{30.27}$$

The scalar λ is called the *forgetting factor* since past data are exponentially weighted less heavily than more recent data. The special case $\lambda = 1$ is known as the *growing memory* case and it was studied in the previous sections. Exponential weighting is one form of data windowing whereby the effective length of the window is $\approx 1/(1-\lambda)$ samples.

Observe that the regularization matrix in (30.26) is chosen as $\lambda^{(N+1)}\Pi$, with the additional scaling factor $\lambda^{(N+1)}$. Since this factor becomes smaller as time progresses, we see that the exponentially-weighted cost function (30.26) is such that it de-emphasizes regularization during the later stages of operation when the data matrix H_N is more likely to have full rank.

Comparing (30.26) with (29.37), and using the identifications $\Pi \leftarrow \lambda^{(N+1)}\Pi$ and $W \leftarrow \Lambda_N$, we find that the solution w_N is obtained by solving

$$\left[\lambda^{(N+1)}\Pi + H_N^*\Lambda_N H_N\right] w_N = H_N^*\Lambda_N y_N \tag{30.28}$$

If we now define the quantities

$$\begin{aligned} P_N &\triangleq \left[\lambda^{(N+1)}\Pi + H_N^*\Lambda_N H_N\right]^{-1} \\ g_N &\triangleq \lambda^{-1}P_{N-1}u_N^*\gamma(N) \\ \gamma(N) &\triangleq 1/(1+\lambda^{-1}u_N P_{N-1}u_N^*) \end{aligned}$$

and repeat the arguments prior to the statement of Alg. 30.1, we arrive at the following statement.

Algorithm 30.2 (Exponentially-weighted RLS) Given $\Pi > 0$, and a forgetting factor $0 \ll \lambda \leq 1$, the solution w_N of the exponentially-weighted regularized least-squares problem (30.27), and the corresponding minimum cost $\xi(N)$, can be computed recursively as follows. Start with $w_{-1} = 0$, $P_{-1} = \Pi^{-1}$, and $\xi(-1) = 0$, and iterate for $i \geq 0$:

$$\begin{aligned} \gamma(i) &= 1/(1+\lambda^{-1}u_i P_{i-1}u_i^*) \\ g_i &= \lambda^{-1}P_{i-1}u_i^*\gamma(i) \\ e(i) &= d(i) - u_i w_{i-1} \\ w_i &= w_{i-1} + g_i e(i) \\ P_i &= \lambda^{-1}P_{i-1} - g_i g_i^*/\gamma(i) \\ \xi(i) &= \lambda\xi(i-1) \;+\; \gamma(i)|e(i)|^2 \end{aligned}$$

At each iteration, P_i has the interpretation $P_i = \left[\lambda^{(i+1)}\Pi + H_i^*\Lambda_i H_i\right]^{-1}$ and w_i is the solution of

$$\min_w \left[\lambda^{(i+1)} w^*\Pi w \;+\; \sum_{j=0}^{i} \lambda^{i-j}|d(j) - u_j w|^2\right]$$

In addition, as was the case with (30.16)–(30.17), the following relations hold:

$$g_i = P_i u_i^*, \qquad \gamma(i) = 1 - u_i P_i u_i^* = 1 - u_i g_i, \qquad r(i) = \gamma(i)e(i)$$

where $r(i) = d(i) - u_i w_i$.

Again, starting from the normal equations (30.28), we can define the quantities

$$\begin{aligned} \Phi_N &\triangleq \lambda^{(N+1)}\Pi + H_N^* \Lambda_N H_N &(30.29)\\ s_N &\triangleq H_N^* \Lambda_N y_N &(30.30)\end{aligned}$$

Then it can be easily verified that they satisfy the recursions

$$\begin{aligned} \Phi_N &= \lambda \Phi_{N-1} + u_N^* u_N, \quad \Phi_{-1} = \Pi &(30.31)\\ s_N &= \lambda s_{N-1} + u_N^* d(N), \quad s_{-1} = 0 &(30.32)\end{aligned}$$

and that w_N can be found by solving the normal equations

$$\Phi_N w_N = s_N \tag{30.33}$$

As mentioned in Sec. 30.3, these equations will be used in Sec. 35.2 to motivate an alternative recursive implementation of the exponentially-weighted RLS algorithm.

Moreover, had we started with a regularized cost function of the form

$$\min_w \left[\lambda^{(N+1)}(w - \bar{w})^* \Pi (w - \bar{w}) \; + \; \sum_{j=0}^{N} \lambda^{N-j} |d(j) - u_j w|^2 \right]$$

with a nonzero $\bar{w}$, then the only modification to the RLS equations of Alg. 30.2 would be in the value of the initial condition, w_{-1}; it becomes $w_{-1} = \bar{w}$. Likewise, equations (30.31)–(30.33) would be replaced by

$$\begin{aligned} \Phi_N &= \lambda \Phi_{N-1} + u_N^* u_N, \quad \Phi_{-1} = \Pi &(30.34)\\ s_N &= \lambda s_{N-1} + u_N^* [d(N) - u_N \bar{w}], \quad s_{-1} = 0 &(30.35)\\ \Phi_N (w_N - \bar{w}) &= s_N &(30.36)\end{aligned}$$

CHAPTER 31

Kalman Filtering and RLS

There is a close relation between regularized least-squares problems, as studied in the previous chapter, and linear least-mean-squares estimation problems, as studied in Part II (*Linear Estimation*). Although the former class of problems deals with deterministic variables and the latter class of problems deals with random variables, both classes turn out to be equivalent in the sense that solving a problem from one class also solves a problem from the other class and vice-versa.

31.1 EQUIVALENCE IN LINEAR ESTIMATION

Stochastic Problem

Let $\boldsymbol{x}$ and $\boldsymbol{y}$ be zero-mean random variables that are related via a linear model of the form

$$\boxed{\boldsymbol{y} = H\boldsymbol{x} + \boldsymbol{v}} \tag{31.1}$$

for some known matrix H and where $\boldsymbol{v}$ denotes a zero-mean random noise vector with known covariance matrix, say, $R_v = \mathsf{E}\,\boldsymbol{v}\boldsymbol{v}^*$. The covariance matrix of $\boldsymbol{x}$ is also known and denoted by $\mathsf{E}\,\boldsymbol{x}\boldsymbol{x}^* = R_x$. Both $\{\boldsymbol{x}, \boldsymbol{v}\}$ are uncorrelated, i.e., $\mathsf{E}\,\boldsymbol{x}\boldsymbol{v}^* = 0$, and we further assume that $R_x > 0$ and $R_v > 0$. We established in Thm. 5.1 that the linear least-mean-squares estimator of $\boldsymbol{x}$ given $\boldsymbol{y}$ is

$$\boxed{\hat{\boldsymbol{x}} = \left[R_x^{-1} + H^* R_v^{-1} H\right]^{-1} H^* R_v^{-1} \boldsymbol{y}} \tag{31.2}$$

and that the resulting minimum mean-square error matrix is

$$\boxed{\text{m.m.s.e.} = \left[R_x^{-1} + H^* R_v^{-1} H\right]^{-1}} \tag{31.3}$$

Deterministic Problem

Now consider instead deterministic variables $\{x, y\}$ and a data matrix H relating them via

$$\boxed{y = Hx + v} \tag{31.4}$$

where v denotes measurement noise. Assume further that we pose the problem of estimating x by solving the weighted regularized least-squares problem:

$$\min_{x} \left[\, x^* \Pi x \;+\; \|y - Hx\|_W^2 \,\right] \tag{31.5}$$

where $\Pi > 0$ is a regularization matrix and $W > 0$ is a weighting matrix. We showed in Thm. 29.5 that the solution $\hat{x}$ is given by

$$\boxed{\hat{x} = \left[\Pi + H^*WH\right]^{-1} H^*Wy} \tag{31.6}$$

and that the resulting minimum cost is

$$\boxed{\xi = y^* \left[W^{-1} + H\Pi^{-1}H^*\right]^{-1} y} \tag{31.7}$$

Equivalence

Expression (31.2) provides the linear least-mean-squares estimator of $\boldsymbol{x}$ in a stochastic framework, while expression (31.6) provides the least-squares estimate of x in the deterministic framework (31.5). Still, it is clear that if we replace the quantities in (31.2) by $R_x \longleftarrow \Pi^{-1}$ and $R_v \longleftarrow W^{-1}$, then the stochastic solution (31.2) would coincide with the deterministic solution (31.6). We therefore say that both problems are equivalent. Such equivalences play a central role in estimation theory since they allow us to move back and forth between deterministic and stochastic formulations, and to determine the solution for one context from the solution to the other. Table 31.1 summarizes the relations between the variables in both contexts. We now illustrate an application of this important result in the context of adaptive filtering.

TABLE 31.1 Equivalence of the stochastic and deterministic frameworks.

Stochastic	Deterministic
Random variables $\{\boldsymbol{x}, \boldsymbol{y}\}$	Deterministic variables $\{x, y\}$
Model $\boldsymbol{y} = H\boldsymbol{x} + \boldsymbol{v}$	Model $y = Hx + v$
Covariance matrix R_x	Inverse regularization matrix Π^{-1}
Noise covariance R_v	Inverse weighting matrix W^{-1}
$\hat{\boldsymbol{x}}$	$\hat{x}$
$\min_K \mathsf{E}\,(\boldsymbol{x} - K\boldsymbol{y})(\boldsymbol{x} - K\boldsymbol{y})^*$	$\min_x \left[x^*\Pi x + \lVert y - Hx\rVert_W^2\right]$
$\hat{\boldsymbol{x}} = \left[R_x^{-1} + H^*R_v^{-1}H\right]^{-1} H^*R_v^{-1}\boldsymbol{y}$	$\hat{x} = \left[\Pi + H^*WH\right]^{-1} H^*Wy$
m.m.s.e. $= \left[R_x^{-1} + H^*R_v^{-1}H\right]^{-1}$	Min. cost $= y^*\left[W^{-1} + H\Pi^{-1}H^*\right]^{-1} y$

31.2 KALMAN FILTERING AND RECURSIVE LEAST-SQUARES

The equivalence established in Table 31.1 between stochastic and deterministic least-squares problems can be used to clarify the relationship that exists between Kalman filtering and RLS algorithms. This relationship is useful for at least two reasons:

1. Kalman filtering theory is well studied and many algorithmic variants have been developed over the years. Therefore, by establishing a connection between Kalman filters and RLS filters, it becomes possible to share ideas and algorithms between both domains.

2. The RLS filter is not equivalent to a full-blown Kalman filter, but only to a special case of it. This fact suggests that extended RLS schemes can be developed with enhanced tracking abilities, and we shall pursue this extension in Sec. 31.A.

In this section, we limit ourselves to showing how RLS is equivalent to a special case of the Kalman filter. The arguments presented here are also applicable to vector-valued observations $\boldsymbol{y}_i$, in which case we would be able to establish the relation between Kalman filtering and the block RLS filter of Prob. VII.36 — see Sec. 31.A. However, for simplicity of presentation, we focus on the case of scalar-valued observations. In addition, other state-space models (i.e., other than (31.8) below) could be used for the same purpose of relating Kalman and RLS filters — see, e.g., (31.41) and (31.40) and the footnote following them.

Thus, consider a collection of zero-mean scalar-valued random variables $\{\boldsymbol{y}(i), 0 \leq i \leq N\}$ that satisfy a special state-space model of the form:

$$\boxed{\begin{array}{rcl} \boldsymbol{x}_{i+1} & = & \lambda^{-1/2}\boldsymbol{x}_i \\ \boldsymbol{y}(i) & = & u_i\boldsymbol{x}_i + \boldsymbol{v}(i) \end{array}} \tag{31.8}$$

with

$$\boxed{\mathsf{E}\,\boldsymbol{x}_0 = 0, \quad \mathsf{E}\,\boldsymbol{x}_0\boldsymbol{x}_0^* = \Pi_0, \quad \mathsf{E}\,\boldsymbol{v}(i)\boldsymbol{v}^*(j) = \delta_{ij}, \quad \mathsf{E}\,\boldsymbol{x}_0\boldsymbol{v}^*(i) = 0} \tag{31.9}$$

This model is a special case of the general state-space model (7.8)–(7.10) and it corresponds to the choices

$$F_i = \lambda^{-1/2}\mathrm{I}, \quad G_i = 0, \quad R_i = 1, \quad H_i = u_i \text{ (a row vector)}$$

where λ is a positive scalar less than or equal to one ($0 \ll \lambda \leq 1$). From Alg. 7.1 we know that the corresponding Kalman filtering equations are given by

$$\begin{array}{rcl} r_e(i) & = & 1 + u_iP_{i|i-1}u_i^* \\ k_{p,i} & = & \lambda^{-1/2}P_{i|i-1}u_i^*/r_e(i) \\ \boldsymbol{\nu}(i) & = & \boldsymbol{y}(i) - u_i\hat{\boldsymbol{x}}_{i|i-1} \\ \hat{\boldsymbol{x}}_{i+1|i} & = & \lambda^{-1/2}\hat{\boldsymbol{x}}_{i|i-1} + k_{p,i}\boldsymbol{\nu}(i), \quad \hat{\boldsymbol{x}}_{0|-1} = 0 \\ P_{i+1|i} & = & \lambda^{-1}\left[P_{i|i-1} - \dfrac{P_{i|i-1}u_i^*u_iP_{i|i-1}}{1 + u_iP_{i|i-1}u_i^*}\right], \quad P_{0|-1} = \Pi_0 \end{array} \tag{31.10}$$

where the notation $\hat{\boldsymbol{x}}_{i+1|i}$ denotes the linear least-mean-squares (l.l.m.s.) estimator of $\boldsymbol{x}_{i+1}$ using the observations $\{\boldsymbol{y}_0, \boldsymbol{y}_1, \ldots, \boldsymbol{y}_i\}$. For convenience of exposition, we are denoting the innovations variable of the Kalman filter by $\boldsymbol{\nu}(i)$, as opposed to the symbol $\boldsymbol{e}(i)$ used in Chapter 7. This is done here in order to avoid a conflict of notation with the symbol $e(i)$ used to denote the output error of RLS.

Actually since, by virtue of (31.8), $\boldsymbol{x}_{i+1}$ is a scaled version of $\boldsymbol{x}_0$, we end up obtaining the estimator of $\boldsymbol{x}_0$ from these same observations. Indeed, observe by iterating (31.8) that $\boldsymbol{x}_0 = \lambda^{(i+1)/2}\ \boldsymbol{x}_{i+1}$ so that

$$\begin{array}{rcl} \hat{\boldsymbol{x}}_{0|i} & = & \lambda^{(i+1)/2}\hat{\boldsymbol{x}}_{i+1|i} \\ & = & \text{l.l.m.s.e. of } \boldsymbol{x}_0 \text{ given } \{\boldsymbol{y}(0), \boldsymbol{y}(1), \ldots, \boldsymbol{y}(i)\} \end{array}$$

We therefore conclude that the Kalman filtering equations (31.10) allow us to solve the stochastic problem of estimating $\boldsymbol{x}_0$ from the observations $\boldsymbol{y} = \text{col}\{\boldsymbol{y}(0), \ldots, \boldsymbol{y}(i)\}$, for any i. If we run recursions (31.10) from $i = 0$ to $i = N$, then we end up with the l.l.m.s. estimator of $\boldsymbol{x}_0$ given the N observations $\{\boldsymbol{y}(0), \boldsymbol{y}(1), \ldots, \boldsymbol{y}(N)\}$:

$$\begin{array}{rcl} \hat{\boldsymbol{x}}_{0|N} & = & \lambda^{(N+1)/2}\hat{\boldsymbol{x}}_{N+1|N} \\ & = & \text{l.l.m.s.e. of } \boldsymbol{x}_0 \text{ given } \boldsymbol{y} = \text{col}\{\boldsymbol{y}(0), \boldsymbol{y}(1), \ldots, \boldsymbol{y}(N)\} \end{array}$$

Observe further that, in view of the state-space model (31.8), the variables $\{\boldsymbol{x}_0, \boldsymbol{y}\}$ are related via the linear model:

$$\underbrace{\begin{bmatrix} \boldsymbol{y}(0) \\ \boldsymbol{y}(1) \\ \boldsymbol{y}(2) \\ \vdots \\ \boldsymbol{y}(N) \end{bmatrix}}_{\triangleq\, \boldsymbol{y}} = \underbrace{\begin{bmatrix} 1 & & & & \\ & \lambda^{-\frac{1}{2}} & & & \\ & & \lambda^{-1} & & \\ & & & \ddots & \\ & & & & [\lambda^{-\frac{1}{2}}]^N \end{bmatrix} \begin{bmatrix} u_0 \\ u_1 \\ u_2 \\ \vdots \\ u_N \end{bmatrix}}_{\triangleq\, H} \boldsymbol{x}_0 + \underbrace{\begin{bmatrix} \boldsymbol{v}(0) \\ \boldsymbol{v}(1) \\ \boldsymbol{v}(2) \\ \vdots \\ \boldsymbol{v}(N) \end{bmatrix}}_{\triangleq\, \boldsymbol{v}} \tag{31.11}$$

that is,

$$\boldsymbol{y} = H\boldsymbol{x}_0 + \boldsymbol{v}$$

where we are denoting the matrix multiplying $\boldsymbol{x}_0$ by H.

Now we can refer to the equivalence result of Table 31.1 in order to characterize the deterministic problem that corresponds to the stochastic problem of estimating $\boldsymbol{x}_0$ from $\boldsymbol{y}$ via recursions (31.10). Indeed, from Table 31.1 we find that the desired deterministic problem is

$$\min_{x_0} \left[x_0^*\Pi_0^{-1}x_0 \;+\; \|y - Hx_0\|^2\right]$$

which we can rewrite as

$$\min_{x_0} \left[x_0^*\Pi_0^{-1}x_0 \;+\; \sum_{j=0}^{N} |y(j) - u_j x_j|^2\right] \tag{31.12}$$

for variables x_j satisfying $x_{j+1} = \lambda^{-1/2}x_j$. In summary, we arrive at the conclusion stated in Lemma 31.1 further ahead.

Relation to Exponentially-Weighted RLS

The point to stress here is that we have arrived at a recursive solution to the deterministic least-squares problem (31.12) by appealing to equivalence with the stochastic solution (31.10)–(31.11) and not by solving it afresh. We can take this argument a step further and show that the recursive solution of Lemma 31.1 is in effect the exponentially-weighted RLS filter of Alg. 30.2. Once this is done, our arguments would have clarified the connection between Kalman filtering and RLS filtering, namely, that the exponentially-weighted RLS problem follows from applying the Kalman filter to the special state-space model (31.8).

To see this, let us consider the regularized least-squares problem (30.27), namely,

$$\min_{w} \left[\lambda^{(N+1)}w^*\Pi w \;+\; \sum_{j=0}^{N} \lambda^{N-j}|d(j) - u_j w|^2\right] \tag{31.13}$$

where $\Pi > 0$ is a regularization matrix. We denote its solution by w_N; it is the estimate for w that is based on the data $\{d(j), u_j\}$ between times $j = 0$ and $j = N$. Although at first sight, this cost function is not of the same form as the cost function appearing in (31.12), it can be reworked into that form with a suitable change of variables. Thus, observe first that solving (31.13) is equivalent to solving

$$\min_{w} \left[w^*(\lambda\Pi)w \;+\; \sum_{j=0}^{N} \lambda^{-j}|d(j) - u_j w|^2\right] \tag{31.14}$$

where we have extracted the constant factor λ^N. Observe further that the cost function in (31.14) can be rewritten as

$$
\begin{aligned}
J(w) &\triangleq w^*(\lambda\Pi)w + \sum_{j=0}^{N}\left|\underbrace{\frac{d(j)}{(\sqrt{\lambda})^j}}_{\triangleq\, y(j)} - u_j \underbrace{\frac{w}{(\sqrt{\lambda})^j}}_{\triangleq\, x_j}\right|^2 \\
&= x_0^*(\lambda\Pi)x_0 + \sum_{j=0}^{N}|y(j) - u_j x_j|^2
\end{aligned}
\tag{31.15}
$$

where we defined the quantities,

$$
y(j) \triangleq \frac{d(j)}{(\sqrt{\lambda})^j}, \qquad x_j \triangleq \frac{w}{(\sqrt{\lambda})^j}, \qquad x_0 = w \tag{31.16}
$$

Now it follows from the definition of x_j in (31.16) that it satisfies $x_{j+1} = \lambda^{-1/2}x_j$. Therefore, solving (31.13) is equivalent to solving

$$
\min_{x_0}\left[x_0^*(\lambda\Pi)x_0 + \sum_{j=0}^{N}|y(j) - u_j x_j|^2\right] \quad \text{subject to} \quad x_{j+1} = \lambda^{-1/2}x_j
$$

This problem is now of the same form as (31.12) with the identification $\Pi_0^{-1} = \lambda\Pi$ and, therefore, its solution can be computed by appealing to the recursions of Lemma 31.1, namely, we start with $P_{0|-1} = \lambda^{-1}\Pi^{-1}$, $\hat{x}_{0|-1} = 0$ and repeat for $i \geq 0$:

$$
\begin{aligned}
r_e(i) &= 1 + u_i P_{i|i-1} u_i^* & (31.17)\\
k_{p,i} &= \lambda^{-1/2} P_{i|i-1} u_i^* / r_e(i) & (31.18)\\
\nu(i) &= y(i) - u_i \hat{x}_{i|i-1} & (31.19)\\
\hat{x}_{i+1|i} &= \lambda^{-\frac{1}{2}}\hat{x}_{i|i-1} + k_{p,i}\nu(i) & (31.20)\\
P_{i+1|i} &= \lambda^{-1}\left[P_{i|i-1} - \frac{P_{i|i-1}u_i^* u_i P_{i|i-1}}{1 + u_i P_{i|i-1} u_i^*}\right] & (31.21)
\end{aligned}
$$

Then at each iteration i, it will hold that

$$
\hat{x}_{0|i} = \lambda^{(i+1)/2}\hat{x}_{i+1|i}
$$

where $\hat{x}_{0|i}$ is the solution of (31.13) using data up to time i, i.e., $\hat{x}_{0|i} = w_i$ where w_i solves

$$
\min_{w}\left[\lambda^{(i+1)}w^*\Pi w + \sum_{j=0}^{i}\lambda^{i-j}|d(j) - u_j w|^2\right]
$$

Lemma 31.1 (Equivalent deterministic problem) Consider a set of $(N+1)$ deterministic data $\{y(i), u_i\}_{i=0}^{N}$, where the $y(i)$ are scalars and the u_i are row vectors. Consider further $n \times 1$ vectors x_i that satisfy $x_{i+1} = \lambda^{-1/2} x_i$, for a positive real scalar $0 \ll \lambda \leq 1$, and let Π_0 be a positive-definite matrix. Then the solution of the regularized least-squares problem:

$$\min_{x_0} \left[x_0^* \Pi_0^{-1} x_0 + \sum_{j=0}^{N} |y(j) - u_j x_j|^2 \right]$$

is equal to $\lambda^{(N+1)/2} \hat{x}_{N+1|N}$, where $\hat{x}_{N+1|N}$ is recursively computed as follows. Start with $P_{0|-1} = \Pi_0$, $\hat{x}_{0|-1} = 0$ and repeat for $i \geq 0$:

$$\begin{aligned}
r_e(i) &= 1 + u_i P_{i|i-1} u_i^* \\
k_{p,i} &= \lambda^{-1/2} P_{i|i-1} u_i^* / r_e(i) \\
\nu(i) &= y(i) - u_i \hat{x}_{i|i-1} \\
\hat{x}_{i+1|i} &= \lambda^{-1/2} \hat{x}_{i|i-1} + k_{p,i} \nu(i) \\
P_{i+1|i} &= \lambda^{-1} \left[P_{i|i-1} - \frac{P_{i|i-1} u_i^* u_i P_{i|i-1}}{1 + u_i P_{i|i-1} u_i^*} \right]
\end{aligned}$$

Moreover, at each iteration i, it holds that $\hat{x}_{0|i} = \lambda^{(i+1)/2} \hat{x}_{i+1|i}$ where $\hat{x}_{0|i}$ is the solution of

$$\min_{x_0} \left[x_0^* \Pi_0^{-1} x_0 + \sum_{j=0}^{i} |y(j) - u_j x_j|^2 \right]$$

using data $\{y(j), u_j\}$ up to time i.

In order to verify that the recursions (31.17)–(31.21) indeed agree with the exponentially-weighted RLS filter of Alg. 30.2, we simply need to rewrite them in terms of the RLS variables $\{d(i), w_i\}$ as opposed to the Kalman variables $\{y(i), \hat{x}_{i|i-1}\}$. Thus, using the substitutions

$$\boxed{y(i) = \frac{d(i)}{(\sqrt{\lambda})^i}, \qquad \hat{x}_{i|i-1} = \frac{w_{i-1}}{(\sqrt{\lambda})^i}} \tag{31.22}$$

in (31.20) leads to

$$\frac{w_i}{\sqrt{\lambda^{i+1}}} = \lambda^{-1/2} \frac{w_{i-1}}{\sqrt{\lambda^i}} + k_{p,i} \left[\frac{d(i)}{\sqrt{\lambda^i}} - u_i \frac{w_{i-1}}{\sqrt{\lambda^i}} \right]$$

or, equivalently, by multiplying both sides by $\sqrt{\lambda^{i+1}}$,

$$w_i = w_{i-1} + \lambda^{1/2} k_{p,i} \left[d(i) - u_i w_{i-1} \right]$$

or, by using the expression for $k_{p,i}$,

$$w_i = w_{i-1} + \frac{P_{i|i-1} u_i^*}{1 + u_i P_{i|i-1} u_i^*} \left[d(i) - u_i w_{i-1} \right] \tag{31.23}$$

Moreover, we already know from (30.19) that the RLS variable P_i satisfies the recursion

$$P_i^{-1} = \lambda P_{i-1}^{-1} + u_i^* u_i, \quad P_{-1}^{-1} = \Pi$$

while by applying the matrix inversion formula (5.4) to recursion (31.21) for $P_{i+1|i}$ we find that

$$P_{i+1|i}^{-1} = \lambda \left[P_{i|i-1}^{-1} + u_i^* u_i \right], \quad P_{0|-1}^{-1} = \lambda \Pi$$

Comparing the recursions for the RLS and Kalman variables $\{P_i^{-1}, P_{i+1|i}^{-1}\}$ we conclude that they are related via

$$\boxed{P_{i|i-1} = \lambda^{-1} P_{i-1}} \tag{31.24}$$

so that (31.23) ends up agreeing with the desired RLS equation from Alg. 30.2, namely

$$w_i = w_{i-1} + \frac{\lambda^{-1} P_{i-1} u_i^*}{1 + \lambda^{-1} u_i P_{i-1} u_i^*} \left[d(i) - u_i w_{i-1} \right]$$

In summary, the above argument shows that the exponentially-weighted RLS solution of Alg. 30.2 is equivalent to the linear least-mean-squares problem of estimating the variable $\boldsymbol{x}_0$ from $\{\boldsymbol{y}(0), \boldsymbol{y}(1), \ldots, \boldsymbol{y}(N)\}$ in the model (31.8) once the Kalman variables are translated into the RLS variables by using the relations (31.16), (31.20) and (31.24).

Actually, there are additional identifications that we can make between the Kalman variables and the RLS variables. Recall, for instance, that with the RLS problem we associate two residuals at each time instant i: the *a priori* error

$$e(i) = d(i) - u_i w_{i-1}$$

and the *a posteriori* error

$$r(i) = d(i) - u_i w_i$$

These residuals can be related to the innovations variable $\nu(i)$ as follows. First note that

$$\nu(i) \;=\; y(i) - u_i \hat{x}_{i|i-1} = \frac{1}{(\sqrt{\lambda})^i} \left[d(i) - u_i w_{i-1} \right] = \frac{1}{(\sqrt{\lambda})^i}\, e(i)$$

while

$$\begin{aligned}
r(i) = d(i) - u_i w_i &= d(i) - (\sqrt{\lambda})^{i+1} u_i \hat{x}_{i+1|i} \\
&= d(i) - (\sqrt{\lambda})^{i+1} u_i \left[\lambda^{-1/2} \hat{x}_{i|i-1} + k_{p,i} \nu(i) \right] \\
&= \left[d(i) - u_i w_{i-1} \right] - \lambda^{1/2} u_i k_{p,i} e(i) \\
&= \left[1 - \sqrt{\lambda} u_i k_{p,i} \right] e(i) \\
&= \left[1 - \frac{u_i P_{i|i-1} u_i^*}{1 + u_i P_{i|i-1} u_i^*} \right] e(i) = e(i)/r_e(i)
\end{aligned}$$

This means that the *conversion factor*, which converts the RLS *a priori* error $e(i)$ to the RLS *a posteriori* error $r(i)$, and which we have denoted by $\gamma(i)$ in Alg. 30.2, is equal to $r_e^{-1}(i)$ (the inverse of the innovations variance). Table 31.2 summarizes the correspondences between the Kalman variables and the RLS variables. This table is useful for the following purpose. By writing down any of the available algorithmic variants for Kalman filtering for model (31.8) (e.g., cf. Apps. 35.A and 37.A), and by using the correspondences from Table 31.2, we can obtain the corresponding RLS variant. Actually, this point of view

TABLE 31.2 Correspondences between Kalman and RLS variables. In the Kalman case, realizations of the random variables are listed (using normal font) rather than the random variables themselves (which would have been described in boldface font).

Description	Kalman variable	RLS variable	Description
Measurement	$y(i)$	$d(i)/(\sqrt{\lambda})^i$	Reference signal
State vector	x_i	$w/(\sqrt{\lambda})^i$	Weight vector
State estimator	$\hat{x}_{i+1\|i}$	$w_i/(\sqrt{\lambda})^{i+1}$	Weight estimate
Error covariance	$P_{i+1\|i}$	$\lambda^{-1}P_i$	Inverse of coefficient matrix
Gain vector	$k_i/r_e(i)$	$\lambda^{-1/2}g_i$	Gain vector
Innovations	$\nu(i)$	$e(i)/(\sqrt{\lambda})^i$	*A priori* error
Innovations	$\nu(i)$	$r(i)r_e(i)/(\sqrt{\lambda})^i$	*A posteriori* error
Innovations variance	$r_e(i)$	$\gamma^{-1}(i)$	Inverse of conversion factor
Initial condition	$\hat{x}_{0\|-1}$	w_{-1}	Initial condition
Initial covariance	Π_0	$\lambda^{-1}\Pi^{-1}$	Regularization matrix

can be used to derive all of the RLS variants described in this book — see Probs. VIII.12 and VIII.13, as well as Prob. IX.16.

31.A APPENDIX: EXTENDED RLS ALGORITHMS

The arguments in Sec. 31.2 show that the exponentially-weighted RLS solution of Alg. 30.2 is equivalent to the linear least-mean-squares problem of estimating $\boldsymbol{x}_0$ from the observations of model (31.8). Now model (31.8) is a special one and, therefore, the RLS filter is equivalent not to a full-blown Kalman filter, but only to a special case of it.

In this section we describe the general deterministic criterion that is equivalent to a full-blown Kalman filter. In so doing, we arrive at extended RLS algorithms that are better suited for tracking the state-vector of general linear state-space models, as opposed to tracking the state-vector of the special model (31.8).

Deterministic Estimation

Consider a collection of $(N+1)$ measurements $\{d_i\}$, possibly column vectors, that satisfy

$$\boxed{d_i = U_i x_i + v_i} \tag{31.25}$$

where the $\{x_i\}$ evolve in time according to the state recursion

$$\boxed{x_{i+1} = F_i x_i + G_i n_i} \tag{31.26}$$

Here the $\{F_i, G_i, U_i\}$ are known matrices and the $\{n_i, v_i\}$ denote disturbances or noises. For generality, we are allowing the observations $\{d_i\}$ to be vectors. Let further Π_0 be a positive-definite regularization matrix, and let $\{Q_i, R_i\}$ be positive-definite weighting matrices. Given the $\{d_i\}$, we pose the problem of estimating the initial state vector x_0 and the signals $\{n_0, n_1, \ldots, n_N\}$ in a regularized least-squares manner by solving

$$\min_{\{x_0, n_0, \ldots, n_N\}} \left[x_0^*\Pi_0^{-1}x_0 + \sum_{i=0}^{N}(d_i - U_i x_i)^* R_i^{-1}(d_i - U_i x_i) + \sum_{i=0}^{N} n_i^* Q_i^{-1} n_i \right] \tag{31.27}$$

subject to the constraint (31.26). We denote the solution by $\{\hat{x}_{0|N}, \hat{n}_{j|N}, 0 \leq j \leq N\}$, and we refer to them as smoothed estimates since they are based on observations beyond the times of occurrence of the respective variables $\{x_0, n_j\}$.

In principle, we could solve (31.27) by using variational (optimization) arguments, e.g., by using a Lagrange multiplier argument. Instead, we shall solve it by appealing to the equivalence result of Table 31.1. In other words, we shall first determine the equivalent stochastic problem and then solve this latter problem to arrive at the solution of (31.27). This method of solving (31.27) not only serves as an illustration of the convenience of equivalence results in estimation theory, but it also shows that sometimes it is easier to solve a deterministic problem in the stochastic domain (or vice-versa). In our case, the problem at hand is more conveniently solved in the stochastic domain.

Define the column vectors

$$z = \text{col}\{x_0, n_0, n_1, \ldots, n_N\}$$

and

$$y = \text{col}\{d_0, d_1, \ldots, d_N\}$$

as well as the block-diagonal matrices

$$W^{-1} \triangleq (R_0 \oplus R_1 \oplus \ldots \oplus R_N), \qquad \Pi^{-1} \triangleq (\Pi_0 \oplus Q_0 \oplus \ldots \oplus Q_N) \tag{31.28}$$

Then the term

$$x_0^* \Pi_0^{-1} x_0 \ + \ \sum_{i=0}^{N} n_i^* Q_i^{-1} n_i$$

that appears in (31.27) is equal to $z^* \Pi z$. Moreover, by using the state equation (31.26) to express each term $U_i x_i$ in terms of combinations of the entries of z, we can verify that

$$\sum_{i=0}^{N} (d_i - U_i x_i)^* R_i^{-1} (d_i - U_i x_i) = (y - Hz)^* W (y - Hz) \ = \ \|y - Hz\|_W^2$$

where the matrix H is block lower-triangular and given by

$$H \triangleq \begin{bmatrix} U_0 & & & & & \\ U_1\Phi(1,0) & U_1 G_0 & & & & \\ U_2\Phi(2,0) & U_2\Phi(2,1)G_0 & U_2 G_1 & & & \\ \vdots & \vdots & \vdots & \ddots & & \\ U_N\Phi(N,0) & U_N\Phi(N,1)G_0 & U_N\Phi(N,2)G_1 & \ldots & U_N G_{N-1} & 0 \end{bmatrix} \tag{31.29}$$

and the matrices $\Phi(i,j)$ are defined by

$$\Phi(i,j) \triangleq \begin{cases} F_{i-1}F_{i-2}\ldots F_j & i > j \\ \text{I} & i = j \end{cases}$$

In other words, we find that we can rewrite the original cost function (31.27) as the regularized least-squares cost function

$$\min_z \ [z^* \Pi z \ + \ (y - Hz)^* W (y - Hz)] \tag{31.30}$$

Let $\hat{z}_{|N}$ denote the solution to (31.30), i.e., $\hat{z}_{|N}$ is a column vector that contains the desired solutions $\{\hat{x}_{0|N}, \hat{n}_{0|N}, \hat{n}_{1|N}, \ldots, \hat{n}_{N|N}\}$. Now, in view of the equivalence result of Sec. 31.1, we know that $\hat{z}_{|N}$ can be obtained by solving an equivalent stochastic estimation problem that is determined as follows.

Stochastic Estimation

We introduce zero-mean random vectors $\{\boldsymbol{z}, \boldsymbol{y}\}$, with the same dimensions and partitioning as the above $\{z, y\}$, and assume that they are related via a linear model of the form

$$\boldsymbol{y} = H\boldsymbol{z} + \boldsymbol{v} \tag{31.31}$$

where H is the same matrix as in (31.29), and where $\boldsymbol{v}$ denotes a zero-mean additive noise vector, uncorrelated with $\boldsymbol{z}$, and partitioned as $\boldsymbol{v} = \text{col}\{\boldsymbol{v}_0, \boldsymbol{v}_1, \ldots, \boldsymbol{v}_N\}$. The dimensions of the $\{\boldsymbol{v}_i\}$ are compatible with those of $\{\boldsymbol{y}_i\}$. We denote the covariance matrices of $\{\boldsymbol{z}, \boldsymbol{v}\}$ by

$$R_z = \mathsf{E}\,\boldsymbol{z}\boldsymbol{z}^*, \qquad R_v = \mathsf{E}\,\boldsymbol{v}\boldsymbol{v}^* \tag{31.32}$$

and we choose them as $R_z = \Pi^{-1}$ and $R_v = W^{-1}$, where $\{\Pi, W\}$ are given by (31.28).

Let $\hat{\boldsymbol{z}}_{|N}$ denote the l.l.m.s. estimator of $\boldsymbol{z}$ given the entries $\{\boldsymbol{d}_0, \boldsymbol{d}_1, \ldots, \boldsymbol{d}_N\}$ in $\boldsymbol{y}$. We further partition $\boldsymbol{z}$ as

$$\boldsymbol{z} = \text{col}\{\boldsymbol{x}_0, \boldsymbol{n}_0, \boldsymbol{n}_1, \ldots, \boldsymbol{n}_N\}$$

Then the equivalence result of Table 31.1 states that the expression for $\hat{\boldsymbol{z}}_{|N}$ in terms of $\boldsymbol{y}$ in the stochastic problem (31.31) is identical to the expression for $\hat{z}_{|N}$ in terms of y in the deterministic problem (31.30).

In order to determine $\hat{\boldsymbol{z}}_{|N}$ or, equivalently, $\{\hat{\boldsymbol{x}}_{0|N}, \hat{\boldsymbol{n}}_{j|N}\}$, we start by noting that the linear model (31.31), coupled with the definitions of $\{R_z, R_v, H\}$ in (31.28), (31.29), and (31.32), show that the stochastic variables $\{\boldsymbol{d}_i, \boldsymbol{v}_i, \boldsymbol{x}_0, \boldsymbol{n}_i\}$ so defined satisfy the following state-space model:

$$\boxed{\begin{aligned} \boldsymbol{x}_{i+1} &= F_i\boldsymbol{x}_i + G_i\boldsymbol{n}_i \\ \boldsymbol{d}_i &= U_i\boldsymbol{x}_i + \boldsymbol{v}_i \end{aligned}} \tag{31.33}$$

with

$$\boxed{\mathsf{E}\begin{bmatrix} \boldsymbol{n}_i \\ \boldsymbol{v}_i \\ \boldsymbol{x}_0 \\ 1 \end{bmatrix}\begin{bmatrix} \boldsymbol{n}_j \\ \boldsymbol{v}_j \\ \boldsymbol{x}_0 \end{bmatrix}^* = \begin{bmatrix} Q_i\delta_{ij} & 0 & 0 \\ 0 & R_i\delta_{ij} & 0 \\ 0 & 0 & \Pi_0 \\ 0 & 0 & 0 \end{bmatrix}} \tag{31.34}$$

We now use this model to derive recursions for estimating $\boldsymbol{z}$ (i.e., for estimating the variables $\{\boldsymbol{x}_0, \boldsymbol{n}_1, \ldots, \boldsymbol{n}_N\}$).

Solving the Stochastic Problem

Let $\hat{\boldsymbol{z}}_{|i}$ denote the linear least-mean-squares (l.l.m.s.) estimator of $\boldsymbol{z}$ given the top entries $\{\boldsymbol{d}_0, \ldots, \boldsymbol{d}_i\}$ in $\boldsymbol{y}$. To determine $\hat{\boldsymbol{z}}_{|i}$, and ultimately $\hat{\boldsymbol{z}}_{|N}$, we can proceed recursively by employing the innovations $\{\boldsymbol{e}_i\}$ of the observations $\{\boldsymbol{y}_i\}$. In this appendix, we stick to our standard notation for the innovations and use $\boldsymbol{e}_i$ instead of $\boldsymbol{\nu}_i$, which was used in the previous appendix. There is no need here to avoid confusion with the error signal for RLS. Using the basic recursive estimation formula (7.6) we have

$$\begin{aligned} \hat{\boldsymbol{z}}_{|i} &= \hat{\boldsymbol{z}}_{|i-1} + (\mathsf{E}\,\boldsymbol{z}\boldsymbol{e}_i^*)\; R_{e,i}^{-1}\,\boldsymbol{e}_i \\ &= \hat{\boldsymbol{z}}_{|i-1} + \left(\mathsf{E}\,\boldsymbol{z}\tilde{\boldsymbol{x}}_{i|i-1}^*\right) U_i^* R_{e,i}^{-1}\,\boldsymbol{e}_i, \quad \hat{\boldsymbol{z}}_{|-1} = 0 \end{aligned} \tag{31.35}$$

where we used in the second equality the innovations equation (cf. (7.21)):

$$\boldsymbol{e}_i \;=\; \boldsymbol{d}_i - U_i\hat{\boldsymbol{x}}_{i|i-1} \;=\; U_i\tilde{\boldsymbol{x}}_{i|i-1} + \boldsymbol{v}_i$$

and the fact that $\mathsf{E}\,\boldsymbol{x}_0\boldsymbol{v}_i^* = 0$ and $\mathsf{E}\,\boldsymbol{n}_j\boldsymbol{v}_i^* = 0$ for all j. Clearly, the entries of $\hat{\boldsymbol{z}}_{|i}$ have the interpretation

$$\hat{\boldsymbol{z}}_{|i} = \text{col}\{\hat{\boldsymbol{x}}_{0|i}, \hat{\boldsymbol{n}}_{0|i}, \hat{\boldsymbol{n}}_{1|i}, \ldots, \hat{\boldsymbol{n}}_{i-1|i}, 0, 0, \ldots, 0\}$$

where the trailing entries of $\hat{\boldsymbol{z}}_{|i}$ are zero since $\hat{\boldsymbol{n}}_{j|i} = 0$ for $j \geq i$.

Let $K_{z,i} = \mathsf{E}\,\boldsymbol{z}\tilde{\boldsymbol{x}}_{i|i-1}^*$. The above recursive construction would be complete, and hence provide the desired quantity $\hat{\boldsymbol{z}}_{|N}$, once we show how to evaluate the gain matrix $K_{z,i}$. For this purpose, we first subtract the equations (from the Kalman filter Alg. 7.1):

$$\boldsymbol{x}_{i+1} = F_i\boldsymbol{x}_i + G_i\boldsymbol{n}_i \quad \text{and} \quad \hat{\boldsymbol{x}}_{i+1|i} = F_i\hat{\boldsymbol{x}}_{i|i-1} + K_{p,i}[U_i\tilde{\boldsymbol{x}}_{i|i-1} + \boldsymbol{v}_i]$$

to obtain

$$\tilde{\boldsymbol{x}}_{i+1|i} = F_{p,i}\tilde{\boldsymbol{x}}_{i|i-1} + G_i\boldsymbol{n}_i - K_{p,i}\boldsymbol{v}_i$$

where $F_{p,i} = F_i - K_{p,i}U_i$. Using this recursion, it is easy to verify that $K_{z,i}$ satisfies the recursion:

$$K_{z,i+1} = \mathsf{E}\, \boldsymbol{z}\tilde{\boldsymbol{x}}^*_{i+1|i} \quad = \quad K_{z,i}F^*_{p,i} \; + \begin{bmatrix} 0 \\ \hline 0 \\ \mathrm{I} \\ 0 \end{bmatrix} Q_i G^*_i, \qquad K_{z,0} = \begin{bmatrix} \Pi_0 \\ 0 \end{bmatrix} \tag{31.36}$$

The identity matrix that appears in the second term of the recursion for $K_{z,i+1}$ occurs at the position that corresponds to the entry $\boldsymbol{n}_i$ in the vector $\boldsymbol{z}$, e.g.,

$$K_{z,1} = \begin{bmatrix} \Pi_0 F^*_{p,0} \\ Q_0 G^*_0 \\ 0 \end{bmatrix}, \qquad K_{z,2} = \begin{bmatrix} \Pi_0 F^*_{p,0} F^*_{p,1} \\ Q_0 G^*_0 F^*_{p,1} \\ Q_1 G^*_1 \\ 0 \end{bmatrix}, \quad \ldots$$

Substituting (31.36) into (31.35) we find that the following recursions hold:

$$\begin{cases} \hat{\boldsymbol{x}}_{0|i} &= \hat{\boldsymbol{x}}_{0|i-1} + \Pi_0 \Phi^*_p(i,0) U^*_i R^{-1}_{e,i} \boldsymbol{e}_i, \quad \hat{\boldsymbol{x}}_{0|-1} = 0 \\ \hat{\boldsymbol{n}}_{j|i} &= \hat{\boldsymbol{n}}_{j|i-1} + Q_j G^*_j \Phi^*_p(i, j+1) U^*_i R^{-1}_{e,i} \boldsymbol{e}_i, \quad j < i \\ \hat{\boldsymbol{n}}_{j|i} &= 0, \;\; j \geq i \end{cases} \tag{31.37}$$

where the matrix $\Phi_p(i,j)$ is defined by

$$\Phi_p(i,j) \triangleq \begin{cases} F_{p,i-1} F_{p,i-2} \ldots F_{p,j} & i > j \\ \mathrm{I} & i = j \end{cases}$$

If we introduce the auxiliary variable

$$\boldsymbol{\rho}_{i|N} \triangleq \sum_{j=i}^{N} \Phi^*_p(j,i) U^*_j R^{-1}_{e,j} \boldsymbol{e}_j$$

then it is easy to verify that recursions (31.37) lead to

$$\begin{cases} \hat{\boldsymbol{x}}_{0|N} &= \Pi_0 \boldsymbol{\rho}_{0|N} \\ \hat{\boldsymbol{x}}_{j+1|j} &= F_{p,j} \hat{\boldsymbol{x}}_{j|j-1} + K_{p,j} \boldsymbol{d}_j, \quad \hat{\boldsymbol{x}}_{0|-1} = 0 \\ \boldsymbol{e}_j &= \boldsymbol{d}_j - U_j \hat{\boldsymbol{x}}_{j|j-1} \\ \hat{\boldsymbol{n}}_{j|N} &= Q_j G^*_j \boldsymbol{\rho}_{j+1|N} \\ \boldsymbol{\rho}_{j|N} &= F^*_{p,j} \boldsymbol{\rho}_{j+1|N} + U^*_i R^{-1}_{e,j} \boldsymbol{e}_j, \quad \boldsymbol{\rho}_{N+1|N} = 0 \end{cases} \tag{31.38}$$

These equations are known as the Bryson-Frazier recursions in the literature of Kalman filtering; the recursions (31.37) evaluate the estimators $\{\hat{\boldsymbol{x}}_{0|i}, \hat{\boldsymbol{n}}_{j|i}\}$ for successive values of i, and not only for $i = N$ as in (31.38). Just like $\{\hat{\boldsymbol{x}}_{0|N}, \hat{\boldsymbol{n}}_{j|N}\}$, the estimators $\{\hat{\boldsymbol{x}}_{0|i}, \hat{\boldsymbol{n}}_{j|i}\}$ can also be related to the solution of a least-squares problem. Indeed, by equivalence, the expressions that provide the solutions $\{\hat{\boldsymbol{x}}_{0|i}, \hat{\boldsymbol{n}}_{j|i}\}$ in (31.37) should coincide with those that provide the solutions $\{\hat{x}_{0|i}, \hat{n}_{j|i}\}$ for the following deterministic problem, with data up to time i (rather than N as in (31.27)):

$$\min_{x_0, n_0, \ldots, n_i} \left[x^*_0 \Pi^{-1}_0 x_0 \; + \sum_{j=0}^{i} (d_j - U_j x_j)^* R^{-1}_j (d_j - U_j x_j) + \sum_{j=0}^{i} n^*_j Q^{-1}_j n_j \right] \tag{31.39}$$

Solving the Deterministic Problem

We know by equivalence that the mappings from $\{\boldsymbol{d}_j\}$ to $\{\hat{\boldsymbol{x}}_{0|N}, \hat{\boldsymbol{n}}_{j|N}\}$ in the stochastic problem (31.31) coincide with the mappings from $\{d_j\}$ to $\{\hat{x}_{0|N}, \hat{n}_{j|N}\}$ in the deterministic problem (31.30). We are therefore led to the following statement.

Algorithm 31.1 (Extended RLS) Consider measurements $\{d_i\}$ that satisfy a state-space model of the form

$$x_{i+1} = F_i x_i + G_i n_i, \quad d_i = U_i x_i + v_i$$

where the $\{F_i, G_i, U_i\}$ are known matrices. The solution $\{\hat{x}_{0|N}, \hat{n}_{j|N}\}$ of

$$\min_{\{x_0, n_0, \ldots, n_N\}} \left[x_0^* \Pi_0^{-1} x_0 + \sum_{i=0}^{N} (d_i - U_i x_i)^* R_i^{-1} (d_i - U_i x_i) + \sum_{i=0}^{N} n_i^* Q_i^{-1} n_i \right]$$

can be determined recursively as follows. Start with $\hat{x}_{0|-1} = 0$, $P_{0|-1} = \Pi_0$, and $\rho_{N+1|N} = 0$, and run the following equations forward in time from $i = 0$ to $i = N$:

$$\begin{aligned} R_{e,i} &= R_i + U_i P_{i|i-1} U_i^* \\ K_{p,i} &= F_i P_{i|i-1} U_i^* R_{e,i}^{-1} \\ e_i &= d_i - U_i \hat{x}_{i|i-1} \\ \hat{x}_{i+1|i} &= F_i \hat{x}_{i|i-1} + K_{p,i} e_i \\ P_{i+1|i} &= F_i P_{i|i-1} F_i^* + G_i Q_i G_i^* - K_{p,i} R_{e,i} K_{p,i}^* \end{aligned}$$

Then run the following recursion backward in time from $i = N$ down to $i = 0$:

$$\rho_{i|N} = F_{p,i}^* \rho_{i+1|N} + U_i^* R_{e,i}^{-1} e_i$$

and set

$$\hat{x}_{0|N} = \Pi_0 \rho_{0|N}$$

and

$$\hat{n}_{i|N} = Q_i G_i^* \rho_{i+1|N} \quad \text{for} \quad 0 \leq i \leq N$$

Special Cases

Algorithm 31.1 is a generalized RLS algorithm and has superior tracking abilities. The generalization is in several respects:

1. To begin with, the measurements $\{d_i\}$ are allowed to be vector-valued. In most of our treatments so far, we have assumed scalar-valued observations arising from a linear model of the form

$$d(i) = u_i w^o + v(i)$$

for some unknown weight vector w^o to be estimated.

2. **Block RLS**. Algorithm 31.1 assumes that the state vectors $\{x_i\}$ that determine the measurements $\{d_i\}$ evolve in time according to the state-space recursion

$$x_{i+1} = F_i x_i + G_i n_i$$

This assumption is more general than assuming $d_i = U_i w^o + v_i$ with an unknown *constant* weight vector w^o. This latter situation can be modeled as $x_{i+1} = x_i$ with $x_0 = w^o$. That is, it corresponds to the special choices $F_i = \mathrm{I}$, $G_i = 0$ and $x_0 = w^o$. In this case, the recursions of Alg. 31.1 reduce to the following. Start with $\hat{x}_{0|-1} = 0$ and $P_{0|-1} = \Pi_0$ and repeat for

$i \geq 0$:

$$\begin{aligned} R_{e,i} &= R_i + U_i P_{i|i-1} U_i^* \\ K_{p,i} &= P_{i|i-1} U_i^* R_{e,i}^{-1} \\ e_i &= d_i - U_i \hat{x}_{i|i-1} \\ \hat{x}_{i+1|i} &= \hat{x}_{i|i-1} + K_{p,i} e_i \\ P_{i+1|i} &= P_{i|i-1} - K_{p,i} R_{e,i} K_{p,i}^* \end{aligned}$$

Now since $x_i = x_0$, the quantities $\hat{x}_{i|i-1}$ correspond to $\hat{x}_{0|i-1}$; i.e., they provide recursive estimates for x_0 and there is no need for the $\rho_{i|N}$ recursion. If we denote $\hat{x}_{0|i}$ by w_i and $P_{i+1|i}$ by P_i, then the above equations will agree with the block RLS algorithm of Prob. VII.36.

3. **RLS**. Clearly, Alg. 31.1 also subsumes the exponentially-weighted RLS filter of Alg. 30.2 as a special case. Comparing (31.14) with (31.39), we see that (31.14) is a special case of the latter if we select

$$R_i = \lambda^i, \qquad \Pi_0^{-1} = \lambda \Pi \tag{31.40}$$

for some scalar $0 \ll \lambda \leq 1$ positive-definite regularization matrix Π. Moreover, we should also select d_i to be a scalar, say, $d(i)$, and U_i to be a row vector, say, u_i, and assume that they are related via

$$x_{i+1} = x_i, \qquad d(i) = u_i x_i + v(i) \tag{31.41}$$

This model corresponds to the special choices $F_i = \mathrm{I}$ and $G_i = 0$. Then the recursions of Alg. 31.1 would reduce to the following. Start with $\hat{x}_{0|-1} = 0$ and $P_{0|-1} = \lambda^{-1}\Pi^{-1}$ and repeat for $i \geq 0$:

$$\begin{aligned} r_e(i) &= \lambda^i + u_i P_{i|i-1} u_i^* \\ k_{p,i} &= P_{i|i-1} u_i^* / r_e(i) \\ e(i) &= d(i) - u_i \hat{x}_{i|i-1} \\ \hat{x}_{i+1|i} &= \hat{x}_{i|i-1} + k_{p,i} e(i) \\ P_{i+1|i} &= P_{i|i-1} - \frac{P_{i|i-1} u_i^* u_i P_{i|i-1}}{\lambda^i + u_i P_{i|i-1} u_i^*} \end{aligned}$$

In this case again, since $x_i = x_0$, the quantities $\hat{x}_{i|i-1}$ correspond to $\hat{x}_{0|i-1}$; i.e., they provide recursive estimates for x_0 and there is no need for the $\rho_{i|N}$ recursion. If we denote $\hat{x}_{0|i}$ by w_i and use the easily verifiable fact that now $P_{i+1|i} = \lambda^i P_i$, then the above equations will agree with the exponentially-weighted RLS filter of Alg. 30.2.

4. **Other state-space models**. In the equivalence argument of App. 31.2 we relied on a different state-space model than (31.41) to arrive at the RLS algorithm. In particular, in model (31.8), we used $F_i = \lambda^{-1/2}\mathrm{I}$, $R(i) = 1$, and scaled the measurements $d(i)$ as in (31.16). The resulting state-space model had a constant R_i whereas (31.40) uses $R_i = \lambda^i$. While in (31.24) we had $P_{i|i-1} = \lambda^{-1} P_{i-1}$, we now have $P_{i|i-1} = \lambda^{i-1} P_{i-1}$. These constructions illustrate that there is freedom in selecting the underlying state-space model. One key advantage of using (31.8) is that this model belongs to the class of *structured* state-space models (cf. defined later by (37.25)), for which fast estimation algorithms can be readily derived. Choosing $R_i = \lambda^i$, on the other hand, leads to a time-variant model, as remarked in the Bibliographic Notes at the end of this part. In this way, model (31.8) becomes useful not only in establishing the connection between plain RLS and Kalman filtering, but also between other RLS and Kalman filtering variants (especially between their fast variants). We shall explain these connections in App. 37.A and Prob. IX.16.

5. **Tracking**. Algorithm 31.1 suggests extensions of RLS in order to track variations in the weight vector that can be captured by linear state-space recursions. Thus, assume, for example, that the measurements $\{d(i)\}$ satisfy

$$\boxed{d(i) = u_i w_i^o + v(i)}$$

and that the unknown weight vector w_i^o evolves in time according to

$$\boxed{w_{i+1}^o = \alpha w_i^o + n_i}$$

for some scalar $|\alpha| \leq 1$ and disturbance n_i. This model is a special case of the state-space model used by Alg. 31.1 if we make the identification $x_i = w_i^o$ and if we set

$$F_i = \alpha \mathrm{I}, \quad G_i = \mathrm{I}$$

Then, again, if we select $\Pi_0^{-1} = \lambda\Pi$, $R_i = \lambda^i$, and $Q_i = q\lambda^i\mathrm{I}$, for some positive scalar q and some scalar $0 \ll \lambda \leq 1$ used to introduce exponential weighting,[14] then the recursions of Alg. 31.1 would reduce to the following. Start with $\hat{x}_{0|-1} = 0$ and $P_{0|-1} = \lambda^{-1}\Pi^{-1}$ and repeat for $i \geq 0$:

$$\begin{aligned}
r_e(i) &= \lambda^i + u_i P_{i|i-1} u_i^* \\
k_{p,i} &= \alpha P_{i|i-1} u_i^* / r_e(i) \\
e(i) &= d(i) - u_i \hat{x}_{i|i-1} \\
\hat{x}_{i+1|i} &= \alpha \hat{x}_{i|i-1} + k_{p,i} e(i) \\
P_{i+1|i} &= |\alpha|^2 \left[P_{i|i-1} - \frac{P_{i|i-1} u_i^* u_i P_{i|i-1}}{\lambda^i + u_i P_{i|i-1} u_i^*} \right] + \lambda^i q \mathrm{I}
\end{aligned}$$

In this case, since $x_i = w_i^o$, the quantity $\hat{x}_{i|i-1}$ serves as an estimate for w_i^o that is based on measurements up to time $i-1$. If we denote $\hat{x}_{i+1|i}$ by w_i and if we introduce the change of variables

$$P_i \stackrel{\Delta}{=} \lambda^{-i} P_{i+1|i}$$

then by multiplying the recursion for $P_{i+1|i}$ by λ^{-i} on both sides, the above equations can be rewritten equivalently in the following form. Start with $w_{-1} = 0$ and $P_{-1} = \Pi^{-1}$ and repeat for $i \geq 0$:

$$\boxed{\begin{aligned}
w_i &= \alpha w_{i-1} + \frac{\lambda^{-1}\alpha P_{i-1} u_i^*}{1 + \lambda^{-1} u_i P_{i-1} u_i^*}[d(i) - u_i w_{i-1}] \\
P_i &= \lambda^{-1}|\alpha|^2 \left[P_{i-1} - \frac{\lambda^{-1} P_{i-1} u_i^* u_i P_{i-1}}{1 + \lambda^{-1} u_i P_{i-1} u_i^*} \right] + q\mathrm{I}
\end{aligned}} \tag{31.42}$$

These recursions collapse to the exponentially-weighted RLS filter of Alg. 30.2 when $\alpha = 1$ and $q = 0$. The associated cost function is

$$\min_{\{w_0^o, n_0, \ldots, n_N\}} \left[\lambda^{(N+1)} w_0^{o*} \Pi w_0^o + \sum_{i=0}^{N} \lambda^{N-i} |d(i) - u_i w_i^o|^2 + q^{-1} \sum_{i=0}^{N} \lambda^{N-i} \|n_i\|^2 \right]$$

subject to $w_{i+1}^o = \alpha w_i^o + n_i$. Note that if we extract a factor λ^N out, then the cost function becomes of the same form shown in Alg. 31.1 for the choices $R_i = \lambda^i$, $Q_i = q\lambda^i\mathrm{I}$, and $\Pi_0^{-1} = \lambda\Pi$.

6. **RLS and Kalman variants**. While in future chapters we focus almost exclusively on the exponentially-weighted RLS scheme of Alg. 30.2, and derive several variants for it, many of these developments can be extended to the more general RLS schemes shown above. These extensions could be pursued afresh by repeating the arguments of the subsequent chapters. Alternatively, once the connection between Kalman filters and extended RLS filters has been established, most of the algorithmic variants that exist for Kalman filters can be applied immediately to extended RLS filters.

[14] Of course, more general choices for R_i and Q_i are allowed by Alg. 31.1. These special choices are for illustration purposes only.

CHAPTER 32

Order and Time-Update Relations

This chapter can be skipped on a first reading. Its results will only be needed in later chapters when we study fast fixed-order and order-recursive least-squares filters in Parts IX (*Fast RLS Algorithms*) and X (*Lattice Filters*).

However, there is a reason why we choose to present the material at this location, and not in later chapters. The reason is that the results we present here are often derived in the literature in the absence of regularization and under the assumption of some structure in the data matrix H (such as requiring its successive rows to be shifted versions of one another, as explained later in the introductory remarks to Chapter 37). In comparison, the derivation given here indicates that the results hold irrespective of data structure, i.e., for any H. In addition, the arguments incorporate regularization and clarify its role in order-update relations. In so doing, the results will allow us to provide later in Chapter 40 a treatment of lattice filters in the presence of regularization. The results also allow the class of lattice filters to be extended to more general data structures, other than the classical tapped-delay-line structure, as was shown in detail in Chapter 16 of Sayed (2003).

For now, it suffices to treat the material in this chapter as an application of the concepts and geometric constructions of the earlier sections.

32.1 BACKWARD ORDER-UPDATE RELATIONS

Consider a weighted regularized least-squares problem of the form

$$\min_{w} \; [\, w^*\Pi w \;+\; (y - Hw)^* W (y - Hw)\,] \tag{32.1}$$

whose optimal solution is given by (cf. Thm. 29.5):

$$\widehat{w} = (\Pi + H^*WH)^{-1}H^*Wy \tag{32.2}$$

Let

$$\boxed{P \stackrel{\Delta}{=} (\Pi + H^*WH)^{-1}} \tag{32.3}$$

In other words, P is the inverse of the coefficient matrix that appears in the normal equations

$$(\Pi + H^*WH)\widehat{w} = H^*Wy$$

Then

$$\widehat{w} = PH^*Wy$$

and the estimate of y is

$$\boxed{\widehat{y} = H\widehat{w} \;=\; HPH^*Wy} \tag{32.4}$$

We refer to $\widehat{y}$ as the regularized projection (or simply projection) of y onto the range space of H. Recall from Sec. 29.5 that, when H has full column rank, the actual projection matrix onto $\mathcal{R}(H)$ is defined by

$$\mathcal{P}_H = H(H^*H)^{-1}H^*$$

For the regularized problem (32.1), we instead have

$$\widehat{y} = H(\Pi + H^*WH)^{-1}H^*Wy \;=\; HPH^*Wy$$

Although the matrix HPH^*W that multiplies y is not an actual projection matrix (cf. (29.19) and (29.20)), we shall still refer to $\widehat{y}$ as the projection of y onto $\mathcal{R}(H)$ for ease of reference. Note that in both cases, with and without regularization, $\widehat{y} \in \mathcal{R}(H)$.

The resulting residual vector is

$$\boxed{\widetilde{y} = y - H\widehat{w}}$$

and the corresponding minimum cost is (cf. Table 29.3):

$$\boxed{\xi = y^*W\widetilde{y}} \tag{32.5}$$

Introduce the scalar

$$\boxed{\gamma \;\stackrel{\Delta}{=}\; 1 - uPu^*} \tag{32.6}$$

where u denotes the last row of H. This scalar plays an important role in least-squares adaptive filtering, and it is called the conversion factor for reasons to be explained later in Sec. 30.4. We shall not employ γ in any of the arguments below, but will only comment on some of its properties whenever appropriate.

Now assume that we extend the data matrix H in (32.1) by adding one column to it, say, h, and consider the extended least-squares problem:

$$\min_{w_z} \left\{ w_z^* \begin{bmatrix} \Pi & 0 \\ 0 & \sigma \end{bmatrix} w_z + \left(y - \begin{bmatrix} H & h \end{bmatrix} w_z\right)^* W \left(y - \begin{bmatrix} H & h \end{bmatrix} w_z\right) \right\} \tag{32.7}$$

We say that the order of the estimation problem has increased by one since we are now estimating y from the column span of the extended data matrix $[H \;\; h]$. Of course, the corresponding weight vector, w_z, has one dimension higher than w and, accordingly, we also extend the regularization matrix, Π, by adding a positive scalar σ to it.[14]

The optimal solution of (32.7) is given by (again cf. Thm. 29.5):

$$\widehat{w}_z = \left(\begin{bmatrix} \Pi & 0 \\ 0 & \sigma \end{bmatrix} + \begin{bmatrix} H^* \\ h^* \end{bmatrix} W \begin{bmatrix} H & h \end{bmatrix} \right)^{-1} \begin{bmatrix} H^* \\ h^* \end{bmatrix} Wy \tag{32.8}$$

Similarly to P in (32.3), we define

$$\boxed{P_z = \left(\begin{bmatrix} \Pi & 0 \\ 0 & \sigma \end{bmatrix} + \begin{bmatrix} H^* \\ h^* \end{bmatrix} W \begin{bmatrix} H & h \end{bmatrix} \right)^{-1}} \tag{32.9}$$

[14] In Prob. VII.18 we show that the results of this section will still apply, with minor modifications, even if the new regularization matrix were not merely a diagonal extension of Π but, more generally, of the form

$$\begin{bmatrix} \Pi & m \\ m^* & \sigma \end{bmatrix}$$

for some row vector m and positive scalar σ.

so that the new estimate of y is given by

$$\boxed{\widehat{y}_z = \begin{bmatrix} H & h \end{bmatrix} \widehat{w}_z = \begin{bmatrix} H & h \end{bmatrix} P_z \begin{bmatrix} H^* \\ h^* \end{bmatrix} Wy} \tag{32.10}$$

We say that $\widehat{y}_z$ is the regularized projection of y onto the range space of the extended data matrix $[H \;\; h]$. The resulting residual vector is

$$\boxed{\widetilde{y}_z = y - \begin{bmatrix} H & h \end{bmatrix} \widehat{w}_z}$$

and the corresponding minimum cost is (cf. Table 29.3):

$$\boxed{\xi_z = y^* W \widetilde{y}_z} \tag{32.11}$$

The associated conversion factor is

$$\boxed{\gamma_z \triangleq 1 - \begin{bmatrix} u & \alpha \end{bmatrix} P_z \begin{bmatrix} u^* \\ \alpha^* \end{bmatrix}} \tag{32.12}$$

where $[u \;\; \alpha]$ denotes the last row of $[H \;\; h]$,

$$\begin{bmatrix} H & h \end{bmatrix} = \begin{bmatrix} \times & \times \\ \times & \times \\ \hline u & \alpha \end{bmatrix}$$

Our objective is to examine the relations between the solutions of the least-squares problems (32.1) and (32.7). We shall use both algebraic and geometric arguments for the sake of illustration. We start with the algebraic argument and later show how the geometry of least-squares theory can be used to arrive at the same conclusions.

Algebraic Argument

The first step toward relating the solution vectors $\{\widehat{w}, \widehat{w}_z\}$ in (32.2) and (32.8) is to relate the coefficient matrices $\{P, P_z\}$ in (32.3) and (32.9). To achieve this, we start by observing from (32.9) that

$$P_z^{-1} = \begin{bmatrix} \Pi & 0 \\ 0 & \sigma \end{bmatrix} + \begin{bmatrix} H^* \\ h^* \end{bmatrix} W \begin{bmatrix} H & h \end{bmatrix} = \begin{bmatrix} P^{-1} & H^*Wh \\ h^*WH & \sigma + h^*Wh \end{bmatrix} \tag{32.13}$$

where the top leftmost corner entry of P_z^{-1} is seen to be P^{-1},

$$P^{-1} = \Pi + H^*WH$$

In order to relate P_z to P we invoke the easily verifiable matrix identity:

$$\begin{bmatrix} A & B \\ C & D \end{bmatrix}^{-1} = \begin{bmatrix} A^{-1} & 0 \\ 0 & 0 \end{bmatrix} + \begin{bmatrix} -A^{-1}B \\ \mathrm{I} \end{bmatrix} (D - CA^{-1}B)^{-1} \begin{bmatrix} -CA^{-1} & \mathrm{I} \end{bmatrix} \tag{32.14}$$

which relates the inverse of a block matrix to the inverse of its top leftmost corner block. Applying this identity to (32.13) we obtain

$$P_z = \begin{bmatrix} P & 0 \\ 0 & 0 \end{bmatrix} + \frac{1}{\nu} \begin{bmatrix} -PH^*Wh \\ 1 \end{bmatrix} \begin{bmatrix} -h^*WHP & 1 \end{bmatrix} \tag{32.15}$$

where the scalar ν is given by

$$\nu \;=\; \sigma + h^*Wh - h^*WHPH^*Wh = \sigma + h^*W[h - HPH^*Wh] \quad (32.16)$$

The quantity PH^*Wh that appears in (32.15) and (32.16) can be interpreted as the weight vector that solves the following regularized least-squares problem:

$$\min_{w^b} \left[w^{b*}\Pi w^b + (h - Hw^b)^*W(h - Hw^b) \right] \quad (32.17)$$

Indeed, the solution of (32.17) is

$$\widehat{w}^b = PH^*Wh$$

and it corresponds to the weight vector that projects (in a regularized manner) the column vector h onto the column span of H. This projection problem is the reason for the title of this subsection, namely, *backward projection*. The terminology refers to the fact that we are projecting h onto the preceding columns of $[H \;\; h]$, as indicated schematically in Fig. 32.1.
We denote the resulting estimate and residual vectors of this projection problem by

$$\begin{aligned} \widehat{h} &= H\widehat{w}^b = HPH^*Wh \\ \widetilde{h} &= h - \widehat{h} = h - H\widehat{w}^b = h - HPH^*Wh \end{aligned}$$

The last entry of $\widetilde{h}$ is $\alpha - u\widehat{w}^b$; it corresponds to the error in estimating α and we denote it by

$$\boxed{\widetilde{\alpha} \triangleq \alpha - u\widehat{w}^b} \quad (32.18)$$

Also, the minimum cost of (32.17) is given by (cf. Table 29.3):

$$\boxed{\xi_h = h^*W\widetilde{h}} \quad (32.19)$$

Using the above definitions of $\{\widehat{w}^b, \widetilde{h}\}$ we can now rewrite (32.15) as

$$\boxed{P_z = \begin{bmatrix} P & 0 \\ 0 & 0 \end{bmatrix} + \frac{1}{\sigma + \xi_h} \begin{bmatrix} -\widehat{w}^b \\ 1 \end{bmatrix} \begin{bmatrix} -\widehat{w}^{b*} & 1 \end{bmatrix}} \quad (32.20)$$

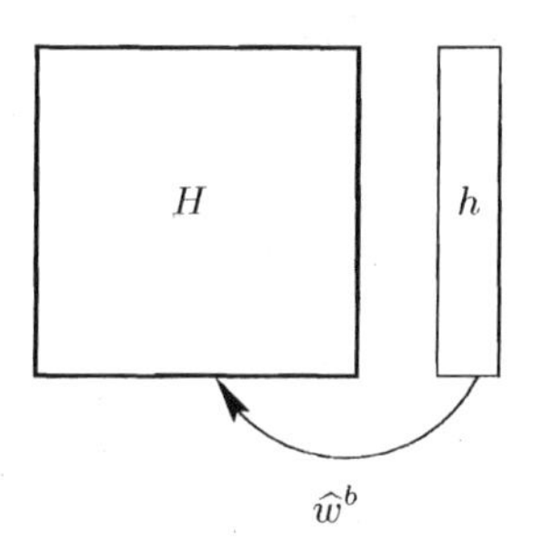

FIGURE 32.1 Backward projection: The last column of $[H \;\; h]$ is projected onto the column span of the matrix H.

which provides the desired relation between $\{P_z, P\}$; the relation is in terms of $\{\sigma, \xi_h, \widehat{w}^b\}$ with the last two quantities arising from the backward projection problem (32.17).

If we multiply both sides of (32.20) from the right by

$$\begin{bmatrix} H^* \\ h^* \end{bmatrix} Wy$$

and use the definitions (32.2) and (32.8) for $\{\widehat{w}_z, \widehat{w}\}$ we find that

$$\widehat{w}_z = \begin{bmatrix} \widehat{w} \\ 0 \end{bmatrix} + \frac{1}{\sigma + \xi_h} \begin{bmatrix} -\widehat{w}^b \\ 1 \end{bmatrix} \widetilde{h}^* Wy$$

If we introduce the scalar quantity

$$\boxed{\kappa = \frac{\widetilde{h}^* Wy}{\sigma + \xi_h}} \tag{32.21}$$

then the above relation between $\{\widehat{w}_z, \widehat{w}\}$ becomes

$$\boxed{\widehat{w}_z = \begin{bmatrix} \widehat{w} \\ 0 \end{bmatrix} + \kappa \begin{bmatrix} -\widehat{w}^b \\ 1 \end{bmatrix} = \begin{bmatrix} \widehat{w} - \kappa \widehat{w}^b \\ \kappa \end{bmatrix}} \tag{32.22}$$

Likewise, the projections $\{\widehat{y}, \widehat{y}_z\}$ are related via

$$\widehat{y}_z = \begin{bmatrix} H & h \end{bmatrix} \widehat{w}_z = \begin{bmatrix} H & h \end{bmatrix} \begin{bmatrix} \widehat{w} - \kappa \widehat{w}^b \\ \kappa \end{bmatrix} = H(\widehat{w} - \kappa \widehat{w}^b) + \kappa h = \widehat{y} + \kappa(h - H\widehat{w}^b)$$

That is,

$$\boxed{\widehat{y}_z = \widehat{y} + \kappa \widetilde{h}} \tag{32.23}$$

and, consequently, the residual vectors $\{\widetilde{y}_z, \widetilde{y}\}$ satisfy

$$\boxed{\widetilde{y}_z = \widetilde{y} - \kappa \widetilde{h}} \tag{32.24}$$

It is also straightforward to see that the corresponding minimum costs $\{\xi, \xi_z\}$ satisfy

$$\xi_z = y^* W \widetilde{y}_z = y^* W [\widetilde{y} - \kappa \widetilde{h}] = y^* W \widetilde{y} - \kappa y^* W \widetilde{h}$$

i.e., using (32.21),

$$\boxed{\xi_z = \xi - \frac{|\rho|^2}{\sigma + \xi_h}} \tag{32.25}$$

where we are defining the scalar

$$\boxed{\rho \stackrel{\Delta}{=} y^* W \widetilde{h}} \tag{32.26}$$

Finally, if we multiply both sides of (32.20) by $[u \;\; \alpha]$ from the left and $\text{col}\{u^*, \alpha^*\}$ from the right, and use the definitions of $\{\gamma_z, \gamma, \widetilde{\alpha}\}$ in (32.6), (32.12), and (32.18), we find that

$$\boxed{\gamma_z = \gamma - \frac{|\widetilde{\alpha}|^2}{\sigma + \xi_h}} \tag{32.27}$$

Before re-deriving and interpreting the above results by means of geometric arguments, we summarize the conclusions in the following statement.

Lemma 32.1 (Backward order-updates) Consider the regularized least-squares problems (32.1) and (32.7), where the data matrix H in the first problem is extended by one column to $[H \;\; h]$ in the second problem. The corresponding solutions $\{\widehat{w}, \widehat{w}_z\}$, coefficient matrices $\{P, P_z\}$, residuals $\{\widetilde{y}, \widetilde{y}_z\}$, minimum costs $\{\xi, \xi_z\}$, and conversion factors $\{\gamma, \gamma_z\}$ are related as follows.

1. Let $\widehat{w}^b$ be the weight vector that projects h onto H, according to (32.17), i.e., $\widehat{w}^b = PH^*Wh$, where $P = (\Pi + H^*WH)^{-1}$. Let $\widetilde{h}$ denote the resulting residual vector, $\widetilde{h} = h - H\widehat{w}^b$, whose last entry is $\widetilde{\alpha}$. Let also ξ_h denote the corresponding minimum cost, $\xi_h = h^*W\widetilde{h}$.
2. Define the scalars $\rho = y^*W\widetilde{h}$ and $\kappa = \rho^*/(\sigma + \xi_h)$.

Then the following relations hold:

$$\begin{aligned}
\widehat{y}_z &= \widehat{y} + \kappa\widetilde{h} \\
\widetilde{y}_z &= \widetilde{y} - \kappa\widetilde{h} \\
\xi_z &= \xi - |\rho|^2/(\sigma + \xi_h) \\
\gamma_z &= \gamma - |\widetilde{\alpha}|^2/(\sigma + \xi_h)
\end{aligned}$$

$$\begin{aligned}
\widehat{w}_z &= \begin{bmatrix} \widehat{w} \\ 0 \end{bmatrix} + \kappa \begin{bmatrix} -\widehat{w}^b \\ 1 \end{bmatrix} \\
P_z &= \begin{bmatrix} P & 0 \\ 0 & 0 \end{bmatrix} + \frac{1}{\sigma + \xi_h} \begin{bmatrix} -\widehat{w}^b \\ 1 \end{bmatrix} \begin{bmatrix} -\widehat{w}^{b*} & 1 \end{bmatrix}
\end{aligned}$$

Special Case: Absence of Regularization

In order to illustrate how the presence of regularization affects the interpretation of the results, let us examine what happens when regularization is ignored. Thus, assume that $\Pi = 0$ and $\sigma = 0$, so that the backward projection problem (32.17) reduces to

$$\min_{w^b} \quad (h - Hw^b)^*W(h - Hw^b) \tag{32.28}$$

Its solution $\widehat{w}^b$ is now such that it satisfies the orthogonality condition

$$H^*W(h - H\widehat{w}^b) = 0 \tag{32.29}$$

and, in this case, we get

$$\widehat{h}^*W\widetilde{h} = 0$$

since $\widehat{h} \in \mathcal{R}(H)$. This fact allows us to rewrite the minimum cost ξ_h in (32.19) as

$$\xi_h = h^*W\widetilde{h} = [\widehat{h} + \widetilde{h}]^*W\widetilde{h} = \widetilde{h}^*W\widetilde{h} \tag{32.30}$$

and expression (32.21) for κ becomes

$$\kappa = \frac{\widetilde{h}^*Wy}{\widetilde{h}^*W\widetilde{h}} \tag{32.31}$$

This expression identifies κ as the coefficient that projects y onto $\widetilde{h}$, i.e., it is the coefficient that solves the least-squares problem:

$$\min_k \ (y-\widetilde{h}k)^*W(y-\widetilde{h}k) \quad \Longrightarrow \quad \kappa$$

In other words, when $\Pi=0$ and $\sigma=0$, the term $\kappa\widetilde{h}$ in (32.23) can be interpreted as the projection of y onto $\widetilde{h}$. The equality (32.31) does *not* hold in the regularized case and, therefore, we cannot interpret κ in that case as the solution to the problem of projecting y onto $\widetilde{h}$. That is, in the regularized case (32.17), it does *not* hold that κ solves

$$\min_k \ \sigma|k|^2 \ + \ (y-\widetilde{h}k)^*W(y-\widetilde{h}k)$$

since the solution to this problem would be $\kappa_\sigma = \widetilde{h}^*Wy/(\sigma+\widetilde{h}^*W\widetilde{h})$, with the term $\widetilde{h}^*W\widetilde{h}$ appearing in the denominator instead of $h^*W\widetilde{h}$, as in expression (32.21) for κ.

Actually, more can be said about κ in (32.31). If we write

$$y=\widehat{y}+\widetilde{y}$$

and recall that $\widehat{y}\in\mathcal{R}(H)$, then by virtue of the orthogonality condition (32.29) we have $\widetilde{h}^*W\widehat{y}=0$, so that

$$\widetilde{h}^*Wy \ = \ \widetilde{h}^*W(\widehat{y}+\widetilde{y}) \ = \ \widetilde{h}^*W\widetilde{y}$$

and (32.31) can be replaced by

$$\kappa=\frac{\widetilde{h}^*W\widetilde{y}}{\widetilde{h}^*W\widetilde{h}} \tag{32.32}$$

That is, in the un-regularized case (32.28), we can also interpret κ as the coefficient that projects $\widetilde{y}$ onto $\widetilde{h}$, i.e., it solves

$$\min_k \ (\widetilde{y}-\widetilde{h}k)^*W(\widetilde{y}-\widetilde{h}k) \quad \Longrightarrow \quad \kappa \tag{32.33}$$

In view of this result, the difference $\widetilde{y}-\kappa\widetilde{h}$ in (32.24) can be interpreted as the residual vector that results from (32.33).

In conclusion, from (32.23)–(32.24) and from (32.31)–(32.32), we have that in the *un-regularized* case the new projection $\widehat{y}_z$ can be obtained from the old projection y as follows:

1. Project h onto $\mathcal{R}(H)$ and find the residual vector $\widetilde{h}$.

2. Project y onto $\widetilde{h}$.

3. Then $\widehat{y}_z=\widehat{y}+\widehat{y}_{\widetilde{h}}$, where $\widehat{y}_{\widetilde{h}}$ denotes the projection of y onto $\widetilde{h}$.

4. Also, $\widetilde{y}_z=\widetilde{y}-\kappa\widetilde{h}$.

That is, we obtain the new projection, $\widehat{y}_z$, by projecting y separately onto H and $\widetilde{h}$ and adding the results. Alternatively, we can obtain the new residual vector $\widetilde{y}_z$ of step 4) as follows:

4.a) Find the residual vector $\widetilde{y}$ that results from projecting y onto $\mathcal{R}(H)$.

4.b) Find the residual vector $\widetilde{h}$ that results from projecting h onto $\mathcal{R}(H)$.

4.c) Then $\widetilde{y}_z$ is the residual vector that results from projecting $\widetilde{y}$ onto $\widetilde{h}$.

This construction of $\widetilde{y}_z$ in the un-regularized case is depicted in Fig. 32.2. In the figure, the arrows indicate that $\widetilde{y}$ results from projecting y onto $\mathcal{R}(H)$ and $\widetilde{h}$ results from projecting h onto $\mathcal{R}(H)$.

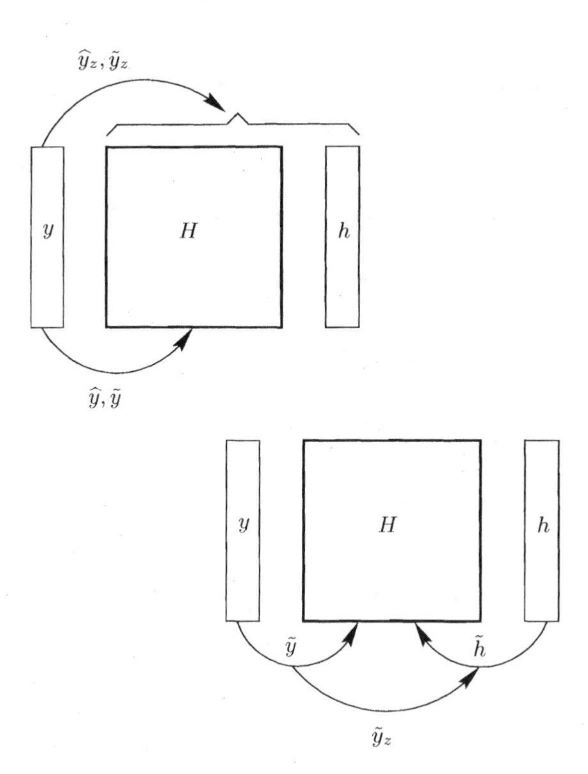

FIGURE 32.2 For unregularized least-squares problems, the order-update of the residual vector $\widetilde{y}$ into the residual vector $\widetilde{y}_z$ (*top figure*) is obtained by projecting $\widetilde{y}$ onto $\widetilde{h}$ (*bottom figure*).

Geometric Argument

We now re-derive the result of Lemma 32.1 by calling upon some of the geometric properties of least-squares problems. To begin with, the major difference between problems (32.1) and (32.7) relates to the data matrices H and $\begin{bmatrix} H & h \end{bmatrix}$. In the first problem, we are projecting y onto the column span of H while in the second problem we are projecting y onto the column span of $[H \;\; h]$; here we mean regularized projections. The column vector h is the additional piece of information that distinguishes the two least-squares problems.

Now assume that we replace $[H \;\; h]$ by a data matrix that spans the *same* column space in the following manner. Let $\widetilde{h}$ denote the residual vector that results from the regularized projection of h onto $\mathcal{R}(H)$, i.e.,

$$\widetilde{h} = h - H\widehat{w}^b$$

where, as before in (32.17), the weight vector $\widehat{w}^b$ is the solution to

$$\min_{w^b} \left[w^{b*}\Pi w^b + (h - Hw^b)^*W(h - Hw^b) \right] \tag{32.34}$$

and is given by

$$\widehat{w}^b = PH^*Wh$$

Then we can write

$$\boxed{\begin{bmatrix} H & h \end{bmatrix} = \begin{bmatrix} H & \widetilde{h} \end{bmatrix}\begin{bmatrix} \mathrm{I} & \widehat{w}^b \\ 0 & 1 \end{bmatrix}} \tag{32.35}$$

which shows that the matrices $[H \;\; h]$ and $[H \;\; \widetilde{h}]$ are related by an invertible transformation, so that they span the same column space, as desired.

If we were dealing with un-regularized least-squares problems, i.e., if $\Pi = 0$ in (32.34), then clearly $\widetilde{h} \perp \mathcal{R}(H)$, i.e., $H^*W\widetilde{h} = 0$. For the general regularized least-squares problem (32.34), however, this orthogonality condition is replaced by (cf. Table 29.2):

$$\boxed{H^*W\widetilde{h} = \Pi\widehat{w}^b} \tag{32.36}$$

In either case, with and without regularization, the orthogonality condition and the transformation (32.35), allow us to re-establish the relations of Lemma 32.1, as we now verify.

Relating $\{P_z, P\}$. In view of (32.36), and using (32.35), we have

$$\begin{aligned}
P_z^{-1} &\triangleq \begin{bmatrix} \Pi & 0 \\ 0 & \sigma \end{bmatrix} + \begin{bmatrix} H^* \\ h^* \end{bmatrix} W \begin{bmatrix} H & h \end{bmatrix} \\
&= \begin{bmatrix} \Pi & 0 \\ 0 & \sigma \end{bmatrix} + \begin{bmatrix} \mathrm{I} & 0 \\ \widehat{w}^{b*} & 1 \end{bmatrix} \begin{bmatrix} H^* \\ \widetilde{h}^* \end{bmatrix} W \begin{bmatrix} H & \widetilde{h} \end{bmatrix} \begin{bmatrix} \mathrm{I} & \widehat{w}^b \\ 0 & 1 \end{bmatrix} \\
&= \begin{bmatrix} \Pi & 0 \\ 0 & \sigma \end{bmatrix} + \begin{bmatrix} \mathrm{I} & 0 \\ \widehat{w}^{b*} & 1 \end{bmatrix} \begin{bmatrix} H^*WH & H^*W\widetilde{h} \\ \widetilde{h}^*WH & \widetilde{h}^*W\widetilde{h} \end{bmatrix} \begin{bmatrix} \mathrm{I} & \widehat{w}^b \\ 0 & 1 \end{bmatrix} \\
&= \begin{bmatrix} \mathrm{I} & 0 \\ \widehat{w}^{b*} & 1 \end{bmatrix} \left\{ \overline{\Pi} + \begin{bmatrix} H^*WH & H^*W\widetilde{h} \\ \widetilde{h}^*WH & \widetilde{h}^*W\widetilde{h} \end{bmatrix} \right\} \begin{bmatrix} \mathrm{I} & \widehat{w}^b \\ 0 & 1 \end{bmatrix}
\end{aligned}$$

where

$$\begin{aligned}
\overline{\Pi} &\triangleq \begin{bmatrix} \mathrm{I} & 0 \\ \widehat{w}^{b*} & 1 \end{bmatrix}^{-1} \begin{bmatrix} \Pi & 0 \\ 0 & \sigma \end{bmatrix} \begin{bmatrix} \mathrm{I} & \widehat{w}^b \\ 0 & 1 \end{bmatrix}^{-1} \\
&= \begin{bmatrix} \mathrm{I} & 0 \\ -\widehat{w}^{b*} & 1 \end{bmatrix} \begin{bmatrix} \Pi & 0 \\ 0 & \sigma \end{bmatrix} \begin{bmatrix} \mathrm{I} & -\widehat{w}^b \\ 0 & 1 \end{bmatrix} \\
&= \begin{bmatrix} \Pi & -\Pi\widehat{w}^b \\ -\widehat{w}^{b*}\Pi & \sigma + \widehat{w}^{b*}\Pi\widehat{w}^b \end{bmatrix}
\end{aligned} \tag{32.37}$$

Therefore, using

$$P^{-1} = \Pi + H^*WH$$

and (32.36), we get

$$P_z^{-1} = \begin{bmatrix} \mathrm{I} & 0 \\ \widehat{w}^{b*} & 1 \end{bmatrix} \begin{bmatrix} P^{-1} & 0 \\ 0 & \sigma + h^*W\widetilde{h} \end{bmatrix} \begin{bmatrix} \mathrm{I} & \widehat{w}^b \\ 0 & 1 \end{bmatrix} \tag{32.38}$$

where we used the fact that

$$\begin{aligned}
\sigma + \widehat{w}_b^*\Pi\widehat{w}^b + \widetilde{h}^*W\widetilde{h} &= \sigma + \widehat{w}_b^*\Pi\widehat{w}^b + (h - H\widehat{w}^b)^*W\widetilde{h} \\
&= \sigma + h^*W\widetilde{h} + \widehat{w}_b^* \underbrace{(\Pi\widehat{w}^b - H^*W\widetilde{h})}_{=0 \text{ by } (32.36)} \\
&= \sigma + h^*W\widetilde{h}
\end{aligned}$$

It follows from (32.38) that

$$\begin{aligned} P_z &= \begin{bmatrix} \text{I} & -\widehat{w}^b \\ 0 & 1 \end{bmatrix} \begin{bmatrix} P & 0 \\ 0 & \dfrac{1}{\sigma + h^* W\widetilde{h}} \end{bmatrix} \begin{bmatrix} \text{I} & 0 \\ -\widehat{w}^{b*} & 1 \end{bmatrix} \\ &= \begin{bmatrix} P & 0 \\ 0 & 0 \end{bmatrix} + \frac{1}{\sigma + h^* W\widetilde{h}} \begin{bmatrix} -\widehat{w}^b \\ 1 \end{bmatrix} \begin{bmatrix} -\widehat{w}^{b*} & 1 \end{bmatrix} \end{aligned}$$

which coincides with expression (32.20). Observe that this argument avoids the need for the algebraic identity (32.14). Moreover, the argument leading to (32.38) becomes more immediate if regularization were not present. Indeed, note that when $\Pi = 0$ and $\sigma = 0$, and using

$$P_z^{-1} = \begin{bmatrix} H^* \\ h^* \end{bmatrix} W \begin{bmatrix} H & h \end{bmatrix}, \quad P^{-1} = H^* W H, \quad h = \widetilde{h} + H\widehat{w}^b, \quad H^* W\widetilde{h} = 0$$

relation (32.38) would immediately follow in the following form

$$P_z^{-1} = \begin{bmatrix} \text{I} & 0 \\ \widehat{w}^{b*} & 1 \end{bmatrix} \begin{bmatrix} P^{-1} & 0 \\ 0 & h^* W\widetilde{h} \end{bmatrix} \begin{bmatrix} \text{I} & \widehat{w}^b \\ 0 & 1 \end{bmatrix}$$

The argument that led to (32.38) assumes regularization. An alternative argument appears in Prob. VII.13, which is based on reformulating a regularized least-squares problem as a standard least-squares problem.

Relating $\{\widehat{w}_z, \widehat{w}\}$. Recall that $\widehat{w}_z$ is the solution to the extended problem (32.7), namely,

$$\min_{w_z} \left\{ w_z^* \begin{bmatrix} \Pi & 0 \\ 0 & \sigma \end{bmatrix} w_z + \|y - \begin{bmatrix} H & h \end{bmatrix} w_z\|_W^2 \right\}$$

Replacing the data matrix $[H \;\; h]$ by (32.35), the above problem becomes

$$\min_{w_z} \left\{ w_z^* \begin{bmatrix} \Pi & 0 \\ 0 & \sigma \end{bmatrix} w_z + \left\| y - \begin{bmatrix} H & \widetilde{h} \end{bmatrix} \begin{bmatrix} \text{I} & \widehat{w}^b \\ 0 & 1 \end{bmatrix} w_z \right\|_W^2 \right\}$$

This alternative rewriting of the cost function suggests that we introduce

$$w_s \stackrel{\Delta}{=} \begin{bmatrix} \text{I} & \widehat{w}^b \\ 0 & 1 \end{bmatrix} w_z$$

and solve instead

$$\min_{w_s} \left\{ w_s^* \overline{\Pi} w_s + \|y - \begin{bmatrix} H & \widetilde{h} \end{bmatrix} w_s\|_W^2 \right\}$$

where $\overline{\Pi}$ is as in (32.37). The solution $\widehat{w}_s$ is given by

$$\widehat{w}_s = \left(\overline{\Pi} + \begin{bmatrix} H^* \\ \widetilde{h} \end{bmatrix} W \begin{bmatrix} H & \widetilde{h} \end{bmatrix} \right)^{-1} \begin{bmatrix} H^* \\ \widetilde{h}^* \end{bmatrix} Wy$$

or, equivalently, using (32.38),

$$\widehat{w}_s = \begin{bmatrix} P & 0 \\ 0 & \dfrac{1}{\sigma + h^* W\widetilde{h}} \end{bmatrix} \begin{bmatrix} H^* Wy \\ \widetilde{h}^* Wy \end{bmatrix} = \begin{bmatrix} \widehat{w} \\ \kappa \end{bmatrix}$$

Then,

$$\widehat{w}_z = \begin{bmatrix} I & -\widehat{w}^b \\ 0 & 1 \end{bmatrix} \widehat{w}_s = \begin{bmatrix} \widehat{w} - \kappa \widehat{w}^b \\ \kappa \end{bmatrix}$$

which is the same expression we derived before for $\widehat{w}_z$ in terms of $\widehat{w}$ in (32.22) — see Prob. VII.14 for an alternative derivation that is based on reformulating a regularized least-squares problem as a standard least-squares problem.

Relating $\{\widehat{y}_z, \widehat{y}\}$. The projection $\widehat{y}_z$ is given by

$$\widehat{y}_z = \begin{bmatrix} H & \widetilde{h} \end{bmatrix} \widehat{w}_s = H\widehat{w} + \kappa \widetilde{h}$$

which is again the same relation we derived before in (32.23).

Relating $\{\gamma_z, \gamma\}$. From the definition (32.12) for γ_z we have:

$$\begin{aligned}
\gamma_z &= 1 - \begin{bmatrix} u & \alpha \end{bmatrix} P_z \begin{bmatrix} u^* \\ \alpha^* \end{bmatrix} \\
&\overset{(32.38)}{=} 1 - \begin{bmatrix} u & \alpha \end{bmatrix} \begin{bmatrix} I & -\widehat{w}^b \\ 0 & 1 \end{bmatrix} \begin{bmatrix} P & 0 \\ 0 & \dfrac{1}{\sigma + h^* W \widetilde{h}} \end{bmatrix} \begin{bmatrix} I & 0 \\ -\widehat{w}^{b*} & 1 \end{bmatrix} \begin{bmatrix} u^* \\ \alpha^* \end{bmatrix} \\
&= 1 - \begin{bmatrix} u & \widetilde{\alpha} \end{bmatrix} \begin{bmatrix} P & 0 \\ 0 & \dfrac{1}{\sigma + h^* W \widetilde{h}} \end{bmatrix} \begin{bmatrix} u^* \\ \widetilde{\alpha}^* \end{bmatrix} \\
&= 1 - uPu^* - \frac{|\widetilde{\alpha}|^2}{\sigma + h^* W \widetilde{h}} \\
&= \gamma - \frac{|\widetilde{\alpha}|^2}{\sigma + h^* W \widetilde{h}}
\end{aligned}$$

where $\{\alpha, \widetilde{\alpha}\}$ are the last entries of $\{h, \widetilde{h}\}$. This expression coincides with (32.27) since $\xi_h = h^* W \widetilde{h}$.

32.2 FORWARD ORDER-UPDATE RELATIONS

In Sec. 32.1 we started from the least-squares problem (32.1) and extended the data matrix H by adding a column h to its right, as in (32.7). Then in Lemma 32.1 we related the solutions to both problems. We can similarly consider the problem of extending H by adding a column to its *left* (rather than to its right), and by extending Π accordingly, say,

$$\min_{w_z} \left\{ w_z^* \begin{bmatrix} \sigma & 0 \\ 0 & \Pi \end{bmatrix} w_z + \left(y - \begin{bmatrix} h & H \end{bmatrix} w_z\right)^* W \left(y - \begin{bmatrix} h & H \end{bmatrix} w_z\right) \right\} \quad (32.39)$$

The optimal solution of this extended problem is now given by

$$\widehat{w}_z = \left(\begin{bmatrix} \sigma & 0 \\ 0 & \Pi \end{bmatrix} + \begin{bmatrix} h^* \\ H^* \end{bmatrix} W \begin{bmatrix} h & H \end{bmatrix} \right)^{-1} \begin{bmatrix} h^* \\ H^* \end{bmatrix} Wy$$

We again denote the inverse of its extended coefficient matrix by

$$\boxed{P_z = \left(\begin{bmatrix} \sigma & 0 \\ 0 & \Pi \end{bmatrix} + \begin{bmatrix} h^* \\ H^* \end{bmatrix} W \begin{bmatrix} h & H \end{bmatrix} \right)^{-1}} \tag{32.40}$$

so that

$$\boxed{\widehat{y}_z = \begin{bmatrix} h & H \end{bmatrix} \widehat{w}_z = \begin{bmatrix} h & H \end{bmatrix} P_z \begin{bmatrix} h^* \\ H^* \end{bmatrix} Wy}$$

The resulting residual vector is

$$\boxed{\widetilde{y}_z = y - \begin{bmatrix} h & H \end{bmatrix} \widehat{w}_z}$$

with the minimum cost of (32.39) equal to (cf. Table 29.3):

$$\boxed{\xi_z = y^* W \widetilde{y}_z}$$

We further associate with (32.39) the conversion factor

$$\boxed{\gamma_z = 1 - \begin{bmatrix} \alpha & u \end{bmatrix} P \begin{bmatrix} \alpha^* \\ u^* \end{bmatrix}}$$

where $[\alpha \;\; u]$ denotes the last row of $[h \;\; H]$.

We can again relate the quantities $\{P, P_z\}$, $\{w, w_z\}$, $\{\widehat{y}, \widehat{y}_z\}$, and $\{\gamma, \gamma_z\}$ of problems (32.1) and (32.39). The arguments (both algebraic and geometric) are identical to those in Sec. 32.1. For this reason, we only state the final results here and leave the details to Probs. VII.15–VII.17.

The relation between problems (32.1) and (32.39) is determined in terms of the solution to the following regularized least-squares problem,

$$\min_{w^f} \left[w^{f*} \Pi w^f + (h - Hw^f)^* W (h - Hw^f) \right] \tag{32.41}$$

namely,

$$\widehat{w}^f = PH^* Wh$$

This weight vector performs the (regularized) projection of h onto the column span of H. And this projection problem is the reason for the title of this subsection, namely, *forward projection*. The terminology refers to the fact that we are projecting h onto the posterior columns of $[h \;\; H]$ — see Fig. 32.3. The minimum cost of (32.41) is given by

$$\boxed{\xi_h = h^* W \widetilde{h}} \tag{32.42}$$

where

$$\widetilde{h} = h - \widehat{h} = h - H\widehat{w}^f = h - HPH^* Wh$$

Lemma 32.2 (Forward order-updates) Consider the regularized problems (32.1) and (32.39), where the data matrix H in the first problem is extended by one column to $[h\ H]$ in the second problem. The corresponding solutions $\{\widehat{w}, \widehat{w}_z\}$, coefficient matrices $\{P, P_z\}$, residuals $\{\widetilde{y}, \widetilde{y}_z\}$, minimum costs $\{\xi, \xi_z\}$, and conversion factors $\{\gamma, \gamma_z\}$ are related as follows.

1. Let $\widehat{w}^f$ be the weight vector that projects h onto H, according to (32.41), i.e., $\widehat{w}^f = PH^*Wh$, where $P = (\Pi + H^*WH)^{-1}$. Let $\widetilde{h}$ denote the resulting residual vector, $\widetilde{h} = h - H\widehat{w}^f$, whose last entry is $\widetilde{\alpha}$. Let also ξ_h denote the corresponding minimum cost, $\xi_h = \widetilde{h}^*W\widetilde{h}$.
2. Define the scalars $\rho = y^*W\widetilde{h}$ and $\kappa = \rho^*/(\sigma + \xi_h)$.

Then the following relations hold

$$
\begin{aligned}
\widehat{y}_z &= \widehat{y} + \kappa\widetilde{h} \\
\widetilde{y}_z &= \widetilde{y} - \kappa\widetilde{h} \\
\xi_z &= \xi - |\rho|^2/(\sigma + \xi_h) \\
\gamma_z &= \gamma - |\widetilde{\alpha}|^2/(\sigma + \xi_h)
\end{aligned}
$$

$$
\begin{aligned}
\widehat{w}_z &= \begin{bmatrix} 0 \\ \widehat{w} \end{bmatrix} + \kappa \begin{bmatrix} 1 \\ -\widehat{w}^f \end{bmatrix} \\
P_z &= \begin{bmatrix} 0 & 0 \\ 0 & P \end{bmatrix} + \frac{1}{\sigma + \xi_h} \begin{bmatrix} 1 \\ -\widehat{w}^f \end{bmatrix} \begin{bmatrix} 1 & -\widehat{w}^{f*} \end{bmatrix}
\end{aligned}
$$

Special Case: Absence of Regularization

Let us examine again what happens when $\Pi = 0$ and $\sigma = 0$ in problem (32.39), which then reduces to

$$
\min_{w^f} \ (h - Hw^f)^*W(h - Hw^f) \tag{32.43}
$$

Its solution is such that it should satisfy the orthogonality condition $H^*W(h - H\widehat{w}^f) = 0$. Therefore, in this case, we also have $\widehat{h}^*W\widetilde{h} = 0$ since $\widehat{h} \in \mathcal{R}(H)$. This fact allows us to rewrite the minimum cost $\xi_h = \widetilde{h}^*W\widetilde{h}$ in (32.42) as

$$
\xi_h = \widetilde{h}^*W\widetilde{h} = [\widehat{h} + \widetilde{h}]^*W\widetilde{h} = h^*W\widetilde{h} \tag{32.44}
$$

so that the expression for κ in Lemma 32.2 becomes

$$
\kappa = \frac{\widetilde{h}^*Wy}{h^*W\widetilde{h}} \tag{32.45}
$$

As explained in the case of backward projection following Lemma 32.1, the above expression allows us to identify κ as the coefficient that projects y onto $\widetilde{h}$, i.e., it solves the least-squares problem

$$
\min_{k} \ (y - \widetilde{h}k)^*W(y - \widetilde{h}k) \quad \Longrightarrow \quad \kappa
$$

In this way, the term $\kappa\widetilde{h}$ in the expression for $\widehat{y}_z$ in Lemma 32.2 can be interpreted as the projection of y onto $\widetilde{h}$. As explained before, the equality (32.44) does *not* hold in the regularized case (32.39).

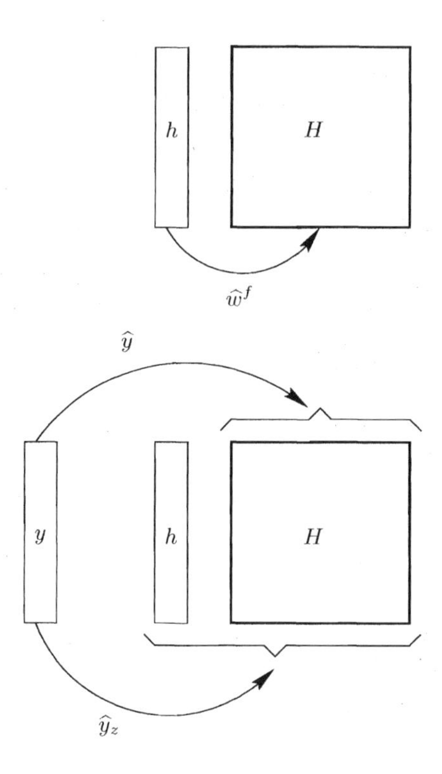

FIGURE 32.3 Forward projection: The leading column of $[h \quad H]$ is projected onto the column span of H (*top figure*). This step helps relate the order-update problems of projecting y onto the column spans of H and $[h \quad H]$ (*bottom figure*).

More can be said about κ in (32.45). If we write

$$y = \widehat{y} + \widetilde{y}$$

and recall that $\widehat{y} \in \mathcal{R}(H)$, then since $\widetilde{h}^* W \widehat{y} = 0$ by the orthogonality condition of weighted least-squares solutions, we get

$$\widetilde{h}^* W y = \widetilde{h}^* W (\widehat{y} + \widetilde{y}) = \widetilde{h}^* W \widetilde{y}$$

and (32.45) can be replaced by

$$\kappa = \frac{\widetilde{h}^* W \widetilde{y}}{\widetilde{h}^* W \widetilde{h}} \tag{32.46}$$

That is, in the un-regularized case (32.43), we can also interpret κ as the coefficient that projects $\widetilde{y}$ onto $\widetilde{h}$, i.e., it solves

$$\min_{k} \; (\widetilde{y} - \widetilde{h}k)^* W (\widetilde{y} - \widetilde{h}k) \quad \Longrightarrow \quad \kappa \tag{32.47}$$

In view of this result, the difference $\widetilde{y} - \kappa \widetilde{h}$ in the expression for $\widetilde{y}_z$ in Lemma 32.2 can be interpreted as the residual vector that results from (32.47).

In conclusion, we find that in the *un-regularized* case the new projection $\widehat{y}_z$ can be obtained from the old projection y as follows:

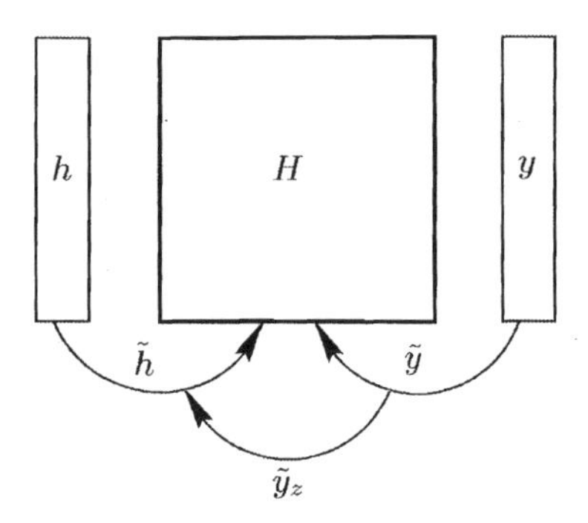

FIGURE 32.4 For un-regularized least-squares problems, the order-update of the residual vector $\widetilde{y}$ into the residual vector $\widetilde{y}_z$ is obtained by projecting $\widetilde{y}$ onto $\widetilde{h}$.

1. Project h onto $\mathcal{R}(H)$ and find the residual vector $\widetilde{h}$.

2. Project y onto $\widetilde{h}$.

3. Then $\widehat{y}_z = \widehat{y} + \widehat{y}_{\widetilde{h}}$, where $\widehat{y}_{\widetilde{h}}$ denotes the projection of y onto $\widetilde{h}$.

4. Also, $\widetilde{y}_z = \widetilde{y} - \kappa\widetilde{h}$.

That is, we obtain the new projection, $\widehat{y}_z$, by projecting y separately onto H and $\widetilde{h}$ and adding the results. Alternatively, we can obtain the new residual vector $\widetilde{y}_z$ of step 4) as follows:

4.a) Find the residual vector $\widetilde{y}$ that results from projecting y onto $\mathcal{R}(H)$.

4.b) Find the residual vector $\widetilde{h}$ that results from projecting h onto $\mathcal{R}(H)$.

4.c) Then $\widetilde{y}_z$ is the residual vector that results from projecting $\widetilde{y}$ onto $\widetilde{h}$.

This construction of $\widetilde{y}_z$ in the un-regularized case is depicted in Fig. 32.4. In the figure, the arrows indicate that $\widetilde{y}$ results from projecting y onto $\mathcal{R}(H)$ and $\widetilde{h}$ results from projecting h onto $\mathcal{R}(H)$.

32.3 TIME-UPDATE RELATION

The discussion in this section complements the one presented in Secs. 32.1 and 32.2. The results derived here will only be used later in Part X (*Lattice Filters*), when we study order-recursive adaptive filters.

Thus, consider an $N \times M$ data matrix H_{N-1} and partition it as

$$H_{N-1} \;\stackrel{\Delta}{=}\; \begin{bmatrix} x_{N-1} & \bar{H}_{N-1} & z_{N-1} \end{bmatrix} \qquad (32.48)$$

where x_{N-1} and z_{N-1} denote the leading and trailing columns of H_{N-1}, and $\bar{H}_{N-1}$ denotes its middle columns. Let $\widehat{z}_{N-1}$ denote the (regularized) projection of z_{N-1} onto $\mathcal{R}(\bar{H}_{N-1})$, i.e.,

$$\widehat{z}_{N-1} = \bar{H}_{N-1} w_{N-1,z}$$

where $w_{N-1,z}$ is obtained by solving the regularized least-squares problem

$$\min_{w} \left[\lambda^N w^* \Pi w + (z_{N-1} - \bar{H}_{N-1} w)^* \Lambda_{N-1} (z_{N-1} - \bar{H}_{N-1} w) \right] \tag{32.49}$$

and Λ_{N-1} is defined as in (30.25). Let further $\widetilde{z}_{N-1}$ denote the resulting residual vector,

$$\widetilde{z}_{N-1} = z_{N-1} - \widehat{z}_{N-1} \;=\; z_{N-1} - \bar{H}_{N-1} w_{N-1,z}$$

and define the weighted inner product

$$\boxed{\Delta(N-1) \;\stackrel{\Delta}{=}\; x_{N-1}^* \Lambda_{N-1} \widetilde{z}_{N-1}}$$

In other words, $\Delta(N-1)$ is the weighted inner product between the first column of H_{N-1} and the residual $\widetilde{z}_{N-1}$ that results from projecting its last column onto the middle columns $\bar{H}_{N-1}$ — see Fig. 32.5.

Now assume that one more row is appended to H_{N-1} in (32.48), say,

$$\boxed{H_N \;\stackrel{\Delta}{=}\; \begin{bmatrix} x_{N-1} & \bar{H}_{N-1} & z_{N-1} \\ \alpha(N) & \bar{h}_N & \beta(N) \end{bmatrix} \;\stackrel{\Delta}{=}\; \begin{bmatrix} x_N & \bar{H}_N & z_N \end{bmatrix}} \tag{32.50}$$

where $\alpha(N)$ and $\beta(N)$ are scalars, while $\bar{h}_N$ is a row vector. That is, H_{N-1} is *time-updated* to H_N.

As above, let $\widehat{z}_N$ denote the (regularized) projection of z_N onto $\mathcal{R}(\bar{H}_N)$, i.e.,

$$\widehat{z}_N = \bar{H}_N w_{N,z}$$

where $w_{N,z}$ is obtained by solving the regularized least-squares problem

$$\min_{w} \left[\lambda^{(N+1)} w^* \Pi w \;+\; (z_N - \bar{H}_N w)^* \Lambda_N (z_N - \bar{H}_N w) \right] \tag{32.51}$$

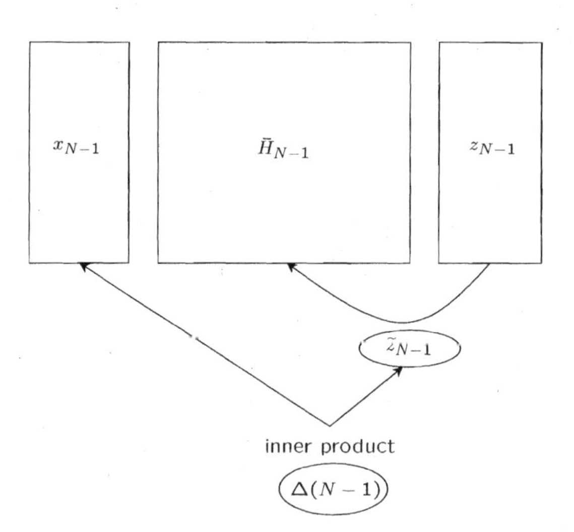

FIGURE 32.5 Inner product between the residual vector $\widetilde{z}_{N-1}$ and x_{N-1}.

Let further $\widetilde{z}_N$ denote the resulting residual vector,

$$\widetilde{z}_N = z_N - \widehat{z}_N = z_N - \bar{H}_N w_{N,z}$$

and define the corresponding weighted inner product

$$\boxed{\Delta(N) \stackrel{\Delta}{=} x_N^* \Lambda_N \widetilde{z}_N} \tag{32.52}$$

Again, $\Delta(N)$ is the weighted inner product between the first column of H_N and the residual vector that results from projecting its last column onto the middle columns $\bar{H}_N$ — see Fig. 32.6.

We would like to relate $\Delta(N)$ and $\Delta(N-1)$, i.e., we would like to determine a time-update relation for the variable Δ. For this purpose, let $\widehat{x}_N$ denote the (regularized) projection of x_N onto $\mathcal{R}(\bar{H}_N)$:

$$\widehat{x}_N = \bar{H}_N w_{N,x}$$

where $w_{N,x}$ is obtained by solving the regularized least-squares problem

$$\min_w \left[\lambda^{(N+1)} w^* \Pi w + (x_N - \bar{H}_N w)^* \Lambda_N (x_N - \bar{H}_N w) \right] \tag{32.53}$$

Introduce the estimation errors

$$\boxed{\begin{array}{rcl} \widetilde{\alpha}(N) & = & \alpha(N) - \bar{h}_N w_{N,x} \\ \widetilde{\beta}(N) & = & \beta(N) - \bar{h}_N w_{N,z} \end{array}} \tag{32.54}$$

Here, $\widetilde{\alpha}(N)$ is the *a posteriori* error in estimating the last entry of x_N, while $\widetilde{\beta}(N)$ is the *a posteriori* error in estimating the last entry of z_N.

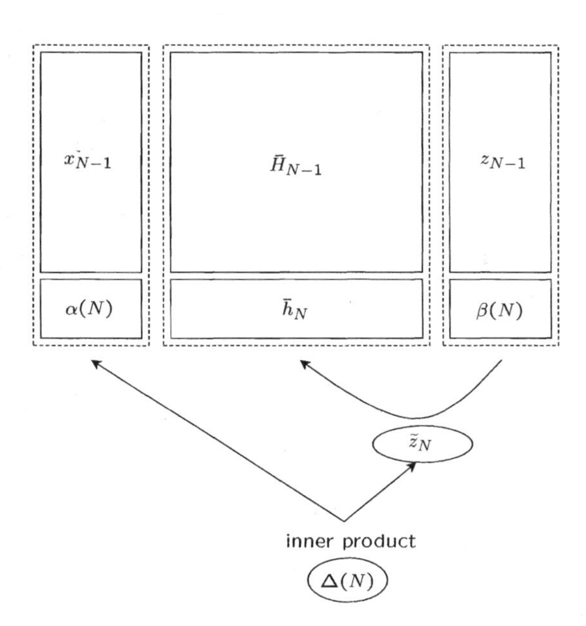

FIGURE 32.6 Inner product between the residual vector $\widetilde{z}_N$ and x_N.

Define further the conversion factor

$$\boxed{\bar{\gamma}(N) \;\stackrel{\Delta}{=}\; 1 - \bar{h}_N \bar{P}_N \bar{h}_N^*} \tag{32.55}$$

where

$$\bar{P}_N \;=\; \left[\lambda^{(N+1)}\Pi + \bar{H}_N^* \Lambda_N \bar{H}_N\right]^{-1}$$

Clearly, $\bar{\gamma}(N)$ is the factor that relates the *a posteriori* error $\widetilde{\beta}(N)$ to its *a priori* version, which is defined by

$$\widetilde{\beta}_a(N) \;=\; \beta(N) - \bar{h}_N w_{N-1,z}$$

with $w_{N-1,z}$ used instead of $w_{N,z}$. That is,

$$\widetilde{\beta}(N) = \bar{\gamma}(N)\widetilde{\beta}_a(N)$$

Actually, $\bar{\gamma}(N)$ is also the factor that relates the *a posteriori* error $\widetilde{\alpha}(N)$ to its *a priori* version, which is defined as

$$\widetilde{\alpha}_a(N) \;=\; \alpha(N) - \bar{h}_N w_{N-1,x}$$

with $w_{N-1,x}$ used instead of $w_{N,x}$, and where $w_{N-1,x}$ is the solution to a problem similar to (32.49) with z_{N-1} replaced by x_{N-1}. That is,

$$\widetilde{\alpha}(N) = \bar{\gamma}(N)\widetilde{\alpha}_a(N)$$

Now from the definition (32.52) for $\Delta(N)$, and using the fact that

$$\Lambda_N = \left[\begin{array}{cc} \lambda\Lambda_{N-1} & 0 \\ 0 & 1 \end{array}\right]$$

we obtain

$$\begin{aligned}
\Delta(N) \;&=\; \left[\begin{array}{cc} x_{N-1}^* & \alpha^*(N) \end{array}\right]\left[\begin{array}{cc} \lambda\Lambda_{N-1} & 0 \\ 0 & 1 \end{array}\right]\left(\left[\begin{array}{c} z_{N-1} \\ \beta(N) \end{array}\right] - \left[\begin{array}{c} \bar{H}_{N-1} \\ \bar{h}_N \end{array}\right] w_{N,z}\right) \\
&=\; \lambda x_{N-1}^*\Lambda_{N-1}z_{N-1} + \alpha^*(N)\beta(N) - (\lambda x_{N-1}^*\Lambda_{N-1}\bar{H}_{N-1} + \alpha^*(N)\bar{h}_N)w_{N,z} \\
&=\; \lambda x_{N-1}^*\Lambda_{N-1}z_{N-1} + \alpha^*(N)\widetilde{\beta}(N) - \lambda x_{N-1}^*\Lambda_{N-1}\bar{H}_{N-1}w_{N,z}
\end{aligned}$$

Moreover, the RLS recursion (30.13) allows us to relate $w_{N,z}$ and $w_{N-1,z}$ as

$$w_{N,z} \;=\; w_{N-1,z} + \bar{P}_N\bar{h}_N^*\left[\widetilde{\beta}(N)/\bar{\gamma}(N)\right]$$

Substituting into the expression for $\Delta(N)$, and using

$$\lambda\bar{\gamma}^{-1}(N)\bar{P}_N\bar{h}_N^* = \bar{P}_{N-1}\bar{h}_N^*$$

we obtain, after grouping terms,

$$\boxed{\Delta(N) = \lambda\Delta(N-1) + \widetilde{\alpha}^*(N)\widetilde{\beta}(N)/\bar{\gamma}(N)} \tag{32.56}$$

This is a useful relation and it plays an important role in the derivation of order-recursive algorithms (see Part X (*Lattice Filters*)). The relation holds for generic regularized least-squares problems and there are no structural restrictions imposed on the data matrices $\{H_{N-1}, H_N\}$.

Lemma 32.3 (Inner product time-update) Consider the data matrix (32.50), namely,

$$H_N \triangleq \begin{bmatrix} x_{N-1} & \bar{H}_{N-1} & z_{N-1} \\ \alpha(N) & \bar{h}_N & \beta(N) \end{bmatrix} = \begin{bmatrix} x_N & \bar{H}_N & z_N \end{bmatrix}$$

Let $\widetilde{z}_{N-1}$ and $\widetilde{z}_N$ denote the residual vectors that result from projecting z_{N-1} onto $\mathcal{R}(\bar{H}_{N-1})$ and z_N onto $\mathcal{R}(\bar{H}_N)$; both projections are meant in the regularized least-squares senses (32.49) and (32.53). Then the weighted inner products

$$\Delta(N) = x_N^* \Lambda_N \widetilde{z}_N, \qquad \Delta(N-1) = x_{N-1}^* \Lambda_{N-1} \widetilde{z}_{N-1}$$

are related via the time-update relation

$$\Delta(N) = \lambda\Delta(N-1) + \widetilde{\alpha}^*(N)\widetilde{\beta}(N)/\bar{\gamma}(N)$$

where $\{\widetilde{\alpha}(N), \widetilde{\beta}(N)\}$ are the *a posteriori* errors in estimating the entries $\{\alpha(N), \beta(N)\}$, as defined by (32.54), and $\bar{\gamma}(N)$ is the conversion factor defined by (32.55).

Summary and Notes

The chapters in this part describe several basic concepts of least-squares and recursive least-squares theory. The main ideas and results are the following.

SUMMARY OF MAIN RESULTS

Least-Squares

1. The standard least-squares criterion seeks the vector $\widehat{y}$ that is closest to a vector y in the column span of a matrix H. It does so by solving

$$\min_{w} \|y - Hw\|^2$$

and by setting $\widehat{y} = H\widehat{w}$, where $\widehat{w}$ is any solution to the normal equations $H^*H\widehat{w} = H^*y$.

2. The normal equations are always consistent and they may even have infinitely many solutions. However, regardless of which solution $\widehat{w}$ we pick, the projection $\widehat{y}$ is unique.

3. Least-squares solutions are characterized by a fundamental orthogonality principle, which states that $\widehat{w}$ is a solution if, and only if, the residual vector $\widetilde{y} = y - H\widehat{w}$ is orthogonal to $\mathcal{R}(H)$.

4. The norms of the vectors $\{y, \widehat{y}, \widetilde{y}\}$ satisfy the relation $\|y\|^2 = \|\widehat{y}\|^2 + \|\widetilde{y}\|^2$, which is an extension to higher dimensions of the famed Pythagorean theorem regarding the sum of the squares of the sides of a right-angle triangle.

5. The matrix that transforms y into $\widehat{y}$ is called the projection matrix and, when H has full-column rank, it is given by $\mathcal{P}_H = H(H^*H)^{-1}H^*$.

6. Often, in practice, we resort to regularized and weighted least-squares problems, which seek the vector $\widehat{w}$ that solves

$$\min_{w} \left[(w - \bar{w})^*\Pi(w - \bar{w}) \;+\; (y - Hw)^*W(y - Hw)\right]$$

where $\Pi > 0$, $W \geq 0$ and $\bar{w}$ is an initial condition (usually zero). The solution is given by

$$\widehat{w} = \bar{w} + (\Pi + H^*WH)^{-1}H^*W(y - H\bar{w})$$

The positive-definiteness of Π, and the nonnegative definiteness of W, guarantee that the coefficient matrix $(\Pi + H^*WH)$ is positive-definite and, hence, invertible.

7. The orthogonality condition of standard least-squares problem becomes $H^*W\widetilde{y} = \Pi(\widehat{w} - \bar{w})$ in the weighted regularized case. For ease of presentation, we continue to refer to $\widehat{y} = H\widehat{w}$ as the projection of y onto $\mathcal{R}(H)$ although, of course, this qualification is not accurate.

8. Regularized and weighted least-squares solutions can be order-updated in a rather elegant fashion. The derivation in the text considers both forward and backward order-updates, whereby the data matrix H is modified by adding a column to its left or to its right. The key conclusion is that order-updating least-squares solutions involves two steps:

(i) We first project the additional column onto the original data matrix by solving the relevant regularized and weighted least-squares problem.

(ii) We then update the original least-squares solution by using the results of this projection.

Such order-updates will play a key role in the derivation of so-called lattice filters later in Part X.

9. There is a fundamental equivalence between deterministic least-squares estimation and linear least-mean-squares estimation. These estimation procedures are equivalent in the sense that solving a problem from one class also solves a problem from the other class and vice-versa. The details are spelled out in Sec. 31.1.

Recursive Least-Squares

1. The recursive least-squares (RLS) algorithm is a recursive procedure that relates the weight vector solution at iteration i to the weight vector solution at iteration $i-1$ of a regularized least-squares problem.
2. Usually, RLS is used with exponential weighting in order to incorporate a forgetting mechanism into the operation of the filter; the forgetting factor gives less weight to data in the remote past and more weight to more recent data.
3. In Prob. VII.33 we explain how to downdate (as opposed to update) least-squares solutions. Then in Prob. VII.34 we combine the update and downdate procedures to derive a sliding window RLS algorithm, whereby the successive least-squares solutions are based on a fixed length of data rather than on growing-memory data.
4. In Prob. VII.36 we derive a block version of RLS, whereby the observation data and the regression data are allowed to be vectors and matrices, respectively (rather than just scalars and rows). The resulting filter equations are similar in structure to classical RLS with the conversion factor now becoming a matrix quantity.
5. In Chapter 31 we exploit the equivalence result of Sec. 31.1 between least-squares and least-mean-squares estimation problems and use it to clarify the relationship that exists between Kalman filtering and RLS algorithms. This relationship is useful for at least two reasons:

   (i) It allows us to share ideas and algorithms between both domains.

   (ii) It allows us to develop extended variants of RLS with enhanced tracking abilities.

BIBLIOGRAPHIC NOTES

Least-Squares

Least-squares and Gauss. The standard least-squares problem of Sec. 29.1 has had an interesting and controversial history since its inception in the late 1700s. It was formulated by Gauss in 1795 at the age of 18 — see Gauss (1809). At that time, there was interest in a claim by a philosopher named Hegel who claimed that he has proven, using pure logic, that there were exactly seven planets. Then on Jan. 1, 1801, an astronomer discovered a moving object in the constellation of Aries, and the location of this celestial body was observed for 41 days before suddenly dropping out of sight. Gauss' contemporaries sought his help in predicting the location of the heavenly body so that they could ascertain whether it was a planet or a comet (see Hall (1970) for an account of this story). With measurements available from the earlier sightings, Gauss formulated and solved a least-squares problem that could predict the location of the body (which turned out to be the planetoid Ceres). Actually, Gauss went further and formulated the recursive-least-squares solution that we shall describe in Chapter 30; this step helped him save the trouble of having to solve a least-squares problem afresh every time a new measurement became available. For some reason, Gauss did not bother to publish his least-squares solution, and controversy erupted in 1805 when Legendre published a book where he independently invented the least-squares method — see Legendre (1805,1810). Since then, the controversy has been settled and credit is nowadays given to Gauss as the sole inventor of the method of least-squares.

Here is how Gauss himself motivated the least-squares problem:

"... if several quantities depending on the same unknown have been determined by inexact observations, we can recover the unknown either from one of the observations or from any of an infinite number of combinations of the observations. Although the value of an unknown determined in this way is always subject to error, there will be less error in some combinations than in others.... One of the most important problems in the application of mathematics to the natural sciences is to choose the best of these many combinations, i.e., the combination that yields values of the unknowns that are least subject to the errors."

Extracted from Stewart (1995, pp. 31,33).

Gauss' choice of the "best" combination was the one that minimizes the least-squares criterion!

Iterative reweighted least-squares algorithm. The least-squares solution (and, more specifically, its weighted version) is useful in solving non-quadratic optimization problems of the form

$$\min_x \left(\sum_{i=0}^{N-1} |d(i) - u_i x|^p \right)$$

for some positive exponent p (usually $1 < p < 2$). This can be seen by reformulating the criterion as a weighted least-squares criterion in the following manner. Define the scalars (assumed nonzero):

$$\alpha(i) \stackrel{\Delta}{=} |d(i) - u_i x|^{p-2}, \quad i = 0, 1, \ldots, N-1$$

and introduce the diagonal weighting matrix $W = \text{diag}\{\alpha(0), \alpha(1), \ldots, \alpha(N-1)\}$. Then the above optimization problem can be rewritten in the form

$$\min_x \left(\sum_{i=0}^{N-1} (y - Hx)^* W (y - Hx) \right)$$

where $y = \text{col}\{d(0), d(1), \ldots, d(N-1)\}$ and $H = \text{col}\{u_0, u_1, \ldots, u_{N-1}\}$. Of course, this reformulation is not truly a weighted least-squares problem of the same form studied in Sec. 29.6 since W is dependent on the unknown vector x. Still, this rewriting of the cost function suggests the following iterative technique for solving it, and which has been proven useful in practice. Given an estimate $\widehat{x}_{k-1}$ at iteration $k-1$ we do the following:

$$\begin{aligned} \text{compute } \alpha_k(i) &= |d(i) - u_i \widehat{x}_{k-1}|^{p-2}, \quad i = 0, 1, \ldots, N-1 \\ \text{set } W_k &= \text{diag}\{\alpha_k(0), \alpha_k(1), \ldots, \alpha_k(N-1)\} \\ \text{compute the new estimate as } \widehat{x}_k &= (H^* W_k H)^{-1} H^* W_k y \\ &\text{and repeat} \end{aligned}$$

This implementation assumes that the successive W_k are invertible, and the algorithm is known as the *iterative reweighted least-squares* (IRLS) algorithm. There are several variations of IRLS with improved stability and convergence performance relative to the above implementation (see, e.g., Osborne (1985) and Björck (1996). See also Fletcher, Grant, and Hebden (1971) and Kahng (1972)). One such variation is to evaluate $\widehat{x}_k$ not directly as above but as a convex combination in terms of the prior iterate $\widehat{x}_{k-1}$ as follows:

$$\begin{aligned} \text{compute } \alpha_k(i) &= |d(i) - u_i \widehat{x}_{k-1}|^{p-2}, \quad i = 0, 1, \ldots, N-1 \\ \text{set } W_k &= \text{diag}\{\alpha_k(0), \alpha_k(1), \ldots, \alpha_k(N-1)\} \\ \text{set } \overline{x}_k &= (H^* W_k H)^{-1} H^* W_k y \\ \text{set } \widehat{x}_k &= \alpha \overline{x}_k + (1-\alpha) \widehat{x}_{k-1} \\ &\text{and repeat} \end{aligned}$$

for some $0 < \alpha \leq 1$.

Maximum likelihood estimation. There is a fundamental connection between least-squares estimation for deterministic quantities and maximum likelihood (ML) and maximum *a posteriori* (MAP) estimation for jointly Gaussian random variables. Maximum likelihood estimation was perhaps the most significant contribution to estimation theory following Gauss' own formulation of the least-squares problem in 1795. It was introduced by Fisher (1912) more than 100 years after Gauss' work. In order to explain the connection between ML and least-squares, we shall briefly review the ML and MAP formulations.

Let $\boldsymbol{y}$ be a random variable with a probability density function $f_{\boldsymbol{y}}(y;x)$, which is parameterized in terms of some unknown constant parameter x. The maximum likelihood estimate of x given a realization y of $\boldsymbol{y}$ is obtained by solving (see, e.g., Van Trees (1968), Scharf (1991), and Kay (1993)):

$$\widehat{x} = \arg\left[\max_{x} f_{\boldsymbol{y}}(y;x)\right]$$

The function $f_{\boldsymbol{y}}(y;x)$ is called the likelihood function, and its logarithm is called the log-likelihood function, $L(x,y) = \ln f_{\boldsymbol{y}}(y;x)$.

The concept of maximum likelihood estimation can be extended to the case where the parameter is a random quantity, $\boldsymbol{x}$, and not a constant, x. In this case, we would maximize the *joint* probability density function of $\boldsymbol{x}$ and $\boldsymbol{y}$ with $\boldsymbol{y}$ fixed at the value of its realization:

$$\widehat{x} = \arg\left[\max_{x} f_{\boldsymbol{x},\boldsymbol{y}}(x,y)\right]$$

However, in view of Bayes' rule, which states that $f_{\boldsymbol{x},\boldsymbol{y}}(x,y) = f_{\boldsymbol{x}|\boldsymbol{y}}(x|y)f_{\boldsymbol{y}}(y)$, we see that maximizing $f_{\boldsymbol{x},\boldsymbol{y}}(x,y)$ over x is equivalent to maximizing $f_{\boldsymbol{x}|\boldsymbol{y}}(x|y)$ over x, i.e.,

$$\widehat{x} = \arg\left[\max_{x}\ f_{\boldsymbol{x}|\boldsymbol{y}}(x|y)\right]$$

This latter formulation is known as maximum *a posteriori* estimation (MAP) and it maximizes the conditional pdf of $\boldsymbol{x}$ given $\boldsymbol{y}$.

Now assume that $\boldsymbol{y} = Hw + \boldsymbol{v}$, where w is an unknown constant vector and where $\boldsymbol{v}$ has a circular Gaussian distribution with zero mean and identity covariance matrix, written as $\boldsymbol{v} \sim \mathcal{N}(0, \mathrm{I})$. Assume also that $\boldsymbol{y}$ is $N-$dimensional. Then $\boldsymbol{y}$ will be normally distributed with mean Hw and identity covariance matrix as well. The probability density function of $\boldsymbol{y}$ will be given by (cf. Lemma A.1):

$$f_{\boldsymbol{y}}(y;w) = \frac{1}{\pi^N}\ \exp^{\{-(y-Hw)^*(y-Hw)\}}$$

and it is parameterized in terms of w. It then follows that the maximum likelihood estimate of w is the value that solves

$$\min_{w} \|y - Hw\|^2$$

which is the least-squares solution. In other words, the ML estimate of w based on a realization of $\boldsymbol{y} = Hw + \boldsymbol{v}$, is equal to the least-squares solution of $\min\limits_{w}\|y - Hw\|^2$.

In a similar vein, we can provide a statistical interpretation for the *regularized* least-squares problem as follows. Assume now that

$$\boldsymbol{y} = H\boldsymbol{w} + \boldsymbol{v}, \qquad \text{with } \boldsymbol{v} \sim \mathcal{N}(0, W^{-1}) \quad \text{and} \quad \boldsymbol{w} \sim \mathcal{N}(0, \Pi^{-1})$$

In other words, the unknown $\boldsymbol{w}$ is now modeled as a random variable. Then we already know from the discussion in Sec. 2.2 that the optimal estimator of $\boldsymbol{w}$ given $\boldsymbol{y}$ is

$$\hat{\boldsymbol{w}} = \mathsf{E}(\boldsymbol{w}|\boldsymbol{y}) = R_{wy}R_y^{-1}\boldsymbol{y}$$

which is also equal to the MAP estimator of $\boldsymbol{w}$ given $\boldsymbol{y}$. We further know from the calculations in Sec. 5.1 that the expression for $\hat{\boldsymbol{w}}$ evaluates to

$$\hat{\boldsymbol{w}} = (\Pi + H^*WH)^{-1}H^*W\boldsymbol{y}$$

which, as we see, has the form of a weighted regularized least-squares solution.

Equivalence. Actually, deterministic and stochastic least-squares problems can be related more directly than above and without the need for the Gaussian assumption on the data and noise. In Sec. 31.1 we show that there is an intimate relation between regularized least-squares problems, as studied in this chapter, and linear least-mean-squares estimation problems, as studied in Part II (*Linear Estimation*). Although the former class of problems deals with deterministic variables while the latter class of problems deals with random variables, both classes turn out to be equivalent in the sense that solving a problem from one class also solves a problem from the other class and vice-versa. Among other results, such equivalence statements allow us, for example, to relate Kalman filtering (which deals with stochastic estimation — cf. Chapter 7) to adaptive RLS filtering (which deals with deterministic estimation) — see Sec. 31.2 and Sayed and Kailath (1994b).

Reliable numerical methods. There is a huge literature on least-squares problems and on reliable numerical methods for their solution — see, e.g., Higham (1996), Lawson and Hanson (1995), and the detailed treatment by Björck (1996). Among the most reliable methods for solving least-squares problems is the QR method, which is described in Prob. VII.12 (and which is studied further in Prob. VII.32 and also in Sec. 35.2 in the context of adaptive least-squares filtering). The origin of the QR method goes back to Householder (1953, pp. 72–73), followed by Golub (1965) and Businger and Golub (1965). Since then, there has been an explosion of interest on solution methods for least-squares problems.

An application to OFDM communications. In a computer project at the end of this part we illustrate how least-squares (and also least-mean-squares) solutions are useful in the design of orthogonal frequency-division multiplexing (OFDM) systems. OFDM is a multi-carrier modulation scheme (MCM) that attempts to achieve the theoretical capacity of a channel. The principle of MCM is to divide the channel into multiple subchannels over which data distortion is negligible, and to divide the data stream into parallel substreams that are used to modulate several carriers. The data rates of the substreams are adjusted in accordance with the spectral properties of the noise over the subchannels. In this way, MCM, and similarly OFDM, can achieve high data rates even over hostile channels. The first works on MCM systems were done by the military in the 1950's, and an early reference on OFDM is the patent by Chang (1970); filed a few years earlier in 1966 and issued in 1970. For further details see, e.g., the books by Bahai and Saltzberg (1999), van Nee and Prasad (2000), and Terry and Heiskala (2002). The works by Tarighat and Sayed (2003,2005) provides further illustration of the use of linear least-squares and least-mean-squares techniques in the design of OFDM receivers for both SISO and MIMO channels.

Recursive Least-Squares

RLS and Gauss. Gauss was also motivated to develop the recursive least-squares algorithm in his work on celestial bodies (ca. 1795). Of course, Gauss' notation and derivation were reminiscent of late 18th century mathematics and, therefore, they do not bear much resemblance with the linear algebraic and matrix arguments used in this chapter; see, e.g., the useful translation of Gauss' original work that appears in Stewart (1995). In modern times, the original work on the RLS algorithm is often credited to Plackett (1950,1972).

Regularization. Regularization has a rich history in numerical analysis and matrix computation methods (see, e.g., Golub and Van Loan (1996)). One of the earliest references on regularization seems to be Tikhonov (1963). In the context of RLS filtering, an examination of the effect of regularization on RLS convergence appears in Moustakides (1997).

Sliding window RLS. A finite-memory RLS algorithm is derived in Prob. VII.34. An efficient implementation as the combination of two pre-windowed RLS solutions is presented in Manolakis, Ling, and Proakis (1987).

Numerical issues. The exponentially-weighted RLS algorithm of Sec. 30.6 can face numerical difficulties in finite word-length implementations. Such difficulties arise mainly from two sources of inaccuracies: loss of Hermitian symmetry and loss of positive-definiteness of P_i due to round-off error accumulation. The loss of symmetry problem was observed in Verhaegen (1989a), and it was suggested by Verhaegen (1989b) and Yang (1994) that this problem could be alleviated by propagating only the lower or upper triangular parts of P_i. In Part VIII (*Array Algorithms*), we shall describe other so-called array implementations of RLS; these array methods are more robust to numerical difficulties and they are also more reliable than plain RLS implementations.

Filter divergence. Since the conversion factor satisfies $0 < \gamma(i) \leq 1$, it can be interpreted as an angle variable (more precisely, as the cosine of an angle) — see Lee, Morf, and Friedlander (1981). In this way, monitoring its value during filter operation can provide useful feedback on the proper operation of the filter. It was proposed by Bottomley and Alexander (1989) that monitoring of the conversion factor is a reasonable mechanism for detecting filter divergence — see Sec. 39.4.

Error propagation. In the adaptive filtering literature, the numerical stability of RLS is usually studied by resorting to a *single-error* propagation model. In other words, a disturbance is assumed to be introduced at some iteration, with no additional disturbances occurring afterwards, and then the evolution of this single error is studied (an example to this effect appears later in Prob. VIII.11). An early study along these lines is that of Ljung and Ljung (1985), where it was shown that RLS is numerically reliable for $\lambda < 1$ and diverges for $\lambda = 1$. However, conclusions based on such simplified error analyses can be misleading. More elaborate models for error propagation in RLS are needed, and progress in this direction appears in the work of Liavas and Regalia (1999). Actually, there have also been many important works in the numerical linear algebra community, most notably by Paige (1979ab,1985) and Paige and Saunders (1985), on stabilized least-squares methods, which are relevant to the adaptive filtering context.

RLS and Kalman filtering. There is a fundamental connection between RLS and Kalman filtering, so much so that solving a problem in one domain amounts to solving a problem in the other domain. The details of this equivalence were developed by Sayed and Kailath (1993,1994b) and they are discussed in Sec. 31.2. The relationship between RLS and Kalman filtering is useful for at least two reasons:

1. It becomes possible to share ideas and algorithms between both domains.
2. Since RLS is equivalent not to a full-blown Kalman filter, but only to a special case of it, it becomes possible to develop extended RLS schemes with enhanced tracking abilities as well as block RLS algorithms (see Sec. 31.A and also Haykin et al. (1997) for additional examples).

One of the earliest mentions of a relation between least-squares and Kalman filtering seems to be Ho (1963); however, this reference considers only a special estimation problem where the successive regression vectors are identical. Later references are Sorenson (1966) and Aström and Wittenmark (1971); these works focus only on the standard (i.e., unregularized) least-squares problem, in which case an exact relationship between least-squares and Kalman filtering does not actually exist, especially during the initial stages of adaptation when the least-squares problem is under-determined. Soon afterwards, in work on channel equalization, Godard (1974) rephrased the growing-memory (i.e., $\lambda = 1$) RLS problem in a stochastic state-space framework, with the unknown state corresponding to the unknown weight vector. Similar constructions also appeared in Willsky (1979, pp. 38–41), Anderson and Moore (1979, pp. 135–136), Ljung (1987, p. 309), Strobach (1990, pp. 331–335), and Söderström (1994, pp. 145–146). In the works by Anderson and Moore (1979), Ljung (1987), and Söderström (1994), the underlying models went a step further and incorporated the case of exponentially decaying memory (i.e., $\lambda < 1$) by formulating state-space models with a time-variant noise variance (as described in (31.40)). Nevertheless, annoying discrepancies persisted that precluded a direct correspondence between the RLS and the Kalman variables. Some of these discrepancies were overcome essentially by fiat (see, e.g., the treatment by Haykin (1991, pp. 502–504)). This lack of a direct correspondence may have inhibited application of the extensive body of Kalman filter results to the adaptive filtering problem until the work of Sayed and Kailath (1993,1994b).

In retrospect, by a simple device, Sayed and Kailath (1994b) were able to obtain a perfectly matched state-space model for the case of exponentially decaying memory, with a direct correspondence between the variables in the exponentially weighted RLS problem and the variables in the state-space estimation problem. The main benefit of this result is that recursive state-space estimation problems have been extensively studied since the early sixties. Besides the celebrated Riccati-equation-based Kalman filtering algorithm (cf. Chapter 7), many algorithmic and implementation alternatives have been studied over the years (see Apps. 35.A and 37.A and, for more details, see the textbook by Kailath, Sayed, and Hassibi (2000)). These include the so-called information filter forms and for certain kinds of time-variant state-space models (including those encountered in adaptive filtering), the Riccati recursions can be replaced by the order-of-magnitude faster Chandrasekhar recursions; moreover, all these variants have certain computationally better square-root (or array) forms. The interesting fact then is that when the exponentially-weighted RLS filtering problem is reformulated in state-space form, the Kalman filtering solutions turn out to be equivalent to the various classes of RLS adaptive filtering algorithms that are going to be introduced in future chapters. The details of this equivalence are spelled out in Chapter 31, in the reference by Sayed and Kailath (1994b), in Apps. 35.A and 37.A, as well as in Probs. VIII.12, VIII.13, and IX.16.

Problems and Computer Projects

PROBLEMS

Problem VII.1 (Rank of a matrix) Consider any $N \times M$ matrix H. Show that its row rank is equal to its column rank. That is, show that the number of independent columns is always equal to the number of independent rows (for any N and M), and hence, we can simply talk about the rank of a matrix.

Problem VII.2 (Frobenius norm of a matrix) Given an $N \times M$ matrix A of rank r, let $A = \sum_{i=1}^{r} \sigma_i u_i v_i^*$ denote its singular value decomposition (cf. Sec. B.6). Show that

$$\|A\|_{\mathrm{F}} = \sqrt{\mathrm{Tr}(A^*A)} = \sqrt{\sum_{i=1}^{r} \sigma_i^2}$$

Problem VII.3 (Projecting onto the orthogonal complement space) Consider an $N \times M$ full-rank matrix H with $N \geq M$, and two column vectors y and z of dimensions $N \times 1$ each. Let $\widetilde{y} = \mathcal{P}_H^{\perp} y$ and $\widetilde{z} = \mathcal{P}_H^{\perp} z$. Are the residual vectors $\widetilde{y}$ and $\widetilde{z}$ collinear in general? If your answer is positive, justify it. If the answer is negative, can you give conditions on N and M for which $\widetilde{y}$ and $\widetilde{z}$ will be collinear?

Problem VII.4 (Orthogonal complement space) Let H be $N \times M$ with full-column rank. Show that any vector in the column span of $\mathcal{P}_H^{\perp}$ is orthogonal to any vector in the column span of H. That is, show that $H^* \mathcal{P}_H^{\perp} = 0$.

Problem VII.5 (Special cases) Consider the least-squares problem $\min_w \|y - Hw\|^2$. Comment on the solution in the following cases: (a) $y \in \mathcal{N}(H)$, (b) $y \in \mathcal{R}(H)$, and (c) $y \in \mathcal{N}(H^*)$.

Problem VII.6 (Minimum-norm solution) Consider the under-determined least-squares problem $\min\limits_{w} \|y - Hw\|^2$, where y is $N \times 1$, H is $N \times M$, and $N < M$. Assume further that H has full rank.

(a) Verify that many solutions $\widehat{w}$ exist.

(b) Show that the minimum norm solution is given by $\widehat{w} = H^*(HH^*)^{-1}y$. Specifically, show first that this $\widehat{w}$ satisfies the normal equations. Then show that any other solution, say $\widehat{w} + r$ for some nonzero r, has Euclidean norm strictly larger than $\widehat{w}$.

Problem VII.7 (Affine projection algorithm) Refer to the discussion in Sec. 13.1. Show that the affine projection algorithm (13.5) can be obtained as the solution to the following regularized least-squares problem $\min\limits_{w_i} \left[\, \epsilon\|w_i - w_{i-1}\|^2 \; + \; \|d_i - U_i w_i\|^2 \right]$, for given $\{w_{i-1}, d_i, U_i\}$.

Problem VII.8 (MIMO least-squares problem) Let $\|A\|_{\mathrm{F}}$ denote the Frobenius norm of A, i.e.,

$$\|A\|_{\mathrm{F}} \;\stackrel{\Delta}{=}\; \sqrt{\sum_{i=0}^{n-1}\sum_{j=0}^{m-1} |a_{ij}|^2}$$

for an $n \times m$ matrix A. Consider an $N \times m$ matrix Y, an $N \times M$ matrix H, and an $M \times m$ matrix X. Show that $\widehat{X}$ is a solution of $\min_X \|Y - HX\|_F$ if, and only if, it satisfies the normal equations $H^*H\widehat{X} = H^*Y$.

Problem VII.9 (Weighted least-squares) Let $W > 0$ be a given weighting matrix and let H be $N \times M$.

(a) Show that the normal equations, $H^*WH\widehat{w} = H^*Wy$, are consistent. More specifically, show that $\mathcal{R}(H^*WH) = \mathcal{R}(H^*W)$.

(b) Show that H^*WH is singular if, and only if, H^*H is singular.

(c) Show that when the normal equations $H^*WH\widehat{w} = H^*Wy$ have many solutions, regardless of which one we pick, the projection vector $\widehat{y} = H\widehat{w}$ remains invariant.

Problem VII.10 (Cholesky method) Consider the normal equations $H^*H\widehat{w} = H^*y$, where H is $N \times M$. Assume H has full column rank (i.e., the rank of H is M) so that H^*H is positive-definite. The normal equations can be solved by the standard method of Gaussian elimination. They can also be solved by appealing to the Cholesky factorization of H^*H.

We showed in Sec. B.3 that every positive-definite matrix admits a *unique* triangular factorization of the form $H^*H = \bar{L}\bar{L}^*$, where $\bar{L}$ is lower-triangular with positive entries on its diagonal. The matrix $\bar{L}$ is called the Cholesky factor of H^*H. Show that the normal equations can be solved by means of the following two steps:

1. Solve the lower triangular system of equations $\bar{L}\widehat{x} = H^*y$ for $\widehat{x}$.
2. Solve the upper triangular system of equations $\bar{L}^*\widehat{w} = \widehat{x}$ for $\widehat{w}$.

Problem VII.11 (Danger of squaring) Solving the normal equations $H^*H\widehat{x} = H^*y$ by forming the matrix H^*H (i.e., by squaring the data) is a bad idea in general. This is because for ill-conditioned matrices H, numerical precision is lost when the matrix product H^*H is formed. Recall that ill-conditioned matrices are those that have a very large ratio of largest to smallest singular values, as defined in Sec. B.6, i.e., they are close to being rank-deficient. Consider the full-rank matrix

$$H = \begin{bmatrix} 1 & 1 \\ 0 & \epsilon \\ 1 & 1 \end{bmatrix}$$

where ϵ is a very small positive number that is of the same order of magnitude as the machine precision. Assuming $2 + \epsilon^2 = 2$ in finite precision, what is the rank of H^*H?

Problem VII.12 (QR method) Consider the same setting of Prob. VII.10. A method to reduce the effects of ill-conditioning of H on the solution of the normal equations is to avoid forming the product H^*H and to determine the Cholesky factor $\bar{L}$ by working directly with H. This can be achieved by appealing to the so-called QR decomposition of H, as explained in Sec. B.5, namely,

$$H = Q\begin{bmatrix} R \\ 0 \end{bmatrix}$$

where Q is $N \times N$ unitary and R is $M \times M$ upper-triangular with positive diagonal entries.

(a) Show that $\bar{L} = R^*$.

Remark. With the $\bar{L}$ so determined, we can proceed to solve the normal equations by using the two-step procedure of Prob. VII.10. Alternatively, we can proceed as below, which is nowadays the preferred way of solving the normal equations due to its numerical reliability.

(b) Let $\text{col}\{z_1, z_2\} = Q^*y$, where z_1 is $M \times 1$. Verify that $\|y - Hw\|^2 = \|z_1 - Rw\|^2 + \|z_2\|^2$. Conclude that the least-squares solution $\widehat{w}$ can be obtained by solving the triangular linear system of equations $R\widehat{w} = z_1$. Conclude further that the resulting minimum cost is $\|z_2\|^2$.

Problem VII.13 (Order-updating P via backward projection) Refer to the geometric argument in Sec. 32.1 and, in particular, to relation (32.35) between $[H \;\; h]$ and $[H \;\; \widetilde{h}]$.

(a) Use the orthogonality condition (32.36) to justify the relation

$$\underbrace{\begin{bmatrix} \Lambda^{1/2}U^* & 0 \\ 0 & \sigma^{1/2} \\ H & h \end{bmatrix}}_{A} = \underbrace{\begin{bmatrix} \Lambda^{1/2}U^* & -\Lambda^{-1/2}U^*H^*W\widetilde{h} \\ 0 & \sigma^{1/2} \\ H & \widetilde{h} \end{bmatrix}}_{B} \underbrace{\begin{bmatrix} \mathrm{I} & \widehat{w}^b \\ 0 & 1 \end{bmatrix}}_{C}$$

which we write as $A = BC$.

(b) Show that

$$A^* \begin{bmatrix} \mathrm{I} & 0 \\ 0 & W \end{bmatrix} A = P_z^{-1}, \quad B^* \begin{bmatrix} I & 0 \\ 0 & W \end{bmatrix} B = \begin{bmatrix} P^{-1} & 0 \\ 0 & \sigma + h^*W\widetilde{h} \end{bmatrix}$$

(c) Use the equality $A^*(\mathrm{I} \oplus W)A = C^*B^*(\mathrm{I} \oplus W)BC$ to re-establish relation (32.20) between $\{P, P_z\}$.

Problem VII.14 (Order-update of weight vector via backward projection) Refer again to the geometric argument in Sec. 32.1 and, in particular, to relation (32.35) between $[H \;\; h]$ and $[H \;\; \widetilde{h}]$.

(a) Argue that determining the vector $\widehat{w}_z$ that solves (32.7) is equivalent to determining the vector $\widehat{w}_s$ that solves

$$\min_{w_s} \left\| \begin{bmatrix} 0 \\ 0 \\ y \end{bmatrix} - \begin{bmatrix} \Lambda^{1/2}U^* & -\Lambda^{-1/2}U^*H^*W\widetilde{h} \\ 0 & \sigma^{1/2} \\ H & \widetilde{h} \end{bmatrix} w_s \right\|^2_{(\mathrm{I} \oplus W)}$$

where $\widehat{w}_s = \begin{bmatrix} \mathrm{I} & \widehat{w}^b \\ 0 & 1 \end{bmatrix} w_z$ and $\widehat{w}^b$ is the solution of (32.17).

(b) Solve the problem of part (a) and re-derive (32.22).

Problem VII.15 (Forward projection) Refer to the discussion in Sec. 32.2 and, in particular, to the definition of P_z in (32.40). Let $P^{-1} = \Pi + H^*WH$.

(a) Establish the validity of the matrix identity:

$$\begin{bmatrix} A & B \\ C & D \end{bmatrix}^{-1} = \begin{bmatrix} 0 & 0 \\ 0 & D^{-1} \end{bmatrix} + \begin{bmatrix} \mathrm{I} \\ -D^{-1}C \end{bmatrix} (A - BD^{-1}C)^{-1} \begin{bmatrix} \mathrm{I} & -BD^{-1} \end{bmatrix}$$

(b) Use the identity of part (a) to show that

$$P_z = \begin{bmatrix} 0 & 0 \\ 0 & P \end{bmatrix} + \frac{1}{\sigma + \xi_h} \begin{bmatrix} 1 \\ -\widehat{w}^f \end{bmatrix} \begin{bmatrix} 1 & -\widehat{w}^{f*} \end{bmatrix}$$

where $\{\widehat{w}^f, \xi_h\}$ are as defined by (32.41) and (32.42).

(c) Multiply both sides of the equality of part (b) from the right by $\begin{bmatrix} h^* \\ H^* \end{bmatrix} Wy$ and show that

$$\widehat{w}_z = \begin{bmatrix} 0 \\ \widehat{w} \end{bmatrix} + \kappa \begin{bmatrix} 1 \\ -\widehat{w}^f \end{bmatrix}$$

where $\kappa = \widetilde{h}^*Wy/(\sigma + \xi_h)$.

Problem VII.16 (Order-updating P via forward projection) Refer again to the discussion in Sec. 32.2.

(a) Verify that

$$\begin{bmatrix} h & H \end{bmatrix} = \begin{bmatrix} \widetilde{h} & H \end{bmatrix} \begin{bmatrix} 1 & 0 \\ \widehat{w}^f & \mathrm{I} \end{bmatrix}$$

(b) Use the orthogonality condition $H^*W\widetilde{h} = \Pi\widehat{w}^f$ to justify the relation

$$\underbrace{\begin{bmatrix} \sigma^{1/2} & 0 \\ 0 & \Lambda^{1/2}U^* \\ h & H \end{bmatrix}}_{A} = \underbrace{\begin{bmatrix} \sigma^{1/2} & 0 \\ -\Lambda^{-1/2}U^*H^*W\widetilde{h} & \Lambda^{1/2}U^* \\ \widetilde{h} & H \end{bmatrix}}_{B} \underbrace{\begin{bmatrix} 1 & 0 \\ \widehat{w}^f & \mathrm{I} \end{bmatrix}}_{C}$$

which we write as $A = BC$.

(c) As in parts (b) and (c) of Prob. VII.13, use the equality $A = BC$ to arrive at

$$P_z = \begin{bmatrix} 0 & 0 \\ 0 & P \end{bmatrix} + \frac{1}{\sigma + h^*W\widetilde{h}} \begin{bmatrix} 1 \\ -\widehat{w}^f \end{bmatrix} \begin{bmatrix} 1 & -\widehat{w}^{f*} \end{bmatrix}$$

Problem VII.17 (Order-update of weight vector via forward projection) Refer again to the discussion in Sec. 32.2.

(a) Argue that determining the vector $\widehat{w}_z$ that solves (32.39) is equivalent to determining the vector $\widehat{w}_s$ that solves

$$\min_{w_s} \left\| \begin{bmatrix} 0 \\ 0 \\ y \end{bmatrix} - \begin{bmatrix} \sigma^{1/2} & 0 \\ -\Lambda^{-1/2}U^*H^*W\widetilde{h} & \Lambda^{1/2}U^* \\ \widetilde{h} & H \end{bmatrix} w_s \right\|^2_{(\mathrm{I}\oplus W)}$$

where $\{w_z, w_s\}$ are related via

$$\widehat{w}_s = \begin{bmatrix} 1 & 0 \\ \widehat{w}^f & \mathrm{I} \end{bmatrix} w_z$$

(b) Solve the problem of part (a) and establish that

$$\widehat{w}_z = \begin{bmatrix} 0 \\ \widehat{w} \end{bmatrix} + \kappa \begin{bmatrix} 1 \\ -\widehat{w}^f \end{bmatrix}$$

where $\kappa = \widetilde{h}^*Wy/(\sigma + \xi_h)$.

Problem VII.18 (Regularized backward projection) Refer to the discussion in Sec. 32.1 and assume that we replace the regularization matrix in (32.7) by the more general choice

$$\begin{bmatrix} \Pi & m \\ m^* & \sigma \end{bmatrix}$$

for some row vector m and positive scalar σ.

(a) Verify that (32.20) becomes

$$\boxed{P_z = \begin{bmatrix} P & 0 \\ 0 & 0 \end{bmatrix} + \frac{1}{\zeta} \begin{bmatrix} -q \\ 1 \end{bmatrix} \begin{bmatrix} -q^* & 1 \end{bmatrix}}$$

where $q = P(H^*Wh + m)$, $\zeta = \sigma + \xi_h - m^*q - \widehat{w}^{b*}m$, and $\widehat{w}^b$ is still the solution to the backward projection problem (32.17). Verify further that $q = \widehat{w}^b + t$, where $t = Pm$.

(b) Define $\bar{h} \triangleq h - Hq$. Show that (32.22) is replaced by

$$\boxed{\widehat{w}_z = \begin{bmatrix} \widehat{w} \\ 0 \end{bmatrix} + \kappa \begin{bmatrix} -q \\ 1 \end{bmatrix} = \begin{bmatrix} \widehat{w} - \kappa q \\ \kappa \end{bmatrix}}$$

where $\kappa = \bar{h}^* W y / \zeta$.

(c) Likewise, show that $\widehat{y}_z = \widehat{y} + \kappa \bar{h}$ and $\widetilde{y}_z = \widetilde{y} - \kappa \bar{h}$, while $\xi_z = \xi - |\rho|^2/\zeta$ where $\rho = y^* W \bar{h}$.

(d) Finally, let $\bar{\alpha}$ denote the last entry of $\bar{h}$. Show that $\gamma_z = \gamma - |\bar{\alpha}|^2/\zeta$.

Problem VII.19 (Stochastic properties of least-squares solutions) Let $\boldsymbol{y} = Hw^o + \boldsymbol{v}$, where w^o is an unknown vector that we wish to estimate and $\boldsymbol{v}$ is a zero-mean random vector with covariance matrix $\mathsf{E}\,\boldsymbol{v}\boldsymbol{v}^* = \sigma_v^2 \mathrm{I}$. Moreover, H is $N \times M$, $N \geq M$, and has full rank. If y is a realization of $\boldsymbol{y}$, then the least-squares estimate of w^o given y is $\widehat{w} = (H^*H)^{-1}H^*y$. The resulting residual vector is $\widetilde{y} = y - H\widehat{w}$.

In order to study the stochastic properties of this least-squares solution, we need to treat it as an estimator and write instead $\hat{\boldsymbol{w}} = (H^*H)^{-1}H^*\boldsymbol{y}$, in terms of the random quantity $\boldsymbol{y}$. We also write $\tilde{\boldsymbol{y}} = \boldsymbol{y} - H\hat{\boldsymbol{w}}$; different realizations for $\boldsymbol{y}$ lead to different realizations for $\{\hat{\boldsymbol{w}}, \tilde{\boldsymbol{y}}\}$.

(a) Verify that $\hat{\boldsymbol{w}} = w^o + (H^*H)^{-1}H^*\boldsymbol{v}$, and conclude that $\hat{\boldsymbol{w}}$ is an unbiased estimator, i.e., $\mathsf{E}\,\hat{\boldsymbol{w}} = w^o$.

(b) Show that $\mathsf{E}\,(w^o - \hat{\boldsymbol{w}})(w^o - \hat{\boldsymbol{w}})^* = \sigma_v^2 (H^*H)^{-1}$.

(c) Assume $N > M$ and let $\hat{\boldsymbol{\sigma}}_v^2 = \|\tilde{\boldsymbol{y}}\|^2/(N-M)$ denote an estimator for the noise variance. Verify that $\mathsf{E}\,\|\tilde{\boldsymbol{y}}\|^2 = \sigma_v^2 \mathrm{Tr}(\mathrm{I} - \mathcal{P}_H)$, where $\mathcal{P}_H$ is the projection matrix onto $\mathcal{R}(H)$. Show further that $\mathrm{Tr}(\mathrm{I} - \mathcal{P}_H) = N - M$ and conclude that $\hat{\boldsymbol{\sigma}}_v^2$ is an unbiased estimator for σ_v^2.

Problem VII.20 (Constrained least-squares) In constrained least-squares problems we want to minimize $\|y - Hw\|^2$ over w subject to the linear constraint $Aw = b$, where H and A are $N \times n$ $(N \geq n)$ and $M \times n$ $(M \leq n)$ full-rank matrices, respectively. Show that the solution is given by $\widehat{w}_c = \widehat{w} - (H^*H)^{-1}A^*\left[A(H^*H)^{-1}A^*\right]^{-1}(A\widehat{w} - b)$, where $\widehat{w}$ is the standard least-squares solution, $\widehat{w} = (H^*H)^{-1}H^*y$.

Problem VII.21 (QR solution of constrained least-squares) Refer to Prob. VII.20 and introduce the QR decomposition of A^*, namely,

$$A^* = Q \begin{bmatrix} R \\ 0 \end{bmatrix}$$

where Q is $n \times n$ unitary and R is $M \times M$ upper triangular. Introduce the change of variables $z = Q^* w$ and $\bar{H} = HQ$ and partition $\{z, \bar{H}\}$ into

$$z = \begin{bmatrix} z_1 \\ z_2 \end{bmatrix}, \qquad \bar{H} = \begin{bmatrix} \bar{H}_1 & \bar{H}_2 \end{bmatrix}$$

where z_1 is $M \times 1$ and $\bar{H}_1$ is $N \times M$. Let $\bar{y}_2 = y - \bar{H}_1 z_1$.

(a) Show that the constrained least-squares formulation of Prob. VII.20 is equivalent to

$$\min_z \|y - \bar{H}z\|^2 \quad \text{subject to } R^* z_1 = b$$

(b) Show that the minimizing solution is given by $\widehat{z} = \mathrm{col}\{\widehat{z}_1, \widehat{z}_2\}$, where $\widehat{z}_1 = R^{-*}b$ while $\widehat{z}_2$ is determined by solving the least-squares problem $\min_{z_2} \|\bar{y}_2 - \bar{H}_2 z_2\|^2$.

Problem VII.22 (Gauss-Markov theorem) Recall the statement of Thm. 6.1. Given a random vector $\boldsymbol{y}$ that is linearly related to an unknown vector w via $\boldsymbol{y} = Hw + \boldsymbol{v}$, where $\boldsymbol{v}$ is zero-mean noise with covariance matrix R_v, the minimum-variance unbiased linear estimator of w is

$\hat{\boldsymbol{w}} = (H^* R_v^{-1} H)^{-1} H^* R_v^{-1} \boldsymbol{y}$ or, equivalently, the estimate of w given an observation y is $\widehat{w} = (H^* R_v^{-1} H)^{-1} H^* R_v^{-1} y$. Verify that this expression can be interpreted as the solution to the weighted least-squares problem $\min_w \; (y - Hw)^* R_v^{-1} (y - Hw)$.

Remark. It follows that the Gauss-Markov theorem suggests that an optimal choice for the weighting matrix W in (29.21) is $W = R_v^{-1}$, i.e., the inverse of the covariance matrix of the noise component in y.

Problem VII.23 (Separating signal from perturbation) Consider the linear model $\boldsymbol{y} = Hx + S\theta + \boldsymbol{v}$, where $\boldsymbol{v}$ is a zero-mean additive noise with unit covariance matrix, and $\{x, \theta\}$ are unknown constant vectors. The matrices H and S are $N \times n$ and $N \times m$, respectively, and such that $[H \;\; S]$ has full rank with $N \geq m + n$. The term $S\theta$ can be interpreted as a structured perturbation that lies in the column span of S, while $s = Hx$ denotes the desired signal that is corrupted by $S\theta$ and by $\boldsymbol{v}$. We wish to estimate Hx from $\boldsymbol{y}$ and, hence, separate the signal component, Hx, from the perturbation, $S\theta$.

(a) Let $z = \text{col}\{x, \theta\}$. Determine the minimum variance unbiased estimator, $\hat{z}$, of z given $\boldsymbol{y}$. *Hint:* Recall Thm. 6.1.

(b) Let $\hat{\boldsymbol{z}} = \text{col}\{\hat{\boldsymbol{x}}, \hat{\boldsymbol{\theta}}\}$ and consider the estimator $\hat{\boldsymbol{s}} = H\hat{\boldsymbol{x}}$ for s. Show that

$$\hat{\boldsymbol{s}} = \mathcal{P}_H \left[\text{I} - S(S^* \mathcal{P}_H^{\perp} S)^{-1} S^* \mathcal{P}_H^{\perp} \right] \boldsymbol{y}$$

with $\mathcal{P}_H^{\perp} = \text{I} - \mathcal{P}_H$, $\mathcal{P}_S^{\perp} = \text{I} - \mathcal{P}_S$ and where $\mathcal{P}_S$ and $\mathcal{P}_H$ denote the orthogonal projectors onto the column spans of S and H, respectively, i.e., $\mathcal{P}_S = S(S^*S)^{-1}S^*$ and $\mathcal{P}_H = H(H^*H)^{-1}H^*$.

(c) Assume instead that $\boldsymbol{x}$ is modeled as a zero-mean random variable with covariance matrix $\Pi^{-1} = \mathsf{E}\,\boldsymbol{x}\boldsymbol{x}^* > 0$. Show that the linear least-mean-squares estimator of $\boldsymbol{s} = H\boldsymbol{x}$ is now given by $\hat{\boldsymbol{s}} = \mathcal{P}_H^{\pi} \left[\text{I} - S(S^* \mathcal{P}_H^{\pi,\perp} S)^{-1} S^* \mathcal{P}_H^{\pi,\perp} \right] \boldsymbol{y}$, where $\mathcal{P}_H^{\pi,\perp} = \text{I} - \mathcal{P}_H^{\pi}$ and $\mathcal{P}_H^{\pi} = H(\Pi + H^*H)^{-1}H^*$. Compute the resulting minimum mean-square error (m.m.s.e.) and compare it with that of part (b).

Problem VII.24 (Property of regularized solutions) Let $\widehat{w}$ denote the unique solution of the regularized least-squares problem

$$\min_w \; \left[\gamma \|w\|^2 + \|y - Hw\|^2 \right]$$

where H is $N \times n$ and has full rank, with $N > n$, and y does not lie in the column span of H. Assume further that $H^*y \neq 0$ so that $\widehat{w}$ is nonzero. Moreover, γ is a finite positive number. Let $\eta^2 = \gamma^2 \|\widehat{w}\|^2 / \|y - H\widehat{w}\|^2$. Show that $\eta^2 < \|H^*y\|^2 / \|y\|^2$.

Remark. This result can be used to show that, under the given conditions on the data $\{H, y\}$, the nonzero solution of a regularized least-squares problem is also the solution to a robust estimation problem of the form studied in Probs. VII.25–VII.26. For more details on such robust formulations, see Sayed, Nascimento, and Chandrasekaran (1998). Further extensions and applications appear in Sayed and Chandrasekaran (2000), Sayed (2001), Sayed, Nascimento, and Cipparrone (2002), and Sayed and Chen (2002).

Problem VII.25 (Zero solution of a robust problem) Consider an $N \times n$ full rank matrix H with $N \geq n$, and an $N \times 1$ vector y that does not belong to the column span of H. Let η be a positive real number and consider the set of all matrices δH whose norm does not exceed η, $\|\delta H\| \leq \eta$. Here, by the norm of a matrix we mean its maximum singular value (see Sec. B.6). For the purposes of this problem, all you need to know about this norm is that it satisfies the property $\|Hx\| \leq \|H\| \, \|x\|$ for any vector x, where $\|x\|$ denotes the Euclidean norm of x. Now consider the following optimization problem

$$\min_w \; \max_{\|\delta H\| \leq \eta} \; \|y - (H + \delta H)w\|$$

That is, we seek a vector w that minimizes the maximum residual over the set $\{\|\delta H\| \leq \eta\}$.

(a) Argue from the conditions of the problem that we must have $N > n$.

(b) Show that the uncertainty set $\{\|\delta H\| \leq \eta\}$ contains a perturbation δH^o such that y is orthogonal to $(H + \delta H^o)$ if, and only if, $\eta \geq \|H^* y\|/\|y\|$.

(c) Show that the above optimization problem has a unique solution at $\widehat{w} = 0$ if, and only if, the condition on η in part (b) holds.

Problem VII.26 (Nonzero solution of a robust problem) Consider an $N \times n$ full rank matrix H with $N \geq n$, and an $N \times 1$ vector y that does not belong to the column span of H.

(a) For any nonzero $n \times 1$ column vector w, show that the following rank-one modification of H (denoted by $H(w)$),

$$H(w) \triangleq \left[H + \eta \frac{Hw - y}{\|Hw - y\|} \frac{w^*}{\|w\|}\right]$$

still has full rank for any positive real number η.

(b) Verify that $\|y - H(w)w\| = \|y - Hw\| + \eta\|w\|$, and that the vectors $y - H(w)w$ and $y - Hw$ are collinear and point in the same direction (that is, one is a positive multiple of the other).

(c) Show that $\|y - H(w)w\| = \max_{\|\delta H\| \leq \eta} \|y - (H + \delta H)w\|$.

(d) Show that the optimization problem $\min_w \max_{\|\delta H\| \leq \eta} \|y - (H + \delta H)w\|$ has a nonzero solution $\widehat{w}$ if, and only if, $\eta < \|H^* y\|/\|y\|$.

(e) Show that $\widehat{w}$ is a nonzero solution of the optimization problem in part (d) if, and only, if $H^*(\widehat{w})[y - H\widehat{w}] = 0$. That is, the residual vector $y - H\widehat{w}$ should be orthogonal to the perturbed matrix $H(\widehat{w})$. Show further that this condition is equivalent to $H^*(\widehat{w})[y - H(\widehat{w})\widehat{w}] = 0$.

(f) Assume two nonzero solutions $\widehat{w}_1$ and $\widehat{w}_2$ exist that satisfy the orthogonality condition of part (e). Argue that $H^*(\widehat{w}_2)[y - H(\widehat{w}_2)\widehat{w}_1] = 0$, and conclude that $\widehat{w}_1 = \widehat{w}_2$ so that the solution is unique.

Problem VII.27 (Circulant matrices and DFT transformations) Refer to the discussion on OFDM receivers in Computer Project VII.1. Use the fact that the DFT matrix F is symmetric to verify that $\Lambda = F^* H^{\mathsf{T}} F$. Conclude from this relation that (VII.2) should hold.

Problem VII.28 (Inner product equality) Consider a data matrix H and partition it as $H = \begin{bmatrix} y & \bar{H} & z \end{bmatrix}$, with y and z denoting its leading and trailing columns, respectively. Let $\widehat{y}$ and $\widehat{z}$ denote the regularized least-squares estimates of y and z given $\bar{H}$, namely, $\widehat{y} = \bar{H}\widehat{x}_y$, $\widehat{z} = \bar{H}\widehat{x}_z$, $\widetilde{y} = y - \widehat{y}$, and $\widetilde{z} = z - \widehat{z}$, where $\widehat{x}_y$ and $\widehat{x}_z$ are the solutions of

$$\min_{x_y} \left(x_y^* \Pi x_y + \|y - \bar{H}x_y\|^2\right) \quad \text{and} \quad \min_{x_z} \left(x_z^* \Pi x_z + \|z - \bar{H}x_z\|^2\right)$$

Show that $\widetilde{y}^* z = y^* \widetilde{z}$. Define $\kappa \triangleq \widetilde{y}^* \widetilde{z}/(\|\widetilde{y}\|\, \|\widetilde{z}\|)$. Show that $|\kappa| \leq 1$.

Problem VII.29 (Minimum cost) Refer to the derivation in Sec. 30.5 for the time-update of the minimum cost. Assume instead that we start from a regularized cost function of the form:

$$\min_w \left[(w - \bar{w})^* \Pi (w - \bar{w}) \; + \; \|y_N - H_N w\|^2\right]$$

with a nonzero $\bar{w}$. Its minimum cost is given by $\xi(N) = (y_N - H_N \bar{w})^* (y_N - H_N w_N)$ — see Table 10.3. Repeat the derivation of that section to show that $\xi(N)$ still satisfies the recursion $\xi(N) = \xi(N-1) + e(N) r^*(N)$, with $\xi(-1) = 0$ and where $e(N) = d(N) - u_N w_{N-1}$ and $r(N) = d(N) - u_N w_N$.

Problem VII.30 (Modified RLS) Refer again to the discussion in Sec. 30.6 on the exponentially weighted least-squares problem but assume now that y_N evolves in time in the following manner:

$$y_N = \begin{bmatrix} a y_{N-1} \\ d(N) \end{bmatrix}$$

for some scalar a. The choice $a = 1$ reduces to the situation studied in Sec. 30.6. Repeat the arguments prior to the statement of Alg. 30.1 and show that the solution w_N, and the corresponding minimum cost, $\xi(N)$, can be computed recursively as follows. Start with $w_{-1} = 0$, $P_{-1} = \Pi^{-1}$, and $\xi(-1) = 0$, and iterate for $i \geq 0$:

$$\begin{aligned}
\gamma(i) &= 1/(1 + \lambda^{-1} u_i P_{i-1} u_i^*) \\
g_i &= \lambda^{-1} P_{i-1} u_i^* \gamma(i) \\
e(i) &= d(i) - a u_i w_{i-1} \\
w_i &= a w_{i-1} + g_i e(i) \\
P_i &= \lambda^{-1} P_{i-1} - g_i g_i^* / \gamma(i) \\
\xi(i) &= \lambda |a|^2 \xi(i-1) \; + \; \gamma(i) |e(i)|^2
\end{aligned}$$

In particular, observe that the scalar a appears in the expressions for $\{w_i, e(i), \xi(i)\}$. Show further that $r(i) = \gamma(i) e(i)$ where $r(i) = d(i) - u_i w_i$.

Problem VII.31 (Inner-product update) Refer to the discussion in Sec. 32.3 and assume the extended matrix H_N in (32.50) is defined instead as follows

$$H_N \;\stackrel{\Delta}{=}\; \begin{bmatrix} a x_{N-1} & \bar{H}_{N-1} & z_{N-1} \\ \alpha(N) & \bar{h}_N & \beta(N) \end{bmatrix} \stackrel{\Delta}{=} \begin{bmatrix} x_N & \bar{H}_N & z_N \end{bmatrix}$$

with a scalar a multiplying x_{N-1}. Repeat the derivation of that section to show that now

$$\Delta(N) = \lambda a^* \Delta(N-1) + \widetilde{\alpha}^*(N) \widetilde{\beta}(N) / \bar{\gamma}(N)$$

where $\{\widetilde{\beta}(N), \widetilde{\alpha}(N)\}$ are defined as before, while $\widetilde{\alpha}_a(N) = \alpha(N) - a \bar{h}_N w_{N-1,x}$.

Problem VII.32 (QR method for updating LS solutions) Refer to the discussion and notation in Sec. 30.2, and consider the two standard least-squares problems:

$$\min_w \|y_{N-1} - H_{N-1} w\|^2 \quad \text{and} \quad \min_w \left\| \begin{bmatrix} y_{N-1} \\ d(N) \end{bmatrix} - \begin{bmatrix} H_{N-1} \\ u_N \end{bmatrix} w \right\|^2$$

whose solutions we denote by w_{N-1} and w_N, respectively. Assume the data matrices $\{H_{N-1}, H_N\}$ have full-column ranks so that $\{w_{N-1}, w_N\}$ are uniquely defined. Suppose that at step $N-1$ we have solved the first problem via the QR method of Prob. VII.12 and obtained w_{N-1}. Let

$$H_{N-1} = Q_{N-1} \begin{bmatrix} R_{N-1} \\ 0 \end{bmatrix}$$

denote the QR factorization of H_{N-1}, and define $Q_{N-1}^* y_{N-1} = \text{col}\{z_{1,N-1}, z_{2,N-1}\}$ where $z_{1,N-1}$ is $M \times 1$. Let also

$$\begin{bmatrix} R_{N-1} \\ u_N \end{bmatrix} = Q \begin{bmatrix} R_N \\ 0 \end{bmatrix}$$

denote the QR factorization of $\text{col}\{R_{N-1}, u_N\}$.

(a) Verify that

$$\left\| \begin{bmatrix} Q_{N-1}^* & 0 \\ 0 & 1 \end{bmatrix} (y_N - H_N w) \right\|^2 = \left\| \begin{bmatrix} z_{1,N-1} \\ d(N) \end{bmatrix} - \begin{bmatrix} R_{N-1} \\ u_N \end{bmatrix} w \right\|^2 + \|z_{2,N-1}\|^2$$

(b) Define the entries $\{z_{1,N}, z_{2,N}\}$ as follows:

$$\begin{bmatrix} z_{1,N} \\ z_{2,N} \end{bmatrix} \stackrel{\Delta}{=} Q^* \begin{bmatrix} z_{1,N-1} \\ d(N) \end{bmatrix}$$

Show that $w_N = R_N^{-1} z_{1,N}$. Show also that the minimum cost $\xi(N)$ at step N satisfies

$$\xi(N) = \xi(N-1) + \|z_{2,N}\|^2$$

Remark. In summary, we find that $\{R_N, z_{1,N}, z_{2,N}\}$ are updated via the equation

$$\boxed{\begin{bmatrix} R_{N-1}^* & u_N^* \\ z_{1,N-1}^* & d^*(N) \end{bmatrix} Q = \begin{bmatrix} R_N^* & 0 \\ z_{1,N}^* & z_{2,N}^* \end{bmatrix}}$$

We shall re-derive this QR method in Sec. 35.2 via a different route and, actually, for more general regularized least-squares problems. At that time, we shall also provide further insight into the meaning of the variables $\{z_{1,N}, z_{2,N}\}$.

Problem VII.33 (Downdating least-squares solutions) We use the notation of Sec. 30.2. In this problem, we consider a new issue, namely, that of removing the effect of earlier data. Thus consider again the data $\{y_N, H_N\}$ from (30.5) and partition them instead as follows:

$$y_N = \begin{bmatrix} d(0) \\ y_{1:N} \end{bmatrix}, \quad H_N = \begin{bmatrix} u_0 \\ H_{1:N} \end{bmatrix}$$

where the notation $y_{1:N}$ denotes the last entries of y_N (i.e., from time 1 to time N). Likewise, $H_{1:N}$ denotes the last rows of H_N. The subscript notation $_{1:N}$ indicates that $H_{1:N}$ consists of rows 1 through N; likewise for $y_{1:N}$.

Suppose now that we wish to remove the effect of the initial data $\{d(0), u_0\}$, i.e., we are interested in the solution of the regularized least-squares problem:

$$\min_w \left[w^* \Pi w + \|y_{1:N} - H_{1:N} w\|^2 \right]$$

We denote its solution by $w_{1:N}$, i.e., $w_{1:N} = (\Pi + H_{1:N}^* H_{1:N})^{-1} H_{1:N}^* y_{1:N}$ (cf. (30.7)). The purpose of this problem is to relate $w_{1:N}$ to w_N in (30.7), which solves (30.6). Let $P_{1:N} = (\Pi + H_{1:N}^* H_{1:N})^{-1}$, and recall from (30.8) that $P_N = (\Pi + H_N^* H_N)^{-1}$. The arguments that follow, for relating $\{w_{1:N}, w_N\}$ and $\{P_{1:N}, P_N\}$, are essentially identical to those in Sec. 30.2 while deriving the RLS algorithm.

(a) Show that $P_{1:N} = P_N - P_N u_0^* u_0 P_N / (-1 + u_0 P_N u_0^*)$.

(b) Show also that $w_{1:N} = w_N + \frac{P_N u_0^*}{-1 + u_0 P_N u_0^*} [d(0) - u_0 w_N]$.

(c) Show that $1 - u_0 P_N u_0^* > 0$.

Problem VII.34 (Sliding window RLS) We can combine the result of Prob. VII.33 with the recursions of Alg. 30.1 to derive a recursive least-squares solution with *finite-memory*. The solution w_N in (30.7) is usually termed a growing-memory solution since it depends on all data prior to and including time N. A finite-memory solution can be developed as follows — see also Prob. IX.14.

Introduce the quantities

$$y_{N-L+1:N} \triangleq \begin{bmatrix} d(N-L+1) \\ \vdots \\ d(N) \end{bmatrix}, \quad H_{N-L+1:N} \triangleq \begin{bmatrix} u_{N-L+1} \\ \vdots \\ u_N \end{bmatrix}$$

The vector $y_{N-L+1:N}$ contains entries over an interval of length L, namely, from time $N-L+1$ up to time N. Likewise, the data matrix $H_{N-L+1:N}$ contains regressors over the same interval of time. Let $w_{N-L+1:N}$ denote the solution to the regularized least-squares problem:

$$\min_w \left[w^* \Pi w + \|y_{N-L+1:N} - H_{N-L+1:N} w\|^2 \right]$$

The purpose of this problem is to derive an algorithm for going from $w_{N-L:N-1}$ to $w_{N-L+1:N}$, as depicted in Fig. VII.1. This can be achieved by combining the downdating recursion $w_{N-L:N-1} \rightarrow$

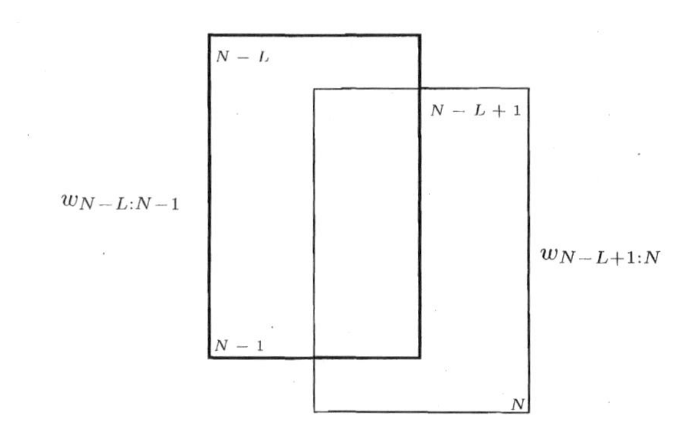

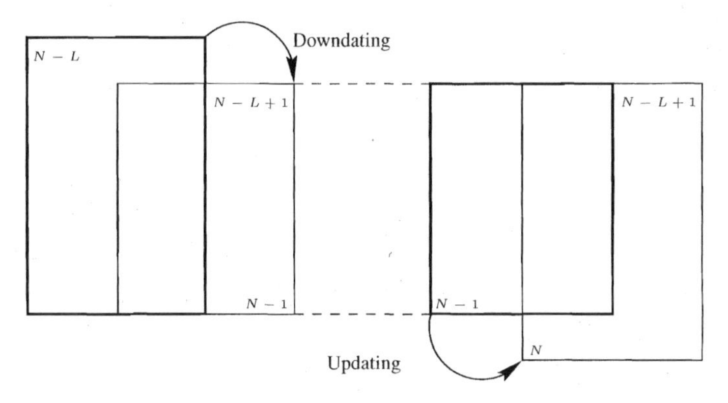

FIGURE VII.1 A procedure for finite-memory recursive least-squares by means of updating and downdating least-squares solutions.

$w_{N-L+1:N-1}$ (which has a similar form to that derived in Prob. VII.33) with the updating recursion $w_{N-L+1:N-1} \rightarrow w_{N-L+1:N}$ (which has a similar form to that in Alg. 30.1), i.e.,

$$\boxed{w_{N-L:N-1}} \xrightarrow{\text{downdating}} w_{N-L+1:N-1} \xrightarrow{\text{updating}} \boxed{w_{N-L+1:N}}$$

Introduce

$$P_{N-L+1:N-1} \triangleq \left(\Pi + H^*_{N-L+1:N} H_{N-L+1:N}\right)^{-1}$$

(a) Argue as in Prob. VII.33 to show that

$$\begin{aligned}
w_{N-L+1:N-1} &= w_{N-L:N-1} + \frac{P_{N-L:N-1}u^*_{N-L}}{-1 + u_{N-L}P_{N-L:N-1}u^*_{N-L}} e_d(N-L) \\
e_d(N-L) &= d(N-L) - u_{N-L}w_{N-L:N-1} \\
P_{N-L+1:N-1} &= P_{N-L:N-1} - \frac{P_{N-L:N-1}u^*_{N-L}u_{N-L}P_{N-L:N-1}}{-1 + u_{N-L}P_{N-L:N-1}u^*_{N-L}}
\end{aligned}$$

(b) Argue as in the steps that led to Alg. 30.1 to show that

$$\begin{aligned}
w_{N-L+1:N} &= w_{N-L+1:N-1} + \frac{P_{N-L+1:N-1}u^*_N}{1 + u_N P_{N-L+1:N-1}u^*_N} e_u(N) \\
e_u(N) &= d(N) - u_N w_{N-L+1:N-1} \\
P_{N-L+1:N} &= P_{N-L+1:N-1} - \frac{P_{N-L+1:N-1}u^*_N u_N P_{N-L+1:N-1}}{1 + u_N P_{N-L+1:N-1}u^*_N}
\end{aligned}$$

(c) For simplicity of notation, denote the vectors $\{w_{N-L:N-1}, w_{N-L+1:N}\}$ by $\{w_{N-1}^L, w_N^L\}$; i.e., they are solutions based on windows of length L and they become available at times $\{N-1, N\}$. Using this notation, conclude that the solution w_N^L can be computed recursively as follows. Start with $w_{-1}^L = 0$ and $P_{-1}^L = \Pi^{-1}$ and repeat for $i \geq 0$:

$$\begin{array}{lcl}
e_d(i-L) & = & d(i-L) - u_{i-L}w_{i-1}^L \\
\gamma_L(i-1) & = & 1/(1 - u_{i-L}P_{i-1}^L u_{i-L}^*) \\
g_{i-1}^L & = & P_{i-1}^L u_{i-L}^* \gamma_L(i-1) \\
w_{i-1}^{L-1} & = & w_{i-1}^L - g_{i-1}^L e_d(i-L) \\
P_{i-1}^{L-1} & = & P_{i-1}^L + g_{i-1}^L \left(g_{i-1}^L\right)^* / \gamma_L(i-1) \\
e_u(i) & = & d(i) - u_i w_{i-1}^{L-1} \\
\gamma_{L-1}(i) & = & 1/(1 + u_i P_{i-1}^{L-1} u_i^*) \\
g_i^{L-1} & = & P_{i-1}^{L-1} u_i^* \gamma_{L-1}(i-1) \\
w_i^L & = & w_{i-1}^{L-1} + g_i^{L-1} e_u(i) \\
P_i^L & = & P_{i-1}^{L-1} - g_i^{L-1} \left(g_i^{L-1}\right)^* / \gamma_{L-1}(i)
\end{array}$$

Problem VII.35 (Subspace constrained RLS) Consider an $M \times m$ full-rank matrix A ($M > m$) and let w be any vector in its range space, i.e., $w \in \mathcal{R}(A)$. Let w_N denote the solution to the following regularized least-squares problem

$$\min_{w \in \mathcal{R}(A)} \left[\lambda^{N+1} w^* \Pi w + \sum_{j=0}^{N} \lambda^{N-j} |d(j) - u_j w|^2 \right]$$

where $\Pi > 0$ and u_j is $1 \times M$. Find a recursion relating w_{N+1} to w_N.

Problem VII.36 (Block RLS) In our treatment of recursive least-squares problems in Secs. 30.2–30.6 we assumed that the new data that are incorporated into $\{y_N, H_N\}$ in (30.5) consist of a *scalar* $d(N)$ and a *row* vector u_N. However, there are applications where block updates are necessary, in which case $d(N)$ is replaced by a vector, say d_N, and u_N is replaced by a matrix, say U_N. In these situations, the same arguments that were employed in Sec. 30.2 can be repeated to develop a block RLS algorithm. The purpose of this problem is to derive this algorithm.

Consider a regularized least-squares problem of the form

$$\min_w \left[(w - \bar{w})^* \Pi (w - \bar{w}) \; + \; (y_{N-1} - H_{N-1}w)^* W_{N-1} (y_{N-1} - H_{N-1}w) \right], \quad \Pi > 0$$

and partition the entries of $\{y_{N-1}, H_{N-1}\}$ into

$$y_{N-1} = \begin{bmatrix} d_0 \\ d_1 \\ \vdots \\ d_{N-1} \end{bmatrix}, \qquad H_{N-1} = \begin{bmatrix} U_0 \\ U_1 \\ \vdots \\ U_{N-1} \end{bmatrix}$$

where each d_j has dimensions $p \times 1$ and each U_j has dimensions $p \times M$. We further assume that the positive-definite weighting matrix W_{N-1} has a block diagonal structure, with $p \times p$ positive-definite diagonal blocks, say $W_{N-1} = \text{diag}\{R_0^{-1}, R_1^{-1}, \ldots, R_{N-1}^{-1}\}$. Let w_{N-1} denote the solution of the above least-squares problem and let $P_{N-1} = (\Pi + H_{N-1}^* W_{N-1} H_{N-1})^{-1}$.

(a) Show that $P_N = P_{N-1} - P_{N-1}U_N^* \Gamma_N U_N P_{N-1}$, with initial condition $P_{-1} = \Pi^{-1}$ and where $\Gamma_N = (R_N + U_N P_{N-1} U_N^*)^{-1}$.

(b) Show that $w_N = w_{N-1} + P_{N-1}U_N^* \Gamma_N [d_N - U_N w_{N-1}]$.

(c) Conclude that w_N can be computed recursively by means of the following block RLS algorithm. Start with $w_{-1} = \bar{w}$ and $P_{-1} = \Pi^{-1}$ and repeat for $i \geq 0$:

$$\begin{array}{|rcl|} \Gamma_i & = & (R_i + U_i P_{i-1} U_i^*)^{-1} \\ G_i & = & P_{i-1} U_i^* \Gamma_i \\ w_i & = & w_{i-1} + G_i[d_i - U_i w_{i-1}] \\ P_i & = & P_{i-1} - G_i \Gamma_i^{-1} G_i^* \end{array}$$

(d) Establish also the equalities $G_N = P_N U_N^* R_N^{-1}$ and $\Gamma_N = R_N^{-1} - R_N^{-1} U_N P_N U_N^* R_N^{-1}$.

(e) Let $\{r_N, e_N\}$ denote the *a posteriori* and *a priori* error vectors, $r_N = d_N - U_N w_N$ and $e_N = d_N - U_N w_{N-1}$. Show that $R_N^{-1} r(N) = \Gamma_N e_N$.

(f) Let $\xi(N-1)$ denote the minimum cost associated with the solution w_{N-1}. Show that it satisfies the time-update relation

$$\boxed{\xi(N) = \xi(N-1) + r_N^* R_N^{-1} e_N \; = \; \xi(N-1) + e_N^* \Gamma_N e_N, \quad \xi(-1) = 0}$$

Conclude that $\xi(N) = \sum_{i=0}^{N} e_i^* \Gamma_i e_i$.

Problem VII.37 (Alternative form of block RLS) Consider the same regularized cost function of Prob. VII.36 and define $\Phi_N = \Pi + H_N^* W_N H_N$ and $s_N = H_N^* W_N (y_N - H_N \bar{w})$. Show that $\{\Phi_i, s_i, w_i\}$ satisfy the following recursions. Start with $\Phi_{-1} = \Pi$, $s_{-1} = 0$, and repeat for $i \geq 0$:

$$\begin{array}{l} \Phi_i = \Phi_{i-1} + U_i^* R_i^{-1} U_i \\ s_i = s_{i-1} + U_i^* R_i^{-1} [d_i - U_i \bar{w}] \\ \text{solve } \Phi_i (w_i - \bar{w}) = s_i \end{array}$$

Problem VII.38 (Exponentially weighted block RLS) Consider a regularized least-squares problem of the form

$$\min_w \left[\lambda^{N+1} (w - \bar{w})^* \Pi (w - \bar{w}) \; + \; \sum_{j=0}^{N} \lambda^{N-j} (d_j - U_j w)^* R_j^{-1} (d_j - U_j w) \right]$$

where each d_j is $p \times 1$, each U_j is $p \times M$, and each R_j is $p \times p$ and positive-definite. Moreover, $0 \ll \lambda \leq 1$ is an exponential forgetting factor and $\Pi > 0$. Let $\xi(N)$ denote the value of the minimum cost associated with the optimal solution w_N.

(a) Repeat the arguments of Prob. VII.36 to show that the solution w_N can be time-updated as follows:

$$\begin{array}{|rcl|} \Gamma_i & = & (R_i + \lambda^{-1} U_i P_{i-1} U_i^*)^{-1} \\ G_i & = & \lambda^{-1} P_{i-1} U_i^* \Gamma_i \\ e_i & = & d_i - U_i w_{i-1} \\ w_i & = & w_{i-1} + G_i[d_i - U_i w_{i-1}], \quad w_{-1} = \bar{w} \\ P_i & = & \lambda^{-1} P_{i-1} - G_i \Gamma_i^{-1} G_i^*, \quad P_{-1} = \Pi^{-1} \\ r_i & = & d_i - U_i w_i \\ \xi(i) & = & \lambda \xi(i-1) + e_i^* \Gamma_i e_i, \quad \xi(-1) = 0 \\ & = & \lambda \xi(i-1) + r_i^* R_i^{-1} e_i \end{array}$$

Verify also that the quantities $\{G_i, \Gamma_i\}$ admit the alternative expressions $G_i = P_i U_i^* R_i^{-1}$ and $\Gamma_i = R_i^{-1} - R_i^{-1} U_i P_i U_i^* R_i^{-1}$.

(b) Repeat the arguments of Prob. VII.37 to show that the solution w_N can also be evaluated as follows:

$$\begin{array}{l} \Phi_i = \lambda \Phi_{i-1} + U_i^* R_i^{-1} U_i, \quad \Phi_{-1} = \Pi \\ s_i = \lambda s_{i-1} + U_i^* R_i^{-1} [d_i - U_i \bar{w}], \quad s_{-1} = 0 \\ \text{solve } \Phi_i (w_i - \bar{w}) = s_i \end{array}$$

In other words, $w_i = \bar{w} + \Phi_i^{-1} s_i$.

Problem VII.39 (Block RLS with singular weighting matrices) Probs. VII.36–VII.38 assume positive-definite weighting factors $\{R_j\}$. This restriction can be relaxed. Thus consider a regularized least-squares problem of the form

$$\min_w \left[\lambda^{N+1}(w-\bar{w})^*\Pi(w-\bar{w}) \; + \; \sum_{j=0}^{N} \lambda^{N-j}(d_j - U_j w)^* A_j (d_j - U_j w) \right]$$

where each d_j is $p \times 1$, each U_j is $p \times M$, and each weighting factor A_j is $p \times p$ and positive semi-definite (hence, possibly singular). Moreover, $0 \ll \lambda \leq 1$ is an exponential forgetting factor and $\Pi > 0$. Repeat the arguments of Prob. VII.37 to show that the solution w_N can be evaluated iteratively as follows:

$$\begin{aligned} &\Phi_i = \lambda\Phi_{i-1} + U_i^* A_i U_i, \quad \Phi_{-1} = \Pi \\ &s_i = \lambda s_{i-1} + U_i^* A_i [d_i - U_i \bar{w}], \quad s_{-1} = 0 \\ &\text{solve } \Phi_i(w_i - \bar{w}) = s_i \end{aligned}$$

Argue that Φ_i is invertible.

Problem VII.40 (Data fusion) At each time $n \geq 0$, a total of M noisy measurements of a scalar unknown variable x are collected across M spatially-distributed sensors, say $\boldsymbol{y}_k(n) = x + \boldsymbol{v}_k(n), k = 0, 1, \ldots, M-1$. The noises $\{\boldsymbol{v}_n(k)\}$ are assumed to be spatially and temporally white with variances $\{\sigma_{v,k}^2(n)\}$.

(a) Show that the minimum-variance unbiased estimator (m.v.u.e) of x given the observation vector $\boldsymbol{y}_n = \text{col}\{\boldsymbol{y}_0(n), \boldsymbol{y}_1(n), \ldots, \boldsymbol{y}_{M-1}(n)\}$ is given by

$$\hat{\boldsymbol{x}}_n = \frac{\sum_{k=0}^{M-1} \boldsymbol{y}_k(n)/\sigma_{v,k}^2(n)}{\sum_{k=0}^{M-1} 1/\sigma_{v,k}^2(n)}$$

(b) More generally, assume x is estimated instead by solving a deterministic least-squares problem of the form:

$$\min_x \sum_{n=0}^{N} \lambda^{N-n} \left(\sum_{k=0}^{M-1} \alpha_k(n) \, |y_k(n) - x|^2 \right) \implies \widehat{x}_N$$

where $0 \ll \lambda \leq 1$ is an exponential forgetting factor and $\alpha_k(n)$ are some nonnegative weighting coefficients (for example, $\alpha_k(j) = 1/\sigma_{v,k}^2(n)$). Show that $\widehat{x}_N$ can be computed recursively as follows:

$$\begin{aligned} \phi(n) &= \lambda\phi(n-1) + \sum_{k=0}^{M-1} \alpha_k(n), \quad \phi(-1) = 0 \\ s(n) &= \lambda s(n-1) + \sum_{k=0}^{M-1} \alpha_k(n) y_k(n), \quad s(-1) = 0 \\ \widehat{x}_n &= s(n)/\phi(n) \end{aligned}$$

Problem VII.41 (Data reweighting) Consider a regularized least-squares problem of the form

$$\min_w \; [\, w^*\Pi w \; + \; (y_{N-1} - H_{N-1}w)^* W_{N-1} (y_{N-1} - H_{N-1}w) \,], \quad \Pi > 0$$

and partition the entries of $\{y_{N-1}, H_{N-1}\}$ into

$$y_{N-1} = \begin{bmatrix} d_0 \\ d_1 \\ \vdots \\ d_{N-1} \end{bmatrix}, \quad H_{N-1} = \begin{bmatrix} U_0 \\ U_1 \\ \vdots \\ U_{N-1} \end{bmatrix}$$

where each d_j has dimensions $p \times 1$ and each U_j has dimensions $p \times M$. We further assume that the positive-definite weighting matrix W_{N-1} has a block diagonal structure, with $p \times p$ positive-definite diagonal blocks, $W_{N-1} = \text{diag}\{R_0^{-1}, R_1^{-1}, \ldots, R_{N-1}^{-1}\}$. Let w_{N-1} denote the solution of the above least-squares problem. Now assume the weighting matrix W_N is related to W_{N-1} as follows

$$W_N = \begin{bmatrix} D_{N-1}W_{N-1} & \\ & R_N^{-1} \end{bmatrix}$$

where the diagonal matrix D has the form $D_{N-1} = \text{diag}\{\mathrm{I}_p, \ldots, \mathrm{I}_p, \beta\mathrm{I}_p, \mathrm{I}_p, \ldots, \mathrm{I}_p\}$, and $\beta > 1$ is a positive scalar. The scalar β appears at the location corresponding to the $k-$th block R_k^{-1}. Find a recursion relating w_N to w_{N-1}.

COMPUTER PROJECTS

Project VII.1 (OFDM receiver) This project illustrates how least-squares and least-mean-squares solutions are useful in the design of orthogonal frequency-division multiplexing (OFDM) receivers.

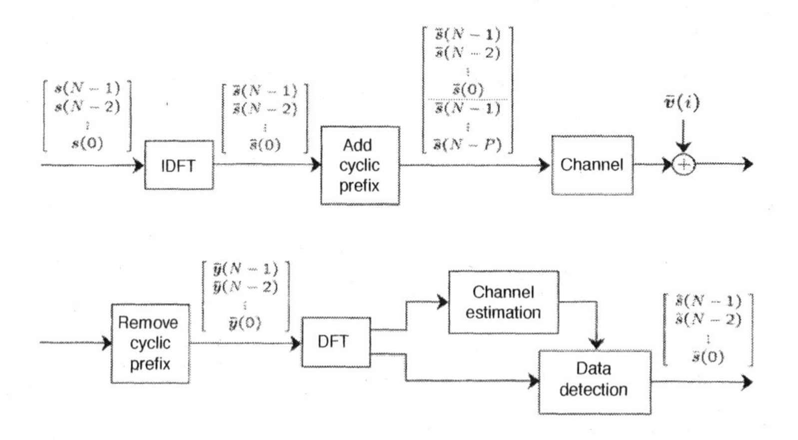

FIGURE VII.2 OFDM transmitter and receiver structures.

In an OFDM system, data are transmitted in blocks, say of size N symbols each, with additional P symbols added for cyclic prefixing purposes (in a manner similar to what we discussed before in Probs. II.27 and II.28). Specifically, consider a block of data of size N,

$$s = \text{col}\{\ s(N-1), s(N-2), \ldots, s(0)\ \}$$

Before transmission, this block of data is transformed by the inverse DFT matrix, i.e., s is transformed into $\bar{s} = F^* s$, where F is the unitary DFT matrix of size N defined by

$$[F]_{ik} \triangleq \frac{1}{\sqrt{N}} e^{\frac{-j2\pi ik}{N}}, \quad i,k = 0,1,\ldots,N-1, \quad j = \sqrt{-1}$$

Then a cyclic prefix of length P is added to the transformed data, so that the transmitted sequence ends up being (see Fig. VII.2):

$$\text{col}\{\ \underbrace{\bar{s}(N-1), \bar{s}(N-2), \ldots, \bar{s}(0)}_{\text{transformed block of size } N},\ \underbrace{\bar{s}(N-1), \bar{s}(N-2), \ldots, \bar{s}(N-P)}_{\text{cyclic prefix of size } P}\}$$

with the rightmost sample transmitted at time instant 1 and the leftmost sample transmitted at time instant $N+P$. At the receiver, these $(N+P)$ samples are observed in the presence of additive noise and collected into a vector, say

$$\text{col}\{\underbrace{\bar{y}(N+P-1),\bar{y}(N+P-2),\ldots,\bar{y}(P)}_{\text{last } N \text{ received samples}},\underbrace{\bar{y}(P-1),\ldots,\bar{y}(2),\bar{y}(1),\bar{y}(0)}_{\text{first } P \text{ received samples}}\}$$

The first P received samples are discarded, while the remaining N received samples are collected into an $N\times 1$ vector, $\bar{y}$.

Data model. Assuming an M-tap FIR model for the channel with impulse response sequence,

$$h=\text{col}\{h(0),h(1),\ldots,h(M-1)\}$$

we can verify that (recall Prob. II.27):

$$\underbrace{\begin{bmatrix}\bar{y}(N+P-1)\\ \bar{y}(N+P-2)\\ \bar{y}(N+P-3)\\ \vdots\\ \bar{y}(P)\end{bmatrix}}_{\triangleq\,\bar{y}}=\underbrace{\begin{bmatrix}\bar{v}(N+P-1)\\ \bar{v}(N+P-2)\\ \bar{v}(N+P-3)\\ \vdots\\ \bar{v}(P)\end{bmatrix}}_{\triangleq\,\bar{v}}+$$

$$\underbrace{\begin{bmatrix}h(0) & h(1) & \cdots & h(M-1) & & & & \\ & h(0) & h(1) & \cdots & h(M-1) & & & \\ & & h(0) & h(1) & \cdots & h(M-1) & & \\ & & & \ddots & & & \ddots & \\ & & & & h(0) & h(1) & \cdots & h(M-1)\\ \ddots & & & & & \ddots & & \vdots\\ h(2) & \cdots & h(M-1) & & & & h(0) & h(1)\\ h(1) & h(2) & \cdots & h(M-1) & & & & h(0)\end{bmatrix}}_{H}\underbrace{\begin{bmatrix}\bar{s}(N-1)\\ \bar{s}(N-2)\\ \bar{s}(N-3)\\ \vdots\\ \bar{s}(0)\end{bmatrix}}_{\triangleq\,\bar{s}}$$

or, more compactly,

$$\boxed{\bar{y}=H\bar{s}+\bar{v}}\qquad\text{(transform domain)}$$

where H is the channel matrix and $\bar{v}$ denotes measurement noise in the transformed domain.

Now observe that H has a circulant structure and, therefore, it can be diagonalized by the DFT matrix. Let $\Lambda=FHF^*$ be the diagonalized channel, and multiply the above equation by F from the left. Then we obtain

$$\boxed{y=\Lambda s+v}\qquad\text{(time domain)}\tag{VII.1}$$

where the time-domain vector quantities $\{y,s,v\}$ are defined by

$$\boxed{y=F\bar{y},\qquad s=F\bar{s},\qquad v=F\bar{v}}$$

Let λ be a column vector with the entries of Λ, i.e., $\lambda=\text{diag}\{\Lambda\}$. Then it holds that (see Prob. VII.27):

$$\lambda=\sqrt{N}F^*\begin{bmatrix}h_{M\times 1}\\ 0_{(N-M)\times 1}\end{bmatrix}\tag{VII.2}$$

That is, λ is the inverse DFT of the channel impulse response with its length extended to N.

Data recovery. Given the linear relation (VII.1) between $\{y,s\}$, and assuming the channel is known, we can invoke the results of Sec. 5.1 to conclude that the linear least-mean-squares estimator of s

given $\boldsymbol{y}$ is

$$\hat{\boldsymbol{s}} = R_s\Lambda^*\left[R_v + \Lambda R_s\Lambda^*\right]^{-1}\boldsymbol{y}$$

where $R_s = \mathsf{E}\,\boldsymbol{s}\boldsymbol{s}^*$ and $R_v = \mathsf{E}\,\boldsymbol{v}\boldsymbol{v}^*$. Assuming that the data $\{\boldsymbol{s}(\cdot)\}$ and the noise $\{\boldsymbol{v}(\cdot)\}$ are i.i.d. with variances σ_s^2 and σ_v^2, respectively, then $R_s = \sigma_s^2\mathrm{I}$ and $R_v = \sigma_v^2\mathrm{I}$. In this way, the above expression for $\hat{\boldsymbol{s}}$ simplifies to

$$\boxed{\hat{\boldsymbol{s}} = \Lambda^*\left[\frac{\sigma_v^2}{\sigma_s^2}\mathrm{I} + \Lambda\Lambda^*\right]^{-1}\boldsymbol{y}} \quad \text{(data recovery)} \qquad \text{(VII.3)}$$

Observe that the matrix being inverted is diagonal, which therefore results in a simple receiver structure as shown in Fig. VII.2. If we denote the entries of $\boldsymbol{y}$ by

$$\boldsymbol{y} = \text{col}\{\,\boldsymbol{y}(N+P-1), \boldsymbol{y}(N+P-2), \ldots, \boldsymbol{y}(P)\,\}$$

then the estimators for the individual symbols are given by

$$\hat{\boldsymbol{s}}(i) = \left(\frac{\lambda_i^*}{\frac{\sigma_v^2}{\sigma_s^2} + |\lambda_i|^2}\right)\boldsymbol{y}(i+P), \quad i = 0, 1, \ldots, N-1$$

We may remark that expression (VII.3) for $\hat{\boldsymbol{s}}$ in terms of $\boldsymbol{y}$ could have also been obtained had we treated $\{\boldsymbol{s}, \boldsymbol{y}\}$ as deterministic quantities $\{s, y\}$ (i.e., as realizations) and formulated the problem of estimating s from y in the regularized least-squares sense

$$\min_s \left[\frac{1}{\mathsf{SNR}}\|s\|^2 + \|y - \Lambda s\|^2\right]$$

with regularization parameter $1/\mathsf{SNR}$, where $\mathsf{SNR} = \sigma_s^2/\sigma_v^2$.

Training. In order to recover the transmitted signals (i.e., in order to estimate $\boldsymbol{s}$) as above, the channel taps are needed (i.e., Λ is needed). Different training schemes can be used to enable the receiver to estimate the channel and, consequently, Λ. The most common training scheme is to allocate some of the tones, i.e., some of the $\{\boldsymbol{s}(i)\}$ in an OFDM symbol, to *known* training data. We shall refer to these known symbols by writing $\{s(i)\}$ (with normal font instead of boldface letters) since they are not random any longer. The channel taps are then estimated as follows. We first rewrite (VII.1) as

$$\boldsymbol{y} = \begin{bmatrix} \boldsymbol{s}(N-1) & & & \\ & \boldsymbol{s}(N-2) & & \\ & & \ddots & \\ & & & \boldsymbol{s}(0) \end{bmatrix}\lambda + \boldsymbol{v} \;\stackrel{\Delta}{=}\; \boldsymbol{S}\lambda + \boldsymbol{v}$$

Let $\{k_1, k_2, \cdots, k_L\}$ denote the indices of the L ($L \geq M$) elements of $\boldsymbol{s}$ that are used as training tones, and are therefore known. We collect these transmitted training tones, which are known and, hence, deterministic quantities, and the corresponding received data into two vectors

$$\begin{aligned} s_t &= \text{col}\{s(k_L), s(k_{L-1}), \ldots, s(k_1)\} \\ y_t &= \text{col}\{y(k_L+P), y(k_{L-1}+P), \ldots, y(k_1+P)\} \end{aligned}$$

Let Q denote the corresponding $M \times L$ submatrix of F,

$$Q = \begin{bmatrix} [F]_{0,k_1} & [F]_{0,k_2} & \cdots & [F]_{0,k_L} \\ [F]_{1,k_1} & [F]_{1,k_2} & \cdots & [F]_{1,k_L} \\ \vdots & & \ddots & \vdots \\ [F]_{M-1,k_1} & [F]_{M-1,k_2} & \cdots & [F]_{M-1,k_L} \end{bmatrix} \qquad (M \times L)$$

and let S_t be the corresponding $L \times L$ submatrix of S, i.e., $S_t = \text{diag}\{s_t\}$. Then

$$\boxed{\underbrace{y_t}_{L\times 1} = \sqrt{N} \cdot \underbrace{S_t Q^*}_{L\times M} \underbrace{h}_{M\times 1} + v_t}$$

where v_t is the corresponding noise vector, $v_t = \text{col}\{v(k_L + P), \ldots, v(k_1 + P)\}$. We can now recover h by solving a least-squares problem, namely, as

$$\widehat{h} = \frac{1}{\sqrt{N}} \cdot \left[QS_t^* S_t Q^*\right]^{-1} QS_t^* y_t$$

Using the fact that S_t is diagonal, and assuming the training data satisfy $|s(i)|^2 = 1$ so that $S_t^* S_t = \text{I}$, then the above least-squares expression simplifies to

$$\boxed{\widehat{h} = \frac{1}{\sqrt{N}} \cdot [QQ^*]^{-1} QS^* \begin{bmatrix} y_t(k_L + P)/s_t(k_L) \\ \vdots \\ y_t(k_2 + P)/s_t(k_2) \\ y_t(k_1 + P)/s_t(k_1) \end{bmatrix}} \quad \text{(channel estimation)}$$

The estimate of Λ is therefore

$$\boxed{\widehat{\Lambda} = \sqrt{N} \cdot \text{diag}(\widehat{\lambda}) = \text{diag}\left(F^* \begin{bmatrix} \widehat{h} \\ 0 \end{bmatrix}\right)}$$

Channel tracking. The above explanation assumes that one OFDM symbol is used to estimate the channel taps. However, in practice, this procedure is not accurate enough due to noise and variations in the channel. Therefore, multiple OFDM symbols are usually used to estimate and track the channel. Thus assume that k OFDM symbols are transmitted with the training tones in the same locations. Then repeating the equality $y_t = S_t Q^* h + v_t$ for the k received symbols we have

$$\begin{bmatrix} y_{t,k} \\ y_{t,k-1} \\ \vdots \\ y_{t,1} \end{bmatrix} = \begin{bmatrix} S_{t,k} \\ S_{t,k-1} \\ \vdots \\ S_{t,1} \end{bmatrix} Q^* h + \begin{bmatrix} v_{t,k} \\ v_{t,k-1} \\ \vdots \\ v_{t,1} \end{bmatrix}$$

where the indices $\{1, \ldots, k\}$ correspond to the received OFDM symbols. Now, if we assume that the training data used in different symbols are the same (which is a valid assumption in practice), then the least-squares estimate of h is simply the average of the channel estimates derived for each OFDM symbol separately, i.e.,

$$\widehat{h} = \frac{1}{k} \sum_{i=1}^{k} \widehat{h}_i$$

where $\widehat{h}_i$ is the channel estimate derived from the i-th received symbol, as in the previous subsection.

Training schemes. Different training schemes can be used in OFDM systems to estimate the channel, as indicated in Fig. VII.3. One scheme is to assign all tones in an OFDM symbol for training, and then use all tones in the subsequent symbols for data transmission. In other words, some symbols are entirely for training purposes and others are purely for data transmission. This scheme is shown in Fig. VII.3(a). An alternative scheme is to allocate certain tone indices for training on all symbols and to use the other tones for data transmission. In this scheme, the training sequence and the data are transmitted simultaneously on all symbols, as shown in Fig. VII.3(b). The advantage of the first scheme is that the matrix inversion $(QQ^*)^{-1}$ becomes mute since $QQ^* = \text{I}$ in this case. Yet, this scheme cannot track channel variations occurring between the symbols assigned for training. The second scheme has a better tracking performance, but requires the matrix inversion $[QQ^*]^{-1}$.

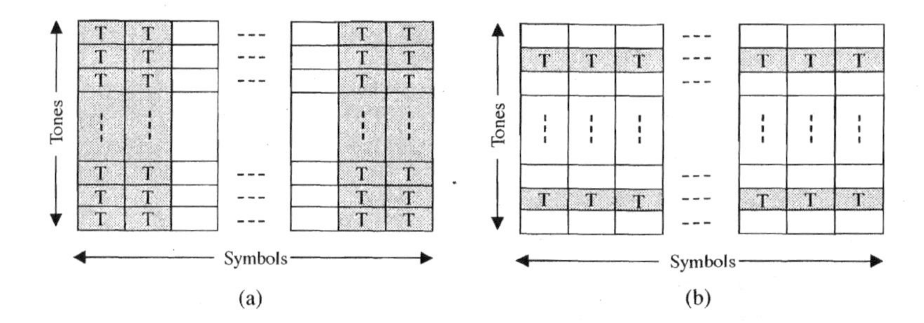

FIGURE VII.3 Two training schemes for an OFDM system: (a) training followed by data transmission; and (b) simultaneous training and data transmission.

Remark. Comparing the structure of an OFDM system with that of a single-carrier frequency-domain-equalization system, as described earlier in Prob. II.28, we find that in the latter scheme a cyclic prefix is added to the raw (time-domain) data before transmission. In OFDM, on the other hand, the data is first transformed to the frequency domain and then a cyclic prefix is added before transmission. As a result, OFDM systems tend to suffer from a peak-to-average ratio (PAR) problem, i.e., large signal peaks can occur due to constructive interferences among the signals in the sub-channels. For a comparison of both modulation schemes (single-carrier-frequency-domain equalization and OFDM) see, e.g., Sari, Karam, and Jeanclaude (1995) and Czylwik (1997).

Simulations. The following are the parameters used in our simulation of an OFDM system: $N = 64$ (block or symbol size), $P = 16$ (cyclic prefix length), $M = 8$ (channel length) and $L = 8$ (training tones in a symbol). The transmitted data are drawn from a QPSK constellation with unit power. The channel taps are complex numbers with Gaussian distribution for the real and imaginary parts.

(a) Write a program that simulates the performance of the two training schemes of Fig. VII.3. Specifically, for the first scheme, use all 64 tones of the first symbol for training and channel estimation. For the second scheme, use only 8 tones corresponding to indices

$$\{4, 12, 20, 28, 36, 44, 52, 60\}$$

for training. In order to perform a fair comparison, in the second scheme, estimate the channel only after receiving 8 consecutive symbols. This is to ensure that both schemes have the same total training overhead. Set the SNR at 10 dB at the input to the receiver and simulate the operation of the OFDM receiver; estimate the channel taps both in the time domain (h) and in the frequency domain (Λ).

(b) For each of the training schemes of Fig. VII.5, generate BER curves versus SNR by averaging over 100 randomly generated channels, i.e., perform training and data recovery over 100 channels and average their bit-error-rate performance. Vary the SNR between 0 and 20 dB. Generate also the BER curve assuming exact knowledge of the channel. Fix the SNR at 25 dB and plot the scatter diagrams of the transmitted sequence, the received sequence, and the recovered data using scheme (a) from Fig. VII.3.

Project VII.2 (Tracking Rayleigh fading channels) We reconsider the problem of tracking a Rayleigh fading channel from Computer Project IV.2. Let w^o denote the tap vector of an $M-$th order multipath channel and assume all taps fade at the same Doppler rate f_D. We mentioned in Computer Project IV.2 that one simple approximation for the time-variation in the channel would be to model it as $w_i^o = \alpha w_{i-1}^o + q_i$, where $\alpha = \mathcal{J}_o(2\pi f_D T_s)$ and q_i is an i.i.d. random vector with covariance matrix $Q = q\text{I} = (1-\alpha^2)\text{I}$. The quantities $\{\alpha, q\}$ are defined in terms of the maximum Doppler frequency f_D, the sampling rate T_s, and the Bessel function

$$\mathcal{J}_o(x) \triangleq \frac{1}{\pi}\int_0^{\pi} \cos(x \sin\theta)d\theta$$

which characterizes the auto-correlation sequence of each of the taps as

$$r(k) \triangleq \mathsf{E}\,\boldsymbol{x}(n)\boldsymbol{x}(n-k) = \mathcal{J}_o\left(2\pi f_D T_s\, k\right), \quad k = \ldots, -1, 0, 1, \ldots$$

(a) Let $M = 5$, $f_D = 10\text{Hz}$, and $T_s = 0.8\mu\text{s}$. Design an adaptive tracker for the channel using 1) ϵ-NLMS with $\mu = 0.25$; 2) RLS with $\lambda = 0.995$; 3) the extended RLS algorithm (31.42) with the above values for α and q and $\lambda = 0.995$. Generate learning curves for the three algorithms by averaging over 200 experiments. Set the SNR level at 30 dB. Compare the steady-state mean-square error values with those predicted by theory.

(b) Increase the Doppler frequency to $f_D = 80\text{Hz}$ and repeat the simulations. What do you observe?

(c) In parts (a) and (b), the value of the parameter α is very close to unity and the value of q is also very small. In order to illustrate more clearly the difference in performance that arises between RLS and its extended version (31.42), consider a model with more distinct values for $\{\alpha, q\}$, say $\alpha = 0.95$ and $q = 0.1$, and repeat the simulations of part (a). Repeat also for $\alpha = 1$ and $q = 0.1$. Start from an initial condition $\boldsymbol{w}_{-1}^o$ that is normally distributed with identity covariance matrix.

PART 

ARRAY ALGORITHMS

CHAPTER 33

Norm and Angle Preservation

Array methods are powerful algorithmic variants that are theoretically equivalent to the recursive least-squares algorithm, but they nevertheless perform the required computations in a more *reliable* manner. In these array forms, an algorithm is described not as an explicit set of equations, but as a sequence of elementary operations on arrays of numbers (or matrices). Usually, a pre-array of numbers has to be triangularized by a sequence of elementary rotations in order to yield a post-array of numbers. The quantities needed to form the next pre-array are read off from the entries of the post-array, and the procedure can be repeated. The explicit forms of the rotation matrices are not needed in most cases, and they can be implemented in a variety of well-known ways, e.g., as a sequence of elementary circular or hyperbolic rotations. The purpose of this chapter is to develop several array-based methods for RLS filtering. In order to motivate such array algorithms, we shall first consider a simple (yet contrived) example that helps highlight some important issues.

33.1 SOME DIFFICULTIES

Thus, consider the update equation (30.12) for the variable P_N in the RLS algorithm, and assume that all variables are real and scalar-valued (and, hence, we shall write $\{u(N), P(N)\}$ instead of $\{u_N, P_N\}$). Assume further that at some iteration n_o, especially during the initial stages of adaptation where $P(N)$ is more likely to assume large values, $u(n_o) = 1$ and $P(n_o - 1)$ is sufficiently large so that, in *finite precision*, the value of $1 + P(n_o - 1)$ evaluates to $P(n_o - 1)$, i.e.,

$$1 + P(n_o - 1) = P(n_o - 1) \qquad \text{(in finite precision)} \tag{33.1}$$

From (30.12) we have that the value of $P(n_o)$ is obtained via the update

$$P(n_o) = P(n_o - 1) - \frac{P^2(n_o - 1)}{1 + P(n_o - 1)} \tag{33.2}$$

which is also equivalent to

$$P(n_o) = \frac{P(n_o - 1)}{1 + P(n_o - 1)} \tag{33.3}$$

Now assume that the term

$$P^2(n_o - 1)/[1 + P(n_o - 1)]$$

in (33.2) is evaluated as follows:

$$\frac{P^2(n_o - 1)}{1 + P(n_o - 1)} = P(n_o - 1) \cdot \frac{P(n_o - 1)}{1 + P(n_o - 1)}$$

That is, we first evaluate the ratio

$$P(n_o - 1)/[1 + P(n_o - 1)]$$

and then multiply the result by $P(n_o - 1)$. Then, because of (33.1), the above ratio evaluates to 1 and, therefore, if we compute $P(n_o)$ using (33.2) we get

$$P(n_o) = P(n_o - 1) - P(n_o - 1) \cdot 1 = 0$$

On the other hand, recursion (33.3) gives $P(n_o) = 1$.

The values obtained for $P(n_o)$ are obviously different and the second one is the desired value since, from (33.3),

$$\lim_{P(n_o-1)\to\infty} P(n_o) = 1$$

We therefore find from the equivalent expressions (33.2) and (33.3) that these two different implementations of the *same* recursion can behave differently in the presence of round-off errors. There is a reason for the failure of the implementation based on (33.2); it evaluates the nonnegative quantity $P(n_o)$ as the difference of two nearly equal and large nonnegative numbers, and the result is an undesired cancellation; in some other more complex scenarios, the variable P_N may even lose its positive-definiteness. Such cancellations are often, but not always, bad phenomena in finite-precision implementations so much so that they are usually called *catastrophic cancellations*!

This simple example shows that there is always merit in considering alternative implementations even for the same algorithm. Another example is presented in Prob. VIII.11 where two theoretically equivalent implementations of the same algorithm are also shown to react differently in response to perturbation in the data.

For these and other reasons, we are motivated to develop *array* methods for recursive least-squares problems. The array methods will have several intrinsic advantages over a plain RLS implementation; for one thing, they will be more reliable in finite-precision implementations (as explained in the sequel and as illustrated later in the computer project at the end of the chapter). We start with a handful of definitions.

33.2 SQUARE-ROOT FACTORS

A key element in array algorithms is the concept of a square-root of a positive-definite matrix.

Definition

Although the concept of square-roots can be defined for nonnegative-definite matrices, it is enough for our purposes here to focus on positive-definite matrices. Thus, let A denote an $n \times n$ positive-definite matrix and introduce its eigen-decomposition

$$A = U\Lambda U^* \tag{33.4}$$

where Λ is an $n \times n$ diagonal matrix with real positive entries, which correspond to the eigenvalues of A, and U is a unitary matrix, namely an $n \times n$ square matrix that satisfies

$$UU^* = U^*U = I$$

The columns of U correspond to the orthonormal eigenvectors of A as can be seen by re-writing (33.4) as $AU = U\Lambda$. Let $\Lambda^{1/2}$ denote a diagonal matrix whose entries are the positive square-roots of the diagonal entries of Λ and, hence,

$$\Lambda = \Lambda^{1/2}\left(\Lambda^{1/2}\right)^*$$

Then we can rewrite (33.4) as

$$A = \left(U\Lambda^{1/2}\right)\cdot\left(U\Lambda^{1/2}\right)^*$$

which expresses A as the product of an $n \times n$ matrix and its conjugate transpose, namely,

$$A = XX^* \qquad \text{with} \qquad X = U\Lambda^{1/2}$$

We say that X is *a* square-root of A.

> **Definition 33.1 (Square-root factors)** A square-root of an $n \times n$ positive-definite matrix A is any $n \times n$ matrix X satisfying $A = XX^*$.

The construction prior to the definition exhibits one possible choice for X, namely, $X = U\Lambda^{1/2}$, in terms of the eigenvectors and eigenvalues of A. However, square-root factors are highly nonunique. This is true even for scalars. For instance, the number 4 has infinitely many square-roots over the field of complex numbers, namely, $2e^{j\phi}$ for any $\phi \in [-\pi, \pi]$. For matrices, if we take the above X and multiply it by any unitary matrix Θ, say, $\bar{X} = X\Theta$ where $\Theta\Theta^* = \mathrm{I}$, then $\bar{X}$ is also a square-root factor of A since

$$\bar{X}\bar{X}^* = X\underbrace{\Theta\Theta^*}_{=\mathrm{I}}X^* = XX^* = A$$

Notation

It is customary to use the notation $A^{1/2}$ to refer to a square-root of a matrix A and, therefore, we write

$$A = A^{1/2}\left(A^{1/2}\right)^*$$

It is also customary to employ the compact notations

$$A^{*/2} \triangleq \left(A^{1/2}\right)^*, \qquad A^{-1/2} \triangleq \left(A^{1/2}\right)^{-1}, \qquad A^{-*/2} \triangleq \left(A^{1/2}\right)^{-*}$$

so that

$$\boxed{A = A^{1/2}A^{*/2}, \qquad A^{-1} = A^{-*/2}A^{-1/2}} \tag{33.5}$$

Cholesky Factor

One of the most widely used square-root factors of a positive-definite matrix is its Cholesky factor. Recall that we showed in Sec. B.3 that every positive-definite matrix A admits a *unique* triangular factorization of the form

$$A = \bar{L}\bar{L}^* \tag{33.6}$$

where $\bar{L}$ is a lower-triangular matrix with positive entries on its diagonal. We could also consider the alternative factorization $A = \bar{U}\bar{U}^*$ in terms of an upper triangular matrix $\bar{U}$. However, the lower triangular form will be the standard form for our discussions. The factor $\bar{L}$ is called the Cholesky factor of A. Comparing (33.6) with the defining relation (33.5), we see that $\bar{L}$ is also a square-root factor of A. For our purposes, whenever we refer to the square-root of a matrix A we shall mean its Cholesky factor.[16] It has two advantages in relation to other square-root factors: it is lower triangular and is uniquely defined (i.e., there is no other lower triangular square-root factor with positive diagonal entries).

Array Algorithms

Now, an array algorithm generally implements transformations of the form

$$\begin{bmatrix} \times & \times & \times & \times \\ \times & \times & \times & \times \\ \times & \times & \times & \times \\ \times & \times & \times & \times \end{bmatrix} \Theta = \begin{bmatrix} \times & 0 & 0 & 0 \\ \times & \times & 0 & 0 \\ \times & \times & \times & 0 \\ \times & \times & \times & \times \end{bmatrix}$$

where Θ is some unitary matrix whose purpose is to transform the pre-array of numbers to some triangular form. There are many ways to implement unitary transformations of this kind. For example, it is explained in Chapter 34 that Θ could be implemented as a sequence of elementary transformations (known as Givens rotations) or reflection transformations (known as Householder reflections). Such array descriptions have many intrinsic advantages:

1. They have better numerical properties for two main reasons. First, unitary transformations are numerically well-behaved since they do not amplify numerical errors. And second, the entries in the pre- and post-arrays are usually square-root factors of certain variables and these entries tend to assume values within smaller dynamic ranges.

2. They are easy to implement as a sequence of elementary rotations or reflections, as we explain in Chapter 34.

3. They admit modular and parallelizable implementations, since each rotation or reflection can be applied simultaneously to all rows of the pre-array.

33.3 PRESERVATION PROPERTIES

Since unitary transformations are at the core of most array methods, it is important to examine some of their most distinctive properties. To begin with, a key property of unitary transformations is that the norms of vectors and the inner products between vectors are preserved by them. To see this, assume that x and y are two row vectors that are related by some unitary transformation Θ, say, $x\Theta = y$, then

$$\|y\|^2 \stackrel{\Delta}{=} yy^* = x\Theta\Theta^* x^* = xx^* \stackrel{\Delta}{=} \|x\|^2$$

That is, the vectors $\{x, y\}$ will have the same Euclidean norm. We therefore say that unitary transformations preserve Euclidean norms:

$$\boxed{x\Theta = y \implies \|x\|^2 = \|y\|^2} \quad \text{(norm preservation)} \qquad (33.7)$$

[16] Accordingly, the square-root of a positive number is taken to be its positive square-root.

In addition, if a and b are two other row vectors that are related by the same transformation, say, $a\Theta = b$, then we have that

$$yb^* = x\Theta\Theta^* a^* = xa^*$$

and we find that unitary transformations also preserve inner products between vectors, i.e.,

$$\boxed{x\Theta = y \text{ and } a\Theta = b \implies xa^* = yb^*} \quad \text{(angle preservation)} \qquad (33.8)$$

The reason why we are referring to the inner-product preservation property as an angle-preservation property is the following. For the case of real data we have

$$xa^{\mathsf{T}} = \|x\| \cdot \|a\| \cdot \cos(\theta)$$

with θ denoting the angle between x and a. Likewise,

$$yb^{\mathsf{T}} = \|y\| \cdot \|b\| \cdot \cos(\alpha)$$

with α denoting the angle between y and b. But since

$$\|a\| = \|b\| \quad \text{and} \quad \|x\| = \|y\|$$

we then conclude from $xa^{\mathsf{T}} = yb^{\mathsf{T}}$ that

$$\cos(\theta) = \cos(\alpha)$$

This latter equality amounts to angle preservation. We thus say that the vectors $\{x, a\}$ are transformed into $\{y, b\}$ in such a way that their norms are preserved as well as the angles between them (see Fig. 33.1).

The conclusions (33.7)–(33.8) can be combined into a stronger (if, and only if) statement, which will be the basis for most of our derivations of array algorithms.

> **Lemma 33.1 (Basis rotation)** Consider two $n \times M$ $(n \leq M)$ matrices A and B. Then $AA^* = BB^*$ if, and only if, there exists an $M \times M$ unitary matrix Θ such that $A = B\Theta$.

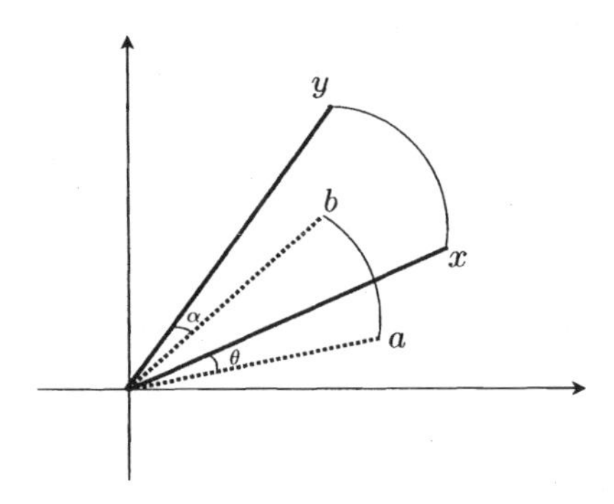

FIGURE 33.1 Rotation of vectors $\{a, x\}$ into $\{b, y\}$ with the vector norms and the angles between them preserved: $\theta = \alpha$, $\|a\| = \|b\|$, and $\|x\| = \|y\|$.

Proof: One direction is obvious. If $A = B\Theta$, for some unitary matrix Θ, then

$$AA^* = (B\Theta)(B\Theta)^* = B(\Theta\Theta^*)B^* = BB^*$$

One proof for the converse implication follows by using the singular value decompositions of A and B (cf. Sec. B.6) — see Prob. VIII.2 for another proof:

$$A = U_A \begin{bmatrix} \Sigma_A & 0 \end{bmatrix} V_A^*, \quad B = U_B \begin{bmatrix} \Sigma_B & 0 \end{bmatrix} V_B{}^*$$

where U_A and U_B are $n \times n$ unitary matrices, V_A and V_B are $M \times M$ unitary matrices, and Σ_A and Σ_B are $n \times n$ diagonal matrices with nonnegative entries. The squares of the diagonal entries of Σ_A (Σ_B) are the eigenvalues of AA^* (BB^*). Moreover, U_A (U_B) are constructed from an orthonormal basis for the right eigenvectors of AA^* (BB^*). Hence, it follows from the identity $AA^* = BB^*$ that $\Sigma_A = \Sigma_B$ and $U_A = U_B$. Let $\Theta = V_B V_A^*$. Then $\Theta\Theta^* = \mathrm{I}$ and $B\Theta = A$.

◇

Our derivation of array algorithms will be based on the result of Lemma 33.1. Actually, the result of the lemma is more than we need. It will be enough for our purposes to use only one direction of the lemma, namely, the statement that

$$\boxed{AA^* = BB^* \implies \text{There exists a unitary } \Theta \text{ such that } A = B\Theta} \tag{33.9}$$

33.4 MOTIVATION FOR ARRAY METHODS

Before plunging into the derivation of RLS array methods, we shall study a simple example in some detail in order to highlight the main ideas underlying the mechanics of array methods. Thus, consider two scalars $\{a, b\}$ and assume that we wish to evaluate the positive scalar c that satisfies

$$|c|^2 = |a|^2 + |b|^2 \tag{33.10}$$

The first method that comes to mind is to evaluate the squares $|a|^2$ and $|b|^2$, add them, and then compute the square-root of the sum to find c.

Preservation of Norms

A less obvious way for determining c, albeit one that will be more useful for our purposes (especially when we deal with matrix quantities $\{A, B, C\}$ as opposed to scalars $\{a, b, c\}$), is to use an array method. It can be motivated as follows. Observe that the right-hand side of (33.10) is the sum of two squares and it can be expressed as an inner product:

$$|a|^2 + |b|^2 = \begin{bmatrix} a & b \end{bmatrix} \begin{bmatrix} a^* \\ b^* \end{bmatrix}$$

Similarly, the left-hand side of (33.10) can be expressed as an inner product:

$$|c|^2 = \begin{bmatrix} c & 0 \end{bmatrix} \begin{bmatrix} c^* \\ 0 \end{bmatrix}$$

In this way, relation (33.10) in effect amounts to an equality of the form

$$\begin{bmatrix} c & 0 \end{bmatrix} \begin{bmatrix} c^* \\ 0 \end{bmatrix} = \begin{bmatrix} a & b \end{bmatrix} \begin{bmatrix} a^* \\ b^* \end{bmatrix} \tag{33.11}$$

which has the same form as $AA^* = BB^*$ with the identifications

$$A \longleftarrow \begin{bmatrix} c & 0 \end{bmatrix}, \qquad B \longleftarrow \begin{bmatrix} a & b \end{bmatrix} \tag{33.12}$$

Therefore, using (33.9), we conclude that there should exist a 2×2 unitary matrix Θ that maps $[a \;\; b]$ to $[c \;\; 0]$,

$$\underbrace{\begin{bmatrix} a & b \end{bmatrix}}_{B} \Theta = \underbrace{\begin{bmatrix} c & 0 \end{bmatrix}}_{A} \tag{33.13}$$

If we can find Θ, then applying it to the pre-array $[a \;\; b]$ would result in the desired value for c.

Now recall that the proof of Lemma 33.1 provides an expression for the unitary matrix Θ that transforms B to A. However, that expression is in terms of the right singular vectors of $\{A, B\}$ and, therefore, it requires that we know beforehand both A and B. Clearly, this construction is not helpful in situations like (33.13) where A is not known. For this reason, the conclusion (33.9) is useful only in that it guarantees the *existence* of a Θ that performs the required transformation (33.13).

In order to find Θ we would argue differently as follows. Choose *any* unitary Θ that transforms the pre-array, $[a \;\; b]$, to the generic form

$$\begin{bmatrix} a & b \end{bmatrix} \Theta = \begin{bmatrix} \times & 0 \end{bmatrix} \tag{33.14}$$

That is, choose any unitary Θ that annihilates the second entry of $[a \;\; b]$ and let $\times$ denote the resulting leading entry of the post-array. We explain in Lemma 34.2, and in the remark following it, how such a Θ can be found for any $[a \;\; b]$, e.g., a Givens rotation could be used, which in this case would be given by

$$\Theta = \begin{cases} \dfrac{e^{-j\phi_a}}{\sqrt{1+|\rho|^2}} \begin{bmatrix} 1 & -\rho \\ \rho^* & 1 \end{bmatrix} & \text{if } a \neq 0 \text{ and where } \rho = b/a \\[2ex] e^{-j\phi_b} \begin{bmatrix} 0 & 1 \\ 1 & 0 \end{bmatrix} & \text{if } a = 0 \end{cases}$$

where $\{\phi_a, \phi_b\}$ are the phases of (the possibly complex numbers) $\{a, b\}$. If we apply this choice of Θ to the pre-array $[a \;\; b]$ in (33.14), then it is easy to see by direct calculation that a post-array of the form $[\times \;\; 0]$ will result. Specifically, as explained in the remark following Lemma 34.2, we get

$$\begin{bmatrix} \times & 0 \end{bmatrix} = \begin{bmatrix} \sqrt{|a|^2 + |b|^2} & 0 \end{bmatrix}$$

which readily identifies $\times$ as the desired c.

Alternatively, the value of $\times$ could have been identified by using the property (33.7) and without assuming any explicit knowledge of Θ. Indeed, (33.7) states that unitary transformations preserve Euclidean norms, so that the norm of the pre-array $[a \;\; b]$ must coincide with the norm of the post-array $[\times \;\; 0]$. Therefore, by "squaring" both sides of (33.14), namely, by writing

$$\begin{bmatrix} a & b \end{bmatrix} \underbrace{\Theta\Theta^*}_{\mathrm{I}} \begin{bmatrix} a^* \\ b^* \end{bmatrix} = \begin{bmatrix} \times & 0 \end{bmatrix} \begin{bmatrix} \times^* \\ 0 \end{bmatrix}$$

we get

$$|a|^2 + |b|^2 = |\times|^2$$

so that $\mathsf{x} = c$.

In summary, this discussion shows that whenever we encounter an equality of the form (33.10), where both sides can be interpreted as sums of squares and can therefore be expressed as inner products of certain vectors with themselves, then we can compute the left-hand side entry in array form as follows. Form a pre-array and annihilate its second entry by means of any unitary rotation that results in a positive leading entry in the post-array, as in (33.14). Then this leading entry should be the desired c.

Preservation of Inner Products

In order to further appreciate the convenience of the array formulation, assume now that in addition to the scalars $\{a, b, c\}$ satisfying (33.10), we are also given scalars $\{d, e\}$ and that we want to evaluate the scalar f that satisfies

$$fc^* = da^* + eb^* \tag{33.15}$$

Of course, given the $\{a, b\}$ we could first determine c as explained above, and then evaluate f by dividing the right-hand side of (33.15) by c^* (or c since c is real in this example). Alternatively, we could evaluate f in array form, just like we did for c, as follows.

We start by noting that the right-hand side of (33.15) can be interpreted as the inner product between two vectors:

$$da^* + eb^* = \begin{bmatrix} d & e \end{bmatrix} \begin{bmatrix} a^* \\ b^* \end{bmatrix}$$

Now we explained above how to find a unitary transformation Θ that takes $[a \;\; b]$ to $[c \;\; 0]$. In addition, we know from our discussions in Sec. 33.3, that any such unitary transformation preserves not only vector norms but also inner products between vectors (recall (33.8)). This second property can be used to our advantage here. Assume we apply Θ to both $[a \;\; b]$ and $[d \;\; e]$. We already know that in the first case we obtain $[c \;\; 0]$ as the post-array, whereas in the second case we would obtain some other post-array that we denote by $[\mathsf{y} \;\; \mathsf{z}]$:

$$\begin{aligned} \begin{bmatrix} a & b \end{bmatrix} \Theta &= \begin{bmatrix} c & 0 \end{bmatrix} \\ \begin{bmatrix} d & e \end{bmatrix} \Theta &= \begin{bmatrix} \mathsf{y} & \mathsf{z} \end{bmatrix} \end{aligned}$$

The preservation of inner products then implies that the inner-product of the pre-array vectors should coincide with the inner-product of the post-array vectors, i.e.,

$$\begin{bmatrix} d & e \end{bmatrix} \begin{bmatrix} a^* \\ b^* \end{bmatrix} = \begin{bmatrix} \mathsf{y} & \mathsf{z} \end{bmatrix} \begin{bmatrix} c^* \\ 0 \end{bmatrix}$$

or, equivalently,

$$da^* + eb^* = \mathsf{y}c^*$$

Comparing with (33.15), we see that we can immediately identify y as the desired f. As for z, while it is not of immediate interest to us here, its value can be identified by noting that $[d \;\; e]$ and $[\mathsf{y} \;\; \mathsf{z}]$ must have identical Euclidean norms, so that

$$|\mathsf{z}|^2 + |f|^2 = |d|^2 + |e|^2$$

Array Description

In conclusion, the above discussion shows that calculations of the type (33.10) and (33.15), aimed at determining $\{c, f\}$ from knowledge of $\{a, b, d, e\}$, can be accomplished in array form as follows. We form the pre-array

$$\mathcal{A} = \begin{bmatrix} a & b \\ d & e \end{bmatrix} \tag{33.16}$$

and choose a unitary matrix Θ that lower triangularizes $\mathcal{A}$, namely, it reduces $\mathcal{A}$ to the form

$$\mathcal{A}\Theta = \begin{bmatrix} \times & 0 \\ \mathsf{y} & \mathsf{z} \end{bmatrix} \tag{33.17}$$

with a positive $\times$. The determination of Θ is solely dependent on the first row of the pre-array $\mathcal{A}$; the entries of the second row of $\mathcal{A}$ are not used to define Θ.

Then the entries $\{\times, \mathsf{y}\}$ can be identified as the desired $\{c, f\}$; this identification follows from the preservation of norms and inner products by unitary matrices. In particular, the identification of $\times$ as c follows from the fact that the top rows of the pre- and post-arrays $\{\mathcal{A}, \mathcal{A}\Theta\}$ must have the same Euclidean norms, while the identification of y as f follows from the fact that the inner product of the rows in the pre- and post-arrays must coincide.

Another way of carrying out this procedure for identifying the entries of the post-array $\mathcal{A}\Theta$ is as follows. Given the pre-array $\mathcal{A}$ as in (33.16), then the entries of the post-array $\mathcal{A}\Theta$ in (33.17) can be identified by "squaring" both sides of (33.17), i.e., by writing

$$\mathcal{A}\underbrace{\Theta\Theta^*}_{\mathrm{I}}\mathcal{A} = \mathcal{A}\mathcal{A}^* = \begin{bmatrix} \times & 0 \\ \mathsf{y} & \mathsf{z} \end{bmatrix}\begin{bmatrix} \times & 0 \\ \mathsf{y} & \mathsf{z} \end{bmatrix}^*$$

and then by comparing terms on both sides of the resulting equality:

$$\begin{bmatrix} a & b \\ d & e \end{bmatrix}\begin{bmatrix} a & b \\ d & e \end{bmatrix}^* = \begin{bmatrix} \times & 0 \\ \mathsf{y} & \mathsf{z} \end{bmatrix}\begin{bmatrix} \times & 0 \\ \mathsf{y} & \mathsf{z} \end{bmatrix}^* \tag{33.18}$$

Doing so results in the relations

$$|\times|^2 = |a|^2 + |b|^2 \quad \text{and} \quad \mathsf{y}\times^* = da^* + eb^*$$

which identify $\times$ as c and y as f.

Vector Case

The above example, with scalar entries $\{a, b, c, d, e, f\}$, illustrates the main ideas behind the derivation of array algorithms. In the context of adaptive filtering, however, we shall encounter vector analogues of relations (33.10) and (33.15), such as determining a lower triangular matrix C, with positive diagonal entries, satisfying

$$CC^* = AA^* + BB^* \tag{33.19}$$

and determining a matrix F satisfying

$$FC^* = DA^* + EB^* \tag{33.20}$$

where $\{A, B, D, E\}$ are generally matrix or vector quantities. The same arguments that we used above will reveal that $\{C, F\}$ can be determined by means of an array method as follows. We form the pre-array (cf. (33.16)):

$$\mathcal{A} = \begin{bmatrix} A & B \\ D & E \end{bmatrix}$$

and reduce it via a unitary transformation Θ to the lower triangular form (cf. (33.17)):

$$\begin{bmatrix} \mathsf{X} & 0 \\ \mathsf{Y} & \mathsf{Z} \end{bmatrix}$$

where X is lower triangular with positive entries along its diagonal; sometimes Z is a square matrix and Θ is also required to generate it in lower-triangular form along with X:

$$\begin{bmatrix} A & B \\ D & E \end{bmatrix} \Theta = \begin{bmatrix} \mathsf{X} & 0 \\ \mathsf{Y} & \mathsf{Z} \end{bmatrix} \tag{33.21}$$

The matrix Θ is not 2×2 any longer; but it can be implemented as a sequence of elementary (Givens) rotations or Householder reflections, as explained in Chapter 34, where we show how to lower triangularize a matrix via a sequence of rotations or reflections. The reader is encouraged to consult Chapter 34 at this stage in order to learn how such unitary transformations Θ are implemented. The subsequent presentation in this chapter assumes that the reader has familiarized himself with the material of Chapter 34.

An explicit expression for Θ in (33.21) is not needed. All we need to do is find the right sequence of rotations that yields the desired triangular post-array. Then, by "squaring" both sides of (33.21) we get

$$\begin{bmatrix} A & B \\ D & E \end{bmatrix} \underbrace{\Theta\Theta^*}_{\mathrm{I}} \begin{bmatrix} A & B \\ D & E \end{bmatrix}^* = \begin{bmatrix} \mathsf{X} & 0 \\ \mathsf{Y} & \mathsf{Z} \end{bmatrix} \begin{bmatrix} \mathsf{X} & 0 \\ \mathsf{Y} & \mathsf{Z} \end{bmatrix}^*$$

so that we must have

$$\mathsf{X}\mathsf{X}^* = AA^* + BB^* \quad \text{and} \quad \mathsf{Y}\mathsf{X}^* = DA^* + EB^*$$

In this way, X can be identified as the lower triangular Cholesky factor of the matrix $AA^* + BB^*$, and since Cholesky factors are unique, we conclude that X must coincide with the desired C. From the second equality above we conclude that $\mathsf{Y} = F$, so that the array algorithm (33.21) enables us to determine $\{C, F\}$. We may remark that we are not restricted to array methods with two (block) rows in the pre-array and post-arrays as in (33.21). If additional relations are available that satisfy certain norm and inner-product preservation properties, then these could be incorporated into the array algorithm as well. A demonstration to this effect is the QR algorithm of Sec. 35.2.

CHAPTER 34

Unitary Transformations

In this chapter we describe two classes of unitary transformations (Givens and Householder) that can be used to annihilate selected entries in a vector or matrix and thereby reduce a matrix to triangular (or similar) form, as is often required by array algorithms. Special care needs to be taken when dealing with complex-valued data as compared to real-valued data.

34.1 GIVENS ROTATIONS

Givens rotations provide an effective way to annihilate specific entries in a vector and it is enough to explain their operation on 2-dimensional row vectors. We consider the case of real-valued data first.

Real data

Consider a 1×2 real-valued vector $z = \begin{bmatrix} a & b \end{bmatrix}$, and assume that we wish to determine a 2×2 matrix Θ that transforms it to the form:

$$\begin{bmatrix} a & b \end{bmatrix} \Theta = \begin{bmatrix} \alpha & 0 \end{bmatrix} \tag{34.1}$$

for some real number α to be determined, and where Θ is required to be orthogonal, i.e., it should satisfy

$$\Theta\Theta^{\mathsf{T}} = \Theta^{\mathsf{T}}\Theta = \mathrm{I}$$

We refer to $[a \;\; b]$ as the pre-array and to $[\alpha \;\; 0]$ as the post-array.

Now any orthogonal matrix Θ has the important property that it preserves vector norms. Indeed, it is easy to see from (34.1) that the following equality must hold:

$$\begin{bmatrix} a & b \end{bmatrix} \underbrace{\Theta\Theta^{\mathsf{T}}}_{\mathrm{I}} \begin{bmatrix} a \\ b \end{bmatrix} = \begin{bmatrix} \alpha & 0 \end{bmatrix} \begin{bmatrix} \alpha \\ 0 \end{bmatrix}$$

or, equivalently, $a^2 + b^2 = \alpha^2$. In this way, no matter which orthogonal transformation Θ we choose to implement the transformation (34.1), it will always hold that the Euclidean norm of the post-array should coincide with the Euclidean norm of the pre-array. This fact allows us to conclude the value of α, namely,

$$\alpha = \pm\sqrt{a^2 + b^2}$$

even before knowing the expression of any Θ that achieves (34.1). Note that there are two choices for α and which one we pick depends on how Θ is implemented, as explained next.

An expression for an orthogonal Θ that achieves the transformation (34.1) is given by

$$\Theta = \frac{1}{\sqrt{1+\rho^2}} \begin{bmatrix} 1 & -\rho \\ \rho & 1 \end{bmatrix} \quad \text{where} \quad \rho = \frac{b}{a}, \quad a \neq 0 \tag{34.2}$$

This choice is known as a Givens or circular rotation. Surely, it can be verified by direct calculation that this Θ is orthogonal and that it leads to

$$\begin{bmatrix} a & b \end{bmatrix} \Theta = \begin{bmatrix} \pm\sqrt{a^2+b^2} & 0 \end{bmatrix}$$

The choice of which sign to pick depends on whether the value of the square root in the expression (34.2) for Θ is chosen to be negative or positive. In general, we choose the positive sign.

The reason for the denomination *circular rotation* for Θ can be seen by expressing Θ in the form

$$\Theta = \begin{bmatrix} c & -s \\ s & c \end{bmatrix}, \quad c = \frac{1}{\sqrt{1+\rho^2}}, \quad s = \frac{\rho}{\sqrt{1+\rho^2}}, \quad \rho = \frac{b}{a}$$

in terms of cosine and sine parameters. In this way, we find that the effect of Θ is to rotate any point (x, y) in the two-dimensional Euclidean space along the *circle* of equation

$$x^2 + y^2 = a^2 + b^2$$

When $a = 0$, we simply select Θ to be the permutation matrix

$$\Theta = \begin{bmatrix} 0 & -1 \\ 1 & 0 \end{bmatrix} \tag{34.3}$$

We could also choose

$$\Theta = \begin{bmatrix} 0 & 1 \\ 1 & 0 \end{bmatrix}$$

with $+1$ instead of -1. The effect of this permutation will be the same; we choose (34.3) with a minus sign so that (34.3) can be regarded as the limit of the Givens rotation (34.2) when $\rho \to \infty$. In summary, we have the following result.

Lemma 34.1 (Real Givens rotation) Consider a 1×2 vector $[a \;\; b]$ with real entries. Then choose Θ as in (34.2) to get

$$\begin{bmatrix} a & b \end{bmatrix} \Theta = \pm\sqrt{a^2+b^2} \begin{bmatrix} 1 & 0 \end{bmatrix}$$

If $a = 0$, then choose Θ as in (34.3) to get $[0 \;\; b]\Theta = [b \;\; 0]$.

Rotations versus Reflections

We can also use elementary reflections as opposed to rotations. These are defined by matrices of the form

$$\Theta = \frac{1}{\sqrt{1+\rho^2}} \begin{bmatrix} 1 & \rho \\ \rho & -1 \end{bmatrix} \tag{34.4}$$

Observe that the determinant of a rotation matrix is equal to $+1$ while the determinant of a reflection matrix is equal to -1.

The reason for the denomination "reflection" is the following. Let $z = \|z\|e^{j\theta}$ denote the polar representation of a point $z = [a\ b]$ in the two-dimensional Euclidean space. Let $\rho = b/a$ and consider the matrix Θ defined by (34.4). The effect of this matrix on z is to align it with $[1\ 0]$. The manner by which this alignment is achieved is by reflecting z across the line that passes through the origin and the point $(\cos(\theta/2), \sin(\theta/2))$ — see Fig. 34.1.

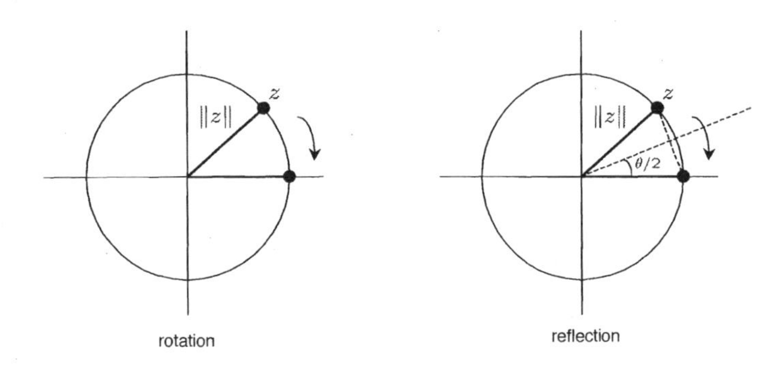

FIGURE 34.1 Aligning a vector z with the first basis vector by means of rotation (*left*) or reflection across the line passing through $(\cos(\theta/2), \sin(\theta/2))$(*right*).

The distinction between rotations and reflections becomes more obvious by examining their effect on other vectors (i.e., other than the vector z that determined them). So let, for example,

$$z = [\sqrt{2}\ \sqrt{2}]$$

so that $\rho = 1$, $\|z\| = 2$, and $\theta = \pi/4$. The corresponding rotation and reflection matrices are given by

$$\Theta^{\text{rot}} = \frac{\sqrt{2}}{2}\begin{bmatrix} 1 & -1 \\ 1 & 1 \end{bmatrix}, \qquad \Theta^{\text{ref}} = \frac{\sqrt{2}}{2}\begin{bmatrix} 1 & 1 \\ 1 & -1 \end{bmatrix}$$

Applying any of these matrices to z will align it with the basis vector $[1\quad 0]$, as shown in see Fig. 34.1. Consider now the vector

$$z' = [0\ \ 2]$$

which lies on the same circle as z. Multiplying z' by Θ^{rot} results in

$$z'\Theta^{\text{rot}} = [\ \sqrt{2}\ \ \sqrt{2}\ ]$$

so that z' is rotated by $\pi/4$ radians in the clockwise direction until it reaches its destination at $[\sqrt{2}\ \sqrt{2}]$ — see Fig. 34.2. On the other hand, multiplying z' by Θ^{ref} results in

$$z'\Theta^{\text{ref}} = [\ \sqrt{2}\ \ -\sqrt{2}\ ]$$

so that, in this case, z' is reflected along the line passing through the origin and the point $(\cos(\pi/8), \sin(\pi/8))$ in order to attain its destination at $[\sqrt{2}\ \ -\sqrt{2}]$ — see again Fig. 34.2.

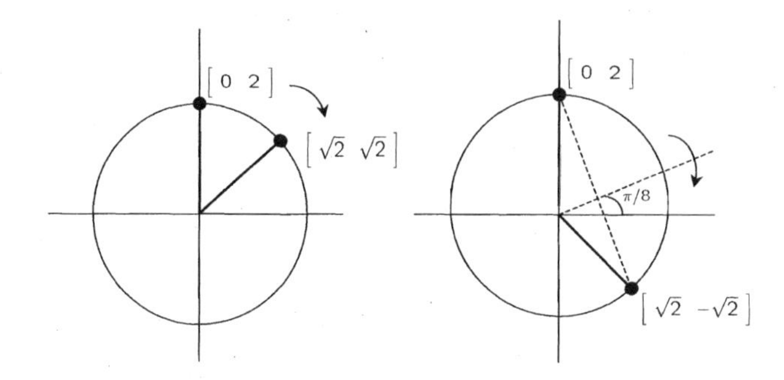

FIGURE 34.2 Rotating the vector $[0\ 2]$ into $[\sqrt{2}\ \sqrt{2}]$ (*left*) vs. reflecting it into $[\sqrt{2}\ -\sqrt{2}]$ across the line passing through the origin at an angle of $\pi/8$ radians (*right*).

Complex data

When the entries of $z = \begin{bmatrix} a & b \end{bmatrix}$ are complex-valued, we should seek a unitary (as opposed to orthogonal) matrix Θ that achieves the transformation (34.1), namely, Θ should now satisfy

$$\Theta\Theta^* = \Theta^*\Theta = \mathrm{I}$$

An expression for such a Θ is

$$\boxed{\Theta = \frac{1}{\sqrt{1+|\rho|^2}}\begin{bmatrix} 1 & -\rho \\ \rho^* & 1 \end{bmatrix} \quad \text{where} \quad \rho = \frac{b}{a}, \quad a \neq 0} \tag{34.5}$$

Let $a = |a|e^{j\phi_a}$ denote the polar representation of a. Then it can be verified that we now obtain

$$\begin{bmatrix} a & b \end{bmatrix}\Theta = \begin{bmatrix} \pm e^{j\phi_a}\sqrt{|a|^2+|b|^2} & 0 \end{bmatrix} \tag{34.6}$$

In other words, the value of α is in general complex and its phase is determined by the phase of a. Again, the choice of the sign depends on whether the value of the square root in (34.5) is chosen to be negative or positive. When $a = 0$, we choose Θ as the permutation matrix (34.3).

Lemma 34.2 (Complex Givens rotation) Consider a 1×2 vector $[a\ b]$ with possibly complex entries. Then choose Θ as in (34.5) to get

$$\begin{bmatrix} a & b \end{bmatrix}\Theta = \pm e^{j\phi_a}\sqrt{|a|^2+|b|^2}\begin{bmatrix} 1 & 0 \end{bmatrix}$$

where ϕ_a denotes the phase of a. If $a = 0$, then choose Θ as in (34.3) to get $[0\ b]\Theta = [b\ 0]$.

Remark 34.1 (Real post-array) Usually, it is desirable to obtain a real-valued (and also positive) α even in the complex case (34.6). This property can be enforced by choosing Θ as

$$\Theta = \frac{e^{-j\phi_a}}{\sqrt{1+|\rho|^2}}\begin{bmatrix} 1 & -\rho \\ \rho^* & 1 \end{bmatrix} \quad \text{where} \quad \rho = \frac{b}{a}, \quad a \neq 0$$

To enforce $\alpha > 0$, we simply choose the plus sign in (34.6). If a already happens to be real-valued, then of course $\phi_a = 0$ and the same Θ as before in (34.5) results in a real α. Note further that, in this case, the diagonal entries of Θ will be real.

When $a = 0$, we choose Θ as

$$\Theta = e^{-j\phi_b}\begin{bmatrix} 0 & -1 \\ 1 & 0 \end{bmatrix}$$

in terms of the phase of b. This remark applies to all our future discussions for complex-valued data; so it will not be repeated again.

◇

Example 34.1 (Using Givens rotations)

Assume we are given a 2×3 pre-array $\mathcal{A}$,

$$\mathcal{A} = \begin{bmatrix} 1 & 0.75 & 0.25 \\ 0.4 & 0.2 & 0.2 \end{bmatrix} \tag{34.7}$$

and that we wish to reduce it to the form

$$\mathcal{A}\Theta = \begin{bmatrix} \times & 0 & 0 \\ \times & \times & 0 \end{bmatrix} \tag{34.8}$$

via a sequence of Givens rotations. This can be obtained, among several possibilities, as follows. We first annihilate the $(1,3)$ entry of $\mathcal{A}$ by pivoting with its $(1,1)$ entry. From the construction (34.2), we know that the orthogonal matrix Θ_1 that achieves this transformation is given by

$$\Theta_1 = \frac{1}{\sqrt{1+\rho_1^2}}\begin{bmatrix} 1 & -\rho_1 \\ \rho_1 & 1 \end{bmatrix} = \begin{bmatrix} 0.9701 & -0.2425 \\ 0.2425 & 0.9701 \end{bmatrix}, \quad \rho_1 = 0.25/1 = 0.25$$

Applying Θ_1 to $\mathcal{A}$, and leaving the second column of $\mathcal{A}$ unchanged, leads to

$$\underbrace{\begin{bmatrix} \boxed{1} & 0.75 & \boxed{0.25} \\ 0.4 & 0.2 & 0.2 \end{bmatrix}}_{\mathcal{A}}\begin{bmatrix} 0.9701 & 0 & -0.2425 \\ 0 & 1 & 0 \\ 0.2425 & 0 & 0.9701 \end{bmatrix} = \underbrace{\begin{bmatrix} 1.0307 & 0.7500 & 0 \\ 0.4365 & 0.2000 & 0.0970 \end{bmatrix}}_{\mathcal{A}_1}$$

We now annihilate the $(1,2)$ entry of the post-array $\mathcal{A}_1$ by pivoting with its $(1,1)$ entry. For this purpose, we choose the orthogonal matrix as

$$\Theta_2 = \frac{1}{\sqrt{1+\rho_2^2}}\begin{bmatrix} 1 & -\rho_2 \\ \rho_2 & 1 \end{bmatrix} = \begin{bmatrix} 0.8086 & -0.5884 \\ 0.5884 & 0.8086 \end{bmatrix}, \quad \rho_2 = \frac{0.7500}{1.0307}$$

Applying Θ_2 to $\mathcal{A}_1$, and leaving the third column of $\mathcal{A}_1$ unchanged, leads to

$$\underbrace{\begin{bmatrix} \boxed{1.0307} & \boxed{0.7500} & 0 \\ 0.4365 & 0.2000 & 0.0970 \end{bmatrix}}_{\mathcal{A}_1}\begin{bmatrix} 0.8086 & -0.5884 & 0 \\ 0.5884 & 0.8086 & 0 \\ 0 & 0 & 1 \end{bmatrix} = \underbrace{\begin{bmatrix} 1.2748 & 0 & 0 \\ 0.4707 & -0.0951 & 0.0970 \end{bmatrix}}_{\mathcal{A}_2}$$

We finally annihilate the $(2,3)$ entry of $\mathcal{A}_2$ by pivoting with its $(2,2)$ entry. One way to achieve this transformation is to use the orthogonal matrix

$$\Theta_3 = \frac{1}{\sqrt{1+\rho_3^2}}\begin{bmatrix} 1 & -\rho_3 \\ \rho_3 & 1 \end{bmatrix} = \begin{bmatrix} 0.7001 & 0.7140 \\ -0.7140 & 0.7001 \end{bmatrix}, \quad \rho_3 = \frac{0.0970}{-0.0951}$$

and to apply it to $\mathcal{A}_2$, without modifying its first column, thus leading to

$$\underbrace{\begin{bmatrix} 1.2748 & 0 & 0 \\ 0.4707 & \boxed{-0.0951} & \boxed{0.0970} \end{bmatrix}}_{\mathcal{A}_2}\begin{bmatrix} 1 & 0 & 0 \\ 0 & 0.7001 & 0.7140 \\ 0 & -0.7140 & 0.7001 \end{bmatrix} = \underbrace{\begin{bmatrix} 1.2748 & 0 & 0 \\ 0.4707 & -0.1359 & 0 \end{bmatrix}}_{\mathcal{A}_3}$$

Clearly, the negative entry -0.1359 could have been replaced by a positive entry 0.1359 had we employed $-\Theta_3$ instead of Θ_3.

A more direct way to achieve the last step is to avoid forming Θ_3 altogether and to simply replace the vector $[-0.0951 \quad 0.0970]$ by $[\alpha \quad 0]$, where α is the norm of the vector, i.e., by $[0.1359 \quad 0]$. In this way, the resulting post-array becomes

$$\begin{bmatrix} 1.2748 & 0 & 0 \\ 0.4707 & 0.1359 & 0 \end{bmatrix}$$

We have therefore determined a succession of three elementary orthogonal transformations that triangularize the original pre-array $\mathcal{A}$. The combined effect of these transformations is to achieve the desired transformation (34.8).

◇

34.2 HOUSEHOLDER TRANSFORMATIONS

In contrast to Givens rotations, Householder transformations can be used to annihilate multiple entries in a row vector at once. We describe them below for both cases of real and complex data.

Real data

Let $e_0 = \begin{bmatrix} 1 & 0 & \dots & 0 \end{bmatrix}$ denote the leading row basis vector in $n-$dimensional Euclidean space, and consider a $1 \times n$ real-valued vector z with entries $\{z(i), i = 0, 1, \dots, n-1\}$. Assume that we wish to transform z to the form

$$\begin{bmatrix} z(0) & z(1) & \dots & z(n-1) \end{bmatrix}\Theta = \alpha\, e_0 \tag{34.9}$$

for some real scalar α to be determined, and where the transformation Θ is required to be both orthogonal and involutary. That is, Θ should satisfy $\Theta\Theta^{\mathsf{T}} = \mathrm{I}$ and $\Theta^2 = \mathrm{I}$ (or, equivalently, $\Theta = \Theta^{\mathsf{T}}$ and $\Theta^2 = \mathrm{I}$). We remark in passing that matrices Q that satisfy $Q^2 = \mathrm{I}$ are called involutary matrices.

Of course, the scalar α cannot be arbitrary and its value can be determined even before determining the expression for a matrix Θ that achieves (34.9). Indeed, note from (34.9) and from the orthogonality of Θ that

$$z\Theta\Theta^{\mathsf{T}}z^{\mathsf{T}} = \|z\|^2 = \alpha^2$$

so that we must have $\alpha = \pm\|z\|$. Both values of α are possible (since if Θ achieves $z\Theta = \|z\|e_0$, then $-\Theta$ is orthogonal and achieves $z\Theta = -\|z\|e_0$). One way to achieve the

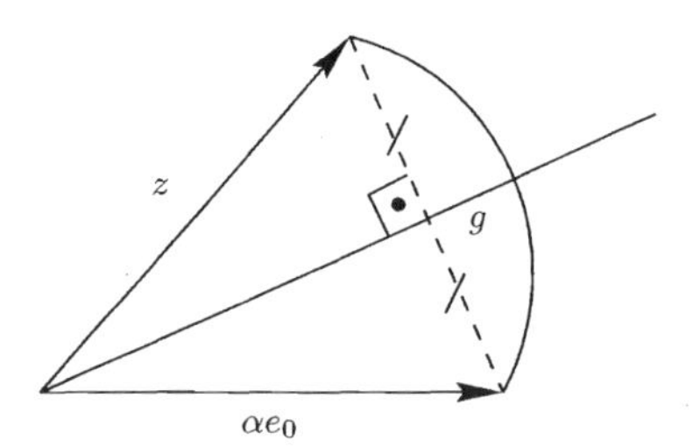

FIGURE 34.3 The vector z is aligned with e_0 by reflecting it across the line that bisects the angle between the sides z and αe_0. This construction provides a geometric interpretation for the Householder transformation.

transformation (34.9) is to employ a Householder reflection. We motivate it by means of a geometric argument.

Thus, refer to Fig. 34.3, which shows the vector z and its destination αe_0. Since $\alpha = \|z\|$, the triangle with sides z and αe_0 is isosceles and we denote its base by

$$g = z - \alpha e_0$$

If we drop a perpendicular from the origin of z to g, it will divide g into two equal parts, with the upper part being the projection of z onto g and is equal to $zg^{\mathsf{T}}\|g\|^{-2}g$.

This means that $g = 2zg^{\mathsf{T}}\|g\|^{-2}g$ and, consequently,

$$\alpha e_0 = z - 2zg^{\mathsf{T}}(gg^{\mathsf{T}})^{-1}g = z \underbrace{\left[\mathrm{I} - 2\frac{g^{\mathsf{T}}g}{gg^{\mathsf{T}}}\right]}_{\Theta} \tag{34.10}$$

where

$$\boxed{\Theta \triangleq \mathrm{I} - 2\frac{g^{\mathsf{T}}g}{gg^{\mathsf{T}}}} \tag{34.11}$$

We thus have a matrix Θ that maps z to αe_0. It is straightforward to verify that this Θ is orthogonal and involutary, as desired. The matrix Θ so defined is called an (elementary) Householder transformation or reflection; it is a reflection since its effect is to reflect z across the line that bisects the angle between the sides z and αe_0 — see Fig. 34.3. Moreover, for the Householder matrix Θ in (34.11), we see that it is a rank-one modification of the identity matrix. Therefore, it has $(n-1)$ eigenvalues at 1 and a single eigenvalue at -1. Therefore, $\det \Theta = -1$, which again confirms that it is a reflector. In summary, we established the following result.

Lemma 34.3 (Real Householder reflection) Consider an $n-$dimensional vector z with real-valued entries. Choose $g = z \pm \|z\|e_0$ and

$$\Theta = \mathrm{I}_n - 2\,\frac{g^{\mathsf{T}}g}{gg^{\mathsf{T}}}$$

to get $z\Theta = \mp\|z\|e_0$. Here, $e_0 = [1\ 0\ 0\ldots\ 0]$ is the first basis vector.

Usually, the sign in the expression for g is chosen to be the same as the sign of the leading entry of z in order to avoid a vector g with small Euclidean norm.

Complex data

More generally, consider a $1 \times n$ vector z with possibly complex entries, and assume that we wish to determine a transformation Θ that transforms it to the form (34.9) with a scalar α that is possibly complex-valued, and where Θ is required to be a Hermitian unitary matrix, i.e., it should satisfy $\Theta = \Theta^*$ and $\Theta^2 = \mathrm{I}$.

Again, the scalar α cannot be arbitrary and its value can be determined even before determining the expression for a matrix Θ that achieves (34.9). Indeed, note from (34.9), and from the unitarity of Θ, that $|\alpha|^2 = \|z\|^2$ so that $|\alpha| = \|z\|$. Moreover, from the equality $z\Theta z^* = \alpha z^*(0)$, and from the fact that $z\Theta z^*$ is real (since Θ is Hermitian), we conclude that $\alpha z^*(0)$ must be real. If we introduce the polar representation of the first entry of z, namely, $z(0) = |a|\ e^{j\phi_a}$, then it follows that α is given by

$$\alpha = \pm\|z\|e^{j\phi_a}$$

The prior geometric construction of Θ can be repeated in the complex case and it leads to the following conclusion.

Lemma 34.4 (Complex Householder reflection) Consider a $1 \times n$ vector z with possibly complex-valued entries. Choose $g = z \pm \|z\|e^{j\phi_a}e_0$ and

$$\Theta = \mathrm{I}_n - 2\,\frac{g^*g}{gg^*}$$

to get $z\Theta = \mp\|z\|e^{j\phi_a}e_0$. Here,$e_0 = [1\ 0\ 0 \ldots\ 0]$ and ϕ_a is the phase of the leading entry of z.

Proof: Besides the geometric argument, we can establish the result of the lemma algebraically as follows. We express g as $g = z + \alpha e_0$, where α is a scalar that satisfies $|\alpha|^2 = \|z\|^2$ and αa^* is real. Then direct calculation shows that

$$\begin{aligned} \|g\|^2 &= 2\|z\|^2 + 2\alpha a^* \\ zg^*g &= z\|z\|^2 + \alpha\|z\|^2 e_0 + \alpha^* az + \alpha(\alpha a^*)e_0 \end{aligned}$$

and, hence,

$$z\Theta = z - \frac{2zg^*g}{\|g\|^2} = -\alpha e_0$$

Example 34.2 (Using Householder transformations)

We re-consider the pre-array of numbers $\mathcal{A}$ in (34.7) and now show how to transform it to the form (34.8) by means of Householder transformations.

Let x_1 denote the top row of $\mathcal{A}$, i.e., $x_1 = \begin{bmatrix} 1 & 0.75 & 0.25 \end{bmatrix}$. Our first step is to annihilate the last two entries of x_1. From expression (34.11), we know that this transformation can be achieved by using the following 3×3 Householder transformation:

$$\Theta_1 = \mathrm{I}_3 - 2\,\frac{g_1^{\mathsf{T}} g_1}{g_1 g_1^{\mathsf{T}}}$$

where $g_1 = x_1 \pm \|x_1\| \begin{bmatrix} 1 & 0 & 0 \end{bmatrix}$. We initially choose the sign in the expression for g_1 to be the same as the sign of the leading entry of x_1, which is positive, so that

$$g_1 = \begin{bmatrix} 1 & 0.75 & 0.25 \end{bmatrix} + \begin{bmatrix} 1.2748 & 0 & 0 \end{bmatrix} = \begin{bmatrix} 2.2748 & 0.1500 & 1.0000 \end{bmatrix}$$

Applying Θ_1 to the two rows of $\mathcal{A}$ gives

$$x_1\Theta_1 = x_1 - 2\frac{x_1 g_1^{\mathsf{T}}}{g_1 g_1^{\mathsf{T}}} g_1 \;=\; \begin{bmatrix} -1.2748 & 0 & 0 \end{bmatrix}$$

and

$$x_2\Theta_1 = x_2 - 2\frac{x_2 g_1^{\mathsf{T}}}{g_1 g_1^{\mathsf{T}}} g_1 \;=\; \begin{bmatrix} -0.4707 & -0.0871 & 0.1043 \end{bmatrix}$$

In other words,

$$\underbrace{\begin{bmatrix} 1 & 0.75 & 0.25 \\ 0.4 & 0.2 & 0.2 \end{bmatrix}}_{\mathcal{A}} \Theta_1 = \underbrace{\begin{bmatrix} -1.2748 & 0 & 0 \\ -0.4707 & -0.0871 & 0.1043 \end{bmatrix}}_{\mathcal{A}_1}$$

Of course, had we chosen the sign in the expression for g_1 to be the negative sign, the signs of all entries in the above post-array would have been switched, say,

$$\underbrace{\begin{bmatrix} 1 & 0.75 & 0.25 \\ 0.4 & 0.2 & 0.2 \end{bmatrix}}_{\mathcal{A}} \Theta_1 = \underbrace{\begin{bmatrix} 1.2748 & 0 & 0 \\ 0.4707 & 0.0871 & -0.1043 \end{bmatrix}}_{\mathcal{A}_1}$$

In order to annihilate the $(2,3)$ entry of $\mathcal{A}_1$ we can replace the vector $[0.0871 \;\; 0.1043]$ by one of the form $[\alpha \;\; 0]$, where α is the norm of the vector, i.e., by $[0.1359 \;\; 0]$. In this way, it is seen that the resulting post-array in (34.8) can be taken as

$$\begin{bmatrix} 1.2748 & 0 & 0 \\ 0.4707 & 0.1359 & 0 \end{bmatrix}$$

CHAPTER **35**

QR and Inverse QR Algorithms

In preparation for our discussions of RLS array methods, we now recall briefly the defining equations of RLS for ease of reference.

Consider a collection of data $\{u_j, d(j)\}_{j=0}^{N}$, where the $\{u_j\}$ are $1 \times M$ and the $\{d(j)\}$ are scalars, in addition to an $M \times 1$ column vector $\bar{w}$, an $M \times M$ positive-definite matrix Π, and a scalar λ satisfying $0 \ll \lambda \leq 1$. Then the solution, w_N, of the regularized least-squares problem

$$\min_{w} \left[\lambda^{(N+1)}(w-\bar{w})^*\Pi(w-\bar{w}) + \sum_{j=0}^{N} \lambda^{N-j}|d(j)-u_j w|^2 \right] \tag{35.1}$$

and the corresponding minimum cost, $\xi(N)$, can be obtained recursively as follows (recall Alg. 30.2). Start with $w_{-1} = \bar{w}$, $P_{-1} = \Pi^{-1}$, $\xi(-1) = 0$, and repeat for $i \geq 0$:

$$\begin{array}{rcl} \gamma(i) & = & 1/(1+\lambda^{-1}u_i P_{i-1} u_i^*) \\ g_i & = & \lambda^{-1}\gamma(i) P_{i-1} u_i^* \\ e(i) & = & d(i) - u_i w_{i-1} \\ w_i & = & w_{i-1} + g_i e(i) \\ P_i & = & \lambda^{-1} P_{i-1} - g_i g_i^*/\gamma(i) \\ r(i) & = & d(i) - u_i w_i \\ \xi(i) & = & \lambda\xi(i-1) \ + \ r(i)e^*(i) \end{array} \tag{35.2}$$

Moreover, the following relations also hold at each iteration i:

$$\gamma(i) = 1 - u_i P_i u_i^*, \qquad g_i = P_i u_i^* \tag{35.3}$$

We further remarked following Alg. 30.2 that the $\{w_i\}$ also satisfy the following construction. Start with $w_{-1} = \bar{w}$, $s_{-1} = 0$, $\Phi_{-1} = \Pi$, and repeat for $i \geq 0$:

$$\Phi_i = \lambda\Phi_{i-1} + u_i^* u_i, \qquad s_i = \lambda s_{i-1} + u_i^*[d(i) - u_i\bar{w}] \tag{35.4}$$

Then, at each iteration i, it holds that

$$\Phi_i[w_i - \bar{w}] = s_i \tag{35.5}$$

$$\gamma(i) = 1 - u_i \Phi_i^{-1} u_i^* \tag{35.6}$$

because, by the definition (30.30), $\Phi_i = P_i^{-1}$. Equations (35.4)–(35.6) will be significant in this chapter in that they will form the basis for one of the most celebrated array variants of RLS (see Sec. 35.2).

35.1 INVERSE QR ALGORITHM

The first array algorithm we derive is known as the inverse QR algorithm (and also as the square-root RLS algorithm). It is based on the observation that the expressions for $\{\gamma^{-1}(i), g_i\}$ in (35.2) can be put into the desirable forms (33.19) and (33.20). Indeed, note that

$$\left\{\begin{array}{lll} \gamma^{-1}(i) & = & 1+\lambda^{-1}u_iP_{i-1}u_i^* \\ g_i\gamma^{-1}(i) & = & \lambda^{-1}P_{i-1}u_i^* \end{array}\right. \tag{35.7}$$

Comparing with (33.19) and (33.20) we see that we can make the identifications:

$$\left\{\begin{array}{lll} C\leftarrow\gamma^{-1/2}(i) & A\leftarrow 1 & B\leftarrow\lambda^{-1/2}u_iP_{i-1}^{1/2} \\ F\leftarrow g_i\gamma^{-1/2}(i) & D\leftarrow 0 & E\leftarrow\lambda^{-1/2}P_{i-1}^{1/2} \end{array}\right.$$

where $P_{i-1}^{1/2}$ denotes the Cholesky factor of P_{i-1}. In other words, the expression for $\gamma^{-1}(i)$ in (35.7) corresponds to the norm preservation identity (remember that $\gamma(i)$ is a real variable even for complex data):

$$\left[\begin{array}{cc} \gamma^{-1/2}(i) & 0 \end{array}\right]\left[\begin{array}{c} \gamma^{-1/2}(i) \\ 0 \end{array}\right] = \left[\begin{array}{cc} 1 & \lambda^{-1/2}u_iP_{i-1}^{1/2} \end{array}\right]\left[\begin{array}{c} 1 \\ \lambda^{-1/2}P_{i-1}^{*/2}u_i^* \end{array}\right]$$

whereas the expression for $g_i\gamma^{-1}(i)$ in (35.7) corresponds to the inner-product preservation identity:

$$\left[\begin{array}{cc} g_i\gamma^{-1/2}(i) & \times \end{array}\right]\left[\begin{array}{c} \gamma^{-1/2}(i) \\ 0 \end{array}\right] = \left[\begin{array}{cc} 0 & \lambda^{-1/2}P_{i-1}^{1/2} \end{array}\right]\left[\begin{array}{c} 1 \\ \lambda^{-1/2}P_{i-1}^{*/2}u_i^* \end{array}\right]$$

Therefore, motivated by the discussion that led to (33.21) from (33.19)–(33.20), we let Θ_i be a unitary matrix that transforms the pre-array

$$\mathcal{A} = \left[\begin{array}{cc} 1 & \lambda^{-1/2}u_iP_{i-1}^{1/2} \\ 0 & \lambda^{-1/2}P_{i-1}^{1/2} \end{array}\right]$$

to lower triangular form, say,

$$\mathcal{B} = \left[\begin{array}{cc} \mathsf{x} & 0 \\ \mathsf{y} & \mathsf{Z} \end{array}\right]$$

for some variables $\{\mathsf{x},\mathsf{y},\mathsf{Z}\}$ to be determined, i.e.,

$$\left[\begin{array}{cc} 1 & \lambda^{-1/2}u_iP_{i-1}^{1/2} \\ 0 & \lambda^{-1/2}P_{i-1}^{1/2} \end{array}\right]\Theta_i = \left[\begin{array}{cc} \mathsf{x} & 0 \\ \mathsf{y} & \mathsf{Z} \end{array}\right] \tag{35.8}$$

where x is a scalar, y is a vector, and Z is a lower-triangular matrix with positive-diagonal entries. Observe that the pre- and post-arrays in (35.8) have the forms (assuming $M=3$):

$$\left[\begin{array}{c|ccc} 1 & \times & \times & \times \\ \hline 0 & \times & 0 & 0 \\ 0 & \times & \times & 0 \\ 0 & \times & \times & \times \end{array}\right] \quad \text{and} \quad \left[\begin{array}{c|ccc} \times & 0 & 0 & 0 \\ \hline \times & \times & 0 & 0 \\ \times & \times & \times & 0 \\ \times & \times & \times & \times \end{array}\right]$$

respectively. We already know from the explanation in Sec. 33.4 that the values of x and y in (35.8) should be

$$\mathsf{x} = \gamma^{-1/2}(i), \qquad \mathsf{y} = g_i\gamma^{-1/2}(i)$$

Alternatively, we can identify the values of $\{\mathsf{x},\mathsf{y}\}$, and also the value of Z, by "squaring" both sides of (35.8) to obtain the equality:

$$\begin{bmatrix} 1 & \lambda^{-1/2}u_iP_{i-1}^{1/2} \\ 0 & \lambda^{-1/2}P_{i-1}^{1/2} \end{bmatrix}\begin{bmatrix} 1 & \lambda^{-1/2}u_iP_{i-1}^{1/2} \\ 0 & \lambda^{-1/2}P_{i-1}^{1/2} \end{bmatrix}^* = \begin{bmatrix} \mathsf{x} & 0 \\ \mathsf{y} & \mathsf{Z} \end{bmatrix}\begin{bmatrix} \mathsf{x} & 0 \\ \mathsf{y} & \mathsf{Z} \end{bmatrix}^*$$

Comparing terms on both sides we find that the following equalities must hold:

$$\begin{cases} |\,\mathsf{x}\,|^2 &= 1+\lambda^{-1}u_iP_{i-1}u_i^* \\ \mathsf{y}\mathsf{x}^* &= \lambda^{-1}P_{i-1}u_i^* \\ \mathsf{y}\mathsf{y}^*+\mathsf{Z}\mathsf{Z}^* &= \lambda^{-1}P_{i-1} \end{cases}$$

From the first equation we get $|\,\mathsf{x}\,|^2 = \gamma^{-1}(i)$ so that $\mathsf{x} = \gamma^{-1/2}(i)$. From the second equation we get $\mathsf{y} = g_i\gamma^{-1/2}(i)$, and from the third equation we get

$$\mathsf{Z}\mathsf{Z}^* \;=\; \lambda^{-1}P_{i-1}-\mathsf{y}\mathsf{y}^* \;=\; \lambda^{-1}P_{i-1}-g_ig_i^*/\gamma(i) \;=\; P_i$$

where the last equality follows from recursion (35.2). Therefore, since Z is lower triangular with positive diagonal entries, we conclude that Z is the Cholesky factor of P_i. We thus find that, in addition to the quantities $\{\gamma^{-1}(i), g_i\gamma^{-1}(i)\}$, the array algorithm (35.8) also provides the square-root factor $P_i^{1/2}$, which is needed to form the pre-array for the next time instant. In this way, we obtain a self-contained array method where the quantities that are needed to form the pre-array are obtained in the post-array and the procedure can be repeated. In summary, we arrive at the following algorithm.

Algorithm 35.1 (Inverse QR) Consider data $\{u_j, d(j)\}_{j=0}^N$, where the u_j are $1\times M$ and the $d(j)$ are scalars. Consider also an $M\times 1$ vector $\bar{w}$, an $M\times M$ positive-definite matrix Π, and a scalar $0 \ll \lambda \le 1$. The solution, w_N, of the least-squares problem (35.1) can be computed recursively as follows.

Let $\Sigma = \Pi^{-1}$ and introduce the Cholesky decomposition $\Sigma = \Sigma^{1/2}\Sigma^{*/2}$, where $\Sigma^{1/2}$ is lower triangular with positive-diagonal entries. Then start with $w_{-1} = \bar{w}$, $P_{-1}^{1/2} = \Sigma^{1/2}$, and repeat for $i \ge 0$.

1. Find a unitary matrix Θ_i that lower triangularizes the pre-array shown below and generates a post-array with positive diagonal entries. Then the entries in the post-array will correspond to

$$\begin{bmatrix} 1 & \lambda^{-1/2}u_iP_{i-1}^{1/2} \\ 0 & \lambda^{-1/2}P_{i-1}^{1/2} \end{bmatrix}\Theta_i = \begin{bmatrix} \gamma^{-1/2}(i) & 0 \\ g_i\gamma^{-1/2}(i) & P_i^{1/2} \end{bmatrix}$$

2. Update the weight vector as

$$w_i = w_{i-1} + \left[g_i\gamma^{-1/2}(i)\right]\left[\gamma^{-1/2}(i)\right]^{-1}[d(i)-u_iw_{i-1}]$$

where the quantities $\{g_i\gamma^{-1/2}(i), \gamma^{-1/2}(i)\}$ are read from the post-array.

The computational complexity of this algorithm is $O(M^2)$ operations per iteration.

Remark 35.1 (Terminology) The above array algorithm is known as the inverse QR method in the adaptive filtering literature for the following reason. The qualification *inverse* refers to the fact that $P_i^{1/2}$ is a square-root factor of the *inverse* of Φ_i in (35.4). Just note that since, by definition, $\Phi_i = P_i^{-1}$ and $P_i = P_i^{1/2}P_i^{*/2}$, we have $\Phi_i^{-1} = P_i^{1/2}P_i^{*/2}$. The qualification QR, on the other hand, is because this array method relates to the QR decomposition of a matrix, as explained in the third remark below. The algorithm is also known as square-root RLS since it propagates the square-root factor $P_i^{1/2}$; this latter terminology is borrowed from Kalman filtering — see Prob. VIII.12.

◇

Remark 35.2 (Reliability) By propagating a square-root factor of P_i, rather than P_i itself as in a plain RLS implementation, the danger of having P_i lose its positive-definiteness due to numerical inaccuracies is essentially eliminated. If needed, P_i can be recovered by squaring, i.e., via $P_i = P_i^{1/2}P_i^{*/2}$. However, this step is not necessary since the array method already evaluates the gain vector g_i and, moreover, the entries of its post-array contain everything we need in order to proceed to the next iteration.

◇

Remark 35.3 (Relation to the QR decomposition) If we conjugate-transpose the array equations of Alg. 35.1 we get

$$\begin{bmatrix} 1 & 0 \\ \lambda^{-1/2}P_{i-1}^{*/2}u_i^* & \lambda^{-1/2}P_{i-1}^{*/2} \end{bmatrix} = \Theta_i \begin{bmatrix} \gamma^{-1/2}(i) & g_i^*\gamma^{-1/2}(i) \\ 0 & P_i^{*/2} \end{bmatrix}$$

where the rightmost array is upper triangular with positive diagonal entries. Since Θ_i is unitary, this way of expressing the array equations amounts to a QR decomposition (cf. Sec. B.5) of the matrix

$$\begin{bmatrix} 1 & 0 \\ \lambda^{-1/2}P_{i-1}^{*/2}u_i^* & \lambda^{-1/2}P_{i-1}^{*/2} \end{bmatrix} \tag{35.9}$$

where the matrix Θ_i plays the role of the Q factor (recall the defining expression (B.9)). This observation shows that an alternative way of implementing the array algorithm is by performing the QR decomposition of the matrix (35.9).

◇

Remark 35.4 (Further motivation) The inverse QR algorithm could have been alternatively derived as follows. Starting from the equations for $\{g_i, \gamma(i), P_i\}$ in (35.2), we note that they can be combined together in factored form as follows:

$$\begin{bmatrix} 1 & \lambda^{-1/2}u_iP_{i-1}^{1/2} \\ 0 & \lambda^{-1/2}P_{i-1}^{1/2} \end{bmatrix} \begin{bmatrix} 1 & 0 \\ \lambda^{-1/2}P_{i-1}^{*/2}u_i^* & \lambda^{-1/2}P_{i-1}^{*/2} \end{bmatrix} =$$

$$\begin{bmatrix} \gamma^{-1/2}(i) & 0 \\ g_i\gamma^{-1/2}(i) & P_i^{1/2} \end{bmatrix} \begin{bmatrix} \gamma^{-1/2}(i) & g_i^*\gamma^{-1/2}(i) \\ 0 & P_i^{*/2} \end{bmatrix}$$

To verify that this is indeed the case, simply expand both sides and compare terms. Now this equality fits precisely into the statement of Lemma 33.1 by choosing

$$A = \begin{bmatrix} 1 & \lambda^{-1/2}u_iP_{i-1}^{1/2} \\ 0 & \lambda^{-1/2}P_{i-1}^{1/2} \end{bmatrix} \quad \text{and} \quad B = \begin{bmatrix} \gamma^{-1/2}(i) & 0 \\ g_i\gamma^{-1/2}(i) & P_i^{1/2} \end{bmatrix}$$

so that there should exist a unitary matrix Θ_i that takes A to B.

35.2 QR ALGORITHM

The second array algorithm we consider is known as the QR algorithm (and also as the square-root information RLS algorithm). It follows from the observation that the recursions for $\{\Phi_i, s_i\}$ in (35.4) can be put into the desirable forms (33.19) and (33.20).

Indeed, let $\Phi_i^{1/2}$ denote the Cholesky factor of Φ_i, and introduce the auxiliary signals

$$\boxed{\bar{d}(i) \stackrel{\Delta}{=} d(i) - u_i \bar{w}, \qquad \bar{w}_i \stackrel{\Delta}{=} w_i - \bar{w}} \tag{35.10}$$

as well as the auxiliary column vector

$$\boxed{q_i \stackrel{\Delta}{=} \Phi_i^{*/2}[w_i - \bar{w}]} \tag{35.11}$$

These auxiliary signals would reduce to $d(i)$ and $\Phi^{*/2} w_i$, respectively, when $\bar{w} = 0$, which is usually the case. We consider $\bar{w}$ for generality.

From the equation for w_i in (35.5), it is seen that q_i satisfies

$$\boxed{\Phi_i^{1/2} q_i = s_i} \tag{35.12}$$

With $\{\bar{d}(i), \bar{w}_i, q_i\}$ so defined, we can rewrite the recursions in (35.4) as

$$\left\{ \begin{array}{lcl} \Phi_i^{1/2}\Phi_i^{*/2} & = & \lambda \Phi_{i-1}^{1/2}\Phi_{i-1}^{*/2} + u_i^* u_i \\ \Phi_i^{1/2} q_i & = & \lambda \Phi_{i-1}^{1/2} q_{i-1} + u_i^* \bar{d}(i) \end{array} \right. \tag{35.13}$$

If we conjugate the second equation, we get

$$\left\{ \begin{array}{lcl} \Phi_i^{1/2}\Phi_i^{*/2} & = & \lambda \Phi_{i-1}^{1/2}\Phi_{i-1}^{*/2} + u_i^* u_i \\ q_i^* \Phi_i^{*/2} & = & \lambda q_{i-1}^* \Phi_{i-1}^{*/2} + \bar{d}^*(i) u_i \end{array} \right. \tag{35.14}$$

Comparing with (33.19) and (33.20) we see that we can make the identifications:

$$\left\{ \begin{array}{lll} C \leftarrow \Phi_i^{1/2} & A \leftarrow \lambda^{1/2}\Phi_{i-1}^{1/2} & B \leftarrow u_i^* \\ F \leftarrow q_i^* & D \leftarrow \lambda^{1/2} q_{i-1}^* & E \leftarrow \bar{d}^*(i) \end{array} \right.$$

In other words, the expression for Φ_i in (35.14) corresponds to the norm preservation identity:

$$\left[\begin{array}{cc} \Phi_i^{1/2} & 0 \end{array} \right] \left[\begin{array}{c} \Phi_i^{*/2} \\ 0 \end{array} \right] = \left[\begin{array}{cc} \lambda^{1/2}\Phi_{i-1}^{1/2} & u_i^* \end{array} \right] \left[\begin{array}{c} \lambda^{1/2}\Phi_{i-1}^{*/2} \\ u_i \end{array} \right] \tag{35.15}$$

whereas the expression for s_i in (35.14) corresponds to the inner-product preservation identity:

$$\left[\begin{array}{cc} q_i^* & \times \end{array} \right] \left[\begin{array}{c} \Phi_i^{*/2} \\ 0 \end{array} \right] = \left[\begin{array}{cc} \lambda^{1/2} q_{i-1}^* & \bar{d}^*(i) \end{array} \right] \left[\begin{array}{c} \lambda^{1/2}\Phi_{i-1}^{*/2} \\ u_i \end{array} \right] \tag{35.16}$$

Therefore, motivated again by the discussion that led to (33.21) from (33.19)–(33.20), we let Θ_i be a unitary matrix that transforms the pre-array

$$\mathcal{A} = \left[\begin{array}{cc} \lambda^{1/2}\Phi_{i-1}^{1/2} & u_i^* \\ \lambda^{1/2} q_{i-1}^* & \bar{d}^*(i) \end{array} \right]$$

to lower triangular form, say,

$$\mathcal{B} = \begin{bmatrix} \mathsf{X} & 0 \\ \mathsf{y} & \mathsf{z} \end{bmatrix}$$

for some variables $\{\mathsf{X}, \mathsf{y}, \mathsf{z}\}$ to be determined, with X lower triangular with positive diagonal entries, y a row vector, and z a scalar, i.e.,

$$\begin{bmatrix} \lambda^{1/2}\Phi_{i-1}^{1/2} & u_i^* \\ \lambda^{1/2}q_{i-1}^* & \bar{d}^*(i) \end{bmatrix} \Theta_i = \begin{bmatrix} \mathsf{X} & 0 \\ \mathsf{y} & \mathsf{z} \end{bmatrix} \tag{35.17}$$

Observe that the pre- and post-arrays in (35.17) have the forms (assuming $M = 3$):

$$\left[\begin{array}{ccc|c} \times & 0 & 0 & \times \\ \times & \times & 0 & \times \\ \times & \times & \times & \times \\ \hline \times & \times & \times & \times \end{array}\right] \quad \text{and} \quad \left[\begin{array}{ccc|c} \times & 0 & 0 & 0 \\ \times & \times & 0 & 0 \\ \times & \times & \times & 0 \\ \hline \times & \times & \times & \times \end{array}\right]$$

respectively. Therefore, the purpose of Θ_i is to introduce the zeros in the last column and to generate a lower triangular post-array with positive diagonal entries. This can be achieved, for example, by using Givens rotations and pivoting with the diagonal entries of $\lambda^{1/2}\Phi_{i-1}^{1/2}$ — as explained in Chapter 34.

Continuing with the array description (35.17), we already know from the explanation in Sec. 33.4 that the values of X and y in (35.17) should be $\mathsf{X} = \Phi_i^{1/2}$ and $\mathsf{y} = q_i^*$. Alternatively, we can identify the values of $\{\mathsf{X}, \mathsf{y}\}$, and also z, by "squaring" both sides of (35.17) to obtain the equality:

$$\begin{bmatrix} \lambda^{1/2}\Phi_{i-1}^{1/2} & u_i^* \\ \lambda^{1/2}q_{i-1}^* & \bar{d}^*(i) \end{bmatrix} \begin{bmatrix} \lambda^{1/2}\Phi_{i-1}^{1/2} & u_i^* \\ \lambda^{1/2}q_{i-1}^* & \bar{d}^*(i) \end{bmatrix}^* = \begin{bmatrix} \mathsf{X} & 0 \\ \mathsf{y} & \mathsf{z} \end{bmatrix} \begin{bmatrix} \mathsf{X} & 0 \\ \mathsf{y} & \mathsf{z} \end{bmatrix}^*$$

Then comparing terms on both sides we find that the following equalities must hold:

$$\left\{ \begin{array}{rcl} \mathsf{XX}^* & = & \lambda\Phi_{i-1} + u_i^* u_i \\ \mathsf{yX}^* & = & \lambda s_{i-1}^* + u_i \bar{d}^*(i) \\ \mathsf{yy}^* + \mathsf{zz}^* & = & \lambda\|q_{i-1}\|^2 + |\bar{d}(i)|^2 \end{array} \right.$$

From the first equation we get $\mathsf{XX}^* = \Phi_i$. But since X is lower triangular with positive diagonal entries, we conclude that X is the Cholesky factor of Φ_i, namely,

$$\mathsf{X} = \Phi_i^{1/2}$$

From the second equation we get $\mathsf{y}\Phi_i^{*/2} = s_i^*$ so that, from (35.12),

$$\mathsf{y} = q_i^*$$

Thus, so far we have established that the array algorithm (35.17) leads to a recursion of the form

$$\begin{bmatrix} \lambda^{1/2}\Phi_{i-1}^{1/2} & u_i^* \\ \lambda^{1/2}q_{i-1}^* & \bar{d}^*(i) \end{bmatrix} \Theta_i = \begin{bmatrix} \Phi_i^{1/2} & 0 \\ q_i^* & \mathsf{z} \end{bmatrix} \tag{35.18}$$

Note in particular that the quantities $\{q_i, \Phi_i^{1/2}\}$ that are needed to form the next pre-array are available in the post-array and, hence, the procedure can be continued. However, it is useful to persist on a complete characterization of the variable z. The identification of z requires more effort.

From the third equation we find that the scalar z satisfies

$$\begin{aligned}\mathsf{zz}^* &= \lambda\|q_{i-1}\|^2 + |\bar{d}(i)|^2 - \|\mathsf{y}\|^2 \\ &= \lambda\|q_{i-1}\|^2 + |\bar{d}(i)|^2 - \|q_i\|^2\end{aligned}$$

Substituting q_i and q_{i-1} by their definitions (35.11), namely,

$$q_i = \Phi_i^{*/2}\bar{w}_i, \qquad q_{i-1} = \Phi_{i-1}^{*/2}\bar{w}_{i-1}$$

we obtain

$$\mathsf{zz}^* = \lambda\bar{w}_{i-1}^*\Phi_{i-1}\bar{w}_{i-1} - \bar{w}_i^*\Phi_i\bar{w}_i + |\bar{d}(i)|^2$$

or, equivalently,

$$\mathsf{zz}^* = \lambda\bar{w}_{i-1}^* s_{i-1} - s_i^*\bar{w}_i + |\bar{d}(i)|^2$$

since, from (35.4), $\Phi_i\bar{w}_i = s_i$ and $\Phi_{i-1}\bar{w}_{i-1} = s_{i-1}$. Using the time-update relation for s_i from (35.4) we find that

$$\begin{aligned}\mathsf{zz}^* &= \lambda(\bar{w}_{i-1}^* s_{i-1} - s_{i-1}^*\bar{w}_i) \;+\; \bar{d}^*(i)[\bar{d}(i) - u_i\bar{w}_i] \\ &= \lambda(\bar{w}_{i-1}^* s_{i-1} - s_{i-1}^*\bar{w}_i) \;+\; \bar{d}^*(i)[d(i) - u_i w_i] \\ &= \lambda(\bar{w}_{i-1}^* s_{i-1} - s_{i-1}^*\bar{w}_i) \;+\; \bar{d}^*(i)r(i)\end{aligned} \tag{35.19}$$

in terms of the *a posteriori* error,

$$r(i) = d(i) - u_i w_i$$

However, from

$$\Phi_{i-1}\bar{w}_{i-1} = s_{i-1}$$

we have

$$\lambda s_{i-1}^*\bar{w}_i = \lambda\bar{w}_{i-1}^*\Phi_{i-1}\bar{w}_i$$

so that

$$\begin{aligned}\lambda(\bar{w}_{i-1}^* s_{i-1} - s_{i-1}^*\bar{w}_i) &= \bar{w}_{i-1}^*[\lambda s_{i-1} - \lambda\Phi_{i-1}\bar{w}_i] \\ &= \bar{w}_{i-1}^*[\lambda s_{i-1} - (\Phi_i - u_i^* u_i)\bar{w}_i] \\ &= \bar{w}_{i-1}^*[\lambda s_{i-1} - s_i + u_i^* u_i\bar{w}_i] \\ &= \bar{w}_{i-1}^*[-u_i^*\bar{d}(i) + u_i^* u_i\bar{w}_i] \\ &= -\bar{w}_{i-1}^* u_i^*[\bar{d}(i) - u_i\bar{w}_i] \\ &= -\bar{w}_{i-1}^* u_i^* r(i)\end{aligned}$$

Substituting into (35.19) we find that

$$\begin{aligned}\mathsf{zz}^* &= [\bar{d}(i) - u_i\bar{w}_{i-1}]^* r(i) \\ &= [d(i) - u_i w_{i-1}]^* r(i)\end{aligned}$$

or, equivalently,

$$\mathsf{zz}^* = |\mathsf{z}|^2 = e^*(i)r(i) = |e(i)|^2\gamma(i) \tag{35.20}$$

in terms of the *a priori* error

$$e(i) = d(i) - u_i w_{i-1}$$

and the conversion factor $\gamma(i)$. Therefore, from the arguments presented to this point, we can only identify the magnitude of z, namely,

$$|\mathsf{z}| = |e(i)|\gamma^{1/2}(i) \tag{35.21}$$

More information is needed in order to identify the phase of z. To do so, we first note that the arrays (35.18) can be expanded in order to allow for the evaluation of the conversion factor as well. Specifically, by incorporating the row vector $[0 \quad 1]$ into the pre-array in (35.18), we obtain an array description of the form:

$$\begin{bmatrix} \lambda^{1/2}\Phi_{i-1}^{1/2} & u_i^* \\ \lambda^{1/2}q_{i-1}^* & \bar{d}^*(i) \\ 0 & 1 \end{bmatrix} \Theta_i = \begin{bmatrix} \Phi_i^{1/2} & 0 \\ q_i^* & \mathsf{z} \\ \mathsf{t} & \mathsf{s} \end{bmatrix} \tag{35.22}$$

for some row t and scalar s to be identified. Clearly, the value of s agrees with the rightmost diagonal entry of Θ_i, and this entry can be enforced to be positive — see Prob. VIII.6.

Equating the inner product of the top and last rows on both sides of (35.22) we get

$$\begin{bmatrix} 0 & 1 \end{bmatrix} \begin{bmatrix} \lambda^{1/2}\Phi_{i-1}^{*/2} \\ u_i \end{bmatrix} = \begin{bmatrix} \mathsf{t} & \mathsf{s} \end{bmatrix} \begin{bmatrix} \Phi_i^{*/2} \\ 0 \end{bmatrix}$$

or, equivalently, $u_i = \mathsf{t}\Phi_i^{*/2}$, so that

$$\mathsf{t} = u_i\Phi_i^{-*/2}$$

Likewise, equating the norms of the last rows on both sides of (35.22) we get

$$\begin{bmatrix} 0 & 1 \end{bmatrix} \begin{bmatrix} 0 \\ 1 \end{bmatrix} = \begin{bmatrix} \mathsf{t} & \mathsf{s} \end{bmatrix} \begin{bmatrix} \mathsf{t}^* \\ \mathsf{s}^* \end{bmatrix}$$

so that, from the just derived value for t,

$$1 = u_i\Phi_i^{-1}u_i^* + |\mathsf{s}|^2$$

Using expression (35.6) for $\gamma(i)$, and the fact that s is positive, we can identify s as

$$\mathsf{s} = \gamma^{1/2}(i)$$

Finally, equating the inner product of the last two rows in (35.22) we get

$$\begin{bmatrix} 0 & 1 \end{bmatrix} \begin{bmatrix} \lambda^{1/2}q_{i-1} \\ \bar{d}(i) \end{bmatrix} = \begin{bmatrix} \mathsf{t} & \mathsf{s} \end{bmatrix} \begin{bmatrix} q_i \\ \mathsf{z}^* \end{bmatrix}$$

i.e.,

$$\begin{aligned} \bar{d}(i) &= \left(u_i\Phi_i^{-*/2}\right) q_i + \mathsf{s}\mathsf{z}^* \\ &\overset{(35.12)}{=} u_i[w_i - \bar{w}] + \gamma^{1/2}(i)\mathsf{z}^* \end{aligned}$$

or, equivalently, by using $\bar{d}(i) = d(i) - u_i\bar{w}$,

$$d(i) - u_iw_i = \gamma^{1/2}(i)\mathsf{z}^* = r(i) \tag{35.23}$$

which allows us to identify z as

$$\mathsf{z} = e^*(i)\gamma^{1/2}(i)$$

This expression is of course consistent with (35.21).

Algorithm 35.2 (QR algorithm) Consider data $\{u_j, d(j)\}_{j=0}^N$, where the u_j are $1 \times M$ and the $d(j)$ are scalars. Consider also an $M \times 1$ vector $\bar{w}$, an $M \times M$ positive-definite matrix Π, and a scalar $0 \ll \lambda \leq 1$. Let $\bar{d}(i) = d(i) - u_i\bar{w}$. The solution, w_N, of the least-squares problem (35.1) can be computed recursively as follows.

Start with $\Phi_{-1}^{1/2} = \Pi^{1/2}$, $q_{-1} = 0$, and repeat for $i \geq 0$.

1. Find a unitary matrix Θ_i that lower triangularizes the pre-array shown below and generates a post-array with positive diagonal entries in $\Phi_i^{1/2}$, as well as a positive rightmost corner entry in the last row of the post-array. The entries in the post-array will then correspond to

$$\begin{bmatrix} \lambda^{1/2}\Phi_{i-1}^{1/2} & u_i^* \\ \lambda^{1/2}q_{i-1}^* & \bar{d}^*(i) \\ 0 & 1 \end{bmatrix} \Theta_i = \begin{bmatrix} \Phi_i^{1/2} & 0 \\ q_i^* & e^*(i)\gamma^{1/2}(i) \\ u_i\Phi_i^{-*/2} & \gamma^{1/2}(i) \end{bmatrix}$$

2. Obtain w_i by solving the triangular system of equations $\Phi_i^{*/2}[w_i - \bar{w}] = q_i$, where the quantities $\{\Phi_i^{*/2}, q_i\}$ are read from the post-array.

The computational complexity of this algorithm is $O(M^2)$ operations per iteration.

This algorithm is also sometimes referred to as the square-root information RLS algorithm; a terminology that is borrowed from Kalman filtering — see Prob. VIII.13.

35.3 EXTENDED QR ALGORITHM

The QR algorithm of the previous section determines the weight vector w_i by solving the triangular linear system of equations $\Phi_i^{*/2}[w_i - \bar{w}] = q_i$. An alternative procedure that avoids this step can be obtained by expanding the array equations even further. Starting from the RLS update equation (cf. (35.2)):

$$w_i = w_{i-1} + g_i e(i) \tag{35.24}$$

and writing it as

$$\Phi_i^{-*/2}q_i = \Phi_{i-1}^{-*/2}q_{i-1} + g_i e(i) \tag{35.25}$$

we find that (35.25) justifies the inner product equality

$$\begin{bmatrix} \lambda^{-1/2}\Phi_{i-1}^{-*/2} & 0 \end{bmatrix} \begin{bmatrix} \lambda^{1/2}q_{i-1} \\ \bar{d}(i) \end{bmatrix} = \begin{bmatrix} \Phi_i^{-*/2} & -g_i\gamma^{-1/2}(i) \end{bmatrix} \begin{bmatrix} q_i \\ e(i)\gamma^{1/2}(i) \end{bmatrix}$$

This equality suggests that we can extend the array equations of Alg. 35.2 as follows:

$$\begin{bmatrix} \lambda^{1/2}\Phi_{i-1}^{1/2} & u_i^* \\ \lambda^{1/2}q_{i-1}^* & d^*(i) \\ 0 & 1 \\ \lambda^{-1/2}\Phi_{i-1}^{-*/2} & 0 \end{bmatrix} \Theta_i = \begin{bmatrix} \Phi_i^{1/2} & 0 \\ q_i^* & e^*(i)\gamma^{1/2}(i) \\ u_i\Phi_i^{-*/2} & \gamma^{1/2}(i) \\ \mathsf{R} & \mathsf{m} \end{bmatrix} \tag{35.26}$$

for some matrix R and column vector m to be determined. As usual, by equating the inner products of the first and last rows on both sides of (35.26) we get $\mathsf{I} = \mathsf{R}\Phi_i^{*/2}$ so that $\mathsf{R} = \Phi_i^{-*/2}$. Likewise, equating the inner product of the third and fourth rows on both sides of (35.26) we see that we must have $0 = \Phi_i^{-1}u_i^* + \mathsf{m}\gamma^{1/2}(i)$. But since $\Phi_i^{-1} = P_i$ and $P_i u_i^* = g_i$, we get

$$\mathsf{m} = -g_i\gamma^{-1/2}(i) \tag{35.27}$$

In summary, we are led to the following algorithm.

Algorithm 35.3 (Extended QR) Consider the same setting of Alg. 35.2. The solution w_N can be recursively computed as follows. Start with $\Phi_{-1}^{1/2} = \Pi^{1/2}$, $\Phi_{-1}^{-*/2} = \Pi^{-*/2}$, $q_{-1} = 0$, and repeat for $i \geq 0$.

1. Find a unitary matrix Θ_i that lower triangularizes the pre-array shown below and generates a post-array with positive diagonal entries in $\Phi_i^{1/2}$, as well as a positive rightmost entry in the third (block) row of the post-array (the entry corresponding to $\gamma^{1/2}(i)$). The entries in the post-array will then correspond to

$$\begin{bmatrix} \lambda^{1/2}\Phi_{i-1}^{1/2} & u_i^* \\ \lambda^{1/2}q_{i-1}^* & d^*(i) \\ 0 & 1 \\ \lambda^{-1/2}\Phi_{i-1}^{-*/2} & 0 \end{bmatrix} \Theta_i = \begin{bmatrix} \Phi_i^{1/2} & 0 \\ q_i^* & e^*(i)\gamma^{1/2}(i) \\ u_i\Phi_i^{-*/2} & \gamma^{1/2}(i) \\ \Phi_i^{-*/2} & -g_i\gamma^{-1/2}(i) \end{bmatrix}$$

2. Obtain w_i recursively via

$$w_i = w_{i-1} - \left(-g_i\gamma^{-1/2}(i)\right) \cdot \left(e^*(i)\gamma^{1/2}(i)\right)^*$$

The computational complexity of this algorithm is still $O(M^2)$ operations per iteration.

Observe that the last row in the above array equations resembles the last row of the inverse QR method of Alg. 35.2. However, since the pre- and post-arrays in the extended QR array algorithm propagate both $\Phi_i^{1/2}$ and its inverse, this method can face numerical difficulties in finite-precision implementations.

35.A APPENDIX: ARRAY KALMAN FILTERS

The idea of using array algorithms to propagate square-root factors, rather than the matrices themselves, has originated in the context of Kalman filtering. The arguments used in the body of the chapter to derive array methods for recursive least-squares are similar to what is needed to derive array methods for Kalman filtering. For this reason, we shall be brief. Once the array methods are presented, the reader will then be able to recognize the similarities between the RLS and Kalman filtering domains (by simply working out Probs. VIII.12 and VIII.13); this conclusion is of course expected in view of the already established equivalence result of Sec. 31.2. Thus, refer to the material on Kalman filtering in Chapter 7 and consider again the standard state-space model (7.8)–(7.10):[17]

$$\boxed{\begin{array}{l} \boldsymbol{x}_{i+1} = F_i\boldsymbol{x}_i + G_i\boldsymbol{n}_i, \quad i \geq 0 \\ \boldsymbol{y}_i \;\;= H_i\boldsymbol{x}_i + \boldsymbol{v}_i \\ \mathsf{E}\begin{bmatrix} \boldsymbol{n}_i \\ \boldsymbol{v}_i \\ \boldsymbol{x}_0 \\ 1 \end{bmatrix}\begin{bmatrix} \boldsymbol{n}_j \\ \boldsymbol{v}_j \\ \boldsymbol{x}_0 \end{bmatrix}^* = \begin{bmatrix} \begin{bmatrix} Q_i & 0 \\ 0 & R_i \end{bmatrix}\delta_{ij} & 0 \\ & 0 \\ 0 \;\; 0 & \Pi_0 \\ 0 \;\; 0 & 0 \end{bmatrix} \end{array}} \tag{35.28}$$

The Kalman filter recursions, in covariance form, for computing the innovations and, subsequently, predicting the state vector are given by (cf. Alg. 7.1):[18]

$$\begin{array}{rcl} R_{e,i} & = & R_i + H_iP_{i|i-1}H_i^* \\ K_{p,i} & = & F_iP_{i|i-1}H_i^*R_{e,i}^{-1} \\ \boldsymbol{\nu}_i & = & \boldsymbol{y}_i - H_i\hat{\boldsymbol{x}}_{i|i-1} \\ \hat{\boldsymbol{x}}_{i+1|i} & = & F_i\hat{\boldsymbol{x}}_{i|i-1} + K_{p,i}\boldsymbol{\nu}_i \\ P_{i+1|i} & = & F_iP_{i|i-1}F_i^* + G_iQ_iG_i^* - K_{p,i}R_{e,i}K_{p,i}^* \end{array} \tag{35.29}$$

with initial conditions $\hat{\boldsymbol{x}}_{0|-1} = 0$ and $P_{0|-1} = \Pi_0$. The time- and measurement update forms of the Kalman filter are similarly given by (cf. Alg. 7.2):

$$\begin{array}{rcl} R_{e,i} & = & R_i + H_iP_{i|i-1}H_i^* \\ K_{f,i} & = & P_{i|i-1}H_i^*R_{e,i}^{-1} \\ \boldsymbol{\nu}_i & = & \boldsymbol{y}_i - H_i\hat{\boldsymbol{x}}_{i|i-1} \\ \hat{\boldsymbol{x}}_{i|i} & = & \hat{\boldsymbol{x}}_{i|i-1} + K_{f,i}\boldsymbol{\nu}_i \\ \hat{\boldsymbol{x}}_{i+1|i} & = & F_i\hat{\boldsymbol{x}}_{i|i} \\ P_{i|i} & = & P_{i|i-1} - P_{i|i-1}H_i^*R_{e,i}^{-1}H_iP_{i|i-1} \\ P_{i+1|i} & = & F_iP_{i|i}F_i^* + G_iQ_iG_i^* \end{array} \tag{35.30}$$

Time-update array form. An array algorithm is evident for the time-update problem. Indeed, from the identity $P_{i+1|i} = F_iP_{i|i}F_i^* + G_iQ_iG_i^*$, we can express $P_{i+1|i}$ as

$$P_{i+1|i} = \begin{bmatrix} F_iP_{i|i}^{1/2} & G_iQ_i^{1/2} \end{bmatrix}\begin{bmatrix} F_iP_{i|i}^{1/2} & G_iQ_i^{1/2} \end{bmatrix}^*$$

This suggests the following array construction. Let Θ_i^{tu} be any unitary matrix that triangularizes the pre-array shown below:

$$\begin{bmatrix} F_iP_{i|i}^{1/2} & G_iQ_i^{1/2} \end{bmatrix}\Theta_i^{tu} = \begin{bmatrix} \mathsf{X} & 0 \end{bmatrix}$$

where X is lower-triangular with positive diagonal entries. Then, by comparing entries on both sides of the squared identity,

$$\begin{bmatrix} F_iP_{i|i}^{1/2} & G_iQ_i^{1/2} \end{bmatrix}\underbrace{\Theta_i^{tu}\Theta_i^{tu*}}_{=\mathrm{I}}\begin{bmatrix} F_iP_{i|i}^{1/2} & G_iQ_i^{1/2} \end{bmatrix}^* = \begin{bmatrix} \mathsf{X} & 0 \end{bmatrix}\begin{bmatrix} \mathsf{X} & 0 \end{bmatrix}^*$$

[17]We shall assume, without loss of generality, that $S_i = 0$, i.e., that the processes $\{\boldsymbol{n}_i, \boldsymbol{v}_i\}$ are uncorrelated. This case is sufficient for our purposes here.

[18]For convenience of exposition, we are denoting the innovations variable of the Kalman filter by $\boldsymbol{\nu}(i)$, as opposed to the symbol $\boldsymbol{e}(i)$ used in Chapter 7. This is in order to avoid a conflict of notation with the symbol $e(i)$ used to denote the output error of RLS.

we conclude that $XX^* = P_{i+1|i}$ so that $X = P_{i+1|i}^{1/2}$. In summary, we are led to the following time-update array form:

$$\boxed{\begin{bmatrix} F_i P_{i|i}^{1/2} & G_i Q_i^{1/2} \end{bmatrix} \Theta_i^{tu} = \begin{bmatrix} P_{i+1|i}^{1/2} & 0 \end{bmatrix}} \qquad (35.31)$$

where Θ_i^{tu} is any unitary matrix that triangularizes the pre-array.

Measurement-update array form. For the measurement-update problem, we choose any unitary matrix Θ_i^{mu} that triangularizes the pre-array below:

$$\begin{bmatrix} R_i^{1/2} & H_i P_{i|i-1}^{1/2} \\ 0 & P_{i|i-1}^{1/2} \end{bmatrix} \Theta_i^{mu} = \begin{bmatrix} X & 0 \\ Y & Z \end{bmatrix}$$

where X and Z are lower triangular with positive diagonal entries. Then, by comparing entries on both sides of the squared identity:

$$\begin{bmatrix} R_i^{1/2} & H_i P_{i|i-1}^{1/2} \\ 0 & P_{i|i-1}^{1/2} \end{bmatrix} \underbrace{\Theta_i^{mu} \Theta_i^{mu*}}_{=I} \begin{bmatrix} R_i^{1/2} & H_i P_{i|i-1}^{1/2} \\ 0 & P_{i|i-1}^{1/2} \end{bmatrix}^* = \begin{bmatrix} X & 0 \\ Y & Z \end{bmatrix} \begin{bmatrix} X & 0 \\ Y & Z \end{bmatrix}^*$$

we arrive at the identifications shown in the following measurement-update array form:

$$\boxed{\begin{bmatrix} R_i^{1/2} & H_i P_{i|i-1}^{1/2} \\ 0 & P_{i|i-1}^{1/2} \end{bmatrix} \Theta_i^{mu} = \begin{bmatrix} R_{e,i}^{1/2} & 0 \\ \bar{K}_{f,i} & P_{i|i}^{1/2} \end{bmatrix}} \qquad (35.32)$$

where $\bar{K}_{f,i} = K_{f,i} R_{e,i}^{1/2}$.

Array covariance form. By combining the just derived array forms for the measurement- and time-update steps we obtain the following array form for the covariance filter, also known as the square-root form,

$$\boxed{\begin{bmatrix} R_i^{1/2} & H_i P_{i|i-1}^{1/2} & 0 \\ 0 & F_i P_{i|i-1}^{1/2} & G_i Q_i^{1/2} \end{bmatrix} \Theta_i = \begin{bmatrix} R_{e,i}^{1/2} & 0 & 0 \\ \bar{K}_{p,i} & P_{i+1|i}^{1/2} & 0 \end{bmatrix}} \qquad (35.33)$$

where $\bar{K}_{p,i} = K_{p,i} R_{e,i}^{1/2}$ and Θ_i is any unitary matrix that triangularizes the pre-array. Alternatively, we can deduce the same algorithm as follows. Let Θ_i be any unitary matrix that triangularizes the pre-array shown below:

$$\begin{bmatrix} R_i^{1/2} & H_i P_{i|i-1}^{1/2} & 0 \\ 0 & F_i P_{i|i-1}^{1/2} & G_i Q_i^{1/2} \end{bmatrix} \Theta_i = \begin{bmatrix} X & 0 & 0 \\ Y & Z & 0 \end{bmatrix}$$

where X and Z are lower triangular with positive diagonal entries. Then, by comparing terms on both sides of the squared identity

$$\begin{bmatrix} R_i^{1/2} & H_i P_{i|i-1}^{1/2} & 0 \\ 0 & F_i P_{i|i-1}^{1/2} & G_i Q_i^{1/2} \end{bmatrix} \underbrace{\Theta_i \Theta_i^*}_{=I} \begin{bmatrix} R_i^{1/2} & H_i P_{i|i-1}^{1/2} & 0 \\ 0 & F_i P_{i|i-1}^{1/2} & G_i Q_i^{1/2} \end{bmatrix}^*$$
$$= \begin{bmatrix} X & 0 & 0 \\ Y & Z & 0 \end{bmatrix} \begin{bmatrix} X & 0 & 0 \\ Y & Z & 0 \end{bmatrix}^*$$

we arrive at the aforementioned array covariance form.

Information forms of the Kalman filter. Starting from the covariance and time- and measurement-update forms (35.28) and (35.30) of the Kalman filter, some algebra will show that, when F_i is invertible, the following forms also hold:

- Measurement-update information form:

$$P_{i|i}^{-1} = P_{i|i-1}^{-1} + H_i^* R_i^{-1} H_i$$
$$P_{i|i}^{-1}\hat{x}_{i|i} = P_{i|i-1}^{-1}\hat{x}_{i|i-1} + H_i^* R_i^{-1} y_i$$

- Time-update information form:

$$\hat{x}_{i+1|i} = F_i \hat{x}_{i|i}$$
$$R_{e,i}^d = Q_i^{-1} + G_i^* F_i^{-*} P_{i|i}^{-1} F_i^{-1} G_i$$
$$K_{p,i}^d = F_i^{-*} P_{i|i}^{-1} F_i^{-1} G_i R_{e,i}^{-d}$$
$$P_{i+1|i}^{-1} = F_i^{-*} P_{i|i}^{-1} F_i^{-1} - K_{p,i}^d R_{e,i}^d K_{p,i}^{d*}, \quad P_{0|-1}^{-1} = \Pi_0^{-1}$$

- Recursion for the inverse Riccati variable:

$$P_{i+1|i}^{-1} = F_i^{-*} P_{i|i-1}^{-1} F_i^{-1} + F_i^{-*} H_i^* R_i^{-1} H_i F_i^{-1} - K_{p,i}^d R_{e,i}^d K_{p,i}^{d*}, \quad P_{0|-1}^{-1} = \Pi_0^{-1}$$

These variants are called information forms since they propagate the inverses of the error covariance matrices, $\{P_{i|i-1}^{-1}, P_{i|i}^{-1}\}$. We can devise array formulations for them as well. The arguments are similar to what we have done so far and, therefore, we shall only state the final recursions; their validity can be verified by simply "squaring" both sides of the corresponding array equations and comparing terms.

Let $\Phi_{i|i-1} = P_{i|i-1}^{-1}$ and $\Phi_{i|i} = P_{i|i}^{-1}$, and introduce their Cholesky factorizations,

$$\Phi_{i|i-1} = \Phi_{i|i-1}^{1/2}\Phi_{i|i-1}^{*/2}, \qquad \Phi_{i|i} = \Phi_{i|i}^{1/2}\Phi_{i|i}^{*/2}$$

where $\{\Phi_{i|i-1}^{1/2}, \Phi_{i|i}^{1/2}\}$ are lower triangular with positive diagonal entries.

Measurement-update information array form.

$$\begin{bmatrix} \Phi_{i|i-1}^{1/2} & H_i^* R_i^{-*/2} \\ \hat{x}_{i|i-1}^* \Phi_{i|i-1}^{1/2} & y_i^* R_i^{-*/2} \end{bmatrix} \Theta_i^{mu} = \begin{bmatrix} \Phi_{i|i}^{1/2} & 0 \\ \hat{x}_{i|i}^* \Phi_{i|i}^{1/2} & \nu_i^* R_{e,i}^{-*/2} \end{bmatrix} \tag{35.34}$$

where Θ_i^{mu} is any unitary matrix that lower triangularizes the pre-array.

Time-update information array form

$$\begin{bmatrix} Q_i^{-*/2} & G_i^* F_i^{-*} \Phi_{i|i}^{1/2} \\ 0 & F_i^{-*}\Phi_{i|i}^{1/2} \end{bmatrix} \Theta_i^{tu} = \begin{bmatrix} V_i^{1/2} & 0 \\ F_i^{-*}\Phi_{i|i} F_i^{-1} G_i V_i^{-*/2} & \Phi_{i+1|i}^{1/2} \end{bmatrix} \tag{35.35}$$

where Θ_i^{tu} is any unitary rotation that triangularizes the pre-array and generates a lower triangular factor $V_i^{1/2}$ with positive diagonal entries. Here, $V_i = Q_i^{-1} + G_i^* A_i G_i$. The combination of (35.34) and (35.35) is referred to as the square-root information filter (SRIF).

Information array form

$$\begin{bmatrix} 0 & F_i^{-*}\Phi_{i|i-1}^{1/2} & F_i^{-*} H_i^* R_i^{-*/2} \\ 0 & \hat{x}_{i|i-1}^* \Phi_{i|i-1}^{1/2} & y_i^* R_i^{-*/2} \\ 0 & 0 & R_i^{-*/2} \end{bmatrix} \Theta_i = \begin{bmatrix} (*) & \Phi_{i+1|i}^{1/2} & 0 \\ (*) & \hat{x}_{i+1|i}^* \Phi_{i+1|i}^{1/2} & \nu_i^* Z_i^{1/2} \\ (*) & K_{p,i}^* \Phi_{i+1|i}^{1/2} & Z_i^{1/2} \end{bmatrix} \tag{35.36}$$

where Θ_i is any unitary rotation that lower triangularizes the pre-array. Moreover, $Z_i^{1/2}$ is a square-root factor for $R_{e,i}^{-1}$, i.e., $R_{e,i}^{-1} = Z_i^{1/2} Z_i^{*/2}$, and the "(*)" notation indicates "don't care" entries.

Summary and Notes

The chapters in this part developed three array variants for recursive least-squares solutions: inverse QR, QR, and extended QR (also known as square-root RLS, square-root information RLS, and extended square-root information RLS, respectively).

SUMMARY OF MAIN RESULTS

1. Array methods are based on transforming a pre-array of numbers into a post-array of numbers by means of unitary transformations. Such transformations are easy to implement as a sequence of elementary rotations or reflections (e.g., as Givens rotations or Householder reflections).
2. Array methods are self-contained in that quantities that are needed to form the pre-array are propagated in the post-array.
3. The variables involved in array methods are usually square-root factors whose entries assume values within smaller dynamic ranges. As a result, array methods are more reliable and have better numerical properties in finite-precision arithmetic than a direct RLS implementation — see, e.g., the computer project at the end of this part.
4. The array variants developed in this part have the same computational complexity as RLS, namely, $O(M^2)$ operations per iteration.

BIBLIOGRAPHIC NOTES

QR methods. It is generally accepted that the QR method for solving and updating least-squares solutions is among the most reliable procedures for finite word-length implementations. The origin of the method can be traced back to the works of Householder (1953, pp. 72–73), Golub (1965), and Businger and Golub (1965) on least-squares problems — see Probs. VII.12 and VII.32 for a description of the QR method as usually derived in works on computational linear algebra.

In the context of adaptive filtering, simulation studies in Yang and Böhme (1992) suggest that the QR method of Sec. 35.2 is reliable numerically for $\lambda < 1$ and can diverge for $\lambda = 1$. This is because the mechanism for the propagation of a single numerical error is exponentially stable for $\lambda < 1$. Further experimental evidence for the numerical stability of the QR method can be found in Ward, Hargrave, and McWhirter (1986). However, as mentioned in the concluding remarks of Part VII (*Least-Squares Methods*), such conclusions on the numerical stability of RLS-type algorithms from single-error propagation models should be interpreted with care.

The extended QR method of Sec. 35.3 was developed by Yang and Böhme (1992), and it also appears in Sayed (1992) and Sayed and Kailath (1994b). Unfortunately, the algorithm can face numerical difficulties in finite word-length implementations since its pre- and post-arrays involve both $\Phi_i^{1/2}$ and its inverse.

Systolic implementation. Array methods can be parallelized in view of the fact that the rotations can be applied simultaneously to all rows in a pre-array. In particular, for the QR method of Sec. 35.2, a systolic array implementation was developed by McWhirter (1983) in the form of a triangular array of processors. Related descriptions can be found in McWhirter and Proudler (1993), Shepherd and McWhirter (1993), Sayed, Lev-Ari, and Kailath (1994), and Haykin (1996,2002). Parallel and systolic array implementations for the inverse QR algorithm of Sec. 35.1 can be found in Pan and Plemmons (1989) and Alexander and Ghirnikar (1993).

Basis rotation. The idea of using the basis rotation result of Lemma 33.1 as a tool for deriving array RLS methods is due to Sayed and Kailath (1994b). The result of the lemma provides a convenient way for motivating and describing array methods (see, e.g., the presentation in Haykin (1996,2000) and also Manolakis, Ingle, and Kogon (2000)).

Array methods in estimation. The idea of using array algorithms to propagate square-root factors, rather than the matrices themselves, dates back to the mid 1960s in works on Kalman filtering by Potter and Stern (1963). The need for introducing such array methods was sparked by the necessity to develop reliable Kalman filtering implementations for precision approach and moon landing systems. However, this initial work was limited to the measurement-update step of the Kalman filter (cf. Chapter 7), and extensive subsequent developments followed that extended array methods to different forms of Kalman filtering implementations (for both filtering and smoothing applications). Among these earlier works we may mention those of Dyer and McReynolds (1969), Hanson and Lawson (1969), Schmidt (1970), Bierman (1974,1977), and Morf and Kailath (1975). The last two references are notable for their generality, with the work by Morf and Kailath (1975) being the closest to the array descriptions in this chapter — see App. 35.A and Probs. VIII.12 and VIII.13.

RLS and Kalman filtering. Since RLS can be regarded as a special case of the Kalman filter for a special state-space model, as was shown in Sec. 31.2, then there should also exist a correspondence between the RLS array algorithms of this chapter and the Kalman filtering array algorithms. This correspondence was developed by Sayed and Kailath (1993,1994b) and it is detailed in Chapter 31, Apps. 35.A and 37.A, and Probs. VIII.12 and VIII.13 — it is also covered in Haykin (1996,2000).

Paige's method. There have been several interesting works in numerical linear algebra on stable array methods for recursively updating least-squares solutions. These methods are relevant to the adaptive filtering context, and most notable among them is an array algorithm developed by Paige (1985) as a result of studies by Paige (1979a,1979b) and Paige and Saunders (1977). Paige's form is useful when dealing with ill-conditioned data. Rather than propagate a square-root factor of P_i, the algorithm propagates factors of $P_i^{1/2}$ itself. In this way, it avoids matrix products while forming the pre-arrays and leads to improved (in fact, stable) numerical performance albeit at some increased computational cost. A description of the algorithm appears in the textbook by Kailath, Sayed, and Hassibi (2000, Chapter 12). Paige proved that his algorithm is "backward" stable, which is a desirable feature. What this means is the following.

A numerical algorithm for solving a linear system of equations $Ax = b$ is said to be *backward stable* if the computed solution $\bar{x}$ can be shown to be the *exact* solution of a slightly perturbed system, namely, $\bar{x}$ satisfies $(A + \delta A)\bar{x} = (b + \delta b)$ for small perturbations $\{\delta A, \delta b\}$.

Problems and Computer Projects

PROBLEMS

Problem VIII.1 (Rank-one modifications) Let $X = \text{I} - \alpha x x^*$, where x is a column vector, α is a scalar, and I is the identity matrix.

(a) Show that $XX^* = \text{I} - \beta x x^*$, for some real number β.

(b) Consider a matrix Y of the form $Y = \text{I} - \beta x x^*$, for some real scalar β. For what condition on β is Y positive semi-definite? When this is the case, show that Y admits a Hermitian square-root factor of the form $Y^{1/2} = \text{I} - \alpha x x^*$ for some real scalar α.

Problem VIII.2 (Basis rotation) In this problem we provide another proof for Lemma 33.1 by resorting to the QR decomposition of a matrix rather than its singular value decomposition. The SVD proof given in the text is more general since the argument in this problem assumes that the matrices A and B have full rank.

Introduce the QR decompositions

$$A^* = Q_A \begin{bmatrix} R_A \\ 0 \end{bmatrix}, \qquad B^* = Q_B \begin{bmatrix} R_B \\ 0 \end{bmatrix}$$

where Q_A and Q_B are $M \times M$ unitary matrices, and R_A and R_B are $n \times n$ upper triangular matrices with positive diagonal entries (due to the full rank assumption on A and B).

(a) Show that $AA^* = R_A^* R_A = R_B^* R_B$.

(b) Conclude, by uniqueness of the Cholesky factorization (cf. Sec. B.3), that $R_A = R_B$. Complete the argument to show that the matrix $\Theta = Q_B Q_A^*$ is unitary and maps B to A.

Problem VIII.3 (Sample covariance matrix) Let $\Phi_i = \sum_{j=0}^{i} \lambda^{i-j} u_j^* u_j$, where the u_j are $1 \times M$ row vectors and $0 \ll \lambda \leq 1$. The matrix Φ_i can have full-rank or it may be rank deficient. Assume its rank is $r \leq M$. Let $\Phi_i^{1/2}$ denote an $M \times r$ full-rank square-root factor, i.e., $\Phi_i^{1/2}$ has rank r and satisfies $\Phi_i^{1/2} \Phi_i^{*/2} = \Phi_i$. Show that u_i^* belongs to the column span of $\Phi_i^{1/2}$.

Problem VIII.4 (Rank-three update) Consider a recursion of the form

$$\Phi_i = \lambda \Phi_{i-1} + a^* a + b^* b + c^* c$$

where λ is a positive scalar and $\{a, b, c\}$ are row vectors.

(a) Let $P_i = \Phi_i^{-1}$. Find a procedure that updates P_{i-1} to P_i and that does not require any matrix inversion; only scalar inversions are allowed.

(b) Derive an array algorithm for updating $\Phi_{i-1}^{1/2}$ to $\Phi_i^{1/2}$.

(c) Derive an array algorithm for updating $P_{i-1}^{1/2}$ to $P_i^{1/2}$.

Problem VIII.5 (QR method for a modified cost) Consider the formulation of Prob. VII.30. Repeat the arguments of Sec. 35.2 to derive the following QR-based implementation. Start with $\Phi_{-1}^{1/2} = \Pi^{1/2}$, $q_{-1} = 0$, and repeat for $i \geq 0$.

1. Find a unitary matrix Θ_i that lower triangularizes the pre-array shown below and generates a post-array with positive diagonal entries in $\Phi_i^{1/2}$, as well as a positive rightmost corner entry in the last row of the post-array. The entries in the post-array will then correspond to

$$\begin{bmatrix} \lambda^{1/2}\Phi_{i-1}^{1/2} & u_i^* \\ \lambda^{1/2}a^*q_{i-1}^* & d^*(i) \\ 0 & 1 \end{bmatrix} \Theta_i = \begin{bmatrix} \Phi_i^{1/2} & 0 \\ q_i^* & e^*(i)\gamma^{1/2}(i) \\ u_i\Phi_i^{-*/2} & \gamma^{1/2}(i) \end{bmatrix}$$

2. Obtain w_i by solving the triangular linear system of equations $\Phi_i^{*/2}w_i = q_i$, where the quantities $\{\Phi_i^{*/2}, q_i\}$ are read from the post-array.

Problem VIII.6 (Rotation matrix for QR algorithm) Refer to the discussion in Sec. 35.2 on the QR algorithm and recall that the purpose of the transformation Θ_i in (35.17) is to perform a transformation of the form (assuming $M = 3$):

$$\begin{bmatrix} \lambda^{1/2}\Phi_{i-1}^{1/2} & u_i^* \\ \lambda^{1/2}q_{i-1}^* & d^*(i) \end{bmatrix} \Theta_i \equiv \left[\begin{array}{ccc|c} \times & 0 & 0 & \times \\ \times & \times & 0 & \times \\ \times & \times & \times & \times \\ \hline \times & \times & \times & \times \end{array}\right] \Theta_i = \left[\begin{array}{ccc|c} \times & 0 & 0 & 0 \\ \times & \times & 0 & 0 \\ \times & \times & \times & 0 \\ \hline \times & \times & \times & \times \end{array}\right]$$

One way to implement Θ_i is via a sequence of three elementary Givens rotations in order to annihilate the three entries of u_i^*, one at a time. Now since, by assumption, the diagonal entries of $\Phi_{i-1}^{1/2}$ are positive, we have that the diagonal entries of the triangular matrix in the pre-array are positive. Recall further that we desire a post-array in (35.8) having a $\Phi_i^{1/2}$ with positive diagonal entries as well. Therefore, in view of the remark following Lemma 34.2, the diagonal entries of the individual Givens rotations will be positive. Use this fact to conclude that the rightmost diagonal entry of Θ_i is positive.

Problem VIII.7 (Block inverse QR) Refer to the block RLS algorithm of Prob. VII.36. Follow the derivation in Sec. 35.1 to derive the following array variant for it.

Let w_N denote the solution of the least-squares problem

$$\min_w \left[(w-\bar{w})^*\Pi(w-\bar{w}) + \sum_{j=0}^{N}(d_j - U_jw)^* R_j^{-1}(d_j - U_jw)\right]$$

Then w_N can be recursively computed as follows. Let $\Sigma = \Pi^{-1}$ and introduce the Cholesky factorization $\Sigma = \Sigma^{1/2}\Sigma^{*/2}$, where $\Sigma^{1/2}$ is lower triangular with positive-diagonal entries. Then start with $w_{-1} = \bar{w}$, $P_{-1}^{1/2} = \Sigma^{1/2}$, and repeat for $i \geq 0$.

1. Find a unitary matrix Θ_i that lower triangularizes the pre-array shown below and generates a post-array with positive diagonal entries. Then the entries in the post-array will correspond to

$$\begin{bmatrix} R_i^{1/2} & U_iP_{i-1}^{1/2} \\ 0 & P_{i-1}^{1/2} \end{bmatrix} \Theta_i = \begin{bmatrix} C_i^{1/2} & 0 \\ G_iC_i^{1/2} & P_i^{1/2} \end{bmatrix}$$

where $R_i = R_i^{1/2}R_i^{*/2}$ and $\Gamma_i^{-1} = C_i^{1/2}C_i^{*/2}$ denote the Cholesky decompositions of R_i and Γ_i^{-1}, respectively.

2. Update the weight vector as $w_i = w_{i-1} + \left[G_iC_i^{1/2}\right]\left[C_i^{1/2}\right]^{-1}[d_i - U_iw_{i-1}]$, where the quantities $\{G_iC_i^{1/2}, C_i^{1/2}\}$ are read from the post-array.

Problem VIII.8 (Block QR) Refer to the statement of the block RLS algorithm in Prob. VII.37. Follow the derivation in Sec. 35.2 to derive the following array variant for it.

Let w_N denote the solution of the least-squares problem

$$\min_w \left[(w-\bar{w})^*\Pi(w-\bar{w}) + \sum_{j=0}^{N}(d_j - U_jw)^* R_j^{-1}(d_j - U_jw)\right]$$

and let $\bar{d}_i = d_i - U_i\bar{w}$. Then w_N can be recursively computed as follows. Start with $\Phi_{-1}^{1/2} = \Pi^{1/2}$, $q_{-1} = 0$, and repeat for $i \geq 0$.

1. Find a unitary matrix Θ_i that lower triangularizes the pre-array shown below and generates a post-array with positive diagonal entries in the lower triangular factors $\Phi_i^{1/2}$ and $\Gamma_i^{1/2}$ in the post-array. The entries in the post-array will then correspond to

$$\begin{bmatrix} \Phi_{i-1}^{1/2} & U_i^* R_i^{-*/2} \\ q_{i-1}^* & \bar{d}_i^* R_i^{-*/2} \\ 0 & R_i^{-*/2} \end{bmatrix} \Theta_i = \begin{bmatrix} \Phi_i^{1/2} & 0 \\ q_i^* & e_i^* \Gamma_i^{1/2} \\ R_i^{-1} U_i \Phi_i^{-*/2} & \Gamma_i^{1/2} \end{bmatrix}$$

where $R_i = R_i^{1/2} R_i^{*/2}$ and $\Gamma_i = \Gamma_i^{1/2}\Gamma_i^{*/2}$ denote the Cholesky decompositions of R_i and Γ_i, respectively.

2. Obtain w_i by solving the triangular linear system of equations $\Phi_i^{*/2}[w_i - \bar{w}] = q_i$, where the quantities $\{\Phi_i^{*/2}, q_i\}$ are read from the post-array.

Problem VIII.9 (Block extended QR) Refer to the statement of the block RLS algorithm in Prob. VII.37. Follow the derivation in Sec. 35.3 to derive the following array variant for it.

Let w_N denote the solution of the least-squares problem

$$\min_w \left[(w - \bar{w})^* \Pi (w - \bar{w}) + \sum_{j=0}^{N} (d_j - U_j w)^* R_j^{-1} (d_j - U_j w) \right]$$

and let $\bar{d}_i = d_i - U_i\bar{w}$. Then w_N can be recursively computed as follows. Start with $\Phi_{-1}^{1/2} = \Pi^{1/2}$, $\Phi_{-1}^{-*/2} = \Pi^{-*/2}$, $q_{-1} = 0$, and repeat for $i \geq 0$.

1. Find a unitary matrix Θ_i that lower triangularizes the pre-array shown below and generates a post-array with positive diagonal entries in $\Phi_i^{1/2}$ and $\Gamma_i^{1/2}$ in the post-array. The entries in the post-array will then correspond to

$$\begin{bmatrix} \Phi_{i-1}^{1/2} & U_i^* R_i^{-*/2} \\ q_{i-1}^* & \bar{d}_i^* R_i^{-*/2} \\ 0 & R_i^{-*/2} \\ \Phi_{i-1}^{-*/2} & 0 \end{bmatrix} \Theta_i = \begin{bmatrix} \Phi_i^{1/2} & 0 \\ q_i^* & e_i^* \Gamma_i^{1/2} \\ R_i^{-1} U_i \Phi_i^{-*/2} & \Gamma_i^{1/2} \\ \Phi_i^{-*/2} & -G_i \Gamma_i^{-*/2} \end{bmatrix}$$

where $R_i = R_i^{1/2} R_i^{*/2}$ and $\Gamma_i = \Gamma_i^{1/2}\Gamma_i^{*/2}$ denote the Cholesky decompositions of R_i and Γ_i, respectively.

2. Update the weight vector as $w_i = w_{i-1} - \left(-G_i \Gamma_i^{-*/2}\right) \cdot \left(e_i^* \Gamma_i^{1/2}\right)^*$.

Problem VIII.10 (Numerical example) Transform the pre-array

$$A = \begin{bmatrix} 0.27 & 0.14 & 0.14 & 0.35 \\ 0.22 & 0.36 & 0.21 & 0.25 \\ -0.22 & 0.28 & -0.14 & 0.30 \end{bmatrix} \quad \text{to the form} \quad \begin{bmatrix} \times & 0 & 0 & 0 \\ \times & \times & 0 & 0 \\ \times & \times & \times & 0 \end{bmatrix}$$

by using (i) Givens rotations, (ii) Householder transformations, and (iii) a mixture of Givens and Householder transformations.

Problem VIII.11 (Effect of perturbations) Refer to the discussion in App. 35.A on Kalman filtering, and consider a one-dimensional model of the form $\boldsymbol{x}(i+1) = a\boldsymbol{x}(i)$ and $\boldsymbol{y}(i) = \boldsymbol{x}(i) + \boldsymbol{v}(i)$ with $a < 1$, $\mathsf{E}\,|\boldsymbol{x}(0)|^2 = 1$, and $\mathsf{E}\,\boldsymbol{v}(i)\boldsymbol{v}^*(j) = \frac{1}{a^2}\delta_{ij}$.

(a) Verify that, for this model, the covariance form and the information form recursions for the Riccati variable are given by

$$p(i+1|i) = \frac{a^2 p(i|i-1)}{1 + a^2 p(i|i-1)} \quad \text{and} \quad p^{-1}(i+1|i) = \frac{1}{a^2} p^{-1}(i|i-1) + 1$$

with initial condition $p(0|-1) = 1$; here $p(i|i-1)$ is a scalar variable.

(b) Assume that at iteration i_o, a perturbation is introduced into $p(i_o|i_o-1)$, say due to numerical errors. No further perturbations are introduced afterwards. Let $\bar{p}(i|i-1)$ denote the Riccati variable that results for $i \geq i_o$ from

$$\bar{p}(i+1|i) = \frac{a^2\bar{p}(i|i-1)}{1+a^2\bar{p}(i|i-1)}, \quad \bar{p}(i_o|i_o-1), \quad i \geq i_o$$

Let also $\hat{p}^{-1}(i|i-1)$ denote the inverse Riccati variable that results for $i \geq i_o$ from

$$\hat{p}^{-1}(i+1|i) = \frac{1}{a^2}\hat{p}^{-1}(i|i-1) + 1, \quad \hat{p}^{-1}(i_o|i_o-1), \quad i \geq i_o$$

Show that $\hat{p}^{-1}(i+1|i) - p^{-1}(i+1|i) = [\hat{p}^{-1}(i|i-1) - p^{-1}(i|i-1)]/a^2$, whereas

$$\bar{p}(i+1|i) - p(i+1|i) = \frac{a^2(\bar{p}(i|i-1) - p(i|i-1))}{1+a^2p(i|i-1) + a^2\bar{p}(i|i-1) + a^4\bar{p}(i|i-1)p(i|i-1)}$$

(c) Conclude that the recursion for $\bar{p}(i|i-1)$ recovers from the perturbation, while the error $\hat{p}(i|i-1) - p(i|i-1)$ grows unbounded!

Remark. This problem shows how different (but mathematically equivalent) filter implementations can behave quite differently under perturbations.

Problem VIII.12 (Inverse QR and array covariance form) In Sec. 31.2 we showed that the exponentially weighted RLS algorithm can be obtained via equivalence to the Kalman filter as follows. We start from the state-space model (31.8) and write down the Kalman recursions for estimating its state variable. Then we translate the Kalman variables into the RLS variables by using the correspondences summarized in Table 31.2. In this problem, as well as in Prob. VIII.13, we want to apply this same procedure to the array variants of the Kalman filter that are described in App. 35.A. In this way, we would be able to recover the array variants of RLS as special cases. Thus consider model (31.8), which is a special case of the state-space model (35.28) with $R_i = 1$, $G_i = 0$ and $Q_i = 0$.

(a) Use (35.33) to verify that the corresponding covariance array form is given by

$$\begin{bmatrix} 1 & u_i P_{i|i-1}^{1/2} \\ 0 & \lambda^{-1/2} P_{i|i-1}^{1/2} \end{bmatrix} \Theta_i = \begin{bmatrix} r_e^{1/2}(i) & 0 \\ k_{p,i} r_e^{1/2}(i) & P_{i+1|i}^{1/2} \end{bmatrix}$$

where Θ_i is any unitary matrix that triangularizes the pre-array. Moreover,

$$\hat{\boldsymbol{x}}_{i+1|i} = \lambda^{-1/2}\hat{\boldsymbol{x}}_i + k_{p,i}\boldsymbol{\nu}(i), \quad \boldsymbol{\nu}(i) = \boldsymbol{y}(i) - u_i\hat{\boldsymbol{x}}_{i|i-1}$$

(b) Use the correspondences from Table 11.1 and replace the Kalman variables

$$\{P_{i|i-1}, r_e(i), k_{p,i}, P_{i+1|i}, \hat{\boldsymbol{x}}_{i|i-1}, \boldsymbol{\nu}(i)\}$$

by the corresponding RLS variables. Verify that the array form of part (a) would then reduce to the inverse QR array of Alg. 35.1, i.e.,

$$\begin{bmatrix} 1 & \lambda^{-1/2} u_i P_{i-1}^{1/2} \\ 0 & \lambda^{-1/2} P_{i-1}^{1/2} \end{bmatrix} \Theta_i = \begin{bmatrix} \gamma^{-1/2}(i) & 0 \\ g_i \gamma^{-1/2}(i) & P_i^{1/2} \end{bmatrix}$$

while the state estimator leads to

$$w_i = w_{i-1} + \left[g_i\gamma^{-1/2}(i)\right]\left[\gamma^{-1/2}(i)\right]^{-1}[d(i) - u_i w_{i-1}]$$

Remark. Problems VIII.12, VIII.13, and later Prob. IX.16, clarify the relation between RLS array algorithms and their Kalman filtering counterparts, following the work of Sayed and Kailath (1994b).

Problem VIII.13 (Array QR and Kalman information form) We follow the procedure of Prob. VIII.12 and relate now the QR array form of RLS (cf. Alg. 35.2) to the array information form (35.36) of the Kalman filter. So consider again model (31.8).

(a) Use (35.36) to verify that the corresponding information array form is given by

$$\begin{bmatrix} \lambda^{1/2}\Phi_{i|i-1}^{1/2} & \lambda^{1/2}u_i^* \\ \hat{x}_{i|i-1}^*\Phi_{i|i-1}^{1/2} & y^*(i) \\ 0 & 1 \end{bmatrix} \Theta_i = \begin{bmatrix} \Phi_{i+1|i}^{1/2} & 0 \\ \hat{x}_{i+1|i}^*\Phi_{i+1|i}^{1/2} & \nu^*(i)r_e^{-1/2}(i) \\ k_{p,i}^*\Phi_{i+1|i}^{1/2} & r_e^{-1/2}(i) \end{bmatrix}$$

where Θ_i is any unitary rotation that lower triangularizes the pre-array and $\Phi_{i|i-1} = P_{i|i-1}^{-1}$.

(b) Use the correspondences from Table 11.1 to verify that the array form of part (a) reduces to the QR array form of Alg. 35.2, i.e.,

$$\begin{bmatrix} \lambda^{1/2}\Phi_{i-1}^{1/2} & u_i^* \\ \lambda^{1/2}q_{i-1}^* & d^*(i) \\ 0 & 1 \end{bmatrix} \Theta_i = \begin{bmatrix} \Phi_i^{1/2} & 0 \\ q_i^* & e^*(i)\gamma^{1/2}(i) \\ u_i\Phi_i^{-*/2} & \gamma^{1/2}(i) \end{bmatrix}$$

with w_i obtained by solving $\Phi_i^{*/2}w_i = q_i$ (assume, for simplicity, $\bar{w} = 0$).

COMPUTER PROJECT

Project VIII.1 (Performance of array implementations in finite precision) The purpose of this project is to compare the performance of RLS and one of its array variants in finite-precision. Refer to the channel estimation application shown in Fig. VIII.1.

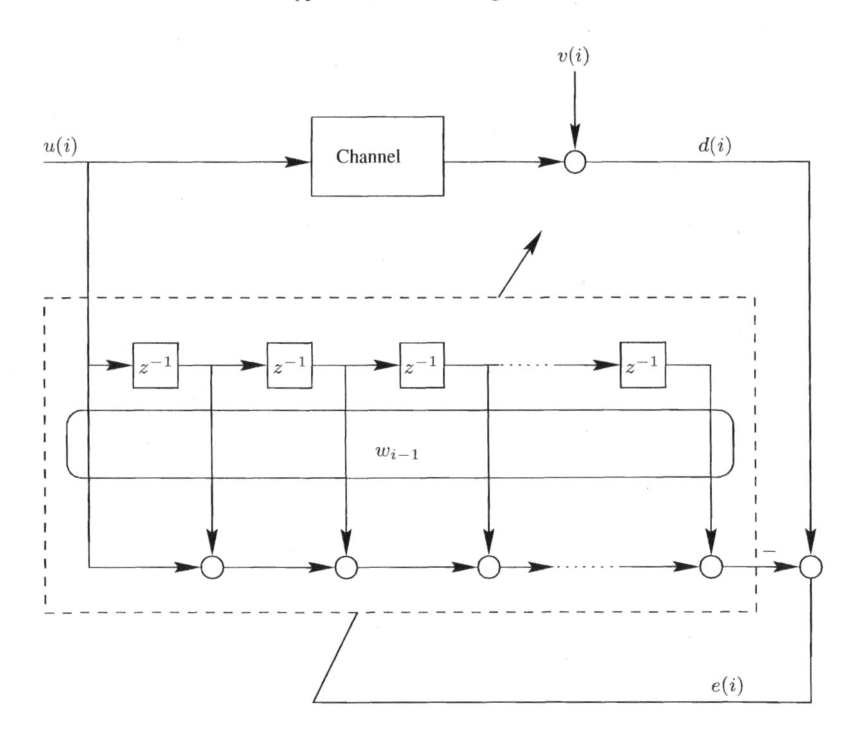

FIGURE VIII.1 Adaptive configuration for channel estimation.

Generate five random samples of a channel impulse response sequence and normalize the norm of this impulse response to unity. Feed unit-variance Gaussian input data through the channel and add Gaussian noise to its output. Set the noise power at 30 dB below the input signal power. Train an adaptive filter for $N = 200$ iterations using RLS and the QR array method of Alg. 35.2. Use $\lambda = 0.995$, $\Pi = 1 \times 10^{-6}\mathrm{I}$ and $\bar{w} = 0$. Assume also a finite-precision implementation is used with B_r bits for signals and B_c bits for coefficients (including the sign bit in both cases). In order to simulate the quantized behavior of the filters, you may use a routine quantize.m. The routine receives as input two parameters: a real number x and the total number of bits, B (including the sign bit), for its quantized representation. The routine then returns a real number that corresponds to the quantized value of x. This value is determined as follows. With B total bits, the largest integers that can be represented are

$$\pm 2^{(B-1)} - 1 .$$

If x exceeds these extreme values then its quantized representation is taken as either one of them, depending on the sign of x. If, on the other hand, x falls within the interval

$$x \in \left(-2^{(B-1)} - 1,\ 2^{(B-1)} - 1 \right)$$

then the routine determines how many bits are needed to represent the integer part of x, and the remaining bits are used to represent the fractional part of x.

For each algorithm (RLS and QR array method), generate an ensemble-average learning curve by averaging over 100 experiments and for the following choices:

1. $B_r = B_c = 30$ bits.
2. $B_r = B_c = 25$ bits.
3. $B_r = B_c = 16$ bits.

PART IX
FAST RLS ALGORITHMS

CHAPTER 36

Hyperbolic Rotations

It is sometimes necessary, especially when deriving fast least-squares algorithms (as we shall discuss in the next chapter), to employ $J-$unitary (also called hyperbolic) transformations, as opposed to unitary transformations, in order to annihilate certain entries in a pre-array of numbers. A $J-$unitary transformation Θ is one that satisfies

$$\boxed{\Theta J \Theta^* = \Theta^* J \Theta = J}$$

for some signature matrix J, i.e., a diagonal matrix with ± 1 entries. The special case $J = I$ corresponds to unitary transformations and was studied in Chapter 34. In this chapter, we extend the results to the $J-$unitary case, starting with Givens rotations and followed by Householder transformations.

36.1 HYPERBOLIC GIVENS ROTATIONS

As in our discussions in Chapter 34, we again distinguish between the cases of real data and complex data.

Real Data

Thus, consider a 1×2 real-valued vector $z = \begin{bmatrix} a & b \end{bmatrix}$, and assume that we wish to determine a 2×2 matrix Θ that transforms it to the form:

$$\begin{bmatrix} a & b \end{bmatrix} \Theta = \begin{bmatrix} \alpha & 0 \end{bmatrix} \tag{36.1}$$

for some nonzero real number α to be determined, and where Θ is required to be hyperbolic, i.e., it should satisfy

$$\Theta J \Theta^{\mathsf{T}} = \Theta^{\mathsf{T}} J \Theta = J \qquad \text{where} \qquad J = \text{diag}\{1, -1\}$$

Unfortunately, and in contrast to the case of orthogonal Givens transformations in Sec. 34.1, the transformation (36.1) is not always possible. To see this, note from (36.1) that

$$\begin{bmatrix} a & b \end{bmatrix} \underbrace{\Theta J \Theta^{\mathsf{T}}}_{J} \begin{bmatrix} a \\ b \end{bmatrix} = \begin{bmatrix} \alpha & 0 \end{bmatrix} J \begin{bmatrix} \alpha \\ 0 \end{bmatrix} = \begin{bmatrix} \alpha & 0 \end{bmatrix} \begin{bmatrix} 1 & 0 \\ 0 & -1 \end{bmatrix} \begin{bmatrix} \alpha \\ 0 \end{bmatrix}$$

i.e., $a^2 - b^2 = \alpha^2$. Now since $\alpha^2 > 0$, no matter what α is, this means that the transformation (36.1) is only possible if $|a| > |b|$. When this is not the case, i.e., if $|a| < |b|$, then we should seek instead a hyperbolic rotation Θ that transforms z into the alternative form

$$\begin{bmatrix} a & b \end{bmatrix} \Theta = \begin{bmatrix} 0 & \alpha \end{bmatrix} \tag{36.2}$$

In this case, the transformation (36.2) will guarantee $a^2 - b^2 = -\alpha^2$, which is consistent with the fact that $a^2 - b^2 < 0$. So let us examine the cases $|a| > |b|$ and $|a| < |b|$ separately.

$|a| > |b|$ **(first entry of the vector is dominant)**

In this case, an expression for Θ that achieves (36.1) is given by

$$\Theta = \frac{1}{\sqrt{1-\rho^2}} \begin{bmatrix} 1 & -\rho \\ -\rho & 1 \end{bmatrix} \quad \text{where} \quad \rho = \frac{b}{a} \tag{36.3}$$

It is a straightforward exercise to verify that

$$\begin{bmatrix} a & b \end{bmatrix} \Theta = \begin{bmatrix} \pm\sqrt{a^2 - b^2} & 0 \end{bmatrix}$$

where the sign of the resulting α depends on whether the value of the square-root in the expression (36.3) for Θ is chosen to be negative or positive.

The reason for the denomination *hyperbolic rotation* for Θ can be seen by expressing Θ in the form

$$\Theta = \begin{bmatrix} \mathsf{ch} & -\mathsf{sh} \\ -\mathsf{sh} & \mathsf{ch} \end{bmatrix}, \quad \mathsf{ch} = \frac{1}{\sqrt{1-\rho^2}}, \quad \mathsf{sh} = \frac{\rho}{\sqrt{1-\rho^2}}.$$

in terms of hyperbolic cosine and sine parameters, ch and sh, respectively. In this way, we find that the effect of Θ is to rotate any point (x, y) in the two-dimensional Euclidean space along the *hyperbola* of equation $x^2 - y^2 = a^2 - b^2$ — see Fig. 36.1.

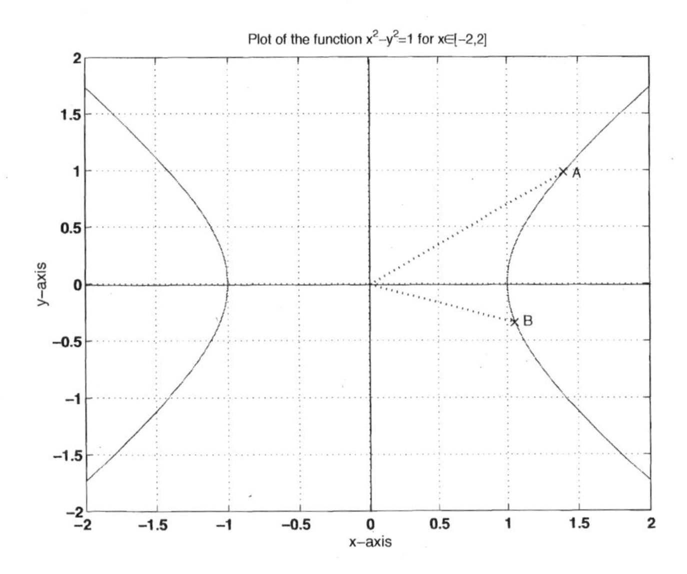

FIGURE 36.1 A hyperbolic rotation in 2−dimensional Euclidean space moves points along the hyperbola of equation $x^2 - y^2 = a^2 - b^2$. In the figure, $a^2 - b^2 = 1$ and point A is moved into point B.

$|b| > |a|$ **(second entry of the vector is dominant)**

In this second case, an expression for Θ that achieves (36.2) is given by

$$\boxed{\Theta = \frac{1}{\sqrt{1-\rho^2}} \begin{bmatrix} 1 & -\rho \\ -\rho & 1 \end{bmatrix} \quad \text{where} \quad \rho = \frac{a}{b}} \tag{36.4}$$

which now leads to the result

$$\begin{bmatrix} a & b \end{bmatrix} \Theta = \begin{bmatrix} 0 & \pm\sqrt{b^2 - a^2} \end{bmatrix}$$

In summary, we arrive at the following conclusion.

Lemma 36.1 (Real hyperbolic Givens) Consider a 1×2 vector $[a\ b]$ with real entries. If $|a| > |b|$, then choose Θ as in (36.3) to get

$$\begin{bmatrix} a & b \end{bmatrix} \Theta = \pm\sqrt{a^2 - b^2} \begin{bmatrix} 1 & 0 \end{bmatrix}$$

If, on the other hand, $|a| < |b|$, then choose Θ as in (36.4) to get

$$\begin{bmatrix} a & b \end{bmatrix} \Theta = \pm\sqrt{b^2 - a^2} \begin{bmatrix} 0 & 1 \end{bmatrix}$$

Complex Data

More generally, consider a 1×2 vector $z = \begin{bmatrix} a & b \end{bmatrix}$ with possibly complex entries, and assume that we wish to determine an elementary 2×2 matrix Θ that transforms it to either forms (36.1) or (36.2) with a possibly complex-valued α, and where Θ is now required to satisfy

$$\Theta J \Theta^* = \Theta^* J \Theta = J \qquad \text{where} \qquad J = \text{diag}\{1, -1\}$$

We again need to distinguish between two cases.

$\boxed{|a| > |b|}$ **(first entry of the vector is dominant)**

In this case, we choose Θ as

$$\boxed{\Theta = \frac{1}{\sqrt{1-|\rho|^2}} \begin{bmatrix} 1 & -\rho \\ -\rho^* & 1 \end{bmatrix} \quad \text{where} \quad \rho = \frac{b}{a}} \tag{36.5}$$

This choice achieves the transformation (36.1). Specifically, it gives

$$\begin{bmatrix} a & b \end{bmatrix} \Theta = \begin{bmatrix} \pm e^{j\phi_a}\sqrt{|a|^2 - |b|^2} & 0 \end{bmatrix}$$

where ϕ_a denotes the phase of a.

$\boxed{|a| < |b|}$ **(second entry of the vector is dominant)**

Now we choose Θ as

$$\boxed{\Theta = \frac{1}{\sqrt{1-|\rho|^2}} \begin{bmatrix} 1 & -\rho \\ -\rho^* & 1 \end{bmatrix} \quad \text{where} \quad \rho = \frac{a^*}{b^*}} \tag{36.6}$$

This choice achieves the transformation (36.2). Specifically, it gives

$$\begin{bmatrix} a & b \end{bmatrix} \Theta = \begin{bmatrix} 0 & \pm e^{j\phi_b}\sqrt{|b|^2 - |a|^2} \end{bmatrix}$$

where ϕ_b denotes the phase of b. In summary, we are led to the following conclusion.

> **Lemma 36.2 (Complex hyperbolic Givens)** Consider a 1×2 vector $[a\ b]$ with possibly complex entries. If $|a| > |b|$, then choose Θ as in (36.5) to get
>
> $$\begin{bmatrix} a & b \end{bmatrix} \Theta = \pm\, e^{j\phi_a}\sqrt{|a|^2 - |b|^2} \begin{bmatrix} 1 & 0 \end{bmatrix}$$
>
> If, on the other hand, $|a| < |b|$, then choose Θ as in (36.6) to get
>
> $$\begin{bmatrix} a & b \end{bmatrix} \Theta = \pm\, e^{j\phi_b}\sqrt{|b|^2 - |a|^2} \begin{bmatrix} 0 & 1 \end{bmatrix}$$

36.2 HYPERBOLIC HOUSEHOLDER TRANSFORMATIONS

In contrast to hyperbolic Givens rotations, Householder transformations can be used to annihilate multiple entries in a row vector at once. We describe them below for both cases of real and complex data.

Real Data

Let e_0 denote the leading row basis vector in n−dimensional Euclidean space,

$$e_0 = \begin{bmatrix} 1 & 0 & \ldots & 0 \end{bmatrix}$$

and consider a $1 \times n$ real-valued vector z with entries $\{z(i), i = 0, 1, \ldots, n-1\}$. Assume that we wish to transform z to the form

$$\begin{bmatrix} z(0) & z(1) & \ldots & z(n-1) \end{bmatrix} \Theta = \alpha\, e_0 \tag{36.7}$$

for some nonzero real scalar α to be determined, and where the transformation Θ is required to be both J-orthogonal and involutary. By a J−orthogonal transformation we mean one that satisfies

$$\Theta J \Theta^{\mathsf{T}} = \Theta^{\mathsf{T}} J \Theta = J$$

for some given signature matrix J with $\pm$ diagonal entries, usually of the form

$$J = (\mathrm{I}_p \oplus -\mathrm{I}_q), \quad p \geq 1, \quad q \geq 1$$

and by an involutary matrix Θ we mean one that satisfies $\Theta^2 = \mathrm{I}$.

Again, and in contrast to the case of orthogonal Householder transformations in Sec. 34.1, the transformation (36.7) is not always possible. To see this, note from (36.7) that

$$z \underbrace{\Theta J \Theta^{\mathsf{T}}}_{J} z^{\mathsf{T}} \;=\; \alpha e_0 J e_0^{\mathsf{T}} \alpha \;=\; \alpha^2$$

i.e., $zJz^{\mathsf{T}} = \alpha^2$. Now since $\alpha^2 > 0$, no matter what the value of α is, this means that the transformation (36.7) is only possible whenever $zJz^{\mathsf{T}} > 0$. If, this is not the case, i.e., if

$zJz^{\mathsf{T}} < 0$, then we should seek instead a $J-$unitary transformation Θ that transforms z to the alternative form

$$\begin{bmatrix} z(0) & z(1) & \dots & z(n-1) \end{bmatrix} \Theta = \alpha\, e_{n-1} \tag{36.8}$$

where e_{n-1} is the last basis vector,

$$e_{n-1} = \begin{bmatrix} 0 & \dots & 0 & 1 \end{bmatrix}$$

In this case, the transformation (36.8) will guarantee $zJz^{\mathsf{T}} = -\alpha^2$, which is consistent with the fact that $zJz^{\mathsf{T}} < 0$. So let us examine the cases $zJz^{\mathsf{T}} > 0$ and $zJz^{\mathsf{T}} < 0$ separately.

$\boxed{zJz^{\mathsf{T}} > 0}$ **(positive value)**

To determine an expression for Θ that meets the requirement (36.7), we can follow the same geometric argument that we used in the orthogonal case in Chapter 34, except that we replace gg^{T} by gJg^{T} and the inner product zg^{T} by zJg^{T}. Therefore, the first step is to choose $\alpha = \pm\sqrt{zJz^{\mathsf{T}}}$ and $g = z - \alpha e_0$, and then to write

$$\alpha e_0 = z - 2zJg^{\mathsf{T}}(gJg^{\mathsf{T}})^{-1}g = z\underbrace{\left[\mathrm{I} - 2J\frac{g^{\mathsf{T}}g}{gJg^{\mathsf{T}}}\right]}_{\triangleq\,\Theta} \tag{36.9}$$

The indicated matrix Θ is called a hyperbolic Householder transformation; it is both J-unitary and involutary.

$\boxed{zJz^{\mathsf{T}} < 0}$ **(negative value)**

In this case, we can also follow the same geometric argument to determine an expression for Θ to achieve (36.8). The first steps are now to choose $\alpha = \pm\sqrt{-zJz^{\mathsf{T}}}$ and $g = z - \alpha e_{n-1}$, and then to write

$$\alpha e_{n-1} = z - 2zJg^{\mathsf{T}}(gJg^{\mathsf{T}})^{-1}g = z\underbrace{\left[\mathrm{I} - 2J\frac{g^{\mathsf{T}}g}{gJg^{\mathsf{T}}}\right]}_{\triangleq\,\Theta} \tag{36.10}$$

The indicated matrix Θ is also a hyperbolic Householder transformation; it is both J-unitary and involutary. In summary we arrive at the following statement.

Lemma 36.3 (Real hyperbolic Householder) Consider an $n-$dimensional vector z with real-valued entries. If $zJz^{\mathsf{T}} > 0$, then choose $g = z \pm \sqrt{zJz^{\mathsf{T}}}e_0$ and Θ as in (36.9) to get

$$z\Theta = \mp\sqrt{zJz^{\mathsf{T}}}e_0$$

If, on the other hand, $zJz^{\mathsf{T}} < 0$, then choose $g = z \pm \sqrt{-zJz^{\mathsf{T}}}e_{n-1}$ and Θ as in (36.10) to get

$$z\Theta = \mp\sqrt{-zJz^{\mathsf{T}}}\, e_{n-1}$$

Complex Data

More generally, consider a $1 \times n$ vector z with possibly complex entries, and assume that we wish to determine a transformation Θ that transforms it to either forms (36.7) or (36.8) with a possibly complex-valued α, and where Θ is now required to satisfy

$$\Theta J \Theta^* = \Theta^* J \Theta = J \quad \text{and} \quad \Theta^2 = \mathrm{I}$$

for some given signature matrix J and with the same involutory condition. We again need to distinguish between two cases.

$\boxed{zJz^* > 0}$ **(positive value)**

Introduce the polar representation of the first entry of z, namely, let $z(0) = |a|e^{j\phi_a}$. Then choose $g = z \pm \sqrt{zJz^*}\ e^{j\phi_a} e_0$ and Θ as

$$\boxed{\Theta \triangleq \mathrm{I} - 2\frac{Jg^*g}{gJg^*}} \tag{36.11}$$

This choice gives

$$z\Theta = \mp\sqrt{zJz^*}\ e^{j\phi_a} e_0$$

$\boxed{zJz^* < 0}$ **(negative value)**

Introduce the polar representation of the last entry of z, namely, let $z(n-1) = |b|e^{j\phi_b}$. Then choose $g = z \pm \sqrt{-zJz^*}\ e^{j\phi_b} e_{n-1}$ and Θ as in (36.11). This leads to

$$z\Theta = \mp\sqrt{-zJz^*}\ e^{j\phi_b} e_{n-1}$$

In summary, we arrive at the following conclusion.

Lemma 36.4 (Complex hyperbolic Householder) Consider a $1 \times n$ vector z with possibly complex-valued entries. If $zJz^* > 0$, then choose $g = z \pm \sqrt{zJz^*}e^{j\phi_a}\ e_0$ and Θ as in (36.11) to get

$$z\Theta = \mp e^{j\phi_a}\sqrt{zJz^*}e_0$$

If, on the other hand, $zJz^* < 0$, then choose $g = z \pm \sqrt{-zJz^*}e^{j\phi_b}e_{n-1}$ and Θ as in (36.11) to get

$$z\Theta = \mp e^{j\phi_b}\sqrt{-zJz^*}\ e_{n-1}$$

Here $\{ae^{j\phi_a}, be^{j\phi_b}\}$ denote the polar representations of the leading and trailing entries of z, respectively.

Remark 36.1 (Improved numerical accuracy) Computations with hyperbolic transformations in finite-precision can face numerical difficulties due to the possible accumulation of roundoff errors. If $p = q\Theta$, and if $\hat{p}$ denotes the computed vector that results from the evaluation of the product $q\Theta$ in finite precision, then it is known that

$$\|p - \hat{p}\| \leq O(\epsilon) \cdot \|q\| \cdot \|\Theta\|$$

where $\|\Theta\|$ denotes the spectral norm of Θ (i.e., its maximum singular value), ϵ denotes the machine precision, and $O(\epsilon)$ denotes a quantity of the order of the machine precision. This result assumes floating-point arithmetic and can be found, e.g., in Golub and Van Loan (1996). Now since hyperbolic rotations Θ can have relatively large norms, we find that, in general, the computed quantity $\hat{p}$ need not be evaluated accurately enough. Still, there are some careful ways for implementing hyperbolic transformations (especially hyperbolic Givens transformations) that help ameliorate numerical problems in finite-precision implementations. Two such methods are described in App. 14.A of Sayed (2003).

◇

36.3 HYPERBOLIC BASIS ROTATIONS

We now extend the result of Lemma 33.1 by replacing the equality $AA^* = BB^*$ by $AJA^* = BJB^*$, for some signature matrix J. We start with the following statement.

Lemma 36.5 (More columns than rows with a full-rank requirement) Let A and B be two $n \times m$ matrices with $n \leq m$ (i.e., the matrices have more columns than rows). Let $J = (\mathrm{I}_p \oplus -\mathrm{I}_q)$ be a signature matrix and assume that AJA^* has full rank. If $AJA^* = BJB^*$, then there should exist a J-unitary matrix Θ that maps B to A, i.e., $A = B\Theta$.

Proof: Let $S^{-1} = AJA^*$. Then S^{-1} is $n \times n$ Hermitian. Moreover, $S^{-1} = BJB^*$. Let $\mathrm{In}(S^{-1}) = \{\alpha, \beta\}$ denote the inertia of S^{-1}, with $\alpha+\beta = n$, and introduce the two block triangular factorizations:[18]

$$\begin{bmatrix} S^{-1} & A \\ A^* & J \end{bmatrix} = \begin{bmatrix} \mathrm{I} & \\ A^*S & \mathrm{I} \end{bmatrix} \begin{bmatrix} S^{-1} & \\ & J - A^*SA \end{bmatrix} \begin{bmatrix} \mathrm{I} & \\ A^*S & \mathrm{I} \end{bmatrix}^*$$

$$= \begin{bmatrix} \mathrm{I} & AJ \\ & \mathrm{I} \end{bmatrix} \begin{bmatrix} \underbrace{S^{-1} - AJA^*}_{=0} & \\ & J \end{bmatrix} \begin{bmatrix} \mathrm{I} & AJ \\ & \mathrm{I} \end{bmatrix}^*$$

Using Sylvester's law of inertia (cf. Lemma B.5), we conclude that the center matrices in the above factorizations must have the same inertia. In other words, it must hold that

$$\mathrm{In}\{J - A^*SA\} = \mathrm{In}\{J\} - \mathrm{In}\{S^{-1}\} = \{p - \alpha, q - \beta, n\}$$

where, from the definition of J, $p + q = m$. Similarly, $\mathrm{In}\{J - B^*SB\} = \{p - \alpha, q - \beta, n\}$. We therefore find that $J - A^*SA$ and $J - B^*SB$ are $m \times m$ matrices with n zero eigenvalues and $m - n$ nonzero eigenvalues. These matrices can then be factored as $J - A^*SA = XJ_1X^*$ and $J - B^*SB = YJ_1Y^*$, where $J_1 = (\mathrm{I}_{p-\alpha} \oplus -\mathrm{I}_{q-\beta})$ and $\{X, Y\}$ are $m \times (m - n)$. Now define the square matrices

$$\Sigma_1 = \begin{bmatrix} A \\ X^* \end{bmatrix}, \qquad \Sigma_2 = \begin{bmatrix} B \\ Y^* \end{bmatrix}$$

Then it follows that

$$\Sigma_1^*(S \oplus J_1)\Sigma_1 = J \qquad \text{and} \qquad \Sigma_2^*(S \oplus J_1)\Sigma_2 = J \tag{36.12}$$

so that Σ_1 and Σ_2 are invertible. Multiplying the first equality by $J\Sigma_1^*$ from the right we get $\Sigma_1^*(S \oplus J_1)\Sigma_1(J\Sigma_1^*) = J(J\Sigma_1^*) = \Sigma_1^*$, so that $\Sigma_1 J\Sigma_1^* = (S^{-1} \oplus J_1)$. Likewise, $\Sigma_2 J\Sigma_2^* = (S^{-1} \oplus J_1)$

[18]This elegant argument was suggested to the author by his late colleague Professor Tiberiu Constantinescu.

and, consequently, $\Sigma_1 J\Sigma_1^* = \Sigma_2 J\Sigma_2^*$. From (36.12) we have that $J\Sigma_1^* = \Sigma_1^{-1}(S^{-1} \oplus J)$ so that $\Sigma_1 = \Sigma_2[J\Sigma_2^*(S \oplus J_1)\Sigma_1]$. If we set $\Theta = [J\Sigma_2^*(S \oplus J_1)\Sigma_1]$, then Θ is J-unitary and, from the equality of the first block row of $\Sigma_1 = \Sigma_2\Theta$, we get $A = B\Theta$.

◇

In the above statement, the matrices A and B were either square or had more columns than rows (since $n \leq m$). We can establish a similar result when $n \geq m$ and A and B have full ranks. For this purpose, we first note that if A is $n \times m$ and has full rank, with $n \geq m$, then its SVD has the form

$$A = U \begin{bmatrix} \Sigma \\ 0 \end{bmatrix} V^*$$

where Σ is $n \times n$ and invertible. The pseudo inverse of A is $A^\dagger \triangleq V \begin{bmatrix} \Sigma^{-1} & 0 \end{bmatrix} U^*$ and it satisfies $A^\dagger A = \mathrm{I}_m$. In other words, the matrix A admits a right inverse. A similar conclusion holds for B.

Lemma 36.6 (More rows than columns) Consider two $n \times m$ matrices A and B with $n \geq m$ (i.e., the matrices have more rows than columns). Let $J = (\mathrm{I}_p \oplus -\mathrm{I}_q)$ be a signature matrix and assume that A and B have full rank. The equality $AJA^* = BJB^*$ holds if, and only if, there exists a $m \times m$ $J-$unitary matrix Θ such that $A = B\Theta$.

Proof: The "if" statement is immediate. If $A = B\Theta$, for some $J-$unitary Θ, then clearly $AJA^* = BJB^*$. For the converse statement, assume $AJA^* = BJB^*$ and define $\Theta = B^\dagger A$. Then this choice of Θ is $J-$unitary and it maps B to A. Indeed, note that

$$\underbrace{B^\dagger A}_{\Theta} J \underbrace{A^* B^{\dagger *}}_{\Theta^*} = \underbrace{B^\dagger B}_{\mathrm{I}_m} J \underbrace{B^\dagger B}_{\mathrm{I}_m} \underbrace{B^\dagger B}_{\mathrm{I}_m} = J$$

Moreover, from the relations $AJA^* = BJB^*$, $B^\dagger B = \mathrm{I}_m$, and $A^\dagger A = \mathrm{I}_m$ we obtain the following equalities:

$$AJA^* = B\underbrace{(B^\dagger B)}_{\mathrm{I}_m} JB^* = BB^\dagger(BJB^*) = BB^\dagger AJA^*$$

which, upon further multiplication from the right by $A^{\dagger *}$, give

$$AJ\underbrace{A^* A^{\dagger *}}_{\mathrm{I}_m} = BB^\dagger AJ\underbrace{A^* A^{\dagger *}}_{\mathrm{I}_m} = B\underbrace{B^\dagger A}_{\Theta} J$$

That is, $AJ = B\Theta J$. But since J is invertible we arrive at $A = B\Theta$, as desired.

CHAPTER 37

Fast Array Algorithm

All least-squares algorithms studied so far, including RLS, inverse QR, QR, and extended QR algorithms, do not assume any structure in the data. As a result, the computational complexity of each of these algorithms is $O(M^2)$ operations per iteration, where M is the order of the filter. However, when data structure is present, more efficient implementations are possible.

Thus, consider a collection of $(N+1)$ data $\{d(j), u_j\}$ where the $\{u_j\}$ are $1 \times M$ and the $d(j)$ are scalars. All the aforementioned algorithms are recursive procedures for determining the solution w_N, and the minimum cost $\xi(N)$, of the regularized least-squares problem:

$$\min_w \left[\lambda^{(N+1)}(w - \bar{w})^* \Pi (w - \bar{w}) + \sum_{j=0}^{N} \lambda^{N-j} |d(j) - u_j w|^2 \right] \tag{37.1}$$

where $\bar{w}$ is $M \times 1$, $\Pi > 0$ is $M \times M$ and $0 \ll \lambda \leq 1$. In particular, RLS evaluates w_N recursively as follows (cf. Alg. 30.2). Start with $w_{-1} = \bar{w}$, $P_{-1} = \Pi^{-1}$, $\xi(-1) = 0$, and repeat for $i \geq 0$:

$$\begin{array}{rcl} \gamma(i) & = & 1/(1 + \lambda^{-1} u_i P_{i-1} u_i^*) \ = \ 1 - u_i P_i u_i^* \\ g_i & = & \lambda^{-1}\gamma(i) P_{i-1} u_i^* \ = \ P_i u_i^* \\ e(i) & = & d(i) - u_i w_{i-1} \\ w_i & = & w_{i-1} + g_i e(i) \\ P_i & = & \lambda^{-1} P_{i-1} - g_i g_i^* / \gamma(i) \\ r(i) & = & d(i) - u_i w_i \\ \xi(i) & = & \lambda \xi(i-1) \ + \ r(i) e^*(i) \end{array} \tag{37.2}$$

These equations hold irrespective of any structure in the $\{u_j\}$.

Now, it is often the case that the regressors $\{u_i\}$ exhibit some form of structure. In the chapters in this part, we study the case in which the $\{u_i\}$ arise as regressors of a tapped-delay-line implementation, as shown in Fig. 37.1. That is, we assume that the entries of u_i are formed from time-delayed samples of an input sequence $\{u(\cdot)\}$, say,

$$\boxed{u_i = \left[\ u(i) \quad u(i-1) \quad \ldots \quad u(i-M+2) \quad u(i-M+1)\ \right]} \tag{37.3}$$

In this case, two successive regressors will share most of their entries since, for example, u_{i-1} will be given by

$$\boxed{u_{i-1} = \left[\ u(i-1) \quad u(i-2) \quad \ldots \quad u(i-M+1) \quad u(i-M)\ \right]} \tag{37.4}$$

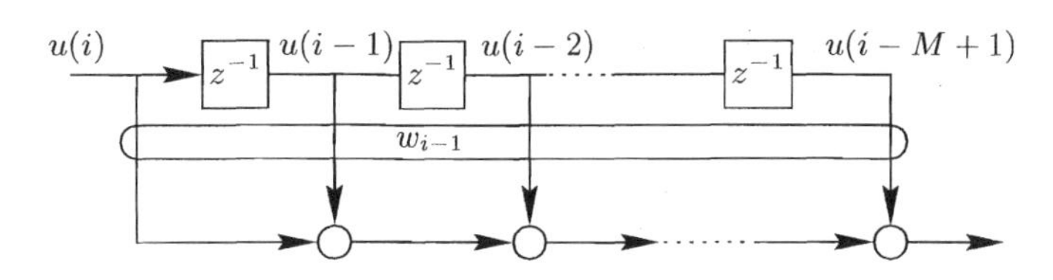

FIGURE 37.1 A tapped-delay-line structure resulting in regressors with shift-structure

Comparing expressions (37.3) and (37.4) for u_i and u_{i-1} we see that u_i is obtained from u_{i-1} by shifting the entries of the latter by one position to the right and introducing a new entry, $u(i)$, at the left. One way to capture this relation is to note that the following equality holds:

$$\boxed{\begin{bmatrix} u(i) & u_{i-1} \end{bmatrix} = \begin{bmatrix} u_i & u(i-M) \end{bmatrix}} \tag{37.5}$$

That is, if we extend u_{i-1} by one entry to the left and u_i by one entry to the right, then the extended vectors will coincide.

The shift structure in the regressors can be exploited to great effect in order to devise efficient recursive least-squares solutions. By efficient we mean algorithms whose computational complexity is $O(M)$ operations per iteration, as opposed to $O(M^2)$; i.e., they are an order of magnitude more efficient than the slower implementations that we studied before (RLS, inverse QR, QR, extended QR). There are several classes of efficient RLS algorithms that can be derived by exploiting data structure. In this part we study fixed-order algorithms, while in Part X (*Lattice Filters*) we study order-recursive algorithms.

We start our discussions by deriving an efficient RLS algorithm in *array* form; we do so since the derivation of the array form is more immediate than other efficient RLS algorithms. Later, in Secs. 39.1–39.3 we derive efficient algorithms in explicit forms.

37.1 TIME-UPDATE OF THE GAIN VECTOR

Consider the RLS update (37.2) for w_i, namely,

$$w_i = w_{i-1} + g_i[d(i) - u_i w_{i-1}] \tag{37.6}$$

and note that RLS requires the gain vector g_i in order to compute w_i. In turn, the evaluation of g_i requires the matrix P_{i-1} (or P_i), and updating $\{P_{i-1} \text{ or } P_i\}$ requires $O(M^2)$ operations per iteration. This update step is the main computational bottleneck in the RLS algorithm. However, when the $\{u_i\}$ have shift structure, as indicated by (37.5), the gain vector g_i can be evaluated more immediately without requiring evaluation of P_{i-1} (or P_i). This will be achieved by developing a time-update for g_i itself, i.e., by showing how to compute g_i from g_{i-1} directly.

From the definition of g_i in (37.2) we have that

$$g_i\gamma^{-1}(i) = \lambda^{-1}P_{i-1}u_i^* \quad \text{and} \quad g_{i-1}\gamma^{-1}(i-1) = \lambda^{-1}P_{i-2}u_{i-1}^* \tag{37.7}$$

Now, because of the shift structure in the regressors $\{u_i, u_{i-1}\}$, we can express the extended vector $\text{col}\{P_{i-1}u_i^*, 0\}$ in the equivalent forms

$$\begin{bmatrix} P_{i-1}u_i^* \\ 0 \end{bmatrix} = \begin{bmatrix} P_{i-1} & 0 \\ 0 & 0 \end{bmatrix}\begin{bmatrix} u_i^* \\ u^*(i-M) \end{bmatrix} = \begin{bmatrix} P_{i-1} & 0 \\ 0 & 0 \end{bmatrix}\begin{bmatrix} u^*(i) \\ u_{i-1}^* \end{bmatrix} \tag{37.8}$$

where in the second equality we used (37.5). Likewise, we can write

$$\begin{bmatrix} 0 \\ P_{i-2}u_{i-1}^* \end{bmatrix} = \begin{bmatrix} 0 & 0 \\ 0 & P_{i-2} \end{bmatrix} \begin{bmatrix} u^*(i) \\ u_{i-1}^* \end{bmatrix} \tag{37.9}$$

Consequently, subtracting (37.8) and (37.9), we get

$$\lambda^{-1}\begin{bmatrix} P_{i-1}u_i^* \\ 0 \end{bmatrix} - \lambda^{-1}\begin{bmatrix} 0 \\ P_{i-2}u_{i-1}^* \end{bmatrix} = \lambda^{-1}\left(\begin{bmatrix} P_{i-1} & 0 \\ 0 & 0 \end{bmatrix} - \begin{bmatrix} 0 & 0 \\ 0 & P_{i-2} \end{bmatrix}\right)\begin{bmatrix} u^*(i) \\ u_{i-1}^* \end{bmatrix}$$

Let δP_{i-1} denote the difference

$$\delta P_{i-1} \stackrel{\Delta}{=} \begin{bmatrix} P_{i-1} & 0 \\ 0 & 0 \end{bmatrix} - \begin{bmatrix} 0 & 0 \\ 0 & P_{i-2} \end{bmatrix} \qquad (M+1)\times(M+1)$$

Using the defining relations (37.7) for $\{g_i, g_{i-1}\}$, we arrive at

$$\boxed{\begin{bmatrix} g_i\gamma^{-1}(i) \\ 0 \end{bmatrix} = \begin{bmatrix} 0 \\ g_{i-1}\gamma^{-1}(i-1) \end{bmatrix} + \lambda^{-1}\delta P_{i-1}\begin{bmatrix} u^*(i) \\ u_{i-1}^* \end{bmatrix}} \tag{37.10}$$

This is a significant result. It shows that in order to time-update the gain vector from $g_{i-1}\gamma^{-1}(i-1)$ to $g_i\gamma^{-1}(i)$, it is only necessary to know what the difference δP_{i-1} is; it is not necessary to know the value of the individual matrices $\{P_i, P_{i-1}\}$ themselves. In this way, it suffices to know how to update δP_{i-1} to δP_i in order to carry out the updates:

$$g_{i-1}\gamma^{-1}(i-1) \xrightarrow{\delta P_{i-1}} g_i\gamma^{-1}(i) \xrightarrow{\delta P_i} g_{i+1}\gamma^{-1}(i+1) \longrightarrow \dots$$

37.2 TIME-UPDATE OF THE CONVERSION FACTOR

Although knowledge of δP_{i-1} is enough to update $g_{i-1}\gamma^{-1}(i-1)$ to $g_i\gamma^{-1}(i)$, we still need to recover g_i itself. In other words, we still need to know how to remove the scaling by $\gamma^{-1}(i)$ from $g_i\gamma^{-1}(i)$. If we were to evaluate $\gamma(i)$ as suggested by the RLS implementation (37.2), in terms of P_{i-1} (or P_i), then we would be back to square one in terms of excessive computational complexity. However, it turns out that the conversion factor $\gamma(i)$ can also be time-updated in a manner that only requires knowledge of δP_{i-1}.

To see this, we use the expression for $\gamma(i)$ in (37.2) to write

$$\gamma^{-1}(i) \;=\; 1+\lambda^{-1}u_iP_{i-1}u_i^*, \quad \gamma^{-1}(i-1) = 1+\lambda^{-1}u_{i-1}P_{i-2}u_{i-1}^*$$

so that, upon subtraction, we get

$$\gamma^{-1}(i) = \gamma^{-1}(i-1) + \lambda^{-1}\left[u_iP_{i-1}u_i^* - u_{i-1}P_{i-2}u_{i-1}^*\right]$$

Again, using the partitioning (37.5), we can express the difference

$$\nabla \stackrel{\Delta}{=} u_iP_{i-1}u_i^* - u_{i-1}P_{i-2}u_{i-1}^*$$

as

$$\begin{aligned} \nabla &= \begin{bmatrix} u(i) & u_{i-1} \end{bmatrix}\left(\begin{bmatrix} P_{i-1} & 0 \\ 0 & 0 \end{bmatrix} - \begin{bmatrix} 0 & 0 \\ 0 & P_{i-2} \end{bmatrix}\right)\begin{bmatrix} u^*(i) \\ u_{i-1}^* \end{bmatrix} \\ &= \begin{bmatrix} u(i) & u_{i-1} \end{bmatrix}\delta P_{i-1}\begin{bmatrix} u^*(i) \\ u_{i-1}^* \end{bmatrix} \end{aligned}$$

and, hence,

$$\boxed{\gamma^{-1}(i) = \gamma^{-1}(i-1) + \lambda^{-1}\begin{bmatrix} u(i) & u_{i-1} \end{bmatrix} \delta P_{i-1} \begin{bmatrix} u^*(i) \\ u^*_{i-1} \end{bmatrix}} \tag{37.11}$$

which shows, as desired, that knowledge of δP_{i-1} is also sufficient for the time-update of the conversion factor.

37.3 INITIAL CONDITIONS

In order to complete the argument, we need to show how to compute the factor δP_{i-1} that appears in (37.10) and (37.11) in an efficient manner. This step requires that the regularization matrix Π be chosen in an appropriate manner, as we now explain.

Assume $u(i) = 0$ for $i < 0$ so that $u_{-1} = 0$ (i.e., the initial state of the filter is zero). Then $g_{-1} = 0, \gamma(-1) = 1, P_{-1} = \Pi^{-1}$, and $P_{-2} = \lambda P_{-1} = \lambda\Pi^{-1}$, so that the difference δP_i at time -1 is given by

$$\delta P_{-1} = \begin{bmatrix} P_{-1} & 0 \\ 0 & 0 \end{bmatrix} - \begin{bmatrix} 0 & 0 \\ 0 & P_{-2} \end{bmatrix} = \begin{bmatrix} \Pi^{-1} & 0 \\ 0 & 0 \end{bmatrix} - \begin{bmatrix} 0 & 0 \\ 0 & \lambda\Pi^{-1} \end{bmatrix} \tag{37.12}$$

The value of this difference depends on our choice of Π. So assume that we choose Π^{-1} as the diagonal matrix

$$\boxed{\Pi^{-1} = \eta \cdot \text{diag}\ \left\{\lambda^2, \lambda^3, \ldots, \lambda^{M+1}\right\}} \tag{37.13}$$

where η is a positive scalar (usually large) and λ is the forgetting factor. Then δP_{-1} becomes

$$\delta P_{-1} = \eta\lambda^2 \cdot \text{diag}\{1, 0, \ldots, 0, -\lambda^M\}$$

That is, δP_{-1} reduces to a rank-two matrix with one positive eigenvalue and one negative eigenvalue. Actually, it is not difficult to verify that δP_{-1} can be factored as

$$\boxed{\delta P_{-1} = \lambda \cdot \bar{L}_{-1} S_{-1} \bar{L}^*_{-1}}$$

where $\bar{L}_{-1}$ is $(M+1) \times 2$ and S_{-1} is 2×2 and given by

$$\boxed{\bar{L}_{-1} = \sqrt{\eta\lambda} \cdot \begin{bmatrix} 1 & 0 \\ 0 & 0 \\ \vdots & \vdots \\ 0 & 0 \\ 0 & \lambda^{M/2} \end{bmatrix}, \qquad S_{-1} = \begin{bmatrix} 1 & 0 \\ 0 & -1 \end{bmatrix}} \tag{37.14}$$

The signature matrix, S_{-1}, indicates that the difference δP_{-1} has one positive eigenvalue and one negative eigenvalue; all other eigenvalues are zero since the rank is 2.

This argument shows that the choice (37.13) for Π leads to a rank 2 difference matrix δP_{-1} with signature $(1 \oplus -1)$. A striking property that is established in the next subsection is that the rank and inertia properties of δP_i remain *invariant* over time; the rank never goes up; it is always 2, with the same signature matrix S_{-1}, except in rare degenerate situations where the rank can only drop below 2. More specifically, we argue ahead that once the low-rank property holds at a certain time instant $i-1$, say,

$$\boxed{\begin{bmatrix} P_{i-1} & 0 \\ 0 & 0 \end{bmatrix} - \begin{bmatrix} 0 & 0 \\ 0 & P_{i-2} \end{bmatrix} = \lambda \cdot \bar{L}_{i-1} S_{i-1} \bar{L}_{i-1}^*} \tag{37.15}$$

for some $\{\bar{L}_{i-1}, S_{i-1}\}$, then three important facts hold:

1. The low-rank property will be valid at time i as well, say,

$$\begin{bmatrix} P_i & 0 \\ 0 & 0 \end{bmatrix} - \begin{bmatrix} 0 & 0 \\ 0 & P_{i-1} \end{bmatrix} = \lambda \cdot \bar{L}_i S_i \bar{L}_i^* \qquad \text{for some } \{\bar{L}_i, S_i\}.$$

2. There will exist an array algorithm that updates $\bar{L}_{i-1}$ to $\bar{L}_i$, and this same algorithm will also provide the gain vector g_i that is needed in (37.6).

3. The signature matrices $\{S_{i-1}, S_i\}$ will coincide.

37.4 ARRAY ALGORITHM

The desired array algorithm for updating $\{\bar{L}_{i-1}, g_{i-1}, \gamma(i-1)\}$ to $\{\bar{L}_i, g_i, \gamma(i)\}$ can be motivated and derived in much the same manner as we did in Sec. 33.4.

Thus, starting from (37.10) and (37.11), namely,

$$\begin{aligned} \gamma^{-1}(i) &= \gamma^{-1}(i-1) + \begin{bmatrix} u(i) & u_{i-1} \end{bmatrix} \bar{L}_{i-1} S_{i-1} \bar{L}_{i-1}^* \begin{bmatrix} u^*(i) \\ u_{i-1}^* \end{bmatrix} \\ \begin{bmatrix} g_i \gamma^{-1}(i) \\ 0 \end{bmatrix} &= \begin{bmatrix} 0 \\ g_{i-1}\gamma^{-1}(i-1) \end{bmatrix} + \bar{L}_{i-1} S_{i-1} \bar{L}_{i-1}^* \begin{bmatrix} u^*(i) \\ u_{i-1}^* \end{bmatrix} \end{aligned}$$

we note that these expressions are of the form

$$CC^* = AA^* + BSB^*, \qquad FC^* = DA^* + ESB^* \tag{37.16}$$

with the following identifications

$$\begin{cases} C \leftarrow \gamma^{-1/2}(i) & A \leftarrow \gamma^{-1/2}(i-1) & B \leftarrow \begin{bmatrix} u(i) & u_{i-1} \end{bmatrix} \bar{L}_{i-1} \\ F \leftarrow \begin{bmatrix} g_i \gamma^{-1/2}(i) \\ 0 \end{bmatrix} & D \leftarrow \begin{bmatrix} 0 \\ g_{i-1}\gamma^{-1/2}(i-1) \end{bmatrix} & E \leftarrow \bar{L}_{i-1} \\ S \leftarrow S_{i-1} & & \end{cases}$$

Generic Array Algorithm

The forms (37.16) play a role similar to the "norm-preserving" and "inner-product preserving" equalities (33.19) and (33.20), which we used in Sec. 33.4 to motivate and derive array algorithms. The main distinction is the presence of the signature matrix S in (37.16). Still, the same arguments that we employed in Sec. 33.4 could be repeated here with the only issue being that we now need to deal with hyperbolic transformations as opposed to unitary transformations.

More specifically, as in (33.21), given general equalities of the form (37.16) where it is desired to evaluate $\{C, F\}$ from knowledge of $\{A, B, D, E, S\}$, we would form the pre-array

$$\mathcal{A} = \begin{bmatrix} A & B \\ D & E \end{bmatrix} \quad \text{and reduce it to the form} \quad \begin{bmatrix} \mathsf{X} & 0 \\ \mathsf{Y} & \mathsf{Z} \end{bmatrix} \tag{37.17}$$

by annihilating B by means of a transformation Θ that is now required to be $(\mathrm{I}\oplus S)$-unitary, i.e., it should satisfy

$$\Theta\begin{bmatrix} \mathrm{I} & \\ & S \end{bmatrix}\Theta^* = \begin{bmatrix} \mathrm{I} & \\ & S \end{bmatrix}$$

The question of course is whether such a Θ exists. While in Sec. 33.4 we started from equality (33.19) and appealed to (33.9) to justify the existence of a Θ that achieves (33.21), this same argument cannot be applied to the equality

$$CC^* = AA^* + BSB^* \tag{37.18}$$

due to the presence of the signature matrix S. However, Lemma 36.5 already extends the result (33.9) to handle such more general cases with signature matrices. Specifically, by writing (37.18) as

$$\begin{bmatrix} C & 0 \end{bmatrix}\begin{bmatrix} C^* \\ 0 \end{bmatrix} = \begin{bmatrix} A & B \end{bmatrix}\begin{bmatrix} \mathrm{I} & \\ & S \end{bmatrix}\begin{bmatrix} A^* \\ B^* \end{bmatrix}$$

we are able to appeal to Lemma 36.5 to conclude that an $(\mathrm{I}\oplus S)-$unitary Θ exists that maps $[A \quad B]$ to $[C \quad 0]$. Therefore, in a manner similar to the explanations that led to (33.21), in order to determine $\{C, F\}$ we would first use one such $(\mathrm{I}\oplus S)-$unitary matrix Θ to perform the transformation

$$\begin{bmatrix} A & B \\ C & D \end{bmatrix}\Theta = \begin{bmatrix} \mathsf{X} & 0 \\ \mathsf{Y} & \mathsf{Z} \end{bmatrix}$$

and then "square" and compare entries on both sides of the equality:

$$\begin{bmatrix} A & B \\ C & D \end{bmatrix}\underbrace{\Theta\begin{bmatrix} \mathrm{I} & \\ & S \end{bmatrix}\Theta^*}_{(\mathrm{I}\oplus S)}\begin{bmatrix} A & B \\ C & D \end{bmatrix}^* = \begin{bmatrix} \mathsf{X} & 0 \\ \mathsf{Y} & \mathsf{Z} \end{bmatrix}\begin{bmatrix} \mathrm{I} & \\ & S \end{bmatrix}\begin{bmatrix} \mathsf{X} & 0 \\ \mathsf{Y} & \mathsf{Z} \end{bmatrix}^*$$

in order to identify X as C and Y as F.

Fast RLS Array Algorithm

Applying this construction to the RLS case (37.16), we would first form the pre-array

$$\mathcal{A} = \begin{bmatrix} \gamma^{-1/2}(i-1) & \begin{bmatrix} u(i) & u_{i-1} \end{bmatrix}\bar{L}_{i-1} \\ \begin{bmatrix} 0 \\ g_{i-1}\gamma^{-1/2}(i-1) \end{bmatrix} & \bar{L}_{i-1} \end{bmatrix}$$

For example, when $M = 3$, this array has the form (recall that $\bar{L}_i$ is $(M+1)\times 2$):

$$\mathcal{A} = \left[\begin{array}{c|cc} \times & \times & \times \\ \hline 0 & \times & \times \\ \times & \times & \times \\ \times & \times & \times \\ \times & \times & \times \end{array}\right]$$

Then we would define the signature matrix $J = (1\oplus S_{i-1})$, and choose a $J-$unitary matrix Θ_i (i.e., $\Theta_i J \Theta_i^* = J$) such that it transforms $\mathcal{A}$ to the form

$$\mathcal{B} = \begin{bmatrix} \times & 0 \\ \mathsf{y} & \mathsf{Z} \end{bmatrix} \tag{37.19}$$

for some quantities $\{\times, \mathsf{y}, \mathsf{Z}\}$ to be determined, with $\times$ a positive scalar, y a column vector, and Z a two column matrix. Again, for $M = 3$, the post-array will be of the form:

$$\mathcal{B} = \left[\begin{array}{c|cc} \times & 0 & 0 \\ \hline \times & \times & \times \\ \times & \times & \times \\ \times & \times & \times \\ \times & \times & \times \end{array}\right]$$

The matrix Θ_i can be implemented in many ways (as described in Chapter 36). For example, we could employ a circular (Givens) rotation that pivots with the left-most entry of the first row and annihilates its second entry. We could then employ a hyperbolic rotation that pivots again with the left-most entry and annihilates the last entry of the same row:

$$\left[\begin{array}{ccc} \boxed{\times} & \boxed{\times} & \times \\ \hline 0 & \times & \times \\ \times & \times & \times \\ \times & \times & \times \\ \times & \times & \times \end{array}\right] \xrightarrow{\text{Givens}} \left[\begin{array}{ccc} \boxed{\times'} & 0 & \boxed{\times} \\ \hline \times' & \times' & \times \\ \times' & \times' & \times \\ \times' & \times' & \times \\ \times' & \times' & \times \end{array}\right] \xrightarrow{\text{hyperbolic}} \left[\begin{array}{ccc} \times'' & 0 & 0 \\ \hline \times'' & \times' & \times'' \\ \times'' & \times' & \times'' \\ \times'' & \times' & \times'' \\ \times'' & \times' & \times'' \end{array}\right]$$

Now, in order to identify the entries $\{\times, \mathsf{y}, \mathsf{Z}\}$ in the post-array (37.19), i.e., in the equality below

$$\underbrace{\left[\begin{array}{cc} \gamma^{-1/2}(i-1) & \left[\begin{array}{cc} u(i) & u_{i-1} \end{array}\right] \bar{L}_{i-1} \\ \left[\begin{array}{c} 0 \\ g_{i-1}\gamma^{-1/2}(i-1) \end{array}\right] & \bar{L}_{i-1} \end{array}\right]}_{\mathcal{A}} \Theta_i = \underbrace{\left[\begin{array}{cc} \times & 0 \\ \mathsf{y} & \mathsf{Z} \end{array}\right]}_{\mathcal{B}}$$

we simply compare entries on both sides of the equality $\mathcal{A}\Theta_i J \Theta_i^* \mathcal{A}^* = \mathcal{B} J \mathcal{B}^*$ to find that

$$\left\{\begin{array}{lcl} |\times|^2 & = & \gamma^{-1}(i-1) + \left[\begin{array}{cc} u(i) & u_{i-1} \end{array}\right] \bar{L}_{i-1} S_{i-1} \bar{L}_{i-1}^* \left[\begin{array}{c} u^*(i) \\ u_{i-1}^* \end{array}\right] \\ \mathsf{y}\times^* & = & \left[\begin{array}{c} 0 \\ g_{i-1}\gamma^{-1}(i-1) \end{array}\right] + \bar{L}_{i-1} S_{i-1} \bar{L}_{i-1}^* \left[\begin{array}{c} u^*(i) \\ u_{i-1}^* \end{array}\right] \\ \mathsf{y}\mathsf{y}^* + \mathsf{Z} S_{i-1} \mathsf{Z}^* & = & \left[\begin{array}{c} 0 \\ g_{i-1}\gamma^{-1/2}(i) \end{array}\right] \left[\begin{array}{c} 0 \\ g_{i-1}\gamma^{-1/2}(i-1) \end{array}\right]^* + \bar{L}_{i-1} S_{i-1} \bar{L}_{i-1}^* \end{array}\right. \tag{37.20}$$

The right-hand side of the first equality coincides with that of (37.11) so that we can identify $\times$ as $\times = \gamma^{-1/2}(i)$. Similarly, the right-hand side of the second equality in (37.20) coincides with that of (37.10), so that we can identify y as

$$\mathsf{y} = \left[\begin{array}{c} g_i \gamma^{-1/2}(i) \\ 0 \end{array}\right]$$

Finally, the last equality in (37.20) leads to

$$\begin{aligned}
\mathsf{Z}S_{i-1}\mathsf{Z}^* &= \begin{bmatrix} 0 \\ g_{i-1}\gamma^{-1/2}(i-1) \end{bmatrix}\begin{bmatrix} 0 \\ g_{i-1}\gamma^{-1/2}(i-1) \end{bmatrix}^* \\
&\quad + \begin{bmatrix} \lambda^{-1}P_{i-1} & 0 \\ 0 & 0 \end{bmatrix} - \begin{bmatrix} 0 & 0 \\ 0 & \lambda^{-1}P_{i-2} \end{bmatrix} - \begin{bmatrix} g_i\gamma^{-1/2}(i) \\ 0 \end{bmatrix}\begin{bmatrix} g_i\gamma^{-1/2}(i) \\ 0 \end{bmatrix}^* \\
&= \begin{bmatrix} 0 & 0 \\ 0 & \lambda^{-1}P_{i-2} - P_{i-1} \end{bmatrix} + \begin{bmatrix} \lambda^{-1}P_{i-1} & 0 \\ 0 & 0 \end{bmatrix} - \begin{bmatrix} 0 & 0 \\ 0 & \lambda^{-1}P_{i-2} \end{bmatrix} \\
&\quad - \begin{bmatrix} \lambda^{-1}P_{i-1} - P_i & 0 \\ 0 & 0 \end{bmatrix} \\
&= \begin{bmatrix} P_i & 0 \\ 0 & 0 \end{bmatrix} - \begin{bmatrix} 0 & 0 \\ 0 & P_{i-1} \end{bmatrix} = \delta P_i
\end{aligned}$$

The difference δP_i is, by definition, $\lambda \bar{L}_i S_i \bar{L}_i^*$, so that

$$\mathsf{Z}S_{i-1}\mathsf{Z}^* = \lambda \bar{L}_i S_i \bar{L}_i^*$$

This result shows that the difference δP_i has the same signature matrix, S_{i-1}, as δP_{i-1} and, consequently, S_i remains invariant for all i. For this reason, we shall drop the time subscript i from S_i and write S instead. Moreover, we can identify Z as $\mathsf{Z} = \sqrt{\lambda} \cdot \bar{L}_i$. In summary, we arrive at the following statement.

Algorithm 37.1 (Fast RLS array algorithm) Consider data $\{u_j, d(j)\}_{j=0}^N$, where the u_j are $1 \times M$ and the $d(j)$ are scalars. Consider also an $M \times 1$ vector $\bar{w}$, a scalar $0 \ll \lambda \leq 1$, and an $M \times M$ positive-definite matrix Π^{-1} of the form

$$\Pi^{-1} = \eta \cdot \text{diagonal}\,\{\lambda^2, \lambda^3, \ldots, \lambda^{M+1}\}, \quad \eta > 0$$

When the $\{u_j\}$ correspond to regressors of a tapped-delay-line implementation, the solution w_N of the least-squares problem (37.1) can be recursively computed as follows. Start with $w_{-1} = \bar{w}$, $\gamma^{-1/2}(-1) = 1$, $g_{-1} = 0$, $\bar{L}_{-1}$ and S as in (37.14), $J = (1 \oplus S)$, and repeat for $i \geq 0$:

1. Find a $J-$unitary matrix Θ_i that annihilates the last two entries in the top row of the post-array below and generates a positive leading entry. Then the entries of the post-array will correspond to

$$\begin{bmatrix} \gamma^{-1/2}(i-1) & \begin{bmatrix} u(i) & u_{i-1} \end{bmatrix}\bar{L}_{i-1} \\ \begin{bmatrix} 0 \\ g_{i-1}\gamma^{-1/2}(i-1) \end{bmatrix} & \bar{L}_{i-1} \end{bmatrix}\Theta_i$$

$$= \begin{bmatrix} \gamma^{-1/2}(i) & \begin{bmatrix} 0 & 0 \end{bmatrix} \\ \begin{bmatrix} g_i\gamma^{-1/2}(i) \\ 0 \end{bmatrix} & \sqrt{\lambda}\,\bar{L}_i \end{bmatrix}$$

2. Update $w_i = w_{i-1} + \left[g_i\gamma^{-1/2}(i)\right]\left[\gamma^{-1/2}(i)\right]^{-1}\left[d(i) - u_i w_{i-1}\right]$, where the quantities $\{g_i\gamma^{-1/2}(i), \gamma^{-1/2}(i)\}$ are read from the post-array.

Observe that this array algorithm computes the gain vector g_i without evaluating the $M \times M$ matrix P_i. Instead, the low-rank factor $\bar{L}_i$, which is $(M+1) \times 2$, is propagated, resulting in a lower computational complexity. Later, in Table 39.1, we shall compare the computational requirements of several fast fixed-order variants.

37.A APPENDIX: CHANDRASEKHAR FILTER

The fast algorithms of this chapter have connections with a fast alternative to the Kalman filter for constant state-space models, known as the Chandrasekhar filter.

Thus, refer to the material on Kalman filtering in Chapter 7 and consider again the standard state-space model (7.8)–(7.10) (say, of order n):

$$\boxed{\begin{array}{l} \boldsymbol{x}_{i+1} = F_i \boldsymbol{x}_i + G_i \boldsymbol{n}_i, \quad i \geq 0 \\ \boldsymbol{y}_i \ = \ H_i \boldsymbol{x}_i + \boldsymbol{v}_i \\ \mathsf{E}\left[\begin{array}{c} \boldsymbol{n}_i \\ \boldsymbol{v}_i \\ \boldsymbol{x}_0 \\ 1 \end{array}\right]\left[\begin{array}{c} \boldsymbol{n}_j \\ \boldsymbol{v}_j \\ \boldsymbol{x}_0 \end{array}\right]^* = \left[\begin{array}{cc} \left[\begin{array}{cc} Q_i & 0 \\ 0 & R_i \end{array}\right]\delta_{ij} & \begin{array}{c} 0 \\ 0 \end{array} \\ \begin{array}{cc} 0 & 0 \end{array} & \Pi_0 \\ \begin{array}{cc} 0 & 0 \end{array} & 0 \end{array}\right] \end{array}} \tag{37.21}$$

We assume, without loss of generality, that $S_i = 0$, i.e., that the processes $\{\boldsymbol{n}_i, \boldsymbol{v}_i\}$ are uncorrelated. This case is sufficient for our purposes here.

The Kalman filter recursions, in covariance form, for computing the innovations and, subsequently, predicting the state vector are given by (cf. Alg. 7.1):

$$\begin{array}{rcl} R_{e,i} & = & R_i + H_i P_{i|i-1} H_i^* \\ K_{p,i} & = & F_i P_{i|i-1} H_i^* R_{e,i}^{-1} \\ \boldsymbol{\nu}_i & = & \boldsymbol{y}_i - H_i \hat{\boldsymbol{x}}_{i|i-1} \\ \hat{\boldsymbol{x}}_{i+1|i} & = & F_i \hat{\boldsymbol{x}}_i + K_{p,i} \boldsymbol{\nu}_i \\ P_{i+1|i} & = & F_i P_{i|i-1} F_i^* + G_i Q_i G_i^* - K_{p,i} R_{e,i} K_{p,i}^* \end{array} \tag{37.22}$$

with initial conditions $\hat{\boldsymbol{x}}_{0|-1} = 0$ and $P_{0|-1} = \Pi_0$. Again, for convenience of exposition, we are denoting the innovations variable of the Kalman filter by $\boldsymbol{\nu}(i)$ in order to avoid conflict of notation with the symbol $e(i)$ used to denote the output error of RLS.

Array Chandrasekhar filter for constant models. We present first the array form of the Chandrasekhar filter. Its derivation relies on the same kind of arguments used in the body of the chapter while deriving the fast array RLS method. For this reason, we shall be brief. Once the array method is presented, the reader will be able to recognize the close similarity between the RLS and Kalman filtering domains (by pursuing Prob. IX.16).

So assume initially that the model parameters are constant, say, $\{F, G, H, Q, R\}$, and introduce the factorization $P_{i+1|i} - P_{i|i-1} = \bar{L}_{i|i-1} S \bar{L}_{i|i-1}^*$, where $\bar{L}_{i|i-1}$ is $n \times \alpha$, S is an $\alpha \times \alpha$ signature matrix with as many $\pm 1's$ as $(P_{i+1|i} - P_{i|i-1})$ has positive and negative eigenvalues. Moreover, $\alpha = \text{rank}(P_{i+1|i} - P_{i|i-1})$. The array algorithm follows by forming the pre-array

$$A = \left[\begin{array}{cc} R_{e,i}^{1/2} & H\bar{L}_{i|i-1} \\ K_{p,i} R_{e,i}^{1/2} & F\bar{L}_{i|i-1} \end{array}\right]$$

and triangularizing it via an $(\mathsf{I} \oplus S)-$unitary matrix Θ_i, i.e.,

$$A\Theta_i = \left[\begin{array}{cc} R_{e,i}^{1/2} & H\bar{L}_{i|i-1} \\ K_{p,i} R_{e,i}^{1/2} & F\bar{L}_{i|i-1} \end{array}\right] \Theta_i = \left[\begin{array}{cc} \mathsf{X} & 0 \\ \mathsf{Y} & \mathsf{Z} \end{array}\right]$$

for some Θ_i such that

$$\Theta_i \begin{bmatrix} \mathrm{I} & 0 \\ 0 & S \end{bmatrix} \Theta_i^* = \begin{bmatrix} \mathrm{I} & 0 \\ 0 & S \end{bmatrix}$$

We can identify the $\{\mathsf{X},\mathsf{Y},\mathsf{Z}\}$ terms by comparing entries on both sides of the equality

$$\begin{bmatrix} R_{e,i}^{1/2} & H\bar{L}_{i|i-1} \\ K_{p,i}R_{e,i}^{1/2} & F\bar{L}_{i|i-1} \end{bmatrix} \underbrace{\Theta_i \begin{bmatrix} \mathrm{I} & 0 \\ 0 & S \end{bmatrix} \Theta_i^*}_{=(\mathrm{I}\oplus S)} \begin{bmatrix} R_{e,i}^{1/2} & H\bar{L}_{i|i-1} \\ K_{p,i}R_{e,i}^{1/2} & F\bar{L}_{i|i-1} \end{bmatrix}^*$$

$$= \begin{bmatrix} \mathsf{X} & 0 \\ \mathsf{Y} & \mathsf{Z} \end{bmatrix} \begin{bmatrix} \mathrm{I} & 0 \\ 0 & S \end{bmatrix} \begin{bmatrix} \mathsf{X} & 0 \\ \mathsf{Y} & \mathsf{Z} \end{bmatrix}^* \tag{37.23}$$

and using the constancy of the model parameters. Doing so leads to the array method (see Prob. IX.15):

$$\boxed{\begin{bmatrix} R_{e,i}^{1/2} & H\bar{L}_{i|i-1} \\ K_{p,i}R_{e,i}^{1/2} & F\bar{L}_{i|i-1} \end{bmatrix} \Theta_i = \begin{bmatrix} R_{e,i+1}^{1/2} & 0 \\ K_{p,i+1}R_{e,i+1}^{1/2} & \bar{L}_{i+1|i} \end{bmatrix}} \tag{37.24}$$

where Θ_i is any $(\mathrm{I}\oplus S)-$unitary matrix that produces the block zero entry in the post-array. In other words, this array form propagates the low-rank factor $\{\bar{L}_{i|i-1}\}$ as well as the gain matrix $\{K_{p,i}\}$. The algorithm does not involve computation or propagation of the Riccati variable $P_{i|i-1}$, which therefore results in an order of magnitude improvement in complexity. While the covariance form of the Kalman filter requires $O(n^3)$ operations per iteration, the above array form requires $O(\alpha n^2)$ operations per iteration and $\alpha \ll n$ usually.

The initial conditions are $R_{e,0} = R + H\Pi_0 H^*$ and $K_{p,0}R_{e,0}^{1/2} = F\Pi_0 H^* R_{e,0}^{-*/2}$ with $(\bar{L}_{0|-1}, S)$ obtained from the factorization

$$P_{1|0} - \Pi_0 = \left[F\Pi_0 F^* + GQG^* - K_0 R_{e,0}^{-1} K_0^* - \Pi_0\right] = \bar{L}_{0|-1} S \bar{L}_{0|-1}^*$$

Array Chandrasekhar filter for structured models. The connection between the Chandrasekhar filter and the fast RLS algorithms is more natural to establish by developing an extended version of the Chandrasekhar recursions; one that applies to a class of non-constant but *structured* state-space models. The need for this extension can be seen by examining the state-space model (31.8) that is associated with RLS. In this model, the regressor u_i plays the role of the matrix H_i in model (37.21). However, although the regressor is time-variant, it nevertheless varies in a structured manner. For example, in tapped-delay-line implementations, the regressors will possess shift structure.

To handle such structured situations, we define a class of structured models as follows. Returning to (37.21), we shall say that the model is *structured* if there exist $n \times n$ matrices $\{\Psi_i\}$ such that the parameters $\{F_i, H_i, G_i\}$ vary according to the rule:

$$\boxed{H_i = H_{i+1}\Psi_i, \qquad F_{i+1}\Psi_i = \Psi_{i+1}F_i, \qquad G_{i+1} = \Psi_{i+1}G_i} \tag{37.25}$$

We continue to assume that the covariance matrices $\{R_i, Q_i\}$ are constant, although extensions are possible. Now introduce the factorization

$$P_{i+1|i} - \Psi_i P_{i|i-1}\Psi_i^* = \bar{L}_{i|i-1} S \bar{L}_{i|i-1}^*$$

Then a fast array algorithm can be derived in much the same way as we did before for (37.24). We omit the details and state the final extended Chandrasekhar recursions:

$$\boxed{\begin{bmatrix} R_{e,i}^{1/2} & H_{i+1}\bar{L}_{i|i-1} \\ \Psi_{i+1}K_{p,i}R_{e,i}^{1/2} & F_{i+1}\bar{L}_{i|i-1} \end{bmatrix} \Theta_{i+1} = \begin{bmatrix} R_{e,i+1}^{1/2} & 0 \\ K_{p,i+1}R_{e,i+1}^{1/2} & \bar{L}_{i+1|i} \end{bmatrix}} \tag{37.26}$$

where Θ_{i+1} is any $(\mathrm{I}\oplus S)$-unitary matrix that lower triangularizes the pre-array. Moreover, $(\bar{L}_{0|-1}, S)$ are found from the factorization

$$P_{1|0} - \Psi_0 P_{0|-1}\Psi_0^* = \left(F_0\Pi_0 F_0^* + G_0 Q G_0^* - K_0 R_{e,0}^{-1} K_0^*\right) - \Psi_0\Pi_0\Psi_0^* = \bar{L}_{0|-1} S \bar{L}_{0|-1}^*$$

Chandrasekhar filter in explicit form. Finally, we shall only state that the Chandrasekhar filter can be expressed in explicit form as follows. We write $K_{p,i} = K_i R_{e,i}^{-1}$, and generate $\{K_i, R_{e,i}\}$ via recursions involving certain auxiliary sequences $\{L_{i|i-1}, R_{r,i}\}$:

$$\begin{aligned} K_{i+1} &= K_i - F L_{i|i-1} R_{r,i}^{-1} L_{i|i-1}^* H^* \\ L_{i+1|i} &= (F - K_i R_{e,i}^{-1} H) L_{i|i-1} \\ R_{e,i+1} &= R_{e,i} - H L_{i|i-1} R_{r,i}^{-1} L_{i|i-1}^* H^* \\ R_{r,i+1} &= R_{r,i} - L_{i|i-1}^* H^* R_{e,i}^{-1} H L_{i|i-1} \end{aligned} \tag{37.27}$$

The variables $\{L_{i|i-1}, R_{r,i}\}$ have the following interpretation:

$$P_{i+1|i} - P_{i|i-1} = -L_{i|i-1} R_{r,i}^{-1} L_{i|i-1}^*$$

For structured state-space models we have

$$\begin{aligned} K_{i+1} &= \Psi_{i+1} K_i - F_{i+1} L_{i|i-1} R_{r,i}^{-1} L_{i|i-1}^* H_{i+1}^* \\ L_{i+1|i} &= F_{i+1} L_{i|i-1} - \Psi_{i+1} K_i R_{e,i}^{-1} H_{i+1} L_{i|i-1} \\ R_{e,i+1} &= R_{e,i} - H_{i+1} L_{i|i-1} R_{r,i}^{-1} L_{i|i-1}^* H_{i+1}^* \\ R_{r,i+1} &= R_{r,i} - L_{i|i-1}^* H_{i+1}^* R_{e,i}^{-1} H_{i+1} L_{i|i-1} \end{aligned} \tag{37.28}$$

where now $\{L_{i|i-1}, R_{r,i}\}$ have the following interpretation:

$$P_{i+1|i} - \Psi_i P_{i|i-1} \Psi_i^* = -L_{i|i-1} R_{r,i}^{-1} L_{i|i-1}^*$$

CHAPTER 38

Regularized Prediction Problems

The derivation of the fast array algorithm in Chapter 37 was based on the realization that, by proper choice of the regularization matrix Π as in (37.13), the successive differences δP_{i-1} in (37.15) will have rank 2 with a constant signature matrix, $S = (1 \oplus -1)$, and with two-column factors $\bar{L}_{i-1}$. In this chapter, we provide an interpretation for the columns of $\bar{L}_{i-1}$. In the process of doing so, we shall arrive at other efficient implementations of RLS. These implementations will *not* be in array form, but in terms of explicit sets of equations; they are known as the fast Kalman filter, the fast *a posteriori* error sequential technique (FAEST), and the fast transversal filter (FTF).

First, however, to facilitate the presentation, we need to adopt a more explicit notation in order to indicate the fact that the $\{P_i, P_{i-1}\}$ are $M \times M$ matrices. For this reason, we shall write $P_{M,i}$ and $P_{M,i-1}$, instead of P_i and P_{i-1}. We shall also write $u_{M,i}$ instead of u_i, Π_M instead of Π, $w_{M,i}$ instead of w_i, $\gamma_M(i)$ instead of $\gamma(i)$, and $g_{M,i}$ instead of g_i. The subscript M in all these variables is used to indicate the order of the underlying estimation problem, i.e., the dimension of the regression vectors $\{u_{M,i}\}$.

Thus, consider again the regularized least-squares problem

$$\min_{w_M} \left[\lambda^{(i+1)} w_M^* \Pi_M w_M + \sum_{j=0}^{i} \lambda^{i-j} |d(j) - u_{M,j} w_M|^2 \right] \tag{38.1}$$

with regularization matrix

$$\Pi_M = \eta^{-1} \cdot \text{diag}\{\lambda^{-2}, \lambda^{-3}, \ldots, \lambda^{-(M+1)}\}, \quad \eta > 0$$

and with data up to time i. We assume $\bar{w} = 0$ from now on. The corresponding data matrix and measurement vector are denoted by

$$H_{M,i} = \begin{bmatrix} u_{M,0} \\ u_{M,1} \\ \vdots \\ u_{M,i} \end{bmatrix}, \qquad y_i = \begin{bmatrix} d(0) \\ d(1) \\ \vdots \\ d(i) \end{bmatrix} \tag{38.2}$$

with a subscript M also attached to the data matrix in order to indicate its column dimension. The weight vector solution is given by (cf. (30.28))

$$w_{M,i} = P_{M,i} H_{M,i}^* \Lambda_i y_i$$

where the inverse coefficient matrix $P_{M,i}$ is given by

$$\boxed{P_{M,i} = [\lambda^{(i+1)} \Pi_M \; + \; H_{M,i}^* \Lambda_i H_{M,i}]^{-1}}$$

and

$$\Lambda_i = \text{diag}\{\lambda^i, \lambda^{i-1}, \ldots, \lambda, 1\}$$

In addition, the conversion factor at time i is given by

$$\gamma_M(i) = 1 - u_{M,i} P_{M,i} u^*_{M,i} \tag{38.3}$$

and the gain vector is

$$g_{M,i} = P_{M,i} u^*_{M,i}$$

Our objective is to examine the meaning of the low-rank factor $\bar{L}_i$. For this purpose, we shall derive an explicit expression for the difference

$$\begin{bmatrix} P_i & 0 \\ 0 & 0 \end{bmatrix} - \begin{bmatrix} 0 & 0 \\ 0 & P_{i-1} \end{bmatrix} \tag{38.4}$$

The derivation requires that we introduce the so-called backward and forward prediction problems.

38.1 REGULARIZED BACKWARD PREDICTION

Problem (38.1) projects, in a regularized and weighted manner, the measurement vector y_i onto $\mathcal{R}(H_{M,i})$, where the data matrix $H_{M,i}$ is given by (say, for $M = 4$):

$$H_{4,i} = \begin{bmatrix} u(0) & 0 & 0 & 0 \\ u(1) & u(0) & 0 & 0 \\ u(2) & u(1) & u(0) & 0 \\ u(3) & u(2) & u(1) & u(0) \\ u(4) & u(3) & u(2) & u(1) \\ \vdots & \vdots & \vdots & \vdots \\ u(i) & u(i-1) & u(i-2) & u(i-3) \end{bmatrix}$$

Now assume that we augment $H_{M,i}$ by one column to the right and define

$$H_{M+1,i} \stackrel{\Delta}{=} \begin{bmatrix} H_{M,i} & h \end{bmatrix} \tag{38.5}$$

e.g., for $M = 4$,

$$H_{5,i} = \left[\begin{array}{cccc|c} u(0) & 0 & 0 & 0 & 0 \\ u(1) & u(0) & 0 & 0 & 0 \\ u(2) & u(1) & u(0) & 0 & 0 \\ u(3) & u(2) & u(1) & u(0) & 0 \\ u(4) & u(3) & u(2) & u(1) & u(0) \\ u(5) & u(4) & u(3) & u(2) & u(1) \\ \vdots & \vdots & \vdots & \vdots & \vdots \\ u(i) & u(i-1) & u(i-2) & u(i-3) & u(i-4) \end{array}\right]$$

where the entries of the additional column h are

$$h = \text{col}\{0, \ldots, 0, u(0), u(1), \ldots, u(i-M)\} \tag{38.6}$$

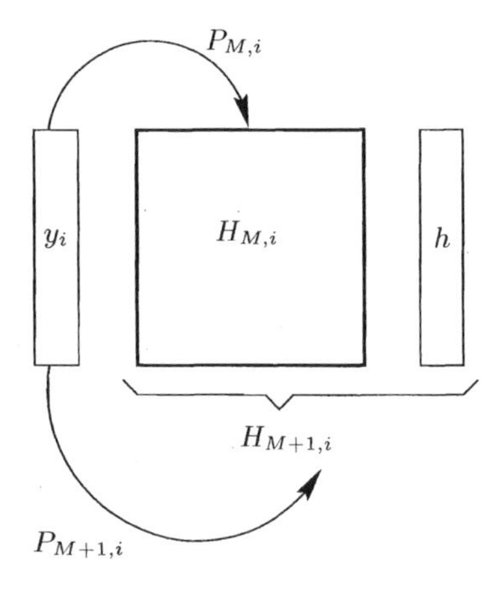

FIGURE 38.1 Projection of y_i onto $\mathcal{R}(H_{M,i})$ and $\mathcal{R}(H_{M+1,i})$ with the corresponding inverse coefficient matrices $\{P_{M,i}, P_{M+1,i}\}$.

We can then consider the problem of projecting y_i onto the extended range space $\mathcal{R}(H_{M+1,i})$, as shown in Fig. 38.1.

This step requires that we solve the following estimation problem of order $M+1$:

$$\min_{w_{M+1}} \left[\lambda^{(i+1)} w_{M+1}^* \Pi_{M+1} w_{M+1} + \sum_{j=0}^{i} \lambda^{i-j} |d(j) - u_{M+1,j} w_{M+1}|^2 \right] \tag{38.7}$$

Its solution is given by

$$w_{M+1,i} = P_{M+1,i} H_{M+1,i}^* \Lambda_i y_i$$

with

$$\boxed{P_{M+1,i} = [\lambda^{(i+1)} \Pi_{M+1} \;+\; H_{M+1,i}^* \Lambda_i H_{M+1,i}]^{-1}} \tag{38.8}$$

Since

$$\Pi_{M+1} = (\Pi_M \oplus \eta^{-1} \lambda^{-(M+2)})$$

the regularization matrix $\lambda^{(i+1)} \Pi_{M+1}$ is seen to satisfy

$$\boxed{\lambda^{(i+1)} \Pi_{M+1} \;=\; \begin{bmatrix} \lambda^{(i+1)} \Pi_M & 0 \\ 0 & \eta^{-1} \lambda^{i-M-1} \end{bmatrix}}$$

Comparing problems (38.1) and (38.7) we see that they are of the same form as the order-update problems (32.1) and (32.7) studied in Sec. 32.1 with the identifications

$$\left\{ \begin{array}{lll} y \leftarrow y_i & H \leftarrow H_{M,i} & W \leftarrow \Lambda_i \\ \Pi \leftarrow \lambda^{(i+1)} \Pi_M & \sigma \leftarrow \eta^{-1} \lambda^{i-M-1} & h \leftarrow h \\ P_z \leftarrow P_{M+1,i} & P \leftarrow P_{M,i} & \end{array} \right.$$

We can therefore invoke the result of Lemma 32.1 to relate $\{w_{M,i}, w_{M+1,i}\}$ as well as $\{P_{M,i}, P_{M+1,i}\}$. For our purposes, we are more interested in the latter relation, which from Lemma 32.1 is given by

$$P_{M+1,i} = \begin{bmatrix} P_{M,i} & 0 \\ 0 & 0 \end{bmatrix} + \frac{1}{\eta^{-1}\lambda^{i-M-1} + \xi^b_M(i)} \begin{bmatrix} -w^b_{M,i} \\ 1 \end{bmatrix} \begin{bmatrix} -w^{b*}_{M,i} & 1 \end{bmatrix} \tag{38.9}$$

where $w^b_{M,i}$ is the weight vector that projects h in (38.6) onto $\mathcal{R}(H_{M,i})$, namely it solves

$$\min_{w^b_M} \left[\lambda^{(i+1)} w^{b*}_M \Pi_M w^b_M + (h - H_{M,i}w^b)^* \Lambda_i (h - H_{M,i}w^b_M) \right] \tag{38.10}$$

and $\xi^b_M(i)$ is the corresponding minimum cost. The cost function (38.10) can be rewritten as

$$\min_{w^b_M} \left[\lambda^{(i+1)} w^{b*}_M \Pi_M w^b_M + \sum_{j=0}^{i} \lambda^{i-j} \left| u(j-M) - u_{M,j}w^b_M \right|^2 \right] \tag{38.11}$$

which can be interpreted as estimating each entry $u(j-M)$ from the future values in $u_{M,j}$,

$$\boxed{u_{M,j} = \begin{bmatrix} u(j) & u(j-1) & \ldots & u(j-M+1) \end{bmatrix}} \quad \boxed{u(j-M)}$$

Hence, the use of the superscript ^{b} to indicate a *backward prediction problem*. Let

$$b_M(i) \overset{\Delta}{=} u(i-M) - u_{M,i}w^b_{M,i}$$

denote the backward error that results from estimating $u(i-M)$ from $u_{M,i}$; it corresponds to the last entry of the residual vector $h - H_{M,i}w^b_{M,i}$. By invoking again Lemma 32.1 with the identifications

$$\gamma_z \leftarrow \gamma_{M+1}(i), \qquad \gamma \leftarrow \gamma_M(i), \qquad \widetilde{\alpha} \leftarrow b_M(i)$$

we can relate the conversion factors $\{\gamma_M(i), \gamma_{M+1}(i)\}$ of problems (38.1) and (38.7) as

$$\gamma_{M+1}(i) = \gamma_M(i) - \frac{|b_M(i)|^2}{\eta^{-1}\lambda^{i-M-1} + \xi^b_M(i)} \tag{38.12}$$

This expression provides an order-update relation for the conversion factor.

If we further multiply (38.9) from the right by

$$u^*_{M+1,i} = \begin{bmatrix} u^*_{M,i} \\ u^*(i-M) \end{bmatrix}$$

then we obtain an order-update relation for the gain vector as well:

$$g_{M+1,i} = \begin{bmatrix} g_{M,i} \\ 0 \end{bmatrix} + \frac{b^*_M(i)}{\eta^{-1}\lambda^{i-M-1} + \xi^b_M(i)} \begin{bmatrix} -w^b_{M,i} \\ 1 \end{bmatrix} \tag{38.13}$$

38.2 REGULARIZED FORWARD PREDICTION

Our second step toward evaluating the difference (38.4) is to derive *both* order- and time-update relations for the variables $\{\gamma_M(i-1), g_{M,i-1}, P_{M,i-1}\}$. That is, we now show how to go from these variables to $\{\gamma_{M+1}(i), g_{M+1,i}, P_{M+1,i}\}$.

Thus, consider again the matrix $H_{M+1,i}$ in (38.5) but now partition it as

$$H_{M+1,i} = \left[\begin{array}{c|c} h & \begin{array}{c} 0 \\ H_{M,i-1} \end{array} \end{array}\right] \tag{38.14}$$

That is, we now separate its leading column (as opposed to its trailing column in (38.5)) from the remaining columns. The remaining columns have a top zero row, which is represented by the zero entry in (38.14). The entries of h are

$$h = \text{col}\{u(0), u(1), u(2), \ldots, u(i)\} \tag{38.15}$$

We then consider the problem of projecting the same vector y_i onto the range space of

$$\begin{bmatrix} 0 \\ H_{M,i-1} \end{bmatrix}$$

as shown in Fig. 38.2. This is a projection problem of order M and it corresponds to solving

$$\min_{w_M} \left[\lambda^i w_M^* \Pi_M w_M + \sum_{j=0}^{i} \lambda^{i-j} |d(j) - u_{M,j-1} w_M|^2\right] \tag{38.16}$$

where $u_{M,-1} = 0$. We denote its solution by $w_{M,i-1}$ and it is given by

$$w_{M,i-1} = P_{M,i-1} H_{M,i-1}^* \Lambda_{i-1} \cdot \text{col}\{d(1), d(2), \ldots, d(i)\}$$

where

$$\boxed{P_{M,i-1} = [\lambda^i \Pi_M \; + \; H_{M,i-1}^* \Lambda_{i-1} H_{M,i-1}]^{-1}} \tag{38.17}$$

Comparing problems (38.7) and (38.16) we see that we face the same situation studied in Sec. 32.2 on order-updates for forward prediction problems; just compare with problems (32.1) and (32.39) with the identifications:

$$\left\{\begin{array}{lll} y \leftarrow y_{i-1} & H \leftarrow H_{M,i-1} & W \leftarrow \Lambda_{i-1} \\ \Pi \leftarrow \lambda^i \Pi_M & \sigma \leftarrow \eta^{-1} \lambda^{i-1} & h \leftarrow h \\ P_z \leftarrow P_{M+1,i} & P \leftarrow P_{M,i-1} & \end{array}\right.$$

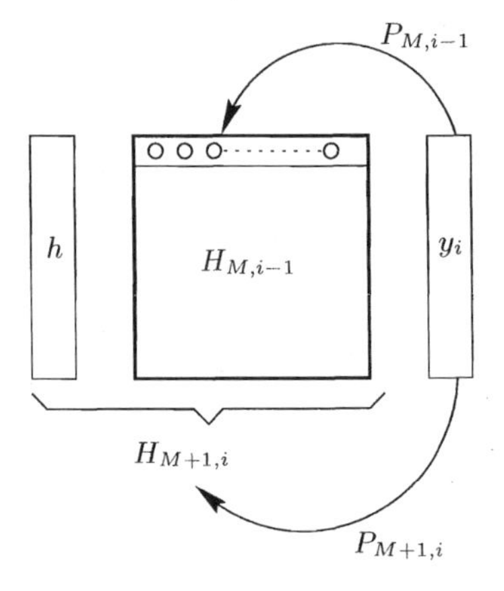

FIGURE 38.2 Projection of y_i onto $\mathcal{R}(\text{col}\{0, H_{M,i-1}\})$ and $\mathcal{R}(H_{M+1,i})$ with the corresponding inverse coefficient matrices $\{P_{M,i-1}, P_{M+1,i}\}$.

We can then invoke the result of Lemma 32.2 to relate $\{P_{M,i-1}, P_{M+1,i}\}$, namely,

$$\boxed{P_{M+1,i} = \begin{bmatrix} 0 & 0 \\ 0 & P_{M,i-1} \end{bmatrix} + \frac{1}{\eta^{-1}\lambda^{i-1} + \xi^f_{M,i}} \begin{bmatrix} 1 \\ -w^f_{M,i} \end{bmatrix} \begin{bmatrix} 1 & -w^{f*}_{M,i} \end{bmatrix}} \tag{38.18}$$

where $w^f_{M,i}$ is the weight vector that projects h in (38.15) onto $\mathcal{R}(\text{col}\{0, H_{M,i-1}\})$, namely it solves

$$\min_{w^f_M} \left[\lambda^i w^{f*}_M \Pi_M w^f_M + \left(h - \begin{bmatrix} 0 \\ H_{M,i-1} \end{bmatrix} w^f_M \right)^* \Lambda_i \left(h - \begin{bmatrix} 0 \\ H_{M,i-1} \end{bmatrix} w^f_M \right) \right] \tag{38.19}$$

and $\xi^f_M(i)$ is the corresponding minimum cost. The cost function (38.19) can be rewritten as

$$\min_{w^f_M} \left[\lambda^i w^{f*}_M \Pi_M w^f_M + \sum_{j=0}^{i} \lambda^{i-j} \left| u(j) - u_{M,j-1} w^f_M \right|^2 \right] \tag{38.20}$$

which, since $u_{M,-1} = 0$, is also equivalent to

$$\min_{w^f_M} \left[\lambda^i w^{f*}_M \Pi_M w^f_M + \sum_{j=0}^{i-1} \lambda^{i-1-j} \left| u(j+1) - u_{M,j} w^f_M \right|^2 \right] \tag{38.21}$$

This problem can be interpreted as estimating each entry $u(j+1)$ from its past values in $u_{M,j}$,

$$\boxed{u(j+1)} \qquad \boxed{\begin{bmatrix} u(j) & u(j-1) & \ldots & u(j-M+1) \end{bmatrix} = u_{M,j}}$$

Hence, the use of the superscript ^{f} to indicate a *forward prediction problem*. Let

$$f_M(i) \triangleq u(i) - u_{M,i-1} w^f_{M,i}$$

denote the forward error that results from estimating $u(i)$ from $u_{M,i-1}$; it corresponds to the last entry of the residual vector

$$h - \begin{bmatrix} 0 \\ H_{M,i-1} \end{bmatrix} w^f_{M,i}$$

By invoking again Lemma 32.2, with the identifications

$$\gamma_z \leftarrow \gamma_{M+1}(i), \qquad \gamma \leftarrow \gamma_M(i-1), \qquad \widetilde{\alpha} \leftarrow f_M(i)$$

we can relate the conversion factors $\{\gamma_{M+1}(i), \gamma_M(i-1)\}$ of problems (38.7) and (38.16) as

$$\boxed{\gamma_{M+1}(i) = \gamma_M(i-1) - \frac{|f_M(i)|^2}{\eta^{-1}\lambda^{i-1} + \xi^f_M(i)}} \tag{38.22}$$

If we further multiply (38.18) from the right by

$$u^*_{M+1,i} = \begin{bmatrix} u^*(i) \\ u^*_{M,i-1} \end{bmatrix}$$

we obtain a time- and order-update relation for the gain vector as well:

$$\boxed{g_{M+1,i} = \begin{bmatrix} 0 \\ g_{M,i-1} \end{bmatrix} + \frac{f_M^*(i)}{\eta^{-1}\lambda^{i-1} + \xi_M^f(i)} \begin{bmatrix} 1 \\ -w_{M,i}^f \end{bmatrix}} \tag{38.23}$$

38.3 LOW-RANK FACTORIZATION

Now subtracting (38.9) and (38.18) we find that

$$\begin{bmatrix} P_{M,i} & 0 \\ 0 & 0 \end{bmatrix} - \begin{bmatrix} 0 & 0 \\ 0 & P_{M,i-1} \end{bmatrix} = \frac{1}{\eta^{-1}\lambda^{i-1} + \xi_M^f(i)} \begin{bmatrix} 1 \\ -w_{M,i}^f \end{bmatrix} \begin{bmatrix} 1 & -w_{M,i}^{f*} \end{bmatrix}$$
$$- \frac{1}{\eta^{-1}\lambda^{i-M-1} + \xi_M^b(i)} \begin{bmatrix} -w_{M,i}^b \\ 1 \end{bmatrix} \begin{bmatrix} -w_{M,i}^{b*} & 1 \end{bmatrix} \tag{38.24}$$

The coefficient matrices $\{P_{M,i}, P_{M,i-1}\}$ that appear in the above expression have the same order M. Dropping M, we can rewrite the above relation as

$$\begin{bmatrix} P_i & 0 \\ 0 & 0 \end{bmatrix} - \begin{bmatrix} 0 & 0 \\ 0 & P_{i-1} \end{bmatrix} =: \lambda L_i T_i L_i^* \tag{38.25}$$

where the factors $\{L_i, T_i\}$ are defined by

$$\begin{aligned} L_i &= \begin{bmatrix} 1 & -w_{M,i}^b \\ -w_{M,i}^f & 1 \end{bmatrix} \\ T_i &= \text{diag}\left\{ \frac{\lambda^{-1}}{\eta^{-1}\lambda^{i-1} + \xi_M^f(i)}, \; -\frac{\lambda^{-1}}{\eta^{-1}\lambda^{i-M-1} + \xi_M^b(i)} \right\} \end{aligned} \tag{38.26}$$

Alternatively, we can write (38.25) as

$$\begin{bmatrix} P_i & 0 \\ 0 & 0 \end{bmatrix} - \begin{bmatrix} 0 & 0 \\ 0 & P_{i-1} \end{bmatrix} = \lambda \bar{L}_i S \bar{L}_i^* \tag{38.27}$$

where $S = (1 \oplus -1)$ and the columns of $\bar{L}_i$ are given by

$$\begin{aligned} \text{first column of } \bar{L}_i &= \sqrt{\frac{1}{\eta^{-1}\lambda^{i} + \lambda\xi_M^f(i)}} \begin{bmatrix} 1 \\ -w_{M,i}^f \end{bmatrix} \\ \text{second column of } \bar{L}_i &= \sqrt{\frac{1}{\eta^{-1}\lambda^{i-M} + \lambda\xi_M^b(i)}} \begin{bmatrix} -w_{M,i}^b \\ 1 \end{bmatrix} \end{aligned}$$

We have therefore achieved our objective of interpreting the columns of $\bar{L}_i$; they are scaled multiples of the forward and backward prediction vectors $\{w_{M,i}^b, w_{M,i}^f\}$, which are solutions to (38.11) and (38.21).

CHAPTER 39

Fast Fixed-Order Filters

The arguments so far do not only provide an interpretation for the columns of $\bar{L}_i$ but they also motivate other efficient RLS implementations. These alternative implementations are not in the form of array algorithms, as was the case in Chapter 37, but in terms of explicit sets of recursions. We start with the Fast Transversal Filter (FTF) form.

39.1 FAST TRANSVERSAL FILTER

Define the normalized gain vector

$$c_{M,i} \stackrel{\Delta}{=} g_{M,i}\gamma_M^{-1}(i)$$

and observe that the RLS recursion (37.2) can be used to time-update the weight vectors $\{w^b_{M,i}, w^f_{M,i}\}$ as follows:

$$w^b_{M,i} = w^b_{M,i-1} + c_{M,i}b_M(i) \tag{39.1}$$

$$w^f_{M,i} = w^f_{M,i-1} + c_{M,i-1}f_M(i) \tag{39.2}$$

where $c_{M,i-1}$ is used in the recursion for $w^f_{M,i}$ instead of $c_{M,i}$, since the forward projection problem (38.19) is based on $H_{M,i-1}$ while the backward projection problem (38.10) is based on $H_{M,i}$.

Moreover, writing expression (37.10) in terms of $\{c_{M,i}, c_{M,i-1}\}$, and using the low-rank factorization (38.25) at time $(i-1)$, we obtain

$$\begin{bmatrix} c_{M,i} \\ 0 \end{bmatrix} = \begin{bmatrix} 0 \\ c_{M,i-1} \end{bmatrix} + \frac{\lambda^{-1}\alpha^*_M(i)}{\eta^{-1}\lambda^{i-2} + \xi^f_M(i-1)} \begin{bmatrix} 1 \\ -w^f_{M,i-1} \end{bmatrix} - \frac{\lambda^{-1}\beta^*_M(i)}{\eta^{-1}\lambda^{i-M-2} + \xi^b_M(i-1)} \begin{bmatrix} -w^b_{M,i-1} \\ 1 \end{bmatrix} \tag{39.3}$$

in terms of the *a priori* estimation errors

$$\alpha_M(i) = u(i) - u_{M,i-1}w^f_{M,i-1}, \quad \beta_M(i) = u(i-M) - u_{M,i}w^b_{M,i-1}$$

which are related to the *a posteriori* errors $\{f_M(i), b_M(i)\}$ via the respective conversion factors, i.e.,

$$b_M(i) = \gamma_M(i)\beta_M(i) \quad \text{and} \quad f_M(i) = \gamma_M(i-1)\alpha_M(i)$$

Now the sum of the first two terms on the right-hand side of (39.3) can be shown to be equal to $c_{M+1,i}$ — see Prob. IX.6. In this way, the update for $c_{M,i}$ can be split into two steps. We first perform the time- and order-update:

$$c_{M+1,i} = \begin{bmatrix} 0 \\ c_{M,i-1} \end{bmatrix} + \frac{\lambda^{-1}\alpha_M^*(i)}{\eta^{-1}\lambda^{i-2} + \xi_M^f(i-1)} \begin{bmatrix} 1 \\ -w_{M,i-1}^f \end{bmatrix} \tag{39.4}$$

followed by the order down-date:

$$\begin{bmatrix} c_{M,i} \\ 0 \end{bmatrix} = c_{M+1,i} - \frac{\lambda^{-1}\beta_M^*(i)}{\eta^{-1}\lambda^{i-M-2} + \xi_M^b(i-1)} \begin{bmatrix} -w_{M,i-1}^b \\ 1 \end{bmatrix} \tag{39.5}$$

Using the fact that the last entry on the left-hand side of the above equality is zero, we find that

$$\text{last entry of } c_{M+1,i} = \frac{\lambda^{-1}\beta_M^*(i)}{\eta^{-1}\lambda^{i-M-2} + \xi_M^b(i-1)}$$

Hence, if we denote this last entry by $\nu_{M+1}(i)$, then $\beta_M(i)$ can be evaluated via the alternative formula:

$$\beta_M(i) = \lambda[\eta^{-1}\lambda^{i-M-2} + \xi_M^b(i-1)]\nu_{M+1}^*(i) \tag{39.6}$$

Moreover, since $\xi_M^f(i)$ denotes the minimum cost of the regularized least-squares problem (38.19), and since $\xi_M^b(i)$ denotes the minimum cost of the regularized least-squares problem (38.10), they both satisfy time-update relations of the form (cf. Alg. 30.2):

$$\begin{aligned} \xi_M^f(i) &= \lambda\xi_M^f(i-1) + \alpha_M(i)f_M^*(i), \quad \xi_M^f(-1) = 0 \\ \xi_M^b(i) &= \lambda\xi_M^b(i-1) + \beta_M(i)b_M^*(i), \quad \xi_M^b(-1) = 0 \end{aligned}$$

By collecting the relations we derived so far, along with two easily derived relations for the conversion factor (see Prob. IX.2), we obtain the so-called fast transversal filter (FTF). In the listing of the algorithm below, we have introduced the auxiliary variables:

$$\zeta_M^f(i) \triangleq \eta^{-1}\lambda^{i-1} + \xi_M^f(i), \qquad \zeta_M^b(i) \triangleq \eta^{-1}\lambda^{i-M-1} + \xi_M^b(i)$$

It is easy to see that these variables satisfy similar recursions to those of $\{\xi_M^f(i), \xi_M^b(i)\}$, namely,

$$\begin{aligned} \zeta_M^f(i) &= \lambda\zeta_M^f(i-1) + \alpha_M^*(i)f_M(i), \quad \zeta_M^f(-1) = \eta^{-1}\lambda^{-2} \\ \zeta_M^b(i) &= \lambda\zeta_M^b(i-1) + \beta_M^*(i)b_M(i), \quad \zeta_M^b(-1) = \eta^{-1}\lambda^{-(M+2)} \end{aligned}$$

albeit with different initial conditions.

Algorithm 39.1 (Fast transversal filter) Consider the same setting as Alg. 37.1 with $\bar{w} = 0$. The solution w_N can be recursively computed as follows. Start with $u_{M,-1} = 0$, $\gamma_M(-1) = 1$, $c_{M,-1} = 0$, $w^f_{M,-1} = 0$, $w^b_{M,-1} = 0$, $w_{M,-1} = 0$, $\zeta^f_M(-1) = \eta^{-1}\lambda^{-2}$, $\zeta^b_M(-1) = \eta^{-1}\lambda^{-(M+2)}$ and repeat for $i \geq 0$:

$$
\begin{aligned}
\alpha_M(i) &= u(i) - u_{M,i-1}w^f_{M,i-1} \\
f_M(i) &= \gamma_M(i-1)\alpha_M(i) \\
\zeta^f_M(i) &= \lambda\zeta^f_M(i-1) + \alpha^*_M(i)f_M(i) \\
\gamma_{M+1}(i) &= \gamma_M(i-1)\,\lambda\zeta^f_M(i-1)/\zeta^f_M(i) \\
w^f_{M,i} &= w^f_{M,i-1} + f_M(i)c_{M,i-1} \\
c_{M+1,i} &= \begin{bmatrix} 0 \\ c_{M,i-1} \end{bmatrix} + \frac{\alpha^*_M(i)}{\lambda\zeta^f_M(i-1)} \begin{bmatrix} 1 \\ -w^f_{M,i-1} \end{bmatrix} \\
\nu_{M+1}(i) &= \text{last entry of } c_{M+1,i} \\
\beta_M(i) &= \lambda\zeta^b_M(i-1)\nu^*_{M+1}(i) \\
\gamma_M(i) &= \gamma_{M+1}(i)/[1 - \beta_M(i)\gamma_{M+1}(i)\nu_{M+1}(i)] \\
\begin{bmatrix} c_{M,i} \\ 0 \end{bmatrix} &= c_{M+1,i} - \nu_{M+1}(i) \begin{bmatrix} -w^b_{M,i-1} \\ 1 \end{bmatrix} \\
b_M(i) &= \gamma_M(i)\beta_M(i) \\
\zeta^b_M(i) &= \lambda\zeta^b_M(i-1) + \beta^*_M(i)b_M(i) \\
w^b_{M,i} &= w^b_{M,i-1} + b_M(i)c_{M,i} \\
e(i) &= d(i) - u_{M,i}w_{i-1} \\
r(i) &= \gamma_M(i)e(i) \\
w_i &= w_{i-1} + r(i)c_{M,i}
\end{aligned}
$$

39.2 FAEST FILTER

The fast *a posteriori* error sequential technique (FAEST) is similar to the FTF implementation we described above except that FAEST relies on propagating the inverses of the conversion factors, rather than the conversion factors themselves.

Thus, observe from (38.22), or from Prob. IX.2, that we can equivalently write

$$
\gamma_{M+1}(i) = \gamma_M(i-1)\left[1 - \frac{\alpha^*_M(i)f_M(i)}{\zeta^f_M(i)}\right] = \gamma_M(i-1)\frac{\lambda\zeta^f_M(i-1)}{\zeta^f_M(i)}
$$

so that

$$
\gamma^{-1}_{M+1}(i) = \gamma^{-1}_M(i-1)\frac{\zeta^f_M(i)}{\lambda\zeta^f_M(i-1)} = \gamma^{-1}_M(i-1)\frac{\lambda\zeta^f_M(i-1) + \alpha^*_M(i)f_M(i)}{\lambda\zeta^f_M(i-1)}
$$

and, hence,

$$\gamma_{M+1}^{-1}(i) = \gamma_M^{-1}(i-1) + \frac{|\alpha_M(i)|^2}{\lambda \zeta_M^f(i-1)} \tag{39.7}$$

Likewise, from the relation in part (c) of Prob. IX.2 we get

$$\gamma_M^{-1}(i) = \gamma_{M+1}^{-1}(i) - \beta_M(i)\nu_{M+1}(i)$$

All other equations in the FTF description remain unchanged.

Algorithm 39.2 (FAEST filter) Consider the same setting as Alg. 37.1 with $\bar{w} = 0$. The solution w_N can be recursively computed as follows. Start with $u_{M,-1} = 0$, $\gamma_M^{-1}(-1) = 1$, $c_{M,-1} = 0$, $w_{M,-1}^f = 0$, $w_{M,-1}^b = 0$, $w_{M,-1} = 0$, $\zeta_M^f(-1) = \eta^{-1}\lambda^{-2}$, $\zeta_M^b(-1) = \eta^{-1}\lambda^{-(M+2)}$ and repeat for $i \geq 0$:

$$\begin{aligned}
\alpha_M(i) &= u(i) - u_{M,i-1}w_{M,i-1}^f \\
f_M(i) &= \alpha_M(i)/\gamma_M^{-1}(i-1) \\
\zeta_M^f(i) &= \lambda\zeta_M^f(i-1) + \alpha_M^*(i)f_M(i) \\
\gamma_{M+1}^{-1}(i) &= \gamma_M^{-1}(i-1) + |\alpha_M(i)|^2/\lambda\zeta_M^f(i-1) \\
w_{M,i}^f &= w_{M,i-1}^f + f_M(i)c_{M,i-1} \\
c_{M+1,i} &= \begin{bmatrix} 0 \\ c_{M,i-1} \end{bmatrix} + \frac{\alpha_M^*(i)}{\lambda\zeta_M^f(i-1)} \begin{bmatrix} 1 \\ -w_{M,i-1}^f \end{bmatrix} \\
\nu_{M+1}(i) &= \text{last entry of } c_{M+1,i} \\
\beta_M(i) &= \lambda\zeta_M^b(i-1)\nu_{M+1}^*(i) \\
\gamma_M^{-1}(i) &= \gamma_{M+1}^{-1}(i) - \beta_M(i)\nu_{M+1}(i) \\
\begin{bmatrix} c_{M,i} \\ 0 \end{bmatrix} &= c_{M+1,i} - \nu_{M+1}(i) \begin{bmatrix} -w_{M,i-1}^b \\ 1 \end{bmatrix} \\
b_M(i) &= \beta_M(i)/\gamma_M^{-1}(i) \\
\zeta_M^b(i) &= \lambda\zeta_M^b(i-1) + \beta_M^*(i)b_M(i) \\
w_{M,i}^b &= w_{M,i-1}^b + b_M(i)c_{M,i} \\
e(i) &= d(i) - u_{M,i}w_{i-1} \\
r(i) &= e(i)/\gamma_M^{-1}(i) \\
w_i &= w_{i-1} + r(i)c_{M,i}
\end{aligned}$$

39.3 FAST KALMAN FILTER

The fast Kalman filter is another efficient RLS implementation (actually the earliest such implementation); its equations are in essence similar to what we have done so far for FTF and FAEST except that they rely on propagating the gain vector $g_{M,i}$ instead of its normalized version $c_{M,i}$. The update of $g_{M,i}$ is based on equations (38.13) and (38.23), as we shall explain below. Also, the fast Kalman filter does not use the conversion factors to evaluate *a posteriori* errors from *a priori* errors. All error quantities are evaluated directly from their definitions. For this reason, FTF and FAEST are more efficient than fast Kalman — see Table 39.1.

The time-update of the gain vector, from $g_{M,i-1}$ to $g_{M,i}$, is performed as follows. First, equation (38.23) is used to update $g_{M,i-1}$ to $g_{M+1,i}$, namely,

$$g_{M+1,i} = \begin{bmatrix} 0 \\ g_{M,i-1} \end{bmatrix} + \frac{f_M^*(i)}{\zeta_M^f(i)} \begin{bmatrix} 1 \\ -w_{M,i}^f \end{bmatrix}$$

Now from (38.13), we see that the last entry of the just computed $g_{M+1,i}$ is equal to $b_M^*(i)/\zeta_M^b(i)$. We denote this last entry by $\sigma_{M+1}(i)$:

$$\boxed{\sigma_{M+1}(i) \stackrel{\Delta}{=} \text{last entry of } g_{M+1,i} = b_M^*(i)/\zeta_M^b(i)}$$

Then $g_{M,i}$ can be recovered from (38.13) as

$$g_{M,i} = g_{M+1,i}[0:M-1] + \sigma_{M+1}(i) w_{M,i}^b$$

where the notation $g_{M+1,i}[0:M-1]$ denotes the top M entries of $g_{M+1,i}$. If we further replace $w_{M,i}^b$ in the above equation by its update

$$w_{M,i}^b = w_{M,i-1}^b + g_{M,i}\beta_M(i)$$

and solve for $g_{M,i}$, we get

$$\boxed{g_{M,i} = \frac{g_{M+1,i}[0:M-1]}{1-\sigma_{M+1}(i)\beta_M(i)}}$$

In this way, we arrive at the statement of the fast Kalman filter.

For comparison purposes, Table 39.1 lists the estimated computational cost per iteration for the fast array method, FTF, FAEST, and fast Kalman assuming real data. The costs are in terms of the number of multiplications, additions, and divisions that are needed for each iteration. It is seen that FTF and FAEST require $O(7M)$ operations per iteration, while fast Kalman requires $O(9M)$ operations per iteration.

TABLE 39.1 Estimated computational cost per iteration for fast array, FTF, FAEST, and fast Kalman, assuming real data.

Algorithm	$\times$	$+$	$\div$	$\sqrt{\cdot}$
FAEST	$7M+6$	$7M+2$	5	–
FTF	$7M+10$	$7M+1$	3	–
fast Kalman	$9M+2$	$8M+1$	2	–
fast array	$6M+4$	$10M+16$	6	2

Algorithm 39.3 (Fast Kalman filter) Consider the same setting as Alg. 37.1 with $\bar{w} = 0$. The solution w_N can be recursively computed as follows. Start with $u_{M,-1} = 0$, $g_{M,-1} = 0$, $w^f_{M,-1} = 0$, $w^b_{M,-1} = 0$, $w_{M,-1} = 0$, $\zeta^f_M(-1) = \eta^{-1}\lambda^{-2}$, and repeat for $i \geq 0$:

$$
\begin{aligned}
\alpha_M(i) &= u(i) - u_{M,i-1}w^f_{M,i-1} \\
w^f_{M,i} &= w^f_{M,i-1} + \alpha_M(i)g_{M,i-1} \\
f_M(i) &= u(i) - u_{M,i-1}w^f_{M,i} \\
\zeta^f_M(i) &= \lambda\zeta^f_M(i-1) + \alpha^*_M(i)f_M(i) \\
g_{M+1,i} &= \begin{bmatrix} 0 \\ g_{M,i-1} \end{bmatrix} + \frac{f^*_M(i)}{\zeta^f_M(i)}\begin{bmatrix} 1 \\ -w^f_{M,i} \end{bmatrix} \\
\sigma_{M+1}(i) &= \text{last entry of } g_{M+1,i} \\
\beta_M(i) &= u(i-M) - u_{M,i}w^b_{M,i-1} \\
g_{M,i} &= \frac{g_{M+1,i}[0:M-1]}{1 - \sigma_{M+1}(i)\beta_M(i)} \\
w^b_{M,i} &= w^b_{M,i-1} + \beta_M(i)g_{M,i} \\
e(i) &= d(i) - u_{M,i}w_{i-1} \\
w_i &= w_{i-1} + g_{M,i}e(i)
\end{aligned}
$$

39.4 STABILITY ISSUES

Unfortunately, the fast least-squares algorithms of this chapter tend to suffer from numerical instabilities when implemented in finite precision. This is because the algebraic equalities that are used to derive the algorithms tend to break down in finite precision. A computer project at the end of this part illustrates some of these instability problems; the project also shows that some implementations are more reliable than others.

Still, there are several ingenious methods that have been proposed over the years in order to combat the instability problem with varied degrees of success. In this section, we describe and comment on some of these techniques.

Array Implementation

We start with the array implementation of Alg. 37.1, namely,

$$
\begin{bmatrix} \gamma^{-1/2}(i-1) & \begin{bmatrix} u(i) & u_{i-1} \end{bmatrix}\bar{L}_{i-1} \\ \begin{bmatrix} 0 \\ g_{i-1}\gamma^{-1/2}(i-1) \end{bmatrix} & \bar{L}_{i-1} \end{bmatrix}\Theta_i = \begin{bmatrix} \gamma^{-1/2}(i) & \begin{bmatrix} 0 & 0 \end{bmatrix} \\ \begin{bmatrix} g_i\gamma^{-1/2}(i) \\ 0 \end{bmatrix} & \sqrt{\lambda}\,\bar{L}_i \end{bmatrix} \tag{39.8}
$$

where Θ_i is J-unitary with

$$
J = (1 \oplus S) \quad \text{and} \quad S = \text{diag}\{1, -1\}
$$

Although from a theoretical point of view, any Θ_i that produces the zero entries in the first row of the post-array in (39.8) will do, different implementations of Θ_i tend to lead to different numerical behavior. To explain this, consider for illustration purposes the case $M = 3$. Then the pre- and post-arrays will have the generic forms:

$$\left[\begin{array}{c|cc} \times & \times & \times \\ \hline \times & \times & \times \\ \times & \times & \times \\ \times & \times & \times \\ \times & \times & \times \end{array}\right] \Theta_i = \left[\begin{array}{c|cc} \times & 0 & 0 \\ \hline \times & \times & \times \\ \times & \times & \times \\ \times & \times & \times \\ 0 & \times & \times \end{array}\right]$$

In order to create the zero pattern in the first row of the post array, the J-unitary transformation Θ_i can be constructed based solely on the first row of the pre-array, which means that only the information that is needed to update $\gamma^{-1/2}(i-1)$ to $\gamma^{-1/2}(i)$ is used to determine Θ_i. In this way, no information from the other equations of the RLS algorithm (37.2) influences the choice of Θ_i.

An alternative way to construct Θ_i is as follows. First, we create a zero entry in the first row of the post-array by means of a circular (Givens) rotation, using instead the entries $(0,0)$ and $(0,1)$ of the pre-array:

$$\begin{bmatrix} \otimes & \otimes & \times \\ \times & \times & \times \\ \times & \times & \times \\ \times & \times & \times \\ \times & \times & \times \end{bmatrix} \xrightarrow{\text{Givens}} \begin{bmatrix} \times' & \boxed{0} & \times \\ \times' & \times' & \times \\ \times' & \times' & \times \\ \times' & \times' & \times \\ \times' & \times' & \times \end{bmatrix}$$

Now note that the additional hyperbolic rotation that is needed to zero out the remaining entry in the first row of the post-array, should also result in a zero entry in the $(M+1,0)$ position of the post-array. Therefore, rather than determine this hyperbolic rotation by using the entries $(0,0)$ and $(0,2)$ of the *first* row of the pre-array, we can determine it by using the entries $(M+1,0)$ and $(M+1,2)$ of the *last* row of the pre-array. That is,

$$\begin{bmatrix} \times' & 0 & \times \\ \times' & \times' & \times \\ \times' & \times' & \times \\ \times' & \times' & \times \\ \otimes & \times' & \otimes \end{bmatrix} \xrightarrow{\text{hyperbolic}} \begin{bmatrix} \times'' & 0 & 0 \\ \times'' & \times' & \times'' \\ \times'' & \times' & \times'' \\ \times'' & \times' & \times'' \\ \boxed{0} & \times' & \times'' \end{bmatrix}$$

This is a reasonable choice since, in this case, the construction of Θ_i is affected by other RLS variables. In the computer project at the end of this part, it is observed that the array method is more robust to numerical effects than the other (explicit) fast least-squares variants.

Rescue Mechanisms

With regards to the FTF, FAEST, and fast Kalman implementations, there are several mechanisms to rescue them from slipping into divergence. The main idea in these methods is to monitor a certain variable whose value is known theoretically to be positive. When the variable becomes negative, the algorithm is restarted as explained below.

The rescue variable is chosen as

$$\boxed{\text{rescue} \triangleq 1 - \beta_M(i)\gamma_{M+1}(i)\nu_{M+1}(i)} \qquad \text{(for FTF)} \tag{39.9}$$

which appears in the expression for evaluating $\gamma_M(i)$ from $\gamma_{M+1}(i)$ in the FTF algorithm. At each iteration, the sign of the rescue variable is checked. If it is positive, the algorithm continues its flow. Otherwise, if the rescue variable becomes negative at some iteration i_o, the algorithm is restarted for iteration $i_o + 1$ as follows:

$$\boxed{\begin{array}{l} w_{i_o} = w_{i_o-1}, \quad w^f_{M,i_o} = 0, \quad w^b_{M,i_o} = 0 \\ u_{M,i_o} = 0, \quad c_{M,i_o} = 0, \quad \gamma_M(i_o) = 1 \\ \zeta^b_M(i_o) = \eta^{-1}\lambda^{-(M+2)}, \quad \zeta^f_M(i_o) = \eta^{-1}\lambda^{-2} \end{array}} \tag{39.10}$$

That is, all variables are set to their original initial values except for the weight vector w_{i_o}, which is set to the current solution. In a similar manner, for FAEST we choose the rescue variable as the quantity

$$\boxed{\text{rescue} \stackrel{\Delta}{=} \gamma^{-1}_{M+1}(i) - \beta_M(i)\nu_{M+1}(i)} \qquad \text{(for FAEST)}$$

whereas for fast Kalman it is

$$\boxed{\text{rescue} \stackrel{\Delta}{=} 1 - \sigma_{M+1}(i)\beta_M(i)} \qquad \text{(for fast Kalman)}$$

It has been observed in simulations that a more reliable rescue mechanism is perhaps to keep $\zeta^f_M(i_o)$ at its current value and to re-initialize $\xi^b_M(i_o)$ by using the expression from Prob. IX.4, namely,

$$\zeta^b_M(i_o) \quad = \quad \frac{\gamma_M(i_o)\zeta^f_M(i_o)}{\lambda^M} \quad = \quad \frac{\zeta^f_M(i_o)}{\lambda^M}$$

This expression enforces the relation that should hold between $\{\xi^b_M(i_o), \xi^f_M(i_o)\}$. In this way, an alternative rescue mechanism is to restart the algorithm for iteration $i_o + 1$ as follows:

$$\boxed{\begin{array}{l} w_{i_o} = w_{i_o-1}, \quad w^f_{M,i_o} = 0, \quad w^b_{M,i_o} = 0 \\ u_{M,i_o} = 0, \quad c_{M,i_o} = 0, \quad \gamma_M(i_o) = 1 \\ \zeta^b_M(i_o) = \zeta^f_M(i_o)/\lambda^M \end{array}} \tag{39.11}$$

Feedback Stabilization

A second approach for addressing the stability problem of FTF and FAEST is based on introducing redundancy into the computation of certain quantities. In so doing, it becomes possible to estimate the numerical errors accumulated in these quantities and to use these errors in a feedback mechanism in order to combat destabilizing effects.

Consider the *a priori* backward prediction error $\beta_M(i)$. In FTF and FAEST it is computed by means of the expression

$$\boxed{\beta'_M(i) = \lambda\zeta^b_M(i-1)\nu^*_{M+1}(i)}$$

However, it could also be computed more directly (but also less efficiently) by using its definition, namely,

$$\boxed{\beta''_M(i) = u(i-M) - u_{M,i}w^b_{M,i-1}}$$

If we now modify the FTF and FAEST recursions to incorporate the evaluation of $\beta_M(i)$ by means of both of these expressions, then a suitable combination of their values could be interpreted as a better value for $\beta_M(i)$. Stabilization by feedback is based on this idea. Specifically, the difference between $\{\beta'_M(i), \beta''_M(i)\}$ is fed back, scaled by a certain gain τ, to evaluate $\beta_M(i)$, i.e.,

$$\begin{aligned} \beta_M(i) &= \beta'_M(i) + \tau[\beta''_M(i) - \beta'_M(i)] \\ &= \beta_M(i) = \tau\beta''_M(i) + (1-\tau)\beta'_M(i) \end{aligned}$$

The resulting equations are listed in Alg. 39.4; their computational complexity is $O(8M)$ operations per iteration; the increase from $O(7M)$ to $O(8M)$ is due to the evaluation of $\beta''_M(i)$. The algorithm is more similar to FAEST than to FTF since it relies on propagating the inverses of the conversion factors. It is distinct from both in that it propagates the inverse of $\xi^f_M(i)$ as well (see Prob. IX.5).

The algorithm actually computes three convex combinations of $\{\beta'_M(i), \beta''_M(i)\}$ and evaluates three *a priori* errors:

$$\boxed{\begin{aligned} \beta^{(1)}_M(i) &= \tau_1\beta''_M(i) + (1-\tau_1)\beta'_M(i) \\ \beta^{(2)}_M(i) &= \tau_2\beta''_M(i) + (1-\tau_2)\beta'_M(i) \\ \beta^{(3)}_M(i) &= \tau_3\beta''_M(i) + (1-\tau_3)\beta'_M(i) \end{aligned}}$$

Each of these values is used at a different location in the algorithm. The challenge lies in selecting suitable values for the coefficients $\{\tau_1, \tau_2, \tau_3\}$. Typical values are

$$\{\tau_1, \tau_2, \tau_3\} = \{1.5, 2.5, 1\} \tag{39.12}$$

but these may need to be adjusted (usually by simulation), because the efficacy of such a "stabilization" procedure is dependent on signal statistics.

Observe further that the algorithm evaluates $\gamma^{-1}_M(i)$ in two different ways:

$$\gamma^{-1}_M(i) = \gamma^{-1}_{M+1}(i) - \beta^{(3)}_M(i)\nu_{M+1}(i)$$

and (cf. Prob. IX.4),

$$\gamma^{-1}_M(i) = \zeta^f_M(i)/\zeta^b_M(i)\lambda^M$$

It also evaluates $\gamma^{-1}_{M+1}(i)$ via

$$\boxed{\gamma^{-1}_{M+1}(i) = \gamma^{-1}_M(i-1) + \theta_{M+1}(i)\alpha_M(i)} \tag{39.13}$$

This recursion follows from (39.7) by noting from the expression for $c_{M+1,i}$ in the statements of FTF or FAEST that its top entry is equal to $\alpha^*_M(i)/\lambda\zeta^f_M(i-1)$:

$$\boxed{\theta_{M+1}(i) \triangleq \text{top entry of } c_{M+1,i} = \alpha^*_M(i)/\lambda\zeta^f_M(i-1)} \tag{39.14}$$

We may finally remark that, despite the qualification "stabilized FTF", this procedure can still suffer from instability problems (see the computer project at the end of the chapter). Care needs to be taken in the choice of the scaling factors $\{\tau_1, \tau_2, \tau_3\}$. In addition, the value of λ needs to be sufficiently close to one, usually

$$1 - \frac{1}{2M} \leq \lambda \leq 1$$

which limits the performance of the algorithm in nonstationary environments. Of course, one could also consider incorporating into the algorithm the rescue procedures (39.10) or (39.11).

Algorithm 39.4 (Stabilized FTF) Consider the same setting as Alg. 39.1 with same initial conditions. Choose combination coefficients $\{\tau_1, \tau_2, \tau_3\}$, for instance, as in (39.12), and repeat for $i \geq 0$:

$$
\begin{aligned}
\alpha_M(i) &= u(i) - u_{M,i-1}w^f_{M,i-1} \\
f_M(i) &= \alpha_M(i)/\gamma_M^{-1}(i-1) \\
c_{M+1,i} &= \begin{bmatrix} 0 \\ c_{M,i-1} \end{bmatrix} + \lambda^{-1}\alpha_M^*(i)\xi_M^{-f}(i-1)\begin{bmatrix} 1 \\ -w^f_{M,i-1} \end{bmatrix} \\
\nu_{M+1}(i) &= \text{last entry of } c_{M+1,i} \\
\theta_{M+1}(i) &= \text{top entry of } c_{M+1,i} \\
\gamma_{M+1}^{-1}(i) &= \gamma_M^{-1}(i-1) + \theta_{M+1}(i)\alpha_M(i) \\
\zeta_M^{-f}(i) &= \lambda^{-1}\zeta_M^{-f}(i-1) - |\theta_{M+1}(i)|^2/\gamma_{M+1}^{-1}(i) \\
w^f_{M,i} &= w^f_{M,i-1} + f_M(i)c_{M,i-1} \\
\beta'_M(i) &= \lambda\zeta^b_M(i-1)\nu^*_{M+1}(i) \\
\beta''_M(i) &= u(i-M) - u_{M,i}w^b_{M,i-1} \\
\beta^{(1)}_M(i) &= \tau_1\beta''_M(i) + (1-\tau_1)\beta'_M(i) \\
\beta^{(2)}_M(i) &= \tau_2\beta''_M(i) + (1-\tau_2)\beta'_M(i) \\
\beta^{(3)}_M(i) &= \tau_3\beta''_M(i) + (1-\tau_3)\beta'_M(i) \\
\gamma_M^{-1}(i) &= \gamma_{M+1}^{-1}(i) - \beta^{(3)}_M(i)\nu_{M+1}(i) \\
b^{(1)}_M(i) &= \beta^{(1)}_M(i)/\gamma_M^{-1}(i) \\
b^{(2)}_M(i) &= \beta^{(2)}_M(i)/\gamma_M^{-1}(i) \\
\zeta^b_M(i) &= \lambda\zeta^b_M(i-1) + \beta^{*(2)}_M(i)b^{(2)}_M(i) \\
\begin{bmatrix} c_{M,i} \\ 0 \end{bmatrix} &= c_{M+1,i} - \nu_{M+1}(i)\begin{bmatrix} -w^b_{M,i-1} \\ 1 \end{bmatrix} \\
w^b_{M,i} &= w^b_{M,i-1} + b^{(1)}_M(i)c_{M,i} \\
\gamma_M^{-1}(i) &= \zeta^f_M(i)\zeta_M^{-b}(i)/\lambda^M \\
e(i) &= d(i) - u_{M,i}w_{i-1} \\
r(i) &= e(i)/\gamma_M^{-1}(i) \\
w_i &= w_{i-1} + r(i)c_{M,i}
\end{aligned}
$$

Incorporating Leakage

Another approach to addressing the stability difficulties of FTF and FAEST is based on inserting leakage correction factors into some of the FTF equations. The motivation for using leakage is similar to what we discussed before for LMS in the concluding remarks of Part IV (*Mean-Square Performance*) and in Prob. IV.39. The complexity of the resulting algorithm becomes $O(11M)$ operations per iteration; but it can be reduced to $8M$ by applying leakage less often.

The procedure consists in first reducing the number of recursive loops that are sensitive to numerical effects. In this regard, it computes $\gamma_M(i)$ directly from its definition in (37.2), namely, $\gamma_M(i) = 1/(1 + u_{M,i}c_{M,i})$, rather than recursively as in FTF and FAEST. Then a leakage factor ϵ is incorporated into the update equations for the quantities $\{c_{M+1,i}, c_{M,i}, w^f_{M,i}, w^b_{M,i}\}$:

$$\begin{aligned}
c_{M+1,i} &= \epsilon\begin{bmatrix} 0 \\ c_{M,i-1} \end{bmatrix} + \frac{\epsilon\alpha^*_M(i)}{\lambda\zeta^f_M(i-1)}\begin{bmatrix} 1 \\ -w^f_{M,i-2} \end{bmatrix} \\
w^f_{M,i} &= \epsilon w^f_{M,i-1} + \epsilon f_M(i)c_{M,i-1} \\
\begin{bmatrix} c_{M,i} \\ 0 \end{bmatrix} &= c_{M+1,i} - \epsilon\nu_{M+1}(i)\begin{bmatrix} -w^b_{M,i-1} \\ 1 \end{bmatrix} \\
w^b_{M,i} &= \epsilon w^b_{M,i-1} + \epsilon b_M(i)c_{M,i}
\end{aligned}$$

where ϵ is suitably chosen to be smaller but close to one, e.g., $\epsilon = 0.98$ or $\epsilon = 0.99$. The "stabilization" of the FTF recursions is achieved, however, at the expense of degradation in performance since the leakage factor introduces bias in the weight vector estimates, especially during the initial stages of adaptation.

Summary and Notes

The chapters in this part describe several efficient implementations of RLS.

SUMMARY OF MAIN RESULTS

1. By exploiting the fact that regressors have shift structure, the computational cost of RLS can be reduced from $O(M^2)$ to $O(M)$ operations per iteration.

2. The key insight is to realize that the gain vector g_i, and the conversion factor $\gamma(i)$, can be time-updated without requiring explicit evaluation of the matrix P_i. Instead, only the low-rank difference

$$\begin{bmatrix} P_{i-1} & 0 \\ 0 & 0 \end{bmatrix} - \begin{bmatrix} 0 & 0 \\ 0 & P_{i-2} \end{bmatrix}$$

   needs to be time-updated.

3. Interestingly enough, the regularization matrix Π can be chosen so as to enforce a rank-two difference right at the first iteration. Once this is done, the rank-two property preserves itself over time.

4. Four efficient implementations were described in this part: (a) fast array filter, (b) fast transversal filter (FTF), (c) fast *a posteriori* error sequential technique (FAEST), and (d) fast Kalman filter. The array form is the simplest to derive and to describe; it also shows reasonable robustness in finite-precision implementation, as illustrated in the computer project at the end of this part. The FTF and FAEST filters are the most efficient but tend to suffer from numerical difficulties. This is because many of the equalities used to derive them tend to break down in finite precision. The fast Kalman filter is the oldest among them.

5. The derivation of FTF, FAEST, and fast Kalman is based on studying forward and backward prediction problems, and on showing that the update of the gain vector can be described in terms of the weight vectors for the forward and backward prediction problems.

6. The derivation in the chapters assumes that regularization is present throughout.

7. Several stabilization procedures are described including: (a) rescue mechanisms, (b) feedback stabilization, and (c) leakage.

8. In Prob. IX.13 we show that fast RLS filters can be derived even for cases involving regressors without shift structure; another important example to this effect is discussed in Chapter 16 of Sayed (2003) in the context of RLS Laguerre filtering.

BIBLIOGRAPHIC NOTES

Regularization. In comparison to conventional derivations of fast fixed-order filters, we have incorporated regularization into our arguments right from the early stages of the derivation.

Conversion factor and original derivations. The variable $\gamma(i)$ plays the role of a conversion factor, which converts an *a priori* error into an *a posteriori* error. This is a useful property since it allows us to evaluate an *a posteriori* error before updating the relevant tap-weight vector itself. This property was exploited to great advantage in our presentation, as well as in the original derivations of the fast *a posteriori* error sequential technique (FAEST) by Carayannis, Manolakis, and Kalouptsidis (1983) and the fast transversal filter (FTF) by Cioffi and Kailath (1984). Further related works on efficient implementations of the least-squares type appear in Moustakides and Theodoridis (1991) and Glentis, Berberidis, and Theodoridis (1999).

Fast Kalman filter. The fast Kalman filter of Sec. 39.3 is the earliest fast RLS filter. It was developed by Ljung, Morf, and Falconer (1978) and was also used in Falconer and Ljung (1978).

Fast array filter. The derivation that led to the array form of Alg. 37.1 is a streamlined version of a derivation given by Sayed and Kailath (1992,1994a,1998). This array form is actually a special case of a more general algorithm known as the extended Chandrasekhar algorithm (37.26), and which was derived by Sayed and Kailath (1994a) as a fast state-space estimation method for models whose parameters vary in a structured manner. This connection with state-space estimation was explained in Sayed and Kailath (1994b) and is also presented in App. 37.A and Prob. IX.16 (see also Prob. IX.13).

Backward consistency. As explained in Chapter 39, fast fixed-order least-squares filters are sensitive to round-off errors and tend to suffer from instability problems in finite word-length implementations (see, e.g., Botto (1988) as well as the computer project at the end of this part). Using the concept of backward consistency, Regalia (1992,1993) and Slock (1992) provided an explanation for the origin of such difficulties. Their analyses indicated that the dynamics of a fast transversal filter (FTF) contains unstable modes that are not excited in exact arithmetic but that become active in finite word-length implementations. Appendix 14.C of Sayed (2003) provides a summary of their results; see also Bunch, Leborne, and Proudler (2001) for a reference dealing with the issue of consistency for adaptive signal processing. It was to address such numerical problems that Lin (1984), Cioffi and Kailath (1984), and Eleftheriou and Falconer (1987) suggested using rescue variables in order to monitor the behavior of the FTF algorithm and to re-initialize it whenever abnormal behavior is observed (as was discussed in Sec. 39.4).

Rescue schemes. Most rescue mechanisms monitor the value of the conversion factor to detect filter divergence. This is because it is known beforehand that the value of the conversion factor should lie inside the interval $(0, 1]$. Motivated by this fact, Lin (1984) suggested monitoring the rescue variable (39.9). Cioffi and Kailath (1984) employed a similar rescue procedure, in addition to adding a small white noise to the input data (a technique known as dithering). The purpose of dithering was to avoid the matrix P_i from becoming close-to-singular in steady-state. This method introduces a small bias in the weight estimate and is only effective if λ is chosen as $\lambda \geq (5M - 1)/(5M + 1)$, where M is the filter order.

Another technique for improving the reliability of fast least-squares filters is the one based on introducing feedback into the operation of the filters in order to influence the propagation of the errors. This technique resulted in the stabilized FTF variant of Alg. 39.4 by Slock and Kailath (1988,1991). The technique was also applied to the fast Kalman filter by Botto and Moustakides (1989). Still, even with "stabilization", the stabilized FTF algorithm can suffer from numerical difficulties especially when the forgetting factor is not sufficiently close to unity. In Slock and Kailath (1991), it was suggested that λ should be chosen within the range $(2M - 1)/2M < \lambda < 1$. Yet another rescue technique is the one based on inserting leakage into some of the FTF recursions, as described in Sec. 39.4. This technique was proposed by Binde (1995). Apparently, the performance of the method does not depend on the input signal statistics. However, the "stabilization" of the FTF recursions is achieved at the expense of degradation in performance due to bias, especially during the initial adaptation phase. The overall complexity of this procedure is $O(11M)$ operations per iteration; nevertheless, the complexity can be reduced to $8M$, by applying leakage less often.

Fast filtering methods. The fast algorithms of this part, array-based and otherwise, have connections with a fast alternative to the Kalman filter for constant state-space models known as the Chandrasekhar filter. The filter is described in App. 37.A and it was originally developed in the early 1970s by Kailath, Morf, and Sidhu (1973) and Morf, Sidhu, and Kailath (1974), as an extension to discrete-time of results derived in continuous time by Kailath (1972,1973). Interestingly, the continuous-time results were motivated by earlier works by Ambartsumian (1943) and Chandrasekhar (1947a,1947b) on radiative transfer theory. Lindquist (1974,1976) also independently obtained a fast algorithm for discrete-time filtering; albeit one that is specific to stationary processes. The array version of the Chandrasekhar filter appeared in Morf and Kailath (1975) and Kailath, Vieira, and Morf (1978).

Extended Chandrasekhar filter. Connections between the original Chandrasekhar filter (37.24) for state-space estimation and fast RLS algorithms have been discussed by Houacine and Demoment (1986), Slock (1989), and Houacine (1991) by formulating time-invariant state-space models. However, the state-space model that arises in the context of adaptive filtering is not constant (as we saw, for example, in model (31.8)); this is because the regressor u_i varies with time. Motivated by this observation, Sayed and Kailath (1992) showed how to extend the original Chandrasekhar filter to a class of *structured* state-space models (cf. (37.25)), a special case of which is model (31.8). They derived the extended Chandrasekhar filter (37.26), which is an efficient estimation procedure for state-space models that vary in a certain structured manner. In this way, the relation between the Chandrasekhar filter and fast RLS methods becomes more natural and also broader (see Probs. IX.13 and IX.16). In addition, the extended Chandrasekhar filter can be used to derive fast RLS methods even for some non-tapped-delay-line structures (as was shown by Merched and Sayed (1999) and as is discussed in detail in Chapter 16 of Sayed (2003)).

Problems and Computer Projects

PROBLEMS

Problem IX.1 (Initial conditions) Refer to the discussion in Sec. 37.3 on initial conditions and let Π be as in (37.13).

(a) Verify that in this case $\delta P_0 = \eta\lambda \cdot \text{diag}\left\{\frac{1}{1+\eta\lambda|u(0)|^2}, 0, \ldots, 0, -\lambda^M\right\}$.

(b) Conclude that

$$\bar{L}_0 = \sqrt{\frac{\eta}{1+\eta\lambda|u(0)|^2}} \cdot \begin{bmatrix} 1 & 0 \\ 0 & 0 \\ \vdots & \vdots \\ 0 & 0 \\ 0 & \lambda^{M/2} \end{bmatrix}, \qquad S_0 = \begin{bmatrix} 1 & 0 \\ 0 & -1 \end{bmatrix}$$

along with $w_0 = \bar{w} + g_0[d(0) - u_0\bar{w}]$ and

$$\gamma^{-1/2}(0) = \sqrt{1+\eta\lambda|u(0)|^2}, \qquad g_0 = \frac{\eta\lambda}{1+\eta\lambda|u(0)|^2} \cdot \text{col}\{u^*(0), 0, \ldots, 0\}$$

Problem IX.2 (Order- and time-update of conversion factor) Start from expression (38.22) for $\gamma_{M+1}(i)$ in terms of $\gamma_M(i-1)$ and rewrite it as

$$\gamma_{M+1}(i) = \gamma_M(i-1)\left[1 - \frac{\alpha_M(i)f_M^*(i)}{\eta^{-1}\lambda^{i-1} + \xi_M^f(i)}\right]$$

(a) Since $\xi_M^f(i)$ denotes the minimum cost of the regularized least-squares problem (38.19), it satisfies $\xi_M^f(i) = \lambda\xi_M^f(i-1) + \alpha_M(i)f_M^*(i)$ (cf. Alg. 30.2). Use this relation to show that the above expression for $\gamma_{M+1}(i)$ leads to the equality

$$\boxed{\gamma_{M+1}(i)/\gamma_M(i-1) \;=\; \lambda\zeta_M^f(i-1)/\zeta_M^f(i)}$$

(b) Likewise, show that

$$\boxed{\gamma_{M+1}(i)/\gamma_M(i) \;=\; \lambda\zeta_M^b(i-1)/\zeta_M^b(i)}$$

(c) Use part (b) to show that $1 - |\beta_M(i)|^2\gamma_{M+1}(i)/\lambda\zeta_M^b(i-1) = 1 - b_M^*(i)\beta_M(i)/\zeta_M^b(i)$, and conclude from (38.12), and from the definition of $\nu_{M+1}(i)$ in (39.6), namely, $\nu_{M+1}(i) = \beta_M^*(i)/\lambda\zeta_M^b(i-1)$, that

$$\boxed{\gamma_M(i) = \frac{\gamma_{M+1}(i)}{1 - \beta_M(i)\gamma_{M+1}(i)\nu_{M+1}(i)}}$$

Problem IX.3 (Relation between conversion factors) Use the result of Prob. IX.2 to show that

$$\boxed{\gamma_{M+1}(i)/\gamma_M(i-1) \leq 1 \quad \text{and} \quad \gamma_{M+1}(i)/\gamma_M(i) \leq 1}$$

Problem IX.4 (Time-update for conversion factor) Use again the result of Prob. IX.2 to derive the time-update relation:

$$\boxed{\gamma_M(i) \;=\; \gamma_M(i-1)\frac{\zeta_M^b(i)}{\zeta_M^b(i-1)}\frac{\zeta_M^f(i-1)}{\zeta_M^f(i)}}$$

Conclude, by iterating this result, that $\gamma_M(i) = \lambda^M \zeta_M^b(i)/\zeta_M^f(i)$.

Problem IX.5 (Updating the inverse cost) Start from $\zeta_M^f(i) = \lambda\zeta_M^f(i-1) + \alpha_M^*(i)f_M(i)$, and the matrix inversion lemma (30.11) to the right-hand side, and use (39.7) to conclude that

$$\boxed{1/\zeta_M^f(i) = \lambda^{-1}/\zeta_M^f(i-1) - \gamma_{M+1}(i)|\theta_{M+1}(i)|^2}$$

where $\theta_{M+1}(i)$ is defined by (39.14).

Problem IX.6 (Order- and time-update of gain vector) In this problem we want to establish relation (39.4).

(a) Use the result of part (a) of Prob. IX.2, and (38.23), to show that

$$g_{M+1,i} = \begin{bmatrix} 0 \\ g_{M,i-1} \end{bmatrix} + \gamma_{M+1}(i)\frac{\lambda^{-1}\alpha_M^*(i)}{\eta^{-1}\lambda^{i-2} + \xi_M^f(i-1)}\begin{bmatrix} 1 \\ -w_{M,i}^f \end{bmatrix}$$

Conclude that

$$c_{M+1,i} = \begin{bmatrix} 0 \\ g_{M,i-1} \end{bmatrix}\frac{1}{\gamma_{M+1}(i)} + \frac{\lambda^{-1}\alpha_M^*(i)}{\eta^{-1}\lambda^{i-2} + \xi_M^f(i-1)}\begin{bmatrix} 1 \\ -w_{M,i}^f \end{bmatrix}$$

(b) Use (39.2), namely, $w_{M,i}^f = w_{M,i-1}^f + g_{M,i-1}\gamma_M^{-1}(i-1)f_M(i)$, and substitute $w_{M,i}^f$ into the expression of part (a) in terms of $w_{M,i-1}^f$ in order to establish (39.4).

Problem IX.7 (Triangular factorization) Introduce the notation

$$\zeta_k^b(i) \;\stackrel{\Delta}{=}\; \eta^{-1}\lambda^{i-k-1} + \xi_k^b(i), \quad k = 0, 1, \ldots, M-1$$

Iterate relation (38.9) to conclude that the matrix $P_{M,i}$ admits the upper-diagonal-lower triangular factorization $P_{M,i} = U_{M,i}D_{M,i}U_{M,i}^*$, where

$$D_{M,i} = \begin{bmatrix} 1/\zeta_0^b(i) & & & \\ & 1/\zeta_1^b(i) & & \\ & & \ddots & \\ & & & 1/\zeta_{M-1}^b(i) \end{bmatrix}, \quad U_{M,i} = \begin{bmatrix} 1 & \times & \times & \ldots & \times \\ & 1 & \times & \ldots & \times \\ & & 1 & & \times \\ & & & \ddots & \vdots \\ & & & & 1 \end{bmatrix}$$

where the columns of $U_{M,i}$ are defined in terms of the backward prediction vectors

$$\begin{bmatrix} -w_{k,i}^b \\ 1 \end{bmatrix}, \quad k = 0, 1, \ldots, M-1$$

Problem IX.8 (Determinant expression) Show that

$$\det P_{M,i} = \prod_{k=0}^{M-1} \left(\frac{1}{\zeta_k^b(i)} \right)$$

Conclude from part (b) of Prob. IX.2 that

$$\gamma_M(i) = \lambda^M \frac{\det P_{M,i}}{\det P_{M,i-1}}$$

Problem IX.9 (Initial iterations of FTF) Refer to the FTF recursions in Alg. 39.1. Verify that the variable $\nu_{M+1}(i)$, and consequently the errors $\{\beta_M(i), b_M(i)\}$, are zero for $i = 0$ to $M-1$.

(a) Conclude that during these initial time instants the FTF recursions are given by:

$$\text{for } i = 0, 1, \ldots, M-1:$$

$$\boxed{\begin{array}{rcl}
\alpha_M(i) & = & u(i) - u_{M,i-1} w^f_{M,i-1} \\
f_M(i) & = & \gamma_M(i-1)\alpha_M(i) \\
\zeta^f_M(i) & = & \lambda\zeta^f_M(i-1) + \alpha^*_M(i) f_M(i) \\
\gamma_M(i) & = & \gamma_M(i-1)\, \lambda\zeta^f_M(i-1)/\zeta^f_M(i) \\
w^f_{M,i} & = & w^f_{M,i-1} + f_M(i) c_{M,i-1} \\
\begin{bmatrix} c_{M,i} \\ 0 \end{bmatrix} & = & \begin{bmatrix} 0 \\ c_{M,i-1} \end{bmatrix} + \dfrac{\alpha^*_M(i)}{\lambda\zeta^f_M(i-1)} \begin{bmatrix} 1 \\ -w^f_{M,i-1} \end{bmatrix} \\
\zeta^b_M(i) & = & \lambda\zeta^b_M(i-1) \\
e(i) & = & d(i) - u_{M,i} w_{i-1} \\
r(i) & = & \gamma_M(i) e(i) \\
w_i & = & w_{i-1} + r(i) c_{M,i}
\end{array}}$$

(b) Use (38.1) to argue that w_{M-1} is the solution to the least-squares problem:

$$\min_w \left[\lambda^M w^* \Pi_M w + (y_{M-1} - H_{M-1} w)^* \Lambda_{M-1} (y_{M-1} - H_{M-1} w) \right]$$

where H_{M-1} is lower triangular Toeplitz with first column $\{u(0), u(1), \ldots, u(M-1)\}$.

Problem IX.10 (Rank-three factorization) Use the recursion for P_i from (37.2) and (38.25) to show that

$$\boxed{\begin{bmatrix} P_i & 0 \\ 0 & 0 \end{bmatrix} - \begin{bmatrix} 0 & 0 \\ 0 & \lambda P_i \end{bmatrix} = \lambda \bar{L}_i S \bar{L}_i^* + \frac{\lambda}{\gamma(i)} \begin{bmatrix} 0 \\ g_i \end{bmatrix} \begin{bmatrix} 0 \\ g_i \end{bmatrix}^*}$$

Problem IX.11 (Generating functions) Introduce the basis function

$$\mathcal{B}_M(z) \triangleq \begin{bmatrix} 1 & z^{-1}\lambda^{-1/2} & z^{-2}\lambda^{-1} & \ldots & z^{-(M-1)}\lambda^{-(M-1)/2} \end{bmatrix}$$

and use it to transform the matrix and vector quantities $\{P_i, w^f_{M,i}, w^b_{M,i}, g_i\}$ into functions of z and s as follows:

$$\begin{array}{rcl}
P(z,s) & = & \mathcal{B}_M(z) P_i \mathcal{B}^*_M(s), \quad g(z) = z^{-1}\gamma^{-1/2}(i)\mathcal{B}_M(z) g_i \\
w^f(z) & = & \sqrt{\dfrac{\lambda}{\zeta^f_M(i)}} \mathcal{B}_{M+1}(z) \begin{bmatrix} 1 \\ -w^f_{M,i} \end{bmatrix}, \quad w^b(z) = \sqrt{\dfrac{\lambda}{\zeta^b_M(i)}} \mathcal{B}_{M+1}(z) \begin{bmatrix} -w^b_{M,i} \\ 1 \end{bmatrix}
\end{array}$$

(a) Show that the equality of Prob. IX.10 reduces to the function equality:

$$\boxed{(1 - 1/zs^*) P(z,s) = w^f(z) w^{f*}(s) - w^b(z) w^{b*}(s) + g(z) g^*(s)}$$

(b) Choose $s = z$ and conclude that because P_i is positive-definite, the following inequalities should hold:

$$w^f(z)w^{f*}(z) + g(z)g^*(z) - w^b(z)w^{b*}(z) \begin{cases} > 0, & |z| > 1 \\ = 0, & |z| = 1 \\ < 0, & |z| < 1 \end{cases}$$

Conclude that $w^b(z)$ cannot have zeros inside the unit circle.

(c) Verify that $w^b(z)w^{b*}(1/z^*) = w^f(z)w^{f*}(1/z^*) + g(z)g^*(1/z^*)$.

Remark. This equation can be used to argue that $w^b(z)$ is uniquely defined by $\{w^f(z), g(z)\}$ (more specifically, it is the spectral factor of the right-hand-side expression). Moreover, the quantity $\zeta^b_M(N)$ can be inferred from the last coefficient of $w^b(z)$ — see Regalia (1992,1993).

Problem IX.12 (Leading and trailing columns of $P_{M+1,i}$) Show that the last column of $P_{M+1,i}$ is proportional to the backward prediction vector col$\{-w^b_{M,i}, 1\}$. Show further that the first column of $P_{M+1,i}$ is proportional to the forward prediction vector col$\{1, -w^f_{M,i}\}$.

Problem IX.13 (General fast array algorithm) Consider data $\{u_j, d(j)\}_{j=0}^N$, where the u_j are $1\times M$ and the $d(j)$ are scalars. Consider also an $M\times 1$ vector $\bar{w}$, a scalar $0 \ll \lambda \leq 1$, and an $M\times M$ positive-definite matrix Π. Assume further that for a particular choice of Π there exists some $M\times M$ matrix Ψ such that the difference $P_{-1} - \Psi P_{-2}\Psi^*$ is low rank, say $P_{-1} - \Psi P_{-2}\Psi^* = \bar{L}_{-1}S\bar{L}^*_{-1}$ for some $\bar{L}_{-1}$ and signature matrix S. It is further assumed that the regressors satisfy $u_i\Psi = u_{i-1}$. Follow the discussion in Chapter 37 to show that the solution w_N of the least-squares problem

$$\min_w \left[\lambda^{(N+1)}(w-\bar{w})^*\Pi(w-\bar{w}) + \sum_{j=0}^{N}\lambda^{N-j}|d(j) - u_jw|^2\right]$$

can be recursively computed as follows. Start with $w_{-1} = \bar{w}$, $\gamma^{-1/2}(-1) = 1$, $g_{-1} = 0$, $\bar{L}_{-1}$ from the above low-rank factorization, and repeat for $i \geq 0$:

$$\begin{bmatrix} \gamma^{-1/2}(i-1) & \lambda^{-1/2}u_i\bar{L}_{i-1} \\ \Psi g_{i-1}\gamma^{-1/2}(i-1) & \lambda^{-1/2}\bar{L}_{i-1} \end{bmatrix}\Theta_i = \begin{bmatrix} \gamma^{-1/2}(i) & 0 \\ g_i\gamma^{-1/2}(i) & \bar{L}_i \end{bmatrix}$$

where Θ_i is a $(1\oplus S)-$unitary matrix that produces the zero entries in the post-array, and where the factor $\bar{L}_i$ satisfies $P_i - \Psi P_{i-1}\Psi^* = \bar{L}_iS\bar{L}^*_i$. *Remark.* These recursions are a special case of the extended Chandrasekhar filter of Sayed and Kailath (1994a,1994b) for structured state-space models — see App. 37.A. They form the basis for the derivation of fast RLS Laguerre adaptive filters — see Sec. 16.3 of Sayed (2003).

Problem IX.14 (Array algorithm for sliding window RLS) Refer to the finite-memory RLS algorithm derived in Prob. VII.34. Follow the arguments in Sec. 35.1 to motivate the following array algorithm. Let $\Sigma = \Pi^{-1}$ and introduce the Cholesky decomposition $\Sigma = \Sigma^{1/2}\Sigma^{*/2}$, where $\Sigma^{1/2}$ is lower triangular with positive-diagonal entries. Then start with $w^L_{-1} = \bar{w}$, $(P^L_{-1})^{1/2} = \Sigma^{1/2}$, and repeat for $i \geq 0$.

1. (downdating step). Find a $(1\oplus -\mathrm{I})-$unitary matrix Θ^d_i that lower triangularizes the pre-array shown below and generates a post-array with positive diagonal entries. Then the entries in the post-array will correspond to

$$\begin{bmatrix} 1 & u_{i-L}\left(P^L_{i-1}\right)^{1/2} \\ 0 & \left(P^L_{i-1}\right)^{1/2} \end{bmatrix}\Theta^d_i = \begin{bmatrix} \gamma_L^{-1/2}(i-1) & 0 \\ -g^L_{i-1}\gamma_L^{-1/2}(i-1) & \left(P^{L-1}_{i-1}\right)^{1/2} \end{bmatrix}$$

2. (updating step). Find a unitary matrix Θ^u_i that lower triangularizes the pre-array shown below and generates a post-array with positive diagonal entries. Then the entries in the post-array will correspond to

$$\begin{bmatrix} 1 & u_i\left(P^{L-1}_{i-1}\right)^{1/2} \\ 0 & \left(P^{L-1}_{i-1}\right)^{1/2} \end{bmatrix}\Theta^u_i = \begin{bmatrix} \gamma_{L-1}^{-1/2}(i) & 0 \\ g^{L-1}_i\gamma_{L-1}^{-1/2}(i) & \left(P^L_i\right)^{1/2} \end{bmatrix}$$

3. Evaluate the errors $e_d(i-L) = d(i-L) - u_{i-L}w_{i-1}^L$ and $e_u(i) = d(i) - u_i w_{i-1}^{L-1}$.

4. Update the weight vector from w_{i-1}^L to w_i^L as follows:

$$\begin{aligned} w_{i-1}^{L-1} &= w_{i-1}^L - \left(g_{i-1}^L \gamma_L^{-1/2}(i-1)\right) \cdot \left(\gamma_L^{-1/2}(i-1)\right)^{-1} e_d(i-L) \\ w_i^L &= w_{i-1}^{L-1} + \left(g_i^{L-1}\gamma_{L-1}^{-1/2}(i)\right) \cdot \left(\gamma_{L-1}^{-1/2}(i)\right)^{-1} e_u(i) \end{aligned}$$

Problem IX.15 (Array Chandrasekhar form) Refer to the discussion on the Chandrasekhar filter in App. 37.A. Compare entries on both sides of (37.23) and identify the terms $\{\mathsf{X}, \mathsf{Y}, \mathsf{Z}\}$ as in (37.24). Likewise, verify the validity of the extended Chandrasekhar recursion (37.26).

Problem IX.16 (Fast array RLS and Chandrasekhar filter) In Sec. 31.2 we showed that the exponentially-weighted RLS algorithm can be obtained via equivalence to the Kalman filter as follows. We start from the state-space model (31.8) and write down the Kalman recursions for estimating its state variable. Then we translate the Kalman variables into the RLS variables by using the correspondences summarized in Table 31.2. We used this procedure in Probs. VIII.12 and VIII.13 in order to examine the relation between RLS array methods and Kalman filtering array methods.

In this problem, we use the same reasoning to explain the connection between the array Chandrasekhar filter of App. 37.A and the fast array RLS method of Alg. 37.1 when u_i has shift structure, i.e., when $u_i = \begin{bmatrix} u(i) & u(i-1) & \ldots & u(i-M+1) \end{bmatrix}$. To do so, we need to replace the time-variant state-space model (31.8) by a structured state-space model of the form defined by (37.25).

Thus consider the least-squares formulation (37.1), with $N+1$ data points, and assume the regressors u_i have shift structure. Introduce the following $(N+1)$–dimensional state-space model (in contrast, model (31.8) is M–dimensional): $\boldsymbol{x}_{i+1} = \lambda^{-1/2}\boldsymbol{x}_i$ and $\boldsymbol{y}(i) = h_i \boldsymbol{x}_i + \boldsymbol{v}(i)$, with h_i defined as the $(N+1)$-long vector $h_i = [u(i)\ u(i-1)\ \ldots\ u(0)\ 0_{1\times N-i-1}]$. That is, h_i has all the input data from time 0 up to and including time i. The remaining entries are zeros. Moreover, the trailing $N-M$ entries of the state vectors $\boldsymbol{x}_i$ are taken to be zero, i.e., $\boldsymbol{x}_i = \text{col}\{\times, 0_{N-M\times 1}\}$, so that $h_i\boldsymbol{x}_i = u_i\boldsymbol{x}_i$ for all i. Let

$$\mathsf{E}\,\boldsymbol{v}(i)\boldsymbol{v}^*(j) = \delta_{ij}, \qquad \mathsf{E}\,\boldsymbol{x}_0\boldsymbol{x}_0^* = \begin{bmatrix} \Pi_0 & 0 \\ 0 & 0_{(N-M)\times(N-M)} \end{bmatrix}$$

(a) Verify that h_i satisfies $h_i = h_{i+1}Z$, where Z is the lower triangular shift matrix with ones on the first subdiagonal and zeros elsewhere. Conclude that the state-space model defined above is structured (i.e., it satisfies (37.25) with $\Psi_i = Z$).

(b) From the Kalman recursions (37.22) for the above model, i.e., from

$$\begin{aligned} \hat{\boldsymbol{x}}_{i+1|i} &= \lambda^{-1/2}\hat{\boldsymbol{x}}_{i|i-1} + k_{p,i}\left[\boldsymbol{y}(i) - h_i\hat{\boldsymbol{x}}_{i|i-1}\right] \\ r_e(i) &= 1 + h_i P_{i|i-1} h_i^* \\ k_{p,i} &= \lambda^{-1/2} P_{i|i-1} h_i^* / r_e(i) \\ P_{i+1|i} &= \lambda^{-1}\left[P_{i|i-1} - P_{i|i-1}h_i^* h_i P_{i|i-1}/r_e(i)\right], \qquad P_{0|-1} = \left(\Pi_0 \oplus 0_{(N-M)\times(N-M)}\right) \end{aligned}$$

verify that the normalized gain vector $k_{p,i}r_e^{1/2}(i)$ has trailing zeros. Specifically, argue that it has the form

$$k_{p,i}r_e^{1/2}(i) = \begin{bmatrix} \bar{c}_i \\ 0_{(N-M)\times 1} \end{bmatrix}$$

for some $M \times 1$ vector $\bar{c}_i$.

(c) Verify that the extended Chandrasekhar recursions (37.26) that correspond to the above model are given by

$$\begin{bmatrix} r_e^{1/2}(i-1) & h_i \bar{L}_{i-1|i-2} \\ Z\begin{bmatrix} \bar{c}_{i-1} \\ 0_{(N-M)\times 1} \end{bmatrix} & \lambda^{-1/2}\bar{L}_{i-1|i-2} \end{bmatrix} \Theta_i = \begin{bmatrix} r_e^{1/2}(i) & \begin{bmatrix} 0 & 0 \end{bmatrix} \\ \begin{bmatrix} \bar{c}_i \\ 0_{(N-M)\times 1} \end{bmatrix} & \bar{L}_{i|i-1} \end{bmatrix}$$

where Θ_i is any $(1 \oplus S)$-unitary matrix that lower triangularizes the pre-array. Moreover, $(\bar{L}_{0|-1}, S)$ are found from the factorization $P_{1|0} - ZP_{0|-1}Z^* = \bar{L}_{0|-1}S\bar{L}_{0|-1}^*$.

(d) Argue that $\bar{L}_{i|i-1}$ also has trailing zeros and write it as

$$\bar{L}_{i|i-1} = \begin{bmatrix} \widetilde{L}_{i|i-1} \\ 0_{(N-M-1)\times\alpha} \end{bmatrix}$$

where $\widetilde{L}_{i|i-1}$ is $(M+1)\times\alpha$. Let $\widetilde{h}_i$ denote the row vector with the first $M+1$ entries of h_i. Verify that the Chandrasekhar recursions of part (c) reduce to

$$\begin{bmatrix} r_e^{1/2}(i-1) & \widetilde{h}_i\widetilde{L}_{i-1|i-2} \\ \begin{bmatrix} 0 \\ \bar{c}_{i-1} \end{bmatrix} & \lambda^{-1/2}\widetilde{L}_{i-1|i-2} \end{bmatrix} \Theta_i = \begin{bmatrix} r_e^{1/2}(i) & \begin{bmatrix} 0 & 0 \end{bmatrix} \\ \begin{bmatrix} \bar{c}_i \\ 0 \end{bmatrix} & \widetilde{L}_{i|i-1} \end{bmatrix}$$

(e) Use the correspondences from Table 11.1 between RLS and Kalman variables to show that the above Chandrasekhar filter leads to the fast array method of Alg. 37.1, namely,

$$\begin{bmatrix} \gamma^{-1/2}(i-1) & \begin{bmatrix} u(i) & u_{i-1} \end{bmatrix} \bar{L}_{i-1} \\ \begin{bmatrix} 0 \\ g_{i-1}\gamma^{-1/2}(i-1) \end{bmatrix} & \bar{L}_{i-1} \end{bmatrix} \Theta_i = \begin{bmatrix} \gamma^{-1/2}(i) & \begin{bmatrix} 0 & 0 \end{bmatrix} \\ \begin{bmatrix} g_i\gamma^{-1/2}(i) \\ 0 \end{bmatrix} & \sqrt{\lambda}\,\bar{L}_i \end{bmatrix}$$

Problem IX.17 (Fast multichannel RLS) Consider N FIR channels with $\{M_k,\ k = 1,\ldots,N\}$ taps each. Let $\{u^{(k)}(i),\ k = 1,\ldots,N\}$ denote the input sequences to these channels with the corresponding regression vectors and tap vectors denoted by

$$u_i^{(k)} \triangleq \begin{bmatrix} u^{(k)}(i) & u^{(k)}(i-1) & \ldots & u^{(k)}(i-M_k+1) \end{bmatrix}, \quad k = 1,\ldots,N$$

and $\{w^{o(k)},\ k = 1,\ldots,N\}$. The outputs of the channels are combined together to yield the noisy measurement $d(i) = \sum_{k=1}^{N} u_i^{(k)} w^{o(k)} + v(i)$ — see Fig. IX.1.

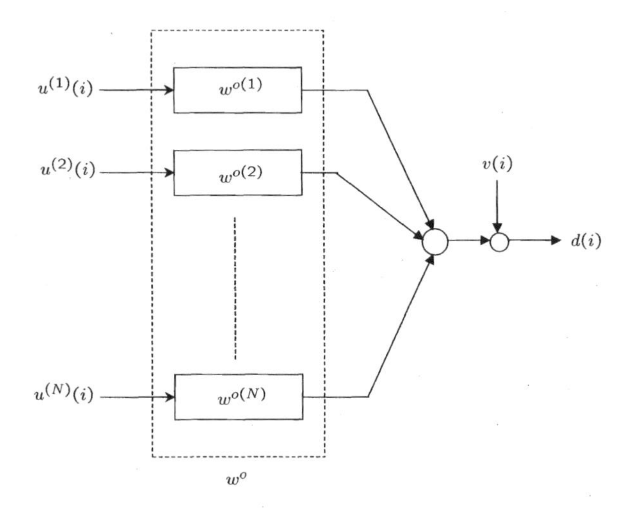

FIGURE IX.1 Data transmission over N FIR channels.

This formulation can be recast in terms of a single-channel description by collecting all regressors and tap vectors into a single extended regressor and a single extended tap vector:

$$u_i \triangleq \left[\begin{array}{cccc} u_i^{(1)} & u_i^{(2)} & \ldots & u_i^{(N)} \end{array}\right], \quad w^o \triangleq \text{col}\left\{w^{o(1)}, w^{o(2)}, \ldots, w^{o(N)}\right\}$$

so that $d(i) = u_i w^o + v(i)$. Therefore, given measurements $\{d(i), u_i\}$ we can estimate the weight vector w^o and, subsequently, the individual channel weight vectors $\{w^{o(k)}\}$, by applying an adaptive filter (of the LMS or RLS type) to the data $\{d(i), u_i\}$, e.g., by using

$$w_i = w_{i-1} + \mu u_i^*[d(i) - u_i w_{i-1}] \qquad \text{(multichannel LMS)}$$

or

$$\begin{array}{rcll} \gamma(i) & = & 1/(1 + \lambda^{-1} u_i P_{i-1} u_i^*) & \\ g_i & = & \lambda^{-1}\gamma(i) P_{i-1} u_i^* & \\ w_i & = & w_{i-1} + g_i[d(i) - u_i w_{i-1}] & \text{(multichannel RLS)} \\ P_i & = & \lambda^{-1} P_{i-1} - g_i g_i^*/\gamma(i) & \end{array}$$

The computational cost of the LMS implementation will be $O(M)$ operations per iteration, while that of RLS will be $O(M^2)$ operations per iteration, where $M = (M_1 + M_2 + \ldots + M_N)$.

In this problem we are interested in showing that in the case of RLS, an efficient $O(M)$ implementation can be pursued by observing the following. Although, strictly speaking, the regressor u_i is not shift-structured, it still consists of individual vectors $\{u_i^k\}$ that are shift-structured themselves. This fact can be exploited to derive fast RLS implementations.

(a) Define

$$\left[\Pi^{(k)}\right]^{-1} \triangleq \eta \cdot \text{diag}\{\lambda^2, \lambda^3, \ldots, \lambda^{M_k+1}\} \qquad (M_k \times M_k)$$

and choose the regularization matrix Π (i.e., P_{-1}) as $\Pi = \text{diag}\{\Pi^{(1)}, \Pi^{(2)}, \ldots, \Pi^{(N)}\}$. Define further the $(M_k + 1) \times 2$ and 2×2 matrices

$$\bar{L}_{-1}^{(k)} = \sqrt{\eta\lambda}\left[\begin{array}{cc} 1 & 0 \\ 0 & 0 \\ \vdots & \vdots \\ 0 & 0 \\ 0 & \lambda^{M_k/2} \end{array}\right], \qquad S^{(k)} = \left[\begin{array}{cc} 1 & 0 \\ 0 & -1 \end{array}\right]$$

as well as the $(M+N) \times 2N$ matrix $\bar{L}_{-1} = \text{diag}\{\ \bar{L}_{-1}^{(1)},\ \bar{L}_{-1}^{(2)},\ \bar{L}_{-1}^{(3)},\ \ldots, \bar{L}_{-1}^{(N)}\ \}$, and the $2N \times 2N$ signature matrix $S = \text{diag}\{S^{(1)}, S^{(2)}, \ldots, S^{(N)}\}$. Partition g_i into

$$g_i = \text{col}\{g_i^{(1)}, g_i^{(2)}, \ldots, g_i^{(N)}\} \qquad (M \times 1)$$

where each $g_i^{(k)}$ is $M_k \times 1$. Show, by either repeating the derivation of the fast array method of Chapter 37 or by appealing to equivalence with Kalman filtering as in Prob. IX.16, that the $\{g_i^{(k)}\}$ can be propagated iteratively as follows:

$$\left[\begin{array}{cc} \gamma^{-1/2}(i-1) & \left[\begin{array}{cccccc} u^{(1)}(i) & u_{i-1}^{(1)} & \ldots & u^{(N)}(i) & u_{i-1}^{(N)} \end{array}\right]\bar{L}_{i-1} \\ \left[\begin{array}{c} 0 \\ g_{i-1}^{(1)}\gamma^{-1/2}(i-1) \\ \vdots \\ 0 \\ g_{i-1}^{(N)}\gamma^{-1/2}(i-1) \end{array}\right] & \text{diag}\left\{\bar{L}_{i-1}^{(1)}, \bar{L}_{i-1}^{(2)}, \ldots, \bar{L}_{i-1}^{(N)}\right\} \end{array}\right]\Theta_i$$

$$= \begin{bmatrix} \gamma^{-1/2}(i) & 0_{1\times 2N} \\ \begin{bmatrix} g_i^{(1)}\gamma^{-1/2}(i) \\ 0 \\ \vdots \\ g_i^{(N)}\gamma^{-1/2}(i) \\ 0 \end{bmatrix} & \sqrt{\lambda}\,\text{diag}\left\{\bar{L}_i^{(1)}, \bar{L}_i^{(2)}, \ldots, \bar{L}_i^{(N)}\right\} \end{bmatrix}$$

where each $\bar{L}_i^{(k)}$ is $(M_k+1)\times 2N$ and Θ_i is any $(1\oplus S)$-unitary transformation that produces the zeros in the top row of the post-array.

(b) The computational cost of the array method of part (a) is $O(2NM)$ operations per iteration. It can be reduced to $O(2M)$ operations per iteration when all channels have the same number of taps, say $M_k = K$. Consider again the $N \times N$ regularization matrix Π and let $\{\Pi^{(kl)}\}$ denote its (k,l) block; each such block is $K \times K$. Rather than select Π as a block diagonal matrix, as was the case in part (a), we now choose its diagonal and off-diagonal blocks as follows:

$$\left[\Pi^{(kl)}\right]^{-1} = \eta \cdot \text{diag}\{\lambda^2, \lambda^3, \ldots, \lambda^{K+1}\} \quad (K \times K)$$

Show that the same array algorithm of part (a) will still hold with $\bar{L}_{-1}$ now defined as the $N(K+1)\times 2$ matrix

$$\bar{L}_{-1} = \begin{bmatrix} \bar{L}_{-1}^{(1)} \\ \bar{L}_{-1}^{(2)} \\ \vdots \\ \bar{L}_{-1}^{(N)} \end{bmatrix}$$

and $S = (1 \oplus -1)$. Moreover, the $0_{1\times 2N}$ block in the post-array should be replaced by $0_{1\times 2}$ and each $\bar{L}_i^{(k)}$ is now $(K+1)\times 2$.

Remark. For a discussion on multichannel least-squares problems and equivalence with Kalman filtering, see Khalaj, Sayed, and Kailath (1993).

Problem IX.18 (Adaptive Volterra filtering) A second-order Volterra filter is described by a nonlinear mapping of the form

$$d(i) = \sum_{j=0}^{M-1} w^o(j)u(i-j) \; + \; \sum_{j=0}^{M-1}\sum_{k=j}^{M-1} w^o(j,k)u(i-j)u(i-k) \; + \; v(i)$$

where $\{u(i)\}$ is the input sequence, $\{v(i)\}$ is the noise sequence, and $\{w^o(j), w^o(j,k)\}$ denote the filter coefficients. Compared with a standard FIR mapping, we find that products of the form $\{u(i-j)u(i-k)\}$ also appear. We can reformulate this problem as a multichannel filtering problem by defining $M+2$ channels with the following regression vectors:

$$u_i^{(1)} \triangleq \begin{bmatrix} u(i) & u(i-1) & \ldots & u(i-M+1) \end{bmatrix} \quad \text{(channel 1; } M \text{ taps)}$$

$$u_i^{(2)} \triangleq \begin{bmatrix} u^2(i) & u^2(i-1) & \ldots & u^2(i-M+1) \end{bmatrix} \quad \text{(channel 2; } M \text{ taps)}$$

$$u_i^{(3)} \triangleq \begin{bmatrix} u(i)u(i-1) & \ldots & u(i-M+2)u(i-M+1) \end{bmatrix} \quad \text{(channel 3; } M-1 \text{ taps)}$$

$$u_i^{(4)} \triangleq \begin{bmatrix} u(i-1)u(i-2) & \ldots & u(i-M+3)u(i-M+1) \end{bmatrix} \quad \text{(channel 4; } M-2 \text{ taps)}$$

$$\vdots$$

$$u_i^{(M+2)} \triangleq \begin{bmatrix} u(i+1)u(i-M+1) \end{bmatrix} \quad \text{(channel } M+2\text{; 1 tap)}$$

and with the following tap vectors:

$$w^{o(1)} \triangleq \left[\begin{array}{ccc} w^o(0) & \ldots & w^o(M-1) \end{array}\right] \quad \text{(channel 1; } M \text{ taps)}$$

$$w^{o(2)} \triangleq \left[\begin{array}{ccc} w^o(0,0) & \ldots & w^o(M-1,M-1) \end{array}\right] \quad \text{(channel 2; } M \text{ taps)}$$

$$w^{o(3)} \triangleq \left[\begin{array}{ccc} w^o(0,1) & \ldots & w^o(M-2,M-1) \end{array}\right] \quad \text{(channel 3; } M-1 \text{ taps)}$$

$$w^{o(4)} \triangleq \left[\begin{array}{ccc} w^o(1,2) & \ldots & w^o(M-3,M-1) \end{array}\right] \quad \text{(channel 4; } M-2 \text{ taps)}$$

$$\vdots$$

$$w^{o(M+2)} \triangleq \left[\begin{array}{c} w^o(0,M-1) \end{array}\right] \quad \text{(channel } M+2\text{; 1 tap)}$$

Again, this formulation can be recast in terms of a single-channel description by collecting all regressors and taps into a single extended regressor and a single extended tap vector:

$$u_i \triangleq \left[\begin{array}{cccc} u_i^{(1)} & u_i^{(2)} & \ldots & u_i^{(M+2)} \end{array}\right], \quad w^o \triangleq \text{col}\{w^{o(1)}, w^{o(2)}, \ldots, w^{o(M+2)}\}$$

so that $d(i) = u_i w^o + v(i)$. Therefore, given measurements $\{d(i), u_i\}$, we can estimate w^o and, subsequently, the individual taps $\{w^o(j), w^o(j,k)\}$, by applying an adaptive filter (of the LMS or RLS type) to the data $\{d(i), u_i\}$ as in Prob. IX.17. In the RLS case, show that the fast array solution of part (a) of Prob. IX.17 still applies in this application.

Remark. For more details on Volterra filters, their modeling abilities, and their use in the context of adaptive filtering, see Schetzen (1980) and Mathews (1991b). See also Khalaj, Sayed, and Kailath (1993) for additional discussion on the results of this problem.

Problem IX.19 (Laguerre adaptive filters) Consider an adaptive transversal filter structure with multiple poles $\{|a_k| < 1\}$, as shown in Fig. IX.2. Let $\theta_k = \sqrt{1 - |a_k|^2}$ and introduce the regression vector $u_i = [u(i,0)\ u(i,1)\ \ldots\ u(i,M-1)]$, where the notation $u(i,j)$ refers to the j-th entry of u_i. Show that two successive regressors satisfy the structural relation $[u(i,0)\ u_{i-1}] = [u_i\ u(i-1,M-1)]\Psi$, where Ψ is the $(M+1) \times (M+1)$ matrix, e.g., for $M = 5$,

$$\Psi = \begin{bmatrix} 1 & a_0^* & 0 & 0 & 0 & 0 \\ 0 & \theta_1\theta_0 & a_1^* & 0 & 0 & 0 \\ 0 & -a_1\theta_2\theta_0 & \theta_2\theta_1 & a_2^* & 0 & 0 \\ 0 & a_1 a_2 \theta_3 \theta_0 & -a_2\theta_3\theta_1 & \theta_3\theta_2 & a_3^* & 0 \\ 0 & -a_1 a_2 a_3 \theta_0/\theta_4 & a_2 a_3 \theta_1/\theta_4 & -a_3\theta_2/\theta_4 & \theta_3/\theta_4 & 0 \\ 0 & a_1 a_2 a_3 a_4 \theta_0/\theta_4 & -a_2 a_3 a_4 \theta_1/\theta_4 & a_3 a_4 \theta_2/\theta_4 & -a_4\theta_3/\theta_4 & 1 \end{bmatrix}$$

Remark. We thus find that the successive regressors of the filter structure of Fig. IX.2 satisfy a certain structural relation. Comparing this relation with the result (37.5) in the shift-structured case, we see that it reduces to (37.5) when the $\{a_k\}$ are all zero since then $\Psi = \mathrm{I}$. As in the case of tapped-delay-line implementations, the result of this problem can be exploited to great effect to derive fast fixed-order (and also order-recursive) least-squares algorithms for Laguerre filters as well; see Chapter 16 of Sayed (2003) and also Merched and Sayed (2000b,2001a,2001b) for details.

COMPUTER PROJECT

Project IX.1 (Stability issues in fast least-squares) The purpose of this project is to illustrate some of the stability problems that arise when dealing with fast fixed-order least-squares algorithms. We do so by considering the same adaptive channel estimation application shown in Fig. VIII.1.

(a) Load the file channel.mat. It contains the 64 samples of a randomly generated channel impulse response sequence. The norm of this impulse response has been normalized to unity.

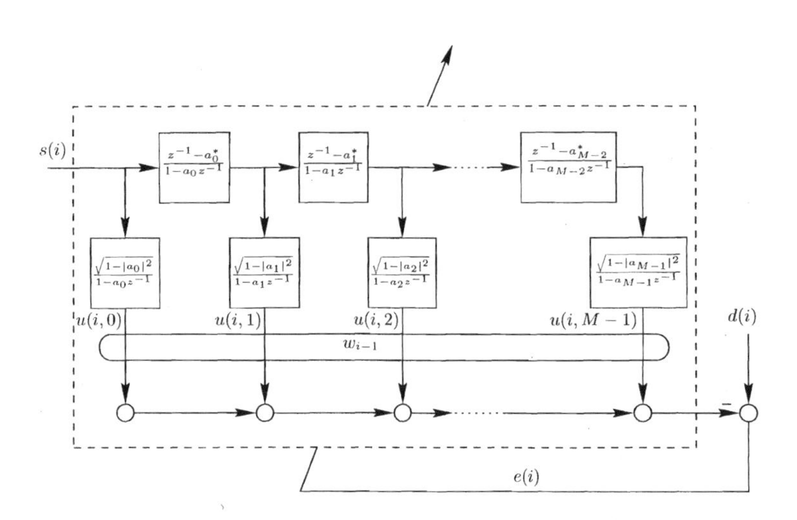

FIGURE IX.2 An adaptive transversal structure with multiple poles.

Feed unit-variance Gaussian input data through the channel and add Gaussian noise to its output. Set the noise power at 30 dB below the input signal power. Train an adaptive filter using:

1. The fast transversal filter of Alg. 39.1;
2. The same FTF algorithm but incorporating the rescue mechanism (39.10);
3. The stabilized FTF implementation of Alg. 39.4;
4. The fast array filter of Alg. 37.1.

For each algorithm, generate an ensemble-average learning curve by averaging over 100 experiments. Compare the performance of the algorithms.

(b) All computations in MATLAB are performed in double precision. In order to illustrate some of the instability problems that arise in finite precision, we repeat the simulation of part (a) in a quantized environment. To do so, you may rely on the routine quantize.m from Computer Project VIII.1. Usually, in practice, the internal variables of a filter implementation are suitably scaled in order to reduce the occurrences of overflow and underflow when operating in finite precision. In order to simplify this project, and thereby avoid the need for incorporating scaling, we shall choose the number of bits to be relatively high. Thus fix the word-length for both data and coefficients at $B = 36$ bits (including the sign bit). Although high, this number of bits will still be useful to illustrate some quantization effects.

Repeat the simulation of part (a) and generate ensemble-average learning curves for each of the algorithms. Compare their performance.

PART 

LATTICE FILTERS

CHAPTER 40

Three Basic Estimation Problems

The recursive least-squares algorithms described so far in Parts VIII (*Least-Squares Methods*) and IX (*Fast RLS Algorithms*), including array variants and fast least-squares variants, are usually qualified as *fixed-order* algorithms. The qualification "fixed-order" means that, from one iteration to another, these implementations propagate quantities that relate to estimation problems of fixed-order.

In this part, we shall study RLS algorithms that are *order-recursive* in nature, as opposed to fixed-order. They are widely known as lattice filters and have several desirable properties such as improved numerical behavior, stability, modularity, in addition to computational efficiency. In these implementations, least-squares problems of increasing orders are solved successively so that, in addition to time-updates, the lattice filters rely heavily on order-updates for various quantities.

Our treatment of least-squares lattice filters has at least three features:

1. First, all relevant order-recursive relations are derived *without* assuming any structure in the regression vectors.

2. Second, and because of the above, the derivation is able to show that it is possible to design efficient lattice filters even for cases where the regressors *do not* possess shift structure. This generalization is achieved by pinpointing the variable whose update is affected by data structure, and by showing what kind of structure enables an efficient order-recursive relation for the variable.

3. Third, all order-recursive relations are derived by solving regularized least-squares problems, as opposed to standard least-squares problems without regularization.

We start our presentation by motivating the need for lattice filters and by introducing the notation that is necessary to describe such filters. The reader will soon realize that, while the derivation of this class of filters relies on familiar concepts, it is the excessive use of subscripts and superscripts that may confuse the uninitiated reader. The unfortunate truth is that the notation is necessary, and the reader needs to become familiar with it. The good news is that the notation is suggestive and very much tied to the context. Moreover, the underlying concepts and arguments are fairly familiar by now.

Notation for Order-Recursive Problems

To study order-recursive problems, it is necessary to adjust the notation in order to be able to indicate *both* the size of a variable and the time instant at which it becomes available. For example, when referring to a weight vector w_i at time i, we shall write $w_{M,i}$ with two subscripts, M and i. The first subscript, M, is used to indicate that the weight vector is of size M or, equivalently, that it is computed as the solution to a least-squares problem of order M, as in (40.1) and (40.3) below. The second subscript, i, is used to indicate that the weight vector is dependent on data up to time i and, therefore, becomes available at time i.

In a similar vein, we shall write $H_{M,i}$ instead of H_i to refer to a data matrix with column dimension M and with data up to time i. Similarly, we shall write Π_M instead of Π to refer to an $M \times M$ regularization matrix. With this notation, we can now provide a brief review of the regularized least-squares problem.

40.1 MOTIVATION FOR LATTICE FILTERS

So consider a collection of $(i+1)$ data $\{d(j), u_{M,j}\}_{j=0}^{i}$ and introduce the observation vector y_i and the data matrix $H_{M,i}$ defined by

$$y_i = \begin{bmatrix} d(0) \\ d(1) \\ \vdots \\ d(i) \end{bmatrix}, \qquad H_{M,i} = \begin{bmatrix} u_{M,0} \\ u_{M,1} \\ \vdots \\ u_{M,i} \end{bmatrix}$$

The exponentially-weighted least-squares problem of order M seeks the $M \times 1$ column vector w that solves (cf. Sec. 30.6):

$$\min_{w_M} \left[\lambda^{i+1} w_M^* \Pi_M w_M \;+\; (y_i - H_{M,i} w_M)^* \Lambda_i (y_i - H_{M,i} w_M)\right] \tag{40.1}$$

where Π_M is an $M \times M$ positive-definite regularization matrix. In the sequel, we shall choose Π_M in a manner similar to the fast array method of Chapter 37 (cf. (37.13)), namely,

$$\boxed{\Pi_M \;=\; \eta^{-1} \text{diag}\{\lambda^{-2}, \lambda^{-3}, \ldots, \lambda^{-(M+1)}\}} \tag{40.2}$$

Moreover,

$$\Lambda_i = \text{diag}\{\lambda^i, \lambda^{i-1}, \ldots, \lambda, 1\}$$

is a diagonal weighting matrix, defined in terms of a forgetting factor λ that satisfies $0 \ll \lambda \leq 1$. It is sometimes convenient to rewrite (40.1) more explicitly in terms of the individual data $\{d(j), u_{M,j}\}$ as follows:

$$\min_{w_M} \left[\lambda^{i+1} w_M^* \Pi_M w_M \;+\; \sum_{j=0}^{i} \lambda^{i-j} |d(j) - u_{M,j} w_M|^2\right] \tag{40.3}$$

We denote the solution of (40.3) by $w_{M,i}$ and we already know that it is given by (cf. Thm. 29.5

$$\boxed{w_{M,i} \;=\; P_{M,i} H_{M,i}^* \Lambda_i y_i} \tag{40.4}$$

where

$$\boxed{P_{M,i} = (\lambda^{i+1} \Pi_M + H_{M,i}^* \Lambda_i H_{M,i})^{-1}} \tag{40.5}$$

The regularization term $\lambda^{i+1}\Pi_M$ guarantees an invertible coefficient matrix, i.e., an invertible $P_{M,i}$. In the absence of regularization (i.e., when $\Pi_M = 0$), we would need to assume that $H_{M,i}$ has full-column rank so that $H_{M,i}^* \Lambda_i H_{M,i}$ is invertible. Observe that the regularization matrix in (40.5) has the form

$$\lambda^{i+1} \Pi_M \;=\; \eta^{-1} \text{diag}\{\lambda^{i-1}, \lambda^{i-2}, \ldots, \lambda^{i-M}\}$$

We further let $\widehat{y}_{M,i}$ denote the estimate of y_i,

$$\boxed{\widehat{y}_{M,i} = H_{M,i}w_{M,i}} \tag{40.6}$$

and we refer to $\widehat{y}_{M,i}$ as the regularized projection (or simply projection) of y_i onto the range space of $H_{M,i}$, written as $\mathcal{R}(H_{M,i})$. Recall from Sec. 29.5 that, when $H_{M,i}$ has full-column rank, the projection matrix onto $\mathcal{R}(H_{M,i})$ is $\mathcal{P}_H = H_{M,i}(H^*_{M,i}H_{M,i})^{-1}H^*_{M,i}$. For the regularized problem (40.1), we have $\widehat{y}_{M,i} = H_{M,i}P_{M,i}H^*_{M,i}\Lambda_i y_i$. Although the matrix $H_{M,i}P_{M,i}H^*_{M,i}$ is not an actual projection matrix, we shall still refer to $\widehat{y}_{M,i}$ as the (regularized) projection of y_i onto $\mathcal{R}(H_{M,i})$ for ease of reference.

We also define two error vectors: the *a posteriori* and *a priori* error vectors:

$$\boxed{r_{M,i} = y_i - H_{M,i}w_{M,i}, \qquad e_{M,i} = y_i - H_{M,i}w_{M,i-1}} \tag{40.7}$$

where $w_{M,i-1}$ is the solution to a least-squares problem similar to (40.1) and (40.3) with data up to time $i-1$ and with λ^{i+1} replaced by λ^i, i.e.,

$$\min_{w_M}\left[\lambda^i w^*_M \Pi_M w_M + \sum_{j=0}^{i-1}\lambda^{i-1-j}|d(j) - u_{M,j}w_M|^2\right] \implies w_{M,i-1}$$

The last entries of the error vectors $\{r_{M,i}, e_{M,i}\}$ at time i are denoted by[20]

$$\begin{cases} r_M(i) = d(i) - u_{M,i}w_{M,i} & (\textit{a posteriori} \text{ error}) \\ e_M(i) = d(i) - u_{M,i}w_{M,i-1} & (\textit{a priori} \text{ error}) \end{cases} \tag{40.8}$$

and they are related by the conversion factor, $\gamma_M(i)$,

$$\boxed{r_M(i) = \gamma_M(i)e_M(i)} \tag{40.9}$$

which is defined by

$$\boxed{\gamma_M(i) = 1 - u_{M,i}P_{M,i}u^*_{M,i} = \frac{1}{1+\lambda^{-1}u_{M,i}P_{M,i-1}u^*_{M,i}}} \tag{40.10}$$

Moreover, the minimum cost of the least-squares problem (40.1) is given by (cf. Thm. 29.5):

$$\boxed{\xi_M(i) = y^*_i\Lambda_i r_{M,i} = y^*_i\Lambda_i[y_i - H_{M,i}w_{M,i}]} \tag{40.11}$$

We also know from Alg. 30.2 that RLS allows us to update $w_{M,i}$ and $\xi_M(i)$ recursively as follows:

$$\begin{cases} \gamma_M^{-1}(i) = 1 + \lambda^{-1}u_{M,i}P_{M,i-1}u^*_{M,i} \\ g_{M,i} = \lambda^{-1}\gamma_M(i)P_{M,i-1}u^*_{M,i} \\ e_M(i) = d(i) - u_{M,i}w_{M,i-1} \\ w_{M,i} = w_{M,i-1} + g_{M,i}e_M(i) \\ P_{M,i} = \lambda^{-1}P_{M,i-1} - g_{M,i}g^*_{M,i}/\gamma_M(i) \\ r_M(i) = d(i) - u_{M,i}w_{M,i} \\ \xi_M(i) = \lambda\xi_M(i-1) + r_M(i)e^*_M(i) \end{cases} \tag{40.12}$$

[20]Recall that we use subscripts to indicate the time index of a vector quantity and parenthesis for a scalar quantity. Thus, we write $e_{M,i}$ and $e_M(i)$. Likewise, we write $r_{M,i}$ and $r_M(i)$.

with initial conditions $w_{M,-1} = 0$, $\xi_M(-1) = 0$, and $P_{M,-1} = \Pi_M^{-1}$. It also holds that $g_{M,i} = P_{M,i}u_{M,i}^*$.

The RLS algorithm (40.12) allows us to update $w_{M,i-1}$ to $w_{M,i}$, i.e., it only performs a time-update of the weight-vector solution. Here, both $w_{M,i-1}$ and $w_{M,i}$ are $M-$dimensional vectors with the former computed from data up to time $i-1$ while the latter is computed from data up to time i.

Now, similar to (40.1), consider a least-squares problem of order $M+1$, i.e.,

$$\min_{w_{M+1}} \left[\lambda^{i+1} w_{M+1}^* \Pi_{M+1} w_{M+1} + \sum_{j=0}^{i} \lambda^{i-j} |d(j) - u_{M+1,j} w_{M+1}|^2 \right]$$

Its solution is an $(M+1)\times 1$ column vector that we denote by $w_{M+1,i}$. Although an order-update relation that takes $w_{M,i}$ to $w_{M+1,i}$ is possible (recall Lemmas 32.1 and 32.2; see also Prob. X.9), the lattice filters of this chapter are concerned with other kinds of order-update relations.

Specifically, lattice filters are not concerned with the weight vectors themselves, but rather with the corresponding projections $\{\widehat{y}_{M,i}, \widehat{y}_{M+1,i}\}$. So let $d_M(i)$ denote the estimate of $d(i)$ of order M; it is the last entry of $\widehat{y}_{M,i}$, i.e., $d_M(i) = u_{M,i}w_{M,i}$. Likewise, let $d_{M+1}(i)$ denote the estimate of $d(i)$ of order $M+1$, which is the last entry of $\widehat{y}_{M+1,i}$,

$$d_{M+1}(i) = u_{M+1,i} w_{M+1,i} \tag{40.13}$$

The corresponding *a posteriori* estimation errors are $r_M(i) = d(i) - d_M(i)$ and $r_{M+1}(i) = d(i) - d_{M+1}(i)$, respectively. It would seem that in order to update $d_M(i)$ to $d_{M+1}(i)$, we may need to order-update $w_{M,i}$ to $w_{M+1,i}$. However, this is not the case. The lattice solutions that we study in this chapter will allow us to update $r_M(i)$ to $r_{M+1}(i)$ *directly* without the need to evaluate the weight vectors $w_{M,i}$ and $w_{M+1,i}$ or even update them. In so doing, the lattice filters will end up being an efficient alternative to RLS; efficient in the sense that their computational cost will be an order of magnitude smaller than that of RLS, namely, $O(M^2)$ vs. $O(M)$ operations per iteration.

40.2 JOINT PROCESS ESTIMATION

We start our derivation of lattice filters by examining the problem of order-updating the projection vector $\widehat{y}_{M,i}$, i.e., of relating $\widehat{y}_{M+1,i}$ to $\widehat{y}_{M,i}$. This problem is known as *joint process estimation*. In order to simplify the presentation, and without loss of generality, we illustrate the arguments and constructions for the case $M=3$. Later, we show how the results extend to generic M.

Thus, assume $M=3$ and consider the data matrix

$$H_{3,i} = \begin{bmatrix} u(0,0) & u(0,1) & u(0,2) \\ u(1,0) & u(1,1) & u(1,2) \\ u(2,0) & u(2,1) & u(2,2) \\ \vdots & \vdots & \vdots \\ u(i,0) & u(i,1) & u(i,2) \end{bmatrix} = \begin{bmatrix} u_{3,0} \\ u_{3,1} \\ u_{3,2} \\ \vdots \\ u_{3,i} \end{bmatrix} \tag{40.14}$$

The subscript 3 refers to the order of the estimation problem (i.e., to the column dimension of the data matrix), and the subscript i indicates that the data matrix contains data up to time i. Observe that we are denoting the individual entries of $H_{3,i}$ and, correspondingly, of the regressors $\{u_{3,i}\}$, by $\{u(i,j)\}$ with the first index referring to time and the second index referring to the column position within the regression vector, namely,

$$u_{3,i} = \begin{bmatrix} u(i,0) & u(i,1) & u(i,2) \end{bmatrix}$$

In other words, we are not assuming shift-structure in $u_{3,i}$, i.e., the entries of $u_{3,i}$ are not assumed to be delayed versions of some input sequence. If this were the case, then $H_{3,i}$ would have been of the form

$$H_{3,i} = \begin{bmatrix} u(0) & & \\ u(1) & u(0) & \\ u(2) & u(1) & u(0) \\ \vdots & \vdots & \vdots \\ u(i) & u(i-1) & u(i-2) \end{bmatrix}$$

However, since all results in the sequel, until Sec. 41.1, will hold irrespective of any structure in $u_{3,i}$, we shall proceed with our arguments by treating the general case (40.14).

The (regularized) projection of y_i onto $\mathcal{R}(H_{3,i})$ is given by (cf. (40.6)):

$$\boxed{\widehat{y}_{3,i} = H_{3,i}P_{3,i}H_{3,i}^*\Lambda_i y_i \;=\; H_{3,i}w_{3,i}} \tag{40.15}$$

where

$$\boxed{P_{3,i} = (\lambda^{i+1}\Pi_3 + H_{3,i}^*\Lambda_i H_{3,i})^{-1}, \qquad w_{3,i} = P_{3,i}H_{3,i}^*\Lambda_i y_i} \tag{40.16}$$

and

$$\boxed{\lambda^{i+1}\Pi_3 \;=\; \eta^{-1}\text{diag}\{\lambda^{i-1}, \lambda^{i-2}, \lambda^{i-3}\}} \tag{40.17}$$

We say that $\widehat{y}_{3,i}$ is the third-order projection of y_i onto $\mathcal{R}(H_{3,i})$. Now suppose that one more column is appended to $H_{3,i}$, which then becomes

$$H_{4,i} = \begin{bmatrix} H_{3,i} & | & x_{3,i} \end{bmatrix} = \left[\begin{array}{ccc|c} u(0,0) & u(0,1) & u(0,2) & u(0,3) \\ u(1,0) & u(1,1) & u(1,2) & u(1,3) \\ u(2,0) & u(2,1) & u(2,2) & u(2,3) \\ \vdots & \vdots & \vdots & \vdots \\ u(i,0) & u(i,1) & u(i,2) & u(i,3) \end{array}\right] \tag{40.18}$$

where we are denoting the last column of $H_{4,i}$ by $x_{3,i}$. The (regularized) projection of the same vector y_i onto the extended range space $\mathcal{R}(H_{4,i})$ is now given by

$$\boxed{\widehat{y}_{4,i} = H_{4,i}P_{4,i}H_{4,i}^*\Lambda_i y_i \;=\; H_{4,i}w_{4,i}} \tag{40.19}$$

where

$$\boxed{P_{4,i} = (\lambda^{i+1}\Pi_4 + H_{4,i}^*\Lambda_i H_{4,i})^{-1}, \qquad w_{4,i} = P_{4,i}H_{4,i}^*\Lambda_i y_i} \tag{40.20}$$

and

$$\boxed{\lambda^{i+1}\Pi_4 \;=\; \eta^{-1}\text{diag}\{\lambda^{i-1}, \lambda^{i-2}, \lambda^{i-3}, \lambda^{i-4}\}} \tag{40.21}$$

Comparing expressions (40.15) and (40.19) for $\{\widehat{y}_{3,i}, \widehat{y}_{4,i}\}$ we see that they differ by virtue of the difference between the data matrices $\{H_{3,i}, H_{4,i}\}$. However, these data matrices are identical except for the last column in $H_{4,i}$. Therefore, it should be possible to relate the projections $\{\widehat{y}_{3,i}, \widehat{y}_{4,i}\}$ and to obtain an order-update relation for them. This process of order-updating the projection of the observation vector is known as *joint process estimation*.

We already studied such order-update problems in Sec. 32.1. Recall that in that section we derived, both algebraically and geometrically, the relations that exist between the (regularized) projection of an observation vector onto a data matrix H and onto its augmented version $[H \;\; h]$, for some column h. More specifically, comparing with the statement of Lemma 32.1, we can make the following identifications

$$\left\{\begin{array}{llll} H \longleftarrow H_{3,i} & h \longleftarrow x_{3,i} & P \longleftarrow P_{3,i} & P_z \longleftarrow P_{4,i} \\ \Pi \longleftarrow \lambda^{i+1}\Pi_3 & \sigma \longleftarrow \eta^{-1}\lambda^{i-4} & \gamma \longleftarrow \gamma_3(i) & \gamma_z \longleftarrow \gamma_4(i) \\ \widehat{w} \longleftarrow w_{3,i} & \widehat{w}_z \longleftarrow w_{4,i} & \widetilde{y} \longleftarrow r_{3,i} & \widetilde{y}_z \longleftarrow r_{4,i} \end{array}\right.$$

Therefore, using the result of Lemma 32.1, we can relate the variables of the projection problems that result in $\{\widehat{y}_{3,i}, \widehat{y}_{4,i}\}$ as follows. Let $w^b_{3,i}$ denote the solution of the least-squares problem:

$$\min_{w^b_3} \left[\, \lambda^{i+1} w^{b*}_3 \Pi_3 w^b_3 \;+\; (x_{3,i} - H_{3,i}w^b_3)^* \Lambda_i (x_{3,i} - H_{3,i}w^b_3) \,\right] \tag{40.22}$$

That is, $w^b_{3,i}$ is the vector that projects $x_{3,i}$ onto $\mathcal{R}(H_{3,i})$,

$$\boxed{w^b_{3,i} = P_{3,i}H^*_{3,i}\Lambda_i x_{3,i}}$$

The subscript 3 refers to an estimation problem of order 3, while the subscript i denotes the use of data up to time i. The superscript b refers to backward projection. The reason for this terminology is that problem (40.22) amounts to estimating the last column of $H_{4,i}$ from its leading columns, $H_{3,i}$. Let $\xi^b_3(i)$ denote the minimum cost of (40.22), i.e.,

$$\boxed{\xi^b_3(i) = x^*_{3,i}\Lambda_i b_{3,i}}$$

where $b_{3,i}$ is the (backward) *a posteriori* error vector that results from projecting $x_{3,i}$ onto $\mathcal{R}(H_{3,i})$,

$$\boxed{b_{3,i} = x_{3,i} - H_{3,i}w^b_{3,i}}$$

We denote the last entry of $b_{3,i}$ by $b_3(i)$ and it refers to the estimation error in estimating the last entry of $x_{3,i}$ from the last row of $H_{3,i}$ (namely, $u_{3,i}$).

Define further the scalar coefficient

$$\boxed{\kappa_3(i) \;\stackrel{\Delta}{=}\; \frac{b^*_{3,i}\Lambda_i y_i}{\eta^{-1}\lambda^{i-4} + \xi^b_3(i)} \;=\; \frac{\rho^*_3(i)}{\eta^{-1}\lambda^{i-4} + \xi^b_3(i)}} \tag{40.23}$$

where

$$\boxed{\rho_3(i) \;\stackrel{\Delta}{=}\; y^*_i\Lambda_i b_{3,i}} \tag{40.24}$$

Then from Lemma 32.1 we conclude that the following order-update relations hold:

$$P_{4,i} = \begin{bmatrix} P_{3,i} & 0 \\ 0 & 0 \end{bmatrix} + \frac{1}{\eta^{-1}\lambda^{i-4} + \xi_3^b(i)} \begin{bmatrix} -w_{3,i}^b \\ 1 \end{bmatrix} \begin{bmatrix} -w_{3,i}^{b*} & 1 \end{bmatrix} \quad (40.25)$$

$$\widehat{y}_{4,i} = \widehat{y}_{3,i} + \kappa_3(i)\, b_{3,i} \quad (40.26)$$

$$r_{4,i} = r_{3,i} - \kappa_3(i) b_{3,i} \quad (40.27)$$

$$\xi_4(i) = \xi_3(i) - \frac{|\rho_3(i)|^2}{\eta^{-1}\lambda^{i-4} + \xi_3^b(i)} \quad (40.28)$$

$$\gamma_4(i) = \gamma_3(i) - \frac{|b_3(i)|^2}{\eta^{-1}\lambda^{i-4} + \xi_3^b(i)} \quad (40.29)$$

$$w_{4,i} = \begin{bmatrix} w_{3,i} \\ 0 \end{bmatrix} + \kappa_3(i) \begin{bmatrix} -w_{3,i}^b \\ 1 \end{bmatrix} \quad (40.30)$$

We therefore arrived at an order-update relation (40.27) for the *a posteriori* error vectors $\{r_{3,i}, r_{4,i}\}$. It tells us that in order to update $r_{3,i}$ to $r_{4,i}$ we need to know $b_{3,i}$. In the same vein, in order to move forward and update $r_{4,i}$ to $r_{5,i}$ we need $b_{4,i}$ and so on. This means that it is necessary to know how to order-update the backward error vectors as well, which motivates us to examine more closely the backward estimation problem.

40.3 BACKWARD ESTIMATION PROBLEM

For this purpose, we return to the data matrix $H_{3,i}$ in (40.18) and partition it as

$$H_{3,i} = \begin{bmatrix} x_{0,i} & | & \bar{H}_{2,i} \end{bmatrix}$$

with $x_{0,i}$ denoting its leading column and $\bar{H}_{2,i}$ denoting the remaining columns. In this way, the extended data matrix $H_{4,i}$ of (40.18) can be partitioned as

$$H_{4,i} = \begin{bmatrix} H_{3,i} & | & x_{3,i} \end{bmatrix} = \begin{bmatrix} x_{0,i} & \bar{H}_{2,i} & x_{3,i} \end{bmatrix} \quad (40.31)$$

with $\{x_{0,i}, x_{3,i}\}$ denoting its leading and trailing columns, and $\bar{H}_{2,i}$ denoting the center columns.

We can then consider two backward estimation problems: one has order 3 and estimates $x_{3,i}$ from $H_{3,i}$, and the other has order 2 and estimates $x_{3,i}$ from $\bar{H}_{2,i}$. The first problem is the one we considered above in (40.22) with regularization matrix $\lambda^{i+1}\Pi_3$ and it leads to the backward residual vector $b_{3,i}$,

$$\boxed{b_{3,i} = x_{3,i} - H_{3,i} w_{3,i}^b} \quad (40.32)$$

with the corresponding coefficient matrix

$$\boxed{P_{3,i} = (\lambda^{i+1}\Pi_3 + H_{3,i}^* \Lambda_i H_{3,i})^{-1}} \quad (40.33)$$

The second problem corresponds to solving the following least-squares problem:

$$\min_{w_2^{\bar{b}}} \left[\lambda^i w_2^{\bar{b}*} \Pi_2 w_2^{\bar{b}} + (x_{3,i} - \bar{H}_{2,i} w_2^{\bar{b}})^* \Lambda_i (x_{3,i} - \bar{H}_{2,i} w_2^{\bar{b}}) \right] \quad (40.34)$$

with regularization matrix chosen as

$$\lambda^i \Pi_2 = \eta^{-1} \text{diag}\{\lambda^{i-2}, \lambda^{i-3}\} \quad (40.35)$$

The optimal solution of (40.34) is denoted by $w_{2,i}^{\bar{b}}$ and is given by

$$\boxed{w_{2,i}^{\bar{b}} = \bar{P}_{2,i}\bar{H}_{2,i}^{*}\Lambda_i x_{3,i}} \tag{40.36}$$

with

$$\boxed{\bar{P}_{2,i} = (\lambda^i\Pi_2 + \bar{H}_{2,i}^{*}\Lambda_i\bar{H}_{2,i})^{-1}} \tag{40.37}$$

and whose residual vector we denote by

$$\boxed{\bar{b}_{2,i} = x_{3,i} - \bar{H}_{2,i}w_{2,i}^{\bar{b}}}$$

The resulting minimum cost of (40.34) is denoted by $\xi_2^{\bar{b}}(i)$. The reason for the notation $\bar{b}_{2,i}$ (with an overbar) as opposed to $b_{2,i}$ is that in our development, $b_{2,i}$ would correspond to the residual vector that results from projecting the third column of $H_{3,i}$ onto the range space of its leading two columns. More specifically, denote the columns of $H_{4,i}$ generically by $H_{4,i} = \begin{bmatrix} m & n & o & p \end{bmatrix}$. Then projecting o onto $[m \;\; n]$ results in the residual vector $b_{2,i}$, while projecting p onto $[m \;\; n \;\; o]$ results in the residual vector $b_{3,i}$. Observe that in both cases we start from the initial column m. In contrast, projecting p onto $[n \;\; o]$ results in the residual vector $\bar{b}_{2,i}$. The initial column now is n and, hence, the use of the bar notation to distinguish between both second-order projections: o onto $[m \;\; n]$ and p onto $[n \;\; o]$ — see Fig. 40.1. We shall study more closely later the relation between $\{\bar{b}_{2,i}, b_{2,i}\}$, e.g., in Sec. 41.1 where we show that, when the regressors have shift structure, it will hold that $\bar{b}_{2,i}$ is related to $b_{2,i-1}$. For now, it suffices to proceed with $\bar{b}_{2,i}$.

The argument that follows for relating $\bar{b}_{2,i}$ and $b_{3,i}$ is similar to the argument we employed in the previous section for relating $r_{3,i}$ and $r_{4,i}$. Thus, note that we are faced with the problem of projecting the same vector $x_{3,i}$ onto the range spaces of two data matrices: one is $\bar{H}_{2,i}$ and the other is $H_{3,i}$, which is obtained from $\bar{H}_{2,i}$ by augmenting it by a column to the left.

We already studied such order-update problems in Sec. 32.2 in some detail. Recall that in that section we derived, both algebraically and geometrically, the relations that exist between the (regularized) projection of an observation vector onto a data matrix H and its augmented version $[h \;\; H]$, for some column h. More specifically, comparing with the

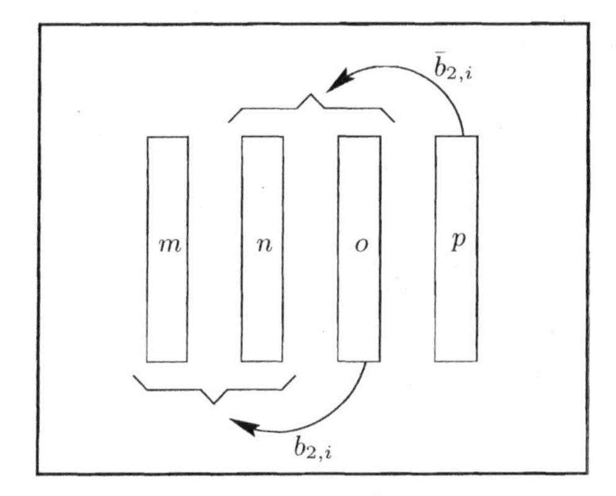

FIGURE 40.1 Two second-order backward projection problems with the corresponding residual vectors.

statement of Lemma 32.2, we can make the following identifications:

$$\left\{\begin{array}{llll} H \longleftarrow \bar{H}_{2,i} & h \longleftarrow x_{0,i} & P \longleftarrow \bar{P}_{2,i} & P_z \longleftarrow P_{3,i} \\ \Pi \longleftarrow \lambda^i \Pi_2 & \sigma \leftarrow \eta^{-1}\lambda^{i-1} & \gamma \longleftarrow \bar{\gamma}_2(i) & \gamma_z \longleftarrow \gamma_3(i) \\ \widehat{w} \longleftarrow w^{\bar{b}}_{2,i} & \widehat{w}_z \longleftarrow w^b_{3,i} & \widetilde{y} \longleftarrow \bar{b}_{2,i} & \widetilde{y}_z \longleftarrow b_{3,i} \end{array}\right.$$

Therefore, using the result of Lemma 32.2, we can relate the variables of the projection problems that result in $\{\bar{b}_{2,i}, b_{3,i}\}$ as follows.

Let $w^f_{2,i}$ denote the solution to the least-squares problem:

$$\min_{w^f_2} \left[\lambda^i w^{f*}_2 \Pi_2 w^f_2 \; + \; (x_{0,i} - \bar{H}_{2,i} w^f_2)^* \Lambda_i (x_{0,i} - \bar{H}_{2,i} w^f_2) \right] \tag{40.38}$$

which projects the leading column $x_{0,i}$ onto $\mathcal{R}(\bar{H}_{2,i})$, namely,

$$\boxed{w^f_{2,i} = \bar{P}_{2,i} \bar{H}^*_{2,i} \Lambda_i x_{0,i}}$$

The subscript 2 in $w^f_{2,i}$ refers to an estimation problem of order 2, while the subscript i denotes the use of data up to time i. The superscript f refers to forward projection. The reason for this terminology is that the above problem amounts to estimating the leading column of $H_{3,i}$ from its trailing columns, $\bar{H}_{2,i}$.

Let $\xi^f_2(i)$ denote the minimum cost of (40.38), i.e.,

$$\boxed{\xi^f_2(i) = x^*_{0,i} \Lambda_i f_{2,i}}$$

where $f_{2,i}$ is the (forward) *a posteriori* error vector that results from projecting $x_{0,i}$ onto $\mathcal{R}(\bar{H}_{2,i})$,

$$\boxed{f_{2,i} = x_{0,i} - \bar{H}_{2,i} w^f_{2,i}}$$

We denote the last entry of $f_{2,i}$ by $f_2(i)$ and it refers to the estimation error in estimating the last entry of $x_{0,i}$ from the last row of $\bar{H}_{2,i}$.

Define further the scalar coefficient

$$\boxed{\kappa^b_2(i) \;\stackrel{\Delta}{=}\; \frac{f^*_{2,i} \Lambda_i x_{3,i}}{\eta^{-1}\lambda^{i-1} + \xi^f_2(i)} = \frac{\delta_2(i)}{\eta^{-1}\lambda^{i-1} + \xi^f_2(i)}} \tag{40.39}$$

where

$$\boxed{\delta_2(i) \;\stackrel{\Delta}{=}\; f^*_{2,i} \Lambda_i x_{3,i}} \tag{40.40}$$

Then from Lemma 32.2 we conclude that the following relations hold:

$$P_{3,i} \;=\; \begin{bmatrix} 0 & 0 \\ 0 & \bar{P}_{2,i} \end{bmatrix} + \frac{1}{\eta^{-1}\lambda^{i-1} + \xi^f_2(i)} \begin{bmatrix} 1 \\ -w^f_{2,i} \end{bmatrix} \begin{bmatrix} 1 & -w^{f*}_{2,i} \end{bmatrix} \tag{40.41}$$

$$b_{3,i} \;=\; \bar{b}_{2,i} - \kappa^b_2(i) f_{2,i} \tag{40.42}$$

$$\xi^b_3(i) \;=\; \xi^{\bar{b}}_2(i) - \frac{|\delta_2(i)|^2}{\eta^{-1}\lambda^{i-1} + \xi^f_2(i)} \tag{40.43}$$

$$\gamma_3(i) \;=\; \bar{\gamma}_2(i) - \frac{|f_2(i)|^2}{\eta^{-1}\lambda^{i-1} + \xi^f_2(i)} \tag{40.44}$$

We therefore arrived at an order-update relation (40.42) for the *a posteriori* backward residual vectors $\{b_{3,i}, \bar{b}_{2,i}\}$. It tells us that in order to update $\bar{b}_{2,i}$ to $b_{3,i}$ we need to know $f_{2,i}$. In the same vein, in order to move forward and update $\bar{b}_{3,i}$ to $b_{4,i}$ we need $f_{3,i}$ and so on. This means that it is necessary to know how to order update the forward error vectors as well, which motivates us to examine more closely the forward estimation problem.

40.4 FORWARD ESTIMATION PROBLEM

To do so, we reconsider the data matrix $H_{4,i}$ in (40.18) and now partition it as

$$H_{4,i} = \left[\begin{array}{c|c} x_{0,i} & \bar{H}_{3,i} \end{array}\right] = \left[\begin{array}{c|cc} x_{0,i} & \bar{H}_{2,i} & x_{3,i} \end{array}\right] \tag{40.45}$$

where $\bar{H}_{3,i}$ denotes its trailing columns. We then consider two forward estimation problems: one has order 2 and estimates $x_{0,i}$ from $\bar{H}_{2,i}$, and the other has order 3 and estimates $x_{0,i}$ from $\bar{H}_{3,i}$. The first problem is the one we considered above in (40.38) with regularization matrix $\lambda^i\Pi_2$ and leads to the forward residual vector $f_{2,i}$,

$$\boxed{f_{2,i} = x_{0,i} - \bar{H}_{2,i}w_{2,i}^f} \tag{40.46}$$

with

$$\boxed{\bar{P}_{2,i} = (\lambda^i\Pi_2 + \bar{H}_{2,i}^*\Lambda_i\bar{H}_{2,i})^{-1}} \tag{40.47}$$

The second problem corresponds to solving the following least-squares problem:

$$\min_{w_3^f} \left[\lambda^i w_3^{f*}\Pi_3 w_3^f \ + \ (x_{0,i} - \bar{H}_{3,i}w_3^f)^*\Lambda_i(x_{0,i} - \bar{H}_{3,i}w_3^f)\right] \tag{40.48}$$

with regularization matrix

$$\lambda^i\Pi_3 \ = \ \eta^{-1}\text{diag}\{\lambda^{i-2}, \lambda^{i-3}, \lambda^{i-4}\} \tag{40.49}$$

The optimal solution of (40.48) is denoted by $w_{3,i}^f$,

$$\boxed{w_{3,i}^f = \bar{P}_{3,i}\bar{H}_{3,i}^*\Lambda_i x_{0,i}} \tag{40.50}$$

with coefficient matrix

$$\boxed{\bar{P}_{3,i} = (\lambda^i\Pi_3 + \bar{H}_{3,i}^*\Lambda_i\bar{H}_{3,i})^{-1}} \tag{40.51}$$

The corresponding residual vector is

$$\boxed{f_{3,i} = x_{0,i} - \bar{H}_{3,i}w_{3,i}^f}$$

and the resulting minimum cost is denoted by $\xi_3^f(i)$. Observe that we are now denoting the residual vectors of problems (40.38) and (40.48) by $f_{2,i}$ and $f_{3,i}$, respectively, without the need for the bar notation. Thus, note that if we again denote the columns of $H_{4,i}$ generically by $H_{4,i} = \left[\begin{array}{cccc} m & n & o & p \end{array}\right]$, then projecting m onto $[n \ \ o]$ results in the residual vector $f_{2,i}$, while projecting m onto $[n \ \ o \ \ p]$ results in the residual vector $f_{3,i}$. In both cases, we start from the same initial column n — see Fig. 40.2.

Again, the argument that follows for relating $f_{2,i}$ and $f_{3,i}$ is similar to the arguments we employed in Secs. 40.2 and 40.3 for relating $r_{3,i}$ and $r_{4,i}$, as well as $\bar{b}_{2,i}$ and $b_{3,i}$. Thus,

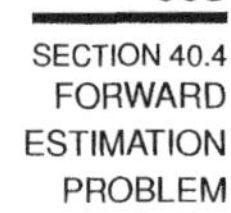

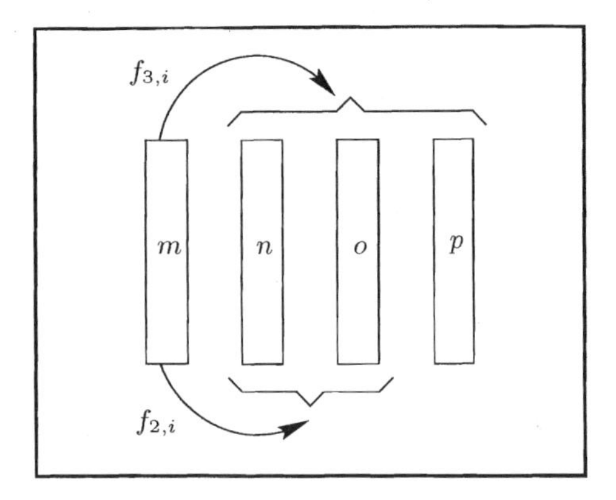

FIGURE 40.2 Two forward projection problems with the corresponding residual vectors.

note that we are faced with the problem of projecting the same column vector, $x_{0,i}$, onto the range spaces of two data matrices: one is $\bar{H}_{2,i}$ and the other is $\bar{H}_{3,i}$, which is obtained from $\bar{H}_{2,i}$ by augmenting it by a column to the right.

We studied such order-update problems in Sec. 32.1. Recall that in that section we derived, both algebraically and geometrically, the relations that exist between the (regularized) projection of an observation vector onto a data matrix H and onto its augmented version $[H \quad h]$, for some column h. More specifically, comparing with the statement of Lemma 32.1, we can make the following identifications:

$$\left\{\begin{array}{llll} H \longleftarrow \bar{H}_{2,i} & h \longleftarrow x_{3,i} & P \longleftarrow \bar{P}_{2,i} & P_z \longleftarrow \bar{P}_{3,i} \\ \Pi \longleftarrow \lambda^i \Pi_2 & \sigma \leftarrow \eta^{-1}\lambda^{i-4} & \gamma \longleftarrow \bar{\gamma}_2(i) & \gamma_z \longleftarrow \bar{\gamma}_3(i) \\ \widehat{w} \longleftarrow w_{2,i}^f & \widehat{w}_z \longleftarrow w_{3,i}^f & \widetilde{y} \longleftarrow f_{2,i} & \widetilde{y}_z \longleftarrow f_{3,i} \end{array}\right.$$

Define further the scalar coefficient

$$\boxed{\kappa_2^f(i) \stackrel{\Delta}{=} \frac{\bar{b}_{2,i}^* \Lambda_i x_{0,i}}{\eta^{-1}\lambda^{i-4} + \xi_2^{\bar{b}}(i)} = \frac{\delta_2^*(i)}{\eta^{-1}\lambda^{i-4} + \xi_2^{\bar{b}}(i)}} \tag{40.52}$$

Note that we are using $\delta_2^*(i)$ in the numerator of $\kappa_2^f(i)$, with $\delta_2(i)$ being the coefficient we used in the numerator of $\kappa_2^b(i)$ in (40.39). This is because

$$\begin{aligned} \bar{b}_{2,i}^* \Lambda_i x_{0,i} = [x_{3,i} - \bar{H}_{2,i} w_{2,i}^{\bar{b}}]^* \Lambda_i x_{0,i} &= [x_{3,i} - \bar{H}_{2,i}\bar{P}_{2,i}\bar{H}_{2,i}^*\Lambda_i x_{3,i}]^* \Lambda_i x_{0,i} \\ &= x_{3,i}^* \Lambda_i [I - \bar{H}_{2,i}\bar{P}_{2,i}\bar{H}_{2,i}^*\Lambda_i] x_{0,i} \\ &= x_{3,i}^* \Lambda_i [x_{0,i} - \bar{H}_{2,i} w_{2,i}^f] \\ &= x_{3,i}^* \Lambda_i f_{2,i} \end{aligned} \tag{40.53}$$

That is, $\delta_2(i)$ is also given by

$$\boxed{\delta_2(i) = x_{0,i}^* \Lambda_i \bar{b}_{2,i}} \tag{40.54}$$

Therefore, using the result of Lemma 32.1, we can relate the variables of the projection problems that result in $\{f_{2,i}, f_{3,i}\}$ as follows:

$$\bar{P}_{3,i} = \begin{bmatrix} \bar{P}_{2,i} & 0 \\ 0 & 0 \end{bmatrix} + \frac{1}{\eta^{-1}\lambda^{i-4} + \xi_2^{\bar{b}}(i)} \begin{bmatrix} -w_{2,i}^{\bar{b}} \\ 1 \end{bmatrix} \begin{bmatrix} -w_{2,i}^{\bar{b}*} & 1 \end{bmatrix} \quad (40.55)$$

$$f_{3,i} = f_{2,i} - \kappa_2^f(i)\bar{b}_{2,i} \quad (40.56)$$

$$\xi_3^f(i) = \xi_2^f(i) - \frac{|\delta_2(i)|^2}{\eta^{-1}\lambda^{i-4} + \xi_2^{\bar{b}}(i)} \quad (40.57)$$

$$\bar{\gamma}_3(i) = \bar{\gamma}_2(i) - \frac{|\bar{b}_2(i)|^2}{\eta^{-1}\lambda^{i-4} + \xi_2^{\bar{b}}(i)} \quad (40.58)$$

40.5 TIME AND ORDER-UPDATE RELATIONS

We summarize the order-update relations derived so far for the case of a generic order M.

Order-Update of Estimation Errors

Consider several equivalent partitionings of the data matrix $H_{M+1,i}$:

$$\begin{aligned} H_{M+1,i} &= \begin{bmatrix} x_{0,i} & x_{1,i} & \dots & x_{M,i} \end{bmatrix} \\ &= \begin{bmatrix} H_{M,i} & x_{M,i} \end{bmatrix} = \begin{bmatrix} x_{0,i} & \bar{H}_{M,i} \end{bmatrix} = \begin{bmatrix} x_{0,i} & \bar{H}_{M-1,i} & x_{M,i} \end{bmatrix} \end{aligned}$$

where $\{x_{j,i}\}$ denote the individual columns of $H_{M+1,i}$. Let also $\{u_{M,i}, \bar{u}_{M,i}\}$ denote the last rows of $H_{M,i}$ and $\bar{H}_{M,i}$.

Let further

$$\begin{cases} r_{M,i} = \textit{a posteriori} \text{ residual from projecting } y_i \text{ onto } H_{M,i} \\ b_{M,i} = \textit{a posteriori} \text{ residual from projecting } x_{M,i} \text{ onto } H_{M,i} \\ f_{M,i} = \textit{a posteriori} \text{ residual from projecting } x_{0,i} \text{ onto } \bar{H}_{M,i} \\ \bar{b}_{M,i} = \textit{a posteriori} \text{ residual from projecting } x_{M+1,i} \text{ onto } \bar{H}_{M,i} \end{cases}$$

where the projection problems for $\{r_{M,i}, b_{M,i}\}$ employ the regularization matrix $\lambda^{i+1}\Pi_M$, while the projection problems for $\{f_{M,i}, \bar{b}_{M,i}\}$ employ the regularization matrix $\lambda^i\Pi_M$. Hence,

$$\begin{cases} r_{M,i} = y_i - H_{M,i}w_{M,i} & b_{M,i} = x_{M,i} - H_{M,i}w_{M,i}^b \\ f_{M,i} = x_{0,i} - \bar{H}_{M,i}w_{M,i}^f & \bar{b}_{M,i} = x_{M+1,i} - \bar{H}_{M,i}w_{M,i}^{\bar{b}} \end{cases} \quad (40.59)$$

where, for example, $w_{M,i}^f$ is the solution to the regularized least-squares problem:

$$\min_{w_M^f} \left[\lambda^i w_M^{f*}\Pi_M w_M^f + (x_{0,i} - \bar{H}_{M,i}w_M^f)^*\Lambda_i(x_{0,i} - \bar{H}_{M,i}w_M^f) \right] \quad (40.60)$$

and similarly for $\{w_{M,i}, w_{M,i}^b, w_{M,i}^{\bar{b}}\}$. These constructions are depicted in Fig. 40.3. The last entries of $\{r_{M,i}, b_{M,i}, f_{M,i}, \bar{b}_{M,i}\}$ are denoted by

$$\{r_M(i), b_M(i), f_M(i), \bar{b}_M(i)\} \quad (40.61)$$

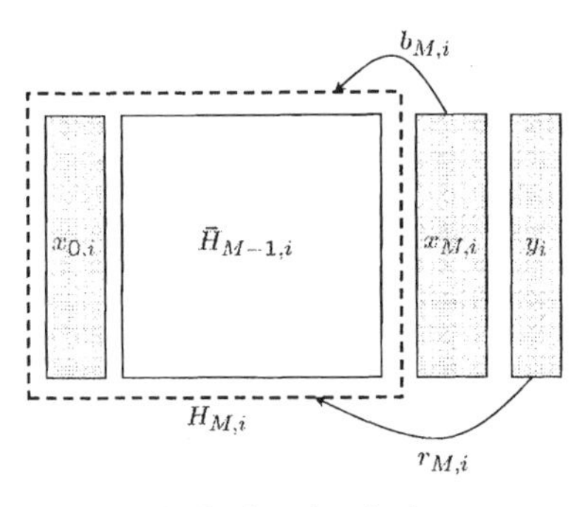

(a) Backward projection

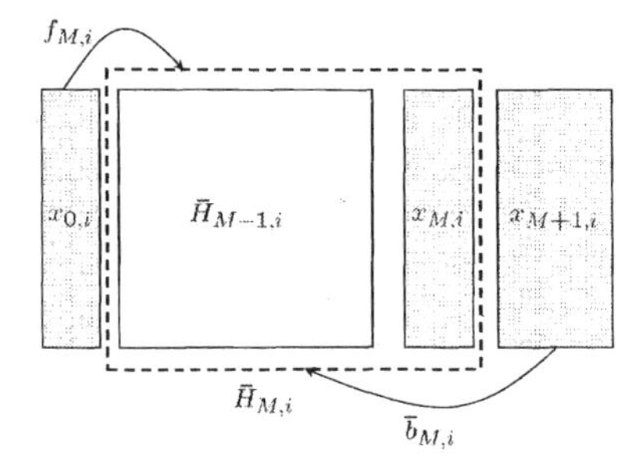

(b) Forward projection

FIGURE 40.3 Projections of $\{x_{0,i}, x_{M,i}, x_{M+1,i}, y_i\}$ onto the relevant data matrices with the resulting residual vectors $\{f_{M,i}, b_{M,i}, \bar{b}_{M,i}, r_{M,i}\}$.

The derivations in the earlier sections show that these residual vectors satisfy updates of the form:

$$r_{M+1,i} = r_{M,i} - \kappa_M(i)b_{M,i}, \quad b_{M+1,i} = \bar{b}_{M,i} - \kappa_M^b(i)f_{M,i}, \quad f_{M+1,i} = f_{M,i} - \kappa_M^f(i)\bar{b}_{M,i} \tag{40.62}$$

where we still need to derive an update for $\bar{b}_{M,i}$. We postpone this discussion to Sec. 41.1 due to its dependence on data structure. From (40.62) we obtain the following relations for the *a posteriori* estimation errors at time i:

$$\left\{\begin{array}{lcl} r_{M+1}(i) & = & r_M(i) - \kappa_M(i)b_M(i) \\ b_{M+1}(i) & = & \bar{b}_M(i) - \kappa_M^b(i)f_M(i) \\ f_{M+1}(i) & = & f_M(i) - \kappa_M^f(i)\bar{b}_M(i) \end{array}\right. \tag{40.63}$$

where the scaling coefficients $\{\kappa_M(i), \kappa_M^b(i), \kappa_M^f(i)\}$, also called reflection coefficients, are defined as the ratios

$$\left\{\begin{array}{lcl} \kappa_M(i) & = & \rho_M^*(i)/(\eta^{-1}\lambda^{i-M-1} + \xi_M^b(i)) \\ \kappa_M^b(i) & = & \delta_M(i)/(\eta^{-1}\lambda^{i-1} + \xi_M^f(i)) \\ \kappa_M^f(i) & = & \delta_M^*(i)/(\eta^{-1}\lambda^{i-M-2} + \xi_M^{\bar{b}}(i)) \end{array}\right. \tag{40.64}$$

and the quantities $\{\delta_M(i), \rho_M(i), \xi_M^b(i), \xi_M^{\bar{b}}(i), \xi_M^f(i)\}$ are defined in terms of the inner products

$$\left\{\begin{array}{lcl} \rho_M(i) & = & y_i^*\Lambda_i b_{M,i} \\ \delta_M(i) & = & x_{0,i}^*\Lambda_i\bar{b}_{M,i}^* \;=\; f_{M,i}^*\Lambda_i x_{M+1,i} \\ \xi_M^f(i) & = & x_{0,i}^*\Lambda_i f_{M,i} \end{array}\right. \qquad \begin{array}{lcl} \xi_M^b(i) & = & x_{M,i}^*\Lambda_i b_{M,i} \\ \xi_M^{\bar{b}}(i) & = & x_{M+1,i}^*\Lambda_i\bar{b}_{M,i} \\ \xi_M(i) & = & y_i^*\Lambda_i r_{M,i} \end{array} \tag{40.65}$$

The quantities $\{\xi_M(i), \xi_M^b(i), \xi_M^f(i), \xi_M^{\bar{b}}(i)\}$ denote the minimum costs of the projection problems that result in $\{r_{M,i}, b_{M,i}, f_{M,i}, \bar{b}_{M,i}\}$. Let further $\{\gamma_M(i), \bar{\gamma}_M(i)\}$ denote the conversion factors associated with the projection problems $\{r_{M,i}, b_{M,i}\}$ and $\{f_{M,i}, \bar{b}_{M,i}\}$ (the first two have the same conversion factor $\gamma_M(i)$, while the last two have the same conversion factor $\bar{\gamma}_M(i)$). That is,

$$\left\{\begin{array}{lcl} \gamma_M(i) & = & 1 - u_{M,i}P_{M,i}u_{M,i}^* \\ \bar{\gamma}_M(i) & = & 1 - \bar{u}_{M,i}\bar{P}_{M,i}\bar{u}_{M,i}^* \end{array}\right. \qquad \begin{array}{lcl} P_{M,i} & = & \left[\lambda^{i+1}\Pi_M + H_{M,i}\Lambda_i H_{M,i}^*\right]^{-1} \\ \bar{P}_{M,i} & = & \left[\lambda^{i}\Pi_M + \bar{H}_{M,i}\Lambda_i \bar{H}_{M,i}^*\right]^{-1} \end{array}$$

Then the earlier discussions also established the following update relations:

$$\left\{\begin{array}{lcl} \xi_{M+1}(i) & = & \xi_M(i) - |\rho_M(i)|^2/(\eta^{-1}\lambda^{i-M-1} + \xi^b_M(i)) \\ \xi^b_{M+1}(i) & = & \xi^b_M(i) - |\delta_M(i)|^2/(\eta^{-1}\lambda^{i-1} + \xi^f_M(i)) \\ \xi^f_{M+1}(i) & = & \xi^f_M(i) - |\delta_M(i)|^2/(\eta^{-1}\lambda^{i-M-2} + \xi^b_M(i)) \\ \\ \gamma_{M+1}(i) & = & \gamma_M(i) - |b_M(i)|^2/(\eta^{-1}\lambda^{i-M-1} + \xi^b_M(i)) \\ \gamma_{M+1}(i) & = & \bar{\gamma}_M(i) - |f_M(i)|^2/(\eta^{-1}\lambda^{i-1} + \xi^f_M(i)) \\ \bar{\gamma}_{M+1}(i) & = & \bar{\gamma}_M(i) - |\bar{b}_M(i)|^2/(\eta^{-1}\lambda^{i-M-2} + \xi^b_M(i)) \end{array}\right. \tag{40.66}$$

as well as (cf. (40.30)):[21]

$$w_{M+1,i} = \left[\begin{array}{c} w_{M,i} \\ 0 \end{array}\right] + \kappa_M(i)\left[\begin{array}{c} -w^b_{M,i} \\ 1 \end{array}\right] \tag{40.67}$$

We still need to show how to update the factors $\{\rho_M(i), \delta_M(i)\}$ in (40.65) in order to arrive at an efficient recursive scheme. The derivation in the next section shows that time-updates for $\{\rho_M(i), \delta_M(i)\}$ are possible regardless of data structure. This is a useful observation; for example, it allows one to extend least-squares lattice algorithms to more general filter structures (other than tapped-delay-line structures) — see, e.g., Chapter 16 of Sayed (2003) on Laguerre lattice filters.

Time-Update Relations

Consider first the quantity $\delta_M(i) = x^*_{0,i}\Lambda_i\bar{b}_{M,i}$, which appears in the numerator of $\kappa^f_M(i)$ in (40.64), and introduce the data matrix

$$H_{M+2,i} = \left[\begin{array}{ccc} x_{0,i} & \bar{H}_{M,i} & x_{M+1,i} \end{array}\right]$$

We partition it as

$$H_{M+2,i} = \left[\begin{array}{c|c|c} x_{0,i-1} & \bar{H}_{M,i-1} & x_{M+1,i-1} \\ \hline u(i,0) & \bar{u}_{M,i} & u(i,M+1) \end{array}\right]$$

where we are denoting the last entries of $\{x_{0,i}, x_{M+1,i}\}$ by $\{u(i,0), u(i,M+1)\}$, and the last row of $\bar{H}_{M,i}$ by $\bar{u}_{M,i}$. Consider further

$$\delta_M(i-1) = x^*_{0,i-1}\Lambda_{i-1}\bar{b}_{M,i-1}$$

Now recall that $\bar{b}_{M,i}$ is the residual vector that results from projecting $x_{M+1,i}$ onto $\mathcal{R}(\bar{H}_{M,i})$ with regularization matrix $\lambda^i\Pi_M$. Likewise, $\bar{b}_{M,i-1}$ is the residual vector that results from projecting $x_{M+1,i-1}$ onto $\mathcal{R}(\bar{H}_{M,i-1})$ with regularization matrix $\lambda^{i-1}\Pi_M$. We are therefore faced with the problem of time-updating the inner product $\delta_M(i)$, which is of the same form as the problem studied in Sec. 32.3. More specifically, comparing with the statement of Lemma 32.3 (or with the data matrix (32.48) and its time-updated version (32.50)), we see that we can make the identifications:

$$\left\{\begin{array}{lll} \bar{H}_{i-1} \longleftarrow \bar{H}_{M,i-1} & \bar{\gamma}(i) \longleftarrow \bar{\gamma}_M(i) & \beta(i) \longleftarrow u(i,M+1) \\ x_{i-1} \longleftarrow x_{0,i-1} & h_i \longleftarrow \bar{u}_{M,i} & \bar{\alpha}(i) \longleftarrow f_M(i) \\ \alpha(i) \longleftarrow u(i,0) & z_{i-1} \longleftarrow x_{M+1,i-1} & \bar{\beta}(i) \longleftarrow b_M(i) \end{array}\right.$$

[21]This last recursion is used in Prob. X.9 to establish a relation between the standard RLS solution (40.12) and lattice filters.

and arrive at the time-update relation (cf. (32.56)):

$$\boxed{\delta_M(i) = \lambda\delta_M(i-1) + \frac{f_M^*(i)\bar{b}_M(i)}{\bar{\gamma}_M(i)}} \tag{40.68}$$

Consider now the inner product $\rho_M(i) = y_i^*\Lambda_i b_{M,i}$, which appears in the numerator of $\kappa_M(i)$ in (40.64), and introduce the matrix

$$\begin{bmatrix} y_i & H_{M,i} & x_{M,i} \end{bmatrix}$$

Let us partition it as

$$\left[\begin{array}{c|c|c} y_{i-1} & H_{M,i-1} & x_{M,i-1} \\ \hline d(i) & u_{M,i} & u(i,M) \end{array}\right]$$

Now recall that $b_{M,i}$ is the residual vector that results from projecting $x_{M,i}$ onto $\mathcal{R}(H_{M,i})$ with regularization matrix $\lambda^{i+1}\Pi_M$, while $b_{M,i-1}$ is the residual vector that results from projecting $x_{M,i-1}$ onto $\mathcal{R}(H_{M,i-1})$ with regularization matrix $\lambda^i\Pi_M$. We are therefore faced with the problem of time-updating the inner product $\rho_M(i)$, which is again of the same form as the problem studied earlier in Sec. 32.3. More specifically, comparing with the statement of Lemma 32.3 (or with the data matrix (32.48) and its time-updated version (32.50)), we see that we can make the identifications:

$$\left\{\begin{array}{lll} \bar{H}_{i-1} \longleftarrow H_{M,i-1} & \gamma(i) \longleftarrow \gamma_M(i) & \beta(i) \longleftarrow u(i,M) \\ x_{i-1} \longleftarrow y_{i-1} & h_i \longleftarrow u_{M,i} & \widetilde{\alpha}(i) \longleftarrow e_M(i) \\ \alpha(i) \longleftarrow d(i) & z_{i-1} \longleftarrow x_{M,i-1} & \widetilde{\beta}(i) \longleftarrow b_M(i) \end{array}\right.$$

and arrive at the time-update relation (cf. (32.56)):

$$\boxed{\rho_M(i) = \lambda\rho_M(i-1) + \frac{r_M^*(i)b_M(i)}{\gamma_M(i)}} \tag{40.69}$$

We can also obtain time-updates for the minimum costs $\{\xi_M^f(i), \xi_M^b(i), \xi_M^{\bar{b}}(i)\}$ in much the same manner as above. Alternatively, since these variables correspond to the minimum costs of regularized least-squares problems, and since we already know how to time-update such minimum costs (cf. Alg. 30.2), we can readily write

$$\left\{\begin{array}{lll} \xi_M^f(i) & = & \lambda\xi_M^f(i-1) + |f_M(i)|^2/\bar{\gamma}_M(i) \\ \xi_M^b(i) & = & \lambda\xi_M^b(i-1) + |b_M(i)|^2/\gamma_M(i) \\ \xi_M^{\bar{b}}(i) & = & \lambda\xi_M^{\bar{b}}(i-1) + |\bar{b}_M(i)|^2/\bar{\gamma}_M(i) \end{array}\right. \tag{40.70}$$

Table 40.1 collects the various order- and time-update relations derived so far in the chapter. We again emphasize that these relations are independent of any data structure.

For convenience of notation, and also in order to save on addition operations, we introduce the *modified* cost variables:

$$\left\{\begin{array}{lll} \zeta_M^b(i) & \triangleq & \eta^{-1}\lambda^{i-M-1} + \xi_M^b(i) \\ \zeta_M^f(i) & \triangleq & \eta^{-1}\lambda^{i-1} + \xi_M^f(i) \\ \zeta_M^{\bar{b}}(i) & \triangleq & \eta^{-1}\lambda^{i-M-2} + \xi_M^{\bar{b}}(i) \end{array}\right. \tag{40.71}$$

TABLE 40.1 A listing of the time and order-update relations derived in Secs. 40.2–40.5. All these updates are independent of data structure.

$$\begin{aligned}
\xi^f_M(i) &= \lambda\xi^f_M(i-1) + |f_M(i)|^2/\bar{\gamma}_M(i)\\
\xi^b_M(i) &= \lambda\xi^b_M(i-1) + |b_M(i)|^2/\gamma_M(i)\\
\xi^{\bar{b}}_M(i) &= \lambda\xi^{\bar{b}}_M(i-1) + |\bar{b}_M(i)|^2/\bar{\gamma}_M(i)\\[1ex]
\xi_{M+1}(i) &= \xi_M(i) - |\rho_M(i)|^2/\zeta^b_M(i)\\
\xi^b_{M+1}(i) &= \xi^{\bar{b}}_M(i) - |\delta_M(i)|^2/\zeta^f_M(i)\\
\xi^f_{M+1}(i) &= \xi^f_M(i) - |\delta_M(i)|^2/\zeta^{\bar{b}}_M(i)\\[1ex]
\rho_M(i) &= \lambda\rho_M(i-1) + r^*_M(i)b_M(i)/\gamma_M(i)\\
\delta_M(i) &= \lambda\delta_M(i-1) + f^*_M(i)\bar{b}_M(i)/\bar{\gamma}_M(i)\\[1ex]
\kappa_M(i) &= \rho^*_M(i)/\zeta^b_M(i)\\
\kappa^b_M(i) &= \delta_M(i)/\zeta^f_M(i)\\
\kappa^f_M(i) &= \delta^*_M(i)/\zeta^{\bar{b}}_M(i)\\[1ex]
r_{M+1}(i) &= r_M(i) - \kappa_M(i)b_M(i)\\
b_{M+1}(i) &= \bar{b}_M(i) - \kappa^b_M(i)f_M(i)\\
f_{M+1}(i) &= f_M(i) - \kappa^f_M(i)\bar{b}_M(i)\\[1ex]
\gamma_{M+1}(i) &= \gamma_M(i) - |b_M(i)|^2/\zeta^b_M(i)\\
\gamma_{M+1}(i) &= \bar{\gamma}_M(i) - |f_M(i)|^2/\zeta^f_M(i)\\
\bar{\gamma}_{M+1}(i) &= \bar{\gamma}_M(i) - |\bar{b}_M(i)|^2/\zeta^{\bar{b}}_M(i)
\end{aligned}$$

It is easy to verify from the time and order-updates for $\{\xi^b_M(i), \xi^f_M(i), \xi^{\bar{b}}_M(i)\}$ that these modified variables satisfy similar updates, namely

$$\left\{\begin{aligned}
\zeta^f_M(i) &= \lambda\zeta^f_M(i-1) + |f_M(i)|^2/\bar{\gamma}_M(i)\\
\zeta^b_M(i) &= \lambda\zeta^b_M(i-1) + |b_M(i)|^2/\gamma_M(i)\\
\zeta^{\bar{b}}_M(i) &= \lambda\zeta^{\bar{b}}_M(i-1) + |\bar{b}_M(i)|^2/\bar{\gamma}_M(i)\\[1ex]
\zeta_{M+1}(i) &= \zeta_M(i) - |\rho_M(i)|^2/\zeta^b_M(i)\\
\zeta^b_{M+1}(i) &= \zeta^{\bar{b}}_M(i) - |\delta_M(i)|^2/\zeta^f_M(i)\\
\zeta^f_{M+1}(i) &= \zeta^f_M(i) - |\delta_M(i)|^2/\zeta^{\bar{b}}_M(i)
\end{aligned}\right. \tag{40.72}$$

albeit with initial conditions

$$\zeta^b_M(-1) = \eta^{-1}\lambda^{-M-2}, \quad \zeta^f_M(-1) = \eta^{-1}\lambda^{-2}, \quad \zeta^{\bar{b}}_M(-1) = \eta^{-1}\lambda^{-M-3} \tag{40.73}$$

while the initial values for the original variables are

$$\xi^b_M(-1) = \xi^f_M(-1) = \xi^{\bar{b}}_M(-1) = 0 \tag{40.74}$$

CHAPTER 41

Lattice Filter Algorithms

Figure 41.1 illustrates how the error variables $\{r_M(i), b_M(i), f_M(i)\}$ are related in terms of the reflection coefficients $\{\kappa_M(i), \kappa_M^b(i), \kappa_M^f(i)\}$, as was described in Chapter 40. It should be noted that the recursions listed in Table 40.1 help characterize *almost* fully the operation of the structure shown in the figure. The only missing piece of information is to know how to update the error sequence $\{\bar{b}_M(i)\}$. This fact is indicated schematically in Fig. 41.1 by the boxes with question marks. It is the update of these variables that is determined by data structure, and figuring out their update is the key to achieving an efficient algorithm; by efficient we mean $O(M)$ operations per iteration for a filter of order M.

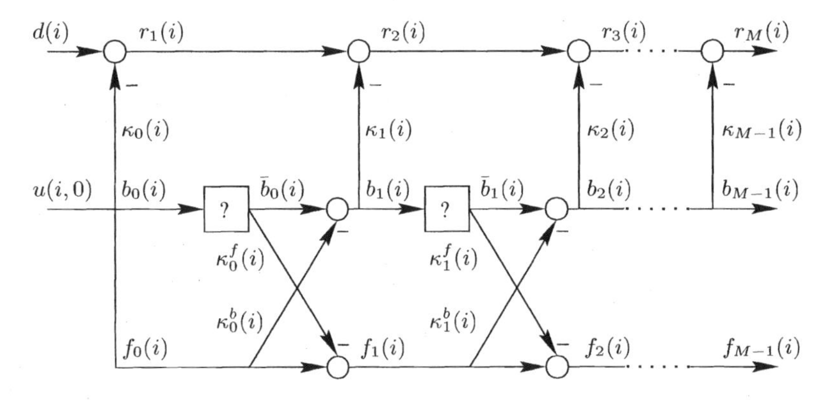

FIGURE 41.1 Relations among the residuals $\{r_M(i), f_M(i), b_M(i)\}$. The boxes with question marks indicate that we still need to develop a relation between $\{b_M(i), \bar{b}_M(i)\}$. This relation turns out to be a function of data structure. For example, if the regressors have shift structure, then the question marks will be replaced by pure delays, z^{-1}. Other data structures would lead to other choices for the blocks with questions marks — see the derivation of Laguerre lattice filters in Chapter 16 of Sayed (2003).

41.1 SIGNIFICANCE OF DATA STRUCTURE

To illustrate how the evaluation of $\bar{b}_M(i)$ is dependent on data structure, we now focus on the case of regressors with shift structure, i.e., we assume that the entries of $u_{M,i}$ are delayed versions of an input sequence $\{u(\cdot)\}$ so that

$$\begin{array}{rcl} u_{M,i} & = & \begin{bmatrix} u(i) & u(i-1) & \ldots & u(i-M+2) & u(i-M+1) \end{bmatrix} \\ u_{M,i+1} & = & \begin{bmatrix} u(i+1) & u(i) & u(i-1) & \ldots & u(i-M+2) \end{bmatrix} \end{array}$$

Comparing the expressions for both $u_{M,i}$ and $u_{M,i+1}$ we see that $u_{M,i+1}$ is obtained from $u_{M,i}$ by shifting the entries of the latter by one position to the right and introducing a new entry, $u(i+1)$, at the left. In this way, the data matrix $H_{M,i}$ will also exhibit structure, e.g., for $M = 4$, it will have the form (compare with (40.18)):

$$H_{4,i} = \begin{bmatrix} u(0) & 0 & 0 & 0 \\ u(1) & u(0) & 0 & 0 \\ u(2) & u(1) & u(0) & 0 \\ u(3) & u(2) & u(1) & u(0) \\ \vdots & \vdots & \vdots & \vdots \\ u(i) & u(i-1) & u(i-2) & u(i-3) \end{bmatrix}$$

where we are assuming $u(j) = 0$ for $j < 0$. Observe that every column of $H_{M,i}$ is a shifted version of the previous column, i.e., every column is obtained from the previous column by shifting its entries downwards by one position and by adding a zero entry. This means that any two successive columns of $H_{M,i}$, say, $\{x_{j,i}, x_{j+1,i}\}$, are related by the lower triangular shift matrix Z, i.e.,

$$\boxed{x_{j+1,i} = Zx_{j,i}} \tag{41.1}$$

where Z is the $(i+1) \times (i+1)$ lower triangular matrix with zeros everywhere except for unit entries on the first sub-diagonal, e.g., for $i = 3$,

$$Z = \begin{bmatrix} 0 & & & \\ 1 & 0 & & \\ & 1 & 0 & \\ & & 1 & 0 \end{bmatrix}$$

Now, as in (40.31) and (40.45), we partition $H_{M+1,i}$ as

$$H_{M+1,i} = \begin{bmatrix} x_{0,i} & \bar{H}_{M,i} \end{bmatrix} = \begin{bmatrix} H_{M,i} & x_{M,i} \end{bmatrix}$$

to find that, in view of (41.1), the following relation holds between $\{H_{M,i}, \bar{H}_{M,i}\}$:

$$\boxed{\bar{H}_{M,i} = ZH_{M,i}} \tag{41.2}$$

In addition, the following relation also holds between $\{H_{M,i}, H_{M,i-1}\}$:

$$ZH_{M,i} = \begin{bmatrix} 0 \\ H_{M,i-1} \end{bmatrix} \tag{41.3}$$

With these relations we can now relate the residual vectors $\bar{b}_{M,i}$ and $b_{M,i}$. Thus, recall their definitions:

$$\bar{b}_{M,i} \;=\; x_{M+1,i} - \bar{H}_{M,i}w^{\bar{b}}_{M,i} \qquad b_{M,i} \;=\; x_{M,i} - H_{M,i}w^{b}_{M,i} \tag{41.4}$$

where

$$\begin{cases} w^{\bar{b}}_{M,i} = \bar{P}_{M,i}\bar{H}^*_{M,i}\Lambda_i x_{M+1,i} & \bar{P}_{M,i} = (\lambda^i\Pi_M + \bar{H}^*_{M,i}\Lambda_i\bar{H}_{M,i})^{-1} \\ w^b_{M,i} = P_{M,i}H^*_{M,i}\Lambda_i x_{M,i} & P_{M,i} = (\lambda^{i+1}\Pi_M + H^*_{M,i}\Lambda_i H_{M,i})^{-1} \end{cases} \quad (41.5)$$

Substituting (41.2) and (41.3) into the expression for $\bar{P}_{M,i}$ we obtain

$$\bar{P}_{M,i} = \left(\lambda^i\Pi_M + \begin{bmatrix} 0 & H^*_{M,i-1} \end{bmatrix}\Lambda_i\begin{bmatrix} 0 \\ H_{M,i-1} \end{bmatrix}\right)^{-1} = (\lambda^i\Pi_M + H^*_{M,i-1}\Lambda_{i-1}H_{M,i-1})^{-1}$$

That is,

$$\boxed{\bar{P}_{M,i} = P_{M,i-1}}$$

Substituting this result, along with (41.1), into expression (41.5) for $w^{\bar{b}}_{M,i}$ we obtain

$$\begin{aligned} w^{\bar{b}}_{M,i} &= P_{M,i-1}H^*_{M,i}Z^*\Lambda_i Z x_{M,i} \\ &= P_{M,i-1}H^*_{M,i-1}\Lambda_{i-1}x_{M,i-1} \\ &= w^b_{M,i-1} \end{aligned}$$

so that

$$\begin{aligned} \bar{b}_{M,i} &= x_{M+1,i} - \bar{H}_{M,i}w^{\bar{b}}_{M,i} \\ &= Zx_{M,i} - ZH_{M,i}w^b_{M,i-1} \\ &= \begin{bmatrix} 0 \\ x_{M,i-1} \end{bmatrix} - \begin{bmatrix} 0 \\ H_{M,i-1} \end{bmatrix} w^b_{M,i-1} \\ &= \begin{bmatrix} 0 \\ b_{M,i-1} \end{bmatrix} \end{aligned}$$

and, consequently, by equating the last entries of both sides,

$$\boxed{\bar{b}_M(i) = b_M(i-1)} \quad (41.6)$$

In other words, we find that in the shift-structured case, the residual errors $\{\bar{b}_M(i)\}$ are time delayed versions of $\{b_M(i)\}$. In a similar vein, it can be verified that

$$\boxed{\xi^{\bar{b}}_M(i) = \xi^b_M(i-1), \quad \bar{\gamma}_M(i) = \gamma_M(i-1)} \quad (41.7)$$

Interpretation of the Estimation Errors

In the shift-structured case, we can provide additional insights into the meaning of the *a posteriori* estimation errors $\{f_M(i), b_M(i), r_M(i)\}$. Indeed, in this scenario, we have from the definitions of these errors,

$$\begin{aligned} f_M(i) &= u(i) - u_{M,i-1}w^f_{M,i} \\ b_M(i) &= u(i-M) - u_{M,i}w^b_{M,i} \\ r_M(i) &= d(i) - u_{M,i}w_{M,i} \end{aligned}$$

where

$$u_{M,i} = \begin{bmatrix} u(i) & u(i-1) & \ldots & u(i-M+1) \end{bmatrix}$$

In this way, $f_M(i)$ can be interpreted as the forward prediction error in estimating $u(i)$ from the M past values, while $b_M(i)$ can be interpreted as the backward prediction error in estimating $u(i-M)$ from the M future values.

41.2 A POSTERIORI-BASED LATTICE FILTER

If we substitute the results (41.6)–(41.7) into Table 40.1, we arrive at Alg. 41.1. This filter is known as the *a posteriori* lattice filter since it relies on the propagation of the *a posteriori* errors $\{f_M(i), b_M(i), r_M(i)\}$; the filter is depicted in Fig. 41.2.

Algorithm 41.1 (*A posteriori* lattice filter) Let $0 \ll \lambda \leq 1$ be a forgetting factor and define $\Pi_M = \eta^{-1}\text{diag}\{\lambda^{-2}, \lambda^{-3}, \ldots, \lambda^{-(M+1)}\}$. Consider a reference sequence $\{d(j)\}$ and a regressor sequence $\{u_{M,j}\}$ with shift structure and of dimension $1 \times M$, say, $u_{M,j} = \left[\begin{array}{cccc} u(j) & u(j-1) & \ldots & u(j-M+1) \end{array}\right]$. For each $i \geq 0$, the $M-$th order *a posteriori* estimation error, $r_M(i) = d(i) - u_{M,i}w_{M,i}$, that results from the solution of the regularized least-squares problem:

$$\min_{w_M} \left[\lambda^{i+1} w_M^* \Pi_M w_M \; + \; \sum_{j=0}^{i} \lambda^{i-j} |d(j) - u_{M,j}w_M|^2 \right]$$

can be computed as follows:

1. Initialization. From $m = 0$ to $m = M-1$ set:

$$\begin{array}{l} \delta_m(-1) = \rho_m(-1) = 0, \quad \gamma_m(-1) = 1, \quad b_m(-1) = 0 \\ \zeta_m^f(-1) = \eta^{-1}\lambda^{-2}, \quad \zeta_m^b(-1) = \eta^{-1}\lambda^{-m-2} \end{array}$$

2. For $i \geq 0$, repeat:

- Set $\gamma_0(i) = 1$, $b_0(i) = f_0(i) = u(i)$; and $r_0(i) = d(i)$
- From $m = 0$ to $m = M-1$, repeat:

$$\begin{array}{rcl}
\zeta_m^f(i) & = & \lambda\zeta_m^f(i-1) + |f_m(i)|^2/\gamma_m(i-1) \\
\zeta_m^b(i) & = & \lambda\zeta_m^b(i-1) + |b_m(i)|^2/\gamma_m(i) \\
\delta_m(i) & = & \lambda\delta_m(i-1) + f_m^*(i)b_m(i-1)/\gamma_m(i-1) \\
\rho_m(i) & = & \lambda\rho_m(i-1) + r_m^*(i)b_m(i)/\gamma_m(i) \\
\gamma_{m+1}(i) & = & \gamma_m(i) - |b_m(i)|^2/\zeta_m^b(i) \\
\kappa_m^b(i) & = & \delta_m(i)/\zeta_m^f(i) \\
\kappa_m^f(i) & = & \delta_m^*(i)/\zeta_m^b(i-1) \\
\kappa_m(i) & = & \rho_m^*(i)/\zeta_m^b(i) \\
b_{m+1}(i) & = & b_m(i-1) - \kappa_m^b(i)f_m(i) \\
f_{m+1}(i) & = & f_m(i) - \kappa_m^f(i)b_m(i-1) \\
r_{m+1}(i) & = & r_m(i) - \kappa_m(i)b_m(i)
\end{array}$$

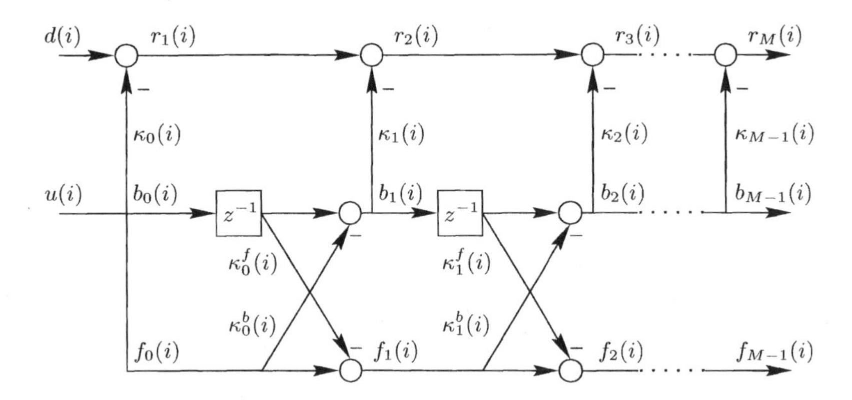

FIGURE 41.2 The *a posteriori*-based lattice filter.

Remark 41.1 (Data structure) What if two successive columns of the data matrix $H_{M,i}$ are not shifted versions of each other as in (41.1), but are instead related by some other matrix Φ? Would it still be possible to derive a lattice algorithm? Interesting enough, the answer is positive — see, e.g., Chapter 16 of Sayed (2003), which deals with Laguerre lattice filters.

41.3 A PRIORI-BASED LATTICE FILTER

Algorithm 41.1 relies on order-updates for the *a posteriori* estimation errors denoted by $\{r_M(i), f_M(i), b_M(i)\}$. A similar algorithm can be obtained by relying instead on *a priori* errors, which are defined as follows.

Refer again to Sec. 40.5 and to the data matrix $H_{M+1,i}$ with its equivalent partitionings:

$$\begin{aligned} H_{M+1,i} = \begin{bmatrix} x_{0,i} & x_{1,i} & \ldots & x_{M,i} \end{bmatrix} &= \begin{bmatrix} H_{M,i} & x_{M,i} \end{bmatrix} \\ &= \begin{bmatrix} x_{0,i} & \bar{H}_{M,i} \end{bmatrix} \\ &= \begin{bmatrix} x_{0,i} & \bar{H}_{M-1,i} & x_{M,i} \end{bmatrix} \end{aligned}$$

We define the *a priori* residual vectors as follows:

$$\left\{ \begin{array}{lcl} e_{M,i} & = & \textit{a priori} \text{ residual from projecting } y_i \text{ onto } H_{M,i} \\ \beta_{M,i} & = & \textit{a priori} \text{ residual from projecting } x_{M,i} \text{ onto } H_{M,i} \\ \alpha_{M,i} & = & \textit{a priori} \text{ residual from projecting } x_{0,i} \text{ onto } \bar{H}_{M,i} \\ \bar{\beta}_{M,i} & = & \textit{a priori} \text{ residual from projecting } x_{M+1,i} \text{ onto } \bar{H}_{M,i} \end{array} \right.$$

where the projection problems for $\{e_{M,i}, \beta_{M,i}\}$ employ the regularization matrix $\lambda^{i+1}\Pi_M$, while the projection problems for $\{\alpha_{M,i}, \bar{\beta}_{M,i}\}$ employ the regularization matrix $\lambda^i\Pi_M$. The term *a priori* in the above definitions means that *prior* weight estimates are used in the definition of the errors. More specifically (compare with (40.59)),

$$\left\{ \begin{array}{lclclcl} e_{M,i} & = & y_i - H_{M,i}w_{M,i-1} & \qquad & \alpha_{M,i} & = & x_{0,i} - \bar{H}_{M,i}w^f_{M,i-1} \\ \beta_{M,i} & = & x_{M+1,i} - H_{M+1,i}w^b_{M,i-1} & \qquad & \bar{\beta}_{M,i} & = & x_{M+1,i} - \bar{H}_{M,i}w^b_{M,i-1} \end{array} \right.$$

where now $w^f_{M,i-1}$, for example, is the solution to a regularized least-squares problem of the form

$$\min_{w^f_M} \left[\lambda^{i-1} w^{f*}_M \Pi_M w^f_M + (x_{0,i-1} - \bar{H}_{M,i-1} w^f_2)^* \Lambda_{i-1} (x_{0,i-1} - \bar{H}_{M,i-1} w^f_M) \right] \tag{41.8}$$

Comparing this cost function with (40.60) we see that i is replaced by $i-1$. The last entries of the above residual vectors are denoted by $\{e_M(i), \alpha_M(i), \beta_M(i), \bar{\beta}_M(i)\}$ and they are referred to as the *a priori* estimation errors. They are, of course, related to the corresponding *a posteriori* errors (40.61) via the associated conversion factors:

$$\begin{aligned} r_M(i) &= e_M(i)\gamma_M(i), \quad b_M(i) = \beta_M(i)\gamma_M(i) \\ f_M(i) &= \alpha_M(i)\bar{\gamma}_M(i), \quad \bar{b}_M(i) = \bar{\beta}_M(i)\bar{\gamma}_M(i) \end{aligned}$$

By following arguments similar to what we did in Secs. 40.2–40.4, and which led to (40.63), it can be verified that these *a priori* errors satisfy the following order-update relations in terms of the same reflection coefficients $\{\kappa_M(i), \kappa^f_M(i), \kappa^b_M(i)\}$:

$$\boxed{\begin{aligned} e_{M+1}(i) &= e_M(i) - \kappa_M(i-1)\beta_M(i) \\ \beta_{M+1}(i) &= \bar{\beta}_M(i) - \kappa^b_M(i-1)\alpha_M(i) \\ \alpha_{M+1}(i) &= \alpha_M(i) - \kappa^f_M(i-1)\bar{\beta}_M(i) \end{aligned}} \tag{41.9}$$

Again, relations (41.9) hold irrespective of data structure. However, as was shown in Sec. 41.1 for the variables $\{\bar{b}_M(i), b_M(i)\}$, when the regressors possess shift structure it will also hold that

$$\boxed{\bar{\beta}_M(i) = \beta_M(i-1)}$$

In addition, the *a priori* estimation errors $\{\alpha_M(i), \beta_M(i), e_M(i)\}$ will admit the following interpretations:

$$\alpha_M(i) = u(i) - u_{M,i-1} w^f_{M,i-1} \tag{41.10}$$

$$\beta_M(i) = u(i-M) - u_{M,i} w^b_{M,i-1} \tag{41.11}$$

$$e_M(i) = d(i) - u_{M,i} w_{M,i-1} \tag{41.12}$$

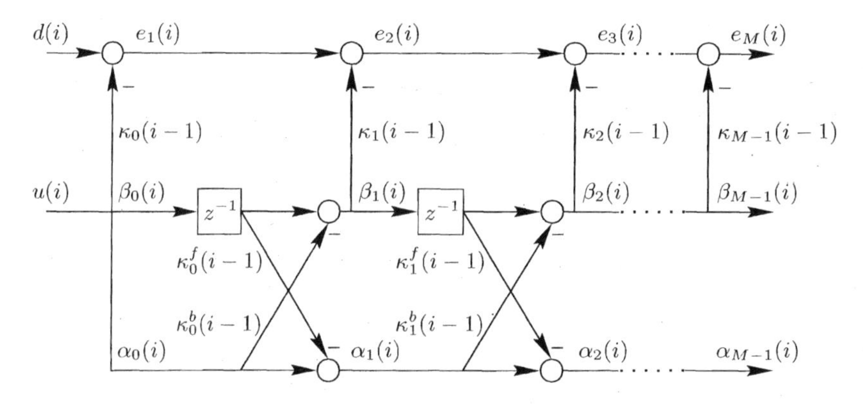

FIGURE 41.3 The *a priori*-based lattice filter.

In other words, $\alpha_M(i)$ denotes the forward prediction error in estimating $u(i)$ from the M past values using the prior weight estimate, $w^f_{M,i-1}$, while $\beta_M(i)$ denotes the backward prediction error in estimating $u(i-M)$ from the M future values using the prior weight estimate, $w^b_{M,i-1}$. The resulting *a priori*-based lattice filter is listed in Alg. 41.2 and shown in Fig. 41.3.

Algorithm 41.2 (*A priori* lattice filter) Consider the same setting of Alg. 41.1. For each $i \geq 0$, the $M-$th order *a priori* estimation error, $e_M(i) = d(i) - u_{M,i}w_{M,i-1}$, that results from the solution of the regularized least-squares problem

$$\min_{w_M}\left[\lambda^i w_M^* \Pi_M w_M + \sum_{j=0}^{i-1}\lambda^{i-1-j}|d(j) - u_{M,j}w_M|^2\right]$$

can be computed as follows:

1. Initialization. From $m = 0$ to $m = M-1$ set:

$$\begin{array}{l} \delta_m(-1) = \rho_m(-1) = 0, \quad \gamma_m(-1) = 1, \quad \beta_m(-1) = 0 \\ \kappa^f_m(-1) = \kappa^b_m(-1) = \kappa_m(-1) = 0 \\ \zeta^f_m(-1) = \eta^{-1}\lambda^{-2}, \quad \zeta^b_m(-1) = \eta^{-1}\lambda^{-m-2} \end{array}$$

2. For $i \geq 0$, repeat:

- Set $\gamma_0(i) = 1$, $\beta_0(i) = \alpha_0(i) = u(i)$, and $e_0(i) = d(i)$
- From $m = 0$ to $m = M-1$, repeat:

$$\begin{array}{rcl} \zeta^f_m(i) &=& \lambda\zeta^f_m(i-1) + |\alpha_m(i)|^2\gamma_m(i-1) \\ \zeta^b_m(i) &=& \lambda\zeta^b_m(i-1) + |\beta_m(i)|^2\gamma_m(i) \\ \delta_m(i) &=& \lambda\delta_m(i-1) + \alpha^*_m(i)\beta_m(i-1)\gamma_m(i-1) \\ \rho_m(i) &=& \lambda\rho_m(i-1) + e^*_m(i)\beta_m(i)\gamma_m(i) \\ \beta_{m+1}(i) &=& \beta_m(i-1) - \kappa^b_m(i-1)\alpha_m(i) \\ \alpha_{m+1}(i) &=& \alpha_m(i) - \kappa^f_m(i-1)\beta_m(i-1) \\ e_{m+1}(i) &=& e_m(i) - \kappa_m(i-1)\beta_m(i) \\ \gamma_{m+1}(i) &=& \gamma_m(i) - |\gamma_m(i)\beta_m(i)|^2/\zeta^b_m(i) \\ \kappa^b_m(i) &=& \delta_m(i)/\zeta^f_m(i) \\ \kappa^f_m(i) &=& \delta^*_m(i)/\zeta^b_m(i-1) \\ \kappa_m(i) &=& \rho^*_m(i)/\zeta^b_m(i) \end{array}$$

CHAPTER 42

Error-Feedback Lattice Filters

Although the *a posteriori*- and *a priori*-based lattice filters of the previous chapter are common lattice forms, several other equivalent implementations exist such as error-feedback forms, array-based forms, and normalized forms. All these variants are, of course, theoretically equivalent. However, they can differ in performance under different operating conditions that arise, for example, in finite-precision implementations or as a result of noise and regularization.

42.1 A PRIORI ERROR-FEEDBACK LATTICE FILTER

In this section we derive the so-called error-feedback form, which tends to exhibit good performance under finite-precision conditions. The algorithm can be motivated as follows. Observe from the *a posteriori* lattice recursions of Alg. 41.1 that a pivotal role is played by the reflection coefficients $\{\kappa_M(i), \kappa_M^b(i), \kappa_M^f(i)\}$. These coefficients are computed as ratios of certain quantities. For example, $\kappa_M^f(i)$ is computed as the ratio $\delta_M^*(i)/\zeta_M^b(i-1)$, with separate recursions used to evaluate the numerator and denominator quantities. Likewise for $\{\kappa_M^b(i), \kappa_M(i)\}$.

The error-feedback lattice form replaces the separate recursions for the numerator and denominator quantities by direct recursions for the reflection coefficients themselves. In principle, we could derive these recursions algebraically as follows. Consider for instance

$$\kappa_M(i) = \rho_M^*(i)/\zeta_M^b(i)$$

where, from the listing of the *a posteriori*-based lattice filter, the numerator and denominator quantities are updated via

$$\left\{\begin{array}{lcl} \rho_M(i) & = & \lambda\rho_M(i-1) + r_M^*(i)b_M(i)/\gamma_M(i) \\ \zeta_M^b(i) & = & \lambda\zeta_M^b(i-1) + |b_M(i)|^2/\gamma_M(i) \end{array}\right. \tag{42.1}$$

Substituting these expressions into the above expression for $\kappa_M(i)$ leads to

$$\kappa_M(i) = \frac{\lambda\rho_M^*(i-1) + r_M(i)b_M^*(i)/\gamma_M(i)}{\lambda\zeta_M^b(i-1) + |b_M(i)|^2/\gamma_M(i)}$$

and some algebra will then result in a relation between $\kappa_M(i)$ and $\kappa_M(i-1)$. We shall not pursue this algebraic route. Instead, we shall employ a more elegant geometric argument and use it to provide an interesting interpretation for the reflection coefficients $\{\kappa_M(i), \kappa_M^b(i), \kappa_M^f(i)\}$. Specifically, we shall show that each one of them can be interpreted as performing the projection of a *normalized* error vector onto another *normalized* error vector. This interpretation allows us to invoke many of the results we have already developed for projection (least-squares) problems and to arrive at recursions for the reflection coefficients almost by inspection.

We start with the reflection coefficient

$$\kappa_M(i) = \rho_M^*(i)/\zeta_M^b(i) \tag{42.2}$$

where from Table 40.1, the numerator and denominator quantities satisfy the time-updates (42.1). Now define the so-called *angle normalized* estimation errors:

$$\boxed{\begin{array}{rcl} b'_M(i) & \triangleq & b_M(i)/\gamma_M^{1/2}(i) \;=\; \beta_M(i)\gamma_M^{1/2}(i) \\ r'_M(i) & \triangleq & r_M(i)/\gamma_M^{1/2}(i) \;=\; e_M(i)\gamma_M^{1/2}(i) \end{array}} \tag{42.3}$$

The reason for the qualification "angle-normalized" is that, since the conversion factor satisfies $0 < \gamma_M(i) \leq 1$, its square-root can be interpreted as the cosine of an angle, say, $\gamma_M^{1/2}(i) = \cos\phi_M(i)$ for some $\phi_M(i)$.

Using these normalized variables, we can rewrite recursions (42.1) as

$$\left\{ \begin{array}{rcl} \rho_M(i) & = & \lambda\rho_M(i-1) + r'^*_M(i)b'_M(i) \\ \zeta_M^b(i) & = & \lambda\zeta_M^b(i-1) + |b'_M(i)|^2 \end{array} \right. \tag{42.4}$$

which, in view of the initial conditions $\rho_M(-1) = 0$ and $\xi_M^b(-1) = 0$, lead to

$$\rho_M(i) \;=\; \sum_{i=0}^{i} \lambda^{i-i} r'^*_M(i)b'_M(i), \qquad \xi_M^b(i) = \sum_{i=0}^{i} \lambda^{i-i}|b'_M(i)|^2$$

In other words, if we introduce the following vectors of angle normalized errors:[22]

$$b'_{M,i} \triangleq \begin{bmatrix} b'_M(0) \\ b'_M(1) \\ \vdots \\ b'_M(i) \end{bmatrix}, \qquad r'_{M,i} \triangleq \begin{bmatrix} r'_M(0) \\ r'_M(1) \\ \vdots \\ r'_M(i) \end{bmatrix}$$

then the quantities $\{\rho_M(i), \xi_M^b(i)\}$ can be recognized as the inner products

$$\rho_M(i) = r'^*_{M,i}\Lambda_i b'_{M,i}, \qquad \xi_M^b(i) = b'^*_{M,i}\Lambda_i b'_{M,i} \tag{42.5}$$

where

$$\Lambda_i = \text{diag}\{\lambda^i, \lambda^{i-1}, \ldots, \lambda, 1\}$$

Expression (42.5) should be compared with the original definitions (40.65) for these variables, namely,

$$\rho_M(i) = y_i^*\Lambda_i b_{M,i} \quad \text{and} \quad \xi_M^b(i) = x^*_{M,i}\Lambda_i b_{M,i}$$

[22]The vectors $b_{M,i}$ and $b'_{M,i}$ differ in a fundamental way; the entries of $b_{M,i}$ are not $\{b_M(0), b_M(1), \ldots, b_M(i)\}$! Likewise for $r_{M,i}$ and $r'_{M,i}$.

The key observation is that while the original definitions involve three vectors, $\{y_i, b_{M,i}, x_{M,i}\}$ the representations (42.5) involve only two vectors, $\{b'_{M,i}, r'_{M,i}\}$. In particular, they allow us to re-express the defining relation (42.2) for $\kappa_M(i)$ in the form:

$$\kappa_M(i) = (\bar{\eta}^{-1}\lambda^{i+1} + b'^*_{M,i}\Lambda_i b'_{M,i})^{-1} b'^*_{M,i}\Lambda_i r'_{M,i} \tag{42.6}$$

where $\bar{\eta} = \eta\lambda^{M+2}$. This expression for $\kappa_M(i)$ has the form of the solution to a regularized least-squares problem! In other words, $\kappa_M(i)$ can be regarded as the solution to the problem that projects (in a regularized manner) the vector $r'_{M,i}$ onto the vector $b'_{M,i}$, namely, it solves

$$\min_{\kappa_M} \left[\bar{\eta}^{-1}\lambda^{i+1}|\kappa_M|^2 + \|r'_{M,i} - \kappa_M b'_{M,i}\|^2_{\Lambda_i} \right] \implies \kappa_M(i) \tag{42.7}$$

This observation readily establishes that $\kappa_M(i)$ can be time-updated via a standard RLS update of the form:

$$\begin{aligned}
\kappa_M(i) &= \kappa_M(i-1) + \frac{b'^*_M(i)}{\zeta^b_M(i)}[r'_M(i) - b'_M(i)\kappa_M(i-1)] \qquad (42.8)\\
&= \kappa_M(i-1) + \frac{b^*_M(i)}{\zeta^b_M(i)}[r_M(i) - \beta_M(i)\kappa_M(i-1)]\\
&= \kappa_M(i-1) + \frac{\beta^*_M(i)\gamma_M(i)}{\zeta^b_M(i)} e_{M+1}(i) \qquad (42.9)
\end{aligned}$$

Comparing the statement (42.7) with that of a generic regularized least-squares problem (40.1), we see that κ_M plays the role of w_M, $b'_{M,i}$ plays the role of $H_{M,i}$, $r'_{M,i}$ plays the role of y_i, $\bar{\eta}^{-1}$ plays the role of Π_M, and $\zeta^b_M(i)$ plays the role of $P^{-1}_{M,i}$; this latter conclusion is because $\zeta^b_M(i)$ can be seen to be the coefficient of the normal equations (42.6), namely, $\zeta^b_M(i)\kappa_M(i) = b'^*_{M,i}\Lambda_i r'_{M,i}$. Then (42.8) follows from the update for $w_{M,i}$ in (40.12). Indeed, the last equation (42.9) is obtained from the order-update (41.9) for $e_{M+1}(i)$. In a similar vein, we can derive updates for $\kappa^f_M(i)$ and $\kappa^b_M(i)$. Thus, define the *angle normalized* estimation errors:

$$\boxed{\begin{aligned}
\bar{b}'_M(i) &\triangleq \bar{b}_M(i)/\bar{\gamma}^{1/2}_M(i) = \beta_M(i)\bar{\gamma}^{1/2}_M(i)\\
f'_M(i) &\triangleq f_M(i)/\bar{\gamma}^{1/2}_M(i) = \alpha_M(i)\bar{\gamma}^{1/2}_M(i)
\end{aligned}} \tag{42.10}$$

and the corresponding vectors

$$\bar{b}'_{M,i} \triangleq \begin{bmatrix} \bar{b}'_M(0)\\ \bar{b}'_M(1)\\ \vdots\\ \bar{b}'_M(i) \end{bmatrix}, \qquad f'_{M,i} \triangleq \begin{bmatrix} f'_M(0)\\ f'_M(1)\\ \vdots\\ f'_M(i) \end{bmatrix}$$

Then the defining expressions for $\{\kappa^f_M(i), \kappa^b_M(i)\}$ from Table 40.1 can be written as

$$\begin{aligned}
\kappa^f_M(i) &= (\eta^{-1}\lambda^{i-M-1} + \bar{b}'^*_{M,i}\Lambda_i\bar{b}'_{M,i})^{-1}\bar{b}'^*_{M,i}\Lambda_i f'_{M,i}\\
\kappa^b_M(i) &= (\eta^{-1}\lambda^{i-1} + f'^*_{M,i}\Lambda_i f'_{M,i})^{-1} f'^*_{M,i}\Lambda_i\bar{b}'_{M,i}
\end{aligned}$$

These expressions again suggest that we can interpret $\{\kappa_M^f(i), \kappa_M^b(i)\}$ as the solution of the problems that project (in a regularized manner) the vectors $\{f'_{M,i}, \bar{b}'_{M,i}\}$ onto the vectors $\{\bar{b}'_{M,i}, f'_{M,i}\}$, namely, they solve

$$\min_{\kappa_M^f} \left[\bar{\eta}^{-1}\lambda^{i+1}|\kappa_M^f|^2 + \left\| f'_{M,i} - \kappa_M^f \bar{b}'_{M,i} \right\|^2_{\Lambda_i} \right] \Longrightarrow \kappa_M^f(i)$$

$$\min_{\kappa_M^b} \left[\breve{\eta}^{-1}\lambda^{i+1}|\kappa_M^b|^2 + \left\| \bar{b}'_{M,i} - \kappa_M^b f'_{M,i} \right\|^2_{\Lambda_i} \right] \Longrightarrow \kappa_M^b(i)$$

where $\breve{\eta} = \eta\lambda^2$. These observations therefore establish that $\{\kappa_M^f(i), \kappa_M^b(i)\}$ can also be time-updated via standard RLS recursions as

$$\kappa_M^f(i) = \kappa_M^f(i-1) + \bar{\beta}_M^*(i)\bar{\gamma}_M(i)\alpha_{M+1}(i)/\zeta_M^{\bar{b}}(i) \tag{42.11}$$

$$\kappa_M^b(i) = \kappa_M^b(i-1) + \alpha_M^*(i)\bar{\gamma}_M(i)\beta_{M+1}(i)/\zeta_M^f(i) \tag{42.12}$$

Table 42.1 collects the various order and time-update relations derived so far for the *a priori* estimation errors, in addition to updates that are obtained immediately from those of Table 40.1 by simply rewriting them in terms of *a priori* errors (like, e.g., the first three relations in Table 42.1). The relations of Table 42.1 are again independent of any data structure. The only update that is missing is one for the error sequence $\{\bar{\beta}_M(i)\}$.

TABLE 42.1 A listing of time and order-update relations for the *a priori* estimation errors; these updates are independent of data structure.

$$\begin{aligned}
\xi_M^f(i) &= \lambda\xi_M^f(i-1) + |\alpha_M(i)|^2\bar{\gamma}_M(i) \\
\xi_M^b(i) &= \lambda\xi_M^b(i-1) + |\beta_M(i)|^2\gamma_M(i) \\
\xi_M^{\bar{b}}(i) &= \lambda\xi_M^{\bar{b}}(i-1) + |\bar{\beta}_M(i)|^2\bar{\gamma}_M(i)
\end{aligned}$$

$$\begin{aligned}
\xi_{M+1}(i) &= \xi_{M+1}(i) - |\rho_M(i)|^2/\zeta_M^b(i) \\
\xi_{M+1}^b(i) &= \xi_M^{\bar{b}}(i) - |\delta_M(i)|^2/\zeta_M^f(i) \\
\xi_{M+1}^f(i) &= \xi_M^f(i) - |\delta_M(i)|^2/\zeta_M^{\bar{b}}(i)
\end{aligned}$$

$$\begin{aligned}
e_{M+1}(i) &= e_M(i) - \kappa_M(i-1)\beta_M(i) \\
\beta_{M+1}(i) &= \bar{\beta}_M(i) - \kappa_M^b(i-1)\alpha_M(i) \\
\alpha_{M+1}(i) &= \alpha_M(i) - \kappa_M^f(i-1)\bar{\beta}_M(i)
\end{aligned}$$

$$\begin{aligned}
\kappa_M(i) &= \kappa_M(i-1) + \beta_M^*(i)\gamma_M(i)e_{M+1}(i)/\zeta_M^b(i) \\
\kappa_M^f(i) &= \kappa_M^f(i-1) + \bar{\beta}_M^*(i)\bar{\gamma}_M(i)\alpha_{M+1}(i)/\zeta_M^{\bar{b}}(i) \\
\kappa_M^b(i) &= \kappa_M^b(i-1) + \alpha_M^*(i)\bar{\gamma}_M(i)\beta_{M+1}(i)/\zeta_M^f(i)
\end{aligned}$$

$$\begin{aligned}
\gamma_{M+1}(i) &= \gamma_M(i) - |\gamma_M(i)\beta_M(i)|^2/\zeta_M^b(i) \\
\gamma_{M+1}(i) &= \bar{\gamma}_M(i) - |\bar{\gamma}_M(i)\alpha_M(i)|^2/\zeta_M^f(i) \\
\bar{\gamma}_{M+1}(i) &= \bar{\gamma}_M(i) - |\bar{\gamma}_M(i)\bar{\beta}_M(i)|^2/\zeta_M^{\bar{b}}(i)
\end{aligned}$$

However, as was shown in Sec. 41.1 for the variables $\{\bar{b}_M(i), b_M(i)\}$, when the regressors possess shift structure it will hold that $\bar{\beta}_M(i) = \beta_M(i-1)$. The resulting *a priori*-based error feedback lattice filter is listed in Alg. 42.1. The reason for the qualification "error-feedback" is that, as seen from the listing of the algorithm, the errors $\{e_{M+1}(i), \alpha_{M+1}(i), \beta_{M+1}(i-1)\}$ are fed into the recursions for the reflection coefficients $\{\kappa_M(i), \kappa_M^f(i), \kappa_M^b(i)\}$.

Algorithm 42.1 (*A priori* error-feedback filter) Let $0 \ll \lambda \leq 1$ be a forgetting factor and define $\Pi_M = \eta^{-1}\text{diag}\{\lambda^{-2}, \lambda^{-3}, \ldots, \lambda^{-(M+1)}\}$. Consider a reference sequence $\{d(j)\}$ and a regressor sequence $\{u_{M,j}\}$ with shift structure and of dimension $1\times M$, say, $u_{M,j} = \begin{bmatrix} u(j) & \ldots & u(j-M+1) \end{bmatrix}$. For each $i \geq 0$, the $M-$th order *a priori* estimation error, $e_M(i) = d(i) - u_{M,i}w_{M,i-1}$, that results from the solution of the regularized least-squares problem:

$$\min_{w_M} \left[\lambda^i w_M^* \Pi_M w_M + \sum_{j=0}^{i-1} \lambda^{i-1-j} |d(j) - u_{M,j} w_M|^2 \right]$$

can be computed as follows:

1. Initialization. From $m = 0$ to $m = M-1$ set:

$$\begin{array}{l} \gamma_m(-1) = 1, \quad \beta_m(-1) = 0 \\ \kappa_m^f(-1) = \kappa_m^b(-1) = \kappa_m(-1) = 0 \\ \zeta_m^f(-1) = \eta^{-1}\lambda^{-2}, \quad \zeta_m^b(-1) = \eta^{-1}\lambda^{-m-2} \end{array}$$

2. For $i \geq 0$, repeat:

   - Set $\gamma_0(i) = 1$, $\beta_0(i) = \alpha_0(i) = u(i)$, and $e_0(i) = d(i)$
   - From $m = 0$ to $m = M-1$, repeat:

$$\begin{array}{rcl} \zeta_m^f(i) & = & \lambda\zeta_m^f(i-1) + |\alpha_m(i)|^2\gamma_m(i-1) \\ \zeta_m^b(i) & = & \lambda\zeta_m^b(i-1) + |\beta_m(i)|^2\gamma_m(i) \\ \beta_{m+1}(i) & = & \beta_m(i-1) - \kappa_m^b(i-1)\alpha_m(i) \\ \alpha_{m+1}(i) & = & \alpha_m(i) - \kappa_m^f(i-1)\beta_m(i-1) \\ e_{m+1}(i) & = & e_m(i) - \kappa_m(i-1)\beta_m(i) \\ \kappa_m(i) & = & \kappa_m(i-1) + \beta_m^*(i)\gamma_m(i)e_{m+1}(i)/\zeta_m^b(i) \\ \kappa_m^f(i) & = & \kappa_m^f(i-1) + \beta_m^*(i-1)\gamma_m(i-1)\alpha_{m+1}(i)/\zeta_m^b(i-1) \\ \kappa_m^b(i) & = & \kappa_m^b(i-1) + \alpha_m^*(i)\gamma_m(i-1)\beta_{m+1}(i)/\zeta_m^f(i) \\ \gamma_{m+1}(i) & = & \gamma_m(i) - |\gamma_m(i)\beta_m(i)|^2/\zeta_m^b(i) \end{array}$$

42.2 A POSTERIORI ERROR-FEEDBACK LATTICE FILTER

An alternative feedback form that relies on *a posteriori* errors, as opposed to *a priori* errors, can also be derived. This can be achieved by simply expressing the updates for the reflection coefficients in terms of *a posteriori* quantities and rearranging terms. For example, consider recursion (42.8) for the reflection coefficient $\kappa_M(i)$, which can be written as

$$\kappa_M(i) = \left(1 - \frac{|b'_M(i)|^2}{\zeta^b_M(i)}\right)\kappa_M(i-1) + \frac{b'^*_M(i)r'_M(i)}{\zeta^b_M(i)} \tag{42.13}$$

Now note from Table 40.1, and from the definition of $b'_M(i)$ in (42.3), that

$$\begin{aligned}\gamma_{M+1}(i) &= \gamma_M(i) - \frac{|b_M(i)|^2}{\zeta^b_M(i)} \\ &= \gamma_M(i)\left[1 - \frac{|b'_M(i)|^2}{\zeta^b_M(i)}\right]\end{aligned}$$

so that

$$\begin{aligned}\kappa_M(i) &= \frac{\gamma_{M+1}(i)}{\gamma_M(i)}\kappa_M(i-1) + \frac{b'^*_M(i)r'_M(i)}{\zeta^b_M(i)} \\ &= \frac{\gamma_{M+1}(i)}{\gamma_M(i)}\left[\kappa_M(i-1) + \frac{b'^*_M(i)r'_M(i)\gamma_M(i)}{\gamma_{M+1}(i)\zeta^b_M(i)}\right]\end{aligned}$$

Note further that

$$\begin{aligned}\gamma_{M+1}(i) &= \gamma_M(i)\left[\frac{\zeta^b_M(i) - |b'_M(i)|^2}{\zeta^b_M(i)}\right] \\ &= \gamma_M(i)\left[\frac{\lambda\zeta^b_M(i-1)}{\zeta^b_M(i)}\right]\end{aligned}$$

so that

$$\gamma_{M+1}(i)\zeta^b_M(i) = \lambda\gamma_M(i)\zeta^b_M(i-1)$$

Substituting this equality into the last expression above for $\kappa_M(i)$ we get

$$\kappa_M(i) = \frac{\gamma_{M+1}(i)}{\gamma_M(i)}\left[\kappa_M(i-1) + \frac{b^*_M(i)r_M(i)}{\lambda\gamma_M(i)\zeta^b_M(i-1)}\right] \tag{42.14}$$

In a similar manner, we can derive time-updates for the other reflection coefficients:

$$\kappa^f_M(i) = \frac{\bar{\gamma}_{M+1}(i)}{\bar{\gamma}_M(i)}\left[\kappa^f_M(i-1) + \frac{\bar{b}^*_M(i)f_M(i)}{\lambda\bar{\gamma}_M(i)\zeta^{\bar{b}}_M(i-1)}\right] \tag{42.15}$$

$$\kappa^b_M(i) = \frac{\gamma_{M+1}(i)}{\bar{\gamma}_M(i)}\left[\kappa^b_M(i-1) + \frac{f^*_M(i)\bar{b}_M(i)}{\lambda\bar{\gamma}_M(i)\zeta^f_M(i-1)}\right] \tag{42.16}$$

As before, when the regression vectors have shift structure, it will hold that

$$\bar{b}_M(i) = b_M(i-1), \quad \bar{\gamma}_M(i) = \gamma_M(i-1), \quad \xi^{\bar{b}}_M(i) = \xi^b_M(i-1)$$

In this case, we arrive at the *a posteriori*-based error feedback lattice filter of Alg. 42.2.

Algorithm 42.2 (*A posteriori* error-feedback filter) Let $0 \ll \lambda \leq 1$ be a forgetting factor and define $\Pi_M = \eta^{-1}\text{diag}\{\lambda^{-2}, \lambda^{-3}, \ldots, \lambda^{-(M+1)}\}$. Consider a reference sequence $\{d(j)\}$ and a regressor sequence $\{u_{M,j}\}$ with shift structure and of dimension $1 \times M$, say, $u_{M,j} = \left[\begin{array}{ccc} u(j) & \ldots & u(j-M+1) \end{array}\right]$. For each $i \geq 0$, the $M-$th order *a posteriori* estimation error, $r_M(i) = d(i) - u_{M,i}w_{M,i}$, that results from the solution of the regularized least-squares problem:

$$\min_{w_M} \left[\lambda^{i+1} w_M^* \Pi_M w_M + \sum_{j=0}^{i} \lambda^{i-j} |d(j) - u_{M,j} w_M|^2 \right]$$

can be computed as follows:

1. Initialization. From $m = 0$ to $m = M - 1$ set:

$$\begin{array}{l} \gamma_m(-1) = 1, \quad b_m(-1) = 0, \quad \kappa_m^f(-1) = \kappa_m^b(-1) = \kappa_m(-1) = 0 \\ \zeta_m^f(-1) = \eta^{-1}\lambda^{-2}, \quad \zeta_m^b(-1) = \eta^{-1}\lambda^{-m-2} \end{array}$$

2. For $i \geq 0$, repeat:

   - Set $\gamma_0(i) = 1$, $b_0(i) = f_0(i) = u(i)$, and $r_0(i) = d(i)$
   - From $m = 0$ to $m = M - 1$, repeat:

$$\begin{aligned}
\zeta_m^f(i) &= \lambda\zeta_m^f(i-1) + |f_m(i)|^2/\gamma_m(i-1) \\
\zeta_m^b(i) &= \lambda\zeta_m^b(i-1) + |b_m(i)|^2/\gamma_m(i) \\
\gamma_{m+1}(i) &= \gamma_m(i) - |b_m(i)|^2/\zeta_m^b(i) \\
\kappa_m(i) &= \frac{\gamma_{m+1}(i)}{\gamma_m(i)} \left[\kappa_m(i-1) + \frac{b_m^*(i) r_m(i)}{\lambda\gamma_m(i)\zeta_m^b(i-1)} \right] \\
\kappa_m^f(i) &= \frac{\gamma_{m+1}(i-1)}{\gamma_m(i-1)} \left[\kappa_m^f(i-1) + \frac{b_m^*(i-1) f_m(i)}{\lambda\gamma_m(i-1)\zeta_m^b(i-2)} \right] \\
\kappa_m^b(i) &= \frac{\gamma_{m+1}(i)}{\gamma_m(i-1)} \left[\kappa_m^b(i-1) + \frac{f_m^*(i) b_m(i-1)}{\lambda\gamma_m(i-1)\zeta_m^f(i-1)} \right] \\
r_{m+1}(i) &= r_m(i) - \kappa_m(i) b_m(i) \\
b_{m+1}(i) &= b_m(i-1) - \kappa_m^b(i) f_m(i) \\
f_{m+1}(i) &= f_m(i) - \kappa_m^f(i) b_m(i-1)
\end{aligned}$$

42.3 NORMALIZED LATTICE FILTER

The lattice filters studied so far require the evaluation of three reflection coefficients,

$$\{\kappa_M^f(i), \kappa_M^b(i), \kappa_M(i)\}$$

There is another equivalent lattice form that employs only two reflection coefficients. We denote these new coefficients by $\{\kappa_M^a(i), \kappa_M^c(i)\}$. The algorithm can be derived as follows.

First, recall the definitions of $\{\kappa_M^f(i), \kappa_M^b(i), \kappa_M(i)\}$ from Table 40.1:

$$\begin{aligned}
\kappa_M(i) &= \rho_M^*(i)/\zeta_M^b(i) \\
\kappa_M^b(i) &= \delta_M(i)/\zeta_M^f(i) \\
\kappa_M^f(i) &= \delta_M^*(i)/\zeta_M^{\bar{b}}(i)
\end{aligned}$$

and introduce the modified cost

$$\zeta_M(i) \triangleq \eta^{-1}\lambda^{i-1} + \xi_M(i)$$

Define also the normalized estimation errors

$$\begin{aligned}
b_M''(i) &\triangleq b_M(i)/\gamma_M^{1/2}(i)\zeta_M^{b/2}(i) \\
\bar{b}_M''(i) &\triangleq \bar{b}_M(i)/\bar{\gamma}_M^{1/2}(i)\zeta_M^{\bar{b}/2}(i) \\
f_M''(i) &\triangleq f_M(i)/\bar{\gamma}_M^{1/2}(i)\zeta_M^{f/2}(i) \\
r_M''(i) &\triangleq r_M(i)/\gamma_M^{1/2}(i)\zeta_M^{1/2}(i)
\end{aligned}$$

Comparing, for example, $b_M''(i)$ above with the angle-normalized variable $b_M'(i)$ in (42.3) we see that we are now further normalizing by the square-root of $\zeta_M^b(i)$.

The normalized reflection coefficients that we are interested in are defined as follows:

$$\boxed{\begin{aligned}
\kappa_M^a(i) &\triangleq \frac{\delta_M^*(i)}{\zeta_M^{\bar{b}/2}(i)\zeta_M^{f/2}(i)} = \kappa_M^f(i)\cdot\frac{\zeta_M^{\bar{b}/2}(i)}{\zeta_M^{f/2}(i)} = \kappa_M^{b*}(i)\cdot\frac{\zeta_M^{f/2}(i)}{\zeta_M^{\bar{b}/2}(i)} \\
\kappa_M^c(i) &\triangleq \frac{\rho_M^*(i)}{\zeta_M^{b/2}(i)\zeta_M^{1/2}(i)} = \kappa_M(i)\cdot\frac{\zeta_M^{b/2}(i)}{\zeta_M^{1/2}(i)}
\end{aligned}}$$

That is, we scale the forward projection coefficient $\kappa_M^f(i)$ by $\zeta_M^{\bar{b}/2}(i)/\zeta_M^{f/2}(i)$. Likewise for $\kappa_M^c(i)$. We now derive updates for these coefficients and show how they lead to a normalized lattice form.

Time-Update for the Normalized Reflection Coefficients

We start with the recursion for $\kappa_M^f(i)$ from Table 42.1:

$$\kappa_M^f(i) = \kappa_M^f(i-1) + \frac{\bar{\beta}_M^*(i)\bar{\gamma}_M(i)}{\zeta_M^{\bar{b}}(i)}\alpha_{M+1}(i)$$

which can be expressed in terms of *a posteriori* errors as follows:

$$\begin{aligned}
\kappa_M^f(i) &= \kappa_M^f(i-1) + \frac{\bar{\beta}_M^*(i)\bar{\gamma}_M(i)}{\zeta_M^{\bar{b}}(i)}[\alpha_M(i) - \kappa_M^f(i-1)\bar{\beta}_M(i)] \\
&= \kappa_M^f(i-1) + \frac{\bar{b}_M^*(i)}{\zeta_M^{\bar{b}}(i)}[\alpha_M(i) - \kappa_M^f(i-1)\bar{\beta}_M(i)] \\
&= \kappa_M^f(i-1) + \frac{\bar{b}_M^*(i)}{\bar{\gamma}_M(i)\zeta_M^{\bar{b}}(i)}[f_M(i) - \kappa_M^f(i-1)\bar{b}_M(i)] \\
&= \left(1 - \frac{|\bar{b}_M^*(i)|^2}{\bar{\gamma}_M(i)\zeta_M^{\bar{b}}(i)}\right)\kappa_M^f(i-1) + \frac{\bar{b}_M^*(i)f_M(i)}{\bar{\gamma}_M(i)\zeta_M^{\bar{b}}(i)} \\
&= \left(1 - |\bar{b}_M''(i)|^2\right)\kappa_M^f(i-1) + \frac{\bar{b}_M^*(i)f_M(i)}{\bar{\gamma}_M(i)\zeta_M^{\bar{b}}(i)}
\end{aligned}$$

Multiplying both sides of the above equality by the ratio $\zeta_M^{\bar{b}/2}(i)/\zeta_M^{f/2}(i)$, and using the definition of the normalized coefficient $\kappa_M^a(i)$, we obtain

$$\kappa_M^a(i) = \frac{\zeta_M^{\bar{b}/2}(i)}{\zeta_M^{f/2}(i)}\kappa_M^f(i-1)(1 - |\bar{b}_M''(i)|^2) + f_M''(i)\bar{b}_M''^*(i) \tag{42.17}$$

However, the modified minimum costs $\zeta_M^{\bar{b}}(i)$ and $\zeta_M^f(i)$ satisfy the time-update relations (cf. (40.72)):

$$\zeta_M^f(i) = \lambda\zeta_M^f(i-1) + |f_M(i)|^2/\bar{\gamma}(i), \quad \zeta_M^{\bar{b}}(i) = \lambda\zeta_M^{\bar{b}}(i-1) + |\bar{b}_M(i)|^2/\bar{\gamma}_M(i)$$

These updates can be re-expressed in terms of the normalized estimation errors $\{f_M''(i), \bar{b}_M''(i)\}$ Indeed, note that

$$\zeta_M^f(i) = \lambda\zeta_M^f(i-1) + |f_M''(i)|^2\zeta_M^f(i) \quad \text{so that} \quad \zeta_M^f(i) = \frac{\lambda\zeta_M^f(i-1)}{1 - |f_M''(i)|^2}$$

Likewise,

$$\zeta_M^{\bar{b}}(i) = \frac{\lambda\zeta_M^{\bar{b}}(i-1)}{1 - |\bar{b}_M''(i)|^2}$$

Now, by taking square-roots, we have

$$\zeta_M^{\bar{b}/2}(i) = \frac{\sqrt{\lambda}\zeta_M^{\bar{b}/2}(i-1)}{\sqrt{1 - |\bar{b}_M''(i)|^2}}, \qquad \zeta_M^{f/2}(i) = \frac{\sqrt{\lambda}\zeta_M^{f/2}(i-1)}{\sqrt{1 - |f_M''(i)|^2}}$$

and substituting these equations into (42.17), we arrive at a time-update recursion for the first reflection coefficient:

$$\boxed{\kappa_M^a(i) = \kappa_M^a(i-1)\sqrt{(1 - |\bar{b}_M''(i)|^2)(1 - |f_M''(i)|^2)} + f_M''(i)\bar{b}_M''^*(i)} \tag{42.18}$$

This recursion is in terms of the normalized errors $\{b_M''(i), f_M''(i)\}$. We thus need to determine order-updates for these errors as well, which we pursue in the next section.

In a similar manner, we can derive a time-update recursion for the second normalized reflection coefficient, $\kappa_M^c(i)$. We start from recursion (42.9) for $\kappa_M(i)$ (or from Table 42.1):

$$\begin{aligned}\kappa_M(i) &= \kappa_M(i-1) + \frac{b_M^*(i)}{\zeta_M^b(i)}[e_M(i) - \beta_M(i)\kappa_M(i-1)] \\ &= \kappa_M(i-1) + \frac{b_M^*(i)}{\zeta_M^b(i)\gamma_M(i)}[r_M(i) - b_M(i)\kappa_M(i-1)] \\ &= \left(1 - \frac{b_M^*(i)}{\zeta_M^b(i)\gamma_M(i)}\right)\kappa_M(i-1) + \frac{b_M^*(i)r_M(i)}{\zeta_M^b(i)\gamma_M(i)} \\ &= \left(1 - |b_M''(i)|^2\right)\kappa_M(i-1) + \frac{b_M^*(i)r_M(i)}{\zeta_M^b(i)\gamma_M(i)}\end{aligned}$$

Multiplying both sides of the above equality by the ratio $\zeta_M^{b/2}(i)/\zeta_M^{1/2}(i)$, and using the definition of the normalized coefficient $\kappa_M^c(i)$, we obtain

$$\kappa_M^c(i) = \frac{\zeta_M^{b/2}(i)}{\zeta_M^{1/2}(i)}\kappa_M(i-1)(1 - |b_M''(i)|^2) + r_M''(i)b_M''^*(i) \tag{42.19}$$

However, the quantities $\zeta_M^b(i)$ and $\zeta_M(i)$ satisfy the time-update relations (cf. (40.72)):

$$\zeta_M^b(i) \;=\; \lambda\zeta_M^b(i-1) \;+\; |b_M(i)|^2/\gamma(i), \qquad \zeta_M(i) = \lambda\zeta_M(i-1) \;+\; |r_M(i)|^2/\gamma_M(i)$$

These updates can be re-expressed in terms of the normalized estimation errors $\{b_M''(i), e_M''(i)\}$. Indeed, note that

$$\zeta_M^b(i) = \lambda\zeta_M^b(i-1) \;+\; |b_M''(i)|^2\zeta_M^b(i) \quad \text{so that} \quad \zeta_M^b(i) = \frac{\lambda\zeta_M^b(i-1)}{1 - |b_M''(i)|^2}$$

Likewise, $\zeta_M(i) = \lambda\zeta_M(i-1)/(1 - |r_M''(i)|^2)$. Taking square-roots, we have

$$\zeta_M^b(i) = \frac{\sqrt{\lambda}\zeta_M^{b/2}(i-1)}{\sqrt{1 - |b_M''(i)|^2}}, \qquad \zeta_M^{1/2}(i) = \frac{\sqrt{\lambda}\zeta_M^{1/2}(i-1)}{\sqrt{1 - |e_M''(i)|^2}} \tag{42.20}$$

Substituting these equations into (42.19), we arrive at a time-update recursion for the second reflection coefficient:

$$\boxed{\kappa_M^c(i) = \kappa_M^c(i-1)\sqrt{(1 - |b_M''(i)|^2)(1 - |r_M''(i)|^2)} \;+\; r_M''(i)b_M''^*(i)} \tag{42.21}$$

This recursion is again in terms of the normalized errors $\{e_M''(i), b_M''(i)\}$. We thus need to determine order-updates for these errors as well.

Order-Update for the Normalized Estimation Errors

We now derive order-updates for the normalized errors $\{f_M''(i),\ b_M''(i),\ \bar{b}_M''(i),\ r_M''(i)\}$. We start with the order-update for $b_{M+1}(i)$ from Table 40.1, namely, $b_{M+1}(i) = \bar{b}_M(i) - \kappa_M^b(i)f_M(i)$. Dividing both sides by $\zeta_{M+1}^{b/2}(i)\gamma_{M+1}^{1/2}(i)$, we obtain

$$b_{M+1}''(i) = \frac{\bar{b}_M(i) - \kappa_M^b(i)f_M(i)}{\zeta_{M+1}^{b/2}(i)\gamma_{M+1}^{1/2}(i)} \tag{42.22}$$

Now note that the quantities $\{\zeta^b_M(i), \gamma_M(i)\}$ satisfy the order-updates (cf. (40.72)):

$$\zeta^b_{M+1}(i) = \zeta^{\bar{b}}_M(i) - |\delta_M(i)|^2/\zeta^f_M(i), \quad \gamma_{M+1}(i) = \bar{\gamma}_M(i) - |f_M(i)|^2/\zeta^f_M(i)$$

Both of these updates can be rewritten in terms of the normalized reflection coefficient $\kappa^a_M(i)$:

$$\zeta^b_{M+1}(i) = \zeta^{\bar{b}}_M(i)(1-|\kappa^a_M(i)|^2), \quad \gamma_{M+1}(i) = \bar{\gamma}_M(i)(1-|f''_M(i)|^2)$$

Substituting these relations into (42.22), we obtain

$$\boxed{b''_{M+1}(i) = \frac{\bar{b}''_M(i) - \kappa^{a*}_M(i)f''_M(i)}{\sqrt{(1-|f''_M(i)|^2)(1-|\kappa^a_M(i)|^2)}}} \tag{42.23}$$

Similarly, using the order-updates for $f_{M+1}(i)$, $\zeta^f_M(i)$, and $\bar{\gamma}_M(i)$ from Table 40.1 we obtain

$$\boxed{f''_{M+1}(i) = \frac{f''_M(i) - \kappa^a_M(i)\bar{b}''_M(i)}{\sqrt{(1-|\bar{b}''_M(i)|^2)(1-|\kappa^a_M(i)|^2)}}} \tag{42.24}$$

Observe that the order-updates for $\{b''_M(i), f''_M(i)\}$ are in terms of a single reflection coefficient, namely $\kappa^a_M(i)$ and its conjugate. This is in contrast to the un-normalized lattice forms, where two separate reflection coefficients, $\{\kappa^b_M(i), \kappa^f_M(i)\}$, are needed for the order-updates of $\{b_M(i), f_M(i)\}$.

Likewise, using the order-update for $\zeta_M(i)$ and $\gamma_M(i)$ (cf. Table 40.1 and (40.72)), we can establish the following recursion:

$$\boxed{r''_{M+1}(i) = \frac{r''_M(i) - \kappa^c_M(i)b''_M(i)}{\sqrt{(1-|b''_M(i)|^2)(1-|\kappa^c_M(i)|^2)}}} \tag{42.25}$$

Although the normalized lattice filter returns the normalized residual $r''_{M+1}(i)$, the estimation error $r_{M+1}(i)$ can be recovered as follows. From the definition of $r''_{M+1}(i)$ we have

$$r_{M+1}(i) = \zeta^{1/2}_{M+1}(i)\gamma^{1/2}_{M+1}(i)r''_{M+1}(i)$$

which indicates that we need to evaluate the scaling factor $\sigma_{M+1}(i) = \zeta^{1/2}_{M+1}(i)\gamma^{1/2}_{M+1}(i)$. Now using the order-update for $\gamma_M(i)$ from Table 40.1, we have that

$$\boxed{\gamma_{M+1}(i) = \gamma_M(i)(1-b''_M(i)|^2)} \tag{42.26}$$

Likewise, using the order-update (40.72), we obtain $\zeta_{M+1}(i) = \zeta_M(i)\left(1-|\kappa^c_M(i)|^2\right)$. Combining the above relations we arrive at the following order-update for the scaling factor:

$$\boxed{\sigma_{M+1}(i) = \sigma_M(i)\sqrt{(1-|\kappa^c_M(i)|^2)(1-b''_M(i)|^2)}} \tag{42.27}$$

We should again indicate that all the relations derived so far for the normalized lattice filter hold independent of data structure. When the regressors have shift structure, we can relate $\{b''_M(i), \bar{b}''_M(i)\}$ via $\bar{b}''_M(i) = b''_M(i-1)$. In this case, the recursions collapse to the following.

Algorithm 42.3 (Normalized lattice filter) Consider the same setting of Alg. 42.2. For each $i \geq 0$, the $M-$th order *a posteriori* and *a priori* errors, $\{e_M(i), r_M(i)\}$ can be computed as follows:

1. Initialization. Set

$$\zeta_0^b(-1) = \eta^{-1}\lambda^{-2},\ \zeta_0(-1) = \eta^{-1}\lambda^{-2},\ b''_m(-1) = 0 \ \text{ for } m = 0, \ldots, M-1$$

2. For $i \geq 0$, repeat:

   - Set

$$\begin{aligned} \zeta_0^b(i) &= \lambda\zeta_0^b(i-1) + |u(i)|^2,\ \zeta_0(i) = \lambda\zeta_0(i-1) + |d(i)|^2,\ \sigma_0(i) = \zeta_0^{1/2}(i) \\ b''_0(i) &= f''_0(i) = u(i)/\zeta_0^{b/2}(i),\ r''_0(i) = d(i)/\zeta_0^{1/2}(i),\ \gamma_0(i) = 1 \end{aligned}$$

   - From $m = 0$ to $m = M-1$, repeat:

$$\begin{aligned} p_m^b(i) &= \sqrt{1 - |b''_m(i)|^2},\ p_m^f(i) = \sqrt{1 - |f''_m(i)|^2} \\ p_m(i) &= \sqrt{1 - |r''_m(i)|^2} \\ \kappa_m^a(i) &= \kappa_m^a(i-1)p_m^b(i-1)p_m^f(i) + f''_m(i)b''^*_m(i-1) \\ \kappa_m^c(i) &= \kappa_m^c(i-1)p_m^b(i)p_m(i) + b''^*_m(i)r''_m(i) \\ p_m^a(i) &= \sqrt{1 - |\kappa_m^a(i)|^2},\ p_m^c(i) = \sqrt{1 - |\kappa_m^c(i)|^2} \\ r''_{m+1}(i) &= \frac{1}{p_m^b(i)p_m^c(i)}\left(r''_m(i) - \kappa_m^c(i)b''_m(i)\right) \\ b''_{m+1}(i) &= \frac{1}{p_m^f(i)p_m^a(i)}\left(b''_m(i-1) - \kappa_m^{a*}(i)f''_m(i)\right) \\ f''_{m+1}(i) &= \frac{1}{p_m^b(i-1)p_m^a(i)}\left(f''_m(i) - \kappa_m^a(i)b''_m(i-1)\right) \\ \sigma_{m+1}(i) &= \sigma_m(i)\, p_m^c(i)p_m^b(i),\ \gamma_{m+1}(i) = \gamma_m(i)\ \left(p_m^b(i)\right)^2 \\ r_{m+1}(i) &= \sigma_{m+1}(i)r''_{m+1}(i) \\ e_{m+1}(i) &= r_{m+1}(i)/\gamma_{m+1}(i) \end{aligned}$$

CHAPTER 43

Array Lattice Filters

The interpretations we provided for the coefficients $\{\kappa_M(i), \kappa_M^f(i), \kappa_M^b(i)\}$ in Sec. 42.1, in terms of solutions to first-order least-squares problems, can be used to motivate yet another lattice implementation in array form. We discussed array methods and their advantages in some detail in Chapter 33. We show here that such array methods can also be developed for order-recursive problems.

Thus, recall that in Sec. 42.1 we introduced the angle-normalized estimation errors

$$\{b'_M(i),\ r'_M(i),\ f'_M(i),\ \bar{b}'_M(i)\}$$

and the corresponding angle-normalized error vectors

$$\{b'_{M,i},\ r'_{M,i},\ f'_{M,i},\ \bar{b}'_{M,i}\}$$

We then argued that the reflection coefficients $\{\kappa_M(i), \kappa_M^f(i), \kappa_M^b(i)\}$ can be interpreted as the solutions to three simple (regularized) projection problems, namely

$$\begin{cases} \kappa_M(i) \text{ projects } r'_{M,i} \text{ onto } b'_{M,i} \\ \kappa_M^f(i) \text{ projects } f'_{M,i} \text{ onto } \bar{b}'_{M,i} \\ \kappa_M^b(i) \text{ projects } \bar{b}'_{M,i} \text{ onto } f'_{M,i} \end{cases}$$

That is, each of these reflection coefficients solves the problem of projecting one angle-normalized error vector onto another. More specifically, they solve the following regularized least-squares problems:

$$\begin{cases} \min_{\kappa_M} \left[\bar{\eta}^{-1}\lambda^{i+1}|\kappa_M|^2 + \left\|r'_{M,i} - \kappa_M b'_{M,i}\right\|^2_{\Lambda_i} \right] \implies \kappa_M(i) \\ \min_{\kappa_M^f} \left[\bar{\eta}^{-1}\lambda^{i+1}|\kappa_M^f|^2 + \left\|f'_{M,i} - \kappa_M^f \bar{b}'_{M,i}\right\|^2_{\Lambda_i} \right] \implies \kappa_M^f(i) \\ \min_{\kappa_M^b} \left[\breve{\eta}^{-1}\lambda^{i+1}|\kappa_M^b|^2 + \left\|\bar{b}'_{M,i} - \kappa_M^b f'_{M,i}\right\|^2_{\Lambda_i} \right] \implies \kappa_M^b(i) \end{cases} \tag{43.1}$$

where

$$\bar{\eta} = \eta\lambda^{M+2} \quad \text{and} \quad \breve{\eta} = \eta\lambda^2$$

The above interpretations were used in Sec. 42.1 to show that the reflection coefficients $\{\kappa_M(i), \kappa_M^f(i), \kappa_M^b(i)\}$ can be time-updated by resorting to the RLS algorithm in each case.

Now in Chapter 33, we argued that least-squares solutions can also be updated in array form, e.g., by using the QR algorithm of Sec. 35.2. The QR method can therefore be used here to develop array methods for updating the reflection coefficients themselves.

43.1 ORDER-UPDATE OF OUTPUT ESTIMATION ERRORS

We start with the reflection coefficient $\kappa_M(i)$. Comparing the cost function for $\kappa_M(i)$ in (43.1) with the one that appears in the statement of the QR method in Alg. 35.2 we see that we can make the following identifications:

$$\left\{\begin{array}{lll} w \longleftarrow \kappa_M & \bar{w} \longleftarrow 0 & \Pi \longleftarrow \bar{\eta}^{-1} \\ d(i) \longleftarrow r'_M(i) & u_i \longleftarrow b'_M(i) & \end{array}\right.$$

and

$$\Phi_i \longleftarrow \bar{\eta}^{-1}\lambda^{i+1} + b'^*_{M,i}\Lambda_i b'_{M,i} = \bar{\eta}^{-1}\lambda^{i+1} + \xi^b_M(i) \stackrel{\Delta}{=} \zeta^b_M(i)$$

If we now write down the QR equations of Alg. 35.2 for these new variables, we arrive at the following statement. Define the normalized reflection coefficient

$$\boxed{q_M(i) \stackrel{\Delta}{=} \zeta_M^{b/2}(i)\kappa_M(i)} \tag{43.2}$$

Then start with $\zeta_M^{b/2}(-1) = \sqrt{\eta^{-1}\lambda^{-M-2}}$ and $q_M(-1) = 0$, and repeat for $i \geq 0$. At each iteration, find a 2×2 unitary matrix $\Theta_{M,i}$ that generates the zero entry in the post-array shown below, along with a leading positive entry in the first row and a positive entry s. The entries in the post-array would then correspond to:

$$\begin{bmatrix} \lambda^{1/2}\zeta_M^{b/2}(i-1) & b'^*_M(i) \\ \lambda^{1/2}q^*_M(i-1) & r'^*_M(i) \\ 0 & 1 \end{bmatrix} \Theta_{M,i} = \begin{bmatrix} \zeta_M^{b/2}(i) & 0 \\ q^*_M(i) & \mathsf{z} \\ b'_M(i)/\zeta_M^{b/2}(i) & \mathsf{s} \end{bmatrix} \tag{43.3}$$

where, as was the case with Alg. 35.2, the scalar quantities $\{\mathsf{s}, \mathsf{z}\}$ can be determined from the identities:

$$\mathsf{s}\mathsf{z}^* = r'_M(i) - b'_M(i)\kappa_M(i), \qquad |\mathsf{s}|^2 = 1 - \frac{|b'_M(i)|^2}{\zeta^b_M(i)} = \frac{\gamma_{M+1}(i)}{\gamma_M(i)}$$

The first identity follows by equating the inner products of the second and third lines of the arrays, while the second identity follows from equating the norms of the last lines of the arrays. It is easy to see that the first expression leads to

$$\begin{aligned} \mathsf{s}\mathsf{z}^* &= r'_M(i) - b'_M(i)\kappa_M(i) \\ &= \frac{1}{\gamma_M^{1/2}(i)}\left[r_M(i) - b_M(i)\kappa_M(i)\right] \\ &= \frac{1}{\gamma_M^{1/2}(i)}\, r_{M+1}(i) \\ &= \frac{\gamma_{M+1}^{1/2}(i)}{\gamma_M^{1/2}(i)}\, r'_{M+1}(i) \end{aligned}$$

whereas the second expression leads to $\mathsf{s} = \gamma_{M+1}^{1/2}(i)/\gamma_M^{1/2}(i)$ and, correspondingly, $\mathsf{z} = r'^*_{M+1}(i)$. The array algorithm then becomes

$$\begin{bmatrix} \lambda^{1/2}\zeta_M^{b/2}(i-1) & b'^*_M(i) \\ \lambda^{1/2}q_M^*(i-1) & r'^*_M(i) \\ 0 & 1 \end{bmatrix} \Theta_{M,i} = \begin{bmatrix} \zeta_M^{b/2}(i) & 0 \\ q_M^*(i) & r'^*_{M+1}(i) \\ b'_M(i)/\zeta_M^{b/2}(i) & \gamma_{M+1}^{1/2}(i)/\gamma_M^{1/2}(i) \end{bmatrix} \tag{43.4}$$

If we further multiply the last rows on both sides of (43.4) by $\gamma_M^{1/2}(i)$ we arrive at the array equation:

$$\boxed{\begin{bmatrix} \lambda^{1/2}\zeta_M^{b/2}(i-1) & b'^*_M(i) \\ \lambda^{1/2}q_M^*(i-1) & r'^*_M(i) \\ 0 & \gamma_M^{1/2}(i) \end{bmatrix} \Theta_{M,i} = \begin{bmatrix} \zeta_M^{b/2}(i) & 0 \\ q_M^*(i) & r'^*_{M+1}(i) \\ b_M(i)/\zeta_M^{b/2}(i) & \gamma_{M+1}^{1/2}(i) \end{bmatrix}} \tag{43.5}$$

This step tells us how to order-update the angle-normalized variable $r'_M(i)$. If desired, the reflection coefficient $\kappa_M(i)$ can be determined from the equality

$$\boxed{\kappa_M(i) = q_M(i)/\zeta_M^{b/2}(i)} \tag{43.6}$$

We now derive array methods for order-updating the angle-normalized variables $\{f'_M(i), b'_M(i)\}$ by applying similar arguments to the other cost functions in (43.1).

43.2 ORDER-UPDATE OF BACKWARD ESTIMATION ERRORS

Consider the reflection coefficient $\kappa_M^b(i)$. Comparing its cost function from (43.1) with the one that appears in the statement of the QR method in Alg. 35.2 we see that we can make the following identifications

$$\begin{cases} w \longleftarrow \kappa_M^b & \bar{w} \longleftarrow 0 & \Pi \longleftarrow \breve{\eta}^{-1} \\ d(i) \longleftarrow b'_M(i) & u_i \longleftarrow f'_M(i) & \end{cases}$$

and

$$\Phi_i \longleftarrow \breve{\eta}^{-1}\lambda^{i+1} + f'^*_{M,i}\Lambda_i f'_{M,i} = \breve{\eta}^{-1}\lambda^{i+1} + \xi_M^f(i) \triangleq \zeta_M^f(i)$$

If we now write down the QR equations of Alg. 35.2 for these new variables, we arrive at the following statement. Define the normalized reflection coefficient

$$\boxed{q_M^b(i) \triangleq \zeta_M^{f/2}(i)\kappa_M^b(i)} \tag{43.7}$$

Then start with $\zeta_M^{f/2}(-1) = \sqrt{\breve{\eta}^{-1}\lambda^{-2}}$ and $q_M^b(-1) = 0$, and repeat for $i \geq 0$. At each iteration, find a 2×2 unitary matrix $\Theta_{M,i}^b$ that generates the zero entry in the post-array below, along with a positive leading entry in the first row and a positive s. The entries in the post-array would then correspond to:

$$\begin{bmatrix} \lambda^{1/2}\zeta_M^{f/2}(i-1) & f'^*_M(i) \\ \lambda^{1/2}q_M^{b*}(i-1) & b'^*_M(i) \\ 0 & 1 \end{bmatrix} \Theta_{M,i}^b = \begin{bmatrix} \zeta_M^{f/2}(i) & 0 \\ q_M^{b*}(i) & \mathsf{z} \\ f'_M(i)/\zeta_M^{f/2}(i) & \mathsf{s} \end{bmatrix} \tag{43.8}$$

where, as was the case with Alg. 35.2, the scalar quantities $\{s, z\}$ can be determined from the identities:

$$sz^* = \bar{b}'_M(i) - f'_M(i)\kappa^b_M(i), \qquad |s|^2 = 1 - \frac{|f'_M(i)|^2}{\zeta^f_M(i)} = \frac{\gamma_{M+1}(i)}{\bar{\gamma}_M(i)}$$

The first identity follows by equating the inner products of the second and third lines of the arrays, while the second identity follows from equating the norms of the last lines of the arrays. It is easy to see that the first expression leads to

$$\begin{aligned} sz^* &= \bar{b}'_M(i) - f'_M(i)\kappa^b_M(i) \\ &= \frac{1}{\bar{\gamma}^{1/2}_M(i)}\left[\bar{b}_M(i) - f_M(i)\kappa^b_M(i)\right] \\ &= \frac{1}{\bar{\gamma}^{1/2}_M(i)}\, b_{M+1}(i) \\ &= \frac{\gamma^{1/2}_{M+1}(i)}{\bar{\gamma}^{1/2}_M(i)}\, b'_{M+1}(i) \end{aligned}$$

whereas the second expression gives $s = \gamma^{1/2}_{M+1}(i)/\bar{\gamma}^{1/2}_M(i)$. In this way, the array algorithm (43.8) becomes

$$\begin{bmatrix} \lambda^{1/2}\zeta^{f/2}_M(i-1) & f'^*_M(i) \\ \lambda^{1/2}q^{b*}_M(i-1) & \bar{b}'^*_M(i) \\ 0 & 1 \end{bmatrix} \Theta^b_{M,i} = \begin{bmatrix} \zeta^{f/2}_M(i) & 0 \\ q^{b*}_M(i) & b'^*_{M+1}(i) \\ f'_M(i)/\zeta^{f/2}_M(i) & \gamma^{1/2}_{M+1}(i)/\bar{\gamma}^{1/2}_M(i) \end{bmatrix} \tag{43.9}$$

If we further multiply the last rows on both sides of (43.9) by $\bar{\gamma}^{1/2}_M(i)$ we arrive at the array equation:

$$\boxed{\begin{bmatrix} \lambda^{1/2}\zeta^{f/2}_M(i-1) & f'^*_M(i) \\ \lambda^{1/2}q^{b*}_M(i-1) & \bar{b}'^*_M(i) \\ 0 & \bar{\gamma}^{1/2}_M(i) \end{bmatrix} \Theta^b_{M,i} = \begin{bmatrix} \zeta^{f/2}_M(i) & 0 \\ q^{b*}_M(i) & b'^*_{M+1}(i) \\ f_M(i)/\zeta^{f/2}_M(i) & \gamma^{1/2}_{M+1}(i) \end{bmatrix}} \tag{43.10}$$

This step tells us how to order-update the angle-normalized variable $b'_M(i)$. If desired, the reflection coefficient $\kappa^b_M(i)$ can be determined from the equality

$$\boxed{\kappa^b_M(i) = q^b_M(i)/\zeta^{f/2}_M(i)} \tag{43.11}$$

43.3 ORDER-UPDATE OF FORWARD ESTIMATION ERRORS

Finally, consider the reflection coefficient $\kappa^f_M(i)$. Comparing its cost function from (43.1) with the one that appears in the statement of the QR method in Alg. 35.2 we see that we can make the following identifications:

$$\begin{cases} w \longleftarrow \kappa^f_M & \bar{w} \longleftarrow 0 & \Pi \longleftarrow \bar{\eta}^{-1} \\ d(i) \longleftarrow f'_M(i) & u_i \longleftarrow \bar{b}'_M(i) & \end{cases}$$

and

$$\Phi_i \longleftarrow \bar{\eta}^{-1}\lambda^{i+1} + \bar{b}'^*_{M,i}\Lambda_i\bar{b}'_{M,i} = \bar{\eta}^{-1}\lambda^{i+1} + \xi^{\bar{b}}_M(i) \triangleq \zeta^{\bar{b}}_M(i)$$

If we now write down the QR equations of Alg. 35.2 for these new variables, we arrive at the following statement. Define the normalized reflection coefficient

$$\boxed{q_M^f(i) \stackrel{\Delta}{=} \zeta_M^{\bar{b}/2}(i)\kappa_M^f(i)} \tag{43.12}$$

Then start with $\zeta_M^{\bar{b}/2}(-1) = \sqrt{\eta^{-1}\lambda^{-M-2}}$ and $q_M^f(-1) = 0$, and repeat for $i \geq 0$. At each iteration find a 2×2 unitary matrix $\Theta_{M,i}^f$ that generates the zero entry in the post-array below, along with a leading positive entry in the first row and a positive s. The entries in the post-array would then correspond to:

$$\begin{bmatrix} \lambda^{1/2}\zeta_M^{\bar{b}/2}(i-1) & \bar{b}_M'^*(i) \\ \lambda^{1/2}q_M^{f*}(i-1) & f_M'^*(i) \\ 0 & 1 \end{bmatrix} \Theta_{M,i}^f = \begin{bmatrix} \zeta_M^{\bar{b}/2}(i) & 0 \\ q_M^{f*}(i) & \mathsf{z} \\ \bar{b}_M'(i)/\zeta_M^{\bar{b}/2}(i) & \mathsf{s} \end{bmatrix} \tag{43.13}$$

where, as was the case with Alg. 35.2, the scalar quantities $\{\mathsf{s}, \mathsf{z}\}$ can be determined from the identities:

$$\mathsf{s}\mathsf{z}^* = f_M'(i) - \bar{b}_M'(i)\kappa_M^f(i), \qquad |\mathsf{s}|^2 = 1 - \frac{|\bar{b}_M'(i)|^2}{\zeta_M^{\bar{b}}(i)} = \frac{\bar{\gamma}_{M+1}(i)}{\bar{\gamma}_M(i)}$$

The first identity follows by equating the inner products of the second and third lines of the arrays, while the second identity follows from equating the norms of the last lines of the arrays. It is easy to see that the first expression leads to

$$\begin{aligned} \mathsf{s}\mathsf{z}^* &= f_M'(i) - \bar{b}_M'(i)\kappa_M^f(i) \\ &= \frac{1}{\bar{\gamma}_M^{1/2}(i)}\left[f_M(i) - \bar{b}_M(i)\kappa_M^f(i)\right] \\ &= \frac{1}{\bar{\gamma}_M^{1/2}(i)} f_{M+1}(i) \\ &= \frac{\bar{\gamma}_{M+1}^{1/2}(i)}{\bar{\gamma}_M^{1/2}(i)} f_{M+1}'(i) \end{aligned}$$

whereas the second expression gives $\mathsf{s} = \bar{\gamma}_{M+1}^{1/2}(i)/\bar{\gamma}_M^{1/2}(i)$ and, consequently, $\mathsf{z} = f_{M+1}'(i)$. In this way, the array algorithm (43.13) becomes

$$\begin{bmatrix} \lambda^{1/2}\zeta_M^{\bar{b}/2}(i-1) & \bar{b}_M'^*(i) \\ \lambda^{1/2}q_M^{f*}(i-1) & f_M'^*(i) \\ 0 & 1 \end{bmatrix} \Theta_{M,i}^f = \begin{bmatrix} \zeta_M^{\bar{b}/2}(i) & 0 \\ q_M^{f*}(i) & f_{M+1}'^*(i) \\ \bar{b}_M'(i)/\zeta_M^{\bar{b}/2}(i) & \bar{\gamma}_{M+1}^{1/2}(i)/\bar{\gamma}_M^{1/2}(i) \end{bmatrix} \tag{43.14}$$

If we further multiply the last rows on both sides of (43.14) by $\bar{\gamma}_M^{1/2}(i)$ we arrive at the array equation:

$$\boxed{\begin{bmatrix} \lambda^{1/2}\zeta_M^{\bar{b}/2}(i-1) & \bar{b}_M'^*(i) \\ \lambda^{1/2}q_M^{f*}(i-1) & f_M'^*(i) \\ 0 & \bar{\gamma}_M^{1/2}(i) \end{bmatrix} \Theta_{M,i}^f = \begin{bmatrix} \zeta_M^{\bar{b}/2}(i) & 0 \\ q_M^{f*}(i) & f_{M+1}'^*(i) \\ \bar{b}_M(i)/\zeta_M^{\bar{b}/2}(i) & \bar{\gamma}_{M+1}^{1/2}(i) \end{bmatrix}} \tag{43.15}$$

This step tells us how to order-update the angle-normalized variable $f'_M(i)$. If desired, the reflection coefficient $\kappa^f_M(i)$ can be determined from the equality

$$\boxed{\kappa^f_M(i) = q^b_M(i)/\zeta^{\bar{b}/2}_M(i)} \qquad (43.16)$$

43.4 SIGNIFICANCE OF DATA STRUCTURE

As we already know, when the successive regressors have shift structure it holds that

$$\bar{b}_M(i) = b_M(i-1), \quad \bar{b}'_M(i) = b'_M(i-1), \quad \bar{\gamma}_M(i) = \gamma_M(i-1), \quad \zeta^{\bar{b}}_M(i) = \zeta^b_M(i-1)$$

and we are led to the array-based lattice algorithm, also known as the QRD-based lattice filter — see Fig. 43.1; the qualification "QRD-based" is used to indicate that the array recursions correspond to QR decompositions of the corresponding pre-arrays (recall the third remark following the statement of Alg. 35.1).

For comparison purposes, Table 43.1 lists the estimated computational cost per iteration for the various lattice filters derived in this chapter assuming real data. The costs are in terms of the number of multiplications, additions, divisions, and square-roots that are needed for each iteration. It is seen that lattice filters generally require $O(20M)$ operations per iteration.

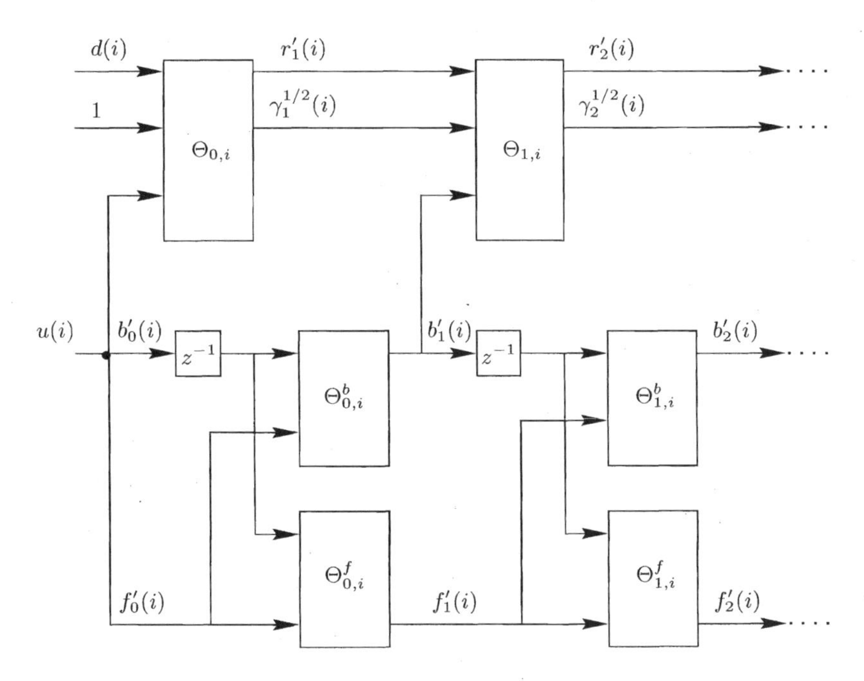

FIGURE 43.1 The QRD-based lattice filter.

Algorithm 43.1 (Array lattice filter) Consider again the same setting of Alg. 42.2. For each $i \geq 0$, the M-th order *a posteriori* estimation error, $r_M(i) = d(i) - u_{M,i}w_{M,i}$, that results from the solution of the regularized least-squares problem

$$\min_{w_M} \left[\lambda^{i+1} w_M^* \Pi_M w_M + \sum_{j=0}^{i} \lambda^{i-j} |d(j) - u_{M,j} w_M|^2 \right]$$

can be computed as follows:

1. Initialization. From $m = 0$ to $m = M-1$ set:

$$\zeta_m^{f/2}(-1) = \sqrt{\eta^{-1}\lambda^{-2}},\ \zeta_m^{b/2}(-1) = \sqrt{\eta^{-1}\lambda^{-m-2}},\ \zeta_m^{b/2}(-2) = \sqrt{\eta^{-1}\lambda^{-m-3}}$$
$$q_m(-1) = 0,\quad q_m^f(-1) = 0,\quad q_m^b(-1) = 0,\quad b_m(-1) = 0$$

2. For $i \geq 0$, repeat:

   - Set $\gamma_0^{1/2}(i) = 1$, $b_0'(i) = f_0'(i) = u(i)$, and $r_0'(i) = d(i)$
   - For $m = 0$ to $m = M-1$, apply 2×2 unitary rotations $\Theta_{m,i}^f$, $\Theta_{m,i}$, and $\Theta_{m,i}^b$, with positive (2,2) entries, in order to annihilate the $(1,2)$ entries of the post-arrays below:

$$\begin{bmatrix} \lambda^{1/2}\zeta_m^{b/2}(i-2) & b_m^{'*}(i-1) \\ \lambda^{1/2}q_m^{f*}(i-1) & f_m^{'*}(i) \end{bmatrix} \Theta_{m,i}^f = \begin{bmatrix} \zeta_m^{b/2}(i-1) & 0 \\ q_m^{f*}(i) & f_{m+1}^{'*}(i) \end{bmatrix}$$

$$\begin{bmatrix} \lambda^{1/2}\zeta_m^{b/2}(i-1) & b_m^{'*}(i) \\ \lambda^{1/2}q_m^{*}(i-1) & r_m^{'*}(i) \\ 0 & \gamma_m^{1/2}(i) \end{bmatrix} \Theta_{m,i} = \begin{bmatrix} \zeta_m^{b/2}(i) & 0 \\ q_m^{*}(i) & r_{m+1}^{'*}(i) \\ b_m(i)/\zeta_m^{b/2}(i) & \gamma_{m+1}^{1/2}(i) \end{bmatrix}$$

$$\begin{bmatrix} \lambda^{1/2}\zeta_m^{f/2}(i-1) & f_m^{'*}(i) \\ \lambda^{1/2}q_m^{b*}(i-1) & b_m^{'*}(i-1) \end{bmatrix} \Theta_{m,i}^b = \begin{bmatrix} \zeta_m^{f/2}(i) & 0 \\ q_m^{b*}(i) & b_{m+1}^{'*}(i) \end{bmatrix}$$

   and set $r_{m+1}(i) = r'_{m+1}(i)\gamma_{m+1}^{1/2}(i)$.

TABLE 43.1 Estimated computational cost per iteration for various lattice filters

Algorithm	$\times$	$+$	$\div$	$\sqrt{\cdot}$	Reference
A posteriori lattice	$16M$	$8M$	$8M$	-	Algorithm 41.1
A priori lattice	$16M$	$8M$	$8M$	-	Algorithm 41.2
A priori error-feedback lattice	$18M$	$8M$	$2M$	-	Algorithm 42.1
A posteriori error-feedback lattice	$16M$	$9M$	$7M$	-	Algorithm 42.2
Normalized lattice	$18M$	$5M$	$3M$	$5M$	Algorithm 42.3
QRD-based lattice	$23M$	$8M$	$6M$	$2M$	Algorithm 43.1
Givens-based lattice	$27M$	$8M$	$4M$	$2M$	Problem X.7

Summary and Notes

The chapters in this part describe several order-recursive (lattice) implementations of RLS.

SUMMARY OF MAIN RESULTS

1. The lattice forms are primarily concerned with order-updating the output estimation error and not the weight vector itself. To do so, forward and backward prediction errors also need to be order-updated.

2. All order-update relations derived in the chapter hold irrespective of data structure. As indicated in Fig. 41.1, the only place where the structure of the regressors is relevant is in knowing how to generate the error quantities $\bar{b}_M(i)$ from the error quantities $b_M(i)$: both these errors are related to backward projection problems.

3. When the regressors possess shift structure, it holds that $\bar{b}_M(i) = b_M(i-1)$. There are situations where the regressors are not shifted versions of each other and yet one can still relate $\{\bar{b}_M(i), b_M(i)\}$ — see, e.g., Chapter 16 of Sayed (2003) on Laguerre lattice filters and Merched and Sayed (2000b,2001a).

4. Seven lattice forms are described in the text: (a) *a posteriori* lattice form, (b) *a priori* lattice form, (c) *a priori* lattice form with error feedback, (d) *a posteriori* lattice form with error feedback, (e) normalized lattice form, (f) array lattice form, and (g) Givens-based lattice form (described in Prob. X.7).

5. The lattice forms with error feedback are based on time-updating the three reflection coefficients $\{\kappa_M(i), \kappa_M^f(i), \kappa_M^b(i)\}$. The standard *a posteriori* and *a priori* lattice forms are based on evaluating the reflection coefficients as ratios. The normalized lattice form, on the other hand, uses only two reflection coefficients. The array form uses three Givens rotations and updates angle-normalized errors. The Givens-based lattice is simply the array form with the equations spelled out.

6. The array lattice form seems to be the most reliable in finite precision along with the *a priori* form with error feedback. The *a posteriori* lattice form with error feedback seems to be the least reliable. A computer project at the end of this part illustrates these behaviors.

7. The derivations of lattice filters in this book account for regularization. Moreover, the derivation of the time-updates for the reflection coefficients exploits the useful observation that these coefficients can be interpreted as solutions to least-squares problems in their own right and, therefore, that RLS updates can be used to time-update them.

BIBLIOGRAPHIC NOTES

Regularization. In comparison to conventional derivations of lattice filters in the literature, we have incorporated regularization into our derivations.

Pre-windowed data. In this book we assume pre-windowed data, which is the case of most interest in practice. In pre-windowing, the input data $\{u(i), d(i)\}$ are assumed to be zero prior to filter operation, i.e., $u(i) = d(i) = 0$ for $i < 0$. Other forms of data windowing are possible (see, e.g., Honig and Messerschmitt (1984) and Alexander (1986)).

Lattice forms. The *a posteriori*-based lattice filter was developed by Morf and Lee (1978) and Lee, Morf, and Friedlander (1981). Further analysis and geometric derivations appear in Friedlander (1982) and Lev-Ari, Kailath, and Cioffi (1984). The idea of updating the reflection coefficients directly in order to improve the numerical properties of finite-word-length implementations, as in the *a priori* and *a posteriori* error-feedback lattice forms, is due to Ling, Manolakis, and Proakis (1985,1986).

Numerical issues. In Levin and Cowan (1994), the performance of several recursive least-squares algorithms are compared in a finite-precision environments. The results suggest that the array lattice form has superior numerical properties; a fact that agrees with the simulation results in the computer project at the end of this part.

Array derivation. The idea of using pre- and post-arrays to derive the array-based lattice filter of Chapter 43, also known as the QRD-lattice filter, is from Sayed and Kailath (1994b).

QR-based lattice. The idea of using the QR method as the backbone for deriving fast recursive-least-squares filters was put forward independently in the late 1980s by a number of authors including Cioffi (1988,1990), Bellanger (1988), Proudler, McWhirter, and Shepherd (1988,1989,1990), and Ling (1989,1991). The papers by Proudler, McWhirter, and Shepherd (1989,1990) and Ling (1989,1991) specifically derive lattice structures; the first work relies on the QR-decomposition while the second work uses the modified Gram-Schmidt orthogonalization procedure; both lattice algorithms are essentially the Givens-based lattice form of Prob. X.7; see also Proakis et al. (1992). The relation between fast QR methods and lattice filters was discussed by Regalia and Bellanger (1991). Related work on fast QR least-squares algorithms can also be found in Rontogiannis and Theodoridis (1998).

Prediction theory. There has been extensive work on lattice structures for least-mean-squares (as opposed to least-squares) estimation and prediction since the late 1940s. These earlier investigations have a rich history and they relate to contributions by Szegö (1939) on orthogonal polynomials and by Levinson (1947) and Durbin (1960) on the linear prediction of stationary time series. The reflection coefficients of the resulting lattice prediction filters are usually referred to as PARCOR (i.e., as partial correlation) coefficients, since they relate to the correlation between forward and backward prediction errors — recall Probs. II.31 and II.32. It is instructive to compare the recursions of Prob. II.31 for the least-mean-squares lattice filter with the recursions of the *a posteriori* lattice filter of Alg. 41.1; compare also with the normalized lattice filter of Alg. 42.3. One of the earliest applications of these lattice least-mean-squares (prediction) filters was in the context of speech analysis and synthesis by Itakura and Saito (1972).

State-space derivation. In Sec. 42.1 we showed that each one of the reflection coefficients, $\{\kappa_M(i), \kappa_M^b(i), \kappa_M^f(i)\}$, can be interpreted as performing the projection of a normalized error vector onto another normalized error vector. In other words, each one of these coefficients can be regarded as the solution to a first-order regularized least-squares problem. In this way, by invoking the equivalence that exists between RLS and Kalman filtering (cf. Sec. 31.2), we should be able to derive lattice filters by equivalently formulating state-space estimation problems of first-order for updating each of these reflection coefficients. This is indeed the case, as was shown by Sayed and Kailath (1994b). This point of view provides another convenient way of deriving lattice filters — see Probs. X.11–X.13.

Batch versus sequential processing. Assume the total number of data (i.e., samples) that is available for processing by a lattice filter is $N + 1$, say $\{d(j), u(j), 0 \leq j \leq N\}$. At each time instant i, the estimate $d_M(i)$ computed by the lattice implementation is an estimate of $d(i)$ of order

M and it is based on regression data up to and including time i. Specifically, $d_M(i) = u_{M,i}w_{M,i}$, where $u_{M,i}$ is the $M-$th order regressor at time i and $w_{M,i}$ is the weight-vector solution at time i as well. All lattice filters described in this part, as well as all other least-squares adaptive filters described in the previous chapters, operate in this *sequential* manner. Namely, the signal estimates that are available at a particular time instant are based on the regression data up to that same time instant. This fact is in contrast to a *batch* least-squares solution (cf. Chapter 29), whereby all $N+1$ regression vectors and reference data are used to generate the signal estimates. For instance, in a batch implementation, the estimate of $d(i)$ would be computed as $\widehat{d}(i) = u_{M,i}w_{M,N}$, where $w_{M,N}$ is the weight estimate that is based on *all* data (i.e., up to time N). There have been several investigations in the literature pertaining to the issue of how to approximate a batch solution from a collection of sequential solutions. Such investigations have been conducted most notably in the coding and prediction theory literature, e.g., by Rissanen (1984,1986), Ryabko (1984,1988), Wax (1988), and others. In later works by Merhav and Feder (1993,1998), Singer and Feder (1999), and Singer, Kozat, and Feder (2002), the authors examined the relation between batch and sequential processing in the context of adaptive filtering. For example, Singer and Feder (1999) showed how to select some weighting coefficients $\{\eta_m(i)\}$ and constructed a so-called universal predictor for $d(i)$, denoted by $d_u(i)$, by combining the sequential order-recursive estimates $\{d_m(i)\}$ that are generated by a lattice implementation as $d_u(i) = \sum_{m=1}^{M} \eta_m(i)d_m(i)$. The performance of this universal solution was then shown to approximate reasonably well that of the batch solution; specifically, the performance of both batch and universal solutions as measured in terms of the energies of the resulting residual vectors (over all data) were shown to be within $O(N^{-1}\ln N)$ of each other.

Problems and Computer Projects

PROBLEMS

Problem X.1 (Forward and backward residuals) Assume the regressors have shift structure and use the following updates for the conversion factor from Table 40.1,

$$\gamma_{M+1}(i) = \gamma_M(i-1) - |f_M(i)|^2/\zeta^f_M(i), \qquad \gamma_{M+1}(i) = \gamma_M(i) - |b_M(i)|^2/\zeta^b_M(i)$$

to establish the identity

$$\boxed{\frac{|f_{M+1}(i)|^2}{\zeta^f_{M+1}(i)} + \frac{|b_M(i-1)|^2}{\zeta^b_M(i-1)} = \frac{|f_M(i)|^2}{\zeta^f_M(i)} + \frac{|b_{M+1}(i)|^2}{\zeta^b_{M+1}(i)}}$$

Problem X.2 (Conversion factor) Show that $\gamma^{1/2}_{M+1}(i) = \left[1 - \sum_{j=0}^{M}\left(|b_j(i)|^2/\zeta^b_j(i)\right)\right]^{1/2}$.

Problem X.3 (Prediction vectors) Assume the regressors $\{u_{M,j}\}$ have shift structure. Use the discussion in Secs. 40.2–40.4, and the results of Lemmas 32.1 and 32.2, to argue that the weight prediction vectors $\{w^f_{M,i}, w^b_{M,i}\}$ satisfy the relations

$$\begin{aligned} w^b_{M+1,i} &= \begin{bmatrix} 0 \\ w^b_{M,i-1} \end{bmatrix} + \kappa^b_M(i)\begin{bmatrix} 1 \\ -w^f_{M,i} \end{bmatrix} \\ w^f_{M+1,i} &= \begin{bmatrix} w^f_{M,i} \\ 0 \end{bmatrix} + \kappa^f_M(i)\begin{bmatrix} -w^b_{M,i-1} \\ 1 \end{bmatrix} \end{aligned}$$

Problem X.4 (Prediction errors) Introduce the shift operator $q^{-1}u(i) = u(i-1)$. Applying an operator $A(q^{-1}) = \sum_{k=0}^{M} a_k(i)q^{-k}$ to a sequence $u(i)$ results in $A(q^{-1})u(i) = \sum_{k=0}^{M} a_k(i)u(i-k)$. Show that the forward and backward prediction errors of order $M+1$ can be generated as follows:

$$\begin{bmatrix} b_{M+1}(i) \\ f_{M+1}(i) \end{bmatrix} = \left(\prod_{m=0}^{M}\begin{bmatrix} q^{-1} & -\kappa^b_m(i) \\ -q^{-1}\kappa^f_m(i) & 1 \end{bmatrix}\right)\begin{bmatrix} u(i) \\ u(i) \end{bmatrix}$$

Problem X.5 (Reflection coefficients) Starting from recursion (42.8), and using the time-update (42.4) for $\zeta^b_M(i)$, show that, when the regressors have shift structure,

$$\kappa_M(i) = \frac{\lambda\zeta^b_M(i-1)}{\zeta^b_M(i)}\kappa_M(i-1) + \frac{b'^*_M(i)}{\zeta^b_M(i)}r'_M(i)$$

Likewise, show that

$$\begin{aligned} \kappa^f_M(i) &= \frac{\lambda\zeta^b_M(i-2)}{\zeta^b_M(i-1)}\kappa^f_M(i-1) + \frac{b'^*_M(i-1)}{\zeta^b_M(i-1)}f'_M(i) \\ \kappa^b_M(i) &= \frac{\lambda\zeta^f_M(i-1)}{\zeta^f_M(i)}\kappa^b_M(i-1) + \frac{f'^*_M(i)}{\zeta^f_M(i)}b'_M(i-1) \end{aligned}$$

Problem X.6 (Givens rotations) Refer to the statement of the QRD-based lattice filter in Alg. 43.1. The unitary matrix $\Theta_{m,i}$ is a Givens rotation of the form

$$\Theta_{m,i} = \frac{1}{\sqrt{1+|\rho|^2}}\begin{bmatrix} 1 & -\rho \\ \rho^* & 1 \end{bmatrix} \quad \text{where} \quad \rho = \frac{b'_m(i)}{\sqrt{\lambda\zeta^b_m(i-1)}}$$

Verify that the entries of $\Theta_{m,i}$ and $\Theta^b_{m,i}$ can be identified as

$$\begin{aligned}
\Theta_{m,i} &= \begin{bmatrix} \sqrt{\lambda\zeta^b_m(i-1)/\zeta^b_m(i)} & -b'_m(i)/\sqrt{\zeta^b_m(i)} \\ b'^*_m(i)/\sqrt{\zeta^b_m(i)} & \sqrt{\lambda\zeta^b_m(i-1)/\zeta^b_m(i)} \end{bmatrix} \triangleq \begin{bmatrix} c^b_m(i) & -s^b_m(i) \\ s^{b*}_m(i) & c^b_m(i) \end{bmatrix} \\
\Theta^b_{m,i} &= \begin{bmatrix} \sqrt{\lambda\zeta^f_m(i-1)/\zeta^f_m(i)} & -f'_m(i)/\sqrt{\zeta^f_m(i)} \\ f'^*_m(i)/\sqrt{\zeta^f_m(i)} & \sqrt{\lambda\zeta^f_m(i-1)/\zeta^f_m(i)} \end{bmatrix} \triangleq \begin{bmatrix} c^f_m(i) & -s^f_m(i) \\ s^{f*}_m(i) & c^f_m(i) \end{bmatrix}
\end{aligned}$$

Problem X.7 (Givens-based lattice) Use the identifications of Prob. X.6 to deduce from the QRD-based lattice filter of Alg. 43.1 the following equivalent form, which propagates angle-normalized errors:

1. Initialization. From $m = 0$ to $m = M - 1$ set:

$$\zeta^{f/2}_m(-1) = \sqrt{\eta^{-1}\lambda^{-2}}, \quad \zeta^{b/2}_m(-1) = \sqrt{\eta^{-1}\lambda^{-m-2}}, \quad q_m(-1) = q^f_m(-1) = q^b_m(-1) = 0$$
$$c^b_m(-1) = 1, \quad s^b_m(-1) = 0, \quad b'_m(-1) = 0$$

2. For $i \geq 0$, repeat:

   - Set $\gamma_0(i) = 1$, $b'_0(i) = f'_0(i) = u(i)$, and $r'_0(i) = d(i)$
   - For $m = 0$ to $m = M - 1$ do:

$$\boxed{\begin{aligned}
\zeta^{f/2}_m(i) &= \left[\lambda\left(\zeta^{f/2}_m(i-1)\right)^2 + |f'_m(i)|^2\right]^{1/2} \\
\zeta^{b/2}_m(i) &= \left[\lambda\left(\zeta^{b/2}_m(i-1)\right)^2 + |b'_m(i)|^2\right]^{1/2} \\
c^b_m(i) &= \lambda^{1/2}\zeta^{b/2}_m(i-1)/\zeta^{b/2}_m(i) \\
s^b_m(i) &= b'_m(i)/\zeta^{b/2}_m(i) \\
c^f_m(i) &= \lambda^{1/2}\zeta^{f/2}_m(i-1)/\zeta^{f/2}_m(i) \\
s^f_m(i) &= f'_m(i)/\zeta^{f/2}_m(i) \\
f'_{m+1}(i) &= c^b_m(i-1)f'_m(i) \; - \; \lambda^{1/2}s^{b*}_m(i-1)q^f_m(i-1) \\
q^f_m(i) &= \lambda^{1/2}c^b_m(i-1)q^f_m(i-1) \; + \; s^b_m(i-1)f'_m(i) \\
b'_{m+1}(i) &= c^f_m(i)b'_m(i-1) \; - \; \lambda^{1/2}s^{f*}_m(i)q^b_m(i-1) \\
q^b_m(i) &= \lambda^{1/2}c^f_m(i)q^b_m(i-1) \; + \; s^f_m(i)b'_m(i-1) \\
\\
r'_{m+1}(i) &= c^b_m(i)r'_m(i) \; - \; \lambda^{1/2}s^{b*}_m(i)q_m(i-1) \\
q_m(i) &= \lambda^{1/2}c^b_m(i)q_m(i-1) \; + \; s^b_m(i)r'_m(i) \\
\gamma^{1/2}_{m+1}(i) &= \gamma^{1/2}_m(i)c^b_m(i) \\
r_{m+1}(i) &= r'_{m+1}(i)\gamma^{1/2}_{m+1}(i)
\end{aligned}}$$

Remark. As can be seen, the Givens-based lattice filter simply amounts to identifying explicitly the parameters of the rotation matrices, and subsequently expanding the equations of the QRD-based lattice filter.

Problem X.8 (Minimum cost updates) In the Givens-based filter of Prob. X.7, explain why it is more convenient to update the quantities $\{\zeta^{f/2}_m(i), \zeta^{b/2}_m(i)\}$ as indicated in that problem rather than use the following updates, which follow from the QRD-based implementation (cf. Alg. 43.1):

$$\begin{aligned}
\zeta^{f/2}_m(i) &= \lambda^{1/2}c^f_m(i)\zeta^{f/2}_m(i-1) + s^{f*}_m(i)f'_m(i) \\
\zeta^{b/2}_m(i) &= \lambda^{1/2}c^b_m(i)\zeta^{b/2}_m(i-1) + s^{b*}_m(i)b'_m(i)
\end{aligned}$$

Problem X.9 (Relation between RLS and lattice filters) The purpose of this problem is to establish a relation between lattice filters and the standard RLS solution (40.12). It turns out that there is a relation between the weight vector $w_{M,i}$ and the backward prediction vectors $\{w_{j,i}^b\}$ of increasing orders $j = 0, 1, \ldots, M-1$. Iterate recursion (40.67) starting from the initial condition $w_{0,i} = 0$ and up to $j = M-1$, to establish that

$$\boxed{w_{M,i} = \begin{bmatrix} 1 & \times & \times & \ldots & \times \\ & 1 & \times & \ldots & \times \\ & & 1 & & \times \\ & & & \ddots & \vdots \\ & & & & 1 \end{bmatrix} \begin{bmatrix} \kappa_0(i) \\ \kappa_1(i) \\ \kappa_2(i) \\ \vdots \\ \kappa_{M-1}(i) \end{bmatrix}}$$

where the columns of the upper-triangular matrix are defined in terms of the backward prediction vectors

$$\begin{bmatrix} -w_{j,i}^b \\ 1 \end{bmatrix}, \quad j = 0, 1, \ldots, M-1$$

Remark. This result shows that $w_{M,i}$ and the vector of reflection coefficients $\{\kappa_j(i)\}$ at time i determine each other uniquely.

Problem X.10 (Backward prediction errors) Show that the backward prediction errors of successive orders (the ones appearing in Fig. 41.2) are related to the input sequence via the relation:

$$\begin{bmatrix} b_0(i) & b_1(i) & \ldots & b_{M-1}(i) \end{bmatrix} = u_{M,i}\, U_{M,i}$$

where $U_{M,i}$ is the same upper triangular matrix appearing in Prob. X.9.

Problem X.11 (State-space formulation of lattice filters) In Sec. 42.1 we explained that the reflection coefficients $\{\kappa_M(i), \kappa_M^f(i), \kappa_M^b(i)\}$ can be interpreted as solutions to first-order least-squares problems. Specifically, we introduced the angle-normalized vectors $\{b'_{M,i}, r'_{M,i}, f'_{M,i}, \bar{b}'_{M,i}\}$ and showed that $\{\kappa_M(i), \kappa_M^f(i), \kappa_M^b(i)\}$ are the solutions to the following three (regularized) projection problems:

$$\begin{array}{lcl}
\min_{\kappa_M} \left[\bar{\eta}^{-1}\lambda^{i+1}|\kappa_M|^2 + \left\|r'_{M,i} - \kappa_M b'_{M,i}\right\|^2_{\Lambda_i} \right] & \Longrightarrow & \kappa_M(i) \\
\min_{\kappa_M^f} \left[\bar{\eta}^{-1}\lambda^{i+1}|\kappa_M^f|^2 + \left\|f'_{M,i} - \kappa_M^f \bar{b}'_{M,i}\right\|^2_{\Lambda_i} \right] & \Longrightarrow & \kappa_M^f(i) \\
\min_{\kappa_M^b} \left[\breve{\eta}^{-1}\lambda^{i+1}|\kappa_M^b|^2 + \left\|\bar{b}'_{M,i} - \kappa_M^b f'_{M,i}\right\|^2_{\Lambda_i} \right] & \Longrightarrow & \kappa_M^b(i)
\end{array}$$

where $\bar{\eta} = \eta\lambda^{M+2}$ and $\breve{\eta} = \eta\lambda^2$. That is,

$$\begin{array}{l}
\kappa_M(i) \text{ projects } r'_{M,i} \text{ onto } b'_{M,i} \\
\kappa_M^f(i) \text{ projects } f'_{M,i} \text{ onto } \bar{b}'_{M,i} \\
\kappa_M^b(i) \text{ projects } \bar{b}'_{M,i} \text{ onto } f'_{M,i}
\end{array}$$

These interpretations were used in Sec. 42.1 and Chapter 43 to derive different forms of lattice filters by solving first-order least-squares problems.

In this problem, as well as in Probs. X.12–X.13, we shall derive the same filters by appealing instead to the equivalence relation that exists between RLS and Kalman filtering. Thus recall from the discussion in App. 31.2 that we showed that the exponentially-weighted RLS algorithm can be obtained via equivalence to the Kalman filter as follows. We start from the state-space model (31.8) and write down the Kalman recursions for estimating its state variable. Then we translate the Kalman variables into the RLS variables by using the correspondences summarized in Table 31.2. In Probs. X.11–X.13, we want to use this equivalence construction in order to re-derive lattice filters. As a result, we shall clarify the connections between lattice filters and Kalman filtering.

(a) Let us first consider the problem of projecting $f'_{M,i}$ onto $\bar{b}'_{M,i}$ in order to evaluate $\kappa_M^f(i)$. Use the equivalence between RLS and Kalman filtering from App. 31.2 to argue that this problem can be solved by introducing the following one-dimensional state-space model:

$$\boxed{\boldsymbol{x}^f(i+1) = \lambda^{-1/2}\boldsymbol{x}^f(i), \quad \boldsymbol{y}^f(i) = \bar{b}'_M(i)\boldsymbol{x}^f(i) + \boldsymbol{v}^f(i)}$$

with $\mathsf{E}\,|\boldsymbol{x}^f(0)|^2 = \lambda^{-1}\bar{\eta}$ and $\mathsf{E}\,\boldsymbol{v}^f(i)\boldsymbol{v}^{f*}(j) = \delta_{ij}$, and where the variables $\{\boldsymbol{x}^f(i), \boldsymbol{y}^f(i)\}$ are identified with the least-squares variables $\{\kappa_M^f, f'_M(i)\}$ as follows:

$$\boxed{\boldsymbol{x}^f(i) \longleftrightarrow \frac{\kappa_M^f}{(\sqrt{\lambda})^i}, \quad \boldsymbol{y}^f(i) \longleftrightarrow \frac{f'_M(i)}{(\sqrt{\lambda})^i}}$$

(b) Consider next the problem of projecting $\bar{b}'_{M,i}$ onto $f'_{M,i}$ in order to evaluate $\kappa_M^b(i)$. Argue that this problem can be solved by introducing the following one-dimensional state-space model:

$$\boxed{\boldsymbol{x}^b(i+1) = \lambda^{-1/2}\boldsymbol{x}^b(i), \quad \boldsymbol{y}^b(i) = f'_M(i)\boldsymbol{x}^b(i) + \boldsymbol{v}^b(i)}$$

with $\mathsf{E}\,|\boldsymbol{x}^b(0)|^2 = \lambda^{-1}\breve{\eta}$ and $\mathsf{E}\,\boldsymbol{v}^b(i)\boldsymbol{v}^{b*}(j) = \delta_{ij}$, and where the variables $\{\boldsymbol{x}^b(i), \boldsymbol{y}^b(i)\}$ are identified with the least-squares variables $\{\kappa_M^b, \bar{b}'_M(i)\}$ as follows:

$$\boxed{\boldsymbol{x}^b(i) \longleftrightarrow \frac{\kappa_M^b}{(\sqrt{\lambda})^i}, \quad \boldsymbol{y}^b(i) \longleftrightarrow \frac{\bar{b}'_M(i)}{(\sqrt{\lambda})^i}}$$

(c) Consider now the problem of projecting $r'_{M,i}$ onto $b'_{M,i}$ in order to evaluate $\kappa_M(i)$. Argue that this problem can be solved by introducing the following one-dimensional state-space model:

$$\boxed{\boldsymbol{x}(i+1) = \lambda^{-1/2}\boldsymbol{x}(i), \quad \boldsymbol{y}(i) = b'_M(i)\boldsymbol{x}(i) + \boldsymbol{v}(i)}$$

with $\mathsf{E}\,|\boldsymbol{x}(0)|^2 = \lambda^{-1}\bar{\eta}$ and $\mathsf{E}\,\boldsymbol{v}(i)\boldsymbol{v}^*(j) = \delta_{ij}$, and where the state-space variables $\{\boldsymbol{x}(i), \boldsymbol{y}(i)\}$ are identified with the least-squares variables $\{\kappa_M, r'_M(i)\}$ as follows:

$$\boxed{\boldsymbol{x}(i) \longleftrightarrow \frac{\kappa_M}{(\sqrt{\lambda})^i}, \quad \boldsymbol{y}(i) \longleftrightarrow \frac{r'_M(i)}{(\sqrt{\lambda})^i}}$$

Remark. For further discussion on the connections between lattice filters and Kalman filters, see the article by Sayed and Kailath (1994b).

Problem X.12 (Array lattice and Kalman filtering) Continue with the same setting as in Prob. X.11.

(a) Consider the state-space model of part (a) of that problem, which relates to projecting $f'_{M,i}$ onto $\bar{b}'_{M,i}$. Use (35.36) to verify that the information array form of the corresponding Kalman filter is given by:

$$\begin{bmatrix} \lambda^{1/2}\phi^{\bar{b}/2}(i|i-1) & \lambda^{1/2}\bar{b}'^*_M(i) \\ \hat{\boldsymbol{x}}^{f*}(i|i-1)\phi^{\bar{b}/2}(i|i-1) & \boldsymbol{y}^{f*}(i) \\ 0 & 1 \end{bmatrix} \Theta^f_{M,i} = \begin{bmatrix} \phi^{\bar{b}/2}(i+1|i) & 0 \\ \hat{\boldsymbol{x}}^{f*}(i+1|i)\phi^{f/2}(i+1|i) & \boldsymbol{\nu}^{f*}(i)r_e^{-\bar{b}/2}(i) \\ k_p^{\bar{b}*}(i)\phi^{\bar{b}/2}(i+1|i) & r_e^{-\bar{b}/2}(i) \end{bmatrix}$$

where $\Theta^f_{M,i}$ is any unitary rotation that introduces the zeros in the post-array and where we are using the notations $\{\phi^{\bar{b}}(i), k_p^{\bar{b}}(i), \boldsymbol{\nu}^f(i), r_e^{\bar{b}}(i)\}$ to denote the variables $\{\Phi_i, K_{p,i}, \boldsymbol{\nu}_i, R_{e,i}\}$ that appear in the general formulation (35.36). Moreover, $\boldsymbol{\nu}^f(i) = \boldsymbol{y}^f(i) - \bar{b}'_M(i)\hat{\boldsymbol{x}}^f(i|i-1)$. Use the correspondences from Table 31.2 to verify that the above array equations lead to the RLS array equations (43.14) for updating the forward estimation error.

(b) Consider now the state-space model of part (b) of Prob. X.11, which relates to the problem of projecting $\bar{b}'_{M,i}$ onto $f'_{M,i}$. Verify that the corresponding information array form is given by:

$$\begin{bmatrix} \lambda^{1/2}\phi^{f/2}(i|i-1) & \lambda^{1/2}f'^*_M(i) \\ \hat{\boldsymbol{x}}^{b*}(i|i-1)\phi^{f/2}(i|i-1) & \boldsymbol{y}^{b*}(i) \\ 0 & 1 \end{bmatrix} \Theta^b_{M,i} = \begin{bmatrix} \phi^{f/2}(i+1|i) & 0 \\ \hat{\boldsymbol{x}}^{b*}(i+1|i)\phi^{f/2}(i+1|i) & \boldsymbol{\nu}^{b*}(i)r_e^{-f/2}(i) \\ k_p^{f*}(i)\phi^{f/2}(i+1|i) & r_e^{-f/2}(i) \end{bmatrix}$$

where $\Theta^b_{M,i}$ is any unitary rotation that introduces the zeros in the post-array, and where $\boldsymbol{\nu}^b(i) = \boldsymbol{y}^b(i) - f'_M(i)\hat{\boldsymbol{x}}^b(i|i-1)$. Use the correspondences from Table 31.2 to verify that the above array equations lead to the RLS array equations (43.9) for updating the backward estimation error.

(c) Finally, consider the state-space model of part (c) of Prob. X.11, which relates to the problem of projecting $r'_{M,i}$ onto $b'_{M,i}$. Verify that the corresponding information array form is given by:

$$\begin{bmatrix} \lambda^{1/2}\phi^{b/2}(i|i-1) & \lambda^{1/2}b'^*_M(i) \\ \hat{\boldsymbol{x}}^*(i|i-1)\phi^{b/2}(i|i-1) & \boldsymbol{y}^*(i) \\ 0 & 1 \end{bmatrix} \Theta_{M,i} = \begin{bmatrix} \phi^{b/2}(i+1|i) & 0 \\ \hat{\boldsymbol{x}}^*(i+1|i)\phi^{b/2}(i+1|i) & \boldsymbol{\nu}^*(i)r_e^{-1/2}(i) \\ k_p^{b*}(i)\phi^{b/2}(i+1|i) & r_e^{-1/2}(i) \end{bmatrix}$$

where $\Theta_{M,i}$ is any unitary rotation that introduces the zeros in the post-array, and where $\boldsymbol{\nu}(i) = \boldsymbol{y}(i) - b'_M(i)\hat{\boldsymbol{x}}(i|i-1)$. Use the correspondences from Table 31.2 to verify that the above array equations lead to the RLS array equations (43.4) for updating the output estimation error.

Remark. In other words, each of the array equations in the QR-based lattice algorithm is a special case of the array information form of the Kalman filter (Sayed and Kailath (1994b)).

Problem X.13 (Error feedback and Kalman filtering) Continue with the same setting as in Prob. X.11.

(a) Consider the state-space model of part (a) of that problem, which relates to projecting $f'_{M,i}$ onto $\bar{b}'_{M,i}$. Verify that the corresponding Kalman recursion for the state estimator is given by:

$$\hat{\boldsymbol{x}}^f(i+1|i) = \lambda^{-1/2}\hat{\boldsymbol{x}}^f(i|i-1) + \lambda^{1/2}p^{\bar{b}}(i+1|i)\bar{b}'^*_M(i)[\boldsymbol{y}^f(i) - \bar{b}'_M(i)\hat{\boldsymbol{x}}^f(i|i-1)]$$

where the inverse of the Riccati variable $p^{\bar{b}}(i)$ satisfies

$$p^{-\bar{b}}(i+1|i) = \lambda p^{-\bar{b}}(i|i-1) + |\bar{b}'_M(i)|^2, \quad p^{-\bar{b}}(0|-1) = \lambda\bar{\eta}^{-1}$$

Now use the correspondences from Table 31.2 between RLS and Kalman filtering variables, as well as the relation between $\{\bar{b}_M(i), \bar{b}_M(i)\}$, to verify that the above state estimator equation leads to the following update for the reflection coefficient from Table 42.1:

$$\kappa^f_M(i) = \kappa^f_M(i-1) + \bar{\beta}^*_M(i)\bar{\gamma}_M(i)\alpha_{M+1}(i)/\zeta^{\bar{b}}_M(i)$$

(b) Consider now the state-space model of part (b) of Prob. X.11, which relates to projecting $\bar{b}'_{M,i}$ onto $f'_{M,i}$. Verify that the corresponding Kalman recursion for the state estimator is given by:

$$\hat{\boldsymbol{x}}^b(i+1|i) = \lambda^{-1/2}\hat{\boldsymbol{x}}^b(i|i-1) + \lambda^{1/2}p^f(i+1|i)f'^*_M(i)[\boldsymbol{y}^b(i) - f'_M(i)\hat{\boldsymbol{x}}^b(i|i-1)]$$

where the inverse of the Riccati variable $p^f(i)$ satisfies the recursion

$$p^{-f}(i+1|i) = \lambda p^{-f}(i|i-1) + |\bar{f}'_M(i)|^2, \quad p^{-f}(0|-1) = \lambda\breve{\eta}^{-1}$$

Now use the correspondences from Table 40.1 between RLS and Kalman filtering variables, as well as the relation between $\{f'_M(i), f_M(i)\}$, to verify that the above state estimator equation leads to the following update for the reflection coefficient from Table 42.1:

$$\kappa^b_M(i) = \kappa^b_M(i-1) + \alpha^*_M(i)\bar{\gamma}_M(i)\beta_{M+1}(i)/\zeta^f_M(i)$$

(c) Finally, consider the state-space model of part (c) of Prob. X.11, which relates to projecting $r'_{M,i}$ onto $b'_{M,i}$. Verify that the corresponding Kalman recursion for the state estimator is given by:

$$\hat{\boldsymbol{x}}(i+1|i) = \lambda^{-1/2}\hat{\boldsymbol{x}}(i|i-1) + \lambda^{1/2}p(i+1|i)b'^*_M(i)[\boldsymbol{y}(i) - b'_M(i)\hat{\boldsymbol{x}}(i|i-1)]$$

where the inverse of the Riccati variable $p(i)$ satisfies the recursion

$$p^{-1}(i+1|i) = \lambda p^{-1}(i|i-1) + |b'_M(i)|^2, \quad p^{-1}(0|-1) = \lambda\bar{\eta}^{-1}$$

Now use the correspondences from Table 40.1 between RLS and Kalman filtering variables, as well as the relation between $\{b'_M(i), b_M(i)\}$, to verify that the above state estimator equation leads to the following update for the reflection coefficient from Table 42.1:

$$\kappa_M(i) = \kappa_M(i-1) \; + \; \beta^*_M(i)\gamma_M(i)e_{M+1}(i)/\zeta^b_M(i)$$

Remark. In other words, each of the update equations for the reflection coefficients in error-feedback lattice forms is simply a special case of the prediction equation of the Kalman filter (cf. Sayed and Kailath (1994b)).

COMPUTER PROJECT

Project X.1 (Performance of lattice filters in finite precision) Although equivalent from a theoretical point of view, the performance of the varied lattice filters differ under finite-precision conditions. The purpose of this computer project is to illustrate these differences, as well as illustrate the recovery mechanism of some of the filters during the occurrence of impulsive interferences.

(a) Generate 10 random coefficients of a channel and normalize its energy to unity. Feed unit-variance Gaussian input data through the channel and add Gaussian noise to its output. Set the noise power at 30 dB below the input signal power. Choose $\lambda = 0.999$ and $\eta = 10^6$ and train the following lattice filters using the input sequence of the channel as input to the lattice implementations and the noisy output of the channel as the reference sequence: 1. *A posteriori* lattice form; 2. *a priori* lattice form; 3. *a priori* lattice form with error feedback; 4. *a posteriori* lattice form with error feedback; 5. normalized lattice form and 6. array lattice form. Assume in your simulations a finite-precision implementation with B bits for signals including the sign bit; use the routine quantize.m from Computer Project IX.1. For each algorithm, generate an ensemble-average learning curve by averaging over 50 experiments of duration $N = 200$ iterations each for the following choices: 1. $B = 35$ bits; 2. $B = 25$ bits; 3. $B = 20$ bits; 4. $B = 16$ bits and 5. $B = 10$ bits. Which lattice forms appear to be most reliable in finite precision?

(b) For this part, assume first a floating-point implementation. Introduce an impulsive interference of unit amplitude to the input sequence at time instant $i = 200$. Generate ensemble-average learning curves for the lattice filters over $N = 500$ iterations and observe whether they recover from the impulsive disturbance.

(c) Repeat the simulations of part (b) in finite precision using $B = 20$ bits and $B = 10$ bits.

PART XI

ROBUST FILTERS

CHAPTER 44

Indefinite Least-Squares

We end our treatment of adaptive filtering in this book by studying the robustness of adaptive filters in the presence of disturbances and uncertainties in the data. A study of this kind requires that we first define what we mean by robustness. For our purposes, and in loose terms, a robust filter will be one for which small disturbances in the data do not degrade the performance of the filter appreciably. The measure of smallness and largeness of a signal will be chosen as its energy, so that a robust filter will be one such that disturbances with small energy cannot lead to estimation errors with large energy and, more generally, the estimation error energy will remain bounded as long as the disturbance energy is bounded.

There are of course other characterizations of robustness. The one described above lends itself to analysis and mathematical manipulations. In particular, its characterization will involve studying quadratic cost functions that share many of the characteristics of regularized least-squares, except for the appearance of indefinite weighting matrices (as opposed to positive-definite weighting matrices). For this reason, many of the features of least-squares solutions will manifest themselves again in this part of the book; albeit in modified forms that result from the presence of indefinite weights.

44.1 INDEFINITE LEAST-SQUARES FORMULATION

The notion of robust adaptive filters will be defined in mathematical terms in Secs. 45.1 and 45.3. At that point, it will become clear that the design of robust filters rests on the minimization of certain regularized quadratic cost functions with indefinite weighting matrices. For this reason, in this section and in Sec. 44.2, we study such cost functions under rather general conditions. Then in Secs. 45.1 and 45.3 we specialize the ensuing theory to the design of robust filters.

The indefinite least-squares problem that we study is a variation of the regularized least-squares problem studied earlier in Sec. 29.8. Thus, given an $N \times 1$ measurement vector y, an $N \times M$ data matrix H, an $N \times N$ Hermitian matrix W, and an $M \times M$ Hermitian matrix Π, we consider the quadratic cost function

$$\boxed{J(w) \triangleq w^*\Pi w + (y - Hw)^*W(y - Hw)} \tag{44.1}$$

where w is $M \times 1$. The matrix W plays the role of a weighting matrix except that now it is not required to be positive-definite or even invertible any longer (it can have both positive and negative eigenvalues). This is in contrast to the weighted least-squares problem (29.37) where W was taken to be positive-definite. Likewise, the matrix Π plays the role of a regularization matrix, except that it too is not required to be positive-definite or even invertible, as was the case with the regularized cost function in (29.37).

In Sec. 29.8, with positive-definite matrices $\{\Pi, W\}$, we found that $J(w)$ always had a unique minimizing solution (cf. Thm. 29.5). However, when the matrices $\{\Pi, W\}$ are not necessarily positive-definite, as is the case under consideration, other scenarios can occur and the minimization problem may not even make sense. To explain what can happen in this general case, we follow the same completion-of-squares argument that we employed in Sec. 29.3. In this way, the reader will be able to appreciate the parallels between both treatments.

We first rewrite $J(w)$ as

$$\boxed{J(w) = \begin{bmatrix} y^* & w^* \end{bmatrix} \begin{bmatrix} W & -WH \\ -H^*W & \Pi + H^*WH \end{bmatrix} \begin{bmatrix} y \\ w \end{bmatrix}} \tag{44.2}$$

and proceed to express it as the sum of two terms: one is dependent on the unknown w and the other is independent of w. The sum will allow us to examine the behavior of $J(w)$ in some detail. However, in order to present the main ideas without much worries about technicalities, we distinguish between two cases. We treat below the case of an invertible coefficient matrix, $\Pi + H^*WH$, which is the case of interest to our treatment of robust filters. Observe that the matrix $\Pi + H^*WH$ need not be invertible even when H has full-column rank and $\{\Pi, W\}$ are both invertible. Consider, for example, the choices $\Pi = \mathrm{I}$, $H = \text{col}\{\mathrm{I}, \mathrm{I}\}$ and $W = \text{diag}\{\mathrm{I}, -2\mathrm{I}\}$, which lead to $\Pi + H^*WH = 0$. Such situations can never arise when $\{\Pi, W\}$ are positive-definite since then $\Pi + H^*WH > 0$. For this reason, there was no need to distinguish between singular and nonsingular coefficient matrices in Sec. 29.8. Appendix 17.A of Sayed (2003) treats the general case of possibly singular $\Pi + H^*WH$.

Invertible Coefficient Matrix

Assume $\Pi + H^*WH$ is invertible. Then the center matrix in (44.2) can be factored as the product of upper-triangular, block-diagonal, and lower-triangular matrices:

$$\begin{bmatrix} W & -WH \\ -H^*W & \Pi + H^*WH \end{bmatrix} = \begin{bmatrix} \mathrm{I} & -WH(\Pi + H^*WH)^{-1} \\ 0 & \mathrm{I} \end{bmatrix} \begin{bmatrix} W - WH(\Pi + H^*WH)^{-1}H^*W & 0 \\ 0 & \Pi + H^*WH \end{bmatrix} \begin{bmatrix} \mathrm{I} & 0 \\ -(\Pi + H^*WH)^{-1}H^*W & \mathrm{I} \end{bmatrix} \tag{44.3}$$

Substituting the right-hand side into (44.2) and expanding leads to the representation:

$$\boxed{J(w) = y^*[W - WH(\Pi + H^*WH)^{-1}H^*W]y \ + \ (w - \widehat{w})^*(\Pi + H^*WH)(w - \widehat{w})} \tag{44.4}$$

where we introduced the column vector

$$\boxed{\widehat{w} \stackrel{\Delta}{=} (\Pi + H^*WH)^{-1}H^*Wy} \tag{44.5}$$

In (44.4), $J(w)$ is expressed as the sum of two factors, and only the rightmost factor depends on the unknown w.

If our aim is to minimize $J(w)$, then it does not follow any longer that the choice $w = \widehat{w}$ minimizes $J(w)$, even though this choice for w cancels the rightmost factor in (44.4). This is because the coefficient matrix $(\Pi + H^*WH)$ is in general indefinite so that the choice

$w = \widehat{w}$ could correspond to a minimizer of $J(w)$, a maximizer of $J(w)$, or neither, as we explain below. Contrast this situation with the one we encountered in (29.15) where the coefficient matrix was H^*H and, hence, nonnegative definite, and the choice $w = \widehat{w}$ could only be a minimizer.

In order to examine the nature of the solution (44.5), we consider the following three possibilities for the coefficient matrix:

(a) $\underline{\Pi + H^*WH > 0}$. In this case, the second term in (44.4) is nonnegative for all w,

$$(w - \widehat{w})^*(\Pi + H^*WH)(w - \widehat{w}) \geq 0$$

and it is zero only when $w = \widehat{w}$. Consequently, from (44.4),

$$J(w) \;\geq\; y^*[W - WH(\Pi + H^*WH)^{-1}H^*W]y$$

with equality only when $w = \widehat{w}$. This means that $J(w)$ attains its global minimum at $w = \widehat{w}$ so that the minimization problem below has a unique solution at $\widehat{w}$:

$$\boxed{\min_{w} J(w) \quad \Longrightarrow \quad \widehat{w} = (\Pi + H^*WH)^{-1}H^*Wy \quad \text{when } \Pi + H^*WH > 0}$$

Of course, saying that $\widehat{w}$ is the minimizing solution of $J(w)$ means that if we start at the point $w = \widehat{w}$ and modify w in any direction, the cost function $J(w)$ can only increase in value. Figure 44.1 shows a typical plot of a quadratic cost function with a global minimum.

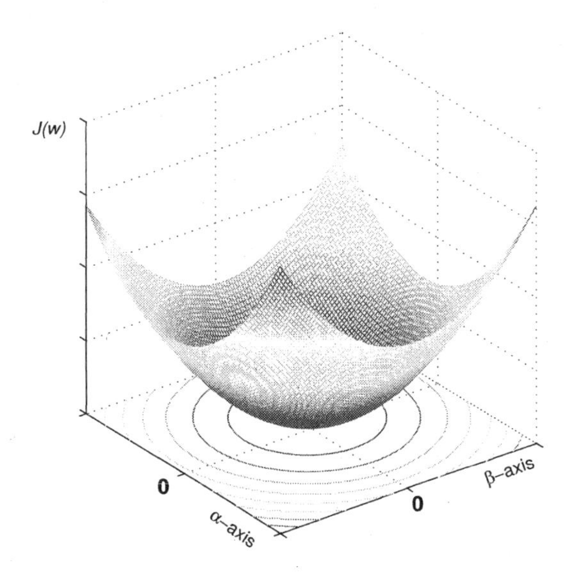

FIGURE 44.1 A typical plot of a quadratic cost function $J(w)$ with a global minimum for the case in which w is two-dimensional, say, $w = \text{col}\{\alpha, \beta\}$. The plot also shows the contour curves of $J(w)$.

(b) $\underline{\Pi + H^*WH < 0}$. In this case, the second term in (44.4) is non-positive for all w,

$$(w - \widehat{w})^*(\Pi + H^*WH)(w - \widehat{w}) \leq 0$$

and it is zero only when $w = \widehat{w}$. Consequently, from (44.4),

$$J(w) \leq y^*[W - WH(\Pi + H^*WH)^{-1}H^*W]y$$

with equality only when $w = \widehat{w}$. This means that $J(w)$ attains its global maximum at $w = \widehat{w}$, so that the maximization problem below has a unique solution at the specified $\widehat{w}$:

$$\boxed{\max_w J(w) \implies \widehat{w} = (\Pi + H^*WH)^{-1}H^*Wy \quad \text{when} \quad (\Pi + H^*WH) < 0}$$

Again, saying that $\widehat{w}$ is the maximizing solution of (44.1) means that if we start at the point $w = \widehat{w}$ and modify w in any direction, the cost function $J(w)$ can only decrease in value. Figure 44.2 shows a typical plot of a quadratic cost function with a global maximum.

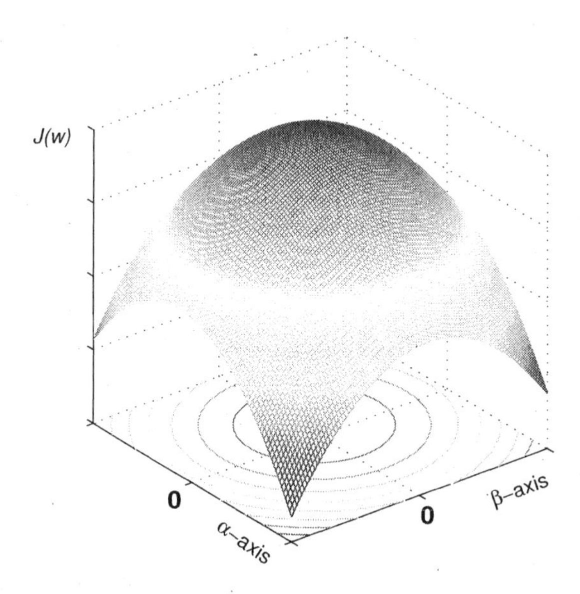

FIGURE 44.2 A typical plot of a quadratic cost function $J(w)$ with a global maximum for the case in which w is two-dimensional, say, $w = \text{col}\{\alpha, \beta\}$. The plot also shows the contour curves of $J(w)$.

(c) $\Pi + H^*WH$ is indefinite. In this case, the second term in (44.4) can be of any sign (positive or negative). The term is still zero when $w = \widehat{w}$ (but it can be zero at other choices for w as well — see Prob. XI.1). We say that $J(w)$ has a *saddle point* at $w = \widehat{w}$. A saddle point is such that if we depart from $w = \widehat{w}$, then the cost function increases along some directions and decreases along others. To explain this behavior, introduce the eigen-decomposition $\Pi + H^*WH = V\Lambda V^*$, where the diagonal entries of Λ, $\{\lambda_1, \lambda_2, \ldots, \lambda_M\}$, can be positive or negative, and V is a unitary matrix, i.e., $VV^* = V^*V = \text{I}$. Assume, for illustration purposes, that the first diagonal entry of Λ is positive with a corresponding eigenvector v_1, while the last diagonal entry of Λ is negative with a corresponding eigenvector v_M. Now suppose we choose w such that $(w - \widehat{w}) = \alpha v_1$, for any nonzero scalar α. This means that starting from $\widehat{w}$ we are modifying w along the direction of v_1. Then the second term in expression (44.4) for $J(w)$ evaluates to a positive value for all $\alpha \neq 0$,

$$(w - \widehat{w})^*(\Pi + H^*WH)(w - \widehat{w}) \;=\; |\alpha|^2\lambda_1 > 0$$

It follows that the cost function will increase along this direction. On the other hand, if we choose w such that $w = \widehat{w} + \alpha v_M$, so that starting from $\widehat{w}$ we are modifying w along the direction of v_M, then

$$(w - \widehat{w})^*(\Pi + H^*WH)(w - \widehat{w}) \;=\; |\alpha|^2\lambda_M < 0$$

which means that the cost function will decrease along this direction. Figure 44.3 provides an example of a quadratic cost function with a saddle point.

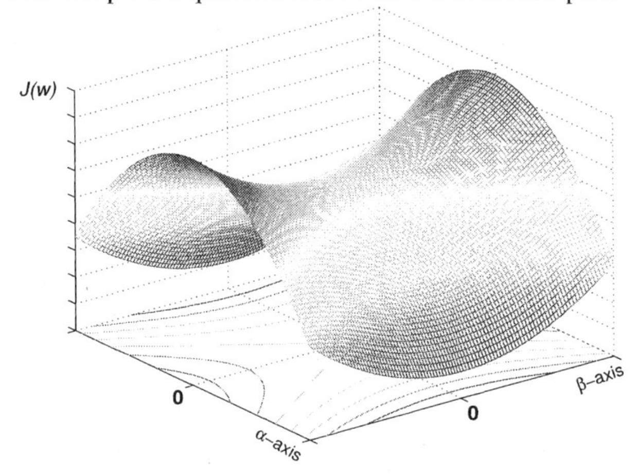

FIGURE 44.3 A typical plot of a quadratic cost function $J(w)$ with a saddle point for the case in which w is two-dimensional, say, $w = \text{col}\{\alpha, \beta\}$. The plot also shows the contour curves of $J(w)$.

In all three cases (a)–(c) considered above, we say that $\widehat{w}$ is the *stationary point*, also called the critical point, of the cost function $J(w)$ (a stationary point is a point at which the gradient of $J(w)$ with respect to w evaluates to zero — see App. 44.A). We therefore refer to the process of determining $\widehat{w}$ as the process of stationarizing $J(w)$. In summary, we arrive at the following statement.

Theorem 44.1 (Invertible coefficient matrix) Consider the problem of stationarizing the quadratic cost $J(w)$ in (44.1), where $\{\Pi, W\}$ are Hermitian matrices such that $\Pi + H^*WH$ is nonsingular. The following facts hold:

1. $J(w)$ has a unique stationary point at $\widehat{w} = (\Pi + H^*WH)^{-1}H^*Wy$, and
$$J(\widehat{w}) = y^*[W - WH(\Pi + H^*WH)^{-1}H^*W]y$$
2. The vector $\widehat{w}$ is a global minimum if, and only if, $\Pi + H^*WH > 0$.
3. The vector $\widehat{w}$ is a global maximum if, and only if, $\Pi + H^*WH < 0$.
4. The stationary point is a saddle point, otherwise.

44.2 RECURSIVE MINIMIZATION ALGORITHM

Theorem 44.1 characterizes the stationarization of quadratic cost functions under mild conditions on the matrices $\{\Pi, W\}$; these matrices are only required to be Hermitian but could be singular. However, it is usually the case that $\{\Pi, W\}$ are invertible, i.e.,

$$\boxed{\{\Pi, W\} \text{ are invertible Hermitian matrices}} \tag{44.6}$$

In addition, since we are mostly interested in the case when the cost function $J(w)$ has a unique minimizing solution, we shall require $\{\Pi, W\}$ to also satisfy

$$\boxed{\Pi + H^*WH > 0} \tag{44.7}$$

For these reasons, from now on, we shall assume that conditions (44.6)–(44.7) hold; the matrices $\{\Pi, W\}$ themselves could still be indefinite.

Our objective now is to develop a recursive algorithm that *time-updates* the minimizing solution of $J(w)$, in a manner that is similar to the recursive least-squares (RLS) algorithm of Sec. 30.2. However, two complications arise in comparison to our treatment of RLS. First, we need to pay particular attention to the existence condition (44.7) in order to guarantee that the successive iterates do indeed correspond to minima of the corresponding cost functions. Second, we need to derive a *multi-channel* version of the recursive algorithm (as we did for RLS in Prob. VII.36), whereby the measurements are taken as vectors rather than scalars. This generalization is needed in order for the resulting algorithm to be useful in the design of robust filters in Secs. 45.1 and 45.3.

Derivation of the Algorithm

So consider the quadratic cost function

$$J(w) = w^*\Pi w \;+\; (y - Hw)^*W(y - Hw) \tag{44.8}$$

with matrices $\{\Pi, W\}$ satisfying (44.6)–(44.7). We already know from Thm. 44.1 that, under condition (44.7), the unique minimizing solution of $J(w)$ is given by

$$\widehat{w} = (\Pi + H^*WH)^{-1}H^*Wy \tag{44.9}$$

We partition the vector y into its individual components, $y = \text{col}\{z_0, z_1, z_2, \ldots, z_{N-1}\}$, where each $\{z_j\}$ could be a scalar or a vector, say, of dimension p. In the former case, y is $N \times 1$, while in the latter case, y is $Np \times 1$. The reason why we allow for vector entries $\{z_j\}$ in y is because in our study of robust filters later in Secs. 45.1 and 45.3, the entries of y will need to be vectors.

Likewise, we partition H into

$$H = \text{col}\{U_0, U_1, U_2, \ldots, U_{N-1}\} \tag{44.10}$$

where each U_j is $p \times M$. When $p = 1$, the $\{U_j\}$ become rows. We further assume that the weighting matrix W has a block diagonal structure, with $p \times p$ invertible diagonal entries,

$$W = \text{diag}\{R_0, R_1, \ldots, R_{N-1}\} \tag{44.11}$$

In recursive minimization, we deal with the issue of increasing N. Therefore, in order to indicate the dependency of the solution on N, we shall denote the vector $\widehat{w}$ in (44.9) by w_{N-1}; this notation is meant to indicate that $\widehat{w}$ is based on data up to time $N-1$. We shall also write $\{y_{N-1}, H_{N-1}, W_{N-1}, J_{N-1}(\cdot)\}$ instead of $\{y, H, W, J(\cdot)\}$, so that the cost function (44.8) becomes

$$\boxed{J_{N-1}(w) = w^*\Pi w \ + \ (y_{N-1} - H_{N-1}w)^* W_{N-1}(y_{N-1} - H_{N-1}w)}$$

and, from (44.9), its minimizing solution is

$$\boxed{w_{N-1} = (\Pi + H_{N-1}^* W_{N-1} H_{N-1})^{-1} H_{N-1}^* W_{N-1} y_{N-1}} \tag{44.12}$$

provided that condition (44.7) holds, i.e.,

$$\boxed{\Pi + H_{N-1}^* W_{N-1} H_{N-1} > 0} \tag{44.13}$$

Now suppose that one more (block) row is added to H_{N-1}, one more (vector) entry is added to y_{N-1}, and one more invertible weighting block is added to W_{N-1} leading to

$$y_N \ \stackrel{\Delta}{=} \ \begin{bmatrix} y_{N-1} \\ z_N \end{bmatrix}, \quad H_N = \begin{bmatrix} H_{N-1} \\ U_N \end{bmatrix}, \quad W_N = \begin{bmatrix} W_{N-1} & \\ & R_N \end{bmatrix} \tag{44.14}$$

Then the minimizing solution of the extended cost function

$$J_N(w) \ \stackrel{\Delta}{=} \ w^*\Pi w \ + \ (y_N - H_N w)^* W_N (y_N - H_N w)$$

is given by

$$w_N = (\Pi + H_N^* W_N H_N)^{-1} H_N^* W_N y_N \tag{44.15}$$

provided that

$$\Pi + H_N^* W_N H_N > 0 \tag{44.16}$$

Our objective is to relate w_N to w_{N-1}, as well as relate the existence conditions (44.13) and (44.16).

To do so, we start by introducing

$$\boxed{P_N \ \stackrel{\Delta}{=} \ (\Pi + H_N^* W_N H_N)^{-1} \qquad \text{with } P_{-1} = \Pi^{-1}} \tag{44.17}$$

Then the existence conditions (44.13) and (44.16) can be restated in terms of $\{P_N, P_{N-1}\}$ as

$$P_N > 0 \quad \text{and} \quad P_{N-1} > 0 \tag{44.18}$$

In addition, the solutions $\{w_N, w_{N-1}\}$ can be expressed more compactly as

$$w_N = P_N H_N^* W_N y_N, \qquad w_{N-1} = P_{N-1} H_{N-1}^* W_{N-1} y_{N-1} \tag{44.19}$$

Now note that the inverse of P_N satisfies

$$P_N^{-1} \ = \ \Pi + H_N^* W_N H_N \ = \ \underbrace{\Pi + H_{N-1}^* W_{N-1} H_{N-1}}_{P_{N-1}^{-1}} + U_N^* R_N U_N$$

i.e.,

$$\boxed{P_N^{-1} = P_{N-1}^{-1} + U_N^* R_N U_N} \tag{44.20}$$

Using the matrix inversion identity

$$(A + BCD)^{-1} = A^{-1} - A^{-1}B(C^{-1} + DA^{-1}B)^{-1}DA^{-1} \tag{44.21}$$

with the identifications $A \leftarrow P_{N-1}^{-1}, B \leftarrow U_N^*, C \leftarrow R_N$, and $D \leftarrow U_N$, we obtain a recursive formula for updating P_N directly rather than its inverse,

$$\boxed{P_N = P_{N-1} - P_{N-1}U_N^*\Gamma_N U_N^* P_{N-1}, \qquad P_{-1} = \Pi^{-1}} \tag{44.22}$$

where

$$\boxed{\Gamma_N \stackrel{\Delta}{=} (R_N^{-1} + U_N P_{N-1} U_N^*)^{-1}} \tag{44.23}$$

The above recursion for P_N also gives one for updating w_{N-1} to w_N. Using the expression (44.19) for w_N, and substituting the recursion for P_N into it, we obtain:

$$\begin{aligned}
w_N &= P_N\left[H_{N-1}^* W_{N-1} y_{N-1} + U_N^* R_N z_N\right] \\
&= (P_{N-1} - P_{N-1}U_N^*\Gamma_N U_N P_{N-1})\left[H_{N-1}^* W_{N-1} y_{N-1} + U_N^* R_N z_N\right] \\
&= \underbrace{P_{N-1}H_{N-1}^* W_{N-1} y_{N-1}}_{w_{N-1}} - P_{N-1}U_N^*\Gamma_N U_N \underbrace{P_{N-1}H_{N-1}^* W_{N-1} y_{N-1}}_{w_{N-1}} + \\
&\quad P_{N-1}U_N^*\left(R_N - \Gamma_N U_N P_{N-1} U_N^* R_N\right) z_N \\
&= w_{N-1} - P_{N-1}U_N^*\Gamma_N U_N w_{N-1} + P_{N-1}U_N^*\Gamma_N \underbrace{\left(\Gamma_N^{-1} R_N - U_N P_{N-1} U_N^* R_N\right)}_{\mathrm{I}} z_N \\
&= w_{N-1} + P_{N-1}U_N^*\Gamma_N[z_N - U_N w_{N-1}] \\
&= w_{N-1} + G_N[z_N - U_N w_{N-1}]
\end{aligned}$$

where we are defining the gain matrix

$$G_N \stackrel{\Delta}{=} P_{N-1}U_N^*\Gamma_N \tag{44.24}$$

In summary, we arrive at the following statement. Observe that when $p = 1$, so that the $\{z_j\}$ are scalars and the $\{U_j\}$ are row vectors, and when $W_N = \mathrm{I}$, the recursions stated below have the same form as the RLS solution of Alg. 30.1.

Algorithm 44.1 (Recursive minimization algorithm) Consider an invertible Hermitian matrix Π, and an invertible block diagonal weighting matrix $W_N = \text{diag}\{R_j\}$ with $p \times p$ block entries. The solution w_N that minimizes the cost

$$J_N(w) = w^*\Pi w + (y_N - H_N w)^* W_N (y_N - H_N w)$$

can be computed recursively as follows. Start with $w_{-1} = 0$ and $P_{-1} = \Pi^{-1}$ and iterate for $i \geq 0$:

$$\begin{aligned}
\Gamma_i &= (R_i^{-1} + U_i P_{i-1} U_i^*)^{-1} \\
G_i &= P_{i-1} U_i^* \Gamma_i \\
w_i &= w_{i-1} + G_i[z_i - U_i w_{i-1}] \\
P_i &= P_{i-1} - G_i \Gamma_i^{-1} G_i^*
\end{aligned}$$

For each iteration $0 \leq i \leq N$, the vector w_i minimizes the cost

$$J_i(w) = w^*\Pi w \;+\; (y_i - H_i w)^* W_i (y_i - H_i w)$$

if, and only if, $P_i > 0$.

Alternative Form

We can derive two alternative expressions for the quantities $\{G_{N+1}, \Gamma_{N+1}\}$ in (44.23) and (44.24). Multiplying the recursion for P_N by U_N^* from the right gives

$$\begin{aligned} P_N U_N^* &= P_{N-1}U_N^* - P_{N-1}U_N^*\Gamma_N U_N P_{N-1}U_N^* \\ &= P_{N-1}U_N^*\Gamma_N\left[\Gamma_N^{-1} - U_N P_{N-1}U_N^*\right] \\ &= P_{N-1}U_N^*\Gamma_N\left[R_N^{-1} + U_N P_{N-1}U_N^* - U_N P_{N-1}U_N^*\right] \\ &= P_{N-1}U_N^*\Gamma_N R_N^{-1} \end{aligned} \tag{44.25}$$

which leads to the following expression for G_N:

$$\boxed{G_N = P_N U_N^* R_N} \tag{44.26}$$

By further multiplying the identity (44.25) by U_N from the left we get

$$U_N P_N U_N^* R_N = U_N P_{N-1} U_N^* \Gamma_N$$

But, from the definition (44.23) for Γ_N,

$$U_N P_{N-1} U_N^* = \Gamma_N^{-1} - R_N^{-1}$$

so that substituting into the previous equation we obtain the following alternative expressions for Γ_N:

$$\boxed{\Gamma_N = R_N - R_N U_N P_N U_N^* R_N} \tag{44.27}$$

Expressions (44.26) and (44.27) are in terms of P_N rather than P_{N-1}, so that the recursion for w_N from Alg. 44.1 can be rewritten as

$$\boxed{w_N = w_{N-1} + P_N U_N^* R_N [z_N - U_N w_{N-1}]} \tag{44.28}$$

44.3 TIME-UPDATE OF THE MINIMUM COST

In addition to relating the solution vectors $\{w_N, w_{N-1}\}$, we can also relate the corresponding minimum costs, $\{J_N(w_N), J_{N-1}(w_{N-1})\}$. The argument we employ here is similar to the one that led to Lemma 30.1 in the RLS case. Thus, let $\xi(N) = J_N(w_N)$ denote the minimum cost at time N. Let also $\{e_N, r_N\}$ denote the $p \times 1$ *a priori* and *a posteriori* error vectors defined by

$$\boxed{e_N = z_N - U_N w_{N-1}, \qquad r_N = z_N - U_N w_N} \tag{44.29}$$

Then from Thm. 44.1 and Alg. 44.1 we have

$$\xi(N) = y_N^* W_N [y_N - H_N w_N] \quad \text{and} \quad w_N = w_{N-1} + G_N e_N$$

so that, using the partitioning (44.14) for $\{y_N, H_N, W_N\}$, we can rework the above expression for $\xi(N)$ and show that either of the following two expressions can be used to time-update the minimum cost:

$$\boxed{\xi(N) = \xi(N-1) + r_N^* R_N r_N} \tag{44.30}$$

$$\boxed{\xi(N) \;=\; \xi(N-1) \;+\; e_N^* \Gamma_N e_N} \tag{44.31}$$

Indeed, we have

$$\begin{aligned}
\xi(N) &= \begin{bmatrix} y_{N-1}^* W_{N-1} & z_N^* R_N \end{bmatrix} \begin{bmatrix} y_{N-1} - H_{N-1}[w_{N-1} + G_N e_N] \\ z_N - U_N[w_{N-1} + G_N e_N] \end{bmatrix} \\
&= \begin{bmatrix} y_{N-1}^* W_{N-1} & z_N^* R_N \end{bmatrix} \begin{bmatrix} (y_{N-1} - H_{N-1} w_{N-1}) - H_{N-1} G_N e_N \\ (\mathrm{I} - U_N G_N) e_N \end{bmatrix} \\
&= \xi(N-1) \;-\; y_{N-1}^* W_{N-1} H_{N-1} G_N e_N + z_N^* R_N [\mathrm{I} - U_N G_N] e_N \\
&= \xi(N-1) + [z_N^* R_N - y_N^* W_N H_N G_N]\, e_N \\
&= \xi(N-1) + \left(z_N^* R_N - \underbrace{y_N^* W_N H_N P_N}_{w_N^*} U_N^* R_N \right) e_N \\
&= \xi(N-1) + (z_N^* - w_N^* U_N^*)\, R_N e_N \\
&= \xi(N-1) + r_N^* R_N e_N
\end{aligned}$$

which establishes (44.30). In order to derive (44.31), we relate r_N and e_N as follows:

$$\begin{aligned}
r_N = z_N - U_N w_N = z_N - U_N[w_{N-1} + G_N e_N] &= [\mathrm{I} - U_N G_N] e_N \\
&= [\mathrm{I} - U_N P_{N-1} U_N^* \Gamma_N] e_N \\
&= [\Gamma_N^{-1} - U_N P_{N-1} U_N^*] \Gamma_N e_N \\
&= R_N^{-1} \Gamma_N e_N
\end{aligned}$$

That is,

$$\boxed{r_N = R_N^{-1} \Gamma_N e_N} \tag{44.32}$$

so that $r_N^* R_N e_N = e_N^* \Gamma_N e_N$. Substituting into (44.30) leads to (44.31).

Lemma 44.1 (Estimation errors) Consider the same setting of Alg. 44.1. At each iteration i, the *a priori* and *a posteriori* errors defined by $e_i = z_i - U_i w_{i-1}$ and $r_i = z_i - U_i w_i$, are related via $r_i = R_i^{-1} \Gamma_i e_i$. In addition, the minimum cost, $\xi(i) = J_i(w_i)$, can be time-updated via either recursion:

$$\xi(i) = \xi(i-1) + r_i^* R_i e_i \;=\; \xi(i-1) + e_i^* \Gamma_i e_i$$

Using the initial condition $\xi(-1) = 0$, it also follows that the minimum cost at time N is given by

$$\xi(N) = \sum_{i=0}^{N} e_i^* \Gamma_i e_i$$

44.4 SINGULAR WEIGHTING MATRICES

Although the derivations so far assumed invertible matrices $\{\Pi, W\}$, with a block diagonal W, the results can be extended in order to accommodate a possibly singular W. We illustrate this possibility by considering one particular special case, which will arise in our study of robust filters in Sec. 45.3.

Assume all matrices R_i in (44.11) are invertible for $0 \leq i \leq N-1$, while R_N in (44.14) is singular, say, of the particular form

$$R_N = \begin{bmatrix} A_N & \\ & 0 \end{bmatrix}$$

with an invertible leading submatrix A_N. In other words, assume W_N has the form

$$W_N = \begin{bmatrix} W_{N-1} & \\ & \begin{bmatrix} A_N & \\ & 0 \end{bmatrix} \end{bmatrix}$$

with invertible $\{W_{N-1}, A_N\}$. The question is how does the new minimizing solution w_N relate to w_{N-1}, and how do the minimum costs relate to each other?

To answer these questions, we partition the entries $\{U_N, z_N\}$ in (44.14) accordingly with R_N, say,

$$U_N = \begin{bmatrix} L_N \\ B_N \end{bmatrix} \qquad z_N = \begin{bmatrix} s_N \\ d_N \end{bmatrix}$$

Then, by repeating the derivation of Sec. 44.2, it is easy to see that the recursions of Alg. 44.1 still hold for the time instants $0 \leq i \leq N-1$, i.e., it still holds that

$$\begin{aligned} \Gamma_i &= (R_i^{-1} + U_i P_{i-1} U_i^*)^{-1} \\ G_i &= P_{i-1} U_i^* \Gamma_i \\ w_i &= w_{i-1} + G_i[z_i - U_i w_{i-1}] \\ P_i &= P_{i-1} - G_i \Gamma_i^{-1} G_i^* \end{aligned}$$

while the iteration for going from time $N-1$ to time N is now given by

$$\begin{aligned} \Gamma_N &= (A_N^{-1} + L_N P_{N-1} L_N^*)^{-1} \\ G_N &= P_{N-1} L_N^* \Gamma_N \\ w_N &= w_{N-1} + G_N[s_N - L_N w_{N-1}] \\ P_N &= P_{N-1} - G_N \Gamma_N^{-1} G_N^* \end{aligned}$$

with z_N replaced by s_N, R_N replaced by A_N, and U_N replaced by L_N. Moreover, for any $0 \leq i \leq N$, it still holds that w_i is a minimum of the corresponding cost function $J_i(w)$ if, and only if, $P_i > 0$. In particular, w_N minimizes $J_N(w)$ if, and only if, $P_N > 0$. But since now

$$P_N^{-1} = P_{N-1}^{-1} + L_N^* A_N L_N$$

we find that the condition at time N can be equivalently stated as requiring

$$\boxed{P_{N-1}^{-1} + L_N^* A_N L_N > 0} \tag{44.33}$$

In addition, the minimum value of $J_N(w)$ is given by either of the following expressions:

$$\xi(N) = \sum_{i=0}^{N} e_i^* \Gamma_i e_i = \sum_{i=0}^{N} r_i^* R_i e_i \tag{44.34}$$

where

$$\begin{cases} e_i = z_i - U_i w_{i-1} & \text{for } 0 \leq i \leq N-1 \\ r_i = z_i - U_i w_i & \text{for } 0 \leq i \leq N-1 \\ e_N = s_N - L_N w_{N-1} \\ r_N = s_N - L_N w_N \end{cases} \quad (44.35)$$

We are now in a position to use the theory of indefinite least-squares, as developed so far, to design robust filters.

44.A APPENDIX: STATIONARY POINTS

Let $J(w)$ denote a function of w that is not necessarily quadratic.

Definition 44.1 (Stationary points) A stationary point (also called a critical point) of $J(w)$ is defined as any $\widehat{w}$ at which the gradient vector is annihilated:

$$\widehat{w} \text{ is a stationary point } \iff \nabla_w J(w)\big|_{w=\widehat{w}} = 0$$

Moreover, the following facts hold:

1. A stationary point $\widehat{w}$ is a local minimum if $\nabla_w^2[J(\widehat{w})] > 0$.
2. A stationary point $\widehat{w}$ is a local maximum if $\nabla_w^2[J(\widehat{w})] < 0$.
3. A stationary point $\widehat{w}$ is a saddle point if $\nabla_w^2[J(\widehat{w})]$ is indefinite and invertible. In this case, the behavior of the function at the saddle point looks similar to the shape of a saddle (hence, the name).
4. If the Hessian matrix is singular, no conclusion can be drawn about the stationary point (i.e., whether it is a minimum, a maximum, or a saddle point). Further analysis is required.

The discussion in Sec. 44.1 shows that when $J(w)$ is *quadratic* in w, as in (44.1), then more can be said about its stationary points.

Definition 44.2 (Stationary points of quadratic functions) When the cost function $J(w)$ is quadratic in w, the following facts hold:

1. A stationary point $\widehat{w}$ is a global minimum if, and only if, $\nabla_w^2[J(\widehat{w})] \geq 0$.
2. A stationary point $\widehat{w}$ is a global maximum if $\nabla_w^2[J(\widehat{w})] \leq 0$.
3. Otherwise, a stationary point $\widehat{w}$ is a saddle point.

44.B APPENDIX: INERTIA CONDITIONS

We stated in Alg. 44.1 that each w_N will be a minimum of $J_N(w)$ only if the corresponding matrix P_N is positive-definite. This positivity condition needs to be checked at each iteration in order to ensure that the successive minimization problems are well-defined. Checking the positive-definiteness of P_N, for every N, is computationally demanding but simplifications are possible as we now verify by induction.

Assume $P_i > 0$ for $0 \leq i \leq N-1$ and let us devise a simplified condition for checking whether $P_N > 0$. We start by recalling that $P_N = P_{N-1} - P_{N-1}U_N^*\Gamma_N U_N P_{N-1}$, which shows that P_N

can be obtained as the Schur complement of the block matrix

$$X \triangleq \begin{bmatrix} P_{N-1} & P_{N-1}U_N^* \\ U_N P_{N-1} & \Gamma_N^{-1} \end{bmatrix}$$

with respect to its lower-right corner entry. This observation can be used as the basis for an alternative method for checking the positive-definiteness of the matrices $\{P_{N-1}, P_N\}$. To see this, we factor X in two ways:

$$X = \begin{bmatrix} \mathrm{I} & G_N \\ & \mathrm{I} \end{bmatrix} \begin{bmatrix} P_N & \\ & \Gamma_N^{-1} \end{bmatrix} \begin{bmatrix} \mathrm{I} & \\ G_N^* & \mathrm{I} \end{bmatrix} \tag{44.36}$$

$$= \begin{bmatrix} \mathrm{I} & \\ U_N & \mathrm{I} \end{bmatrix} \begin{bmatrix} P_{N-1} & \\ & R_N^{-1} \end{bmatrix} \begin{bmatrix} \mathrm{I} & U_N^* \\ & \mathrm{I} \end{bmatrix} \tag{44.37}$$

Each of the triangular factorizations (44.36) and (44.37) has the form of a congruence relation. In other words, each one of them has the form $X = CTC^*$ for some nonsingular matrix C and center matrix T.

Now recall that Sylvester's law of inertia (cf. Lemma B.5) states that the inertia of a Hermitian matrix is preserved under congruence, where the inertia of a Hermitian matrix T is defined as the triplet of integers $\{I_+(T), I_-(T), I_0(T)\}$, such that

$$I_{(+,-,0)}(T) \triangleq \text{number of (positive,negative,zero) eigenvalues of } T \tag{44.38}$$

Applying this result to the factorizations (44.36) and (44.37) we find that the matrices $\text{diag}\{P_N, \Gamma_N^{-1}\}$ and $\text{diag}\{P_{N-1}, R_N^{-1}\}$ should have the same inertia, or more explicitly,

$$\begin{aligned} I_+(P_N \oplus \Gamma_N^{-1}) &= I_+(P_{N-1} \oplus R_N^{-1}) \\ I_-(P_N \oplus \Gamma_N^{-1}) &= I_-(P_{N-1} \oplus R_N^{-1}) \end{aligned} \tag{44.39}$$

where the notation $a \oplus b$ denotes a block diagonal matrix with entries $\{a, b\}$, i.e., $a \oplus b = \text{diag}\{a, b\}$. Using the rather obvious fact that the inertia of a block diagonal matrix is the sum of the inertias of its individual diagonal blocks, it follows from (44.39) that the following inertia equalities must hold:

$$\begin{aligned} I_+(P_N) + I_+(\Gamma_N^{-1}) &= I_+(P_{N-1}) + I_+(R_N^{-1}) \\ I_-(P_N) + I_-(\Gamma_N^{-1}) &= I_-(P_{N-1}) + I_-(R_N^{-1}) \end{aligned}$$

From this result, and from $P_{N-1} > 0$, we find that $P_N > 0$ will hold if, and only if,

$$I_+(\Gamma_N^{-1}) = I_+(R_N^{-1}) \quad \text{and} \quad I_-(\Gamma_N^{-1}) = I_-(R_N^{-1})$$

In other words, since a matrix and its inverse have the same inertia, we find that $P_{N-1} > 0$ and $P_N > 0$ will hold if, and only if, the matrices

$$\boxed{\{\Gamma_N, R_N\} \quad \text{have the same inertia}} \tag{44.40}$$

Lemma 44.2 (Minimization conditions) Consider the same setting of Alg. 44.1. Then each w_i is a minimizer of the corresponding cost function $J_i(\cdot)$, for $i = 0, 1, \ldots, N$, if, and only if, $P_i > 0$ for $i = 0, 1, \ldots, N$ or, equivalently, if and only if, $\{\Gamma_i, R_i\}$ have the same inertia for $i = 0, 1, \ldots, N$.

The second condition in the lemma is easier to check since the matrices $\{\Gamma_i\}$ are smaller in size than the $\{P_i\}$; the former is $p \times p$ while the latter is $M \times M$ and usually $p \ll M$. For example, in the special case $p = 1$, we have that P_i is $M \times M$ while $\{\Gamma_i, R_i\}$ are scalars. In the scalar case, requiring the $\{\Gamma_i, R_i\}$ to have the same inertia is equivalent to requiring them to have the same sign.

CHAPTER 45

Robust Adaptive Filters

We now formulate a robust design criterion and proceed to devise adaptive filters that meet the desired robustness performance. The derivations and arguments in this chapter are similar in nature to the arguments we employed in Chapters 29 and 30 while studying recursive least-squares problems. The main distinction will be in the use of quadratic cost functions with *indefinite* weighting matrices, as opposed to positive-definite weights. A key conclusion from the discussion here will be that some of the adaptive filters that we encountered before, e.g., LMS and $\epsilon-$NLMS, and which were derived in Chapters 10 and 11 by appealing to stochastic-gradient approximations, will now be shown to satisfy the adopted robustness measure. Actually, the arguments in the current chapter will allow us to motivate and derive these algorithms as *optimal*, as opposed to approximate, recursive solutions to well-defined optimization problems, in much the same way as the RLS algorithm was derived in Chapter 30 as the optimal recursive solution to a regularized least-squares problem.

45.1 A POSTERIORI-BASED ROBUST FILTERS

Thus, consider measurements $\{d(i)\}$ that are related to an unknown vector w^o via

$$\boxed{d(i) = u_i w^o + v(i)} \tag{45.1}$$

where $v(i)$ denotes an unknown disturbance and u_i is a row vector. The disturbance sequence is assumed to have finite energy, so that

$$\sum_{j=0}^{i} |v(j)|^2 < \infty \quad \text{for all } i \tag{45.2}$$

Given the $\{d(\cdot)\}$, we are interested in estimating some linear transformation of w^o, say,

$$\boxed{s_i \triangleq L_i w^o} \tag{45.3}$$

for some given matrices L_i. For example, if L_i is chosen as the identity matrix, $L_i = \mathrm{I}$, then the problem amounts to that of estimating w^o itself. If, on the other hand, $L_i = u_i$, then the problem amounts to estimating $u_i w^o$, which is the uncorrupted part of $d(i)$. Different choices for L_i lead to different interpretations for s_i. The derivation that follows is for arbitrary L_i.

Clearly, whenever we say that the observations $\{d(i)\}$ satisfy a model of the form (45.1), there is the implicit assumption that such a w^o exists and, furthermore, that we know its dimension. However, it is not hard to imagine that the assumption need not be valid in general and, therefore, modeling mismatches are bound to occur. Consider, for example, a first-order auto-regressive model with transfer function

$$H(z) = \frac{1}{1 - \alpha z^{-1}} \quad \text{with } \alpha \text{ real and } |\alpha| < 1$$

so that

$$H(z) = 1 + \alpha z^{-1} + \alpha^2 z^{-2} + \alpha^3 z^{-3} + \ldots \tag{45.4}$$

Assume further that we feed a sequence $\{u(i)\}$ into this model and measure its output sequence, say, $\{o(i)\}$, in the presence of additive noise $\{n(i)\}$, i.e., assume we measure

$$d(i) = o(i) + n(i)$$

If we choose some FIR model of order M to approximate $H(z)$, then we are in effect assuming that $o(i) \approx u_i w^o$ where, for this example,

$$u_i = \begin{bmatrix} u(i) & u(i-1) & \ldots & u(i-M+1) \end{bmatrix}$$

and w^o could consist of the first M coefficients of the expansion (45.4), i.e.,

$$w^o = \text{col}\{1, \alpha, \alpha^2, \ldots, \alpha^{M-1}\}$$

The mismatch between $o(i)$ and $u_i w^o$ is due to the terms that are ignored from the expansion (45.4). If we incorporate the effect of these ignored terms into $n(i)$, then we end up with a relation of the same form as in (45.1), i.e.,

$$d(i) = u_i w^o + v(i)$$

with the term $v(i)$ accounting for both measurement noise and modeling uncertainties or errors. It is for this reason that we sometimes state that $v(i)$ in the model (45.1) represents not only measurement noise but also model uncertainties.

Besides model uncertainties, a second source of uncertainty that is relevant in the context of adaptive filtering pertains to the choice of the initial condition w_{-1} of an adaptive algorithm. The value of w_{-1} can interfere with the performance of the filter, e.g., it may delay convergence or even cause divergence. Usually, we choose $w_{-1} = 0$ and, therefore, the squared Euclidean norm of w^o, $\|w^o\|^2$, serves as a measure of how far this initial guess is from w^o. In our formulation of robust filters we shall also account for the effect of the initial condition on filter performance.

Robustness Criterion

Now let $\widehat{s}_{i|i}$ denote an estimate for s_i that is based *causally* on the measurements $\{d(j)\}$, i.e., $\widehat{s}_{i|i}$ can only depend on $\{d(0), d(1), \ldots, d(i)\}$ and not on any future value of $d(\cdot)$. Our objective is to compute estimates $\{\widehat{s}_{i|i}\}$ that are robust according to the following criterion.

Let γ be a given positive number and let Π be a given positive-definite matrix. We want to determine estimates

$$\{\widehat{s}_{0|0}, \widehat{s}_{1|1}, \widehat{s}_{2|2}, \ldots, \widehat{s}_{N|N}\}$$

at times $i = 0, 1, \ldots, N$ such that, regardless of the disturbance $\{v(\cdot)\}$ and the unknown w^o, the estimates satisfy the performance measure:

$$\frac{\sum_{j=0}^{i} \|\widehat{s}_{j|j} - s_j\|^2}{w^{o*}\Pi w^o + \sum_{j=0}^{i} |v(j)|^2} < \gamma^2 \quad \text{for all } i = 0, 1, \ldots, N \tag{45.5}$$

This requirement has the following interpretation. For every time instant i, the numerator is a measure of the estimation-error energy up to time i, while the denominator consists of two terms:

1. One term is the energy of the disturbance, $v(j)$, over the same period of time.

2. The other term, $w^{*o}\Pi w^o$, is the weighted energy of the error in estimating w^o by using, for the lack of any other information, a zero initial guess for it. That is, $w^{o*}\Pi w^o = \widetilde{w}^*\Pi\widetilde{w}$ where $\widetilde{w} = w^o - w_{-1} = w^o$.

We therefore say that the denominator in (45.5) is a measure of the energies of the disturbances in the problem, namely, $\{v(\cdot), \widetilde{w}\}$, while the numerator is a measure of the resulting estimation error energies. In this way, the criterion (45.5) requires that we determine estimates $\{\widehat{s}_{j|j}\}$ such that the ratio of energies, from disturbances to estimation errors, does not exceed γ^2 for any $\{v(\cdot), w^o\}$. When this property holds, we shall say that the resulting estimates are robust in the sense that bounded disturbance energies lead to bounded estimation-error energies and, in a similar vein, small disturbance energies lead to small estimation-error energies. Of course, the smaller the value of γ, the more robust the solution is.[23] Observe further that no statistical assumptions on the data are being made. In this way, the robustness studies that are carried out in this chapter allow us to highlight some features of adaptive algorithms that hold regardless of statistical considerations.

Relation to Indefinite Least-Squares

The condition (45.5) can be equivalently stated as requiring, for $i = 0, 1, \ldots, N$,

$$w^{o*}\Pi w^o + \sum_{j=0}^{i} |v(j)|^2 - \gamma^{-2}\sum_{j=0}^{i} \|\widehat{s}_{j|j} - s_j\|^2 > 0$$

no matter what the value of the unknown w^o is, as well as the value of the disturbance sequence $\{v(j)\}$. Using the relation

$$d(j) = u_j w^o + v(j)$$

we can equivalently rewrite the above requirement as

$$w^{o*}\Pi w^o + \sum_{j=0}^{i}\left(\begin{bmatrix} \widehat{s}_{j|j} \\ d(j) \end{bmatrix} - \begin{bmatrix} L_j \\ u_j \end{bmatrix} w^o\right)^* \begin{bmatrix} -\gamma^{-2}\mathrm{I} & \\ & 1 \end{bmatrix}\left(\begin{bmatrix} \widehat{s}_{j|j} \\ d(j) \end{bmatrix} - \begin{bmatrix} L_j \\ u_j \end{bmatrix} w^o\right) > 0 \tag{45.6}$$

[23]However, the value of γ cannot be reduced at will by the designer. As we are going to see, its value needs to meet certain existence conditions (see, e.g., (45.15)).

regardless of w^o. In other words, the problem of determining estimates $\{\widehat{s}_{j|j}\}$ in order to satisfy (45.5), for any w^o and for any $\{v(\cdot)\}$ satisfying (45.2), is analogous to the problem of determining $\{\widehat{s}_{j|j}\}$ in order to satisfy (45.6) for any w^o.

If we now introduce the vector and matrix quantities

$$y_i \triangleq \begin{bmatrix} \widehat{s}_{0|0} \\ d(0) \\ \hline \widehat{s}_{1|1} \\ d(1) \\ \hline \vdots \\ \hline \widehat{s}_{i|i} \\ d(i) \end{bmatrix}, \qquad H_i \triangleq \begin{bmatrix} L_0 \\ u_0 \\ \hline L_1 \\ u_1 \\ \hline \vdots \\ \hline L_i \\ u_i \end{bmatrix} \tag{45.7}$$

as well as the block-diagonal weighting matrix

$$W_i = \text{diag}\left\{ \begin{bmatrix} -\gamma^{-2} & \\ & 1 \end{bmatrix}, \begin{bmatrix} -\gamma^{-2} & \\ & 1 \end{bmatrix}, \ldots, \begin{bmatrix} -\gamma^{-2} & \\ & 1 \end{bmatrix} \right\} \tag{45.8}$$

then problem (45.6) amounts to determining estimates $\{\widehat{s}_{j|j}\}$ that guarantee $J_i(w) > 0$ for $i = 0, 1, \ldots, N$ and for any w, where the cost function $J_i(w)$ is defined by

$$\boxed{J_i(w) \;=\; w^*\Pi w \;+\; (y_i - H_i w)^* W_i (y_i - H_i w)} \tag{45.9}$$

and where we are denoting the indeterminate variable more generally by w instead of w^o.

The cost function $J_i(w)$ so defined is quadratic in w and has an indefinite weighting matrix W_i; it has the same form as the cost function $J(w)$ studied in (44.1). Therefore, the results developed in Secs. 44.1–44.2 are immediately applicable. Specifically, in order for the *quadratic* function $J_i(w)$ to be positive for all w, it is necessary and sufficient that both of the following conditions hold:

1. $J_i(w)$ should have a minimum with respect to w.
2. The estimates $\{\widehat{s}_{j|j}\}$ should be chosen such that the value of $J_i(w)$ at its minimum is positive.

Solving the Minimization Step

Let us first show how $J_i(w)$ can be guaranteed to have a minimum for each $i = 0, 1, \ldots, N$, and, in addition, how to update the minimizing solutions. Comparing expression (45.9) for $J_i(w)$ with the expression for $J_i(w)$ from Alg. 44.1, we can make the following identifications:

$$z_i \longleftarrow \begin{bmatrix} \widehat{s}_{i|i} \\ d(i) \end{bmatrix}, \quad U_i \longleftarrow \begin{bmatrix} L_i \\ u_i \end{bmatrix}, \quad R_i \longleftarrow \begin{bmatrix} -\gamma^{-2}\mathrm{I} & \\ & 1 \end{bmatrix} \tag{45.10}$$

so that the algorithm that updates the successive minima is given by the following equations. Start with $w_{-1} = 0$ and $P_{-1} = \Pi^{-1}$ and iterate

$$\begin{aligned}
\Gamma_i &= \left(\begin{bmatrix} -\gamma^2\mathrm{I} & \\ & 1 \end{bmatrix} + \begin{bmatrix} L_i \\ u_i \end{bmatrix} P_{i-1} \begin{bmatrix} L_i^* & u_i^* \end{bmatrix} \right)^{-1} && (45.11) \\
G_i &= P_{i-1} \begin{bmatrix} L_i^* & u_i^* \end{bmatrix} \Gamma_i && (45.12) \\
w_i &= w_{i-1} + G_i \left(\begin{bmatrix} \widehat{s}_{i|i} \\ d(i) \end{bmatrix} - \begin{bmatrix} L_i \\ u_i \end{bmatrix} w_{i-1} \right) && (45.13) \\
P_i &= P_{i-1} - G_i \Gamma_i^{-1} G_i^* && (45.14)
\end{aligned}$$

Each w_i is a minimum of the corresponding cost function $J_i(w)$ if, and only if, $P_i > 0$, or, equivalently,

$$\Gamma_i \text{ and } \begin{bmatrix} -\gamma^2 \mathrm{I} & \\ & 1 \end{bmatrix} \text{ have the same inertia.} \tag{45.15}$$

Enforcing Positivity

We still need to show how to determine estimates $\{\widehat{s}_{j|j}\}$ such that the values of the successive $J_i(w)$ are positive at their minima. We pursue the construction of the $\{\widehat{s}_{j|j}\}$ by induction. So assume that estimates $\{\widehat{s}_{0|0}, \ldots, \widehat{s}_{i-1|i-1}\}$ have been chosen such that the values of the corresponding cost functions $\{J_0, J_1, \ldots, J_{i-1}\}$ are positive at their respective minima:

$$J_0(w_0) > 0, \;\; J_1(w_1) > 0, \;\; \ldots, \;\; J_{i-1}(w_{i-1}) > 0$$

Using the result of Lemma 44.1, we know that

$$J_i(w_i) \;=\; J_{i-1}(w_{i-1}) \;+\; e_i^* \Gamma_i e_i$$

where, in view of the identifications (45.10),

$$e_i \;\stackrel{\Delta}{=}\; \begin{bmatrix} \widehat{s}_{i|i} \\ d(i) \end{bmatrix} - \begin{bmatrix} L_i \\ u_i \end{bmatrix} w_{i-1}$$

But since $J_{i-1}(w_{i-1}) > 0$, we find that in order to guarantee $J_i(w_i) > 0$ it is *sufficient* to choose $\widehat{s}_{i|i}$ such that the term below is positive,

$$e_i^* \Gamma_i e_i > 0 \tag{45.16}$$

This construction provides *one* possibility for enforcing $J_i(w_i) > 0$, and it will lead to what is known as the central solution. Other constructions for $\widehat{s}_{i|i}$ are possible but are less immediate. We now explain how (45.16) can be achieved. For this purpose, we partition the vector e_i as

$$e_i \;=\; \begin{bmatrix} \widehat{s}_{i|i} \\ d(i) \end{bmatrix} - \begin{bmatrix} L_i \\ u_i \end{bmatrix} w_{i-1} \;\stackrel{\Delta}{=}\; \begin{bmatrix} e_{s,i} \\ \hline e(i) \end{bmatrix}$$

where its bottom entry is a scalar, i.e.,

$$e(i) = d(i) - u_i w_{i-1}$$

and its top entry is dependent on $\widehat{s}_{i|i}$, i.e.,

$$e_{s,i} = \widehat{s}_{i|i} - L_i w_{i-1}$$

Using the defining expression (45.11) for Γ_i, condition (45.16) is equivalent to choosing $\widehat{s}_{i|i}$ such that

$$\begin{bmatrix} e_{s,i}^* & e^*(i) \end{bmatrix} \begin{bmatrix} -\gamma^2 \mathrm{I} + L_i P_{i-1} L_i^* & L_i P_{i-1} u_i^* \\ u_i P_{i-1} L_i^* & 1 + u_i P_{i-1} u_i^* \end{bmatrix}^{-1} \begin{bmatrix} e_{s,i} \\ e(i) \end{bmatrix} > 0 \tag{45.17}$$

The expression on the left-hand side of (45.17) is quadratic in $\widehat{s}_{i|i}$; it can be rewritten as the sum of two squares by resorting to a completion-of-squares argument.

Introduce the upper-diagonal-lower factorization of the inverse of the central matrix in (45.17), namely,

$$\Gamma_i^{-1} \triangleq \begin{bmatrix} -\gamma^2 \mathrm{I} + L_i P_{i-1} L_i^* & L_i P_{i-1} u_i^* \\ u_i P_{i-1} L_i^* & 1 + u_i P_{i-1} u_i^* \end{bmatrix} \tag{45.18}$$

$$= \begin{bmatrix} \mathrm{I} & \dfrac{L_i P_{i-1} u_i^*}{1 + u_i P_{i-1} u_i^*} \\ 0 & 1 \end{bmatrix} \begin{bmatrix} \Delta & 0 \\ 0 & 1 + u_i P_{i-1} u_i^* \end{bmatrix} \begin{bmatrix} \mathrm{I} & 0 \\ \dfrac{u_i P_{i-1} L_i^*}{1 + u_i P_{i-1} u_i^*} & 1 \end{bmatrix}$$

where

$$\Delta \triangleq (-\gamma^2 \mathrm{I} + L_i P_{i-1} L_i^*) - L_i P_{i-1} u_i^* (1 + u_i P_{i-1} u_i^*)^{-1} u_i P_{i-1} L_i^* \tag{45.19}$$

Although unnecessary for our present argument, the matrix Δ can be shown to be negative-definite — see Prob. XI.2. Inverting both sides of (45.18) we obtain a similar, albeit lower-diagonal-upper, factorization for the center matrix in (45.17),

$$\Gamma_i \triangleq \begin{bmatrix} -\gamma^2 \mathrm{I} + L_i P_{i-1} L_i^* & L_i P_{i-1} u_i^* \\ u_i P_{i-1} L_i^* & 1 + u_i P_{i-1} u_i^* \end{bmatrix}^{-1} \tag{45.20}$$

$$= \begin{bmatrix} \mathrm{I} & 0 \\ -\dfrac{u_i P_{i-1} L_i^*}{1 + u_i P_{i-1} u_i^*} & 1 \end{bmatrix} \begin{bmatrix} \Delta^{-1} & 0 \\ 0 & (1 + u_i P_{i-1} u_i^*)^{-1} \end{bmatrix} \begin{bmatrix} \mathrm{I} & 0 \\ -\dfrac{u_i P_{i-1} L_i^*}{1 + u_i P_{i-1} u_i^*} & 1 \end{bmatrix}^*$$

Substituting into (45.17) leads to

$$\begin{bmatrix} e_{s,i}^* - \dfrac{e^*(i) u_i P_{i-1} L_i^*}{1 + u_i P_{i-1} u_i^*} & e^*(i) \end{bmatrix} \begin{bmatrix} \Delta^{-1} & 0 \\ 0 & \dfrac{1}{1 + u_i P_{i-1} u_i^*} \end{bmatrix} \begin{bmatrix} e_{s,i} - \dfrac{L_i P_{i-1} u_i^* e(i)}{1 + u_i P_{i-1} u_i^*} \\ e(i) \end{bmatrix} > 0 \tag{45.21}$$

Now since

$$(1 + u_i P_{i-1} u_i^*) > 0$$

(in view of $P_{i-1} > 0$), the positivity condition can be met by setting

$$e_{s,i} - e(i) \frac{L_i P_{i-1} u_i^*}{1 + u_i P_{i-1} u_i^*} = 0 \tag{45.22}$$

or, equivalently, using $e_{s,i} = \widehat{s}_{i|i} - L_i w_{i-1}$,

$$\begin{aligned} \widehat{s}_{i|i} &= L_i w_{i-1} + \frac{L_i P_{i-1} u_i^*}{1 + u_i P_{i-1} u_i^*} [d(i) - u_i w_{i-1}] \\ &= L_i \left[w_{i-1} + \frac{P_{i-1} u_i^*}{1 + u_i P_{i-1} u_i^*} (d(i) - u_i w_{i-1}) \right] \end{aligned}$$

This choice for $\widehat{s}_{i|i}$ allows us to simplify the recursion for w_i in (45.13). Indeed, substituting the expression for $e_{s,i}$ from (45.22) into (45.13) leads to

$$w_i = w_{i-1} + G_i \begin{bmatrix} \frac{L_i P_{i-1} u_i^*}{1 + u_i P_{i-1} u_i^*} \\ 1 \end{bmatrix} e(i)$$

Using the factorization (45.20) for Γ_i in the expression (45.12) for G_i gives

$$G_i \begin{bmatrix} \frac{L_i P_{i-1} u_i^*}{1+u_i P_{i-1} u_i^*} \\ 1 \end{bmatrix} = \frac{P_{i-1} u_i^*}{1 + u_i P_{i-1} u_i^*}$$

so that the recursion for w_i reduces to the form shown in the statement below.Comparing with the above expression for $\widehat{s}_{i|i}$ we conclude that $\widehat{s}_{i|i}$ can be written more compactly as $\widehat{s}_{i|i} = L_i w_i$.

Algorithm 45.1 (*A posteriori* robust filtering) Consider data $\{d(i), u_i\}$ that satisfy the model $d(i) = u_i w^o + v(i)$, for some unknown vector w^o and finite energy noise sequence $\{v(i)\}$. Consider further a positive scalar γ and a positive-definite matrix Π. Let $s_i = L_i w^o$ denote some linear transformation of w^o that we wish to estimate causally from the $\{d(j)\}$ such that

$$\frac{\sum_{j=0}^{i} \|\widehat{s}_{j|j} - s_j\|^2}{w^{o*}\Pi w^o \; + \; \sum_{j=0}^{i} |v(j)|^2} < \gamma^2 \quad \text{for all } i = 0, 1, \ldots, N \tag{45.23}$$

where $\widehat{s}_{j|j}$ denotes an estimate for s_j using the data $\{d(0), d(1), \ldots, d(j)\}$. One construction for the desired estimates can be computed as follows. Start with $w_{-1} = 0$ and $P_{-1} = \Pi^{-1}$ and iterate

$$\begin{aligned}
w_i &= w_{i-1} + \frac{P_{i-1} u_i^*}{1 + u_i P_{i-1} u_i^*}[d(i) - u_i w_{i-1}] \\
\widehat{s}_{i|i} &= L_i w_i \\
\Gamma_i &= \left(\begin{bmatrix} -\gamma^2 \mathrm{I} & \\ & 1 \end{bmatrix} + \begin{bmatrix} L_i \\ u_i \end{bmatrix} P_{i-1} \begin{bmatrix} L_i^* & u_i^* \end{bmatrix} \right)^{-1} \\
G_i &= P_{i-1} \begin{bmatrix} L_i^* & u_i^* \end{bmatrix} \Gamma_i \\
P_i &= P_{i-1} - G_i \Gamma_i^{-1} G_i^*
\end{aligned}$$

This solution satisfies (45.23) if, and only if, the $\{P_i\}$ are positive-definite, or, equivalently, if and only if, Γ_i and $\text{diag}\{-\gamma^2 \mathrm{I}, 1\}$ have the same inertia for $0 \leq i \leq N$.

It turns out that some of the adaptive algorithms that we encountered earlier in Chapter 10, and which were motivated there as stochastic-gradient methods, can now be re-examined in light of the robust solution of Alg. 45.1 with some useful conclusions.

45.2 ϵ-NLMS ALGORITHM

Let $L_i = u_i$ so that $s_i = u_i w^o$, i.e., s_i is now a scalar and we rewrite it as $s(i)$. This choice of L_i corresponds to a situation in which we are interested in estimating the undisturbed part of $d(i)$. In this case, the expressions for Γ_i and G_i from Alg. 45.1 become

$$\begin{aligned}\Gamma_i &= \left(\begin{bmatrix} -\gamma^2 & \\ & 1\end{bmatrix} + \begin{bmatrix} u_i \\ u_i \end{bmatrix} P_{i-1} \begin{bmatrix} u_i^* & u_i^* \end{bmatrix}\right)^{-1} \\ G_i &= P_{i-1} \begin{bmatrix} u_i^* & u_i^* \end{bmatrix} \Gamma_i\end{aligned}$$

so that the recursion for P_i becomes

$$P_i = P_{i-1} - P_{i-1} \begin{bmatrix} u_i^* & u_i^* \end{bmatrix} \left(\begin{bmatrix} -\gamma^2 & \\ & 1\end{bmatrix} + \begin{bmatrix} u_i \\ u_i \end{bmatrix} P_{i-1} \begin{bmatrix} u_i^* & u_i^* \end{bmatrix}\right)^{-1} \begin{bmatrix} u_i \\ u_i \end{bmatrix} P_{i-1}$$

This recursion can be simplified. Inverting both sides and using the matrix inversion lemma (44.21) we find that

$$P_i^{-1} = P_{i-1}^{-1} + \begin{bmatrix} u_i^* & u_i^* \end{bmatrix} \begin{bmatrix} -\gamma^{-2} & \\ & 1\end{bmatrix} \begin{bmatrix} u_i \\ u_i \end{bmatrix}$$

or, equivalently,

$$\boxed{P_i^{-1} = P_{i-1}^{-1} + (1-\gamma^{-2}) u_i^* u_i} \tag{45.24}$$

Inverting both sides of (45.24) allows us to re-express the recursion for P_i in the form

$$P_i = P_{i-1} - \frac{P_{i-1} u_i^* u_i P_{i-1}}{(1-\gamma^{-2})^{-1} + u_i P_{i-1} u_i^*}$$

Observe from (45.24) that starting from $P_{-1}^{-1} = \Pi > 0$, all successive P_i^{-1} will be positive-definite for any $\gamma \geq 1$. In other words, the robust filtering problem that corresponds to the choice $L_i = u_i$ is guaranteed to have a solution for any $\gamma \geq 1$. But can it have a solution for $\gamma < 1$? The answer is in general negative.

To see this, assume that the regressors $\{u_i\}$ are *sufficiently exciting*, i.e.,

$$\lim_{N\to\infty} \left(\sum_{k=0}^{N} u_k^* u_k\right) = \infty \tag{45.25}$$

This is a mild condition on the regressors since it amounts to requiring them not to "vanish quickly". Now iterating (45.24) gives

$$P_i^{-1} = \Pi \; + \; (1-\gamma^{-2}) \sum_{k=0}^{i} u_k^* u_k \tag{45.26}$$

For any $\gamma < 1$, and because of (45.25), it is easy to conclude that for sufficiently large i, at least one diagonal element of the matrix

$$\Pi \; + \; (1-\gamma^{-2}) \sum_{k=0}^{i} u_k^* u_k$$

must become negative. This fact violates the required positive-definiteness of P_i so that $\gamma < 1$ is not possible when the regressors are sufficiently exciting. In Prob. XI.4 it is shown that, when the regressors do not satisfy (45.25), choices of $\gamma < 1$ are possible.

Assume now we set $\gamma = 1$ and choose $\Pi = \epsilon \mathrm{I}$, for some small $\epsilon > 0$. Then recursion (45.24) implies that

$$P_i^{-1} = P_{i-1}^{-1} \quad \text{for all } i$$

and, hence, $P_i = \epsilon^{-1}\mathrm{I}$. In this case, the recursions of Alg. 45.1 collapse to the ϵ-NLMS algorithm with unit step-size ($\mu = 1$ — see Alg. 11.1).

Algorithm 45.2 (ϵ–NLMS algorithm) Consider data $\{d(i), u_i\}$ that satisfy the model $d(i) = u_i w^o + v(i)$, for some unknown vector w^o and finite energy noise sequence $\{v(i)\}$. Let $s(i) = u_i w^o$ denote the uncorrupted part of $d(i)$ and assume we wish to estimate $s(i)$ causally from the $\{d(j)\}$ such that

$$\frac{\sum_{j=0}^{i} |\widehat{s}(j|j) - s(j)|^2}{\epsilon \|w^o\|^2 + \sum_{j=0}^{i} |v(j)|^2} < 1 \quad \text{for all } i = 0, 1, \ldots, N$$

where $\widehat{s}(j|j)$ denotes an estimate for $s(j)$ using the data $\{d(0), d(1), \ldots, d(j)\}$. One construction for the desired estimates can be computed as follows. Start with $w_{-1} = 0$ and iterate

$$\begin{aligned} w_i &= w_{i-1} + \frac{u_i^*}{\epsilon + \|u_i\|^2}[d(i) - u_i w_{i-1}] \\ \widehat{s}(i|i) &= u_i w_i \end{aligned}$$

When the regressors $\{u_i\}$ are sufficiently exciting, as in (45.25), then it can be verified that that ϵ-NLMS, with unit step-size, is a solution to the following min-max problem over all finite-energy noise sequences (see Prob. XI.5):

$$\gamma_{\text{opt}}^2 \stackrel{\Delta}{=} \inf_{\{\widehat{s}(j|j)\}} \sup_{\{w^o, v(j)\}} \left(\frac{\sum_{j=0}^{\infty} |\widehat{s}(j|j) - u_j w^o|^2}{\epsilon \|w^o\|^2 + \sum_{j=0}^{\infty} |v(j)|^2} \right) \tag{45.27}$$

In other words, the recursions of the algorithm minimize, through the choice of the estimates $\{\widehat{s}(j|j)\}$, the largest possible value of the energy gain from the disturbance sequence $\{w^o, v(j)\}$ to the error sequence $\{\widetilde{s}(j|j)\}$ and, moreover, $\gamma_{\text{opt}}^2 = 1$.

We therefore find that ϵ–NLMS, which was derived in Chapter 10 by appealing to stochastic approximations, is in fact an optimal recursive solution to the optimization problem (45.27), in much the same way as RLS is the optimal recursive solution to the regularized least-squares problem.

45.3 A PRIORI-BASED ROBUST FILTERS

In the robust formulation of Sec. 45.1, the estimate $\widehat{s}_{i|i}$ was required to depend on measurements $\{d(j)\}$ up to *and including* time i. In this section, we shall show how to solve a similar problem where the estimate of $s_i = L_i w^o$ is now required to depend on the $\{d(j)\}$

in a *strictly* causal manner. Specifically, we shall show how to determine an estimate $\widehat{s}_{i|i-1}$ that is based on the data $\{d(j)\}$ up to and including time $i-1$; hence, the notation $\widehat{s}_{i|i-1}$ as opposed to $\widehat{s}_{i|i}$.

Let again $\{v(i)\}$ denote any disturbance sequence with finite energy as in (45.2). We wish to determine estimates

$$\{\widehat{s}_{0|-1}, \widehat{s}_{1|0}, \widehat{s}_{2|1}, \ldots, \widehat{s}_{N|N-1}\}$$

at times $i = 0, 1, \ldots, N$ such that, regardless of the disturbance sequence $\{v(i)\}$ and the unknown w^o, the estimates should satisfy the performance measure:

$$\boxed{\frac{\sum_{j=0}^{i} \|\widehat{s}_{j|j-1} - s_j\|^2}{w^{o*}\Pi w^o + \sum_{j=0}^{i-1} |v(j)|^2} < \gamma^2 \quad \text{for all } i = 0, 1, \ldots, N} \tag{45.28}$$

Comparing with (45.5), we observe that the upper limit on the sum involving $v(i)$ is now $i-1$ rather than i. The condition (45.28) can be equivalently stated as requiring for $i = 0, 1, \ldots, N$,

$$w^{o*}\Pi w^o + \sum_{j=0}^{i-1} |v(j)|^2 - \gamma^{-2} \sum_{j=0}^{i} \|\widehat{s}_{j|j-1} - s_j\|^2 > 0 \tag{45.29}$$

or, equivalently, by using the relation $d(j) = u_j w^o + v(j)$,

$$\begin{array}{c} w^{o*}\Pi w^o - \gamma^{-2}\|\widehat{s}_{i|i-1} - s_i\|^2 \\ + \sum_{j=0}^{i-1} \left(\begin{bmatrix} \widehat{s}_{j|j-1} \\ d(j) \end{bmatrix} - \begin{bmatrix} L_j \\ u_j \end{bmatrix} w^o \right)^* \begin{bmatrix} -\gamma^{-2}\mathrm{I} & \\ & 1 \end{bmatrix} \left(\begin{bmatrix} \widehat{s}_{j|j-1} \\ d(j) \end{bmatrix} - \begin{bmatrix} L_j \\ u_j \end{bmatrix} w^o \right) > 0 \end{array} \tag{45.30}$$

no matter what w^o is. In other words, the problem of determining estimates $\{\widehat{s}_{j|j-1}\}$ in order to satisfy (45.28), for any w^o and for any disturbance sequence $\{v(j)\}$ satisfying (45.2), is equivalent to the problem of determining the $\{\widehat{s}_{j|j-1}\}$ in order to satisfy (45.30) for any w^o.

Observe that the index for the sum involving $\widehat{s}_{j|j-1}$ in (45.29) and (45.30) runs from o up to i, while the index for the sum involving $v(j)$ (and, consequently, $d(j)$) runs from 0 to $i-1$. In order to treat both sums uniformly, with the indices running from 0 to i in both cases, we define the weighting matrix

$$W_i = \text{diag}\left\{ \begin{bmatrix} -\gamma^{-2} & \\ & 1 \end{bmatrix}, \begin{bmatrix} -\gamma^{-2} & \\ & 1 \end{bmatrix}, \ldots, \begin{bmatrix} -\gamma^{-2} & \\ & 1 \end{bmatrix}, \begin{bmatrix} -\gamma^{-2} & \\ & 0 \end{bmatrix} \right\} \tag{45.31}$$

with a last block entry that is singular. Likewise, we define the vector and matrix quantities

$$y_i \triangleq \begin{bmatrix} \widehat{s}_{0|-1} \\ d(0) \\ \hline \widehat{s}_{1|0} \\ d(1) \\ \hline \vdots \\ \hline \widehat{s}_{i|i-1} \\ d(i) \end{bmatrix}, \qquad H_i \triangleq \begin{bmatrix} L_0 \\ u_0 \\ \hline L_1 \\ u_1 \\ \hline \vdots \\ \hline L_i \\ u_i \end{bmatrix} \tag{45.32}$$

Then we can rewrite (45.30) as requiring us to determine estimates $\{\widehat{s}_{j|j-1}\}$ that guarantee $J_i(w) > 0$ for $i = 0, 1, \ldots, N$ and for any w, where the cost function $J_i(w)$ is defined by

$$\boxed{J_i(w) \;=\; w^*\Pi w \;+\; (y_i - H_i w)^* W_i (y_i - H_i w)}$$

and where we are now denoting the indeterminate variable more generally by w.

Since the last block entry of the weighting matrix W_i is singular, while all other blocks in it are invertible, we are therefore faced with the situation discussed in Sec. 44.4. Using the results of that section, with i playing the role of N, we find that the minimizing argument w_i of $J_i(w)$ can be determined recursively as follows. Start with $w_{-1} = 0$ and $P_{-1} = \Pi^{-1}$ and iterate for $0 \leq j \leq i-1$:

$$\begin{aligned}
\Gamma_j &= \left(\begin{bmatrix} -\gamma^2 \mathrm{I} & \\ & 1 \end{bmatrix} + \begin{bmatrix} L_j \\ u_j \end{bmatrix} P_{j-1} \begin{bmatrix} L_j^* & u_j^* \end{bmatrix}\right)^{-1} && (45.33)\\
G_j &= P_{j-1} \begin{bmatrix} L_j^* & u_j^* \end{bmatrix} \Gamma_j && (45.34)\\
w_j &= w_{j-1} + G_j \left(\begin{bmatrix} \widehat{s}_{j|j-1} \\ d(j) \end{bmatrix} - \begin{bmatrix} L_j \\ u_j \end{bmatrix} w_{j-1}\right) && (45.35)\\
P_j &= P_{j-1} - G_j \Gamma_j^{-1} G_j^* && (45.36)
\end{aligned}$$

At iteration i, the vector w_i is found via

$$\begin{aligned}
\Gamma_i &= \left(-\gamma^2 \mathrm{I} + L_i P_{i-1} L_i^*\right)^{-1} && (45.37)\\
G_i &= P_{i-1} L_i^* \Gamma_i && (45.38)\\
w_i &= w_{i-1} + G_i \left(\widehat{s}_{i|i-1} - L_i w_{i-1}\right) && (45.39)\\
P_i &= P_{i-1} - G_i \Gamma_i^{-1} G_i^* && (45.40)
\end{aligned}$$

The resulting w_i minimizes $J_i(w)$ if, and only if,

$$P_{i-1}^{-1} - \gamma^{-2} L_i^* L_i > 0 \qquad (45.41)$$

Moreover, the resulting minimum cost at time i is given by

$$J_i(w_i) \;=\; \sum_{j=0}^{i-1} e_j^* \Gamma_j e_j \;+\; e_i^* \Gamma_i e_i$$

where

$$\begin{aligned}
e_j &= \begin{bmatrix} \widehat{s}_{j|j-1} \\ d(j) \end{bmatrix} - \begin{bmatrix} L_j \\ u_j \end{bmatrix} w_{j-1} \;\stackrel{\Delta}{=}\; \begin{bmatrix} e_{s,j} \\ e(j) \end{bmatrix}, \qquad \text{for } \; 0 \leq j \leq i-1\\
e_i &= \widehat{s}_{i|i-1} - L_i w_{i-1} \;\stackrel{\Delta}{=}\; e_{s,i}
\end{aligned}$$

In other words, by using the expressions for $\{\Gamma_j, \Gamma_i\}$,

$$\begin{aligned}
J_i(w_i) \;=\; & \sum_{j=0}^{i-1} \begin{bmatrix} e_{s,j}^* & e^*(j) \end{bmatrix} \left(\begin{bmatrix} -\gamma^2 \mathrm{I} & \\ & 1 \end{bmatrix} + \begin{bmatrix} L_j \\ u_j \end{bmatrix} P_{j-1} \begin{bmatrix} L_j^* & u_j^* \end{bmatrix}\right)^{-1} \begin{bmatrix} e_{s,j} \\ e(j) \end{bmatrix}\\
& + e_{s,i}^* \left(-\gamma^2 \mathrm{I} + L_i P_{i-1} L_i^*\right)^{-1} e_{s,i}
\end{aligned}$$

This expression for the minimum cost can be simplified as follows. Introduce the upper-diagonal-lower factorization for Γ_j:

$$\begin{bmatrix} -\gamma^2 \mathrm{I} + L_j P_{j-1} L_j^* & L_j P_{j-1} u_j^* \\ u_j P_{j-1} L_j^* & 1 + u_j P_{j-1} u_j^* \end{bmatrix}^{-1} = \begin{bmatrix} \mathrm{I} & -(-\gamma^2 \mathrm{I} + L_j P_{j-1} L_j^*)^{-1} L_j P_{j-1} u_j^* \\ 0 & 1 \end{bmatrix} \begin{bmatrix} (-\gamma^2 \mathrm{I} + L_j P_{j-1} L_j^*)^{-1} & \\ & \Delta_j^{-1} \end{bmatrix} \begin{bmatrix} \mathrm{I} & 0 \\ -u_j P_{j-1} L_j^* (-\gamma^2 \mathrm{I} + L_j P_{j-1} L_j^*)^{-1} & 1 \end{bmatrix} \tag{45.42}$$

where

$$\begin{aligned} \Delta_j &\stackrel{\Delta}{=} 1 + u_j P_{j-1} u_j^* - u_j P_{j-1} L_j^* (-\gamma^2 \mathrm{I} + L_j P_{j-1} L_j^*)^{-1} L_j P_{j-1} u_j^* \\ &= 1 + u_j \left[P_{j-1} - P_{j-1} L_j^* (-\gamma^2 \mathrm{I} + L_j P_{j-1} L_j^*)^{-1} L_j P_{j-1} \right] u_j^* \\ &= 1 + u_j (P_{j-1}^{-1} - \gamma^{-2} L_j^* L_j)^{-1} u_j^* \\ &\stackrel{\Delta}{=} 1 + u_j \widetilde{P}_{j-1} u_j^* \end{aligned}$$

and

$$\boxed{\widetilde{P}_{j-1}^{-1} \stackrel{\Delta}{=} P_{j-1}^{-1} - \gamma^{-2} L_j^* L_j} \tag{45.43}$$

We know from condition (45.41) that $\widetilde{P}_{i-1}$ is positive-definite, so that Δ_i itself is positive-definite. Then we can write

$$J_i(w_i) = \sum_{j=0}^{i} e_{s,j}^* (-\gamma^2 \mathrm{I} + L_j P_{j-1} L_j^*)^{-1} e_{s,j} \; + \; \sum_{j=0}^{i-1} [d(j) - \bar{d}(j)]^* \Delta_j^{-1} [d(j) - \bar{d}(j)]$$

where we introduced

$$\bar{d}(j) \stackrel{\Delta}{=} u_j w_{j-1} \; + \; u_j P_{j-1} L_j^* (-\gamma^2 \mathrm{I} + L_j P_{j-1} L_j^*)^{-1} e_{s,j}$$

It is now clear that $J_i(w_i)$ can be made positive by choosing $e_{s,j} = 0$ or, equivalently,

$$\widehat{s}_{j|j-1} = L_j w_{j-1}$$

Substituting into recursion (45.35) for w_j we get

$$\begin{aligned} w_j &= w_{j-1} + P_{j-1} \begin{bmatrix} L_j^* & u_j^* \end{bmatrix} \Gamma_j \begin{bmatrix} 0 \\ e(j) \end{bmatrix} \\ &= w_{j-1} + \frac{\widetilde{P}_{j-1} u_j^*}{1 + u_i \widetilde{P}_{i-1} u_i^*} e(j) \end{aligned}$$

where the second equality follows by using the factorization (45.42) for Γ_j. On the other hand, from (45.39) we get $w_i = w_{i-1}$. We therefore arrive at the following statement.

Algorithm 45.3 (*A priori* robust filtering) Consider data $\{d(i), u_i\}$ that satisfy the model $d(i) = u_i w^o + v(i)$, for some unknown vector w^o and finite energy noise sequence $\{v(i)\}$. Consider further a positive scalar γ and a positive-definite matrix Π. Let $s_i = L_i w^o$ denote some linear transformation of w^o that we wish to estimate in a strictly causal manner from the $\{d(j)\}$ such that

$$\frac{\sum_{j=0}^{i} \|\widehat{s}_{j|j-1} - s_j\|^2}{w^{o*}\Pi w^o \ + \ \sum_{j=0}^{i-1} |v(j)|^2} < \gamma^2 \quad \text{for all } i = 0, 1, \ldots, N \tag{45.44}$$

where $\widehat{s}_{j|j-1}$ denotes an estimate for s_j using the data $\{d(0), d(1), \ldots, d(j-1)\}$. One construction for the desired estimates can be computed as follows. Start with $w_{-1} = 0$ and $P_{-1} = \Pi^{-1}$ and iterate for $0 \leq i < N$:

$$\begin{aligned}
\widehat{s}_{i|i-1} &= L_i w_{i-1} \\
\widetilde{P}_{i-1}^{-1} &= P_{i-1}^{-1} - \gamma^{-2} L_i^* L_i \\
w_i &= w_{i-1} + \frac{\widetilde{P}_{i-1} u_i^*}{1 + u_i \widetilde{P}_{i-1} u_i^*}[d(i) - u_i w_{i-1}] \\
\Gamma_i &= \left(\begin{bmatrix} -\gamma^2 \mathrm{I} & \\ & 1 \end{bmatrix} + \begin{bmatrix} L_i \\ u_i \end{bmatrix} P_{i-1} \begin{bmatrix} L_i^* & u_i^* \end{bmatrix} \right)^{-1} \\
G_i &= P_{i-1} \begin{bmatrix} L_i^* & u_i^* \end{bmatrix} \Gamma_i \\
P_i &= P_{i-1} - G_i \Gamma_i^{-1} G_i^*
\end{aligned}$$

while $\widehat{s}_{N|N-1} = L_N w_{N-1}$ and $w_N = w_{N-1}$. This solution satisfies (45.44) if, and only if, the $\{\widetilde{P}_i\}$ are positive-definite for $0 \leq i < N$.

45.4 LMS ALGORITHM

Consider again the choice $L_i = u_i$ so that $s_i = u_i w^o$, i.e., s_i is now a scalar, $s(i)$. In this case, we know from Sec. 45.2 that the recursion for P_i reduces to (45.24). We can then pose the same question with regards to the value of γ in (45.44) when $L_i = u_i$. That is, can it be smaller than one? As before, the answer is negative for sufficiently exciting regressors as in (45.25). To see this, we use expression (45.24) for P_i^{-1}, along with

$$\widetilde{P}_{i-1}^{-1} = P_{i-1}^{-1} - \gamma^{-2} u_i^* u_i$$

to write

$$\widetilde{P}_{i-1}^{-1} = \Pi \ + \ (1 - \gamma^{-2}) \sum_{k=0}^{i-1} u_k^* u_k \ - \ \gamma^{-2} u_i^* u_i \tag{45.45}$$

For any $\gamma < 1$, it is easy to conclude that for sufficiently large i, at least one diagonal element of the matrix

$$\Pi \ + \ (1 - \gamma^{-2}) \sum_{k=0}^{i-1} u_k^* u_k$$

must become negative. This fact violates the required positive-definiteness of $\{\widetilde{P}_{i-1}\}$ so that $\gamma < 1$ is not possible when the regressors are sufficiently exciting. What about $\gamma = 1$? If we set $\gamma = 1$ and choose $\Pi = \mu^{-1}\mathrm{I}$, for some small $\mu > 0$, then it follows from (45.24) that

$$P_i = \mu\mathrm{I}$$

Moreover, the following equalities hold:

$$\widetilde{P}_{i-1}u_i^*(1 + u_i\widetilde{P}_{i-1}u_i^*)^{-1} \;=\; (\widetilde{P}_{i-1}^{-1} + u_i^*u_i)^{-1}u_i^* \;=\; P_{i-1}u_i^* \;=\; \mu u_i^*$$

so that the recursions of Alg. 45.3 collapse to the LMS algorithm with step-size μ — see Alg. 10.1.

Moreover, when the regressors $\{u_i\}$ are sufficiently exciting, as in (45.25), it can be verified that LMS provides a solution to the following min-max problem over all finite-energy noise sequences (see Prob. XI.5):

$$\gamma_{\rm opt}^2 \;\stackrel{\Delta}{=}\; \inf_{\{\widehat{s}(j|j-1)\}} \;\sup_{\{w^o, v(j)\}} \left(\frac{\sum_{j=0}^{\infty} |\widehat{s}(j|j-1) - s(j)|^2}{\mu^{-1}\|w^o\|^2 \;+\; \sum_{j=0}^{\infty}|v(j)|^2} \right) \tag{45.46}$$

In other words, the recursions of the algorithm minimize, through the choice of the strictly causal estimates $\{\widehat{s}(j|j-1)\}$, the largest possible value of the energy gain from the disturbance sequence $\{w^o, v(j)\}$ to the error sequence $\{\widetilde{s}(j|j-1)\}$ and, moreover, $\gamma_{\rm opt}^2 = 1$. We therefore find that LMS, which was derived in Sec. 10.2 by appealing to stochastic-gradient approximations, is in fact an optimal recursive solution to the optimization problem (45.46).

Algorithm 45.4 (LMS algorithm) Consider data $\{d(i), u_i\}$ that satisfy the model $d(i) = u_i w^o + v(i)$, for some unknown vector w^o and finite energy noise sequence $\{v(i)\}$. Let $s(i) = u_i w^o$ denote the uncorrupted part of $d(i)$ and assume we wish to estimate $s(i)$ in a strictly causal manner from the $\{d(j)\}$ such that

$$\frac{\sum_{j=0}^{i} |\widehat{s}(j|j-1) - s(j)|^2}{\mu^{-1}\|w^o\|^2 \;+\; \sum_{j=0}^{i-1}|v(j)|^2} < 1 \quad \text{for all } i = 0, 1, \ldots, N \tag{45.47}$$

where $\widehat{s}(j|j-1)$ denotes an estimate for $s(j)$ using the data $\{d(0), \ldots, d(j-1)\}$. One construction for the desired estimates can be computed as follows. Start with $w_{-1} = 0$ and iterate for $0 \leq i < N$:

$$\begin{aligned} \widehat{s}(i|i-1) &= u_i w_{i-1} \\ w_i &= w_{i-1} + \mu u_i^*[d(i) - u_i w_{i-1}] \end{aligned}$$

while $\widehat{s}(N|N-1) = u_N w_{N-1}$. This solution satisfies the robustness condition (45.47) if, and only if, the matrices $\{\mu^{-1}\mathrm{I} - u_i^* u_i\}$ are positive-definite for $0 \leq i < N$.

Observe that the existence condition in the statement of the algorithm is in terms of the positive-definiteness of the rank-one matrices $\{\mu^{-1}\mathrm{I} - u_i^* u_i\}$. The eigenvalues of every such matrix are given by $\{\mu^{-1}, \mu^{-1}, \ldots, \mu^{-1}, \mu^{-1} - \|u_i\|^2\}$, with $(M-1)$ eigenvalues at μ^{-1} and one eigenvalue at $\mu^{-1} - \|u_i\|^2$ (assuming regressors of size $1 \times M$). Therefore, the matrices $\{\mu^{-1}\mathrm{I} - u_i^* u_i\}$ will be positive-definite for $i = 0, 1, \ldots, N$ if the step-size μ is chosen to satisfy

$$\sup_i \mu \|u_i\|^2 < 1 \tag{45.48}$$

In other words, the step-size μ needs to be sufficiently small in order for the LMS filter to satisfy the robustness condition (45.47).

45.A APPENDIX: $\mathcal{H}^\infty$ FILTERS

In this appendix we explain the general form of $\mathcal{H}^\infty$ filters for linear state-space estimation, and then show how the adaptive robust filters derived in the body of the chapter can be obtained as special cases (in much the same way as RLS itself was seen to be a special case of the Kalman filter in App. 31.2). For more details, the reader is referred to the monograph by Hassibi, Sayed, and Kailath (1999, Chapter 4) and to the article by Sayed, Hassibi, and Kailath (1996a); the inertia (existence) conditions listed below follow the latter reference where they are derived in a manner similar to the arguments used in App. 44.B.

Consider a state-space model of the form

$$x_{i+1} = F_i x_i + G_i u_i, \qquad y_i = H_i x_i + v_i, \qquad i \geq 0 \tag{45.49}$$

with initial condition x_0, and where $\{F_i, G_i, H_i\}$ are known $n \times n$, $n \times m$, and $p \times n$ matrices, respectively. Moreover, the $\{u_i, v_i\}$ denote disturbances, which along with x_0, are assumed to be unknown. Using the observations $\{y_j\}$, we would like to estimate some linear combination of the state entries, say, $s_i = L_i x_i$ for some given $q \times n$ matrices L_i. Let $\widehat{s}_{i|i}$ denote an estimate of s_i that is based on the observations $\{y_j\}$ from time 0 up to and including time i. Likewise, let $\widehat{s}_{i|i-1}$ denote an estimate of s_i that is based on the observations $\{y_j\}$ from time 0 up to time $i-1$. In other words, $\widehat{s}_{i|i}$ depends on the observations in a causal manner, while $\widehat{s}_{i|i-1}$ depends on the observations in a strictly causal manner. We refer to $\widehat{s}_{i|i}$ as a *filtered* estimate and to $\widehat{s}_{i|i-1}$ as a *predicted* estimate. Define further the filtered and predicted errors $\widetilde{s}_{i|i} = s_i - \widehat{s}_{i|i}$ and $\widetilde{s}_{i|i-1} = s_i - \widehat{s}_{i|i-1}$.

$\mathcal{H}^\infty$ A Posteriori Filters

$\mathcal{H}^\infty$ *a posteriori* filters are concerned with the computation of filtered estimates. In this context, we want to determine estimates $\{\widehat{s}_{0|0}, \widehat{s}_{1|1}, \widehat{s}_{2|2}, \ldots, \widehat{s}_{N|N}\}$ at times $i = 0, 1, \ldots, N$ such that, regardless of the disturbances $\{u_j, v_j\}$ and the unknown x_0, the estimates guarantee

$$\left(\frac{\sum_{j=0}^{i} \|\widetilde{s}_{j|j}\|^2}{x_0^* \Pi_0^{-1} x_0 + \sum_{j=0}^{i} \|u_j\|^2 + \sum_{j=0}^{i} \|v_j\|^2} \right) < \gamma_f^2 \quad \text{for } i = 0, 1, \ldots, N \tag{45.50}$$

for some given positive scalar γ_f^2 and positive-definite weighting matrix Π_0. This requirement has the following interpretation. For every time instant i, the numerator is a measure of the estimation error energy up to time i, while the denominator consists of two terms:

1. One term is the energy of the disturbances, $\{u_j, v_j\}$, over the same period of time.
2. The other term, $x_0^* \Pi_0^{-1} x_0$, is the weighted energy of the error in estimating x_0 by using, for the lack of any other information, a zero initial guess for it.

We therefore say that the denominator in (45.50) is a measure of the energies of the disturbances in the problem, while the numerator is a measure of the resulting estimation error energies. In this way, criterion (45.50) requires that we determine estimates $\{\widehat{s}_{j|j}\}$ such that the ratio of energies, from disturbances to estimation errors, does not exceed γ_f^2 for any $\{v_j, u_j, w\}$. When this property holds, we say that the resulting estimates are robust in the sense that bounded disturbance energies lead to bounded estimation error energies.

It turns out that a recursive solution exists for state-space models of the form (45.49); in a manner similar to the solution of Alg. 45.1.

Algorithm 45.5 ($\mathcal{H}^\infty$ *a posteriori* filter) One construction of an *a posteriori* $\mathcal{H}^\infty$ filter is found as follows:

$$\begin{aligned}
\widehat{x}_{i|i} &= F_{i-1}\widehat{x}_{i-1|i-1} + P_{i|i-1}H_i^*(\mathrm{I} + H_iP_{i|i-1}H_i^*)^{-1}(y_i - H_iF_{i-1}\widehat{x}_{i-1|i-1}) \\
\widehat{s}_{i|i} &= L_i\widehat{x}_{i|i} \\
R_i &= \begin{bmatrix} -\gamma_f^2\mathrm{I} & 0 \\ 0 & \mathrm{I} \end{bmatrix}, \quad R_{e,i} = R_i + \begin{bmatrix} L_i \\ H_i \end{bmatrix} P_{i|i-1} \begin{bmatrix} L_i^* & H_i^* \end{bmatrix} \\
P_{i+1|i} &= F_iP_{i|i-1}F_i^* + G_iG_i^* - F_iP_{i|i-1}\begin{bmatrix} L_i^* & H_i^* \end{bmatrix} R_{e,i}^{-1} \begin{bmatrix} L_i \\ H_i \end{bmatrix} P_{i|i-1}F_i^*
\end{aligned}$$

with initial conditions $\widehat{x}_{-1|-1} = 0$ and $P_{0|-1} = \Pi_0$. This construction satisfies (45.50) if, and only if, R_i and $R_{e,i}$ have the same inertia for all $0 \le i \le N$ or, equivalently, if and only if, $\mathrm{I} + H_iP_{i|i-1}H_i^* > 0$ and $-\gamma_f^2\mathrm{I} + L_i(P_{i|i-1}^{-1} + H_i^*H_i)^{-1}L_i^* < 0$ for $0 \le i \le N$.

$\mathcal{H}^\infty$ *A Priori* Filters

$\mathcal{H}^\infty$ *a priori* filters are concerned with the computation of predicted estimates. In this context, we want to determine estimates $\{\widehat{s}_{0|-1}, \widehat{s}_{1|0}, \widehat{s}_{2|1}, \ldots, \widehat{s}_{N|N-1}\}$ at times $i = 0, 1, \ldots, N$ such that, regardless of the disturbances $\{u_j, v_j\}$ and the unknown x_0, the estimates guarantee

$$\left(\frac{\sum_{j=0}^{i} \|\widetilde{s}_{j|j-1}\|^2}{x_0^*\Pi_0^{-1}x_0 + \sum_{j=0}^{i-1}\|u_j\|^2 + \sum_{j=0}^{i-1}\|v_j\|^2} \right) < \gamma_p^2 \quad \text{for } i = 0, 1, \ldots, N \tag{45.51}$$

for some given positive scalar γ_p^2 and positive-definite weighting matrix Π_0. Compared with (45.50), we see that the sums over $\|v_j\|^2$ and $\|u_j\|^2$ run only up to time instant $i-1$ due to the strict causality constraint. Again, it turns out that a recursive solution exists for state-space models of the form (45.49); in a manner similar to the solution of Alg. 45.3.

Algorithm 45.6 ($\mathcal{H}^\infty$ *a priori* filter) One construction of an *a priori* $\mathcal{H}^\infty$ filter is as follows:

$$\begin{aligned}
\widehat{s}_{i|i-1} &= L_i\widehat{x}_{i|i-1}, \quad \widetilde{P}_{i|i-1}^{-1} = P_{i|i-1}^{-1} - \gamma_p^{-2}L_i^*L_i \\
\hat{x}_{i+1|i} &= F_i\hat{x}_{i|i-1} + F_i\widetilde{P}_{i|i-1}H_i^*(\mathrm{I} + H_i\widetilde{P}_{i|i-1}H_i^*)^{-1}(y_i - H_i\hat{x}_{i|i-1}) \\
R_i &= \begin{bmatrix} -\gamma_p^2\mathrm{I} & 0 \\ 0 & \mathrm{I} \end{bmatrix}, \quad R_{e,i} = R_i + \begin{bmatrix} L_i \\ H_i \end{bmatrix} P_{i|i-1} \begin{bmatrix} L_i^* & H_i^* \end{bmatrix} \\
P_{i+1|i} &= F_iP_{i|i-1}F_i^* + G_iG_i^* - F_iP_{i|i-1}\begin{bmatrix} L_i^* & H_i^* \end{bmatrix} R_{e,i}^{-1} \begin{bmatrix} L_i \\ H_i \end{bmatrix} P_{i|i-1}F_i^*
\end{aligned}$$

with initial conditions $\widehat{x}_{0|-1} = 0$ and $P_{0|-1} = \Pi_0$. This construction satisfies (45.51) if, and only if, $\widetilde{P}_{i|i-1} > 0$ for all $0 \le i \le N$.

Special Case of Adaptive Filters

It is now straightforward to verify that the *a posteriori* and *a priori* $\mathcal{H}^\infty$ filters of Algs. 45.5 and 45.6 reduce to the robust adaptive filters of Algs. 45.1 and 45.3 when the state-space model (45.49) is specialized to $x_{i+1} = x_i$ and $y(i) = u_i x_i + v(i)$. That is, when F_i is set equal to the identity matrix and H_i is set equal to a row vector u_i. Moreover, y_i becomes a scalar $y(i)$. Substituting these choices into the *a posteriori* $\mathcal{H}^\infty$ filter of Alg. 45.5, we find that it collapses to

$$\begin{aligned}
\widehat{x}_{i|i} &= \widehat{x}_{i-1|i-1} + \frac{P_{i|i-1}u_i^*}{1+u_iP_{i|i-1}u_i^*}(y(i) - u_i\widehat{x}_{i-1|i-1}) \\
\widehat{s}_{i|i} &= u_i\widehat{x}_{i|i} \\
R_i &= \begin{bmatrix} -\gamma_f^2 \mathrm{I} & 0 \\ 0 & 1 \end{bmatrix}, \quad R_{e,i} = R_i + \begin{bmatrix} L_i \\ u_i \end{bmatrix} P_{i|i-1} \begin{bmatrix} L_i^* & u_i^* \end{bmatrix} \\
P_{i+1|i} &= P_{i|i-1} - P_{i|i-1}\begin{bmatrix} L_i^* & u_i^* \end{bmatrix} R_{e,i}^{-1} \begin{bmatrix} L_i \\ u_i \end{bmatrix} P_{i|i-1}
\end{aligned}$$

Comparing with the statement of the *a posteriori* adaptive robust filter in Alg. 45.1, we find that the equations are identical once the identifications shown in Table 45.1 below are made between the $\mathcal{H}^\infty$ variables and the adaptive filter variables. A similar conclusion holds for the *a priori* $\mathcal{H}^\infty$ filter. Its equations reduce to

$$\begin{aligned}
\widehat{s}_{i|i-1} &= L_i\widehat{x}_{i|i-1} \\
\widetilde{P}_{i|i-1}^{-1} &= P_{i|i-1}^{-1} - \gamma_p^{-2} L_i^* L_i \\
\hat{x}_{i+1|i} &= \hat{x}_{i|i-1} + \frac{\widetilde{P}_{i|i-1}u_i^*}{1+u_i\widetilde{P}_{i|i-1}u_i^*}(y(i) - u_i\hat{x}_{i|i-1}) \\
R_i &= \begin{bmatrix} -\gamma_p^2 \mathrm{I} & 0 \\ 0 & 1 \end{bmatrix}, \quad R_{e,i} = R_i + \begin{bmatrix} L_i \\ u_i \end{bmatrix} P_{i|i-1} \begin{bmatrix} L_i^* & u_i^* \end{bmatrix} \\
P_{i+1|i} &= P_{i|i-1} - P_{i|i-1}\begin{bmatrix} L_i^* & u_i^* \end{bmatrix} R_{e,i}^{-1} \begin{bmatrix} L_i \\ u_i \end{bmatrix} P_{i|i-1}
\end{aligned}$$

which agree with the equations of Alg. 45.3.

TABLE 45.1 Correspondences between $\mathcal{H}^\infty$ and robust adaptive variables.

Description	$\mathcal{H}^\infty$ variable	Adaptive variable	Description
Measurement	$y(i)$	$d(i)$	Reference signal
State vector	x_i	w	Weight vector
State estimator	$\hat{x}_{i\vert i}$	w_i	Weight estimate
State estimator	$\hat{x}_{i+1\vert i}$	w_i	Weight estimate
Riccati variable	$P_{i+1\vert i}$	P_i	Inverse of coefficient matrix
Coefficient matrix	$\widetilde{P}_{i\vert i-1}$	$\widetilde{P}_{i-1}$	Inverse of coefficient matrix
Initial condition	$\hat{x}_{-1\vert -1}$	w_{-1}	Initial condition
Initial covariance	Π_0	Π^{-1}	Regularization matrix

CHAPTER 46

Robustness Properties

This chapter deals with robustness analysis, as opposed to robust filter design. While the presentation in Chapter 45 was concerned with a framework for designing robust filters and applying it to the study of LMS and ϵ-NLMS, the results obtained therein are not immediately applicable to studying the robustness performance of other adaptive filters. To do so, in this chapter we resort to the same energy-conservation arguments that we employed in Parts IV (*Mean-Square Performance*) and V (*Transient Performance*) while studying the performance of adaptive filters. As a byproduct, we shall gain further insights into the robustness performance not only of LMS and $\epsilon-$NLMS, but of other adaptive filters as well. Besides providing a more intuitive route to the robustness results of Chapter 45, the energy arguments also lead to tighter robustness bounds. The discussion in this chapter is self-contained and it approaches the subject of robustness from first principles; the presentation does not rely on the indefinite least-squares theory developed in the previous two chapters.

46.1 ROBUSTNESS OF LMS

Let us start with the LMS algorithm in order to illustrate the main ideas. Consider again measurements $\{d(i)\}$ that arise from a model of the form

$$\boxed{d(i) = u_i w^o + v(i)} \tag{46.1}$$

for some unknown weight vector w^o and unknown disturbance $v(i)$. The LMS algorithm estimates w^o recursively according to the rule:

$$\boxed{w_i = w_{i-1} \ + \ \mu u_i^*[d(i) - u_i w_{i-1}], \qquad w_{-1} \ = \ \text{initial condition}} \tag{46.2}$$

where μ is the step-size parameter. Introduce the error quantities:

$$\widetilde{w}_i = w^o - w_i, \qquad e_a(i) = u_i \widetilde{w}_{i-1}, \qquad e_p(i) = u_i \widetilde{w}_i$$

They can be related as follows. Subtracting both sides of (46.2) from w^o leads to

$$\widetilde{w}_i = \widetilde{w}_{i-1} - \mu u_i^*[e_a(i) + v(i)] \tag{46.3}$$

since

$$d(i) - u_i w_{i-1} = e_a(i) + v(i)$$

Squaring both sides of (46.3) and rearranging terms we obtain

$$\boxed{\|\widetilde{w}_i\|^2 - \|\widetilde{w}_{i-1}\|^2 + \mu|e_a(i)|^2 - \mu|v(i)|^2 = \mu \cdot |e_a(i) + v(i)|^2 \cdot (\mu\|u_i\|^2 - 1)} \tag{46.4}$$

This equality relates the energies of several quantities in an LMS implementation. The identity is equivalent to the energy-conservation identity that we encountered before in Thm. 15.1. Nevertheless, the energy relation is now written in a different form, in terms of $\{e_a(i), v(i)\}$ and not $\{e_a(i), e_p(i)\}$, which is more convenient to our present purposes.

To begin with, observe that the right-hand side in (46.4) is the product of three terms. Two of these terms, μ and $|e_a(i) + v(i)|^2$, are nonnegative, whereas the last term, $(\mu \cdot \|u_i\|^2 - 1)$, could be positive, negative, or zero depending on the relative sizes of μ and $\|u_i\|^2$. It follows that the quantities $\{\widetilde{w}_i, \widetilde{w}_{i-1}, e_a(i), v(i)\}$ always satisfy the following inequalities in terms of their squared norms:

$$\begin{cases} \|\widetilde{w}_i\|^2 + \mu\,|e_a(i)|^2 \;\leq\; \|\widetilde{w}_{i-1}\|^2 + \mu\,|v(i)|^2 & \text{if} \;\; \mu\|u_i\|^2 \leq 1 \\ \|\widetilde{w}_i\|^2 + \mu\,|e_a(i)|^2 \;=\; \|\widetilde{w}_{i-1}\|^2 + \mu\,|v(i)|^2 & \text{if} \;\; \mu\|u_i\|^2 = 1 \\ \|\widetilde{w}_i\|^2 + \mu\,|e_a(i)|^2 \;\geq\; \|\widetilde{w}_{i-1}\|^2 + \mu\,|v(i)|^2 & \text{if} \;\; \mu\|u_i\|^2 \geq 1 \end{cases} \tag{46.5}$$

or, more compactly, assuming a nonzero denominator,

$$\frac{\|\widetilde{w}_i\|^2 + \mu\,|e_a(i)|^2}{\|\widetilde{w}_{i-1}\|^2 + \mu\,|v(i)|^2} \quad \begin{cases} \leq 1 & \text{if} \;\; \mu\|u_i\|^2 \leq 1 \\ = 1 & \text{if} \;\; \mu\|u_i\|^2 = 1 \\ \geq 1 & \text{if} \;\; \mu\|u_i\|^2 \geq 1 \end{cases} \tag{46.6}$$

This representation in the form of a ratio is more compact and is adopted only for convenience of presentation. Of course, we can avoid the assumption of nonzero denominators by working with the differences (46.5) rather than the ratio (46.6).

The first inequality in (46.6), which corresponds to the case $\mu\|u_i\|^2 \leq 1$, has an interesting interpretation. It states that, for any step-size μ satisfying

$$\mu\|u_i\|^2 \leq 1$$

and no matter what $v(i)$ is, we have

$$\boxed{\mu^{-1}\|\widetilde{w}_i\|^2 \;+\; |e_a(i)|^2 \;\leq\; \mu^{-1}\|\widetilde{w}_{i-1}\|^2 \;+\; |v(i)|^2} \tag{46.7}$$

Summing over $0 \leq i \leq N$ we get

$$\sum_{i=0}^{N} \left(\mu^{-1}\|\widetilde{w}_i\|^2 \;+\; |e_a(i)|^2\right) \;\leq\; \sum_{i=0}^{N} \left(\mu^{-1}\|\widetilde{w}_{i-1}\|^2 \;+\; |v(i)|^2\right) \tag{46.8}$$

But since several of the terms $\|\widetilde{w}_j\|^2$ get cancelled from both sides of the inequality (46.8), we find that it simplifies to

$$\boxed{\mu^{-1}\|\widetilde{w}_N\|^2 \;+\; \sum_{i=0}^{N}|e_a(i)|^2 \;\leq\; \mu^{-1}\|\widetilde{w}_{-1}\|^2 \;+\; \sum_{i=0}^{N}|v(i)|^2} \tag{46.9}$$

or, in ratio format again,

$$\boxed{\frac{\mu^{-1}\|\widetilde{w}_N\|^2 \;+\; \sum_{i=0}^{N}|e_a(i)|^2}{\mu^{-1}\|\widetilde{w}_{-1}\|^2 \;+\; \sum_{i=0}^{N}|v(i)|^2} \;\leq\; 1} \tag{46.10}$$

This relation holds for all N whenever

$$\boxed{\mu\|u_i\|^2 \leq 1 \quad \text{for} \quad 0 \leq i \leq N} \tag{46.11}$$

We have therefore derived a robustness result for LMS that is consistent with the result of Alg. 45.4. Indeed, note that Alg. 45.4 indicates, by setting $\widehat{s}(j|j-1) = u_j w_{j-1}$, that LMS satisfies

$$\frac{\sum_{j=0}^{N} |e_a(j)|^2}{\mu^{-1}\|w^o\|^2 + \sum_{j=0}^{N} |v(j)|^2} \leq 1 \tag{46.12}$$

where $w^o = \widetilde{w}_{-1}$ since $w_{-1} = 0$. Relation (46.10) is a tighter result, with the additional factor $\mu^{-1}\|\widetilde{w}_N\|^2$ appearing in the numerator.

Cauchy-Schwartz Interpretation

Before proceeding to examine, in a similar manner, the robustness of other adaptive filters, it is of value to step back and to take a closer look at the above robustness analysis of LMS. In the previous section, we simply squared both sides of the weight-error recursion (46.3) to conclude that inequality (46.7) holds for any μ satisfying $\mu\|u_i\|^2 \leq 1$. By summing both sides of (46.7) over $0 \leq i \leq N$, we were then led to inequality (46.10), which establishes the robustness of LMS. Given the simplicity of this energy argument, it is worth examining the origin of inequality (46.7) from first principles.

To do so, let w_{i-1} denote *any* generic estimate at iteration $i-1$ of some unknown vector w^o. This estimate could have been generated by the LMS filter or by any other filter. Now given any vector u_i, pick an arbitrary positive number μ satisfying

$$\boxed{\mu\|u_i\|^2 \leq 1} \tag{46.13}$$

Then it *always* holds that

$$\boxed{|u_i w^o - u_i w_{i-1}|^2 \leq \mu^{-1} \; \|w^o - w_{i-1}\|^2} \tag{46.14}$$

This is because condition (46.13) and the Cauchy-Schwartz inequality guarantee that[24]

$$|u_i w^o - u_i w_{i-1}|^2 \;\leq\; \|u_i\|^2 \cdot \|w^o - w_{i-1}\|^2 \;\leq\; \mu^{-1} \; \|w^o - w_{i-1}\|^2$$

Note that the quantity on the left-hand side of (46.14) is the energy of the estimation error $e_a(i) = u_i \widetilde{w}_{i-1}$. Likewise, the quantity on the right-hand side of (46.14) is the energy of the weight-error vector, $\widetilde{w}_{i-1}$ (weighted by μ^{-1}). Note further that if the right-hand side of (46.14) is increased by any nonnegative value, the inequality will continue to hold. Therefore, assume we add $|v(i)|^2$ to the right-hand side of (46.14), for any disturbance value $v(i)$, then it always holds that

$$\boxed{|e_a(i)|^2 \;\leq\; \mu^{-1}\|\widetilde{w}_{i-1}\|^2 \;+\; |v(i)|^2} \tag{46.15}$$

no matter how w_{i-1} is generated! In other words, every adaptive filter should satisfy this inequality.

[24]The Cauchy-Schwartz inequality states that for any two column vectors x and y, it holds that $|x^* y| \;\leq\; \|x\| \cdot \|y\|$.

Lemma 46.1 (Trivial inequality) For any μ satisfying $\mu\|u_i\|^2 \leq 1$, for any $\{v(i), u_i\}$, and for any estimate w_{i-1} of an unknown vector w^o, it always holds that

$$|e_a(i)|^2 \leq \mu^{-1}\|\widetilde{w}_{i-1}\|^2 + |v(i)|^2$$

This inequality holds for any adaptive filter.

The natural question is how does inequality (46.15) change when the estimate w_{i-1} is generated by LMS? That is, how can knowledge of the specific algorithm that generates w_{i-1} be used to improve the inequality? To see this, assume that w_{i-1} is obtained from the LMS recursion (46.2). Then we already know from (46.7) that a tighter inequality holds, namely,

$$\boxed{\mu^{-1}\|\widetilde{w}_i\|^2 + |e_a(i)|^2 \leq \mu^{-1}\|\widetilde{w}_{i-1}\|^2 + |v(i)|^2} \tag{46.16}$$

with an additional factor $\mu^{-1}\|\widetilde{w}_i\|^2$ added to the left hand-side, when compared with (46.15). In other words, although the left-hand side is now larger than before, the inequality still holds

Robustness Interpretation

It is also useful to examine the robustness result (46.10) from the perspective of bounded mappings. Specifically, it turns out that the result (46.10) can be interpreted as bounding the norm of the mapping induced by LMS between its *a priori* estimation errors and disturbances. To see this, introduce the column vectors:

$$\underline{\text{dist}} = \begin{bmatrix} \frac{1}{\sqrt{\mu}}\widetilde{w}_{-1} \\ v(0) \\ v(1) \\ \vdots \\ v(N) \end{bmatrix} \quad \text{and} \quad \underline{\text{error}} = \begin{bmatrix} e_a(0) \\ e_a(1) \\ \vdots \\ e_a(N) \\ \frac{1}{\sqrt{\mu}}\widetilde{w}_N \end{bmatrix}$$

The vector dist contains the disturbances that affect the performance of the filter; its energy is the quantity that appears in the denominator of (46.10). Likewise, the vector error contains the *a priori* errors and the final weight-error vector; its energy is the quantity that appears in the numerator of (46.10). Now the LMS update (46.2) allows us to relate the entries of both vectors in a straightforward manner. For example, we have

$$e_a(0) = u_0\widetilde{w}_{-1} = (\sqrt{\mu}\,u_0)\left(\frac{1}{\sqrt{\mu}}\,\widetilde{w}_{-1}\right)$$

which shows how the first entry of error relates to the first entry of dist. Similarly, for $e_a(1) = u_1\widetilde{w}_0$ we obtain

$$e_a(1) \;=\; (\sqrt{\mu}u_1[\mathrm{I} - \mu u_0^* u_0])\,\frac{1}{\sqrt{\mu}}\widetilde{w}_{-1} \;-\; (\mu u_1 u_0^*)\,v(0)$$

which relates $e_a(1)$ to the first two entries of the vector dist. Continuing in this manner, we can relate $e_a(2)$ to the first three entries of dist, $e_a(3)$ to the first four entries of dist, and so on. We can express these relations as:

$$\underbrace{\begin{bmatrix} e_a(0) \\ e_a(1) \\ \vdots \\ e_a(N) \\ \hline \frac{1}{\sqrt{\mu}}\widetilde{w}_N \end{bmatrix}}_{\underline{\text{error}}} = \underbrace{\begin{bmatrix} \times & & & & & \\ \times & \times & & & & \\ \vdots & & \ddots & & & \\ \times & \times & \times & \times & \times & \times \end{bmatrix}}_{\mathcal{T}} \underbrace{\begin{bmatrix} \frac{1}{\sqrt{\mu}}\widetilde{w}_{-1} \\ \hline v(0) \\ v(1) \\ \vdots \\ v(N) \end{bmatrix}}_{\underline{\text{dist}}} \tag{46.17}$$

with a lower-triangular mapping $\mathcal{T}$ relating dist to error. The symbol $\times$ is used to denote the generic entries of $\mathcal{T}$. The causal nature of the adaptive algorithm results in a lower triangular mapping $\mathcal{T}$.

The ratio (46.10) can now be seen to be equal to the ratio between the energies of the input and output vectors of $\mathcal{T}$. However, we know from the discussion in Sec. B.6 that the maximum singular value of $\mathcal{T}$ admits the interpretation:

$$\bar{\sigma}(\mathcal{T}) = \max_{x \neq 0} \frac{\|\mathcal{T}x\|}{\|x\|}$$

In other words, $\bar{\sigma}(\mathcal{T})$ measures the maximum energy gain from the input vector x to the resulting vector $\mathcal{T}x$. It then follows that relation (46.10) amounts to saying that

$$\max_{\underline{\text{dist}} \neq 0} \frac{\| \mathcal{T}\underline{\text{dist}} \|}{\| \underline{\text{dist}} \|} \leq 1$$

so that the maximum singular value of $\mathcal{T}$ must be bounded by 1. We therefore say that the robustness of LMS guarantees that the singular values of the mapping $\mathcal{T}$ between dist and error are bounded by unity.

46.2 ROBUSTNESS OF ϵ−NLMS

We now examine the robustness of other adaptive filters, by employing energy arguments similar to what we have just done for LMS.

Consider the ϵ−NLMS recursion with unit step-size,

$$\boxed{w_i = w_{i-1} + \frac{u_i^*}{\epsilon + \|u_i\|^2}[d(i) - u_i w_{i-1}]} \tag{46.18}$$

Subtracting w^o from both sides we find

$$\widetilde{w}_i = \widetilde{w}_{i-1} - \frac{u_i^*}{\epsilon + \|u_i\|^2}[e_a(i) + v(i)]$$

Multiplying both sides by u_i from the left allows us to relate $\{e_p(i), e_a(i)\}$ as follows:

$$\boxed{e_p(i) = \frac{\epsilon}{\epsilon + \|u_i\|^2}e_a(i) \; - \; \frac{\|u_i\|^2}{\epsilon + \|u_i\|^2}v(i)}$$

so that

$$e_a(i) = \frac{(\epsilon + \|u_i\|^2)}{\epsilon}e_p(i) \; + \; \frac{\|u_i\|^2}{\epsilon}v(i)$$

Using this expression for $e_a(i)$ we get

$$d(i) - u_i w_{i-1} = e_a(i) + v(i) = \frac{(\epsilon + \|u_i\|^2)}{\epsilon}[e_p(i) + v(i)] = \frac{(\epsilon + \|u_i\|^2)}{\epsilon}[d(i) - u_i w_i]$$

and substituting back into (46.18) we conclude that the $\epsilon-$NLMS recursion can be rewritten in the equivalent form

$$\boxed{w_{i-1} = w_i - \epsilon^{-1} u_i^* [d(i) - u_i w_i]} \tag{46.19}$$

with w_i appearing on the right-hand side as well. This form looks similar to the LMS update (46.2), so that most of the derivations from Sec. 46.1 should extend almost literally to the present case.

Indeed, subtracting w^o from both sides of (46.18) now leads to

$$\widetilde{w}_{i-1} = \widetilde{w}_i + \epsilon^{-1} u_i^* [e_p(i) + v(i)] \tag{46.20}$$

with $e_p(i)$ on the right-hand side, as opposed to $e_a(i)$ as in the LMS case (46.3). After squaring both sides of (46.20) we get

$$\|\widetilde{w}_{i-1}\|^2 - \|\widetilde{w}_i\|^2 - \epsilon^{-1}|e_p(i)|^2 + \epsilon^{-1}|v(i)|^2 = \epsilon^{-1} \cdot |e_p(i) + v(i)|^2 \cdot (\epsilon^{-1}\|u_i\|^2 + 1)$$

where the right-hand side is now seen to be always nonnegative! Therefore, it always holds that

$$\epsilon\|\widetilde{w}_i\|^2 + |e_p(i)|^2 \ \leq \ \epsilon\|\widetilde{w}_{i-1}\|^2 + |v(i)|^2$$

Summing over $0 \leq i \leq N$ we get, in ratio form,

$$\boxed{\frac{\epsilon\|\widetilde{w}_N\|^2 \ + \ \sum_{i=0}^{N}|e_p(i)|^2}{\epsilon\|\widetilde{w}_{-1}\|^2 \ + \ \sum_{i=0}^{N}|v(i)|^2} \ \leq \ 1} \tag{46.21}$$

This relation holds for all N. In this way, we have derived a robustness result for ϵ-NLMS that is consistent with the result from Alg. 45.2. Recall from the statement of Alg. 45.2, by using $\widehat{s}(j|j) = u_j w_j$, that ϵ-NLMS satisfies

$$\frac{\sum_{j=0}^{N}|e_p(j)|^2}{\epsilon\|w^o\|^2 \ + \ \sum_{j=0}^{N}|v(j)|^2} \ \leq \ 1$$

where $w^o = \widetilde{w}_{-1}$ since $w_{-1} = 0$. Relation (46.21) provides a tighter bound with the additional term $\epsilon\|\widetilde{w}_N\|^2$ appearing in the numerator.

46.3 ROBUSTNESS OF RLS

We now examine the robustness of the RLS filter as stated in Alg. 30.1, namely,

$$w_i = w_{i-1} + \frac{P_{i-1}u_i^*}{1 + u_i P_{i-1} u_i^*}[d(i) - u_i w_{i-1}]$$

where, in addition, we know from (30.10) that

$$\boxed{P_i^{-1} = P_{i-1}^{-1} + u_i^* u_i} \tag{46.22}$$

Using a derivation similar to the one used above for $\epsilon-$NLMS, it can be verified that the update equation for RLS can be rewritten in the equivalent form (see Prob. XI.7):

$$\boxed{w_{i-1} = w_i - P_{i-1}u_i^*[d(i) - u_i w_i]}$$

Subtracting w^o from both sides leads to

$$\widetilde{w}_{i-1} = \widetilde{w}_i + P_{i-1}u_i^*[e_p(i) + v(i)] \tag{46.23}$$

so that computing the weighted norm of both sides, by using P_{i-1}^{-1} as a weighting matrix, we get

$$\widetilde{w}_{i-1}^* P_{i-1}^{-1}\widetilde{w}_{i-1} = (\ \widetilde{w}_i + P_{i-1}^{-1}u_i^*[e_p(i) + v(i)]\ )^* \cdot P_{i-1}^{-1} \cdot (\ \widetilde{w}_i + P_{i-1}u_i^*[e_p(i) + v(i)]\ )$$

which, upon expansion and using (46.22), provides the equality

$$\begin{aligned} &\widetilde{w}_{i-1}^* P_{i-1}^{-1}\widetilde{w}_{i-1} - \widetilde{w}_i^* P_i^{-1}\widetilde{w}_i - |e_p(i)|^2 + |v(i)|^2 \\ &= (e_p(i) + v(i))^*[u_i P_{i-1}u_i^* + 1](e_p(i) + v(i)) \ - \ |e_p(i)|^2 \end{aligned} \tag{46.24}$$

It is observed now that the right-hand side in the above equality is *not* guaranteed to be nonnegative, in general. However, observe the following. For any $\{e_p(i), v(i)\}$ it holds that

$$|e_p(i) + v(i)|^2 \ \geq \ \frac{1}{2}|e_p(i)|^2 \ - \ |v(i)|^2$$

as can be easily checked from the trivial inequality

$$\left|\frac{1}{\sqrt{2}}e_p(i) + \sqrt{2}v(i)\right|^2 \geq 0$$

Therefore, adding $\frac{1}{2}|e_p(i)|^2 + |v(i)|^2$ to both sides of (46.24) we get

$$\widetilde{w}_{i-1}^* P_{i-1}^{-1}\widetilde{w}_{i-1} - \widetilde{w}_i^* P_i^{-1}\widetilde{w}_i - \frac{1}{2}|e_p(i)|^2 + 2|v(i)|^2$$

$$= (e_p(i) + v(i))^*[u_i P_{i-1}u_i^* + 1](e_p(i) + v(i)) \ - \ \left(\frac{1}{2}|e_p(i)|^2 - |v(i)|^2\right)$$

where the right-hand side is now guaranteed to be nonnegative since

$$\begin{aligned} (e_p(i) + v(i))^*[u_i P_{i-1}u_i^* + 1](e_p(i) + v(i)) \ &= \ (e_p(i) + v(i))^*(u_i P_{i-1}u_i^*)(e_p(i) + v(i)) \\ &\quad + |e_p(i) + v(i)|^2 \\ &\geq \ (e_p(i) + v(i))^*(u_i P_{i-1}u_i^*)(e_p(i) + v(i)) \\ &\quad + \frac{1}{2}|e_p(i)|^2 \ - \ |v(i)|^2 \end{aligned}$$

It follows that

$$\widetilde{w}_i^* P_i^{-1}\widetilde{w}_i + \frac{1}{2}|e_p(i)|^2 \leq \widetilde{w}_{i-1}^* P_{i-1}^{-1}\widetilde{w}_{i-1} + 2|v(i)|^2$$

Summing over $0 \leq i \leq N$ we get, in ratio form,

$$\boxed{\frac{\widetilde{w}_N^* P_N^{-1} \widetilde{w}_N + \frac{1}{2}\sum_{i=0}^{N} |e_p(i)|^2}{\widetilde{w}_{-1}^* \Pi \widetilde{w}_{-1} \; + \; 2\sum_{i=0}^{N} |v(i)|^2} \leq 1} \tag{46.25}$$

This relation holds for all N. In this way, we arrive at a robustness result for RLS as well. This result can be reworked into a more familiar form as follows. The inequality (46.25) would still hold if we increase the denominator by adding $\widetilde{w}_{-1}^* \Pi \widetilde{w}_{-1}$ to it, so that

$$\frac{\widetilde{w}_N^* P_N^{-1} \widetilde{w}_N + \frac{1}{2}\sum_{i=0}^{N} |e_p(i)|^2}{\widetilde{w}_{-1}^* \Pi \widetilde{w}_{-1} \; + \; \sum_{i=0}^{N} |v(i)|^2} \leq 2$$

from which we conclude that

$$\frac{\sum_{i=0}^{N} |e_p(i)|^2}{\widetilde{w}_{-1}^* \Pi \widetilde{w}_{-1} \; + \; \sum_{i=0}^{N} |v(i)|^2} \leq 4 \tag{46.26}$$

The result (46.25) gives a tighter bound than (46.26). This last inequality shows that for RLS, the ratio of the energies from disturbances $\{\widetilde{w}_{-1}, v(i)\}$ to estimation errors $\{e_p(i)\}$ never exceeds 4. In Prob. XI.17, a similar bound is derived using the *a priori* estimation errors $\{e_a(\cdot)\}$; this latter bound, however, will be data dependent.

Summary and Notes

The chapters in this part describe a notion of robustness and a procedure for designing robust filters by relying on indefinite-least-squares theory.

SUMMARY OF MAIN RESULTS

1. A recursive procedure similar to RLS is derived for time-updating the solution of an indefinite least-squares problem. Algorithm 44.1 does so for the general case in which the observation data and the regression data are allowed to be vectors and matrices, respectively. In contrast to the block RLS solution of Prob. VII.36, for example, we now require the successive P_i to be positive-definite in order for the successive weight vectors to correspond to global minima.

2. There are many notions of robustness. The one described in Chapter 45 requires that we bound the energy gain from disturbances to estimation errors at each iteration. This formulation reduces the design of robust filters to the solution of indefinite least-squares problems.

3. Two kinds of robust filters are designed: *a posteriori* filters and *a priori* filters. It turns out that ϵ-NLMS is a special case of the former while LMS is a special case of the latter. It is further argued in the text and in Prob. XI.5 that LMS and $\epsilon-$NLMS are optimal robust filters.

4. Chapter 46 develops a framework for robustness analysis by using the same energy conservation arguments that we employed earlier in Parts IV (*Mean-Square Performance*) and V (*Transient Performance*). This framework allows us to examine the robustness of a larger class of adaptive filters (e.g., RLS, filtered-error LMS, Perceptron, etc.), and to bring forth connections with some system-theoretic concepts such as passivity relations, small gain conditions, and l_2-stability — see, e.g., Probs. XI.6–XI.12.

BIBLIOGRAPHIC NOTES

Derivation of robust filters. The *a posteriori* and *a priori* robust adaptive filters studied in Secs. 45.1 and 45.3 are special cases of a broader family of robust filters known as $\mathcal{H}^\infty$ filters (in much the same way as RLS itself is a special case of the Kalman filter, as was shown in App. 31.2). An overview of $\mathcal{H}^\infty$ filters is given in App. 45.A. A more detailed treatment of $\mathcal{H}^\infty$ filters, as well as $\mathcal{H}^\infty$ controllers, can be found in the textbook by Green and Limebeer (1995) and in the monograph by Hassibi, Sayed, and Kailath (1999). Several approaches can be used to derive $\mathcal{H}^\infty$ filters, especially for general state-space models, such as completion-of-squares arguments, game-theoretic arguments, and Krein space arguments. In Secs. 45.1 and 45.3 we chose instead to follow the presentation of the least-squares theory in Chapter 30. In this way, readers will be able to realize more immediately the connections between robust adaptive filters and least-squares adaptive filters. Further discussions on the connections between robust adaptive filters and $\mathcal{H}^\infty$ filters can be found in Hassibi, Sayed, and Kailath (1996,1999), Sayed, Hassibi, and Kailath (1996a), and Sayed and Rupp (1997).

$\mathcal{H}^\infty$ framework. Research on $\mathcal{H}^\infty$ designs was pursued rather systematically during the 1980's and 1990's. Most works in this area were primarily concerned with the analysis and design of robust filters and controllers; robust in the sense that they limit the effect of modeling uncertainties on the performance and stability of the resulting filters and controllers. The original work in the field is that of Zames (1981) on sensitivity issues in feedback control. Since then many ingenious approaches and viewpoints have been put forward with applications over a wide range of areas. See, for example, the books by Green and Limebeer (1995), Basar and Bernhard (1995), Dahleh and Diaz-Bobillo (1995), Zhou, Doyle, and Glover (1996), and Helton and Merlino (1998), which focus mostly on control problems, and the monograph by Hassibi, Sayed, and Kailath (1999), which takes an estimation perspective to the solution of both filtering and control problems. Some earlier references are the articles by Khargonekar and Nagpal (1991), Yaesh and Shaked (1991), and Shaked and Theodor (1992), which deal with filtering issues as well.

$\mathcal{H}^\infty$-optimality of LMS. As we saw in Chapter 10, the LMS algorithm is usually derived as a stochastic-gradient approximation to a steepest-descent method. In Sec. 45.4, we argued that LMS can be derived as an *optimal* solution to a well-defined robustness criterion (cf. the statement of Alg. 45.4). This result was established by Hassibi, Sayed, and Kailath (1993,1996) using connections with $\mathcal{H}^\infty$ filtering theory (cf. App. 45.A). Extensions of this conclusion to the backpropagation algorithm for neural network training can also be found in Hassibi, Sayed, and Kailath (1994,1999). Further extensions to time-variant step-sizes, Gauss-Newton recursions, and filtered-error algorithms can be found in Sayed and Rupp (1994,1996,1997,1998) and Rupp and Sayed (1996ab).

Energy conservation. The energy-conservation relation (46.4) (or, equivalently, (15.32)) is due to Sayed and Rupp (1995). It was used by the authors in a series of works, including Rupp and Sayed (1996ab,1997,1998,2000) and Sayed and Rupp (1996,1997,1998), in their studies on the robustness and small gain analysis of adaptive filters. Several of their robustness results are discussed in the problems at the end of this part. The energy arguments of Chapter 46 are due to Sayed and Rupp (1994,1996), as well as the Cauchy-Schwartz argument of Sec. 46.1; this latter argument also appears in Sayed and Kailath (1994b). An extension of the energy conservation argument of Chapter 46 to continuous-time LMS appears in Sayed and Kokotovic (1995).

Comparison of RLS and LMS. It is instructive to compare the robustness performance of LMS and RLS for the case of single-tap adaptive filters (in which case the regressor is a scalar). The following example is extracted from Hassibi, Sayed and Kailath (1996). Assume that the regression signal randomly assumes the values ± 1 with probability $1/2$ and let $w^o = 0.25$ be the unknown weight that we wish to estimate. The noise signal is assumed to be zero-mean and Gaussian. We first employ LMS, i.e.,

$$w(i) = w(i-1) + \mu u(i)[d(i) - u(i)w(i-1)]$$

and compute the initial $N = 100$ weight estimates starting from $w(-1) = 0$ and using $\mu = 0.97$. We also evaluate the entries of the resulting mapping $\mathcal{T}$ from (46.17), which we now denote by $\mathcal{T}_{lms}$. [Observe that the chosen value for μ satisfies the requirement $\mu\|u(i)\|^2 \leq 1$ for all i.]

We then employ RLS, i.e.,

$$w(i) = w(i-1) + \frac{p(i-1)u(i)}{1+p(i-1)}[d(i) - u(i)w(i-1)], \quad p(i) = \frac{p(i-1)}{1+p(i-1)}$$

with initial conditions $p(-1) = \mu$ and $w(-1) = 0$. We again compute the initial 100 weight estimates using $\mu = 0.97$ and evaluate the entries of the corresponding mapping $\mathcal{T}$, now denoted by $\mathcal{T}_{rls}$ (in this case, it can be verified that $p(i) = \mu/(1+(i+1)\mu)$ for all $i \geq -1$).

Figure XI.1 shows a plot of the 100 singular values of the mappings $\mathcal{T}_{lms}$ and $\mathcal{T}_{rls}$. As expected from the robustness analysis in the chapter, we find that the singular values of $\mathcal{T}_{lms}$ (indicated by an almost horizontal line at 1) are all bounded by one, whereas the maximum singular value of $\mathcal{T}_{rls}$ is approximately 1.65. Observe, however, that most of the singular values of $\mathcal{T}_{rls}$ are considerably smaller than one while the singular values of $\mathcal{T}_{lms}$ are clustered around one. This fact has an interesting interpretation. If the disturbance vector $\underline{\text{dist}}$ happens to lie in the range space of the right singular vectors of $\mathcal{T}_{rls}$ that are associated with the smaller singular values, then its effect will be significantly attenuated by RLS. This fact indicates that while LMS has the best worst-case performance (in the

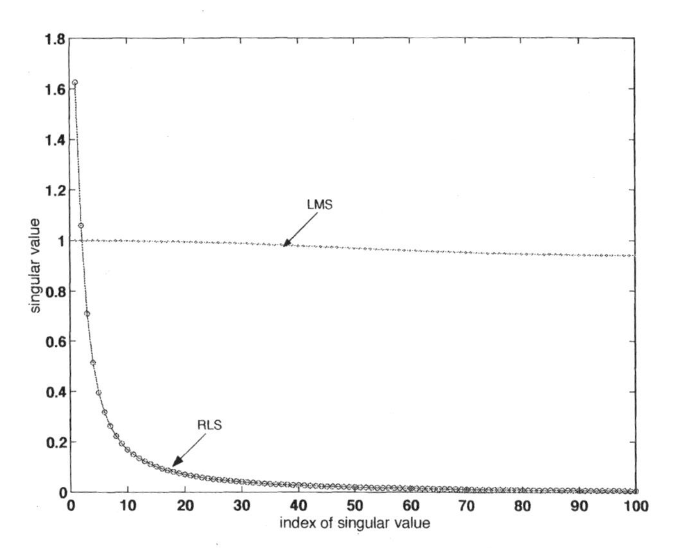

FIGURE XI.1 Plot of the first 100 singular values associated with the LMS and RLS mappings.

sense that it guards against worst-case disturbances), the RLS algorithm is expected to have a better performance on average.

Feedback structure. In Fig. 15.5 we showed that an intrinsic feedback structure can be associated with every adaptive filter update in the energy domain (cf. Sayed and Rupp (1995) and Rupp and Sayed (1996a)). The feedback structure was motivated in these references, and in Chapter 15, using energy arguments. It is seen to consist of two distinctive blocks: a time-variant lossless (i.e., energy preserving) feedforward mapping and a time-variant feedback path. In Probs. XI.9–XI.13 we examine this configuration more closely and show that it lends itself to analysis via a so-called small gain theorem from system theory (see, e.g., Vidyasagar (1993) and Khalil (1996)). Such analyses lead to stability and robustness conditions that require the contractivity of certain operators. Results along these lines can be found in Rupp and Sayed (1996a,1997) and Sayed and Rupp (1997).

There have been several earlier interesting works in the literature on feedback structures for adaptive filtering, e.g., by Ljung (1977b) and Landau (1979,1984), as well as by Popov (1973) on hyperstability or passivity results — see also Anderson et al. (1986). These earlier works were of a different nature than the structure of Fig. 15.5; they were concerned with the fact that the update equations of adaptive filters can be represented in a recursive manner. The feedback structure of Fig. 15.5 is instead concerned with energy propagation and, among other features, it enforces a lossless mapping in the feedforward path and allows for a time-variant mapping in the feedback (in comparison, hyperstability analyses require one of the paths to be time-invariant (see Landau (1979, p. 381)).

TABLE XI.1 Three FxLMS variants.

Algorithm	Complexity	Memory	Convergence
FxLMS	$2M + M_F$	$3M$	poor
mFxLMS Bjarnason (1992), Kim et al. (1994)	$3M + 2M_F$	$3M + M_F$	good
mFxLMS Rupp and Sayed (1998)	$2M + 2M_F$	$3M + M_F$	good

Robustness of FxLMS and active noise control. A widely used algorithm in active noise control is the so-called filtered-x least-mean-squares (FxLMS) algorithm (see, e.g., Widrow and Stearns

(1985), Sethares, Anderson and Johnson (1989), and Elliott and Nelson (1993)). The algorithm is discussed in the computer project at the end of this part. In its standard form, the algorithm has LMS complexity, i.e., it requires $2M$ operations per sample. However, it exhibits poor convergence performance. Modifications, referred to as modified FxLMS, were proposed by Bjarnason (1992), Kim et al. (1994), and Rupp and Sayed (1998), in order to ameliorate the convergence problem — see Table XI.1, which compares the performance of these variants in terms of complexity, memory requirement, and convergence. In the table, M is the length of the adaptive filter and M_F is the length of the error filter (as described in the computer project).

Problems and Computer Projects

PROBLEMS

Problem XI.1 (Indefinite quadratic cost) Consider the quadratic cost function (44.1) and choose $\Pi = \mathrm{I}, H = \mathrm{I}, W = (0 \oplus -2), y = \mathrm{col}\{\alpha, \beta\}$.

(a) What is the signature of the coefficient matrix $\Pi + H^* W H$?

(b) Verify that the gradient vector of $J(w)$ evaluates to zero at $\widehat{w} = \mathrm{col}\{0, 2\beta\}$, and that $J(\widehat{w}) = 2|\beta|^2$. Show that $\widehat{w}$ is the only vector where the gradient of $J(w)$ is zero.

(c) Are there any other choices for w at which $J(w)$ evaluates to $2|\beta|^2$?

(d) Plot the contour curves of $J(w)$.

Problem XI.2 (Negative-definite coefficient) Refer to (45.19).

(a) Conclude from (45.20) that the matrices $\{\Delta^{-1} \oplus (1 + u_i P_{i-1} u_i^*)^{-1}\}$ and Γ_i are congruent.

(b) Use the existence condition (45.15) to conclude that $\Delta < 0$.

Problem XI.3 (Robust formulation with weighted noise energy) Refer to the discussion in Sec. 45.1. Assume again that we are given measurements $d(i) = u_i w^o + v(i)$ and that we wish to estimate $s(i) = u_i w^o$ in order to satisfy the criterion

$$\frac{\sum_{j=0}^{i} |\widehat{s}(j|j) - s(j)|^2}{w^{o*}\Pi w^o \ + \ \alpha \sum_{j=0}^{i} |v(j)|^2} < \gamma^2 \quad \text{for all } \ i = 0, 1, \ldots, N$$

for some $\alpha > 0$. Repeat the arguments of that section to show that one construction for the desired estimates is as follows. Start with $w_{-1} = 0$ and $P_{-1} = \Pi^{-1}$ and iterate:

$$\begin{aligned} w_i &= w_{i-1} + \frac{P_{i-1} u_i^*}{\alpha^{-1} + u_i P_{i-1} u_i^*}[d(i) - u_i w_{i-1}], \quad \widehat{s}(i|i) = u_i w_i \\ P_i &= P_{i-1} - \frac{P_{i-1} u_i^* u_i P_{i-1}}{(\alpha - \gamma^{-2})^{-1} + u_i P_{i-1} u_i^*} \end{aligned}$$

Argue that this solution satisfies the above robustness criterion for all $\gamma > 1/\sqrt{\alpha}$.

Problem XI.4 (Non-sufficiently exciting regressors) Assume that condition (45.25) does not hold so that $\lim_{N\to\infty} \left(\sum_{k=0}^{N} u_k^* u_k\right) \leq \rho \mathrm{I}$, for some $\rho < \infty$. Choose $\Pi = \epsilon \mathrm{I}$, $\epsilon > 0$. Use (45.26) to argue that $P_i > 0$ for any γ satisfying $\gamma^2 > \rho/(\epsilon + \rho)$.

Problem XI.5 (Min-max interpretation) Refer to the discussion in Sec. 45.2 and assume the regressors are sufficiently exciting as in (45.25).

(a) Argue that, for any $\gamma \geq 1$, the estimates $\{\widehat{s}(j|j)\}$ that are constructed according to the ϵ-NLMS algorithm guarantee the following bound, for all finite-energy noise sequences $\{v(i)\}$:

$$\sup_{\{w^o, v(j)\}} \left(\frac{\sum_{j=0}^{\infty} |\widehat{s}(j|j) - s(j)|^2}{\epsilon \|w^o\|^2 + \sum_{j=0}^{\infty} |v(j)|^2} \right) < \gamma^2$$

(b) Conclude that the ϵ−NLMS recursion of Alg. 45.2 is a solution to the following min-max problem over all finite-energy noise sequences $\{v(i)\}$:

$$\gamma_{\text{opt}}^2 \stackrel{\Delta}{=} \inf_{\{\widehat{s}(j|j)\}} \sup_{\{w^o, v(j)\}} \left(\frac{\sum_{j=0}^{\infty} |\widehat{s}(j|j) - s(j)|^2}{\epsilon \|w^o\|^2 + \sum_{j=0}^{\infty} |v(j)|^2} \right)$$

Argue further that $\gamma_{\text{opt}}^2 = 1$.

(c) Repeat the derivation for the LMS algorithm of Sec. 45.4. More specifically, show that LMS is a solution to the following min-max problem over all finite-energy noise sequences $\{v(i)\}$:

$$\gamma_{\text{opt}}^2 \stackrel{\Delta}{=} \inf_{\{\widehat{s}(j|j-1)\}} \sup_{\{w^o, v(j)\}} \left(\frac{\sum_{j=0}^{\infty} |\widehat{s}(j|j-1) - s(j)|^2}{\mu^{-1} \|w^o\|^2 + \sum_{j=0}^{\infty} |v(j)|^2} \right)$$

Argue further that $\gamma_{\text{opt}}^2 = 1$.

Problem XI.6 (Passivity relation) Consider LMS-type updates with time-variant step-sizes of the form $w_i = w_{i-1} + \mu(i) u_i^* [d(i) - u_i w_{i-1}]$ with $\mu(i) > 0$.

(a) Show that

$$\begin{cases} \|\widetilde{w}_i\|^2 + \mu(i)\, |e_a(i)|^2 \leq \|\widetilde{w}_{i-1}\|^2 + \mu(i)\, |v(i)|^2 & \text{if } \mu(i)\|u_i\|^2 \leq 1 \\ \|\widetilde{w}_i\|^2 + \mu(i)\, |e_a(i)|^2 = \|\widetilde{w}_{i-1}\|^2 + \mu(i)\, |v(i)|^2 & \text{if } \mu(i)\|u_i\|^2 = 1 \\ \|\widetilde{w}_i\|^2 + \mu(i)\, |e_a(i)|^2 \geq \|\widetilde{w}_{i-1}\|^2 + \mu(i)\, |v(i)|^2 & \text{if } \mu(i)\|u_i\|^2 \geq 1 \end{cases}$$

(b) Define the normalized quantities $\bar{e}_a(i) = \sqrt{\mu(i)} e_a(i)$ and $\bar{v}(i) = \sqrt{\mu(i)} v(i)$ and assume $\mu(i)\|u_i\|^2 \leq 1$. Conclude that

$$\boxed{\|\widetilde{w}_N\|^2 + \sum_{i=0}^{N} |\bar{e}_a(i)|^2 \leq \|\widetilde{w}_{-1}\|^2 + \sum_{i=0}^{N} |\bar{v}(i)|^2}$$

(c) What does the relation of part (b) collapse to when $\mu(i) = \mu$?

(d) What does the relation of part (b) collapse to when $\mu(i) = 1/(\epsilon + \|u_i\|^2)$?

Remark. For more details see Sayed and Rupp (1996).

Problem XI.7 (RLS recursion) Consider the RLS recursion of Sec. 46.3. Follow the argument of Sec. 46.2 to show that the recursion for w_i can be rewritten in the equivalent form $w_{i-1} = w_i - P_{i-1} u_i^* [d(i) - u_i w_i]$.

Problem XI.8 (Convergence result) Consider the LMS recursion

$$w_i = w_{i-1} + \mu u_i^* [d(i) - u_i w_{i-1}], \quad w_{-1} = 0$$

and assume the regressors $\{u_i\}$ are bounded and that the step-size μ is positive and bounded from above, say $\mu < c < \inf_i \left(1/\|u_i\|^2\right)$. Assume further that the data $\{d(i), u_i\}$ satisfy $d(i) = u_i w^o +$

$v(i)$ for some unknown $M \times 1$ bounded vector w^o, and for some unknown noise sequence $v(i)$ with finite energy, i.e., $\lim_{N\to\infty}\left(\sum_{i=0}^{N}|v(i)|^2\right) < \infty$.

(a) Use the contraction relation (46.9) to conclude that $e_a(i) \to 0$ as $i \to \infty$.

(b) Assume the regressors are *persistently exciting*, namely, there exists a finite integer $L \geq M$ such that the matrix $\text{col}\{u_i, u_{i+1}, \ldots, u_{i+L}\}$ has full rank for sufficiently large i. Conclude from part (a) that $w_i \to w^o$ as $i \to \infty$.

(c) Assume instead that the noise sequence has finite power, as opposed to finite energy, i.e.,

$$\lim_{N\to\infty}\left(\frac{1}{N}\sum_{i=0}^{N}|v(i)|^2\right) < \infty$$

Show that the sequence $\{e_a(i)\}$ will also have finite power.

Problem XI.9 (Small gain condition) Consider the feedback structure of Fig. XI.2 It has a finite-dimensional lossless mapping $\mathcal{T}$ in the feedforward path, i.e., $\mathcal{T}$ satisfies $\mathcal{T}\mathcal{T}^* = \mathcal{T}^*\mathcal{T} = \text{I}$. It also has an arbitrary finite-dimensional mapping $\mathcal{F}$ in the feedback path. The input and output signals of interest are denoted by $\{x, y, r, v, e\}$. In this system, the signals x, v play the role of disturbances.

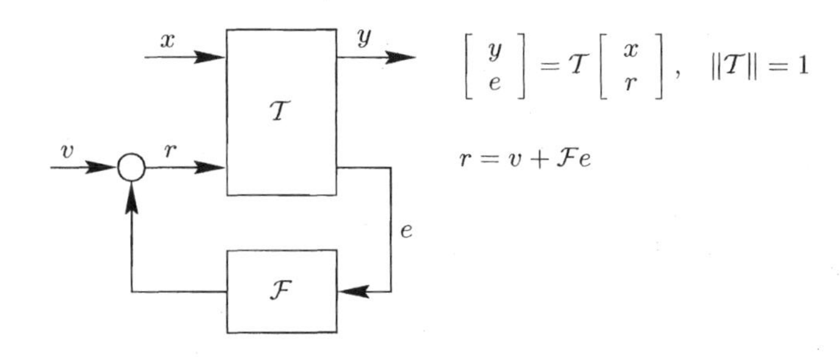

FIGURE XI.2 A feedback structure with a lossless feedforward mapping.

(a) Establish the energy conservation relation $\|y\|^2 + \|e\|^2 = \|x\|^2 + \|r\|^2$, and conclude that $\|e\| \leq \|x\| + \|r\|$. Here $\|\cdot\|$ denotes the Euclidean norm of its vector argument.

(b) Let $\|\mathcal{F}\|$ denote the maximum singular value of $\mathcal{F}$. Use the triangle inequality of norms to show that $\|r\| \leq \|v\| + \|\mathcal{F}\| \cdot \|e\|$.

(c) Conclude that if $\mathcal{F}$ satisfies the small gain condition $\|\mathcal{F}\| < 1$ (in which case we say that $\mathcal{F}$ is contractive), then

$$\boxed{\|e\| \leq \frac{1}{1-\|\mathcal{F}\|}\cdot[\,\|x\| + \|v\|\,]}$$

Remark. This result indicates that a contractive feedback mapping $\mathcal{F}$ guarantees a certain robustness performance from the disturbances $\{x, v\}$ to the output $\{e\}$. In this case, we say that the mapping from $\{x, v\}$ to $\{e\}$ is l_2-stable.

(d) Plot $1/(1 - \|\mathcal{F}\|)$ for $0 < \|\mathcal{F}\| < 1$. Conclude that the smaller the value of $\|\mathcal{F}\|$:

(d.1) The smaller the effect of $\{x, v\}$ on $\{e\}$.

(d.2) The smaller the upper bound on $\|e\|$.

Problem XI.10 (l_2-stability of adaptive filters) Consider adaptive filters of the LMS-type with time-dependent step-sizes

$$\boxed{w_i = w_{i-1} + \mu(i)u_i^*[d(i) - u_i w_{i-1}]}$$

and assume the data $\{d(i), u_i\}$ satisfy $d(i) = u_i w^o + v(i)$ for some unknown vector w^o and disturbance $v(i)$. Define the weight-error vector and the *a priori* and *a posteriori* estimation errors $\widetilde{w}_i = w^o - w_i$, $e_a(i) = u_i \widetilde{w}_{i-1}$ and $e_p(i) = u_i \widetilde{w}_i$. Let also $\bar{\mu}(i) = (\|u_i\|^2)^\dagger$.

(a) Show that $e_p(i) = \left(1 - \mu(i)\|u_i\|^2\right) e_a(i) \; - \; \mu(i)\|u_i\|^2 v(i)$.

(b) Establish the energy conservation relation

$$\boxed{\|\widetilde{w}_i\|^2 + \bar{\mu}(i)\,|e_a(i)|^2 \;=\; \|\widetilde{w}_{i-1}\|^2 + \bar{\mu}(i)\,|e_p(i)|^2}$$

Conclude that the mapping from

$$\{\widetilde{w}_N, \sqrt{\bar{\mu}(0)}\, e_a(0), \sqrt{\bar{\mu}(1)}\, e_a(1), \ldots, \sqrt{\bar{\mu}(N)}\, e_a(N)\}$$

to

$$\{\widetilde{w}_{-1}, \sqrt{\bar{\mu}(0)}\, e_p(0), \sqrt{\bar{\mu}(1)}\, e_p(1), \ldots, \sqrt{\bar{\mu}(N)}\, e_p(N)\}$$

is lossless.

(c) Assume $u_i \neq 0$. Define

$$\eta(N) = \max_{0 \leq i \leq N} \left|1 - \mu(i)\|u_i\|^2\right| \qquad \text{and} \qquad \kappa(N) = \max_{0 \leq i \leq N} \mu(i)\|u_i\|^2$$

Use the small gain condition of Prob. XI.9 to establish that if $\eta(N) < 1$, i.e., if $\mu(i)\|u_i\|^2 < 2$ for $i = 0, 1, \ldots, N$, then the following inequalities hold:

$$\sqrt{\sum_{i=0}^{N} \bar{\mu}(i)\,|e_a(i)|^2} \;\leq\; \frac{1}{1-\eta(N)} \left[\|\widetilde{w}_{-1}\| + \kappa(N)\sqrt{\sum_{i=0}^{N} \bar{\mu}(i)|v(i)|^2} \right]$$

$$\sqrt{\sum_{i=0}^{N} \mu(i)\,|e_a(i)|^2} \;\leq\; \frac{\kappa^{1/2}(N)}{1-\eta(N)} \left[\|\widetilde{w}_{-1}\| + \kappa^{1/2}(N)\sqrt{\sum_{i=0}^{N} \mu(i)|v(i)|^2} \right]$$

How would these results change if $u_i = 0$ for some i, say at $i = i_o$?

(d) Assume $\eta(N) < c < 1$ and $\kappa(N) < b < \infty$ for all N. That is, $\{\eta(N), \kappa(N)\}$ are uniformly bounded as indicated. Assume further that $\|u_i\|^{-2} > \epsilon > 0$ for all i. If the normalized noise sequence $\{\sqrt{\mu(i)}v(i)\}$ has finite energy, conclude that $e_a(i) \to 0$ as $i \to \infty$. If, in addition, the regressors $\{u_i\}$ are persistently exciting, as defined in Prob. XI.8, conclude that $w_i \to w^o$. *Remark.* Compared with Prob. XI.8, the convergence result is now obtained under the condition of a finite-energy sequence $\{\sqrt{\mu(i)}v(i)\}$, which requires $\sqrt{\mu(i)}v(i) \to 0$ as opposed to $v(i) \to 0$ (see Sayed and Rupp (1996) and Rupp and Sayed (1996a)).

Problem XI.11 (Energy propagation in feedback) Consider the same setting as in Prob. XI.10.

(a) Explain that the results of parts (a) and (b) of that problem can be represented in diagram form as shown in Fig. XI.3, where the lossless mapping from $\{\widetilde{w}_{i-1}, \sqrt{\mu(i)}e_a(i)\}$ to $\{\widetilde{w}_i, \sqrt{\mu(i)}e_p(i)\}$ is indicated by $\mathcal{L}$.

(b) Assume $v(i) = 0$. From parts (a) and (b) of Prob. XI.10 show that

$$\boxed{\|\widetilde{w}_i\|^2 \;=\; \|\widetilde{w}_{i-1}\|^2 - \mu(i)\left[2 - \mu(i)\|u_i\|^2\right]|e_a(i)|^2}$$

Conclude that $\|\widetilde{w}_i\|^2 \leq \|\widetilde{w}_{i-1}\|^2$ if, and only if, $0 < \mu(i)\|u_i\|^2 < 2$.

(c) Show that choosing $\mu(i)$ such that $\mu(i)\|u_i\|^2 = 1$ results in the largest energy decrease in going from $\|\widetilde{w}_{i-1}\|^2$ to $\|\widetilde{w}_i\|^2$.

(d) Which algorithm corresponds to the choice $\mu(i)\|u_i\|^2 = 1$?

Remark. Observe that for the choice of step-size in part (c), the feedback path in Fig. XI.4 is disconnected, so that there is no energy flowing back from the lower output of the lossless map into its lower input.

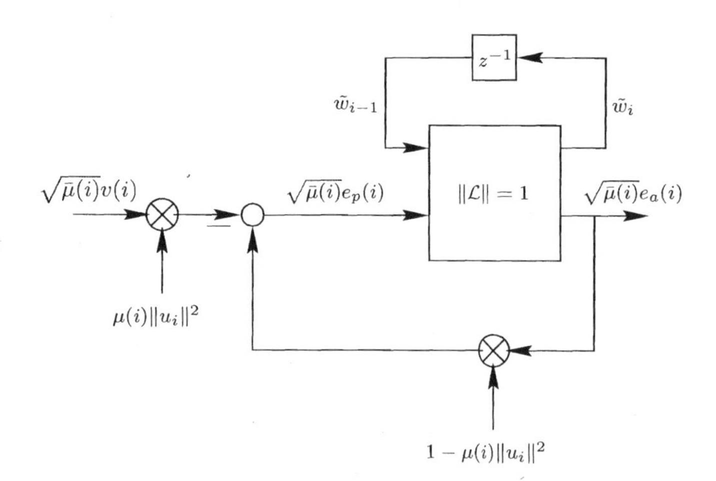

FIGURE XI.3 A feedback structure representing each iteration of an LMS-type filter with time-dependent step-sizes.

Problem XI.12 (Filtered-error adaptive filter) In applications such as active noise control and vibration control, adaptive filters that rely on filtered errors become necessary, i.e., adaptive updates of the form

$$\boxed{w_i = w_{i-1} \;+\; \mu(i)u_i^* F[d(i) - u_i w_{i-1}]}$$

where the notation $F[\cdot]$ denotes some filter that operates on the error signal $d(i) - u_i w_{i-1}$. In these applications, filtered versions of the error are more readily available than the error signal itself, so that the adaptive filter update would need to rely on $F[d(i) - u_i w_{i-1}]$ rather than on $d(i) - u_i w_{i-1}$. For our purposes here, we shall assume that $F[\cdot]$ is an FIR filter of order M_F with transfer function $F(z) = \sum_{j=0}^{M_F-1} f(j)z^{-j}$.

(a) Let $d(i) = u_i w^o + v(i)$. Introduce the estimation errors $e_a(i) = u_i\widetilde{w}_{i-1}$ and $e_p(i) = u_i\widetilde{w}_i$, where $\widetilde{w}_i = w^o - w_i$. Show that

$$\boxed{e_p(i) = \left(e_a(i) - \mu(i)\|u_i\|^2 F[e_a(i)]\right) \;-\; \mu(i)\|u_i\|^2 F[v(i)]}$$

and

$$\boxed{\|\widetilde{w}_i\|^2 + \bar{\mu}(i)\,|e_a(i)|^2 \;=\; \|\widetilde{w}_{i-1}\|^2 + \bar{\mu}(i)\,|e_p(i)|^2}$$

where $\bar{\mu}(i) = 1/\|u_i\|^2$ if $u_i \neq 0$ and $\bar{\mu}(i) = 0$ otherwise. *Remark.* Of course, this is the same energy-conservation relation that we derived earlier in Thm. 15.1 for general functions $g(\cdot)$ of $d(i) - u_i w_{i-1}$.

(b) Assume $u_i \neq 0$, $v(i) = 0$, and choose $\mu(i) = \alpha/\|u_i\|^2$ for some $\alpha > 0$. Show that the mapping from $\{\sqrt{\bar{\mu}(0)}e_a(0), \ldots, \sqrt{\bar{\mu}(N)}e_a(N)\}$ to $\{\sqrt{\bar{\mu}(0)}e_p(0), \ldots, \sqrt{\bar{\mu}(N)}e_p(N)\}$ is described by the $(N+1) \times (N+1)$ lower triangular matrix

$$\mathcal{F}_N = \begin{bmatrix} 1-\alpha f(0) & & & \\ -\alpha\frac{\sqrt{\bar{\mu}(1)}}{\sqrt{\bar{\mu}(0)}}f(1) & 1-\alpha f(0) & & \\ -\alpha\frac{\sqrt{\bar{\mu}(2)}}{\sqrt{\bar{\mu}(0)}}f(2) & -\alpha\frac{\sqrt{\bar{\mu}(2)}}{\sqrt{\bar{\mu}(1)}}f(1) & 1-\alpha f(0) & \\ \vdots & & & \ddots \end{bmatrix}$$

in terms of the coefficients $\{f(j)\}$ of the filter $F(z)$. Since usually $N+1 > M_F$, argue that $\mathcal{F}_N$ is banded with only M_F diagonals.

(c) Define the filtered-noise sequence $\bar{v}(i) = F[v(i)]$ and the coefficients $\eta(N) = \|\mathcal{F}_N\|$ and $\kappa(N) = \max_{0 \le i \le N} \mu(i)\|u_i\|^2$. That is, $\eta(N)$ is the largest singular value of $\mathcal{F}_N$. If $\eta(N) < 1$, show that relations similar to those in part (c) of Prob. XI.10 hold with $v(i)$ replaced by $\bar{v}(i)$.

(d) Usually, the length M_F is much smaller than the length of the regression vector u_i. Thus assume that the energy of the regression vectors does not change appreciably over M_F successive iterations, so that $\bar{\mu}(i) \approx \bar{\mu}(i-1) \approx \ldots \approx \bar{\mu}(i-M_F)$. Verify that, in this case, $\mathcal{F}_N$ can be approximated by $\mathcal{F}_N \approx \mathrm{I} - \alpha \mathcal{C}_N$, where $\mathcal{C}_N$ is the $(N+1) \times (N+1)$ convolution matrix associated with the filter $F(z)$, i.e., $\mathcal{C}$ is a banded Toeplitz matrix, e.g., for $M_F = 3$,

$$\mathcal{C}_N = \begin{bmatrix} f(0) & & & & \\ f(1) & f(0) & & & \\ f(2) & f(1) & f(0) & & \\ & f(2) & f(1) & f(0) & \\ & & \ddots & \ddots & \ddots \end{bmatrix}$$

Argue that the contractivity requirement $\|\mathcal{F}_N\| < 1$ can be met if the step-size parameter α satisfies $\max_\omega \left|1 - \alpha F(e^{j\omega})\right| < 1$, where $F(e^{j\omega})$ denotes the frequency response of $F(z)$.

(e) Assume $v(i) = 0$ and let $e = \{\sqrt{\bar{\mu}(0)}e_a(0), \sqrt{\bar{\mu}(1)}e_a(1), \ldots, \sqrt{\bar{\mu}(N)}e_a(N)\}$ which, as seen from part (b), is the input to the mapping $\mathcal{F}_N$. Under the same conditions of part (d) show that $\|\widetilde{w}_N\|^2 \le \|\widetilde{w}_{-1}\|^2 + \|(\mathrm{I} - \alpha\mathcal{C}_N)\|^2\|e\|^2 - \|e\|^2$. Argue that an approximate choice for α that maximizes the decrease in energy from $\|\widetilde{w}_{-1}\|^2$ to $\|\widetilde{w}_N\|^2$ can be obtained by solving $\min_\alpha \max_\omega \left|1 - \alpha F(e^{j\omega})\right|$ over all values of α that guarantee $\max\limits_\omega \left|1 - \alpha F(e^{j\omega})\right| < 1$.

Remark. For more details see Rupp and Sayed (1996a).

Problem XI.13 (Transfer function analysis) Consider the LMS recursion

$$w_i = w_{i-1} + \mu u_i^* e(i), \qquad e(i) = d(i) - u_i w_{i-1}$$

and assume u_i has shift structure, i.e., $u_i = \begin{bmatrix} u(i) & u(i-1) & \ldots & u(i-M+1) \end{bmatrix}$. Assume further that $u(i)$ is sinusoidal, say $u(i) = C\cos(\omega_o i)$ for some $\{C, \omega_o\}$. Let $d(i) = u_i w^o + v(i)$ and define the weight-error vector $\widetilde{w}_i = w^o - w_i$.

(a) Verify that the update for the $k-$th entry of the weight-error vector is given by

$$[\widetilde{w}_i]_k = [\widetilde{w}_{i-1}]_k - \frac{\mu C}{2}\left[e^{j(i-k)\omega_o} + e^{-j(i-k)\omega_o}\right] e(i)$$

Let $\widetilde{W}_k(z)$ denote the $z-$transform of the sequence $\{[\widetilde{w}_i]_k\}$. Ignoring initial conditions, verify that

$$\widetilde{W}_k(z) = -\frac{z}{z-1}\frac{\mu C}{2}\left[E(ze^{-j\omega_o})e^{-jk\omega_o} + E(ze^{j\omega_o})e^{jk\omega_o}\right]$$

(b) Likewise, using $e_a(i) = u_i \widetilde{w}_{i-1}$ show that

$$E_a(z) = z^{-1}\frac{C}{2}\sum_{k=0}^{M-1}\left[\widetilde{W}_k(ze^{-j\omega_o})e^{j\omega_o(1-k)} + \widetilde{W}_k(ze^{j\omega_o})e^{j\omega_o(k-1)}\right]$$

(c) Substitute the expression for $\widetilde{W}_k(z)$ into $E_a(z)$ and ignore the effects of the mixing terms $E(ze^{2j\omega_o})$ and $E(ze^{-2j\omega_o})$ to show that

$$E_a(z) \quad \approx \quad \frac{\mu C^2 M}{2} \frac{1 - z\cos(\omega_o)}{z^2 - 2z\cos(\omega_o) + 1} E(z)$$

Using $E(z) = E_a(z) + V(z)$, conclude that the transfer function from $v(\cdot)$ to $e_a(\cdot)$ is approximately

$$\boxed{\frac{E_a(z)}{V(z)} \approx \frac{\frac{\mu}{\bar{\mu}}\left[1 - z\cos(\omega_o)\right]}{z^2 - 2z\cos(\omega_o)\left(1 - \frac{\mu}{2\bar{\mu}}\right) + \left(1 - \frac{\mu}{\bar{\mu}}\right)}}$$

where $\bar{\mu} = 2/C^2 M$.

(d) Define $\overline{V}(z) = \frac{\mu}{\bar{\mu}}V(z) - \left(1 - \frac{\mu}{\bar{\mu}}\right) E_a(z)$. Show that $E_a(z)/\overline{V}(z) = (z^{-1} - \cos(\omega_o))/(z - \cos(\omega_o))$. Conclude that the transfer function from $\overline{v}(\cdot)$ to $e_a(\cdot)$ is all-pass. Conclude further that the transfer function from v to e_a can be represented as a feedback structure, as shown in Fig. XI.4, with an all-pass filter in the forward path and a constant gain in the feedback loop.

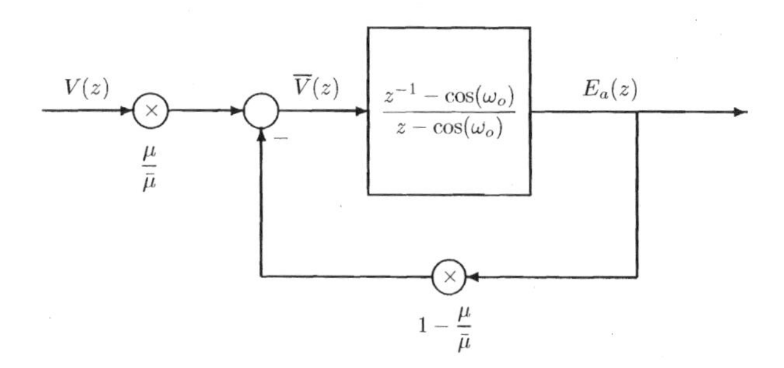

FIGURE XI.4 A transfer function description for LMS in terms of an all-pass forward mapping and a feedback gain.

Remark. There are some analogies between the transfer function representation of Fig. XI.4 and the feedback structure of Fig. 15.5 derived earlier using energy conservation arguments. However, the arguments that led to Fig. 15.5 did not use any approximation and were carried out in the time domain. The transfer function derivation, on the other hand, assumes regressors with shift structure, uses a sinusoidal input sequence (although some other classes of signals can be used), and ignores the effect of initial conditions and nonlinear mixing terms. Still, the transfer function approach offers useful insights into the performance of LMS-type algorithms. It was first described by Glover (1977) and later extended by Clarkson and White (1987).

Problem XI.14 (Filtered-error LMS) Consider the filtered-error LMS recursion from Prob. XI.12,

$$w_i = w_{i-1} + \mu u_i^* F[e(i)], \qquad e(i) = d(i) - u_i w_{i-1}$$

Repeat the transfer function derivation of Prob. XI.13 and show that it leads to

$$\frac{E_a(z)}{V(z)} \approx \frac{\frac{\mu C^2 M}{2} F(z)(1 - z\cos(\omega_o))}{z^2 - 2z\cos(\omega_o)\left(1 - \frac{\mu C^2 M}{4}F(z)\right) + \left(1 - \frac{\mu C^2 M}{2}F(z)\right)}$$

Conclude that the transfer function from $v(\cdot)$ to $e_a(\cdot)$ can again be described in terms of an all-pass feedforward path and a dynamic feedback loop given by $1 - \frac{\mu}{\bar{\mu}}F(z)$, where $\bar{\mu} = 2/C^2 M$.

Problem XI.15 (Two inequalities) Consider any column vectors $\{x, y\}$, and any positive-definite matrix P.

(a) Show that for any $\alpha > 0$, $(x+y)^* P(x+y) \geq \left(1 - \frac{1}{\alpha}\right) x^* P x + (1 - \alpha) y^* P y$.

(b) Likewise, show that for any $\alpha > 0$, $(x+y)^* P(x+y) \leq \left(1 + \frac{1}{\alpha}\right) x^* P x + (1 + \alpha) y^* P y$.

Problem XI.16 (Two optimization problems) Establish the following conclusions.

(a) For any $\beta > 0$,

$$\min_{\alpha>1} \left(\frac{\alpha^2 + (\beta - 1)\alpha}{\alpha - 1} \right) = (1 + \sqrt{\beta})^2$$

and the minimum occurs at $\alpha_o = 1 + \sqrt{\beta}$.

(b) For any $\beta > 1$,

$$\max_{\alpha>0} \left(\frac{-\alpha^2 + (\beta - 1)\alpha}{\alpha + 1} \right) = (\sqrt{\beta} - 1)^2$$

and the maximum occurs at $\alpha_o = \sqrt{\beta} - 1$.

Problem XI.17 (RLS algorithm) Let $d(i) = u_i w^o + v(i)$ and consider the RLS filter:

$$\begin{aligned} w_i &= w_{i-1} + \frac{P_{i-1}u_i}{1 + u_i P_{i-1} u_i^*}[d(i) - u_i w_{i-1}], \quad i = 0, 1, \ldots, N \\ P_i &= P_{i-1} - \frac{P_{i-1}u_i u_i^* P_{i-1}}{1 + u_i P_{i-1} u_i^*}, \quad P_{-1} = \Pi^{-1} \end{aligned}$$

which we know minimizes, over w, the cost function $J(w) = w^*\Pi w + \sum_{i=0}^{N} |d(i) - u_i w|^2$. We also know from Lemma 30.1 that the resulting minimum cost is given by either of the following expressions:

$$\xi(N) = \sum_{i=0}^{N} \frac{|e(i)|^2}{1 + u_i P_{i-1} u_i^*} = \sum_{i=0}^{N} |r(i)|^2 (1 + u_i P_{i-1} u_i^*)$$

where $e(i) = d(i) - u_i w_{i-1}$ and $r(i) = d(i) - u_i w_i$. Let $\widetilde{w}_i = w^o - w_i$. Then it also holds that $e(i) = e_a(i) + v(i)$ and $r(i) = e_p(i) + v(i)$.

(a) Since $J(w) \geq \xi(N)$ for any w and $\{v(\cdot)\}$, conclude that the following inequality holds:

$$w^{o*}\Pi w^o + \sum_{i=0}^{N} |v(i)|^2 \geq \left\{ \min_{0 \leq i \leq N} \left(\frac{1}{1 + u_i P_{i-1} u_i^*} \right) \right\} \cdot \left(\sum_{i=0}^{N} |e_a(i) + v(i)|^2 \right)$$

(b) Define $\overline{r} = \max_{0 \leq i \leq N} (1 + u_i P_{i-1} u_i^*)$. Use the result of part (a) from Prob. XI.15 to conclude that, for any $\alpha > 1$,

$$\sum_{i=0}^{N} |e_a(i)|^2 \leq \frac{\overline{r}}{1 - 1/\alpha} w^{o*}\Pi w^o + \left(\frac{\alpha^2 + \alpha(\overline{r} - 1)}{\alpha - 1} \right) \sum_{i=0}^{N} |v(i)|^2$$

(c) Use the results of Prob. XI.16 to conclude that

$$\boxed{\frac{\sum_{i=0}^{N} |e_a(i)|^2}{w^{o*}\Pi w^o + \sum_{i=0}^{N} |v(i)|^2} \leq (1 + \sqrt{\overline{r}})^2 \quad \text{and} \quad \frac{\sum_{i=0}^{N} |e_p(i)|^2}{w^{o*}\Pi w^o + \sum_{i=0}^{N} |v(i)|^2} \leq (1 + \sqrt{1/\underline{r}})^2}$$

where $\underline{r} = \min_{0 \leq i \leq N} (1 + u_i P_{i-1} u_i^*)$, and conclude that

$$\boxed{\frac{\sum_{i=0}^{\infty} |e_p(i)|^2}{w^{o*}\Pi w^o + \sum_{i=0}^{\infty} |v(i)|^2} \leq 4}$$

Remark. This result agrees with (46.26). The energy argument that led to (46.26) provides a tighter bound. For more details on the results of Probs. XI.15–XI.17 see Hassibi and Kailath (2001) and also the monograph by Hassibi, Sayed, and Kailath (1999).

COMPUTER PROJECT

Project XI.1 (Active noise control) Consider the setting of Fig. XI.5. The noise from an engine, usually in an enclosure such as a duct, travels towards the rightmost microphone location. In order to diminish the noise level at that location, the incoming noise wave is measured by the leftmost microphone and used to generate a noise-like signal by a loudspeaker. The purpose of this replica is to cancel the arriving noise signal at the rightmost microphone.

Primary source　Microphone　Secondary source　Microphone
$d(i)$　$-$
Noise signal　Noise-like signal　$\hat{d}(i)$
Processing

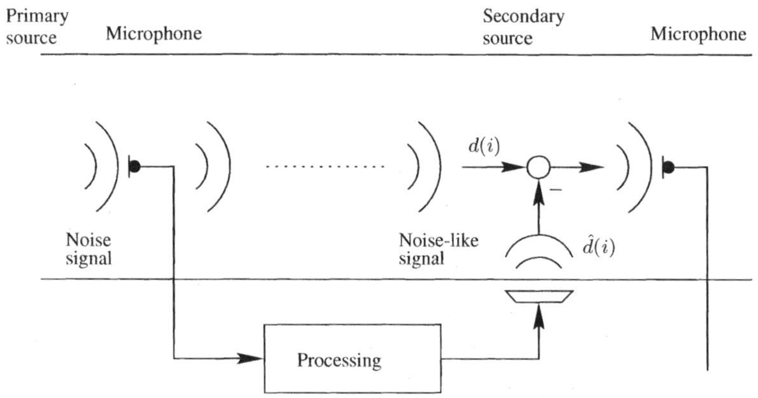

FIGURE XI.5 An active noise control system in a duct.

In an adaptive implementation of an active noise control system, the structure of Fig. XI.6 is used. The figure shows the input noise signal $u(i)$ and a filtered version of it, denoted by $d(i)$, which corresponds to $u(i)$ travelling down the enclosure until it reaches the secondary source.

In Fig. XI.6, an anti-noise (or noise-like) sequence, $\hat{d}(i)$, is generated by an adaptive FIR filter of length M at the secondary source with the intent of cancelling $d(i)$. The difference between both signals $d(i)$ and $\hat{d}(i)$ cannot be measured directly but only a filtered version of it, which we shall denote by $e_f(i) = F[d(i) - \hat{d}(i)]$. The filter F is assumed to be FIR and its presence is due to the fact that the signals $\{d(i), \hat{d}(i)\}$ have to further travel a path before reaching the rightmost microphone. This path is usually unknown, and the objective is to update the adaptive filter weights in order to reduce or cancel the filtered error $e_f(i)$.

One adaptive algorithm that has been developed for such purposes is the so-called filtered-x least-mean-squares (FxLMS) algorithm. The algorithm requires filtering the regression vector by the same filter F, namely, it relies on the update

$$\boxed{w_i = w_{i-1} + \mu(i)\,(F[u_i])^*\, F[e(i)], \quad e(i) = d(i) - \hat{d}(i), \quad \hat{d}(i) = u_i w_{i-1}} \qquad \text{(FxLMS)}$$

Here, as usual, u_i denotes the regressor of the tapped-delay-line adaptive filter. Moreover, the notation $F[x]$ is used to denote the filter F applied to the sequence x.

The FxLMS algorithm as such requires knowledge of F in order to process the regression data by it. In addition, it tends to exhibit poor convergence performance. In this project, we shall focus mostly on a filtered-error least-mean-squares variant (FeLMS), which does not require knowledge of F. Its update equation has the form:

$$\boxed{w_i = w_{i-1} + \mu(i) u_i^* F[e(i)], \quad e(i) = d(i) - u_i w_{i-1}} \qquad \text{(FeLMS)}$$

That is, it is an LMS update with $e(i)$ replaced by $e_f(i)$. This is the same algorithm studied in Prob. XI.12. We shall assume that $F(z) = 1 - 1.2\, z^{-1} + 0.72\, z^{-2}$. Let $\mu(i) = \alpha/\|u_i\|^2$ for some parameter $\alpha > 0$ and fix the filter order at $M = 10$.

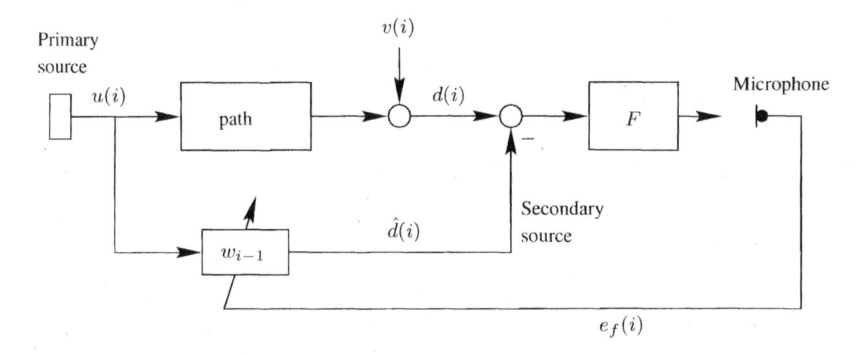

FIGURE XI.6 Structure for adaptive active noise control.

(a) Consider the maximization $g(\alpha) = \max_{\omega} \left|1 - \alpha F(e^{j\omega})\right|$. Plot $g(\alpha)$. For what range of α is $g(\alpha) < 1$? Determine the value of α that minimizes $g(\alpha)$.

(b) Load the file path.mat; it contains 10 samples of the signal path impulse response from $u(i)$ to $d(i)$. Assume that an additive zero-mean white Gaussian noise $v(i)$ corrupts the signal $d(i)$ and set its variance level at 60 dB below the input signal power. Train the filtered-error LMS filter and generate ensemble-average mean-square deviation curves for the choices $\alpha \in \{0.01, 0.02, 0.03, 0.05, 0.1\}$. Observe how the steady-state values of $\mathsf{E}\,\|\tilde{w}_N\|^2$ vary with α. Use binary data ± 1 with equal probability as input data $\{u(i)\}$. Run each experiment for 30000 iterations and average the results over 50 experiments.

(c) Train again filtered-error LMS and generate ensemble-average learning curves for the choices $\alpha \in \{0.05, 0.1, 0.2, 0.5, 0.6\}$. Which value of α results in fastest convergence of the learning curve? Run each experiment for 5000 iterations and average the results over 50 experiments. Generate learning curves for both cases:

   1. input $u(i)$ is ± 1 with equal probability.
   2. input $u(i)$ is white Gaussian with unit variance.

(d) When the variations in the weight estimates are slow over the length of the filter F, say $w_i \approx w_{i-1} \approx \ldots \approx w_{i-M_F}$, we can approximate $F[u_i w_{i-1}]$ by $F[u_i]w_{i-1}$, so that the FxLMS update can be approximated by

$$w_i = w_{i-1} + \mu(i)\,(F[u_i])^*\,(F[d(i)] - F[u_i]w_{i-1})$$

This update now has the same form as a standard LMS update and, correspondingly, similar performance, with the signals $\{u_i, d(i)\}$ replaced by their filtered versions $\{F[u_i], F[d(i)]\}$. A modification to FxLMS that also makes its update become similar to that of an LMS update, without the need for the slow adaptation assumption, can be obtained by adding two terms to the update equation as follows:

$$\boxed{w_i = w_{i-1} + \mu(i)\,(F[u_i])^*\,(F[e(i)] \;+\; F[u_i w_{i-1}] - F[u_i]w_{i-1})} \quad \text{(mFxLMS)}$$

This variation shows improved convergence performance over standard FxLMS.

Set $\alpha = 0.15$ and generate learning curves for the following four algorithms: a) NLMS, b) filtered-error LMS (FeLMS), c) filtered-x LMS (FxLMS), and d) modified filtered-x LMS (mFxLMS). Run each experiment for 5000 iterations and average the results over 50 experiments. Use white Gaussian input data with unit variance.

REFERENCES AND INDICES

References

T. ABOULNASR AND K. MAYYAS (1997), "A robust variable step-size LMS-type algorithm: Analysis and simulations," *IEEE Trans. Signal Process.*, vol. 45, no. 3, pp. 631–639.

S. ALAMOUTI (1998), "A simple transmit diversity technique for wireless communications," *IEEE J. Selected Areas Commun.*, vol. 16, no. 8, pp. 1451–1458.

A. E. ALBERT AND L. S. GARDNER, JR. (1967), *Stochastic Approximation and Nonlinear Regression*, MIT Press, Cambridge, MA.

N. AL-DHAHIR (2001), "Single-carrier frequency-domain equalization for space-time block-coded transmissions over frequency-selective fading channels," *IEEE Commun. Lett.*, vol. 5, no. 7, pp. 304–306.

N. AL-DHAHIR AND J. CIOFFI (1995), "MMSE decision feedback equalizers and coding: Finite-length results,"*IEEE Trans. Inform. Theory*, pp. 961–976.

N. AL-DHAHIR AND A. H. SAYED (2000), "The finite-length multi-input multi-output MMSE-DFE," *IEEE Trans. Signal Process.*, vol. 48, no. 10, pp. 2921–2936.

L. V. AHLFORS (1979), *Complex Analysis*, 3rd edition, McGraw Hill, New York.

S. T. ALEXANDER (1986), *Adaptive Signal Processing: Theory and Applications*, Springer-Verlag, New York.

S. T. ALEXANDER (1987), "Transient weight misadjustment properties for the finite precision LMS algorithm," *IEEE Trans. Acoust. Speech Signal Process.*, vol. 35, no. 9, pp. 1250–1258.

S. T. ALEXANDER AND A. L. GHIRNIKAR (1993), "A method for recursive least-squares filtering based upon an inverse QR decomposition," *IEEE Trans. Signal Process.*, vol. 41, pp. 20–30.

T. Y. AL-NAFFOURI AND A. H. SAYED (2001a), "Transient analysis of adaptive filters," *Proc. ICASSP*, Salt Lake City, UT, vol. 6, pp. 3869–3872.

T. Y. AL-NAFFOURI AND A. H. SAYED (2001b), "Adaptive filters with error nonlinearities: Mean-square analysis and optimum design," *EURASIP Journal on Applied Signal Processing*, special issue on Nonlinear Signal Processing and Applications, no. 4, pp. 192–205.

T. Y. AL-NAFFOURI AND A. H. SAYED (2003a), "Transient analysis of data-normalized adaptive filters," *IEEE Trans. Signal Process.*, vol. 51, no. 3, pp. 639–652.

T. Y. AL-NAFFOURI AND A. H. SAYED (2003b), "Transient analysis of adaptive filters with error nonlinearities," *IEEE Trans. Signal Process.*, vol. 51, no. 3, pp. 653–663.

V. A. AMBARTSUMIAN (1943), "Diffuse reflection of light by a foggy medium," *Dokl. Akad. Sci. SSSR*, vol. 38, pp. 229–322.

P. E. AN, M. BROWN, AND C. J. HARRIS (1997), "On the convergence rate performance of the normalized least-mean-square adaptation," *IEEE Trans. Neural Networks*, vol. 8, no. 5, pp. 1211–1214.

P. ANDERSON (1985), "Adaptive forgetting in recursive identification through multiple models," *Int. J. Control*, vol. 42, pp. 1175–1193.

B. D. O. ANDERSON AND J. B. MOORE (1979), *Optimal Filtering*, Prentice Hall, Englewood Cliffs, NJ.

B. D. O. ANDERSON, R. R. BITMEAD, C. R. JOHNSON, P. V. KOKOTOVIC, R. L. KOSUT, I. M. Y. MAREELS, L. PRALY, AND B. D. RIEDLE (1986), *Stability of Adaptive Systems: Passivity and Averaging Analysis*, MIT Press, Cambridge, MA.

J. APOLINÁRIO, JR., M. L. R. CAMPOS, AND P. S. R. DINIZ (2000), "Convergence analysis of the binormalized data-reusing LMS algorithm," *IEEE Trans. Signal Process.*, vol. 48, no. 11, pp. 3235–3242.

S. P. APPLEBAUM (1966), *Adaptive Arrays*, Rep. SPLTR 66-1, Syracuse University Research Corporation.

J. ARENAS-GARCIA, A. R. FIGUEIRAS-VIDAL, AND A. H. SAYED (2006), "Mean-square performance of a convex combination of two adaptive filters," *IEEE Trans. Signal Process.*, vol. 54, no. 3, pp. 1078–1090.

M. R. ASHARIF, F. AMANO, S. UNAGAMI, AND K. MURAMO (1986a), "A new structure of echo canceller based on frequency bin adaptive filtering (FBAF)," *Proc. DSP Symp.*, Japan, pp. 165–169.

M. R. ASHARIF, T. TAKEBAYASHI, T. CHUGO, AND K. MURAMO (1986b), "Frequency domain noise canceller: Frequency-bin adaptive filtering (FBAF)," *Proc. ICASSP*, Tokyo, Japan, pp. 41.22.1–41.22.4.

K. J. ASTRÖM (1970), *Introduction to Stochastic Control Theory*, Academic Press, New York.

K. J. ASTRÖM AND B. WITTENMARK (1971), "Problems of identification and control," *J. Math. Anal. App.*, vol. 34, pp. 90–113.

M. E. AUSTIN (1967), *Decision-Feedback Equalization for Digital Communication Over Dispersive Channels*, Tech. Rep. 437, MIT Lincoln Laboratory, Cambridge, MA.

A. R. S. BAHAI AND B. R. SALTZBERG (1999), *Multi-Carrier Digital Communications - Theory and Applications of* OFDM, Kluwer Academic, MA.

A. R. S. BAHAI AND M. SARRAF (1997), "A frequency offset estimation technique for nonstationary channels," *Proc. ICASSP*, Munich, Germany, vol. 5, pp. 21–24.

T. BASAR AND P. BERNHARD (1995), *$\mathcal{H}^\infty$-Optimal Control and Related Minimax Design Problems: A Dynamic Games Approach*, Birkhauser, Boston.

F. BEAUFAYS (1995), "Transform-domain adaptive filters: An analytical approach," *IEEE Trans. Signal Process.*, vol. 43, pp. 422–431.

F. BEAUFAYS AND B. WIDROW (1994), "Two-layer linear structures for fast adaptive filtering," *Proc. World Congress on Neural Networks*, San Diego, vol. III, pp. 87–93.

R. T. BEHRENS AND L. L. SCHARF (1994), "Signal processing applications of oblique projection operators," *IEEE Trans. Signal Process.*, vol. 42, pp. 1413–1424.

M. G. BELLANGER (1988), "The FLS-QR algorithm for adaptive filtering," *Signal Processing*, vol. 17, pp. 291–304.

M. G. BELLANGER (1987), *Adaptive Digital Filters and Signal Analysis,* Marcel Dekker, New York.

R. E. BELLMAN (1953), *Stability Theory of Differential Equations*, McGraw Hill, New York.

R. E. BELLMAN (1957), *Dynamic Programming*, Princeton University Press, Princeton, NJ.

R. E. BELLMAN (1970), *Introduction to Matrix Analysis*, 2nd edition, McGraw Hill, New York.

J. BENESTY, T. GÄNSLER, D. R. MORGAN, M. M. SONDHI, AND S. L. GAY (2001), *Advances in Network and Acoustic Echo Cancellation*, Springer-Verlag, Berlin.

A. BENVENISTE (1987), "Design of adaptive algorithms for the tracking of time-varying systems," *Int. J. Adap. Contr. Signal Process.*, pp. 3–27.

A. BENVENISTE AND M. GOURSAT (1984), "Blind equalizer," *IEEE Trans. Commun.*, vol. 32, pp. 871–883.

A. BENVENISTE, M. MÉTIVIER, AND P. PRIOURET (1987), *Adaptive Algorithms and Stochastic Approximations*, Springer-Verlag, New York.

A. BENVENISTE AND G. RUGET (1982), "A measure of the tracking capability of recursive stochastic algorithms with constant gains," *IEEE Trans. Automat. Contr.*, vol. 27, pp. 639–649.

J. C. M. BERMUDEZ AND N. J. BERSHAD (1996a), "A nonlinear analytical model for the quantized LMS algorithm – The arbitrary step-size case," *IEEE Trans. Signal Process.,* vol. 44, no. 5, pp. 1175–1183.

J. C. M. BERMUDEZ AND N. J. BERSHAD (1996b), "Transient and tracking performance analysis of the quantized LMS algorithm for time-varying system identification," *IEEE Trans. Signal Process.*, vol. 44, no. 8, pp. 1990–1997.

N. J. BERSHAD (1986a), "On the optimum data nonlinearity in LMS adaptation," *IEEE Trans. Acoust. Speech Signal Process.*, vol. 34, no. 1, pp. 69–76.

N. J. BERSHAD (1986b), "Analysis of the normalized LMS algorithm with Gaussian inputs," *IEEE Trans. Acoust. Speech Signal Process.*, vol. 34, pp. 793–806.

N. J. BERSHAD (1987), "Behavior of the ϵ-normalized LMS algorithm with Gaussian inputs," *IEEE Trans. Acoust. Speech Signal Process.,* vol. 35, pp. 636–644.

N. J. BERSHAD (1988), "On error saturation nonlinearities in LMS adaptation," *IEEE Trans. Acoust. Speech Signal Process.*, vol. 36, no. 4, pp. 440–452.

N. J. BERSHAD AND M. BONNET (1990), "Saturation effects in LMS adaptive echo cancellation for binary data," *IEEE Trans. Acoust. Speech Signal Process.,* vol. 38, no. 10, pp. 1687–1696.

N. J. BERSHAD, P. FEINTUCH, A. REED, AND B. FISHER (1980), "Tracking characteristics of the LMS adaptive line enhancer: Response to a linear chirp signal in noise," *IEEE Trans. Acoust. Speech Signal Process.*, vol. 28, pp. 504–516.

N. J. BERSHAD, D. LINEBARGER, AND S. MCLAUGHLIN (2001), "A stochastic analysis of the affine projection algorithm for Gaussian autoregressive inputs," *Proc. ICASSP*, Salt Lake City, UT, pp. 3837–3840.

N. J. BERSHAD AND S. ROY (1990), "Performance of the 2-2 constant modulus (CM) adaptive algorithm for Rayleigh fading sinusoids in Gaussian noise," *Proc. ICASSP*, Albuquerque, NM, pp. 1675–1678.

G. J. BIERMAN (1974), "Sequential square-root filtering and smoothing for discrete linear systems," *Automatica*, vol. 10, pp. 147–158.

G. J. BIERMAN (1977), *Factorization Methods for Discrete Sequential Estimation*, Academic Press, New York.

S. BINDE (1995), "Numerically stable fast transversal filter with leakage correction," *IEEE Signal Process. Lett.*, vol. 2, no. 6, pp. 114–116.

D. BINI AND P. FAVATI (1993), "On a matrix algebra related to the discrete Hartley transform," *SIAM J. Matrix Anal. Appl.*, vol. 14, no. 2, pp. 500–507.

R. R. BITMEAD AND B. D. O. ANDERSON (1980), "Performance of adaptive estimation algorithms in dependent random environments," *IEEE Trans. Automat. Contr.*, vol. 25, pp. 788–794.

R. R. BITMEAD AND B. D. O. ANDERSON (1981), "Adaptive frequency sampling filters," *IEEE Trans. Circuits Syst.*, vol. 28, no. 6, pp. 524–533.

R. R. BITMEAD, B. D. O. ANDERSON, AND T. S. NG (1986), "Convergence rate determination for gradient-based adaptive estimators," *Automatica*, vol. 22, pp. 185–191.

E. BJARNASON (1992), "Noise cancellation using a modified form of the filtered–XLMS algorithm," *Proc. Eusipco Signal Processing V*, Brussels.

A. BJÖRCK (1996), *Numerical Methods for Least-Squares Methods*, SIAM, PA.

J. R. BLUM (1954), "Multidimensional stochastic approximation methods," *Ann. Math. Stat.*, vol. 25, pp. 737–744.

J. M. P. BORALLO AND M. G. OTERO (1992), "On the implementation of a partitioned block frequency domain adaptive filter (PBFDAF) for long acoustic echo cancellation," *Signal Processing*, vol. 27, pp. 301–315.

J. L. BOTTO (1988), "Numerical divergence of fast recursive least-squares transversal filters," *RAIRO-APII*, vol. 22, pp. 231–254.

J. L. BOTTO AND G. V. MOUSTAKIDES (1989), "Stabilizing the fast Kalman algorithms," *IEEE Trans. Acoust. Speech Signal Process.*, vol. 37, pp. 1342–1348.

G. E. BOTTOMLEY AND S. T. ALEXANDER (1989), "A theoretical basis for the divergence of conventional recursive least-squares filters," *Proc. ICASSP*, Glasgow, pp. 908–911.

G. E. P. BOX AND G. C. TIAO (1973), *The Bayesian Inference in Statistical Analysis*, Addison-Wesley, Reading, MA.

D. M. BRADY (1970), "An adaptive coherent diversity receiver for data transmission through dispersive media," *Conf. Rec. ICC*, pp. 21.35–21.40.

D. G. BRENNAN (1959), "Linear diversity combining techniques," *Proc. IRE*, vol. 47, pp. 1075–1102. [Reproduced in *Proc. IEEE*, vol. 91, no. 2, pp. 331–356, 2003.]

J. R. BUNCH, R. C. LEBORNE, AND I. K. PROUDLER (2001), "A conceptual framework for consistency, conditioning, and stability issues in signal processing," *IEEE Trans. Signal Process.*, vol. 49, no. 9, pp. 1971–1981.

P. BUSINGER AND G. H. GOLUB (1965), "Linear least-squares solution by Householder transformations," *Numer. Math.*, vol. 7, pp. 269–276.

J. J. BUSSGANG (1952), *Cross-Correlation Functions of Amplitude Distorted Gaussian Signals*, Tech. Report 216, MIT Research Laboratory of Electronics, Cambridge, MA.

H. J. BUTTERWECK (1995), "A steady-state analysis of the LMS adaptive algorithm without use of the independence assumption," *Proc. ICASSP*, Detroit, MI, pp. 1404–1407.

H. J. BUTTERWECK (2001), "A wave theory of long adaptive filters," *IEEE Trans. Circuits Syst. I: Fundamental Theory and Applications*, vol. 84, pp. 739–747.

P. E. CAINES (1988), *Linear Stochastic Systems*, Wiley, New York.

C. CARAISCOS AND B. LIU (1984), "A roundoff error analysis of the LMS adaptive algorithm," *IEEE Trans. Acoust. Speech Signal Process.*, vol. 32, no. 1, pp. 34–41.

F. CATTIVELLI, C. G. LOPES, AND A. H. SAYED (2008), "Diffusion recursive least-squares for distributed estimation over adaptive networks," *IEEE Trans. Signal Processing*, vol. 56.

J. A. CHAMBERS, O. TANRIKULU, AND A. G. CONSTANTINIDES (1994), "Least mean mixed-norm adaptive filtering," *Electron. Lett.*, vol. 30, no. 19, pp. 1574–1575.

J. A. CHAMBERS, O. TANRIKULU, AND A. G. CONSTANTINIDES (1994), "Least mean mixed-norm adaptive filtering," *Electron. Lett.*, vol. 30, no. 19, pp. 1574–1575.

C. K. CHAN AND J. J. SHYNK (1990), "Stationary points of the constant modulus algorithm for real Gaussian signals," *IEEE Trans. Acoust. Speech Signal Process.*, vol. 38, no. 12, pp. 2176–2181.

S. CHANDRASEKHAR (1947a), "On the radiative equilibrium of a stellar atmosphere – Pt XXI," *Astrophys. J.*, vol. 106, pp. 152–216.

S. CHANDRASEKHAR (1947b), "On the radiative equilibrium of a stellar atmosphere — Pt XXII," *Astrophys. J.*, vol. 107, pp. 48–72.

R. W. CHANG (1970), *Orthogonal Frequency Division Multiplexing*, U.S. Patent 3,488,445.

P. S. CHANG AND A. N. WILLSON (1995), "A roundoff error analysis of the normalized LMS algorithm," *Proc. Asilomar Conference on Signals, Systems and Computers*, vol. 2, Pacific Grove, CA, pp. 1337–1341.

J. CHEN, H. BES, J. VANDEWALLE, AND P. JANSSENS (1988), "A new structure for subband acoustic echo canceller," *Proc. ICASSP*, New York, pp. 2574–2577.

S. CHO AND V. J. MATHEWS (1990), "Tracking analysis of sign algorithm in nonstationary environments," *IEEE Trans. Acoust. Speech Signal Process.*, vol. 38, no. 12, pp. 2046–2057.

J. M. CIOFFI (1987), "Limited-precision effects in adaptive filtering," *IEEE Trans. Circuits Syst.*, vol. 34, no. 7, pp. 821–833.

J. M. CIOFFI (1988), "High speed systolic implementation of the fast QR adaptive filters," *Proc. ICASSP*, New York, pp. 1584–1588.

J. M. CIOFFI (1990), "The fast adaptive rotor's RLS algorithm," *IEEE Trans. Acoust. Speech Signal Process.*, vol. 38, pp. 631–653.

J. M. CIOFFI AND T. KAILATH (1984), "Fast recursive least-squares transversal filters for adaptive filtering," *IEEE Trans. Acoust. Speech Signal Process.*, vol. 32, pp. 304–337.

J. M. CIOFFI AND J. J. WERNER (1985), "The tap-drifting problem in digitally implemented data-driven echo cancellers," *Bell Syst. Tech. J.*, vol. 64, no. 1, pp. 115–138.

T. CLAASEN AND W. MECKLENBRÄUKER (1981), "Comparison of the convergence of two algorithms for adaptive FIR digital filters," *IEEE Trans. Circuits Syst.*, vol. 28, no. 6, pp. 510–518.

M. V. CLARK (1998), "Adaptive frequency-domain equalization and diversity combining for broadband wireless communications," *IEEE J. Selected Areas Commun.*, vol. 16, no. 8, pp. 1385–1395.

G. A. CLARK, S. K. MITRA, AND S. R. PARKER (1981), "Block implementation of adaptive digital filters," *IEEE Trans. Circuits Syst.*, vol. 28, pp. 584–592.

P. M. CLARKSON AND P. R. WHITE (1987), "Simplified analysis of the LMS adaptive filter using a transfer function approximation," *IEEE Trans. Acoust. Speech Signal Process.*, vol. 35, pp. 987–993.

D. R. COX AND D. V. HINKLEY (1974), *Theoretical Statistics*, Chapman and Hall, New York.

A. CZYLWIK (1997), "Comparison between adaptive OFDM and single carrier modulation with frequency domain equalization," *Proc. VTC*, pp. 865–869.

M. A. DAHLEH AND I. J. DIAZ-BOBILLO (1995), *Control of Uncertain Systems: A Linear Programming Approach*, Prentice-Hall, Englewood Cliffs, NJ.

P. DEWILDE AND H. DYM (1981), "Schur recursions, error formulas, and convergence of rational estimators for stationary stochastic sequences," *IEEE Trans. Inform. Theory*, vol. 27, no. 4, pp. 446–461.

P. S. R. DINIZ (2002), *Adaptive Filtering: Algorithms and Practical Implementation*, 2nd edition, Kluwer Academic Publisher, Norwell, MA.

M. J. DITORO (1965), "A new method for high speed adaptive signal communication through any time variable and dispersive transmission medium," *Proc. 1st IEEE Annu. Commun. Conf.* , pp. 763–767.

S. C. DOUGLAS AND T. H. -Y MENG (1994a), "Stochastic gradient adaptation under general error criterion," *IEEE Trans. Signal Processing*, vol. 42, no. 6, pp. 1335–1351.

S. C. DOUGLAS AND T. H. -Y MENG (1994b), "Normalized data nonlinearities for LMS adaptation," *IEEE Trans. Signal Process.*, vol. 42, no. 6, pp. 1352–1365.

S. C. DOUGLAS AND W. PAN (1995), "Exact expectation analysis of the LMS adaptive filter," *IEEE Trans. Signal Process.*, vol. 43, no. 12, pp. 2863–2871.

J. DURBIN (1960), "The fitting of time series models," *Rev. L'Institut Intl. de Statistique*, vol. 28, pp. 233–244.

R. DURRETT (1996), *Probability: Theory and Examples*, Duxbury Press, 2nd edition.

D. L. DUTTWEILER (1982), "Adaptive filter performance with nonlinearities in the correlation multiplier," *IEEE Trans. Acoust. Speech Signal Process.*, vol. 30, no. 4, pp. 578–586.

P. DYER AND S. R. MCREYNOLDS (1969), "Extension of square-root filtering to include process noise," *J. Optimiz. Theory Appl.*, vol. 3, pp. 444–459.

E. ELEFTHERIOU AND D. D. FALCONER (1986), "Tracking properties and steady-state performance of RLS adaptive filter algorithms," *IEEE Trans. Acoust. Speech Signal Process.*, vol. 34, pp. 1097–1110.

E. ELEFTHERIOU AND D. D. FALCONER (1987), "Adaptive equalization techniques for HF channels," *IEEE J. Selected Areas in Commun.*, vol. 5, pp. 238–247.

S. J. ELLIOTT AND P. A. NELSON (1993), "Active noise control," *IEEE Signal Process. Magaz.*, vol. 1, no. 4, pp. 12–35.

O. K. ERSOY (1997), *Fourier-Related Transforms, Fast Algorithms, and Applications*, Prentice-Hall, Englewood Cliffs, NJ.

E. EWEDA (1990a), "Analysis and design of signed regressor LMS algorithm for stationary and nonstationary adaptive filtering with correlated Gaussian data," *IEEE Trans. Circuits Syst.*, vol. 37, no. 11, pp. 1367–1374.

E. EWEDA (1990b), "Optimum step size of the sign algorithm for nonstationary adaptive filtering," *IEEE Trans. Acoust. Speech Signal Process.*, vol. 38, no. 11, pp. 1897–1901.

E. EWEDA (1992), "Convergence analysis and design of an adaptive filter with finite-bit-power-of-two quantizer error," *IEEE Trans. Circuits Syst. II: Analog and Digital Signal Processing*, vol. 39, pp. 113–115.

E. EWEDA (1994), "Comparison of RLS, LMS, and sign algorithms for tracking randomly time-varying channels," *IEEE Trans. Signal Process.*, vol. 42, no. 11, pp. 2937–2944.

E. EWEDA (1997), "A quantitative and tracking comparison of two adaptive echo cancellation algorithms," *Signal Processing*, vol. 59, pp. 285–290.

E. EWEDA AND O. MACCHI (1985), "Tracking error bounds of adaptive nonstationary filtering," *Automatica*, pp. 296–302.

E. EWEDA, W. M. YOUNIS, AND S. H. EL-RAMLY (1998), "Tracking performance of a quantized adaptive filter equipped with the sign algorithm," *Signal Processing*, vol. 69, no. 2, pp. 157–162.

E. EWEDA, N. R. YOUSEF, AND S. H. EL-RAMLY (1998), "Reducing the effect of finite wordlength on the performance of an LMS adaptive filter," *Proc. ICC*, Atlanta, GA, vol. 2, pp. 688–692.

D. K. FADDEEV AND V. N. FADDEEVA (1963), *Computational Methods of Linear Algebra*, Freeman, San Francisco, CA.

D. D. FALCONER AND L. LJUNG (1978), "Application of fast Kalman estimation to adaptive equalization," *IEEE Trans. Commun.*, vol. 26, no. 10, pp. 1439–1446.

D. FARDEN (1981), "Tracking properties of adaptive signal processing algorithms," *IEEE Trans. Acoust. Speech Signal Process.*, vol. 29, pp. 439–446.

B. FARHANG-BOROUJENY (1999), *Adaptive Filters: Theory and Applications*, Wiley, New York.

E. R. FERRARA, JR. (1980), "Fast implementation of LMS adaptive filters," *IEEE Trans. Acoust. Speech Signal Process.*, vol. 28, pp. 474–475.

A. FEUER AND E. WEINSTEIN (1985), "Convergence analysis of LMS filters with uncorrelated Gaussian data," *IEEE Trans. Acoust. Speech Signal Process.*, vol. 33, no. 1, pp. 222–229.

I. FIJALKOW, C. E. MANLOVE, AND C. R. JOHNSON (1998), "Adaptive fractionally spaced blind CMA equalization: Excess MSE," *IEEE Trans. Signal Process.*, vol. 46. no. 1, pp. 227–231.

R. A. FISHER (1912), "On an absolute criterion for fitting frequency curves," *Messenger of Math.*, vol. 41, p. 155.

R. FLETCHER (1987), *Practical Methods of Optimization*, 2nd edition, Wiley, New York.

R. FLETCHER, J. A. GRANT, AND M. D. HEBDEN (1971), "The calculation of linear best L_p approximations," *Comput. J.*, vol. 14, pp. 276–279.

S. FLORIAN AND A. FEUER (1986), "Performance analysis of the LMS algorithm with a tapped delay line (two-dimensional case)," *IEEE Trans. Acoust. Speech Signal Process.*, vol. 34, no. 6, pp. 1542–1549.

J. B. FOLEY AND F. M. BOLAND (1988), "A note on the convergence analysis of LMS adaptive filters with Gaussian data," *IEEE Trans. Acoust. Speech Signal Process.*, vol. 36, pp. 1087–1089.

G. J. FOSCHINI (1985), "Equalizing without altering or detecting data (digital radio signals)," *AT&T Technical Journal*, vol. 64, pp. 1885–1911.

B. FRIEDLANDER (1982), "Lattice filters for adaptive processing," *Proc. IEEE*, vol. 27, pp. 829–867.

I. FURUKAWA (1984), "A design of canceller of broadband acoustic echo," *Proc. Int. Teleconference Symposium*, Tokyo, pp. 1/8–8/8.

D. GABOR, W. P. Z. WILBY, AND R. WOODCOCK (1961), "An universal nonlinear filter, predictor, and simulator which optimizes itself by a learning process," Proc. IEE, vol. 108, no. 40.

F. R. GANTMACHER (1959), *The Theory of Matrices*, Chelsea Publishing Company, New York.

W. A. GARDNER (1984), "Learning characteristic of stochastic-descent algorithms: A general study, analysis, and critique," *Signal Processing*, vol. 6, no. 2, pp. 113–133.

C. F. GAUSS (1809), *Theory of the Motion of Heavenly Bodies*, Dover, New York. [English translation of *Theoria Motus Corporum Coelestium, 1809*.]

D. GEORGE, R. BOWEN, AND J. STOREY (1971), "An adaptive decision feedback equalizer," *IEEE Trans. Commun. Tech.*, vol. 19, pp. 281–292.

A. GERSHO (1969), "Adaptive equalization of highly dispersive channels for data transmission," *Bell Syst. Tech. J.*, vol. 48, pp. 55–70.

A. GERSHO (1984), "Adaptive filtering with binary reinforcement," *IEEE Trans. Inform. Theory*, vol. 30, pp. 191–199.

M. GHOSH (1999), "Blind decision feedback equalization for terrestrial television receivers," *Proc. Asilomar Conference on Signals, Systems, and Computers*, Pacific Grove, CA, vol. 1, pp. 1159–1163.

J. GIBSON AND S. GRAY (1988), "MVSE adaptive filtering subject to a constraint on MSE," *IEEE Trans. Circuits Syst.*, vol. 35, no. 5, pp. 603–608.

A. GILLOIRE AND M. VETTERLI (1988), "Adaptive filtering in subbands," *Proc. ICASSP*, New York, pp. 1572–1575.

A. GILLOIRE AND M. VETTERLI (1992), "Adaptive filtering in subbands with critical sampling: Analysis, experiments, and applications to acoustic echo cancellation," *IEEE Trans. Sig. Process.*, vol. 40, pp. 1862–1875.

R. D. GITLIN, J. S. HAYES, AND S. B. WEINSTEIN (1992), *Data Communications Principles*, Plenum Press, New York.

R. D. GITLIN, J. E. MAZO, AND M. G. TAYLOR (1973), "On the design of gradient algorithms for digitally implemented adaptive filters," *IEEE Trans. Circuit Theory*, vol. 20, pp. 125–136.

R. D. GITLIN, H. C. MEADORS, JR., AND S. B. WEINSTEIN (1982), "The tap-leakage algorithm: An algorithm for the stable operation of a digitally implemented, fractionally spaced adaptive equalizer," *Bell Syst. Tech. J.*, vol. 61, no. 8, pp. 1817–1839.

R. D. GITLIN AND S. B. WEINSTEIN (1981), "Fractionally-spaced equalization: An improved digital transversal filter," *Bell Syst. Tech. J.*, vol. 60, pp. 275–296.

G. O. GLENTIS, K. BERBERIDIS, AND S. THEODORIDIS (1999), "Efficient least-squares adaptive algorithms for FIR transversal filtering," *IEEE Signal Process. Mag.*, vol. 16, pp. 13–41.

J. R. GLOVER, JR. (1977), "Adaptive noise cancelling applied to sinusoidal interferences," *IEEE Trans. Acoust. Speech Signal Process.*, vol. 25, pp. 484–491.

D. N. GODARD (1974), "Channel equalization using a Kalman filter for fast data transmission," *IBM J. Res. Develop.*, vol. 18, pp. 267–273.

D. N. GODARD (1980), "Self-recovering equalization and carrier tracking in two-dimensional data communication systems," *IEEE Trans. Commun.*, vol. 28, no. 11, pp. 1867–1875.

G. H. GOLUB (1965), "Numerical methods for solving linear least-squares problems," *Numer. Math.*, vol. 7, pp. 206–216.

G. H. GOLUB AND C. F. VAN LOAN (1996), *Matrix Computations*, 3rd edition, The Johns Hopkins University Press, Baltimore.

N. GOODMAN (1963), "Statistical analysis based on a certain multivariate complex Gaussian distribution," *Ann. Math. Stat.*, vol. 34, pp. 152–177.

G. C. GOODWIN AND K. S. SIN (1984), *Adaptive Filtering, Prediction, and Control*, Prentice-Hall, Englewood Cliffs, NJ.

M. GREEN AND D. J. N. LIMEBEER (1995), *Linear Robust Control*, Prentice-Hall, Englewood Cliffs, NJ.

L. J. GRIFFITHS (1967), "A simple adaptive algorithm for real-time processing in antenna arrays," *Proc. IEEE*, vol. 57, pp. 1696–1704.

L. GUO (1994), "Stability of recursive stochastic tracking algorithms," *SIAM J. Contr. Optim.*, vol. 32, no. 5, pp. 1195–1225.

M. HAJIVANDI AND W. A. GARDNER (1990), "Measures of tracking performance for the LMS algorithm," *IEEE Trans. Acoust. Speech Signal Process.*, vol. 38, no. 11, pp. 1953–1958.

T. HALL (1970), *Carl Friedrich Gauss: A Biography*, MIT Press, Cambridge, MA.

R. J. HANSON AND C. L. LAWSON (1969), "Extensions and applications of the Householder algorithm for solving linear least-squares problems," *Math. Comput.*, vol. 23, pp. 787–812.

R. W. HARRIS, D. M. CHABRIES, AND F. A. BISHOP (1986), "A variable step (VS) adaptive filter algorithm," *IEEE Trans. Acoust. Speech Signal Process.* vol. 34, no. 2, pp. 309–316.

B. HASSIBI AND T. KAILATH (2001), "$\mathcal{H}^\infty$ bounds for least-squares estimators," *IEEE Trans. Automat. Contr.*, vol. 46, pp. 309–314.

B. HASSIBI, A. H. SAYED, AND T. KAILATH (1993), "LMS is $\mathcal{H}^\infty$ optimal," *Proc. CDC*, vol. 1, pp. 74–79.

B. HASSIBI, A. H. SAYED, AND T. KAILATH (1994), "LMS and backpropagation are minimax filters," in *Neural Computation and Learning*, V. Roychowdhurys, K. Y. Siu, and A. Orlitsky, Eds., Kluwer Academic Publishers, pp. 425–447.

B. HASSIBI, A. H. SAYED, AND T. KAILATH (1996), "$\mathcal{H}^\infty$ optimality of the LMS algorithm," *IEEE Trans. Signal Process.*, vol. 44, no. 2, pp. 267–280.

B. HASSIBI, A. H. SAYED, AND T. KAILATH (1999), *Indefinite Quadratic Estimation and Control: A Unified Approach to $\mathcal{H}^2$ and $\mathcal{H}^\infty$ Theories*, SIAM, PA.

S. HAYKIN (1991), *Adaptive Filter Theory*, 2nd edition, Prentice-Hall, Englewood Cliffs, NJ.

S. HAYKIN (1996), *Adaptive Filter Theory*, 3rd edition, Prentice-Hall, Englewood Cliffs, NJ.

S. HAYKIN (2000), *Adaptive Filter Theory*, 4th edition, Prentice-Hall, Englewood Cliffs, NJ.

S. HAYKIN, A. H. SAYED, J. R. ZEIDLER, P. WEI, AND P. YEE (1997), "Adaptive tracking of linear time-variant systems by extended RLS algorithms," *IEEE Trans. Signal Process.*, vol. 45, no. 5, pp. 1118–1128.

G. HEINIG AND K. ROST (1998), "Representations of Toeplitz-plus-Hankel matrices using trigonometric transformations with application to fast matrix-vector multiplication," *Linear Algebra and Its Applications*, vols. 275/276, pp. 225–248.

J. W. HELTON AND O. MERLINO (1998), *Classical Control Using $\mathcal{H}^\infty$ Methods*, SIAM, PA.

H. V. HENDERSON AND S. R. SEARLE (1981), "On deriving the inverse of a sum of matrices," *SIAM Review*, vol. 23, pp. 53–60.

M. R. HESTENES (1975), *Optimization Theory*, Wiley, New York.

N. J. HIGHAM (1996), *Accuracy and Stability of Numerical Algorithms*, SIAM, PA.

D. HIRSCH AND W. J. WOLF (1970), "A simple adaptive equalizer for efficient data transmission," *IEEE Trans. Commun. Tech.*, vol. 18, no. 1, pp. 5–12.

Y. C. HO (1963), "On the stochastic approximation method and optimal filter theory," *J. Math. Anal. Appl.*, vol. 6, pp. 152–154.

M. L. HONIG AND D. G. MESSERSCHMITT (1984), *Adaptive Filters: Structures, Algorithms, and Applications*, Kluwer Academic, Boston, MA.

R. A. HORN AND C. R. JOHNSON (1987), *Matrix Analysis*, Cambridge University Press, Cambridge, England.

R. A. HORN AND C. R. JOHNSON (1994), *Topics in Matrix Analysis*, Cambridge University Press, Cambridge, England.

L. L. HOROWITZ AND K. D. SENNE (1981), "Performance advantage of complex LMS for controlling narrowband adaptive arrays," *IEEE Trans. Acoust. Speech Signal Process.*, vol. 29, no. 3, pp. 722–736.

A. HOUACINE (1991), "Regularized fast recursive least squares algorithms for adaptive filtering," *IEEE Trans. Signal Process.*, vol. 39, pp. 860–870.

A. HOUACINE AND G. DEMOMENT (1986), "Chandrasekhar adaptive regularizer for adaptive filtering," *Proc. ICASSP*, Tokyo, Japan, pp. 2967–2970.

A. S. HOUSEHOLDER (1953), *Principles of Numerical Analysis*, McGraw Hill, New York.

T. C. HSIA (1983), "Convergence analysis of LMS and NLMS adaptive algorithms," *Proc. ICASSP*, Boston, MA, vol. 1, pp. 667–670.

F. ITAKURA AND S. SAITO (1972), "On the optimum quantization of feature parameters in the PARCOR speech synthesizer," *Proc. IEEE Conf. Speech Commun. Process.*, New York, pp. 434–437.

D. H. JOHNSON AND D. E. DUDGEON (1993), *Array Signal Processing: Concepts and Techniques*, Prentice-Hall, Englewood Cliffs, NJ.

C. R. JOHNSON, P. SCHNITER, I. FIJALKOW, L. TONG, J. D. BEHM, M. G. LARIMORE, D. R. BROWN, R. A. CASAS, T. J. ENDERS, S. LAMBOTHARAN, A. TOUZNI, H. H. ZENG, M. GREEN, AND J. R. TREICHLER (1998), "Blind equalization using the constant modulus criterion: A review," *Proc. IEEE*, vol. 86, no. 10, pp. 1927–1950.

S. K. JONES, R. K. CAVIN, AND W. M. REED (1982), "Analysis of error-gradient adaptive linear estimators for a class of stationary dependent processes," *IEEE Trans. Inform. Theory*, vol. 28, pp. 318–329.

S. W. KAHNG (1972), "Best L_p approximation," *Math. Comput.*, vol. 26, pp. 505–508.

T. KAILATH (1960a), "Estimating filters for linear time-invariant channels," *Quarterly Progress Report 58*, MIT Research Laboratory of Electronics, Cambridge, MA, pp. 185–197.

T. KAILATH (1960b), "Correlation detection of signals perturbed by a random channel," *IRE Trans. Inform. Theory*, vol. 6, pp. 361–366.

T. KAILATH (1961), "Channel characterization: Time-variant dispersive channels," in *Lectures on Communication System Theory*, E. Baghdady, Ed., McGraw Hill, New York.

T. KAILATH (1968), "An innovations approach to least-squares estimation, part I: Linear filtering in additive white noise," *IEEE Trans. Automat. Contr.* vol. 13, pp. 646–655.

T. KAILATH (1972), "Some Chandrasekhar-type algorithms for quadratic regulator problems," *Proc. Conference on Decision and Control*, New Orleans, pp. 219–223.

T. KAILATH (1973), "Some new algorithms for recursive estimation in constant linear systems," *IEEE Trans. Inform. Theory*, vol. 19, pp. 750–760.

T. KAILATH (1974), "A view of three decades of linear filtering theory," *IEEE Trans. Inform. Theory*, vol. 20, pp. 146–181.

T. KAILATH (1980), *Linear Systems*, Prentice-Hall, Englewood Cliffs, NJ.

T. KAILATH, M. MORF, AND G. S. SIDHU (1973), "Some new algorithms for recursive estimation in constant linear discrete-time systems," *Proc. Seventh Princeton Conf. Inform. Sci. Systems*, Princeton, NJ, pp. 344–352.

T. KAILATH AND A. H. SAYED (1995), "Displacement structure: Theory and applications," *SIAM Review*, vol. 37, no. 3, pp. 297–386.

T. KAILATH, A. H. SAYED, AND B. HASSIBI (2000), *Linear Estimation*, Prentice-Hall, Englewood Cliffs, NJ.

T. KAILATH, A. C. VIEIRA, AND M. MORF (1978), "Orthogonal transformation (square–root) implementations of the generalized Chandrasekhar and generalized Levinson algorithms," in *Lecture Notes in Control and Information Sciences*, A. Bensoussan and J. Lions, Eds., Springer-Verlag, New York, vol. 32, pp. 81–91.

R. E. KALMAN (1960), "A new approach to linear filtering and prediction problems," *Trans. ASME J. Basic Eng.*, vol. 82, pp. 34–45.

R. E. KALMAN AND R. S. BUCY (1961), "New results in linear filtering and prediction theory," *Trans. ASME J. Basic Eng.*, vol. 83, pp. 95–107.

S. KAYALAR AND H. L. WEINERT (1989), "Oblique projections: Formulas, algorithms, and error bounds," *Math. Contr. Signals Syst.*, vol. 2, pp. 33–45.

S. M. KAY (1993), *Fundamentals of Statistical Signal Processing*, Prentice-Hall, Englewood Cliffs, NJ.

A. V. KEERTHI, A. MATHUR, AND J. J. SHYNK (1998), "Misadjustment and tracking analysis of the constant modulus array," *IEEE Trans. Signal Process.*, vol. 46, no. 1, pp. 51–58.

W. KELLERMANN (1984), "Kompensation akustischer echos in frequenzteilbändern," *Aachener Kolloquium*, Aachen, Germany, pp. 322–325.

W. KELLERMANN (1985), "Kompensation akustischer echos in frequenzteilbändern," *Frequenz*, vol. 39, no. 7/8, pp. 209–215.

J. L. KELLY, JR. AND R. F. LOGAN (1970), *Self-Adaptive Echo Canceller*, U.S. Patent 3,500,000.

M. KENDALL AND A. STUART (1976–1979), *The Advanced Theory of Statistics*, vols. 1–3, Macmillan, New York.

B. H. KHALAJ, A. H. SAYED, AND T. KAILATH (1993), "A unified approach to multichannel least-squares algorithms," *Proc. ICASSP*, MN, vol. 5, pp. 523–526.

H. K. KHALIL (1996), *Nonlinear Systems*, 2nd edition, Macmillan, New York.

P. P. KHARGONEKAR AND K. M. NAGPAL (1991), "Filtering and smoothing in an $\mathcal{H}^\infty$ setting," *IEEE Trans. Automat. Contr.*, vol. 36, pp. 151–166.

I. KIM, H. NA, K. KIM, AND Y. PARK (1994), "Constraint filtered-x and filtered-u algorithms for the active control of noise in a duct," *J. Acoust. Soc. Am.*, vol. 95, no. 6, pp. 3397–3389.

S. KOIKE (1995), "Convergence analysis of a data echo canceler with a stochastic gradient adaptive FIR filter using the sign algorithm," *IEEE Trans. Signal Process.*, vol. 43, no. 12, pp. 2852–2861.

A. N. Kolmogorov (1939), "Sur l'interpolation et extrapolation des suites stationnaires," *C. R. Acad. Sci.*, vol. 208, p. 2043.

A. N. Kolmogorov (1941a), "Stationary sequences in Hilbert space (in Russian)," *Bull. Math. Univ. Moscow*, vol. 2.

A. N. Kolmogorov (1941b), "Interpolation and extrapolation of stationary random processes," *Bull. Acad. Sci. USSR*, vol. 5. [A translation has been published by the RAND Corp., Santa Monica, CA, as Memo. RM-3090-PR, Apr. 1962.]

C. Komninakis, C. Fragouli, A. H. Sayed, and R. D. Wesel (2002), "Multi-input multi-output fading channel tracking and equalization using Kalman estimation," *IEEE Trans. Signal Process.*, vol. 50, no. 5, pp. 1065–1076.

F. Kozin (1969), "A survey of stability of stochastic systems," *Automatica*, vol. 5, pp. 95–112.

S. G. Kratzer and D. R. Morgan (1985), "The partial-rank algorithm for adaptive beamforming," *Proc. SPIE*, vol. 564, pp. 9–14.

H. Krim and M. Viberg (1996), "Two decades of signal processing: The parametric approach," *IEEE Signal Process. Mag.*, vol. 13, no. 4, pp. 67–94.

H. J. Kushner (1984), *Approximation and Weak Convergence Methods for Random Processes, with Applications to Stochastic System Theory*, MIT Press, Cambridge, MA.

H. J. Kushner and F. J. Vázquez-Abad (1996), "Stochastic approximation methods for systems over an infinite horizon," *SIAM J. Contr. Optim.*, vol. 34, no. 2, pp. 712–756.

H. J. Kushner and G. G. Yin (1997), *Stochastic Approximation Algorithms and Applications*, Springer, New York.

O. W. Kwon, C. K. Un, and J. C. Lee (1992), "Performance of constant modulus adaptive digital filters for interference cancellation," *Signal Processing*, vol. 26, no. 2, pp. 185–196.

R. H. Kwong and E. W. Johnston (1992), "A variable step size LMS algorithm," *IEEE Trans. Signal Process.*, vol. 40, no. 7, pp. 1633–1642.

I. D. Landau (1979), *Adaptive Control: The Model Reference Approach*, Marcel Dekker, New York.

I. D. Landau (1984), "A feedback system approach to adaptive filtering," *IEEE Trans. Inform. Theory*, vol. 30, no. 2, pp. 251–262.

C. L. Lawson and R. J. Hanson (1995), *Solving Least-Squares Problems*, SIAM, PA.

P. Lax (1997), *Linear Algebra*, Wiley, New York.

D. Lay (1994), *Linear Algebra and Its Applications*, Addison-Wesley, Reading, MA.

Y. W. Lee (1933), "Synthesis of electric networks by means of Fourier transforms of Laguerre's functions," *Journal of Mathematics and Physics*, vol. XI, pp. 83–113.

Y. W. Lee (1960), *Statistical Theory of Communication*, Wiley, New York.

D. T. L. Lee, M. Morf, and B. Friedlander (1981), "Recursive least-squares ladder estimation algorithms," *IEEE Trans. Circuits Syst.*, vol. 28, pp. 467–481.

J. C. Lee and C. K. Un (1989), "Performance analysis of frequency-domain block LMS adaptive digital filters," *IEEE Trans. Circuits Syst.*, vol. 36, pp. 173–189.

A. M. Legendre (1805), *Nouvelles Méthodes pour la Détermination des Orbites de Comètes*, Courcier, Paris.

A. M. Legendre (1810), "Méthode de moindres quarres, pour trouver le milieu de plus probable entre les résultats des différentes observations," *Mem. Inst. France*, pp. 149–154.

A. Leon-Garcia (1994), *Probability and Random Processes for Electrical Engineering*, 2nd edition, Addison-Wesley, Reading, MA.

H. Lev-Ari, T. Kailath, and J. Cioffi (1984), "Least-squares adaptive lattice and transversal filters: A unified geometrical theory," *IEEE Trans. Inform. Theory*, vol. 30, pp. 222–236.

M. D. Levin and C. F. N. Cowan (1994), "The performance of eight recursive-least-squares adaptive filtering algorithms in a limited precision environment," *Proc. European Signal Process. Conf.*, Edinburgh, pp. 1261–1264.

N. Levinson (1947), "The Wiener r.m.s. (root-mean-square) error criterion in filter design and prediction," *J. Math. Phys.*, vol. 25, pp. 261–278.

Y. Li and Z. Ding (1996), "Global convergence of fractionally spaced Godard (CMA) adaptive equalizers," *IEEE Trans. Signal Process.*, vol. 44, pp. 818–826.

A. P. Liavas and P. A. Regalia (1999), "On the numerical stability and accuracy of the conventional least-squares algorithm," *IEEE Trans. Signal Process.*, vol. 47, pp. 88–96.

D. W. Lin (1984), "On the digital implementation of the fast Kalman algorithm," *IEEE Trans. Acoust. Speech Signal Process.*, vol. 32, pp. 998–1005.

I. S. Lin and S. K. Mitra (1996), "Overlapped block digital filtering," *IEEE Trans. Circuits Syst. II: Analog and Digital Signal Process.*, vol. 43, no. 8, pp. 586–596.

A. Lindquist (1974), "A new algorithm for optimal filtering of discrete-time stationary processes," *SIAM J. Control*, vol. 12, pp. 736–746.

F. Ling (1988), "Convergence characteristics of LMS and LS adaptive algorithms for signals with rank-deficient correlation matrices," *Proc. ICASSP*, New York, pp. 1499–1502.

F. LING (1989), "Efficient least-squares lattice algorithms based on Givens rotation with systolic array implementations," *Proc. ICASSP*, Glasgow, pp. 1290–1293.

F. LING (1991), "Givens rotation based least-squares lattice and related algorithms," *IEEE Trans. Signal Processing*, vol. 39, pp. 1541–1551.

F. LING, D. G. MANOLAKIS, AND J. G. PROAKIS (1985), "New forms of LS lattice algorithms and an analysis of their round-off error characteristics," *Proc. ICASSP*, Tampa, FL, pp. 1739–1742.

F. LING, D. G. MANOLAKIS, AND J. G. PROAKIS (1985), "Numerically robust least-squares lattice-ladder algorithm with direct updating of the reflection coefficients," *IEEE Trans. Acoust. Speech Signal Process.*, vol. 34, pp. 837–845.

L. LJUNG (1977a), "Analysis of recursive stochastic algorithms," *IEEE Trans. Automat. Contr.*, vol. 22, pp. 551–575.

L. LJUNG (1977b), "On positive real transfer functions and the convergence of some recursive schemes," *IEEE Trans. Automat. Contr.*, vol. 22, pp. 539–551.

L. LJUNG (1987), *System Identification: Theory for the User*, 1st edition, Prentice-Hall, Englewood Cliffs, NJ.

S. LJUNG AND L. LJUNG (1985), "Error propagation properties of recursive least-squares adaptation algorithms," *Automatica*, vol. 21, pp. 157–167.

L. LJUNG, M. MORF, AND D. D. FALCONER (1978), "Fast calculation of gain matrices for recursive estimation schemes," *Int. J. Control*, vol. 27, pp. 1–19.

C. G. LOPES AND A. H. SAYED, (2006), "Distributed adaptive incremental strategies: Formulation and performance analysis," *Proc.ICASSP*, Toulouse, France, vol. 3, pp. 584–587.

C. G. LOPES AND A. H. SAYED, (2007a), "Incremental adaptive strategies over distributed networks," *IEEE Trans. Signal Processing*, vol. 55, no. 8, pp. 4064–4077.

C. G. LOPES AND A. H. SAYED, (2007b), "Diffusion least-mean-squares over adaptive networks," *Proc. ICASSP*, Honolulu, Hawaii, vol. 3, pp. 917–920.

C. G. LOPES AND A. H. SAYED, (2008), "Diffusion least-mean squares over adaptive networks: Formulation and performance analysis," *IEEE Trans. Signal Processing*, vol. 56.

R. LUCKY (1965), "Automatic equalization for digital communication," *Bell Syst. Tech. J.*, vol. 44, pp. 547–588.

O. MACCHI (1995), *Adaptive Processing: The* LMS *Approach with Applications in Transmission*, Wiley, New York.

O. MACCHI AND E. EWEDA (1983), "Second-order convergence analysis of stochastic adaptive linear filtering," *IEEE Trans. Automat. Control*, vol. 28, no. 1, pp. 76–85.

O. MACCHI AND E. EWEDA (1984), "Convergence analysis of self-adaptive equalizers," *IEEE Trans. Inform. Theory*, vol. 30, pp. 161–176.

J. MAI AND A. H. SAYED (2000), "A feedback approach to the steady-state performance of fractionally-spaced blind adaptive equalizers," *IEEE Trans. Signal Process.*, vol. 48, no. 1, pp. 80–91.

D. G. MANOLAKIS, V. K. INGLE, AND S. M. KOGON (2000), *Statistical and Adaptive Signal Processing*, McGraw Hill, New York.

D. G. MANOLAKIS, F. LING, AND J. G. PROAKIS (1987), "Efficient time-recursive least-squares algorithms for finite-memory adaptive filtering," *IEEE Trans. Circuits Syst.*, vol. 34, no. 4, pp. 400–408.

D. MANSOUR AND A. H. GRAY, JR. (1982), "Unconstrained frequency-domain adaptive filter," *IEEE Trans. Acoust. Speech Signal Process.*, vol. 30, no. 5, pp. 726–734.

S. MARCOS AND O. MACCHI (1987), "Tracking capability of the least mean square algorithm: Application to an asynchronous echo canceler," *IEEE Trans. Acoustics Speech Signal Process.*, vol. 35, no. 11, pp. 1570–1578.

D. F. MARSHALL, W. K. JENKINS, AND J. J. MURPHY (1989), "The use of orthogonal transforms for improving performance of adaptive filters," *IEEE Trans. Circuits Syst.*, vol. 36, no. 4, pp. 474–483.

E. MASRY AND F. BULLO (1995), "Convergence analysis of the sign algorithm for adaptive filtering," *IEEE Trans. Inform. Theory*, vol. 41, pp. 489-495.

V. J. MATHEWS (1991a), "Performance analysis of adaptive filters equipped with dual sign algorithm," *IEEE Trans. Signal Process.*, vol. 39, pp. 85–91.

V. J. MATHEWS (1991b), "Adaptive polynomial filters," *IEEE Signal Process. Mag.*, vol. 8, no. 3, pp. 10–26.

V. J. MATHEWS AND S. CHO (1987), "Improved convergence analysis of stochastic gradient adaptive filters using the sign algorithm," *IEEE Trans. Acoustics Speech Signal Process.*, vol. 35, no. 4, pp. 450–454.

V. J. MATHEWS AND Z. XIE (1993) "Stochastic gradient adaptive filters with gradient adaptive step size," *IEEE Trans. Signal Process.*, vol. 41, no. 6, pp. 2075–2087.

K. MAYYAS AND T. ABOULNASR (1997), "The leaky LMS algorithm: MSE analysis for Gaussian data," *IEEE Trans. Signal Process.*, vol. 45, no. 4, pp. 927–934.

J. E. MAZO (1979), "On the independence theory of equalizer convergence," *Bell Syst. Tech. J.*, vol 58, pp. 963–993.

W. F. MCGEE (1969), "Circularly complex Gaussian noise – A Price theorem and a Mehler expansion," *IEEE Trans. Inform. Theory*, vol. 15, pp. 317–319.

J. G. McWhirter and I. K. Proudler (1993), "The QR family," in *Adaptive System Identification and Signal Processing Algorithms*, N. Kalouptsidis and S. Theodoridis, Eds., Prentice-Hall, Englewood Cliffs, NJ, pp. 260–321.

R. Merched and A. H. Sayed (1999), "Fast RLS Laguerre adaptive filtering," *Proc. Allerton Conf. Commun. Contr. and Computing*, pp. 338–347, Allerton, IL.

R. Merched and A. H. Sayed (2000a) "An embedding approach to frequency-domain and subband adaptive filtering," *IEEE Trans. Signal Process.*, vol. 48, no. 9, pp. 2607–2619.

R. Merched and A. H. Sayed (2000b), "Order-recursive RLS Laguerre adaptive filtering," *IEEE Trans. Signal Process.*, vol. 48, no. 11, pp. 3000–3010.

R. Merched and A. H. Sayed (2001a), " RLS-Laguerre lattice adaptive filtering: Error-feedback, normalized and array-based algorithms," *IEEE Trans. Signal Process.*, vol. 49, no. 11, pp. 2565–2576.

R. Merched and A. H. Sayed (2001b), "Extended fast fixed order RLS adaptive filters," *IEEE Trans. Signal Process.*, vol. 49, no. 12, pp. 3015–3031.

N. Merhav and M. Feder (1993), "Universal schemes for sequential decision from individual sequences," *IEEE Trans. Inform. Theory*, vol. 39, pp. 1280-1292.

N. Merhav and M. Feder (1998), "Universal prediction," *IEEE Trans. Inform. Theory*, vol. 44, pp. 2124–2147.

W. B. Mikhael, F. H. Wu, L. G. Kazovsky, G. S. Kang, and L. J. Fransen (1986), "Adaptive filters with individual adaptation of parameters," *IEEE Trans. Circuits Syst.*, vol. 33, no. 7, pp. 677–686.

K. Miller (1974), *Complex Stochastic Processes*, Addison-Wesley, Reading, MA.

P. Monsen (1971), "Feedback equalization for fading dispersive channels," *IEEE Trans. Inform. Theory*, vol. 17, pp. 56–64.

M. Morf and T. Kailath (1975), "Square root algorithms for least squares estimation," *IEEE Trans. Automat. Contr.*, vol. 20, pp. 487–497.

M. Morf and D. T. L. Lee (1978), "Recursive least-squares ladder forms for fast parameter tracking," *Proc. Conference on Decision and Control*, San Diego, pp. 1362–1367.

M. Morf, G. S. Sidhu, and T. Kailath (1974), "Some new algorithms for recursive estimation in constant, linear, discrete-time systems," *IEEE Trans. Automat. Contr.*, vol. 19, pp. 315–323.

D. R. Morgan and S. G. Kratzer (1996), "On a class of computationally efficient, rapidly converging, generalized NLMS algorithms," *IEEE Signal Process. Lett.*, vol. 3, no. 8, pp. 245–247, Aug. 1996.

D. R. Morgan and M. J. C. Thi (1995), "A delayless subband adaptive filter architecture," *IEEE Trans. Signal Process.*, vol. 43, pp. 1818–1830.

G. V. Moustakides (1997), "Study of the transient phase of the forgetting factor RLS," *IEEE Trans. Signal Process.*, vol. 45, pp. 2468–2476.

G. V. Moustakides and S. Theodoridis (1991), "Fast Newton transversal filters – A new class of adaptive estimation algorithms," *IEEE Trans. Signal Process.*, vol. 39, pp. 2184–2193.

J. I. Nagumo and A. Noda (1967), "A learning method for system identification," *IEEE Trans. Automat. Contr.*, vol. 12, pp. 282–287.

S. S. Narayan and A. M. Peterson (1981), "Frequency domain LMS algorithm," *Proc. IEEE*, vol. 69, no. 1, pp. 124–126.

S. S. Narayan, A. M. Peterson, and M. J. Narasimha (1983), "Transform domain LMS algorithm," *IEEE Trans. Acoust. Speech Signal Process.*, vol. 31, no. 3, pp. 609–615.

V. H. Nascimento and A. H. Sayed (1998), "Stability of the LMS adaptive filter by means of a state equation," *Proc. Allerton Conference on Communication, Control, and Computing*, Allerton, IL, pp. 242–251.

V. H. Nascimento and A. H. Sayed (1999), "Unbiased and stable leakage-based adaptive filters," *IEEE Trans. Signal Process.*, vol. 47, no. 12, pp. 3261–3276.

V. H. Nascimento and A. H. Sayed (2000), "On the learning mechanism of adaptive filters," *IEEE Trans. Signal Process.*, vol. 48, no. 6, pp. 1609–1625

M. Niedźwiecki (1992), "Multiple-model approach to finite memory adaptive filtering," *IEEE Trans. Signal Process.*, vol. 40, pp. 470–473.

R. Nitzberg (1985), "Application of the normalized LMS algorithm to MSLC," *IEEE Trans. Aero. Electronic Syst.*, vol. 21, no. 1, pp. 24–31.

D. O. North (1943), "An analysis of the factors which determine signal/noise discrimination in pulse-carrier systems," RCA Tech. Rep. PTR-6C.

M. R. Osborne (1985), *Finite Algorithms in Optimization and Data Analysis*, Wiley, New York.

K. Ozeki and T. Umeda (1984), "An adaptive filtering algorithm using an orthogonal projection to an affine subspace and its properties," *Electron. Commun. Jpn.*, vol. 67-A, no. 5, pp. 19–27.

C. C. Paige (1979a), "Computer solution and perturbation analysis of generalized linear least-squares problems," *Math. Comput.*, vol. 33, pp. 171–183.

C. C. Paige (1979b), "Fast numerically stable computation for generalized linear least-squares problems," *SIAM J. Numer. Anal.*, vol. 16, pp. 165–171.

C. C. Paige (1985), "Covariance matrix representation in linear filtering," *Proc. Joint Summer Research Conference on Linear Algebra and Its Role in Systems Theory*, B. Datta, Ed.

C. C. PAIGE AND M. A. SAUNDERS (1977), "Least-squares estimation of discrete linear dynamic systems using orthogonal transformations," *SIAM J. Numer. Anal.*, vol. 14, pp. 180–193.

C. T. PAN AND R. J. PLEMMONS (1989), "Least-squares modifications with inverse factorizations: Parallel implications," *Comput. Appl. Math.*, vol. 27, pp. 109–127.

A. PAPOULIS (1991), *Probability, Random Variables, and Stochastic Processes*, 3rd edition, McGraw Hill, New York.

A. PARTHASARATHY AND R. M. EVAN-IWANOWSKI (1978), "On the almost sure stability of linear stochastic systems," *SIAM J. Appl. Math.*, vol. 34, no. 4, pp. 643–656.

D. I. PAZAITIS AND A. G. CONSTANTINIDES (1999), "A novel kurtosis driven variable step-size adaptive algorithm," *IEEE Trans. Signal Process.*, vol. 47, no. 3, pp. 864–872.

M. R. PETRAGLIA AND S. K. MITRA (1991), "Generalized fast convolution implementations of adaptive filters," *Proc. ISCAS*, vol. 5, pp. 2916–2919.

M. R. PETRAGLIA AND S. K. MITRA (1993), "Adaptive FIR filter structure based on the generalized subband decomposition of FIR filters," *IEEE Trans. Circuits Syst. II: Analog and Digital Signal Process.*, vol. 40, no. 6, pp. 354–362.

M. R. PETRAGLIA, R. G. ALVES, AND P. S. R. DINIZ (2000), "New structures for adaptive filtering in subbands with critical sampling," *IEEE Trans. Signal Process.*, vol. 48, pp. 3316–3327.

E. PFANN AND R. STEWART (1998), "LMS adaptive filtering with $\Sigma - \Delta$ modulated input signals," *IEEE Signal Process. Lett.*, vol. 5, no. 4, pp. 95–97.

B. PICINBONO (1993), *Random Signals and Systems*, Prentice-Hall, Englewood Cliffs, NJ.

G. PICCHI AND G. PRATI (1987), "Blind equalization and carrier recovery using a stop-and-go decision-direction algorithm," *IEEE Trans. Commun.*, vol. 35, pp. 877–887.

R. L. PLACKETT (1950), "Some theorems in least-squares," *Biometrika*, vol. 37, pp. 149.

R. L. PLACKETT (1972), "The discovery of the method of least-squares," *Biometrika*, vol. 59, pp. 239–251.

V. M. POPOV (1973), *Hyperstability of Control Systems*, Springer-Verlag, New York. [Translation of Romanian edition, 1966.]

J. E. POTTER AND R. G. STERN (1963), "Statistical filtering of space navigation measurements," *Proc. AIAA Guidance Contr. Conference*.

S. S. PRADHAN AND V. U. REDDY (1999), "A new approach to subband adaptive filtering," *IEEE Trans. Signal Process.*, vol. 47, pp. 655–664.

R. PRICE (1954), "The detection of signals perturbed by scatter and noise," *IRE Trans. Inform. Theory*, vol. 4, pp. 163–170.

R. PRICE (1955), "Optimum detection of random signals in noise with application to scatter-multipath communication," *IRE Trans. Inform. Theory*, vol. 2, pp. 125–135.

R. PRICE (1958), "A useful theorem for nonlinear devices having Gaussian inputs," *IRE Trans. Inform. Theory*, vol. 4, pp. 69–72.

R. PRICE (1972), "Nonlinearly feedback-equalized PAM vs. capacity for noisy filter channels," *Record IEEE Int. Conf. Commun.*, Philadelphia, pp. 22.12–22.17

R. PRICE AND P. E. GREEN, JR. (1958), "A communication technique for multipath channels," *Proc. IRE*, vol. 46, pp. 555–570.

J. G. PROAKIS (2000), *Digital Communications*, McGraw Hill, New York, 4th edition.

J. G. PROAKIS AND J. H. MILLER (1969), "Adaptive receiver for digital signaling through channels with intersymbol interference," *IEEE. Trans. Inform. Theory*, vol. 15, pp. 484–497.

J. G. PROAKIS, C. M. RADER, F. LING, AND C. L. NIKIAS (1992), *Advanced Digital Signal Processing*, Macmillan, New York.

I. K. PROUDLER, J. G. MCWHIRTER, AND T. J. SHEPHERD (1988), "Fast QRD-based algorithms for least-squares linear prediction," *Proc. IMA Conf. Math. Signal Process.*, Warwick, England.

I. K. PROUDLER, J. G. MCWHIRTER, AND T. J. SHEPHERD (1989), "QRD-based lattice filter algorithms," *Proc. SPIE*, vol. 1152, pp. 56–67.

I. K. PROUDLER, J. G. MCWHIRTER, AND T. J. SHEPERD (1990), "Computationally efficient QR decomposition approach to least-squares adaptive filtering," *IEE Proceedings*, vol. 138, no. 4, pp. 341–353.

V. S. PUGACHEV (1958), "The determination of an optimal system by some arbitrary criterion," *Automation and Remote Control*, vol. 19, pp. 513–532.

S. U. H. QURESHI (1985), "Adaptive equalization," *Proc. IEEE*, vol. 53, pp. 1349–1387.

S. U. H. QURESHI AND G. D. FORNEY (1977), "Performance and properties of a $T/2$ equalizer," *Nat.l. Telecom. Conf. Record*, Los Angeles, CA, pp. 11:1.1–4.

T. S. RAPPAPORT (1996), *Wireless Communications*, Prentice-Hall, Englewood Cliffs, NJ.

C. R. RAO (1973), *Linear Statistical Inference and Its Applications*, Wiley, New York.

P. A. REGALIA (1992), "Numerical stability issues in fast least-squares adaptation algorithms," *Optical Engineering*, vol. 31, pp. 1144–1152.

P. A. REGALIA (1993), "Numerical stability properties of a QR-based fast least squares algorithm," *Optical Engineering*, vol. 41, pp. 2006–2109.

P. A. REGALIA (1995), *Adaptive IIR Filtering in Signal Processing and Control*, Marcel Dekker, New York.

P. A. REGALIA AND G. BELLANGER (1991), "On the duality between fast QR methods and lattice methods in least-squares adaptive filtering," *IEEE Trans. Signal Process.*, vol. 39, pp. 879–891.

M. REUTER AND J. R. ZEIDLER (1999), "Nonlinear effects in LMS adaptive equalizers," *IEEE Trans. Signal Process.*, vol. 47, pp. 1570–1579.

J. F. RICCATI (1724),"Animadversationnes in aequationes differentiales secundi gradus," *Eruditorum Quae Lipside Publicantur Supplementa*, vol. VIII, pp. 66–73.

J. T. RICKARD AND J. R. ZEIDLER (1979), "Second-order output statistics of the adaptive line enhancer," *IEEE Trans. Acoust. Speech Signal Process.*, vol. 27, pp. 31–39.

J. RISSANEN (1984), "Universal coding, information, prediction, and estimation," *IEEE Trans. Inform. Theory*, vol. 30, pp. 629–636.

J. RISSANEN (1986), "A predictive least-squares principle," *IMA J. Math. Contr. Inform.*, vol. 3, no. 2–3, pp. 221–222.

H. ROBBINS AND S. MONRO (1951), "A stochastic approximation method," *Ann. Math. Stat.*, vol. 22, pp. 400–407.

A. A. RONTOGIANNIS AND S. THEODORIDIS (1998), "New fast QR decomposition least-squares adaptive algorithms," *IEEE Trans. Signal Process.*, vol. 46, pp. 2113–2121.

S. ROY AND J. J. SHYNK (1989), "Analysis of the data-reusing LMS algorithm," *Proc. Midwest Symp. Circuits Syst.*, Urbana, IL, pp. 1127–1130.

M. RUPP (1993), "The behavior of LMS and NLMS algorithms in the presence of spherically invariant processes," *IEEE Trans. Signal Process.*, vol. 41, no. 3, pp. 1149–1160.

M. RUPP (1995), "Bursting in the LMS algorithm," *IEEE Trans. Signal Process.*, vol. 43, no. 10, pp. 2414–2417.

M. RUPP (1998a), "A family of adaptive filter algorithms with decorrelating properties," *IEEE Trans. Signal Process.*, vol. 46, no. 3, pp. 771–775.

M. RUPP (1998b), "LMS tracking behavior under periodically changing systems," *Proc. European Signal Processing Conference*, Island of Rhodes, Greece.

M. RUPP AND A. H. SAYED (1996a), "A time-domain feedback analysis of filtered-error adaptive gradient algorithms," *IEEE Trans. Signal Process.*, vol. 44, no. 6, pp. 1428–1439.

M. RUPP AND A. H. SAYED (1996b), "Robustness of Gauss-Newton recursive methods: A deterministic feedback analysis," *Signal Processing*, vol. 50, no. 3, pp. 165–188.

M. RUPP AND A. H. SAYED (1997), "Supervised learning of Perceptron and output feedback dynamic networks: A feedback analysis via the small gain theorem," *IEEE Trans. Neural Networks*, vol. 8, no. 3, pp. 612–622.

M. RUPP AND A. H. SAYED (1998), "Robust FxLMS algorithms with improved convergence performance," *IEEE Trans. Speech Audio Process.*, vol. 6, no. 1, pp. 78–85.

M. RUPP AND A. H. SAYED (2000), "On the convergence of blind adaptive equalizers for constant-modulus signals," *IEEE Trans. Commun.*, vol. 48, no. 5, pp. 795–803.

B. Y. RYABKO (1984), "Twice-universal coding," *Probl. Inform. Transm.*, vol. 20, pp. 173–177.

B. Y. RYABKO (1988), "Prediction of random sequences and universal coding," *Probl. Inform. Transm.*, vol. 24, pp. 87–96.

J. SALZ (1973), "Optimum mean-square decision feedback equalization," *Bell Syst. Tech. J.*, vol. 50, no. 8, pp. 1341–1373.

S. G. SANKARAN AND A. A. (LOUIS) BEEX (1997), "Normalized LMS algorithm with orthogonal correction factors," *Proc. Asilomar Conf. on Signals, Systems, and Computers*, Pacific Grove, CA, pp. 1670–1673.

S. G. SANKARAN AND A. A. (LOUIS) BEEX (2000), "Convergence behavior of affine projection algorithms," *IEEE Trans. Signal Process.*, vol. 48, no. 4, pp. 1086–1096.

H. SARI, G. KARAM, AND I. JEANCLAUDE (1995), "Transmission techniques for digital terrestrial TV broadcasting," *IEEE Commun. Mag.*, pp. 100–109.

Y. SATO (1975), "A method of self-recovering equalization for multilevel amplitude modulation," *IEEE Trans. Commun.*, vol. 23, pp. 679–682.

A. H. SAYED (1992), *Displacement Structure in Signal Processing and Mathematics*, Ph.D. dissertation, Department of Electrical Engineering, Stanford University, Stanford, CA.

A. H. SAYED (2001), "A framework for state-space estimation with uncertain models," *IEEE Trans. Autom. Contr.*, vol. 46, no. 7, pp. 998–1013.

A. H. SAYED (2003), *Fundamentals of Adaptive Filtering*, Wiley, NJ.

A. H. SAYED AND T. Y. AL-NAFFOURI (2001), "Mean-square analysis of normalized leaky adaptive filters," *Proc. ICASSP*, Salt Lake City, UT, vol. 6, pp. 3873–3876.

A. H. SAYED, T. Y. AL-NAFFOURI, AND V. H. NASCIMENTO (2003), "Energy conservation in adaptive filtering," in *Nonlinear Signal and Image Processing: Theory, Methods, and Applications*, K. E. Barner and G. Arce, Eds., CRC Press, pp. 1–35.

A. H. SAYED AND S. CHANDRASEKARAN (2000), "Parameter estimation with multiple sources and levels of uncertainties," *IEEE Trans. Signal Process.*, vol. 48, no. 3, pp. 680–692.

A. H. SAYED AND H. CHEN (2002), "A uniqueness result concerning a robust regularized least-squares solution," *Syst. & Contr. Lett.*, vol. 46, pp. 361–369.

A. H. SAYED, B. HASSIBI, AND T. KAILATH (1996a), "Inertia conditions for the minimization of quadratic forms in indefinite metric spaces," in vol. 87 of *Operator Theory: Advances and Applications*, I. Gohberg, P. Lancaster, and P. N. Shivakumar, Eds., Birkhauser, pp. 309–347.

A. H. SAYED, B. HASSIBI, AND T. KAILATH (1996b), "Inertia properties of indefinite quadratic forms," *IEEE Signal Process. Letters*, vol. 3, no. 2, pp. 57–59.

A. H. SAYED AND T. KAILATH (1992), "Structured matrices and fast RLS adaptive filtering," *Proc. 2nd IFAC Workshop on Algorithms and Architectures for Real-Time Control*, Seoul, Korea, pp. 211–216.

A. H. SAYED AND T. KAILATH (1993), "A state-space approach to adaptive filtering," *Proc. ICASSP*, MN, vol. 3, pp. 559–562.

A. H. SAYED AND T. KAILATH (1994a) "Extended Chandrasekhar recursions," *IEEE Trans. Automat. Control*, vol. 39, no. 3, pp. 619–623.

A. H. SAYED AND T. KAILATH (1994b), "A state-space approach to adaptive RLS filtering," *IEEE Signal Process. Mag.*, vol. 11, pp. 18–60.

A. H. SAYED AND T. KAILATH (1995), "Oblique state-space estimation algorithms," *Proc. American Control Conference*, Seattle, WA, vol. 3, pp. 1969–1973.

A. H. SAYED AND T. KAILATH (1998), "Recursive least-squares adaptive filters," *DSP Handbook*, Chapter 21, CRC Press, 1998.

A. H. SAYED AND P. V. KOKOTOVIC (1995), "A note on a Lyapunov argument for stochastic gradient methods in the presence of noise," *Proc. ACC*, Seattle, WA, vol. 1, pp. 587–588.

A. H. SAYED, H. LEV-ARI, AND T. KAILATH (1994), "Time-variant displacement structure and triangular arrays," *IEEE Trans. Signal Process.*, vol. 42, pp. 1052–1062.

A. H. SAYED AND C. G. LOPES (2007), "Adaptive processing over distributed networks," *IEICE Transactions on Fundamentals of Electronics, Communications and Computer Sciences*, vol. E90-A, no. 8, pp. 1504-1510.

A. H. SAYED AND V. H. NASCIMENTO (2003), "Energy conservation and the learning ability of LMS adaptive filters," in *Least-Mean-Square Adaptive Filters: New Insights*, S. Haykin and B. Widrow, Eds., Wiley, NY, pp. 79–104.

A. H. SAYED, V. H. NASCIMENTO, AND S. CHANDRASEKARAN (1998), "Estimation and control with bounded data uncertainties," *Linear Alg. Appl.*, vol. 284, pp. 259–306.

A. H. SAYED, V. H. NASCIMENTO, AND F. A. M. CIPPARRONE (2002), "A regularized robust design criterion for uncertain data," *SIAM J. Matrix Anal. Appl.*, vol. 23, no. 4, pp. 1120–1142.

A. H. SAYED AND M. RUPP (1994), "On the robustness, convergence, and minimax performance of instantaneous-gradient adaptive filters," *Proc. Asilomar Conference on Signals, Systems, and Computers*, pp. 592–596.

A. H. SAYED AND M. RUPP (1995), "A time-domain feedback analysis of adaptive algorithms via the small gain theorem," *Proc. SPIE*, vol. 2563, San Diego, CA, pp. 458–469.

A. H. SAYED AND M. RUPP (1996), "Error energy bounds for adaptive gradient algorithms," *IEEE Trans. Signal Process.*, vol. 44, no. 8, pp. 1982–1989.

A. H. SAYED AND M. RUPP (1997), "An l_2−stable feedback structure for nonlinear adaptive filtering and identification," *Automatica*, vol. 33, no. 1, pp. 13–30.

A. H. SAYED AND M. RUPP (1998), "Robustness issues in adaptive filtering," *DSP Handbook*, Chapter 20, CRC Press.

L. L. SCHARF (1991), *Statistical Signal Processing: Detection, Estimation and Time-Series Analysis*, Addison–Wesley, Reading, MA.

M. SCHETZEN (1980), *The Volterra and Wiener Theory of Nonlinear Systems,*, Wiley, New York.

L. SCHMETTERER (1961), "Stochastic approximation," *Proc. Berkeley Symp. Math. Statist. Probab.*, pp. 587–609.

S. F. SCHMIDT (1970), "Computational techniques in Kalman filtering," in *Theory and Applications of Kalman Filtering*, AGARDograph 139, Tech. Rep., NATO Advisory Group for Aerospace Research and Development.

O. SEFL (1960), "Filters and predictors which adapt their values to unknown parameters of the input process," *Transactions of the Second Prague Conference on Information Theory,* Czechoslovak Academy of Sciences, Prague.

W. A. SETHARES (1992), "Adaptive algorithms with nonlinear data and error functions," *IEEE Trans. Signal Process.*, vol. 40, no. 9, pp. 2199–2206.

W. A. SETHARES, B. D. O. ANDERSON, AND C. R. JOHNSON, JR. (1989), "Adaptive algorithms with filtered regressor and filtered error," *Mathematics of Control, Signals, and Systems*, vol. 2, pp. 381–403.

W. A. SETHARES AND C. R. JOHNSON, JR. (1989), "A comparison of two quantized state adaptive algorithms," *IEEE Trans. Acoust. Speech Signal Process.*, vol. 37, pp. 138–143.

W. A. SETHARES, D. A. LAWRENCE, C. R. JOHNSON, JR., AND R. R. BITMEAD (1986), "Parameter drift in LMS adaptive filters," *IEEE Trans. Acoust. Speech Signal Process.*, vol. 34, pp. 868–878.

W. A. SETHARES, I. M. Y. MAREELS, B. D. O. ANDERSON, C. R. JOHNSON, JR., AND R. R. BITMEAD (1988), "Excitation conditions for signed regressor least-mean-square adaptation," *IEEE Trans. Circuits. Syst.*, vol. 35, pp. 613–624.

U. SHAKED AND Y. THEODOR (1992), "$\mathcal{H}^\infty$-optimal estimation: A tutorial," *Proc. CDC*, Tucson, AZ, pp. 2278–2286.

O. SHALVI AND E. WEINSTEIN (1990), "New criteria for blind equalization of nonminimum phase systems channels," *IEEE Trans. Inform. Theory*, vol. 36, pp. 312–321.

R. SHARMA, W. SETHARES, AND J. BUCKLEW (1996), "Asymptotic analysis of stochastic gradient-based adaptive filtering algorithms with general cost functions," *IEEE Trans. Signal Process.*, vol. 44, no. 9, pp. 2186–2194.

T. J. SHEPHERD AND J. G. MCWHIRTER (1993), "Systolic adaptive beamforming," in *Radar Array Processing*, S. Haykin and T. J. Shepherd, Eds., Springer-Verlag, New York, pp. 153–243.

D. T. SHERWOOD AND N. J. BERSHAD (1987), "Quantization effects in the complex LMS adaptive algorithm: Linearization using dither-theory," *IEEE Trans. Circuits Syst.*, vol. 34, pp. 848–854.

O. B. SHEYNIN (1977), "Laplace's theory of errors", *Archive for History of Exact Sciences*, vol. 17, pp. 1–61.

H-C. SHIN AND A. H. SAYED (2004), "Mean-square performance of a family of affine projection algorithms," *IEEE Trans. Signal Process*, vol. 52, no. 1, pp. 90–102.

H-C. SHIN, A. H. SAYED, AND W-J. SONG (2004), "Variable step-size NLMS and affine projection algorithms," *IEEE Signal Processing Letters*, vol. 11, no. 2, part II, pp. 132-135.

H-C. SHIN, A. H. SAYED, AND W-J. SONG (2004b), "Mean-square performance of adaptive filters using averaging theory," *Proc. 38th Asilomar Conference on Signals, Systems and Computers,* Pacific Grove, pp. 229–234.

J. J. SHYNK (1989), "Adaptive IIR filtering," *IEEE ASSP Mag.*, vol. 6, pp. 4–21.

J. J. SHYNK (1992), "Frequency-domain and multirate adaptive filtering," *IEEE Signal Process. Mag.*, vol. 9, no. 1, pp. 14–37.

J. J. SHYNK, R. P. GOOCH, G. KRISNAMURTHY, AND C. K. CHAN (1991), "A comparative performance study of several blind equalization algorithms," *Proc. SPIE Conf. Adv. Signal Process.*, vol. 1565, pp. 102–117.

A. C. SINGER AND M. FEDER (1999), "Universal linear prediction by model order weighting," *IEEE Trans. Signal Process.*, vol. 47, no. 10, pp. 2685–2699.

A. C. SINGER, S. S. KOZAT, AND M. FEDER (2002), "Universal linear least-squares prediction: Upper and lower bounds," *IEEE Trans. Inform. Theory*, vol. 48, no. 8, pp. 2354–2362.

D. T. M. SLOCK (1989), "Fast Algorithms for Fixed-Order Recursive Least-Squares Parameter Estimation," Ph.D. dissertation, Department of Electrical Engineering, Stanford University, Stanford, CA.

D. T. M. SLOCK (1992), "The backward consistency concept and roundoff error propagation dynamics in RLS algorithms," *Optical Engineering*, vol. 31, pp. 1153–1169.

D. T. M. SLOCK (1993), "On the convergence behavior of the LMS and the normalized LMS algorithms," *IEEE Trans. Signal Process.*, vol. 41, no. 9, pp. 2811–2825.

D. T. M. SLOCK AND T. KAILATH (1988), "Numerically stable fast recursive least-squares transversal filters," *Proc. ICASSP*, vol. 3, pp. 1365–1368.

D. T. M. SLOCK AND T. KAILATH (1991), "Numerically stable fast transversal filters for recursive least-squares adaptive filtering," *IEEE Trans. Signal Process.*, vol. 39, no. 1, pp. 92–114.

T. SÖDERSTRÖM (1994), *Discrete-Time Stochastic Systems: Estimation and Control*, Prentice-Hall International, UK.

T. SÖDERSTRÖM AND P. G. STOICA (1983), *Instrumental Variable Methods for System Identification*, Springer-Verlag, Berlin, Heidelberg.

V. SOLO (1997), "The stability of LMS," *IEEE Trans. Signal Process.*, vol. 45, no. 12, pp. 3017–3026.

V. SOLO AND X. KONG (1995), *Adaptive Signal Processing Algorithms*, Prentice-Hall, Englewood Cliffs, NJ.

P. C. W. SOMMEN (1989), "Partitioned frequency domain adaptive filters," *Proc. Asilomar Conf. Signals, Syst. and Computers*, Pacific Grove, CA, pp. 677–681.

P. C. W. SOMMEN, P. J. VAN GERWEN, H. J. KOTMANS, AND A. E. J. M. JANSEN (1987), "Convergence analysis of a frequency-domain adaptive filter with exponential power averaging and generalized window function," *IEEE Trans. Circuits Syst.*, vol. 34, pp. 788–798.

M. M. SONDHI (1967), "An adaptive echo canceller," *Bell Syst. Tech. J.*, vol. 46, no. 3.

M. M. SONDHI AND D. A. BERKLEY (1980), "Silencing echoes in the telephone network," *Proc. IEEE*, vol. 68, pp. 948–963.

J. S. SOO AND K. K. PANG (1987), "A new structure for block FIR adaptive digital filters," *IREECON. Int. Dig. Papers,* Sydney, Australia, pp. 364–367.

J. S. SOO AND K. K. PANG (1990), "Multidelay block frequency domain adaptive filters," *IEEE Trans. Acoust. Speech Signal Process.*, vol. 38, no. 2, pp. 373–376.

H. W. SORENSON, (1966), "Kalman filtering techniques," in *Advances in Control Systems Theory and Applications*, C. T. Leondes, Ed., vol. 3, Academic Press, pp. 219–292.

H. W. SORENSON, (1970), "Least-squares estimation: From Gauss to Kalman," *IEEE Spectrum*, pp. 63–68.

H. W. SORENSON, Ed., (1985), *Kalman Filtering: Theory and Application*, IEEE Press, New York.

H. STARK AND J. W. WOODS (1994), *Probability, Random Processes, and Estimation Theory for Engineers*, 2nd edition, Prentice-Hall, Englewood Cliffs, NJ.

G. W. STEWART (1995), *Theory of the Combination of Observations Least Subject to Errors*, Classics in Applied Mathematics, SIAM, PA. [Translation of original works by C. F. Gauss under the title *Theoria Combinationis Observationum Erroribus Minimis Obnoxiae*.]

K. STEIGLITZ AND L. E. MCBRIDE (1965), "A technique for the identification of linear systems," *IEEE Trans. Automat. Control*, vol. 10, pp. 461–464.

G. STRANG (1988), *Linear Algebra and Its Applications*, 3rd edition, Academic Press, New York.

G. STRANG (1993), *Introduction to Linear Algebra*, Wellesley-Cambridge Press, Wellesley, MA.

P. STROBACH (1990), *Linear Prediction Theory*, Springer-Verlag, Berlin Heidelberg.

G. SZEGÖ (1939), *Orthogonal Polynomials*, American Mathematical Society Colloquium Publishing, vol. 23, Providence, RI.

G. SZEGÖ (1959), *Orthogonal Functions, Interscience Publishers*, New York.

O. TANRIKULU AND J. A. CHAMBERS (1996), "Convergence and steady-state properties of the least-mean mixed-norm (LMMN) adaptive algorithm," *IEE Proceedings – Vision, Image Signal Processing,* vol. 143, no. 3, pp. 137–142.

A. TARIGHAT AND A. H. SAYED (2003), "An optimum OFDM receiver exploiting cyclic prefix for improved data estimation," *Proc. ICASSP*, vol. IV, pp. 217–220, Hong Kong.

A. TARIGHAT AND A. H. SAYED (2004), "Least-mean-phase adaptive filters with application to communications systems," *IEEE Signal Process. Lett.*, vol. 11, no. 2, pp. 220–223.

A. TARIGHAT AND A. H. SAYED (2005), "MIMO OFDM receivers for systems with IQ imbalances," *IEEE Trans. Signal Process.*, vol. 53, no. 9, pp. 3583–3596.

M. TARRAB AND A. FEUER (1988), "Convergence and performance analysis of the normalized LMS algorithm with uncorrelated Gaussian data," *IEEE Trans. Inform. Theory,* vol. 34, no. 4, pp. 680–690.

J. TERRY AND J. HEISKALA (2002), OFDM *Multicarrier Wireless Networks: A Practical Approach*, Sams Publishing.

A. N. TIKHONOV (1963), "On solving incorrectly posed problems and method of regularization," *Doklady Akademii Nauk USSR*, vol. 151, pp. 501–504.

J. T. TRAUB (1965), *Iterative Methods for the Solution of Equations*, Prentice-Hall, Englewood Cliffs, NJ.

J. R. TREICHLER (1979), "Transient and convergent behavior of the adaptive line enhancer," *IEEE Trans. Acoust. Speech Signal Process.*, vol. 27, pp. 53–62.

J. R. TREICHLER AND B. G. AGEE (1983), "A new approach to multipath correction of constant modulus signals," *IEEE Trans. Acoust. Speech Signal Process.*, vol. 31, pp. 459–472.

J. R. TREICHLER, C. R. JOHNSON, JR., AND M. G. LARIMORE (2001), *Theory and Design of Adaptive Filters*, Prentice-Hall, Englewood Cliffs, NJ.

J. R. TREICHLER, M. G. LARIMORE, AND J. C. HARP (1998), "Practical blind demodulators for high-order QAM signals," *Proc. IEEE*, vol. 86, no. 10, pp. 1927–1950.

T. K. TSATSANIS, G. B. GIANNAKIS, AND G. ZHOU (1996), "Estimation and equalization of fading channels with random coefficients," *Signal Processing*, vol. 53, no. 2–3, pp. 211–229.

YA. Z. TSYPKIN (1971), *Adaptation and Learning in Automatic Systems*, Academic Press, New York.

G. L. TURIN (1960), "An introduction to matched filters," *IRE Trans. Inform. Theory*, vol. 6, pp. 311–329.

G. UNGERBOECK (1972), "Theory on the speed of convergence in adaptive equalizers for digital communication," *IBM J. Res. Dev.*, vol. 16, pp. 546–555.

G. UNGERBOECK (1976), "Fractional tap-spacing equalizer and consequences for clock recovery in data modems," *IEEE Trans. Commun.*, vol. 24, no. 8, pp. 856–864.

P. P. VAIDYANATHAN (1993), *Multirate Systems and Filter Banks*, Prentice-Hall, Englewood Cliffs, NJ.

A. VAN DEN BOS (1996) "Price's theorem for complex variables," *IEEE Trans. Inform. Theory*, vol. 42, no. 1, pp. 286–287.

R. D. J. VAN NEE AND R. PRASAD (2000), OFDM *for Wireless Multimedia Communications*, Artech House Publishers.

H. L. VAN TREES (1968), *Detection, Estimation, and Modulation Theory*, Wiley, New York.

B. D. VAN VEEN AND K. M. BUCKLEY (1988), "Beamforming: A versatile approach to spatial filtering," *IEEE ASSP Mag.*, vol. 5, pp. 4–24.

S. VERDU (1998), *Multiuser Detection*, Cambridge University Press, Cambridge, England.

M. H. VERHAEGEN (1989a), "Improved understanding of the loss of symmetry phenomenon in the conventional Kalman filter," *IEEE Trans. Automat. Control*, vol. 34, no. 3, pp. 331–333.

M. H. VERHAEGEN (1989b), "Round-off error propagation in four generally-applicable, recursive least-squares estimation schemes," *Automatica*, vol. 25, no. 3, pp. 437–444.

M. VETTERLI AND H. J. NUSSBAUMER (1984), "Simple FFT and DCT algorithms with reduced number of operations," *Signal Processing*, vol. 6, no. 4, pp. 267–278.

M. VIDYASAGAR (1993), *Nonlinear Systems Analysis*, 2nd edition, Prentice-Hall, Englewood Cliffs, NJ.

A. J. VITERBI (1995), CDMA*: Principles of Spread-Spectrum Communications*, Addison-Wesley, Reading, MA.

R. von Mises and H. Pollaczek-Geiringer (1929), "Praktische verfahren der gleichungs-auflösung," *Z. Agnew. Math. Mech.*, vol. 9.

E. Walach and B. Widrow (1984), "The least-mean fourth (LMF) adaptive algorithm and its family," *IEEE Trans. Inform. Theory*, vol. 30, no. 2, pp. 275–283.

J. L. Walsh (1935), "Interpolation and Approximation by Rational Functions in the Complex Domain," vol. XX of Colloquium Publications, American Mathematical Society.

T. Walzman and M. Schwartz (1973), "Automatic equalization using the discrete frequency domain," *IEEE Trans. Inform. Theory,* pp. 59–68.

C. R. Ward, P. H. Hargrave, and J. G. McWhirter (1986), "A novel algorithm and architecture for adaptive digital beamforming," *IEEE Trans. Antennas Propag.*, vol. 34, pp. 338–346.

M. Wax (1988), "Order selection for AR models by predictive least squares," *IEEE Trans. Acoust. Speech Signal Process.*, vol. 36, pp. 581–588.

A. Weiss and D. Mitra (1979), "Digital adaptive filters: Conditions for convergence, rates of convergence, effects of noise and errors arising from the implementation," *IEEE Trans. Inform. Theory*, vol. 25, pp. 637–652.

J. J. Werner, J. Yang, D. D. Harman, and G. A. Dumont (1999), "Blind equalization for broadband access," *IEEE Comm. Mag.*, pp. 87–93.

G. B. Wetherhill (1966), *Sequential Methods in Statistics,* Methuen, London.

B. Widrow (1966), *Adaptive Filters, I: Fundamentals,* Tech. Rep. 6764-6, Stanford Electronics Laboratory, Stanford, CA.

B. Widrow, J. R. Glover, J. M. McCool, J. Kaunitz, C. S. Williams, R. H. Hearn, J. R. Zeidler, E. Dong, and R. C. Goodlin (1975), "Adaptive noise cancelling: Principles and applications," *Proc. IEEE*, vol. 63, no. 12, pp. 1692–1716.

B. Widrow and M. E. Hoff, Jr. (1960), "Adaptive switching circuits," *IRE WESCON Conv. Rec.*, Pt. 4, pp. 96–104.

B. Widrow and M. A. Lehr (1990), "30 years of adaptive neural networks: Perceptron, Madaline, and backpropagation," *Proc. IEEE*, vol. 78, no. 9, pp. 1415–1441.

B. Widrow, P. Mantey, L. J. Griffiths, and B. Goode (1967), "Adaptive antenna systems," *Proc. IEEE*, vol. 55, no. 12, pp. 2143–2159.

B. Widrow, J. M. McCool, M. G. Larimore, and C. R. Johnson (1976), "Stationary and nonstationary learning characteristics of the LMS adaptive filter," *Proc. IEEE*, vol. 46. no. 8, pp. 1151–1162.

B. Widrow and S. D. Stearns (1985), *Adaptive Signal Processing*, Prentice-Hall, Englewood Cliffs, NJ.

N. Wiener (1933), *The Fourier Integral and Certain of Its Applications*, Cambridge University Press, Cambridge, England.

N. Wiener (1949), *Extrapolation, Interpolation and Smoothing of Stationary Time Series*, Technology Press and Wiley, New York. [Originally published in 1942 as a classified Nat. Defense Res. Council Report. Also published under the title *Time Series Analysis* by MIT Press, Cambridge, MA.]

N. Wiener and E. Hopf (1931), "On a class of singular integral equations," *Proc. Prussian Acad. Math. — Phys. Ser.*, p. 696.

D. J. Wilde (1964), *Optimal Seeking Methods*, Prentice-Hall, Englewood Cliffs, NJ.

D. J. Wilde and C. S. Beightler (1967), *Foundations of Optimization*, Prentice-Hall, Englewood Cliffs, NJ.

A. S. Willsky (1979), *Digital Signal Processing and Control and Estimation Theory*, MIT Press, Cambridge, MA.

H. Wold (1938), *A Study in the Analysis of Stationary Time Series*, Almqvist & Wiksell, Upssala, Sweden.

M. Woodbury (1950), *Inverting Modified Matrices*, Mem. Rep. 42, Statistical Research Group, Princeton University, Princeton, NJ.

R. Wooding (1956), "The multivariate distribution of complex normal variables," *Biometrika*, vol. 43, pp. 212–215.

P. Xue and B. Liu (1986), "Adaptive equalizer using finite-bit power-of-two quantizer," *IEEE Trans. Acoust. Speech Signal Process.*, vol. 34, pp. 1603–1611.

M. Xu and Y. Grenier (1989), "Time-frequency domain adaptive filter," *Proc. ICASSP*, pp. 1154-1157.

I. Yaesh and U. Shaked (1991), "$\mathcal{H}^\infty$ –optimal estimation: The discrete time case," *Proc. Inter. Symp. MTNS*, Kobe, Japan, pp. 261–267.

B. Yang (1994), "A note on error propagation analysis of recursive least-squares algorithms," *IEEE Trans. Signal Process.*, vol. 42, no. 12, pp. 3523–3525.

B. Yang and J. F. Böhme (1992), "Rotation-based RLS algorithms: Unified derivations, numerical properties, and parallel implementations," *IEEE Trans. Signal Process.*, vol. 40, pp. 1151–1167.

J. Yang, J. J. Werner, and G. A. Dumont (1997), "The multi-modulus blind equalization algorithm," *Proc. Int. Conf. Digital Signal Process.*, Santorini, Greece, pp. 127–130.

H. Yasukawa and S. Shimada (1987), "Acoustic echo canceler with high speed quality," *Proc. ICASSP*, Dallas, TX, pp. 2125–2128.

H. YASUKAWA AND S. SHIMADA (1993), "An acoustic echo canceller using subband sampling and decorrelation methods," *IEEE Trans. Sig. Process.*, vol. 41, pp. 926–930.

W. M. YOUNIS, N. AL-DHAHIR, AND A. H. SAYED (2002), "Adaptive frequency-domain equalization of space-time block-coded transmissions," *Proc. ICASSP*, Orlando, FL, vol. 3, pp. 2353–2356.

W. M. YOUNIS, A. H. SAYED, AND N. AL-DHAHIR (2003), "Efficient adaptive receivers for joint equalization and interference cancellation in multi-user space-time block-coded systems," *IEEE Trans. Sig. Process.*, vol. 51, no. 9, pp. 2849–2862.

N. R. YOUSEF AND A. H. SAYED (1999a), "A unified approach to the steady-state and tracking analyses of adaptive filtering algorithms," *Proc. 4th IEEE-EURASIP Workshop on Nonlinear Signal and Image Processing*, Antalya, Turkey, vol. 2, pp. 699–703.

N. R. YOUSEF AND A. H. SAYED (1999b), "A feedback analysis of the tracking performance of blind adaptive equalization algorithms," *Proc. Conference on Decision and Control*, vol. 1, Phoenix, AZ, pp. 174–179.

N. R. YOUSEF AND A. H. SAYED (1999c), "Tracking analysis of the LMF and LMMN adaptive algorithms," *Proc. Asilomar Conference on Signals, Systems, and Computers*, vol. 1, Pacific Grove, CA, pp. 786–790.

N. R. YOUSEF AND A. H. SAYED (2000a), "Steady-state and tracking analyses of the sign algorithm without the explicit use of the independence assumption," *IEEE Signal Process. Lett.*, vol. 7, no. 11, pp. 307–309.

N. R. YOUSEF AND A. H. SAYED (2000b), "A unified approach to the steady-state analysis of quantized adaptive filtering algorithms," *Proc. ISCAS*, vol. 2, Geneva, Switzerland, pp. 341–344.

N. R. YOUSEF AND A. H. SAYED (2001), "A unified approach to the steady-state and tracking analyses of adaptive filters," *IEEE Trans. Signal Process.*, vol. 49, no. 2, pp. 314–324.

N. R. YOUSEF AND A. H. SAYED (2002), "Ability of adaptive filters to track carrier offsets and random channel nonstationarities," *IEEE Trans. Signal Process.*, vol. 50, no. 7, pp. 1533–1544.

N. R. YOUSEF AND A. H. SAYED (2003) "Fixed-point steady-state analysis of adaptive filters,"*Int. J. Adap. Contr. Signal Process.*, vol. 17, no. 3, pp. 237-258.

C. L. ZAHM (1973), "Application of adaptive arrays to suppress strong jammers in the presence of weak signals," *IEEE Trans. Aero. Electronic Syst.*, vol. 9, no. 2, pp. 260–271.

M. ZAKAI (1964), "General error criteria," *IEEE Trans. Inform. Theory*, vol. 10, pp. 94–95.

G. ZAMES (1981), "Feedback and optimal sensitivity: Model reference transformations, multiplicative seminorms, and approximate inverses," *IEEE Trans. Automat. Contr.*, vol. 26, pp. 301–320.

J. R. ZEIDLER (1990), "Performance analysis of LMS adaptive prediction filters," *Proc. IEEE*, vol. 78, pp. 1781–1806.

J. R. ZEIDLER, E. H. SATORIUS, D. M. CHABRIES, AND H. T. WEXLER (1978), "Adaptive enhancement of multiple sinusoids in uncorrelated noise," *IEEE Trans. Acoust. Speech Signal Process.*, vol. 26, pp. 240–254.

H. H. ZENG AND L. TONG (1997), "The MSE performance of constant modulus receivers," *Proc. ICASSP*, Munich, Germany, pp. 3577–3580.

K. ZHOU, J. C. DOYLE, AND K. GLOVER (1996), *Robust and Optimal Control*, Prentice-Hall, Englewood Cliffs, NJ.

Author Index

Subject Index

Printed in the USA/Agawam, MA
January 23, 2013
572226.034